The Ebbsfleet Elephant

Excavations at Southfleet Road, Swanscombe in advance of High Speed 1, 2003–4

The Ebbsfleet Elephant

Excavations at Southfleet Road, Swanscombe in advance of High Speed 1, 2003-4

Edited by Francis Wenban-Smith

with other contributions by

Peter Allen, Richard Allen, Martin Bates, Richard Bates, Silvia M. Bello, David Bridgland, Nigel Cameron, Russell Coope, John Crowther, Denise Druce, Victoria L. Herridge, Richard J. Hewitt, David J. Horne, John Hutchinson, Stuart Foreman, Richard Macphail, Victoria Morris, Simon A. Parfitt, Kirsty Penkman, Richard C. Preece, Barbara Silva, Jean-Luc Schwenninger, John R. Stewart, Antony J. Sutcliffe, Charles Turner, Tom S. White and John E. Whittaker

Principal illustrator
Leo Heatley

Lithic line drawings
Barbara McNee

Other illustrations by
Magdalena Wachnik

Oxford Archaeology Monograph No. 20
2013

This book is one of a series of monographs by Oxford Archaeology (OA)

This publication has been generously funded by High Speed 1

Prepared for publication by Rebecca Nicholson, Oxford Archaeology

Published by Oxford Archaeology

ISBN: 978-0-904220-73-5

Oxford Archaeology, Janus House, Osney Mead, Oxford OX2 0ES,
Registered Charity No. 285627

Typeset by Production Line, Oxford
Printed and bound in Great Britain by Berforts Information Press, Eynsham, Oxford

Contents

Chapter 1. Introduction *by Francis Wenban-Smith*

Chapter 2. Background *by Francis Wenban-Smith*

Chapter 3. Details of excavation, site layout and approaches to analysis
by Francis Wenban-Smith

Chapter 4. Geology, stratigraphy and site phasing *by Francis Wenban-Smith, Martin Bates, Peter Allen, Richard Bates and John Hutchinson*

Chapter 5. Soil micromorphology, loss-on-ignition, phosphate-p concentrations and magnetic susceptibility analyses *by Richard Macphail, John Crowther and Francis Wenban-Smith*

Chapter 6. Clast lithology *by David Bridgland, Francis Wenban-Smith, Tom S. White and Peter Allen*

Chapter 7. The vertebrate remains from Southfleet Road: introduction, taphonomy and palaeoecology *by Simon A. Parfitt*

Chapter 8. The elephant skeleton and the question of human exploitation *by Simon A. Parfitt, Silvia M. Bello, John R. Stewart, Richard J. Hewitt, Victoria L. Herridge and Francis Wenban-Smith*

Chapter 9. Mammalian biostratigraphy *by Simon A. Parfitt*

Chapter 10. Molluscan analyses *by Tom S. White, Richard C. Preece and Francis Wenban-Smith*

Chapter 11. Ostracods and other microfossils *by John E. Whittaker, David J. Horne and Francis Wenban-Smith*

Chapter 12. Pollen *by Charles Turner, Barbara Silva and Francis Wenban-Smith*

Chapter 13. Amino acid dating *by Kirsty Penkman and Francis Wenban-Smith*

Chapter 14. Optically stimulated luminescence (OSL) dating *by Jean-Luc Schwenninger and Francis Wenban-Smith*

Chapter 15. Lithic artefacts: overview and approach to analysis *by Francis Wenban-Smith*

Chapter 16. Lithic artefacts: Phases 1, 2, 3 and 5 *by Francis Wenban-Smith*

Chapter 17. Lithic artefacts: Phase 6, the elephant area *by Francis Wenban-Smith*

Chapter 18. Lithic artefacts: Phase 6, the concentration south of Trench D and other stray pieces *by Francis Wenban-Smith*

Chapter 19. Lithic artefacts: Phase 7, the syncline infill *by Francis Wenban-Smith*

Chapter 20. Lithic artefacts: Phase 8, the overlying fluvial gravel *by Francis Wenban-Smith*

Chapter 21. Lithic artefacts, miscellaneous collections from outside the main sequence: Phases T–1, 9, 9–10 and 11 *by Francis Wenban-Smith*

Chapter 22. Discussion and conclusions: Clactonian elephant hunters, north-west European colonisation and the Acheulian invasion of Britain? *by Francis Wenban-Smith*

Specialist Appendices

List of Digital Appendices

List of Figures

Chapter 4

List of Tables

Chapter 7

Chapter 8

Chapter 9

Summary

This monograph is one of six volumes resulting from archaeological excavations in the Ebbsfleet Valley ahead of the construction of High Speed 1 and the Ebbsfleet International station. It provides the full account of the discovery, excavation and subsequent analysis of remains from a sequence of rich archaeological horizons found late in the construction programme and dating to early in the Palaeolithic period, or Old Stone Age, associated with the Hoxnian interglacial between approximately 425,000 and 375,000 years ago.

The highlight of this work was the identification of the remains of the carcass of an extinct straight-tusked elephant *Palaeoloxodon antiquus*, surrounded by the undisturbed scatter of flint tools used for its butchery, made and abandoned at the spot. Rich fossil palaeo-environmental remains (including pollen, snails and a wide range of vertebrates: rhinoceros, deer, beaver, rabbit, fish, mice, voles and rare specimens of Daubenton's bat and Barbary macaque) from deposits around the elephant skeleton provide a remarkable record of the climate and environment. They show that the elephant lived and died at a time of peak interglacial warmth, when the Ebbsfleet Valley was a lush densely-wooded tributary of the Thames, containing a quiet, almost stagnant swamp. There is no direct evidence of how the elephant met its end, but it is suggested here that it may well have been hunted and killed by the early hominins of this period, whose survival would have depended upon the nutrition provided by large herbivores such as deer, elephants and rhinos.

As well as about 80 flint artefacts around the elephant skeleton, a much larger concentration of about 1900 artefacts was recovered from the same horizon some 30m away. This was from what may have been a higher and drier spot above the swamp containing the elephant carcass, likely to have been more favoured for activity and where occupational evidence may have been more prone to accumulate. All the lithic remains from the elephant horizon show the same technological approach. Namely the manufacture of flakes from simple cores, and then the selection of some sharp-edged flakes for use, either without further modification, or following a minimal amount of further flaking to facilitate handling or to form simple notched cutting-edges. This approach is known from other sites of the same period in south-east England, and has been called 'Clactonian' after Clacton-on-Sea, Essex, where similar remains have been found. The evidence from this new site provides the best record yet of Clactonian remains from this period, establishing that there was a period early in the Hoxnian interglacial when Britain was re-settled (after local extinction due to the great Anglian glaciation) by hominins who did not make handaxes, generally the typical artefact of the earlier Palaeolithic.

The elephant horizon is overlain by a higher level rich in handaxes of various forms, including sharply pointed specimens, bluntly pointed sub-cordates and twisted-profile cordates and ovates. There are various possible interpretations for this difference, discussed in detail in the volume. The possibility is raised that this great contrast reflects different hominin populations, with the appearance of handaxes later in the Hoxnian relating to a second wave of settlement, possibly even by a different hominin lineage. However, on balance it is regarded as more likely that the development of handaxes later in the Hoxnian reflects *in situ* technological development of one south-east hominin group.

Finally, this monograph provides a fascinating case-study of Palaeolithic excavation methods, and how archaeological work is carried out in conjunction with major infrastructure developments such as High Speed 1.

Résumé

L'éléphant d'Ebbsfleet:
Les fouilles de Southfleet Road, à Swanscombe en amont de la construction de la High Speed 1, 2003–4

Cette monographie constitue l'un des six volumes, résultat des fouilles archéologiques ayant eu lieu dans la Vallée d'Ebbsfleet en amont de la construction de la ligne à grande vitesse High Speed 1 et de la Gare de Ebbsfleet International. Cet ouvrage présente le compte-rendu exhaustif de la découverte, des fouilles et de l'analyse ultérieure des vestiges issus d'une séquence d'horizons archéologiques riches, mis au jour vers la fin du programme de construction. Cette séquence date du Paléolithique inférieur, ou *âge de la pierre taillée*, lié à l'interglaciaire Hoxnian il y a environ 425,000 à 375,000 ans.

Le temps fort de ces travaux est marqué par l'identification des restes d'une carcasse d'un éléphant d'une espèce disparue, le *Palaeoloxodon antiquus* aux défenses droites, découvert au milieu d'un ensemble d'outils en silex disséminés, non déplacés et utilisés pour le massacre du mammifère, objets fabriqués et abandonnés sur place. Des vestiges paléoenvironnementaux riches en fossiles (parmi lesquels du pollen, des escargots et une large gamme de vertébrés : rhinocéros, cerf, lapin, poisson, souris, campagnol et de rares specimens du Vespertilion de Daubenton et du Macaque de Barbarie), livrés par les couches encerclant le squelette de l'éléphant, procurent un remarquable témoignage du climat et de l'environnement d'alors. Ces restes démontrent que l'éléphant vécut et s'éteignit à un moment de chaleur interglaciaire maximale, lorsque la Vallée d'Ebbsfleet était un affluent de la Tamise densément boisé et luxuriant, constitué d'un marais à faible courant, quasi-stagnant. Il n'existe pas de témoignage direct sur la manière dont l'éléphant a trouvé la mort, mais tout porte à croire qu'il aurait été chassé et tué par les premiers hominimes de cette période, dont la survie aurait dépendu de la nutrition que procurent les grands herbivores comme le cerf, l'éléphant et le rhinocéros.

En plus des 80 objets en silex retrouvés non loin du squelette de l'éléphant, une concentration beaucoup plus importante d'environ 1900 artéfacts a été mise au jour dans le même horizon à quelque 30 mètres. Ils provenaient sans doute d'un endroit plus haut et sec au-dessus du marais qui a livré la carcasse du mammifère, emplacement privilégié pour une activité et où des indices d'occupation ont eu une plus grande tendance à s'accumuler.

Tous les restes lithiques recueillis dans l'horizon renfermant les restes de l'éléphant témoignent d'une même approche technologique, à savoir la fabrication d'éclats à partir de blocs uniques, puis la sélection de certains éclats tranchants pour leur utilisation, soit sans autre modification, soit après une légère retouche par débitage pour faciliter leur manipulation ou pour former de simples éclats à coches.

Cette approche est connue sur d'autres sites de la même période dans le sud-est de l'Angleterre et est dénommée « Clactonienne » après Clacton-on-Sea dans l'Essex, où des vestiges analogues ont été mis au jour. Les témoins récupérés sur ce nouveau site forme le meilleur assemblage de restes Clactoniens jamais retrouvé pour cette période, permettant d'établir l'existence d'une période au début de l'interglaciaire Hoxnian qui a vu le repeuplement de l'Angleterre (après l'extinction locale due à la grande glaciation Anglienne) par des hominimes qui ne confectionnaient pas de bifaces, mobilier généralement typique du Paléolithique inférieur.

L'horizon où fut découvert l'éléphant était surmonté d'une couche riche en bifaces de formes variées, dont des spécimens très pointus, des subcordiformes pointus émoussés et des cordiformes et ovalaires à profil torse. Diverses interprétations peuvent expliquer cette disparité et sont détaillées dans ce volume. Il est possible que ce fort contraste entre les approches révèle des populations hominimes différentes, avec l'apparition des bifaces plus tard à la période holsteinienne apparentée à une seconde vague d'occupation, peut-être même par une lignée d'hominimes distincte. Toutefois, tout bien considéré, il y aurait plus de chance que le développement tardif des bifaces dans l'Holsteinien traduise le développement technologique *in situ* d'un groupe d'hominimes du sud-est.

Enfin, cette monographie offre une fascinante étude de cas des méthodes de fouilles du Paléolithique et de la manière dont les travaux archéologiques sont exécutés conjointement au développement d'infrastructures majeures tel que la *High Speed 1*.

Zusammenfassung

Der elephant von Ebbsfleet:

Ausgrabungen an der Southfleet Road, Swanscombe, in Vorbereitung des Baus von High Speed 1, 2003–4

Die vorliegende Monographie ist der sechste und letzte Band einer Reihe, die aus den Ausgrabungen im Tal von Ebbsfleet in Vorbereitung des Baus der Bahnstrecke High Speed 1 und des Bahnhofs Ebbsfleet International hervorgegangen ist. Die Monographie ist der vollständige Bericht der Entdeckung, Grabung und Befundanalyse reichhaltiger archäologischer Horizonte, die in einem späten Stadium der Baumaßnahme ergraben wurden und in das frühe Paläolithikum (Altsteinzeit), der Zeit des Hoxnian-Interglazial vor c 425.000 bis 375.000 Jahren, einzuordnen sind.

Herausragender Bestandteil der Arbeit ist die Identifizierung der Überreste eines ausgestorbenen Europäischen Waldelefanten, Palaeoloxodon antiquus, der von einer unberührten Streuung von Feuersteinwerkzeugen umgeben war, welche zum Schlachten des Elefanten benutzt worden waren. Diese Werkzeuge sind an Ort und Stelle hergestellt und zurückgelassen worden. Reichhaltige fossile paläoökologische Hinterlassenschaften (inklusive Pollen, Schnecken und eine breite Auswahl an Wirbeltieren: Nashörner, Rotwild, Bieber, Hasen, Fische, Mäuse, Wühlmäuse, seltene Exemplare von Wasserfledermäusen, Myotis daubentonii und Berberaffen) aus Schichten um das Elefantenskelett lassen einen bemerkenswerten Einblick in das Klima und die Umwelt zu. Es zeigt sich, dass der Elefant während des Höhepunkts einer zwischeneiszeitlichen Wärmeperiode lebte und starb. Zu dieser Zeit war das Tal von Ebbsfleet ein üppig bewaldetes Einzugsgebiet der Themse, mit einem ruhigen, fast stillstehenden Sumpf. Es gibt keine direkten Hinweise darauf wie der Elefant sein Ende gefunden hat, doch wird hier angenommen dass er von frühen Homininen gejagt und getötet wurde, deren Ernährung und Überleben vom Erlegen großer Pflanzenfresser, wie Rotwild, Elefanten und Nashörnern, abhing.

Um das Elefantenskelett verteilt wurden c 80 Feuersteinartefakte gefunden. Eine sehr viel höhere Konzentration von ungefähr 1900 Artefakten wurde im selben Grabungshorizont, etwa 30m entfernt, entdeckt. Dies war vermutlich um eine höher gelegene und damit auch trockenere Stelle oberhalb des Sumpfes, welcher die Elefantenüberreste enthielt, und wurde anscheinend für bestimmte Aktivitäten bevorzugt. Beschäftigungsspuren haben sich hier vermutlich einfacher angesammelt. Die technische Vorgehensweise bei der Steinbearbeitung ist im gesamten Elefantenhorizont sehr einheitlich. Es handelt sich hierbei um die Herstellung von Abschlägen von einfachen Feuersteinknollen. Von diesen wurden scharfkantige Abschläge ausgewählt, welche entweder unbearbeitet genutzt wurden oder mit minimalem Aufwand weiterbehauen wurden, um eine bessere Handhabung zu gewährleisten oder um gekerbte Schnittkanten zu schaffen. Dieses Verfahren ist von weiteren Fundstellen in Südost-England bekannt und wird als 'Clactonian' bezeichnet, nach Clacton-on-Sea (Essex), wo ähnliche Hinterlassenschaften entdeckt wurden. Die Funde der vorliegenden Grabung stellen die bisher besten Ansammlungen von Clactonian Artefakten dieser Periode dar. Daraus lässt sich ableiten, dass es eine Periode in der Hoxnian-Zwischeneiszeit gab, während der Britannien erneut (nach lokalem Aussterben aufgrund der großen Anglischen Vereisung) von Homininen besiedelt wurde, die keine Faustkeile herstellten – das allgemein typische Artfakt des frühen Paläolithikums.

Der Elefantenhorizont von einer höher gelegenen Schicht überlagert, die mit Faustkeilen verschiedenster Formen angereichert ist. Darunter gibt es spitz zulaufende Stücke, stumpfe sub-kordiale und gewundene kordiale Formen sowie ovale Exemplare. Es liegen diverse Interpretationen für diese Unterschiede vor, die in diesem Band ausführlich diskutiert werden. Mit hoher Wahrscheinlichkeit spiegeln diese großen Unterschiede verschiedene Populationen von Homininen wider. Das Auftreten von Faustkeilen im späteren Hoxnian steht vermutlich mit einer zweiten Siedlungswelle in Verbindung, möglicherweise sogar von Homininen anderer Abstammung. Es ist allerdings wahrscheinlicher, dass die Entwicklung von Faustkeilen im späteren Hoxnian die Entwicklung einer einzelnen Homininengruppe vor Ort darstellt.

Letztendlich ist dieser Band ein faszinierendes Fallbeispiel für paläolithische Ausgrabungsmethoden und für die Einbindung archäologischer Arbeiten in große Infrastrukturprojekte, wie z. B. dem Bau der High Speed 1.

Acknowledgements

Bringing to fruition a project on the scale of the Ebbsfleet Elephant excavation has involved literally hundreds of people. Thanks are due to numerous organisations and individuals for their work on the project. It is simply impossible to thank everyone individually, but we are truly grateful to all who have in large and small ways played some role in the project, those named below being only the most directly involved.

Union Railways (North) – latterly High Speed 1 – provided generous funding and support to the archaeological programme, without which the investigation could not have taken place. Josie Murray advised Union Railways on heritage matters and Rail Link Engineering was responsible for managing the design and construction of the railway, including the archaeological programme. Members of the Environment Team, led by Ted Allett and Paul Johnson deserve particular recognition for recognising the importance of the site immediately and negotiating the necessary time and resources for its investigation, in spite of huge pressures to complete the construction programme. Among the archaeological team at Rail Link Engineering, particular thanks are extended to Helen Glass and Steve Haynes for their unstinting support during the fieldwork and post-excavation stages and to Mark Roberts (Institute of Archaeology, UCL) whose role as RLE's specialist external advisor on Palaeolithic matters was also critical in validating the importance of the discovery at an early stage. On the construction side, we are particularly grateful to Jon Gold, the RLE engineer with whom the archaeological team mostly liaised, who had to re-arrange the HS1 and Ebbsfleet International station work programme to accommodate the 6-month excavation programme. Also, to all working on the landscape construction and services in the elephant area, particularly the crew who helped sling and lift the bovid skull found at the very end of the watching brief following the main excavation. Rachel Starling (HS1 Environmental Manager) has subsequently been very patient in supervising completion of the post-excavation analysis and publication on behalf of HS1, while also driving the project firmly to a successful conclusion. As she has pointed out on a number of occasions, this archaeological programme is the last of over a thousand HS1 Project Undertakings to be completed. It was also, to be fair, the last to start.

On the curatorial side, English Heritage (EH) and Kent County Council (KCC) played a vital role as Statutory Consultees on heritage matters, ensuring that Palaeolithic objectives were central to the HS1 research framework for the Ebbsfleet Valley. In particular the project would never have begun without the commitment of Lis Dyson (KCC Heritage Conservation Group) and would never have developed to a successful conclusion without her determined advocacy in the field and post-excavation stages, and thanks are also given to Sharon Thompson and Chris Waite of KCC for their support of the project. Peter Kendall (EH Inspector of Ancient Monuments, SE Region) and Dominique de Moulins (EH Regional Science Advisor) also provided much valuable support and advice throughout the HS1 archaeological project. When the elephant was discovered, early advice indicated that the find was of national significance. Consequently, representations were made by Union Railways, English Heritage and Kent County Council to the Secretary of State for Transport and the Department of Culture, Media and Sport to extend the period available for excavation using the powers available in such circumstances. Accordingly the excavation period was extended and the construction programme amended to accommodate this work. This was the only time that this power was exercised during the whole of the HS1 archaeology programme, in light of the finds' significance. Support from Dartford and Gravesham Borough Councils, Clive Gilbert and Sonia Bunn is also acknowledged in this context.

Oxford Archaeology provided the core field team, as well as project management in the field and post-excavation. From the excellent field team Darko Maricevic is singled out for special mention as the OA site supervisor throughout the excavation and watching brief, and as the first to spot the elephant skeleton when it was uncovered during mechanical excavation. Mary Saunders also deserves a special mention for carrying out so much muddy 'part-sieving' on-site (see Fig. 3.20). We are very grateful to all the Oxford Archaeology excavators, surveyors, finds processors and logistics staff for their hard work on the elephant site: A.Ainsworth, K.Anker, R.Bailey, W.Bedford, R.Blackburn, J.Bolderson, C.Boston, C.Breeden, D.Casey, I.Cook, M.Copley, L.Derry, S.Dobinson, N.Dransfield, N.Gaskell, C.Gerson, M.Gibson, E.Glass, R.Grant, S.Greenslade, D.Harris, R.Hatfield, L.Heatley, R.James, A.Kilgour, A.Kirkpatrick, N.Lambert, S.Laurie-Lynch, M.Littlewood, J.Lord, J.Marchant, L.Martin, A.Mayes, S.Milby, P.Murray, C.Naisbitt, L.Newton, E. Noyce, J.O'Brien, L.Offord, L.O'Gorman, N.Pankhurst, J.Patrick, J.Payne, W.Perkins, S.Pickstone, M.Planas, B.Powell, K.Proctor, R.Radford, C.Rawlings, C.Richardson, C.Sampson, M.Saunders, W.Sawtell, P.Schofield, L.Sikking, M.Sims, M.Sowerby, M.Spalding, A.Stone, R.Tannahill, G.Thacker, D.Thomason, J.Tibber, J.Tierney, G.Walton, R.Whalley, D.Wheeler, M.Wood and M.Wooldridge.

The work of the specialist team on site and in the laboratory is gratefully acknowledged. Martin Bates (Dept. of Geography, University of Wales Trinity St David), Richard Bates (School of Geography and Geosciences, University of St.Andrews), Peter Allen and the late Professor John Hutchinson contributed greatly to the stratigraphic interpretation of the site, Richard being largely responsible for the impressive 3D deposit model. Richard Macphail (Institute of Archaeology, UCL) and John Crowther (Dept. of Geography, University of Wales Trinity St David) worked on the soil micromorphological and soil chemistry studies. Specialist reports were also provided by David Bridgland (University of Durham), Nigel Cameron (Dept. of Geography, University College London), the late Professor Russell Coope, Denise Druce (Oxford Archaeology North), Jean-Luc Schwenninger (RLAHA, University of Oxford), John Stewart (University of Bournemouth), Tom White (Dept. of Zoology, University of Cambridge), John Whittaker (Natural History Museum) and Kirsty Penkman, Victoria Morris and Richard Allen (Dept. of Chemistry, University of York). Simon Parfitt (University College London and the Natural History Museum) led the huge and delicate task of conserving and analysing the faunal remains with crucial assistance from others including: John Stewart, who spent much time on site organising the recovery of fragile faunal remains and bulk environmental samples; Gilbert Marshall, who was involved in excavation of the clay around the elephant and played a major role in the subsequent watching brief; Nigel Larkin who came out at short notice to help conserve and lift the aurochs skull found during the watching brief, and did excellent subsequent conservation work on many of the most important faunal remains. Thanks also to Silvia Bello, Victoria Herridge and the late A.J.Sutcliffe (Natural History Museum) and Richard Hewitt (University of Alcalá (Spain) for their contributions to this report. Glenys Salter (Natural History Museum) is thanked for her invaluable assistance with the day-to-day running of the faunal analysis programme. Silvia Bello and Simon Parfitt acknowledge the support of the Leverhulme Trust and the Calleva Foundation. Silvia Bello's work was part of the Human Behaviour in 3-D project funded by the Calleva Foundation.

Francis Wenban-Smith would particularly like to thank all at the University of Southampton involved in supporting the Centre for Applied Human Origins Research in the Department of Archaeology. During the post-excavation work, key roles in the work carried out in CAHOR were played by: firstly Alison Moore, who, despite being an expert in the Roman period, carried out a huge amount of work washing and marking Palaeolithic flints, entering various records into digital format and typing great tracts of text; secondly James Cole, who constructed the GIS model with the site's faunal and lithic data, and provided a range of graphical outputs and other illustrations; and thirdly Barbara McNee, who produced numerous lithic line drawings of exceptional quality. Various individuals from outside these organisations were involved in the off-site work, in particular Christina Cheung and Christian Lewis, who carried out vast amounts of sample processing.

The British Museum kindly provided working space in the Department of Prehistory and Europe at Franks House for Francis Wenban-Smith during the lithics analysis and have agreed to retain the artefact and paper archives in their collection. Particular thanks are due to Nick Ashton for facilitating this, and to Frank Beresford for his help in curating the lithic collection. Richard Sabin is thanked for providing working space at the natural History's storage facility at Wandsworth. Staff at the Natural History Museum are thanked for greatly facilitating the faunal remains study and kindly agreeing to take the HS1 Ebbsfleet Valley faunal remains into their collection. Digital archive components will be lodged with the Archaeology Data Service.

The specialist team would all like to thank OA South staff including Elizabeth Stafford, (Head of Geo-archaeology), who coordinated the myriad environmental samples, with help from Laura Strafford, variously processing, logging, sub-sampling and distributing them to specialists in an orderly manner. Leigh Allen (Head of Finds) arranged the processing and distribution of artefacts to the specialist team. Other OA staff heavily involved in the post-excavation stages include Leo Heatley who was first involved with the site as an excavator and surveyor, subsequently digitised many of the records and created the initial site archive GIS, and finally turned most of the original illustration briefs into finished figures. Magdalena Wachnik (Head of Graphics) digitised the lithic line drawings and managed the typesetting and printing; Rebecca Nicholson (Head of Environmental Archaeology) produced and copy-edited the monograph in a very short time. Chris Hayden kindly prepared the index. Nathalie Haudecour-Wilks provided the French translation and Markus Dylewski the German translation, with additional input from Susanne Hakenbeck. Bob Williams (OA Chief Operating Officer) and David Jennings (CEO) provided much-needed support and directorial oversight throughout the HS1 project.

Francis Wenban-Smith would like to express especial thanks to the two OA managers with whom he worked throughout the project, firstly Richard Brown who was the project manager during the excavation phase of work, and secondly Stuart Foreman, who managed the project through the subsequent post-excavation analysis and writing-up phases. FW-S would also like to thank family and friends who provided much-needed support throughout the nine year period from the start of this project to its finish, in particular Martin Bates, Laura Coleman, Lis Dyson, Susanne Hakenbeck, Siân Jones, Claire Jowitt, and Mark Roberts.

The managers and staff of Oxford Archaeology, in addition to the people and organisations thanked above, would like to express particular appreciation for the work of Francis Wenban-Smith (CAHOR, University of Southampton) in his various roles as director of HS1 Palaeolithic investigations in the Ebbsfleet Valley, lithics specialist, lead author and general editor of this volume.

In a project with a great many other exciting archaeological distractions, Francis brought to the table an unwavering commitment to, and focus on, the Palaeolithic, without which the elephant site would not have been discovered and this report would never have been completed.

Finally, particular thanks are due to Mark Roberts (UCL Institute of Archaeology), Lis Dyson (Kent County Council), David Bridgland (Dept.of Geography, University of Durham) and Simon Parfitt for their careful reviews of the monograph contents and many pertinent comments. Any remaining errors and opinions are, as is customary, entirely the responsibility of the authors.

Preface

High Speed 1 (HS1) connects Britain to the European High Speed rail network. When Her Majesty the Queen opened High Speed 1 on 6 November 2007 it marked the culmination of Britain's largest construction project, completed on time and within budget. It generated the country's largest archaeological project and created an unprecedented opportunity to excavate along one of the busiest historic corridors between Britain and the Continent, from London and through Kent.

Considerable effort was made in the planning stages of the route to identify important archaeological sites. Where possible they were avoided or preserved in situ. Geophysical survey, field walking and trial trenches were commissioned to provide further detail where there was uncertainty. For sites of interest, an extensive programme of archaeological investigations, analysis and reporting was implemented. Over seventy sites were investigated in this way. A remarkable wealth of information has been gained about the archaeological character and development of the ancient landscapes of Greater London and Kent. The detailed results of the work have been published online through the Archaeology Data Services and subsequently in eight academic books. In addition a book summarising the key sites has also been published, aimed at a non-specialist audience: 'Tracks and Traces, The Archaeology of High Speed 1' (High Speed 1 2011).

The scale of the work required an innovative approach: The RLE archaeology team (HS1's project manager) oversaw all aspects of the archaeology programme. English Heritage, County Archaeologists and university academics were closely involved in setting the High Speed 1 (formerly known as the Channel Tunnel Rail Link) academic research strategy which set the scene for the work. This was implemented within the framework of The Channel Tunnel Rail Link Act 1996 and the project's Environmental Minimum Requirements.

Project managers, planners, design and site engineers, construction, archaeological and historic building contractors, English Heritage, county archaeologists and historic buildings officers came together as one team. It is testament to this team that the fieldwork was undertaken within exacting construction time-scales, whilst ensuring that best practice was achieved. Teamwork has been fundamental to the achievements of the project in general and the Southfleet Road Elephant site in particular.

The programme of archaeological fieldwork extended from 1996 until 2004. It was right at the end of this programme, during a watching brief on earthworks associated with the access road to Ebbsfleet International railway station, that the elephant site was discovered. As the HS1 archaeological programme was nearing its end, the subsequent excavation and analysis of the elephant required significant additional resources and adjustment to the access road construction programme to accommodate appropriate excavation and analysis.

This volume is the penultimate report in the series of archaeological monographs describing the HS1 results and records this significant, unexpected and quite remarkable find. It provides a detailed account of the site and its context, followed by thematic analyses of the elephant and associated finds. The detailed analysis presented here has deepened our understanding of the Palaeolithic/Pleistocene in Britain and of the Ebbsfleet Valley in particular. This is clearly an exceptional and major contribution to our understanding of the past within the context of Kent and south-east England, with significance for academic debates at a national and international level.

The extensive assemblage of artefacts and paper records has been deposited at the British Museum and the faunal remains at the Natural History Museum, for future reference. It gives me great pleasure to thank those involved in this latest research and to commend this book to you.

The archaeological programme for HS1 has been recognised nationally in industry awards for setting exemplary standards of archaeological practice. Thank you to all who have contributed to this significant achievement, from the earliest stages of project planning, through the construction programme to final delivery of the monographs.

Rachel Starling,
HS1 Ltd Environment Manager

Foreword

Palaeolithic archaeologists are brought up to expect the unexpected. We take the discovery of hobbit-size hominins and undreamt of genetic ancestors from Siberia in our stride. We puzzle over their significance for deep human history. We expect the unexpected because we know that the task of charting the variety of hominin forms and behaviour has barely begun. At a global scale the picture is constantly changing and the results of research challenging. It is therefore re-assuring to dip into one of the best known archives of Palaeolithic archaeology contained in the ancient terraces that once fed into the Thames. The excellence of Quaternary research in southern England has in the last thirty years given us a robust chronology and an environmental framework for studying changing hominin life-ways 400,000 years ago.

This was a period of great importance for deep human history. Hominins now had large brains comparable in size to ours. Yet the products of such brain growth are not readily apparent in new technologies, works of art or even the extension of settlement into inhospitable lands. Instead there seem to be disconnects between brains and behaviour.

In this major addition to the British Palaeolithic archive Francis Wenban-Smith and his multi-disciplinary team of Quaternary scientists show us how to interpret the unexpected. The Southfleet Road elephant site so impressively reported on here is important for three main reasons. It points to the co-operative skills of hominins at this time. The information confirms the Southfleet hominins as top-predators, indeed the only predator able to take down a 45 year old male elephant in its prime and without being sneaky about it by immobilising it in boggy ground. As a result the research challenges the time-honoured link between brains, advances in technology and killing-at-will. It achieves this by verifying the independent chronological status of that most un-remarkable of all lithic technologies, the British Clactonian. And all of this was possible due to excellent preservation, a dedicated team and a well-tried series of development controls that made the work possible.

The results of this work will be discussed and no doubt re-interpreted over many years to come. This volume continues the British tradition of publishing primary Palaeolithic data in full so that this key activity of evaluation and re-evaluation can take place. But the Southfleet Road elephant site has major implications for the protection and future investigation of our deep heritage. The opportunity arose as the result of the major High Speed 1 infrastructure project. No grant body would have funded such a huge speculative trench through the Kent countryside. Previous work had singled out the Swanscombe area as potentially important. But even so, the discovery of this 400,000 year old elephant with flint tools so clearly associated together with rich environmental remains was unexpected and remains remarkable. So remarkable that 2013s much-trumpeted find of a minor English monarch in a car park in Leicester is overblown by comparison.

There are great treasures buried deep in the Pleistocene landscapes of southern England. Sometimes they can be predicted, while in other cases they arise from patient watching briefs in the most unexpected places. I congratulate Wenban-Smith, Oxford Archaeology and the specialist team on a magnificent project, brought to fruition. Its legacy will be to make us all aware of the deep archaeology beneath our feet and inspire us to see more of it in the future.

Professor Clive Gamble,
Centre for the Archaeology of Human Origins,
University of Southampton

Chapter 1

Introduction

by Francis Wenban-Smith

ARCHAEOLOGY AND HIGH SPEED 1

High Speed 1 (formerly known as the Channel Tunnel Rail Link) was built by London and Continental Railways Ltd between 1996 and 2007, after securing the necessary parliamentary enabling bill, the Channel Tunnel Rail Link Act 1996. This required due account to be taken of its environmental impact, including appropriate mitigation of any archaeological impact. This new high-speed line, henceforth 'High Speed 1' or HS1, extends for 109km across south-east England between St Pancras station in London and the entrance to the Channel Tunnel near Folkestone (Fig. 1.1a). It was built in two sections: Section 1 lies entirely within Kent and extends from Folkestone to Fawkham Junction (Gravesham), just south of the new Ebbsfleet International station, located in the Ebbsfleet Valley, north-west Kent (Fig. 1.1b). Section 2 extends from the Fawkham Junction to London St Pancras, entering the Ebbsfleet Valley under Pepper Hill, crossing under the Thames between Swanscombe and Purfleet, and then passing through Essex and East London.

A major programme of archaeological investigation was undertaken to mitigate the impact of engineering and construction work on the archaeological resource along the HS1 route. Desk-based assessment and route planning commenced in the early 1990s, followed by an extensive programme of evaluation comprising field-walking, trial-trenching, deep test-pitting and borehole investigations, largely undertaken between 1997 and 2001. Archaeological sites that were revealed, and which could not be avoided or preserved *in situ*, were then excavated in advance of construction. In addition to targeted excavations, watching briefs were maintained during construction along the full route of the line, and on any associated works. Any additional remains thus revealed were also recorded, and incorporated in the subsequent analysis and reporting programme.

The scale of the archaeological programme was so great that results were not quickly produced, but, at the time of writing (in October 2012) many of the research archives from individual investigations have been lodged with the Archaeology Data Service, where they are directly available for on-line perusal by all interested parties. In addition several monographs have already been published or are in final stages of production, covering significant discoveries along the HS1 route, particularly *On Track: the Archaeology of High Speed 1 in Kent* (Booth *et al.* 2011). This includes major discoveries such as a Neolithic house at Whitehorse Stone, a Roman villa at Thurnham and an Anglo-Saxon cemetery at Saltwood Tunnel. Also recently published is *Settling the Ebbsfleet Valley* (Andrews *et al.* 2011a; 2011b; Biddulph *et al.* 2011; Barnett *et al.* 2011), a 4-volume series covering the discovery of an Anglo-Saxon watermill, further investigation of a Roman villa/industrial complex and further investigation of the late Iron Age and Romano British temple complex at the top of the Ebbsfleet Valley at Springhead. *Prehistoric Ebbsfleet* (Wenban-Smith *et al.* forthcoming) covers Palaeolithic, Neolithic and Bronze Age investigations in the Ebbsfleet Valley, apart from the Lower Palaeolithic discovery reported on in this volume. A paper covering some additional investigations of the MIS 9 deposits at Purfleet on the north side of the HS1 Thames crossing is in preparation (Bridgland *et al.* 2012).

This volume marks the final major piece of archaeological reporting resulting from the construction of High Speed 1, covering an unexpected discovery made in late 2003 after the remainder of the archaeological programme had been completed. Consequently, the programme of analysis and reporting has lagged behind the other archaeological work. It has also, unlike the other works that were carried out as a joint venture between Oxford Archaeology and Wessex Archaeology, been solely the product of work by Oxford Archaeology in collaboration with the various external specialists involved.

THE SOUTHFLEET ROAD ELEPHANT SITE: AN UNEXPECTED DISCOVERY

The Ebbsfleet Valley, a tiny south-bank tributary of the Thames in north-west Kent near Dartford (Fig. 1.1b), was a major focus of construction work for HS1. Besides the route itself, additional major impacts were made by construction of the Ebbsfleet International station, its connecting link to the existing North Kent line and associated landscape remodelling and station access routes. Although the Ebbsfleet Valley had previously been subject to extensive aggregate extraction, mostly chalk quarrying, it nonetheless still contained a range of important archaeological remains, from Palaeolithic to Saxon eras, which required mitigating excavations. These investigations took place between April 2001 and March 2003, after which targeted archaeological

Figure 1.1   (a) Location of the High Speed 1 railway line within south-east England; (b) High Speed 1 Sections 1 and 2, and the location of the Ebbsfleet Valley

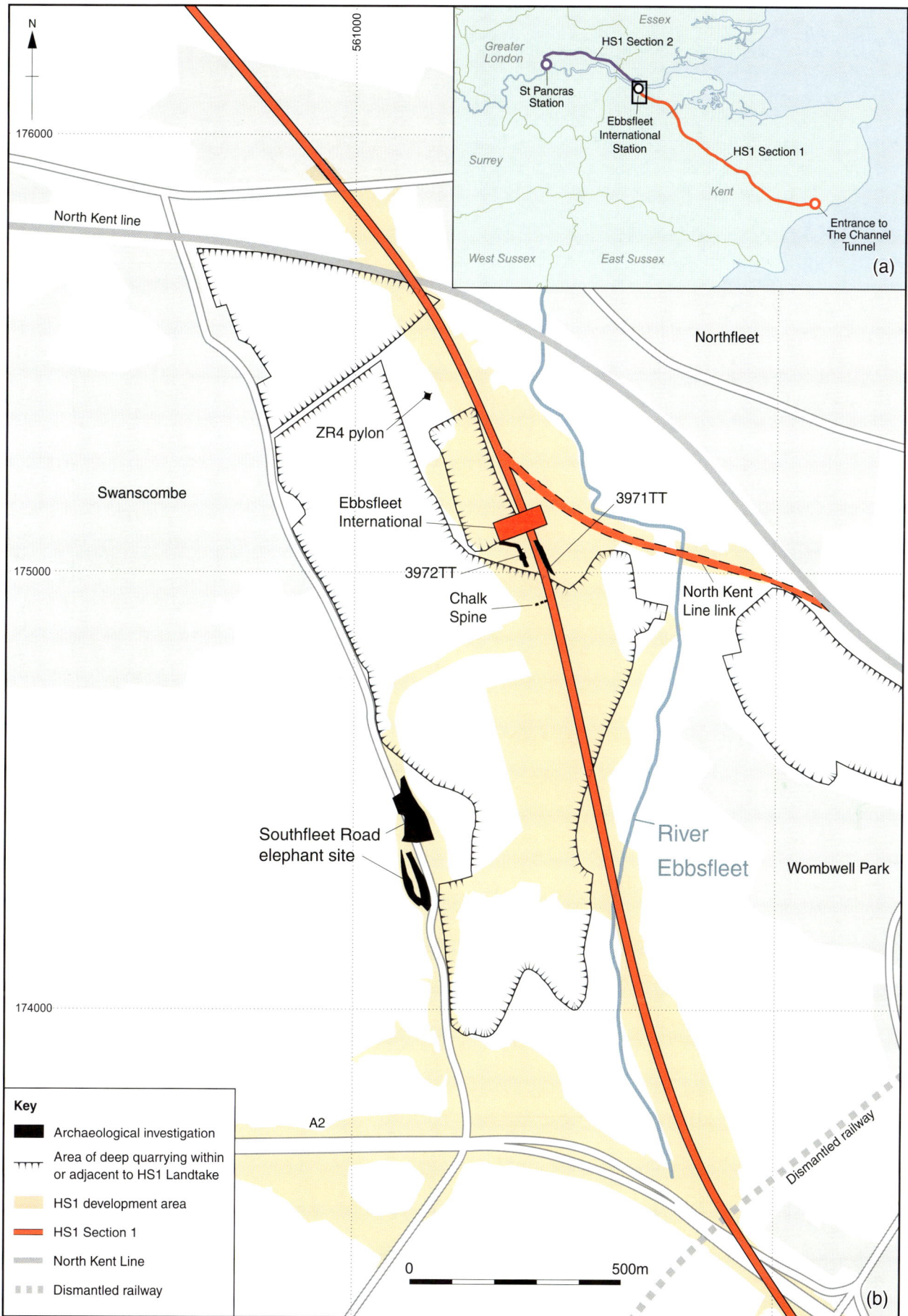

Figure 1.2 (a) High Speed 1, Sections 1 and 2; (b) Location of the Southfleet Road elephant site within HS1 development areas in the Ebbsfleet Valley

fieldwork was considered completed, although a watching brief presence was maintained.

In September 2003, groundwork in the south-west part of the Ebbsfleet Valley, where the access road to the new station was linked with the existing Southfleet Road B259 heading into Swanscombe, exposed Pleistocene gravels in an area not previously considered as of high archaeological potential (Fig. 1.2). Resulting investigation, described in more detail in Chapter 3, led to the discovery not only of the partial skeleton of an extinct straight-tusked elephant *Palaeoloxodon antiquus* associated with an undisturbed scatter of flint artefacts, but also revealed what appeared to be a wider palaeolandsurface at the same level, containing undisturbed concentrations of lithic and faunal remains. Further investigations then revealed that this horizon was merely one part of a deeper sequence containing artefactual and palaeo-environmental evidence at several different levels.

The find was immediately recognised as of such importance that, even though the archaeological programme was already thought to have been completed, extra resources were made available to carry out a thorough excavation, and the construction programme was rearranged to allow time for this to happen. Excavation took place between February and August 2004, followed by a watching brief through to early November 2004. Consideration was given to preservation *in situ*, but this proved an impossible option in view of the extent to which bulk ground extraction and surrounding construction had already taken place.

SCOPE AND OUTLINE OF THE VOLUME

This volume focuses specifically on the Southfleet Road elephant site. Other archaeological investigations in the Ebbsfleet Valley in advance of HS1 are reported on in the separate *Prehistoric Ebbsfleet* volume (Wenban-Smith *et al.* forthcoming), which covers Upper Palaeolithic and later prehistoric investigations and the intensive geo-archaeological and palaeo-environmental studies aimed at (a) improving understanding of marine isotope stage (MIS) 7 deposits in the Ebbsfleet Valley, and (b) contextualising within them the prolific evidence of Levalloisian occupation recovered by previous workers since the 1880s.

Although the Southfleet Road elephant site, henceforth 'the site', is less than 1km from the other significant Lower/Middle Palaeolithic archaeological and geoarchaeological locales of the Ebbsfleet Valley, it is a world away conceptually. The site is linked not with the late Middle Pleistocene and Late Pleistocene deposits that fill the central part of the Ebbsfleet Valley and form the focus of the investigations reported in the *Prehistoric Ebbsfleet* volume, but with significantly earlier Middle Pleistocene sediments lining the upper western flank of the Ebbsfleet Valley. These are associated with a wholly different stage of the Pleistocene, and contain Lower/Middle Palaeolithic remains of far greater age, dating over 150,000 years earlier (see Chapter 2).

Furthermore, the nature of the investigation carried out at the site was quite different to the work at the other Palaeolithic/Pleistocene locales in the central Ebbsfleet Valley. Discussed in more detail subsequently (Chapter 3), the work at the site involved substantial open-area excavation and the exposure, recording and sampling of major deposit sequences at a single location. In contrast, the work in the central Ebbsfleet Valley involved excavation and environmental sampling of numerous isolated test pits and stepped trenches, for which subsequent analysis and interpretation required a quite different approach. Thus, despite superficially all coming under the umbrella of 'Palaeolithic and Pleistocene', it makes sense for reporting of the site to be published as a distinct monograph, separate from the other Palaeolithic/Pleistocene investigations carried out in the Ebbsfleet Valley in advance of HS1.

A preliminary, interim report on the site was produced shortly after its discovery (Wenban-Smith *et al.* 2006), which established the initial biostratigraphical basis for attributing the main archaeological horizon with the elephant and associated artefactual remains to MIS 11, the Hoxnian interglacial. This interim report also provided an overview of the stratigraphic sequence, and summarised the technological and typological characteristics of the lithic industry associated with the elephant horizon, attributing it to the core/flake-tool Clactonian.

This volume provides a much more detailed multidisciplinary report on all aspects of the site. The stratigraphic sequence is described in more detail (Chapter 4), and a revised 11-phase sequence is established, superseding the 6-unit framework presented in the interim report. Results of a range of sedimentary analyses are also presented, namely: micromorphology, loss-on-ignition and magnetic susceptibility (Chapter 5) and clast lithology (Chapter 6).

Faunal and botanical remains including pollen, molluscs, ostracods, small vertebrates and larger mammals were recovered from various horizons throughout the sequence, not only the elephant horizon. More detailed analyses of these remains not only provide a more secure biostratigraphical attribution of the elephant horizon and its associated artefactual remains, but also contribute to a revised reconstruction of the climate and local environment throughout development of the sequence (see Chapters 7-12). This facilitates its integration into the wider chrono-stratigraphic and climatic framework of the Middle Pleistocene.

Two chronometric dating approaches were used to support the biostratigraphical and climatic interpretations. In the first place, recent developments in protocols for amino acid dating, concentrating on opercula of the fluvial gastropod *Bithynia tentaculata*, have provided a new benchmark for dating Middle and Late Pleistocene deposits (Penkman *et al.* 2007 and 2011). Fortunately, *Bithynia* opercula proved to be abundant both at the same horizon as the elephant remains and in another deposit towards the base of the stratigraphic sequence. Thus numerous analyses were carried out on opercula from both horizons (Chapter 13), the results confirming the

MIS 11/Hoxnian attribution for at least the main part of the sequence and also allowing correlations with specific horizons from key comparator sites in other parts of Britain. Deposits above and below these well-dated horizons lacked good dating evidence, so their dates rest on consideration of geomorphological correlations and whether or not there are significant chronological hiatuses above and below the dated middle part of the sequence.

Optically stimulated luminescence (OSL) dating was also applied to a number of horizons in the sequence that lacked any other evidence by which they might be dated. It was not thought that this technique was particularly suitable for deposits broadly attributable to the earlier Middle Pleistocene, but it was intended to investigate whether the upper part of the sequence contained significantly younger deposits above depositional hiatuses. The initial results suggested that this was indeed the case, leading to a second phase of OSL work including analysis of control samples from lower down in the sequence that were confidently attributable to MIS 11. The overall results of this work suggested that, whether solely due to their antiquity or for some chemical or sedimentary reason, the sediments at the site were not providing reliable OSL results. Nonetheless, this work is reported on fully here as a useful case-study for the application of OSL dating to Middle Pleistocene sediments (Chapter 14).

For the lithic evidence of early hominin activity (Chapters 15-21), it is attempted to present not just a technological and typological description leading to an attribution of cultural/industrial affinities, but, where possible, a more holistic approach to lithic analysis. The depositional and post-depositional taphonomic history of the rich assemblage from the elephant horizon is investigated by a combination of spatial distribution, refitting, microdebitage analysis and artefact condition. This leads to interpretation of this level as a palimpsest, with the majority of lithic evidence being a slightly disturbed accumulation, co-occurring with rarer evidence from undisturbed activity episodes, including butchery activity associated with the elephant carcass. Despite a slight degree of disturbance for the majority of the lithic assemblage, it nonetheless has a high degree of integrity. This allows the technological/typological production at

the site to be understood as part of a wider lithic *chaîne opératoire*, within the context of raw material nature and availability, and the manufacture, use and discard of lithic artefacts around the wider landscape in what was evidently a successful adaptation.

Although the great majority of the lithic collection is represented by the rich assemblage from the elephant horizon (Chapters 17-18), lithic artefacts were also recovered from other horizons throughout the sequence, both above and below the elephant horizon. Analysis of the material from these other levels (Chapter 16; Chapters 19-21) establishes a longer history of hominin occupation at the site through MIS 11, as well as perhaps earlier and later, and also demonstrates significant changes in material cultural production and behavioural adaptation during MIS 11.

The final chapter (Chapter 22) presents an integrated summary of the main conclusions reached from the various specialist analyses, with discussion of wider implications within the context of the MIS 11 occupation of Britain and northern Europe. This is followed by some thoughts on Palaeolithic archaeology and mega-development projects, following from what has now been an incredible 20 years of work in the Ebbsfleet Valley since the decision was made in the early 1990s to route HS1 through what was then a quiet post-industrial backwater, with landfilling of the old quarries taking place beside the Blue Circle sports ground. The final section considers the value of this archaeological project, and its legacy.

ARCHIVES

The artefacts and the primary paper archive from this project will be deposited with the British Museum and the faunal remains with the Natural History Museum, London. The digital archive, including the digital appendices listed under 'Contents' will be deposited with the Archaeology Data Service (ADS, 2013), which uses the Digital Object Identifier (DOI) System for uniquely identifying its digital content. The HS1 Ebbsfleet Elephant archive has the following DOI: http://dx.doi.org/10.5284/1018062

Chapter 2

Background

by Francis Wenban-Smith

PERIOD BACKGROUND AND GEOLOGICAL FRAMEWORK

This volume covers the earliest parts of the prehistoric past investigated during construction of HS1, not only in the Ebbsfleet Valley but also along the whole of the route. *Settling the Ebbsfleet Valley* (Andrews *et al.* 2011a; 2011b; Biddulph *et al.* 2011; Barnett *et al.* 2011) covers a relatively restricted burst of settlement activity and political/cultural upheaval associated with the pre-Roman Iron Age, the Roman occupation of southern Britain and the subsequent Saxon settlement. Each of these settlement phases covers at most a few centuries of activity in the general period within a millennium either side of 0 AD, set against the backdrop of a climate and landscape broadly similar to today. In contrast, the earlier prehistory of the Ebbsfleet Valley, covered in the other HS1 Ebbsfleet volume *Prehistoric Ebbsfleet* (Wenban-Smith *et al.* forthcoming), stretches deep into the past, spanning several hundred thousand years and embracing major climatic oscillations and landscape change. *Prehistoric Ebbsfleet* covers both the later Prehistoric evidence, including final Palaeolithic, Mesolithic, Neolithic and Bronze Age, from the end of the last ice age *c* 10,000 years ago and the subsequent Holocene epoch, to earlier evidence of hominin presence and landscape development between *c* 250,000 BP [years Before Present] and the end of the last ice age. This *Ebbsfleet Elephant* (Southfleet Road) volume stretches even further into the prehistoric past, focussing on the period between about 425,000 and 360,000 BP, associated with one particular episode of climatic warmth within the wider framework of Pleistocene climatic change, the Hoxnian interglacial of marine isotope stage (MIS) 11. This framework is recapped here, since it provides the essential background to discussion of the period and interpretation of the remains found at the Southfleet Road site.

The initial Palaeolithic occupation and subsequent settlement of Britain occurred during the Quaternary, a period of time characterised by the onset and recurrence of a series of alternating cold–warm/glacial–interglacial climatic cycles (Lowe and Walker 1997). Over 60 cycles have been identified during the last 1.8 million years, corresponding with fluctuations in proportions of the Oxygen isotopes O^{16} and O^{18} in selected foraminifera from deep-sea sediment sequences. These marine isotope stages (MIS) have been numbered by counting back from the present-day interglacial, or Holocene epoch

(MIS 1), with interglacial peaks having odd numbers and glacial peaks even numbers (Fig. 2.1). Peaks and troughs representing specific stages have been dated by a combination of radiometric dating and tuning to the astronomical timescale of orbital variations, which are now regarded as the fundamental causative agent of the climatic fluctuations represented in the MIS sequence (Hays *et al.* 1976; Martinson *et al.* 1987). These stages are now the yard-stick by which Quaternary scientists (and Palaeolithic archaeologists) consider the evidence and contemplate correlations between sites.

The Quaternary is divided into two epochs: the Holocene and the Pleistocene. The Holocene represents the present-day interglacial, covering the warm period since the end of the last ice age *c* 10,000 BP. The Pleistocene represents the remainder of the Quaternary, and is divided into early, middle and late parts. The great difficulty and fundamental challenge in Pleistocene geology is to match the discontinuous terrestrial sequence, represented in sparse and isolated outcrops of surviving sediment, with the MIS record derived from continuous deep-sea sediment sequences. On the British mainland, surviving Pleistocene sediments have been (where possible) collated into a sequence of named glacial and interglacial periods, and these have so far as possible been tied in with the MIS framework (Fig 2.1). It is generally agreed that deposits attributed to the last glacial (Devensian) are represented in MI Stages 2–5d, dating from *c* 10,000–115,000 BP and that deposits of the preceding last interglacial (the Ipswichian) correlate with the short-lived peak warmth of MIS 5e, dating from *c* 115,000–125,000 BP (eg see Bowen 1999). Beyond that disagreement increases (eg, compare Gibbard 1994 with Bridgland 1994). However most British workers currently feel confident in accepting that the widespread till deposits of the major Anglian glaciation, when ice-sheets reached as far south as the northern outskirts of London, correlate with MIS 12 which ended abruptly *c* 425,000 BP (Shackleton 1987; Bridgland 1994).

The Palaeolithic in Britain covers the timespan from initial colonisation in the late Lower or early Middle Pleistocene, possibly as long ago as MIS 21 *c* 850,000 BP based on artefacts from deposits at Happisburgh on the Norfolk coast (Parfitt *et al.* 2010a, but see Westaway 2011 and response by Preece and Parfitt 2012), to the end of the Late Pleistocene, corresponding with the end of the last ice age some 10,000 years ago. Thus the

Palaeolithic occupies over 800,000 years and includes at least ten major glacial–interglacial cycles and numerous minor cycles which nonetheless represent significant swings of climate for sustained periods (Fig. 2.1). These climatic cycles would have been accompanied by dramatic changes in landscape and environmental resources. At the cold peak of glacial periods, ice-sheets 100s of metres thick would have covered most of Britain, reaching on occasion as far south as London, and the country must have been uninhabitable. At the warm peak of interglacials, the climate was broadly similar to the present day, although sometimes a little warmer based on study of fossil faunal assemblages – in particular insects, molluscs and mammals (Candy *et al.* 2010) – which show an increased presence of what are now slightly more southerly species. For the majority of the time, however, the climate would have been somewhere between these extremes.

The early evidence at Happisburgh consists of a very simple core and flake industry made from locally available flint nodules and pebbles. It was presumably made by descendants of *Homo erectus/ergaster*, or their descendants, known to be present in Africa and central Europe between 2 and 1.5 million years BP (Gabunia *et al.* 2000). Given

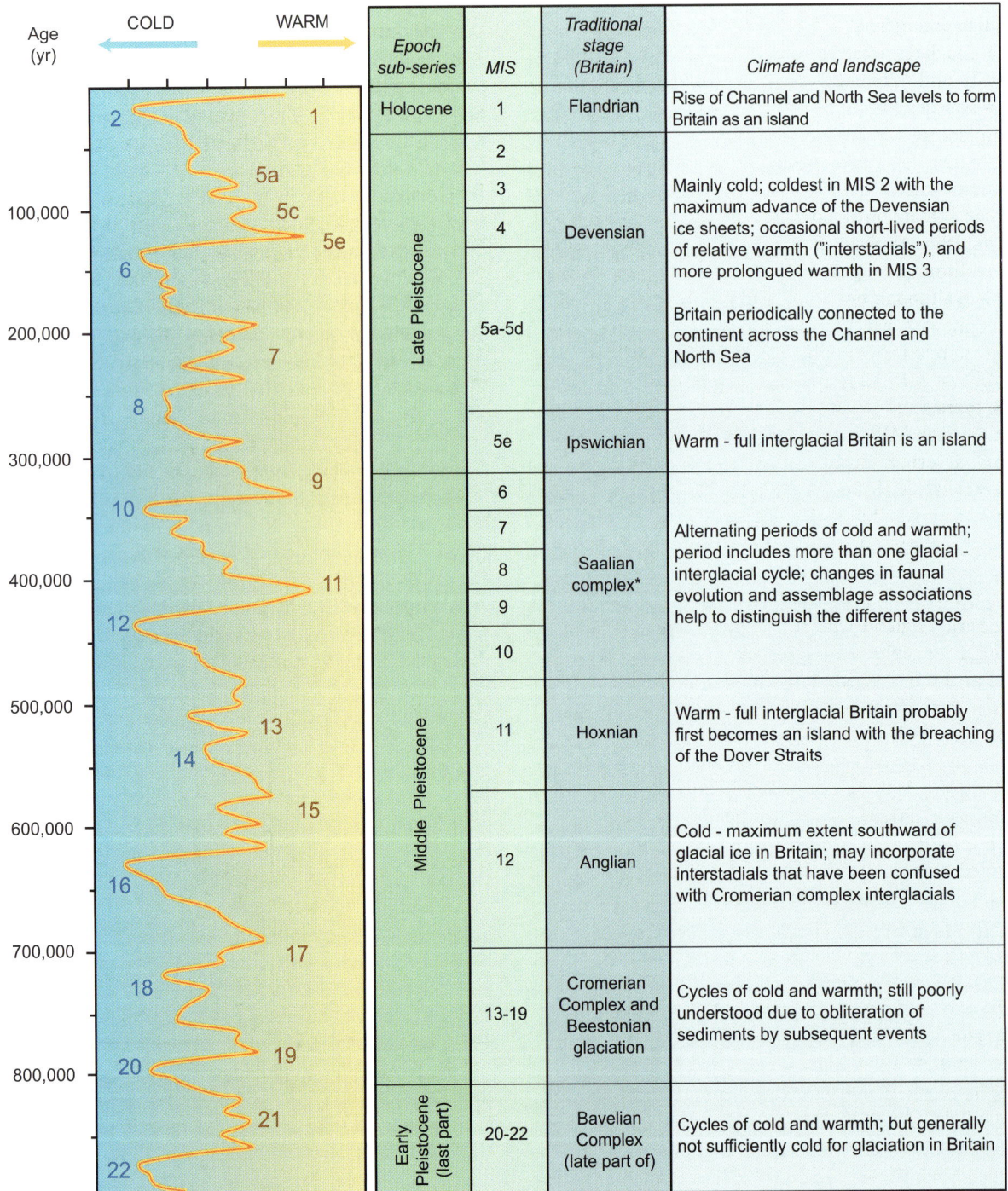

* Netherlands terminology - undefined in Britain

Figure 2.1 British Quaternary and Marine Isotope Stage framework

the lack of hominin remains from Britain and north-west Europe at this time, it is not possible to identify the species involved at Happisburgh, or to establish whether it descends from the *erectus/ergaster* line, or is related to *Homo antecessor*, known to be present in Spain *c* 1 million to 800,000 BP (Klein 2009).

Similar evidence from Pakefield, another early East Anglian interglacial site perhaps dating to MIS 17 to 19 (Parfitt *et al.* 2005), shows the sustained occupation in Europe of this early population, probably expanding their range northwards as climate warmed, but dying out in their northern range as climate cooled (Dennell *et al.* 2011). Following these early occurrences, there are a number of UK sites dating from the immediate pre-Anglian interglacial MIS 13, *c* 500,000 BP associated with the later western European *Homo heidelbergensis* (Pettitt and White 2012). At Boxgrove in Sussex an extensive area of undisturbed evidence from handaxe manufacture and faunal exploitation is associated with a rich range of other palaeo-environmental indicators (Roberts and Parfitt 1999).

The climate must have been too harsh for occupation during the Anglian glacial of MIS 12, but after it Palaeolithic occupation becomes more frequent in Britain, although not continuous. Numerous sites, some of them with exceptional quantities of lithic remains, attest to relatively prolific occupation in the period from the end of MIS 12 to MIS 8 (Wymer 1988; 1999). After this, there seems to have been a decline of activity in the UK through MIS 7, and it appears that Britain may have been deserted during MIS 6 and the Ipswichian interglacial, MIS 5e (Wymer 1988; Ashton and Lewis 2002; Stringer 2006). Until recently, the earliest post-Ipswichian presence in the UK was thought to have occurred in MIS 3, between *c* 60,000 and 40,000 BP. At this time there are a number of sites in Wales and southern England with distinctive *bout coupé* handaxes, thought to represent a late Neanderthal population, as well as the East Anglian site of Lynford (Boismier *et al.* 2012). A newly discovered site at Junction 2 of the M25, near Dartford, Kent suggests, however, that there were probably also earlier Neanderthal incursions into the UK during MIS 5d-5a, around 100,000 BP (Wenban-Smith *et al.* 2010).

The British Palaeolithic has for a long time been divided into three broad, chronologically successive stages, Lower, Middle and Upper, based primarily on changing types of stone tool (see for example Wymer 1968; 1982; Roe 1981) (Table 2.1). This framework has its origins in the 19th century (de Mortillet 1869; 1872), developed before any knowledge of the types of human ancestor associated with the evidence of each period, and without much knowledge of the timescale. This tripartite division has (with minor modifications) nonetheless broadly stood the test of time, proving, at least across Britain and north-west Europe, both to reflect a general chronological succession of lithic technology, and to correspond with the evolution of different ancestral hominin species. Typical Lower and Middle Palaeolithic remains have been shown to date

before 40,000 BP and to be associated with the extinct Neanderthal lineage and their ancestors ('Archaic' *Homo*). Upper Palaeolithic remains date from after about 40,000 BP, after which no Neanderthal remains are known from northern Europe, and are associated with the appearance of anatomically modern humans.

It has, however, become clear in recent decades, with improved dating and lithic analysis of several key sites, that the definition and distinction of Lower and Middle Palaeolithic are less clear-cut in Britain than has hitherto been thought. Earlier 'Lower Palaeolithic' sites embrace a variety of lithic technologies besides handaxe manufacture. Typically 'Middle Palaeolithic' Levalloisian technology has its origins much earlier than previously realised, occurring alongside (typically 'Lower' Palaeolithic) handaxe manufacture, for example at Red Barns, in Hampshire (Wenban-Smith *et al.* 2000), and even at the classic 'Lower' Palaeolithic locality of Swanscombe where a recently investigated site dominated by handaxe-making evidence includes a very Levalloisian-looking core (Wessex Archaeology 2006a). It also seems that later handaxe industries persist alongside fully developed Levalloisian technology in Britain in the period MIS 8-7, for instance at Harnham, Wilts (Whittaker *et al.* 2004; Bates *et al.* in prep.) and Cuxton, Kent (Wenban-Smith *et al.* 2007; Wenban-Smith *et al.* forthcoming). On current evidence, both these sites date to *c* 250,000 BP, or younger, contemporary with Levalloisian activity in the Ebbsfleet Valley (Wenban-Smith *et al.* forthcoming) and Purfleet, Essex (Schreve *et al.* 2002).

In light of these problems, it is perhaps better to talk about a combined Lower/Middle Palaeolithic for the post-Anglian and pre-Ipswichian period, and to reserve the term 'Lower Palaeolithic' for pre-Anglian phases of occupation. After the Ipswichian absence, it seems that *bout coupé* handaxes are specifically associated with occupation from MIS 3 in the middle of the last (Devensian) glaciation (White and Jacobi 2002). So, whether or not labelled 'Middle', they genuinely represent a distinct post-Ipswichian phase of later Neanderthal occupation, prior to the Neanderthal demise and the advent of modern humans, taken here as 'British Mousterian'. These suggested nomenclatural revisions are summarised in the accompanying table (Table 2.1), which also outlines the correspondence of these cultural stages of the British Palaeolithic with the geological and MIS framework.

SITE AREA: LANDSCAPE, TOPOGRAPHY AND GEOLOGY

The Southfleet Road site (national grid reference TQ 6115 7355) is located 2.5km south of the Thames, on the south-east outskirts of Swanscombe village, north-west Kent, at *c* 30m OD [before excavation] on the upper flanks of the western side of the Ebbsfleet Valley (Fig. 2.2). As shown by geological mapping (British Geological Survey 1998), at a macro-scale the site is situated above an east-west ridge of Chalk that divides

Table 2.1 British Prehistoric periods [occupation was almost certainly not continuous within these periods, for instance there is no current evidence for occupation persisting into MIS 6 at the end of the Lower/Middle Palaeolithic and occupation is also unlikely to have persisted through MIS 8 and 10]

Traditional period	Updated period	Hominim species	Lithic artefacts and other material culture	MI Stage	Date (BC)	UK geo-logical stage
Bronze Age	Bronze Age	Anatomically modern humans (*Homo sapiens*)	Some ceremonial lithic artefacts, barbed-and- tanged arrowheads; Beaker pottery	1	700-2,300	Flandrian
Neolithic	Neolithic		Polished stone axes, leaf-shaped arrowheads; pottery		2,300-4,000	
Mesolithic	Mesolithic		Unpolished tranchet axes, blade-based microlithic and scraper industry		4,000-9,500	
Upper Palaeolithic	Upper Palaeolithic		Blade technology and tools made on blade blanks; personal adornment, cave art, bone/antler points	2–3	9,500-35,000	Late Devensian
Middle Palaeolithic	British Mousterian	Neanderthals (*Homo neanderthalensis*)	The appearance of *bout coupé* handaxes; discoidal flake/core reduction strategies	3–5d	35,000-115,000	Early/Middle Devensian
	-	-	Britain uninhabited	5e	115,000-125,000	Ipswichian
	Lower/Middle Palaeolithic	Early pre-Neanderthals, evolving into *Homo neanderthalensis*	Still handaxe-dominated sites (Red Barns, Cuxton and Harnham), but growth of standardised (Levalloisian) production techniques (eg Crayford; Baker's Hole)	6–10 11a-b	125,000-375,000	Saalian Complex
Lower Palaeolithic			Handaxe-dominated (eg Swanscombe), occasional appearance of proto-Levalloisian techniques; early industry without handaxes (Clactonian)	11	375,000-425,000	Hoxnian
	-	-	Britain uninhabited	12	425,000-480,000	Anglian
	Lower Palaeolithic	*Homo heidelbergensis?*	Handaxe-dominated (eg Boxgrove), with unstandardised flake production techniques and simple flake-tools; occasional flake-tool industries without handaxes (High Lodge)	13	480,000-650,000	Cromerian Complex IV
		Homo heidelbergensis? Homo antecessor? Homo ergaster?	Simple flake/core industries (Pakefield)	13-17	500,000-650,000	Cromerian Complex I-III
			Simple flake/core industries (Happisburgh 3)	18–21?	650,000-850,000	Bavelian Complex (late part of)

the London Basin from the Weald, forming the southern boundary of the main axis of the Lower Thames valley. A minor synclinal fold in the Chalk bedrock under the site contains a sequence of Palaeocene and Eocene deposits (Thanet Sand, Woolwich Beds, Blackheath Beds and, in places, patches of London Clay) overlying the Cretaceous Chalk. It forms higher ground to the south-west of the site, where the clayey capping of Woolwich Beds and London Clay has resisted erosion.

The local landscape has been heavily quarried since the 19th century, primarily for chalk for cement manufacture, although the overlying clays and gravels were also exploited for brick manufacture and general building ballast. The site itself survives on an unquarried outcrop between the large quarried areas of Eastern Quarry (to its west) and Baker's Hole (to the east). Pre-quarrying geological and topographic mapping (Fig. 2.3) shows the ground rising steeply to the west of the

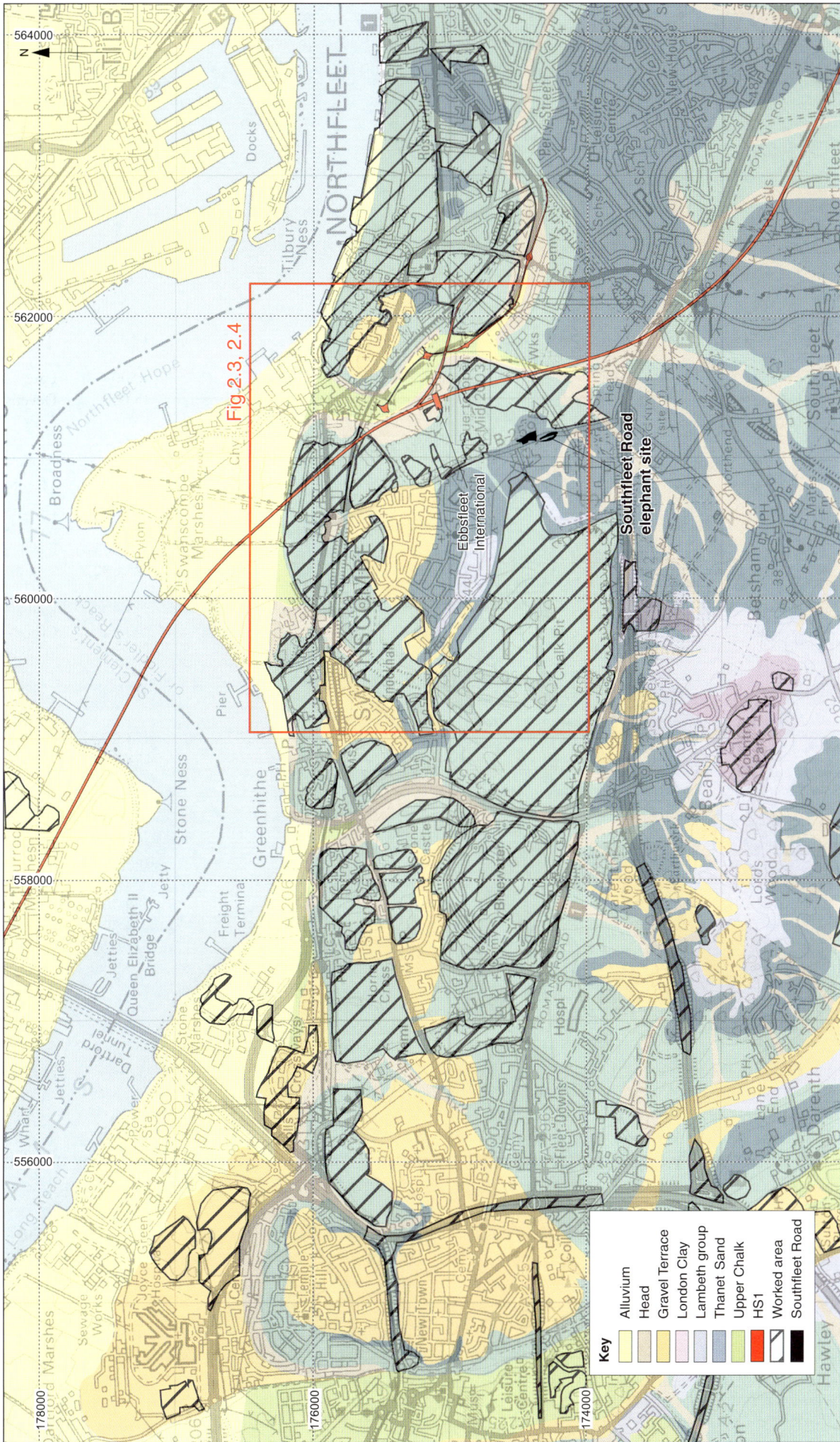

Figure 2.2 Landscape and geological setting of the Southfleet Road elephant site

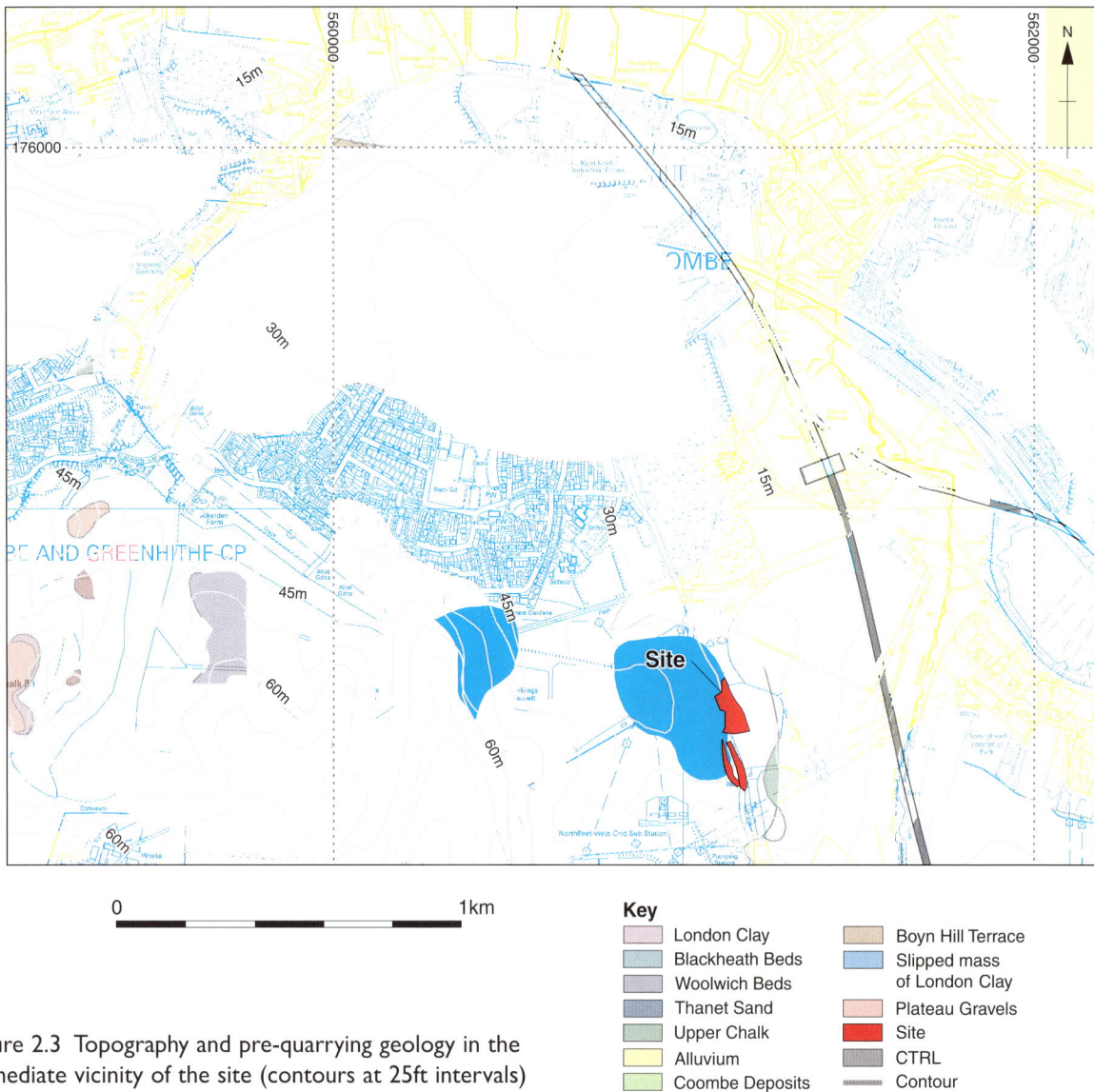

Figure 2.3 Topography and pre-quarrying geology in the immediate vicinity of the site (contours at 25ft intervals)

Key

London Clay		Boyn Hill Terrace	
Blackheath Beds		Slipped mass of London Clay	
Woolwich Beds		Plateau Gravels	
Thanet Sand		Site	
Upper Chalk		CTRL	
Alluvium		Contour	
Coombe Deposits			

site to form what was known in the 19th century as Swanscombe Hill, which was a loosely landscaped wooded park also known as Swanscombe Wood or Swanscombe Park. This park had a high point of about 90m OD formed of London Clay, capping shelly clay and gravel deposits of the Woolwich and Blackheath Beds, overlying a thick body of Thanet Sand. A clayey mass fans out towards the site down the west slope of Swanscombe Hill, mapped in the early 20th century as 'slipped mass, mainly London Clay' (Geological Survey 6' series, edition of 1910, sheet X-NW). The site itself is, according to both early and the most recent geological mapping, situated at the eastern tail-end of the slipped clayey mass, which is underlain by a sand body attributed (wrongly, as it later proves, see Chapter 4) as Thanet Sand. To the east of the site, the pre-quarrying Ebbsfleet Valley contained a continuation of this 'Thanet Sand' outcrop, fading into a north-south spur of Chalk as the ground surface dipped westward. This truncated what was evidently regarded as the broadly horizontal junction between the base of 'Thanet Sand' and the underlying Chalk

SITE AND VICINITY: PALAEOLITHIC AND PLEISTOCENE BACKGROUND

Pleistocene overview

The Lower/Middle Palaeolithic background in the area of the site is inextricably linked with the Pleistocene geology (Fig. 2.2). This is also true for the recovery of flint artefacts from Pleistocene deposits that underpins our recognition of their age and provides contextual information on climate, environment and depositional processes associated with their burial. Patches of high-level gravel (interpreted as terrace deposits of uncertain, but early, Pleistocene age) overlie the Palaeocene high ground to the south-west of the site, and these extended into the now-quarried Eastern Quarry, shown as patches of 'Plateau Gravel' on the pre-quarrying geological mapping (Fig. 2.3). However the main Pleistocene formation in the vicinity is the Middle Pleistocene Boyn Hill/Orsett Heath Formation that underlies much of Swanscombe. Initially recognised in the late 19th and early 20th century as the 'Swanscombe 100-ft terrace' and attributed to the 2nd highest terrace of the Lower Thames sequence (Hinton and Kennard 1905), these deposits were attributed to 'Boyn Hill Gravel' in Geological Survey mapping of the 1920s following nomenclature from the Middle Thames terrace sequence (Dewey *et al.* 1924). More recently, the Swanscombe 100-ft terrace/'Boyn Hill Gravel' deposits have been formally included in the Orsett Heath Formation of the Lower Thames (Bridgland 1994). This Formation is preserved on the south side of the Lower Thames as an intermittent east–west trending series of deposits from Dartford Heath through Dartford, Stone, Greenhithe and Swanscombe to Northfleet. The deposits mostly occur between about 22 and 40m OD in this stretch. They consist of predominantly fluviatile loam, sand and gravel units laid down by the ancient Thames in the immediate post-Anglian interglacial period MIS 11, otherwise generally called the Hoxnian, between *c* 430,000 and 350,000 BP, between MIS 12 and MIS 10 (ibid.). At Swanscombe, the Boyn Hill/Orsett/Heath deposits have mostly been attributed (Bridgland 1995: 43) to a basal 'Orsett Heath Lower Gravel' Member and a middle 'Swanscombe interglacial deposits' Member, despite the former also containing interglacial faunal assemblages. The term 'Swanscombe 100-ft terrace' is used throughout the remainder of this volume to refer to the specific outcrop of the Boyn Hill/Orsett Heath Formation that is present less than 1km to the north of the site, underlying much of Swanscombe, and the nomenclature 'Boyn Hill/Orsett Heath Formation' retained for discussion of the wider Formation of which the Swanscombe 100-ft terrace outcrop is a part. As discussed further below, this Swanscombe outcrop is especially rich in Lower/Middle Palaeolithic remains of high importance, and has in particular been intensively investigated at the site of Barnfield Pit.

Besides the Boyn Hill/Orsett Heath Formation (Swanscombe 100-ft terrace deposits), the patches of older/higher undifferentiated terrace and plateau gravels and the clayey landslipped mass, the only other Pleistocene deposits mapped in the vicinity of the site are Coombe 'Head' deposits. These fill both larger Thames south-bank tributary valleys such as the Ebbsfleet Valley, and smaller dry valley systems feeding into these tributary valleys. The Ebbsfleet Valley cuts northward towards the Thames to the east of the site, through the Swanscombe 100-ft terrace deposits, and consequently is filled with younger sediments at progressively lower levels, down to the Holocene alluvial floodplain and its underlying early post-glacial gravel-filled channel. The major spread of Head deposits on the west side of the Ebbsfleet Valley bury outcrops of silts, sands and gravels occurring between about 0 and 15m OD, which mostly represent different phases of fluvial (and at lower levels perhaps estuarine) deposition between MIS 8 and MIS 5e (Bridgland 1994; Wenban-Smith 1995a; Wenban-Smith *et al.* forthcoming). As with the Swanscombe 100-ft terrace deposits, the Head and fluvial deposits of the Ebbsfleet Valley, which have undergone repeated investigation since the later 19th century, contain rich and well-

investigated Lower/Middle Palaeolithic remains, likewise discussed further below.

The lesser trails of Head deposits filling the numerous minor dry valleys that feed into the Ebbsfleet Valley and towards the Thames and Darent basins are not known to conceal any fluvial outcrops nor to have produced any Lower/Middle Palaeolithic remains. Larger dry valleys with deep and well-developed Head deposits may conceal remnants of fluvial terrace systems or palaeo-landsurfaces, as at the M25/A2 junction where a landsurface of Neanderthal occupation dating to MIS 5 has been identified (Wenban-Smith *et al.* 2010). Small dry valleys probably mostly only contain colluvial slopewash deposits from the last glacial and the Holocene, dating between *c* 110,000 and 3,000 BP, as has been established for those feeding down into the Ebbsfleet Valley from the west by recent investigations in Eastern Quarry (Wessex Archaeology 2006a) and the central Ebsfleet Valley (Wenban-Smith *et al.* forthcoming). Any Lower/Middle Palaeolithic remains found in them are likely to be reworked from earlier Pleistocene deposits. These remains are hence of little importance (unless of very distinctive type, such as a *bout coupé* handaxe) beyond attesting a general distribution of the geographical range of Lower/Middle Palaeolithic presence beyond the surviving outcrops, where the majority of evidence has been recovered.

The Swanscombe 100-ft terrace (Boyn Hill/Orsett Heath Formation)

The Swanscombe 100-ft terrace outcrop of the Boyn Hill/Orsett Heath Formation is rich in Lower Palaeolithic archaeological remains, with numerous locations having produced flint artefacts, faunal remains and biological evidence relating to climate and environment (Wymer 1968; Wessex Archaeology 1993; Table 2.2). The best-investigated site is Barnfield Pit (Fig. 2.4, site 1; Ovey 1964; Conway *et al.* 1996), 2km to the north-west of the HS1 Southfleet Rd elephant site. The deposits at Barnfield Pit contained abundant lithic and faunal remains incorporated in stratified fluvial sand and gravel units, accompanied by biological remains and palaeo-environmental evidence (Table 2.3). The lower levels of the sequence (Phase I, Lower Gravel and Lower Loam) are characterised by a non-handaxe industry, identified as Clactonian since the 1920s (Breuil 1926; Wymer 1968; Roe 1981). This is preserved as undisturbed horizons with intact scatters of flint artefacts representing hominin activity in the Lower Loam (Conway *et al.* 1996). The middle levels (Phase II, Lower Middle Gravel and Upper Middle Gravel) are characterised by a handaxe-dominated industry with a strong emphasis on pointed and sub-cordate handaxe forms with a thick, often only partly worked, butt (Wymer 1968, 338-343). One horizon within the middle phase of the Barnfield Pit sequence, at the base of the Upper Middle Gravel, produced an early human fossil skull (the Swanscombe Skull), as well as copious flint artefacts (Ovey 1964). This makes it one of only two sites in

Table 2.2 Swanscombe 100-ft terrace: key Lower/Middle Palaeolithic sites

Site # (Fig. 2.4)	Name	NGR	Acc.	Summary of finds	Reference/s	Notes/comments
1	Barnfield Pit, Swanscombe	TQ 598 745	A	Classic sequence of Clactonian under Acheulian, along with Swanscombe skull; also good faunal preservation, and mollusc-rich in some parts	Smith & Dewey 1913, 1914; Swanscombe Committee 1938	Prolific source of material 1890s to 1960s; mostly quarried out, but deposits preserved in southern parts
a	Skull site, Wymer excavations 1955-1960	TQ 59800 74230	A	Third skull part discovered at channelled junction where Upper Middle Gravel overlay Lower Middle Gravel	Ovey *et al.* 1964; Wymer 1968	-
b	Waechter excavations 1968-1972	TQ 59840 74250	A	Investigation of Lower Gravel and Lower Loam, *c* 50m NE of skull site area	Conway *et al.* 1996	-
2	Globe Pit, Greenhithe	TQ 58850 74600	A	Numerous handaxes, many in fresh condition; well-made, and with varied typology, but provenance uncertain	Dewey 1932; Wymer 1968; Wenban-Smith 2004a	Large collection made by H Stopes (site 758)
3	Dierden's Pit/Yard	TQ 59450 74750	A	Varied artefact collections and rich faunal and molluscan preservation in places; uncertain association of collections with various faunal/molluscan records	Stopes 1900; Newton 1901; Smith & Dewey 1914; Kerney 1971; Wenban-Smith 2004a & 2009	H Stopes (site 65); over 100 handaxes from site held at NMGW in Cardiff
4	Craylands Lane Pit (New/East)	TQ 60150 74700	A	Deep Pleistocene sequence pit recorded by Smith & Dewey (1914) thought by them to be equivalent to upper part of Barnfield Pit sequence, and containing: (a) an assemblage of twisted ovates at one horizon; and (b) a capping clayey gravel containing Levalloisian-looking flakes	Smith & Dewey 1914; Wymer 1968; Wenban-Smith 1999	Fluvial deposits recorded in 1999 along N side of old pit
5	Rickson's/ Barracks Pit	TQ 60900 74250	A	Abundant Clactonian, handaxe and Levalloisian remains recovered, but mostly not with good provenance	Dewey 1932; Wymer 1968	Unclear how sequence relates to Barnfield Pit sequence
6	Swan Valley Community School	TQ 60750 73750	A	Pleistocene river deposits (Boyn Hill/Orsett Heath, Swanscombe Middle Gravels) with abundant lithic artefacts (handaxes, cores and flakes) and some faunal remains	Wenban-Smith & Bridgland 2001	Discovered 1997; excavated 1997-2001
7	Sweyne County Primary School	TQ 60650 73800	A	Clay deposits interpreted as Upper Loam	Wenban-Smith & Bridgland 2001	Watching brief in 1997
8	Eastern Quarry, Area B	TQ 60850 73700	A	Fluvial deposits of Thames and Ebbsfleet with abundant flint artefactual remains; thick sequences of undisturbed material, deeply buried	Wessex Archaeology 2006a & 2009a	

Table 2.2 (continued)

Site # (Fig. 2.4)	Name	NGR	Acc.	Summary of finds	Reference/s	Notes/comments
12	Bevans Wash-Pit	TQ 610 735	A	22 handaxes and 4 debitage	Spurrell 1890; Wenban-Smith 2004a [Stopes Catalogue, sites 14, 27, 593 & 598]	Two large and fresh pointed sub-cordate handaxes '17ft from the surface'
16	Carreck's ferruginous loam	TQ 6110 7350	A	Has produced occasional very fine mint condition handaxes	Wessex Archaeology 2006b	

Acc: A = accurately known location; E = estimated location; G = general location

Table 2.3 Stratigraphic and archaeological summary of Barnfield Pit sequence, Swanscombe

Phase	MI Stage	Date BP	Stratigraphic unit	Height OD	Palaeolithic archaeology
III	11b–10/10/10–8?	300,000–375,000?	Upper Gravel Upper Loam	c 33–34m c 32–33m	Uncertain; no reliably provenanced material
II	11c–11b	375,000?–400,000?	Upper Middle Gravel	c 28.5–32m	Mostly pointed and sub-cordate handaxes with thick partly trimmed butts (often large and well-made, but also small and crude); also occasional cores, debitage and ad hoc flake-tools – 'Acheulian' [Swanscombe Skull was found at the junction between the Upper and Lower Middle Gravels]
			Lower Middle Gravel	c 26.5–28.5m	
I	11c	400,000–425,000	Lower Loam	c 25–26.5m	Cores, debitage, ad hoc flake tools (often single notches), and very occasional crude 'proto-handaxes' – 'Clactonian'
			Lower Gravel	c 22–26.5m	

England with Lower or Lower/Middle Palaeolithic hominid skeletal evidence, the other being Boxgrove in West Sussex (Roberts and Parfitt 1999). The upper levels (Phase III, Upper Loam and Upper Gravel) are not rich in archaeological remains. None is known with secure provenance, although there are anecdotal reports of a white-patinated ovate-dominated industry from the base of the Upper Loam (Smith and Dewey 1913).

Other important Palaeolithic sites nearby within the Swanscombe 100-ft terrace outcrop of the Boyn Hill/Orsett Heath Formation are the New Craylands Lane Pit (Fig. 2.4, site 4), Dierden's Pit/Yard at Knockhall (Fig 2.4, site 3) and the Globe Pit, Greenhithe (Fig. 2.4, site 2), details of the material from which are tabulated (Table 2.2). At the east side of Swanscombe, the basal terrace deposits (Phase I of the Barnfield Pit sequence) are cut through by the Ebbsfleet Valley and the quarry of Rickson's Pit, also known as Barracks Pit (Fig. 2.4, site 5). The latter exposed the terrace deposits in its west face (Dewey 1932), although it probably contained younger (Phase II, or even younger) sediments in its

main part. At all of these sites there are reliable records of abundant Palaeolithic flint artefacts, often recovered in association with biological remains and from different stratigraphic horizons (Wymer 1968). However, the provenance of much of the material is insufficiently precise to make much contribution to present-day research, although it highlights the potential importance of these locales for further investigations.

The Swanscombe 100-ft terrace outcrop of the Boyn Hill/Orsett Heath Formation is currently mapped by the British Geological Survey (1998) as having its southern boundary between the north edge of the large quarried area of Eastern Quarry and the south edge of the Globe Pit, Greenhithe (Fig. 2.4, site 2). This boundary is then shown as continuing broadly eastward through the centre of Swanscombe, passing to the north of the Swan Valley School (Fig. 2.4, site 6). However, fieldwork from 1997 to 2001 at the school, and between 2003 and 2009 in Eastern Quarry, Area B (Fig. 2.4, site 8), has established the presence of Phase II deposits of the Barnfield Pit sequence significantly further south than

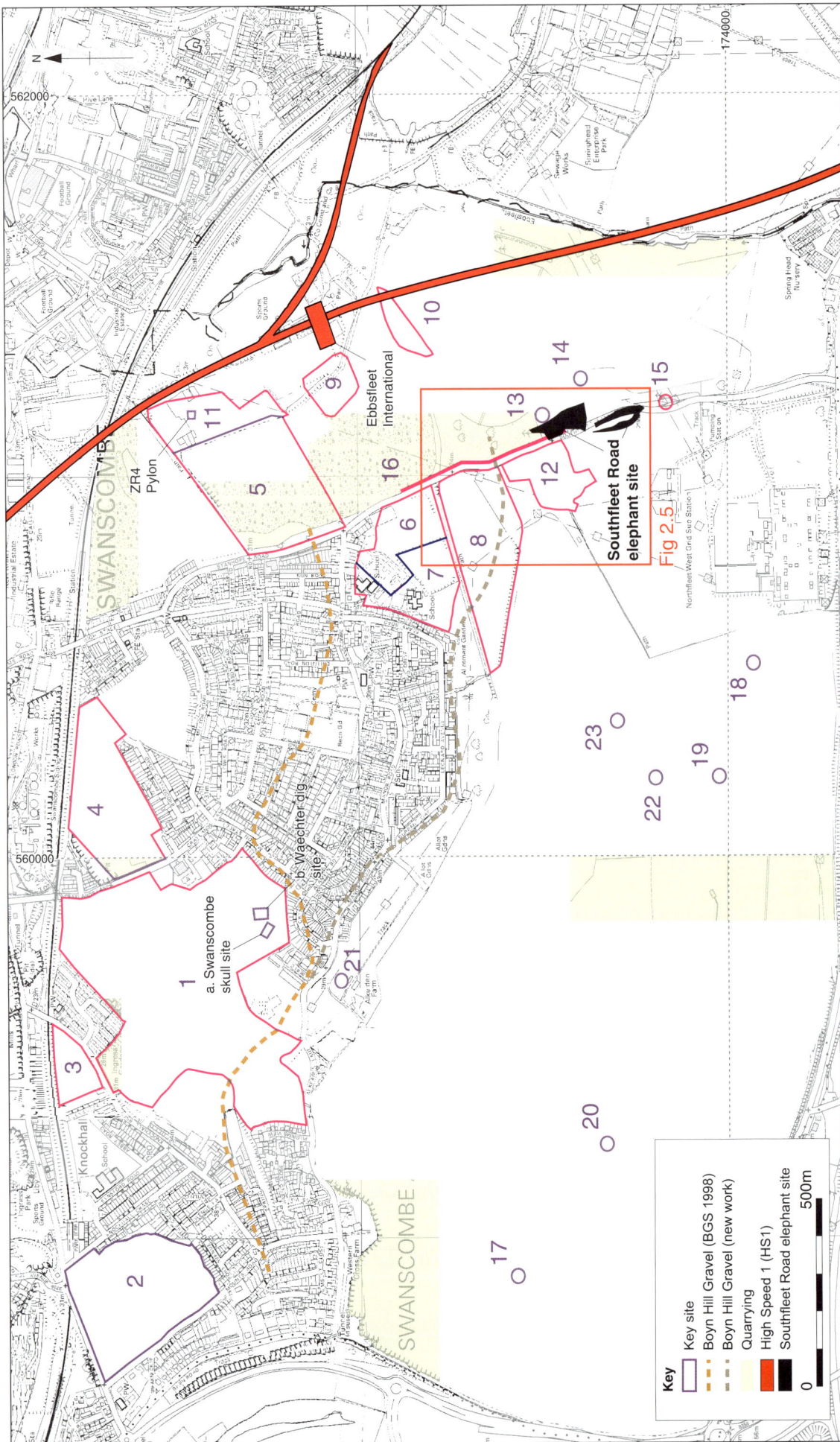

Figure 2.4 Sites in the Swanscombe area: 1 - Barnfield Pit; 2 - Globe Pit, Greenhithe; 3 - Dierden's Pit; 4 - New Craylands Lane Pit; 5 Ricksons Pit (aka Barracks Pit); 6 - Swan Valley Community School; 7 - Sweyne County Primary School; 8 - Eastern Quarry, Area B; 9 - Burchell's Ebbsfleet Channel (Temperate Bed) site [SAM Kent 267b]; 10 - RA Smith's Southfleet Pit Levallois site; 11 - Northfleet Allotments [Baker's Hole SAM Kent 267a]; 12 - Stopes' site 748A 'New Barn Farm'; 14 - Stopes' site 33 'Treadwell's Farm'; 15 - Stopes' site 19 'Treadwell's Hop Ground'; 16 - Carreck's 'Ferruginous Loam'; 18 - Stopes' site 25 'Caerberlarber Hole'; 19 - Stopes' site 29 'Swanscombe Wood'; 20 - Stopes' site 31 'Bartholomew's Hill'; 21 - Stopes' site 588 'Chamber's Farm'; 22 - Swanscombe Wood clay pit; 23 - Swanscombe Hill

mapped (Wenban-Smith and Bridgland 2001; Wessex Archaeology 2009a) in the area mapped as Thanet Sand to the south of the Swanscombe 100-ft terrace outcrop (Fig 2.5). At both sites, fluvial gravels have been found in a number of test pits that are at similar levels to the Swanscombe Lower Middle Gravel and contain non-Wealden clast lithologies indicative of a mainstream Thames origin (analyses carried out by D. R. Bridgland, University of Durham, and T. S. White, University of Cambridge). The gravels also produced handaxes and debitage comparable to material from the Lower Middle Gravel, further cementing their correlation, and consequently extending the southern margin of the Thames channel of MIS 11 significantly further south than recognised in current geological mapping.

An additional relevant record results from rediscovery of the location of the site of 'Bevan's Wash-pit' (Fig. 2.4, site 12), one of numerous sites in the Swanscombe vicinity from which Henry Stopes collected Palaeolithic material in the 1890s. Stopes amassed a massive collection of lithic material, over 100,000 items at its peak. The surviving parts of it have languished unpublished and little-known since the early 20th century in the basement of the National Museum and Galleries of Wales, in Cardiff, due to his untimely early death and the consequent need of his widow to sell his flint collection

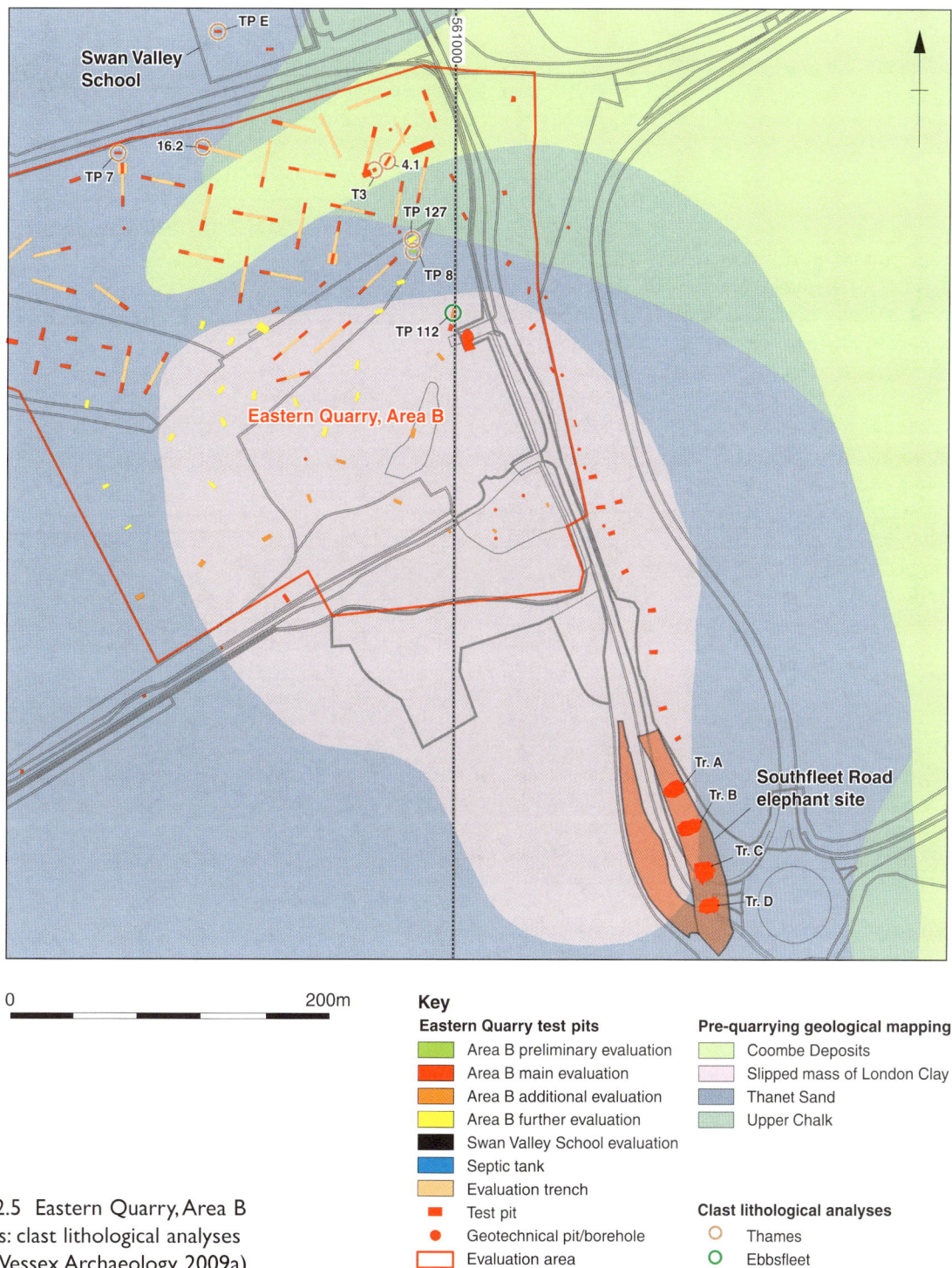

Figure 2.5 Eastern Quarry, Area B test pits: clast lithological analyses (from Wessex Archaeology 2009a)

(Walker 2001). A study was made of this collection between 2002 and 2004, leading to identification of the locations of most of Stopes' sites and analysis of the surviving finds (Wenban-Smith 2004a and 2009). Stopes recovered more than 20 handaxes from his 'Bevans Wash-pit' site. The location of this site can be pinpointed from the ancillary information in Stopes' records ('opposite New Barn Farm') as the brickearth quarry in the slipped clayey mass the other side of Southfleet Road, a short distance to the north-east of the site (Fig. 2.4, site 12). This quarry was also obliquely referenced by Spurrell (1890, cxlv). He provides a description of the stratigraphy as 'masses of brickearth lying on gravel' and also states that 'implements' were found in the brickearth and 'teeth of *Elephas primigenius*' [mammoth; although it is possible that Spurrell has failed to correctly distinguish between mammoth and the extinct straight-tusked elephant *Palaeoloxodon antiquus*, remains of which were, and are, relatively abundant in the Swanscombe area, particularly in the 100-ft terrace deposits to the north-west of Southfleet Road]. Spurrell also states that the gravel under the brickearth can be equated with 'the Dartford Gravel' [ie the Boyn Hill/Orsett Heath Formation in present-day terminology], and can be traced around Swanscombe Hill into the 100-ft terrace outcrop underlying Swanscombe, where it was well exposed in 'the extensive cutting' at Milton Street [ie, Barnfield Pit]. The majority of Stopes' artefacts from Bevan's Wash-pit lack stratigraphic provenance. However two large and fresh condition pointed sub-cordate handaxes are recorded by him as found *in situ* '17ft from

the surface' (Stopes Catalogue, # 598), and therefore come either from towards the base of the brickearth, or from the gravel reported by Spurrell as underlying the brickearth in the pit.

Complementing these records, Carreck (1972, 61) reported a body of 'ferruginous loam' up to 5m deep and extending 365m northward from New Barn Farm in the quarry section along the east side of Southfleet Road (Fig. 2.4, site 16). He suggested, on the basis of height above OD, that it was probably associated with the Barnfield Pit 'Boyn Hill Terrace' sequence, although acknowledging it was not easily equated with any of the known beds.

This accumulation of background records indicates, alongside the more recent research at Swan Valley School and Eastern Quarry, that Pleistocene deposits representing, or broadly contemporary with, the Swanscombe 100-ft terrace extend towards the vicinity of the site. Therefore it is not so unexpected that it has now been shown to contain significant Lower/Middle Palaeolithic remains. They also provide points of reference to contextualise and correlate the deposits later found at the site, as discussed in the remainder of this volume.

The Ebbsfleet Valley: Head and fluvial deposits

To the east of Swanscombe, and to the north-east of the site, the Ebbsfleet Valley is filled (or at least, was once filled, before extensive quarrying) with a varied and complex array of late Middle and Late Pleistocene

Table 2.4 Ebbsfleet Valley: key Lower/Middle Palaeolithic sites

Site # (Fig 2.4)	Name	NGR	Acc.*	Summary of finds	Reference/s	Notes/comments
5	Rickson's/ Barracks Pit	TQ 60900 74250	A	Abundant Clactonian, handaxe and Levalloisian remains recovered, but mostly not with good provenance	Dewey 1932; Wymer 1968	Unclear how stratigraphy relates to Barnfield Pit sequence; probably also contained later deposits
9	Burchell's Ebbsfleet Channel and Temperate Bed site	TQ 61210 74060	A	Middle Palaeolithic artefacts (Levallois flakes and cores) in Pleistocene fluvial deposits with faunal remains and other palaeo-environmental evidence	Burchell 1935 & 1957; Wenban-Smith 1995	SAM Kent 267b
10	RA Smith's 'Baker's Hole' Levallois site (Southfleet/ New Barn Pit)	TQ 61370 73850	A	Abundant Levallois flakes and cores, plus range of fossil fauna including rhino, horse and mammoth	Smith 1911; Wenban-Smith 1995	Site now quarried away
11	Northfleet Allotments Pleistocene site	TQ 61150 74360	A	Pleistocene fluvial deposits with faunal remains and other palaeo-environmental evidence; excavation at 'Site A' in 1970 by British Museum; further work at 'ZR4 pylon' site in 1998 and 2000, for HS1	Kerney & Sieveking 1977; Wenban-Smith 1995; Wenban-Smith *et al.* 2013	SAM Kent 267a; contains CTRL site of ZR4 pylon

*Acc: A = accurately known location; E = estimated location; G = general location

Table 2.5 General Lower/Middle Palaeolithic sites and find-spots near the Southfleet Road elephant site

Site # (Fig. 2.4)	Name	NGR	Acc.*	Summary of finds	Reference/s	Notes/comments
12	Bevans Wash-Pit	TQ 610 735	A	22 handaxes and 4 debitage	Spurrell 1890; Wenban-Smith 2004a [Stopes Catalogue, sites 14, 27, 593 & 598]	Two large and fresh pointed sub-cordate handaxes '17ft from the surface'
13	New Barn Farm	TQ 61260 73550	A	Handaxe (ovate)	Wenban-Smith 2004a [Stopes Catalogue, site 748A]	Found 12ft down in clear brickearth
14	Treadwell's Farm	TQ 61240 73440	E	2 handaxes and 2 debitage	Wenban-Smith 2004a [Stopes Catalogue, site 33]	Surface finds; Treadwell was evidently at New Barn Farmhouse
15	Treadwell's Hop Ground	TQ 61180 73160	E	2 handaxes and 9 flakes	Wenban-Smith 2004a [Stopes Catalogue, site 19]	Surface finds
16	Carreck's ferruginous loam	TQ 6110 7350	A	Has produced occasional very fine mint condition handaxes	Wessex Archaeology 2006b	
17	The Mounts	TQ 58900 73450	G	11 handaxes and 2 debitage	Wenban-Smith 2004a [Stopes Catalogue, site 5]	Probably mostly residual surface finds, now quarried away
18	'Clabber-labber Hole' [aka Caerberlarber Hole]	TQ 60545 72810	A	27 handaxes, 3 flake-tools and 38 flakes	Bull 1990; Wenban-Smith 2004a [Stopes Catalogue, sites 25, 25A-C]	An underground cavern complex entered through a narrow denehole in SE corner of Eastern Quarry, possibly parts of which still survive; finds made on the fields around the entrance
19	Swanscombe Wood	TQ 60300 73000	E	3 handaxes and 9 debitage	Spurrell 1890; Wenban-Smith 2004a [Stopes Catalogue, site 29]	Finds from 'top gravel turned up in planting young trees'
20	Bartholo-mew's Hill	TQ 59500 73200	E	12 handaxes and 3 debitage	Wenban-Smith 2004a [Stopes Catalogue, site 31]	Surface finds
21	Chamber's Farm, Alker-dene [sic]	TQ 59800 73950	G	2 handaxes and 2 debitage	Wenban-Smith 2004a [Stopes Catalogue, site 588]	Surface finds
22	Swanscombe Wood clay pit	Prob. c. TQ 60350 73000	E	Handaxe	Wymer 1968, 352	Various clay pits at different times; other possible locations include TQ 60500 73690, or TQ 59700 73000; apparently 'found 15ft below surface'
23	Swanscombe Hill	TQ 6025 7325	G	4 handaxes	Roe 1968, 185	Prob. surface finds

*Acc: A = accurately known location; E = estimated location; G = general location

sediments. These resulted from fluvial deposition by early northward-flowing channels of the Ebbsfleet and colluvial/solifluction deposition down the sides of the Ebbsfleet Valley (Bridgland 1994; Wenban-Smith 1995a; Wenban-Smith *et al.* forthcoming). The sediments occur at lower levels than the Swanscombe suite of deposits, and mostly date to the younger periods MIS 8 through to MIS 2 (250,000-10,000 BP). These deposits, isolated patches of which still survive in places despite the history of quarrying and the HS1 and Ebbsfleet International station developments, have produced rich Palaeolithic and faunal remains, including prolific Levalloisian flint artefacts and some key fossiliferous locations for MIS 7 (Wymer 1968; Wenban-Smith *et al.* forthcoming). Particularly important locations, shown on Fig. 2.4 and summarised in Table 2.4, include: RA Smith's Baker's Hole Levallois site (Fig. 2.4, site 10); the Northfleet Allotments site, now Scheduled Ancient Monument Kent 267a (Fig. 2.4, site 11), which includes the ZR4 pylon site investigated in 2000 as part of the pre-HS1 programme (Wenban-Smith *et al.* forthcoming); and JPT Burchell's Ebbsfleet Channel Temperate Bed site, Scheduled Ancient Monument Kent 267b (Fig. 2.4, site 9).

Other sites and findspots

As well as these important well-known and well-investigated sites, where artefactual and environmental remains have been recovered from known Pleistocene deposits in the Swanscombe 100-ft terrace and the Ebbsfleet Valley, there are also numerous records of Palaeolithic artefacts (mostly handaxes) recovered as surface finds and/or from forgotten or unknown sites (Table 2.5). Many of these result from the activities of Henry Stopes in the late 19th century, a particularly vigorous collector in the Swanscombe area (Wenban-Smith 2004a and 2009). As well as Carreck's (1972, 61) exposure of 'ferruginous loam' immediately to the north of the site and Stopes' 'Bevan's Wash-pit' site, previously mentioned above, there are three of Stopes' other findspots in the immediate vicinity (Table 2.5 and Fig. 2.4, sites 13, 14 and 15). Two of these were surface findspots, but at one of them (site 13, 'New Barn Farm') Stopes records a handaxe found *in situ* in brickearth, further attesting to the site's Lower/Middle Palaeolithic potential.

There are two more of Stopes' handaxe findspots close to the west of the site (Fig. 2.4 and Table 2.5, sites 18 and 19). These probably represent residual material from activity contemporary with the occupation reflected in the prolific evidence from the Swanscombe 100-ft terrace gravels, discarded on the surface of what would have been the high ground to the south of the Thames at that time. It is, however, also possible that they (all, or partly) represent material of far older age, derived from the earlier Pleistocene gravel outcrops that were once present on the higher ground within Eastern Quarry, and specifically recorded by Spurrell (1890) as occurring at the 'top of Swanscombe Hill'. In addition to these are four other records representing similar surface finds from the general area above the site to its west (Fig. 2.4 and Table 2.5, site 23). Also, three other records, including two more resulting from Stopes' work, from a slightly wider area in the vicinity of the site (Fig. 2.4 and Table 2.5, sites 17, 20 and 21). Likewise, these remains may represent residual evidence of activity on the higher ground above the Thames of the Boyn Hill/Orsett Heath Formation. Alternatively they (or some them) may represent much earlier material derived from the higher level gravel outcrops, possibly of Early Pleistocene age, that capped the high ground in Eastern Quarry. This latter possibility could still be investigated in the vicinity of TQ 575 718, where there are still surviving outcrop of these high level gravels, with a record of at least one nearby handaxe find (Wessex Archaeology 1993, map NWK 4, findspot # 10).

Finally, there is a record (Fig. 2.4 and Table 2.5, site 22) of a handaxe apparently 'found 15ft below surface' from a clay pit in 'Swanscombe Wood'. This would be of more import if it could be determined which of the various clay pits was the source of this find. Various clay pits were opened between the first edition of the OS survey in 1865 and the 1960s, after which Eastern Quarry expanded to engulf the area. These included pits into the London Clay capping the high ground, which would make this a very curious discovery, and, more likely, pits into slipped clayey fans down the sides of Swanscombe Hill, such as Bevan's Wash-pit (Fig. 2.4 and Table 2.5, site 12), already discussed.

Lower/Middle Palaeolithic background overview

It was therefore clear even before work began at the site that it was not in the archaeologically sterile (from the Lower/Middle Palaeolithic viewpoint) area suggested by BGS mapping. Rather, the site was in close proximity to unmapped and poorly investigated Pleistocene deposits that nonetheless had previously produced Lower/Middle Palaeolithic remains, including numerous handaxes and elephant or mammoth remains from Bevan's Wash-pit, a short distance to the north-west. Furthermore, it was also clear that in interpreting any Pleistocene deposits and Palaeolithic remains found at the site it would be necessary to bear in mind the considerable alterations to the topography of the local landscape caused by quarrying. In particular this included the previous existence of a significant area of high ground rising immediately to the west of the site, which was liable to have influenced Pleistocene deposition in the vicinity of the site, on its eastern slopes.

LOWER/MIDDLE PALAEOLITHIC RESEARCH FRAMEWORK

Since the 1990s and the growth of archaeology as a material consideration in the planning process resulting from *Planning Policy Guidance Note 16* (Department of the Environment 1990), the need to curate and manage the archaeological resource in the face of the impact of building and other infrastructural development has stimulated the development of an increasingly formalised framework of research agendas. These specify priorities against which the importance of sites is measured. They also guide decision-making over preservation *in situ* and the allocation of resources for investigation in advance of development.

For the Palaeolithic, the seminal English Heritage publication *Exploring Our Past* identifies three main themes for Lower/Middle Palaeolithic research: physical evolution, cultural development and global colonisation (English Heritage 1991). This strategy document did not go into much detail on how these themes might be addressed and the nature of the relevant evidence. It did, however, echoing Roe (1980), emphasise the importance of undisturbed *in situ* occupational evidence, especially

when in association with biological remains, as the key type of evidence for investigating these questions. This was followed up in the later 1990s by two further documents: (1) *Research Frameworks for the Palaeolithic and Mesolithic of Britain and Ireland* (English Heritage/Prehistoric Society 1999); and (2) *Identifying and Protecting Palaeolithic Remains* (English Heritage 1998).

In the former of these later documents, three revised strategic themes were presented: colonisation, settlement and social organisation, alongside, for each of these major themes, a subsidiary list of research questions and priorities (Table 2.6). Again, however, there was no discussion about the methods by which these issues could be addressed or, critically for curatorial purposes, the most relevant archaeological remains. In the latter document, by contrast, there was a list of 11 criteria for the selection of particularly important

Table 2.6 National primary strategic themes and research priorities for the Palaeolithic in Britain (English Heritage/Prehistoric Society 1999)

Strategic themes	Research priorities
Colonisation and recolonisation	Patterns of interaction with, and impact on, fauna and flora Determination of earliest occupation, and relating this to other well dated deposits across the region and in Northern Europe Examining patterns of recolonisations Tracing relations between Britain, Ireland and NW Europe through archaeological remains, physical anthropology and bio-molecular evidence
Settlement patterns and settlement histories	Establishing when Britain and Ireland were occupied through the Pleistocene Investigating how this settlement history relates to cycles of climatic change through the Pleistocene Investigating changes in landscape use and the organisation of technology in relation to lithic raw materials
Social organisation and belief systems	Application of the chaîne opératoire concept to the analysis of a social technology, rather than just the mechanics of lithic artefact manufacture Investigation of the relationship between social organisation and the spatial distribution of archaeological remains Investigation of regional scale of social systems and social territories, measured through artefact studies

Table 2.7 English Heritage (1998) criteria for recognition of national importance in Palaeolithic sites

Criterion	Details
Human bone	If any human bone is present in relevant deposits
Primary context	The remains* are in an undisturbed, primary context
Period/area rare	The remains* belong to a period or geographic area where evidence of human presence is particularly rare or was previously unknown
Organic artefacts	If any organic artefacts, such as the wooden spear from Clacton-on-Sea, are present
Associated bio-evidence	If there are well-preserved indicators of the contemporary environment (eg. floral, faunal, sedimentological) which can be directly related to remains *
Evidence of lifestyle	There is evidence of lifestyle (such as interference with animal remains)
Stratigraphic relationships	One deposit containing Palaeolithic remains has a clear stratigraphic relationship with another
Artistic evidence	If any artistic representation is present, no matter how simple
Hearths or structures	If any structure, such as a hearth, shelter, floor or securing device survives
Resource exploitation	If the site can be related to the exploitation of a resource, such as a raw material
Artefact abundance	If artefacts are particularly abundant within a particular horizon at a site

* Throughout this table, the word 'remains' generally refers to lithic artefacts; however, in principle it can also cover other evidence of human behaviour, such as: tools made on other materials such as bone, scatters of organic refuse, and structural remains such as paved floor areas or walls

Lower/Middle and Upper Palaeolithic remains (Table 2.7). Even these criteria were primarily aimed at identifying sites of national importance worthy of preservation and protection, although one might reasonably suppose that they also correspond with the types of site most useful for addressing the contemporarily designated research priorities. There also persists a lack of open discussion both on how these remains feed into addressing research priorities, and also most importantly, on the potential of the more abundant instances of less individually important remains to contribute to research. Consequently, a key role in the curatorial process, particularly in relation to Palaeolithic remains, is now played by specialists. These individuals assess the nature and quality of evidence at development sites and attempt to explain to the satisfaction of curators and consultants how the evidence contributes to national and regional research agendas.

Alongside these national curatorial initiatives, the HS1 programme was carried out within the context of its own parallel *Archaeological Research Strategy* (Drewett 1997). This defined five broad landscape zones which the route passes through, and identified five broad archaeological periods. For each period a number of key research objectives was also specified. It was then considered for each of the HS1 landscape zones what the priorities were for archaeological investigation in terms of the period/s and quality of surviving remains represented. This overall HS1 framework was later supplemented for the Section 2 works, between the Ebbsfleet Valley and the St. Pancras terminal, by a more detailed *Research Strategy for Palaeolithic Archaeology and Pleistocene Geology* (Roberts 2000). This provided more specific details of research objectives for different areas of Palaeolithic/ Pleistocene remains along Section 2 between Pepper Hill and the St. Pancras terminus, focusing on remains already known to exist in the central

Ebbsfleet Valley and at Purfleet, in Essex on the other side of the Thames. Under the combination of these two research strategy frameworks, three main research aims P1-P3 were defined for Palaeolithic archaeology and Pleistocene geology in the Ebbsfleet Valley, each with a number of specific subsidiary objectives (Table 2.8).

In contrast to the national framework, these subsidiary objectives incorporated more detail on the nature of remains most relevant to the designated research priorities, and suitable methods of investigation. They provided a vital context within which surviving evidence, as revealed either by targeted evaluations or by unforeseen discovery, could be assessed by specialists and a case made for its potential importance. In the case of the Southfleet Road elephant site, it was not immediately clear that the evidence there was indisputably of high national importance and worthy of significant attention. However there was sufficient evidence to merit some attention within these frameworks. It ultimately became clear not only that the evidence there *was* of high national importance within the context of English Heritage's specified criteria, with potential to contribute significantly to addressing many of the established research priorities in both the national and the RLE frameworks, but also that it was directly relevant to addressing, and perhaps resolving, one specific and long-standing debate in the British Palaeolithic: the so-called 'Clactonian question'.

This debate concerns the interpretation of the technological/typological contrasts between lithic remains from several early Hoxnian sites, and those from several later ones. These contrasts are exemplified in the sequence from nearby Barnfield Pit (Fig. 2.4, site 1), where the lithic remains from Phase I (the Lower Gravel and Lower Loam) comprise nothing but flakes, cores and simple flake-tools. In contrast, those from the directly overlying Phase II deposits (the Lower Middle

Table 2.8 High Speed I research objectives for Palaeolithic archaeology and Pleistocene geology in the Ebbsfleet Valley

Landscape zone research priorities	Site objectives	Details
P1. Investigation of Pleistocene landscape history and evolution	P1a	Clarification of the sequence, lithostratigraphic relationships and geometry of Pleistocene units present
	P1b	Recovery of faunal remains and biological palaeo-environmental evidence from well-provenanced Pleistocene contexts
	P1c	Interpretation of the mode of formation of Pleistocene units
	P1d	Integration/correlation of surviving Pleistocene units with those known from earlier work
P2. Investigation of the range and locations of early hominid activity	P2a	Identification and investigation of undisturbed occupation horizons
	P2b	Recovery of archaeological artefacts from well-provenanced Pleistocene contexts
	P2c	Recovery of faunal remains and biological palaeo-environmental evidence associated with lithic artefacts
P3. Investigation of the effect of climatic and environmental changes on early hominid lifeways and adaptive strategies	P3a	Integration/correlation of previously investigated artefact-bearing horizons and surviving Pleistocene sediments into overall regional framework
	P3b	Integration/correlation of newly discovered artefact-bearing horizons and Pleistocene sediments into overall regional framework

Gravel and Upper Middle Gravel) comprise very numerous handaxes, often well-made with careful shaping and sharp points, with only rare instances of flake/core production (Table 2.3).

The core/flake dominated material from Phase I, subsequently christened Clactonian by Breuil (1926), was interpreted initially by Smith and Dewey (1913) as a culturally distinct non-handaxe industry, occurring earlier within the same interglacial than the subsequent hand-axe dominated industry characteristic of the Phase II deposits. This initial interpretation was subsequently reinforced by further discoveries of Clactonian material from the same period, in particular at the East Anglian site of Barnham where the Clactonian layers were also sealed beneath deposits containing handaxes (Paterson 1937). Since the 1970s, this 'culture-traditional' (or indeed, 'traditional cultural') explanation has been challenged by a series of alternative suggestions. Singer *et al.* (1973) were the first to specifically suggest the possibilities: (a) that the Clactonian was an early precursor of the subsequent Acheulian industry within the Hoxnian interglacial, the latter industry being the product of the descendants of the manufacturers of the former, or (b) that both industries were essentially contemporary, being complementary facies of a single cultural tradition, that incorporated a differing emphasis on handaxe manufacture and flake/core production in different parts of the landscape. A third possibility was later raised in the late 1970s, that (c) the Clactonian was an integral part of the Acheulian, with the former merely representing large and undiagnostic flakes from the early stages of handaxe manufacture, with the apparent distinction resulting from a spatial separation between the early stages of handaxe manufacture and their subsequent finishing and discard (Ohel 1979). Under continuing contention since the 1970s (Ashton and McNabb 1994; Wenban-Smith 1998; White 2000), this debate has such traction because it revolves around fundamental ideas on the nature of the Lower/Middle Palaeolithic record, and the behavioural processes behind its formation. This is subsequently discussed more fully in Chapter 22.

Chapter 3

Excavation methods, site layout and approaches to analysis

by Francis Wenban-Smith

INTRODUCTION

It is worth putting on record the story of how the site came to be discovered. This tale demonstrates that significant archaeological discoveries are not always the inevitable outcome of the slow grinding of the mills of the heritage management process, but can result from a chancy combination of serendipity, error and the pro-active agency of a tenuous chain of interested and engaged individuals. It also shows how close such an important site came to being overlooked altogether, and thus should also be taken as a report of a 'near-miss' incident, from which various lessons could be learnt for future benefit.

The narrative of the subsequent excavation is then presented here as experienced at the time, acknowledging some conflicts between the various stake-holding parties and detailing how the aims and methods of the work evolved in course of the project as new discoveries were made. Some may not find this to their taste, finding it perhaps overly subjective and insufficiently sterile in its explication of what methods were applied in service of which objectives to collect data to feed interpretive conclusions. While I regret this, for those for whom this is the case, I make no apologies; in fact this is one of the key points of this chapter – to demonstrate that the archaeological project is, and this archaeological project was, socially constructed and socially enabled. I hope this serves as a case-study of how work of this nature is not a dehumanised 'scientific' enterprise, following a somehow inevitable path of discovery and recovery, and presents some of the messy truth of how the elephant site investigation was not tidily conceived and delivered (however much this was in principle strived for). Rather, it careered along a compromise path, buffeted by the conflicting wishes of the various parties involved, revising methods and developing new questions as work progressed and new discoveries were made. This critique is not meant to suggest that good scientific work has not been carried out, or that the results of the project are therefore somehow suspect. In fact I believe important and rigorous work has been done, and good results produced with a sound empirical basis. It is merely meant to throw a more realistic light on how projects such as this progress, and to emphasise the role of individual agency and sociality in the rational business of archaeological science.

For those whose hackles are raised, I appeal to you nonetheless to hold your nose and read it. In the first place, it explains why the things that were done were done, even though they were not always ideal. Secondly, it will introduce you to the layout and stratigraphy of the site, and to the major discoveries made, in the order that they were made. By engaging with the narrative of the project, it will genuinely help you understand, and navigate through, the rest of the contents of the volume, which attempts to present as complete an analysis as possible of what has been described as 'possibly the most geologically complex Quaternary site in Britain seen to date' (Peter Allen, pers. comm.). And thirdly, I would request you to ponder where has been the project that has proceeded rationally from start to finish, and that has not developed, if not changed, its goals and methods as work progressed, and not been influenced by the web of human interaction around and within it. I would hazard a confident guess that similar human stories lie behind the dry output of temples of hard physical science such as Fermilab and the CERN Large Hadron Collider. Much of the same site information is covered in 240 words in the published interim report (Wenban-Smith *et al.* 2006), and many might find it instructive to compare and contrast the impression received from this source with that from this chapter.

Thus some interim analytical results and interpretations presented in this chapter were subsequently revised, or in plainer words, shown to be entirely wrong, although they had a crucial bearing in their original form on the priorities and progression of on-site work. Stratigraphic phasing also follows initial interpretation through this chapter, except where otherwise indicated. The eventual overall interpretation of phasing across the site and its immediate surrounds is presented subsequently (Chapter 4), where there is also a table (Table 4.1) cross-referencing the final version of phasing with the two preliminary versions developed in course of the project, one of which was published in the interim report (Wenban-Smith *et al.* 2006).

DISCOVERY AND PRELIMINARY INVESTIGATIONS: SEPTEMBER–DECEMBER 2003

By early March 2003, targeted fieldwork in advance of HS1 in the Ebbsfleet Valley had been completed, and the archaeological programme had moved to its watching

30.7m

Ward Bdy

Southfleet Road
(pre-2003)

SOUTHFLEET ROAD (New Link)

main E-facing
Section

2003 diversion

Den

main W-facing
Section

29.0m

Key

	Excavation area
	Modern spoil
	Temporary pond
	New road layout
	Old road layout
	2003 diversion
	Top of bank
	Toe of bank

0   50m

N

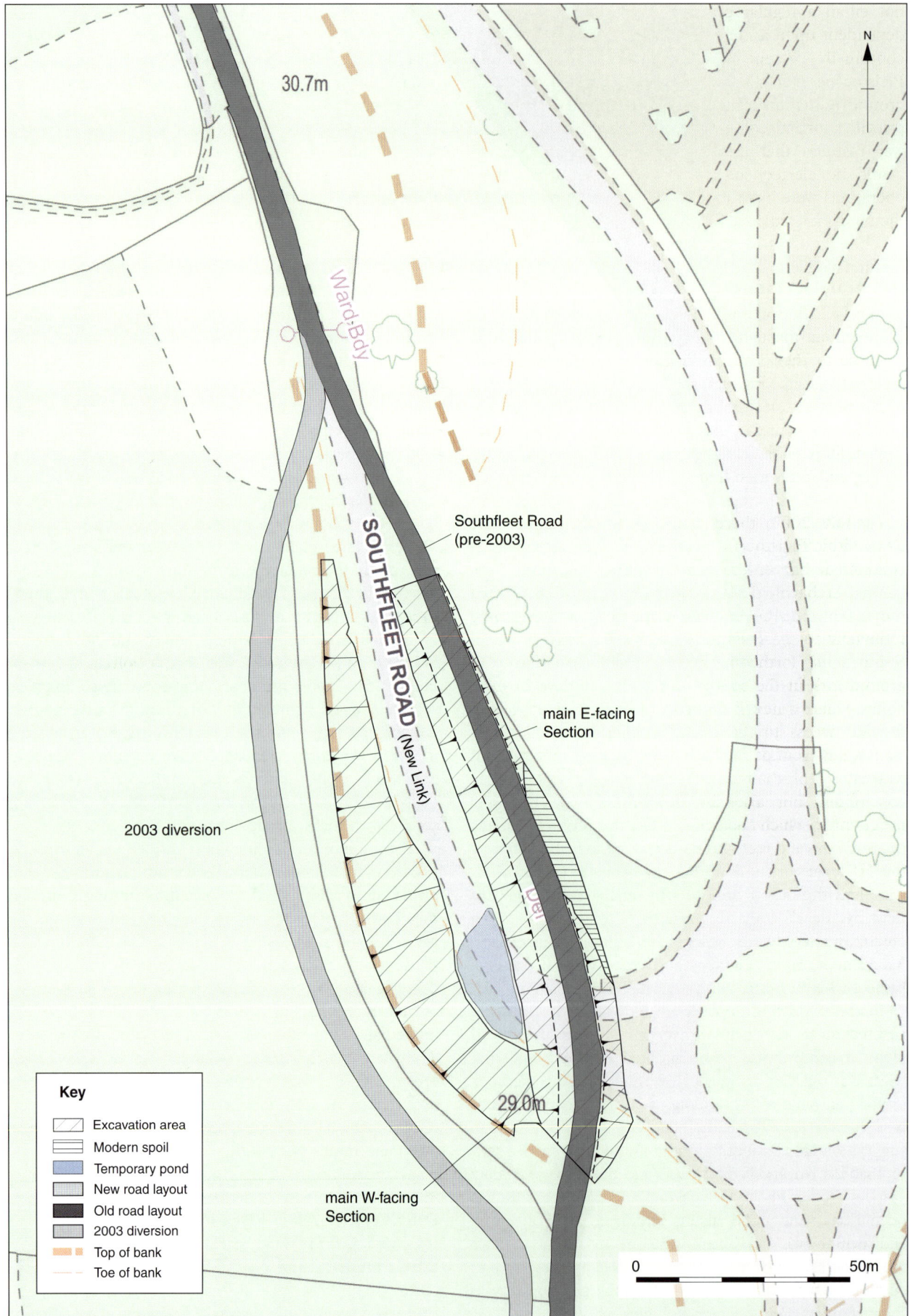

Figure 3.1 Site layout (initial): road diversion and bulk ground extraction at the start of the excavation

brief phase. From the Palaeolithic point of view, this did not entail a regular presence at the site, and it was dependent upon a more general archaeologist who was continually present to identify when a Palaeolithic/ Pleistocene specialist was required. Two specific bulk ground reduction tasks had also been pre-arranged as priorities for Palaeolithic/Pleistocene monitoring, and it was planned that the appropriate specialist (myself) would be alerted shortly in advance of when these operations were likely to take place, to attend and make the necessary records.

The two areas identified for Palaeolithic/Pleistocene monitoring were:

1. The Chalk Spine where the main route of HS1 was being cut through surviving Coombe Rock deposits, close to the location of R.A. Smith's Baker's Hole Levallois site (Fig. 2.4, site 10)
2. An area a short distance to the north of this, where the HS1 route cut through between the face of the Jayflex landfill remediation site and the south end of the large stepped excavation trench 3972 TT (Fig. 1.2b).

By July 2003, these works had taken place, and Palaeolithic/Pleistocene work was now regarded as completed, bar unforeseen discoveries resulting from general archaeological monitoring, and thoughts were moving towards the post-excavation programme, leading ultimately to the *Prehistoric Ebbsfleet* volume (Wenban-Smith *et al.* forthcoming). At this point the area of groundwork in the south-west corner of the Ebbsfleet Valley, where a new link road was being constructed to provide access to the Ebbsfleet International Station from a new roundabout on Southfleet Road (Fig. 3.1), was entirely off the radar. This was true for not only the Palaeolithic but also for the general archaeological programme, which had been focussed since its inception on work more directly associated with the footprint of the HS1 route, the link with the existing North Kent Line and the new station, and also the erection of two new pylons ZR3A and ZR4. The latter were to be erected in part of the Baker's Hole Scheduled Ancient Monument Kent 267a, thus requiring a signficant project (Wenban-Smith *et al.* forthcoming).

In the middle of September 2003 I was monitoring geo-technical test pits in the so-called 'Springhead Quarter' on the high ground on the east side of the HS1 route through the Ebbsfleet Valley, where a new housing estate was shortly to be constructed. This provided a clear view of the HS1 and related operations to the west. In particular I noted what appeared to be a dark reddish/brown gravel body dipping to the north, in an east-facing section revealed by bulk ground extraction on the east side of Southfleet Road (Fig. 3.1).

Previous work at Swan Valley School in 1997 and 1998 and in Eastern Quarry Area B in 2002 had suggested that the Lower Middle Gravel of the Swanscombe 100-ft terrace extended significantly further south than indicated by geological mapping (see Chapter 2) (Wenban-Smith and Bridgland 2001;

Wenban-Smith and CgMs Consulting 2002). It seemed possible, therefore, that this new exposure might represent the southern valley-side edge of the Lower Middle Gravel, and thus define the extent of this archaeologically and geologically significant deposit, and represent an area of high Palaeolithic potential meriting further investigation. However, it was not my remit to instigate this work and I was not optimistic about a suggestion for further Palaeolithic work in advance of HS1 in the Ebbsfleet Valley receiving an enthusiastic response, having recently come to the end of a 5-year programme (as recorded in the *Prehistoric Ebbsfleet* volume, Wenban-Smith *et al.* forthcoming). I was also extremely busy with a number of other projects, at the same time as enduring a bout of ill health, so I did nothing about this exposure.

Two months later, in mid-November, I was carrying out a further phase of investigation in the Springhead Quarter and noted that the afore-mentioned gravel exposure was still extant. Despite the persistence of the same issues that had led to my ignoring it first time round, I suggested to the Kent County Council archaeologist Lis Dyson, who was visiting the Springhead Quarter site, that this gravel should be examined more carefully. Lis had also been one of the Statutory Consultees Group for the HS1 archaeological programme, and was at that time co-directing the Aggregates Levy Sustainability Fund project *Archaeological Survey of Mineral Extraction sites around the Thames Estuary* (Essex County Council and Kent County Council 2004) with which I was also involved.

Lis was suitably impressed that something should be done and liased with Helen Glass, one of the HS1 archaeological team, to allow myself and Peter Allen access to the HS1 works to carry out a closer examination under the aegis of the ALSF mineral extraction sites survey. Permission was granted, and hence on 21st November 2003 Peter and I visited the site for the first time and made a preliminary inspection of the exposed deposits.

Firstly, it was clear that the distant impression of a significant gravel body dipping to the north was correct. A relatively complex sequence of deposits under the gravel was observed, including a yellowish-brown fine-medium sand, with clayey laminations and gravel trails under its highest southern part. This did not appear to be Thanet Sand (as it should have been according to geological mapping). Various other variably clayey and gravelly deposits were also seen. It was also at this point that it was first realised that there were further exposures on the west site of the old Southfleet Road, where bulk extraction had taken place along the route of the proposed new link road from Southfleet Road towards the new roundabout (Fig. 3.1). It had not been possible to extract the deposits from directly under the old Southfleet Road as there were services comprising an old cast-iron water main and a newer plastic gas pipeline running along its sides. Permissions and equipment to dig through and remove these had not yet been obtained. Therefore the site was preserved as a prominent spine of sediment under the old Southfleet Road, about 80m long north-south, 8m wide at the top and 5m high, with sloping sides widening out to about 15m wide at the base (Fig. 3.1). However, significant areas of the exposed sections were obscured by dumped material, and the southern end of the main west-facing section was inaccessible due to partial collapse of the clayey sediments and the presence of a small lake in the cul-de-sac formed by the uncompleted ground extraction there.

Following from this visit, it was agreed that a limited further investigation of the gravels should be carried out as part of the HS1 watching brief programme. Five fieldwork objectives were defined, to:

- Establish the geometry of the exposed gravel deposits in the landscape.
- Retrieve any archaeological remains from them.
- Investigate for the presence/potential of Pleistocene faunal/palaeo-environmental remains.
- Determine the origin and mode of formation of the gravel deposits.
- Establish their date and correlation within the regional framework, in particular whether they are part of the Swanscombe 100-ft terrace, and if so, with which deposit of the classic Barnfield Pit sequence they correlate.

In order to address these objectives five main methods were applied:

- Surveying of the site layout and major stratigraphic boundaries with a Total Station;
- Field recording of sediments, and logging of a representative selection of cleared vertical sections

through the sequence on both sides of the preserved spine of sediments;
- Field examination for artefacts and large faunal remains;
- Sampling for other palaeo-environmental remains;
- Clast lithological sampling of the gravel.

Fieldwork (carried out by Peter and myself in conjunction with Oxford Archaeology, with visiting input from Martin Bates) took place between 9th and 17th December 2003. As described above, the focus at the outset was entirely on the gravel. There was some discussion between ourselves and a visiting Lis Dyson about where the base of the Pleistocene sequence occurred, with suggestions ranging from different depths within the gravel to 'don't know, insufficient information,' which I'm proud to say in the light of later results was my own contribution. The December 2003 fieldwork confirmed that there was much more to the site than just the gravel, which proved to be near the top of a deep and complex Pleistocene sequence (Fig. 3.2; Fig. 3.3; Table 3.1). The gravel was underlain by a sequence of deposits with, at the base, a structureless clayey/silty sand rich in chalk and flint pebbles that was interpreted as a solifluction deposit (Fig. 3.2, deposit 2). This (presumed) evidence of cold climate at the base of the sequence was taken as an indication of Pleistocene age, thus expanding the depth of the sequence of potential Palaeolithic interest below the gravel.

These basal 'solifluction' deposits were overlain by a deposit of sand with wavy sub-parallel clayey lamina-

Table 3.1 Stratigraphic sequence and initial interpretations established during December 2003 fieldwork

Initial deposit group	Sediment	Description and preliminary interpretation
9	Brickearth	Reddish-brown sandy clay-silt, probably of mixed colluvial/alluvial origin, possibly changing from more alluvial at base to more colluvial at top
8	Sand	Parallel-bedded brownish-yellow fine-medium sands — fluvial/alluvial/slopewash?
7	Gravel	Medium/coarse gravel with sand bars from a small northward-flowing river, probably a south-bank tributary of the Thames contemporary with the Lower Middle Gravel of the Swanscombe 100-ft terrace (Boyn Hill/Orsett Heath Formation), known to be present as a major west–east flowing channel *c* 300m to the north of the site
6	Mixed Clay/ Sand/Gravel	Variably pebbly/silty/sandy clay of uncertain origin with some well-developed gravel and sand layers
5	Clay with dark, brecciated upper part	Clay with a dark brown/purple brecciated upper part, representing lake-infill deposits with a palaeo-landsurface developed at the top
4	Clay-laminated Sands	Fine-medium sand with sub-parallel wavy and contorted clay laminae 1– 40mm thick, dipping steeply down to the east in Log 40011 with gravity folding; representing inter mittent fluvial flow and standing water with slopewash influx
3	Clay, clay-silt	Bands of clay, silty in parts, stained grey to brownish-yellow, contorted, and interbedded with sand and gravel horizons
2	Structureless grey clayey/ silty sand with flint and chalk pebbles and Tertiary shell fragments	Interpreted as cold-climate slopewash/solifluction deposits during initial fieldwork, later revised
1	Chalk bedrock	Not seen in the section, but considered as present in the extracted area immediately to the west of the main W-facing section

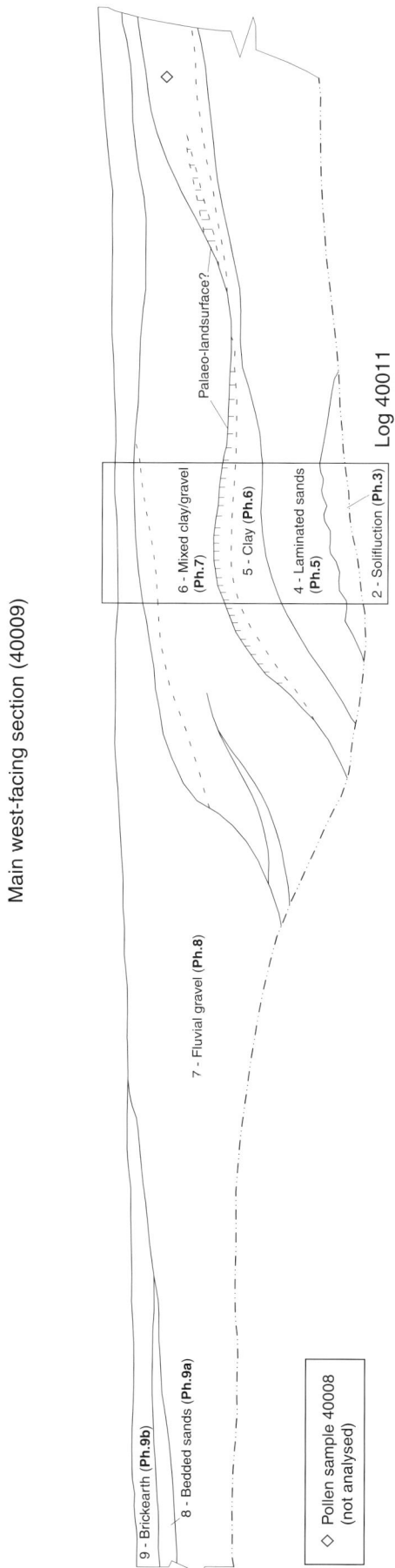

Main west-facing section (40009)

Log 40011

Palaeo-landsurface?

6 - Mixed clay/gravel (**Ph.7**)

5 - Clay (**Ph.6**)

4 - Laminated sands (**Ph.5**)

2 - Solifluction (**Ph.3**)

7 - Fluvial gravel (**Ph.8**)

9 - Brickearth (**Ph.9b**)

8 - Bedded sands (**Ph.9a**)

◇ Pollen sample 40008 (not analysed)

Figure 3.2 Diagrammatic sequence of deposits in the main west-facing Section 40009 as initially recorded (subsequently revised phasing in brackets)

tions dipping steeply to the east with gravity-fold structures (Fig. 3.2, deposit 4). This deposit was interpreted as overbank influx of deposits into a small lake basin. The 'lake-basin' deposit was then overlain by a grey clay (Fig. 3.2, deposit 5), with its top 20-30cm brecciated and coloured very dark brown, almost purple or black in places. This deposit was interpreted as a palaeo-landsurface, rising to the south in the cleared part of the section. The palaeo-landsurface was overlain by a mixed gravelly clay deposit (Fig. 3.2, deposit 6), that in turn was unconformably truncated by the same gravel that had been the original focus of interest (Fig. 3.2, deposit 7). The gravel showed various bedding structures, supporting its interpretation as fluvial. Clast lithological sampling was carried out, although the results were not received until substantially later, in January 2006 (Chapter 6). However, clast orientation studies done in the field indicated a northward fluvial flow direction, and careful field examination of the extensive exposures did not reveal any exotic lithologies characteristic of mainstream post-Anglian Thames deposits. At this point it was therefore considered that these fluvial gravels were most likely to be a palaeo-Ebbsfleet channel heading north to join the Lower Middle Gravel of the Swanscombe 100-ft terrace a short distance to the north, where it was present at a similar height OD in Eastern Quarry and the Swan Valley School (see Chapter 2). The overall sequence was interpreted, purely on these lithological and geo-strati-graphical grounds, as relating to the Hoxnian interglacial, covering from initial warming at the end of the Anglian to more temperate conditions contemporary with the Lower Middle Gravel.

At the northern end of the site, where the palaeo-Ebbsfleet gravel dipped, it was overlain conformably by two further deposits: firstly, a relatively thin bed of brownish-yellow sand (Fig. 3.2, deposit 8) and secondly, a homogenous and structureless reddish-brown slightly sandy clay-silt (Fig. 3.2, deposit 9 – 'brickearth'). The latter could be seen to continue further to the north of the site where the ground surface had been stripped and formed a bank rising up westward towards Southfleet Road. This 'brickearth' deposit was equated with the southern end of Carreck's ferruginous loam, reported by him in exactly this location (Chapter 2).

The far west side of the site was formed by a sloping bank (section 40012) on the west side of the cul-de-sac formed by the cutting for the new Southfleet Road (Fig. 3.1). Having been smoothed and graded by a bulldozer, the sediments in this bank were partly obscured. At the northern end the full exposed sequence was evidently a southerly continuation of Carreck's ferruginous loam, equivalent to the 'Brickearth' of the main site spine. In the southern half, the sequence was more complicated with three main deposits visible, from the base: sand with clayey laminae and occasional thin gravel beds (equivalent to deposit 4 'Laminated sands' of the main site sequence); grey clay with intermittent black staining in its upper part (equivalent to deposit 5 'Clay' of the main site sequence); and this was in turn capped by a

Figure 3.3 Main west-facing Section 40009 as first revealed in December 2003

Table 3.2 Initial pollen evaluation results from December 2003 fieldwork (Oxford Archaeology 2004a, Appendix I)

Deposit group*	Context	Sample <>	Sampling notes	Results**
6	40043	40006	-	A single pollen grain (Poaceae)
5	40039	40007	From upper dark brown/purple brecciated level, the presumed palaeo-landsurface	Eight grass pollen grains (Poaceae) and 29 indeterminate grains
	40029 [later 40100]	40008	From purer, thicker blue-grey clay to south of Log 40011	Not analysed
4	40025	40009	From olive-grey clay lamination	Eighty six pollen grains were identified on the two slides from this sample. The pollen preservation was excellent. Pollen from herbaceous plants dominated the assemblage and included grass (Poaceae), nettle (*Urtica*), goosefoot family (Chenopodiaceae) and stitchwort family (Caryophyllaceae) grains. Some birch (*Betula*), alder (*Alnus*) and hazel (*Corylus*) pollen
		40010	From olive-grey clay lamination	Not analysed
3	40026	40011	Top of grey/orange clay	Not analysed
		40012	Middle of grey/orange clay	Nineteen pollen grains, mainly from herbaceous taxa, were identified on the two slides from this sample. These included pollen from grasses (Poaceae), nettles (*Urtica*), plantains (*Plantago*) and Ericales. Single grains of birch (*Betula*) and oak (*Quercus*) were also recorded. Pollen preservation was again excellent
		40013	Bottom of grey/orange clay	Not analysed
	40027	40014	Middle of lower, more sandy/pebbly clay	Not analysed
		40015	Bottom of lower, more sandy/pebbly clay	Fifty two pollen grains, mainly from herbaceous taxa, were identified on the two slides from this sample. They included pollen from grasses (Poaceae), nettles (*Urtica*), dead nettle family (Lamiaceae) and rock-rose (*Helianthemum*). Some birch (*Betula*), elm (*Ulmus*) and pine (*Pinus*) pollen was also recorded. Pollen preservation was again excellent

* Deposit groups as in initial December 2003 fieldwork   ** All counts based on examination of two slides

Log 40011

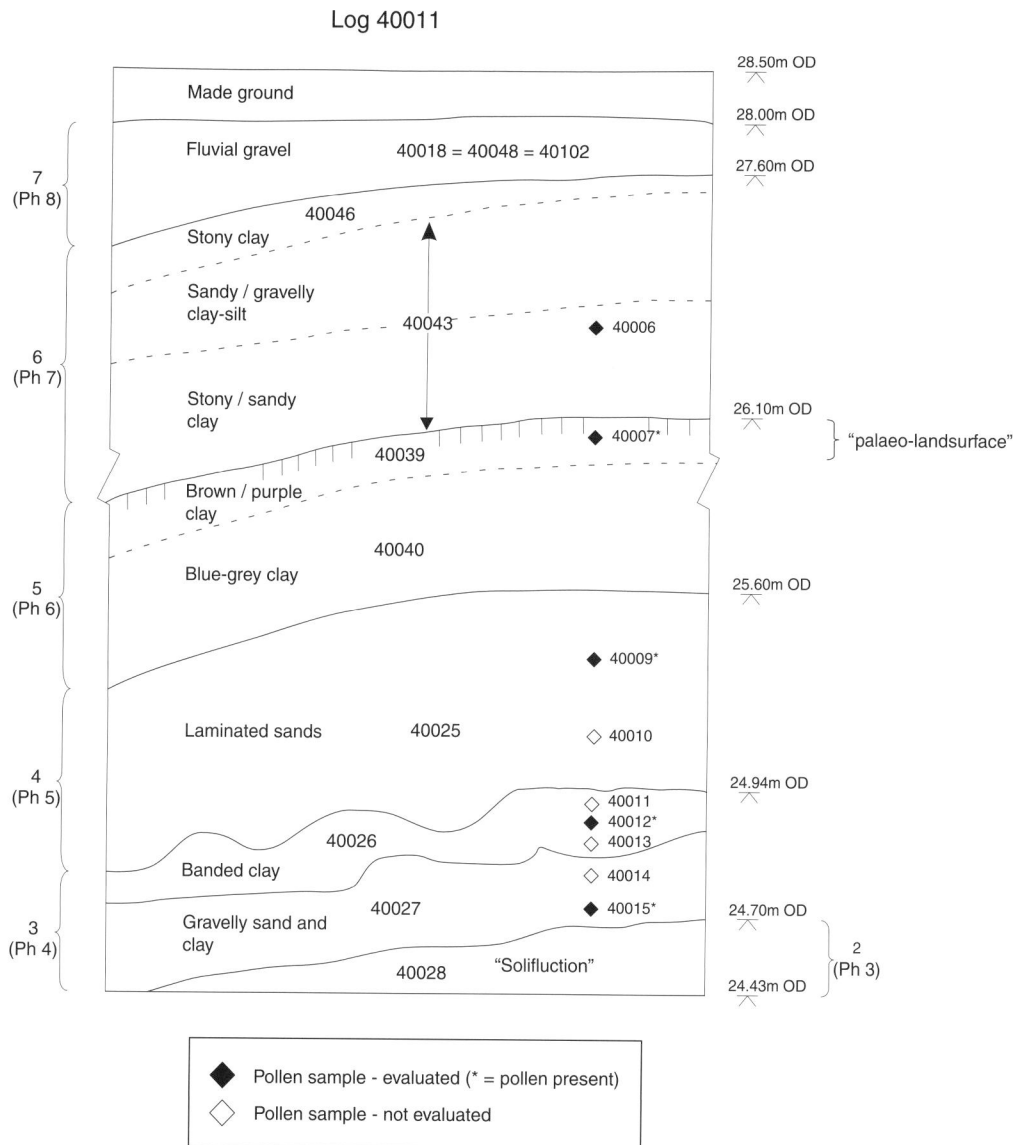

Figure 3.4 Stratigraphic log and pollen sampling in the middle of Section 40009 [Log 40011]

structureles and variably gravelly clay, equivalent to deposit 6 'Mixed clay/gravel' of the main site sequence. Four narrow strips (Fig 3.5, Strips A-D) had been already cleared by machine down the bank, presumably to facilitate drainage. The exposed deposits in these strips were recorded in more detail for the site archive (Section 40012). This sloping bank was unaffected by further works and still survives in the present-day, forming the outside north-east-facing curve of the new Southfleet link road before it joins the roundabout (Fig. 3.11) and is thus still available for study. Therefore no further attention was paid to its examination during the rescue excavation of the rest of the site.

Several artefactual and faunal remains were also recovered. Two large and crude, hard-hammerstruck, mint condition flint flakes were recovered from the grey clay exposed in the sloping bank at the far west side of the site, from a horizon thought to be equivalent to just below the level of the presumed palaeo-landsurface (Fig. 3.5). Although two flakes are clearly inadequate to identify a lithic industry, it was suspected that they

might be associated with Clactonian occupation. This was in light of firstly, their technological characteristics and secondly, their context in a clay underlying a palaeo-Ebbsfleet gravel thought likely to be equivalent to the Lower Middle Gravel, echoing the sequence at Barnfield Pit. Two abraded flint flakes were also recovered from the palaeo-Ebbsfleet gravel (Fig. 3.2, deposit 7). An abraded and white-patinated ovate handaxe was recovered from the ground surface, where the brickearth (Fig. 3.2, deposit 9) that capped the sequence had been disturbed by excavation of a trench for a new water pipeline. Finally, a piece of unidentifiable large mammal bone was found in the basal chalky solifluction deposits (Fig. 3.2, deposit 2).

For environmental sampling, two sets of 75g grab samples were taken through the sequence to investigate for pollen and ostracod remains, focussing on the clayey laminations in the presumed lake infill deposits and the overlying clay deposit with the palaeo-landsurface at its top (Fig. 3.4). No mollusc-bearing sediments were observed, and there was insufficient time at this stage of the fieldwork to systematically investigate for small vertebrate remains; this was, however, identified as a priority for further investigation.

Even before the results of the palaeo-environmental sampling were received, the results of this initial investigation were sufficient to argue that the site was of high Palaeolithic potential, with:

- Evidence of a preserved palaeo-landsurface, with some associated lithic evidence of *in situ* Palaeolithic activity.
- New information on the evolution of the Pleistocene landscape in the Hoxnian interglacial in the nationally important Swanscombe and Ebbsfleet Valley area.
- The possibility of clarifying the context of the rare white-patinated ovate handaxes reported from the Swanscombe area (Smith and Dewey 1913, 193).

The potential importance of the site was then further enhanced in mid-January 2004, when the results of the palaeo-environmental sampling were received. No ostracod or other micro-faunal remains were found, but a number of the samples contained significant numbers of well-preserved pollen grains (Table 3.2). The pollen evidence was regarded as of particular importance for its potential to provide information on the prevailing climate and local environment, to confirm the Hoxnian attribution of the sequence and to provide more precise integration of the sequence with the pollen-based Hoxnian framework established by Turner (1975a). Consequently, during January and early February a 7-week programme of archaeological work was planned, in advance of the site's proposed bulk removal and the continuation of the construction of the new link road.

EXCAVATION PLANNING, OBJECTIVES AND METHODS: JANUARY–FEBRUARY 2004

Six excavation objectives (SO 1-6) were defined (Table 3.3), within the context of the existing framework of national research priorities (Table 2.6) and the HS1 landscape zone Palaeolithic/Pleistocene research objectives for the Ebbsfleet Valley (Table 2.8). To address these objectives, seven elements of fieldwork were initially planned, outlined below together with the recording protocols applied throughout the project. These elements were later modified as new discoveries were made, and working methods were also altered to increase the pace of progress to ensure maximum investigation in the restricted time available. Even though fieldwork eventually lasted until November 2004, this was not anticipated at the outset, and the majority of the excavation took place in an atmosphere of time-pressure, with the worry that work would be imminently curtailed. Unknown to the team on-site, 28 days notice was served to Kent County Council halfway

through the excavation (on 10th of June, retrospectively starting on 21st of May) that work would cease, under paragraph 33 of the Planning Memorandum of the Channel Tunnel Rail Link Act concerning the amount of time allowed to deal with unexpected discoveries. Representations were then made directly to the Secretary of State for Transport by the Secretary of State for Culture, Media and Sport, advised by the statutory curatorial authorities (Kent County Council, Dartford Borough Council and English Heritage), that the site was of exceptional national importance and requesting additional time for recording and excavation. This was fortunately granted, and it is due to these curatorial representations that work on the site was able to be completed on a more extended timescale.

Due to the time-pressure, particularly in the first half of the project, some methods of recovery and recording were adopted in a bid to make quicker progress that it was recognised at the time were not ideal, and which consequently may have compromised some aspects of the resulting archive. The methods applied resulted from series of discussions between myself, Oxford Archaeology, the Statutory Consultees Group, and the HS1 archaeological team advised by their own specialist (Mark Roberts). While these discussions often involved developing a compromise between archaeological recovery and a desire to keep costs and time on site down, there was also substantive support and beneficial input from the HS1 team and their specialist advisor; otherwise the excavation could not have taken place at all. Thus method statements at every stage of the project underwent several iterations, the changing details of which are not itemised here. Despite some complications and compromises, the archive stands comparison with any of the major British Lower/Middle Palaeolithic projects of recent decades, such as: Swanscombe (Conway *et al.* 1996); Hoxne (Singer *et al.* 1993); High Lodge (Ashton *et al.* 1992); Barnham (Ashton *et al.* 1998); and Lynford (Boismier *et al.* 2012) – with the possible exception of Boxgrove, the product of over 10 years of major field seasons between 1983 and 1996 with a cohort of full-time specialists and a substantial excavation team (Pitts and Roberts 1997; Roberts and Parfitt 1999).

The seven elements of excavation work initially planned were:

Element 1 – Main sections, cleaning and recording. It was necessary at the outset to pump dry the pond developed in the southern drainage cul-de-sac to the west of the central site spine. Then, both the east and west faces of the central spine were to be cleared of banked spoil and collapsed talus heaps by mechanical excavator, cleaned by hand and more detailed section drawings made. It was not possible at the outset to relate stratigraphic boundaries between the west and east faces with confidence. The recording of both faces would also provide important information on the overall geometry of the stratigraphic units, once correlations had been made by means of transverse linking trenches (see

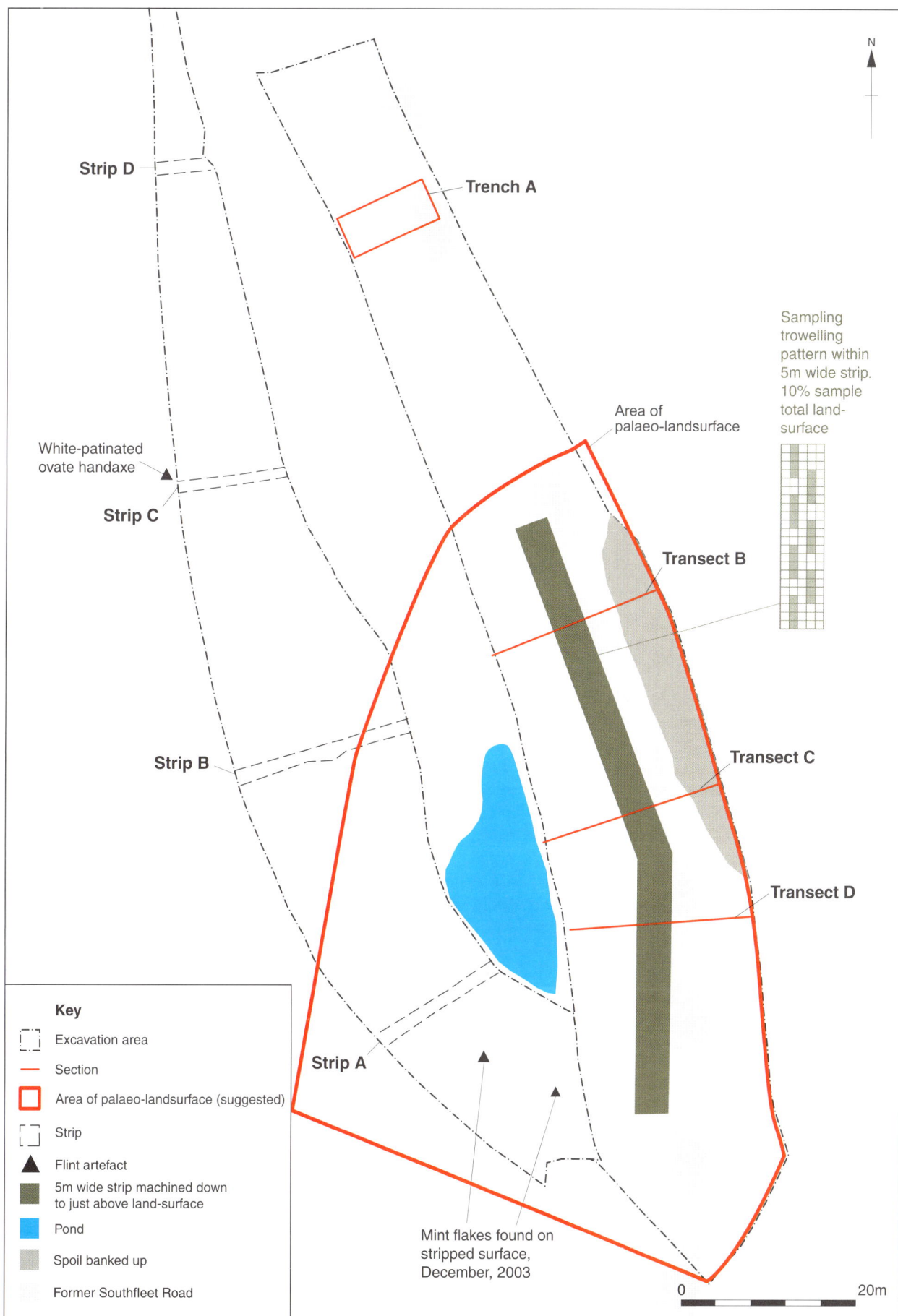

Strip D

Trench A

Sampling
trowelling
pattern within
5m wide strip.
10% sample
total land-
surface

Area of
palaeo-landsurface

White-patinated
ovate handaxe

Strip C

Transect B

Strip B

Transect C

Transect D

Key

Excavation area

Section

Area of palaeo-landsurface (suggested)

Strip

Flint artefact

5m wide strip machined down
to just above land-surface

Pond

Spoil banked up

Former Southfleet Road

Strip A

Mint flakes found on
stripped surface,
December, 2003

0   20m

Figure 3.5 Site layout as initially proposed, with Trench A, Transects B-D and pattern of test pitting for evaluation of palaeo-landsurface

Table 3.3 Site-specific excavation aims cross-referenced with national research priorities (see Table 2.6) and HS1 Palaeolithic/Pleistocene objectives for the Ebbsfleet Valley (see Table 2.8)

Site aims	Details	National priorities	HS1 Ebbsfleet Valley objectives
SO 1	Establish the stratigraphic sequence of deposits and the geometry of major units	2	1a / 1c–d
SO 2	Determine the origin and mode of formation of major stratigraphic units	7	1c–d
SO 3	Establish the date and correlation of major stratigraphic units within the regional and national framework, in particular with the Swanscombe Boyn Hill/Orsett Heath formation and the Hoxnian pollen profile	1–2 6–7	1d 3a–b
SO 4	Investigate changes in climate and local environment through the sequence of deposits at the site	1–2 7	1b–d 3a–b
SO 5	Investigate evidence of human activity on the undisturbed palaeo-landsurface	1 3 5–7 9–10	2a–c 3b
SO 6	Document the presence, distribution and cultural characteristics of artefactual material in all stratigraphic units	1 3 5–7	2b–c

Elements 2, 4 and 6). The faces were cleaned whilst retaining their existing slope for stability, and drawn by hand at a scale of 1:50. The section was drawn by putting a horizontal datum line across the section and measuring up and down from this at regular intervals, to significant stratigraphic boundaries. The drawn record was supported by a digital, colour slide and black-and-white print photographic record. Different stratigraphic units on each section were given unique context numbers, following on from numbers already allocated in the preliminary evaluation.

Element 2 – Palaeo-environmental evaluation. Only those deposits that were recognised and accessible in the preliminary fieldwork had so far been investigated for biological/palaeo-environmental evidence. As a result, some deposits had been shown to contain a small amount of well-preserved pollen and to lack molluscan and ostracods. However, the potential for small vertebrate preservation in the sequence had yet to be established and there also remained several horizons for which pollen (and other palaeo-environmental) potential was unknown. After the main sections had been cleaned and recorded, it was possible to identify the full range of deposits present and to identify those that still had uninvestigated potential for biological/palaeo-environmental evidence. These were then sampled and immediately assessed, to inform the extent and locations of further, mitigating, sampling. Results of the various phases of sampling for different categories of biological remains are discussed in the relevant specialist sections (Chapters 7-12), and a full summary of all environmental sampling is given as an appendix (Appendix 1).

Element 3 – Trench A: machine excavation and bulk sampling in the northern part of the site. A trench with stepped sides (Fig. 3.5, Trench A) would be excavated transversely across the northern end of the central spine

of deposits, linking the east and west faces in the area where bedded sands (deposit 8) and the brickearth (deposit 9) were present. Mechanical excavation would involve taking bulk samples of consistent volume through the sequence for on-site sieving for artefacts and mammalian remains. For the fluvial sands and gravels, a sampling intensity of 500L for every 0.25m thickness of gravels was specified as desirable, although only 100L per 0.25m spit were actually sieved (see Chapter 20). The sections of the trench would be cleaned by hand and drawn at 1:20, and all contexts numbered and described. Optically stimulated luminescence (OSL) sampling of the bedded sands would also be carried out (see Chapter 14).

Element 4 – Transects/Trenches B, C and D: machine excavation and bulk sampling of upper parts, down to palaeo-landsurface. Mechanical excavation was initially intended to proceed from north to south, reducing deposits down to approximately 0.1m above the palaeo-landsurface identified at the top of deposit 5. Bulk sampling and on-site sieving for artefacts and faunal remains was to be as for Trench A, at three transect locations: B, C and D across the main spine of the site, where transverse east-west sections would be temporarily created and recorded (Fig. 3.5). However, this was quickly modified in the field to the more practical (and useful) excavation of three further transverse stepped trenches: B, C and D (Fig. 3.6). This would both create proper sections for recording through the full sequence and allow more accurate attribution of spit bulk samples to horizons in the sections. The upper parts of these trenches would be dug first, down to the level of the clay above the presumed palaeo-landsurface. Then the lower parts would be dug substantially later (see below, Element 6), after the excavation of the palaeo-landsurface horizon (see below, Element 5), which extended from Trench B to the southern edge of the site (Fig. 3.5).

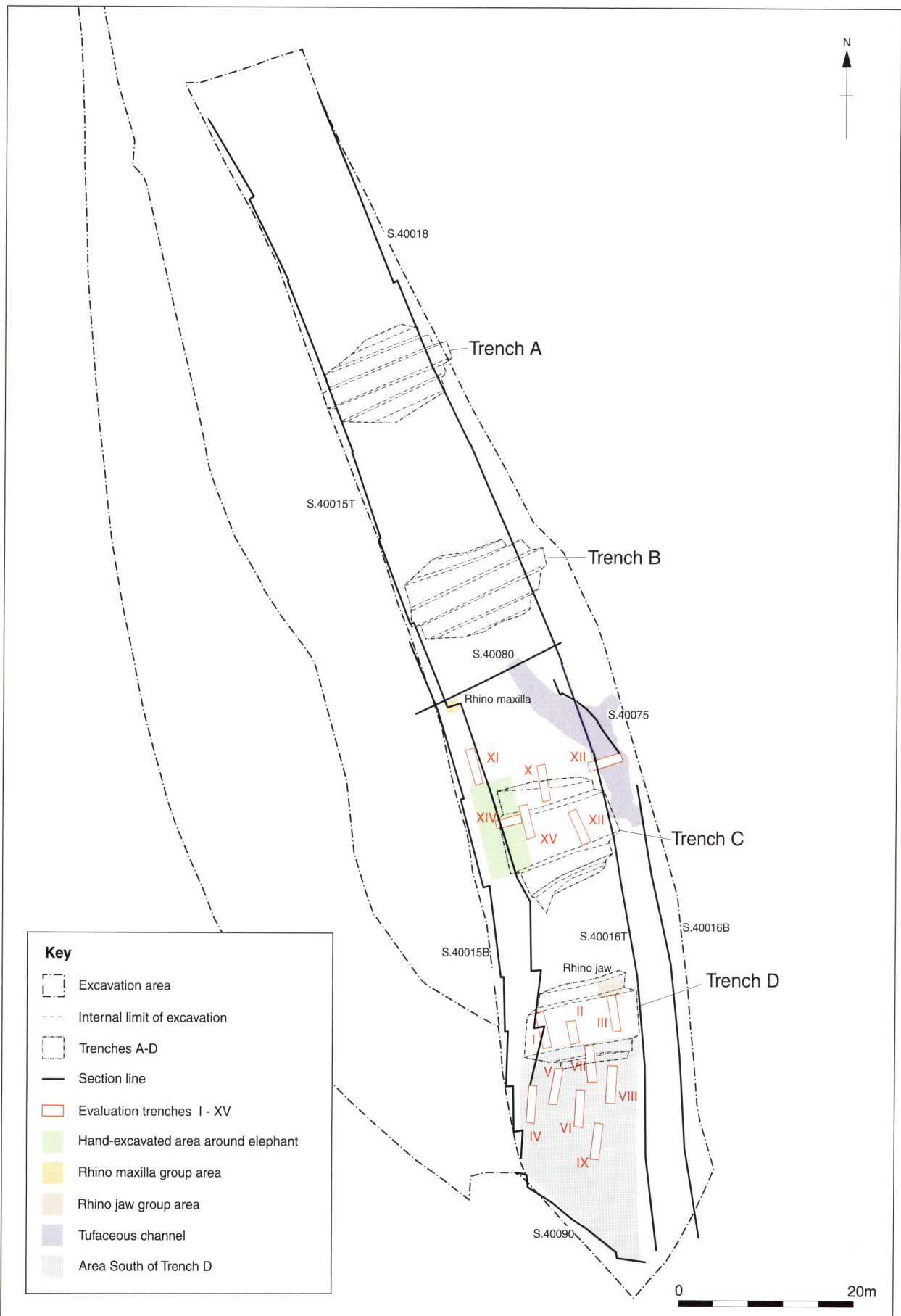

Key

⌐¬ Excavation area

- - - Internal limit of excavation

⌐¬ Trenches A-D

— Section line

▭ Evaluation trenches I - XV

▨ Hand-excavated area around elephant

▨ Rhino maxilla group area

▨ Rhino jaw group area

▨ Tufaceous channel

▨ Area South of Trench D

Figure 3.6 Site layout (main area of excavation) showing: *Paleoloxodon* skeleton and other mega-faunal remains, tufaceous channel, transverse Trenches A–D, evaluation Trenches I–XV and key recorded sections

Element 5 – Hand excavation of palaeo-landsurface. The initial plan was that a grid should be established on the machined level immediately above the palaeo-landsurface. An alternating and overlapping series of north-south test pits 4 x 1m wide would be delicately mattocked across the width of a 5m strip to reveal and evaluate the land surface (see Fig. 3.5), representing an estimated 10% sample of its area. Where artefacts were found, 3-dimensional recording of each artefact would take place and excavation would switch to trowelling, with implementation of a sampling strategy for micro-debitage recovery. In areas where artefacts were not present, delicate mattock excavation would continue downward to at least 0.2m beneath the dark-stained zone marking the presumed palaeo-landsurface. Further open-area trowel excavation would then take place in areas adjacent to artefact concentrations. If no artefacts were revealed, the 10% sample would be regarded as sufficient evidence of the absence of archaeological remains related to the landsurface.

Element 6 – Transects/Trenches B, C and D: machine excavation and bulk sampling of lower parts, below palaeo-landsurface. It was thought at the outset that the sequence of deposits underneath the palaeo-landsurface was where pollen was present and best preserved. After the palaeo-landsurface had been evaluated, and if necessary further excavated, machine excavation would continue down at Transects/Trenches B, C and D, as described above (Element 4) with further bulk sampling and sieving for artefacts and faunal remains at controlled vertical intervals. Sampling for OSL, small vertebrates, pollen and soil micro-morphology would also take place as appropriate. Particular attention would be paid: (a) to sampling for soil micro-morphology, to allow investigation of the extent of soil development and the length of landsurface exposure, and (b) recovery of a continuous sequence of pollen samples (by alternating monoliths) down from the top of the palaeo-landsurface (top of deposit 5) to the base of the Pleistocene sequence (bottom of deposit 2).

Element 7 – Excavation of remaining palaeo-landsurface. Once excavation and grid sampling was completed in the central strip (Element 5), further excavation of the palaeo-landsurface in the adjacent unexcavated areas would only be carried out if/where archaeological remains had been revealed. Initially this was with careful use of mattock/shovel, followed by trowelling and 3-D recording of each individual artefact in areas where they were present.

Site recording protocols

The official site code for the project was ARC 342 W02. Every aspect of the site archive, covering all plans and section drawings, finds and survey data, is tied in with this code. As is normal archaeological practice, different aspects of the site archive have been given unique identifying numbers in sequence, with parallel sequences for different categories of the archive: for instance, sequences for section drawings, environmental samples and finds all began at the same number. Flint and bone finds were not given separate sequences, but were integrated into a single sequence. Almost all site identifying numbers are five-figure numbers in sequences beginning from 40000, although some material from the watching brief phase is '50nnn' or '60nnn' (Table 3.4).

Most flint and bone remains were given an individual small find number (recorded on site within a triangle), with the context identified by being surrounded with an oval, sometimes written on paper with brackets as '(40nnn)'. All individual small finds should have had precise XYZ 3-D positions recorded, using a fixed survey station, although survey records do not survive in some cases. Surveying was not done in relation to the UK Ordnance Survey national grid framework, but the RLE 100 x 100km grid framework unique to the Channel Tunnel and HS1 projects, which nonetheless shared the same Ordnance Datum for height. Consequently, a special Microsoft Excel macro was required to integrate site survey data with other landscape and non-HS1 project data. Some particular clusters of remains were given a group find number, as well as the individual find numbers given to separate remains within a group. Group numbers generally do not have an individual XYZ survey record, although they are the main link with the photographic archive, paper plans and digital photos and survey data taken for subsequent geo-rectification.

Table 3.4 Numbering and quantities of key aspects of the site archive

Aspect	Number sequence in site records	Quantity
Finds Δ.	40001-43943 (main dig)	Flint - *c* 2660
	50001-50189 (watching brief)	Bone - *c* 1350
Sections	40001-40097 (main dig)	115
	40098-40115 (watching brief)	
Samples <>	40001-40422	422; includes full gamut of samples (OSL, bulk sediment, spits for sieving on site, sequences of spot samples, etc.)
Site plans	40001-40036 (main dig)	42
	40037-40042 (watching brief)	
Contexts ()	40001-40178	178 (including several cuts and fills of Holocene archaeological features)

Samples were identified by being surrounded by a diamond, and written on paper as '<40nnn>'.

PROGRESSION OF EXCAVATION: FEBRUARY–AUGUST 2004

Excavations formally began on 23rd February 2004, with the first tasks being to clean and record the main east-facing and west-facing sections, establish a more detailed record of the stratigraphic sequence and carry out sampling for small vertebrates and other palaeo-environmental remains. At this point, the main west-facing section was re-christened no. 40015, and the main east-facing section was broken down into two stretches, the more southerly being no. 40016 and the more northerly no. 40018 (Fig. 3.6). Rather than wait for excavation of Trenches B-D, the parts of the sequence thought to contain pollen (deposit groups 3, 4 and 5 as initially interpreted, equivalent to Phases 5 and 6 of the eventual sequence, see Table 4.1) were sampled by series of overlapping monoliths at various points along the cleaned east-facing and west-facing sections (Fig. 3.7). Likewise, rather than wait for excavation of Trench A, OSL sampling of the bedded sands (deposit 8) was carried out at the northern end of the main west-facing section (see Chapter 14). Progress was initially slow due to a lack of machine-time, since we were relying on the HS1 contractors to provide a machine with driver.

One of the earliest discoveries resulting from the initial section-cleaning was the recognition in the southern part of the main east-facing section (Fig. 4.6) of a 'Tilted block' of sediments. This underlay the rest of the sequence unconformably and appears to represent a detached block of Cretaceous/Tertiary bedrock with a sequence of, from the base, Chalk/Bullhead flint bed/Thanet Sand (Fig. 3.8a). This curiosity is discussed later in this volume (Chapter 4), but is never satisfactorily resolved and it remains unclear whether it is of Pleistocene date or significantly earlier. It must be a clue to structural, and perhaps locally catastrophic, events to the landscape in the site area that may have a significant bearing on its subsequent Quaternary evolution. Perhaps a later worker will one day integrate this isolated and puzzling observation into a wider picture.

Another early discovery was the recognition in the central part of the main east-facing section of some very contorted pockets in the bottom part of the grey clay (Fig. 3.2, deposit 5) filled with a very pale brown calcareous silt/sand (40070) rich in visible molluscan remains (Fig. 3.9). Samples were taken to evaluate for small vertebrate and ostracod remains. These revealed that this deposit, subsequently identified as fill from near the edge of 'the tufaceous channel' (Chapter 4), was also extremely rich in well-preserved small vertebrate remains of all types (see Chapters 7 and 9). Curiously, the main mollusc-rich channel-fill lacked ostracod remains, but it was overlain by white silt (40143) that did later prove to contain ostracods (Chapter 11). This area of the site was then investigated more thoroughly later in the project, with evaluation trenching (particularly Trench XIII), further ostracod sampling, and

Figure 3.7 Overlapping monolith series for pollen sampling (looking south-east)

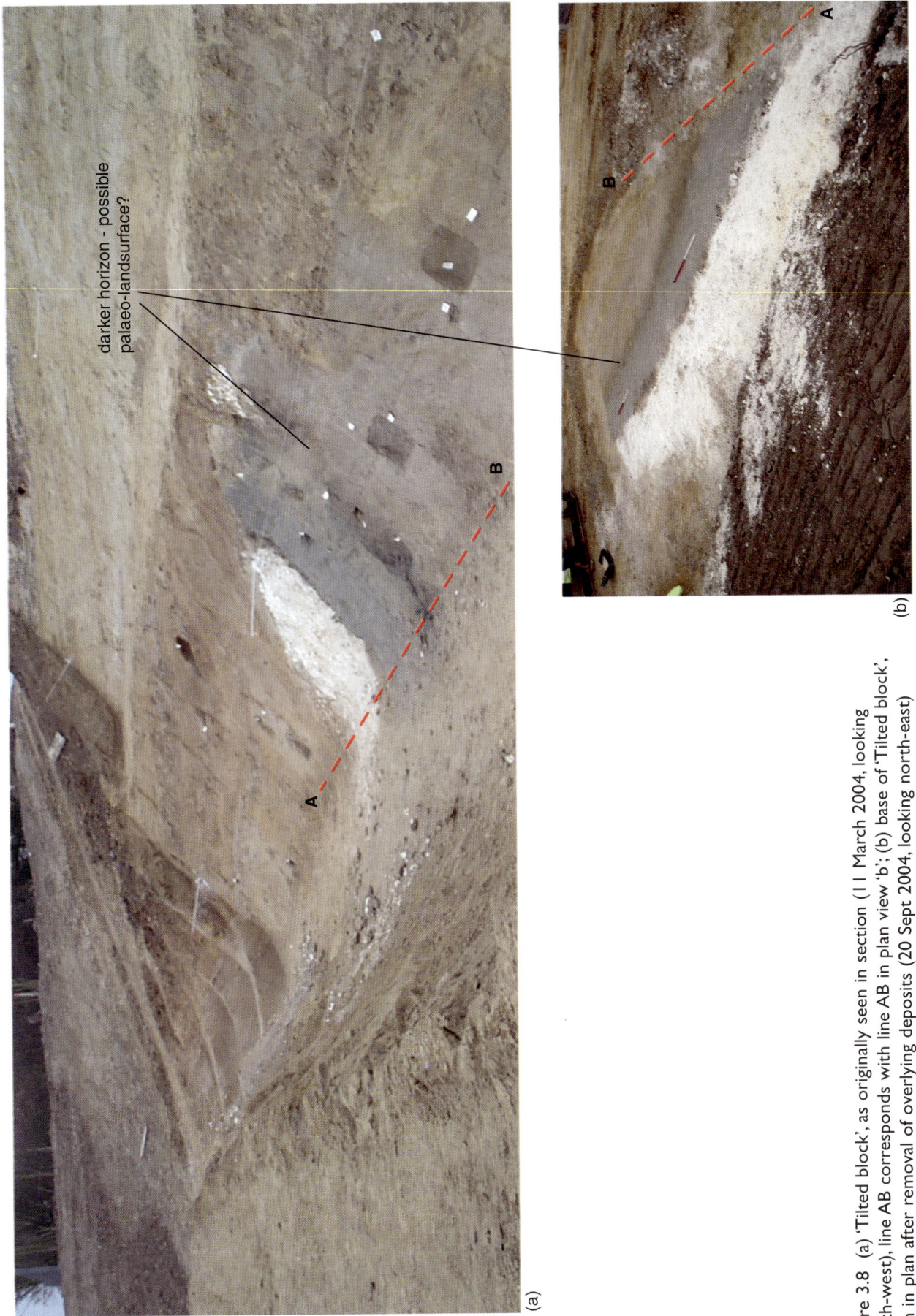

darker horizon - possible palaeo-landsurface?

(a)

(b)

Figure 3.8 (a) 'Tilted block', as originally seen in section (11 March 2004, looking south-west), line AB corresponds with line AB in plan view 'b'; (b) base of 'Tilted block', seen in plan after removal of overlying deposits (20 Sept 2004, looking north-east)

(a)

(b)

Figure 3.9 Tufaceous pockets (context 40070) in main east-facing section, rich in mollusc and small vertebrate remains: (a) general view (looking north-west); (b) closer view of pocket at east side of tufaceous channel (looking west) [50cm divisions on ranging rods]

creation of a longitudinal north-south section through the deepest part of the tufaceous channel. A vertical series of mollusc and small vertebrate samples was taken through the fill sequence.

An additional feature of the early work was the discovery of more handaxes at various locations. One was found in 'made ground' capping the sequence at the south end of the site. This presaged numerous further discoveries as the 'made ground' was removed and undisturbed deposits revealed. Another handaxe, large and in fresh condition (Fig. 3.10), was found on the machine-stripped surface of the sloping brickearth bank to the north of the site. A swift 'fieldwalk' in this area produced another fourteen flint artefacts from nearby, mostly in fresh condition, raising the possibility that another palaeo-landsurface had been discovered. This drew attention to this area as another threatened area of potential Palaeolithic interest. After discussion on-site with the representatives of the Statutory Consultees Group and HS1 archaeologists, it was agreed to investigate this area by machine excavation of three long transects across the stripped surface, which had been obscured by the passage of plant and the effect of surface-water run-off (Fig. 3.11, Transects 1-3). The sequence exposed at the higher west end of Transect 2, next to where the handaxe was found, contained parallel, sub-horizontal beds of varying sand and clay-silt content, possibly commensurate with the presence of palaeo-landsurfaces. The engineering construction plans were then altered to avoid further impact in this area, and therefore no further archaeological investigations

were made here. It subsequently proved impossible to relate most of the deposits seen in the lower, eastern parts of Transects 1 and 2 to those seen at the main site, although they were broadly of similar character to the most important Hoxnian horizons, Phases 3-6 in the final sequencing (Table 4.1). Nonetheless, they did contain deposits of potential interest that also produced a few artefactual and faunal remains (Chapter 21), so this should not be forgotten if further development work is planned in this area.

By the middle of the third week of the project it was clear that it was necessary to hire a separate machine and driver for the archaeological work, as there was insufficient spare capacity for this to be provided by the HS1 contractors. Machine clearing of the main west-facing section, no. 40015, did not, therefore, properly begin until the start of the fourth week of the project, Monday 15th March. The sloping face was being stepped to provide direct access to the full height of the deposits for recording, when, as recorded in the excavation daily journal: "last adjustments revealed a spread of large bones at the level of the step where collided with 'the land surface'. Some of the overlying clay rolled off the step pulling with it some of the remains, including a huge elephant/mammoth tooth...".

This area was then cleaned up by trowel, and the loose sediment with bone remains bagged as a bulk sample. Simon Parfitt (of the Natural History Museum and University College London) and I visited on the following day, and Simon identified that we probably had a skeleton of the extinct straight-tusked elephant

Figure 3.10 Handaxe Δ. 40022 from brickearth bank to north of main site

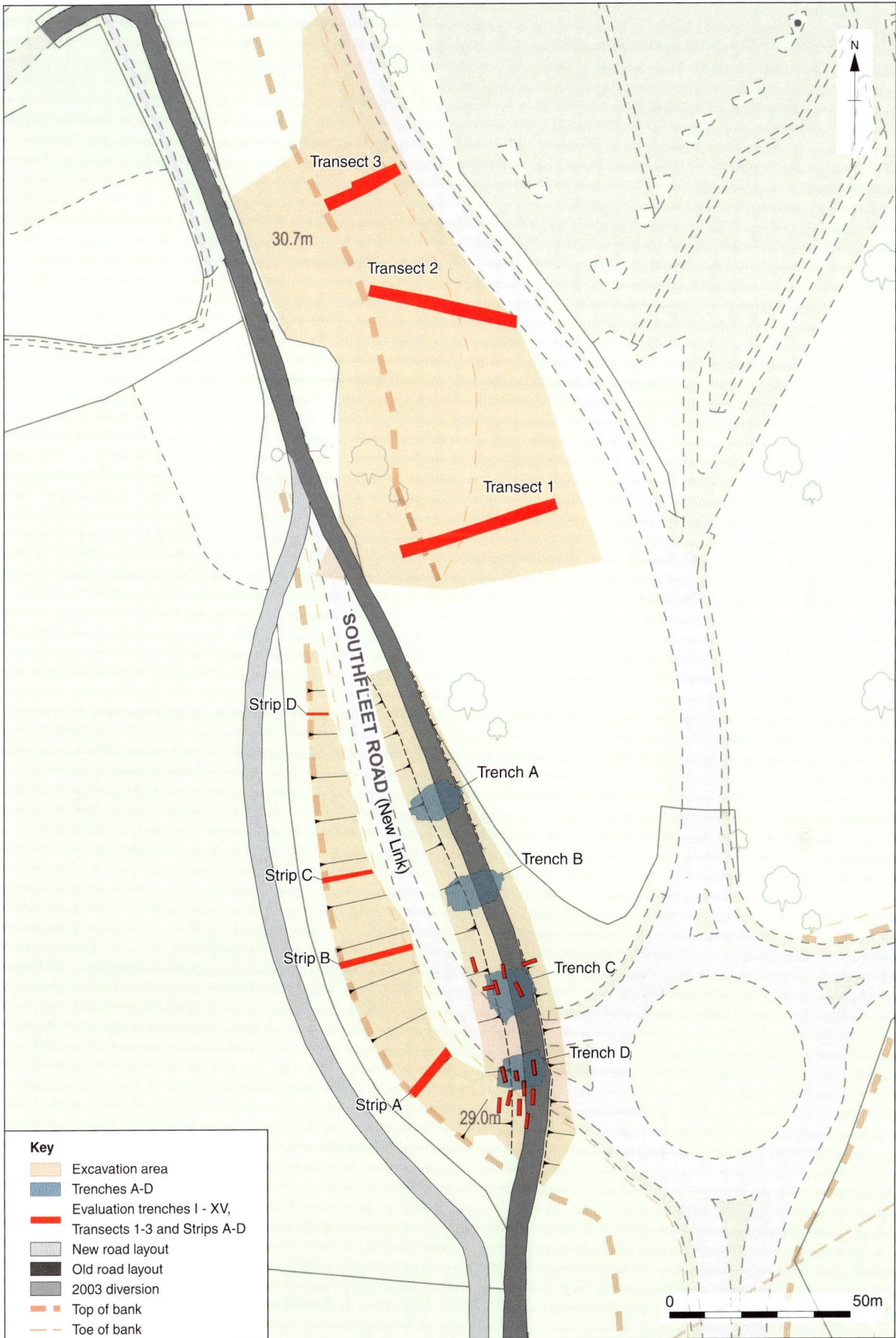

Figure 3.11 Site layout (full extent of investigations): northern Transects 1-3, main spine of site and transverse Trenches A-D

Palaeoloxodon antiquus. One of its notable features was the surviving presence of two very rotted tusks, still parallel with each other (Fig. 3.12). This discovery, even though at this point lacking any evidence of associated hominin activity, bought home the high potential of the palaeo-landsurface and led to identification of the need to evaluate it for further remains and artefactual evidence as a high on-site priority. However, before this could be done it was necessary to deal with the overlying deposits, namely the fluvial gravel (Fig. 3.2, deposit 7/Phase 8) and the mixed clay/gravel (Fig. 3.2, deposit 6/Phase 7).

As described above, these deposits were to be investigated by four transverse trenches, A-D (Fig. 3.6). Trench A having already been completed, with the recovery of numerous artefacts throughout the fluvial gravels including several handaxes, two of them from the trench's basal context 40098 (see Chapter 20; Fig. 20.2b), the upper steps of Trenches B, C and D were now dug. Trench B was located so as to intersect the northern-most end of the palaeo-landsurface as it appeared in the main west-facing section 40015 (Fig. 3.13). Trench C was located further to the south, just above where the *Palaeoloxodon* had been discovered, with the intention of both facilitating exposure of the rest of the skeleton and creating a transverse east-west section near to its location. Trench D was located even further south along the main spine of the site, where the horizon between Trenches B and C interpreted as a

palaeo-landsurface (context 40039) rose up gently and became covered by an increasing thickness of grey clay (context 40100).

As with Trench A, samples for on-site sieving for lithic artefacts were taken at regular intervals down through the fluvial gravel in Trenches B, C and D (variously assigned context numbers 40048 and 40102; see Chapter 4). Artefacts found during machining, but not included in the bulk spit samples, were given unique find ID numbers and their exact 3-D coordinates recorded. The fluvial gravels produced numerous artefacts in all three trenches, with a high proportion of handaxes (see Chapter 20).

The north-facing section of Trench B (Fig. 3.14) gave the first indications of the unusual geometry of the presumed palaeo-landsurface between Trenches B and C, subsequently christened 'the skateboard ramp' (Chapter 4). It also showed the extra thickness and variety of deposits here between the base of the fluvial gravel and the presumed palaeo-landsurface. At the west end of the Trench B section, the palaeo-landsurface dipped almost vertically down, disappearing into the floor of the trench. This was mirrored at the east end of the section, where it re-appeared at a similarly steep angle. The base of the fluvial gravel was underlain in places by another gravelly deposit (initially regarded as part of context 40043, but later re-attributed to context 40167), equally gravelly, but clearly distinct from it with a sharp junction, and being more poorly sorted and

Figure 3.12 *Paleoloxodon* skeleton *in situ* shortly after discovery; paler patches are rotted tusk remnants

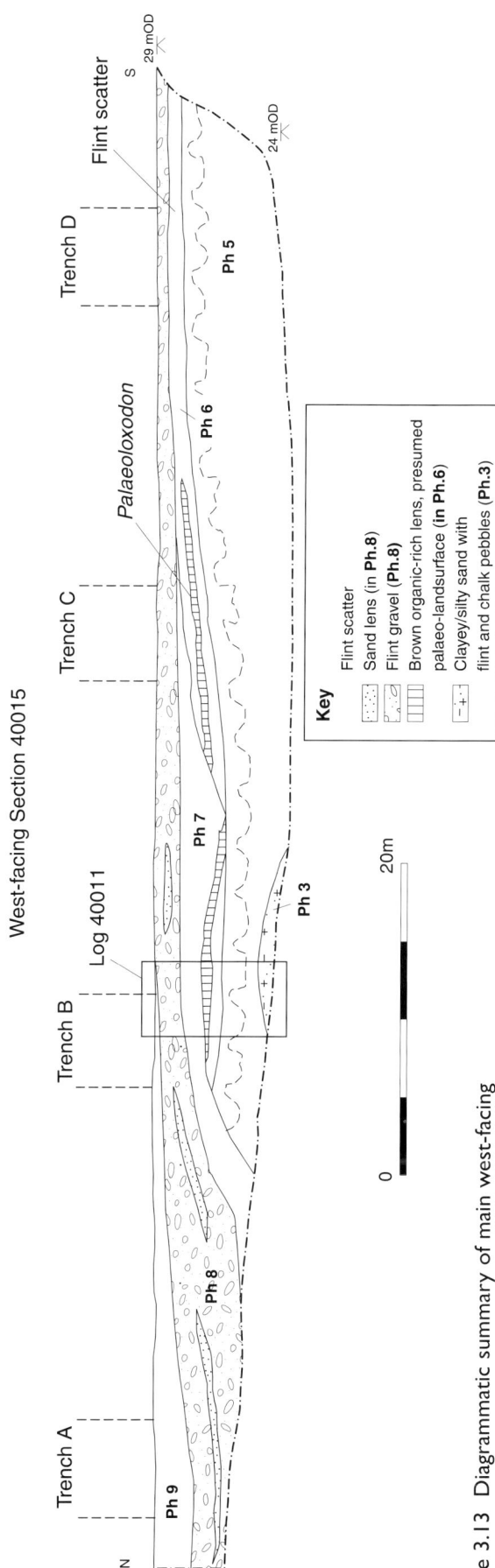

Figure 3.13 Diagrammatic summary of main west-facing Section 40015, showing position of Trenches A, B, C and D

lacking bedding structures. This in turn was underlain by a thick yellowish-brown silty sand (later attributed to context 40166), the upper part of which interdigitated with the base of 40167. Once it became clear that Trench B would have to be dug much deeper than anticipated to approach the presumed landsurface, excavation was halted approximately two metres below the base of the fluvial gravel, and Trench C was commenced.

At Trench C, the presumed palaeo-landsurface (which here contained the *Palaeoloxodon* skeleton) was buried by grey clay (40100), which thickened to the east. The floor of the trench was dug down through the base of the fluvial gravel into the underlying clay, to a level a little above the elephant horizon as exposed in the main west-facing Section 40015 (Fig. 3.13). No artefacts (or faunal remains) were found in the clay during this stage of excavation in Trench C.

Once the upper levels of Trenches B and C were completed, with recording of representative sections and sampling of the fluvial gravel sequence, the area between them was excavated by machine. It was decided that the spit sampling carried out to date was sufficient to provide a controlled sample of the artefactual content of the fluvial gravels, so they were removed in shallow spits, with immediate recovery and 3-D recording of any artefacts found. Several more handaxes and flakes were found in this way, as well as several from freshly excavated spoil for which their precise position was uncertain, but which could still be confidently attributed to the fluvial gravel (Chapter 20).

After the gravel had been removed, machine excavation proceeded through the underlying deposits, with the intention of approaching the palaeo-landsurface so that it could then be evaluated for undisturbed Palaeolithic evidence by means of 4 x 1m evaluation trenches. However, as mentioned above, there was a greater thickness and variety of deposits between the base of the fluvial gravels and the palaeo-landsurface than was apparent in the main west-facing section. As machine excavation progressed through the gravelly deposit (40167) under the base of the fluvial gravel just to the south of Trench B, it also began to produce lithic artefacts. Therefore, since this deposit had not so far been sampled in a controlled manner, a 500L bulk sample, <40197>, was taken and sieved on-site. This produced numerous artefacts; however, unlike in the previous bulk artefact samples from Trenches A-D there were no handaxes, but a range of cores, medium-large flakes and 'flake-flakes' from making notched flake-tools, which appeared decidedly Clactonian as a group.

Since this horizon was at approximately the same height (*c* 25m OD) as the basal hand-axe-producing horizon context 40098 of Trench A, which was likewise clearly stratified below the main fluvial gravel, it then became important to clarify the stratigraphic relationship between context 40098 (in the base of Trench A) and 40167 (under the base of the fluvial gravel in Trench B). This was followed up later in the excavation, when the lower parts of Trenches A and B were dug and the relationship of their lower deposits traced in the main

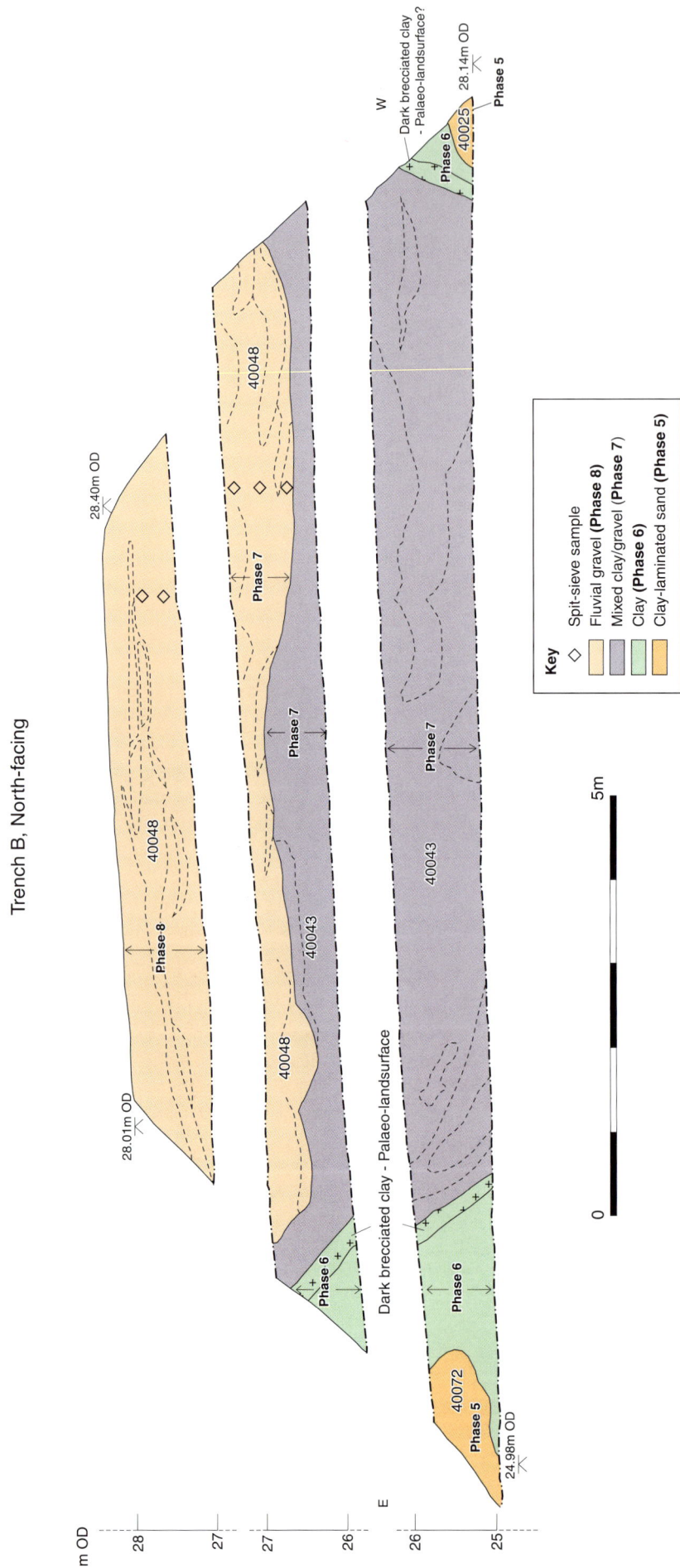

Figure 3.14 Trench B, north-facing Section 40021

west-facing section (Fig. 4.5). It became clear that 40098 was the basal level of the fluvial gravel complex (later categorised as Phase 8, see Chapter 4). It was clearly unconformably stratified above 40167, which was the upper context of a complex of sediments (later categorised as Phase 7, see Chapter 4) filling the synclinal hollow of the 'skateboard ramp'.

At the same time as this work was being carried out, the southern ends of the main east-facing and west-facing sections were being cleaned and recorded in advance of excavation of Trench D. The syncline-infill group of deposits was absent in this southern part of the site. The base of the fluvial gravel was only separated by a grey clay deposit (context 40100) from the horizon thought to represent a continuation of the land surface with the elephant skeleton (context 40039). In contrast to results from further north, this grey clay produced: (a) occasional scraps of large vertebrate bone in reasonably good condition at various depths; and (b) quite numerous, and mint condition, lithic finds in the southern part of the main west-facing section (Fig. 3.13).

At Trench D, a few lithic finds in mint condition were encountered in the grey clay below the fluvial gravel, so the first 4 x 1m evaluation trenches (I-III, Fig. 3.6) were dug here in the base of Trench D, in the grey clay down to the presumed palaeo-landsurface. This landsurface is here considered as context 40039, and manifesting as a yellowish-brown horizon at the base of the grey clay, dividing it from the underlying clay-laminated sands (Fig. 3.13, Phase 5). Trenches I-III produced very few lithics and no faunal remains, so a decision was taken to continue with machine excavation of the area to the south of Trench D. After careful machining of the overlying fluvial gravel, with 3-D recording of any artefacts found (see Chapter 20), the clay was excavated by machine down to about 0. 75m above its base. At this level, which was some 0.4m higher than the level of maximum artefact density in the south end of the west-facing section (Fig. 3.13; Fig. 4.5), relatively numerous medium-large mint condition flint artefacts began to be recovered, so downward machining ceased. These were scattered fairly evenly across the area without any concentration obviously representing an undisturbed knapping scatter and firmly embedded within the grey clay, which lacked any sign of fine stratigraphy related to the presence of the artefacts.

There was concern from the HS1 team about the time implications of open-area excavation through the depth of sediment remaining across the area: about 240m^2 to the south of Trench D. This was recognised as an issue, but I contended that machining could not have been continued down in light of the quantity of finds being recovered. Six evaluation trenches of 4 x 1m were then laid out (Fig. 3.6, Trenches IV-IX), and these were excavated downward to the base of the clay in a series of 0.5m spits (Table 3.5). Half of them were to be done by trowel, and the other half by mattock (with care and reverting to trowelling, if artefact concentrations were found), with unique numbering and full 3-D recording (by Total Station) of any artefacts found. This was intended both to evaluate the artefactual content of the area south of Trench D, and to establish the most appropriate means of excavating it.

Now that an area rich in artefacts had been identified, how to investigate microdebitage had to be considered, in order to help establish how the site was formed and whether artefacts were *in situ* on an undisturbed palaeo-landsurface (or surfaces). It was initially decided to take two of the 4 x 1m sample trenches (Trench IV should be trowelled and Trench V to be mattocked). From each spit a microdebitage sample of 500g from each 0.25 x 0.25m square was to be recovered in areas where larger artefacts were present, but only from each 0.5 x 0.5m square in areas where larger artefacts were absent. The value and efficacy of this was to be reviewed as excavation progressed. It quickly became clear that this desired intensive microdebitage sampling programme was impractically slow, even when applied only within two of the evaluation test pits, so various alterations were made. Rather than sampling the full footprint of each trench, sampling was concentrated in a narrower north-south 0.5m strip down the middle of the trench. It was intended initially to extend these north-south strips beyond the evaluation trenches and to complement them with a continuous east-west strip. However, it was later decided that there was not enough time for this to take place. Consequently the only east-west sampling was within Trench VII which, as well as having a 0.5m wide north-south sampling strip down the middle, was sampled for microdebitage across the full width of its northern end. Even with this curtailed programme, almost 800 separate microdebitage samples were taken from Trenches IV, V and VII (Table 3.5), and they ended up absorbing a perhaps disproportionate amount of post-excavation processing and analytical time in relation to their importance in the grand scheme of the site's results. The results of this microdebitage sampling are discussed in more detail in the relevant lithics chapter (Chapter 18).

Evaluation Trenches IV-IX in the area south of Trench D produced substantial quantities of lithic artefacts, all in mint or very fresh condition, through the full surviving thickness of the clay. This included both the upper grey clay (40100) and the brown basal clay horizon (40039) thought to perhaps be a palaeo-landsurface. Artefacts were, however, most abundant in the grey clay 0.2-0.5m above the presumed palaeo-landsurface (Table 3.6). The artefacts found reinforced the original supposition that the clay might contain a Clactonian industry. Globular cores of varying size and notched flake-tools were present, together with numerous medium-large waste flakes, but without any evidence of handaxes or debitage from their manufacture. Even though it was substantially slower to excavate by trowel, it was thought that this was desirable for the remainder of this area despite the time implications, as use of mattocks would inevitably damage artefacts and lead to less recovery of smaller artefacts.

Consequently, an initial attempt was made to proceed with trowel-excavation of the whole of the area south of Trench D, working southwards (Fig. 3.15). This

Table 3.5 Lithic artefact quantities and microdebitage sampling from evaluation Trenches IV-IX, south of Trench D

Trench	Max depth (m)	No. spits	Excavation method	Lithic recovery (n)	microdebitage sampling	Additional comments
IV	0.60	11	Trowelled	59	Spits 3-11, (total samples = 216)	Uneven transition to contexts 40039, and then 40025, below spit 6
V	0.70	10	Mattocked	41	Spits 3-10, (total samples = 248)	Even transition to context 40039, below spit 9
VI	0.52	10	Trowelled (spits 1-3)	25	-	-
			Mattocked (spits 4-10)	43	-	Uneven transition to contexts 40039, and then 40025, below spit 6
VII	0.83	11	Trowelled/mattocked? *	68	Spits 3-11, (total samples = 312)	Uneven transition to contexts 40039, and then 40025, below spit 6
VIII	0.73	15	Mattocked	44	-	Uneven transition to contexts 40039, and then 40025, below spit 10
IX	0.50	10	Trowelled/mattocked? *	43	-	Uneven transition to contexts 40039, and then 40025, below spit 5

* No record was made of whether these trenches were trowelled or mattocked; it is most likely that they were wholly or mostly mattocked, with any trowelling restricted to the upper spits, as for Tr VI

Table 3.6 Trenches IV-IX, south of Trench D: lithic recovery through clay (contexts 40100 and 40039) by context and spit

Spit	IV		V		VI		VII		VIII		IX	
	40100	40039	40100	40039	40100	40039	40100	40039	40100	40039	40100	40039
1	26	-	6	-	15	-	5	-	1	-	3	-
2		-	1	-	5	-	2	-	-	-	5	-
3		-	5	-	5	-	5	-	-	-	4	-
4	5	-	2	-	11	-	15	-	2	-	4	-
5	4	-	13	-	11	-	8	-	-	-	2	1
6	13	-	3	-	10	-	12	-	4	-	6	1
7	5	-	6	-	5	4	8	1	2	-	-	7
8	-	2	4	-	-	1	3	1	7	-	-	5
9	-	2	-	-	-	1	-	4	14	-	-	4
10	-	-	-	-	-	-	-	3	3	-	Base of clay	
11	-	1*	Base of clay		Base of clay		Base of clay		6	-		
12	Base of clay								2	-		
13									2	1		
Base of clay									Base of clay			
Spit?	-	-	1	-	-	-	1	-	-	-	1	-
Total	58*		41		68		68		44		43	

* Excluding artefact Δ.40818 from Tr IV, from context 40025

included full 3-D recording by Total Station of all lithic artefacts found, albeit without attempting to record finer details such as the orientation of artefacts in the ground. However, this proved problematic for two main reasons. Firstly, there was a substantial amount of clay to deal with (c 175m^3), which would require significant time to excavate in this manner. Secondly, once the clay had been exposed by removal of the overlying gravel, it began to bake in the early summer sunshine to a brick-like hardness. Even once increased staffing was given to this task, it was decided that trowelling was too slow, so the southern two-thirds of this area was excavated by mattock, rather than by trowel. This had inevitable consequences, in that many of the artefacts found by mattock ended up with some impact damage, which caused problems for the subsequent lithic analysis in distinguishing mattock-damage from macro-wear from use and secondary working of flakes into flake tools. This switch from trowelling to mattocking probably also led to reduced recovery of smaller artefacts below c 20mm long. However, even trowelling cannot be relied upon for full recovery of smaller artefacts, which can only be effected by sieving, however meticulous the trowelling. Subsequent analysis of the distribution of lithics in this area (Chapter 18) did not suggest a significant bias in recovery caused by the switch to mattocking.

No lithic finds were encountered in the main east-facing and west-facing sections between Trench C and

Figure 3.15 Trowel excavation in progress, south of Trench D

Figure 3.16 Crushed remnants of rhino jaw (Merck's rhinoceros *Stephanorhinus kirchbergensis*) found to north of evaluation Trench III (Group Δ.40843) [scale bar divisions 10cm]

Trench D, either whilst cleaning the grey clay under the gravel and above the presumed palaeo-landsurface or while machining into the grey clay below the fluvial gravel in the floor of Trench C. Consequently, it was decided not to excavate any 4 x 1m evaluation trenches between Trenches C and D, but to proceed with very delicate machine excavation. This would include careful monitoring and 3-D recording of any finds. Hand excavation would only be used if artefact or faunal find concentrations began to appear. One cluster of rhinoceros teeth and bones (Fig. 3.16), later identified as part of a jaw of Merck's rhinoceros *Stephanorhinus kirchbergensis* (Chapter 7), was found just to the north of Trench D, immediately beyond the end of Evaluation Trench III. An area 2 x 3m in size centred on this cluster was then excavated by hand down to the base of the clay, which proved to be about 0.40m down. Sampling for microdebitage (n=24) was carried out in the 2 x 3m grid around the rhino jaw cluster, which also produced a number of lithic finds.

Otherwise, although tiny unidentifiable scraps of large mammal bone were moderately common, lithic or faunal finds were never again sufficiently concentrated or of sufficiently good quality in the area between Trenches C and D for a reversion to hand trowelling. Therefore the area was excavated by machine down to the base of the clay, with immediate lifting and 3-D recording of all finds as work progressed. Consequently, even more than for the lithic material recovered by mattock in the area to the south of Trench D, this method probably led to some lithic and faunal finds not been recovered and caused some damage to those that were. When faunal remains were sufficiently robust, they were usually uncovered by trowel (after having been initially revealed by machine, which was always operated with a toothless ditching bucket for this work) and placed in a small cardboard box or plastic finds bag, cushioned if necessary by acid-free tissue paper. When less robust, they were given on-site first aid, being consolidated with a dilute solution of PVA (polyvinyl acetate) glue and then, if necessary, encased in plaster-of-Paris, before being lifted as a block including varying amounts of sediment. This latter recovery method was also applied to small concentrations of faunal fragments, which were not individually recovered but rather lifted as a group under a single find number for more detailed subsequent examination off-site.

Meanwhile, back in the elephant area, Trench C had been dug down to a little above the elephant level, and it was possible to begin excavation of the elephant skeleton and to evaluate the surrounding area for its extent and for further lithic and faunal remains. To the north of Trench C, the wider area around the skeleton was excavated by machine to the same level as the base of Trench C, using the same recording methods as between Trench C and Trench D. Once this had been done, six further evaluation trenches (X-XV, Fig. 3.6) were excavated by trowel (reverting to mattock if finds were rare or absent). The aim was to look for further concentrations of artefacts and/or faunal remains in the wider area around the elephant (Trenches X, XII and XIII). It was also intended to define more precisely the extent of the spread of the elephant skeleton and any associated lithic remains (Trenches XI, XIV and XV). Trench XIII was also positioned so as to cross the tufaceous channel transversely, and to allow recovery of two large bulk-samples (sample <40237> of 190L and sample <40241> of 300L) of the calcareous channel-fill sediment, which had been shown to be rich in small vertebrate remains. A series of ostracod samples was also taken from the deposits directly overlying the main tufaceous channel-fill in the section of Trench XIII (Chapter 11), following up the earlier positive results of the ostracod evaluation of these deposits, when exposed in the nearby main east-facing section (Fig. 3.9).

Very few faunal or lithic finds were found in any of these trenches, although they did reveal a good series of sections through the sedimentary sequence in the bottom part of the clay containing the elephant horizon. Therefore an area of *c* 70m^2 (5 x 14m) was defined for hand-excavation around the elephant skeleton (Fig. 3.6). Several lithic artefacts were recovered from directly

Figure 3.17 *Paleoloxodon* remains during excavation, with flint core Δ.40494

Figure 3.18 Topography of the 'skateboard ramp' revealed: the presumed palaeo-landsurface uncovered to the south of Trench B (looking north)

amongst the elephant bone spread (Fig. 3.17), and there was also a clear concentration of artefacts immediately to the east of the northern part of the spread, which was broadly orientated NNW-SSE with the tusks pointing south. A local grid was superimposed on the area of the elephant skeleton. The exposed bones and flint artefacts were cleaned by trowel and their outlines drawn on a plan at a scale of 1:10, before being numbered and lifted. This process was then repeated until no more layers of finds were present. Each individually numbered bone or flint was surveyed in with a Total Station. As with the faunal recording between Trenches C and D (described above), bone pieces were treated on-site with PVA adhesive and/or plaster-jacketed if necessary, and some clusters of small fragments were not individually recorded but block-lifted as a group. Sampling for microdebitage was also carried out in the vicinity of the *Palaeoloxodon* skeleton. A series of 20 contiguous samples were taken in an L-shaped series, to explore variation in microdebitage concentration along two orthogonal axes.

In addition to this standard 3-D recording, after each layer of the elephant skeleton area had been cleaned and planned it was digitally photographed with a grid of geo-referencing targets which were then themselves 3-D surveyed by Total Station. This allowed each layer of the excavation to be reconstructed off-site as a digital model, where for instance it could be represented as a true 2-D vertical view or integrated into a 3-D terrain model. This approach was also used in other parts of the site where more important concentrations of finds were thought to have been found, for instance for the Merck's rhinoceros jaw mentioned above. It was also used for many of the more important stratigraphic cross-sections, including the main east-facing and west-facing sections (nos. 40015, 40016 and 40018) and several of the transverse east-west sections across the main spine of the site.

By mid May 2004 it was clear that the timescale of the original programme was grossly inadequate, even with the various time-saving methodological compromises that had been implemented. This followed the discovery of numerous extra aspects of the site such as the elephant skeleton, the tufaceous channel, the flint concentration

south of Trench D and the widespread distribution of poorly preserved faunal remains requiring on-site conservation and recording throughout the clay overlying the presumed palaeo-landsurface. A revised programme was consequently drawn up (Oxford Archaeology 2004b) with expanded staffing and a more realistic time-scale to tackle the excavation of the elephant area and the flint concentrations south of Trench D.

While these tasks were in progress, the machine was used to remove the remainder of the syncline-infill group of sediments (Phase 7, see Chapter 4) from above the palaeo-landsurface and grey clay group (Phase 6, see Chapter 4) in the area to the north and north-east of the elephant skeleton. This finally revealed the amazing 'skateboard-ramp' topography of the presumed palaeo-landsurface (Fig. 3.18), as well as the actual correlation of the presumed palaeo-landsurface, as originally identified in section 40009 (Fig. 3.3) and log 40011 (Fig. 3.4), with the elephant horizon. It transpired (Fig. 3.1; Fig. 4.5) that these were not direct lateral equivalents, although at broadly the same stratigraphical level and both associated with the grey-clay group of deposits, eventually attributed to Phase 6 (Chapter 4). At the northern end of the exposure of the grey clay, in the vicinity of Trench B, its full thickness was conflated to the black/brown/dark purple horizon around 0.3m thick originally recorded as (40039) in Log 40011. Further south, near Trench C, the grey clay both rose and thickened, containing in its bottom parts various dark-brown horizons rich in fragments of rotted organic material, one of which contained the elephant skeleton. Even further south, towards Trench D, the grey clay continued to rise but lost the dark-brown organic-rich horizons that characterised the area of the elephant skeleton. One of the problems thus created for interpretation of the sequence, and the site as a whole, was when and how this unusual topography formed, and what bearing it might (or might not) have on the recovered archaeological remains.

The machine excavation of the syncline-infill (Chapter 4, Phase 7) also revealed that the deposits within it were sedimentologically very variable, both laterally and vertically. They also contained occasional concentrations

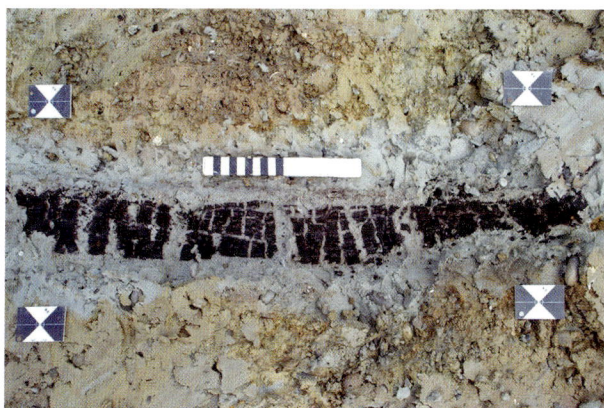

Figure 3.19 Burnt (?) branch from syncline infill [scale bar divisions 1cm]

of what looked like fragments of burnt or rotted plant or wood remains, often crushed into very thin contorted laminations a few mm thick. These proved very difficult to deal with, as they were mostly destroyed by the process of discovery. A number of samples were taken for archiving and later examination, and some of the larger pieces were cleaned up and photographed. One of these in particular was strongly reminiscent of a charred branch (Fig. 3.19), and brought to mind the possibility that burning or fire might play some part in the interpretation of the syncline-infill deposits. This could conceivably be related to the forest-fire postulated by Turner (1970) as being associated with the high non-tree pollen phase in Hoxnian sub-zone HO-IIc. However this remains highly speculative and no strong evidence was found to validate this line of enquiry.

Once it was established (by Evaluation Trenches X, XII and XIII, Fig. 3.6) that the area to the east of the elephant skeleton did not have further mega-faunal scatters in the grey clay requiring open-area hand excavation, this area was reduced carefully by machine, following the same methods as used between Trenches C and D. This was the area where the tufaceous channel had been identified and Trench XIII had been carefully positioned for the tufaceous infill to be bulk sampled and to create an east-west section through the tufaceous channel-fill sequence (Fig. 4.15). This work revealed that the tufaceous channel-fill was contained entirely within the grey clay towards its base, and stratigraphically in the same position as the elephant horizon, although lacking any direct litho-stratigraphical relationship to it (Fig. 4.31). The channel was revealed as a small feature running SSE-NNW, approximately 15m long by 5m wide (at its widest) and never more than 0.7m deep. Its base was highly contorted, although it is likely that these contortions, often forming self-contained pockets at the channel edge, are due to post-depositional deformation. A decision was taken to create a vertical section along the middle of the long axis of the channel (Fig. 3.6, Section 40075), to record its geometry in this direction and to create a face where columns of vertical series of samples (for molluscan and small vertebrate remains) could be taken through the channel-fill sequence (Fig. 10.1a).

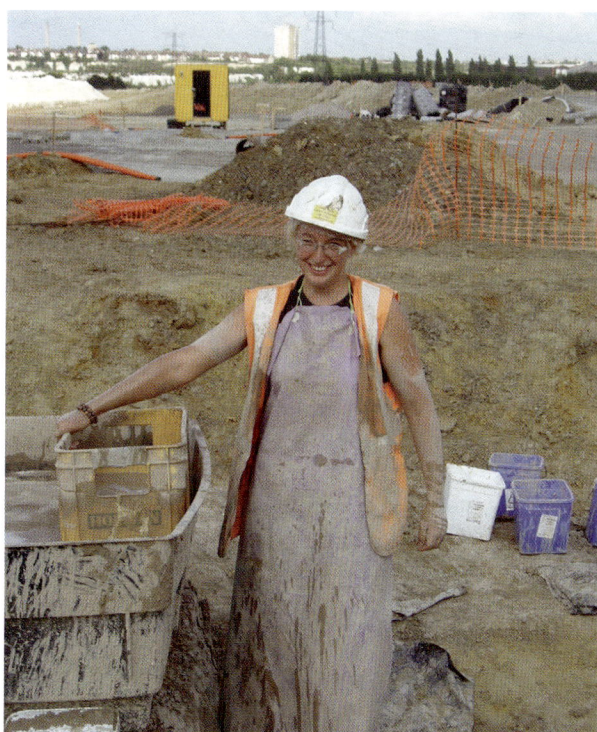

Figure 3.20 Part-sieving of tufaceous channel-fill samples on site

Although the tufaceous channel-fill had now been comprehensively sampled for small vertebrate and other remains, its surviving parts were still visibly rich in small-medium sized vertebrate remains. It was decided initially to use the mechanical excavator to take 500L bulk samples from three consecutive spits through the tufaceous channel-fill. These were to be sieved on-site through a 10mm mesh to recover a representative selection of larger mammalian remains, before continuing with machine excavation through the remainder of the tufaceous channel in the normal manner with recovery of any artefacts and larger and better preserved mammalian remains. However, it was then thought that the tufaceous channel-fill was such a rich resource that as much as possible of it should be investigated, and all faunal remains of all sizes should be recovered as far as possible.

Therefore, as well as recovery of any larger faunal remains found during machine excavation, the tufaceous spoil began to be saved as bulk samples for subsequent processing. The volumes involved quickly led to the accumulation of a very large stack of filled sample buckets, so this approach was modified by coarse-mesh wet-sieving on-site to remove the finer sediment from the samples and thus reduce their volume. This proved a distinctly messy business (Fig. 3.20), and it proved hard to record the depth-range and locations of the material contributing to these 'part-sieved samples'. The mesh-size used for these samples varied between 1 and 4 mm, without a record being kept of the mesh size used for the individual samples, reducing their comparative value. Nonetheless, they produced a huge quantity of identifiable faunal specimens that would otherwise have not have been recovered, and which now form part of the site

Figure 3.21 Trench C, north-facing Section 40085

Figure 3.22 Trench B, south-facing Section 40091, at end of main excavation (05 August 2004)

archive available to all for further research. In general, however, this episode emphasises the value of retrieving series of raw sediment samples from carefully controlled horizons, which are all processed in a similar manner.

The penultimate stage of the main excavation involved completing the lower parts of Trenches A-D, and drawing, and where necessary sampling, their lower transverse east-west sections. In course of this, several lithic artefacts were found in the sediments towards the base of the sequence, from Phases 3 and 5 (see Chapter 16). The north-facing section of Trench C was recut and deepened, to establish the relationship of the elephant horizon with the tufaceous channel and to gain information on the lower lying parts of the sequence (Fig. 3.21; Fig. 4.9). This section also revealed how the brown organic-rich horizon (40078) that contained the elephant skeleton was in fact one of a number of similar beds that developed and rose to the east and south, fading away as they did so. Since the elephant horizon had been initially shown to contain pollen, a result confirmed by Charles Turner while the excavation was in progress (see Chapter 12), a series of overlapping monoliths was also taken from this part of the sequence for subsequent pollen analysis.

In addition to the stratigraphical grid formed by the main east-facing and west-facing sections 40015 and 40016-40018, and the transverse sections of Trenches A-D, various other sections were cleaned and drawn as the excavation progressed. This provided as complete a record as possible of the complex sedimentary geometry and stratigraphy across the site, including an additional east-west transverse section 40080 between Trenches B and C (Fig. 3.6; Fig. 4.12).

The final stage of the main excavation comprised the completion of both the machine excavation of the grey clay between Trenches B and D and the hand excavation of the grey clay south of Trench D. The clay to the north of Trench C formed the synclinal 'skateboard ramp' and was mostly very dark brown or black. This corresponded with the layer originally identified as context 40039 in Log 40011 (Fig. 3.4). Relatively few scattered flint and faunal finds were recovered between Trenches B and D. One particularly important discovery was, however, a

rhinoceros maxilla found *c* 15m to the north of the elephant skeleton (Fig. 3.6), later identified as narrow-nosed rhinoceros *Stephanorhinus hemitoechus* (Chapter 7). It was initially hoped that this maxilla would represent the same individual as the jaw found near Trench D, but it was later established that these two finds represented different individuals, from different species.

The area with the flint scatter south of Trench D was regarded as finished once the clay had been excavated down to below its base. The excavation therefore extended into the top of the underlying clay-laminated sand, and southward beyond the edge of the area that was due to be removed to complete the new Southfleet Road link. When excavation ceased, the vertical section defining the south-western edge was drawn (Fig. 3.6, Section 40090; Fig. 4.17). The edge of the lithic concentration was not reached, however, and numerous lithic artefacts were present in Section 40090, indicating that deposits rich in lithic artefacts survive beyond the edge of the site. The final task carried out was to clean and record the full stepped sequence of the north face of Trench B (Fig. 3.22, Section 40091), and to collect from it various samples for palaeo-environmental work.

WATCHING BRIEF: SEPTEMBER-NOVEMBER 2004

Once all targeted investigations were completed, the archaeological programme switched to a watching brief. The construction works that were to due to be carried out were mostly to reduce the ground to the north of Trench B by about 2m and to reduce the area south of Trench D, clearing the route of the new Southfleet Road link. Also, they included the excavation of an extensive network of service pipelines down both sides of the new Southfleet Road link.

Since the fluvial gravels surviving in the area north of Trench B were known to contain quite numerous lithic artefacts, notably moderately abundant handaxes, this ground reduction was carried out under archaeological control, using the archaeological machine. The machine driver was by that time well-experienced in archaeological methods and in the recognition of lithic artefacts,

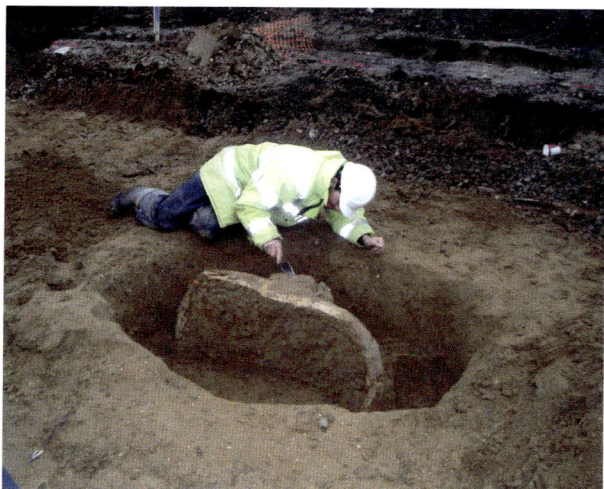

Figure 3.23 Aurochs skull under area south of Trench D, found during Watching Brief

and handaxes in particular, from his cab. A new series of find numbers (commencing with 50,000) were used for this phase of the work, and more than 150 artefacts were recovered, including 18 handaxes and numerous flakes from the fluvial gravels, and various flakes, flake-tools and cores from the underlying deposits.

The archaeologically excavated area between Trenches B and D was then mostly backfilled with spoil from the HS1 works, to bring it back up to the level needed for the HS1 groundworks plan. The deposits below the archaeologically rich level south of Trench D were then removed by the HS1 contractors, with archaeological monitoring to observe the underlying stratigraphy and to recover any archaeological remains revealed. This phase of work cut down into sediments of Phase 1 of the overall site sequence (see Table 4.1), the 'tilted block' (Fig. 3.8a). This was keenly anticipated, to discover if it was perhaps a loose small block, or whether it was part of a large block, perhaps extending into less disturbed bedrock. The base of the block was uncovered and cleaned (Fig. 3.8b), showing that it did not extend very far to the west, and revealing the orientation of its internal sedimentological junctions as steeply dipping to ENE. A few possible lithic artefacts were recovered from sediments from the overlying deposits, Phases 2 and 3 of the overall site sequence (Table 4.1). Those from Phase 2 were later considered as due to machine damage rather than hominin knapping, although those from Phase 3 are of indisputable hominin origin (see Chapter 16). The main discovery here was that of a substantial bovid skull, with its horns mostly intact (Fig. 3.23), from the chalk-rich clayey/pebbly sand towards the base of the sequence (Phase 3 of the overall site sequence, initially attributed as 'deposit 2', Fig. 3.3).

The final phase of the watching brief was to monitor the excavation of the new drainage network around the bottom of the new Southfleet Road link (Fig. 3.24). The drain trenches were mostly 2-3m deep and about 1m wide, and had to be monitored without access into their base. This was problematic as the sides of the trenches were generally smeared by the mechanical

excavation process, making it hard to observe the stratigraphy. Nonetheless, scaled sketches of the stratigraphic sequences were made, and tie-in points along the tops of these drawings were surveyed in at regular intervals, to allow integration of these records with the main site archive. Few archaeological remains were found during this process, probably partly because there was no access to the trenches but mostly because most of the deposits being dug into were from the lower parts of the sequence, where archaeological remains were scarce or absent.

The watching brief work was finally completed on Thursday 4th November 2004, 37 weeks after fieldwork began.

POST-EXCAVATION ASSESSMENT AND ANALYSIS

Introduction to assessment and analysis

After fieldwork was completed, the first stage of the post-excavation process was a preliminary assessment report (Oxford Archaeology 2005). This summarised the results as understood at the time, reviewed the quantity and contents of the site archive (particularly material archaeological remains and associated records), and assessed its importance and potential for further analysis. It also presented a preliminary programme and budget for analysis and reporting. Happily, in contrast to the conflicts and discussions that characterised the early parts of the fieldwork programme, the site was now accepted as of high importance and the proposed programme of analysis was accepted without major modification. Furthermore, of critical importance for its subsequent delivery, it was kept outside the main bulk of HS1 Section 2 post-excavation archaeological work, which was concurrently being undertaken by the Oxford Wessex Archaeology joint venture. This led to a flatter and more effective management structure whereby the post-excavation programme could be directed by a combination of myself and Stuart Foreman (for Oxford Archaeology), within the context of a known and protected budget.

Summary statement of potential

Prior to analysis, the immediate post-excavation assessment of the site was that it had produced varied artefactual and palaeo-environmental Palaeolithic remains from a complex and deep sequence of Middle Pleistocene deposits (Table 3.7), including at least one undisturbed horizon dating to the Hoxnian interglacial. Headline elements comprised:

1. An undisturbed lake-side occupation site with mint condition refitting flint artefacts, megafaunal remains (including extinct straight-tusked elephant, aurochs, rhinoceros and deer) and associated palaeo-environmental indicators (namely pollen, fish, bird, small-

The Ebbsfleet Elephant

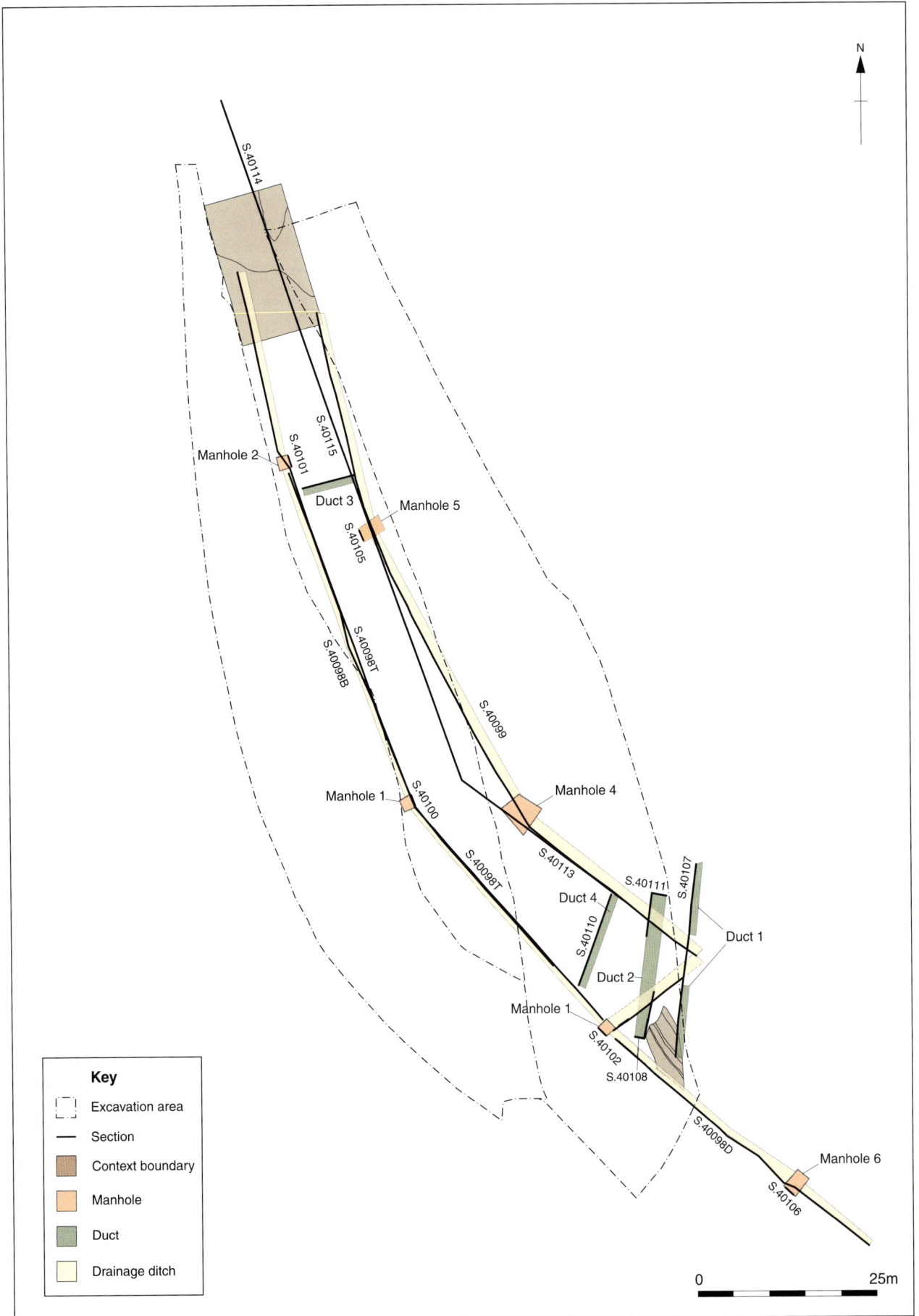

Figure 3.24 Watching Brief drainage layout

mammal, reptile, amphibian and molluscan remains).

2. Megafaunal and artefactual remains from stratigraphic levels above and below the main lake-side occupation horizon.
3. A sudden change of material cultural expression

within the sequence, from a flake/core-based Clactonian industry to a handaxe-based Acheulian industry; this transition is recognised at a number of sites in south-eastern England and its explanation is a subject of controversy.

4. Evidence of dramatic and possibly catastrophic

Table 3.7 Post-excavation assessment: archaeological and palaeo-environmental summary

Stratigraphic group*	Archaeological finds	Lithic industry	Biological remains and sampling
6 – Brickearth	Handaxes and debitage	Acheulian	-
5 – Gravel	Abundant handaxes, plus debitage and a few flake-tools	Acheulian	-
4 – mixed Clay/Sand/Gravel	Cores, flakes	Clactonian – derived?	Wood pieces – burnt?
3 – Clay, landsurface/s and tufaceous deposits	Cores, flakes, notched tools		
	Large mammalian remains – some, or many, dietary	Clactonian	Large mammals
			Small vertebrates
			- rodents
			- reptiles
			- fish
			- amphibians
			Pollen
			Molluscs
			Ostracods
2 – Clay-laminated sands	Cores, flakes, notched tools	Clactonian	Large mammals (top part)
			Fish (isolated undecalcified beds)
1 – Chalk/flint solifluction sands/gravels	Cores, flakes?	Clactonian?	Large mammals

* Stratigraphic groups as determined in December 2005; see Table 4.1 for correlation with eventual site phasing

Table 3.8 Site archive: quantities of finds and main categories of sediment samples recovered

Item	Sub-group, if applicable	Quantity	Notes
Lithic finds	-	2662	Individually recorded lithic finds from site, including finds from spit sample sieving
Faunal finds	-	1348	Individually recorded faunal finds from site
Sediment monoliths	Palaeo-environmental	93	Several series of overlapping monoliths *c* 0.70m long through major sediment units – useful for more detailed sediment descriptions and palaeo-environmental analyses
	Kubiena tins for soil micromorphology	16	Short monoliths *c* 0.15m long taken across a number of possible landsurface horizons
Sediment samples	Unprocessed bulk samples	113	Total volume of *c* 4500L, recovered in 453 x 10L boxes, with typical sample size between 10 and 50L, although sometimes much larger; mostly from tufaceous channel-fill, but other sediments also
	Part-sieved bulk samples	14	Total volume of almost 750L, recovered in 74 x 10L boxes; all from tufaceous channel fill
	microdebitage	820	Recovered as separate 1L sub-samples from 30 numbered samples
	Clast lithology	6	Five of 10L from Dec 03 fieldwork, one of which was lost; the sixth, of 50L, from 2004 fieldwork
	Various isolated spot samples	49	Typical sediment volume between 50cc and 1L; from a range of sediments
	Incremental series	53	Groups of sediment samples recovered as vertical incremental series through various sediment groups; typically between 100cc and 20L
OSL samples	-	24	Taken by J-L Schwenninger from Oxford RLAHA, with background dosimetry readings

landscape evolution in the Swanscombe region in the Middle Pleistocene.

This evidence was recognised as directly relevant to a number of national research themes (Table 2.6) and HS 1 Palaeolithic research priorities (Table 2.8). In particular, it was hoped that the rich evidence from the clay, thought to be undisturbed, would provide an important opportunity to explore the behaviour of hominids of this period. Although isolated finds of elephantid and mammoth bones, and especially molars and tusks, are not that unusual, discovery of relatively complete skeletons of individuals is extremely rare. Four *Palaeoloxodon* skeletons have previously been found in Britain: at Upnor (Andrews 1928), Selsey (Parfitt

1998a), Aveley (Stuart 1982) and Deeping St. James (Langford 1981). None of the four previous finds are reliably associated with evidence of human activity, although some artefacts are reported in the vicinity of the Aveley and Selsey elephants (Aveley – MJ White, pers. comm.; Selsey – Parfitt 1998a). Furthermore, the Southfleet Road specimen is far older than the others, which are thought to date to MIS 7, apart from the Deeping St James find which dates to MIS 5e. Thus the Southfleet Road (Ebbsfleet) elephant is unique in the UK archaeological record for its earlier Middle Pleistocene date, its undisturbed situation and its clear association with human exploitation. A number of other Lower and Middle Pleistocene elephant finds associated with early human exploitation have been made in Spain,

Table 3.9 Site archive: records

Record Group	Comments	Notes
Site Diary		
Daily journal - main excavation	43 A4 sheets	23rd Feb - 23rd April 2004
Daily journal - Watching Brief	3 A4 sheets	27th Sept - 7th Oct 2004
Primary Context records		
Note on levels	1 sheet	
Levels registers	6 sheets	
Context checklists	6 sheets	
Context record sheets	178 sheets	
Trench and spit record sheets	98 sheets	
Survey Records		
Contour survey records	3 sheets	Huge quantity of associated digital data, held and curated by Oxford Archaeology
Daily job records	233 sheets	
Control station location plans	7 sheets	
Catalogue of drawings		
Plan register sheets	4 sheets (plus 1 WB)	
Section register sheets	4 sheets (plus 1 WB)	
Primary drawings		
Plans	8 A1 sheets; 28 A4 sheets	Mostly in pencil on drafting film; some in biro on paper
Sections	25 A1 sheets; 72 A4 sheets	
Primary finds data		
Small finds record sheets	126 sheets	Most finds have digitally surveyed 3-D record
Finds-by-context checklist	19 sheets	
Catalogue of photographs		
Black & white photo record sheets	40 sheets	Each sheet representing a 35mm film of *c* 36 shots
Colour photo record sheets	40 sheets	
Digital photo record sheets	24 sheets	Representing an incrementally numbered series of 1026 images
Geo-rectified photo sheets	c 250 sheets	Sketches showing target ID references for each digital image
Primary environmental records		
Sample collecting sheets	74 sheets	Most samples have digitally surveyed 3-D record
Sample transfer forms	19 sheets	
Photographic record		
Colour slides	c 1440 slides	
Black & white negatives	c 1440 shots	Each film has associated contact sheets
Digital images	1026 images	Each image a JPG of *c* 600 KB

Italy, Africa and the Middle East (Surovell *et al.* 2005; Delagnes *et al.* 2006; Yravedra *et al.* 2012).

The Southfleet Road elephant site (comprising both the elephant area itself and the contemporary scatter south of Trench D) has thus provided an important opportunity not only to investigate Lower Palaeolithic behaviour in the UK, but also to situate this within the wider context of Old World hominin adaptations and megafaunal exploitation. The contemporaneity of the site with nearby Barnfield Pit also gives it additional value. Both sites contain a transition from a flake/core based lithic technology ('Clactonian') to a handaxe-dominant technology ('Acheulian'). As discussed above (Chapter 2), explanation of this transition has since the 1970s been a topic of hot debate. A key element in resolving this debate is identification of contemporary remains in varying landscape situations. The combined evidence from Southfleet Road and Barnfield Pit thus provides an invaluable opportunity to address this problem. Themes identified as priorities for post-excava-

Table 3.10 Assessment and analysis workstreams

Work stream	*Specialist/s, organisation*	*Chapter, Appendix*
ASSESSMENT		
Site review and archive curation	F. Wenban-Smith (University of Southampton); Oxford Archaeology	–
Stratigraphic phasing, sample review, determination of environmental assessment programme	F. Wenban-Smith	Ch 4 Appx 1
Sample processing and logging, sub-sampling and distribution to specialists	Oxford Archaeology (Geo-archaeology and Environmental section)	Appx D1a,b
ANALYSIS		
Stratigraphy and Pleistocene landscape	F. Wenban-Smith; Martin Bates (University of Wales, Trinity St David); Peter Allen; Richard Bates (University of St. Andrews); John Hutchinson	Ch 4; Appx 8
Sediment micro-morphology	Richard Macphail (University College London)	Ch 5; Appx 7
Clast lithology	David Bridgland (University of Durham); Tom White (Dept. of Zoology, University of Cambridge)	Ch 6
Large vertebrates	Simon Parfitt (University College London and Natural History Museum), with input from Silvia Bello on cut-mark identification	Ch 7-9
Small vertebrates	Simon Parfitt (University College London and Natural History Museum), with input from John Stewart (University of Bournemouth) on bird bone identification	Ch 7-9, Appx 10
Molluscs	Tom White, Richard Preece (Dept. of Zoology, University of Cambridge)	Ch 10
Ostracods	John Whittaker (Natural History Museum)	Ch 11
Pollen	Charles Turner, Barbara Silva (University of London, Royal Holloway)	Ch 12
Amino acid dating	Kirsty Penkman (University of York)	Ch 13
OSL dating	Jean-Luc Schwenninger (University of Oxford)	Ch 14
Lithic artefacts, finds and microdebitage	F. Wenban-Smith (University of Southampton), with research assistance from James Cole and Alison Moore	Ch. 15-20; Appx 6
Loss-on-ignition and magnetic susceptibility	John Crowther (University of Wales, Trinity St David)	Ch 5 Appx 2
Diatoms	Nigel Cameron (University College London)	Appx 3
Plant macrofossils	Denise Druce (Oxford Archaeology North)	Appx 4
Insects	Russell Coope	Appx 5

tion analysis were, therefore:

1. Landscape evolution and palaeo-climatic/environmental history.

2. Dating, and in particular more precise correlation of the Southfleet Road sequence with sequences at other Hoxnian sites such as Barnfield Pit, Clacton-on-Sea, Hoxne, Barnham and Beeches Pit.

3. Site formation, depositional and post-depositional processes.

4. Reconstruction of human activity both at the site and also in the wider landscape context, particularly in relation to contemporary evidence from Barnfield Pit.

5. Putting the behavioural and artefactual evidence of the site in the wider context of other UK Lower/Middle Palaeolithic sites and global early Palaeolithic occupation.

Archive summary

The site archive contains a combination of material remains: lithic and faunal finds, and samples of various kinds (Table 3.8) together with ancillary records, comprising various drawn, paper, photographic and digital survey records (Table 3.9). The original site archive is lodged at British Museum, along with the lithic finds. A copy of the site archive is lodged at the Natural History Museum, along with the faunal finds. Copies of key elements of the paper and digital archive are also lodged with the Archaeology Data Service, along with various project reports, digital material and other paperwork created in course of the project.

Assessment and analysis programme

A separate phase of assessment was not carried out. Rather, an integrated assessment and analysis programme was developed, under which the presence and potential of different categories of evidence were assessed separately, feeding into separate analysis workstreams (Table 3.10). The first phase of this programme involved review and curation of the site archive, producing spreadsheets of the finds, samples and section-line drawing-point digital 3-D data, and of the finds and samples registers. After this, a revised stratigraphic phasing model was developed for the site (Table 4.1), and a review was carried out of the stratigraphic range and variety of environmental sampling in different parts of the site, leading to determination of a programme of assessment. Once this programme had been developed, monolith logging and sub-sampling, bulk sample processing and sub-sampling, and microdebitage sample processing had to take place before processed remains could be distributed to relevant specialists for assessment and, if necessary, subsequent analysis.

Four series of sub-samples: A, B, C and D were taken from the primary collection of monoliths and bulk samples, for respectively: A (pollen), B (ostracods), C (molluscs) and D (diatoms). Sub-samples from bulk or spot sediment samples were given a new number retaining the number of their original sample, but suffixed with /A, /B, /C or /D as appropriate. Sub-samples from monoliths were given a new number retaining the number of the original monolith, suffixed as above, and also suffixed with the depth range (in cm) of the sub-sample, measured down from the top of the monolith, eg <NNNNN/A/nn-nn> for a pollen sub-sample. In addition, some monolith samples were amalgamated for bulk-sieving, and these were given a new amalgamated sample number reflecting the constituent samples, ie. two samples <XXXXX> and <YYYYY> were amalgamated to <XXXXX-YYYYY>.

The assessment programme was mostly carried out in mid-late 2010, with any subsequent analysis mostly carried out in 2011, followed by the specialist work being collated into this monograph in 2012.

Chapter 4

Geology, stratigraphy and site phasing

by Francis Wenban-Smith, Martin Bates, Peter Allen, Richard Bates and John Hutchinson

INTRODUCTION

In the first part of this chapter the landscape setting and geomorphological context of the site are described in more detail than in the general background chapter, including use of borehole records to investigate the subsurface geology. This is followed by details of the deposits in the excavated site area. The disparate sequences recorded in various exposures during the excavation have been integrated into a sequence of eleven phases, 1-11 (Table 4.1). Each phase is discussed in turn, with lithological description and interpretation of its depositional origin. At this stage, depositional interpretation is based primarily on sedimentological character, in conjunction with selected results of micro-morphological and clast lithological analyses, taken from the full specialist reports which are presented separately (Chapters 5 and 6). Results of specialist palaeo-environmental investigations, for deposits where such remains were present, are not taken account of here. Nor indeed are the results of the lithic artefact investigations: refitting, microdebitage distribution and assessment of condition, which likewise have some bearing on deposit formation processes for Phases 6-7, which contained the bulk of the lithic artefactual remains. These are presented subsequently, in Chapters 7 to 12 for palaeo-environmental investigations and Chapters 17 to 19 for the lithic work. Any relevant results are then incorporated into the final discussion and conclusions (Chapter 22).

One of the difficulties, and curiosities, of the site is the extreme geometry and topography of various sediment bodies, as exemplified in the 'skateboard ramp' across its central part (Fig. 3.18). Attempting to understand the causes of this geometry is essential both for developing interpretations of formation processes of the sediments concerned and also for wider understanding of how the sequence of deposits at the site relates to key sequences in the surrounding landscape, such as the Swanscombe 100-ft terrace at Barnfield Pit and the Members of the wider Boyn Hill/Orsett Heath Formation. If the observed geometries relate to the original deposition of the sediments, then they may contribute to interpretation of how the sediments were formed; conversely, if the present-day geometries are post-depositional deformations, then they have wholly different implications (see below).

The final two sections of this chapter then consider the Pleistocene sequence at the site in conjunction with data

from borehole and test pit investigations in adjacent areas (Fig. 4.1), identifying correlations with the site phasing sequence and expanding interpretation where possible. This ultimately leads to an overview of the Pleistocene development of the site in its landscape setting.

LANDSCAPE AND GEOLOGICAL CONTEXT

The first key point to keep in mind about the landscape situation and geological context of the site is its drastic alteration over the last 150 years due to the intensity of quarrying around it. Then one must consider how great the geomorphological alterations to the local landscape might have been between the era of its main early Palaeolithic occupation, in the Hoxnian interglacial (see Chapter 22 for a review of dating evidence) and the pre-quarrying landscape of the 19th century. Finally, one must try and incorporate ideas about the palaeo-landscape, its topography and the distribution of resources such as lithic raw material into understanding of the hominin occupational evidence.

In the early 19th century the Ebbsfleet was a small tidal stream less than 3km long, running northward from Springhead to join the Thames, passing between the villages of Swanscombe and Northfleet located on the high ground above the valley sides. As shown on the pre-quarrying geological mapping (Fig. 2.3), the Ebbsfleet passed to the east of a now-quarried spur of chalk (the Associated Portland Cement Manufacturers' Southfleet Pit, excavated between 1907 and 1912) before making a sharp right-angle turn to the west and then curving round to resume a northward course as it enters the Thames. The Southfleet Road elephant site was located at *c* 29m OD (prior to the removal of several metres of deposits by groundworks and archaeological excavation) on the western side of the southern end of this chalk spur. Prior to quarrying this would have been about one third way up the western flank of this part of the Ebbsfleet Valley, which rose westwards for about 1km to the summit of Swanscombe Park at *c* 90m OD. The northern side of the Swanscombe Park hill was quarried away in the second half of the 20th century, contained within Blue Circle's Eastern Quarry as it expanded eastwards and southward from the tunnel linking it with the older Barnfield Pit (Fig. 2.4).

Table 4.1 Stratigraphy and phasing: final version cross-referenced with initial post-evaluation scheme and published interim report (Wenban-Smith *et al.* 2006)

Phase - final	Sub-phase	Interim report (Jan 2006)	Initial phasing (Dec 2003)	Description	Main contexts
11 - Not *in situ*	-	-	-	Out-of-context, or in made ground	0, 40001, 40012
10 - Holocene	-	-	-	Holocene features and layers	-
9 - Brickearth bank	9b	Unit 6	9 - Brickearth	Reddish-brown brickearth, with occ. sandy/gravelly beds	40053, 40075, 40087
	9a	Unit 6	8 - Sand	Sand/gravel bed at base of main brickearth	40051, 40052
8 - Sandy gravel	8c	Unit 5	7 - Gravel	Bedded sand/gravel	40018, 40048, 40049, 40050, 40071, 40102
	8b	Unit 5	7 - Gravel	Bedded sand/gravel, with sandy/clayey patches	40047
	8a	Unit 5	7 - Gravel	Bedded sand/gravel, with sand beds	40098
7-8 Transition	-	-	-	Remnant, truncated clayey/sandy gravel beds that interdigitate with base of overlying Phase 8 gravels	40044, 40045, 40046
7 - Mixed clay/gravel (syncline infill)	-	Unit 4	6 - Mixed Clay/Sand/Gravel	Silts, sands, massive structureless gravelly clay and sandy/clayey gravel	40041, 40042, 40043, 40101, 40164, 40166, 40167
6 - Grey clay, with brown organic-rich beds and tufaceous channel-fill deposits	6	Unit 3	5 - Clay with dark, brecciated upper part	Grey clay, with reddish-brown stained horizons, including v dark humic horizons in north central part of site	40039 (to N of elephant), 40068, 40074, 40078, 40100, 40158
	6b	Unit 3	5 - Clay	Lower grey clay, (central part of site) Tufaceous channel fill	40099, 40160, 40161, 40162 40070, 40143, 40144
	6a	Unit 3	5 - Clay	Basal clay horizons	40039 (in central and southern parts of site), 40040 (taken as transitional to Ph 6), 40103
5 - Clay-laminated sand	-	Unit 2	4 - Clay-laminated sands	Fluvially bedded yellow sands with clay-silt laminae, very wavy and contorted in places, and occ. gravel-rich beds	40025, 40066, 40067, 40072, 40163
4 – Sandy/gravelly clay	-	-	3 - Clay, clay-silt	Clay/silty clay beds, stained in yellowish, brownish and greyish bands, gravel/sand rich in places	40026, 40027
3 – Chalky/silty / gravelly sand	-	Unit 1	2 - Clayey/silty sand with flint and chalk pebbles	Grey clay-silty sand with chalk and flint pebbles, rich in derived Tertiary shell fragments	40028, 40061, 40062, 40063 (decalcified), 40159
2 – Parallel-bedded sand/clay	2b	-	-	Alternating beds of silty clay and sand, dipping to west	40060, 40064, 40065
	2a	-	-	Sandy chalk/flint rubble	40077
1 – Tilted block	-	-	-	Block of translocated/disrupted bedrock with steeply dipping beds of Chalk, Bullhead flint bed and Thanet Sand, then darker zone in upper part of Thanet Sand grading into structureless silty sand	40054, 40055, 40056, 40057, 40058, 40059

Key

London Clay		Alluvium
Blackheath Beds		Coombe Deposit
Woolwich Beds		Boyn Hill Terrace
Thanet Sand		Slipped masses of London Clay
Upper Chalk		

SV   Swan Valley School
EQ.B Eastern Quarry Area B
NW   Northfleet West Sub-Station
SQS   Station Quarter South

0 ——— 250m

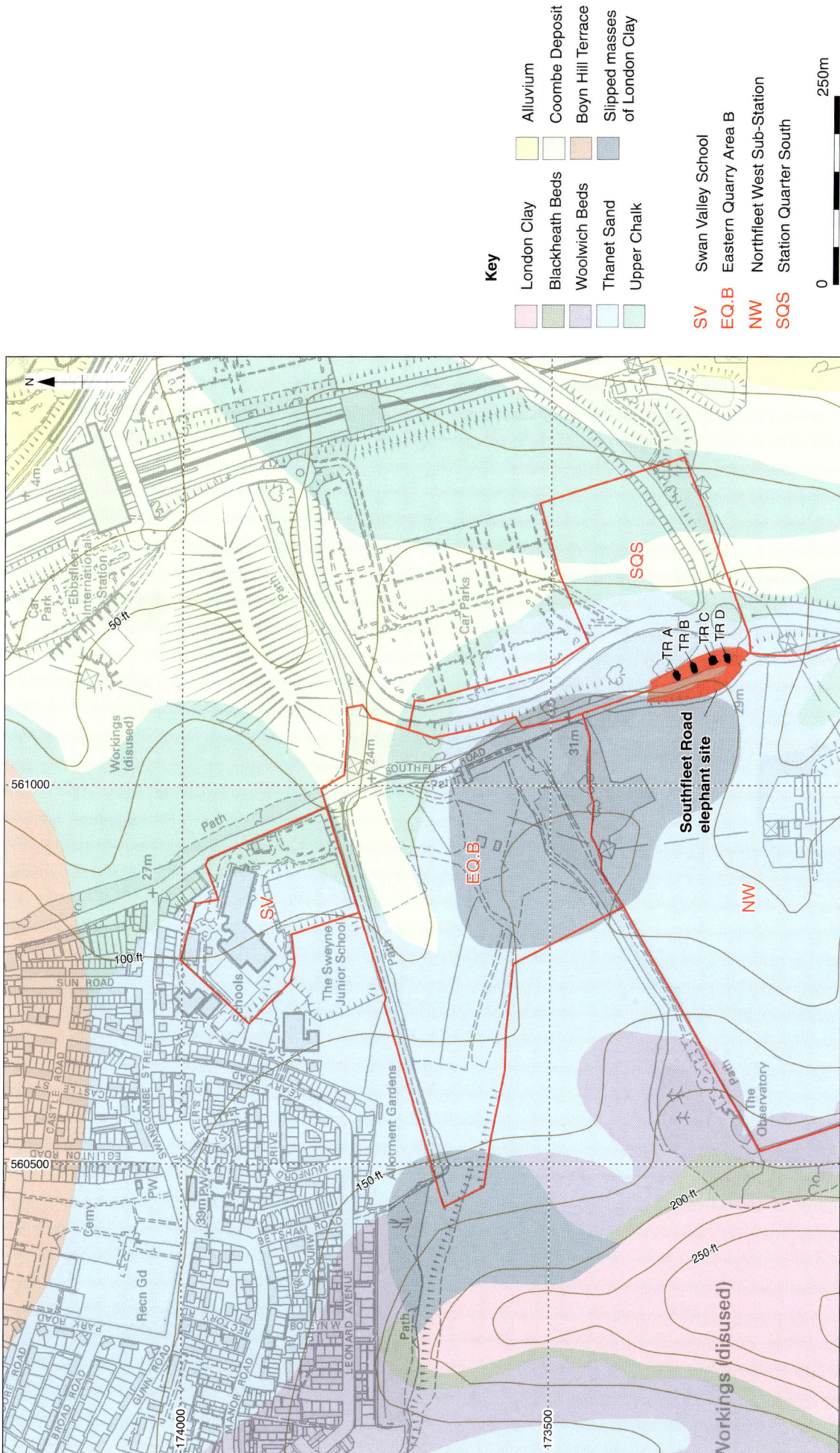

Figure 4.1 Location of non-HS1 interventions around the site: Northfleet West Sub-station; Station Quarter South; Eastern Quarry Area B and Swan Valley Community School [pre-quarrying base geology follows 1922 Geological Survey mapping, 6" sheet X, NW]

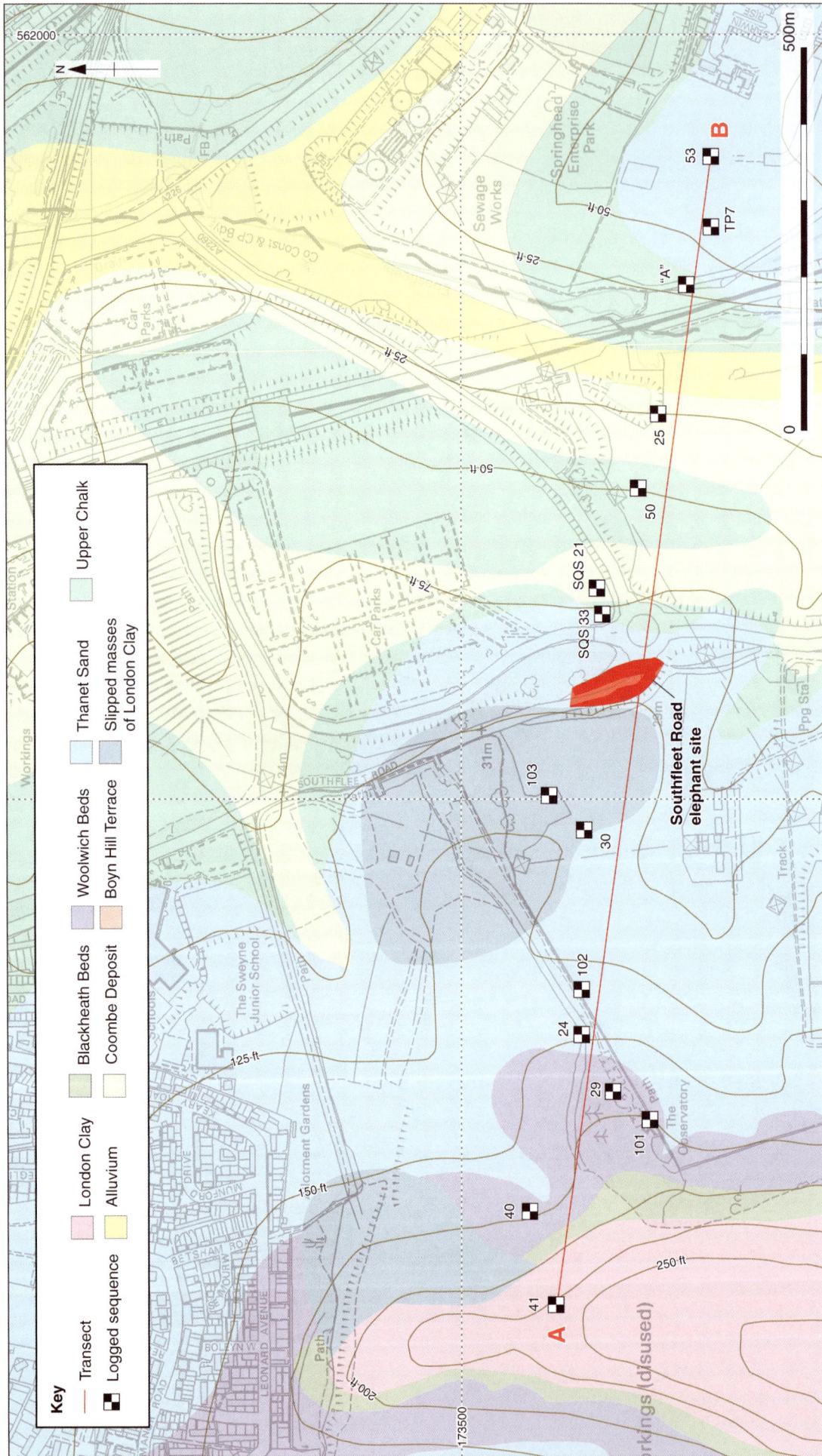

Figure 4.2 Site in geomorphological context, with east-west transect AB points, see Figure 4.3 [pre-quarrying base geology follows 1922 Geological Survey mapping, 6" sheet X, NW]

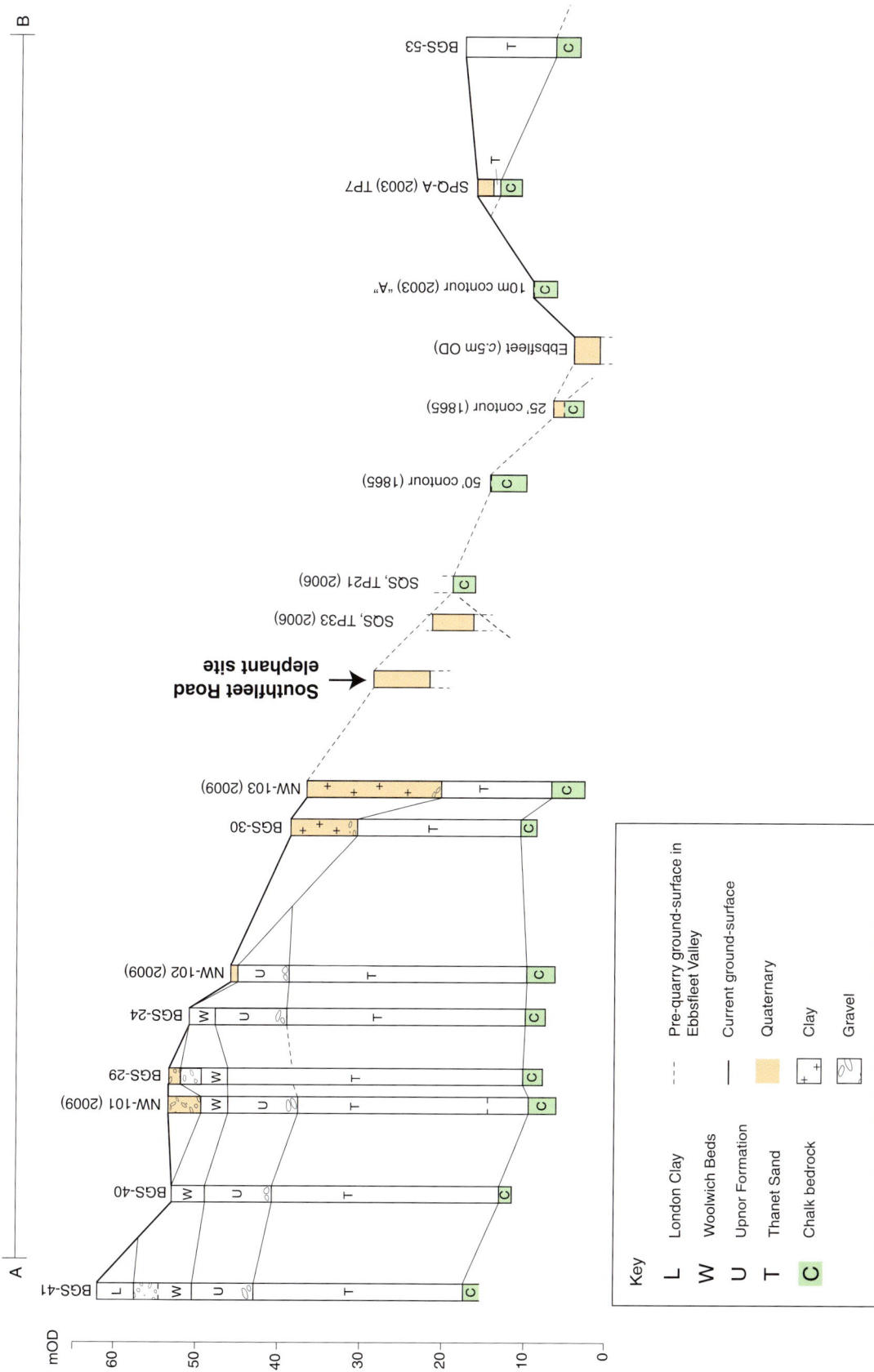

Figure 4.3 East-west geomorphological transect AB across Ebbsfleet Valley and through the site

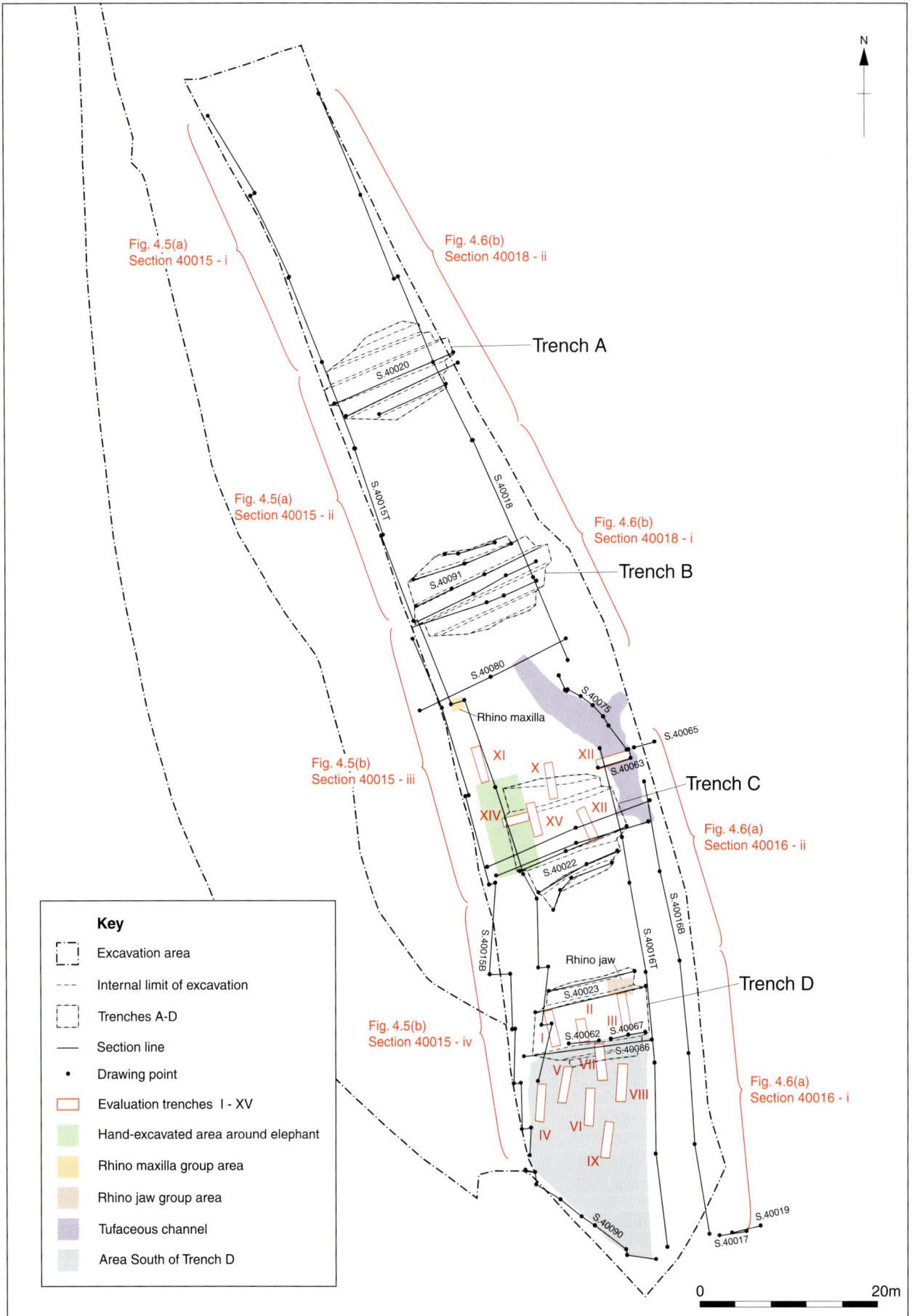

Figure 4.4 Site layout with key section-line drawing points

Pre-quarrying borehole records (Fig. 4.3, borehole BGS-41) show that the Chalk that was the main target of quarrying was overlain by up to 40m of relatively worthless fine-grained deposits, which provides an insight into just how desirable a mineral resource the deeper lying Chalk was. Geological mapping of the late 19th century shows that the Swanscombe Park hill was capped by London Clay, and then progressively older Palaeocene deposits (Woolwich Beds, Upnor Formation and Thanet Sand) were exposed down the eastern slope towards the site (Fig. 2.3 and Fig. 4.2). The geological mapping also identified two 'slipped masses' of mainly London Clay, one close to the summit of the hill and the other half way down the eastern side, with its eastern edge just encroaching over the site's location. This latter 'slipped mass' is at the foot of a spur that extends eastward downslope from the summit of Swanscombe Hill towards the base of the Ebbsfleet Valley. It widens slightly as it descends, suggesting a fan of clay-rich deposits having slipped downslope to the east, originating from the Woolwich Beds and London Clay capping Swanscombe Hill.

A plot of land approximately 30ha in area immediately to the west of the site, bounded to the south by the A2, remains unquarried and currently contains the Northfleet West electricity substation, surrounded by arable fields (Fig. 4.1, area 'NW'). The ground surface within this area is uneven, with steep-sloping dry valleys incised into the soft Tertiary sands that outcrop across most of the area. This plot has recently undergone a range of geotechnical and geo-archaeological investigations (Museum of London Archaeology 2011). These new records, combined with pre-existing records held by the British Geological Survey and results of other recent geotechnical investigations in Station Quarter South to the east of the site (Fig. 4.1, area 'SQS'; Wessex Archaeology 2006b) and Springhead Quarter, further to the east of the Ebbsfleet (CgMs Consulting and Wenban-Smith 2003) allow an east-west transect to be constructed. This shows the site within the context of ground-surface topography and subsurface bedrock geology (Fig. 4.3).

This east-west transect shows that the Chalk bedrock surface is approximately horizontal at *c*10m OD to the west of the site, but rises sharply to over 20m OD immediately to its east, forming the southern end of the spur of Chalk excavated as Southfleet Pit (Fig. 4.3, SQS, TP 21). The eastern side of this chalk spur is then erosionally truncated by the current ground surface of the Ebbsfleet Valley, dipping towards the Ebbsfleet floodplain. The next eastward conformable boundary with the overlying Thanet Sand occurs at *c* 14m OD in the Springhead Quarter (Fig. 4.3, SPQ-A, TP 7). There is thus, immediately to the east of the site, a sharp and anomalous rise in the upper surface of the Chalk bedrock, which then dips steeply eastward, as reflected in the most easterly borehole record of the transect (Fig. 4.3, borehole BGS-53).

The Pleistocene deposits at the site and in its immediate vicinity are discussed in more detail below. It

is, however, apparent from this east-west bedrock transect that they fill a pronounced trough to the west of this Southfleet Pit chalk spur, halfway up the west slope of the Ebbsfleet Valley.

Investigation of the wider geometry of the Chalk bedrock surface, where conformably overlain by the Bullhead (flint) Bed and Thanet Sand rather than erosionally truncated, shows that it rises steadily north-west towards Swanscombe. Here it outcrops beneath the Boyn Hill/Orsett Heath Formation deposits of the Swanscombe 100-ft terrace at *c* 27m OD at the Swan Valley Community School almost 1km away (Wenban-Smith and Bridgland 2001). Apart from the anomalous Southfleet Pit Chalk-high, the site is located in a minor east-west trending depression of the Chalk surface, which also rises to the south, as shown by British Geological Survey mapping (Fig. 2.2).

DEPOSITS AT THE SITE

The full sequence and geometry of deposits at the site was pieced together from section exposures recorded throughout the excavation process. The main sections contributing to this synthesis are the extended north-south sections along each side of the rectangular spine of deposits that formed the main site (see Fig. 4.4 for their locations). These include Section 40015 on its west side (Fig. 4.5) and Section 40016-40018 on its east side (Fig. 4.6) as well as the evenly spaced east-west sections transversely across the site associated with Trenches A-D (Fig. 4.7 – Tr A, Section 40020; Fig. 4.8 – Tr B, Section 40091; Fig. 4.9 – Tr C, Sections 40022 and 40085; Fig. 4.10 – Tr D, Section 40023; and Fig. 4.11 – Tr D, Sections 40062, 40067 and 40086) which provide lithological correlations between the two main north-south sections.

Other important sections include an additional transverse east-west section between Trenches B and C (Fig. 4.12 – Section 40080) and sections associated with the tufaceous channel, namely Section 40064 (Fig. 4.13) and Section 40075 (Fig. 4.14) parallel with its main axis, as well as those recorded transversely across it in and around Trench XIII, the north-facing Section 40063 (Fig. 4.15a). Also, the south-facing composite of Section 40065, 40068 and 40074 (Fig. 4.15b), a small off-set section through the basal phase of deposits at the south-east corner of the site (Fig. 4.16 – Composite Section 40017-40019). Finally, the main north-east-facing Section 40090 (Fig. 4.17), marking the southern boundary of the hand-excavated area south of Trench D.

As will become clear from the descriptions below, and also as is shown in many of the photos in the preceding chapter on excavation progress and methods (Chapter 3), many deposits were extremely uneven in their 3-dimensional topography. In order to try and represent this and to investigate relationships across the site, all sections were integrated as panels into a 3-D site model using the software package Fledermaus (Fig. 4.18) and excerpts from this model are also used as figures where appropriate below.

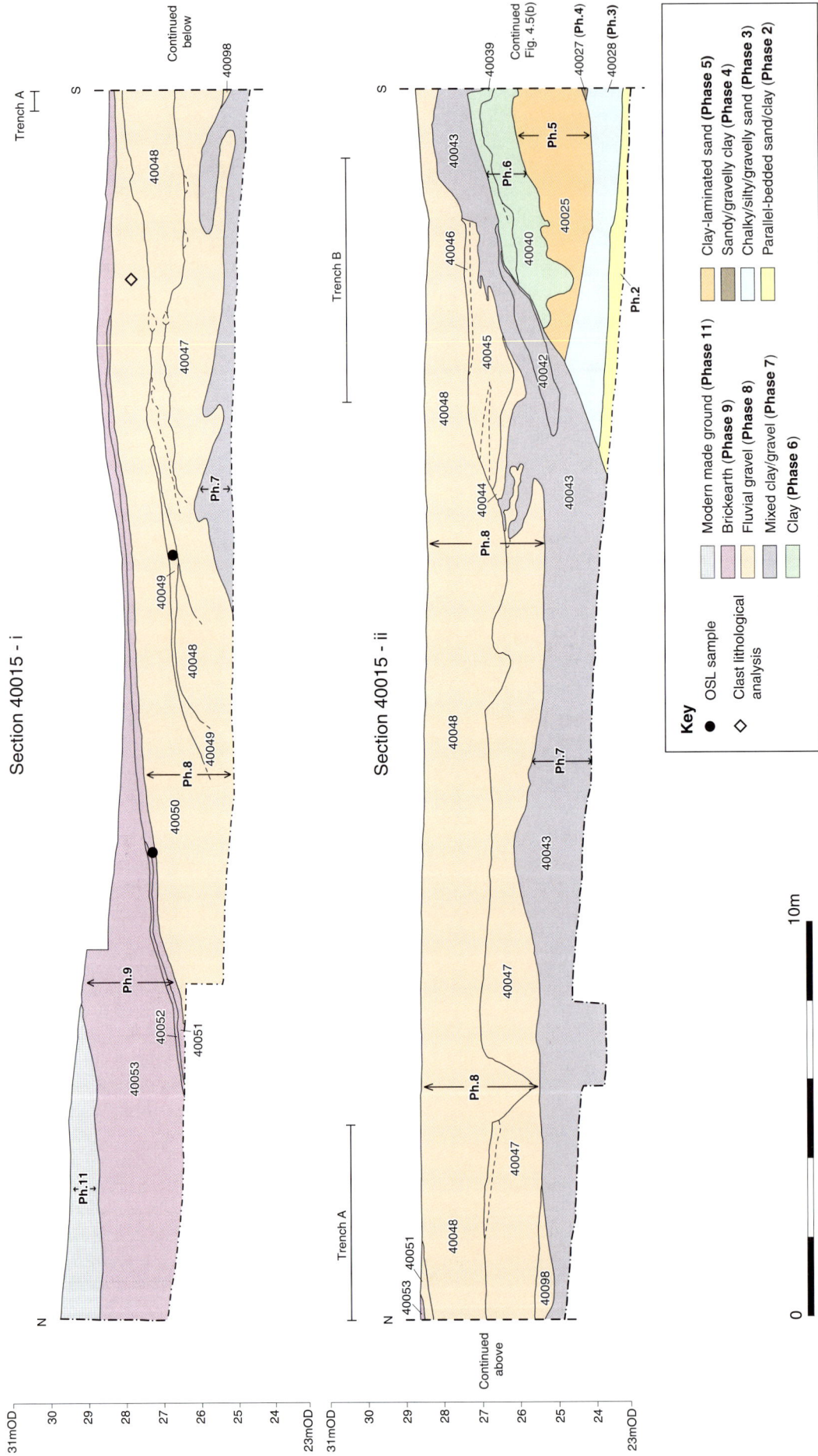

Figure 4.5a Main west-facing section (no. 40015), northern end

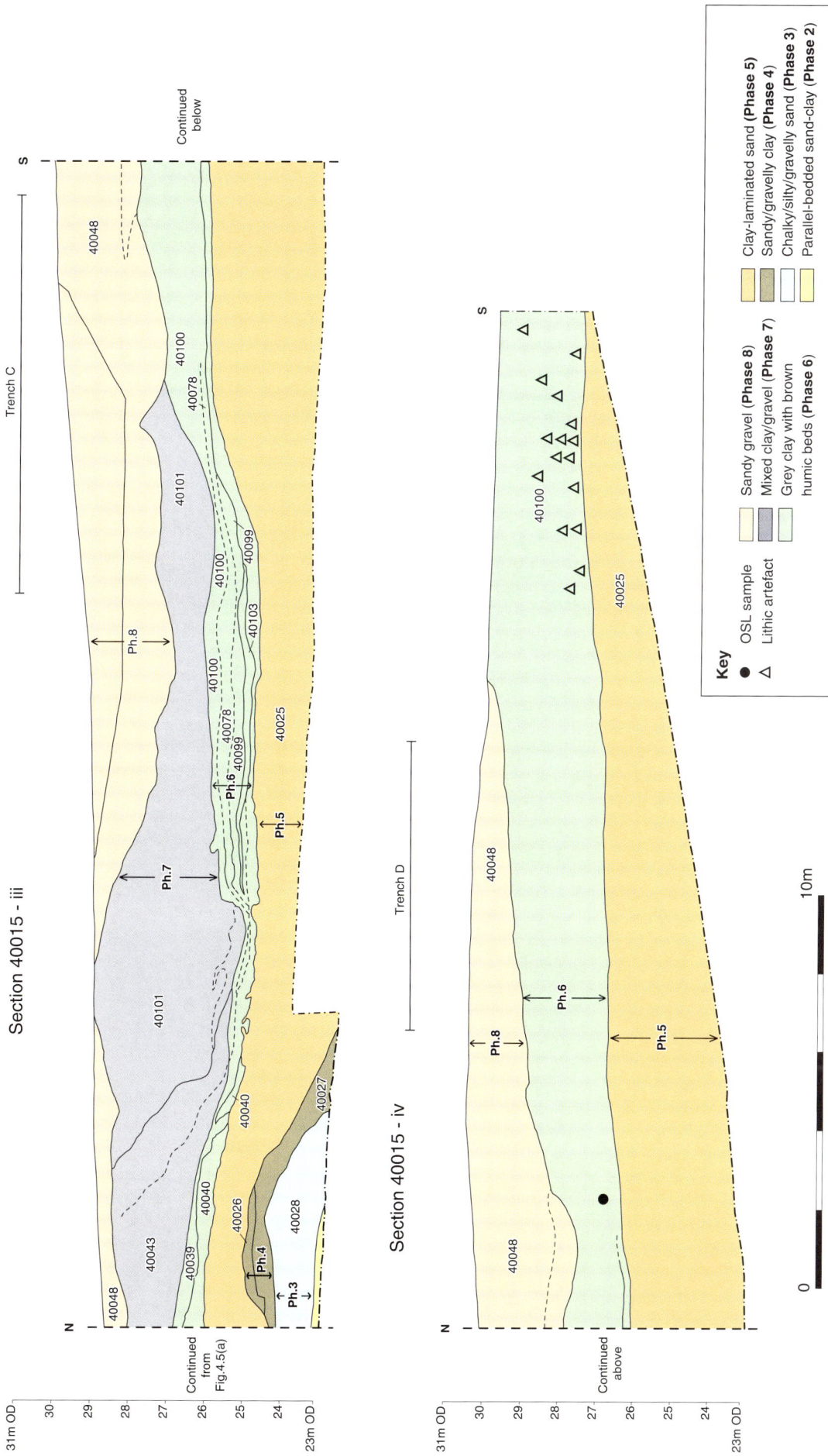

Figure 4.5b Main west-facing section (no. 40015), southern end

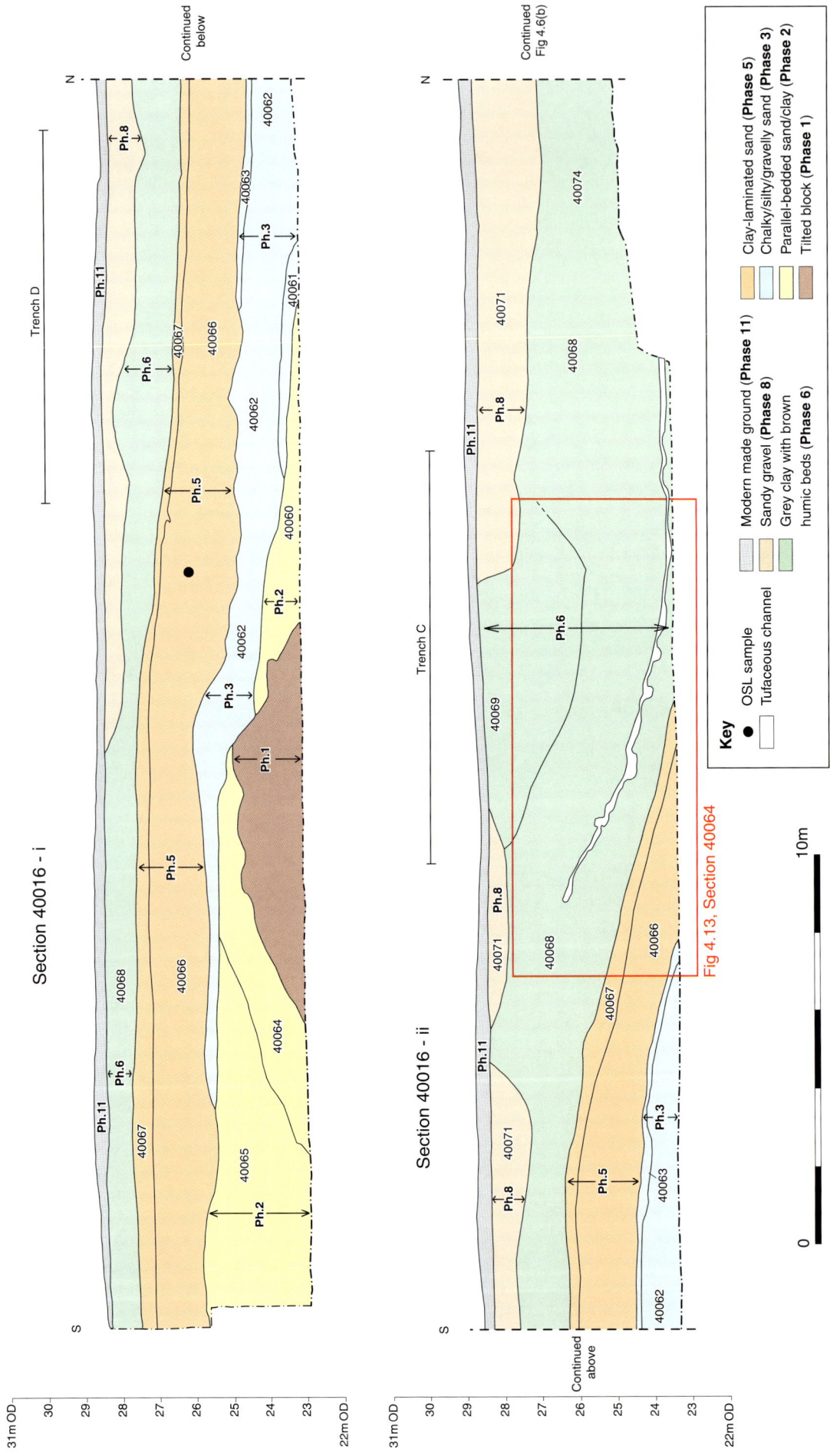

Figure 4.6a Main east-facing section (no. 40016), southern end

Section 40018 - i

Trench B

Continued below

Continued from Fig.4.6(a)

40075

Ph.9

Ph.8

Ph.5

40071

40072

Ph.11

Ph.7

40071

40158

Ph.8

Ph.6

40074

Ph.11

31m OD
30
29
28
27
26
25
24m OD

S

N

Key

◇ Clast lithological sample

Modern made ground (**Phase 11**)

Brickearth (**Phase 9**)

Sandy gravel (**Phase 8**)

Mixed clay/gravel (**Phase 7**)

Grey clay with brown humic beds (**Phase 6**)

Clay-laminated sand (**Phase 5**)

0

10m

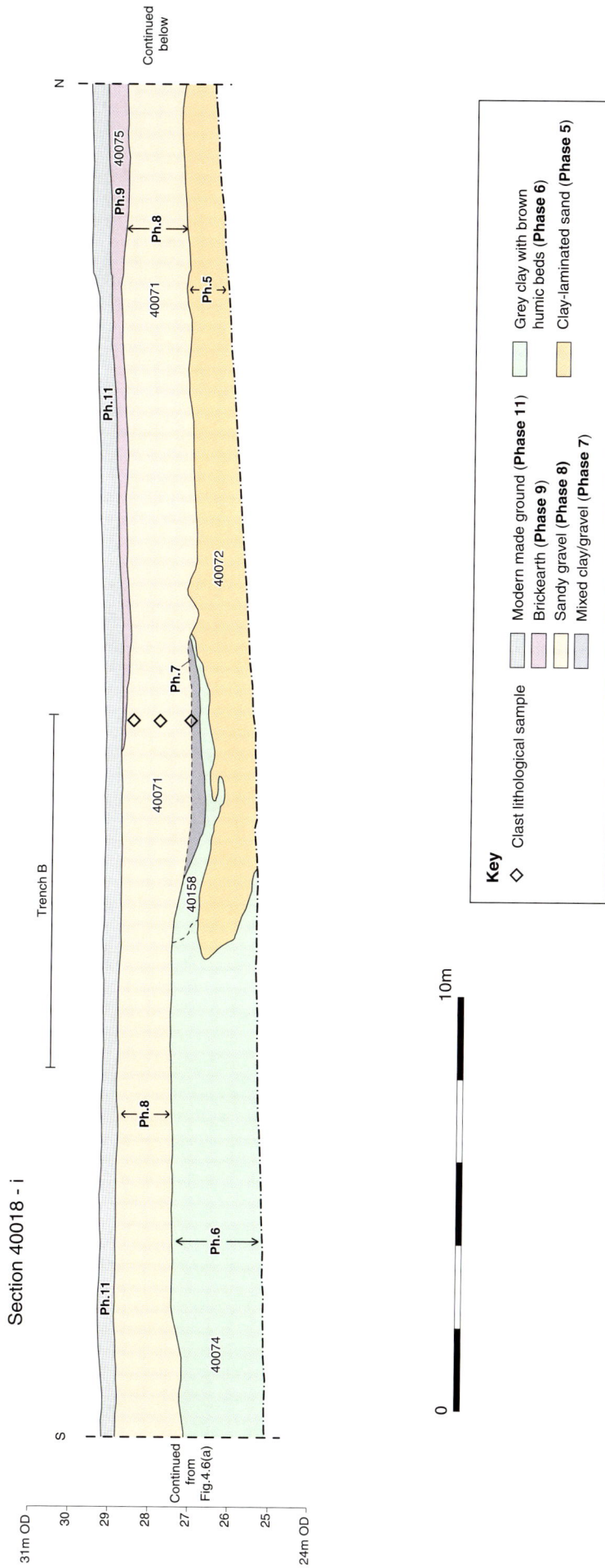

Figure 4.6b Main east-facing section (no. 40018), middle

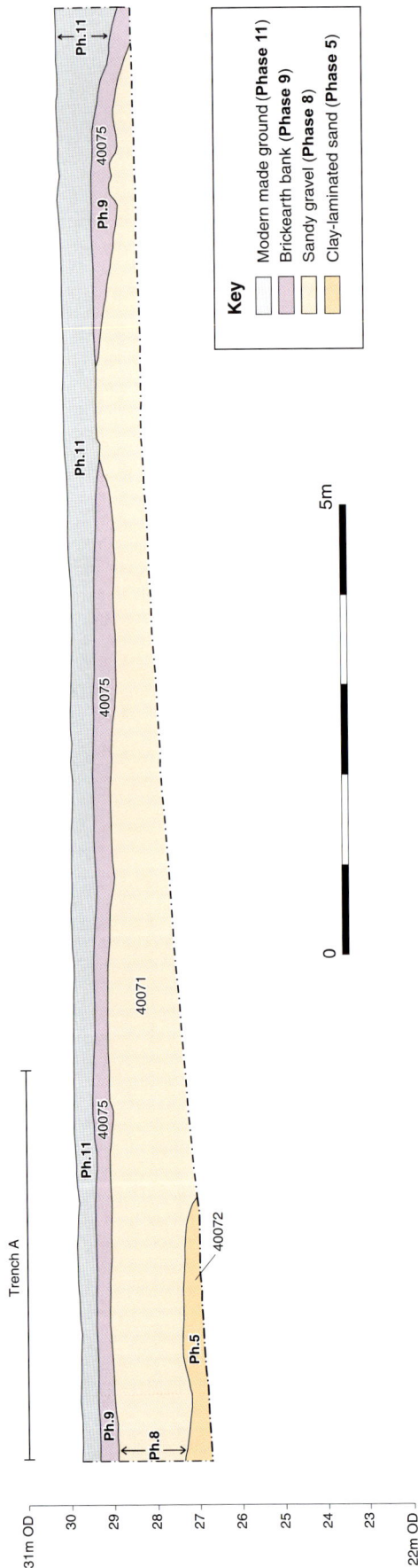

Figure 4.6c Main east-facing section (no. 40018), northern end

Eleven deposit phases were recognised (Table 4.1), from Phase 1 at the base of the sequence to Phase 11 at the top, representing modern made ground and out-of-context finds. This phasing represents an expansion of, and evolution from, previous stratigraphic summaries, mostly presented in unpublished internal site documents (ie Oxford Archaeology 2004a; b and 2005), but also published in the preliminary report (Wenban-Smith *et al.* 2006). This new phasing summary table shows the final version, cross-referenced with (a) the initial phasing summary from the geo-archaeological field evaluation in December 2003 (Oxford Archaeology 2004a) and (b) the published preliminary version of 2006. It also specifies sub-divisions of phases, where relevant, and lists the main archaeological context numbers associated.

No solid bedrock was encountered in course of the project, so the sequence at the site is not securely bottomed. Deposits from the two basal phases, Phases 1 and 2, are of uncertain date and not even necessarily Pleistocene. Those from subsequent phases, Phases 3 through to 8, are unequivocally Middle Pleistocene, and, as explained throughout the remainder of this volume, can be reliably associated with the Hoxnian interglacial, in MIS 11. Phase 9 is of uncertain date; it may follow closely on from Phase 8, or be from the later Middle or Late Pleistocene, or may even include multiple post-MIS 11 depositional phases. Phase 10, the final pre-modern phase, represents anthropogenic deposits associated with late prehistoric (or later) occupation layers and archaeological features from the later Holocene. These last deposits (Phase 10), and associated later prehistoric archaeological material, are not considered further in this volume, which focuses upon the earlier Palaeolithic evidence from the site. They are, however, incorporated in the wider analysis of later prehistoric activity in the Ebbsfleet Valley in the companion *Prehistoric Ebbsfleet* volume (Wenban-Smith *et al.* forthcoming). Phase 11 represents out-of-context artefacts and modern made ground. This chapter, and the remainder of this book, focus therefore upon Phases 1-9, and material recovered from these deposits, with one addition, namely the curious presence of an 18th century gunflint industry from the made ground capping the sequence (Chapter 21).

Phase 1. Tilted Block

An isolated series of steeply dipping sedimentary units christened the 'tilted block' was identified at the base of the sequence, in the southern end of the main east-facing section no. 40016 (Fig. 4.6; Fig. 3.8). These continued south-east beyond the site area, as seen in the lower section forming the southern site boundary (Fig. 4.16). In the main section, the internal unit boundaries appeared to dip vertically, but a better understanding of their geometry resulting from wider exposures (Fig. 4.19) revealed that they in fact dipped steeply to the ENE, *c* 60° down from horizontal. The Phase 1 sediments can be divided into two groups consisting of:

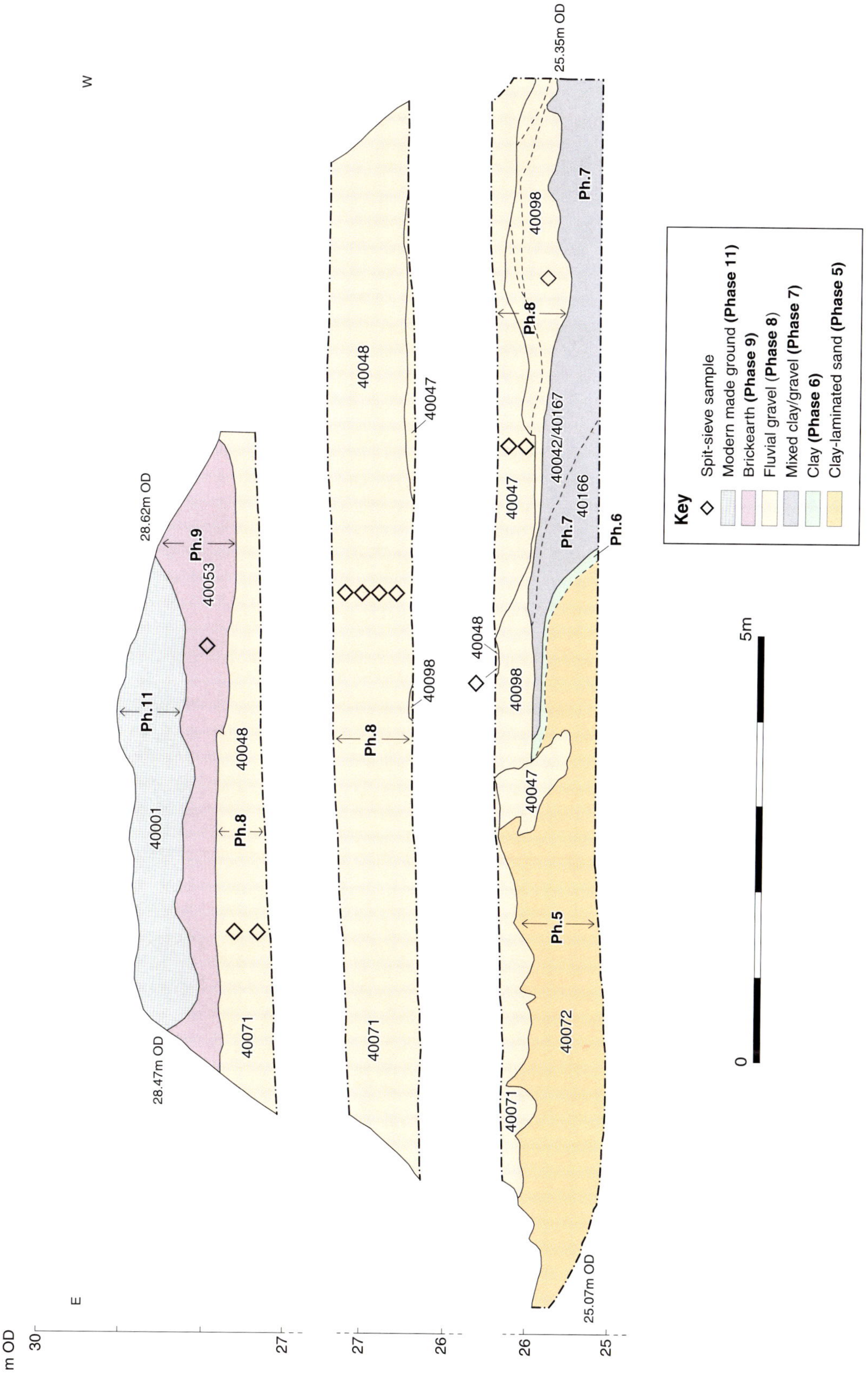

Figure 4.7 Trench A, north-facing section (no. 40020)

The Ebbsfleet Elephant

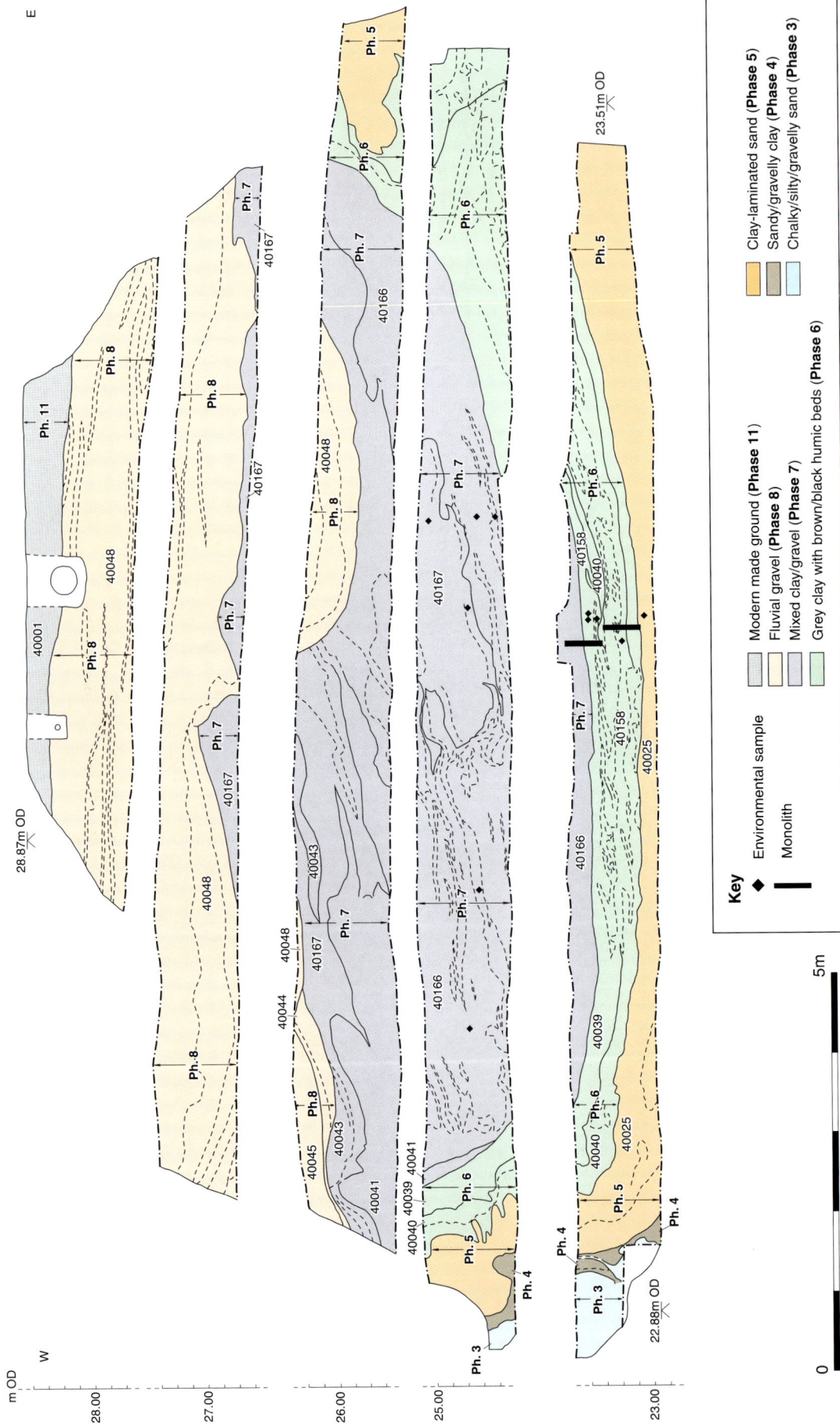

Figure 4.8 Trench B, south-facing section (no. 40091)

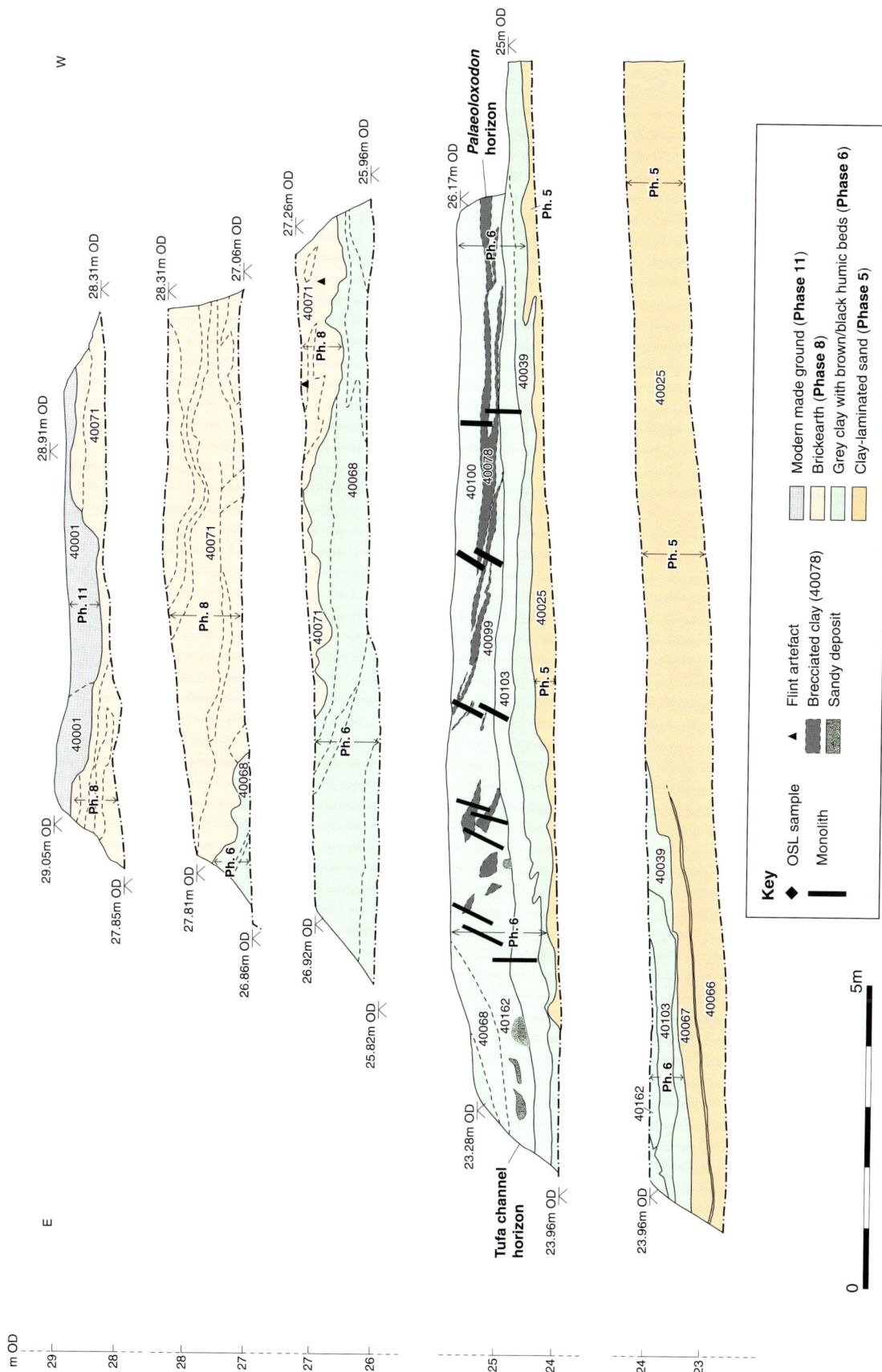

Figure 4.9 Trench C, north-facing sections (nos 40085, 40022)

Section 40023

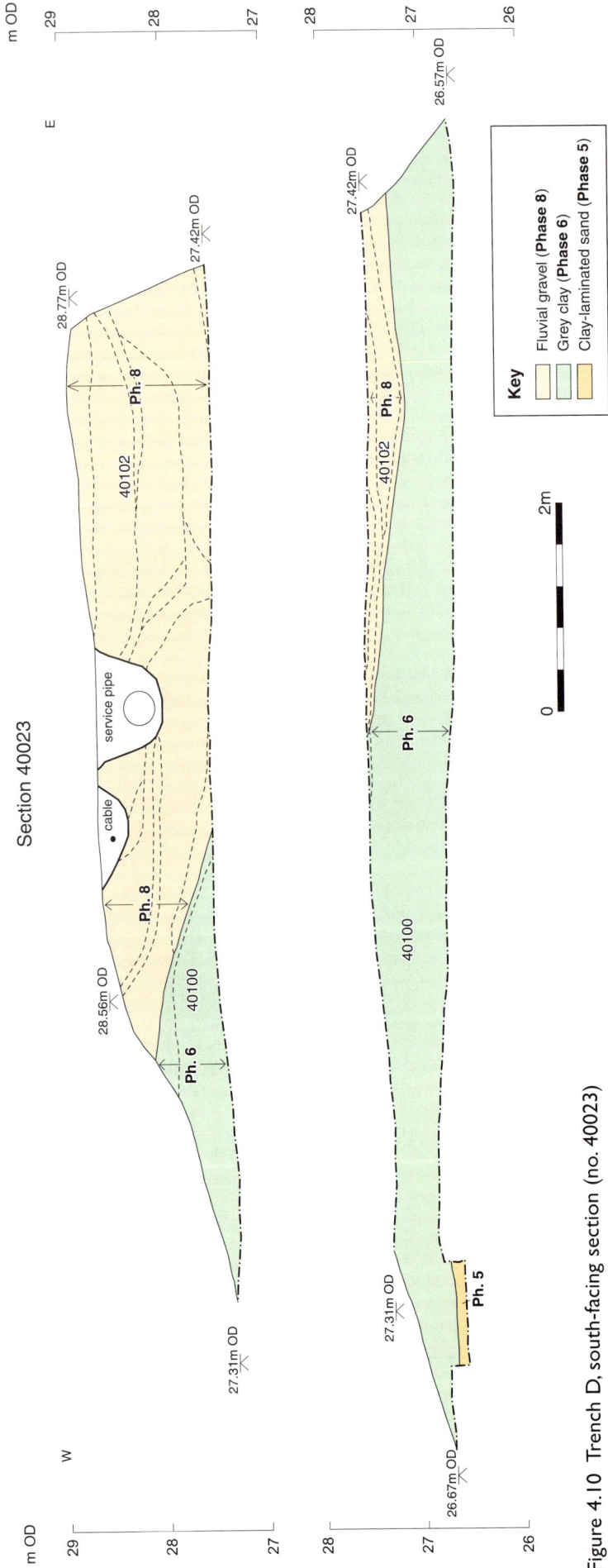

Figure 4.10 Trench D, south-facing section (no. 40023)

Key

Fluvial gravel (**Phase 8**)

Grey clay (**Phase 6**)

Clay-laminated sand (**Phase 5**)

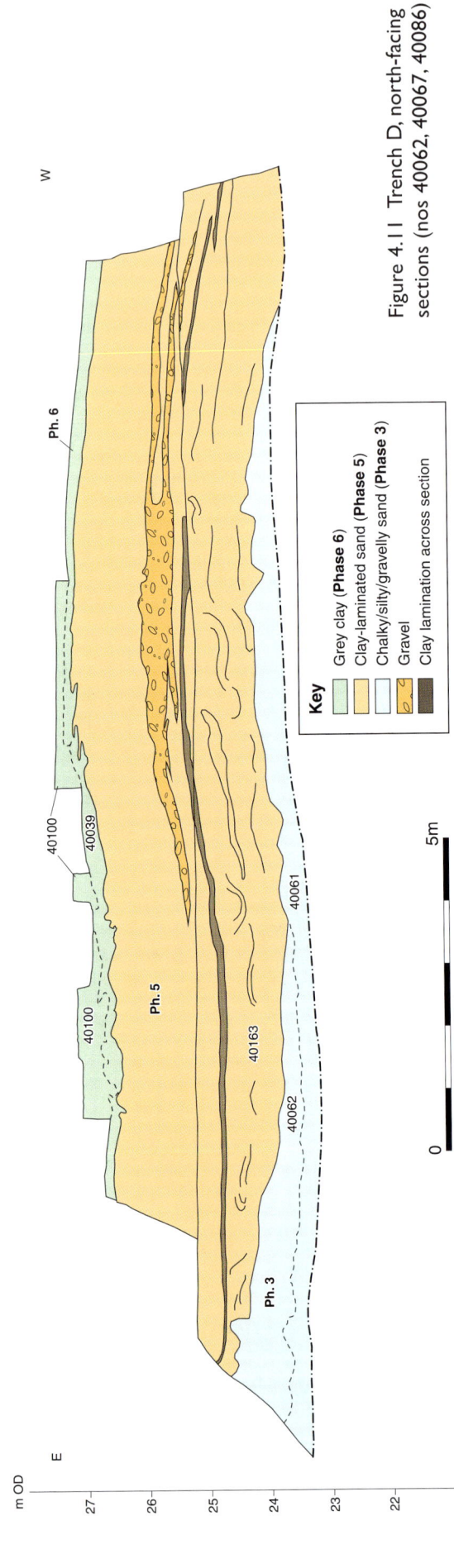

Figure 4.11 Trench D, north-facing sections (nos 40062, 40067, 40086)

Key

Grey clay (**Phase 6**)

Clay-laminated sand (**Phase 5**)

Chalky/silty/gravelly sand (**Phase 3**)

Gravel

Clay lamination across section

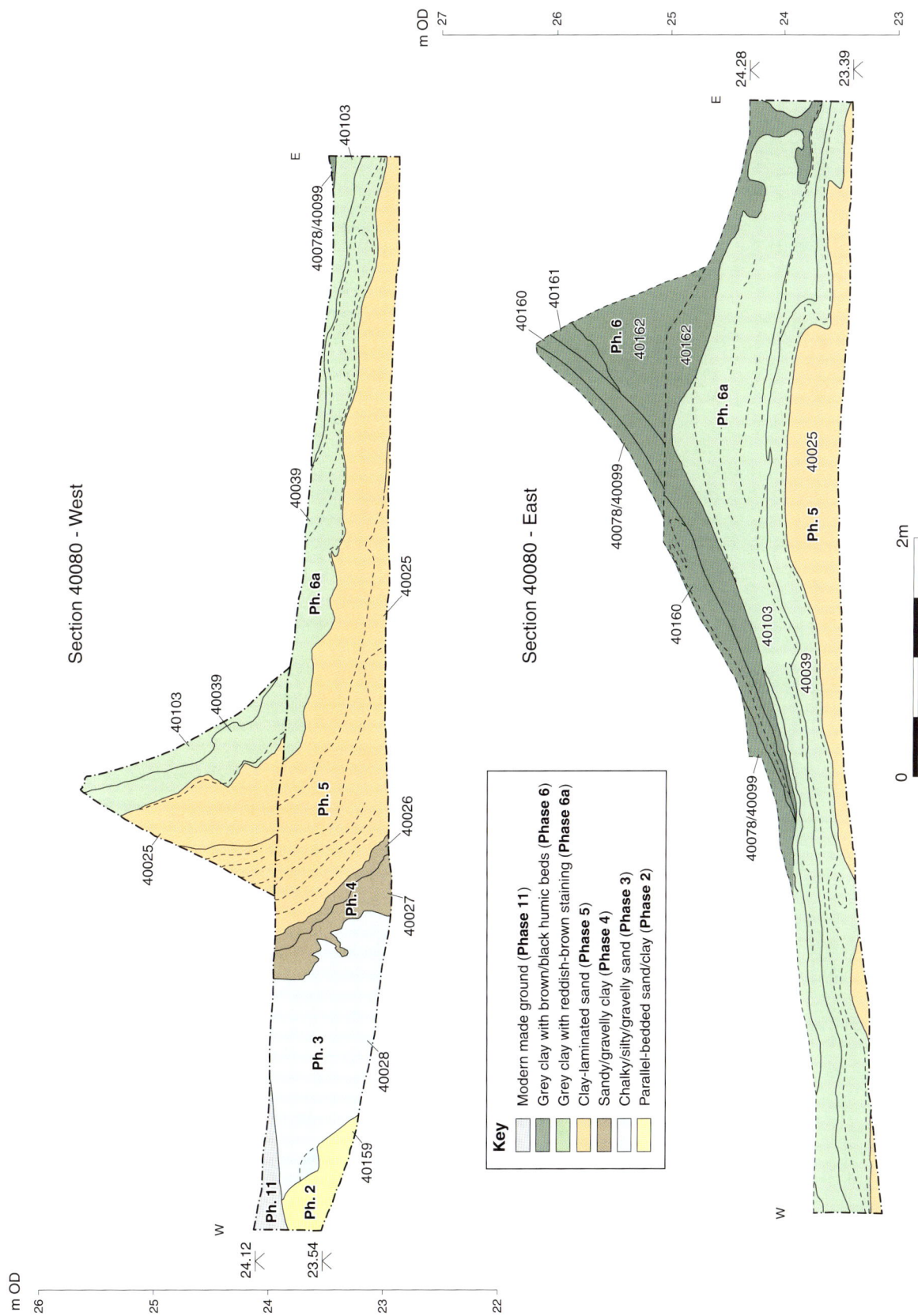

Figure 4.12 Section 40080, south-facing ('skateboard ramp')

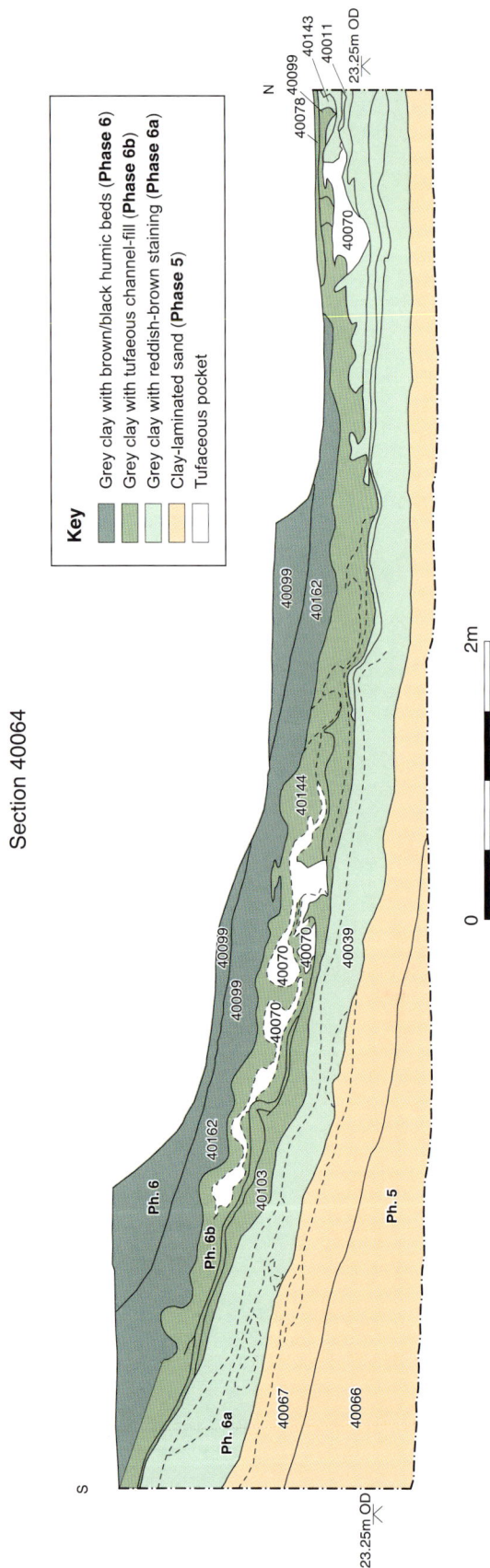

Section 40064

Figure 4.13 Section 40064, showing tufaceous pockets (context 40070) rich in mollusc and small vertebrate remains

Key

Grey clay with brown/black humic beds (**Phase 6**)

Grey clay with tufaeous channel-fill (**Phase 6b**)

Grey clay with reddish-brown staining (**Phase 6a**)

Clay-laminated sand (**Phase 5**)

Tufaceous pocket

- At the base, fractured Chalk (40054), clayey greenish-grey sand with dark-green stained flint nodules with an orange-stained band just beneath their cortical surface (Bullhead flint bed) (40055) and a massive, slightly clayey greenish/greyish-brown sand (Thanet Beds) (40056).

- A series of three overlying sand-dominated sediments, contexts 40057-40059. The deposits within this block are unconformably overlain by the Phase 2 sequence.

The basal three contexts, 40054-40056, appear to represent a body of bedrock moved and re-orientated from the local nearly sub-horizontal junction of the Chalk with the overlying Tertiary beds. Particularly puzzling is the fact that the surface of the Chalk in this tilted block dips steeply in precisely the direction where it is known from previous records that Chalk was present close-by in an anomalously high position in Southfleet Pit (Fig. 4.3). These lower units of the 'tilted block' have clearly been detached *en masse* from regional bedrock; but how this has happened, and from where the block originates, remain entirely unknown. This evidence of disruption may, however, be related to the larger scale disruption reflected in the anomalous chalk-high of Southfleet Pit. Whatever the explanation, the observed effect is that Chalk outcrops, tilted or otherwise, are present in the local landscape in anomalous positions. This has implications for hominin activity, since these Chalk outcrops, or detached masses, provide a source of flint raw material; this matter is returned to subsequently (Chapter 22).

The overlying series of sediments is dominated by sands and clays. Context 40057 consists of a fine to medium sand with a grey black colour possibly associated with weathering and pedogenic activity. Flecking parallel with the lower boundary suggests the possibility that this context was once laminated and the presence of root tubules supports the notion of localised pedogenic activity. The overlying clayey sands (40058, 40059) are characterised by brecciation and shearing, possibly indicative of deposition in low-energy conditions followed by drying out and sub-aerial weathering. The shearing may have occurred in conjunction with overturning of the full block sequence. Although the stratigraphic relationship between the two groups of sediments within the tilted block is clear, the temporal relationship between the two is difficult to resolve, and no environmental or dating evidence was recovered to address this further.

Phase 1-2 transition. Chalk/flint rubble

A single bed (context 40077) exists at the junction between the 'tilted block' sequence and the overlying alternating beds of sand/clay-silt that constitute Phase 2 of the site sequence (Fig. 4.16, section composite 40017-40019). This bed consists of very poorly sorted chalk and flint rubble in a sandy/clayey matrix with solid lumps of greenish-greyish sand clearly derived from context 40056. The bed dips very steeply (*c* 50°) west, therefore

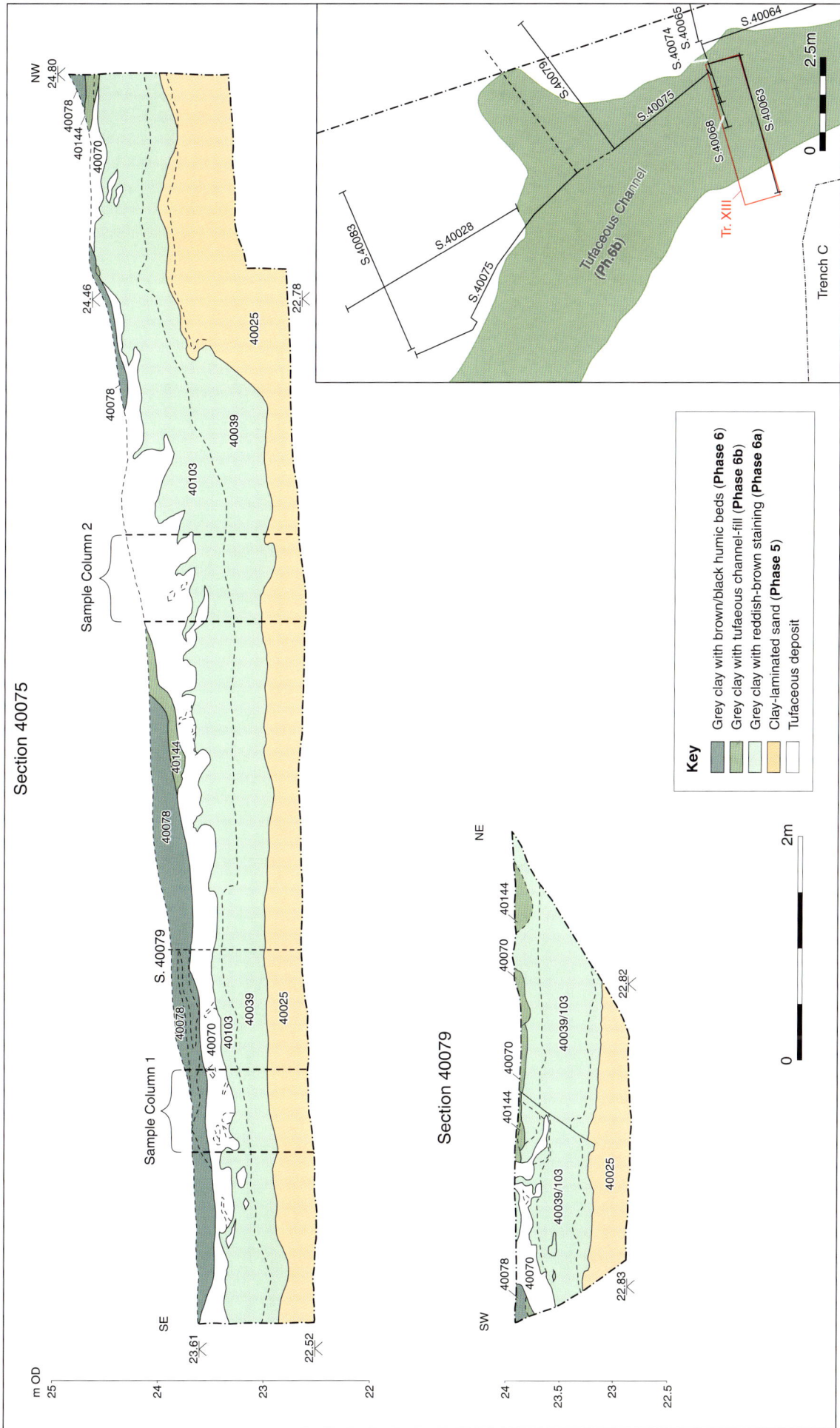

Figure 4.14 Tufaceous channel, main longitudinal section, east-facing (nos 40075, 40082)

Figure 4.15 Tufaceous channel, cross-sections

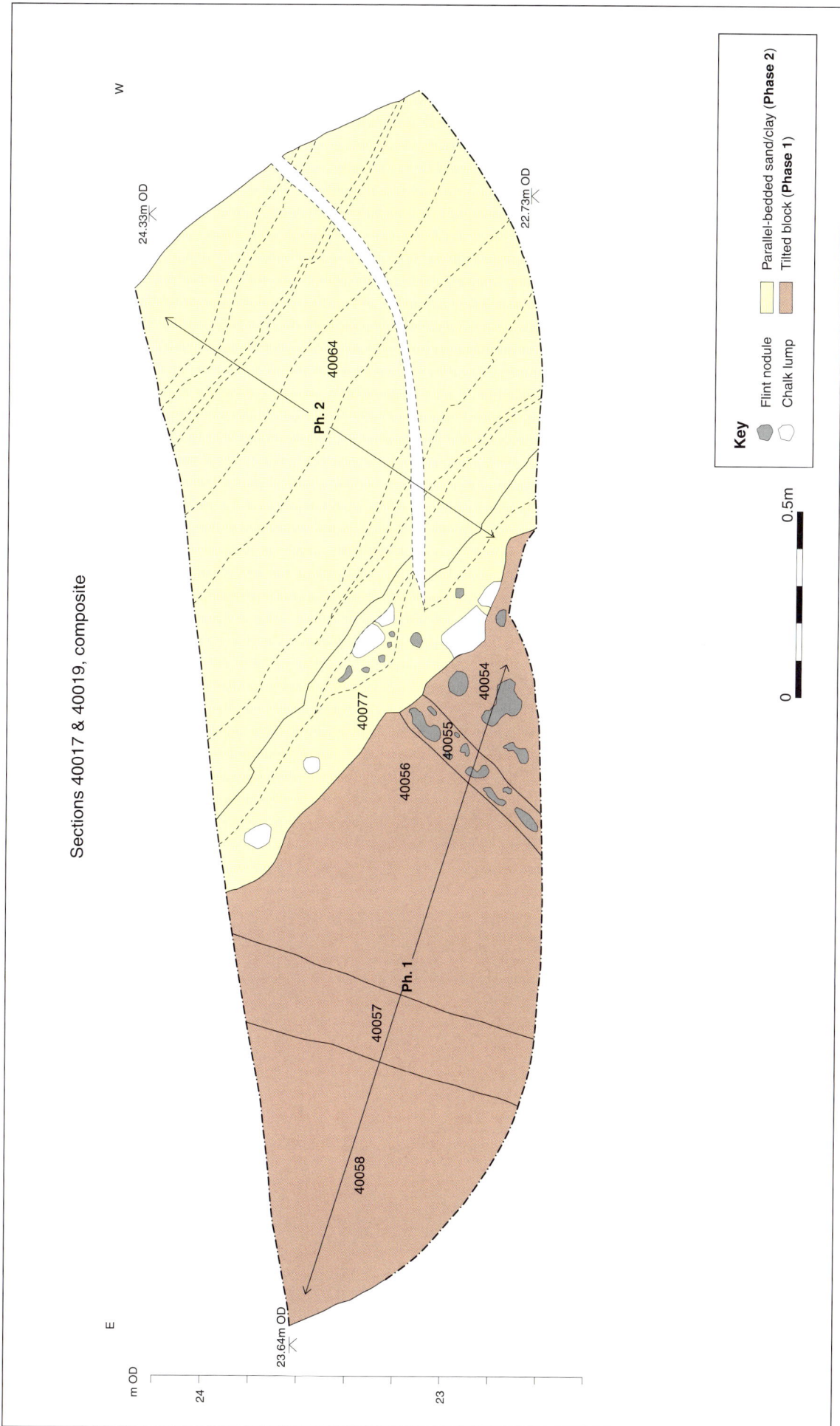

Sections 40017 & 40019, composite

Key

Flint nodule   Parallel-bedded sand/clay (**Phase 2**)
Chalk lump   Tilted block (**Phase 1**)

Figure 4.16 Composite section (40017-40019) showing Phase 1-2 junction in south-east of site

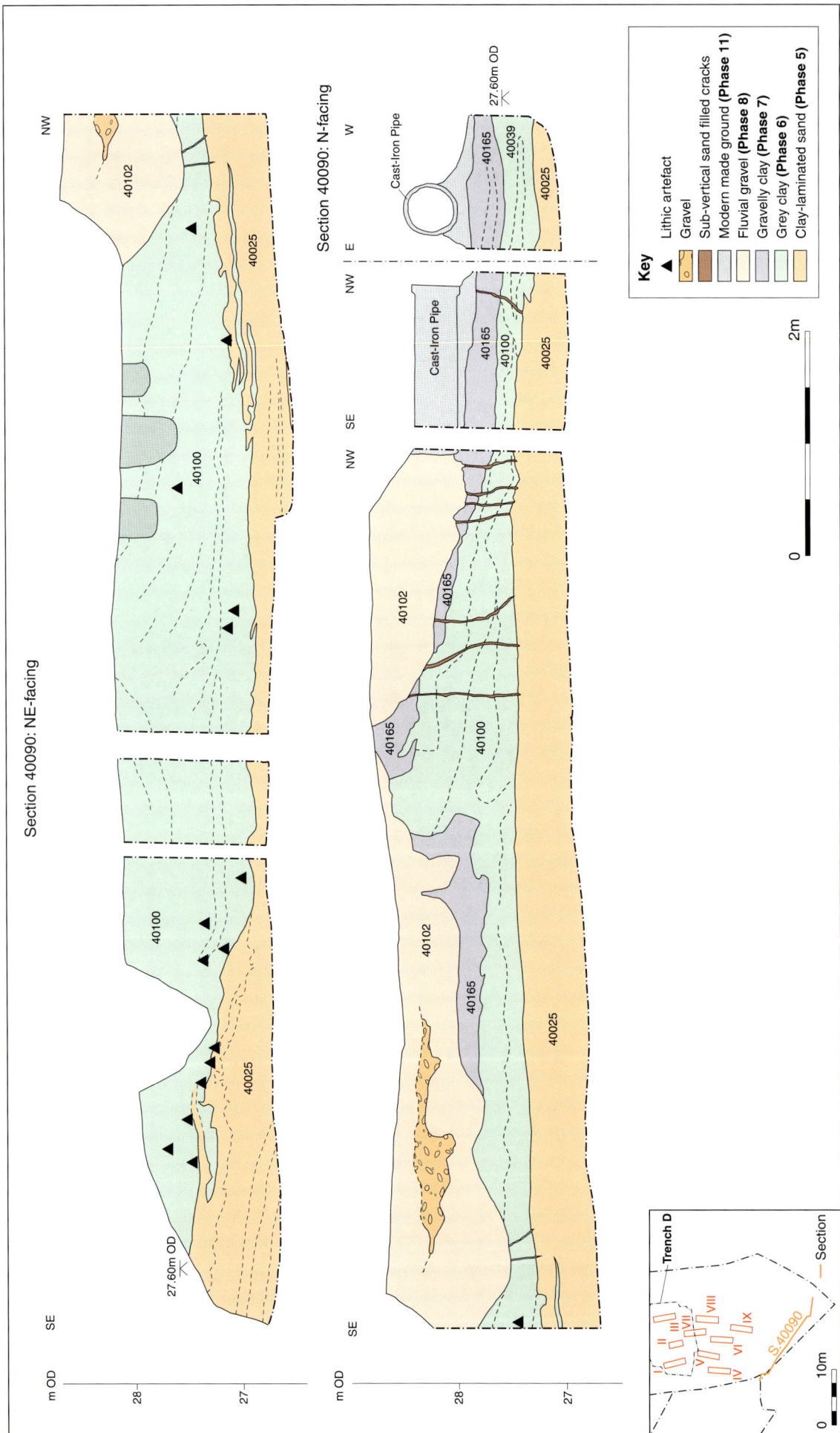

Figure 4.17 Section 40090, north-east facing (southern end of excavated area south of Trench D)

unconformably cuts across the strata of the 'tilted block', while providing a conformable base for the overlying Phase 2 deposits. The chalk clasts vary from small pebbles to cobbles, between angular and sub-angular in shape, and slightly to moderately abraded. Lateral variation is noted in the context, with greater sorting in places. This appears to be either a slopewash or solifluction deposit developed over the tilted block, or a gravel erosion lag. Crucial to its understanding is consideration of whether its current geometry corresponds with the situation when it was formed, or whether there has been subsequent reorientation of deposits *in situ*. The former seems more likely, as subsequent reorientation would have affected overlying deposits that, however, retain undisrupted parallel bedding. Therefore this transitional bed can best be explained as slopewash deposits descending down a steep slope to the west, filling a depression in that direction, with infilling continued with the Phase 2 deposits.

Phase 2. Parallel-bedded sand and clay

Deposits of this phase are only present in the lower levels of the south-east quadrant of the site, at the southern end of the main east-facing section, no. 40016 (Fig. 4.6a) and in the north-facing sections, nos. 40017 and 40019 (Fig. 4.16). The deposits occur in two lithostratigraphically unconnected groups, one on the northern side of the 'tilted block' (context 40060) and the other to its south (contexts 40064 and 40065). The more northerly unit (40060) is dominated by alternating beds of fine sand and clayey fine sand becoming coarser upwards. The bedding is essentially planar and dipping to the north-west. The more southerly units (contexts 40064 and 40065) likewise exhibit parallel sand and silty-clay interbeds, dipping steeply west and conformably overlying the chalk/flint rubble of the Phase 1-2 transitional context 40077. This can be seen at the southern end of site, in the background of the photo of the 'tilted block' in section (Fig. 3.8a).

Phase 3. Clayey/silty sand with flint and chalk gravel

The upper parts of sediments of this phase outcropped at the base of exposures in various locations, namely:

- At the southern end of the main east-facing section, above the more northerly unit (context 40060) of the Phase 2 deposits (Fig. 4.6a, contexts 40061, 40062 and 40063).
- In the central part of the main west-facing section, at the base of the sequence (Fig. 4.5b, context 40028; Fig. 3.4, Log 40011).
- In the stripped surface to the west of the main west-facing section, where they formed (in plan view) a pale chalk-rich elongated patch oriented north-west to south-east (Fig. 4.20).

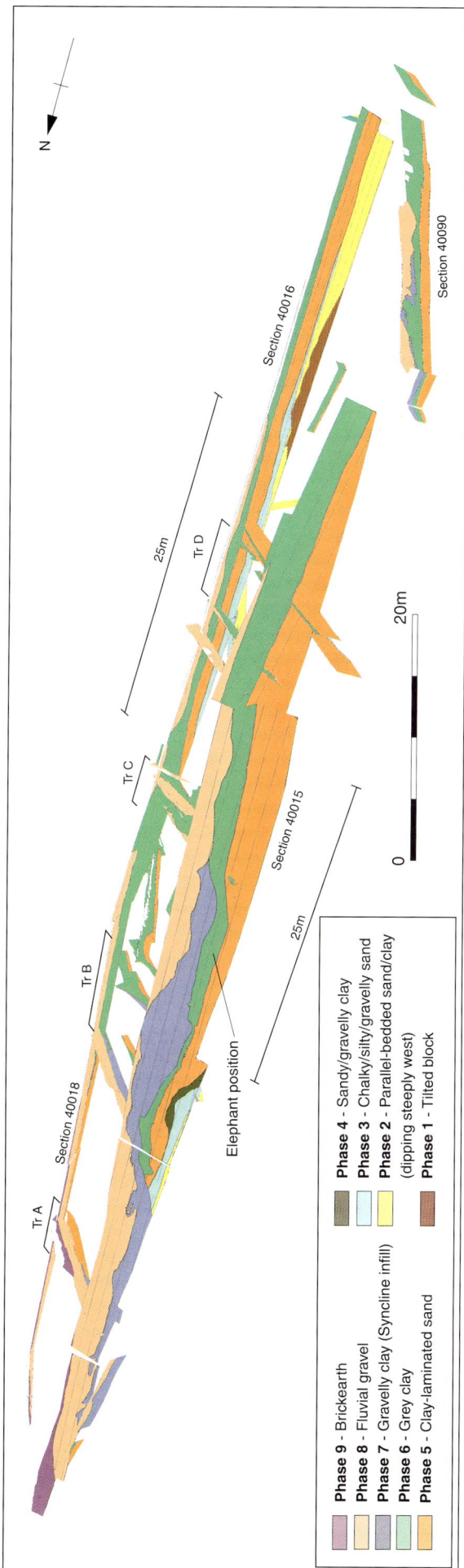

Figure 4.18 Overview of site 3D model, using section drawings as integrated panels coloured by sequence phase (looking north-east)

Phase 9 - Brickearth
Phase 8 - Fluvial gravel
Phase 7 - Gravelly clay (Syncline infill)
Phase 6 - Grey clay
Phase 5 - Clay-laminated sand

Phase 4 - Sandy/gravelly clay
Phase 3 - Chalky/silty/gravelly sand
Phase 2 - Parallel-bedded sand/clay (dipping steeply west)
Phase 1 - Tilted block

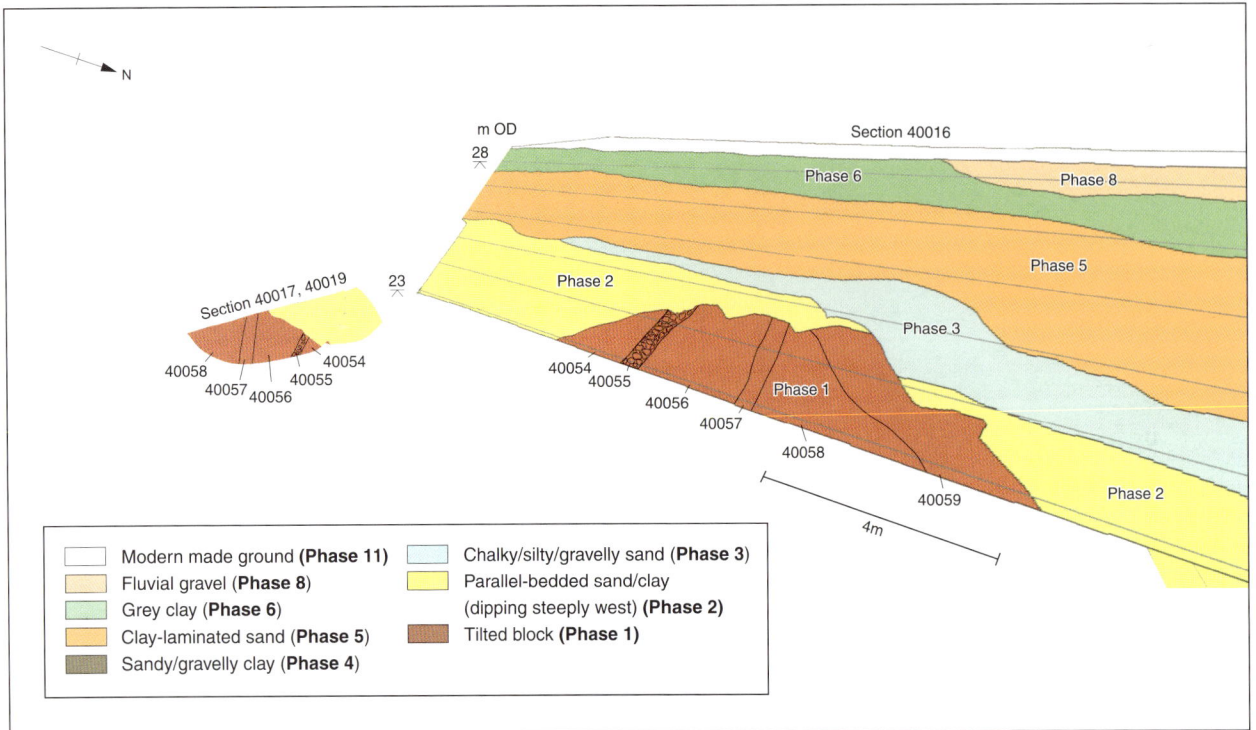

Figure 4.19 Phase 1: 'Tilted block' (looking south-west)

Figure 4.20 Stripped surface to the west of the main west-facing Section 40015 (looking north)

Although not directly recorded in a continuous section, it became clear during the deeper excavations between Trenches B and D that there was a direct sedimentary connection between context 40028 on the east side of the site and 40062 on the west side (Fig. 4.21). This allowed these deposits to be confidently grouped in the same phase, on lithostratigraphic grounds as well as lithological similarities. Phase 3 deposits were also observed at the western ends of the east-west sections 40080 and 40091, between Trenches B and C (Fig. 4.12; Fig. 4.8).

The upper surface of the Phase 3 deposits was observed to pass sub-horizontally east-west across the site at the bottom of Trench D (Fig. 4.11). However, this was probably an unconformable erosional boundary formed by the base of the Phase 5 deposits, and so not indicative of the geometry of the Phase 3 group. The base of the Phase 3 sediments was rarely seen and may also have been affected by the same post-depositional

deformation in the central part of the site as the overlying sediments of Phases 4, 5 and 6 (discussed below). Hence their original geometry remains uncertain, as does their conformability with the underlying Phase 2 sediments. At the southern end of the site, it appears in the bottom part of the main east-facing section that context 40061 at the base of Phase 3 is near-conformable with the top of the Phase 2 sequence (Fig. 4.6a). Further west, however, the Phase 3 deposits of contexts 40062 and 40063, higher up the Phase 3 sequence, lack bedding structures and are unconformable with the well-bedded contexts 40064 and 40065 of Phase 2.

When excavated, the Phase 3 sediments in the central part of the site, between Trenches B and D, formed the basal horizons at the west side of the markedly U-shaped synclinal 'skateboard ramp' feature (Fig. 3.18; Fig. 4.8; Fig. 4.22). Further west, the Phase 3 sediments were exposed in plan, along the route of the cutting for the

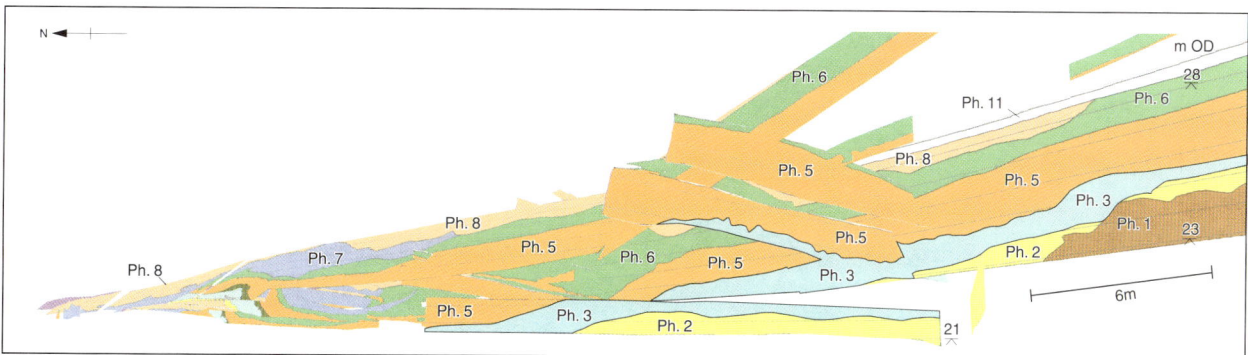

Figure 4.21 Phase 3 (looking up, north-east)

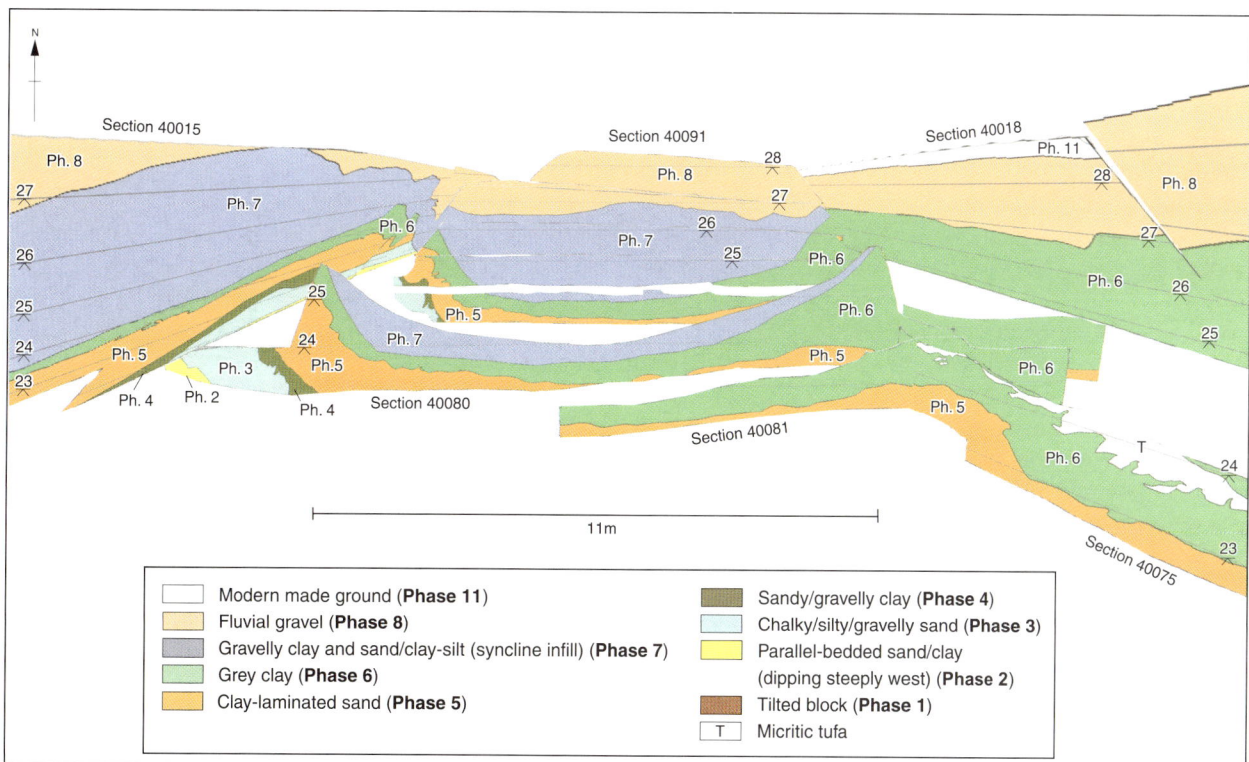

Figure 4.22 Syncline infill, aka 'skateboard ramp' (looking north)

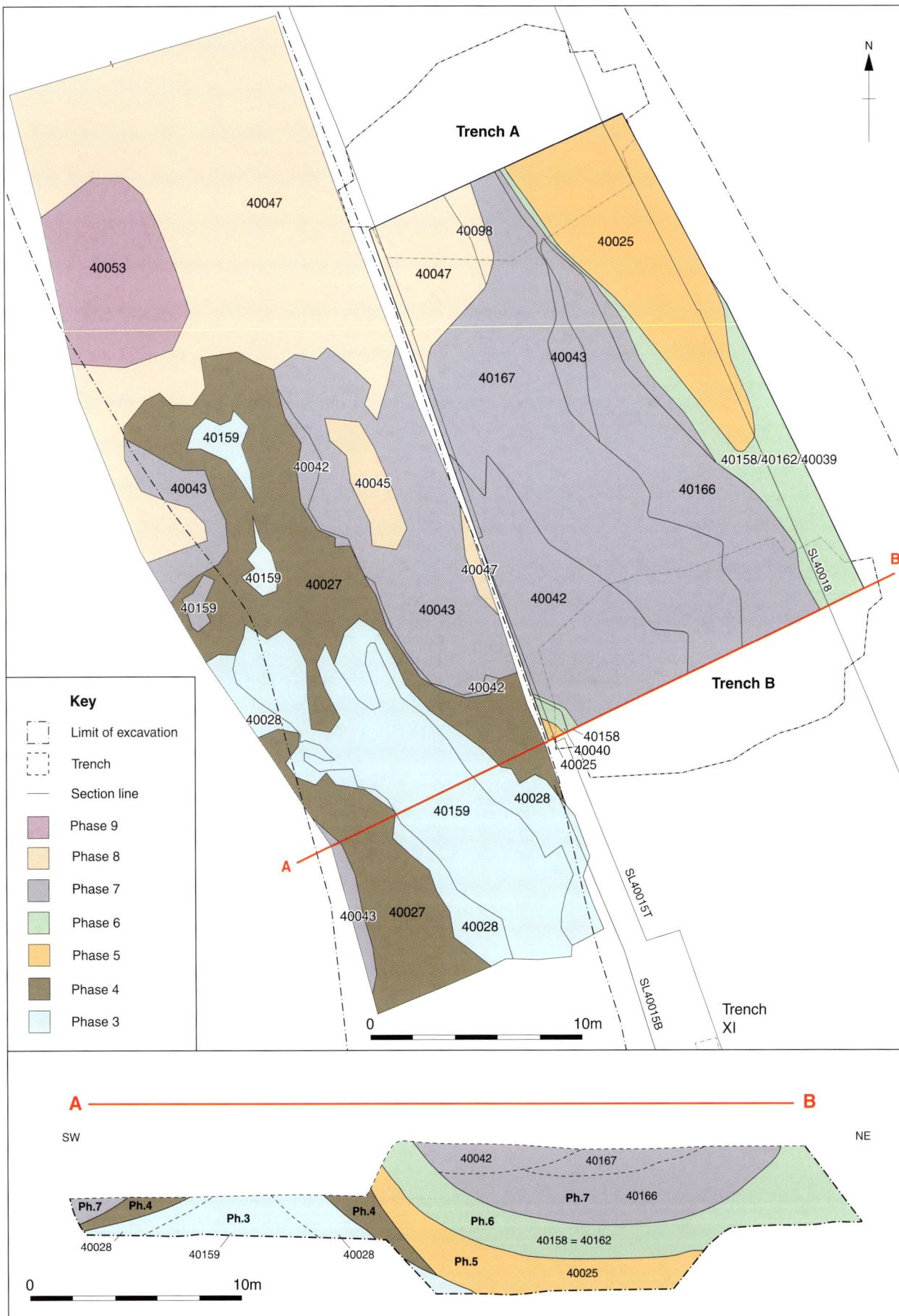

Figure 4.23 Stripped surface to the west of the main west-facing section: plan view showing phasing and diagrammatic cross-section

new Southfleet Road link. Here they formed an elongated tongue oriented roughly north-west to south-east, with the upper surface dipping away again on the west side (Fig. 4.23).

There were two groups of Phase 3 sediments. The upper group (contexts 40028, 40062 and 40063) typically consisted of grey-brown structureless clayey/silty sand with flint and chalk pebbles, the latter decreasing in size and less common eastward across the site. These deposits mostly had a moderately high presence of Tertiary shell fragments, apart from in their upper parts in the main east-facing section, where in places the sediments were decalcified and also lacking chalk clasts (Fig. 4.6, context 40063). The lower group (context 40061) comprised flint gravel beds with common chalk clasts in a silty/sandy matrix rich in Tertiary shell fragments.

Evidence of frost-pitting and sub-angular spalls was noted to be common amongst the flint clasts. Although not bedded, some contexts (for example 40061, which was more gravel-rich) contained evidence of the sub-horizontal alignment of shell fragment and flint pebble long-axes. Minor erosional unconformities were noted in places, for example at the base of 40062. Patches of included sand were noted in some contexts (eg 40063). This phase of deposition is interpreted as an inter-mixing of overbank and slope deposits, forming clay-sand-silt sediments rich in flint and chalk clasts and derived Tertiary shell fragments. The coarse lower bed (40061) may represent a short-lived high-energy fluvial event at the base of the sequence or a narrow torrent of gravelly material washing down the valley side from the west.

Sediments regarded as probably equivalent to Phase 3 were also recorded in subsequent test pits to the east of the site, excavated as part of the field evaluation of Station Quarter South (Wessex Archaeology 2006b). They were also seen to the west of the site as part of the field evaluation of Northfleet West Sub-station (Museum of London Archaeology 2011) (Fig. 4.1). These provided important additional information on the distribution and geometry of Phase 3 deposits, discussed in the following section, below. Furthermore, some important faunal evidence was recovered from Phase 3 deposits, including large vertebrate specimens, *Bithynia* opercula and a rich ostracod fauna. These are discussed subsequently (Chapters 7, 9, 11 and 13) and have additional bearing on the interpretation of the sediments (Chapter 22).

Phase 4. Sandy-gravelly clay

Deposits of this phase were very restricted at the main site, being only present towards the base of the central part of the main west-facing section (Fig. 4.5a, b), shown more closely in Log 40011 (Fig. 3.4). The basal part of this phase consists of a fine clayey sand with some flints (context 40027) passing upwards to a massive grey/brown clay (context 40026). A possible bimodal grain size in the basal unit may be interpreted as a marginal fluvial sand with occasional stronger flows importing clasts giving way up-profile to a very low

energy deposition of clays in a small water body with possible admixtures of colluvially derived material. Mixing across the boundary between the two contexts indicates the sediments were probably saturated after deposition. These sediments dip steeply eastwards, being towards the base, and at the western side, of the series of deposits appearing in a marked synclinal U-shape between Trenches B and C – the 'skateboard ramp' (Fig 3.18; Fig. 4.22). This geometry is regarded as post-depositional (see discussion below), and so has no bearing on interpretation of formation processes with regard to the Phase 4 sediments. Phase 4 sediments did not re-appear on the east side of the skateboard ramp', but a much greater thickness of probably equivalent sediments was observed further east in test pits dug in 2006 (Fig. 4.1) during field evaluation of the Station Quarter South area (Wessex Archaeology 2006b). These are discussed below.

Phase 5. Clay-laminated sands

These sediments (contexts 40025, 40066, 40067 and 40072) were widespread across the site, being the main basal deposit, almost three metres thick, in the southern side of the main west-facing section (Fig. 4.5a, b), and a significant presence in the southern and central parts of the main east-facing section (Fig. 4.6a, b). The links across the site between these two exposures were also directly seen in the transverse east-west sections of Trenches B (Fig. 4.8) and C (Fig. 4.9) and section 40080 (Fig. 4.12; Fig. 4.24). The sediments linked to this phase varied somewhat across the site but typically were brownish-yellow sand with grey clay or clay-silt interbeds, with some reworked Tertiary shell fragments and occasional thin beds of flint pebbles. The sand was planar-bedded, dipping to the north parallel with the junction of their upper surface with the base of the Phase 6 clay deposits, suggesting a broadly conformable relationship between these two sets of deposits. Small ripples were noted in places, along with a number of reactivation surfaces indicating breaks in sedimentation. The silty-clay interbeds were thicker and more frequent higher in the sequence (40067), where they were often highly deformed by loading, suggesting saturated conditions. Gravity folding (*boudinage*), representing soft-sediment deformation under the influence of gravity, was also observed in the central part of the main west-facing section (Fig. 4.5; Fig. 3.4) where the sediments dipped steeply east, forming the western end of the U-shaped 'skateboard ramp' (Fig. 4.22), suggesting post-depositional deformation in a saturated, fluid state.

This sequence appears to reflect local deltaic conditions with alternating sediment supply sources at the edge of a water body. The planar bedding (with occasional ripples) and reactivation surfaces suggest alluvial fan or delta aggradation in a fluctuating fluvial environment with input from overbank colluvial slopewash and surface run-off. The flint pebble beds reflect short-lived episodes of higher fluvial energy alternating with periods of quiet water. Progression

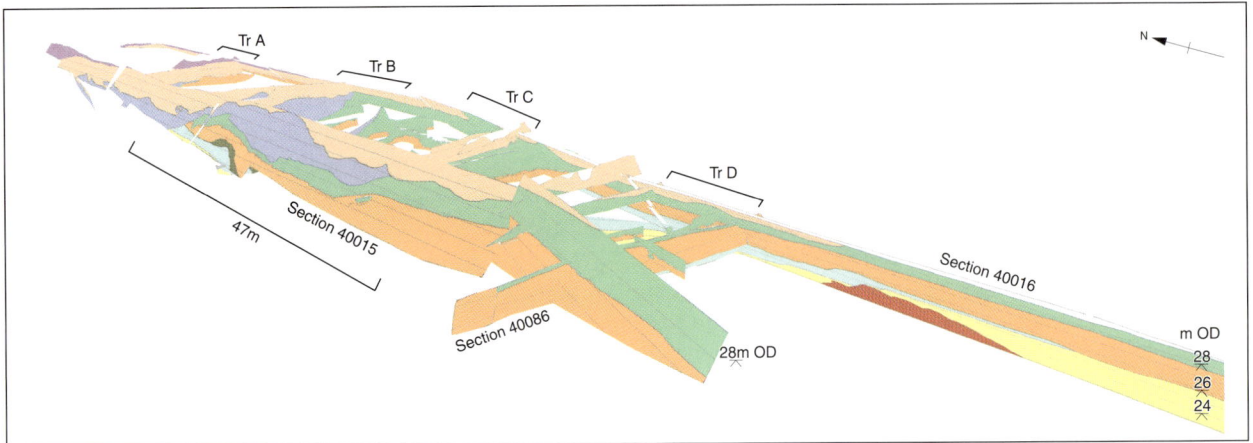

Figure 4.24 Phase 5 sediments (looking down, north-east)

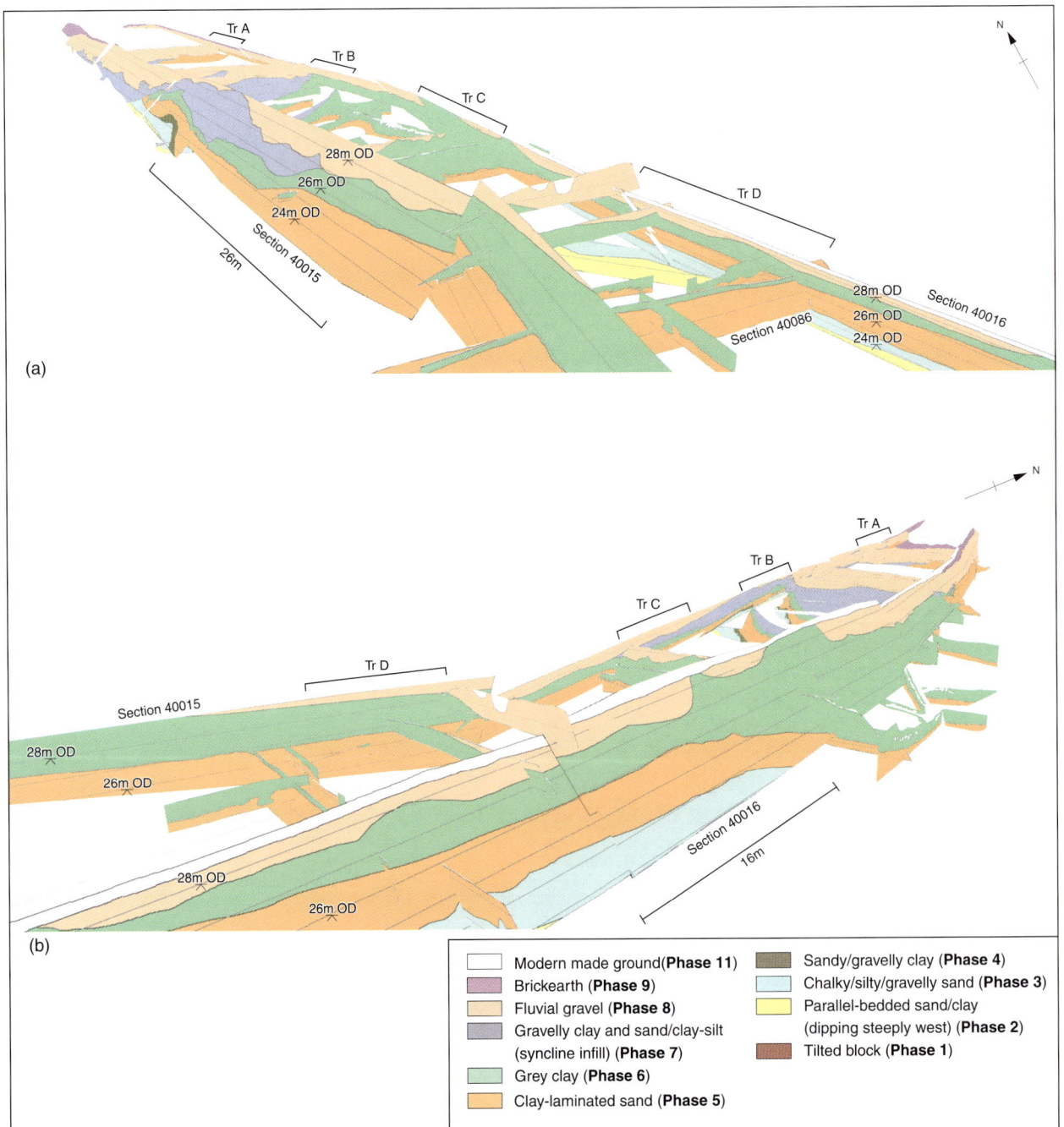

Figure 4.25 Phase 6 sediments: (a) looking down, north-east; (b) looking down, north-west

towards the increased occurrence of low energy conditions are indicated by the presence of the thicker, more frequent fine-grained units upwards. This may also suggest the development of a more extensive low energy water feature in the landscape, such as a very quiet river, or even a lake.

Phase 6. Grey clay with brown humic beds

The sediments associated with this phase (Table 4.1; Table 4.2) are geometrically complex (Fig. 4.25a,b) and also vary laterally. They predominantly consist of grey brecciated clay, often with areas of minor slicken-side faulting, and with undulating, intersecting thin oxidised iron-rich silty bands around 10mm thick in their lower parts in the centre of the site. The clay includes various sandier and siltier facies. These sediments are often associated with colour variations from pale grey to reddish and brownish, almost black in places. The more brownish and black beds are concentrated in the lower parts of the sediments in the centre of the site, where the elephant skeleton was found and associated with organic-rich beds with visible fragments of rotted plant macrofossils. The Phase 6 clay also included (besides concentrations of lithic artefacts) a range of natural flint clasts. These were very rare through most of the deposit, but increasingly common (and generally larger) in the area of the lithic concentration south of Trench D, varying in size from fine gravel to small cobbles, and generally considerably abraded. The depositional and site-formation implications of this are discussed more fully below, in conjunction with the refitting and lithic artefact distributional data (Chapters 17 and 18). The presence of large natural clasts clearly reflects at least some colluvial and/or slopewash input, although, as discussed below, the Phase 6 deposits are thought predominantly to represent deposition in quiet or standing water, with periods of desiccation and the regular exposure of palaeo-landsurfaces.

The Phase 6 sediments form a significant element in the southern parts of the main east- and west-facing sections (Fig. 4.5; Fig. 4.6; Fig. 4.27; Fig. 4.28), and in the transverse east-west sections of Trenches B, C and D between these faces (Fig. 4.8; Fig. 4.9; Fig. 4.10). In general, the Phase 6 sediments wedge out from south to north between Trench D and A. They become significantly deformed in their thinner, northern part into the U-shaped 'skateboard ramp' between Trench C and B, and fade away further north to become a vestigial layer about 10mm thick at Trench A (Fig. 4.7). These sediments contained many of the most important remains found at the site, including the *Palaeoloxodon* skeleton and its associated flint artefacts, the tufaceous channel and its rich micro-faunal remains (Fig. 3.9; Fig. 4.29) and the dense concentration of (mostly) mint condition artefacts found south of Trench D (Fig. 4.26). Therefore their internal stratigraphy and the use of various context numbers within Phase 6 are presented here in some detail.

Description and interpretation of the internal sequence of contexts within Phase 6 is complicated by the varying thickness of the deposits and the use of different numbers for the same sediment in different parts of the site. Additionally, local sedimentary variations were given their own context numbers within the great clayey mass of the Phase 6 deposits (Table 4.2). Context 40039 was used for the full thickness of the Phase 6 deposits when first encountered; a relatively thin (*c* 250mm) dark brown brecciated clay bed in the central part of the main west-facing section (Fig. 4.5a, b). This number was also used, probably erroneously, to represent the strong brown clay at the base of the much thicker grey clay (context 40100) that constituted the Phase 6 deposits in the southern part of the site in the west-facing section. Different context numbers (40068 and 40069) were then used for the same grey clay in the main east-facing section (Fig. 4.6a), and for finer stratigraphic sub-division of the even thicker grey clay with intervening brown humic beds where the *Palaeoloxodon* was found (upwards through the sequence: contexts 40103, 40099, 40078 and 40100 – see Fig. 4.9, west end and Fig. 4.30).

A suite of additional numbers (from the base: 40070, 40144 and 40143) was then used for the sequence of deposits associated with the tufaceous channel, which was contained within the lower part of the Phase 6 grey clay (40068) on the eastern side of the site (Fig. 4.13, section 40064). As shown in Table 4.2, Fig. 4.9 and Fig. 4.22, it is important to note that, while there is no direct lithostratigraphic link between the elephant horizon and the tufaceous channel sequence, they both occurred in equivalent stratigraphic positions, within the bottom part of the Phase 6 clay. Most importantly, recovery of one of the elephant's foot bones from within the tufaceous channel sequence (Chapter 8) also established true contemporaneity of the elephant (and its associated hominin activity) with the rich environmental remains from the tufaceous channel.

Towards the end of the excavation, it was found that the dark brown brecciated clay originally identified as 40039 thinned northward and became almost black, and here it was given a new context number (40158), as seen in the final south-facing section of Trench B (Fig. 4.8: section 40091). Finally, three further context numbers (40160, 40161 and 40162) were later applied to distinct sandy and silty beds within the Phase 6 grey clay in the central eastern part of the site (Fig. 4.12, section 40080).

Two subsidiary phases were identified within Phase 6 (Table 4.2). These were:

1. A basal Phase 6a, representing 40039 (in the central and southern parts of the site where this context did not represent the full thickness of Phase 6), 40040 and 40103.

2. Phase 6b, representing the three contexts of the tufaceous channel-fill sequence (40070, 40144 and 40143).

Table 4.2 Phase 6 contexts, showing relationships and internal phasing across site

Phase	Trenches A-B		Trenches C-D				South of Trench D
6			40100		40160 40161	40069	40100
6 – Elephant			40078	40144			
6b				40143	40162	40068	
			40099	40070			
6a	40158	40039	40103 40039				

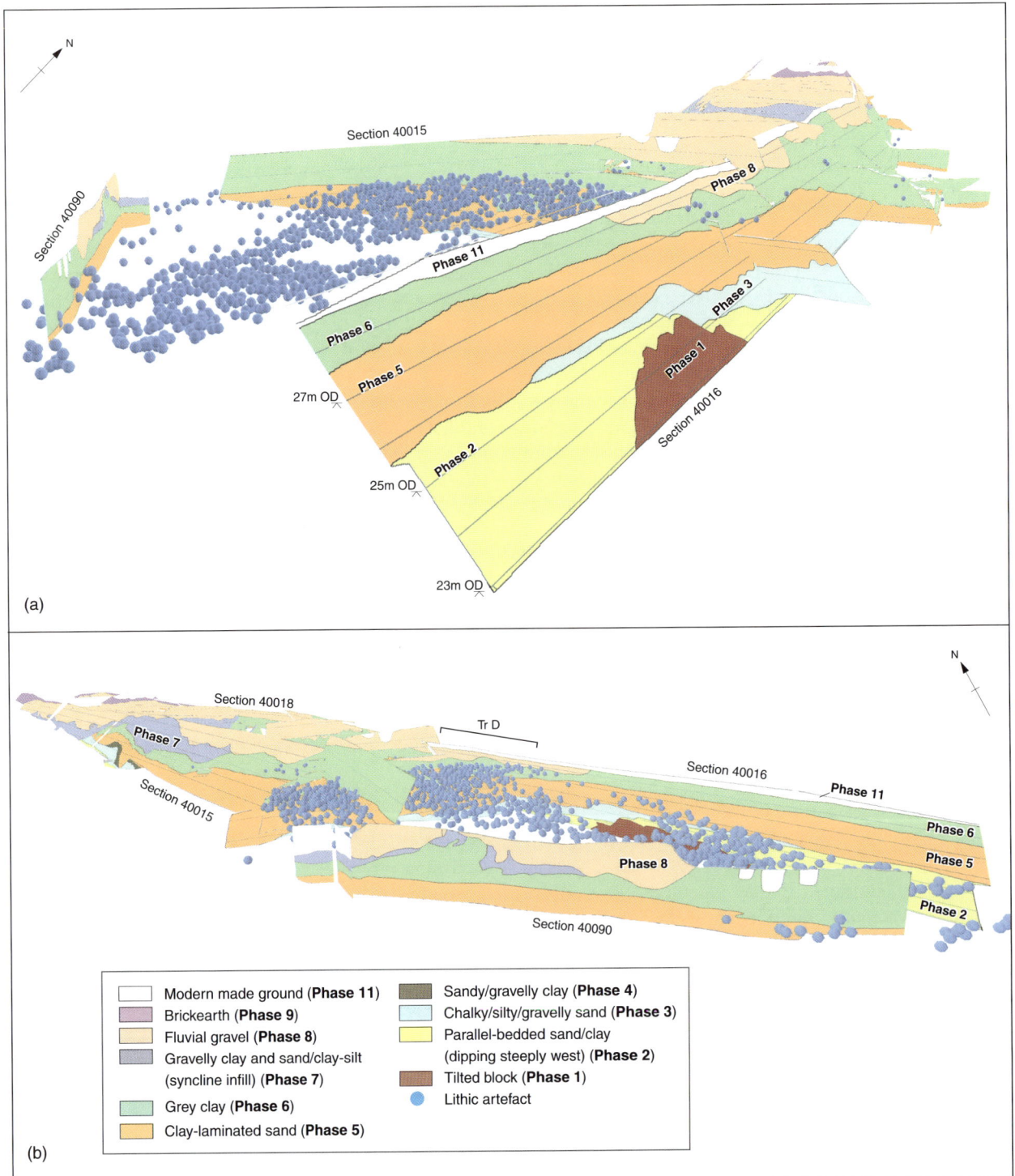

Figure 4.26 Lithic concentration south of Trench D: (a) looking down, north-west; (b) looking down, north-east

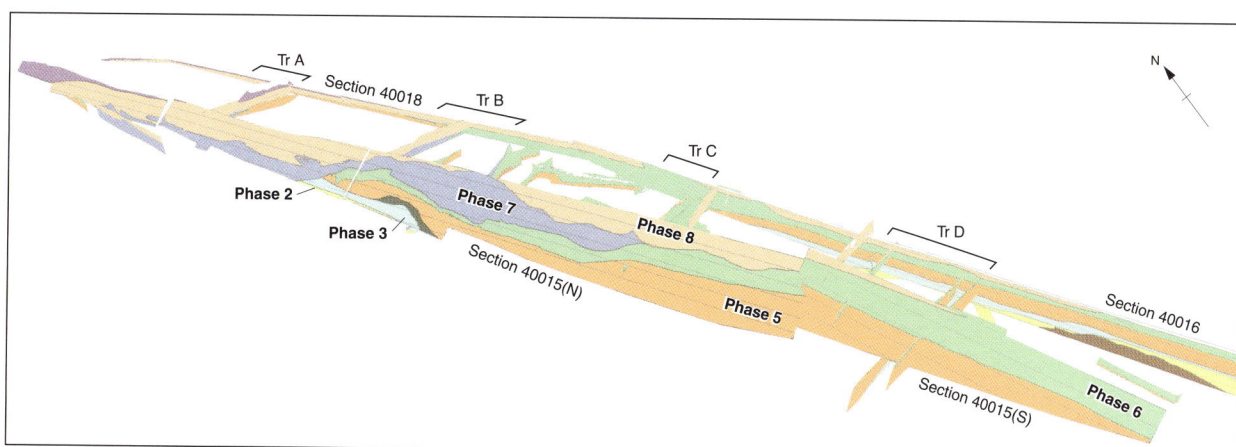

Figure 4.27 Main west-facing section, no. 40015 (looking north-east)

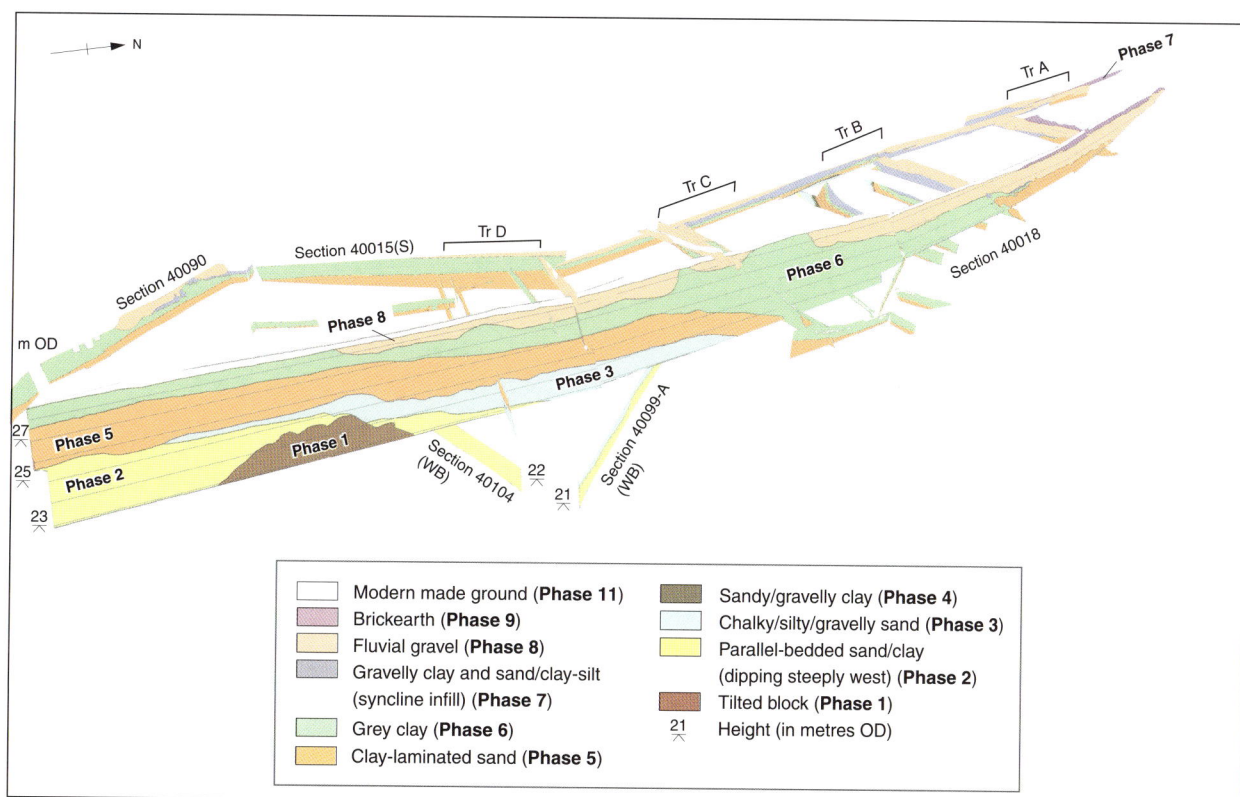

Legend:
- Modern made ground (**Phase 11**)
- Brickearth (**Phase 9**)
- Fluvial gravel (**Phase 8**)
- Gravelly clay and sand/clay-silt (syncline infill) (**Phase 7**)
- Grey clay (**Phase 6**)
- Clay-laminated sand (**Phase 5**)
- Sandy/gravelly clay (**Phase 4**)
- Chalky/silty/gravelly sand (**Phase 3**)
- Parallel-bedded sand/clay (dipping steeply west) (**Phase 2**)
- Tilted block (**Phase 1**)
- 21 Height (in metres OD)

Figure 4.28 Main east-facing section, nos. 40016 and 40018 (looking north-west)

The base of Phase 6 is broadly conformable with, and continues the increasingly clayey upward trend of, the underlying Phase 5 deposits. The basal context (40040) is a mid to dark grey sand and clayey silt with occasional small flint clasts suggesting continuation of fluvial conditions, but with a marked decrease in energy from the underlying Phase 5 sands. As discussed above, the overlying Phase 6 deposits varied in nature and thickness across the site. It was originally thought that the dark brown, almost black in places, brecciated clay (40039; 40158) that overlay this in the central part of the site represented *in situ* pedogenesis and a palaeo-landsurface. Subsequent sedimentary and thin section micromorphological investigations (Chapter 5) indicated that, rather than there just being one land surface, these darker sediments represent regular influxes of sediment rich in organic remains by a combination of alluvial and

colluvial processes interspersed with episodes of desiccation and oxidation. This resulted in the brecciation of the deposit and the periodic exposure of short-lived palaeo-landsurfaces. These processes were evident upwards through the sequence in the central part of the site, near the elephant, with deposition of the grey clay (context 40103). There were sharply-defined undulating sub-horizontal bands of reddish-brown (iron-enriched) silt, with the top of this context being defined by a particularly well-defined reddish iron-rich band that extended across the site (Fig. 4.9; Fig. 4.30). This layer showed greater signs of soil development and may represent a longer-lived palaeo-landsurface, although no particular archaeological remains are associated with it.

Overlying this, and still in the central part of the site, occur the grey clays with browner, organic-rich beds associated with the area of the elephant skeleton. These

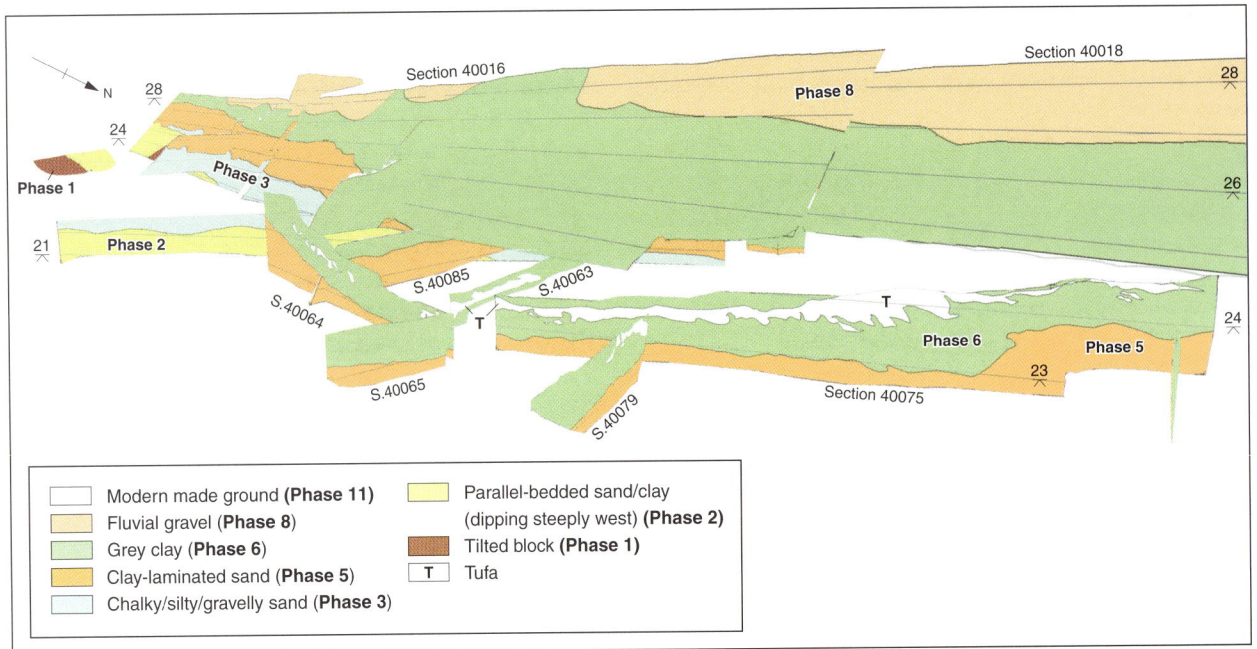

Figure 4.29 Tufaceous channel towards base of Phase 6 (looking south-west)

were best recorded in the west-facing section 40043 beside the elephant skeleton (Fig. 4.30), and at the west end of Trench C in the north-facing section 40085 (Fig. 4.9). The elephant skeleton was mostly contained in one particular brown organic-rich bed (context 40078), which dipped gently north, thickening from 20mm thick at its southernmost manifestation (in the west end of Trench C, Fig. 4.9) to about 150mm thick to the north of the elephant skeleton. At the elephant skeleton, 40078 was divided from the top of the underlying context (40103) by a thin pale grey clay bed (40099), and was overlain by the main body of the Phase 6 grey clay (40100), which rose and thinned southwards (Fig. 4.5). The flints associated with the elephant skeleton were also mostly recovered from the same context (40078) (Fig. 3.17), although several (and also several of the elephant bones) were embedded at their base in the underlying grey clay (40099).

After excavation of the elephant was finished, Trench C was deepened through the deposits to the east of the elephant. It was found that the lower Phase 6 deposits in the vicinity of the elephant were characterised by numerous different brown organic-rich beds, which rose and thinned to the south-east, often merging together as they did so (Fig. 4.9; Fig. 3.21). The apparent presence within these beds of rotted plant macrofossil remains was later complemented by the demonstration of increased loss-on-ignition (Chapter 5) and the recovery of pollen and an *Azolla* spore (Chapter 12), confirming they were associated with enhanced organic preservation. The soil micromorphological studies (Chapter 5) indicated a combination of alluvial and colluvial deposition, perhaps often into standing water, with the formation of peaty beds, now substantially decayed, and periodic desiccation.

Taken together, the evidence suggests that the elephant died (or was killed? – see Chapter 22) in a zone near the

edge of a quiet or still water-body, where fluctuating water levels led to the exposure of dry land alternating with the presence of shallow water and muddy, marshy/swampy conditions. There was some sedimentary input by colluvial slopewash processes from high ground capped by Tertiary sands, clays and gravels to the west. The sediments under the elephant skeleton are not disrupted and its bones, which are often weathered (Chapter 8), were all found within a relatively narrow horizon, certainly when compared to the thickness an intact skeleton could occupy. Consequently it seems likely that the elephant died on a firm palaeo-landsurface that was then shallowly inundated for long enough for the skeleton to be buried by fine-grained clays/silts with various plant remains and peat development.

Another important feature seen in the central part of the site, on its east side, was a small channel (Fig. 3.6; Fig. 4.13, section 40064). This was filled at its base with a very pale brown mollusc-rich sandy/silt (context 40070) overlain by a brownish-grey sandy silt (context 40144), which was capped in places by a thin, very intermittent white silt layer (context 40143). This feature (Fig. 3.9; Fig. 4.29) became known as the 'tufaceous channel' and is allocated to Phase 6b of the site sequence. The main, basal context (40070) consisted of fragments of tufa and micritic tufa, mixed with pale grey silts. Its base was very uneven, forming contorted pockets in places, and interdigitating with the underlying clay, indicating post-depositional deformation. The slightly coarser nature of the sediment in this feature, compared to the remainder of Phase 6, suggests higher energy conditions and that a small stream channel was developed across a dried out surface developed in the underlying sediments. The origin of the source of the fragmented tufa was probably a spring system, perhaps in cascades to the south, with subsequent erosion and transportation of the detritus from primary context and

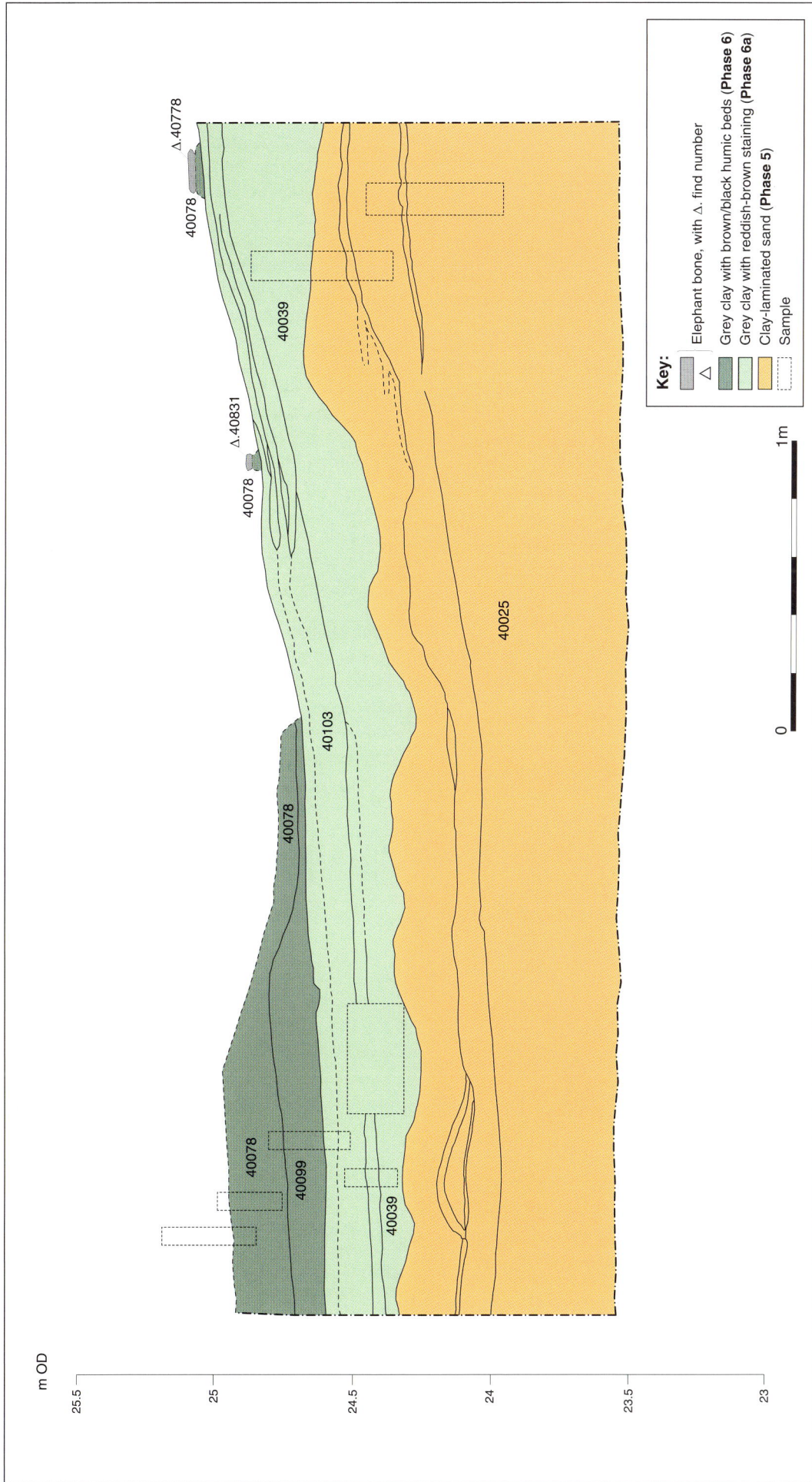

Figure 4.30 Section 40043, closer view of stratigraphy at *Palaeoloxodon* skeleton findspot

its redeposition in the channel. The tufaceous channel deposits were rich in molluscan and small vertebrate remains (Chapters 7, 9 and 10) and there were also ostracods in the overlying white silt of context 40143 (Chapter 11), making it of high importance for biostratigraphic, palaeo-environmental and palaeo-climatic interpretation. There was no direct lithostratigraphic relationship with context 40078 within which the elephant skeleton was found, but both context 40078 and the tufaceous channel occurred in equivalent levels in the bottom part of Phase 6 (Fig. 4.9). Both were above the palaeo-landsurface at the top of context 40103, and both were buried by the main body of clay (contexts 40068; 40100), so they were confidently regarded as broadly contemporary from a Pleistocene stratigraphical viewpoint. Subsequent identification of one of the elephant's well-preserved foot bones from within the tufaceous channel-fill sequence then established true landscape contemporaneity, suggesting the channel-fill was forming at the same time as the elephant carcass was decaying (Chapter 8).

The upper parts of the Phase 6 clays that bury the tufaceous channel and the elephant skeleton rise and extend south across the remainder of the site, thinning in their southern part where they contain the dense lithic concentration south of Trench D (Fig. 4.26). The sedimentological variations that give rise to the complex context sequences in the central part of the site are absent further south. Here there is a very simple sequence of a basal strong-brown-stained context (allocated context number 40039, as it initially appeared in the main west-facing section to be a direct lateral

continuation of the original 40039, in Log 40011, Fig. 3.4) overlain by a homogenous brecciated grey clay (40100). This latter context included some slightly sandier beds and some very faint reddish and yellowish stained sub-horizontal bands, which were not allocated individual context numbers. This suggests build-up in a muddy lacustrine or quiet fluvial backwater environment, with regular colluvial input and frequent desiccation indicated by textural variation and fine brecciation throughout the sequence. The presence of dipping browner beds associated with more abundant larger, coarser natural clasts in section 40090 at the south-west end of the site perhaps indicates that this part of the site is further up the valley-side bank. It may therefore perhaps be more prone to a higher colluvial/slopewash input as well as being slightly more amenable for hominin occupation. As discussed subsequently (Chapter 18), the evidence of the lithic refitting and distribution in this area suggests some minor spatial disturbance and movement, with artefacts having slumped in, or been re-arranged, *en masse* by colluvial/slopewash input from the south/south-west.

Phase 7. Mixed clays and gravels

Although there is generally a sharp boundary between the top of Phase 6 and the base of Phase 7, this is nowhere erosive and unconformable, suggesting no major hiatus between deposition of Phases 6 and 7. The sediments associated with Phase 7 were predominantly represented in the central part of the main west-facing section, no. 40015 (Fig. 4.5; Fig. 4.27) and the south-

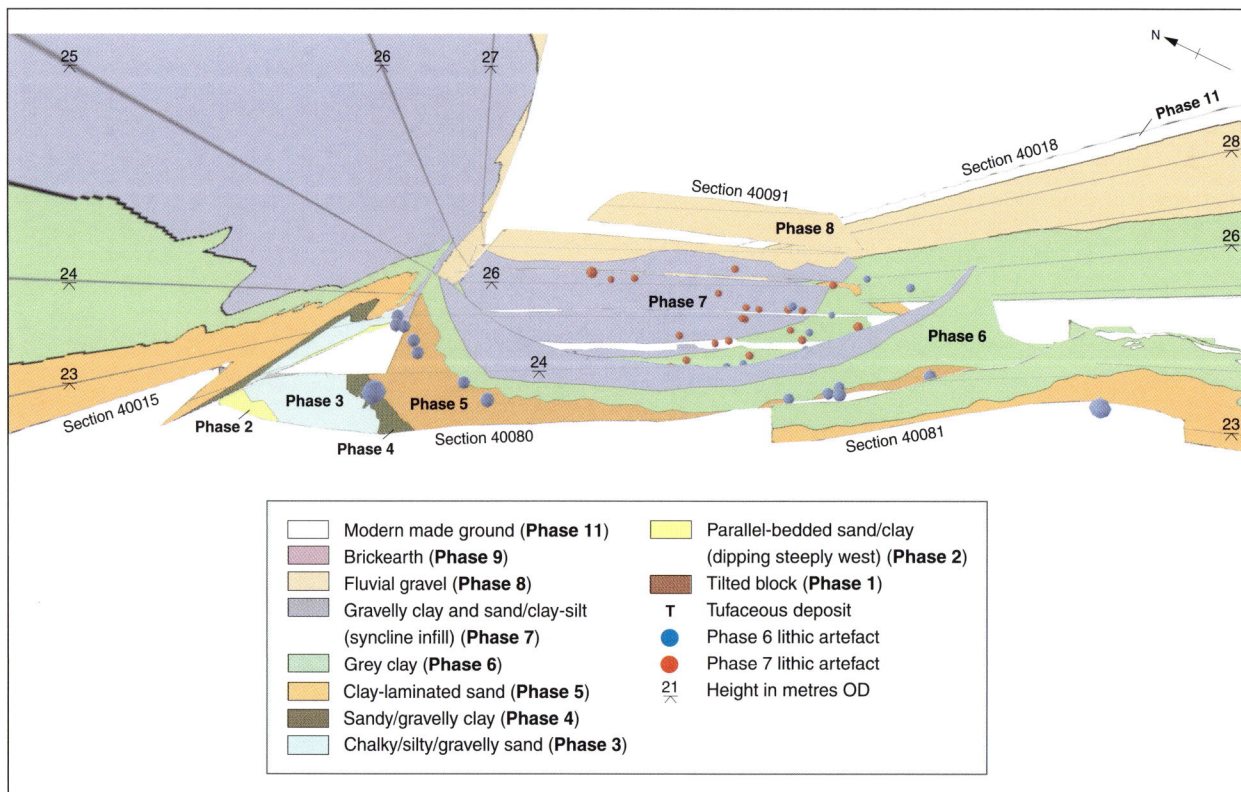

Figure 4.31 Trench B, south-facing Section 40091, also showing flints from Phase 7 (looking north)

facing section, no. 40091 of Trench B (Fig.4.8; Fig. 4.31). In Section 40015 they mostly consisted of variably gravelly greenish/greyish-brown brecciated clay (contexts 40043 and 40101), with minor gravelly and silty beds in places at its base (contexts 40042 and 40041). These deposits were well represented and several metres thick in the central part of the section, but were absent at its northern and southern ends, where the Phase 8 gravel that unconformably overlay them had cut right down to the Phase 6 clay. In the east-west transverse section, 40091 of Trench B, it can be seen that the infill of the synclinal 'skateboard ramp' is dominated by Phase 7 sediments. Here they mostly consisted of a distinct lower facies (context 40166) that is a fine yellowish-brown sandy/clayey silt with contorted clay beds in places, equivalent to context 40041 in the west-facing section 40015. Context 40166 is overlain by contorted clayey gravel (40167), which is equivalent to contexts 40042 and 40043 of Section 40015. As discussed previously (Chapter 3) context 40167 contained occasional concentrations of what looked like rotted and charred plant macro-remains and produced an assemblage of variably stained/abraded lithic artefacts (Chapter 19).

As discussed below, sediments of this phase continue to the west and north-west of the site, extending upslope as a substantial clayey/gravelly mass. This is interpreted as a local mass movement deposit, originating from west or north-west of the site.

Phase 8. Sandy gravel

Passing unconformably across the underlying sequence was a deposit of moderately well-sorted medium-to-coarse gravel with sand lenses, best represented in the main west-facing section no. 40015 (Fig. 4.5; Fig. 4.27) and in the 3-D Fledermaus model (Fig. 4.32a, b). The gravel deposits comprising Phase 8 were given different context numbers in different parts of the site at different stages of the excavation (Table 4.3). Although some might complain that this is typically awkward, in line with the variety of different context numbers used for the same horizons in other phases of the sequence (eg Phase 6, see Table 4.2), this was in fact important and beneficial. It meant that subsidiary stratigraphic phasing

within the gravels that only became apparent when wider continuous exposures were seen later in the excavation could be reconstructed and preserved with suitable amalgamations of context numbers. This enabled the meaningful attribution of clast lithological samples and artefact assemblages to specific subsidiary phases within the build-up of gravel. Three sub-phases of the gravel were identified: 8a, 8b and 8c. The main gravel body was represented by Phase 8c (contexts 40048=40071=40102) which extended all across the site, and the two earlier Phases 8a and 8b were only present in the northern half of the site.

The base level of the main Phase 8c gravel beds was at *c* 27.2m OD at the south end of the site, where it cut directly into Phase 6 clays south of Trench D. It then dipped gently down to the north as it passed over the central part of the site, where its base was at *c* 26.5m OD. It truncated the Phase 7 deposits that featured here, and which in turn here overlay the Phase 6 sediments containing the *Palaeoloxodon* skeleton (Fig. 4.5). Further north, however, although the Phase 8c gravel continued its gentle northward dip, it was underlain by additional fluvial gravel beds (from the base, contexts 40098 and 40047, attributed to Phase 8a and 8b respectively) with the base of the Phase 8a bed seen at 25.2m OD at Trench A. After that it disappeared beneath the ground surface which rose up north of this point. Clast lithological analyses (Chapter 6) and orientation studies established that the gravel was a fluvial deposit laid down by a northward-flowing palaeo-Ebbsfleet.

The Phase 8a gravel (40098) was sandier than the other gravel beds, with the thicker sand beds (for which the grain size was fine-medium) often displaying foresets dipping north/north-west (indicating flow in that direction). Overlying this, the Phase 8b gravel (40047) consisted mostly of well-rounded dark grey gravel clasts (derived from Tertiary deposits) with a medium sand matrix. Occasional clasts of sub-angular flint from Chalk were also noted. The sandy matrix became less clayey from south to north. Clasts were imbricated and dipping to the south, confirming the south-to-north flow direction indicated by the foresets in the Phase 8a gravel. The upper surface of 40047 was channelled and infilled with sediment from 40048, the basal deposit of Phase

Table 4.3 Phase 8 contexts, showing numbering correlations, internal phasing and clast analyses

Phase	Main site sequence	Clast analyses	Logs 40001-40004	Logs 40005-40006
8c	40050	<40004> - lithology	-	40018
	40049 - sand lens	-	-	
	40048=40071=40102 [=40002]	<40001> - lithology	40006	
		<40002> - lithology	40007	
		<40016> - orientation	40008	
		<40017> - orientation	40009 - upper	
8b	40047	-	-	
8a	40098	-	-	-

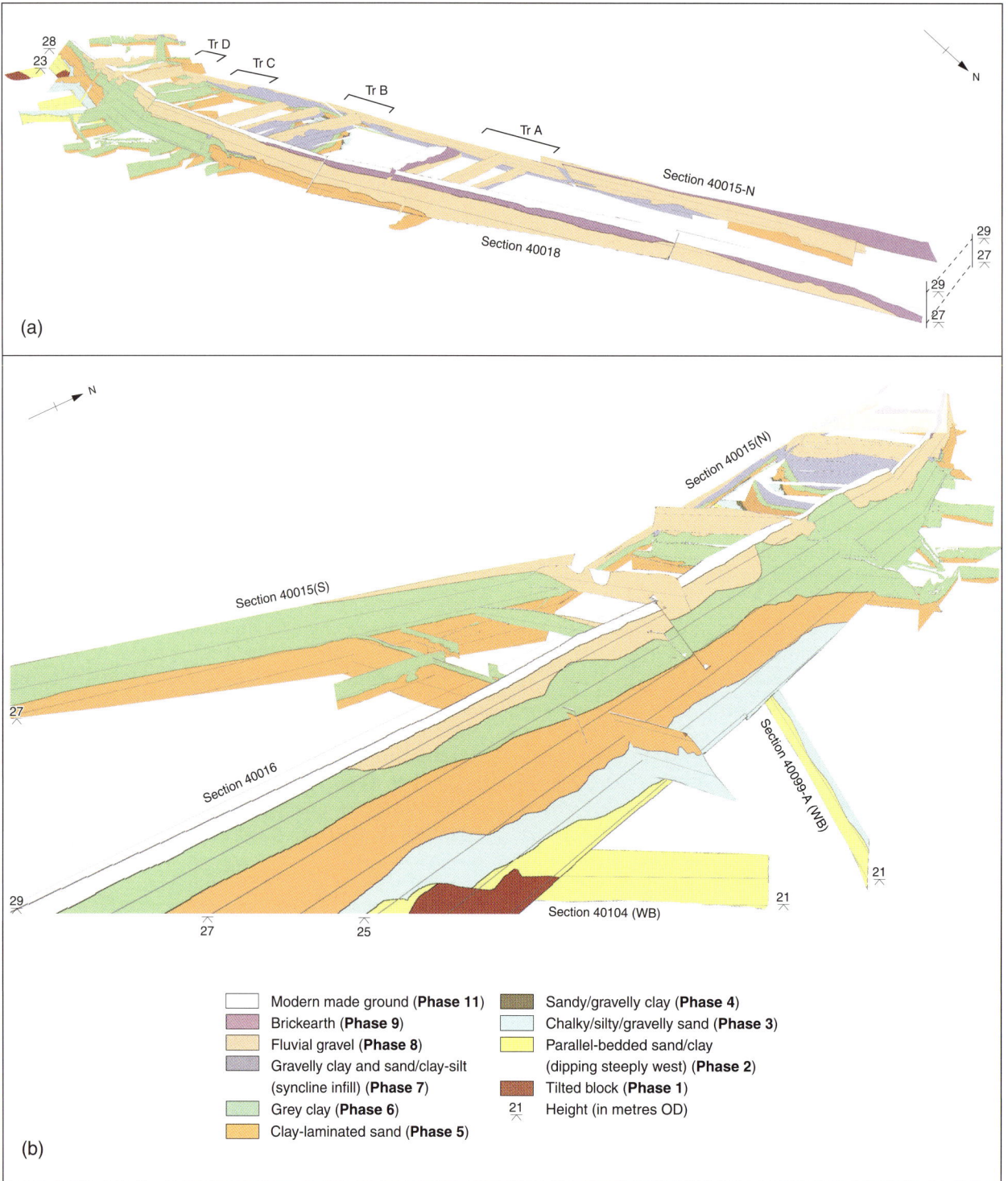

Figure 4.32 Phase 8 sediments: (a) looking south-west; (b) looking north-west

8c, and this junction was also characterised by occasional quite large patches, about 1m long by 0.75m high in section (Fig. 4.5) of clayey sand. Context 40048 was similar to the underlying context 40047, but contained a greater quantity of Chalk-derived flint. Additionally the gravel appeared to exhibit a series of packages of fining upwards sequences that sometimes ended in sands. In the southern part of the site, bedding was sub-horizontal and poorly developed with some bar-edge dipping beds noted. Large scale cross-bedding was noted further to the north.

The Phase 8 sequence appears to show an initial phase of rapid deposition being superseded by the development of a series of superimposed braid bars. Two macro-fabric studies (samples <40016> and <40017>) from the west side of the main site showed a predominant dip to the south (120–230°) with a minor mode to the north-west (285-335°), considered to represent imbricate deposition, with flow to the north. The clasts were predominantly flint, from both the Chalk and the Thanet Sand. There were only a few southern Greensand lithologies (such as chert or sandstone) and too few exotics associable with Thames deposits (such as vein quartz and quartzite) for a Thames origin to be feasible. The top bed of the gravel sequence (context 40050) was characterised by a flint gravel with a medium sand matrix dominated by Tertiary-derived flint with a lesser component of uncertain origin and occasional Chalk-derived flint. Again the clasts exhibit imbrications to the south. This is interpreted as braided fluvial deposition, perhaps prograding on a gentle deltaic front as the channel approaches the point of confluence with the much larger Thames, a short distance to the north.

Phase 9. Reddish-brown clay-silts, 'Brickearth'

The northern part of the gravels, as their upper surface dipped to the north at the north end of section 40015 (Fig. 4.5), was overlain by moderately firm, cohesive and pliable yellowish to reddish-brown sandy/silty clay, locally recognised as 'brickearth'. At its base, this brickearth has a thin bed (context 40051) with thin, continuous, parallel sand/silt laminae. These become coarser upward, grading into a thin reddish-brown gravel bed with poorly sorted flint clasts (context 40052). This is in turn was overlain by the main body of the brickearth, a thick sandy clay-silt (context 40053, also equivalent to 40075), which contained occasional faint sandier beds and gravel trails, as well as patches of intrusive recent rooting. Occasional isolated small gravel clasts were also present. These deposits are likely to predominantly represent slopewash and sheet-wash deposits, including reworked aeolian sediment from upslope.

These sediments are without doubt equivalent to the 'ferruginous loam' reported by Carreck (1972, 61) exposed in the quarry faces to the east of Southfleet Road over a stretch of several hundred metres to the north of the site (Fig. 2.4, site 16). As well as the exposures seen in Transects 1-3 to the north of the main site, the northward continuation of the brickearth was investigated in numerous test pits in Station Quarter South and Eastern Quarry Area B between 2005 and 2008, and this work is discussed below.

GEOMETRY OF THE SEQUENCE AT THE SITE

The sediments exhibited a complex 3-dimensional geometry. The primary bedding structures in the main fluvially lain deposits of Phases 5 and 8 dipped northwards suggesting an underlying trend of progradation from south to north. These sedimentary phases were interspersed with sediments from Phases 3, 4, 6 and 7, which are thought to have been laid down by a combination of colluvial slopewash and surface run-off processes, feeding into fluctuating (and periodically drying out) bodies of standing or very quiet water. However, superimposed upon this, a number of other features and trends can be seen. The sandwich of Phases 3-6 formed an asymmetric north-south trending synclinal basin in the central part of the site, about 18m

across east-west and with a preserved depth of at least three metres (Fig. 3.18; Fig. 4.12). This was flanked to the west by a corresponding anticlinal fold of the same sediments, with a core formed by the north-west to south-east trending Phase 3 chalk-rich sediments, recorded in plan (Fig. 4.20-21). The eastward continuation of the site had been excavated away by HS1 groundworks and so could not be observed. However, faint bedding in the upper Phase 6 sediments in the upper east end of the main north-facing section of Trench C dipped to the east (Fig. 4.9), and it was thought that the main east-facing section of the site contained faulting parallel with its sloping face, suggesting that the upward curve of sediments at the eastern side of the synclinal basin was reversed downward a short distance further east. North of Trench B, the sediments of Phase 6 within the main synclinal basin continued their northward thinning trend in Trench A, persisting here as a vestigial bed about 10mm thick, but the synclinal structure remained, with its eastern limb recorded in plan once the overlying sediments had been machined away (Fig. 4.33).

Synclinal folding was also observed at the east end of Transect 2, *c* 150m to the north of Trench B (Fig. 4.34). This group of contexts could not be tied in with the main phased site sequence, although there were superficial similarities of some beds with Phases 3 (context 40093) and 6 (contexts 40089 and 40091). It is uncertain whether this is the northward continuation of the same synclinal structure as seen at the site, or a different one; it does, however, demonstrate the persistence of sedimentary folding beyond the immediate main site area.

The explanation for these basin features is problematic. Dips in the Solid geology of the central and eastern parts of the London Basin are very low, generally not exceeding a few degrees. Against this background, the high dips observed in some of the Southfleet Road excavations, both in the Chalk and in the overlying Pleistocene deposits, are remarkable and apparently anomalous. A number of possible explanations exist for these features:

1. *Tectonic activity.* Dips in the Chalk as high as 55° occur on the southern limb of the London Basin at the Hog's Back (Sumbler 1996) and, further south, approach the vertical on the south side of the Hampshire Tertiary Basin in the Isle of Wight. This is however an unlikely explanation because of its location in an area of sub-horizontal dip, the very local nature of the feature and the apparent absence of hinge structures between the two basins. Also, there has been no suggestion that any folding has been active during the Pleistocene. As Pleistocene sediments are involved, the deformation must be post-Tertiary and must post-date these regional tectonic structures.

2. *Superficial valley disturbances.* A wide variety of superficial valley disturbances exist, which have been reviewed by Hutchinson (1991 and 1992). These can

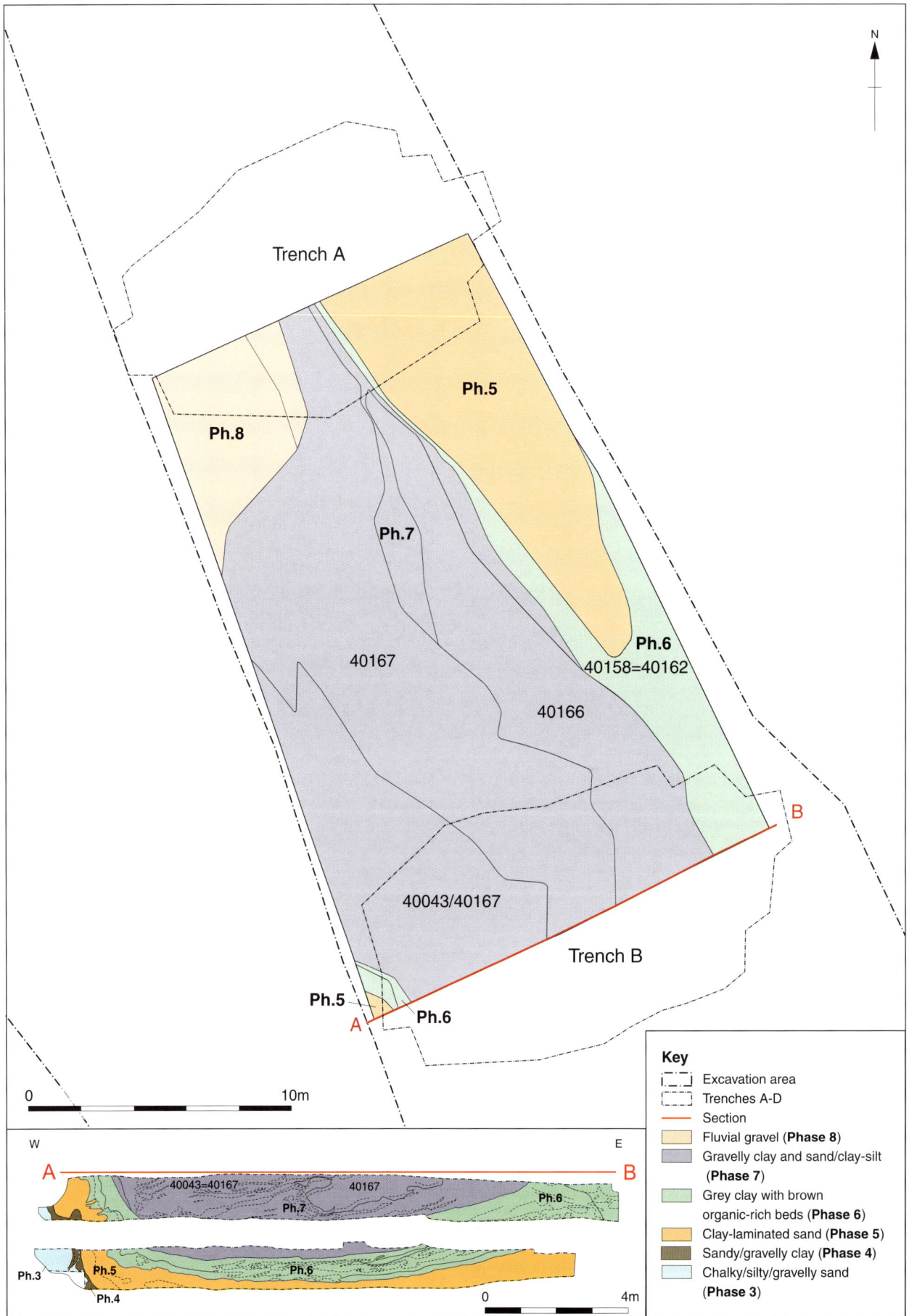

Figure 4.33 Stripped surface between Trenches A and B, showing east side of synclinal basin in plan

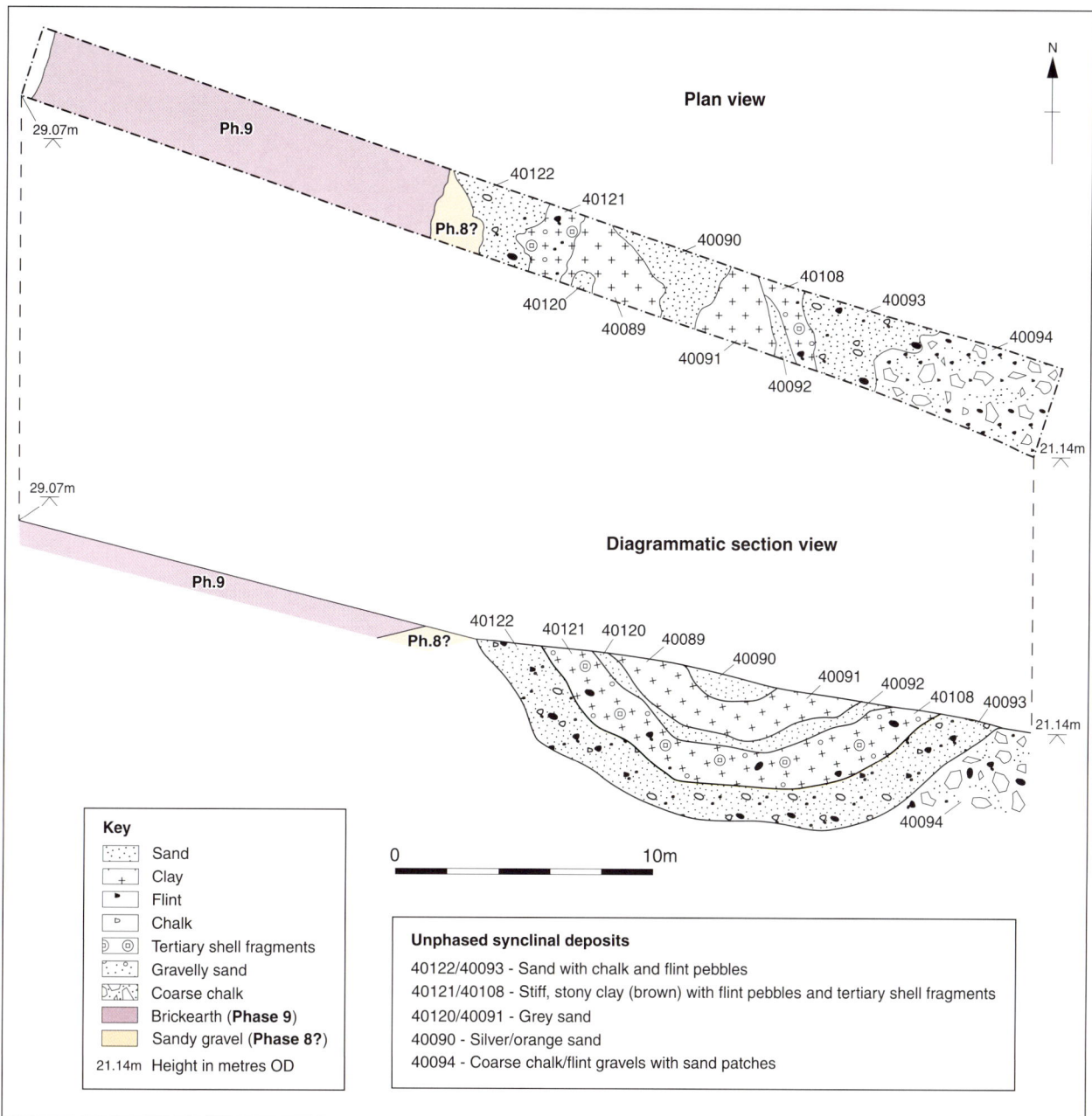

Figure 4.34 Synclinal folding at the east end of Transect 2

produce local dips of 70° or more in valley bulges and are widely distributed in the Jurassic outcrop of central England and in the Weald. However, they require the valley to be underlain by a thick argillaceous stratum. The absence of such a stratum beneath the Southfleet Road site, its location away from the axis of the local valley and the very local nature of the observed anomalous features rule out an explanation depending upon the above mechanisms.

3. *Pingos.* Relict pingos, chiefly of open type and generally (but not always) more or less circular or oval in plan, are common in southern England and Wales. Some are reported to have diapiric structures in their lower parts that can result in steep dips. Pingos are frequently found in clusters towards the

foot of slopes, particularly if artesian groundwater pressure is present. While some of these criteria may be fulfilled at Southfleet Road, the continuity of the beds across the basins and the absence of boggy, peaty infills argue strongly against a pingo origin.

4. *Frost heave and periglacial solifluction.* While both these processes can result in considerable deformation of the ground, they are too shallow to explain the Southfleet Road structures.

5. *Solution of the Chalk.* Solution pipes, swallow holes and solution dolines are numerous in the Chalk, especially around the margin of London Clay capped hills (the umbrella effect). The work of Higginbottom and Fookes (1970) and Jones (1981) indicate that

the extent and magnitude of solution effects in the Chalk have been underestimated. Both this background and the detailed local geology point to the solution of Chalk as a possible mechanism producing the features at the site. Indeed it is not abnormal to have two solution features adjoining each other or for them to have a linear tendency. However, this does not explain the folding of the surrounding beds, nor the marked asymmetry of the basin, which seems suggestive of lateral pressure from the west.

6. *Landsliding*. British Geological Survey mapping shows the site to lie at the lower edge of a landslipped area. Pressure exerted downslope or by the toe of a landslide can produce deformation, folding and faulting of its lower parts and of the resisting ground, although faulting would be less likely to occur in the case of saturated sediments,which would behave in a more fluid manner (Allen 1991). An example at Lyme Regis is described by Hutchinson and Hight (1987), the effects extending no more than about 5m below ground level, which is compatible with the sequences seen here at the Southfleet Road site. The presence of gravity folding, elongation of folds and *boudinage* in Phase 6 especially in section 40091 (Fig. 4.8) and 40085/400285/2 (Fig. 4.9), indicates deformation of sediments in a saturated, fluid state, due to pressure from the west, which lends support to landsliding as a deformative process.

On present evidence, the deformation of the sediments is interpreted as a result of lateral pressure caused by downslope movement of a substantial mass of clay-dominated colluvial deposits from the west, perhaps in conjunction with underlying subsidence caused by solution of the Chalk bedrock. This deformation and landsliding post-dated Phase 6, and pre-dated Phase 8, the sediments of which overlie the deformed sediments unconformably and are not affected. Although the sediments of Phase 7 conformably overlie those of Phase 6 and show some signs of internal deformation, they do not themselves show evidence of having been entirely in place and deformed together with the underlying sequence. Rather, it is suggested that the Phase 7 sediments may themselves have been the source of this pressure, slipping down from the west and both deforming and over-riding the sequence already in place in the valley floor. As is seen in the next section below, the Phase 7 sediments at the site can be correlated with a major body of similar clayey and gravelly sediments that extend upslope to the west and north-west.

Finally an explanation for the nature of the basal parts of the sequence remains elusive. The Phase 1 sediments, ie the intact 'tilted block' of bedrock, and the overlying sediments of Phase 2, remain difficult to explain and insufficient evidence exists to address this convincingly. However, it is likely that the slope failures postulated as responsible for the creation of the synclinal basins discussed above would have had a long history in the Ebbsfleet Valley and that such processes may also be partly responsible for the formation of this earlier part of the sequence. The 'skateboard ramp', Phase 6 (Fig. 4.12), seems to represent an extreme response to lateral pressures from the west, given the scale of the deformations in sections 40091, 40020, 40085/40022; possibly the steeply dipping sediments of Phases 1 and 2 are also related to this.

DEPOSITS IN THE SITE VICINITY

Since 2004 when the elephant excavation took place, the neighbouring areas have been subject to numerous field investigations that have provided new information on the Pleistocene sequences around the elephant site (Fig. 4.1). In particular:

1. Around 70 deep test pits (TP) and 5m window-sample boreholes have been excavated in the field in the north-east quadrant of the Northfleet West electricity substation plot, immediately to the west of the site (Museum of London Archaeology 2011).

2. Some 35 test pits and boreholes have been excavated in the Station Quarter South development plot, immediately to the north and east of the site (Wessex Archaeology 2006b).

3. Approximately 100 deep test pits have been excavated in the north-east unquarried quadrant ('Area B') of Eastern Quarry to the north-west of the site (Wessex Archaeology 2006a; Wessex Archaeology 2009a). Amongst this huge body of work, several interventions have revealed deposit sequences, and on occasion faunal and artefactual remains, that complement the work at the elephant site. These works both expand our understanding of local Pleistocene lithostratigraphy and geometry beyond the relatively restricted area of the elephant site and enhance the scope of the site's interpretation. Key interventions relevant to the elephant site are listed (Table 4.4), with descriptions and photographs of their deposit sequences provided as digital appendices (D1-D4).

Three Transects EF, GH and IJK were constructed using some of the most informative investigations in the closer vicinity of the site (Fig. 4.35). Two of these, EF and GH (Fig. 4.36a; b), cross east-west through the site, and the third, IJK (Fig. 4.36c) originates to the south-west of the site, then passes northward through it.

The west ends of Transects EF and GH both show Thanet Sand bedrock occurring much higher than at the site, at above 30m OD, from which its (erosionally truncated) surface dips steeply eastward to an unknown level (but below 20m OD) less than 100m away at the site. To the west of the site (Northfleet sub-station TPs 235-237, 242-244, 255 and window sample WS 106) the basal group of Pleistocene deposits consist of chalk rubble and chalk-rich sands and flint gravels. These

Table 4.4 Key field interventions in project areas in the vicinity of the elephant site

Project area	Project field code	Test Pits	Section logs	Boreholes, window samples	Report reference
Northfleet West Sub-station	KT-SFL 03	227, 235, 236, 237, 242, 243, 244, 255, 256-B	-	WS 106, WS 163	Museum of London Archaeology (MoLA) 2011
Station Quarter South	63542	1, 3, 5, 11, 12, 16, 17, 20, 21, 23, 25, 27, 30, 31, 32, 33	-	-	Wessex Archaeology 2006b
	63544	-	-	BH 201, WS 3	Wessex Archaeology 2007
Eastern Quarry, Area B	61040	2.2, 3.1, 3.2, 4.1, 6.2, 7.2, 16.2, 55.1, 59, 60, 61	-	-	Wessex Archaeology 2006a
	61041	111, 112	-	-	Wessex Archaeology 2008a
	61042	127	-	-	Wessex Archaeology 2008b
Swan Valley School	SCS 97	D, K, E	19/40	-	Wenban-Smith & Bridgland 2001

deposits dip and thicken towards the site, becoming less chalk-rich and with diminishing chalk-clast size as they do so, and are here interpreted as equivalent to Phase 3 of the site sequence.

These deposits thus both underlie, and also form the western valley-side bank of, deposits of Phase 5-9 at the site, which contain the principal archaeological remains. An important, and very puzzling aspect of these deposits is their abundant chalk content, and by association their nodular flint content, which is probably fundamental to the prolific lithic knapping remains found at the site. Flint nodules would probably have been exposed both in the west side of the (periodically dry) stream/marsh/lake environment of the valley floor, and also in colluvial/slopewash fans slumping down the valley sides towards the valley floor, providing a ready source of flint raw material for hominins at this time.

It is, however, difficult to see, when considering the local geological mapping (Fig. 4.2) and sub-surface solid geology (Fig. 2.2; Fig. 2.3), where the Chalk bedrock source was from which the chalk and flint content of the western valley-side originated. It does not appear feasible that these could have originated from north, south or west, leaving only (however improbable) east as the most likely source direction. As discussed above, there is 'something funny going on' with respect to the chalk bedrock, as evidenced by both the 'tilted block' of the Phase 1 sediment, and the anomalous Chalk high at Southfleet Pit to the east of the site (Fig. 4.3). The most feasible scenario is perhaps that there was originally a pinnacle or cliff of chalk to the east of the site, the westward collapse of which fed chalk rubble and flint nodules into the area that later became the western side of the valley.

There is then a significant hiatus in the depositional sequence to the west of the site, with the Phase 3 sediments there being directly overlain by gravelly clays identical, and interpreted as equivalent, to the Phase 7 deposits at the site. These gravelly clay deposits extend further north and west, and correspond with the protruding lump of higher ground mapped as 'slipped clay' in the early 20th century (Fig. 4.2). Thus it seems clear that the Phase 7 sediments at the site directly originated from the Tertiary deposits capping the high ground to the west, and that these have slipped down to the east, over-riding the Phase 6 clays. As discussed above, it is suggested here that this landsliding has also caused deformation of the underlying sequence at the toe of the landslip, where it over-rides the site area, forming the synclinal basin of the 'skateboard ramp'.

These slipped gravelly clays are in turn overlain by very homogenous reddish-brown clayey silts (present in the top half of the sequence in Northfleet West Sub-station TPs 244 and 255). These pass laterally into the Phase 9 deposits preserved at the top of the Pleistocene sequence in the northern half of the site. Here they overlie the Phase 8 fluvial gravels. The northward continuation of these two deposits is discussed in more detail further below (and see Fig. 4.41).

To the east of the site, chalk-rich sands and gravels interpreted as equivalent to Phase 3 of the site sequence occur at the base of the sequence in Station Quarter South TPs 31-33 at the east end of the transect GH (Fig. 4.36 b), between *c* 17 and 20m OD (with their base not reached). Clast lithological sampling from a gravelly bed of the Phase 3 deposits in the base of TP 32 indicated that here they included fluvially deposited Ebbsfleet gravels (Chapter 6). A distinctive interglacial ostracod fauna, similar to that from the Phase 3 deposits at the site, was also recovered from sand/silt beds (3304 and 3306) at the top of deposits regarded as equivalent to Phase 3 in TP 33 (Chapter 11).

In contrast, only a short distance to the north, at the eastern end of Transect EF (Fig. 4.36a), the Phase 3 chalky sands/gravels were thinner, and were underlain (in Station Quarter South TPs 16 and 17) by a firm laminated sand/silt deposit. This was interpreted as most likely equivalent to Phase 2 in the site sequence, seen

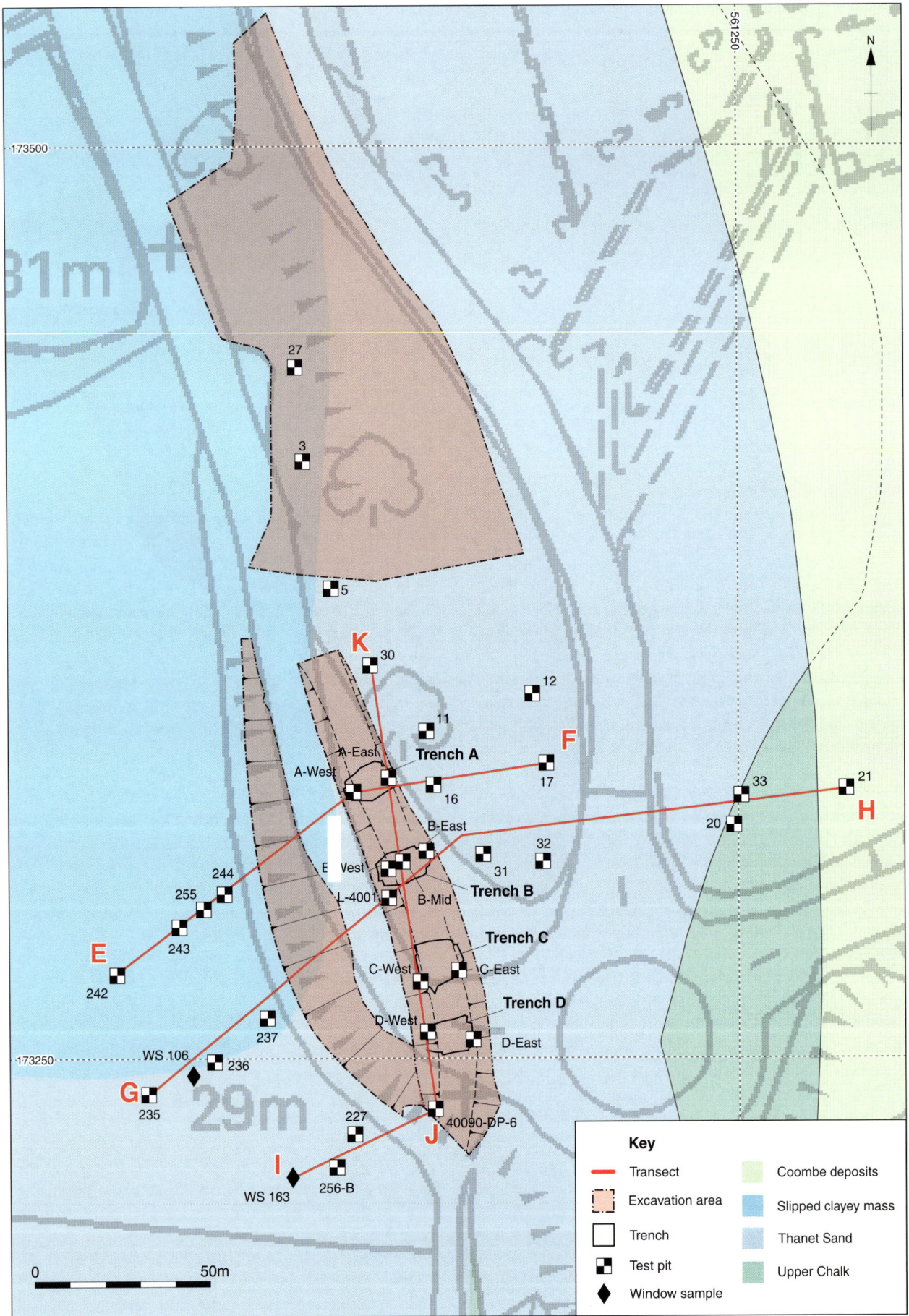

Figure 4.35 Location plan of area around the site, with key interventions from non-HS1 fieldwork and locations of representative Transects EF, GH and IJK

outcropping at the base of the southern end of the main east-facing section (Fig. 4.6).

At the eastern ends of both Transects EF and GH, the Phase 3 deposits were overlain by fine homogenous clayey silts, interpreted as standing lacustrine deposits prone to drying out. This is indicated by the clear polygonal cracking, preserved as an oxidised staining pattern (Fig. 4.37) and equivalent to Phase 4 of the site sequence. These deposits were only thinly present in TP 17 at the east end of Transect EF, but were well-developed and 2-3m thick in TPs 31 and 32, where they occurred between c 20 and 23m OD only 15m east of the site (Fig. 4.36b). The same distinctive ostracod fauna as was found in the Phase 3 deposits at the site and in TP 33 was recovered from the top of the Phase 4 deposits (3107) in TP 31 (Chapter 11). Pleistocene Mollusca were also present in context 3107, in the same horizon as the ostracod-bearing sample, including shells and opercula of *Bithynia tentaculata* (and possibly *B. troschelii*), *Valvata piscinalis* and *Pisidium nitidum* (RC

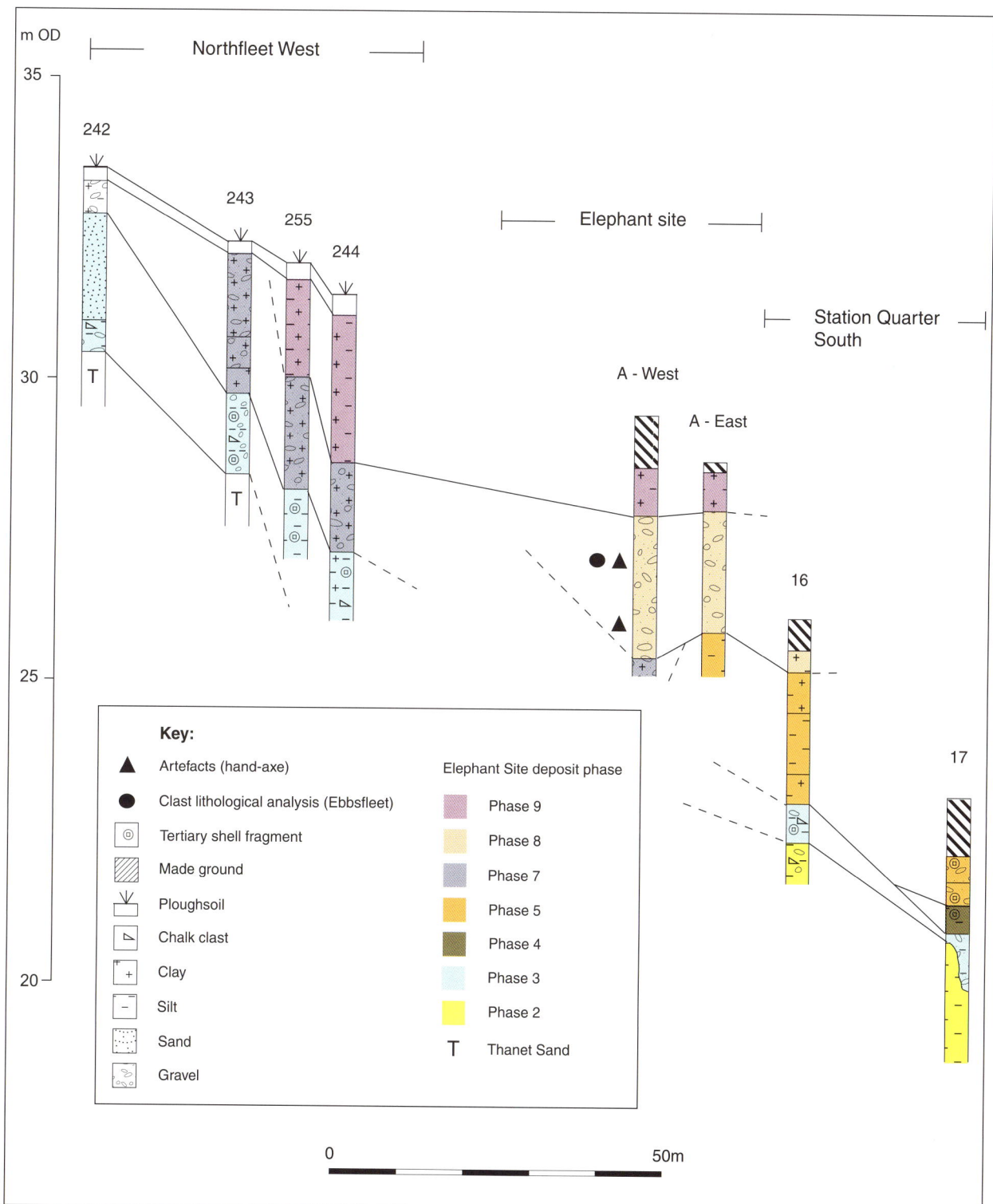

Figure 4.36 Lithostratigraphic transects in the vicinity of the site: (a) EF

The Ebbsfleet Elephant

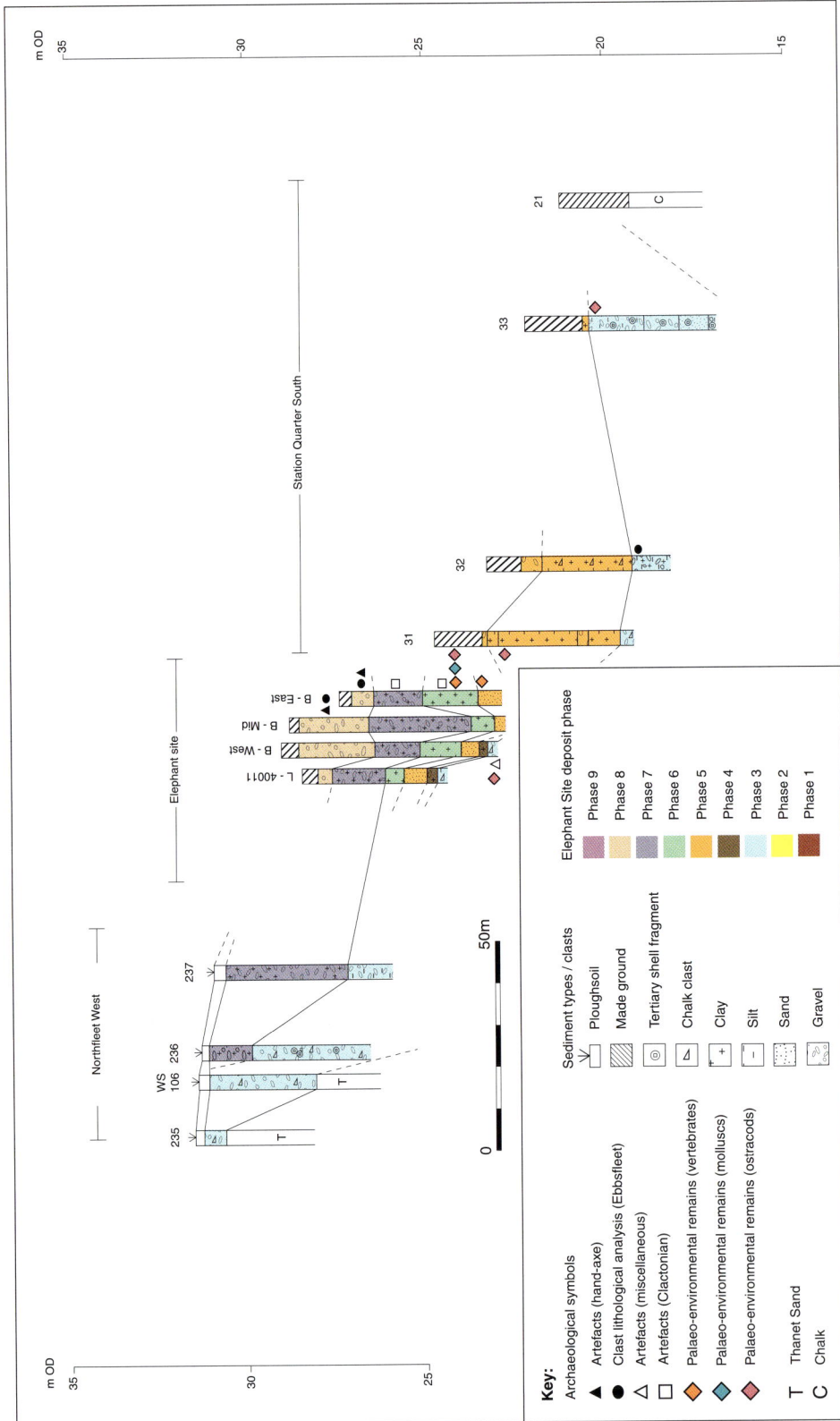

Figure 4.36 Lithostratigraphic transects in the vicinity of the site: (b) GH

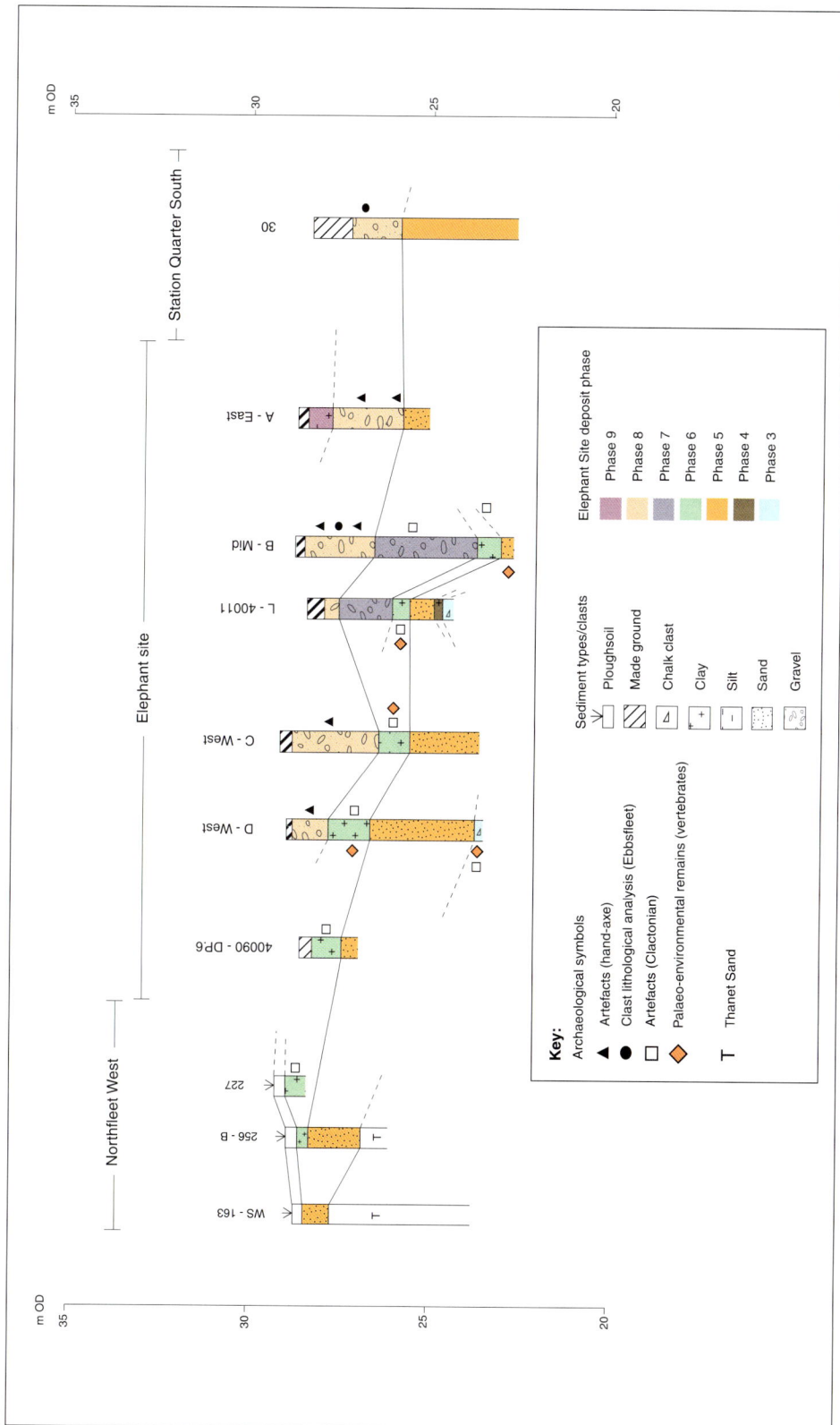

Figure 4.36 Lithostratigraphic transects in the vicinity of the site: (c) IJK

Figure 4.37 (right) Polygonal cracking indicative of desiccated lake-bed sediments (Phase 4) in Station Quarter South, Test Pit 32 [photo Francis Wenban-Smith]

Figure 4.38 (below) Clactonian artefacts *in situ* at Northfleet West Sub-station TP 227: (a) find Δ.1, large flake 0.55m below ground surface; (b) close-up view find Δ.1, showing ventral features and cortical striking platform; (c) close-up view refitting finds Δ.4 and Δ.5, 0.85m below ground surface – note machine smear marks suggesting artefacts were not broken apart by machine bucket [photos Francis Wenban-Smith]

(a)

(b)

(c)

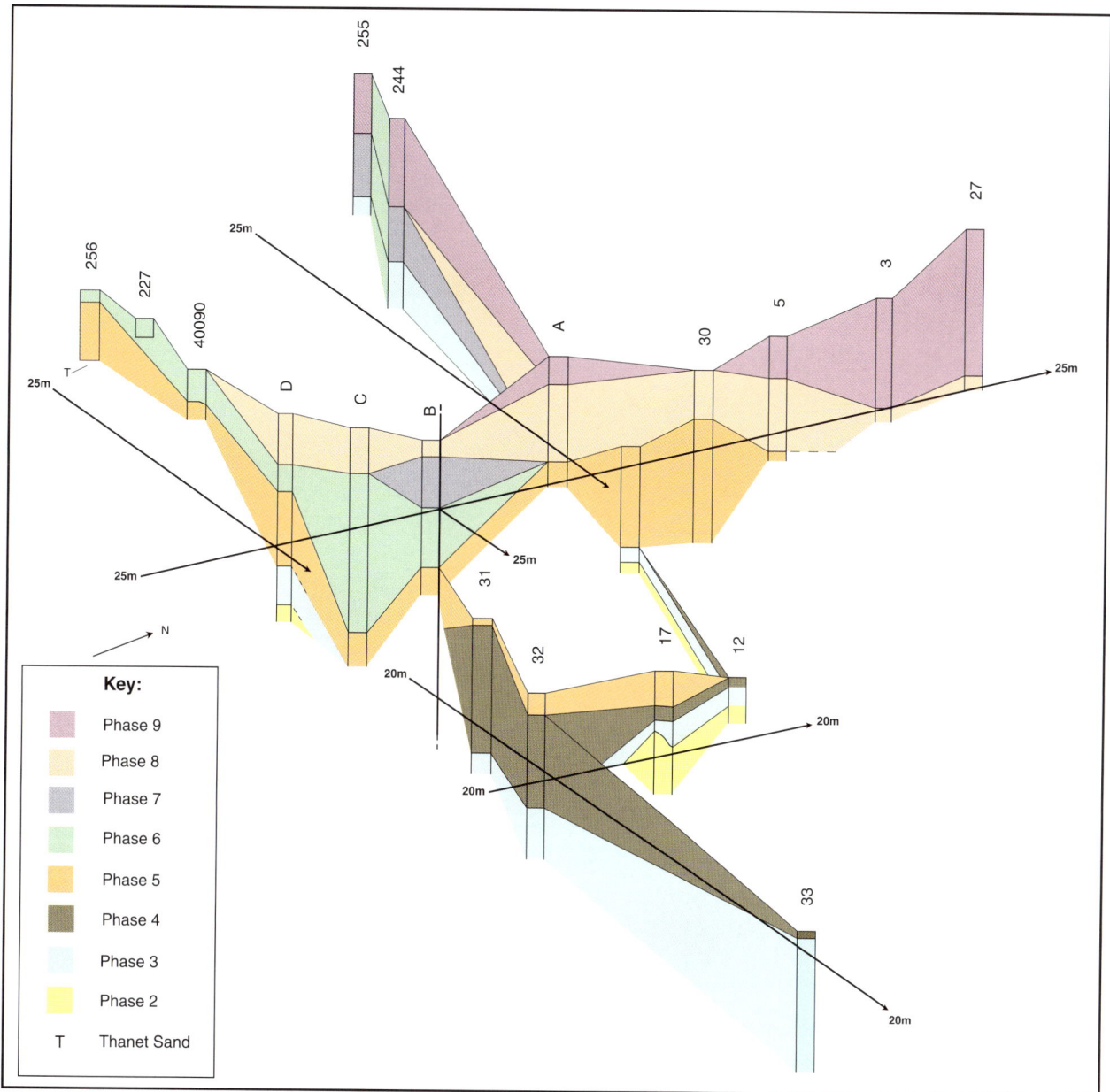

Figure 4.39 Fence diagram showing deposit phases at the site and in the vicinity

Preece, Appendix 3 in Wessex Archaeology 2006b) suggesting a low-energy freshwater habitat.

The Phase 4 sediments are overlain at the east end of Transects EF and GH by clay-laminated sands interpreted as equivalent to Phase 5 of the site sequence (Fig. 4.36a; b). These are mostly present as thin layers at the truncated top of the Pleistocene sequence, overlain by thick deposits of made-up ground, deposited during the HS1 groundworks of 2003-2004. However, thicker Phase 5 deposits were present in TP 16. It can be seen in TP 30 at the northern end of Transect IJK (Fig. 4.36c) that these persist as a well-developed body 2-3m thick immediately to the north and north-east of the site between *c* 23 and 26m OD. Here deposits of Phase 6 and Phase 7 are absent and the Phase 5 sediments are overlain directly by the Phase 8 fluvial gravels, which are in turn overlain by the Phase 9 brickearth. The northward continuation of the Phase 8 and 9 deposits, and their correlation with the

Swanscombe 100-ft terrace deposits at Swan Valley School, Barnfield Pit and other deposits in Eastern Quarry Area B, is discussed further below.

To the south-west of the site, as seen in Transect IJK (Fig. 4.36c), the fluvial clay-laminated sands of Phase 5 were seen to wedge out directly onto Thanet Sand, and were overlain by a continuation of the Phase 6 clay. The base of this clay rose steadily from *c* 27.5m OD at the south-west edge of the site (Section 40090, DP.6) to *c* 28.5m OD at the Northfleet West TP 256-B, where it wedged out beneath the present-day plough-soil. This extension of the Phase 6 clay was, like its counterpoint at the south-west side of the site, still rich in mint condition flint artefacts, several of which were found close beneath the plough-soil in TP 227 (Fig. 4.38).

The relationships, geometry and phasing of deposits in the vicinity of the site, as discussed above, is here integrated with those at the site in a fence diagram (Fig. 4.39).

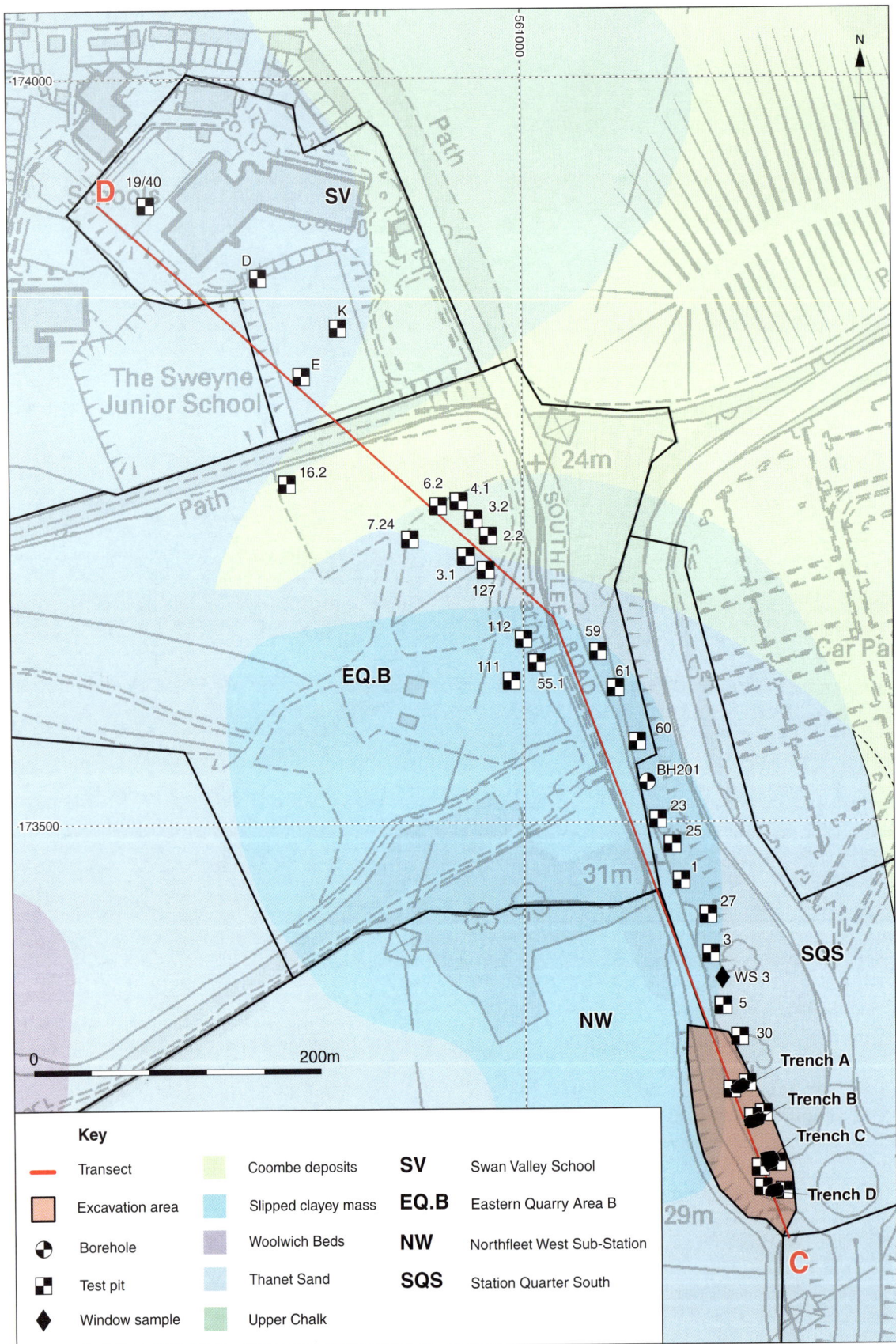

Figure 4.40 Location of north-south Transect CD, Pleistocene sequence between site and Swan Valley School, through Eastern Quarry Area B

As seen in Fig. 4.36c and in the stripped area exposed to the north of the site, deposits of Phases 5, 8 and 9 were seen to extend northward beyond the site. As mentioned previously (Chapter 3), deposits of Phase 9 had earlier been reported by Carreck (1972, 61) as extending several hundred metres further north from the site. There have been a large number of field investigations in various areas to the north and north-west of the site between 2004 and 2008 (Table 4.4; Fig. 4.1). This has provided a wealth of new data that allows correlations of these deposits to be extended northwards towards the proven outcrops of the Swanscombe 100-ft terrace sequence at Swan Valley School and Barnfield Pit. Furthermore, this new work has been accompanied by clast lithological analyses that make a significant addition to our understanding of the course of the Lower Thames and the evolution of the palaeo-landscape during the period represented in these sequences.

From this wealth of new data, the sequences from a relatively small selection (n=26) of field investigations are presented here, distributed along a transect CD that runs broadly north-west from the site to the Swan Valley School, about 700m away (Fig. 4.40). In the southern half of this transect, the dominant deposit is the Phase 9 brickearth (ie Carreck's 'ferruginous loam') which can be traced as far north as TP 112 in Eastern Quarry Area B (Fig. 4.41). This deposit reaches as high as 30m OD in several places, with its base dipping eastward towards the axis of the palaeo-Ebbsfleet valley floor. It is mostly a very homogenous reddish-brown slightly sandy and clayey silt, but contains occasional sandier and fine gravel lenses, also dipping to the east, and more substantial gravel-rich beds as it progresses northward and in its more easterly downslope exposures. This geometry and sedimentology support the notion of the deposits as of colluvial/aeolian origin, although incorporation of a denatured estuarine alluvium perhaps cannot be ruled out. Although mostly lacking in archaeological contents, two mint condition handaxes were recovered *in situ* from this brickearth, including one particularly fine example from TP 25 in the Station Quarter South project area (Fig. 4.42). This find complements the handaxe and accompanying scatter of debitage recovered from the stripped brickearth bank at the west end of Transect 2, and raises the possibility that the Phase 9 brickearth may contain undisturbed, although sparsely distributed, artefactual remains.

The base of the brickearth is underlain along this stretch by a sharp junction with the top of a gravel deposit that is presumed to represent the northward continuation of the Phase 8 gravel. At the southern end of the transect, the gravel was seen at the north end of the site as *c* 2m thick and dipping northward. Further north, the full thickness of the gravel was likewise seen in TPs 30 and 5, where it continued the same northward dip. Between TP 5 and borehole BH 201, deposits thought to be the same gravel were only observable in the inaccessible bases of a series of deep test pits, where its top part was reached beneath deep brickearth deposits. Then, in borehole BH 201, a substantial gravel body was proved between *c* 22 and 25m OD, which

continued the steady downward trend from the north end of the site. North of borehole BH 201, the surface of the gravel rises again, being present at 29m OD in the base of TP 112, in Eastern Quarry Area B. Clast lithological analysis of this gravel bed has established that at this point it remained an Ebbsfleet fluvial gravel (Chapter 6). No artefactual remains were recovered along this stretch of gravel. However this is not indicative of any absence of content, but merely reflects the fact that the gravel was unable to be investigated properly due to its inaccessibility, with only its top parts being exposed at the base of deep test pits.

At the north end of Transect CD, the sequences in the Swan Valley School identified and investigated in the late 1990s (in Trenches D, K and E) have been securely confirmed as containing deposits equivalent to the Lower Middle Gravel (and also, in places, the Upper Middle Gravel) of the Swanscombe 100-ft terrace Barnfield Pit sequence, on the basis of both clast lithological analysis and geomorphology (Wenban-Smith and Bridgland 2001). At the time of this work, part of its significance was extending the southern bank of the Thames of this time significantly further south than previously recognised. This work is now superseded by the new results presented here, which extend the southern valley-side bank even further south, well into Eastern Quarry Area B.

Just to the south of the Swan Valley School plot, TP 16.2 in Eastern Quarry Area B contained what is clearly a slightly further southerly extension of the same Lower Middle Gravel. This result is confirmed by clast lithological analysis and reinforced by its appropriate level above OD and its similar archaeological content, namely handaxes and debitage. TP 16.2 is about 150m to the north of TP 112, where, as discussed above, Ebbsfleet fluvial deposits interpreted as a continuation of the Phase 8 gravels at the elephant site are present at a very similar altitude. However, as discussed below, the situation in between these two locations is more complex, and less clear cut.

Immediately to the south of TP 16.2 there is a south-west to north-east trending dry valley that drops down into the Ebbsfleet valley, filled with last glacial and Holocene sediments. This has cut through and removed any earlier sediments and divides TP 16.2 from a cluster of test pits on its southern side, the sequences from several of which are presented here: TPs 2.2, 3.1, 3.2, 4.1, 6.2, 7.2 and 127 (Fig.4.40; Fig.4.41).

These test pits contain a plethora of sandy gravel deposits, fluvial in appearance and rich in artefactual remains, with recovery of numerous small pointed handaxes and debitage, the handaxes often being fairly crude, but sometimes very neatly made (Fig. 4.43). The deposits occur at a variety of levels, extending to an unknown depth below 23m OD in TP 3.2, and with their highest surface level being nearly 29m OD in TP 127. Clast lithological analysis at TPs 4.1 and 127 has firmly established that the gravels in these test pits are of Thames origin, providing a new maximum southern extent of the Boyn Hill/Orsett Heath formation as represented in the Swanscombe 100-ft terrace, very

The Ebbsfleet Elephant

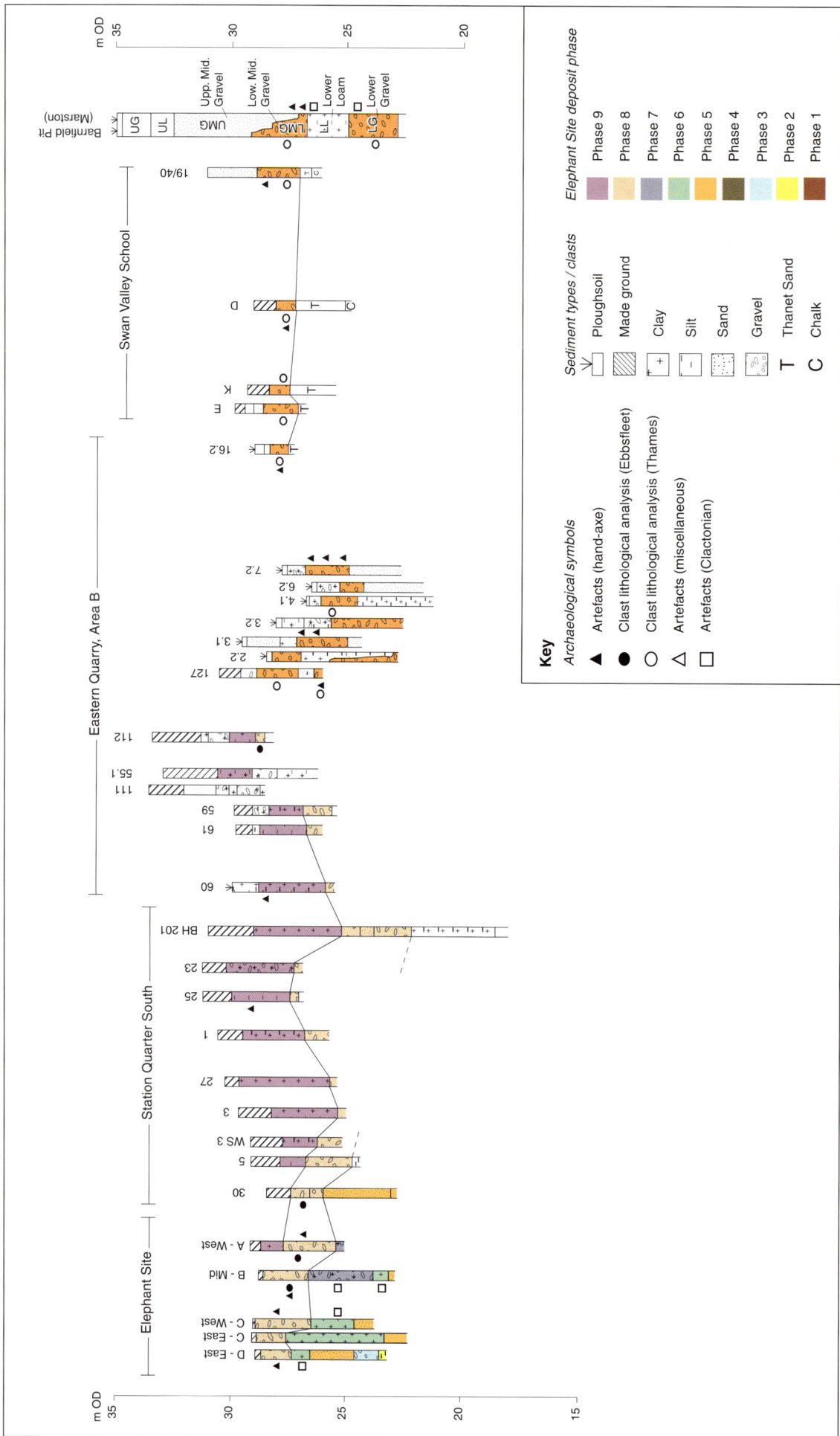

Figure 4.41 (a) North-south Transect CD, lithostratigraphy between the site and the Swanscombe 100-ft terrace at Swan Valley School

(a)

(b)

0   50mm

Figure 4.42 Handaxe from Station Quarter South, Test Pit 25 [photos Francis Wenban-Smith]: (a) as newly discovered in upper part of Phase 9 brickearth; (b) close-up view of handaxe – note imprint in sediment [fine divisions in cm on scale bar]

Figure 4.43 Handaxes from Eastern Quarry Area B, Test Pit 127

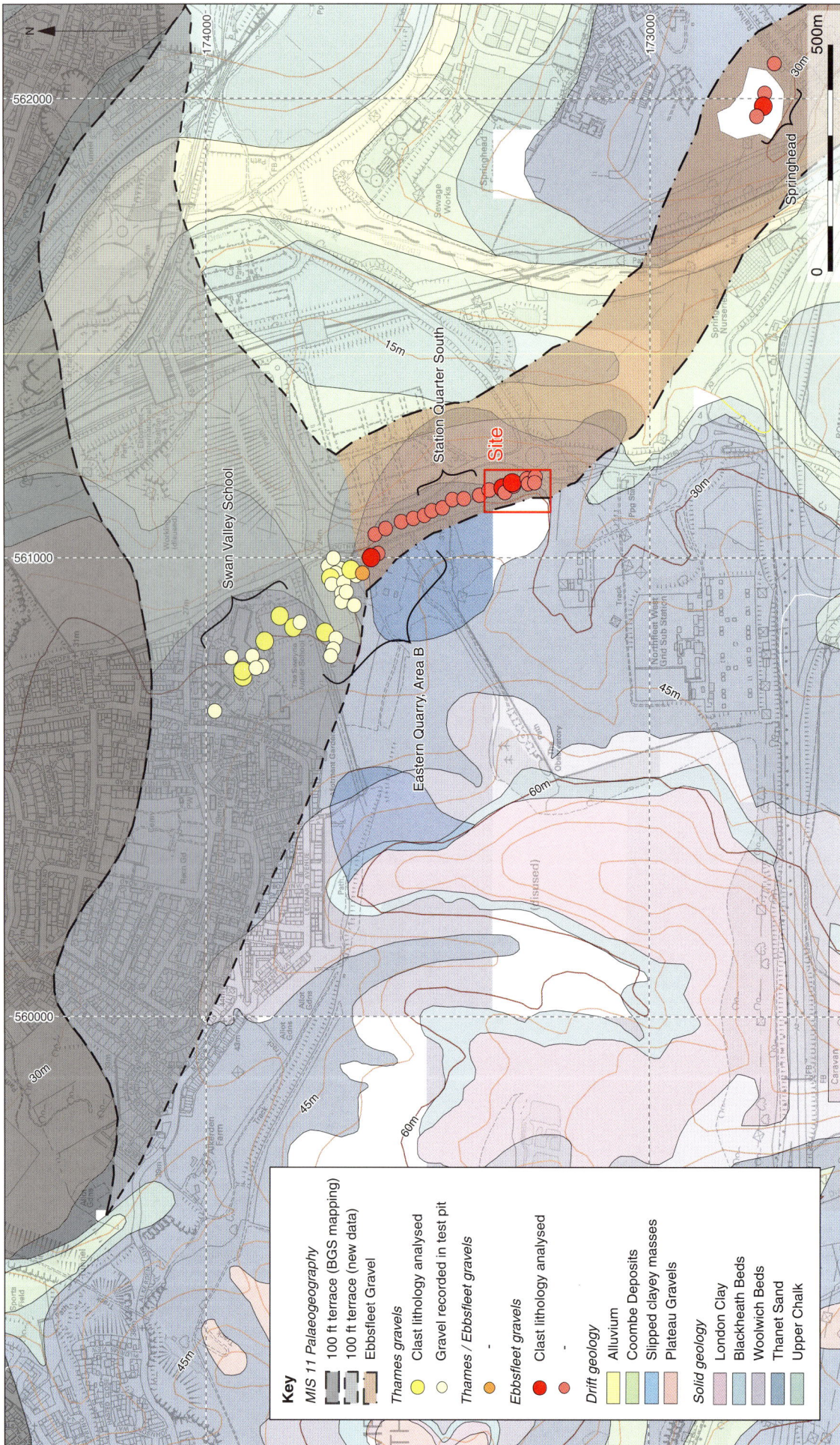

Figure 4.44 Revised palaeo-landscape model of Boyn Hill/Orsett Heath formation (Swanscombe 100-ft terrace outcrop, Phase II) and confluence with the contemporary palaeo-Ebbsfleet (Phase 8)

likely broadly equivalent to the Lower Middle Gravel of the Barnfield Pit sequence.

It was seen in most of these test pits that the base and surface of the gravel beds were not usually horizontal, but dipped in various directions. For example, its upper surface apparently dipped very steeply, declining by several metres to the west in TP 2.2, although it was unclear from the limited exposure whether this was a limited local deformation, caused for instance by underlying solution, or whether indicative of more widespread sediment geometry. This area seems to represent the point of confluence of the palaeo-Ebbsfleet of Phase 8 with the Thames of the Lower Middle Gravel, which would thus on purely geological grounds suggest an MIS II date for the Phase 8 fluvial gravel at the site. The uneven geometry of the gravel, and intervening finer-grain sand/silt beds, at this point is entirely consistent with disturbance relating to a tributary confluence, and with this locale being the furthest outer bend of a large loop of the Thames (Fig. 4.44). There is also, when attempting to interpret the apparent geomorphological geometry of these gravel deposits, the same problem to consider as in interpreting the complex sedimentological geometry at the site, namely: to what extent the present-day geometry represents the original Pleistocene internal position, or whether there have been substantive post-depositional disturbances/events that have altered the deposits, posing extra factors to take account of in their interpretation.

In general, it is suggested here that much of the present-day geometry of the deep Pleistocene sequences, both at the site and along Transect CD between the site and the Thames, has been affected by post-depositional downslope movement (towards the valley floor), and resultant compressional deformation, with possible additional effects from localised chalk solution and vertical collapse. Thus, although broad horizontal correlations remain tenable, particularly when supported by other lines of evidence (such as clast lithological analysis), we would question an undue reliance on more detailed correlation based on very specific altitudes relative to OD and/or interpretation based on apparent sedimentary structures observed in limited exposures, as previously discussed in this chapter.

GEOMORPHOLOGICAL EVOLUTION OF THE SITE LOCALITY

The history of local geomorphological evolution reflected in the sediments in and around the site, discussed above, forms but a small part of the overall geological evolution of the Ebbsfleet Valley from the Middle Pleistocene onwards and elements of this history remain difficult to determine. It is, however, clear that the sediments preserved at the site are the oldest relatively well-understood deposits in the valley and lie at elevations considerably above the recently investigated sequences of deposits associated with the Ebbsfleet International station and its associated works closer to the modern valley floor (Wenban-Smith *et al.* forthcoming).

The overall structure of the sediments, and their fluvial, lacustrine or colluvial origins, suggest that the site would have been close to the valley floor when most of them were deposited, at the foot of the slope of the valley side rising to the west. The contemporary valley side to the east is now missing as a result of the downcutting and eastward shift of the valley axis through subsequent periods of downcutting. The geometry of the sediments indicates progradation of many units from south to north. The associated sedimentological evidence from clast lithology indicates that all fluvial deposits at the site are essentially Ebbsfleet in origin continuing north towards a confluence with the Thames in Eastern Quarry Area B, now pinned down to a narrowly defined area between TPs 112 and 127 (Fig.4.41; Fig. 4.44; Chapter 6). Downcutting and erosion of the valley floor is only indicated with deposition of the final sequence of Pleistocene deposits associated with Phase 9 (the brickearths) that represent valley-side colluvial sequences in the main, perhaps with some aeolian input.

Leaving aside the mysterious earlier history associated with Phase 1 (the 'Tilted Block') and Phase 2 (the 'Primary Sump Infill'), the subsequent history of geomorphological evolution and sediment aggradation seems to reflect the site's position at the junction between valley-side and valley-floor, with alternating cycles of fluvial and colluvial/slopewash deposition. Phase 3 deposits higher up the western valley-side dip and thicken eastward, reflecting slopewash processes. Further east, in the lowest parts of the sequence seen at the site and, especially, in the cluster of Station Quarter South test pits to its east, they contain sand and gravel-rich beds suggesting moderately high energy fluvial deposition. Above this, prior to the deposition of the Phase 8 gravels, the history of landscape development and associated environments of deposition appear to be associated with lower energy water bodies in conjunction with periods of stagnation and drying-out, with variably intense overbank slopewash input.

Although water is implicated in many of the depositional processes associated with Phases 2 to 6, it is probable that deposition occurred in relatively localised bodies of water close to the valley margins at the foot of the valley side rising to the west. Thus local ponds receiving sediment from downslope wash as well as moving bodies of water are likely to have contributed sediments at times. Deltaic conditions have been indicated at times, especially in the well-developed clay-laminated sands of Phase 5, as well as water-edge conditions. However in all cases these are likely to be small-scale local features, probably reflecting minor topographic variations in valley-side fan and valley-floor ponding of water bodies. It is difficult to relate any of these depositional phases to climatic conditions on purely sedimentological grounds, for instance postulating fluvial gravel deposition as a cold/warming or warm/cooling process (cf. Bridgland 2001). Gravel (where present) might easily have been moved/deposited locally by short-lived spate events of little palaeo-climatic significance.

The environment of deposition of the main Phase 6 events associated with the faunal and archaeological remains also reflects this marginal situation between valley side foot and wetter environments on the floodplain. The fine sediments of many of the deposits in this phase clearly indicate deposition under very low energy fluvial, or stagnant marsh/lacustrine, conditions; perhaps in some form of small pond or lake, but where the sediment might have been introduced through downslope washing of sediment into the water body. Drying out of this body is attested to by the brecciation of sediments and various micromorphological features (Chapter 5), as well as the small-channel infilled by reworked tufa (Phase 6b).

After the phase of deposition associated with the Phase 6 sediments, an episode of landsliding occurred, associated with deposition of the major body of gravelly clays that extends upslope to the north-west. This indicates that the site was over-ridden by the toe-end of a major fan of deposits derived from Tertiary deposits capping the high ground to the west. It is suggested here that one possibility for the creation of the asymmetric synclinal 'skateboard ramp' structure between Trenches B and C might be lateral compression by this fan while in a saturated condition, which would have prevented the expected faulting. Other explanations for this structure include localised solution of the underlying Chalk bedrock along north-south joints and the local collapse of organic-rich Phase 6 sediments, although this latter explanation wouldn't address the conformable deformation of the deeper-lying sediments. The massive landslip of Phase 7, along with the nature of the sediments infilling the feature, may indicate that changing local conditions were taking place at this time. Alternatively, it may have been a more unpredictably catastrophic event triggered by an external factor, or by the reaching of an erosional tipping point in the evolutionary history of the local landscape. The Phase 7 sediments are predominantly clay with gravelly patches, although the basal Phase 7 sediments within the heart of the syncline are silts and fine sands (Fig. 4.8). These may be indicative of cold-climate conditions, and the possible inception of minor valley-floor incision at the transition from warm to cold conditions might have destabilised the slope sufficiently to encourage the mass movement of bodies of sediment downslope. The contemporaneous creation and infilling of the syncline would have resulted in conditions necessary for preservation of the feature.

The Phase 7 sediments are unconformably truncated by higher energy fluvial gravels (with sand bars reflecting quieter episodes) of the overlying Phase 8 sequence. Although such gravels are often regarded as reflecting a braided channel environment during cold climatic conditions, this is not necessarily the case. Both the Swanscombe Lower Gravel and Lower Middle Gravel, with the latter of which the Phase 8 gravels are putatively correlated, are both generally accepted as MIS 11 interglacial deposits (eg, Bridgland 1994). The presence of fluvial gravels at this point in the sequence is indicative that any downcutting associated with valley-side destabilisation would only have been of minor, local significance. The cessation of fluvial deposition was followed by the development of the thick sequence of valley-slope brickearths across the site; at this point the fluvial channel may have shifted east, and further downcutting occurred. While one might postulate that this is associated with significant climatic cooling, there is no direct sedimentary structural evidence of this, such as ice-wedge casts. Rather, climatic cooling is only implicit in off-stage downcutting, a small degree of which might be expected to be associated with deterioration of climate at the end of MIS 11 interglacial and the following cold stage MIS 10, followed by more pronounced downcutting at the start of the MIS 10-9-8 cycle associated with formation of the Lynch Hill/Corbets Tey Formation (Bridgland 2006).

Chapter 5

Soil micromorphology, loss-on-ignition, phosphate-P concentrations and magnetic susceptibility analyses

by Richard Macphail, John Crowther and Francis Wenban-Smith

INTRODUCTION

A range of sediment analytical techniques was applied to help understand formation processes associated with the main archaeological horizons, and to perhaps explain some of the apparent stratigraphic breaks and discolorations in the sequence. Analysis was mostly focused on sediments from Phase 6, which contained the richest archaeological remains, including the elephant skeleton and the flint concentration south of Trench D, but also included investigation of the Phase 1 sequence in the 'Tilted Block'. A selection of 16 monoliths was initially examined (Fig. 5.1), leading to identification of 19 target areas for thin section sediment micromorphology, which was complemented by analysis of 'loss-on-ignition' (LOI) and Phosphate-P for 12 associated bulk samples (Table 5.1).

This work was subsequently reinforced by much more detailed LOI and magnetic susceptibility (χ) analyses of 79 more closely spaced samples through the same sequences (Table 5.2), in the hope of identifying soils and/or palaeolandsurfaces. As the sediments accumulated, it would be anticipated that former soils/land surfaces would have had a relatively high organic matter content (as estimated by LOI) as a result of plant growth and inputs of organic litter. An enhanced magnetic susceptibility would also be anticipated as a consequence of natural fermentation processes within soils (Le Borgne 1955). It should be noted, however, that both properties may have been significantly affected by post-depositional processes. Organic matter content is likely to have diminished as a result of decomposition processes and magnetic susceptibility may have been affected by the mobilisation (through gleying), leaching and reprecipitation of iron (Fe) compounds as a result of waterlogging. Also, magnetic susceptibility is affected both by the degree of enhancement and the Fe content, and where (as is likely to be the case in these sedimentary sequences) the latter is quite variable then may poorly reflect the levels of enhancement. The LOI and χ data therefore need to be interpreted with caution.

SAMPLES AND METHODS

Loss-on-ignition (LOI) and phosphate-P

Loss-on-ignition (LOI) and phosphate-P analysis was undertaken on the fine earth fraction (ie < 2 mm) of the bulk samples. LOI was determined by ignition at $375^{\circ}C$ for 16 hrs (Ball 1964). Phosphate-P concentrations were determined colorimetrically following alkaline oxidation with NaOBr, using NH_2SO_4 as extractant (Dick and Tabatabai 1977).

A broad overview of the analytical data from the initial selection of bulk samples, with key features relating to individual contexts highlighted, is presented here (Table 5.3). Full results from more detailed analyses through the different sequences studied are given as an appendix (Appendix 2); significant results from key parts of the site sequence are presented below, in conjunction with the thin section analyses from different parts of the site sequence. All the samples analysed are largely minerogenic; nine of the initial samples have LOI values of < 3.00%; two (both from 40100) have a somewhat higher LOI of 3.42 and 3.63%; and one (from 40158) has a much higher LOI (6.68%). It should also be noted that a number of samples which appeared (by virtue of their darker colour) to be more organic, did not necessarily have relatively high LOI values. LOI in these predominantly minerogenic sediments may be picking up the influence of textural variations between different contexts, with finer-textured contexts having a generally higher LOI. In interpreting these data it should also be borne in mind that organic matter in the various contexts will have been subject to post-depositional decomposition, in other words the LOI recorded therefore underestimates the original organic matter content. The two samples with somewhat higher LOI values may therefore be potentially quite significant within the sequence, perhaps representing periods of soil development. Indeed, the sample from context 40158 could well be derived from what was originally quite a humic, possibly peaty, deposit (see below).

The phosphate-P concentrations exhibit quite marked variability (range: 0.124–1.77mg g-1), though nine of the samples have low concentrations of ≤ 0.346 mg g-1. Somewhat surprisingly, the most organic-rich sample (from context 40158) has the lowest phosphate-P concentration. This is certainly counter-intuitive and tends to suggest that there has been significant post-depositional depletion and movement of phosphate within the sequence. Such movement within the permanently or temporarily waterlogged deposits could be associated with the mobilisation, leaching and deposition of iron. Such a process would seem to be supported

Table 5.1 Overview of source monoliths for thin sections and initial LOI and Phosphate-P bulk samples

Site provenance Site area	Section	Monolith no. <>	Phase	Thin section no.	Depth (cm)	Context/s	Bulk sample no.	Depth (cm)	Context
Central (N)	40091	40418	7/6	M40418A	4.5-12.5	40166/40158	-	-	-
			6	M40418A	7-11(11-12.5)	40158	x40158	7-9	40158
			6/6a	M40418B	17-25	40158/40040	-	-	-
			6a-5	M40418C	40-45	40040-40025	-	-	-
			6-6a-5	M40418C	45-48	40158-40040-40025	-	-	-
Central (EI)	40015	40149	6	M40149	35-43	40078-40099	x40099	30-35	40099, brownish
Central (EI)	40015	40150	6	M40150A	22-24(27)	40099	-	-	-
			6/6a	M40150A	24(27)-38	40099/40103	x40103-Fe pan	24-25	40099/40103 boundary
			6a	M40150B	22-38	40103	x40103-Fe mottling	33-35	40103
Central (EI)	40015	40151	6a	M40151A	7-15	40103/40039	-	-	-
			6a/5	M40151B	32-41	40039/40025	x40039	35-38	40039
Central (EI)	40085	40365	6	M40365A	5-13	40078/40099	x40078	5-7	40078
			6	M40365B	22-30	40099, brownish	x40099	28-30	40099, brownish
			6	M40365C	31-39	40099, brownish	-	-	-
Central (E)	40082	40323	6a	M40323A	0-7	40039	-	-	-
			6a	M40323B	7-13	40039	-	-	-
South (W)	40015 (S end)	40196	6	M40196	8.5-16.5	40100	x40100-a	10-16	40100, brownish
							x40100-b	10-16	40100, greyish
South (W)	40015 (S end)	40195	6	M40195	13-21	40100 (sandy lenses)	-	-	-
South (W)	40015 (S end)	40194	6	M40194	32-40	40100 (stones/artefact inclusions?)	-	-	-
South (W)	40015 (S end)	40193	6/6a	M40193A	29-37	40100/40039	40039-top	30-32	40039, top part, strongly mottled
				M40193B	37-45	40039	40039	34-38	40039
			6a	M40193B			-	-	-
South (E)	40016	40082	1	M40082	29-37	40057	x40057	35-37	40057

Table 5.2 Expanded LOI and magnetic susceptibility (χ) sampling through selected sequences

Figs 5.1 and 5.2 reference	Phase	Context	Monolith <>	Sampling depth (cm)	Sequence depth (cm)	Notes
(f)	7	40166	40418	2-3	2.5	
	6	40158	40418	12-13	12.5	
	6	40158	40418	25-26	25.5	
	6a	40040	40418	36-37	36.5	Stratigraphy complex here; 40160 used for fine yellowish-brown sand wedging out to W, thickening and interdigitating to E with Phase 5
	6a/5	40160	40419	4-5	43	
	6	40158	40418	44-45	44.5	Stratigraphy complex here; 40160 used for fine yellowish-brown sand wedging out to W, thickening and interdigitating to E
	6	40158	40419	11-12	50	Context 40158 (here, in Section 40091, Trench B) is equivalent to the particularly dark facies of 40039 in the central part of the main west facing section in the vicinity of Log 40011
	6	40158	40419	21-22	60	
	6a	40040	40419	31-32	70	
	5	40025	40419	42-43	81	Transitional to 40025
(e)	6	40100	40148	4-5	4.5	
	6	40100	40148	11-12	11.5	
	6	40078	40148	21-22	21.5	
	6	40078	40148	30-31	30.5	
	6	40078	40148	43-44	43.5	
	6	40078	40149	15-16	40.5	
	6	40078	40149	22-23	47.5	
	6	40078	40149	32-33	57.5	
	6	40099	40149	44-45	69.5	
	6	40099	40150	7-8	77.5	
	6	40099	40150	17-18	87.5	Overlap
	6a	40103	40150	32-33	102.5	
	6a	40103	40150	42-43	112.5	
	6a	40039	40151	8-9	111	
	6a	40039	40151	17-18	120	
	6a	40039	40151	24-25	127	
	6a	40039	40151	34-35	137	
	6a	40039	40151	44-45	147	
	5	40025	40152	8-9	148.5	
	5	40025	40152	15-16	155.5	
	5	40025	40152	24-25	164.5	
	5	40025	40152	33-34	173.5	
	5	40025	40152	43-44	183.5	
(d)	6	40078	40365	4-5	4.5	
	6	40099	40365	9-10	9.5	
	6	40099	40365	15-16	15.5	
	6	40099	40365	23-24	23.5	
	6a	40099	40365	30-31	30.5	
	6a	40103	40365	40-41	40.5	
	6a	40103	40365	50-51	50.5	
(c)	6b	40070	40322	4-5	4.5	
	6a	40103	40322	11-12	11.5	
	6a	40039	40323	2-3	22.5	
	6a	40039	40323	11-12	31.5	

continued overleaf

Table 5.2 (continued)

Figs 5.1 and 5.2 reference	Phase	Context	Monolith <>	Sampling depth (cm)	Sequence depth (cm)	Notes
(b)	6	40100	40196	4-5	4.5	
	6	40100	40196	14-15	14.5	
	6	40100	40196	24-25	24.5	
	6	40100	40196	34-35	34.5	
	6	40100	40196	44-45	44.5	
	6	40100	40195	6-7	44	
	6	40100	40195	16-17	54.0	
	6	40100	40195	26-27	64.0	
	6	40100	40195	36-37	74.0	
	6	40100	40195	46-47	84.0	
	6	40100	40194	6-7	81.5	
	6	40100	40194	9-10	84.5	
	6	40100	40194	16-17	91.5	
	6	40100	40194	26-27	101.5	
	6	40100	40194	36-37	111.5	
	6	40100	40194	46-47	121.5	
	6	40100	40193	7-8	122.5	
	6	40100	40193	17-18	132.5	
	6	40100	40193	27-28	142.5	
	6	40039	40193	37-38	152.5	
	6	40039	40193	47-48	162.5	Transitional from context 40039 to 40025?
(a)	1	40058	40082	8-9	8.5	
	1	40058	40082	18-19	18.5	
	1	40058	40082	28-29	28.5	
	1	40057	40082	38-39	38.5	
	1	40057	40082	51-52	51.5	
	1	40057	40081	5-6	5.5	
	1	40057	40081	8-9	8.5	
	1	40057	40081	15-16	15.5	
	1	40057	40081	18-19	18.5	
	1	40057	40081	25-26	25.5	
	1	40056	40081	28-29	28.5	
	1	40056	40081	35-36	35.5	
	1	40056	40081	38-39	38.5	
	1	40056	40081	45-46	45.5	
	1	40056	40081	48-49	48.5	

Table 5.3 Analytical data from bulk samples

Phase	Context	Bulk sample no. <>	Thin section	Bulk sample depth (cm)	LOI -%	Phosphate P (mg g⁻¹)
6	40158	x40158	M40418A	7-9	6.68**	0.124
6	40099, brownish	x40099	M40149	30-35	2.16	0.194
6/6a	40099/40103 boundary	x40103-Fe pan	M40150A	24-25	3.63*	1.77**
6a	40103	x40103-Fe mottling	M40150B	33-35	3.42*	0.538*
6a	40039	x40039	M40151B	35-38	1.88	0.556*
6	40078	x40078	M40365A	5-7	2.44	0.226
	40099, brownish	x40099	M40365B	28-30	2.57	0.346
6	40100, brownish	x40100-a	M40196	10-16	2.52	0.276
	40100, greyish	x40100-b		10-16	1.9	0.127
6a	x40039, strong Fe mottles	40039-top	M40193A	30-32	1.69	0.15
	x40039, Fe mottles	40039		34-38	1.73	0.179
1	40057	x40057	M40082	35-37	1.22	0.156

LOI: samples all largely minerogenic, higher values are highlighted (* = 2.50–4.99%, ** ≥5.00%)   Phosphate-P: samples showing likely signs of enrichment are highlighted (* = slightly enriched 0.500–0.999 mg/g⁻¹), ** = enriched 1.00–1.99 mg/g⁻¹)

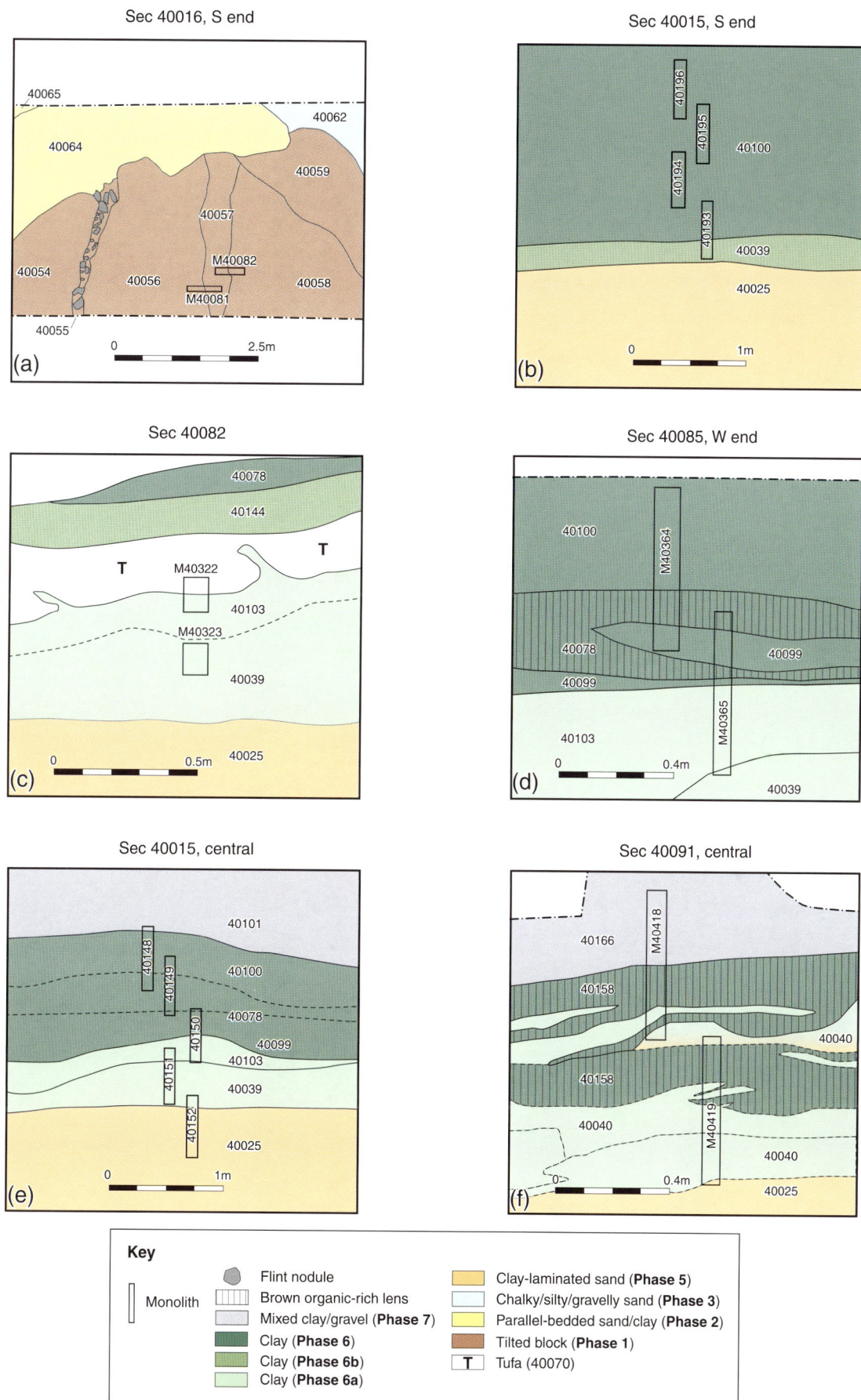

Figure 5.1 Locations of analysed monoliths within local stratigraphic sequences: (a) Phase 1, 'Tilted block';
(b) Phase 6/6a, south of Trench D (west side of site); (c) Phase 6b/6a, tufaceous channel (central east side of site);
(d) Phase 6/6a, west end of Trench C, near elephant; (e) Phase 6/6a/5, central west side of site, near elephant; and
(f) Phase 7/6/6a, north side of Trench B, central part of site

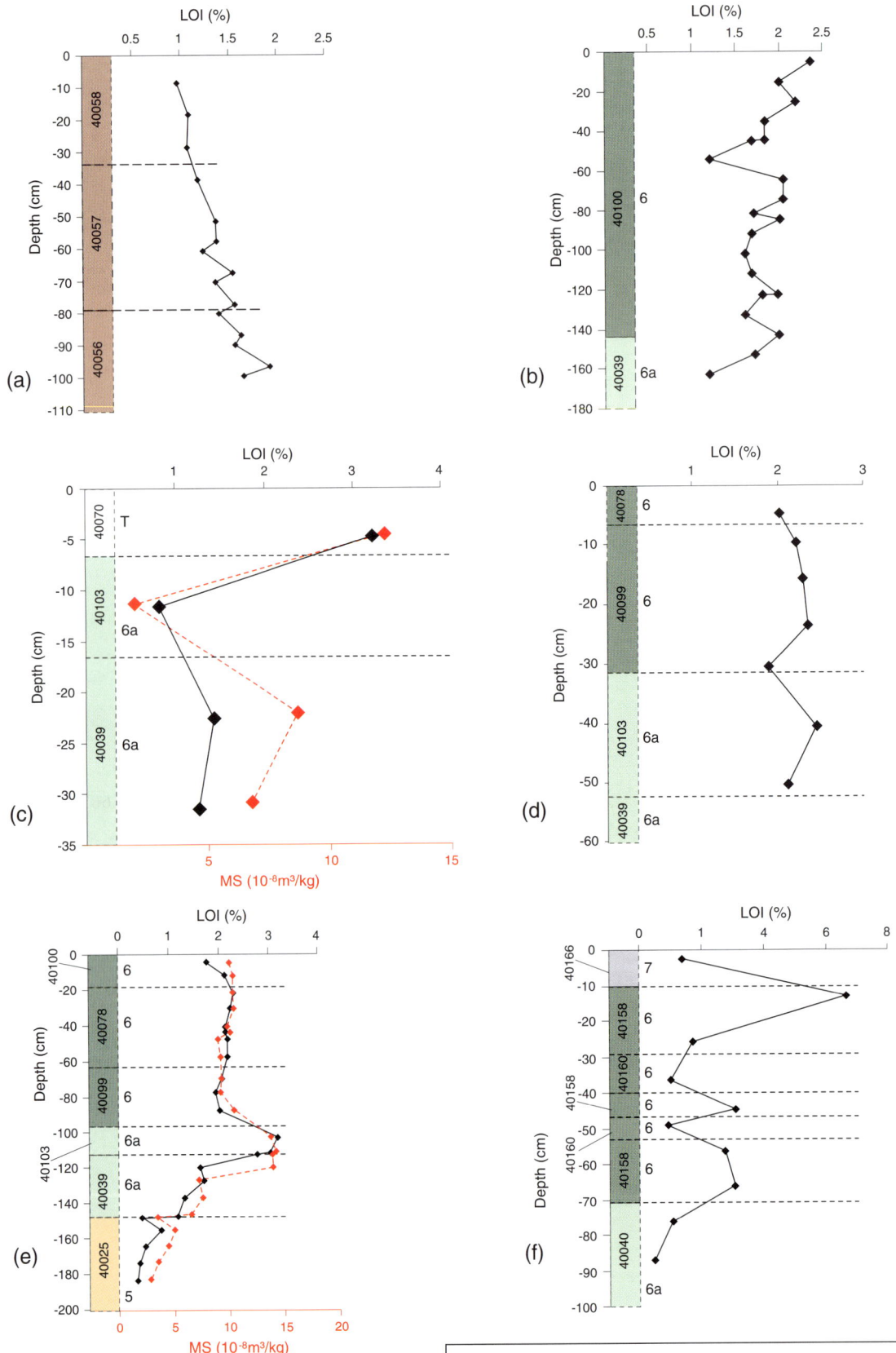

Figure 5.2 Variations in LOI (%) and magnetic susceptibility (χ) through analysed sequences shown in Fig. 5.1: (a) Phase 1, 'Tilted block' through monoliths <40082> and <40081>; (b) Phase 6/6a, south of Trench D (west side of site) through monoliths <40196-40193>; (c) Phase 6b/6a, tufaceous channel (central east side of site) through monoliths <40322> and <40323>; (d) Phase 6/6a, west end of Trench C, near elephant through monolith <40365>; (e) Phase 6/6a/5, central west side of site, near elephant through monoliths <40148> and <40152>; and (f) Phase 7/6/6a, north side of Trench B, central part of site, through monoliths <40418> and <40419>

by the fact that the highest concentration (1.77mg g⁻¹) occurs in the Fe-pan of the sample from context 40103 (discussed further below). In these circumstances, variations in phosphate-P concentrations through the sequence are more likely to reflect patterns of post-depositional redistribution than the original phosphate content of individual contexts, so little further reference is made to the phosphate-P data below.

Magnetic susceptibility (χ)

In addition to χ, determinations were made of χ_{max} (maximum potential magnetic susceptibility, which generally closely reflects the iron content) on 20 samples, representative of the range of χ values recorded, by subjecting a sample to optimum conditions for susceptibility enhancement in the laboratory. χ_{conv} (fractional conversion), which is expressed as a percentage, is a measure of the extent to which the potential susceptibility has been achieved in the original sample, viz: (χ/χ_{max}) x 100.0 (Tite 1972; Scollar *et al.* 1990). In many respects this is a better indicator of magnetic susceptibility enhancement than raw data, particularly in cases where sediments have widely differing χ_{max} values (Crowther and Barker 1995; Crowther 2003). A Bartington MS2 meter was used for magnetic susceptibility measurements. χ_{max} was achieved by heating samples at 650°C in reducing, followed by oxidising conditions. The method used broadly follows that of Tite and Mullins (1971), except that household flour was mixed with the soils and

Table 5.4   Thin section SEM/EDS analyses: (a): M40151B, see Fig. 5.3a, Fig. 5.4a and Fig. 5.13; and (b) M40150A, see Fig. 5.3b and Fig. 5.4b (%; analysed areas and spots)

	Mg	Al	Si	P	K	Ca	Ti	Fe
Area 1								
Matrix	0.93	10.9	31.8		1.73	1.07	0.73	3.90
Fe-staining	0.82	8.18	17.2		1.46	0.55	0.98	32.8
Fe-P-staining 1		6.23	12.5	0.65	0.60	0.53	0.90	43.7
Fe-P- staining 2	0.84	7.43	12.1	0.76	1.01	0.83	0.65	41.6
Fe-P- staining 3	1.16	9.95	21.0	0.55	1.47	0.79	0.73	22.5
Area 2								
Void coating	1.63	7.08	31.0		3.59	0.75		9.54
Matrix	1.16	10.9	31.4		2.39	0.84	0.66	3.98
Pale infill	1.37	16.0	25.0		8.18		0.55	2.41
Area 3								
Fe-P-nodule 1		1.71	10.4	1.07		0.57		55.3
Fe-P-nodule 2		3.82	4.86	1.10		0.86		61.1
Fe-P-nodule 3	0.93	7.09	14.4	0.97	1.05	0.71		38.7
Fe-P-nodule 4		1.86	2.76	1.33		0.85		67.1
Fe-P-staining		5.71	9.81	1.04	0.70	0.76	0.65	48.9
Fe-staining	1.13	20.2	20.8		2.10	0.84		23.8
Fe-void hypocoating	1.04	9.88	26.6		2.07	1.07	0.51	13.8
Matrix	1.37	11.5	25.7		2.35	0.78	0.56	12.7
Pale infill	1.23	12.1	29.9		2.61	0.92	0.60	4.43

(a) M40151B

	Mg	Al	Si	P	K	Ca	Ti	Fe
Area 1 Iron pan 40103								
Marked iron impregnated sediment	0.79	6.74	15.9	1.35	1.12	1.01	0.33	35.4
Ditto	0.47	5.69	11.3	1.75	0.70	1.09	0.29	44.6
Background matrix	1.26	10.6	25.8	0.37	2.22	1.04	0.40	13.2
Area 2 Overlying depleted sediment 40099								
Sediment	1.06	11.4	29.8		2.34	0.99	0.75	5.80
Ditto	1.45	11.9	29.5		2.58	0.93	0.63	5.02
Area 3 Iron pan								
Marked iron impregnated sediment	0.62	6.34	13.9	1.35	1.00	0.84	0.36	39.8
Ditto	0.41	8.12	8.50	1.29	0.80	0.92	0.08	47.0
Background matrix	1.33	10.8	26.8	0.19	2.34	0.83	0.65	11.4
Area 4 Iron pan 40103								
Iron void hypocoating	0.78	5.99	10.8	1.25	1.06	0.99	0.33	45.2
Area 5 Overlying depleted sediment 40099								
Sediment	1.13	12.3	29.5		2.19	0.88	0.55	5.32

(b) M40150A

lids placed on the crucibles to create the reducing environment (after Graham and Scollar 1976; Crowther and Barker 1995).

The most striking feature of the magnetic susceptibility data was the extremely high variability in χ_{max} with values ranging from 22.0–7390 x 10^{-8} m^3 kg^{-1} (Appendix 2). Given this exceptionally high variability, and the relatively

low variability in χ (range, 2.0–27.4 x 10^{-8} m^3 kg^{-1}), it is highly unlikely that there will be a strong relationship between χ and χ_{conv} (Crowther 2003). Furthermore, since such variability in χ_{max} is likely to be largely attributable to variations in Fe content, which in these sedimentary sequences could well have been subject to post-depositional change through gleying and associated

(a): Scan of M40151B, showing mixing of clayey and coarse silt-fine sandy microfacies, location of ferruginised, earlier-formed wood(?) peat, and iron (and phosphate) staining - this and mixing perhaps related to large animal trampling (see Figs 5.13(a)-(h)); note SEM/EDS study area (see T5.4 and Figs 5.13(g)-(h) and Fig 5.4a). Width is ~50mm.

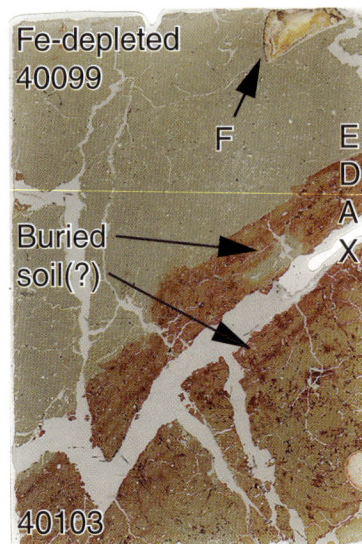

(b): Scan of M40150A; junction of buried soil (upper 40103) and inundation clay (lower 40099); flint (F) and SEM/EDS study are located (see Table 5.4 and Fig 5.4(b)).

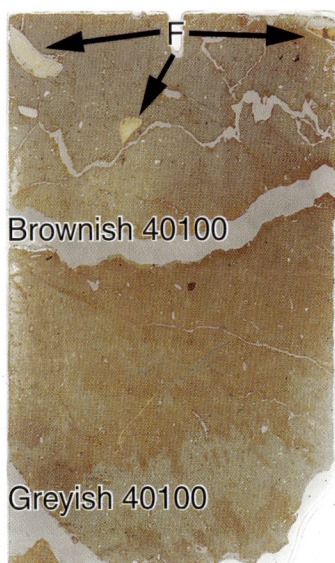

(c) Scan of M40196 (Context 40100), showing greyish and brownish microfacies types; three flint flakes occur in the uppermost layer. Width is ~50mm.

(d): Scan of M40418B showing sandy and peaty microfacies variants in Context 40158 (see Figs 5.19(a)-(d)).

Figure 5.3 Thin Sections after polishing: (a)–(d)

leaching/reprecipitation, the χ and χ_{max} data may poorly reflect the characteristics of the sediments at the time of deposition. Thus, little reliance can be placed on the magnetic susceptibility data, with samples with a high χ_{conv} (maximum, 37.7%) not necessarily being indicative of significant enhancement in the original sediments. Therefore, although the full analyses are presented in Appendix 2, little reference to the magnetic susceptibility data is made in the discussion below.

Soil micromorphology and SEM/EDS analysis

The nineteen thin section subsamples (designated Mxxx, based on the monolith sample number <xxx>) were impregnated with a clear polyester resin-acetone mixture; samples were then topped up with resin, ahead of curing and slabbing for 75x50mm-size thin section manufacture by Spectrum Petrographics, Vancouver, Washington, USA (Goldberg and Macphail 2006; Murphy 1986). Thin sections were further polished with 1,000 grit papers (Fig. 5.3) and analysed using a petrological microscope under plane polarised light (PPL), crossed polarised light (XPL), oblique incident light (OIL) and

using fluorescent microscopy (blue light – BL), at magnifications ranging from x1 to x200/400. Thin sections M40150A and M40151B were also analysed employing quantitative SEM/EDS analyses (Table 5.4; Fig. 5.4). Thin sections were described, ascribed soil microfabric types (MFTs) and microfacies types (MFTs), and counted according to established methods (Bullock *et al.* 1985; Courty *et al.* 1989; Courty 2001; Macphail and Cruise 2001; Stoops 2003; Goldberg and Macphail 2006). An overview of the soil micromorphological results for each thin section is presented here (Table 5.5), and the more detailed data is presented as an appendix (Appendix 7).

PHASE I, 'TILTED BLOCK'

Analysis was carried out on samples from two monoliths, samples <40081> and <40082>, covering the transition through contexts 40056, 40057 and 40058. Context 40056 was interpreted as Thanet Sand, and the overlying, much darker (40057) was thought to possibly be a palaeo-landsurface (Fig. 5.1a).

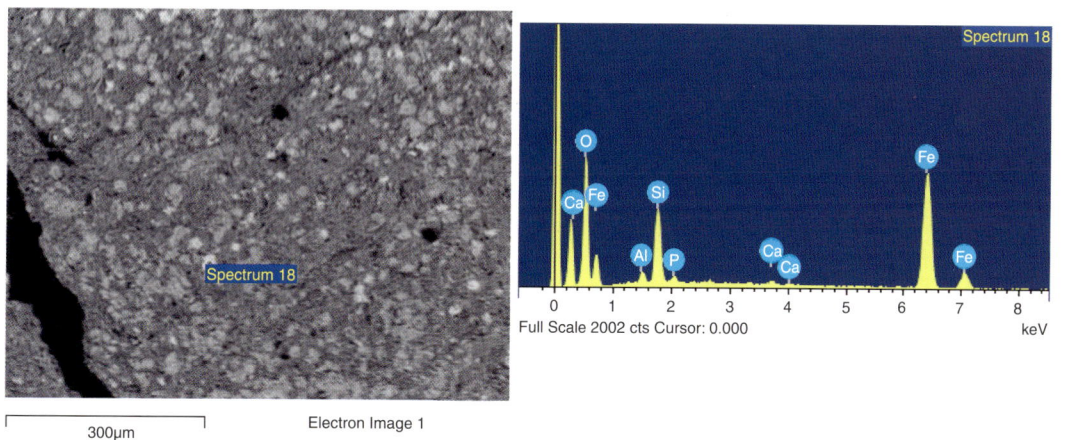

(a): Example of SEM/EDS (M40151B); Backscatter X-ray image and Spectrum of very fine iron nodules (e.g., 55.3% Fe, 1.07% P; see Table 5.4; Figs 5.3(a), 5.13(g)-(h)).

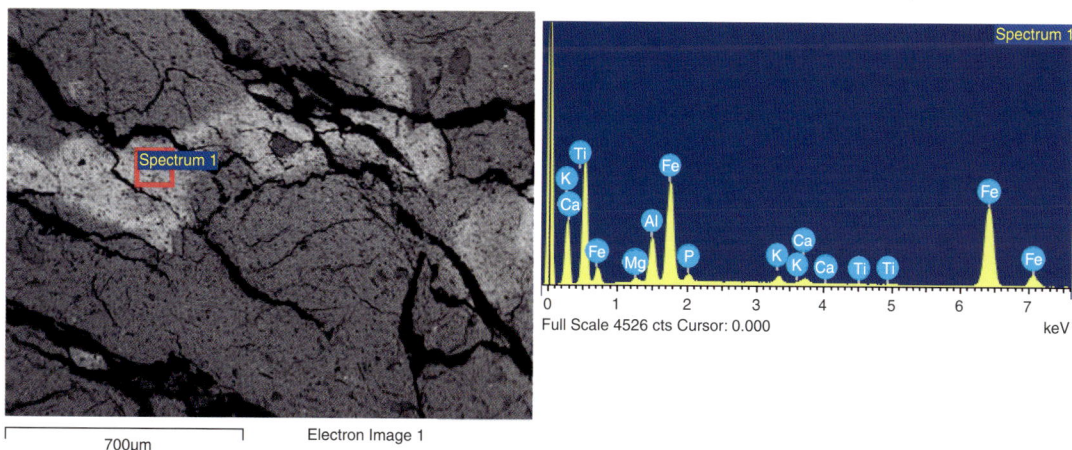

(b): Example of SEM/EDS (M40150A); Backscatter X-ray image and Spectrum of iron pan (e.g., 39.8% Fe, 1.35% P; see Table 5.4; Figs 5.3(b), 5.16(a)-(d)).

Figure 5.4 SEM/EDAX analyses: (a)–(b)

Table 5.5 Soil micro-morphology counts

Thin section	Depth (cm)	Context – rev	Phase	Bulk sample no. <>	MFT	SMT	Voids	Gravel	Clay clasts (soil?)	Glauc.	Root traces	Organic traces	Ferrug. peat	Charcoal
M40418A	4.5-12.5	40166/40158	7/6	-	E4	3a3, 3a1	10%	-	a* papules	f	a?			
M40418A	7-11(11-12.5)	40158	6	x40158	C1	5a1(3a3)	25%	-		*	aa(aaaa)	aaa	aaa frags	
M40418B	17-25	40158/40040	6/6a	-	E3	6a2/3a3	15%			*		aaaaa		aaa
M40418C	40-45	40040-40025	6a-5	-	E2	3a3	40%		?	*		a?	a?	
M40418C	45-48	40158-40040-40025	6-6a-5	-	E1	3a3, 3a4	35%	(a-1)	?	*		a	a	a*
M40149	35-43	40099, brownish	6	x40099	C3b	6a1	30%	(a-1)		*	a*	a		a*
M40150A	22-24(27)	40099	6	-	C3a	6a1	15%	a-1		*	aa	a		a*
M40150A	24(27)-38	40099/40103	6/6a	x40103-Fe pan	C2	5a1	25%		(aa)	*	aaa(aaaa)	aaa		
M40150B	22-38	40103	6a	x40103-Fe mottling	C1	5a1	25%			*	aa(aaaa)	aa		
M40151A	7-15	40103/40039	6a	-	B2	3a1(3a3)	30%			*	a*	a*		
M40151B	32-41	40039/40025	6a/5	x40039	B1	3a1, 3a2, 3a3	20-30%		(aa)	*	a*	aa	aaaa	a(aa)
M40365A	5-13	40078/40099	6	x40078	D3	3a1, 3a2, 3a3	25%	a-1		*	aaa	aaa	(a: org peat)	
M40365B	22-30	40099, brownish	6	x40099	B1	3a1, 3a3	25%			*	aaa	a		
M40365C	31-39	40099, brownish	6	-	B1	3a1, 3a2	10%			*	aaa	a		
M40323A	0-7	40039	6a	-	D2	4a2	10%			*	aa	aa?		
M40323B	7-13	40039	6a	-	D1	4a1	20%	*		*	(a)			a*
M40196	8.5-16.	40100	6	x40100-a; x40100-b	D2a	10a, 10b	10%	(a*)		*	a*			
M40195	13-21	40100 (sandy lenses)	6	-	D1a/D1b	9a1, 9a2	20%	f	(aaa)	f	a*?	a*		a*
M40194	32-40	40100 (stones/ artefact inclusions?)	6	-	C4a	7a1,7a2,7a3	10%	a-1		*		aa(?)		
M40193A	29-37	40100/40039	6/6a	x40039-top; x40039	C4b	8a1, 8a2	25%	a-1		*	a*	a*		
M40193B	37-45	40039	6a	-	C4b	8a1, 8a2	15%			*		?		
M40082	29-37	40057	1	x40057	A1	1a1, 1a2, 2a1	25%			fff	a	a		

Table 5.5 (continued)

Thin section	Depth (cm)	Context - rev	Burned mineral??	Flint chip?	Clay pans	Intercal.	Void clay coats/infill	Secondary Fe	Fine impreg Fe oxide/pyr.	Secondary FeP	Thin burrows	Broad burrows	V. broad burrows
M40418A	4.5-12.5	40166/40158						a					aa
M40418A	7-11(11-12.5)	40158				aa		aaaaa		aa?			
M40418B	17-25	40158/40040			aaaaa	aaa		aaa			aaa		
M40418C	40-45	40040-40025					a	aaaa			aaa		
M40418C	45-48	40158-40040-40025	a-1?				aa	a					a
M40149	35-43	40099, brownish		a-1?		aaaa							
M40150A	22-24(27)	40099		a-1?		aa							
M40150A	24(27)-38	40099/40103				aa	a*	aaaaa	?		aa?		
M40150B	22-38	40103				aa		aaaaa		aa			
M40151A	7-15	40103/40039						aaaa			a*?	a*	
M40151B	32-41	40039/40025				aaa		aaaaa	aaaaa	a			
M40365A	5-13	40078/40099				aaaaa	aa	aa					
M40365B	22-30	40099, brownish			aaaaa			aaaa	(a:ex-pyr)				
M40365C	31-39	40099, brownish				aaa		aaa					
M40323A	0-7	40039			aa	aa		aaaa					
M40323B	7-13	40039	(a*)		aa	aa		a					(aaaa)
M40196	8.5-16.5	40100	a	a-2	aa	aaaaa	aa	(aaaaa)					
M40195	13-21	40100 (sandy lenses)		(a*?)	aa	aaaa	aa	aaaa					
M40194	32-40	40100 (stones/ artefact inclusions?)	a*			aaaaa	aa	aaa	aaa				
M40193A	29-37	40100/40039				aaaaa	aa	aaa					
M40193B	37-45	40039	a-2	a-1?		aaaaa	aaa	aaaaa		?			
M40082	29-37	40057						aa				a	

* – very few 0-5%, f – few 5-15%, ff – frequent 15-30%, fff – common 30-50%, ffff – dominant 50-70%, fffff – very dominant >70%; a – rare <2% (a*1%; a-1, single occurrence), aa – occasional 2-5%, aaa – many 5-10%, aaaa – abundant 10-20%, aaaaa – very abundant >20%

Loss-on-ignition

The sequence through contexts 40058 and 40057 was highly minerogenic, increasing slightly down through the sequence, but with the lowest LOI values in the analysis (Table 5.3; Fig. 5.2a). This trend broadly continued into the underlying sediments, through the bottom part of (40057) and into (40056).

Soil micromorphology

M40082 (context 40057)

As Figure 5.5 shows, within context 40057 there is a boundary zone between firstly, non-calcareous moderately poorly sorted, coarse silty and fine and medium sandy and secondly, glauconitic medium sand-rich sediments with little fine material. The latter became, upwards, mixed with moderately sorted coarse silts and fine sands, sometimes with fewer glauconite, and mixed with clasts, and burrowed-in and washed-in weakly humic clay (LOI=1.22%). Glauconite is moderately weathered with some partially weathered grains (Loveland and Findlay, 1982); opaques include haematite. A relict thin (1.5-2mm) probable root channel is marked by iron hypocoatings. Iron staining also affects traces of organic matter.

This thin section records a moderately mixed junction between fine and medium non-calcareous glauconitic sands, and overlying silty glauconitic sands, and clayey silts, probably deposited as alluvium and recording diminishing energy. There has probably been some burrowing and rooting, the latter affected by secondary iron staining, which marks a minor amount of sediment ripening/soil formation here. Clayey deposits were probably weakly humic originally.

PHASE 6/6a, SOUTH OF TRENCH D (WEST SIDE OF SITE)

Analysis was carried out on samples from the sequence of four monoliths, samples <40196>, <40195>, <40194> and <40193>, down through the Phase 6 grey clay (40100) and into the underlying Phase 6a reddish-brown clay (40039) (Fig. 5.1b). Context 40100 included various variations in colour and texture, with some horizontal sandier horizons and some horizontal bands of faint reddish-brown staining. This area of the site was associated with the dense flint concentration south of Trench D.

Loss-on-ignition

The sequence down through contexts 40100 and 40039 was generally minerogenic, oscillating slightly through the sequence, but with a general downward trend from *c* 2.5% at the top of the sampled sequence, down to *c* 1% at the base of 40039 (Table 5.3; Fig. 5.2b). The slightly higher value (2.52%) of the brownish bulk sample from <40196> in (40100), as opposed to 1.9%

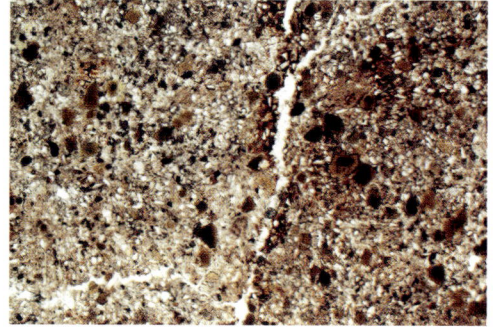

(a): Photomicrograph of M40082 (Context 40057); fine and medium glauconite fluvial sands. Note greenish to brownish weathered glauconite (Loveland and Findlay, 1982). Plane polarised light (PPL), frame width is ~4.62mm.

(b): As (a), under oblique incident light (OIL). A vegetated surface is recorded by iron staining relict of rooting. Glauconite is greenish. Opaque minerals seem to include red haematite.

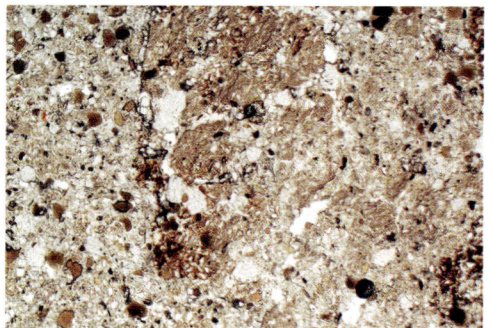

(c): As (a); upwards 40057 has iron-stained roots and burrows; the latter are affected by weak clay inwash (from overlying 40058?). PPL, frame width is ~4.62mm.

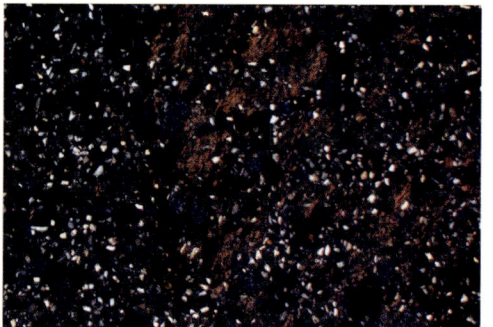

(d): As (c), under crossed polarised light (XPL). Note inwashed clayey burrow fills.

Figure 5.5 Thin Section M40082: (a)–(d)

from the adjacent greyish level, possibly suggests that the more reddish-brown layers within context 40100 in this part of the site may reflect a slightly higher remnant organic content. However, this proposed colour association evidently does not hold for the strongly reddish-brown underlying context (40039).

Soil micromorphology

M40193B (context 40039, lower)

This is similar to M40194 (see below) but more sandy, with three *c* 20-25mm thick layers of clayey fine sand (40% fine sand). There are textural clayey intercalations and iron staining throughout, but these are concentrated

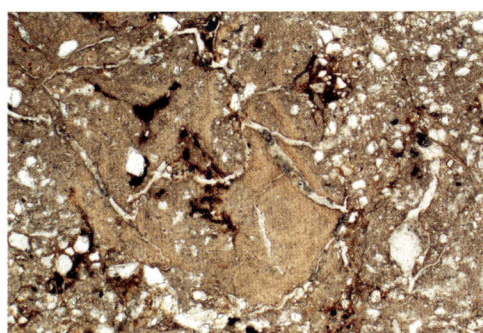

(a): Photomicrograph of M40193B (Context 40100); mixed clayey sands and clays, with clayey inwash feature. PPL, frame width is 4.62mm.

(b): As (a), under XPL; note orientated clay associated with this clayey infill.

(c): As (a), under OIL. Sediment is essentially iron-depleted (gleyed/waterlogged conditions), but affected by later ferruginisation of probably organic staining.

Figure 5.6 Thin Section M40193B: (a)–(c)

in the central layer where there are clay infills up to 3mm wide (Fig. 5.6) and an irregular boundary associated with clayey and sandy clay sediment mixing. Burrows are also iron stained in places. A 400μm-size patinated/calcined flint fragment and an 8mm-size angular flint flake (with possible traces of rubefication) were noted.

This sediment was originally a layered/laminated muddy fine sand (cf. M40151B, Fig. 5.3a), with sediment mixing, clayey intercalations and clayey infills with, for example, marked iron staining, all possibly indicative of trampling. This was possibly again related to large animals affecting an ephemeral surface (or surfaces), and/or by hominins, as the artefact inclusions may imply a hominin presence.

M40193A (context 40100)

The sediment covered by this thin section is similar to that from below (M40193B), but contains more poorly sorted fine to coarse sands and sandy concentrations. There are very abundant textural intercalations, with associated closed vugh formation. Fewer ('many') and more weakly iron stained fabrics occur, however. There is an example of fine gravel and a root trace with ferruginised, once-organic very thin excrements present. It also seems to record muddy, possibly trampled, water-saturated sediments containing higher amounts of poorly sorted sands, compared to below; perhaps as the result of slightly increasing episodic fluvial activity.

M40194 (context 40100, stones/artefact inclusions?)

M40194 covers a heterogeneous, massive, compact iron-depleted clay and iron-stained clay (containing coarse silt and fine sand), with a sharp 2mm-size flint chip (Fig. 5.7 a; b). It is characterised by textural intercalations (associated with uni- and grano-striate b-fabric), which are sometimes silty clay in nature, and patches and infills of yellow-stained clay with fine (5-10μm) ferruginised amorphous organic material (very fine iron nodules/possible relict pyrite pseudomorphs?) (Fig. 5.7c; d). Trace amounts of very fine charcoal, blackened detrital organic matter and phytoliths are present, as well as examples of likely burnt mineral grains. There is fabric mixing of two major microfabric types.

It is a muddy and probably physically disturbed sediment, which was generally iron-depleted (gleyed) but which originally had organic sediment mixed in. The latter was originally affected by possible pyrite framboid formation (associated with decay of relict organic matter under hydromorphic conditions (Miedema *et al.* 1974; Wiltshire *et al.* 1994)) and oxidation/ferruginisation of this material and/or the fine organic material associated with it. Again, as in M40151B (40039) where phosphate concentrations occurred, wet sediment mixing in 40100 may have occurred alongside inputs of possible dung in an animal wallow (cf. Unit 4u at the Boxgrove freshwater pond, Macphail 1999). The occurrence of possible burnt mineral inclusions may indicate local contemporary fire, whether of hominin or natural origin is uncertain.

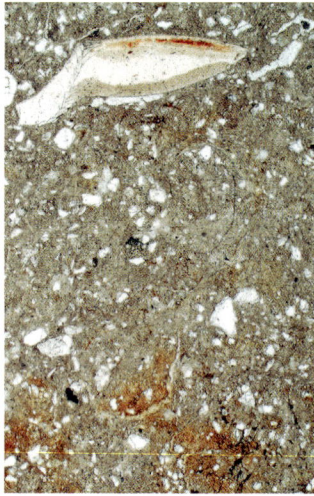

(a): Photomicrograph of M40194 (Context 40100), showing cracked flint flake and iron staining of generally iron-depleted sediment. PPL, frame height is ~4.62mm.

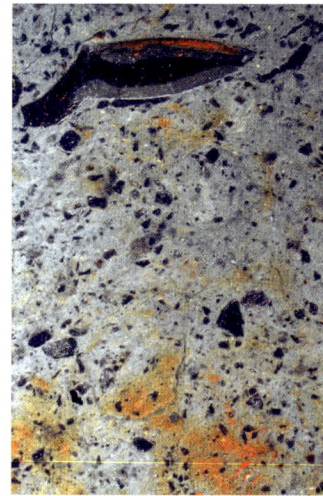

(b): As (a), under OIL. Note iron-staining on flint and generally iron-depleted waterlogged sediment.

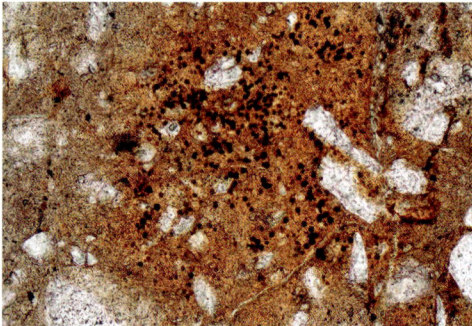

(c): Detail of (a), with very fine (5-10µm) iron nodules formed in iron-stained matrix. PPL, frame width is ~0.90mm.

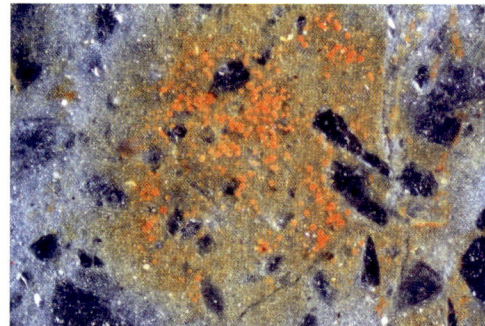

(d): As (c), under OIL. EDS on larger but perhaps similar nodules identified the presence of both Fe and P (see Figs 5.13(g)-(h)); these may also be possibly pseudomorphic of pyrite framboids.

Figure 5.7 Thin Section M40194: (a)–(d)

M40195 (context 40100, sandy lens boundaries)

This thin section is located along layered junctions between sandy microfacies (coarse silt, fine to very coarse sands with fine [max 5mm] flint gravel) and overlying a moderately, sandy clayey layer containing many fine gravel-size brownish clay clasts (as in M40196, see below) (Fig. 5.8). Upwards, is a greyish clayey layer where textural intercalations, matrix void coatings and clayey pans are abundant. There is weak to moderate iron staining throughout, possibly sometimes picking out relict amorphous organic matter. Both weathered and moderately fresh glauconite is present. Rare fine blackened/charred very fine organic matter occurs.

The thin section records a relatively high energy event/wash compared to the clayey deposits generally. First sands and fine gravels were deposited, followed by sandy clays containing gravel-size clay clasts (eroded 40100-type material from elsewhere, probably Tertiary deposits from high ground to the west; see Chapter 2), suggesting a lowering of energy and more muddy, possibly colluvial, deposition upwards. Lastly, muddy clayey sediments are deposited which are characterised by textural intercalations, matrix void coatings and clayey pans. Generally, this appears to be an upward-fining/decreasing energy sequence, the sand lens marking a localised relatively high energy deposition event, perhaps a fan/slopewash deposit.

M40196 (brownish over greyish)

This is a brownish clay loam, partially mixed with grey clay loam, and becoming more dominantly grey downwards (Fig. 5.3c). The microfabric is characterised by very abundant textural pedofeatures (intercalations, pans and matrix void coatings – and associated closed vughs) and iron staining. Fine channelling and fissuring affected the massive soil-sediment. Two flint chips (5mm and 10mm-in size, with a distinct outer patinated/stained zone), and an enigmatic embedded 'reddish' clay clast (c 1mm) occur (Fig. 5.9). Flints are also embedded in the soil-sediment matrix.

The sediment is a muddy mixed clay loam, with mainly iron-stained brownish sediment evident in the upper half of the slide and iron-depleted clay loam

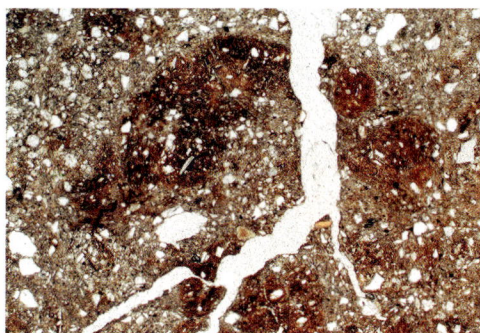

(a): Photomicrograph of M40195 (Context 40100); mixed sandy loam sediments contain rounded very coarse sand and gravel-size dark brown clay (as in M40195, Context 40100, above).PPL, frame width is ~4.62mm.

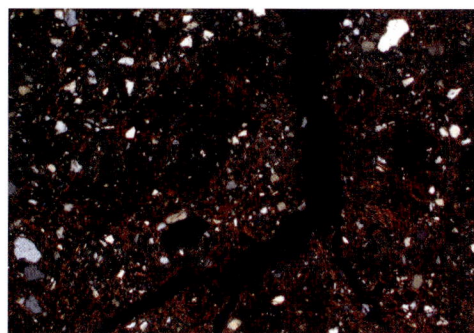

(b): As (a), under XPL; clay clasts act as embedded grains, indicating muddy slurry-like conditions of deposition, probably due to erosion and colluviation.

Figure 5.8 Thin Section M40195: (a)–(b)

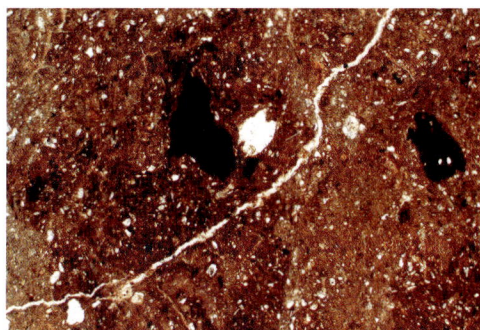

(a): Photomicrograph of M40196 (Context 40100) (see Fig 5.9(c)); highly heterogenous mixture of brown clay, iron depleted grey clay, and including opaque clay fragments. PPL, frame width is 4.62mm.

(b): As (a), under OIL. Note red (rubefied) colour of opaque clay clasts and other fine red mineral material. These could be fine fragments from a combustion feature.

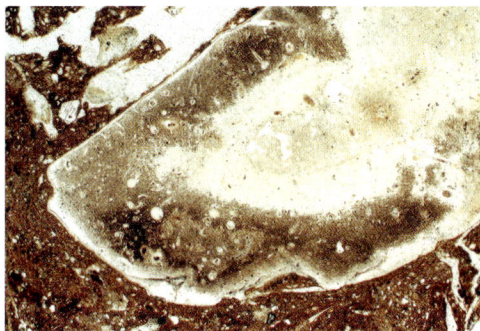

(c): As (a); fragment of 100mm-size fire-cracked flint. PPL, frame width is 4.62mm.

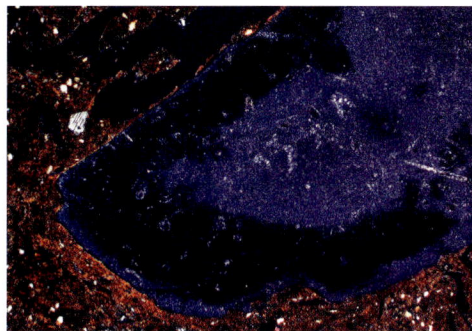

(d): As (c), under XPL. Flint occurs as an 'embedded grain' (aligned matrix clay around flint) within these once-muddy sediments.

(e): As (c), under OIL, illustrating patination? rather than calcination (from burning) and cracked nature of the flint fragment.

Figure 5.9 Thin Section M40196: (a)–(e)

below. The upper part also shows mixed grey and brown microfabrics, and the inclusion of angular flint chips, and possible rubefied clay material, as two ~1mm-size fragments and as a scatter of very fine rubefied material. These materials may have been possibly a reworked relict of a combustion episode, but eroded from upslope and fragmented during downslope transport by colluvial processes. Thus, if this is evidence of contemporary burning, it would have been on higher ground to the west, rather than at the site, and, as above, it is uncertain whether it was of hominin or natural origin.

PHASE 6b/6a, TUFACEOUS CHANNEL (CENTRAL EAST SIDE OF SITE)

Loss-on-ignition analysis was carried out on samples from two short monoliths in the main longitudinal section through the tufaceous channel (Fig. 5.1c). The higher of these, monolith <40322>, covered the base of the mollusc-rich tufa (40070) and the top of the underlying grey clay (context 40103). The lower, monolith <40323>, covered the upper two-thirds of

context 40039, the lowest bed of the Phase 6 clay, which was here stained strong faint yellowish-brown in patches, with occasional strong brown mottles. Two samples were taken from each of these monoliths, to form a sequence down through contexts 40070, 40103 and 40039 (Table 5.2). Thin sections were only analysed from two sub-samples from the lower monolith <40323>. The upper of the two thin section sub-samples, no. M40323A, was formed of the top 7cm of the monolith and covered the top part of context 40039, just below its junction with context 40103. The lower of the two thin section sub-samples was formed of the lower 6cm of the monolith, and covered roughly the middle part of 40039. It was also intended to examine thin sections from monolith <40322>, but this was unfortunately found to be unusable on delivery.

Loss-on-ignition

This sequence (Fig. 5.2c) reveals a close correlation between LOI and magnetic susceptibility χ, and a clear distinction between (40070), which has a higher LOI and χ, and the underlying contexts (40103 and 40039).

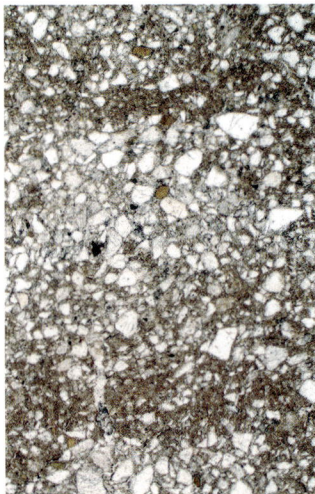

(a): Photomicrograph of M40323B (Context 40039 middle); relict clayey and sandy laminae. PPL, frame height is ~4.62mm.

(b): As (a), under XPL. Note fine and medium subangular and angular quartz sand, and grano-striate b-fabric formed by muddy clay deposition.

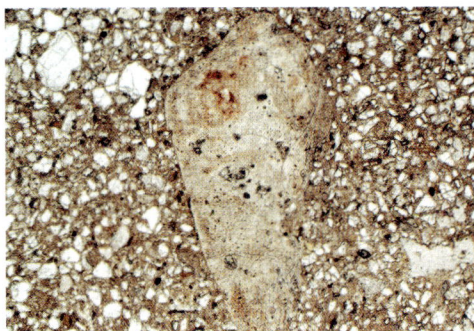

(c): As (a); gravel-size flint inclusion. PPL, frame width is ~4.62mm.

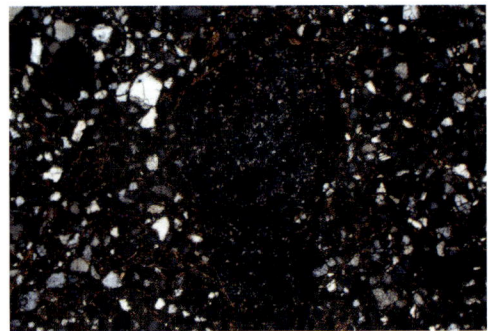

(d): As (c), under XPL; note birefringent clay around flint ('embedded grain'), suggesting inclusion in muddy sediment.

Figure 5.10 Thin Section M40323B: (a)–(d)

Soil micromorphology

M40323B (context 40039, middle)

As seen in this thin section, context 40039 is a massive, moderately well sorted clayey fine and medium sand. It also contains very rare fine gravel clasts, and relict sedimentary laminae (Fig. 5.10). Traces of very broad burrows and penecontemporaneous clayey inwash; intercalation and grano-striate formation were also noted. Clayey intercalations may reflect rooting and associated preferential channel formation. Rare iron staining (for example along the edge of very broad burrow?) and sedimentary laminae hypocoatings were found.

The thin section records moderately low energy muddy fine and medium sandy colluvial deposition, with penecontemporaneous very broad burrowing and clayey infilling. Minor secondary iron staining also affected the sediment.

M40323A (context 40039, upper)

Here, 40039 is a massive, moderately well sorted clayey fine and medium sand, with in the main a strongly iron-depleted fine fabric, which is generally very weakly humic, but appears, upwards, to have a series of thin (<1mm) once-humic (now-iron-replaced) pans. There are trace amounts of phytoliths present, probably derived from contemporary wetland vegetation. In the uppermost 5mm a scatter of 0.5mm-size clay clasts occur (Fig. 5.11). Iron also picks out many traces of probable fine rooting and voids around these clasts.

The sediment characteristics suggest moderately low, becoming lower, energy fine and medium sandy colluvial deposition, marked by thin rooting and periodic organic matter accumulation, possibly seasonal very thin peat formation. Lastly, inwash of eroded clay clasts is recorded. This all occurred under waterlogged conditions (hence iron depletion and organic matter panning), perhaps as the channel silted up. The sediment was affected by secondary but penecontemporaneous iron staining.

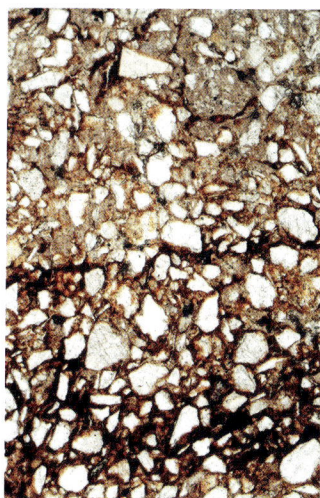

(a): Photomicrograph of M40323A (Context 40039 upper) showing 1mm-thick ironpans and overlying sands containing subrounded clasts of iron-depleted clay. PPL, frame height is ~4.62mm.

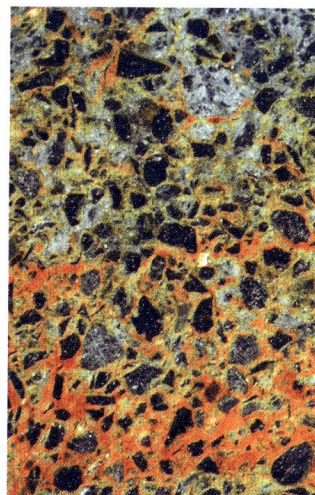

(b): As (a), under OIL. Note ferruginised ironpans. Iron appears to replace very thin once-humic laminae.

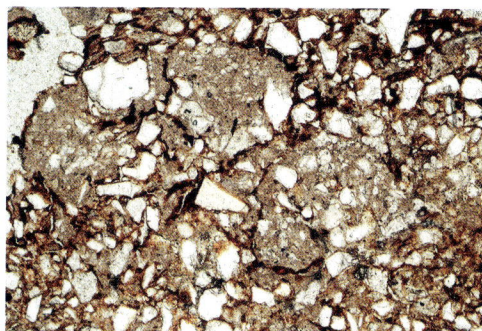

(c): As (a); detail of iron-depleted clay soil-sediment clasts, probably indicating colluvial deposition of eroded soil-sediments from upslope. PPL, frame width is ~2.38mm.

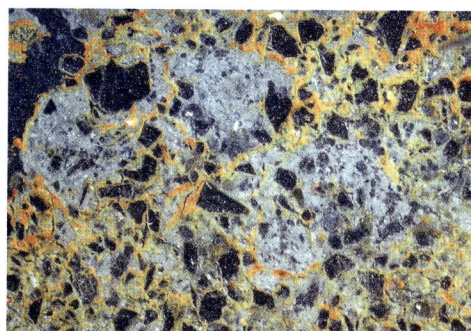

(d): As (c), under OIL. Soil-sediment clasts have become cemented by secondary iron deposition, possibly associated with relict fine rooting in places.

Figure 5.11 Thin Section M40323A: (a)–(d)

PHASE 6/6a, WEST END OF TRENCH C, NEAR ELEPHANT

Analysis was carried out on samples from one monolith, sample <40365>, covering the transition through contexts 40078, 40099 and 40103. This sample was taken at the west end of the south face of Trench C (Fig. 4.9), being the nearest vertical section to the elephant skeleton that could be obtained (Fig. 5.1d). By this stage, it was realised that organic-rich brown-stained horizons, such as contained the elephant (40078), were multiply present in this part of the site towards the base of the Phase 6 clays, and often faded away laterally, or merged with and separated from, each other. Nonetheless, the upper brown-stained horizon in sample <40365> was a lateral continuation of the same bed that contained the majority of the elephant bones and the associated flint artefacts. So this sample was not only subjected to sedimentological analyses, but also relatively intensive pollen analysis (Chapter 12).

Loss-on-ignition

The LOI values recorded in this sequence were quite low, with the values generally between *c* 2.00 and 2.50% (Fig. 5.2d; Table 5.3), although there was a minor peak towards the base of 40103 (at 405mm).

Soil micromorphology

M40365C (context 40103, with reddish-brown staining)

Recorded in this thin section is, generally, an iron depleted, very weakly humic clay, but with many ferruginised traces of roots, some possibly once-woody and up to 3mm in diameter. It is a waterlogged lacustrine clay, with sediments having acted as rooting medium for wetland, and possibly woodland, plants.

M40365B (context 40099, with reddish-brown staining)

This is a mainly iron-depleted clay, characterised by *c* 1mm-thick clay panning (sedimentation), with silt-rich clay loam infilling along possible relict root channel-fills; the latter are marked by iron staining and relict likely hypocoatings. Relict root features are 3-5mm wide; some show ferruginised traces/oxidised traces of probable pyrite framboids associated with roots (Fig. 5.12). These are waterlogged clay sediments deposited as muddy pans, perhaps associated with very low energy alluvial events/flooding. Sediments were rooted by plants, possibly shrubs/woodland, and more silty clay loam infilled these on decay.

M40365A (context 40099-40078 transition)

This is a weakly humic clayey sediment with fine organic fragments and (possibly monocotyledonous) root traces and common patches of partially bedded silty clay loam (40078) over iron-depleted clayey (40099). Context 40078 includes blackened relict plant/root fragments, up to 1mm in size (red and black under PPL); overall organic matter content is reflected in 2.44% LOI. Textural intercalations and associated matrix coatings, and fabric mixing associated with relict channels and fissures.

Layered, weakly humic remains of a putative junction between monocotyledonous minerogenic peat and minerogenic sediments are present, with evidence for flood wash bringing in silts; hence mixing down profile along old root channels. Rooting down through waterlogged clayey sediment caused fabric disruption, mixing and intercalations, typical of shallow lakes that dry out (Cruise *et al*. 2009). This sediment records a period of moderate stasis, possibly with flooding of inwash silts and peat growth.

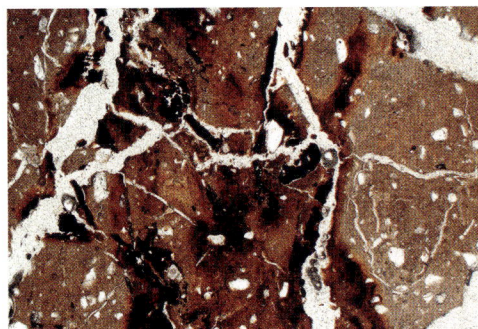

(a): Photomicrograph of M40365B (Context 40098); root traces in waterlogged clayey sediment. PPL frame width is ~4.62mm.

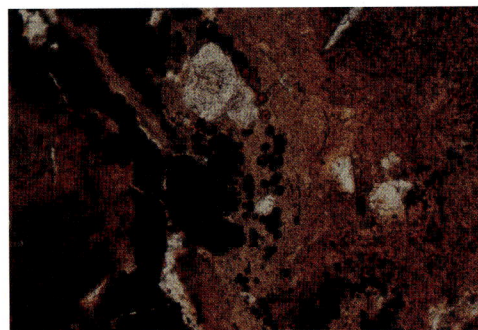

(b): Detail of (a), showing ferruginised pyrite (framboid) pseudomorphs associated with root decay. PPL, frame width is ~0.90mm.

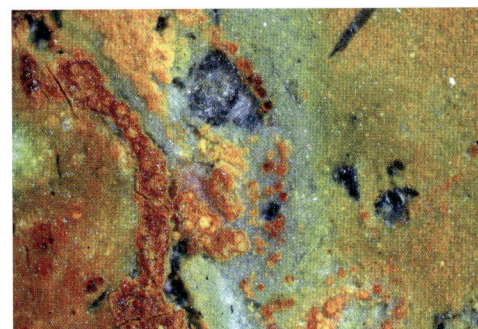

(c): As (b), under OIL.

Figure 5.12 Thin Section M40365B: (a)–(c)

PHASE 6/6a/5, CENTRAL WEST SIDE OF SITE, NEAR ELEPHANT

Analysis was carried out on samples from the sequence of five monoliths down through the Phase 6 clay in the central part of the main west-facing section (Fig. 4.5), and into the underlying Phase 5 clay-laminated sands: monoliths <40148>, <40149>, <40150>, <40151> and <40152> (Fig. 5.1e). This monolith sequence was likewise located reasonably close to the elephant, and preserved the full range of contexts through the lower Phase 6 stratigraphic sequence (apart from Phase 6b, the tufaceous channel, which was not present in this part of the site).

Loss-on-ignition

In this sequence, there is a clear peak in LOI and magnetic susceptibility χ at 102.5–111cm (corresponding with the top and upper part of context 40103 (Fig. 5.2e; Table 5.3). It should also be borne in mind that organic matter in the various contexts will have been subject to post-depositional decomposition; the LOI recorded therefore underestimates the original organic matter content, so this peak is likely to be associated with soil development and surface exposure. Sediments become increasingly minerogenic towards base of sequence (contexts 40039 and, especially, 40025).

There was, however, no peak in LOI or magnetic susceptibility associated with the brown-stained and apparently humic-rich context 40078, although the values were marginally higher than for the grey clay above and below.

Soil micromorphology

M40151B (context 40025/40039 transition)

This thin section includes a partially layered, partially coarsely mixed non-calcareous sediment composed of clays, silty clay, fine and medium sands and ferruginised once-organic sediment (Fig. 5.3a). Silty clays are composed of mixed silt and clay fine laminae. Charcoal as both rare fine (max 800μm-size monocotyledonous charcoal) and occasional very fine material occurs alongside very fine blackened (detrital) organic material, and phytoliths are present (Fig. 5.13a; b).

Patches of totally ferruginised organic matter including plant fragment pseudomorphs occur (current LOI is 1.88%) (Fig. 5.13c; d). Sediments are also characterised by textural intercalations and associated matrix embedded grains and matrix void coatings, some stained yellow. Sediments have evidence of mottling: iron-depleted and iron-stained areas. Rare fine sand-size yellow isotropic nodules/infills, and fine ferruginous nodules are iron phosphate formations, as confirmed by SEM/EDS (Table 5.4; Fig. 5.13e; f). Generally iron-depleted sediments (2.41-3.90% Fe); patches of iron staining also includes phosphate (22.5-43.7% Fe, 0.55-0.76% P, *n*=3). In iron-stained sediments strongly

developed very fine (*c* 30μm) iron nodules are more phosphate-rich (38.7-67.1% Fe, 0.97-1.33% P, *n*=4; Fig. 5.13g; h). The sediment overall is characterised by slight phosphate enrichment, 0.556 mg g^{-1} phosphate-P (Table 5.3).

This deposit was originally a bedded and laminated sediment composed of weakly humic clays, silts, and occasionally medium sands, with patches (possibly inclusions) of peaty material, which are now ferruginised and include plant pseudomorphs. The presence of these organic traces, and charcoal including likely monocotyledonous material, suggests the local presence of wetland (marsh) and, possibly, local incidences of burnt vegetation, and later possible rooting by woody plants: possibly a hydroseral plant succession. Contemporary fabric mixing and associated intercalations, matrix coatings, and yellow phosphate staining and very fine nodule formations that are enriched in P, alongside overall slight phosphate enrichment may infer animal trampling and defecation, possibly in a wallow. Such phosphate enrichment was attributed to large animal concentrations at the Q1B waterhole site at Boxgrove (Macphail *et al.* 2001).

M40151A (context 40039, upper)

Upwards, 40039 continues as massive non-calcareous clay sediments (few sand inclusions), with marked iron-staining overlying 40103. This occurs partially as strongly iron-depleted clay in uppermost 20mm of the thin section. There are infill features (30mm deep by 4.5mm wide curved 'fill'), burrows, once humic peds and excrements, demarcated by strong iron-staining. Sands are only present in rare broad burrow fills. Clays are seemingly less humic compared to lower down, and only very fine charred/blackened detrital organic matter noted.

Minerogenic and sometimes peaty context 40039 develops upwards into a ripened minerogenic wetland clayey sediment which appears to have developed as a topsoil, possibly a hydroseral succession. Here, there are burrows, peds, roots and excrements of relatively humic soil showing marked iron impregnation (Fig. 5.14). Overlying iron-depleted context 40103 is probably a flood clay, recording inundation of the site. The transformation of the buried topsoil through ferruginisation is common to other examples of flood inundated soils, for example Upper Palaeolithic and Early Mesolithic Three Ways Wharf, Uxbridge; Boxgrove Unit 5c and newly inundated coastal analogue sites (Lewis *et al.* 1992; Macphail 1999 and 2009; Lewis and Rackham 2010; Macphail *et al.* 2010).

M40150B (context 40103)

The sediment in this thin section is a massive clay containing small amounts of coarse silt, fine and medium sand. The clay has fissures and relict channels, some with probably ferruginised (woody) root pseudomorphs (0.5-2.5mm; with traces of very thin ferruginised excrements) (Fig. 5.15). The sediment is mottled with iron depleted and iron stained areas and other

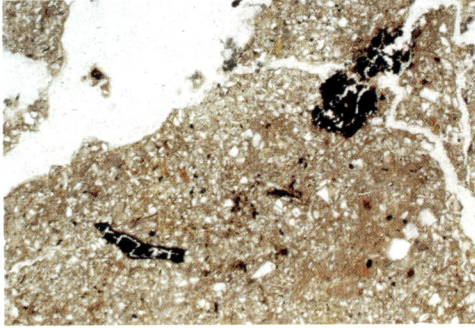

(a): Photomicrograph of M40151B (Context 40039) (see Fig 5.3(a)); mixed weakly humic clays and coarse silts, with very fine and fine charcoal. PPL, frame is ~4.62mm.

(b): Detail of (a), showing possibly inclusion of monocotyledonous charcoal. PPL, frame width is ~2.38mm.

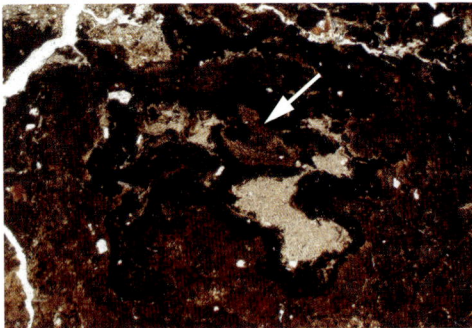

(c): As (a); iron stained area, relict of peat(?). Note possible ferruginised woody(?) root and plant cells (arrow). PPL, frame width is ~4.62mm.

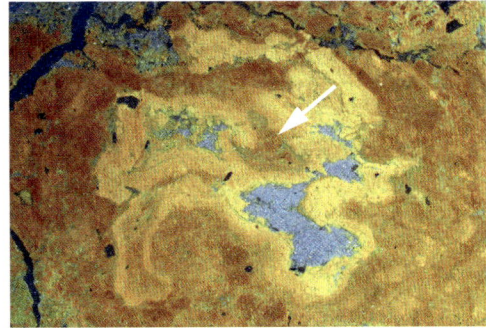

(d): As (c), under OIL.

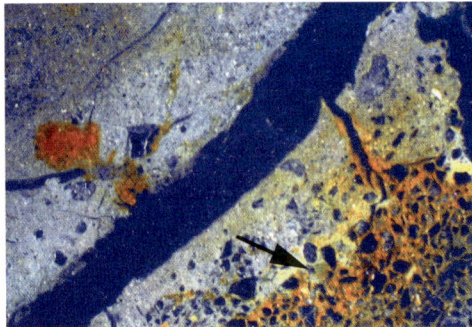

(e): As (a); iron-depleted and iron-stained layers. Some iron staining is yellowish and infills voids (arrows). OIL, frame width is ~4.62mm.

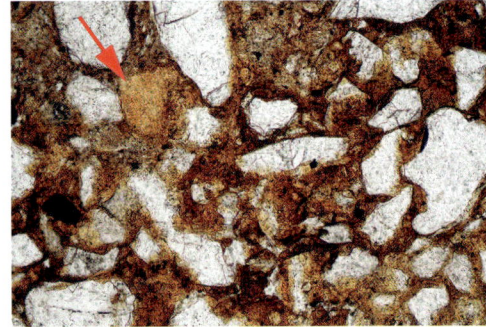

(f): As (e), detail of amorphous Fe-P void infill (arrow) and grain coatings. PPL, frame width is ~0.90mm.

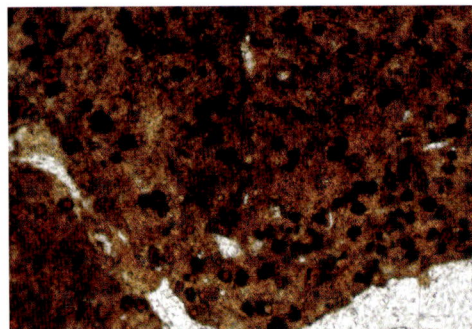

(g): As (a); very fine (~30µm), strongly-formed iron nodules (38.7-67.1% Fe) in iron-stained area; nodules also include 0.97-1.33% P (Table 4, Fig 5.4(a)). PPL, frame width is ~0.90mm.

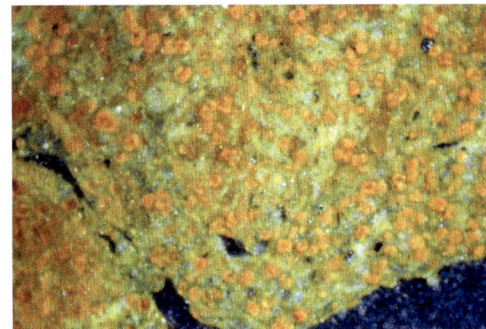

(h): As (g);Fe-P nodules under OIL.

Figure 5.13 Thin Section M40151B: (a)–(h)

possible root stains. Some channels have iron-depleted hypocoatings. There is an example of a channel with a leached/iron-depleted clay fill. These appear to be 'basal' lacustrine clay sediments, containing scattered small amounts of coarse silt, fine and medium sand, possibly locally blown-in/washed-in. This clay acted as a rooting substrate, possibly for woodland, as is suggested by the presence of woody roots (some being dominated by small invertebrate mesofauna). The sediment was also affected by shrinking and swelling (possibly an

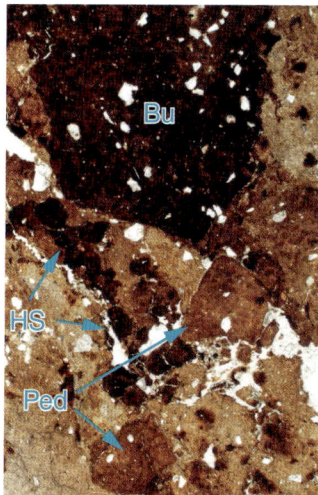

(a): Photomicrograph of M40151A (uppermost 40039); iron-stained remains of flooded soil, with ferruginised once-humic very broad (earthworm?) burrow (Bu), humic soil excrements (HS) and bio-mixed peds (Peds). PPL, frame height is ~4.62mm.

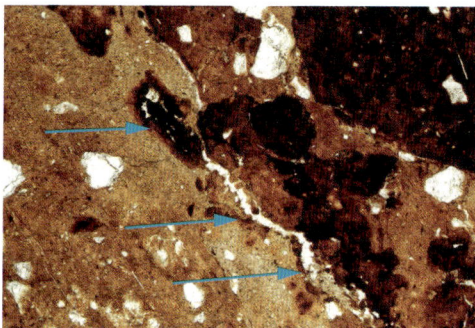

(b): Detail of (a); humic soil excrements associated with ferruginised root trace (arrows). PPL, frame width is ~2.38mm.

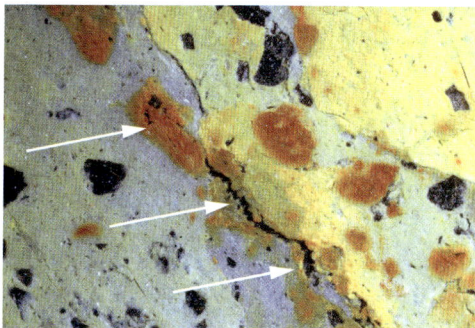

(c): As (b), under OIL.

Figure 5.14 Thin Section M40151A: (a)–(c)

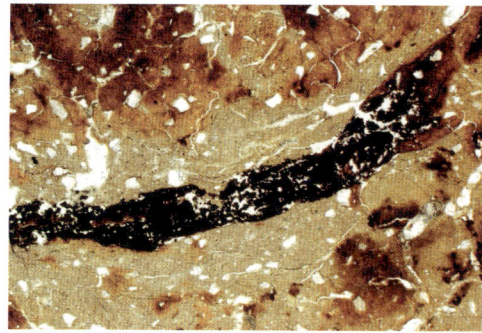

(a): Photomicrograph of M40150B (Context 40103); iron-mottled relict soil, with strongly ferruginised root. PPL, frame width is ~4.62mm.

(b): As (a), under XPL; note clayey textural pedofeatures from inwash around this root and infilling the root channel.

(c): As (a), detail of ferruginised root remains and likely very thin and once-organic excrements of mesofauna. PPL, frame width is ~0.90mm.

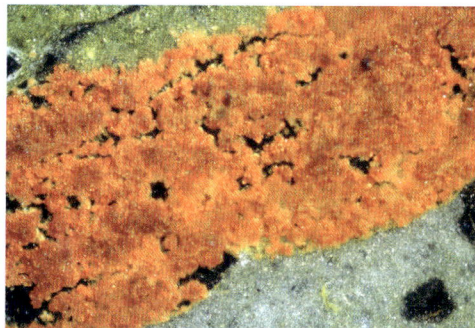

(d): As (c), under OIL.

Figure 5.15 Thin Section M40150B: (a)–(d)

effect of woodland growth/evapotranspiration), marked iron depletion (and possibly clay break-down at times) and iron staining, all hydromorphic/gleying effects (Bouma *et al.* 1990). Enigmatic high measurements of LOI possibly relate to woody root traces, while enhanced P is probably secondary and linked to iron-staining (Thirly *et al.* 2006). This context could represent a woodland subsoil gley horizon (Bg/Cg horizon).

M40150A (context 40103/40099 transition)

Context 40103

This is similar to context 40103 in M40150B, below (see Fig. 5.1e), but with a (sloping?) concentration of iron staining along the uppermost 10mm of this context, forming an iron pan (Fig. 5.16a-d). This is partially made up of a concentration once-humic burrows and channel (fills), and in places 1mm-size 'soil' clasts

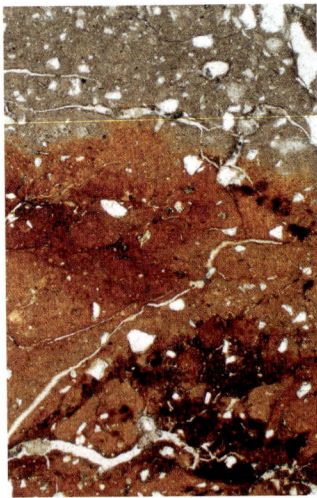

(a): Photomicrograph of M40150A (junction of contexts: iron-depleted [5.32-5.80% Fe] 40099 and iron-stained [35.4-47.0% Fe; 1.29-1.75% P] buried soil 40103)(Table 5.4, Figs 5.3(b) and 5.4 (b). PPL, height is ~4.62mm.

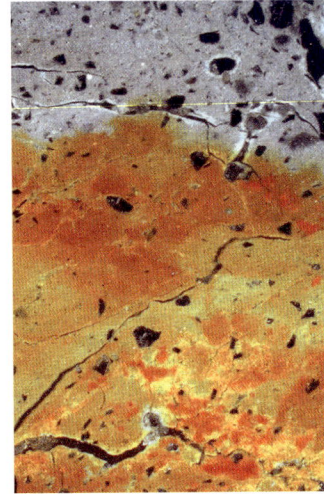

(b): As (a), under OIL; note general iron staining and concentrations where ferruginised once-humic soil existed.

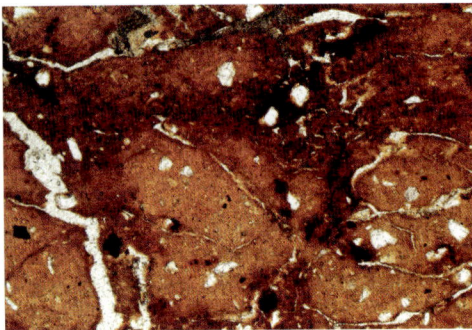

(c): Detail of (a); remains of humic, biologically worked buried topsoil. PPL, frame width is ~2.38mm.

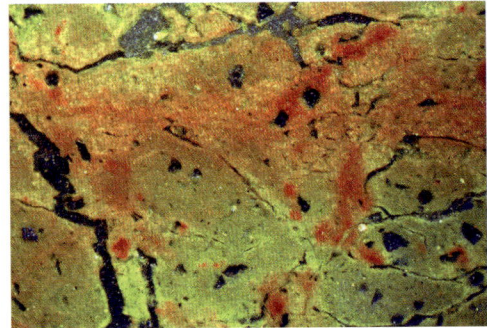

(d): As (c), under OIL, showing panning formation brought about by partial soil slaking due to inundation and 40099 sedimentation.

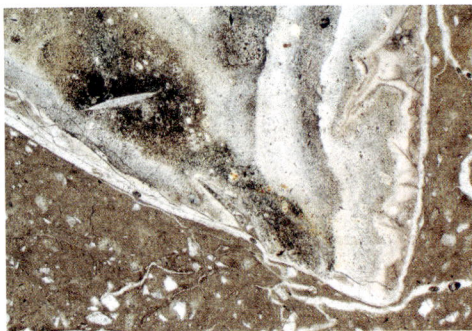

(e): As (a), edge of 12mm-size angular flint flake (see Fig 5.3(b)). PPL, frame width is ~4.62mm.

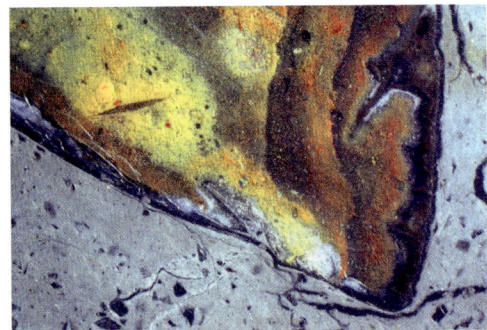

(f): As (e), under OIL; note iron-stained flint and iron-depleted sediment matrix.

Figure 5.16 Thin Section M40150A: (a)–(f)

cemented by infills of sometimes strongly ferruginised (once-humic) fine soil (cf. <40151>). Trace amounts of very fine charcoal and a 6mm clast of rounded flint gravel were noted. A marked high LOI (3.63%) and enriched phosphate-content (1.77mg g[-1] phosphate-P) were found in the initial bulk analysis (Table 5.3). SEM/EDS studies measured 35.4-47.0% Fe and 1.25-1.75% P in the iron-pan.

The uppermost part of 40103 appears to be a possible ripened sediment/soil surface, showing: firstly, slight truncation in places; secondly, traces of rooting and small mesofauna burrowing; thirdly, very locally eroded and transported ripened soil/sediment clasts; and fourthly, local inwash of humic fine soil. This may be a possible buried ripened soil that may have had short-lived woodland cover, before being affected by inundation and renewed sedimentation (context 40099). Gentle inundation caused slaking and structural collapse/formation of intercalations and void matrans, and erosion of local soil/sediments, and inwash (see references employed for M40151, cf. Holocene inundation sites along the Thames (Macphail and Crowther 2009)). This buried 'topsoil' interpretation is consistent with LOI and enriched phosphate probably also reflects this and 'geogenic' concentration of P due to ground-water movement (Thirly *et al.* 2006).

Context 40099

The lowermost part of this context is an iron depleted, very weakly humic clay containing rare detrital fine organic material, blackened (organic, possibly monocotyledonous) root traces and traces of charcoal. It includes a sharp flint chip (12mm) (Fig. 5.3b; Fig. 5.16a;b; Fig. 5.16e; f). Occasional intercalations and matrix infills, and strongly leached fissure and channel hypocoatings were noted. Bulk analyses record 2.26% LOI, 0.194 mg g[-1] phosphate-P (Table 5.3). Iron-depletion was confirmed by SEM/EDS (Table 5.4).

These massive iron-depleted, possibly lacustrine clay sediments, show rooting (by wetland plants: blackened organic monocotyledonous root traces), and inclusions of detrital organic material including trace amounts of very fine charcoal. An anomalous 12mm size flint chip is very likely of anthropogenic origin, and, more specifi-

cally, is almost certainly part of the knapping activity associated with the nearby elephant skeleton. We are here at the north-west edge of the recovered lithic concentration around the elephant (see Chapter 17), and also at the same stratigraphic horizon. Strong leaching has affected the sediment, hence moderate organic matter preservation and markedly low phosphate content.

M40149 (context 40099/40078 transition)

This is an iron depleted clay, as below, with abundant sedimentary clayey intercalations, including a 4mm-thick layer/fill at 410mm depth. A 2mm flint chip is present, as an 'embedded grain' (Fig. 5.17a; b). Near the top, trace amounts of amorphous organic matter inclusions occur in channels. Very weak trace amounts of iron hypocoatings occur.

This thin section records continued muddy lacustrine/wetland sediment accumulation, with another flint chip from the knapping activity around the elephant. Inclusions of preserved amorphous organic matter (40078) may stem from inwash from adjacent areas.

PHASE 7/6/6a, NORTH SIDE OF TRENCH B, CENTRAL PART OF SITE

Analysis was carried out on samples from two monoliths, samples <40418> and <40419>, covering the transition from the base of Phase 7 (context 40166) through to the top of Phase 5 (context 40025), where Phase 6 was reduced in thickness to a particularly black, humic-looking upper bed (context 40158) and a pale grey interdigitating sandy/clayey lower bed (context 40040) (Fig. 5.1f).

Loss-on-ignition

The LOI data show a series of peaks corresponding with the dark context 40158, with a particularly high value of 6.68% in its upper part (Fig. 5.2f; Table 5.3). This very high peak is almost certainly associated with soil (and possibly peat) formation, and the secondary peaks at 44.5 and 56-66cm also seem likely to be associated with periods of soil development and/or surface exposure. As

(a): Photomicrograph of M40149 (Context 40099/40078), containing angular flint fragment. PPL, frame width is ~4.62mm.

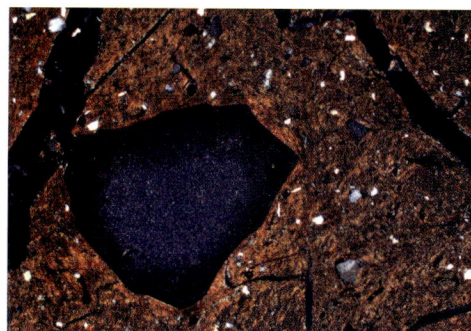

(b): As (a), under XPL. Note oriented clay around this flint fragment (embedded grain), caused by its deposition into a muddy water-saturated clay.

Figure 5.17 Thin Section M40149: (a)–(b)

mentioned above, organic matter will have been subject to post-depositional decomposition, so this context would originally have been significantly enriched in organic content.

Soil micromorphology

M40418C (contexts 40040-40025/40158)

Contexts 40040-40025/40158
As seen here, these contexts are composed of laminated moderately well sorted coarse silts and very fine sands, with patchy 'layer'/infills of clay with occasional ferruginised very fine organic matter and phytoliths present (Fig. 5.18a,b). Very thin iron pans associated with sandy laminae appear to be associated with relict amorphous organic matter and involved with pseudomorphs of very thin relict excrement.

These appear to be fluvial coarse silts and very fine sands, perhaps resulting from seasonal (spring) alluviation, with winter periods of low energy clay deposition, forming infills. Sand laminae also sometimes associated with thin peat formation and its partial working by small invertebrates, again suggestive of seasonality. It may represent relatively cool climate, near-channel alluviation.

Lower context 40158
This is moderately well sorted, massive sediment, which is sometimes laminated with very fine to fine sands that also includes coarse silt and medium sand. Many thin burrows, examples of clayey inwash down fine (1mm) channels (possibly decayed rootlets) and very thin stringers (possibly of ferruginised organic matter) inwash were also recorded. These channel mainly fine sands, were burrowed and rooted, and thus were probably episodically/seasonally exposed. Possibly humic matter from above (see M40418A and M40418B) filtered down and became ferruginised. Clayey sediments above were

introduced down-profile along empty relict root channels.

M40418B (context 40158/40040)

This thin section is mainly composed of laminated very humic clayey fine sands and humic clays, with abundant included fine amorphous organic matter and fine charcoal (100-250µm), and few fragments of amorphous peat; it is increasingly humic upwards. An example of a horizontally oriented 5mm long very thin blackened monocotyledonous leaf/leaves fragment occurs (Fig. 5.19a-d). Humic clays sometimes occur as clayey pans and infills. Minor thin burrowing has occurred. The context has markedly high LOI as discussed above (Fig. 5.2f; Table 5.3).

These are laminated peaty clays and peaty sands, horizontally deposited under low to moderately low energy conditions, allowing horizontal deposition of detrital leaves, fragments of pure peat and ubiquitous fine and very fine charcoal. Organic matter shows no sign of being ferruginised. These are perhaps low energy (seasonal?) minerogenic peats associated with a burned landscape – wildfires and /or fire-managed.

M40418A (contexts 40158/40166)

Upwards, context 40158 sediments become less humic and contain less fine detrital organic material, being clayey deposits, but here also have developed marked (many) iron-staining of organic traces.

These are laminated peaty clays and peaty sands, horizontally deposited in low to moderately low energy conditions, allowing horizontal deposition of detrital leaves, fragments of pure peat and ubiquitous fine and very fine charcoal. The humic content seems rather less, and shows much ferruginisation compared to this context below in M40418B. Again, these may be the result of low energy (seasonal?) minerogenic peat formation associated with burned landscape, most likely wildfires, but here have been affected by penecontempo-

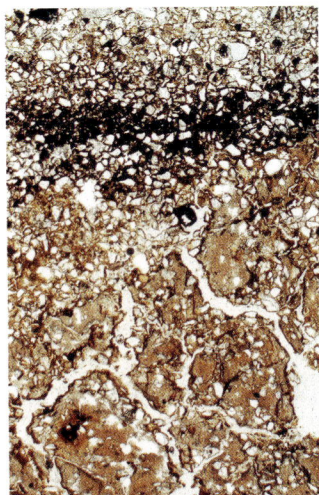

(a): Photomicrograph of M40418C, Context 40025/40040 transition. Sands with clayey infills sealed by ironpan that is relict of a thin peat layer. PPL, frame height is ~4.62mm.

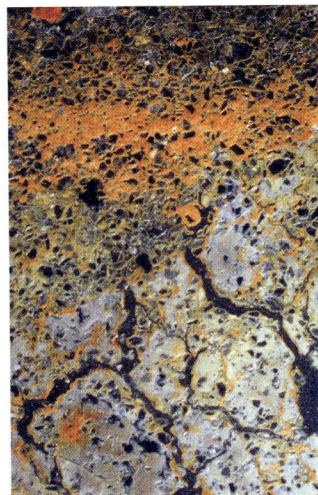

(b): As (a), under OIL; note iron-depleted clay and later ferruginisation, and ironpan formation.

Figure 5.18 Thin Section M40418C: (a)–(b)

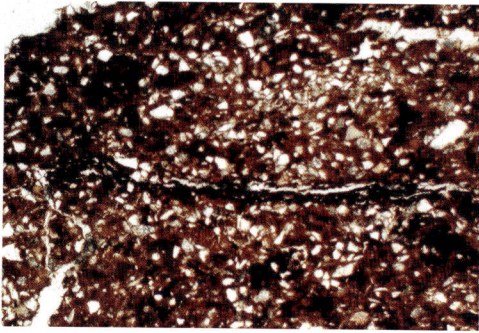

(a): Photomicrograph of M40418B, Context 40158, clayey and peaty fine sands, with 6mm-long horizontally oriented blackened monocotyledonous leaf(?). PPL, frame width is ~4.62mm.

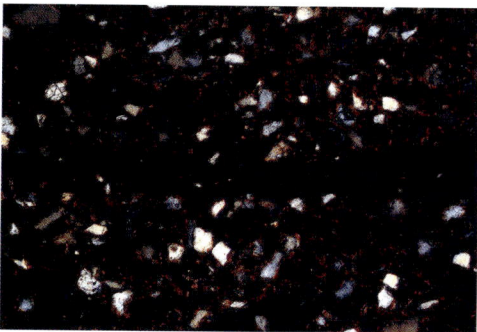

(b): Detail of (a), under XPL, showing grano-striate birefringence of (muddy) clay. Frame width is ~2.38mm.

(c): As (b), under OIL.

(d): As (a); example of charcoal (fragmented on sediment drying); OIL, frame width is ~2.38mm.

Figure 5.19 Thin Section M40418B: (a)–(d)

raneous iron movement and staining affecting this upper part of the context.

Context 40166

In contrast to 40158, context 40166 is highly minerogenic. It is composed of massive and finely laminated, poorly sorted sands with coarsely mixed clay fragments (clay papules), and only shows a trace amount of iron staining, and occasional broad burrows.

Minerogenic poorly sorted laminated sediments record a cessation of 'peat' formation here and probably locally, and show instead slight variations in energy and erosion of sands and more clayey sediments. Possible iron depletion here led to ferruginisation of organic fraction of sediments below.

DISCUSSION

Post-depositional processes

The chief post-depositional processes of interest at this site are: a) iron depletion, b) oxidation of organic matter (often alongside its ferruginisation), c) iron migration and d) phosphate migration. Other processes, such as compaction, possible decalcification etc., which often occur in Pleistocene soil-sediments (cf. Boxgrove), are not specifically relevant to interpretations of the micromorphology here.

a. Iron-depletion is a very important phenomenon at Southfleet Road and is probably the chief process resulting in 'grey' layers. Such gleyed layers occur through hydromorphism where water saturation (and associated extant water tables) led to iron reduction (Fe^{3+} to Fe^{2+}), and the loss of mobile reduced iron (Duchaufour 1982; Bouma *et al.* 1990). This process was probably penecontemporaneous with MIS 11 site formation, and continued sedimentation; the soil associated with the palaeo-landsurface at the top of the Lower Loam at Barnfield Pit itself records hydromorphic effects on soils buried by ensuing alluvial sedimentation and the associated rise in water table (Kemp 1985). Some of the effects of mobile iron at the Southfleet Road site are noted below.

b. Likely oxidation of organic matter and plant remains at the site is noted from the chemistry (see above; Crowther, Appendix 2), and even the most humic context (40158 in thin section M40418B), only recorded a 6.68% LOI. It can be noted that ferruginisation appears to have affected *some* probably once-humic contexts, for example context 40103 (M40150A) which records LOI of 3.63% (40158, in M40418B), seems anomalously unaffected by ferruginisation. In addition, roots for instance show poor pseudomorphic replacement in sometimes otherwise iron-depleted sediments. In relation to this, ferruginised root traces show relict remains of probable pyrite spheroids, where pyrite had formed in association with organic materials under original waterlogged conditions (eg root decay) (Miedema *et al.* 1974; Wiltshire *et al.* 1994), and then become transformed to amorphous iron oxide. It is possible that concentrations of amorphous organic matter may also have been ferruginised (see below). As a comparison, the iron pan at Boxgrove most commonly occurs as a strongly oxidised once-humic minerogenic peat (0.80-0.90% organic C), and only under exceptional conditions of burial has it retained some of its original organic character (5.30% LOI at BH5: Macphail *et al.* 2010). Context 40158 (M40418B) thus must also have experienced a similarly unusual burial history; perhaps it remained water saturated until deep burial sealed it. This is suggested because ferruginisation of relict plant material and other organic remains can be a penecontemporaneous process, as observed at the analogue site of Wallasea Island, Essex (Macphail *et al.* 2010) and in hydromorphic soils in general (Bouma *et al.* 1990).

c. Iron migration, as noted above, is related to hydromorphic iron-depletion (gleying), and led to the reprecipitation of iron as 'ironpans' and iron mottled zones often seemingly associated with once-organic sediments. In addition, ironpans can develop along hydrological boundaries. It is likely that iron-panning (max 47.0% Fe) occurs at context 40103 (<40150A>) below iron-depleted (max 5.80% Fe) context 40099, because (1) 40103 appears to be a relict 'topsoil' (3.63% LOI) and (2) there appears to be a hydraulic barrier between this soil and inundation sediment 40099 (see below).

d. Phosphate migration and deposition, is often associated with (c), and possibly sometimes with (b). In the first instance, mobilised phosphate within the soil-sediment groundwater system has apparently migrated from its original location/source and been reprecipitated along with iron in ironpans and iron mottles (max 1.75% P). This migration of phosphate has been noted previously in waterlogged sediments (Cruise *et al.* 2009; Thirly *et al.* 2006). In the case of 40039, thin section M40151B, however, there appear to be probably localised phosphate concentrations, staining the sediment matrix and infilling voids with Fe-P (see Fig. 5.13e; f). Finely punctuated iron nodules in essentially the same layer (see Fig. 5.3a; Fig. 5.13g; h; Fig. 5.4a) are also enriched in phosphate (eg 1.07%P). It is possible here that iron and phosphate are replacing amorphous organic matter, and that these are not simply FeP migration features, but related to localised inputs of organic matter and phosphate, as faecal waste (cf. Fe-P enriched Unit 4u at the Boxgrove Q1B waterhole site, associated with likely animal trampling, Macphail *et al.* 2001). At a butchery site perhaps animal remains are also a possibility. Associated textural pedofeatures that are indicative of physical mixing of wet sediments at context 40039, may tentatively suggest that

phosphate and once-organic inputs are broadly related to large animal activity (see below).

Phase I deposits ('Tilted Block')

Context 40057 alluvium (coarse silts, fine sands) is characterised by the inclusion of moderately fresh glauconite (Loveland and Findlay 1982), suggesting inclusion of Tertiary sediments. Although weathered glauconite is present throughout the sediments above, fresh glauconite was little in evidence, indicating a change in provenance for Units Phases 6 and 6a, for example.

Phases 6 and 6a

Sediments, soils and vegetation

Sediments are often clayey, but may contain silts and fine sands at times. As noted above, they are strongly characterised by iron depletion and related iron panning and mottling, including the ferruginisation of roots. Clayey laminae can be interdigitated in sands as in the middle part of context 40039 (thin section M40323B, Fig. 5.10), suggesting seasonal variations in sedimentation, while in some coarse silty-fine sandy facies (for example, the upper part of context 40039, thin section M40323A, Fig. 5.11) very thin seasonal peat growth is perhaps recorded as ferruginised laminae (see also Fig. 5.18a; b). Fragments of iron-depleted clay in the same context, eroded from earlier-deposited clays, testify to sediment exposure and erosion related to renewed fluvial activity. It is also possible that cool climate erosion and colluvial processes could be involved, as in Unit 6a at Boxgrove (Macphail *et al.* 2001).

In addition to these variations in sedimentation, ephemeral sediment ripening took place (Avery 1990), when it was seasonally exposed to subaerial weathering. Plant growth is also recorded, and may have involved monocotyledonous wetland plants (see Fig. 5.13a; b; Fig 5.19a-d) and hence the sediment phytolith content, and later rooting by trees (see Fig. 5.13c; d), speculatively in a form of a hydroseral succession (Rodwell 1991) (cf. Pilgrim's School, Winchester, Macphail *et al.* 2009). Pollen evidence of woodland was previously noted (Wenban-Smith *et al.* 2006). The activity of rooting is one mechanism for mixing silts and sand into clayey sediments, alongside localised formation of textural pedofeatures associated with rooting into water-saturated sediments.

There is evidence of soil formation, which is presumably associated with this suggested hydroseral succession, such as in context 40103 (M40150A) and which is now seen as an 'ironpan' (see Fig. 5.3b, Fig. 5.16a-d and Fig. 5.4b). Differentially iron-stained soil and once organic materials are evidence of soil structure formation, activity by small invertebrate mesofauna and roots. When this soil was buried by renewed clayey alluviation (context 40099), the buried soil lost structure (becoming slaked), as is typical of inundated soils (cf. Unit 4c at Boxgrove).

Hominid and large animal activity

It is quite clear that inclusions and microfeatures associated with the presumed presence of hominids and other large animals are reflected in features of the soil-sediments. For example, putative artefacts (flint chips in the size range 4-20mm) occur as embedded clasts in muddy sediments and sediment clay is oriented around them; these small artefacts thus probably sank into the sediments (Fig. 5.10c; d; Fig. 5.16e; f; Fig. 5.17a; b) along with any other heavy objects. In addition, the amount of textural pedofeatures (clayey intercalations, void coatings) and associated closed vughs, infer more marked physical disturbance of water-saturated sediments than can be simply explained by rooting turbation (see above); large animal (and possibly also hominin) trampling may need to be envisaged. Equally, the presence of angular flint chips is anomalous in clayey sediments, and is best interpreted as related to anthropogenic activity.

Even though there are possible traces of rubefication and calcination, suggestive of burning, the latter most likely merely reflects patination and staining as typical fire-cracking is not in evidence (see Fig. 5.7a; b; Fig. 5.9c-e). There does, however, appear to be traces of what may be red burnt clay in the same context (40100; see Fig. 5.9a; b) and regular occurrences of burnt mineral grains. Charcoal is also found in the sediments (Fig. 5.13a; b; Fig. 5.19d). There are therefore weak but consistent indications of fire, but it remains to be established whether this may be derived from Tertiary sediments, or, if regarded as contemporary, whether it is merely natural wildfires periodically occurring, or whether of hominin origin. No hearth layers or other evidence of specifically hominin fire use were noted. Finally, the presence of sediments showing turbation, localised concentrations of relict organic matter and phosphate in the Phase 6 sediments in the vicinity of the elephant skeleton, and the associated flint artefact concentrations (thin sections M40365A, B and C; thin sections M40149 and M40150B) is unlikely to be a coincidence. As well as reflecting the more peaty organic-rich sediments associated with preservation of the skeleton, these characteristics may reflect hominin movement around the elephant skeleton, decay products of the elephant carcass and/or the presence of other living animals in the soft, waterlogged sediments around the elephant.

CONCLUSIONS

Nineteen thin sections (sediment micromorphology and SEM/EDS), 12 associated bulk samples (LOI and P) and 105 further bulk samples (LOI and magnetic susceptibility) were analysed, mostly from Phase 6, the main level of archaeological interest. Deposit formation processes included penecontemporaneous and post-depositional iron-depletion, organic matter oxidation, and iron and phosphate migration. Clayey and coarse silt-fine sandy alluvial and wetland sediments are present, and show probable seasonal variations in sedimentation and thin peat formation, for example. Ephemeral and longer lived sub-aerial weathering and soil formation is in evidence, including possible hydroseral succession(s) from, speculatively, monocotyledonous wetland to woodland and associated topsoil formation. Processes related to water table fluctuations, such as renewed alluviation and site inundation, are recorded generally. Hominid and large animal activity is probably recognisable from water-saturated sediment mixing. Most lithological evidence of burning is most likely of derived Tertiary origin; likewise, pieces of charcoal in the sediments are probably the result of contemporary wildfires, although a hominin role cannot be ruled out. Lastly, the presence of a butchered elephant carcass and artefacts in association with sediments showing turbation and localised concentrations of relict organic matter and phosphate may not simply be a coincidence. These latter observations are most likely to represent hominin movement, and possibly other activity, in the vicinity of the elephant carcass.

Chapter 6

Clast lithology

by David R. Bridgland, Francis Wenban-Smith, Tom S. White and Peter Allen

INTRODUCTION

The objectives of clast lithological analysis at the site were two-fold: first, to complement the sedimentological data in developing interpretations of deposit formation, and in particular to support interpretation of the Phase 8 gravels as fluvial, and second, in the case of confirmed fluvial deposits, to identify their provenance and river system; in particular to distinguish between the Thames and its tributary gravels. For the first, roundness/angularity of the flint clast component was analysed and the resulting distributional profile compared with various data of known depositional origin. For the latter, the variety and proportions of different clast lithologies were recorded, to distinguish between post-Anglian Thames gravels, which have a higher proportion of exotic lithologies, and the deposits of south-bank Thames tributaries, with more restricted local and Wealden composition. This second line of enquiry was particularly salient at the Southfleet Road site, it being located close to where the current Ebbsfleet Valley intersects the Swanscombe Boyn Hill/Orsett Heath Formation (100-ft terrace) outcrop. As recapped earlier (Chapter 3), one of the primary objectives driving initial investigation of the site was to establish whether the Phase 8 gravels that capped the sequence were of fluvial Thames origin, and if so, whether they formed an extension of the Lower Middle Gravel of the Orsett Heath Formation.

Clast lithological analysis was carried out on six gravel samples from the site (Table 6.1). Three of these, samples <40001>, <40002> and <40004>, were from the main body of gravel (Phase 8) that capped much of the site. One sample, <40003>, came from the junction between this gravel and the underlying context 40167 (Phase 7, upper part). Two samples, <40361> and <40420>, came from different levels within the Phase 7 synclinal infill, one entirely from context 40167, and the other from 40164, a thin gravel bed at the very base of the syncline infill sequence (Fig. 6.1). The results of these analyses were then considered in conjunction with analyses carried out at various locations in the surrounding area (Table 6.1; Fig. 6.2).

METHODS

Samples were separated, by wet sieving, into 16–32 and 11.2–16mm fractions for clast analysis. Clast-litholog-

ical analysis was applied to both size fractions (as recommended in the appropriate QRA Technical Guide, Bridgland 1986) and, as a separate procedure, the angularity/roundness characteristics of the flint component of the coarser fraction was also assessed. The latter analysis used a modified version of the Powers (1953) method, adapted for gravel-sized clasts (Fisher and Bridgland 1986) and using the categories defined here (Table 6.2).

ELEPHANT SITE RESULTS

Angularity/roundness

The principal purpose of this analysis was to determine environment of deposition (see Fisher and Bridgland 1986; Bridgland 1999). Unbroken Tertiary flints were excluded, as their extreme roundness is clearly derived from their original marine environment of deposition. The results show the angular class to be modal in all six samples, followed by the very angular (Table 6.3). Very little of the broken flint has edges that have been smoothed even to the subangular condition. The three samples from the Phase 8 gravel (samples <40001>, <40002> and <40004>), thought very likely to be of fluvial origin and all analysed by DRB, show very similar profiles. However the fourth analysed by DRB (sample <40003>), from the basal junction of the Phase 8 gravel with the underlying context (40167), thought more likely to be emplaced by mass movement, shows a slight increase of clasts in the very angular category. The two other samples (both analysed by TW), both from the Phase 7 synclinal infill and one of them thought likely to be from a very similar level to sample <40003>, showed a much stronger modality in the angular category than any of the other samples, including those thought much more likely on independent sedimentological grounds to be fluvial origin. This is most likely to result from variability between classification by the individual analysts, since other instances where DRB and TSW have provided independent analyses of the same gravel in Eastern Quarry (eg samples <4> and <3>) show that TSW generally allocates less material to the 'very angular' in favour of the 'angular' class. It is well established that this technique, which employs mere visual comparison, suffers from subjectivity (and therefore operator bias), despite the use of objective

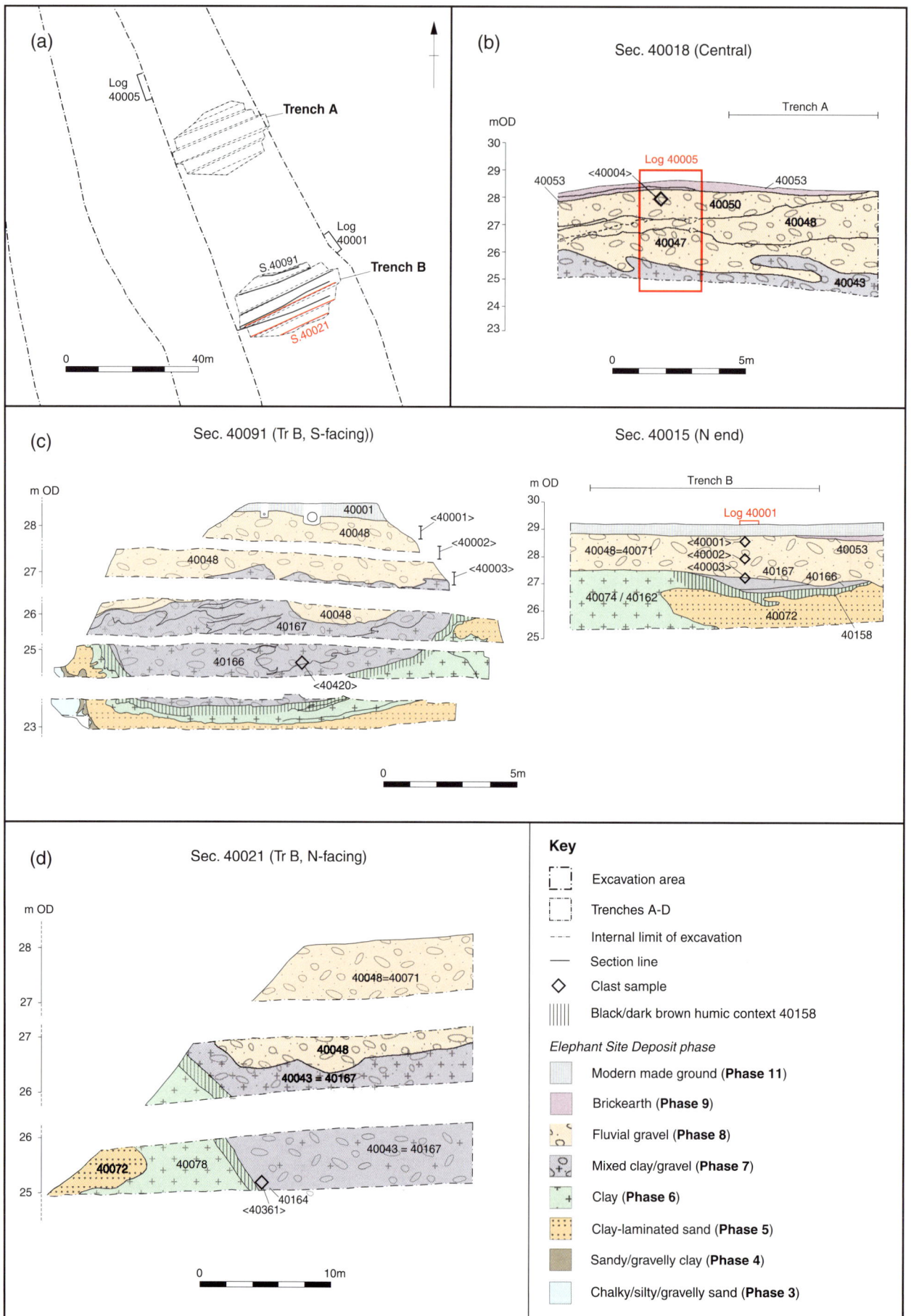

Figure 6.1 Stratigraphic contexts of clast lithological sampling at the Southfleet Road site: (a) thumbnail sketch of sampling locations; (b) Log 40005, sample <40004>; (c) Section 40091 and Log 40001, samples <40001>, <40002>, <40003> and <40420>; (d) Section 40021, sample <40361>

Table 6.1 Clast lithological sampling at the Southfleet Road site (ARC 342 W02) and other projects in the surrounding areas

Project area	Location	Sample <>	Context	Site deposit (phase)	Clast interpretation	Report reference
Southfleet Road site (ARC 342 W02)	Log 40001 (Sec 40018)	40001	40006 [=40071]	Fluvial gravel (8c)	Palaeo-Ebbsfleet gravel	-
		40002	40008 [=40071]	Fluvial gravel (8c)	Palaeo-Ebbsfleet gravel	-
		40003	40009-lower [=40071/40167?]	Fluvial gravel/ Syncline infill? (8c/7?)	Palaeo-Ebbsfleet gravel, reworked by solifluction	-
	Log 40005 (Sec 40015)	40004	40018 [=40050]	Fluvial gravel (8c)	Palaeo-Ebbsfleet gravel	-
	Trench B	40361	40164	Syncline infill, base (7)	Palaeo-Ebbsfleet gravel, reworked by solifluction?	-
		40420	40167	Syncline infill (7)	Palaeo-Ebbsfleet gravel, reworked by solifluction?	-
Springhead (54924)	TP 1121	1121-1	112105	-	Palaeo-Ebbsfleet gravel	Wessex Archaeology 2004
		1121-2	112108	-	Palaeo-Ebbsfleet gravel	Wessex Archaeology 2004
Station Quarter South (63542)	TP 32	8	3207	-	Palaeo-Ebbsfleet gravel	Wessex Archaeology 2006a,b
	TP 30	10	3003	-	Palaeo-Ebbsfleet gravel	Wessex Archaeology 2006a,b
Eastern Quarry (61040)	TP 16.2	2	16.2.03	-	Thames gravel (post-Anglian)	Wessex Archaeology 2006b
	TP 4.1	4	4.1.03	-	Thames gravel (post-Anglian)	Wessex Archaeology 2006b
Eastern Quarry (61041)	TP 104	50	7403	-	Palaeo-Ebbsfleet gravel, reworked by solifluction?	Wessex Archaeology 2008a
		51	7403	-	Palaeo-Ebbsfleet gravel, reworked by solifluction?	Wessex Archaeology 2008a
	TP 112	60	8205	-	Palaeo-Ebbsfleet gravel	Wessex Archaeology 2008a
Eastern Quarry (61042)	TP 127	96	12706	-	Thames gravel (post-Anglian)	Wessex Archaeology 2008b
		101	12708	-	Thames gravel (post-Anglian)	Wessex Archaeology 2008b
Eastern Quarry, septic tank (68990)	Septic Tank, Tr 3	3	11-b	-	Thames gravel (post-Anglian)	Wessex Archaeology 2009

Table 6.2 Angularity/roundness categories used in Table 6.3. These are based on verbal descriptions by Schneiderhöhn (1954, in Pryor 1971) of the categories devised by Powers (1953). Simplified from Fisher and Bridgland (1986)

Category	Characteristic features
WELL-ROUNDED— wr	No flat faces, corners or re-entrants discernible; a uniform convex clast outline
ROUNDED— r	Few remnants of flat faces, with corners all gently rounded
SUBROUNDED— sr	Poorly to moderately developed flat faces with corners well rounded
SUBANGULAR— sa	Strongly developed flat faces with incipient rounding of corners
ANGULAR— a	Strongly developed faces with sharp corners
VERY ANGULAR— va	As angular, but corners and edges very sharp, with no discernible blunting

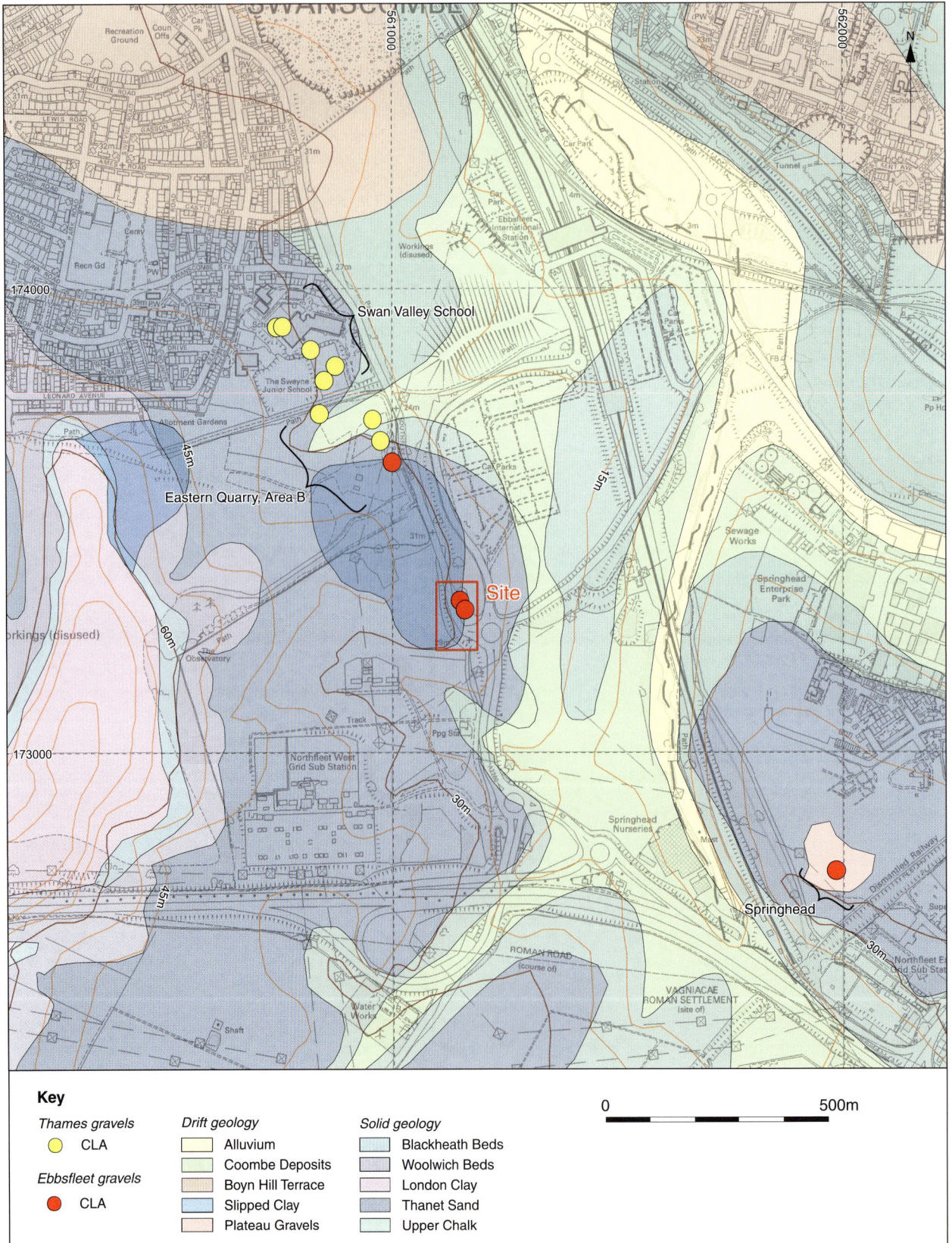

Figure 6.2 Locations of clast lithological sampling (CLA) in project areas surrounding the site [pre-quarrying base geology follows 1922 Geological Survey mapping, 6" sheet 12.1, NW]

Table 6.3 Angularity/roundness of Southfleet Road gravel samples, and comparisons with data from other nearby locales

Locality	Site Phase	wr	r	sr	sa	a	va	Total	Notes	Analyst
					Category (see Table 6.2)			*Total*	*Notes*	
Southfleet Rd, Elephant site										
Southfleet Road <40004>	8.3 - top				7.1	65.7	23.9	67		DRB
Southfleet Road <40001>	8.3			1.1	9.7	66.7	22.6	93		DRB
Southfleet Road <40002>	8.3				9.0	67.0	24.0	100		DRB
Southfleet Road <40003>	8.3/7				8.5	59.8	31.6	117		DRB
Southfleet Road <40420> (40167)	7 - top				6.7	75.5	17.8	279	context 40167	TW
Southfleet Road <40361> (40164)	7 - base				7.3	79.2	13.5	82	context 40164	TW
Comparative material										
Other surrounding locales										
Springhead 1121-1					5.6	59.0	35.4	178		DRB
Springhead 1121-3					7.8	72.0	20.3	232		DRB
Eastern Q (61040) <2>			0.6	1.7	36.3	42.9	18.4	347		DRB
Eastern Q (61040) <4>				2.7	26.4	52.1	18.8	261		DRB
Eastern Q (61041) <50>					4.0	79.8	16.1	124		DRB
Eastern Q (61041) <51>					9.8	73.2	17.1	82		DRB
Eastern Q (61041) <60>				0.5	4.9	85.4	9.3	205		DRB
Eastern Q (61042) <96>				3.5	30.0	61.1	5.7	283		TW
Eastern Q (61042) <101>				2.0	43.3	48.4	6.3	252		TW
Eastern Q (68990) <3>				1.9	23.1	73.1	1.4	216		TW
Station Quarter South (63542) <10>					5.3	67.5	27.2	206		DRB
Station Quarter South (63542) <8>					6.9	48.3	44.8	58		DRB
Northfleet Cement <5> (602)				1.1	55.2	37.9	5.8	377		TW
Northfleet Cement <9> (1003)				0.6	61.0	32.8	5.5	344		TW
Crossways Business Park <18> (1104)				2.0	56.9	29.4	11.7	197		TW
Crossways Business Park <19> (1008)					7.5	72.6	19.9	146		TW
Pleistocene beaches										
Boxgrove 1		1.9	5.8	23.0	29.9	21.0	18.0	618		DRB
Boxgrove 2		1.4	7.4	38.5	28.8	19.1	4.8	351		DRB
Bembridge 1		9.6	21.0	30.5	24.6	11.4	2.9	509		DRB
Bembridge 2		4.6	11.7	30.0	35.9	13.6	4.3	582		DRB
Southwold 1		37.7	27.1	16.9	10.7	3.2	4.4	591	(Westleton Beds)	DRB
Pleistocene fluvial gravels										
Barvills Farm 1		24.8	7.2	3.1	24.1	21.2	19.6	638	(Lower Thames)	DRB
Barvills Farm 1.				1.0	36.8	32.3	29.9	418	(Lower Thames)	DRB
Shakespeare Pit 2A		24.1	6.9	1.3	18.5	22.3	26.8	622	(Lower Medway)	DRB
Shakespeare Pit 2A.				0.7	27.1	32.8	39.4	424	(Lower Medway)	DRB
Aylesford 1				0.8	31.1	17.6	50.4	119	(Middle Medway)	DRB
Aylesford 2			0.7	0.7	26.8	28.9	43.0	142	(Middle Medway)	DRB
Little Hayes 1				0.6	26.7	34.8	37.9	546	(R. Crouch)	DRB
Little Hayes 2				0.6	30.5	41.0	28.6	466	(R. Crouch)	DRB
Rampart Field 4					18.3	54.5	27.2	226	(Ingham River)	DRB
Knettishall 2				1.5	14.1	52.0	32.3	474	(Glacial outwash)	DRB
Solifluction gravels										
Great Fanton Hall 1				0.6	35.2	34.4	29.8	540		DRB
St. Mary's Marshes 1				0.6	15.7	32.9	50.9	540	(TQ 8413 9812)	DRB
Skinners Wick 1				0.6	18.9	74.8		222	(TQ 8106 7804)	DRB
Lodge Hill 1				0.7	14.6	27.2	57.6	151	(TQ 7566 7389)	DRB

parameters in the category definitions (see Table 6.2). Nonetheless, these problems rarely affect interpretations arising from the analyses.

Comparison with previous analyses from known depositional environments shows no perfect match (Table 6.3). Counts with modal angular flint are known only from fluvial gravels, however. The paucity of subangular flint in comparison with the angular class is almost unprecedented, although it can substantially be explained by the fact that the deposits are probably the product of a very small tributary stream, rather than the larger rivers from which most comparative data have been obtained. Furthermore, as discussed above (Chapter 4, *Deposits at the site*), what is suggested here is substantial input by downslope sediment movement of pre-existing Tertiary gravel from the high ground to the west, so this is probably entering the putative river channel in a very angular state. Although many fluvial gravels have more subangular than angular flint, these tend to be from large river systems in which the flint has probably been transported significant distances and/or reworked on multiple occasions. In the lower reaches of the Thames system, for example, where the flint has been carried considerable distances in a large river, the subangular class is generally modal. Solifluction gravels typically peak in the very angular category (Table 6.3), due to the prevalence of frost shattering and pitting in the periglacial environment required for the solifluction process. Even minimal fluvial transport is generally sufficient to dull the edges of fractured flint and result in its classification as angular.

The moderate % very angular in the Southfleet Road samples is typical of fluvial gravels, most of which have probably been deposited under cold (periglacial) climatic conditions, ensuring a rich supply of frost-shattered flint to the bedload, although this is not necessarily the case here, as there is no palaeo-environmental evidence. The gravels are tentatively (although by no means confidently) correlated with the Swanscombe Lower Middle Gravel, which are thought to have been laid down under interglacial conditions. Indeed, the angularity characteristics of the Southfleet Road gravel suggest short-distance fluvial transport in a small stream. The dominance of angular material could result either from modest fluvial discharge or from close proximity to the flint source. These results most closely resemble data from smaller rivers, nearer to flint sources, such as the Medway gravels of north and west Kent and the gravels of the Crouch in Essex (Table 6.3). Comparison with Ingham/Bytham River gravels from Rampart Field, Icklingham, Suffolk is also pertinent (Table 6.3). This was by all accounts a very large river (Rose 1987 and 1991), yet the flint is angular. This is thought to be because the river has only reached a flint source in the close proximity of the Icklingham area and therefore the flint has not been transported far (Bridgland *et al.* 1995). In summary, the data from Southfleet Road suggest that the majority of the flint has been frost-shattered and then transported a short distance before emplacement at the site. Note that the subangular material in the Southfleet Road gravel, like

the Greensand chert, may well have been reworked from older (Darent) gravels.

It is worth noting too, that the data for the Phase 8 gravel are very similar to the results from Springhead (Table 6.3). It seems quite likely that these two gravels represent the same system, if not the same gravel body. The Springhead samples are slightly richer in Tertiary flint and poorer in other material, perhaps reflecting the relative positions of these two locations in the tributary valley, into which the other material was being introduced (probably from high-level Darent gravels such as those centred on Darenth Woods).

Clast lithology

All the Southfleet Road gravel samples contain only flint (96 – >99%) and Greensand chert (0.6–1.1%), with a few ironstones and sandstones of probable southern, Wealden provenance (Table 6.4). The great majority (75–95%) of the flint is of Tertiary origin, in the form of unbroken rounded or broken marine pebbles. Flint from other sources is found in the Southfleet Road deposits; there is 0.5–4.5% nodular flint, as well as weathered and broken flint (not shown in the table), much of which may have come from the Tertiary, although this cannot now be determined. The only other components of the gravel are ironstones of various types (0.2–1.4%) and occasional sandstones of Greensand or (more probably) Tertiary origin (0–0.3%). The ironstones include argillaceous (clayey) and arenaceous (sandy) types, as well as mixtures of the two. They may come from the Greensand, the Tertiary or from iron-cemented Quaternary deposits. Some of the flints show evidence of an origin in the Bullhead Beds (counted with nodular, since these are invariably weathered nodules).

Several samples contained flint pebbles with fire-crazing and red/grey discoloration indicative of burning. This was quantified (Table 6.4) to investigate whether the Phase 7 deposits were associated with a mass movement (landslip) event, and whether this was perhaps associated with a wild-fire event, and indeed perhaps the proposed fire identified as a non-arboreal pollen phase by Turner (1970) in the Hoxnian pollen sequence at Marks Tey. No particular pattern was found; burnt pebbles were generally rare, and were present in samples from all three phases covered by the investigations (Ph 8.3, Ph 8.3/7 and Ph 7). The presence of fire-crazed material is likely to indicate occasional natural conflagrations, Tertiary material having had much longer exposure to successive natural fire events or lightning strikes.

The lithology of the Phase 8 gravels at Southfleet Road demonstrates unequivocally that they are not of Thames origin, and thus not part of the Swanscombe Orsett Heath Formation, which crops out a short distance to the north-west, in Eastern Quarry and at Swan Valley School (Chapter 2). As the comparative data (Table 6.5) show (see also Bridgland 1988 and 1994; Bridgland and D'Olier 1995), Lower Thames gravels, including those at Swanscombe, contain a typical mixture of flint (85–98%),

Greensand chert from southern tributary valleys (0.5-5.0%) and 'exotic' material, derived from outside the London Basin (1.5–12%). The dominance of Tertiary flint pebbles is also informative; these are common in the Thames gravels, but rarely account for more than 70% of the total flint (Table 6.5). The composition of the Phase 8 gravel is suggestive of a south-bank Thames tributary, in other words a palaeo-Ebbsfleet. The range of lithologies includes material that could reflect headwaters draining the Lower Greensand, ie from a catchment basin extending south into the Weald. However, this is not particularly likely, since some of the hills to the south of the Dartford – Swanscombe area are capped with Greensand chert-bearing gravels of probable Darent origin (for example at Darenth Wood). Reworking from such sources is a much more likely origin of the chert in the Southfleet Road gravel. The dominance of Tertiary flint suggests a fairly local source, within the Tertiary outcrops capping the North Downs, and most likely from the now-quarried high ground to the west of the site, where gravel-rich Tertiary beds were mapped (Fig. 2.3). The complete absence of exotics, as well as ruling out a Thames origin, indicates deposition either before the Thames was diverted into its present lower valley (see Bridgland 1988 and 1994) or in a location upstream from any pre-existing Thames deposits, from which occasional exotic material would have been reworked. The supposed correlation between the Southfleet Road Phase 8 gravel and the Swanscombe Lower Middle Gravel deposits, based on their altitudes (Chapter 4, *deposits in the site vicinity*), would mean that they were deposited at a time when there were no pre-existing Thames higher-level deposits for slopewash processes and secondary tributary streams to rework upstream into the Ebbsfleet stream, as is often seen in advance of confluences. The lithology of the other gravel samples, from Phase 7 and the Phase 8/7 junctions, is identical to that from the Phase 8 gravel, suggesting an identical source.

DISCUSSION AND CONCLUSIONS

The results from the Southfleet Road site confirm that the Phase 8 gravels are fluvially deposited, and that they represent an early course of the Ebbsfleet, rather than the main Thames. As discussed above (Chapter 4), they

Table 6.4 Clast lithologies from Southfleet Road gravel samples

Sample <>	Clast size	Flint				Southern			Exotic									TOTAL	Fire-crazed Tertiary pebbles
		Tertiary	Nodular	Weathered	Broken	Greensand chert	Hastings Beds sandstones, siltstones and ironstones	Ironstone	Vein Quartz	Metaquartzite	Orthoquartzite	Carboniferous Chert	Schorl	Rhaxella Chert	Arkose	Igneous	Clinker		
<40004>	16-32mm	136	7	6	1	5		1										156	
	%	87.18	4.49	3.846	0.64	3.205		0.64											
	11.2-16mm	257	3	9	2	6		4										281	
	%	91.46	1.07	3.203	0.71	2.135		1.42											
<40001>	16-32mm	168	7	22	3	13												213	
	%	78.87	3.29	10.33	1.41	6.103													
	11.2-16mm	406	4	27	6	9	2	2										456	
	%	89.0	0.9	5.9	1.3	2.0	0.4	0.4											
<40002>	16-32mm	225	6	10	3	5		1										250	1
	%	90.0	2.4	4.0	1.2	2.0		0.4											
	11.2-16mm	414	7	9	4	3		1										438	6
	%	94.52	1.6	2.055	0.91	0.685		0.23											
<40003>	16-32mm	234	10	2		2												248	
	%	94.35	4.03	0.806		0.806													
	11.2-16mm	579	3	24	4	9	2											621	
	%	93.24	0.48	3.865	0.64	1.449	0.32												
<40420> (40167)	16-32mm	554	23	25	4	7												613	5
	%	90.4	3.8	4.1	0.7	1.1													
	11.2-16mm	1360	22	56	9	38	5	1										1491	13
	%	91.2	1.5	3.8	0.6	2.5	0.3	0.1											
<40361> (40164)	16-32mm	151	6	7	4	1												169	
	%	89.3	3.6	4.1	2.4	0.6													
	11.2-16mm	386	11	14	11	9	3											434	1
	%	88.9	2.5	3.2	2.5	2.1	0.7												

Table 6.5 Comparisons of Southfleet Road clast lithologies with data from nearby locales

Gravel	Site	Sample Identifier	Flint — Tertiary	Flint — Nodular	Flint — Total	Chalk — *Chalk	Southern — Greensand chert	Southern — Total	Quartz	Quartzite	Exotics — Carboniferous Chert	Exotics — Rhaxella Chert	Exotics — Igneous	Exotics — Total	Total count
Swanscombe, Lower Middle Gravel	Barnfield Pit	1 D	58.2	9.8	93.9		0.9	1.2	2.4	1.8	0.5			4.8	1081
	11.2-16	1 D	50.9	5.3	89.9		2.1	2.3	4.4	2.0	0.8		0.1	7.7	1730
		2 D	48.5	12.7	92.7		1.9	2.0	1.9	1.8	0.5	0.1	0.2	5.0	992
	11.2-16	2 D	41.6	5.5	89.7		3.0	3.1	3.5	1.5	0.5	0.2	0.2	6.8	1785
Swan Valley School	LMG	1 D	63.9	7.4	94.3		1.5	1.5	0.6	2.7	0.6	0.2	0.2	4.2	474
	11.2-16	1 D	52.3	4.7	89.2		2.6	3.0	2.1	3.9	0.6	0.1	0.4	7.5	1085
	Trench D	D	66.4	7.3	95.7		1.8	1.8	0.3	1.0	0.6	0.3	0.2	2.5	672
	11.2-16	D D	51.5	6.3	90.5		2.8	3.0	1.4	3.6	0.8		0.1	6.5	1055
	Trench E	D	46.0	13.2	93.1		3.6	3.6	1.1	0.8	0.5	0.3		3.3	889
	11.2-16	E D	42.0	7.1	91.6		3.3	3.4	0.9	2.8	0.8	0.5	0.1	5.0	1089
	Trench K	D	71.6	8.4	95.0		2.0	2.5	0.2	1.5	0.3		0.2	2.3	641
	11.2-16	K D	60.6	3.7	91.4		1.8	1.9	1.8	3.1	0.6	0.3	0.1	6.5	791
	Trench P	D	51.9	13.9	93.5		1.7	1.7	0.2	3.2	0.2	0.3	0.2	4.8	584
	11.2-16	P D	42.1	6.5	89.8		3.7	3.8	1.7	3.4	0.4	0.3	0.2	6.4	999
Springhead	1121/1		92.6	1.1	96.8		1.1	1.1							380
	11.2-16		92.6	0.7	96.6		0.8	0.8							1101
	1121/3		89.5	3.1	96.8		1.1	1.1							551
	11.2-16		95.3	0.8	99.2		0.6	0.6							1331
Eastern Quarry (61040)	TP 4.1	4	51.8	14.6	92.6		4.3	4.5	1.3	0.9	0.7	-	-	2.9	446
	11.2-16	11.2-16	44.1	5.6	90.9		3.3	3.5	1.1	3	1.2	0.2	0.1	5.6	1087
	TP 16.2	2	47.1	13.4	93.2		3.5	3.5	0.9	2	0.4	-	-	3.3	546
	11.2-16	11.2-16	43.8	7.3	90.2		4.4	4.5	1.6	2.5	0.7	0.1	0.2	5.2	1144
Eastern Quarry (61041)	TP 104	50	71.5	18.5	92.7		5.8	5.8							260
	11.2-16	11.2-16	76.5	15	99		2.5	3							200
	TP 104	51	70.7	16.2	94.2		5.2	5.8							154
	11.2-16	11.2-16	78.6	9.3	96.5		2.6	2.6							150
	TP 112	60	90.6	5.6	98.5		0.4	0.6		0.1					461
	11.2-16	11.2-16	91.7	3	98.3		0.8	1.4	0.1						1042
Eastern Q (61042)	TP 127	<96>	80.0	4.1	94.3		4.8	4.8	0.2	0.4				0.7	459
	11.2-16	11.2-16	81.4	2.0	95.5		3.1	3.7	0.1	0.3	0.3		0.1	0.8	1027

Site	Sample												Count
TP 127	<101>	60.1	7.5	93.4	3.4	3.4	2.0	1.0	0.2			3.2	411
	11.2-16	67.0	3.6	93.6	3.9	4.1	1.5	0.5	0.1			2.4	1234
Eastern Q, Septic tank (68990)	<3> (68990)	57.4	7.8	93.3	4.7	4.7	1.0	1.6	0.3	0.3	0.2		387
	11.2-16	52.2	3.8	89.3	4.6	4.6	2.5	2.3	0.1	0.1			853
Station Q South (63542)	10	91.8	2.2	97.6	1.8	2.4							490
	11.2-16	91.8	0.5	98.1	1.2	1.2		0.1			0.1	0.2	1215
	8	94.8	0.9	99.1	0.9	0.9			0.3				116
	11.2-16	90.8	2.4	96.3	0.3	0.3					0.3		295
Northfleet Cement Works	<5> (602)	61.2	7.0	94.3	4.0	4.0	0.7	0.4	0.3			1.6	670
	11.2-16	62.4	3.1	94.8	3.0	3.0	0.9	0.1	0.1		0.1	1.3	1293
	<9> (1003)	68.1	4.9	95.3	3.2	3.2	0.7	0.1				1.1	758
	11.2-16	72.2	2.0	96.3	2.0	2.0	0.8	0.1	0.2	0.1		1.3	1203
Crossways Park	<19> 1003	65.1	17.4	96.9	3.1	3.1	1.6	0.3	0.9			3.1	287
	11.2-16	66.3	18.1	98.4	2.4	2.7	0.8	0.1	0.1		0.4	1.6	1069
Crossways Park	<18> (1104)	48.5	9.3	91.9	5.7	5.7	0.6	0.3	0.3	0.3	0.2	1.8	332
	11.2-16	50.8	4.2	91.9	4.8	4.8	1.5	0.6	0.4	0.1	0.1	3.0	826

can be interpreted on topographic grounds as being broadly equivalent to the Swanscombe Lower Middle Gravel. The Southfleet Road results can be integrated with recent lithological analyses in the surrounding area (Table 6.5) to develop an updated model of the palaeo-landscape, showing the extent of the Swanscombe Lower Middle Gravel channel, the course of the contemporary palaeo-Ebbsfleet and their point of confluence. These new analyses have confirmed the southward extension of Lower Middle Gravel deposits, not only so far as the Swan Valley School (Wenban-Smith and Bridgland 2001), but now well into Eastern Quarry, more than 600m further south than the currently mapped boundary (Fig. 6.2). These new data show the Ebbsfleet passing over the elephant site and entering the Thames from the south-east, with the point of confluence about 400m to the NNW at c TQ 61000 73650 (Fig. 4.44).

Chapter 7

The vertebrate remains from Southfleet Road: introduction, taphonomy and palaeoecology

by Simon A. Parfitt

INTRODUCTION

Vertebrate fossils are preserved in the majority of sedimentary beds containing Clactonian archaeology, but not in the upper part of the sequence, where the deposits with Acheulian artefacts are decalcified and burial conditions did not favour bone preservation. Lower Palaeolithic artefacts and vertebrate fossils were first discovered at Southfleet Road in 2003, during construction of the HS1 international railway station at Ebbsfleet. In 2004, the remaining undisturbed sediments were the focus of a nine-month archaeological investigation, which produced a large collection of vertebrate remains. The artefact assemblage with two distinct superposed industries (ie Acheulian and Clactonian), invites correlation of the Southfleet Road deposits with the internationally famous site at Swanscombe, some 2km to the north-west. A more complete faunal succession is present in the Thames Valley sequences at Barnfield Pit, Swanscombe and a nearby pit at Ingress Vale, whereas vertebrate remains from its tributary at Southfleet Road come solely from the Clactonian levels. Most of the vertebrate fossils from the Swanscombe area do not come from controlled archaeological excavations, but were instead recovered from working faces of pits and quarries as they were being dug for sand, gravel and chalk. As a consequence, these collections include many specimens lacking stratigraphical information. The vertebrate assemblages are inevitably biased in favour of larger specimens and animals, resulting in a fragmented and tantalisingly incomplete picture of the contemporary fauna and confounding efforts to understand human-animal interactions at Swanscombe (cf. Binford 1985). With this in mind, one of the primary objectives of the fieldwork at Southfleet Road was to maximise the recovery of biological remains, in particular those of small mammals, birds, amphibians, reptiles and fishes, which are so important for palaeoecological reconstructions and dating. The combined hand-retrieved and sieved vertebrate assemblage is impressive and the identified sample comprises 18,146 vertebrate specimens. These have been assigned to at least 46 taxa, ranging from isolated teeth of pygmy shrews and bats to the scattered remains of a straight-tusked elephant skeleton. Primary aims of the post-excavation research were to characterise and reconstruct the landscape and environment setting of the

human occupation based on the vertebrate fauna, and to contribute to the dating of the deposits using evidence from mammalian biostratigraphy.

The analysis of the vertebrate remains falls naturally into three chapters. This chapter outlines the methods employed during the analysis of the animal bones and explores the interpretation of the assemblages in terms of depositional history and past environmental conditions. Evidence for faunal change through the sequence is combined with information from bone taphonomy to unravel the origin of the Southfleet Road vertebrate assemblages, providing the basis for understanding the landscape setting and climate at the time of the Clactonian occupation of the site. Chapter 8 explores the question of human exploitation of large mammals, focussing on a description and interpretation of taphonomy of the straight-tusked elephant from Phase 6. Cut marked bones from other contexts are also described. The correlation and age of the deposits are explored from a mammalian biostratigraphical perspective in Chapter 9. The key elements of the biostratigraphic evidence are the recognition of peak interglacial conditions and detailed biometrical comparisons of the large mammal remains with those from other interglacial sites in the Thames valley and the British sequence in general. The mammalian evidence provides strong grounds for linking the Phase 5-6 deposits in the Ebbsfleet Valley at Southfleet Road with the Clactonian levels in the Thames Valley at Swanscombe (Kent) and Clacton (Essex). Taken together, these sites form a transect along a Lower Palaeolithic riverine landscape, from the lower reaches of the Thames in the Swanscombe area to its estuary at Clacton-on-Sea. The discussion in this chapter concludes with discussion of the mammalian evidence for the landscape setting and environmental context of early humans in southern England during the Hoxnian interglacial, some 400,000 years ago.

MATERIAL AND METHODS

The excavated assemblage comprised approximately 1,370 individually-numbered large mammal bones (Table 7.1). Of these specimens, the vast majority comprised unidentifiable, or minimally identifiable fragments of bone and teeth. Most of the material recovered by hand

Table 7.1 Counts and stratigraphical occurrence of hand-collected vertebrate remains from Southfleet Road

Phase	*3*	*3*	*3*	*5*	*5*	*6a*	*6a*	*6a*
Context	*40028*	*40062*	*40159*	*40025*	*40072*	*40039*	*40103*	*40103/039*
No. frags (large mammal)	*1*	*2*	*1*	*15*	*1*	*236*	*20*	*1*
AMPHIBIA								
Anuran indet.								
AVES								
Indet. bird								
MAMMALIA								
Lipotyphla								
Neomys sp., water shrew								
Talpa minor, extinct mole								
Rodentia								
Castor fiber, beaver								
Arvicola cantianus, water vole								
Indet. vole						1		
Lagomorpha								
Oryctolagus cuniculus, rabbit				2		2		
Carnivora								
Panthera leo, lion				1				
Proboscidae								
Palaeoloxodon antiquus, straight-tusked elephant				1				
Elephantidae gen. et sp. indet., indeterminate elephant								
Perissodactyla								
Stephanorhinus hemitoechus, narrow-nosed rhinoceros						23		
Stephanorhinus kirchbergensis, Merck's rhinoceros								
Artiodactyla								
Sus scrofa, wild boar						1		
Dama dama, fallow deer					(1 cf)	1, (3 cf)	1	
Cervus elaphus, red deer						3, (1 cf)	2	(1 cf)
D. dama or *C. elaphus*, fallow or red deer				4		29	5	
Capreolus capreolus, roe deer								
Cervidae gen. et sp. indet, indeterminate deer						2		
Bos primigenius, aurochs		1		1				
Bos primigenius or *Bison priscus*, aurochs or bison				2		2		
Large mammal (fox size)								
Large mammal (Roe deer size)							1	
Large mammal (smaller than fallow deer)						2		
Large mammal (Red/fallow deer size)		1		2		6		
Large mammal (larger than red deer)	1					64		
Large mammal (*Bos* or larger)						19		
Large mammal (rhino size)						1		
Large mammal (rhino or larger)						1		
Large mammal (elephant size)								
Indet large mammal			1	1	1	74	3	
Indet small vertebrate								
Indet vertebrate						1		

Counts refer to the number of identified specimens (NISP); where two values are given, the second refers to specimens tentatively identified; bones that could not be identified to taxon were assigned to a size category

6a/6b 40039/070 2	6b 40070 165	6b 40144 42	6 40099 2	6 40160 3	6 40162 5	6 40078 508	6 40100 293	6 40158 6	6 40068 1	7 40167 1	8 40048 3	8 40049 1	? 40028 1	? 40089 1	Unstratified 25
	2														
	1														
						1									
							1								
							1								
	5					1	1								
	1					1	2								
		1				2									
						61									
		(2 cf)				3	3								
						1	6								
							1								
	2						2, (1 cf)								
	8	1				6	9, (2 cf)								
2	74	23		2	2	50	57	2							9
	1	1					1								
	15	3				18	5	1							
						1									
						2		1		1				1	
							1								
	1	2				3	5								
	4	3		1	1	13	18					1			3
	14					11		1					1		3
	2	1				46	40				2				1
	1						1								
					1	1	38								
			1			172	5								
	40	4	1		1	117	89	1	1		1				9
	1														
	2	1				1									

Table 7.2 Counts and stratigraphical occurrence of vertebrate taxa in sieved samples

Phase	5	5	5	5	5	5	5	5
Context	40025	40025	40025	40025	40025	40025	40025	40025
Sample	40145	40286	40303	40343	40348	40380	40382	40411
Volume (L.)	<1	60	19	-	-	-	-	200
Sieved on site				SOS	SOS	SOS	SOS	
Mollusc sample								
Part sorted								
Lipotyphla								
Sorex minutus, pygmy shrew					+	+		
Sorex sp(p)., shrew				1	+			
Neomys sp., water shrew				1	+			
Soricidae gen. et sp. indet., shrew								
Talpa minor, extinct mole								
Chiroptera								
Myotis daubentonii, Daubenton's bat								
Primates								
Macaca sylvanus, Barbary macaque								
Lagomorpha								
Oryctolagus cuniculus, rabbit				1	1	1		
Rodentia								
Sciurus sp., squirrel								
Spermophilus sp., ground squirrel								+
Clethrionomys glareolus, bank vole				+	+	+		+
Arvicola cantianus, water vole				+	2	1		+
Microtus (Terricola) cf. *subterraneus*, common pine vole						2		
Microtus agrestis★, field vole								
Microtus agrestis/M. arvalis★★, field or common vole				1	1	2		
Microtus oeconomus, northern vole					1	1		
Microtus sp., vole	+			+	+			+
Apodemus maastrichtiensis, mouse (extinct)								
Apodemus sylvaticus, wood mouse								+
Apodemus sp., mouse					+			+
Carnivora								
Mustela cf. *putorius*, polecat								
Mustela sp., mustelid								
Artiodactyla								
Dama dama, fallow deer								
Cervus elaphus, red deer								
D. dama or *C. elaphus*, fallow deer or red deer								
Capreolus capreolus, roe deer								
Small vertebrates (NISP)								
Fish	5		1	190	207	141		129
Amphibian		2		108	161	77		70
Reptile								
Bird						1		
Small mammal	1			106	107	69	1	70
Pre-Pleistocene fish	3			76	72	109	28	79

+ — present; counts for small mammals refer to the minimum number of individuals (MNI); counts for fishes, amphibians, reptiles, birds, and pre-Pleistocene fish remains (principally shark teeth) refer to the number of elements identified

★★ Identifications of *M. agrestis* are based on the presence of diagnostic M^2s

★ *Microtus agrestis* and *M. arvalis* are difficult to separate using M_1s and are therefore grouped

6a	6a	6a	6a	6a	6a	6a	6a	6a	6b	6b	6b	6b
40039	40039	40039	40103	40103	40103/039	40103	40103/039	40103/039	40070	40070	40070	40070
40238	40261	40278	40300	40300c	40301	40312	40320	40325	40035	40162	40252	40277
40	60	0.5	17	-	33	17	20	19	30	50	1	10
				Y								
									+			+
			+			+			1	4		+
			+							1		1
		+										
		+						+		1		+
1		+	1		+	1			2	3	+	3
1	1	+	1			+	1		3	3		2
									1			
									+			
		1							1	3	1	1
+	+			+		+						+
1			2			+			1	2		2
		1	+				+					
										1		
												1
		3	9	1		1	1		48	148	2	38
4		14	56	5	1	49	3	1	135	284	15	100
						2			3	9		1
									3	9		
165	69	28	95	3	11	55	12	17	208	454	7	291
								1	1	8		4

Table 7.2 (continued 1)

Phase	6b	6b	6b	6b	6b	6b
Context	40070	40070	40070	40070	40070	40070
Sample	40282c (8-15 cm)	40282c (15-22 cm)	40296	40298	40298c	40299
Volume (L.)	-	-	3	20	-	16
Sieved on site						
Mollusc sample	Y	Y			Y	
Part sorted						
Lipotyphla						
Sorex minutus, pygmy shrew						
Sorex sp(p)., shrew				3		1
Neomys sp., water shrew				1		1
Soricidae gen. et sp. indet., shrew						
Talpa minor, extinct mole			+	+	+	+
Chiroptera						
Myotis daubentonii, Daubenton's bat						
Primates						
Macaca sylvanus, Barbary macaque						
Lagomorpha						
Oryctolagus cuniculus, rabbit						
Rodentia						
Sciurus sp., squirrel						
Spermophilus sp., ground squirrel						
Clethrionomys glareolus, bank vole	+		+	3	2	2
Arvicola cantianus, water vole	+	+	+	4		1
Microtus (Terricola) cf. subterraneus, common pine vole				1		
Microtus agrestis*, field vole						
Microtus agrestis/M. arvalis**, field or common vole				3		
Microtus oeconomus, northern vole						
Microtus sp., vole	+		+			+
Apodemus maastrichtiensis, mouse (extinct)						
Apodemus sylvaticus, wood mouse			1	1		2
Apodemus sp., mouse						
Carnivora						
Mustela cf. putorius, polecat						
Mustela sp., mustelid						
Artiodactyla						
Dama dama, fallow deer						
Cervus elaphus, red deer						
D. dama or C. elaphus, fallow deer or red deer						
Capreolus capreolus, roe deer						
Small vertebrates (NISP)						
Fish	9		23	137	3	57
Amphibian	6	1	22	224	20	96
Reptile				2		5
Bird						
Small mammal	6	7	22	340	7	151
Pre-Pleistocene fish				2		

+ — present; counts for small mammals refer to the minimum number of individuals (MNI); counts for fishes, amphibians, reptiles, birds, and pre-Pleistocene fish remains (principally shark teeth) refer to the number of elements identified

** Identifications of M. agrestis are based on the presence of diagnostic M^2s

* Microtus agrestis and M. arvalis are difficult to separate using M_1s and are therefore grouped

6b 40070 40299c	6b 40070 40306	6b 40070 40306c	6b 40070 40307	6b 40070 40307c	6b 40070 40308	6b 40070 40308c	6b 40070 40309	6b 40070 40309c	6b 40070 40310	6b 40070 40313	6b 40070 40314	6b 40070 40314c
-	16	-	17	-	11	-	27	-	7	10	28	
Y		Y		Y		Y		Y				Y
											1	
	+	+	1				1					
	1		1				2					
	+		+				+					
2	+		1		+		1			+	1	
+	1		+		+		5	+		1	1	
							1					
							1				+	
							1					
	+				+							
	1		1		+							
			1									
4	51	11	81		27		66		43	4	15	7
10	151	6	71	1	47	4	181	4	14	12	11	16
		1					1		1	1	2	
			3		1						4	
10	96	2	85	1	43	2	215	7	13	11	47	4

Table 7.2 (continued 2)

Phase	6b	6b	6b	6b	6b	6b	6b
Context	40070	40070	40070	40070	40070	40070	40070
Sample	40315	40315c	40316	40316c	40317	40317c	40318
Volume (L.)	28	-	28	-	26	-	28
Sieved on site							
Mollusc sample		Y		Y		Y	
Part sorted							
Lipotyphla							
Sorex minutus, pygmy shrew					+		
Sorex sp(p)., shrew	1	+	1	+	3		1
Neomys sp., water shrew	1		+		2		
Soricidae gen. et sp. indet., shrew							
Talpa minor, extinct mole	+		+		+		+
Chiroptera							
Myotis daubentonii, Daubenton's bat							
Primates							
Macaca sylvanus, Barbary macaque							
Lagomorpha							
Oryctolagus cuniculus, rabbit							
Rodentia							
Sciurus sp., squirrel							
Spermophilus sp., ground squirrel							
Clethrionomys glareolus, bank vole	2		1		2	+	1
Arvicola cantianus, water vole	2		2		2		2
Microtus (Terricola) cf. *subterraneus*, common pine vole							
*Microtus agrestis**, field vole							
*Microtus agrestis/M. arvalis***, field or common vole	1				2		1
Microtus oeconomus, northern vole	1						
Microtus sp., vole							
Apodemus maastrichtiensis, mouse (extinct)					+		
Apodemus sylvaticus, wood mouse	1	2	3				
Apodemus sp., mouse							
Carnivora							
Mustela cf. *putorius*, polecat							
Mustela sp., mustelid	+						
Artiodactyla							
Dama dama, fallow deer							
Cervus elaphus, red deer							
D. dama or *C. elaphus*, fallow deer or red deer							
Capreolus capreolus, roe deer							
Small vertebrates (NISP)							
Fish	244	6	88	1	170	1	1
Amphibian	404	14	216	8	379	4	76
Reptile	11		6		9		6
Bird	2				3		
Small mammal	249	8	174	7	294	6	61
Pre-Pleistocene fish	1						1

+ — present; counts for small mammals refer to the minimum number of individuals (MNI); counts for fishes, amphibians, reptiles, birds, and pre-Pleistocene fish remains (principally shark teeth) refer to the number of elements identified

** Identifications of *M. agrestis* are based on the presence of diagnostic M^2s

* *Microtus agrestis* and *M. arvalis* are difficult to separate using M_1s and are therefore grouped

6b	6b	6b	6b	6b	6b	6b	6b	6b	6b	6b	6b	6b
40070	40070	40070	40070	40070	40070	40070	40070	40070	40070	40070	40070	40070
40318c	40319	40319c	40339	40347	40351	40381	40329	40330	40331	40332	40335	40337
-	13	-	10	-	-	-	-	-	-	-	-	-
			SOS	SOS	SOS	SOS	SOS	SOS	SOS	SOS	SOS	SOS
Y	*Y*	*Y*					*Y*	*Y*	*Y*	*Y*	*Y*	*Y*
						+						
				1			1	+	+	+	2	+
				1		1	2	4	1	1	3	
	+				+							
	+			+	+	+	+	+	+	+	+	
							1					
							1	1				
				1		+					1	
+												
				3	+	2	4	2		1	6	
+	1			7	+	1	7	8	5	1	4	1
							1				1	
				6		2	2	3	3	2	5	
					+							
				2		3	2				5	
				1								
											1	
											3	
1	3	1	5	108	11	78	★	★	★	★	★	★
3	35	3	2	522	64	237	★	★	★	★	★	★
			2		1	9	★	★	★	★	★	★
	1			8		5	★	★	★	★	★	★
4	18	1		525	39	248	★	★	★	★	★	★
				3			★	★	★	★	★	★

Table 7.2 (continued 3)

Phase	6b	6b	6b	6b	6b	6b
Context	40070	40070	40144	40144	40144	40144
Sample	40338	4033x	40282c (2-8 cm)	40294	40297	40305c
Volume (L.)	-	-	-	10	20	-
Sieved on site	SOS	SOS				
Mollusc sample			Y			Y
Part sorted	Y	Y				

Lipotyphla

Sorex minutus, pygmy shrew				+		
Sorex sp(p)., shrew				1	2	
Neomys sp., water shrew					+	
Soricidae gen. et sp. indet., shrew						
Talpa minor, extinct mole					1	

Chiroptera

Myotis daubentonii, Daubenton's bat

Primates

Macaca sylvanus, Barbary macaque

Lagomorpha

Oryctolagus cuniculus, rabbit

Rodentia

Sciurus sp., squirrel						
Spermophilus sp., ground squirrel						
Clethrionomys glareolus, bank vole	+			1	1	
Arvicola cantianus, water vole	+				1	
Microtus (Terricola) cf. subterraneus, common pine vole						
Microtus agrestis*, field vole						
Microtus agrestis/M. arvalis**, field or common vole						
Microtus oeconomus, northern vole						
Microtus sp., vole	+			+		+
Apodemus maastrichtiensis, mouse (extinct)						
Apodemus sylvaticus, wood mouse				1	3	
Apodemus sp., mouse						

Carnivora

Mustela cf. putorius, polecat
Mustela sp., mustelid

Artiodactyla

Dama dama, fallow deer		1				
Cervus elaphus, red deer						
D. dama or C. elaphus, fallow deer or red deer		1				
Capreolus capreolus, roe deer						

Small vertebrates (NISP)

Fish	★		4	6	157	1
Amphibian	★		5	15	269	9
Reptile	★				3	
Bird	★		2	26	228	5
Small mammal	★					
Pre-Pleistocene fish	★					

+ — present; counts for small mammals refer to the minimum number of individuals (MNI); counts for fishes, amphibians, reptiles, birds, and pre-Pleistocene fish remains (principally shark teeth) refer to the number of elements identified

** Identifications of M. agrestis are based on the presence of diagnostic M^2s

* Microtus agrestis and M. arvalis are difficult to separate using M_1s and are therefore grouped

6b	6b	6b	6b	6b	6b	6b	6b	6b	6b	6b	6
40144	40144	40144	40144	40144	40144	40143	40143	40143	40143	40078	40008
40311	40333	40336a&b	40336a	40336b	?	40253	40267	40282c (0-2 cm)	40284c	40293	40327
10	-	-	-	-	-	1	100	-	-	30	20
	SOS	SOS	SOS	SOS							
								Y	Y		
			4	1			1		1		
1	2		2				+				
1	1		2	1			3				
+	1		+	+			+				
			1								
			1								
+	2		6	5			1				
+	3		7	4			1				
	1			1			2				
	cf		3	1							
			1								
+	+										
			cf								
2	3		4	4			3				
			1								
	1										
	1				2						
8	138	1026				15	226	5	104		
38	309	1116				45	778		30		1
	3	21				21	9		1		
24	300	869				2	269		9	1	3
	1	12									

excavation was assigned a unique find number in the field and their positions were recorded using a 'total station'. Dense clusters of large mammal bone and the larger elephant remains were recorded on plans and with vertical photographs. Some of the bones were in good condition, and because they were buried in sandy sediment could be excavated without any problem. However, many of the fossils, chiefly those buried in clayey sediments, were poorly preserved. Particularly fragile pieces were consolidated on site with a polyvinyl acetate (PVA) solution and lifted in blocks of sediment for transport to the laboratory. Some of the larger bones were too large to lift in sediment blocks and had to be plaster-jacketed before removal. In addition to the hand-retrieved sample, large mammal material was also recovered from wet-sieved sediment samples (see Table 7.2).

One of the major objectives of the fieldwork was to obtain small vertebrates remains to aid reconstruction of the depositional setting and palaeoecology of the human occupation. The small vertebrates were recovered from bulk samples that were wet-sieved to 0.5mm and 1mm. Samples totalling more than 5,700L of sediment were collected as spot samples or in columns and transported to the laboratory for processing. Test samples (usually 10L) were sieved to assess small vertebrate abundance and to identify promising samples for further processing. The resulting residues were air-dried and the graded through a nest of sieves, with sorting aided by binocular microscope to recover the smallest bones and teeth. The presence of other biological remains was noted, and ostracods and molluscs were picked to boost samples in contexts where these remains were uncommon (see Chapters 10 and 11). No artefacts or microdebitage were recovered from the sieved samples. In total, 2,643L of sediment was wet-sieved and sorted under laboratory conditions. On-site sieving concentrated on the tufaceous-channel fill deposits (40070) and the underlying sands (40025), which were sieved through coarser meshes. This processing regime was successful in extending the sample of the rarer small vertebrate taxa and adding significantly to the sample of medium-sized vertebrates, which were otherwise under-represented in the hand-excavated assemblage. Some of the tufa samples proved to be spectacularly rich in small vertebrate material, which included bones of fish, amphibians (newts, frogs, toads) snakes, lizards and birds, as well a diverse range of small mammal taxa (principally voles, mice and shrews). Exceptionally rare elements include a small carnivore, provisionally identified as European polecat (*Mustela* cf. *putorius*) and Daubentons' bat (*Myotis daubentonii*), both of which are first records for the British Middle Pleistocene. In all, about 16,800 small vertebrate remains were identified and counted (Table 7.2), making this one of the largest stratified assemblages of late Middle Pleistocene small vertebrates from Britain.

After cleaning and conservation, the bones were identified by direct comparison with modern skeletal material at the Natural History Museum (London) as well as Pleistocene fossils in the NHM palaeontology collections. Detailed comparisons were made with other Hoxnian samples, including large mammals from Swanscombe and Clacton, and small mammals from Beeches Pit and Barnham (Suffolk). Specific details of the large mammal identifications are given in Chapter 9. Bones that were too fragmentary for positive species attribution were assigned to a size category (for example, rhino-sized, bovid-sized, red/fallow deer-sized) based on the presence of known large mammal taxa in the assemblage (Table 7.1). Bone fragments assigned to the 'elephant-sized' category can only have come from elephants, based on large size and their characteristic cortical structure.

A variety of attributes was recorded for each specimen, these included (in addition to standard anatomical and taxonomic identification), bone portion, fusion-state in postcranial bones and tooth wear (to determine ontogenetic age), as well as standard anatomical measurements (see Chapter 9). Taphonomic analysis of the large mammal remains included investigation of skeletal element representation, breakage patterns and bone surface modifications that occurred between the death of the animal and burial (such as climatically induced weathering, trampling, chewing marks from predator activity and grooves). Also recorded were modifications that occurred during burial, such as root-etching, manganese and iron staining, fracturing through soil movements and compression. Each bone was examined under a variable magnification binocular microscope and illuminated with a strong oblique light from a fibre-optic source. This revealed occasional striations and incisions, some of which are interpreted as cut-marks. To aid in the interpretation of these incisions, high-resolution images were recorded with an Alicona variation focus microscope, which generates three-dimensional models of bone surfaces (see Bello and Parfitt, Chapter 8). The assemblage was also searched for joins between fragments of the same bone and articulation between elements. A taphonomic study was conducted on the small mammal remains to determine the accumulating mechanisms that influenced and potentially biased the assemblage and to provide additional information on site formation processes. The methods of small mammal taphonomic analysis as applied to fossil assemblages are fully described by Andrews (1991).

Quantification of the vertebrate remains uses the number of identifiable specimens (NISP). Quantification of the small mammal abundance is based on NISP and the minimum numbers of individuals (MNI) calculated from counts of the most common element (usually upper or lower first molars for rodents, mandibles for soricids and postcrania (humeri) for moles). Ecological interpretations are based on the habitats, ecology and geographical ranges of extant mammal species (Corbet and Southern 1977; Bjarväll and Ullström 1986; Mitchell-Jones *et al.* 1999; ICUN Red List of Threatened Species).

Material and data archives are housed at the Natural History Museum, London and copies of data lists have additionally been lodged with Oxford Archaeology.

Table 7.3 Stratigraphical distribution of mammal taxa from Southfleet Road by phase. Shrews from the tufa (Phase 6b, context 40070) include at least four species of *Sorex*: pygmy shrew (*Sorex minutus*), two medium-sized species (the largest of which is similar in size to common shrew *S. araneus*) and a large species represented by a mandible fragment that is indistinguishable from that of the Early-Middle Pleistocene *Sorex (Drepanosorex)*.

Phase	3	5	6a	6b	6	7
PISCES						
Anguilla anguilla, eel		+		+		
Esox lucius, pike		+		+		
Salmonidae, salmon family		+		+		
Rutilus rutilus, roach				+		
Scardinius erythrophthalmus, rudd				+		
cf. *Phoxinus phoxinus*, minnow				+		
Tinca tinca, tench		+		+		
Cyprinidae indet., carp family		+	+	+		
Pungitius pungitius / Gasterosteus aculeatus, nine-spined or three-spined stickleback		+	+	+		
Gymnocephalus cernuus / Perca fluviatilis, ruffe or perch		+				
AMPHIBIA						
Triturus sp., newt		+	+	+		
Bufo sp., toad		+	+	+		
Hyla sp., tree frog				+		
Rana sp., frog		+	+	+		
REPTILIA						
Lacertidae, undetermined lizard				+		
Anguis fragilis, slow worm				+		
Ophidia indet., snake				+		
AVES						
Anatidae sp., undetermined duck				+		
Turdus philomelos / T. iliacus, song thrush or redwing				+		
Aves indet., bird		+		+		
MAMMALIA						
Lipotyphla						
Sorex minutus, pygmy shrew		+		+		
Sorex sp(p)., shrew		+	+	+		
Neomys sp., water shrew		+	+	+	+	
Soricidae gen. et sp. indet., shrew			+	+		
Talpa minor, extinct mole			+	+	+	
Chiroptera						
Myotis daubentonii, Daubenton's bat				+		
Primates						
Macaca sylvanus, Barbary macaque				+		
Homo sp., hominin (C = Clactonian)	+	C?	C	C	C	C
Lagomorpha						
Oryctolagus cuniculus, rabbit		+	+	+		
Rodentia						
Sciurus sp., squirrel				+		
Spermophilus sp., ground squirrel		+				
Clethrionomys glareolus, bank vole		+	+	+		
Castor fiber, beaver					+	
Arvicola cantianus, water vole		+	+	+	+	
Microtus (Terricola) cf. *subterraneus*, common pine vole		+		+		
Microtus agrestis, field vole				+		
Microtus agrestis / M. arvalis, field or common vole		+	+	+		
Microtus oeconomus, northern vole		+		+		
Microtus sp., vole		+	+	+		
Apodemus maastrichtiensis, extinct mouse				+		
Apodemus sylvaticus, wood mouse		+	+	+		
Apodemus sp., mouse		+	+	+		

Table 7.3 (continued)

Phase	3	5	6a	6b	6	7
Carnivora						
Mustela cf. *putorius*, polecat				+		
Mustela sp., mustelid				+		
Panthera leo, lion		+				
Proboscidae						
Palaeoloxodon antiquus, straight-tusked elephant		+		+	+	
Elephantidae gen. et sp. indet., indeterminate elephant					+	
Perissodactyla						
Stephanorhinus hemitoechus, narrow-nosed rhinoceros			+	cf	+	
Stephanorhinus kirchbergensis, Merck's rhinoceros					+	
Artiodactyla						
Sus scrofa, wild boar			+		+	
Dama dama, fallow deer		cf	+	+	+	
Cervus elaphus, red deer		+	+	+	+	
D. dama or *C. elaphus*, fallow deer or red deer		+	+	+	+	
Capreolus capreolus, roe deer				+	+	
Cervidae gen. et sp. indet, indeterminate deer			+	+	+	
Bos primigenius, aurochs	+	+			+	
Bos primigenius or *Bison priscus*, aurochs or bison		+	+		+	+

RESULTS

The following account provides a summary of the bone assemblages recorded within the stratigraphical framework of the site (Table 7.3). The Minimum Number of Individuals (MNI) was estimated for large mammal assemblages for the main phases. For post-crania, MNIs calculations were based on most numerous non-reproducible skeletal element (excluding shed antlers), with pairing for age (cf. Chaplin 1971; Klein and Cruz-Uribe 1984).

The only species with an MNI of more than one per context is red deer *Cervus elephus* from context 40100, which has two individuals based on scapulae. Specimens not identified to taxon were placed into animal size-groups using bone size and cortical thickness.

Phase 3

Very few vertebrate fossils were recovered from the gravelly sands and silts (Phases 1-4) at the base of the succession. The only identifiable specimen is the posterior part of an aurochs *Bos primigenius* skull with both horncores, from context 40062 (Fig. 7.1; Fig. 9.17). The occurrence of aurochs, which is only known from temperate periods in the British Pleistocene, implies that the prevailing climate was not unduly severe. Its favoured habitats probably included parkland with scattered trees, grassy meadows and wetland habitats, supporting rich herbaceous vegetation (Legge 2010). This provides the only indication from the mammalian evidence of local vegetation conditions during the deposition of this early phase of the site sequence.

Phase 5 (context 40025)

The sediments of Phase 5 are bedded sands with silty clay laminae, which show clear evidence of water transport. Numerous Tertiary shark teeth (Table 7.2) and mollusc fragments were present in most of the sieved residues, supporting the conclusion that the deposit had received substantial input from Tertiary deposits (particularly Woolwich Beds) from high ground to the west (Chapter 4). Hand-collected large mammal material comprises a low density of larger mammal bones, which were dispersed through the deposit (Fig. 7.2). All but one of these was excavated from the area between Trenches B and C. There are no obvious concentrations or anatomically associated (refitting or articulating) specimens. Surfaces of some of the large mammal bones show cracking and splitting (37%) indicating that they were exposed on a land surface. The condition of the bones is variable; some are well preserved (for example the complete lion *Panthera leo* astragalus, Fig. 9.3), but most are fragmentary or weathered (for example the aurochs maxilla and teeth, Fig. 9.18). More than half are iron-stained, but none show evidence of water transport.

Bulk sieving yielded a relatively rich assemblage of small-vertebrate remains (Table 7.2). These appear to be concentrated in localised patches with immediately adjacent samples yielding much lower concentrations. Other samples were totally devoid of Pleistocene faunal material. Fish bones were by far the most abundant small vertebrate specimens, comprising almost 46% of the identifiable remains (Fig. 7.3). Taxa represented included pike *Esox lucius*, eel *Anguilla anguilla*, salmonid (Salmonidae), stickleback (Gasterostidae), ruffe *Gymnocephalus cernuus* or perch *Perca fluviatilis*, and undetermined cyprinids (Cyprinidae: carp family). The assemblage reflects a freshwater environment, probably a large slow-moving watercourse. A fully temperate climate is indicated by the presence of thermophilous fish species (for example tench *Tinca tinca*). Aquatic environments, marsh and other damp habitats are indicated by anurans

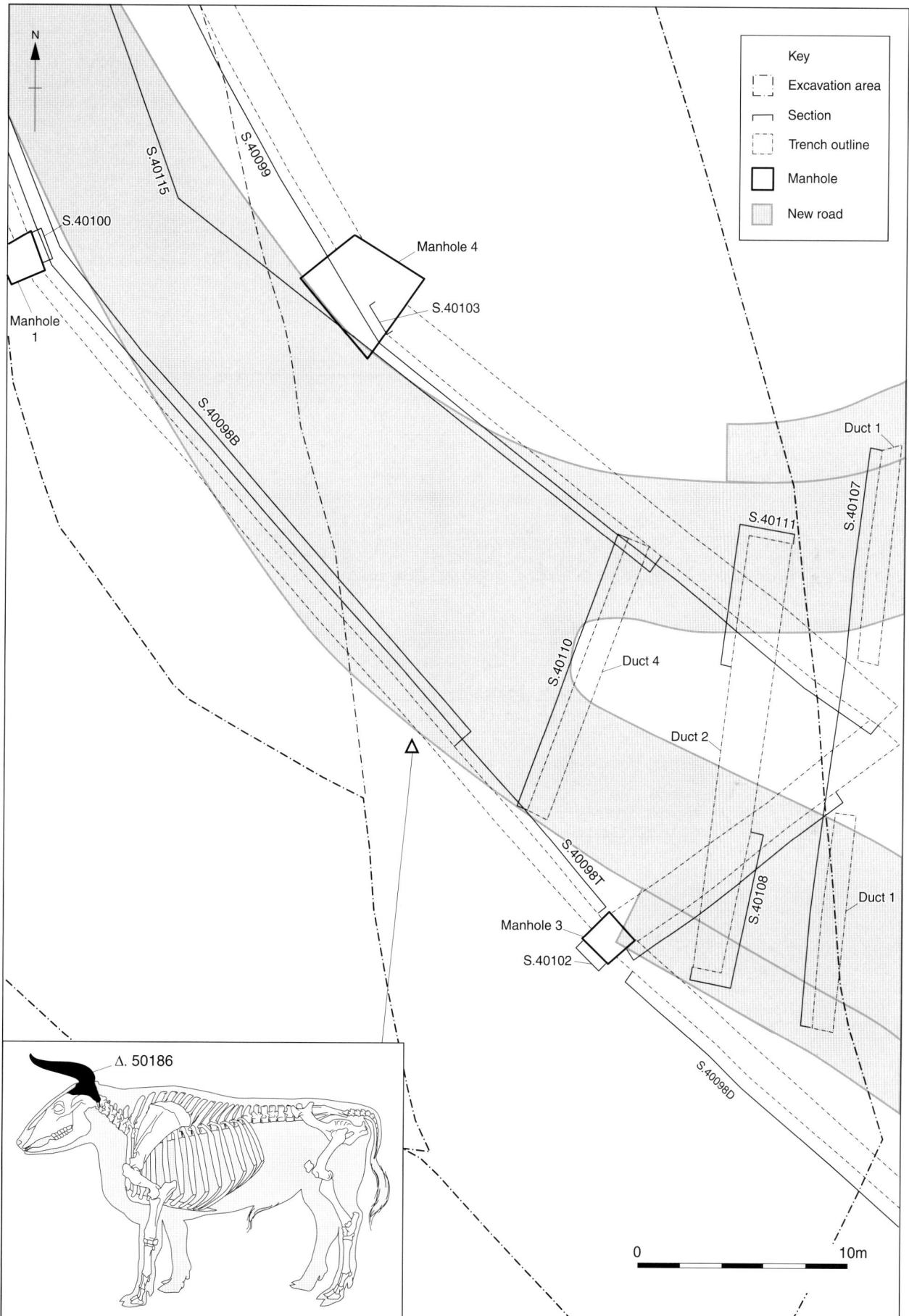

Figure 7.1 Plan showing the location of the aurochs frontal found during machining south of the road cutting near the roundabout at c 23m OD. Inset shows the outline of an auroch's skeleton with the frontal (Δ. 50186) indicated

Trench A

Trench B

Trench C

Key

Excavation area

Trench

Evaluation trench

Bovid

Cervid

Large mammal

Rabbit

Elephant

0   10m

N

XI

X

XIII

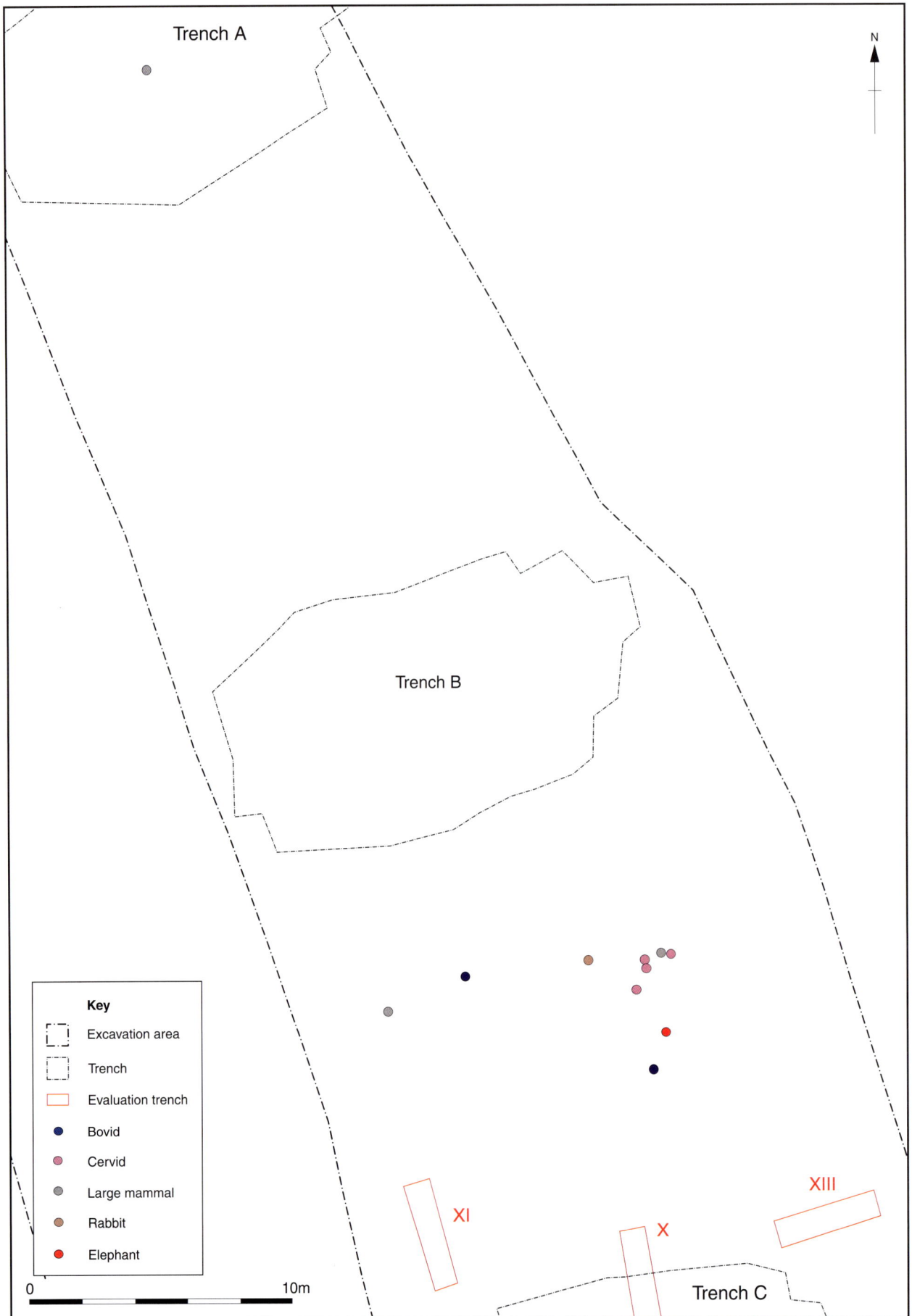

Figure 7.2 Distribution of large mammal bones in Phase 5

(frogs and toads) and newts, which make up 29% of the small vertebrate assemblage (Fig 7.3).

The small number of identifiable large mammal remains (Table 7.4) precludes any meaningful study of taxonomic abundance and ecology, therefore counts are given for identifiable specimens (NISP), rather than minimum number of individuals (MNI). The assemblage does, however, include several species that have specific environmental requirements. These include aurochs and fallow deer *Dama dama*, both of which probably favoured open deciduous or mixed woodland, grassy meadows and wetland habitats. The medium and large-sized ungulates would have provided prey for lions, which prefer to hunt in open grassland or parkland. Turner (1997) has suggested that the greater size of the Pleistocene lions may mean that large bovids, such as aurochs, were also hunted. Rabbits *Oryctolagus cuniculus* build warrens in soft, well-drained, soils. They prefer mixed habitats of grassland with scrubby woodland providing shelter from predators, but avoid coniferous woodland and cold or humid habitats. Dry grassland is also indicated by ground squirrel, *Spermophilus* sp., which is represented by a fragmentary upper cheek tooth. This constitutes the only record to date from the Hoxnian of a species that is more commonly found in cold stage contexts.

The tusk of a straight-tusked elephant *Palaeoloxodon antiquus* was also recovered (Fig. 9.4). Most of the bones are crushed and stained with iron or manganese. The surface layers of the tusk have longitudinal and circum-ferential flaking from exposure to weathering, but the surfaces of the other bones are generally well preserved. Crushing from sediment pressure had deformed the tusk, giving it an apparent curvature that was the basis for its erroneous preliminary identification as *Mammuthus* (Wenban-Smith *et al.* 2006).

The small mammal assemblage includes voles, mice and insectivores and is notable for the relatively high proportion of rodents that prefer damp grassland and other waterside habitats (Fig. 7.3b). Of these the water vole, *Arvicola* , is generally a semi-aquatic species, which burrows into the banks of ponds and slow rivers wherever there is dense vegetation to provide food and cover. Marshes, reedbeds and damp grassland are also a favoured habitat of the northern vole *Microtus oeconomus*, while bank vole *Clethrionomys glareolus* would have preferred scrub or woodland. Rabbits favour open grassland, specifically areas with drier soils and warm microclimates. Overall, the fauna is consistent with extensive open floodplain grassland in a temperate climate.

The mammal assemblage appears unusual as it consists of thermophiles (rabbit and fallow deer) together with ground squirrel and northern vole which are generally thought of as indicators of cool/cold climates in a north-western European context. This association has not previously been found in the British Pleistocene. Today, the natural ranges of rabbit, northern vole and ground squirrels do not overlap, although areas of congruence result from human intervention (Mitchell-

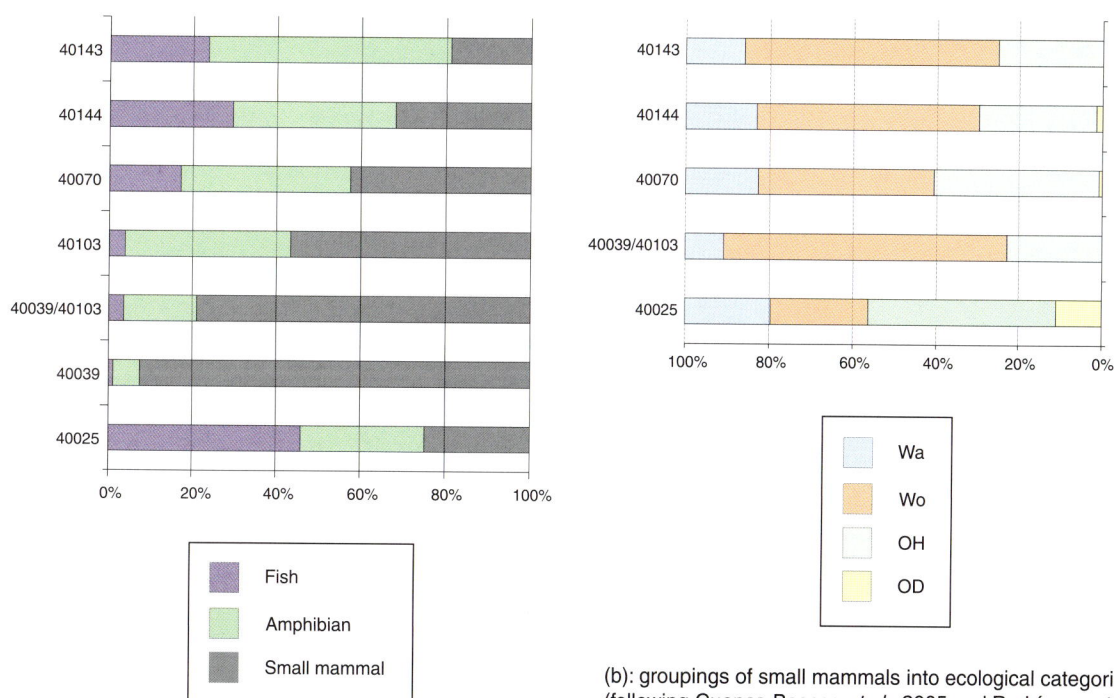

(a): number of identified fragments for fish, amphibians and small mammals as a percentage of the total small vertebrate assemblage.

(b): groupings of small mammals into ecological categories (following Cuenca-Bescos *et al.*, 2005 and Rodríguez *et al.* 2011):
Wa - semi-aquatic favouring waterside habitats,
Wo - woodland including margins and glades with ground cover,
OH - open humid grassland,
OD - open dry grassland

Figure 7.3 Summary diagram of the vertebrate succession at Southfleet Road, showing changes in the composition and ecology of the small vertebrate assemblages

Table 7.4 Phase 5 (context 40025). Taxonomic list of large mammals with body-part data and number of fragments (NISP)

Taxon	NISP	Body parts
Oryctolagus cuniculus	2	Upper molar, L ischium frag
cf. *Dama dama*	1	L antler frag (shed)
Cervidae gen. et sp. indet. (red/fallow deer sized)	4	3 antler frags., R innominate frag
Panthera leo	1	R astragalus
Palaeoloxodon antiquus	1	Tusk
Bos primigenius	1	R premaxilla, maxilla with $M^{1\text{-}3}$ (mid-wear)
Bovidae gen. et sp. indet.	2	L nasal, skull frag
Indet. large mammal (red/fallow deer size)	2	Sacrum frag, long bone shaft frag
Indet. large mammal	1	Indet. bone frag

Table 7.5 Phase 6a. Taxonomic list of large mammals with body-part data and number of fragments (NISP) from contexts 40039 and 40103

Taxon	NISP	Body parts
Context 40039		
Oryctolagus cuniculus	1	L $M_{1\text{-}2}$
S. hemitoechus & *S.* cf. *hemitoechus*	23	Teeth (R P^3, R P^4, R M^1 frag, L P^4, 3 L M^1 frags., R M^2, R M^3, maxilla frags, 2 L P^4 frags, L M^2, L M^3, L. upper premolar frag, 5 cheek tooth frags), L & R petrosal, occipital condyle & skull frags, R lunar
Sus scrofa	1	L upper canine tip (unworn)
Dama dama & cf. *Dama dama*	4	Teeth (L$M^{1\ or\ 2}$, L M^3 – mid wear), L astragalus, L distal tibia
Cervus elaphus & cf. *Cervus elaphus*	4	Axis frag, R distal scapula, R astragalus, L astragalus frag
Cervidae gen. et sp. indet. (red/fallow deer sized)	33	R occipital condyle, frontal with part of antler below base, antler base frag (shed), 15 antler frags, teeth (L M^3, upper molar & maxilla frag, lower molar frag, molar frag, 3 cheek tooth frags), L distal humerus, distal humerus frag, R femur proximal epiphysis (fused), L metatarsal prox & shaft, L metatarsal shaft frag, metapodial shaft frag, metapodial distal condyle
Bovidae gen. et sp. indet.	3	L scaphoid, pisiform, lunate
Indet. large mammal (rhinoceros or larger)	3	2 tooth frags, indet. bone frag
Indet. large mammal (*Bos* or larger)	18	4 skull frags, condyle, long bone shaft frag, 12 indet. bone frags
Indet. large mammal (red deer or larger)	64	Tooth frag, 2 skull frags, long bone shaft frag, 60 indet. bone frags
Indet. large mammal (red/fallow deer size)	5	Petrosal, vertebra frag, 3 long bone shaft frags
Indet. large mammal (smaller than fallow deer)	2	2 long bone shaft frags
Indet. large mammal	71	2 long bone shaft frags, 69 indet. bone frags
Indet. vertebrate	1	Indet bone frag
Context 40103		
Dama dama	1	L antler base with brow tine (shed)
Cervus elaphus	1	R antler base with brow tine (shed), L metatarsal (prox. & shaft) frag
Cervidae gen. et sp. indet. (red/fallow deer sized)	5	Upper cheek tooth frag, molar frag, 2 antler frags, L humerus shaft frag
Indet. large mammal (roe deer size)	1	Rib frag
Indet. large mammal	2	2 indet. bone frags
Context 40103/039		
cf. *Cervus elaphus*	1	L scapula frag

Jones *et al.* 1999). Ground squirrels are restricted to steppe grassland in continental parts of Eurasia (Mitchell-Jones *et al.* 1999). Although common in cold stage assemblages from Britain, ground squirrel also occurs during warm periods in the Thames Valley in the Late Middle Pleistocene. Similarly, northern vole (a common member of cold stage assemblages in Britain) is also found in association with peak interglacial conditions (Stuart 1982) during the Last Interglacial (MIS 5e) and during temperate peaks of the previous interglacial (MIS 7) in the Ebbsfleet Valley (Wenban-Smith *et al.* forthcoming).

Phase 6a (Contexts: 40039, 40103)

Overlying the sand sequence of Phase 3 is a thick accumulation of non-calcareous clay with an interbedded silty tufaceous deposit that makes up Phase 6. The basal clay

Figure 7.4(a) Distribution of large mammal bones in Phase 6a: context 40039

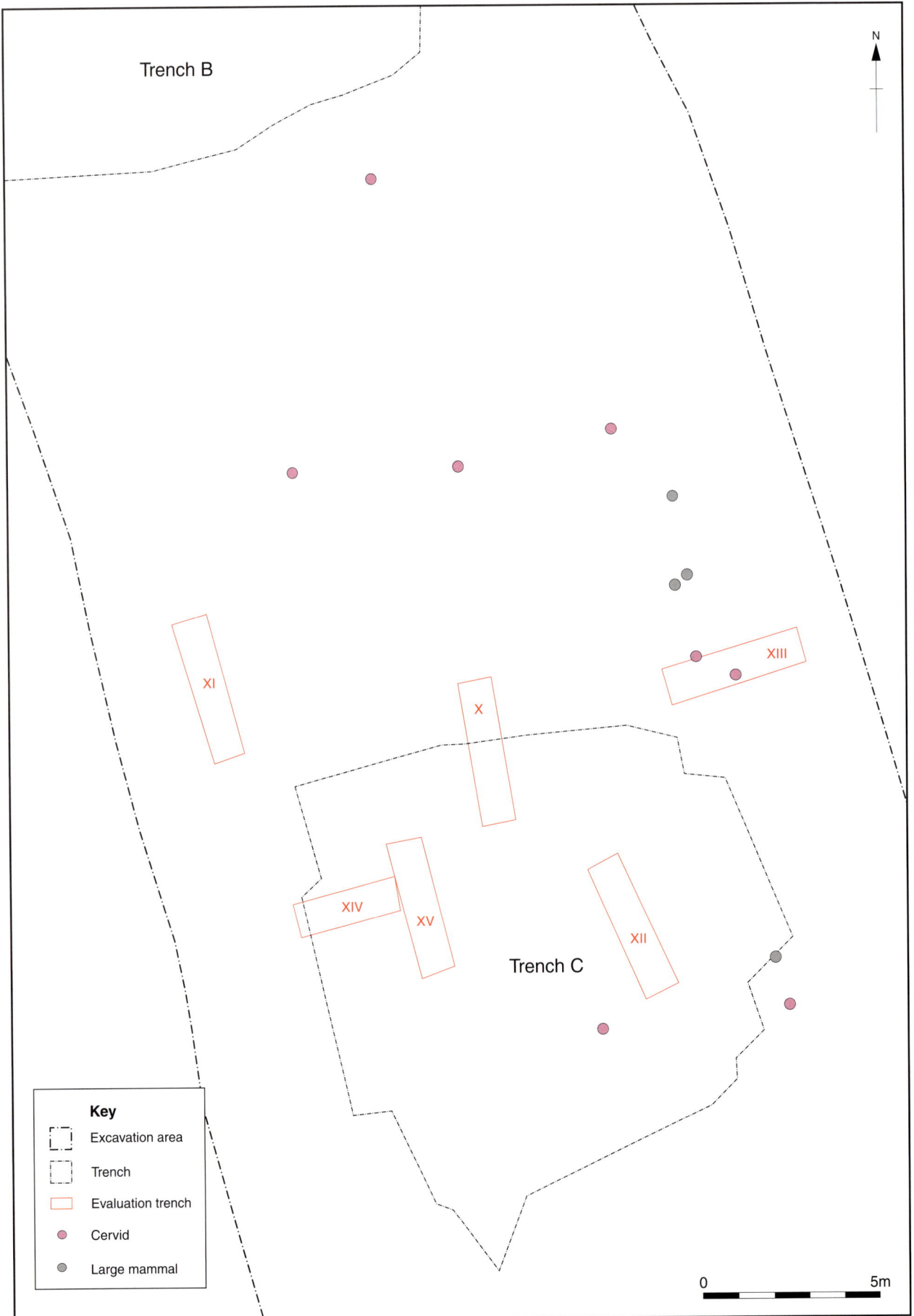

Figure 7.4(b) Distribution of large mammal bones in Phase 6a: context 40103

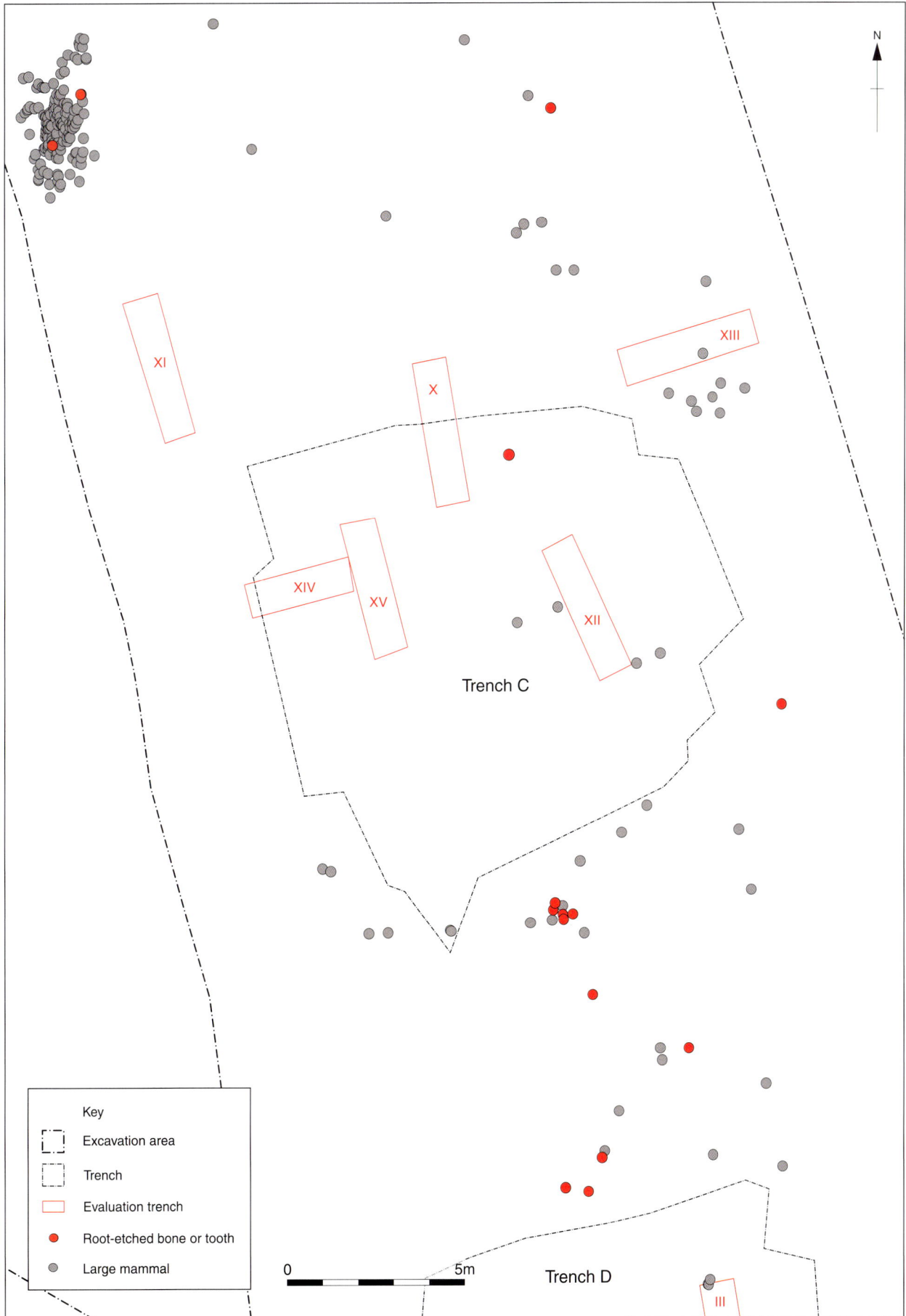

Figure 7.5 Distribution of root-etched bones from context 40039

Figure 7.6 Distribution of large mammal bones from context 40039, showing locations of associated bones from the same individual. For detailed plots see Figs 7.7–7.10

unit (40039) in this sequence is characterised by iron staining and micromorphological features that indicate woodland soil formation. This unit is overlain by mottled clay (interpreted as a flood deposit), which has similarly indications of pedogenesis (Chapter 5 and Appendix 7).

Large mammals are infrequent in context 40103 (NISP = 13), but the assemblage from 40039 is more substantial (NISP = 234) (Table 7.5). Small vertebrates occur in nearly equal numbers in both units (context 40039: NISP = 283; context 40103: NISP = 276). The assemblages from the two contexts are described separately below and the spatial distribution of the large mammal bones is plotted in Figure 7.4a-b.

Bone material from context 40039 is poorly preserved due to decalcification and compaction of the sediments and in general is hard to identify (27.5% identified to taxon). Most fragments are smaller than 50mm in length (68.2%) and often disintegrating and corroded with pitted surfaces as a result of chemical leaching of bone. These corrosive processes have removed much of the outer cortical bone in many specimens (only 40% have more than 75% of the original outer surface preserved). In these cases, all but the most recent stages of surface modifications will have been removed by post-depositional corrosion and rooting, which has resulted in pitting of the bone surfaces. There are, however, indica-

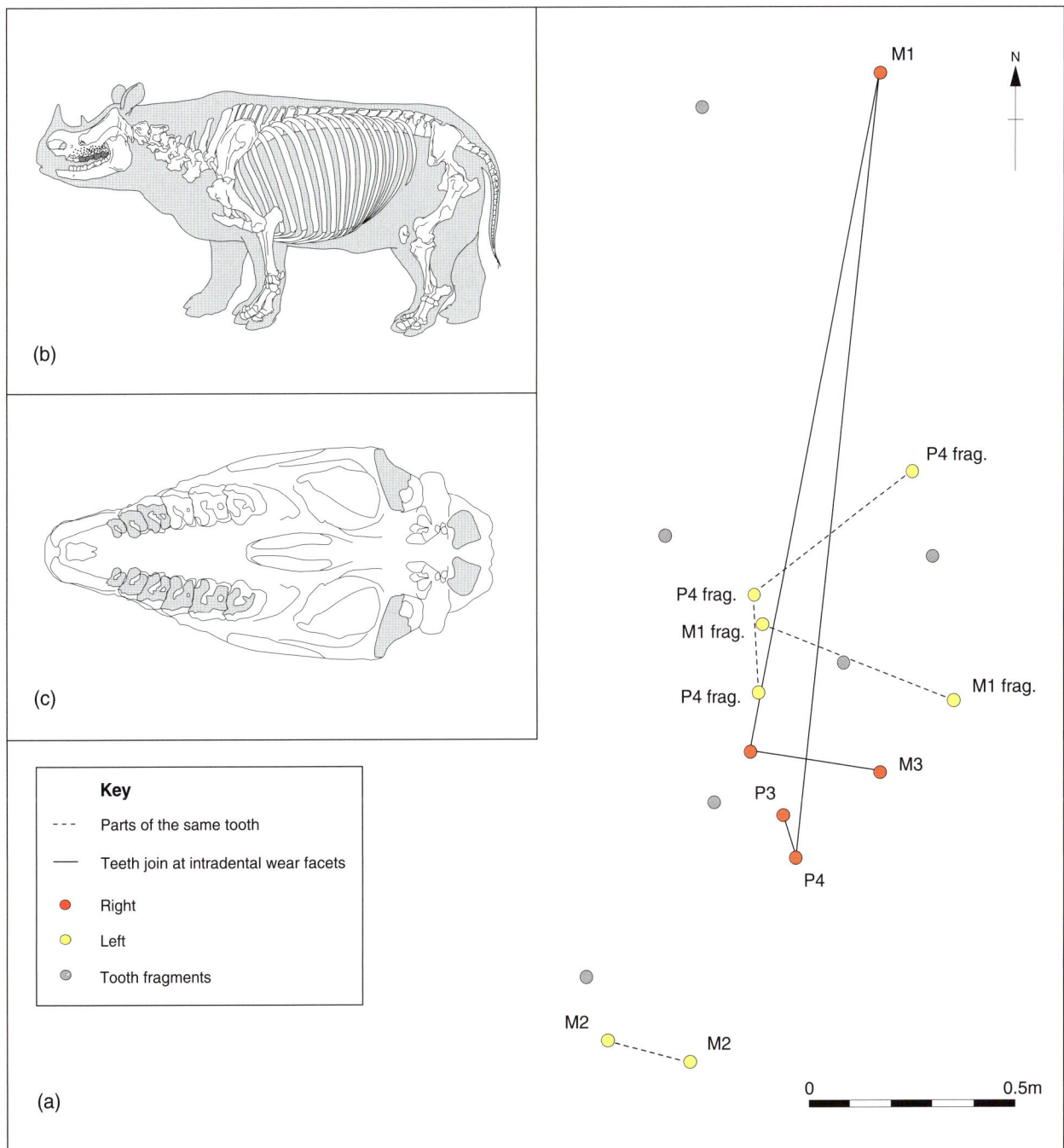

Figure 7.7 (a) distribution of rhinoceros teeth and refits; (b) outline of rhinoceros skeleton with pieces recovered (shaded) from the 'rhinoceros scatter'; (c) outline of rhinoceros skull (occlusal view), showing associated teeth and tooth fragments (shaded)

tions that the bones were modified by carnivores (1.7%) and exposed to weathering (25.1%). Root-etching (6.9%) indicates vegetation cover and burial in a soil (Fig. 7.5). The colour of most of the bones is reddish brown (iron-stained), but colour ranges from white to dark brown. Over half of the bones are iron-stained, but less than one percent have any surface deposits of manganese.

In some cases it has been possible to identify refitting and associated skeletal remains from the same individual. There are four groups of spatially and stratigraphically associated remains (Fig. 7.6). These are a crushed and poorly preserved cranium of narrow-nosed rhinoceros *Stephanorhinus hemitoechus* with most of the upper dentition, two articulating bovid foot bones, the hind-limb bones of a fallow deer and parts of a broken fallow deer antler. The identifiable parts of the rhinoceros skull include the crushed left maxilla with associated M^2 and M^3, an occipital condyle, both perotic bones and several complete and broken cheek teeth (Fig. 7.7 a-c). The associated scatter of comminuted bones includes pieces that are almost certainly parts of the rhinoceros cranium, but they are too fragmentary to identify with certainty. A rhinoceros lunate was found 5m to the east, but it is not possible to say whether this bone is from the same individual. The other cranial pieces and teeth form a dispersed scatter covering about 7 m², (Figs 7.7a and 7.8) and the longest refit distance is 0.5m. The state of wear of the teeth (early-mid wear) implies a prime-aged individual. The cluster also includes the axis vertebra of a red deer, a third upper molar of fallow deer and the tip of a wild boar canine; these are likely to represent part of the background scatter of bones. One curious aspect is breakage of the teeth that occurred before they were buried. This breakage includes chipping and flaking of the enamel, and several of the teeth have been broken into pieces. The latter group includes pieces that can be refitted, but the some of the smaller pieces are missing.

The second group of associated bones includes a metatarsal, tibia and astragalus from the left hind-leg of a fallow deer (Fig. 7.9). The pieces form a tight cluster (*c* 1 m²), with large carnivore gnawing on one of the metatarsal fragments. Studies of carcass decay in natural situations shows that lower hind-limbs are often the last parts to separate into individual bones. This is because the foot bones are tightly bound by tendons and ligaments, which are themselves attached by strong tissue-attachments to the long bones. The third associated group is a cluster of five antler fragments surrounding a frontal bone with the attached basal part of the antler (Fig. 7.10). One of the antler fragments can be positively identified as fallow deer, and the other pieces are not inconsistent with this identification. This cluster covers about 2 m². The final associated group is a bovid left scaphoid that articulates with the lunate; these bones are 15m apart (Fig. 7.6).

The associated groups from context 40039 capture moments in the taphonomic history of the assemblage and help to answer questions about the dispersal and destruction of bones at the site. The main processes operating on the bones were clearly destructive, through the actions of mammalian predators, combined with weathering and trampling. These pre-burial processes acted to disperse and fragment the disarticulated bones. The general low-density of bone finds throughout this unit supports this interpretation and shows that there no significant accumulating agencies (eg, denning carnivores). Post-depositional processes were responsible for further fragmentation and probably also account for the loss of bones through soil corrosion. All of these processes are indicative of taphonomic alterations occurring on a stable land surface.

The small vertebrate faunas of contexts 40103 and 40039 differ from Phase 5 in a number of respects. The most conspicuous difference is the rarity of fish remains, accounting for no more than 1% of the total small vertebrate assemblage in context 40039 and about 4% in context 40103. This is substantially lower than in Phase 5, where fish account for about 46% of the total (Fig. 7.3a). The abrupt decline in fish remains implies that the waterbody had dried-up or that the river had migrated away from the site. The proportion of amphibian (newt, frog and toad) bones increases steadily from context 40039 to 40103 (Fig. 7.3a). This may indicate a return to damper (marshy) conditions, possibly heralding the inundation of the area by flushes and spring-fed streams in Phase 6b.

The small mammal faunas differ from that of Phase 5 in the abundance of woodland taxa (bank vole, wood mouse *Apodemus sylvaticus*) and in the much lower frequency of open-ground and semi-aquatic small mammals (Fig. 7.3b). The later group includes water shrew *Neomys* sp., but the small mole *Talpa minor* would have favoured dry habitats with rich soils supporting abundant invertebrates (earthworms, insects and insect larvae). Areas of herbaceous vegetation would have supported the field/common vole and water vole.

The large mammal fauna is also consistent with the presence of a mixture of woodland and water-edge habitats. Wild boar *Sus scrofa* occurs throughout steppe and broadleaved zones of the Palaearctic. In the western part of its range it favours temperate woodland bordered by open marshy habitats and grassland. It avoids taiga and areas with persistent winter snow cover. Most British Pleistocene records are from interglacials, generally in association with deciduous forest, but aurochs is also recorded from interstadials with open birch woodland (Currant 1986a). In the early Holocene aurochs was present in association with open birch and pine woodland. Fallow deer (found in both 40039 and 40103) is also an indicator species for interglacial conditions. It would have lived in sparse deciduous woodland, grassland and marshes. Deciduous woodland also provides the optimum habitat for red deer, although they can also thrive in open conditions. Open temperate grassland with areas of cover provided by trees or scrub is indicated by the presence of rabbit in context 40039. Finally the narrow-nosed rhino would have grazed in woodland glades or grassland.

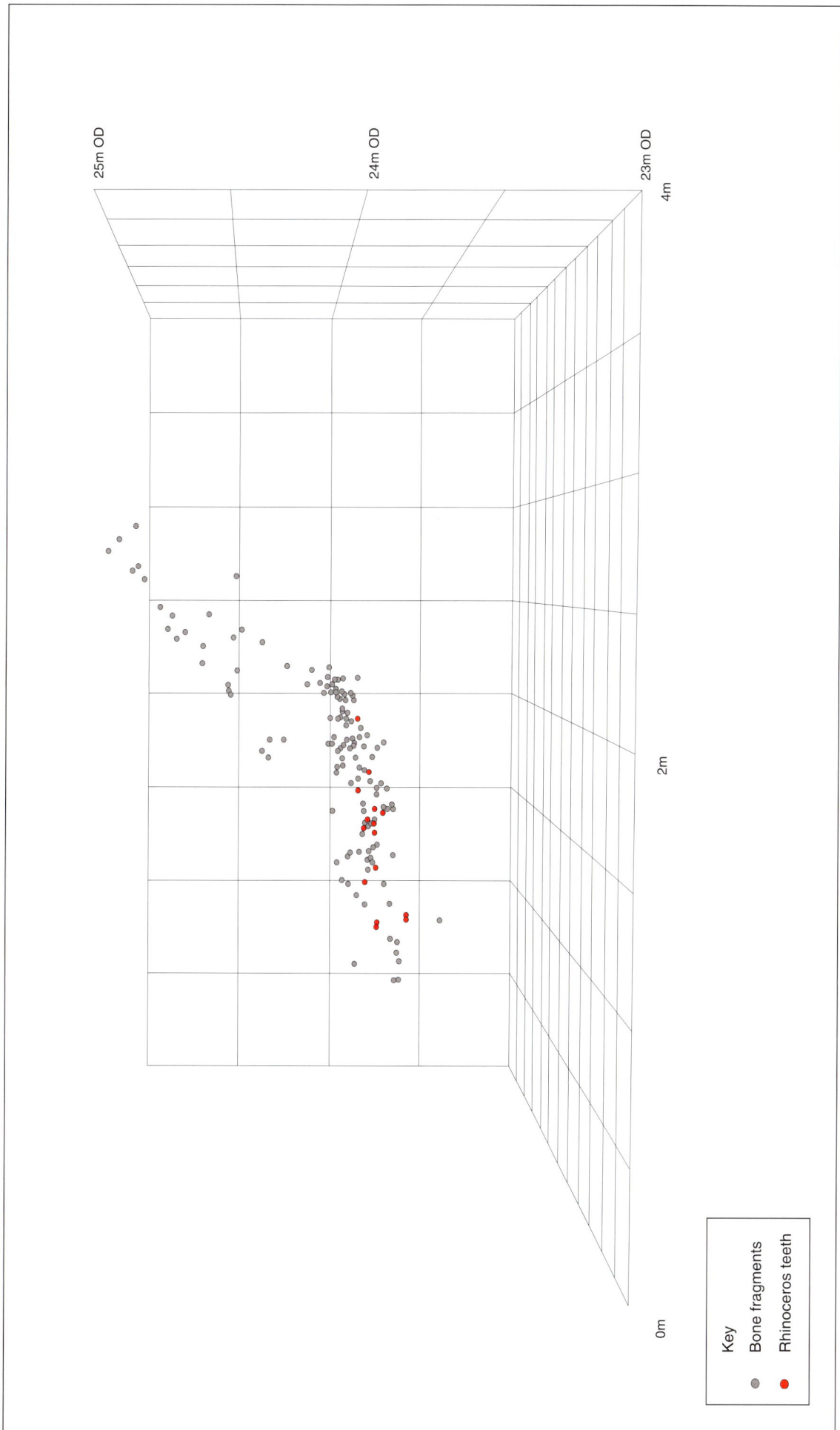

Figure 7.8 3-D plot of rhinoceros teeth (red) and associated bone fragments from context 40039

Key
Bone fragments
Rhinoceros teeth

25m OD
24m OD
23m OD
4m
2m
0m

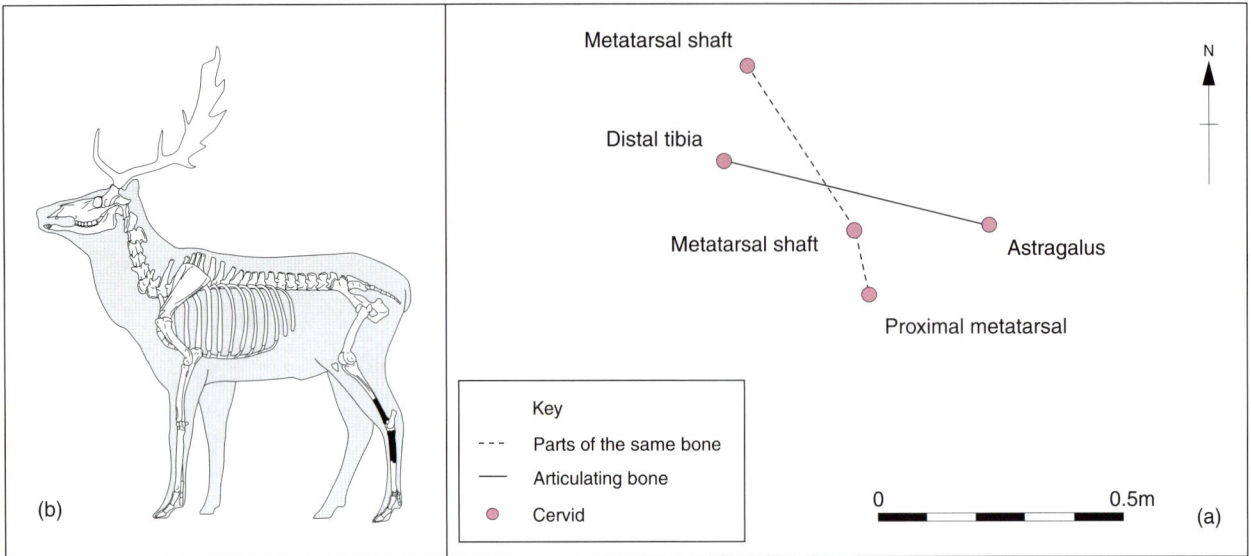

Figure 7.9 Distribution of articulating and refitting hind-limb bones of fallow deer from context 40039. Inset (b) is an outline of a fallow deer skeleton with pieces recovered (shaded)

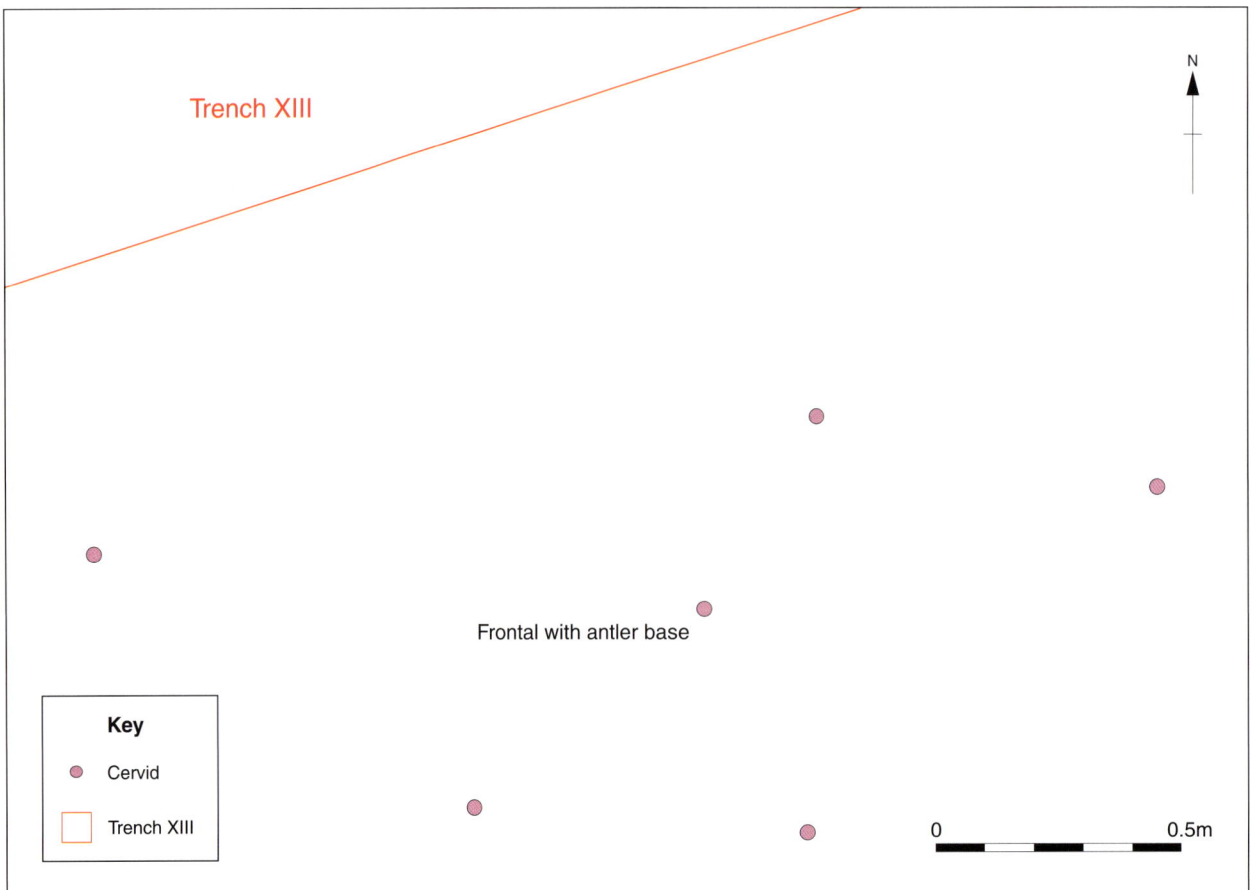

Figure 7.10 Distribution of antler fragments associated with the fallow deer frontal and antler base from context 40039

In conclusion, the mammal fauna indicate a drier more wooded habitat than that of Phase 5. Locally, the environment would have included a mixture of grasslands and woodland. Climatic conditions were probably similar to the present day.

Phase 6b (Contexts: 40070, 40143, 40144)

Immediately above the clays of Phase 6a is a calcareous silt (40070) with tufa oncoliths, rich in shell material and small vertebrates. These sediments fill a linear channel and are overlain by a thin horizon of fine sand (40144), with lenticular patches of white silt (40143) at the top of this succession. As well as being extremely rich in small vertebrate remains, the sediments are also notable for the excellent preservation of most of the vertebrate bones and teeth. Another notable feature of this sequence is the clear ecological succession represented by changes in the composition of the small vertebrates through the sequence. The fact that the same pattern is recorded in different parts of the channel rules out any possibility that the deposit has been mixed or disturbed after deposition.

Context 40070

Although relatively limited in extent, this tufaceous deposit is particularly rich in well-preserved vertebrate fossils (Table 7.6). Large mammal remains are scattered throughout the deposit, with no apparent differentiation (Fig. 7.11). The assemblage consists for the most part of extremely comminuted antlers, but includes teeth and

fragments of limb bones and other postcranial elements (Fig. 7.12). Nearly all of the bone fragments are smaller than 50mm. Bone surfaces are well preserved and cortical bone surfaces are generally intact, with about 60% of the pieces retaining 75% or more of the original outer surface. Surface damage is from a combination of subaerial exposure (5%), root-etching (8.6%) and post-depositional corrosion. In addition, edge-rounding was noted on eight of the pieces, but it was not possible to determine whether this was from fluvial transport or some other taphonomic process. Large mammal bones and teeth are typically purple-brown in colour and about 20% exhibit staining by manganese (10%) or iron (9.8%).

Two bones from the tufa have probable artefact-induced cut marks on them; these are the only specimens with linear striations that could possibly indicate butchery of large mammal carcasses at the site (see Chapter 8).

Table 7.6 Phase 6b. Taxonomic list of large mammals with body-part data and number of fragments (NISP) from contexts 40070 and 40144

Taxon	NISP	Body parts
Context 40070		
Macaca sylvanus	1	1st phalanx
Capreolus capreolus	1	R astraglus
Dama dama & cf. *Dama dama*	2	L antler base (shed), L antler base with brow tine and beam (shed)
Cervus elaphus & cf. *Cervus elaphus*	8	R P^2 (early wear), R M$^{1\,or\,2}$ (early wear), L P$_2$ (early mid-wear), L P$_4$ (mid-late wear), L antler base with brow and bez tines (shed), R lunate, L scaphoid, 2nd phalanx
Cervidae gen. et sp. indet. (red/fallow deer sized)	75	L maxilla with deciduous P$^{3-4}$ (early wear), 3 upper molar frags. (early wear), upper molar frag (mid wear), cheek tooth frag, molar frag, tooth frag, L petrosal, occipital condyle, temporal with antler base, 2 antler bases (shed), 51 antler frags, rib frag, R scapula frag, R scapula frag (distal fused), L scapula frag (dist. fused), R humerus shaft (dist. Fused), R humerus (unfused prox. epiphysis), innominate frag, R tibia shaft, 2 left metatarsal frags, 2nd phalanx (unfused proximal epiphysis)
Cervidae gen. et sp. indet.	14	14 antler frags
Indet. large mammal (rhinoceros or larger)	1	Thoracic vertebra frag
Indet. large mammal (*Bos* or larger)	2	Vertebral centrum (cranial and caudal epiphyses fused), skull frag
Indet. large mammal (red deer or larger)	14	14 indet. bone frags
Indet. large mammal (red/fallow deer size)	4	2 skull frags, rib frag, long bone shaft frag
Indet. large mammal (smaller than fallow deer)	1	Long bone shaft frag
Indet. large mammal	39	39 indet. bone frags
Indet vertebrate	2	2 indet. bone frags
Context 40144		
cf. *Palaeoloxodon antiquus*	1	R cuneiform
Stephanorhinus cf. *hemitoechus*	2	R innominate frag (acetabulum), L innominate frag (pubis)
Capreolus capreolus	1	Metacarpal shaft frag
Cervidae gen. et sp. indet. (red/fallow deer sized)	8	R. deciduous P$_3$ (mid wear), 3 antler bases (shed), R radius (unfused distal epiphysis), L pisiform, R unciform (juvenile), metacarpal shaft frag (juvenile)
Cervidae gen. et sp. indet.	19	19 antler frags
Indet. large mammal (*Bos* or larger)	1	Vertebral centrum (cranial and caudal epiphyses unfused)
Indet. large mammal (red/fallow deer size)	3	Thoracic vertebra frag (cranial and caudal epiphyses unfused), 2 long bone shaft frags
Indet. large mammal (smaller than fallow deer)	2	Thoracic vertebra frag, vertebral centrum (cranial epiphysis unfused)
Indet. large mammal	3	3 indet. bone frags
Indet. vertebrate	1	4 indet. bone frags

Figure 7.11 Distribution of large mammal bones in Phase 6b, context 40070

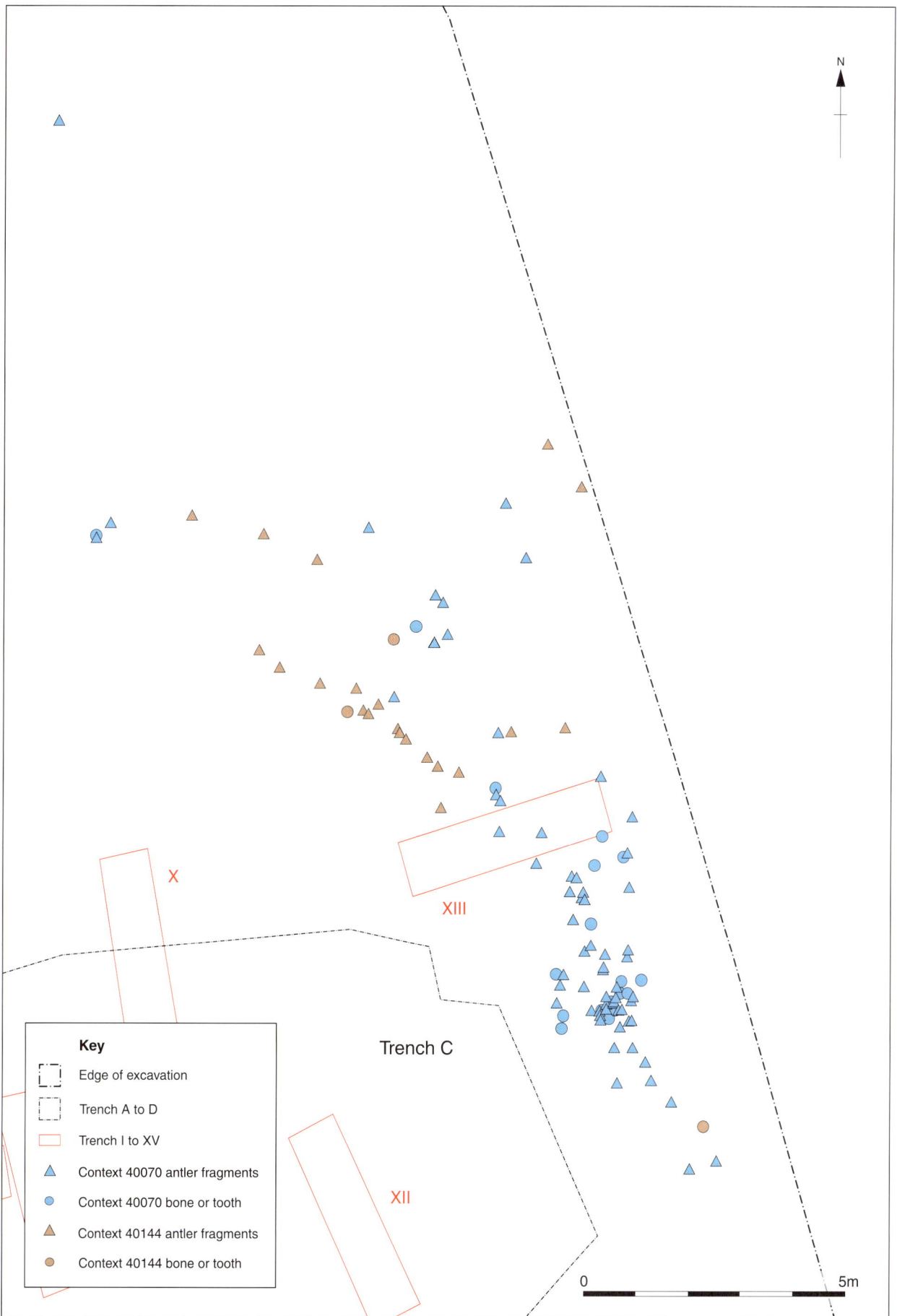

Figure 7.12 Distribution of antler fragments from context 40070 and 40144. No refits were found between vertebrate remains in the channel

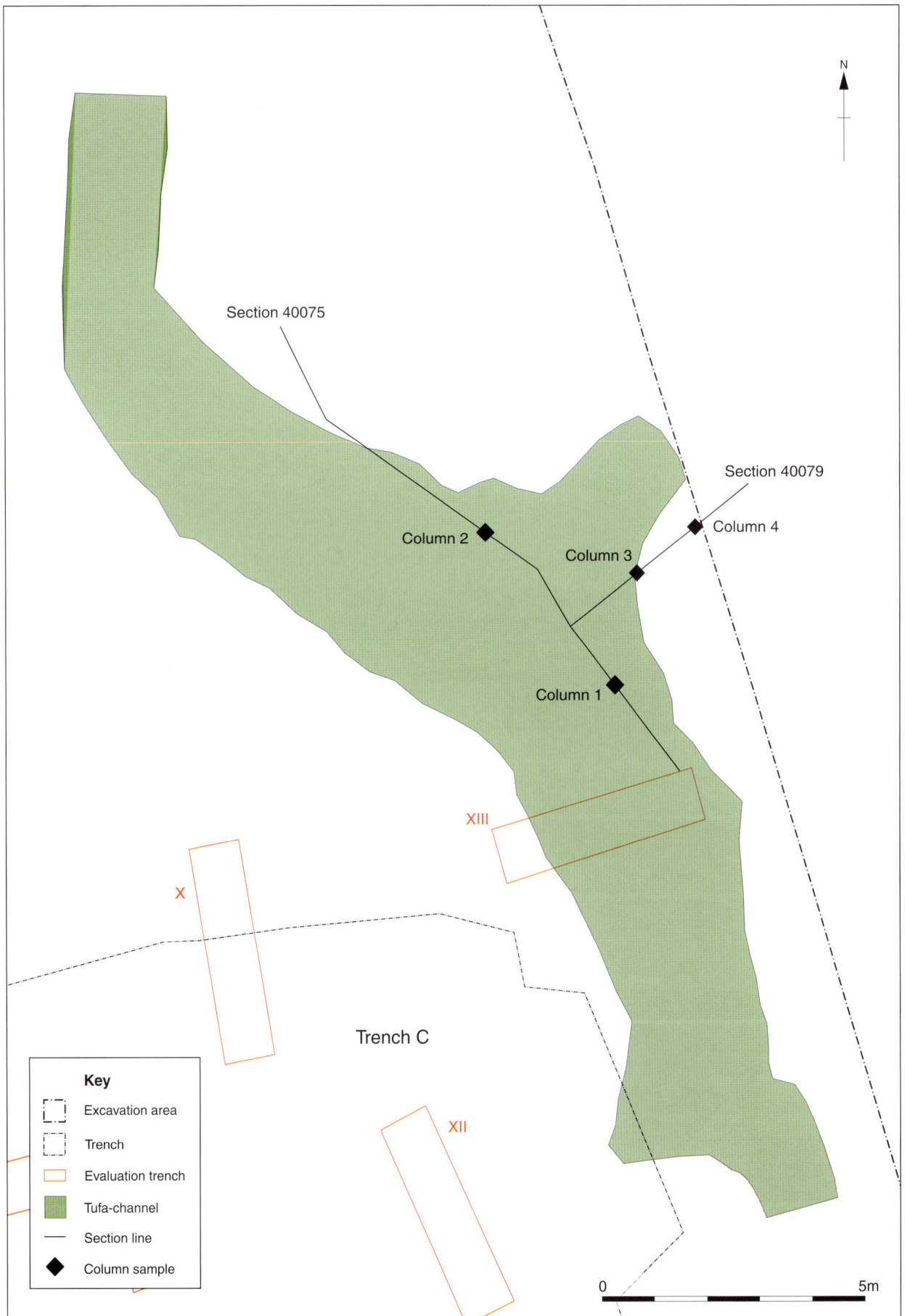

Figure 7.13 Plan of the tufaceous channel (shaded), showing the locations of sections and column samples

Despite the relatively high concentration of large mammal fossils, the hand-excavated large mammal fauna is not particularly diverse (Table 7.1). Of the 100 identifiable specimens, all but one belong to cervids. At least three species are represented, in increasing order of size these are: roe deer (NISP = 1); fallow deer (NISP =2); and red deer (NISP = 8). The remainder cannot be identified to species, but all come from medium-sized deer. In terms of body part representation, most of the cervid remains are highly comminuted pieces of antler beams and tines (n = 70), but five shed antler bases were also identified. The presence of three species of deer is consistent with woodland. Both fallow deer and red deer are to be found in woodland as well as parkland, but roe deer *Capreolus capreolus* is more closely associated with dense woodland. Additional large mammal remains were recovered from wet-sieved samples, including the particularly noteworthy record of macaque *Macaca sylvanus* identified from a single first phalanx. Fossil remains of macaque are exceptionally rare in northern Europe and only one other Hoxnian specimen is known from the Lower Loam at Swanscombe (see Chapter 9). Today, Barbary macaques live in relict populations in North Africa, where they are found among sparse forests of cedar, pine and oak. In the British Pleistocene, they are associated with temperate (deciduous and mixed) woodland phases and local habitats that included a mixture of open woodland and grassland on river floodplains.

The rich small vertebrate faunas from the channel sequence allow a detailed interpretation of the palaeoecology during the infilling of the channel, based on a quantitative study of the faunal succession and taphonomy. Most of the samples analysed were collected from open trench sections from four locations (see Fig. 7.13 for position of samples and Fig. 10.1 for a drawing of the sections showing where the samples were taken). One of the profiles (Column 2) sampled the deepest part of the channel, a second column (Column 1) recovered a somewhat shallower sequence, and spot samples (Columns 3 and 4) were collected from discrete units at the edge of the tufa spread. Samples in Columns 3 and 4 from the margin of the tufa-channel were taken as spot samples from discrete units. Although the spot samples were not taken as a column in a sequence, the samples can nevertheless be arranged in stratigraphical order within each column. The results are summarised as percentage frequency histograms showing the faunal spectra from the four profiles (Figs 7.14-17).

Column 2 (Fig. 7.14)
Six samples were taken at 100mm intervals through the full thickness of the shelly tufaceous sediments in Column 2. Five of the samples were about 28L, but the thinner basal sample was less than half the volume of the other samples. The small vertebrate spectra shows a steady increase in the number of fish remains from about 5% at the base to about 25% in sample <40315>. This is followed by a dramatic decrease in fish remains in the top sample (<40314>). Small mammal abun-

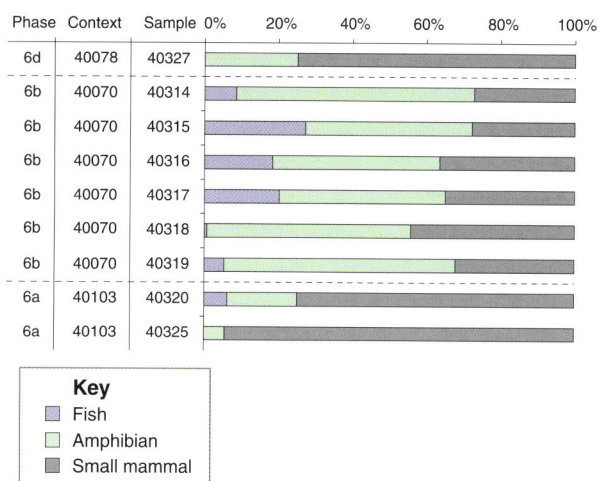

Figure 7.14 Column 2 small-vertebrate diagram

dance decreases slightly through the upper sequence, but amphibians (frogs, toads and newts) are dominant throughout, with the highest number of amphibian bones (*c* 60 %) occurring in the basal and top samples respectively. The overwhelming dominance of amphibians in the top sample, together with the decline in fish remains, argues for a shallowing of the waterbody and a change to more marshy conditions.

Column 1 (Fig. 7.15)
Four samples (from 3 to 20L) were taken at 100mm intervals through shelly tufaceous sediments in Column 1. The sampling included a less tufaceous, sandy/clayey deposit (assigned to context 40144) near the top of the sequence. The percentage of fish in the basal samples is low but rises through the sequence to maximum of about 35%. The representation of amphibians is constant throughout the sequence (*c* 35%), but small mammals show a regular fall in successive samples from 50% at the base to 30% at the top of the profile.

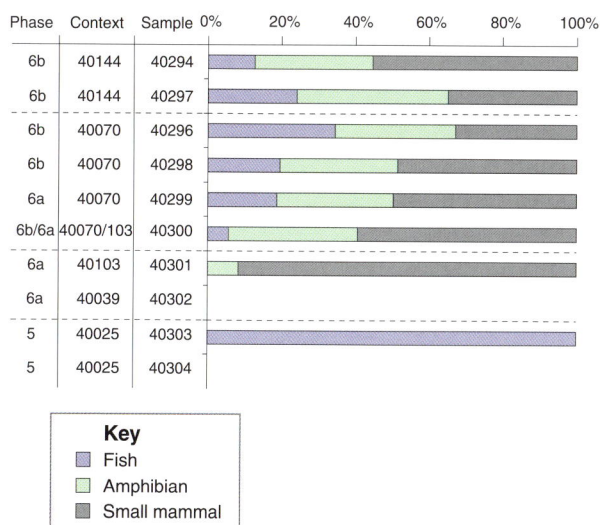

Figure 7.15 Column 1 small-vertebrate diagram. Sampling of the tufa respected the sedimentary boundaries; however contortion and inclusion of discontinuous lenses of clay made some mixing unavoidable

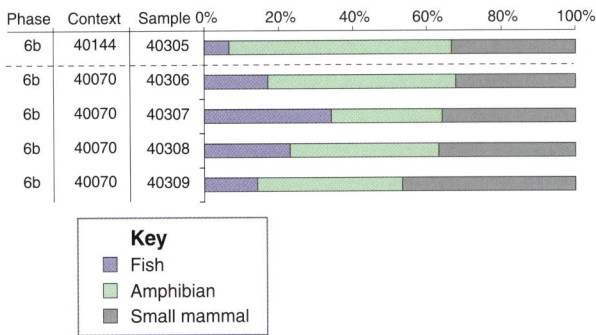

Figure 7.16 Column 3 small-vertebrate diagram

Column 3 (Fig. 7.16)
Four samples (11 to 29L) were collected from successive tufa units. These minor facies differences were identified in the field from differences in particle size (coarser at base) and colour. The faunal succession is similar to that of Column 2, with the top sample being characterised by higher numbers of amphibian remains and a relatively low representation of fish remains.

Column 4 (Fig. 7.17)
The tufa samples that make up Column 4 come from a much shallower (marginal) succession than those from the other columns and they are consequently limited to a two units: a clay-rich tufa (sample <40313>, 10L) and an overlying deposit of fine sandy tufa (sample <40310>, 7L). The spectrum of fish, amphibians and small mammals in sample <40313> resembles that of sample <40308> in Column 3, but <40310> differs from the preceding level in the unusually high representation of fish remains, which now account for over 60% of the small vertebrate assemblage.

It is clear from comparison of the columns that they illustrate approximately the same faunal trends. As has been suggested above, there are two main phases; in the first there is a convincing increase in the frequency of fish at the same time as a decrease in the frequency of small mammal remains. The second phase is characterised by low numbers of fish and small mammals and shift to a dominance of amphibians in Columns 2 and 3. The environment during the first phase is clearly one of a small stream, which was followed by a shallowing phase and the development of marshy conditions.

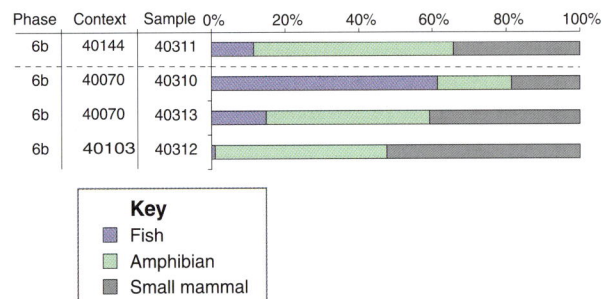

Figure 7.17 Column 4 small-vertebrate diagram

The diversity and abundance of amphibians, which includes newts, toads and frogs, indicates the prevalence of aquatic habitats bordered by marsh. A notable record is tree frog (*Hyla* sp.), which is represented by an ilium from sample <40332>. Two species of tree frog (common tree frog *H. arborea* and stripeless tree frog *H. meridionalis*) occur in Europe; both species have been recorded from the British Pleistocene (Parfitt 2008). European tree frogs favour well-vegetated habitats, preferring areas with trees and bushes or reed-beds. A waterbody with still or slow-flowing water and shallows with good weed growth would have provided a suitable breeding site for the newts, toads and frogs. The predominance of small fishes, together with the presence of minnow *Phoxinus phoxinus*, indicates a smaller and better-oxygenated water body than that represented by context 40025. No estuarine fishes were identified and the abundance of amphibians indicates freshwater conditions.

While the fish and amphibian remains are from animals that lived and died in the immediate vicinity of the site, the small mammal remains were mostly accumulated by avian and mammalian predators and were probably captured from a wider area around the channel. A clear indication of mammalian predator involvement in the accumulation of the assemblage is provided by gastric etching, small tooth punctures and gnawing marks on several of the amphibian and small mammal bones. These traces show that the bones had passed through the gut of a carnivore and were deposited in scats (droppings) close to the depositional site. Indeed, small mustelids are present in the faunal assemblage from context 40070, with polecat *Mustela* cf. *putorius* represented by a mandible and by isolated teeth from other channel samples. The habitat envisaged for the Phase 6b would have been ideally suited for polecat. Today, polecats favour woodland, particularly in damp areas where they can find the frogs that form a large part of their diet. Small rodents, rabbits and birds also form an important part of their diet. Like most carnivores, mustelids mark their territory with faeces (scats) left in prominent places (known as latrines). The scats contain the indigestible prey remains and considerable quantities of bone can accumulate at latrine sites. The bones were probably washed into the channel from both scats and owl pellets that were deposited near the margins of the waterbody.

The small mammal fauna is the most diverse from the site, with at least 18 taxa represented. The insectivore assemblage is remarkable for the high diversity of shrews. This includes at least four species of *Sorex* from the tufa alone. In addition to pygmy shrew *Sorex minutus* and water shrew *Neomys* sp., there is at least one more (and possibly two) medium-sized species, as well as a much larger shrew that is indistinguishable from the Early-Middle Pleistocene *Sorex* (*Drepanosorex*). Amongst the medium-sized forms, mandibles and isolated teeth of a relatively smaller species are common; this form resembles Laxmann's shrew *S. caecutiens*, the most common soricid found at many Middle Pleistocene sites in Britain. Remains of a somewhat larger species resemble common shrew *S. araneus*, but these are infrequent.

Figure 7.18 Distribution of bones from context 40144

In terms of ecology, the small mammal assemblage with abundant bank voles and wood mice is indicative of dense ground-level vegetation, shrubs and woodland, an interpretation which is corroborated by evidence from molluscs (see Chapter 12). There is also a significant open-ground component, indicated by grassland voles such a field vole *Microtus agrestis* and northern vole. Mammals of waterside habitats, including water vole and water shrew, are notably common. Of particular ecological significance is squirrel (*Sciurus* sp.). Squirrels are semi-arboreal rodents that prefer large expanses of mature woodland with continuous canopy. Coniferous woodland is favoured, but they are also found in deciduous woodland (especially beech) or more sparsely wooded areas when numerous (Corbet and Southern 1977). None of the squirrel teeth are waterworn, nor do they show signs of digestion, thereby suggesting that squirrels were living in trees close to the depositional site. The presence of Daubenton's bat *Myotis daubentonii* is also significant as it is closely associated with open water, where it hunts for insects. Daubenton's bat prefers open woodland and would have roosted in hollow trees, since there are no rocky outcrops near the site. The climate was probably as warm, if not warmer, than the present day.

Context 40144 and 40143

The large mammal remains from context 40144 are again dominated by antler fragments (Table 7.6), 16 (72.2%) of which are smaller than 50mm in length. Rhinoceros was identified from two pelvis fragments and straight-tusked elephant from a cuneiform (Fig.7.18). This foot bone is almost certainly part of the elephant carpus, represented by other carpal bones, metapodials and phalanx, found near the western edge of the channel (see Chapter 8). The presence of the cuneiform in (40144) implies that the tufaceous deposits were deposited contemporaneously with clay-silts (40078) that contained the dispersed bones of the straight-tusked elephant. No identifiable large mammal bones were found in the thin white silty layer (40143), but small vertebrates were common in the bulk samples from both contexts. Although all of the small mammal taxa found in contexts 40144 and 40143 were also present in context 40070, there is a greater representation of woodland small mammals in the upper part of the succession.

The small mammal faunas from Phase 6b differ from those of Phase 6a in the nearly equal numbers of woodland and humid grassland animals and the higher numbers of semi-aquatic small mammals (Fig. 7.3b). Overall, the vertebrate evidence is consistent with initial development of a small shallow body of slow-flowing freshwater bordered by open herbaceous vegetation (possibly reeds with grassland patches) in an otherwise forest-dominated landscape. At the top of the sequence, there are indications for drying up of the waterbody, and local development of more marshy conditions. Although the contribution of aquatic and semi-aquatic taxa is rather high, the local area must have included areas of well-drained soils to support both rabbit and mole. The vertebrate fauna does not provide any indication that the climate was significantly different to that of today.

The bird remains *by John R. Stewart*

The bird remains from the site are not very numerous and all come from tufaceous context 40070, Phase 6b (see Appendix 10). Specimens that have been identified to any significant taxonomic level include two bones of a thrush that conform in size with those of song thrush *Turdus philomelos* or redwing *T. iliacus* and the remains of mallard-sized and smaller ducks. As for the rest of the bird remains, they largely represent undetermined small species, most likely various passerines (songbirds) ranging from blackbird to smaller than house sparrow in size.

The presence of song thrush would indicate boreal to warm temperate woodland, whereas redwing lives in more open wooded environments during the breeding season. Areas of woodland or scrub may also be inferred from the relatively high number of passerine remains, as song birds are often abundant in fossil assemblages from such closed environments (Harrison and Stewart 1999). Another, albeit less numerous component of the assemblage are waterbirds, represented by at least two species of duck. The presence of ducks supports the sedimentology and associated molluscan evidence for deposition in slow-flowing freshwater.

The palaeoecological indications provided by the Southfleet Road bird assemblage are therefore consistent with a freshwater course running through temperate woodland. This assemblage is similar in many respects to that from Barnham (Stewart 1998), where bird bones were recovered from waterlain deposits, dating to the same (Hoxnian) interglacial.

Phase 6 (contexts 40099, 40160, 40162)

Three contexts from the lower part of the grey silty clay in the central part of the site yielded large vertebrate material. All of the material is fragmentary and none could be identified to species or genus (Table 7.7). The samples are so small it is not possible to draw any conclusions concerning taphonomy and environment.

Phase 6 (contexts 40078, 40100, 40158)

The main fossil horizon at Southfleet Road is located in the middle-upper part of the grey clay sequence. Stone tools are abundant in this horizon and include many refitting pieces associated with the scattered and poorly-preserved bones and teeth of a straight-tusked elephant. The sediment containing the elephant remains has a blocky ('brecciated') structure with prominent dark brown 'humic' horizons. The brecciation is indicative of cycles of desiccation and wetting, causing the sediment to expand and contract; this is the principal reason for the fractured condition of the elephant bones. In the field, bones were mostly assigned to two contexts: 40078 for those associated with the dense cluster of elephant remains (Fig. 7.19; Table 7.8), and 40100 for finds from the surrounding area of grey clay (Fig. 7.20; Table 7.9). Although distinguished by different context numbers,

Table 7.7 Phase 6 (context 40158, 4099, 40160, 40162). Taxonomic list of large mammals with body-part data and number of fragments (NISP)

Taxon	NISP	Body parts
Context 40158		
Cervidae gen. et sp. indet. (red-fallow deer sized)	2	Antler frag (beam & tine), metapodial shaft frag
Bovidae gen. et sp. indet.	1	R tibia (distal & shaft, dist. fused)
Indet. large mammal (red deer or larger)	1	Indet. bone frag
Indet. large mammal (smaller than fallow deer)	1	Long bone shaft frag
Indet. large mammal	1	Indet. bone frag
Context 40099		
Indet. large mammal (probably elephant)	1	Indet. bone frag
Indet. large mammal	1	Indet. bone frag
Context 40160		
Cervidae gen. et sp. indet. (red-fallow deer sized)	2	Antler base frag (shed), antler tine
Indet. large mammal (red-fallow deer size)	1	Carpal/tarsal frag
Context 40162		
Cervidae gen. et sp. indet. (red-fallow deer sized)	3	Antler tine, antler frag, long bone shaft frag
Indet. large mammal (rhino or larger)	1	Indet. bone frag
Indet. large mammal	1	Indet. bone frag

Table 7.8 Phase 6 (context 40078). Taxonomic list of large mammals with body-part data and number of fragments (NISP)

Taxon	NISP	Body parts
Palaeoloxodon antiquus	2	L & R M³
Elephantidae gen. et sp. indet.	61	29 tusk frags (2 large pieces of L & R tusk), 4 cervical vertebra frag, lumbar vertebra centrum (cranial epiphysis unfused, caudal epiphysis fused), 4 thoracic vertebrae (1 with cranial and caudal epiphyses unfused), lumbar vertebra (cranial epiphysis unfused, caudal epiphysis fusing), 6 vertebra frags, 4 rib frags, 2 L scapula frags, R magnum, R trapezium, R unciform, R metacarpal II, R metacarpal IV, metacarpal frag, metapodial frag, 2nd phalanx IV
Stephanorhinus cf. *hemitoechus*	3	L scapula frag (dist. fused), L humerus (dist. & shaft, fused), L femur (prox. fused)
Stephanorhinus kirchbergensis	1	L ulna (dist.)
Cervus elaphus & cf. *Cervus elaphus*		Antler base. brow & bez tines, beam (shed), R antler base, brow & bez tines, beam (shed), L antler base, brow & bez tines, beam (shed), L antler (shed), L antler trez tine & beam, L tibia frag (dist & shaft, dist fused)
Cervidae gen. et sp. indet. (red/fallow deer sized)	6	R mandible frag with M₃ (mid wear), lower molar frag, molar frag, lower frag, cheek tooth frag, 6 antler base frags (shed), cheek tooth antler base & tine frags (shed), 41 antler frags, vertebral centrum frag, L radius (prox. & shaft, prox fused), R metacarpal (prox. & shaft), L metacarpal (dist, fused), L metacarpal (dist unfused), L metacarpal (prox & shaft), metacarpal shaft frag, metacarpal shaft frag (juvenile), L innominate (unfused), L femur (dist. & shaft), R tibia (prox & dist fused), R tibia frag (dist & shaft, fused), L tibia frag, metatarsal frag, L calcaneus frag (*tuber calcis* unfused)
Cervidae gen. et sp. indet.	2	2 antler frags
Bos primigenius	1	R metatarsal (prox.)
Bovidae gen. et sp. indet.	2	Metacarpal (dist. fused), metapodial condyle
Indet. large mammal (elephant size)	169	4 ?skull frags, 2 vertebra frags, 7 rib frags, 156 indet. bone frags
Indet. large mammal (rhinoceros or larger)	1	Indet. bone frag
Indet. large mammal (*Bos* or larger)	45	Skull & tooth frag, ? vertebra, carpal/tarsal frag, 42 indet. bone frag
Indet. large mammal (red deer or larger)	11	Rib frag, long bone shaft frag, 9 indet. bone frags
Indet. large mammal (red/fallow deer size)	13	Vertebra frag, rib frag, L femur frag (dist. & shaft, dist. fused), 10 long bone shaft frags
Indet. large mammal (smaller than fallow deer)	2	Rib frag, long bone shaft frag
Indet. large mammal	107	103 indet. bone frags. ? skull frag, ? rib frag, 2 long bone shaft frags
Indet. vertebrate	1	Indet. bone frag

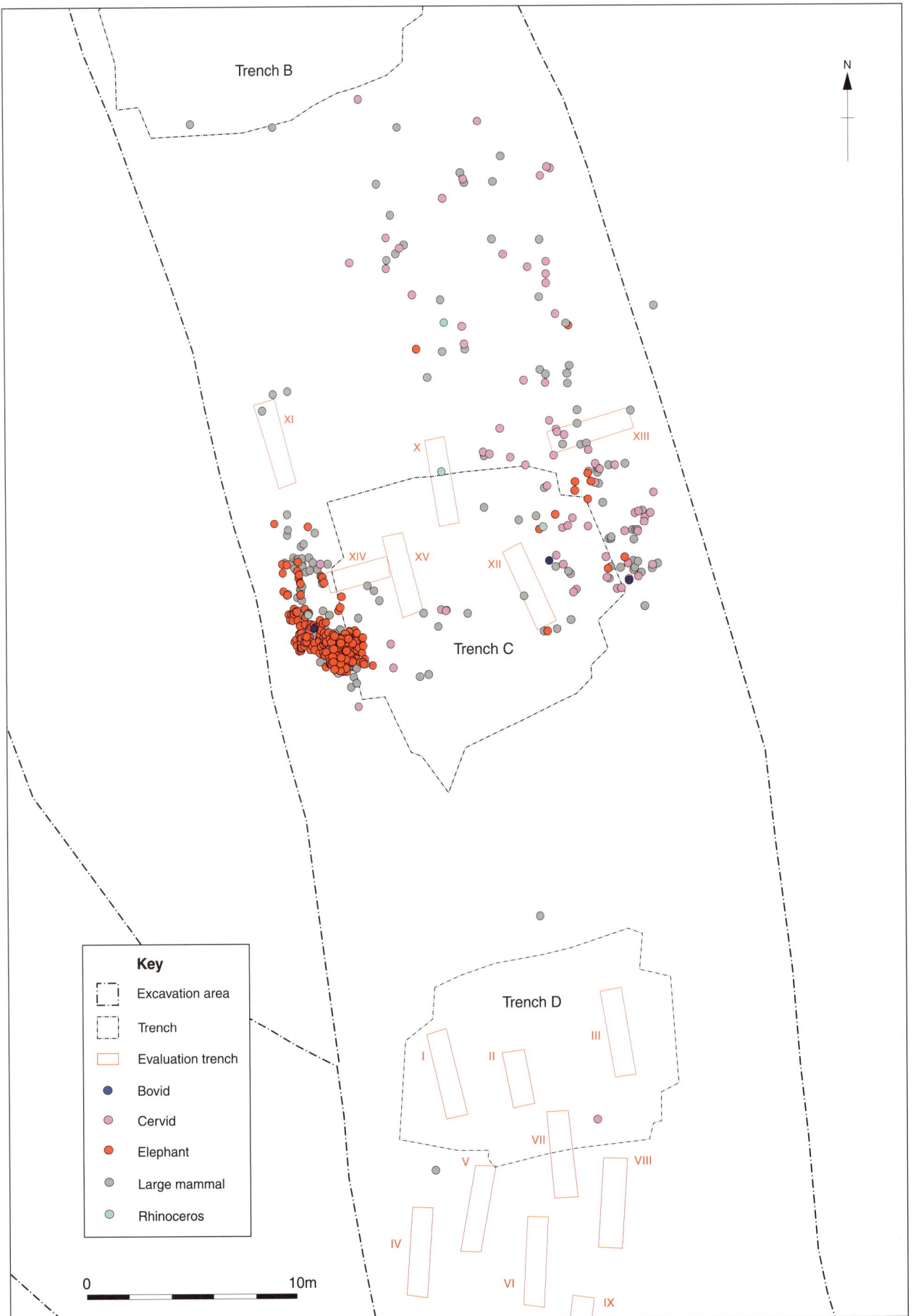

Figure 7.19 Distribution of bones from context 40078

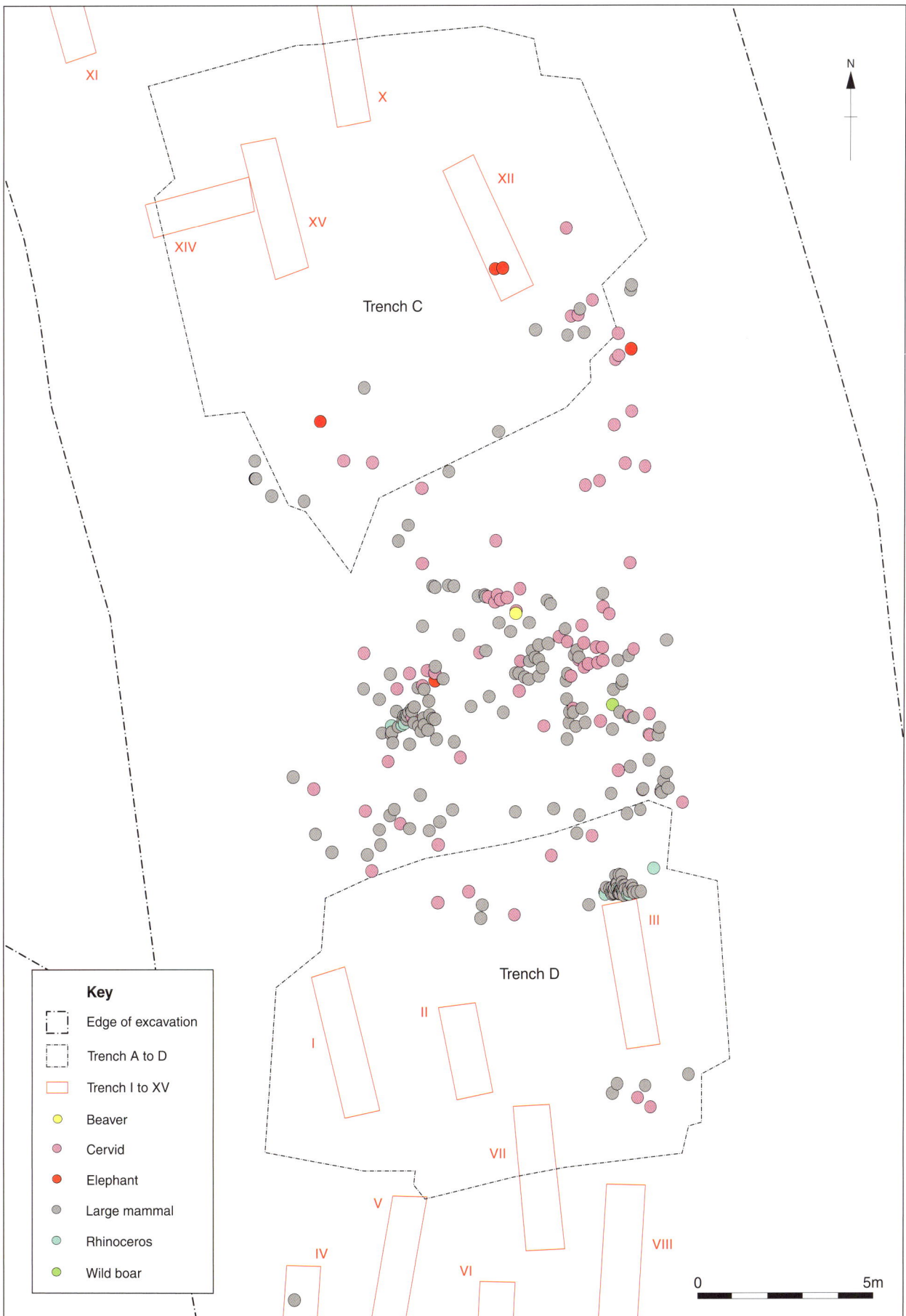

Figure 7.20 Distribution of bones from context 40100

Figure 7.21 Distribution of bones from contexts 40100 and 40078 according to condition. Condition ranges from excellent (1) to extremely poor (4)

40078 and 40100 were not stratigraphically distinct and the two assemblages have been combined in some analyses. There are, however, important differences between the two assemblages, not only in species composition but also in taphonomy, which appear to reflect contrasting depositional conditions. Bulk samples were processed for small vertebrates, but these yielded virtually no bones. Some identifiable small mammal material was recovered by hand excavation, but the sample is scanty and it is not possible to draw any conclusions concerning taphonomy and environment from these remains.

Clear differences between the two large mammal assemblages are in fragment size. The bones from 4100 are mostly smaller than 50mm in length, whereas those from 40078 include 17 bones that are larger than 300mm; only 29% are smaller than 50mm. These differences reflect the predominance of elephant bones in 40078, whereas smaller herbivores form the greater part of the assemblage from 40100. The bones in 40078 are mid- to dark-brown in colour. About twice as many are manganese (5%) stained than are iron stained (10.3%). In contrast more of the bones from 4100 are stained (16.8 % iron stained, 5.9% manganese stained), but this does not account for the lighter colour (fawn to white) of the bones from this context. Overall, surfaces of the bones are better preserved in 40078, with 68% bones having more than three quarters of the original cortical surface preserved, compared with only 35.5% in 40100 (Fig. 7.21). Only four specimens from both contexts are weathered, indicating that most of the bones were either buried quickly or protected from weathering by dense vegetation. Fewer than 2% of the bones from 40078 are root-marked compared with 27% of the bones from 40100 (Fig. 7.22). The high proportion of root-marked bones from the latter context indicates burial in a soil; the distribution of root-etched pieces across the site is likely to reflect the distribution of vegetation shortly after the bones were buried. The low number of root-etched elephant bones may indicate that they were buried in an aquatic environment, whereas those from 40100 were buried in a vegetated terrestrial setting.

The most notable component of the large mammal assemblage from this phase is the dispersed skeletal remains of a large straight-tusked elephant from context 40078 (see Chapter 8). This individual is represented by as many as 237 bones, only 78 of which could be identified to skeletal element. There is a substantial concentration of fragmentary and crushed elephant bones located near the truncated western edge of the site. This cluster is associated with a spatially discrete scatter of flint artefacts; the association of the bones and artefacts in the same horizon, together with limited vertical dispersal, suggests contemporaneity. Almost all of the bones in the cluster are from the axial skeleton and include fragments of the cranium (skull, third upper molar, two large tusks and many tusk fragments), vertebrae (cervical, thoracic and lumbar) and ribs. The spatial disposition of these elements suggests the elephant carcass was lying on its left-hand side, with its head to the south. Many of the smaller bones were probably buried rather rapidly, but weathering and the poor condition of the larger bones and tusks suggest that they had been exposed to prolonged subaerial weathering. An indication of burial processes is provided by the preferred orientation of the smaller bone fragments, which have a strong east-west alignment. The fact that easily transported bones, such as vertebrae, are present suggests that slow-flowing water, possibly slopewash was responsible for depositing the silt in which the bones were buried. The bones were probably subjected several times to flowing water that orientated the bones to the direction of flow without transporting them away from the carcass site. The only appendicular element identified in the main scatter is the left scapula represented by two refitting pieces. Parts of the appendicular elements may have been destroyed during the road construction work, which truncated the western part of the scatter, while other elements may have been transported away from this cluster. This is indicated by the diffuse scatter of elephant remains that extends to the north and east of the main cluster of elephant bones. This scatter includes parts of the left forefoot located some 20–30m to the north-east of the cluster and at the same horizon. Some of these elements can be articulated and their spatial distribution shows that they have been dispersed for short distances in a north-south direction. The elements include carpals, metapodials and a phalanx, which are all light cancellous bones that are easily transported by flowing water. These bones must have been transported to the edge of the tufa-channel before the foot had decayed. This may be explained by the natural sequence of decay in mammalian carcasses, in which the major forelimb elements detach from the body early in the decay sequence. The foot bones, which are held together by strong ligaments and tendons, could have been transported as a unit by a large carnivore or by an inquisitive elephant. As none of the bones are gnawed, it seems more likely that elephants were responsible for displacing some of the bones. African elephants react strongly to the carcasses of other elephants and will carry or kick the bones, sometimes smashing them in the process. Pleistocene elephants probably displayed similar behaviour. Although this is a supposition, the presence of trampling marks on several of the elephant bones in the main cluster suggest that the carcass may have been visited by elephants.

Although the elephant bones and tusks are the most conspicuous element of the large mammal fauna, the deposit has yielded a relatively high diversity of other large mammal species, each represented by a small number of identifiable bones. Ungulates are dominant in the assemblage and point to a rich and varied environment. Some of the taxa are important indicators of the existence of very specific habitats. For example Merck's rhinoceros *Stephanorhinus kirchbergensis* implies a wooded environment, whereas the narrow-nosed rhinoceros would have inhabited more open habitats with rich herbaceous vegetation. Wild boar is generally associated with denser woodland or shrubs and marshy environments. A mixture of woodland and grassland would have favoured red deer,

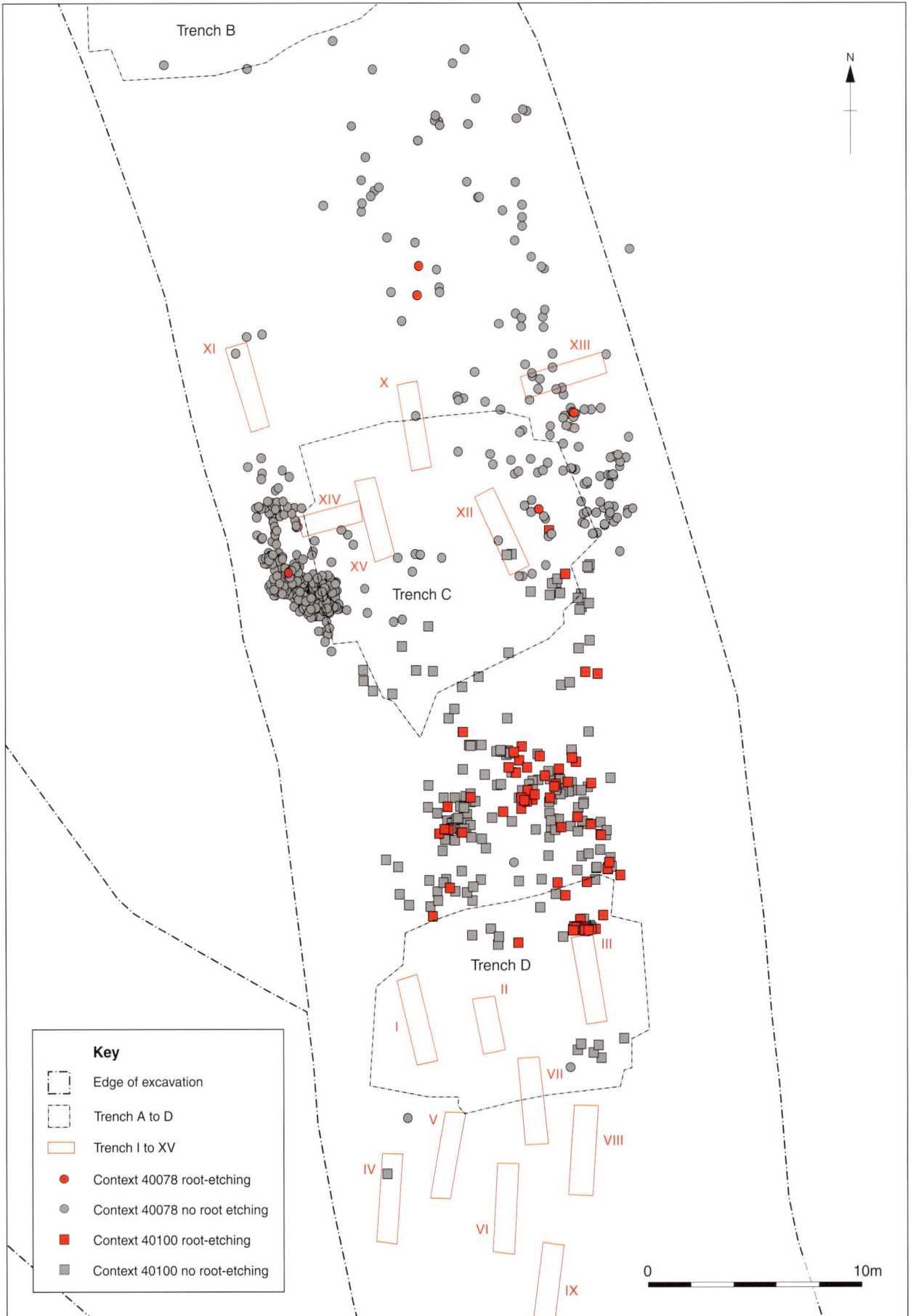

Figure 7.22 Distribution of root-etched bones from contexts 40100 and 40078

Table 7.9 Phase 6 (context 40100). Taxonomic list of large mammals with body-part data and number of fragments (NISP)

Taxon	NISP	Body parts
Castor fiber	1	R P^4
Stephanorhinus hemitoechus & S. cf. hemitoechus	3	3 L radius frags (2 distal, 1 prox. & shaft)
Stephanorhinus kirchbergensis & S. cf. kirchbergensis	6	2 mandible frags, M$_3$ (mid-late wear), R lower cheek tooth & mandible frags, cheek tooth frag, tooth frag
Sus scrofa	1	R I^2 (mid wear)
Dama dama & cf. Dama dama	3	R mandible with M$_{1-3}$ (late-mid wear), L P$_3$ (late wear), R astragalus
Cervus elaphus & cf. Cervus elaphus	11	L antler (shed), L antler base frag (shed), L P$_3$, R M$_3$ (early mid-wear), L scapula frag (dist. fused), L scapula (dist. fused), L radius (prox. & shaft, prox. fused), L tibia (distal & shaft, dist. fused), R distal fibula, R astragalus, 1st phalanx (prox. fused)
Capreolus capreolus	1	Metapodial shaft frag (juvenile)
Cervidae gen. et sp. indet. (red/fallow deer sized)	58	28 antler frags, L M$^{1 \text{ or } 2}$, L M$_{1 \text{ or } 2}$, 3 upper molar frags, upper cheek tooth frag, 3 cheek tooth frags, molar frag, tooth frag, axis frag, ? lumbar vertebral centrum (cranial epiphysis fused), R dist. humerus frag, L radius (prox. & shaft, prox. fused), prox. radius frag, L ulna (prox.), L lunate, L unciform, R metacarpal frag (prox.), L metacarpal frag (prox.), metacarpal frag (dist. & shaft, dist. fused), R innominate frag, 2 astragalus frags, R navicular-cuboid frag, metatarsal shaft frag (juvenile), 2 metapodial distal condyles, metapodial (badly crushed)
Cervidae gen. et sp. indet.	4	4 antler frags
Artiodactyl?	1	Tooth frag
Indet. large mammal (elephant size)	5	Rib frag, 4 indet. bone frags
Indet. large mammal (rhinoceros or larger)	1	L & R occipital condyles
Indet. large mammal (Bos or larger)	40	4 long bone shaft frags, 36 indet. bone frags
Indet. large mammal (red deer or larger)	38	? mandible frag, tooth root frag, 3 long bone shaft frags, 33 indet. bone frags
Indet. large mammal (red/fallow deer size)	18	2 petrosal frags, 3 vertebral centrum frags, 10 long bone shaft frags, 3 indet. bone frags
Indet. large mammal (smaller than fallow deer)	5	5 long bone shaft frag
Indet. large mammal (fox size)	1	? thoracic vertebra frag
Indet. large mammal	85	84 indet. bone frags, tooth frag

fallow deer and aurochs. Nearby freshwater environments are indicated by European beaver *Castor fiber*, which is represented by single tooth from context 40100. The beaver is a valuable environmental indicator species as it is closely tied to large streams or lakes with adjacent deciduous woodland. Beavers feed mostly on herbaceous plants, such as stems and rhizomes of rushes and the leaves and twiggy branches of trees and shrubs. They may have a significant impact on their immediate surroundings by felling trees, which opens-up the forest and creates patches of dense shrubby cover, and by building dams to create ponds (Coles 2010).

The large mammals provide a clear indication for a locally wooded landscape with dry areas and some open herbaceous vegetation close to a large body of freshwater. Overall, the large mammal fauna is entirely consistent with a fully temperate climate.

Phase 7 (context 40167)

A bovid cheek tooth is the only identifiable specimen from this phase (40167). The remnants of the cementum and dentine are heavily corroded, showing that extreme processes must have been operating at some stage during burial and that post-depositional loss is severe.

Phase 8b (context 40048, 40049)

This extensive deposit of bedded clast-supported gravel appears to be largely unfossiliferous. Contexts 40048 and 40049 yielded indeterminate large mammal remains dispersed through the deposit. Two of these pieces are heavily rounded and abraded, with edge-rounding consistent with fluvial transport.

DISCUSSION

Environmental context of the Clactonian occupation

The vertebrate faunas yield a consistent and detailed picture of the contemporary environmental and climatic conditions associated with the Clactonian occupation at the site (Table 7.10). Understanding the taphonomy of the faunal remains is, however, central to attempts at reconstructing site formation processes and the ecological succession at the site. The taphonomic analysis also provides a basis for making broader faunal comparisons that allow the Clactonian archaeology at Southfleet Road to be placed in the wider context of the Lower Palaeolithic occupation in Britain. The following section

Table 7.10 Summary of the Hoxnian environmental history of Southfleet Road, based on evidence from the vertebrate fauna

Phase	Context (Large mammal NISP)	Vertebrates	Depositional context
9		None	?
8b	40048 (3)	Rare large mammals	Fluvial
	40049 (1)		
7	40167 (1)	Rare large mammals (decalcified)	?
6d	40158 (6)	Rare large mammals	?
	40100 (282)	High-diversity ungulate fauna	Vegetated landsurface
	40078 (491)	Straight-tusked elephant skeleton	Landsurface with episodic slopewash
6c	40099 (2)	Rare large mammals	?
	40160 (3)		
	40162 (3)		
6b	40143 (0)	Amphibians & fish dominant	Shallow, slow-flowing stream
	40144 (42)	Fish, amphibian & small mammals present in equal abundance	
	40070 (165)	Fish increase up-sequence. Amphibians dominant in upper part	Shallow, slow-flowing stream. Temporary drying-out of waterbody at top of tufa sequence
6a	40103 (13)	Large mammals & fish (4%) rare, amphibians common (40%)	Landsurface & marsh
	40039 (234)	Mammals common, amphibians & fish rare (1%)	Landsurface (weathering, rootlet corrosion & post-depositional decalcification indicative of burial in a soil)
5	40025 (14)	Rare large mammals, abundant fish (46%); amphibians & small mammals common	Fluvial, deep water
	40072 (1)		
4		None	?
3	40028 (1)	Rare large mammals	?
	40061 (1)		
	40062 (2)		
	40159 (1)		
2		None	?
1		None	?

summarises the taphonomy of the Southfleet Road vertebrates as a prelude to discussions on broader issues concerning the environmental context of the Clactonian.

The sediments at Southfleet Road were deposited within and on the lateral margins of a fluvial channel. In this depositional setting, accumulation and burial of the bones took place on the floodplain, in fluvial channels, near springs and seeps and in colluvium. There is little indication of human involvement in the accumulation of any of the assemblages. Only two possible cut marks were identified, in spite of the presence of artefacts throughout the fossiliferous sequence. Although none of the elephant bones were cut-marked, their intimate association with a cluster of mint-condition refitting artefacts is, however, suggestive of human involvement (see Chapters 17 and 22). Most of the large mammal bones probably accumulated from natural deaths or as carnivore kills. Determining the carnivore or carnivores responsible for accumulating these bones relies on the identification of their skeletal remains and chewing damage. At Southfleet Road, lion is the only large mammalian carnivore represented. The much larger mammal assemblage from contemporaneous deposits at Swanscombe includes, in addition to lion, a smaller jaguar-sized cat, cave bear *Ursus spelaeus*, wolf *Canis lupus*, wild cat *Felis sylvestris* and pine

marten *Martes* cf. *martes*. At Swanscombe, further evidence for carnivore activity is provided by characteristic chewing marks on some of the bones. Carnivores are therefore implicated in taphonomic modification of the large mammal remains and the destruction of bones through chewing at both Swanscombe and at Southfleet Road. This is highlighted by the skeletal element analysis which shows that preservation is strongly correlated with structural density of the bone, with the more dense bones being preferentially preserved. The feeding and scavenging activities of predators was also probably partly responsible for the disarticulated and dispersed distribution of the bones. Bones from most of the contexts were deposited on a land surface and were exposed to such destructive forces as desiccation, frost and damaged by plant growth. Other bones were subsequently modified by transport and weathering, resulting in further breakage or loss of material.

The small vertebrate bones also accumulated on the same landsurfaces and fluvial deposits over a similar period of time. Many of the small vertebrates probably lived and died close to the depositional site, but the remains of other animals show signs of predation by birds of prey and mammalian carnivores. The remains of these hunted small vertebrates were transported to the site,

Environmental interpretation	Climate	Archaeology	Correlation with Swanscomb
?	?		
?	?	Acheulian	Stage II?
?	?	Clactonian (dervived?)	?
?	?		
Mosaic of woodland, grassland and wetland	Fully temperate		
	Fully temperate		
?	?		
Dense woodland with some wetland & grassland areas bordering a stream			
Mosaic of woodland, wetland & grassland habitats bordering a stream			
Wetland habitats, areas of herbaceous vegetation (reed-beds) & closed woodland bordering a stream	Fully temperate	Clactonian	Stage I
Marsh or damp habitats with a mix of woodland & grassland			
Woodland with grassland & distant water-edge habitats			
Large stream or river bordered by marsh and extensive grassland. Drier areas with light soils supporting scrub or woodland			
?	?		
?	Temperate		
?	?		?
?	?		?

most likely from within the confines of the valley itself and probably within a radius of less than 1km from where they were deposited. The incorporation of allochthonous elements (ie transported to the site from various habitats) combined with autochthonous elements from natural deaths at the site needs to be borne in mind in their interpretation. This mixture of elements from different sources accounts for the wide faunal spectrum indicating varied local environments represented, ranging from aquatic to dry grassland and woodland, as well as the presence of fish remains in the colluvial and landsurface deposits.

The taphonomy of the large mammal bones records a dynamic local environment characterised by high fragmentation, dispersal and destruction of individual large mammal bones, with burial of mainly scattered and isolated elements. Some of the bones clearly derive from partial skeletons of animals that died at the site, whereas other bones could have been transported from elsewhere by mammalian carnivores. Few of the bones had been exposed to prolonged weathering, indicating burial was relatively rapid in each of the phases of fluvial, alluvial and colluvial deposition. Studies of modern analogues for fossil assemblages preserved in open sites shows that they can provide sensitive records of population shifts in the dominant herbivore species (Behrensmeyer 1993).

There is no reason to suggest that the large mammals from Southfleet Road do not reflect the composition and abundance of the various large mammal species in the living community.

Summary of faunal change through the sequence

The basal sediments were largely devoid of vertebrate material, with sieved samples from Phases 1, 2 and 3 being totally barren. The presence of aurochs in sediments assigned to Phase 3 provides the only mammalian indication of the prevailing conditions at the time. Locally, the landscape would have supported rich grazing and probably also deciduous trees. In terms of climate, aurochs is known only from interglacial and interstadial periods in Britain, implying that the prevailing climate was not unduly severe. The Phase 3 ostracod fauna (Chapter 11) indicates a fully interglacial climate.

The extensive deposit of bedded fluvial sands of Phase 5 yielded a vertebrate fauna composed of rather few taxa but with a mixture of dry ground, marsh and aquatic species. The water-body, which was of sufficient depth to support relatively large fishes, was probably slow flowing, although gravel lenses suggest occasional episodes of higher energy. Amphibians inhabited marshland or damp terrestrial habitats on the floodplain and the associated mammals are consistent with extensive open grassland that included drier areas. Areas of open woodland supported herds of aurochs and fallow deer, which were predated by lions and other large carnivores. A fully temperate climate is indicated by the presence of thermophilous fish species (eg tench) and mammals, such as rabbit, aurochs and fallow deer. Other elements of the vertebrate fauna (northern vole and ground squirrel) are generally thought of as more typical of cold climates in the British Pleistocene. However, both the ground squirrel and the northern vole have been found in temperate contexts during the Middle Pleistocene in Britain, possibly occurring as survivors from a preceding cold stage.

The vertebrates confirm the sedimentological interpretation that the deposits assigned to Phase 5 are clearly fluvial in origin and deposited by a large stream or river. The sediments were characterised predominantly by sands with some gravel bands and finer sediments. There was a change, with the transition into Phase 6, to predominantly finer-grained silts and clays. This coincided with a change to drier conditions and implies a major change in the character of the depositional environment. The basal sediments of this alluvial sequence were weathered and decalcified. Intense iron-staining reflects chemical alteration during formation of a soil, with drab clays accumulated during periods of flooding, when burial was rapid enough to prevent development of mature palaeosols (Chapter 5). Micromorphological evidence from the iron-stained layers indicates that the soils formed in a woodland setting. The condition of the large mammal bones is also consistent with accumulation on a ground surface and burial in a biologically active soil. Common alterations include corrosion and pitting as a result of chemical leaching and root-etching during burial. These destructive processes have removed the outer layer of cortical bone on many specimens, but several of the better-preserved pieces show carnivore chew marks and splitting from weathering. In combination, these processes account for the fragmentary and disarticulated nature of the remains recovered from these deposits. This transition from predominantly sandy to silty sediments also coincides with abrupt changes in the composition of the small vertebrate fauna. A near absence of fish and amphibians implies a relatively dry terrestrial local environment, which supported abundant woodland mammals. Locally, the landscape appears to have been a drier and more wooded habitat than during Phase 5, although the climate was also fully temperate at the time.

The mammalian assemblage from the tufa-filled channel (Phase 6b) is consistent with deposition near spring-seeps and in shallow, flowing water. Freshwater fish bones, which are particularly common in some of the sampled levels, are mainly from relatively small fishes, indicating a more restricted waterbody than that of Phase 5. The water-body appears to have been relatively shallow but well-oxygenated. Several columns were sampled for small vertebrates and these all show a dominance of amphibians and a steady increase in fish remains with a complementary decrease in small mammal. In localised areas, the upper part of the tufaceous sequence is characterised by low numbers of fish and small mammals with abundant amphibians. This appears to reflect a change in the local environment from that of a small stream, which was steadily encroached by marsh. The terrestrial vertebrate fauna from the tufaceous deposit indicates a mixture of grassland, marsh and aquatic habitats in a generally wooded environment. The most striking aspect of the fauna is the presence of squirrel, which provides a clear indication of local closed-canopy woodland. Aquatic conditions diminish in the upper part of the sequence, with an even greater representation of woodland species in the small mammal fauna, and with a decline in fish and an increase in amphibians. Other elements of the vertebrate assemblage point to a local mosaic of habitats that included woodland, grassland and wetland habitats bordering a stream.

The sediments immediately overlying the tufaceous channel are grey brecciated silty clays with localised darker brown ('humic') horizons; the grey clays extended across most of the site, but the darker brown organic-rich horizons were restricted to the central part of the site to the west of the tufaceous channel. Small vertebrate remains are scarce, but large mammals are relatively common and include the dispersed remains of a straight-tusked elephant. The elephant carcass seems to have been situated in a landscape with significant woodland, possibly of alder-carr type. Open herbaceous vegetation occurred nearby. There is no indication for significant climatic difference, with the vertebrates indicating fully temperate conditions; the presence of a foot bone from the elephant in the tufaceous channel fill establishes its contemporaneity with context 40078, and suggests that much of the vertebrate material from context 40100 could likewise be contemporary.

Landscape setting of the site

Ashton *et al.* (2006) have argued that human occupation in Britain during the Hoxnian was focussed in river valleys. They point out that the valleys were areas of greater biodiversity, with extensive tracts of grassland supporting a higher biomass of grazing herbivores than the more heavily wooded interfluves. The rivers would also have provided drinking water for large mammals, as well easy access to raw materials for making stone tools in the form of gravel bars. The location of the Southfleet Road site in a relatively small tributary valley was somewhat different to most of the sites studied by Ashton *et al.* (ibid.) which were mostly associated with much larger river valleys. Differences in the ecology and taphonomy of the mammal remains in these different landscape settings appears to be reflected in the composition of the mammal faunas at the site. The Southfleet Road site was also situated close to the edge of a shallow valley which opened out into the Thames Valley near Swanscombe. Clast lithological evidence (Chapter 6) indicates sediments were deposited by a small south-bank Thames tributary with a limited catchment that did not extend beyond the chalk high-ground to the south of the site. Sedimentological analysis shows that there were

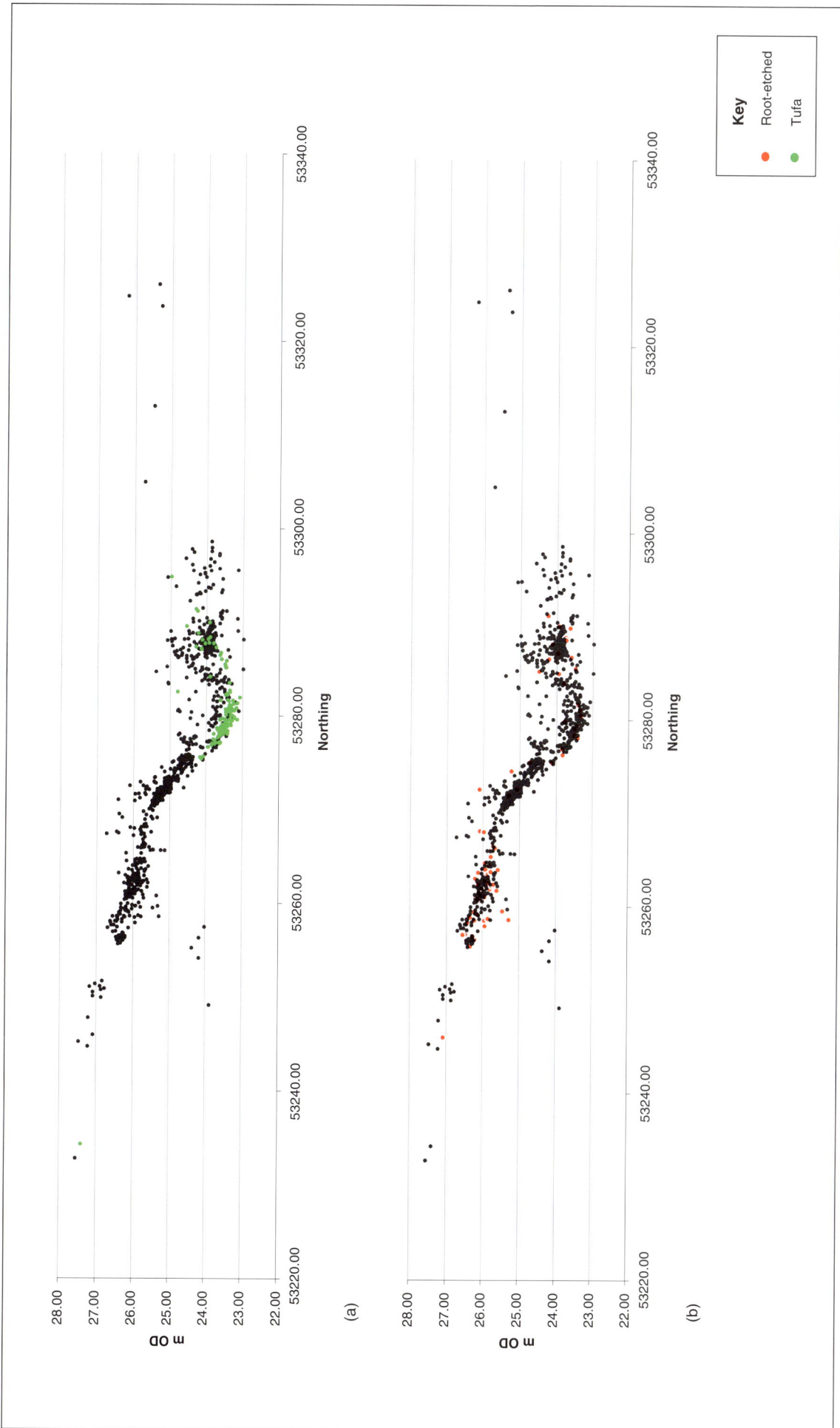

Figure 7.23 Sections of plotted bones from the tufaceous channel (a) compared with root-etched bones (b), the latter suggestive of drier conditions associated with areas of higher topography

changes in hydrological conditions and water-level that gave rise to varying conditions from fluvial channels, wetlands, spring seeps, and dry land.

Interpreting the relationship of the archaeology and faunal remains to these environments is complicated by post-depositional displacement, slumping and deformation of the sediments, which has distorted the original topography (Chapter 4). Nevertheless, the distribution of the bones in Phase 6 gives an indication of a sloping channel-edge with waterlain tufaceous sediments in a shallow stream at the foot of the slope (Fig. 7.23a). These tufaceous deposits have yielded a rich fish assemblage indicative of slow-flowing water. Most of the artefacts, however, are from the area of highest topography at the southern end of the site. In contrast, the major concentration of bones is to the north of the main artefact concentration, on the sloping surface and in the area of lower topography associated with the tufa channel. An outlying concentration of artefacts is associated with the elephant skeleton, but elsewhere artefact density is relatively low. Similarly, root-marked bone is not distributed randomly across the site. In Phase 6, for example, there is a high degree of clustering of bones with root marks coinciding with the area of higher topography (Fig. 7.23b). The concentration of root-marked bones may suggest that this was a dry land surface with well-developed cover of vegetation, abutting the wetter and topographically lower ground to the south. The human occupation therefore took place in a valley setting at the transition between dry land and wetland habitats. This would have offered a variety of environments for early humans, but whether any one of these environments was the preferred habitat is difficult to determine from the evidence at Southfleet Road alone. The vertebrate evidence does however indicate that the local environment was not static and that there was a succession of small-scale changes that appear to relate to local conditions rather than regional climate change (Table 7.10).

The role of herbivores in maintaining the habitat patchwork

Records of Hoxnian vegetational change are preserved in lake sequences, such as those at Hoxne, Suffolk and Marks Tey, Essex (Turner 1970). These reveal a progressive invasion of forest that covered most of the landscape of southern England after the retreat of the Anglian ice sheet. By the early temperate substage an abundance of tree pollen is taken to indicate closed-canopy deciduous forest. Low counts of non-arboreal pollen indicate local herb communities associated with lakes and bogs or glades in the forest. However, the palynological evidence for closed forest at these sites is contradicted by the mammalian evidence, which suggests more open conditions with areas of extensive grassland. This apparent contradiction may be explained by invoking spatial differences in woodland cover, with more open conditions of grassland or open parkland associated with the mammalian evidence from floodplain situations and denser woodland on the interfluves associated with the

lake sites (Ashton et al. 2006). At Southfleet Road there can be little doubt that there were trees nearby when these deposits accumulated (see Chapter 12), but the location and density of the woodland is more difficult to assess. Similarly there are indications for fluctuations in tree density which require explanation.

The mammalian evidence from Southfleet Road suggests that the local vegetation was a patchwork of dense woodland with grassy glades, interspersed with more extensive open grassland, marsh and other water-edge habitats. Changes in faunal composition show the vegetation structure of the immediate area was not static and that there were periods when open grassland was dominant (Phase 5) and other times when dense woodland covered the site (contexts 40070, 40144). Several factors, such as: geomorphic change, fire, fluctuating water levels, floods, edaphic factors and climate change may be implicated in modifying the vegetation. However, climate change is unlikely to have been one of them, given that there are no strong indications that there were major changes in the climate associated with the main faunal horizons (see below). The availability of freshwater at the site evidently attracted large herds of herbivorous mammals and provided a home for European beaver. Large herbivores are an important natural agent in shaping habitats and the activities of these animals are likely to have played an important role in woodland dynamics. For example, the European beaver is an important agent in altering landscapes and ecosystems as a result of their tree-cutting and dam-building behaviour (Coles 2010). They crop small deciduous trees along the edge of water bodies, creating breaks in the canopy that encourage herbaceous vegetation. Large grazing mammals also maintain clearings that are maintained in that condition by intensive grazing and browsing. The feeding activities of elephants are critical in opening-up wooded environments as both African and Asian elephants browse mainly on leaves, twigs and branches and roots. In the process they can cause considerable damage to woodland. It is likely that the straight-tusked elephant would have had a similar impact on the environment at Southfleet Road. Browsing and trampling by mega-herbivores inhibits scrub and woodland regeneration and would have helped to maintain the open grassland areas within the forest. These open areas would have attracted other grazing herbivores to feed (Turner 1975). Turner (ibid.) has shown that the expansion of grassland over what had been woodland can be a direct result of large herbivore activity. Conversely, any reduction in grazing pressure would lead to woodland regeneration; changes to woodland domination may be remarkably fast, due to the spread of already existing trees that had hitherto been confined by grazing.

Comparisons with large mammal assemblages from Clactonian levels at Swanscombe and Clacton-on-Sea

The foregoing discussion has outlined the environmental setting of the Clactonian occupation at Southfleet Road, interpreted from the vertebrate data. The vertebrate faunas from Southfleet Road can now be placed into a wider context of Early Palaeolithic Hoxnian environments by comparison with faunas associated with the Clactonian at Swanscombe and Clacton in the Thames Valley. These sites provide interesting comparisons between localities where Clactonian occupation occurred in different geomorphological settings. At Southfleet Road, the occupation took place in a small tributary valley that supported a patchwork of vegetation, including woodland, marsh and open areas with herbaceous vegetation. At Swanscombe, the sediments were deposited in the channel and on the floodplain of the Thames, which in this area had a wide floodplain flanked by chalk hills. At Clacton, some 45km downstream from Swanscombe, Clactonian artefacts are found in Thames river gravels and silts deposited in the lower reaches of the river, close to its estuary. The floodplain at Clacton was probably more extensive and beyond the floodplain the surrounding landscape was one of generally low-lying topography. The limited relief was formed of fluvial sands, gravels and clays deposited during the Anglian glacial stage. In contrast, Clactonian activity at Southfleet Road took place in a minor tributary valley of Thames. This narrow valley was cut into fine-grained Tertiary sediments and Chalk. Although both Clacton and Swanscombe have yielded substantial quantities of fossil vertebrate material, most of the material from earlier collections was not recovered during archaeological excavations but from searching

foreshore exposures at Clacton and during quarrying at Swanscombe (Currant 1996). As well as producing a bias in favour of larger specimens and animals, many of the fossils collected in this way lack precise stratigraphical information. With the Clacton material there is an added complication as older museum collections incorporate specimens from several sites associated with different channels, one of which post-dates the Hoxnian (Bridgland et al. 1999).

More directly comparable with the Southfleet Road assemblages in terms of recovery techniques are the large mammal assemblages from controlled archaeological excavations undertaken by Wymer at Clacton (Singer et al. 1973) and by Waechter at Swanscombe (Conway et al. 1996). Although bulk wet-sieving was not undertaken at either site, the deposits were nevertheless excavated carefully and all of the retrieved bone fragments were kept. The Swanscombe assemblage even includes rodent teeth and other microfauna recovered during trowelling. The excavations at Swanscombe between 1968 and 1972 were focused on newly recognised Clactonian archaeology in the Lower Loam; the immediately underlying Lower Gravels were also sampled. The Lower Gravels and Lower Loam at Swanscombe represent a single depositional unit reflecting a change from fast flowing-water to quieter conditions of a meander channel to a dry landsurface at the top of the Lower Loam succession (Table 7.11). The excavation recovered refitting scatters of Clactonian knapping debris and abundant large mammal remains from several different horizons

Table 7.11 Swanscombe: correlation of sediments and summary of the environment from Waechter's 1968-72 excavations at Barnfield Pit, based on evidence from the vertebrates, molluscs and sediments (Conway et al. 1996)

Stage	Horizons with mammals	Environment (Conway 1996)
Lower Middle Gravel IIa	Lower Middle Gravel	High-energy fluvial conditions, appearance of Rhenish species
	Base of Lower Middle Gravel	
Lower Loam Ie	Lower Loam surface Weathered Lower Loam	Landsurface with soil formation in dry, open grassland conditions (large mammal footprints). Fully temperate
		Deeper quiet water bordered by dry grassland
Lower Loam Id	Lower Loam main body Lower Loam sandy horizon Base of Lower Loam	Meander channel. Low-energy regime, virtually still if not stagnant, interrupted by phases of channel cutting and infilling and temporary relatively dry landsurface with dessication. In-situ knapping scatters. Molluscs indicate reed swamp and fen with dry grassland in open woodland
	Lower Loam/Lower Gravel junction	Erosional surface, marking boundary between high-energy regime below and low-energy regime above
Midden Horizon Ic	Lower Gravel (midden)	Mollusc fauna comparable with that from upper part of Lower Gravel
Lower Gravel Ib	Lower Gravel unit 4 Lower Gravel unit 3 Lower Gravel unit 2 Lower Gravel unit 1	Fluvial, fully temperate. Molluscs dominated by aquatic species (90%), terrestrial molluscs indicate rather open (marsh) conditions, woodland/scrub species moderately common

The Ebbsfleet Elephant

Table 7.12 Numbers of identified larger mammal specimens (including lagomorphs) from Wymer's (1969–70) excavation at Clacton-on-Sea (Singer *et al.* 1973), Waechter's (1968–1972) excavation at Swanscombe (based on Schreve 1996, with corrections) and Southfleet Road.

	Clacton Sandy gravel Layer 4	Swanscombe Lower Gravel 1	2	3	Lower Loam 4	5
Primates						
Macaca sylvanus, Barbary macaque						
Lagomorpha						
Oryctolagus cuniculus, rabbit			1	5		
Rodentia						
Trogontherium cuvieri, beaver-like rodent	1					
Castor fiber, beaver						
Carnivora						
Canis lupus, wolf			1			
Ursus spelaeus, cave bear		1				
Ursus sp., bear		2	1	1		
Felis sylvestris, wild cat			1			
Panthera leo, lion						
Proboscidae						
Palaeoloxodon antiquus, straight-tusked elephant			6			
Elephantidae gen. et sp. indet., indeterminate elephant	5	2	2			
Perissodactyla						
Equus ferus, horse	6	2				
Stephanorhinus kirchbergensis, Merck's rhinoceros		1	1			1
Stephanorhinus hemitoechus, narrow-nosed rhinoceros	2		2			
Stephanorhinus sp., rhinoceros	3		3	1		
Artiodactyla						
Sus scrofa, wild boar				1		
Megaloceros giganteus, giant deer						
Dama dama clactoniana or *D. dama* ssp. indet., fallow deer	2	10	31	3	2	4
Cervus elaphus, red deer	3	3	3	2		
Capreolus capreolus, roe deer						
Cervidae gen. et sp. indet, indeterminate deer	8	31	53	23		4
Bos primigenius, aurochs		5		1		
Bison priscus, bison		1				
Bos primigenius or *Bison priscus*, aurochs or bison	10	3	9	5		1
Total (NISP)	40	61	114	42	2	10

Clacton (Wymer excavation. Singer *et al.* 1973)
Swanscombe (Waechter excavation. Conway *et al.* 1996)

Swanscombe
1. Lower Gravel
2. Lower Gravel midden
3. Lower Loam - Lower Gravel junction
4. Base of Lower Loam
5. Lower Loam sandy horizon
6. Lower Loam main body
7. Weathered Lower Loam
8. Lower Loam surface / weathered surface

Southfleet Road
1. Phase 5
2. Phase 6a (Context 40039)
3. Phase 6a (Context 40103)
4. Phase 6b (Context 40070)
5. Phase 6b (Context 40144)
6. Phase 6 (Context 40078)
7. Phase 6 (Context 40100)

Note – spatially associated specimens that are likely to have come from the same individual are counted as one specimen.

| | Swanscombe Lower Loam | | | | Southfleet Road | | | | | |
	6	7	8	1	2	3	4	5	6	7
	1						1			
	10	2	2	2	1					
										1
	3									
	3									
	1	1		1						
	3			1				1	2	
									61	
			2							
	1	2			23			2	1	6
									3	3
	1	4			1					1
		1								
	28	5	5	1	4	1	2			3
	6	2	1		4	1	8		6	11
							1	1		1
	59	10	16	4	33	5	88	11	70	62
	2			1					1	
	3	1	3	2	3				2	
	121	28	29	11	69	8	100	15	146	88

(Conway *et al.* 1996). A much smaller vertebrate assemblage was recovered from Wymer's trenches at Clacton (Singer *et al.* 1973). Most of the bones were recovered from a sandy gravel (layer 4) containing Clactonian artefacts. The deposits at Swanscombe and Clacton have been correlated using several lines of evidence, with palynology at Clacton showing that the Clactonian activity took place during the early part of the Hoxnian interglacial (Kerney 1971; Bridgland *et al.* 1999). The Clactonian at Southfleet Road is argued to date from the same period, based on vertebrate palaeontology (Chapter 9), the pollen record (Chapter 12) and amino acid dating (Chapter 13).

Comparisons of the larger mammal faunas from the three excavations are given in Table 7.12. Here, the relative and absolute abundance are assessed by a simple count of the number of identifiable specimens (NISP, excluding indeterminate fragments). The counts shows the expected relationship between the size of the assemblages and the number of taxa; the largest samples are from Southfleet Road (NISP = 437) and Swanscombe (NISP = 387), with 12 and 15 taxa respectively. The much smaller assemblage from Clacton (NISP = 40) has seven taxa represented. Deer, bovids, proboscidians and rhinos are present in all three assemblages, with red deer and fallow deer fossils by far the most common identifiable specimens represented at Swanscombe and Southfleet Road. At the latter two sites, cervids account for more than half of the identified larger mammal specimens. In keeping with trophic structure, carnivores

Table 7.13 The relative abundance of larger mammals at Clacton, Swanscombe and Southfleet Road based on counting a limited set of homologous (non-reproducible) elements: third lower molar, axis, distal humerus, proximal metacarpal, pelvic acetabulum, distal tibia, proximal metatarsal.

	Clacton Sandy gravel Layer 4	*Swanscombe* Lower Gravel 1	2	Lower Loam 3	4	5
Primates						
Macaca sylvanus, Barbary macaque						
Lagomorpha						
Oryctolagus cuniculus, rabbit						
Rodentia						
Trogontherium cuvieri, beaver-like rodent	+					
Castor fiber, beaver						
Carnivora						
Canis lupus, wolf			+			
Ursus spelaeus, cave bear		+				
Ursus sp., bear		+	+	+		
Felis sylvestris, wild cat			1			
Panthera leo, lion						
Proboscidae						
Palaeoloxodon antiquus, straight-tusked elephant			+			
Elephantidae gen. et sp. indet., indeterminate elephant	+	+	1			
Perissodactyla						
Equus ferus, horse	2	1				
Stephanorhinus kirchbergensis, Merck's rhinoceros	1	1	1			+
Stephanorhinus hemitoechus, narrow-nosed rhinoceros	+		2			
Stephanorhinus sp., rhinoceros	1		+	+		
Artiodactyla						
Sus scrofa, wild boar				+		
Megaloceros giganteus, giant deer						
Dama dama clactoniana or *D. dama* ssp. indet., fallow deer	+	2	6	1	+	+
Cervus elaphus, red deer	+	2	2	1		
Capreolus capreolus, roe deer						
Cervidae gen. et sp. indet, indeterminate deer	1	2	7	2		+
Bos primigenius, aurochs		1	+			
Bison priscus, bison		1				
Bos primigenius or *Bison priscus*, aurochs or bison	4	2	+	+		+
Total	8	12	20	4		

Clacton (Wymer excavation. Singer *et al.*, 1973)
Swanscombe (Waechter excavation. Conway *et al.*, 1996)

Swanscombe
1. Lower Gravel
2. Lower Gravel midden
3. Lower Loam - Lower Gravel junction
4. Base of Lower Loam
5. Lower Loam sandy horizon
6. Lower Loam main body
7. Weathered Lower Loam
8. Lower Loam surface / weathered surface

Southfleet Road
1. Phase 5
2. Phase 6a (Context 40039)
3. Phase 6a (Context 40103)
4. Phase 6b (Context 40070)
5. Phase 6b (Context 40144)
6. Phase 6 (Context 40078)
7. Phase 6 (Context 40100)

+ indicates presence of taxa not otherwise represented by a 'countable' element.

	Swanscombe Lower Loam			Southfleet Road						
	6	7	8	1	2	3	4	5	6	7
	+						+			
				1	+					
										+
	+									
	+									
	1	+		+						
				+				+	+	
	+								1	
			+							
	1	+			+			1	+ 2	1 +
	+	+ +			+					+
	4	3	1	+	1	+	+			1
	3	1	+		2	1	+		1	3
							+	+		+
	9	3	1	1	2	+	4	+	6	4
	+			+					1	
	1	+	+	+	+				+	
	19	7	2	2	5	1	4	1	11	9

are rare, although they are better represented at Swanscombe than at Southfleet Road, where lion is joined by wolf, wild cat and the omnivorous cave bear. These differences aside, the Swanscombe and Southfleet Road faunas share many species in common, but there are differences either in qualitative presence/absence or in relative frequency that require explanation. Taxa missing from Southfleet Road assemblages included giant deer, bison (both represented in the Waechter collection by one specimen each) and horse. These differences could result simply from chance, circumstances of preservation, small sample size or inability to identify fragmentary remains (for example the bovid bones from Southfleet Road may include bison but the specimens are generally too fragmentary to make a definitive identifica-

tion). The absence of horse in the Southfleet Road assemblage is more intriguing as it occurs in the both the Lower Loam and Lower Gravels at Swanscombe and in the much smaller Clacton assemblage, where it is represented by more than one individual (Singer et al. 1973). A plausible explanation for the occurrence of horse at Swanscombe and Clacton is that the floodplain bordering the Thames had more extensive areas of open grassland suitable for herds of grazing horses, whereas Southfleet Road was more heavily wooded, especially during the deposition of Phase 6. The narrow wooded valley of the proto-Ebbsfleet may simply have been less suited to horses.

Turning now to the relative proportion of deer species, the NISP counts show that fallow deer specimens are much more common than red deer in both the Lower Gravel and Lower Loam assemblages, whereas the opposite relationship holds in the Southfleet Road assemblages. This could simply result from differences in taphonomy between the two sites, either from differential fragmentation of the bones and the incorporation of shed antlers of red and fallow deer skewing the counts. Both factors could act together to inflate the counts in favour of one species over the other. The problem of differential fragmentation was tackled by counting only a limited set of non-reproducible elements (cf. Davis 1992; O'Connor 2003) and by making separate counts for the antlers. It is clear, both from the NISP (Table 7.12), limited elements (Table 7.13) and the antler counts (Table 7.14), that red deer remains are about three times as common as those of fallow deer in the Southfleet Road assemblage and that fallow deer are consistently more abundant at Swanscombe. These differences would appear to reflect differences in the living population, implying that the local environment at Southfleet Road was better suited to red deer, whereas conditions at Swanscombe provided a more suitable habitat for fallow deer. An indication of the differences in ecology is provided by the molluscan fauna from Swanscombe, which contains a significant component of open ground and marsh taxa, as well as others that prefer woodland or scrub. This mixture of local habitats would have provided ideal conditions for fallow deer, which today are closely associated with parkland, sparse forests, marshland and grassland. Although red deer are found in a similar range of habitats, optimal conditions are provided by dense deciduous forest and forest margins. It may be significant that roe deer is present (although uncommon) in several horizons at Southfleet Road, compared with Swanscombe, where it is represented by only one specimen in the entire collection (Lister 1986). The ecology of roe deer would support other indications of dense woodland at Southfleet Road, which is the preferred habitat of this species, especially in the western part of its range.

More detailed comparisons of the ecology of the Clactonian levels at Swanscombe and Southfleet Road are hampered by the unevenness of the vertebrate records. At Southfleet Road, the vertebrate fauna includes a significant small vertebrate component, which is largely missing in the Swanscombe collections. Although some sieving was undertaken to recover molluscs at Swanscombe (Kerney 1971), the samples were too small to recover a useful sample of the smaller mammals, birds, reptiles, amphibians and fishes. This information gap is unfortunate, as at Southfleet Road the smaller vertebrates have enabled detailed bed-by-bed reconstructions of the local ecological conditions. At Swanscombe, such information is provided by the molluscan fauna, which is much more extensive than at Southfleet Road (Chapter 10). Although the larger vertebrate faunas from the Thames Clactonian sites are clearly very incomplete samples of the once-living populations that they represent, they nevertheless show interesting points of similarity as well as differences that are probably the result of local ecological conditions.

Table 7.14 A comparison of antler counts from Swanscombe (Waechter's excavation) and Southfleet Road.

	Unshed	Shed	Frag	Total
Swanscombe (Lower Gravel)				
Fallow deer	6	9	10	25
Red deer			1	1
Deer		1	26	27
Total	6	10	37	53
Swanscombe (Lower Loam)				
Fallow deer	3	9	7	19
Red deer	2	1	1	4
Deer		1	15	16
Total	5	11	23	39
Southfleet Road (Phase 6)				
Fallow deer		3		3
Red deer		8	1	9
Deer	2	15	178	195
Total	2	26	179	207

Climatic implications of the vertebrate fauna

The vertebrate fauna provides important information about the climatic conditions prevailing during the Clactonian occupation at the site. The climatic conditions can be reconstructed by analogy to the present-day ecology of extant taxa (or their nearest living relatives), combined with the analysis of the whole mammalian fauna using taxonomic habitat indices and the climatic preferences of fossil mammals inferred from associated biological proxies (Andrews 1990; Candy et al. 2010). A comparative framework is provided (Table 7.15), with extant taxa divided into climatological groups according to the northern boundary of their ranges: (1) species whose northern-most boundary extends above the Arctic Circle into the tundra zone; (2) species whose northern-most boundary just reaches the Arctic Circle; and (3) species who only range to southern Scandinavia. In relation to the Pleistocene record, most of the Southfleet Road taxa are known only from warm stages in Britain (Ig = interglacial; Is = interstadial), but a few had wider climatic ranges, and were present in both cold (C) and warm stages.

Mammals are strongly influenced by climatic variation, temperature and humidity. Many of the extant mammals represented in the Southfleet Road assemblage are tolerant of a wide range of conditions and occur widely from the Mediterranean coast and into northern Europe, with some having ranges extending into the tundra above the Arctic Circle (climatological group 3 in Table 7.15). The broad latitudinal distribution of mammals in this group implies that they have wide ecological and climatic tolerances (but see Candy et al. 2010 and Polly and Eronen 2011) for a discussion of the problems of inferring climate from warm-blooded animals). The remaining taxa can be divided into two further climatological groups according to the northern boundary of their ranges. The first group includes

Table 7.15 Ecology and climatic preferences of mammalian taxa from Southfleet Road. Extant taxa are divided into climatological groups according to the northern boundary of their ranges: 1 - Species whose northern-most boundary extends above the Arctic Circle into the tundra zone; 2 - species whose northern-most boundary just reaches the Arctic Circle; 3 - species who only range to southern Scandinavia. Most of the taxa are known only from warm stages in Britain (Ig = interglacial, Is = interstadial) but a few had wider climatic ranges and occur in both cold (C) and warm stages.

	Climatalogical zone			Pleistocene occurrence			Phase					
	1	2	3	Ig	Is	C	3	5	6a	6b	6	7
Extant in Europe												
Oryctolagus cuniculus, rabbit			✓	✓				+	+	+		
Microtus (Terricola) cf. *subterraneus*, common pine vole			✓	✓	✓			+		+		
Apodemus sylvaticus, wood mouse			✓	✓	✓			+	+	+		
Dama dama, fallow deer			✓	✓	✓			cf	+	+	+	
Myotis daubentonii, Daubenton's bat		✓		✓						+		
Mustela cf. *putorius*, polecat		✓		✓						+		
Sus scrofa, wild boar		✓		✓	✓				+		+	
Sciurus sp., squirrel	✓			✓						+		
Neomys sp., water shrew	✓			✓	✓			+	+	+	+	
Castor fiber, beaver	✓			✓	✓					+	+	
Clethrionomys glareolus, bank vole	✓			✓	✓			+	+	+		
Capreolus capreolus, roe deer	✓			✓	✓					+	+	
Sorex minutus, pygmy shrew	✓			✓	✓	✓		+		+		
Spermophilus sp., ground squirrel	✓			✓	✓	✓		+				
Microtus agrestis, field vole	✓			✓	✓	✓				+		
Microtus oeconomus, northern vole	✓			✓	✓	✓		+		+		
Cervus elaphus, red deer	✓			✓	✓	✓		+	+	+	+	
Extinct in Europe												
Macaca sylvanus, Barbary macaque				✓	✓					+		
Panthera leo, lion				✓	✓	✓		+				
Extinct												
Stephanorhinus kirchbergensis, Merck's rhinoceros				✓								+
Apodemus maastrichtiensis, mouse				✓						+		
Bos primigenius, aurochs				✓	✓		+	+				+
Talpa minor, mole				✓	✓				+	+		+
Palaeoloxodon antiquus, straight-tusked elephant				✓	✓			+		+		+
Arvicola cantianus, water vole				✓	✓	✓		+	+	+		+
Stephanorhinus hemitoechus, narrow-nosed rhinoceros				✓	✓	✓			+	cf		+

Daubenton's bat, polecat, and wild boar, whose northern-most boundaries just reach the Arctic Circle. The final group includes species intolerant of cold climate, with ranges extending no further than southern Scandinavia. Included in this group are rabbit, pine vole, wood mouse and fallow deer. Rabbit and fallow deer, in particular, are generally accepted as strong indicators of fully interglacial conditions in north European contexts (Stuart 1982). As a group the mammals from Phase 5 and 6 indicate deciduous woodland and by inference fully temperate conditions.

Of note is the presence of northern vole in the interglacial deposits. This small rodent has been extinct in Britain since the end of the last cold stage; today it has a largely northern Palaearctic distribution, favouring tundra and taiga habitats. There are, however, British Pleistocene records from temperate stages, including the warmest part of the interglacial (MIS 5e), when temperatures were 1 or 2°C warmer than the present day (Stuart 1982) and in fully temperate parts of MIS 7 in the Ebbsfleet Valley (Wenban-Smith *et al.* forthcoming). Similarly, the presence of ground squirrel in Phase 5 is intriguing as most British Pleistocene records are from cold stage contexts (Stuart 1982). As with northern vole, ground squirrels have also been found in more temperate contexts during the late Middle Pleistocene, at Crayford and Erith (Stuart 1982). Consequently, the presence of northern vole and ground squirrel do not necessarily indicate cooler conditions than present day. It seems likely that both species survived from the preceding cold stage in favourable microhabitats of marshland and dry grassland, respectively.

Temperature appears to be the main factor governing the present and past distributions of cold-blooded vertebrates (Stuart 1982). Although the amphibians and reptiles from Southfleet Road have not been identified in detail, the presence of tree frog is climatically informative. Today, *Hyla arborea* (common tree frog) is the most northerly tree frog. It has an extensive range throughout much of southern and central Europe, where it is found as far north as central Denmark and southern Scandinavia. In the east, its range does not extend far into European Russia, where the winters are too severe.

Other cold-blooded fauna are also strongly influenced by climate. For example, water temperature has been identified as a key factor controlling the distribution of cyprinid fishes (Cyprinidae: carp family) across different climatic zones (Wheeler 1977). Amongst the most theromophilic of these are tench *Tinca tinca* and rudd *Scardinius erythrophthalmus*, both of which are present in Phase 5 and 6 at Southfleet Road. Neither occurs farther north than central Fennoscandia and thus provide a further indicator of warm conditions.

Table 7.15 also summarises the inferred climatic preferences of the fossil mammalian taxa. Although some of these mammals had wide ecological tolerances, being found in both temperate and cold stages (for example pygmy shrew, field vole), the majority appear to have been restricted to temperate phases. The most significant of these is Merck's rhinoceros which, although frequently recorded in Middle Pleistocene interglacials, was absent from the intervening intersta-dials and cold stages in northern Europe (Pushkina 2007). The presence of Merck's rhinoceros therefore supports other indications for interglacial conditions.

Another approach to determining palaeoclimate by analysis of the whole mammalian fauna is the Taxonomic Habitat Index (THI) developed by Andrews (1990). The taxonomic habitat index approach adopted for the analysis of the Southfleet Road assemblage makes an assessment of the habitat preferences for all the micromammals in the fauna. The method uses presence or absence of taxa without any reference to their relative abundance, thereby reducing the bias arising from taphonomy (for example predator prey selection). It has the added advantage that it does not rely on single 'indicator species' but combines multiple species and makes allowance for fossil taxa. The habitats were divided into nine types: tundra, boreal forest, deciduous forest, Mediterranean, steppe, forest steppe, arid, tropical and montane. Each species was given a maximum possible score of 1.00 which was broken down according to the habitat preference of that species, so that if an animal occurs in more than one habitat type, it was scored proportionally according to its habitat preference (Andrews 1990). As an example, the hazel dormouse *Muscardinus avellanarius* that lives mainly in deciduous woodland, was scored 0.7 for 'Deciduous'. As it is also widespread in the Mediterranean bioclimatic zone, hazel dormouse was scored 0.3 for 'Mediterranean'. The scores for each habitat for all species are then added together and divided by the number of species to give an average weighed score for each habitat for that fauna. The spectrum produced is presented in the form of a histogram.

The THI spectra for different phases of the Southfleet Road sequence are plotted in Fig. 7.24, compared with the peak MIS 11 interglacial assemblage from Barnham (Parfitt 1998b). The fossil spectra have

been compared with those of 71 modern small-mammal faunas from locations representing the key bio-climatic zones throughout western Eurasia (Parfitt, unpublished data). The THI spectrum of the species in Phase 5 is most similar to those from northern central Europe. Phase 6a is comparable with those of modern central-southern European deciduous woodlands,

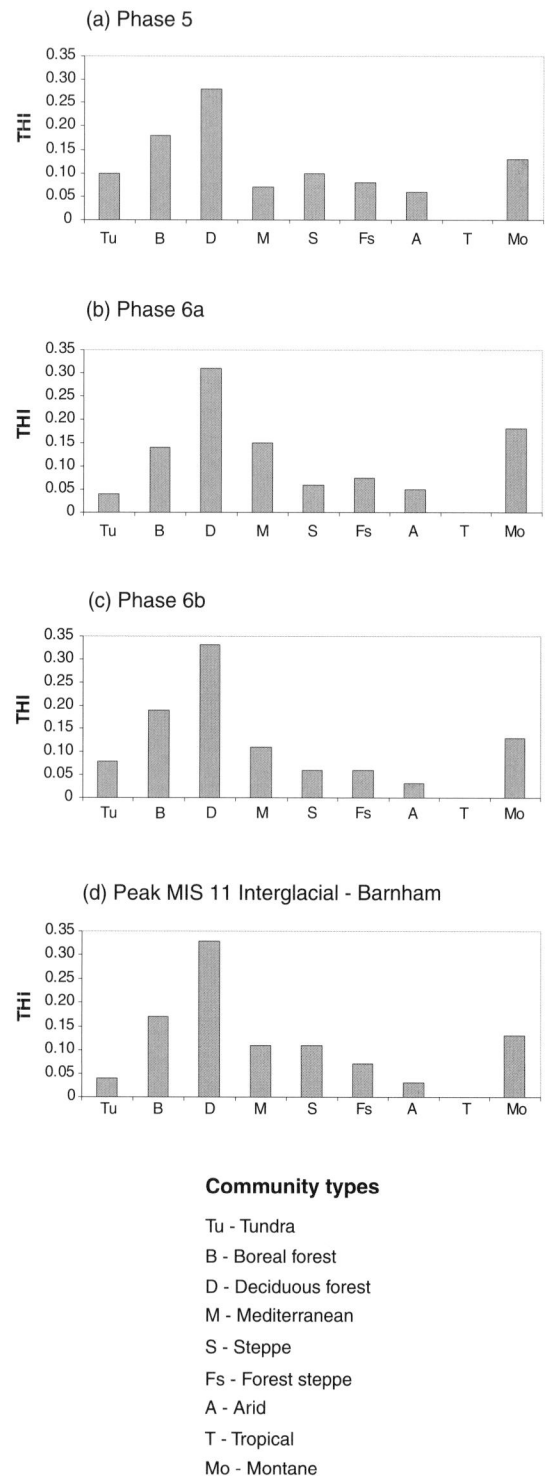

Community types

Tu - Tundra
B - Boreal forest
D - Deciduous forest
M - Mediterranean
S - Steppe
Fs - Forest steppe
A - Arid
T - Tropical
Mo - Montane

Figure 7.24 Taxonomic habitat indices for the micro-mammals from successive phases at Southfleet Road: (a) Phase 5; (b) Phase 6a; (c) Phase 6b, compared to the spectrum of a peak interglacial (MIS 11) assemblage from Barnham, Suffolk

whereas those calculated from the species present in Phase 6b are much more similar to those in the region encompassing southern England and the southern tip of Scandinavia. In Britain, such spectra are characteristic of the climatic optima of Pleistocene interglacials. This is illustrated by comparison with the THI spectrum from the Hoxnian fauna of Barnham (Suffolk). The Barnham fauna includes the thermophilous European pond terrapin *Emys orbicularis* and Aesculapian snake *Zamenis longissimus*. The former requires mean July temperature exceeding 17-18°C for hatching successfully; similar summer temperatures are indicated by Aesculapian snake.

The analysis shows that ecological differences between the spectra are slight, with high values for deciduous elements and somewhat lower scores for boreal elements in the three phases analysed. The spectrum from Phase 5 differs in having a slightly higher score for steppic categories and a somewhat lower Mediterranean component. Overall, the spectra are consistent with the deciduous woodland bioclimatic zone, encompassing central France to the southern tip of Scandinavia. Phase 5 appears to have a more continental aspect, whereas Phase 6a has a closer affinity with central France and 6b with more northerly localities. Overall, the spectra are entirely consistent with interglacial conditions (cf. Andrews 1990).

The vertebrate evidence provides a clear insight into the climatic conditions during the Clactonian occupation at Southfleet Road. A combination of community structure (using the Taxonomic Habitat Method) and climatic preferences of both the extinct and extant taxa yields a consistent picture, indicating that during Phases 5 and 6, the deposits accumulated during an interglacial period. Conditions during the deposition of Phase 3 were probably also temperate. The presence of warmth-loving vertebrates and the complete absence of any exclusively boreal species, suggests that the climate was as warm as that of southern England at the present day and may have been somewhat warmer.

Chapter 8

The elephant skeleton and the question of human exploitation

by Simon A. Parfitt, Silvia M. Bello, John R. Stewart, Richard J. Hewitt,
Victoria L. Herridge and Francis Wenban-Smith

INTRODUCTION

This chapter explores the question of whether humans were involved in the accumulation and modification of the large mammal remains. It follows two lines of evidence; firstly the association of the elephant skeleton with refitting Clactonian artefacts and, secondly, the evidence from cut marks. The focus of the first part of the chapter is the elephant skeleton itself, which has received a great deal of attention because of its preliminary interpretation as a rare example of Lower Palaeolithic elephant butchery. A detailed account of the taphonomy of the elephant skeleton aims at documenting the sequence of taphonomic events from the death of the elephant up to the chance discovery and excavation of its disarticulated and poorly preserved skeleton in 2004. The depositional and environmental context of the bones is also discussed. Age at death, gender and body-size are inferred from an analysis of the bones. These results are integrated with the stone tool evidence in Chapter 17. The second part of the chapter outlines the results from a careful search for cut marks and evidence for marrow extraction. Given the abundance of stone tools at the site, it is surprising that only two cut-marked bones were recorded. The chapter concludes with a discussion of the scant evidence for Clactonian butchery practices and a review of elephant butchery in a wider European Palaeolithic context.

Figure 8.1 The Ebbsfleet elephant remains during excavation: the poor condition of the tusks is evident. [Photo MR Bates]

Table 8.1 List of elephant bones from Southfleet Road, contexts 40078 and 40144.

Context	Find no. Δ	Element	Comments
40078			
	40060	L M³	
	40499	R M³	
	40362	Tusk frag	
	40363	Tusk frag	
	40402	Tusk frag	
	40411	Tusk frag	
	40412	Tusk frag	
	40526	Tusk frag	
	40527	Tusk frag	
	40529	Tusk frag	
	40573	Tusk frag	
	40574	Tusk frag	
	40685	Tusk frag	
	40691	Tusk frag	
	40925	Tusk frag	
	40927	Tusk frag	
	40933	Tusk frag	
	40998	Tusk frag	
	41002	Tusk frag	
	41004	Tusk frag	
	41082	Tusk frag	
	41356	Tusk frag	
	41439	Tusk frag	
	41442	Tusk frag	
	41448	Tusk frag	
	41457	Tusk frag	
	41494	Tusk frag	
	41948	Tusk frag (tip)	
	41444b	Tusk frag	
	40406	? Skull frag	
	40524	? Skull frag	
	40535	? Skull frag	
	41197	? Skull frag	
	40931	Cervical vertebra frag	Articulates with Δ. 41490
	41490	Cervical vertebra frag	Articulates with Δ. 40931
	40990	Cervical vertebra frag	
	40942	Cervical vertebra frag	
	40080	Thoracic vertebra centrum	Cranial & caudal unfused
	40773	Thoracic vertebra	
	40887	Lumbar vertebra	Almost complete (spine missing). Cranial unfused, caudal fusing
	40781	Lumbar vertebra centrum	Cranial unfused, caudal fused
	40999	Thoracic vertebra	
	40780	Thoracic vertebra	
	40079	Vertebra frag	
	40707	Vertebra frag	
	40779	Vertebra frag	
	40798	Vertebra frag	
	40938	Vertebra frag	
	40769	Vertebra frag	
	40074	? Vertebra frag	
	40784	? Vertebra frag	
	40598	Rib frag	
	40684	Rib frag	
	40703	Rib frag (proximal end)	
	40704	Rib frag	
	40596	Rib frag	
	40765	Rib frag (with ?skull frag)	
	42926	Rib frag	
	40516	Rib frag	
	40533	Rib frag	
	43493	? Rib frag	
	41021	? Rib frag	
	41033	? Rib frag	

THE EBBSFLEET ELEPHANT

The excavation took place under rescue conditions and the dry clayey nature of the subsoil contributed to the difficulty of excavating the fragile and fractured elephant remains (Fig. 8.1). Most of the bones were missing and only 47 fragmentary identifiable elephant elements, two molars, two substantial portions of the left and right tusks and 27 tusk fragments were recovered (Table 8.1). These include both cranial and postcranial portions, but notably absent are the pelvis, long bones as well as most of the other limb bones. Nearly all of the elephant bones were found in a cluster covering an area of approximately 25m² (Fig. 8.2), with a separate scatter of articulating foot bones some 20-30m to the north-east. In addition, 159 specimens are too fragmentary to identify to element, but can only have come from an elephant due to their size and cortical thickness. There is no duplication of bone elements and all are consistent in size and ontogenetic age, implying that all of the bones pertain to a single extremely large adult individual (Appendix D5).

The elephant skeleton was largely disarticulated and the remains were found in the central western part of the site, within the lower part of the grey silty clay (Phase 6 deposits) approximately 5m below the modern ground surface. The Phase 6 clay was up to *c* 3m thick, but most of the elephant bones were found in a single brown organic-rich horizon *c* 100mm thick (40078) at the base of the unit. This unit was brecciated with iron-mottled horizons and other darker organic-rich bands in the area of the elephant discovery. The brecciation and iron-mottling indicates fluctuating water levels with periods of desiccation and oxidation. The bone-bearing horizon slopes gently down towards the north, although the extent to which this is the result of post-depositional deformation is unclear. Although pollen was poorly preserved, traces of wood were noted during the excavation and numerous *Alnus* (alder) sieve plates (derived from the breakdown of wood) were observed in the palynological preparations (Chapter 12). It seems likely therefore that the elephant died in or immediately adjacent to a swampy alder carr. Associated vertebrate and palynological evidence suggests that the elephant

lived during a period of fully temperate climatic conditions of the early-temperate substage of the Hoxnian interglacial.

Taxonomy

The identification of elephant skeletal remains is best determined from its cranium, tusks or cheek teeth. The tusks from Southfleet are badly crushed and deformed, but the upper third molars are more-or-less intact. The right molar was dislodged during section cleaning (see Chapter 3) and was damaged by the mechanical excavator. The morphology of the molars corresponds in all respects to the extinct straight-tusked elephant *Palaeoloxodon antiquus* (Fig. 8.3). Features diagnostic for *P. antiquus* include a narrow crown with well-spaced, lozenge-shaped lamellae (plates) that are broader in the middle of the tooth than their extremities and thick enamel. Taxonomic identification of elephant postcranial remains is less straightforward (Lister and Stuart 2010). Nonetheless, the exceptionally large size of the Ebbsfleet elephant bones (including the dispersed carpals and metacarpals), their lack of duplication and their proximity to the identifiable molars suggests that they are almost certainly from the same *P. antiquus* individual.

Age at death

The age of death of elephant can be determined from the state of fusion of the postcranial skeleton and from the eruption and wear state of the cheek teeth (Haynes 1991). The third upper molars are in early wear indicating a prime-aged individual of about 43-49 years old (Appendix D5). Fusion state could be recorded for the metacarpals and second phalanx, which all have fused distal epiphyses, whereas some of the vertebral plates were either unfused or only partially fused at the time of death (Table 8.1, Fig. 8.4). This indicates that animal was not quite skeletally mature. Given the typical pattern of elongated skeletal growth period in elephants, the age estimates derived from teeth are not inconsistent with the fact vertebral epiphyses were not all fully fused.

Table 8.1 (continued)

Context	Find no. Δ	Element	Comments
	40534	L. scapula frag	Conjoins with Δ. 40824
	40824	L. scapula frag	Conjoins with Δ. 40534
	42255	R. magnum	Articulates with: Δ. 41947b, Δ. 42411, Δ. 42412
	42411	R. trapezium	Articulates with: Δ. 42412, Δ. 42255
	41947b	R. unciform	Articulates with: Δ. 42255 & Δ. 42148
	42148	R. metacarpal IV	Articulates with Δ. 41947b
	42975	Metacarpal (crushed)	
	42412	R. metacarpal II	Articulates with: Δ. 42255, Δ. 42411
	42974	Metapodial frag (distal end)	
	42863	2nd Phalanx IV	
40144			
	43378	R. cuneiform	

Abbreviations: frag – fragment; L. – left; R. – right

Gender and body size

The gender of elephant skeletons is best determined from its pelvic bones, skull or tusks (Lister and Agenbroad 1994), but in the absence of these some indication can be gleaned from body size. Skeletons of African and Indian elephants exhibit a considerable size range, related to age, sexual dimorphism and individual variation. Kroll (1991) has shown that the straight-tusked elephants were also sexually dimorphic in body

(a)

size, with a clear size difference between bones of adult male and female *P. antiquus*. Based on this work, estimates of shoulder height suggest that males may have reached 4.3m (range 3.80-4.30m, n=6), whereas adult females were generally smaller than 3.40m (n=3).

The bones of the Ebbsfleet elephant indicate an animal of extremely large body size, although comparisons and estimates of shoulder height are complicated by poor preservation of the material. The only complete measurable bones are from the carpus, which include the second metacarpal with a length of about 230mm. Of the published metacarpal measurements (Table 8.2), only those of the male individuals from Fonte Campanile and Upnor (Kent), elephants are larger

(Andrews 1928). Shoulder height has been estimated for the Crumstadt elephant (with the shorter metacarpal) and for the Upnor elephant. Following these criteria, the subadult individual from Crumstadt had a shoulder height of some 2.90m (Kroll, 1991), whereas the massive Upnor elephant has an estimated height at the shoulder of about 4m. The Ebbsfleet elephant was probably of comparable size, although somewhat smaller than the Upnor elephant (Fig. 8.5a,b). The bones of the Selsey (Sussex) elephant (a probable female) are considerably smaller, as shown in Figure 8.5c,d. Such size differences are almost certainly due to sexual dimorphism. Based on its extremely large size, the Ebbsfleet elephant was almost certainly a male.

Figure 8.2 *(left, above and overleaf)* Geo-rectified vertical photographs of the main cluster of elephant bones (outlined in red) at Southfleet Road during successive stages of excavation (latest–earliest a–c). The poor condition of the larger bones is due to a combination of weathering before burial and soil corrosion, decalcification and mechanical damage during burial.

(c)

Figure 8.2c (left – caption on page 209)

Table 8.2 Measurements of the second metacarpal from the Southfleet Elephant, in comparison with those from other *Palaeoloxodon* skeletons

	Length (mm)	
Crumstadt, Germany[a]	160	Female, subadult
Gröbern II, Germany[b]	179	Female, adult
Riano, Italy	175	Male, adult
Ciechanow, Poland[b]	189	Female, adult
Warschau, Poland[b]	206, 209	Male, subadult
Gröbern I, Germany[b]	217	Male, adult
Jozwin, Poland[b]	226	Male, adult
Southfleet Road	~230	Male, adult
Upnor[c]	233 c	Male, adult
Fonte Campanile, Italy[d]	237 d	

[a] Kroll (1991), Table 15
[b] Kroll (1991), Table 49
[c] Andrews (1928) gives a slightly smaller measurement of 227mm for this specimen.
[d] Estimated from illustration in Trevisan (1947 fig. 25)

Figure 8.3 *(above)* Third upper molars (M^3) of the Ebbsfleet elephant (a-b): Side view (a) right M^3 (Δ.40060); (b) left M^3 (Δ.40499); (c) occlusal view (Δ.40499)

Figure 8.4 *(right)* Lumbar vertebra (Δ..40887), in posterior view. Note partial fusion of caudal bony plate with the body

Taphonomy

Skeletal element representation

Although bones from the skull, fore limbs and torso are represented, the hind limbs were entirely missing (Fig. 8.6). Other bones that were absent include the sacrum, caudal vertebrae and the mandible. Long bones are also conspicuous by their absence, although several of the pieces identified as indeterminate 'elephant-sized' bones appear to be diaphysis fragments. The skeletal elements identified are:

Skull and teeth
Two substantially complete tusks (up to 2m long) were too fragile to recover intact. Many smaller comminuted

Figure 8.5   Metacarpals of straight-tusked elephant from Southfleet Road, Upnor (Kent) and Selsey (West Sussex): (a) Second metacarpal (Δ.42412) of the Ebbsfleet elephant; (b) second metacarpal from the Upnor elephant, (M.11156); (c) third metacarpal from Selsey; (d) third metacarpal from Upnor. The photograph illustrates the considerable size variation, which is almost certainly due to sexual dimorphism

Figure 8.6　Metacarpals of the Ebbsfleet elephant: (a, c) second metacarpal (Δ. 42412); (b) (d) fourth metacarpal (Δ. 42148); in medial (a, b) and (c, d) anterior views

tusk fragments were found nearby and the tip of a tusk (not necessarily from the same individual) was recovered some 15m to the north-east. A small number of fragmentary bones were identified as possible cranial fragments. These were found together with pieces of tusk, and appear to be from the alveoli. The third upper molars are in good preservation state, although both are somewhat damaged at their ends.

Vertebrae and ribs
Elements from the vertebral column are represented by fragments of cervical (mostly articular processes), thoracic and lumbar vertebrae. Only one rib retained its proximal end, the other fragments consisting of entirely of blade fragments.

Scapula
A left scapula is represented by two substantial refitting pieces from part of the blade near the scapular spine.

Figure 8.7　First phalanx (Δ. 42863) of the Ebbsfleet elephant

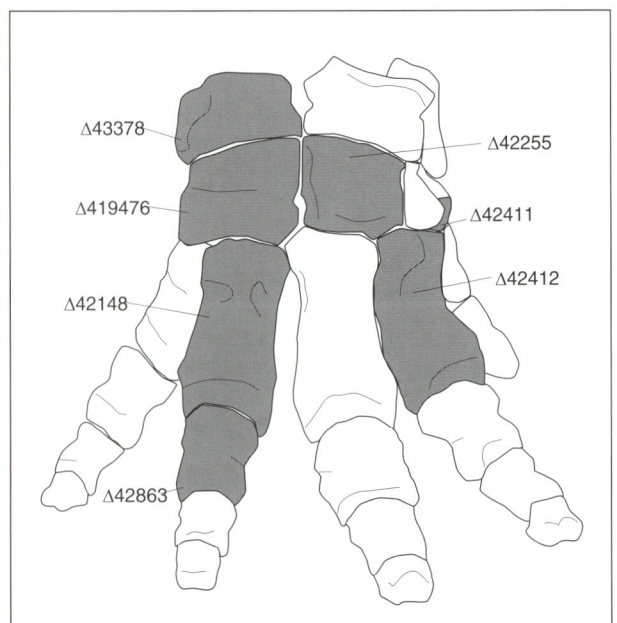

Figure 8.8　Elephant carpus, showing the bones present at Southfleet Road. With the exception of the cuneiform with damaged articular surface, the bones clearly articulate with one another in anatomical sequence

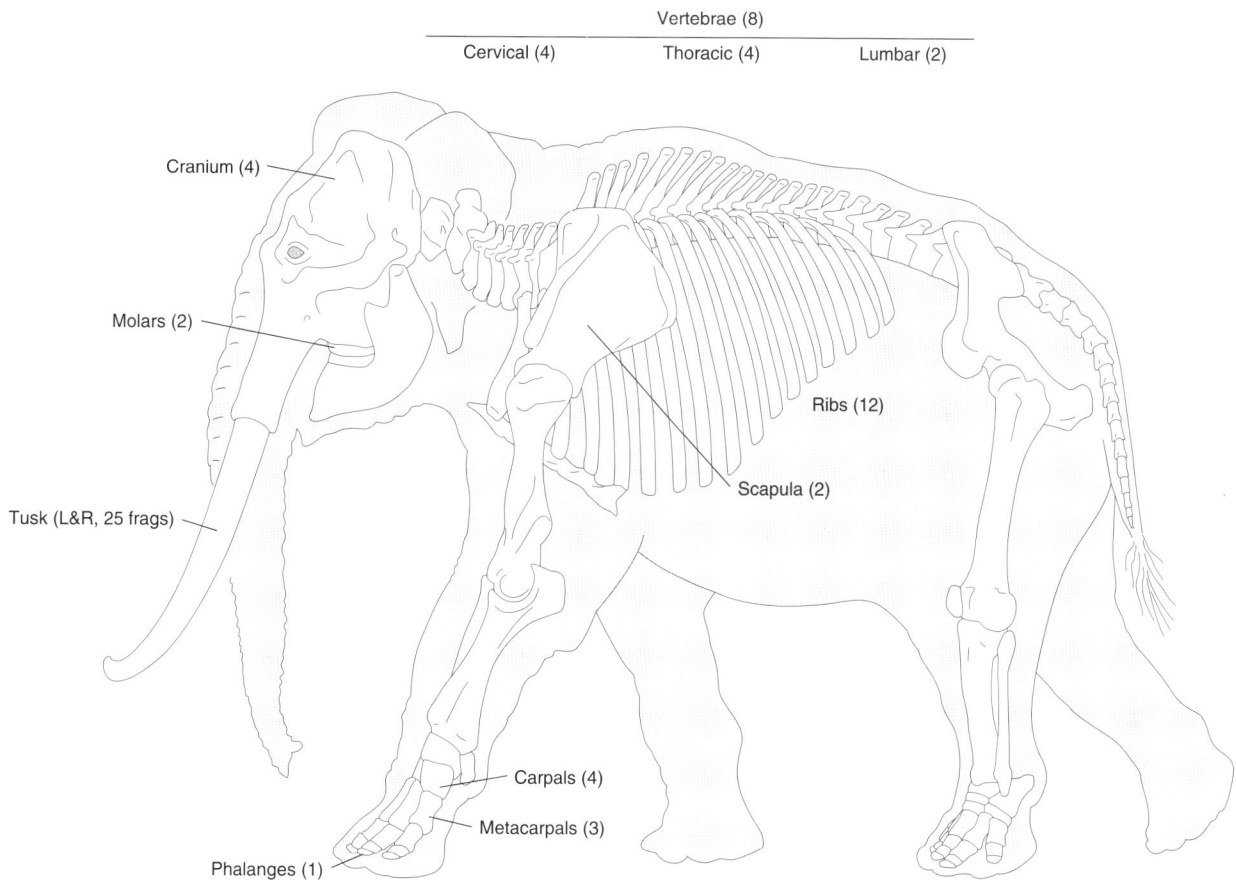

Figure 8.9   Numbers of bone fragments identified to skeletal element. The precise anatomical position of many of the pieces is uncertain due to the considerable fragmentation of the bones

Foot bones
Elements of the right carpus include unciform, magnum, trapezoid, three metacarpals (Fig. 8.7) and a phalanx (Fig. 8.8) from context 40078. The carpals and metacarpals articulate at their articular facets showing that they are from a single individual. The articular facets of the right cuneiform from context 40144 are damaged, but it is most certainly part of the same foot, given its large size and proximity to the other foot bones (Fig. 8.9).

The preceding catalogue highlights the fact that the collection represents no more than about 5% of the elephant skeleton (Fig. 8.6). This raises questions about the processes that led to the loss of the bulk of the skeleton.

Preservation and anatomical distribution of bones

Plots of the bone distribution show that the vast majority of elephant remains were confined to a relatively tight cluster located at the very western edge of the site (Fig. 8.10). Very few elephant bones were found beyond this area, and the distribution of bones to the west of the main cluster is unknown as this part of the site was truncated by mechanical excavation. It is certain that some of the elephant skeleton was lost to mechanical excavation, as some of the surviving bones

were cut in half by the excavator blade (Fig. 3.17). Bones in the main cluster include parts of the axial skeleton, ribs, scapula and the fragments of the cranium (Fig. 8.11). Two large portions of the left and right tusks and two upper molars were also present, but the mandible and lower molars were not recovered. The two largely complete, but badly crushed, tusks mark the southern end of the cluster. Overall, the scatter has a linear form, which is orientated approximately north-west to north-east. To the north and east of the core group is a more dispersed scatter of elephant bones that includes several rib fragments, the tip of a tusk and bones of the right carpus. The carpus bones were deposited along the edge of a shallow channel feature filled with calcareous tufaceous sediments; one of the foot bones (a right cuneiform, Δ.43378) was found in the upper fill (40144) of the tufaceous channel.

To assess the extent to which post-depositional and burial conditions contributed to the loss of bone material, the quality of bone preservation was analysed. This assessment adopted a four point scale with grading, from good (1) to poor (4). Specimens assigned to condition category 4 were denatured due to localised decalcification of the sediments and include pieces that have survived as little more than 'stains'. Preservation of

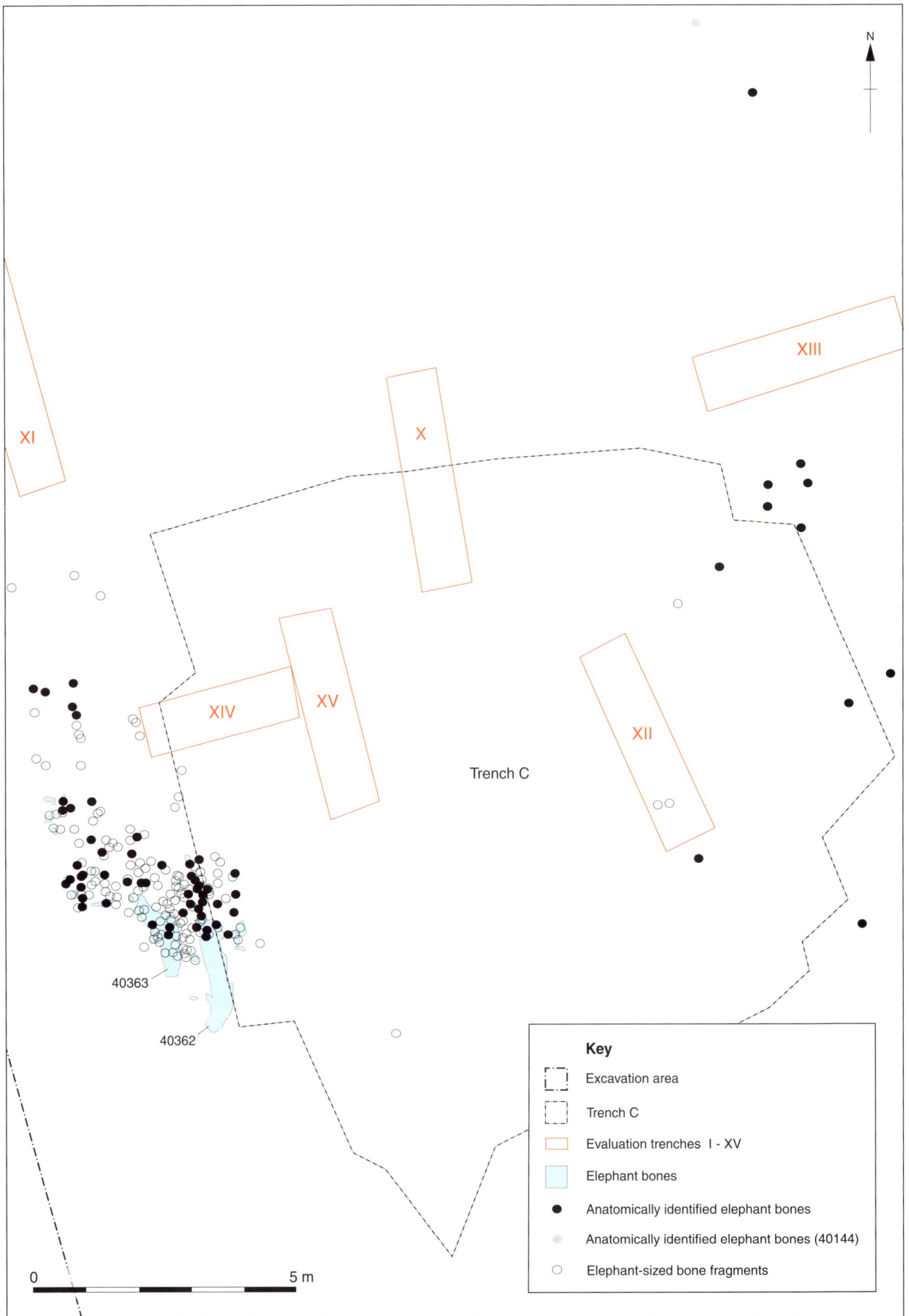

Figure 8.10 Plan of Southfleet Road, showing the distribution of elephant bones identified to element and non-diagnostic (elephant-sized) bone fragments

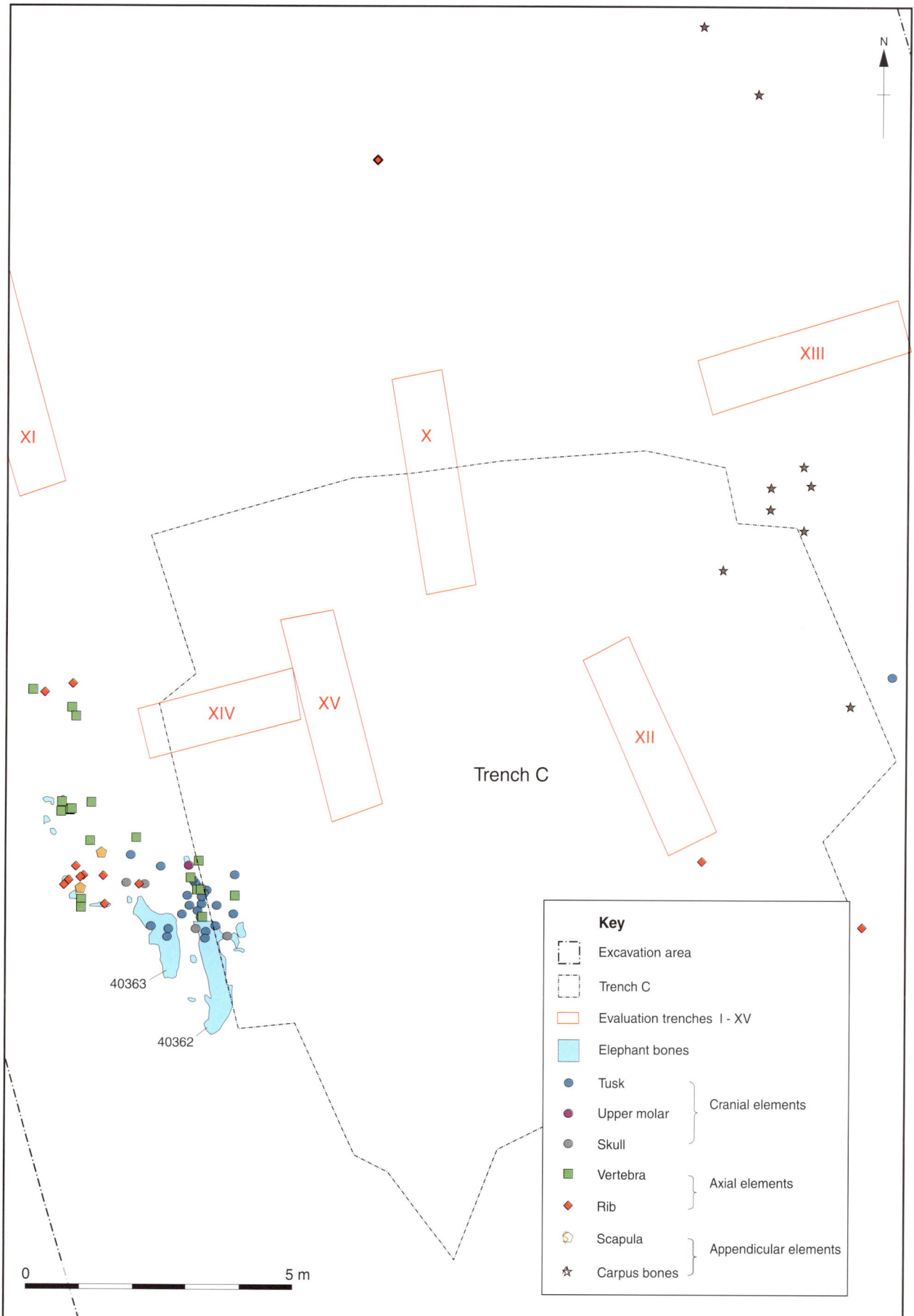

Figure 8.11 Plan of the Southfleet Road excavation showing the distribution of elephant remains according to body part

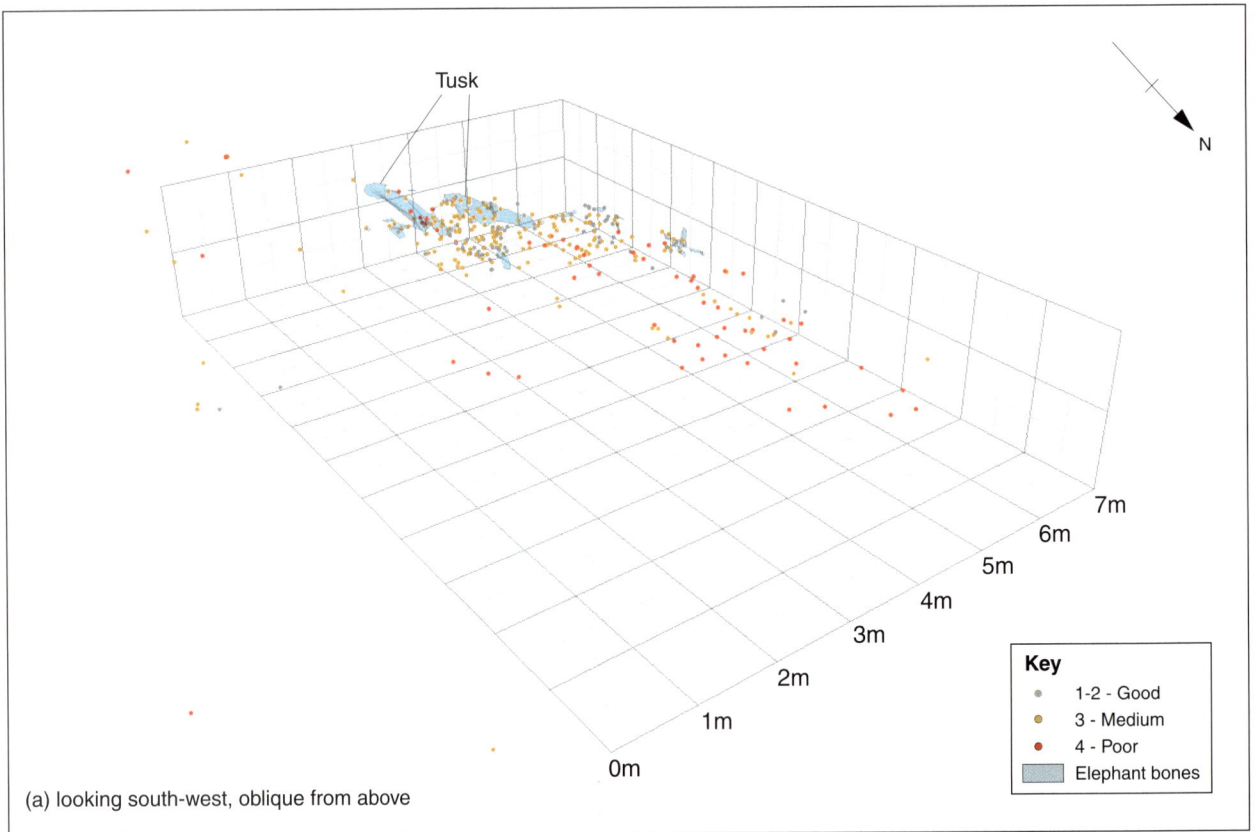

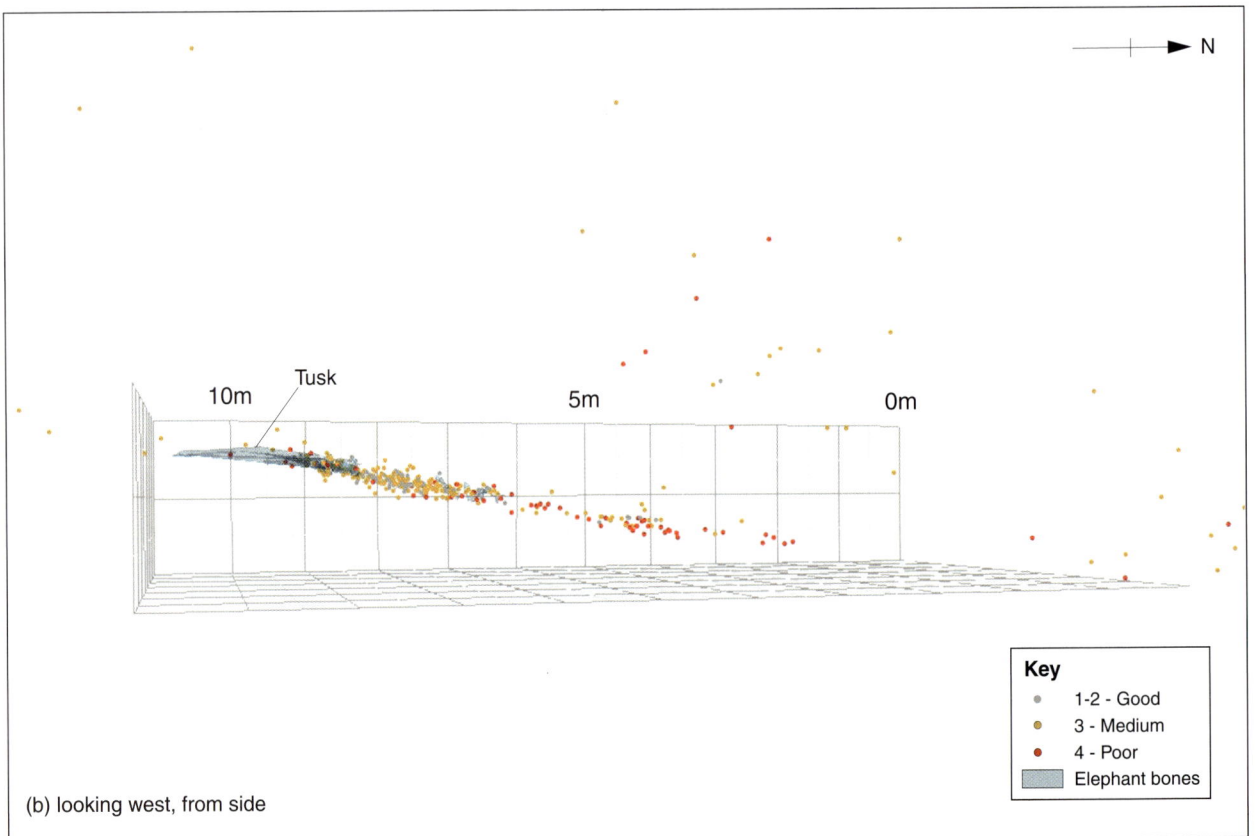

Figure 8.12 3-D plots of the main elephant-bone cluster showing variations in bone condition (1-4): (a) looking from south-west; (b) showing vertical distribution

the elephant bones is extremely variable; a few are not altered or corroded in any way, but most are in poor condition (Fig. 8.12). Fragmentation of the bones is often severe, with most of the bones sustaining fractures as a result of compression during burial. The tusks were in particularly poor condition, being crushed into splinters, although broadly retaining their original shape. The larger bone fragments were generally in an advanced state of disintegration having been subjected to compression and minor lateral movement within the clay. Lateral movement of the sediment body is indicated by slickenside striations in the surrounding sediment, with lateral movement of up to 50mm. Breakage from sediment compaction was compounded by repeated cycles of wetting and desiccation of the clay-rich matrix surrounding the bones, causing the peds to expand and contract. The effect of this process upon the bones was to create sediment-filled split lines and fissures, which has further weakened the bones. In addition, some of the large bones were damaged by mechanical excavation before their discovery. The foot bones close to the edge of the channel were in better condition, possibly due to the higher calcium carbonate content of the sediments associated with the tufaceous channel fill. Heavy post-depositional corrosion has affected fourteen of the bones, which survive as little more than stains or decomposed and crumbly bone (Fig. 8.12).

The state of preservation of bone surfaces is also extremely variable. This spectrum includes pieces with fresh, well-preserved surfaces, while others have lost much of the outer layer of cortical bone. The bones with fibrous, pitted or corroded surfaces may have been exposed to weathering before burial, but much of the corrosion appears to have occurred after burial. This type of surface degradation arises from chemical processes below ground chiefly due to soil conditions, temperature and moisture fluctuations in a biologically active soil and leaching of calcium carbonate (decalcification). Of the bones with well-preserved surface, none bear obvious traces of rounding due to water transport or signs of surface weathering. The smaller pieces of bone are unlikely to have been exposed for any considerable period of time, and were probably rapidly covered by sediment soon after death. In contrast, the largest bones, such as the skull and limb bones were probably exposed on the ground surface for some years, becoming weathered and fractured in the process. Polishing and fine parallel scratches (trample marks) were observed on a small number of the elephant bones. The polishing is likely to be from post-depositional processes and caused by the expansion and contraction of the matrix and slickensiding. Manganese dendrites were present on 42 pieces.

Carcass decay, disarticulation and bone breakage

An examination of the spatial distribution of the skeletal elements has proved invaluable for interpreting the sequence of carcass decay, disarticulation and movement of body parts away from the main cluster of elephant bones. Critical evidence is provided by the degree of bone disarticulation and scattering together with the spatial distribution of bones from different 'zones' of the body. The distribution of carcass components is presented in a series of schematic plans (Figs 8.13-16). Certain general patterns emerge. Firstly, there is a degree of clustering of different skeletal elements that reflects anatomical position within the skeleton (Fig. 8.13). This patterning is best illustrated by the distribution the tusks, vertebrae and ribs, which are in close spatial proximity and retain some resemblance of anatomical order (Fig. 8.14). Also notable is the linear arrangement of the vertebral fragments, which would appear to mark the position of the vertebral column (Fig. 8.14). From the general distribution of bone elements it is possible to infer that the elephant was lying on its left-hand-side. Before the bones were scattered, the skeleton was probably orientated north–south, with the head at the southern end. Much of the rear part of the skeleton would therefore have been located in the area that was dug-away during road construction.

Studies of elephant death sites have shown that some of the bones were not scattered but remain close to where the animal died. Bones remaining at the death site typically include the skull, vertebrae and ribs (Haynes 1988; 1993). These are the elements represented in the main scatter at Southfleet Road giving credence to the view that this area marks the location where the animal died. Observations of carcass decay in natural situations have also shown that the disarticulation of large herbivore skeletons is non-random and predictable (Hill and Behrensmeyer 1984). For example, fore limbs commonly separate off as a single unit and generally disarticulate sooner than the hind limbs. The scapula detaches from the body as the muscles decay, whereas the distal limb bones are held together by stronger ligamentous attachments. The distal elements and feet are often removed at an early stage of carcass decay. It is a commonly observed stage in the dispersal of mammalian skeletons for the proximal hind limbs, axial skeleton and cranium to be preserved at the death site with the feet and fore limbs removed. The diffuse scatter of elephant bone fragments to the east of the main cluster comprises mostly small bone fragments. Whether these were initially complete and broken after transport is unclear. Some of the bones missing from the main cluster were dispersed up to 30m from the carcass (Figs 8.13 and 8.16). Several mechanisms may be invoked to explain the movement and breakage patterns. For example, the dispersal of the bones may have involved scavenging carnivores, the trampling activity of large mammals or transport by flowing water and down-slope movement. Carnivores such as lion, bear and wolf would have been attracted to the carcass, although no unequiv-ocal traces of carnivore chewing were observed. This is somewhat surprising as the assemblage includes bones with well-preserved cortical surfaces and spongy bones that would have been attractive to carnivores. The latter group includes the spongy foot bones, which would have been favoured by scavenging carnivores (Stuart and

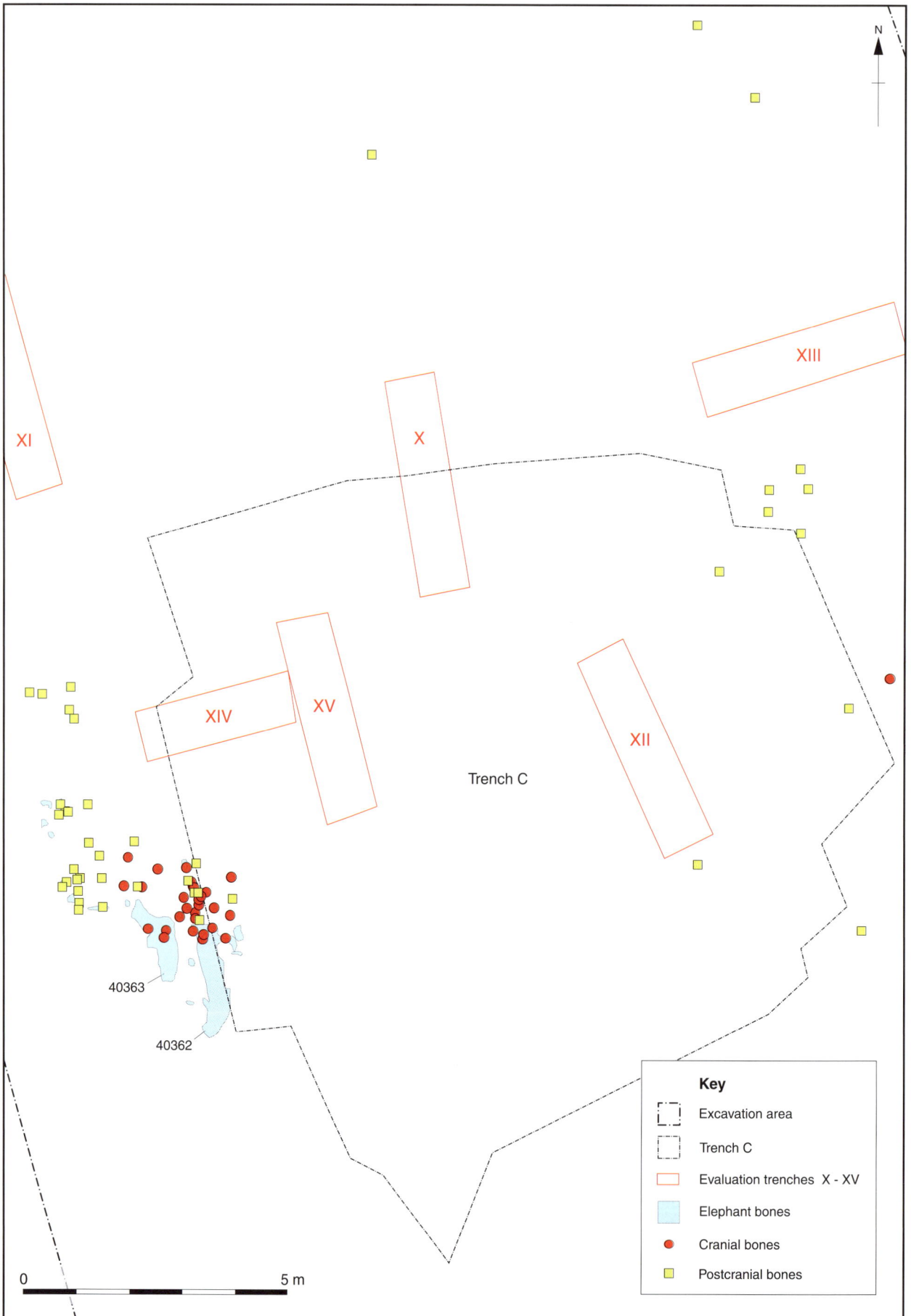

Figure 8.13 Plan of elephant bones, showing the distribution of cranial and postcranial elements

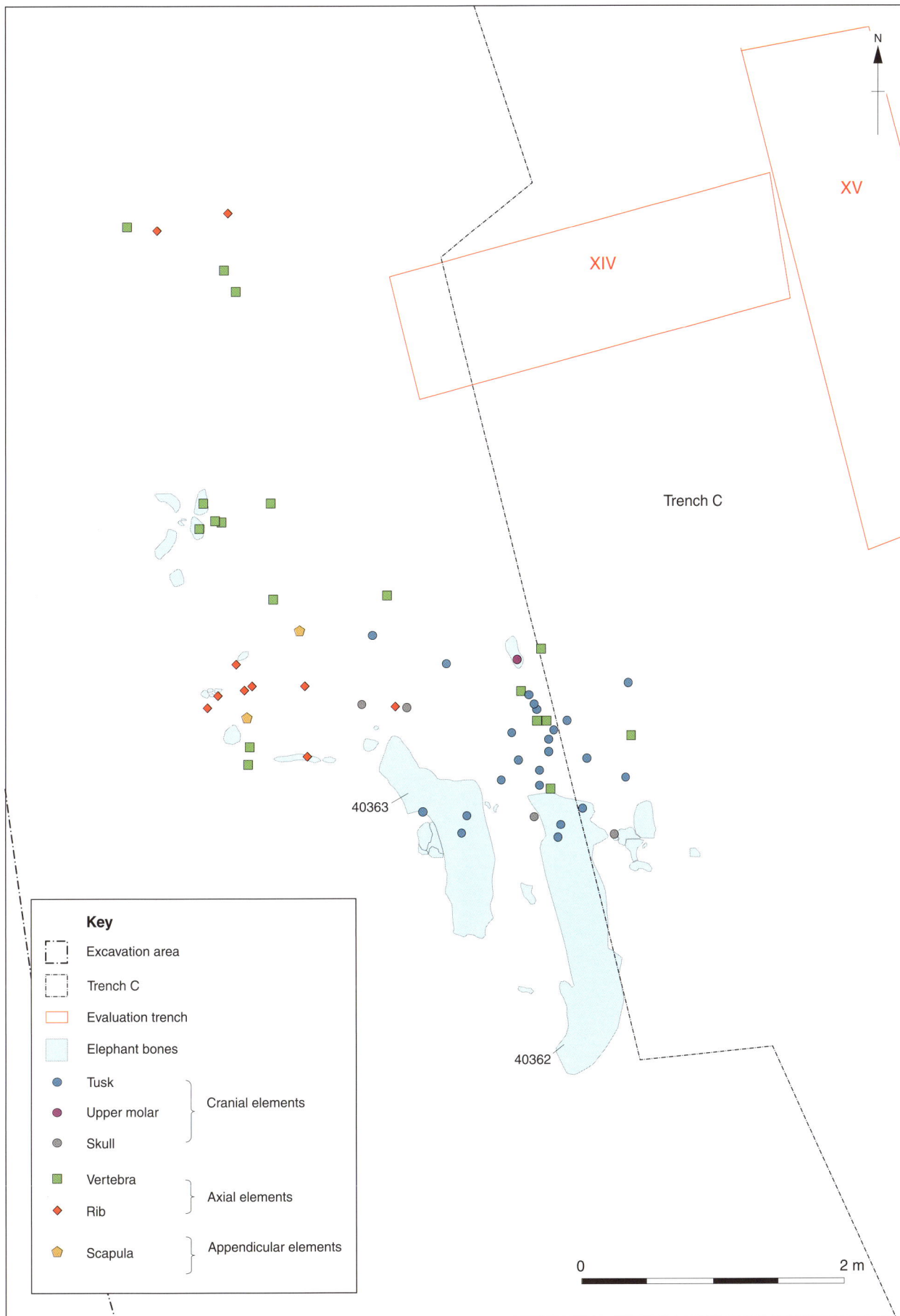

Figure 8.14 The main cluster of elephant bones, showing the distribution of bones according to body part

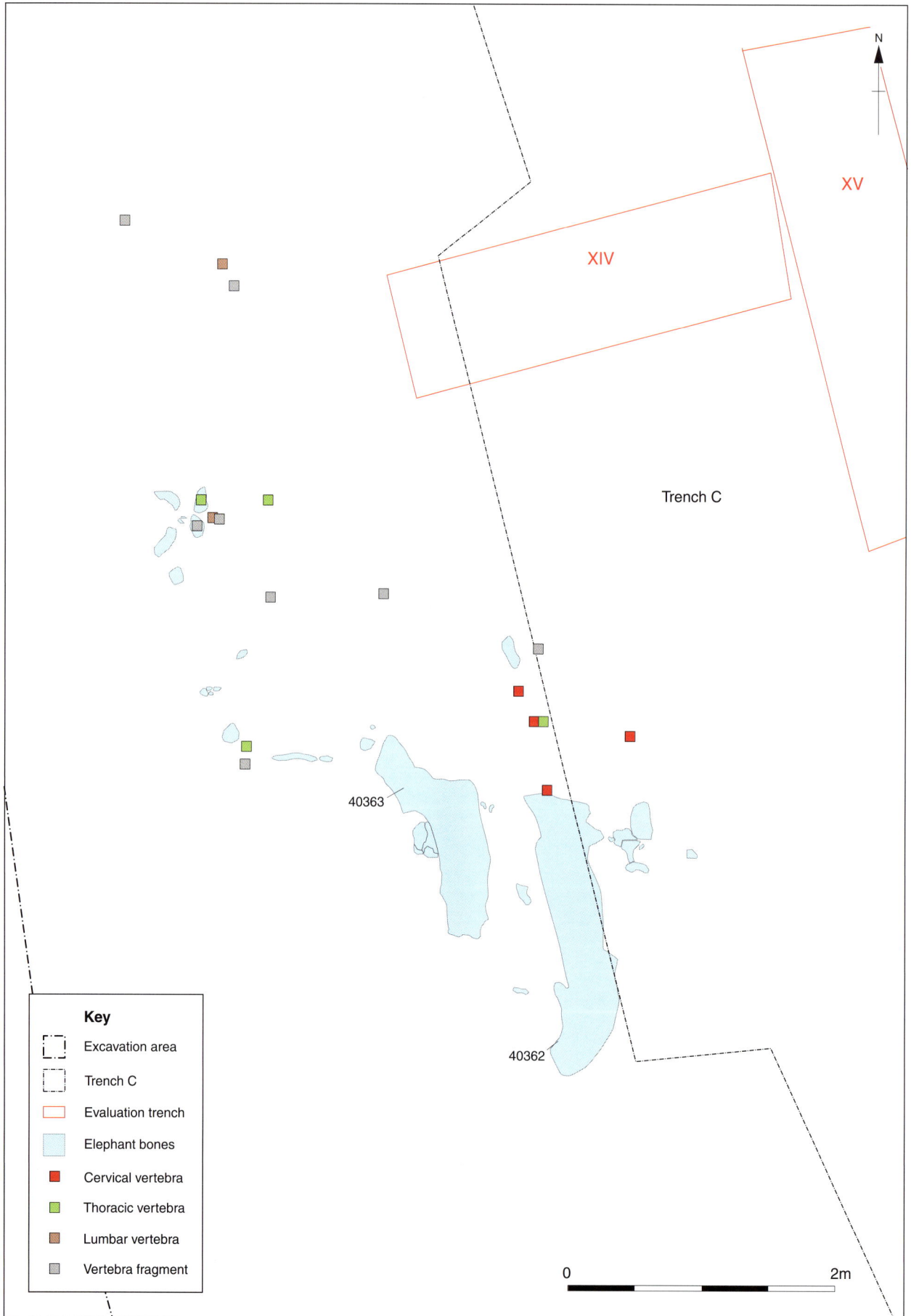

Figure 8.15 Distribution of vertebrae in the main cluster in relation to the tusks (Δ.40363, Δ.40362)

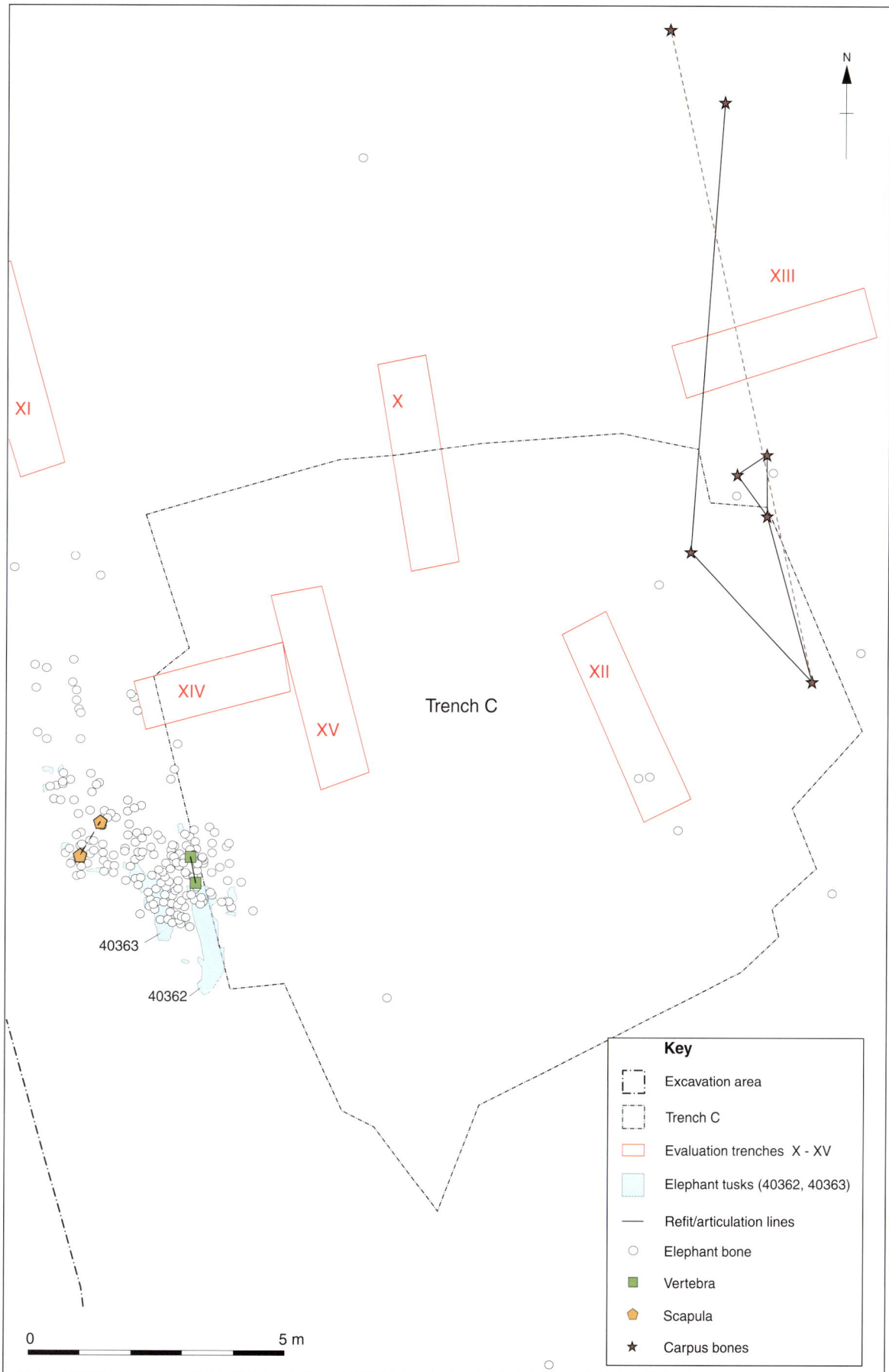

Figure 8.16 Distribution of refitting and articulating elephant bones at Southfleet Road

Larkin 2010), but not necessarily marked by carnivore chewing damage (see Haynes 1988, 139). At Southfleet Road, these bones are in an almost pristine condition, but nevertheless lack any evidence for tooth punctures or other chewing damage. The lack of any evidence for carnivore involvement suggests that the carcass was not heavily scavenged after death and that scavengers did not play a key significant role in dispersing the bones. Evidence for water transport of the bones is equivocal. Although the carcass site is located on the edge of a river valley next to a stream, there is no sedimentological evidence of high-energy water flow (eg channelling and bedded sediments). This is supported by the presence of elements with different transport properties (vertebrae, molars and tusks), which is inconsistent with sorting by fluvial processes (Frison and Todd 1986). Nevertheless, the long axis orientation of the elongated bone

fragments does exhibit a clear preferred orientation (Fig. 8.17), with most of the bone long axis measurements falling between 40 and 120 from north (Fig. 8.12). The most parsimonious explanation is that these bones, which are mostly small pieces less than 50mm long, were aligned by slow-flowing water, possibly from sheet-flow during flooding events. The refit orientations of the articulating foot bones may indicate faster flowing water associated with the tufa channel. Here, the refit orientations are aligned parallel to the channel (Fig. 8.16), implying that stream flow was sufficient to entrain the foot bones and carry them in a north-south direction. The sequence of carcass decay and disarticulation observed in natural situations conforms to the distribution of elephant skeletal elements at Southfleet Road (Fig. 8.16), suggesting that natural processes alone could account for some aspects of the bone distribution and preservation patterns observed at the site.

Most of the elephant bones from Southfleet Road are broken. A notable feature of this assemblage is the high degree of breakage reflected in fragment size distributions (Table 8.3). In the absence of any evidence for carnivore breakage or marrow fracture and other processing activities undertaken by humans, the breakage is likely to have been the result of weathering and trampling. Trampling by large mammals seems to have been an important agent of pre-depositional breakage. This is indicated by patches of sub-parallel, shallow scratch marks on several bones (Fig. 8.18); these are interpreted as trampling abrasions (cf. Andrews and Cook 1985; Behrensmeyer et al. 1986). It is likely that much of the breakage was due to trampling, but because the sediment is fine-grained, the trampling did not always leave visible surface marks. The fact that none of the bones were found lying at a steeply inclined angle suggests that the bones were trampled on a relatively hard substrate. The larger bones may also have been in a weakened condition as a result of prolonged exposure to weathering. This is suggested by the poorest preservation, which is observed in the largest fossils, for example the skull and tusks. These would have been exposed to longer periods of sub-aerial weathering than the smaller more easily buried bones that show little or no evidence for exposure (see above). Clearly, weathering could have been a contributory factor in the breakdown and destruction of thinner-walled bones, such as the skull, which probably disintegrated before sediments built up around them. Elephants may also have had a role in modifying the carcass by breaking and dispersing the bones. Living elephants react strongly to the carcasses of other elephants and there are numerous observations of modern African elephants revisiting sites of elephant skeletons and carrying and kicking bones around. These bones are often smashed and trampled underfoot during their investigations (Haynes 1991). There is no reason to believe that extinct relatives differed in their reaction to elephant carcasses and the actions of inquisitive straight-tusked elephants may provide an explanation for some of the patterns observed at Southfleet Road.

Table 8.3 Summary of elephant bone fragment size

Element	Fragment length				
	<5cm	6-10cm	11-20cm	21-30cm	>30cm
Skull		3			1
Molar				1	1
Tusk	4	1	5	3	4
Vertebra		3	9	4	2
Rib			7	3	2
Scapula				1	1
Carpus bones			6	1	1
Indet.	17	75	49	8	3

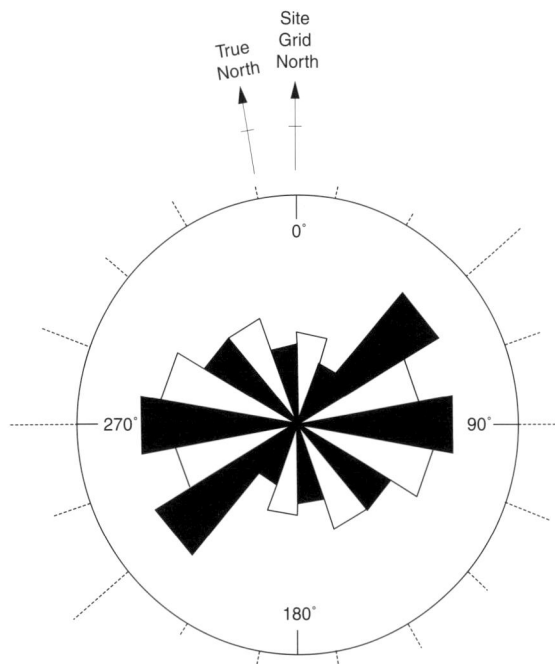

Figure 8.17 Orientations of elephant bone fragments from context 40078 plotted as a bidirectional rose diagram. The long axis orientations (measured from site plans) exhibit a strong preferred orientation

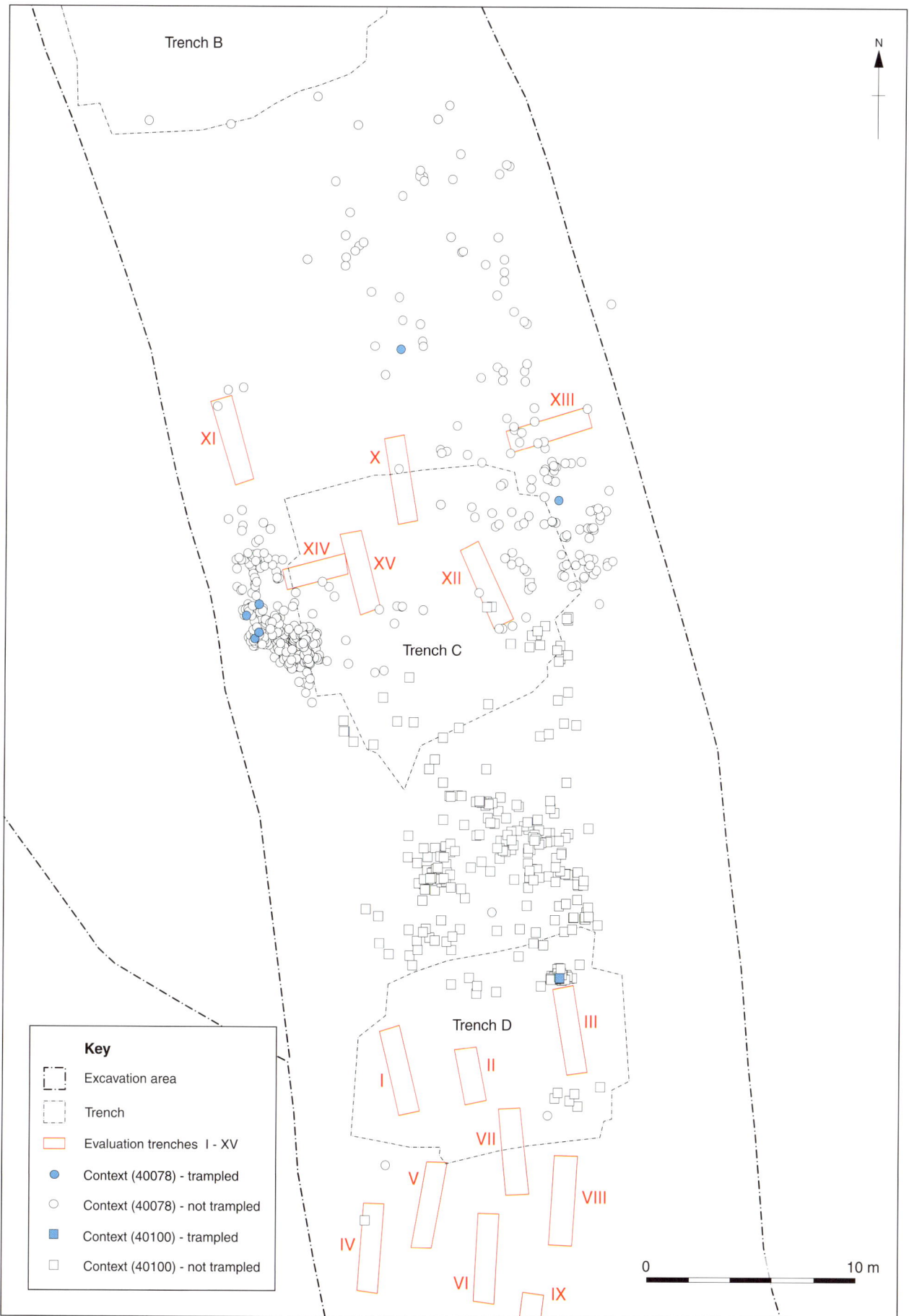

Figure 8.18 (a) Distribution of bones from contexts 40078 and 40100 with surface marks attributed to trampling by large mammals

N

XI

X

43764   XIII
43143

XIV

XV

XII

Trench C

40363
40362

Trench D

III

I

II

VII

V

VIII

IV

VI

IX

Key

Excavation area

Trench

Evaluation trenches I - XV

Possible cutmark

Trample mark

0   10 m

Figure 8.18 (b) Plan showing the location of cut-marked deer bones

Summary

Although less than 5% of the skeleton was recovered and many of the bones are poorly preserved, it has been possible to reconstruct a reasonably coherent picture of the environment in which the elephant lived and died and to infer aspects of its taphonomic history. Although many of the bones were fragmentary and difficult to identify, there appears to be no duplication of skeletal elements in the assemblage. Moreover, the size and ontogenetic age are conformable, supporting the interpretation that the teeth and bones in the main scatter belong to the same individual as the dispersed carpus bones. Wear on the third upper molars indicates that the elephant was about 45 years old at time of death, with no obvious skeletal pathology (Appendix D5). Size comparisons with more complete and better-preserved skeletons indicate that the Ebbsfleet elephant was an enormous animal, probably approaching 4m at the shoulder (based on comparison with the almost complete skeleton of similar size from Upnor, Andrews 1928) and probably weighing *c* 9 tonnes, making male gender very likely (Fig. 8.19). This is considerably larger than any modern African bull elephant, which typically weigh no more than 6 tonnes. The Ebbsfleet elephant is one of the largest straight-tusked elephant skeletons known, although not quite as large as the exceptional Gröbern I individual (Davies 2002). The carcass would have provided a huge source of food and raw materials had it been utilised by early humans.

Aspects of the mode of life of the straight-tusked elephant can be deduced from its skeletal and dental morphology, combined with floral and faunal evidence (Stuart 1982). In northern Europe, the straight-tusked elephant was strongly tied to temperate climates with wooded or mixed vegetation (Stuart 1982), but it is generally absent from intervening cold stage faunas (Lister 2004; Stuart 2005; Mol *et al.* 2007). It almost certainly fed on a mix of herbaceous vegetation and browse. These environmental and climatic associations are fully in keeping with the ecological and climatic context of the Ebbsfleet elephant, with associated environmental evidence suggesting a period of fully temperate climatic conditions and a mixture of open ground and deciduous woodland. The cause of death cannot be determined, but mortality of prime-aged bulls is an unusual event in the wild (Haynes 1991; Conybeare and Haynes 1984). Modern elephants are relatively invulnerable to predation (Haynes 1991) and large bull African elephants have no natural predators (other than man). The extremely large Ebbsfleet elephant is unlikely to have been troubled by lions, which were the largest predator in Britain at that time. There were no obvious bony pathologies, but many fatal diseases do affect the skeleton, and even hunting by humans and butchery may not leave any traces on the bones. One cause of death that can be safely excluded is miring in muddy sediments. Miring has been implicated in the death of the Aveley elephants from the intact foot bones found in anatomical position beneath the disarticulated bones. This contrasts with the situation at Southfleet Road, where the scattered carpus bones were dispersed from the primary carcass location.

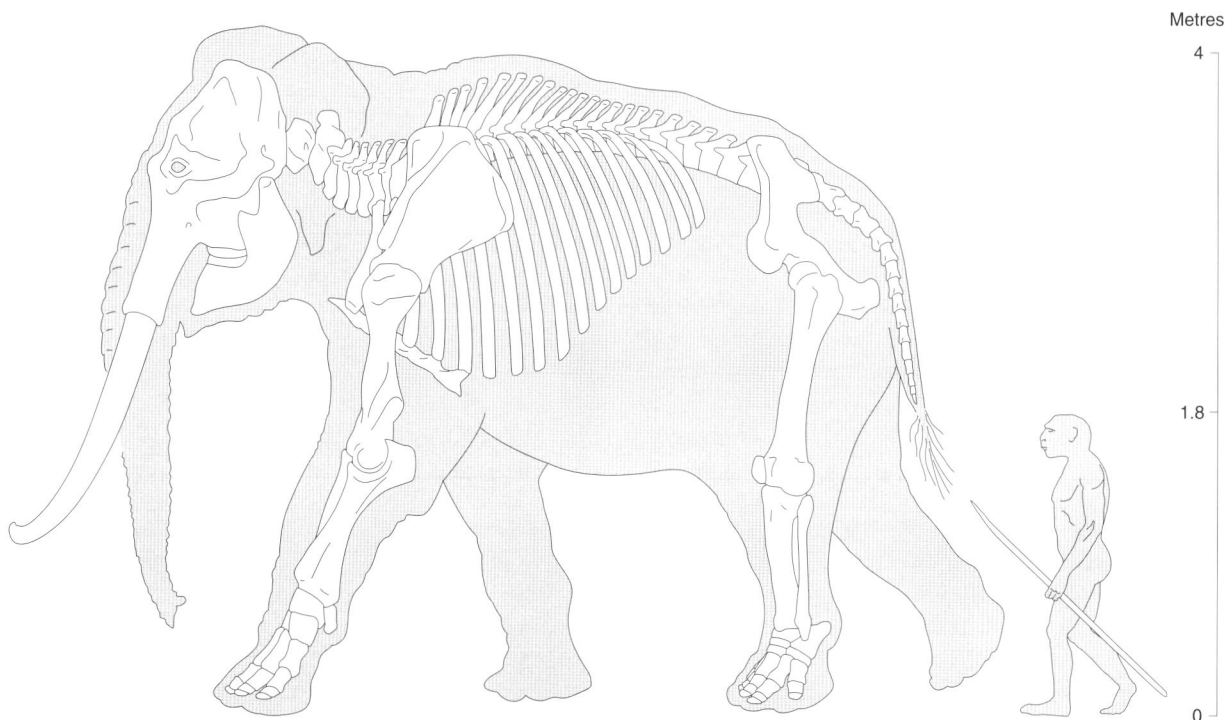

Figure 8.19 Reconstruction of the Ebbsfleet elephant (shoulder height of 4m), compared with estimated size of a Middle Pleistocene male hominin

The elephant appears to have died close to the edge of the valley in an alder carr swamp. The death site is represented by the main cluster of bones and this is where skeletonisation and primary disarticulation occurred. Although a wide range of skeletal elements from the axial skeleton is represented (cranium, tusks, upper molars, vertebrae and ribs), they are all from the anterior part of the animal and the only limb bone represented is the scapula. Missing elements include the sacrum, caudal vertebrae and the hind limbs. These parts of the skeleton were probably destroyed during road construction, which cut through the area of greatest bone concentration. The general distribution of elements suggests the carcass was originally orientated north–south and lying on its left hand side. According to Haynes (1993), this is the 'normal' death posture in the wild.

Shortly after death the carcass would have bloated and started to decompose, initially by autolysis as stomach and intestinal acids and enzymes escaped into the body cavity, and then more-rapidly as putrefaction and maggot infestation took hold. Scavenging carnivores, such as bear, lion and wolf, would have been attracted to the carcass at this stage, although here is no direct evidence for carnivore action from marks on the bones. These processes could have reduced the carcass to a pile of bones in a matter of months, somewhat longer if the elephant had died during the winter, or if the hide had become desiccated during a hot dry summer (Coe 1978). Over time, the skeleton would have completely disarticulated and individual bones become scattered and broken. However, some of the bones were buried close to their original anatomical position, with a linear arrangement of the vertebra and most of the rib fragments located behind the tusks. This suggests that an initial phase of bone breakage occurred shortly after the muscles, ligaments and tendons had decomposed and that some of the smaller bone fragments were buried rather rapidly. The bones in the main scatter also appear to have been inundated by flowing water, possibly at times of high-flood or as sheet-wash from torrential rainfall. This has aligned some of the smaller bones, but the water did not have sufficient power to cut channels, let alone transport the larger bones.

Because of their large size, some of the elephant bones would have remained unburied for many years and exposed to the destructive actions of weathering, lichen and microbial attack and trampling. Weathering is likely to have been the main destructive process. Numerous studies have shown that weathering is capable of reducing robust elephant bones into fragile, exfoliated and unidentifiable splinters (Haynes 1991). Elephants visiting the skeleton may also have contributed to the dispersal of the bones. There are numerous observations of African elephants displacing or rearranging or carrying such bones considerable distances from the carcass sites (Haynes 1991; Stuart and Larkin 2010). Elephants are also implicated in bone breakage, through trampling or by smashing the bones. Evidence for trampling at Southfleet Road is from abrasion marks on the elephant bones, which are typical of surface damage from large mammal trampling (Andrews and Cook 1985; Behrensmeyer et al. 1986; Haynes 1988).

Trampling, kicking and scuffing may also account for the diffuse scatter of rib fragments, the tip of a tusk and numerous unidentifiable elephant bone splinters found to the north and east of the main cluster. The diffuse scatter includes the articulating bones from the right carpus. Observation of modern elephant carcass disarticulation has shown that the foot bones are invariably amongst the first elements of the skeleton to separate from the carcass (Hill and Behrensmeyer 1984) and to be scattered and removed by carnivores (Haynes 1988). Chewing of the thin-walled spongy foot bones can result in puncture marks, grooves and breakage (Stuart and Larkin 2010), although Haynes (1988) has observed that in modern situations these bones 'rarely show signs of carnivore gnaw damage (ibid., 139). The foot of the Ebbsfleet elephant was probably dragged away from the carcass by large carnivores. Subsequently, the foot bones separated when the tissues connecting the bones decayed. The disarticulated bones were then dispersed by flowing water. Unlike many of the elephant bones in the main scatter, the foot bones are not crushed and deformed. The better preservation was probably due to their burial in calcium carbonate-rich sediments close to the edge of the tufa channel. Although the foot bones are well preserved, most of the bones in the main cluster were in poor condition. This is largely the result of unfavourable burial conditions. Geological evidence indicates compaction and down-slope sliding of the sediments containing the bones. The sediments are also leached and decalcified. These processes have combined to weaken the bones in situ. Finally, wetting and drying of the clayey matrix contributed to the poor condition of the bone with the repeating cycles of expansion and contraction of the sediments leading to internal cracking and fracturing.

The Ebbsfleet elephant makes an interesting point of comparison with the four other more-or-less complete straight-tusked elephant skeletons currently known from the British Pleistocene. The youngest of these is from Last Interglacial deposits at Deeping St. James, Lincolnshire (Davies 2002). The majority of this skeleton is missing and the foot bones chewed by spotted hyaena. It entirely lacks any associated artefactual remains. The famous 'Upnor Elephant' skeleton was found in 1911 in fluvial deposits of the River Medway in Kent (Andrews 1928). This skeleton, which is largely complete, was mounted for museum display without the skull, since it was too fragile to conserve. The skeleton represents an extremely large male and dates to the late Middle Pleistocene. As with the former specimen, it entirely lacks archaeological associations. Bones of several individuals have been recovered from late Middle Pleistocene (MIS 7) deposits at Sandy Lane Pit in the Lower Thames Valley at Aveley. The most complete skeleton was excavated by palaeontologists from the British Museum (Natural History) from beneath a mammoth skeleton (Bridgland 1994; Sutcliffe 1995). The presence of two elephants buried in clays and silts close to the margin of a river channel suggests that they had become mired in the soft channel sediments having fallen down the steep and slippery bank of London Clay (Bridgland 1994; Sutcliffe 1995; Davies 2002). There are anecdotal records of lithic material being found in the vicinity (M. J. White, pers. comm.), but there are no provenance records confirming the exact locations and context of the surviving flint artefacts. At Selsey, West Sussex, a partial skeleton of straight-tusked elephant has been excavated from interglacial channel deposits under the beach. A few flint artefacts were recovered in association with the Selsey skeleton. These comprise one small proto-Levallois core, one flake and a piece of irregular waste. As with the Ebbsfleet elephant, no traces of butchery have been observed on the Selsey elephant bones.

MICROSCOPIC EXAMINATION OF CUT MARKS AND TROWEL DAMAGE ON LARGE MAMMAL BONES
by Silvia M. Bello and Simon A. Parfitt

All of the large mammal bones and teeth were scanned for cut and percussion marks using a variable-magnification binocular microscope under low-angle illumination. No unequivocal evidence for marrow processing was found, but two deer bones have linear incisions that are attributed to cuts made during defleshing and disarticulation.

The identification of cut marks is not always straightforward as natural taphonomic processes, such as sediment abrasion and trampling, can create marks on bones that closely resemble cuts, chops and scrapes produced during butchery with stone tools (Shipman 1981; Behrensmeyer *et al.* 1986; Boulestin 1999; Domínguez-Rodrigo *et al.* 2009). In most cases, however, it is possible to identify butchery marks from microscopic features (Shipman and Rose 1983b; Bromage and Boyde 1984; Blumenschine and Selvaggio 1988; Villa and Mahieu 1991; Greenfield 1999; Pickering and Hensley-Marschand 2008) and anatomical position (Binford 1981). Butchery of large mammal carcasses usually involves skinning, disarticulation, defleshing and marrow extraction, with each process leaving a characteristic pattern of marks on the bones, linked to cutting of ligaments and tendons (disarticulation), muscle attachments (defleshing) and skinning and impact damage from marrow bone breakage (Binford 1981). The bone-bearing sediments at Southfleet Road are predominantly fine-grained and natural abrasion is uncommon (see observations on trample marks on elephant bones). More common are grooves, scratches and scrapes inflicted on the bones by metal tools during the excavation. Although the bones were excavated with great care, damage was unavoidable due to the fragile condition of the bones and the hardened clayey matrix in which they were buried. Trowel marks were nevertheless easily to recognise under the binocular microscope as they were generally lighter in colour than the surrounding bone.

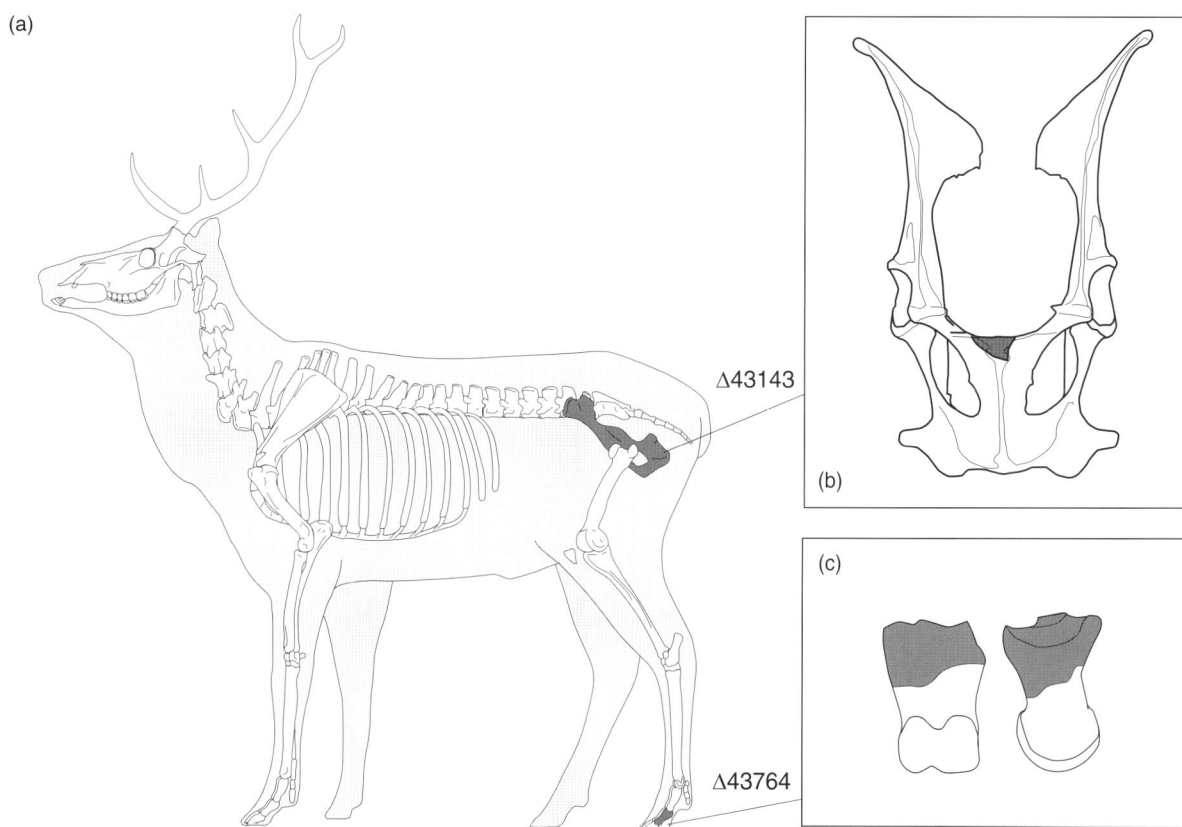

Figure 8.20 Anatomical position of deer remains with probable cut marks

(a)

(b)

(c)

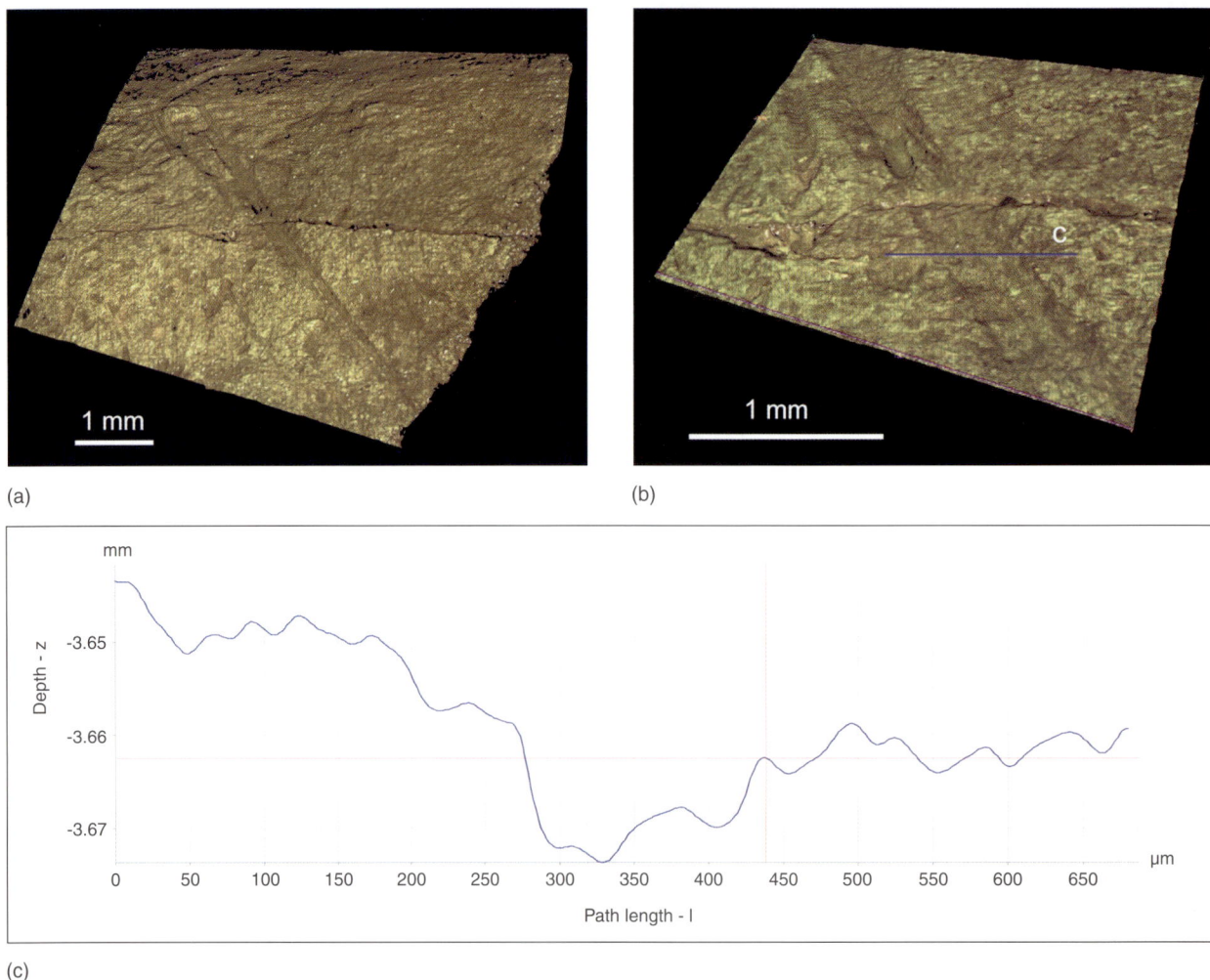

Figure 8.21 Cut mark on the pubis of a medium-sized deer (Δ.43143): (a) oblique view Alicona image of probable cut mark. Note probable 'shoulder effect' (magnification 2.5x, vertical resolution 9.46μm, lateral resolution 19.50μm); (b) close-up of the incision (magnification 10x, vertical resolution 407nm, lateral resolution 1.46μm), showing internal microstriations; (c) cross-sectional profile at the mid-point of the cut mark. Note internal microstriations and asymmetric V-shaped cross-section

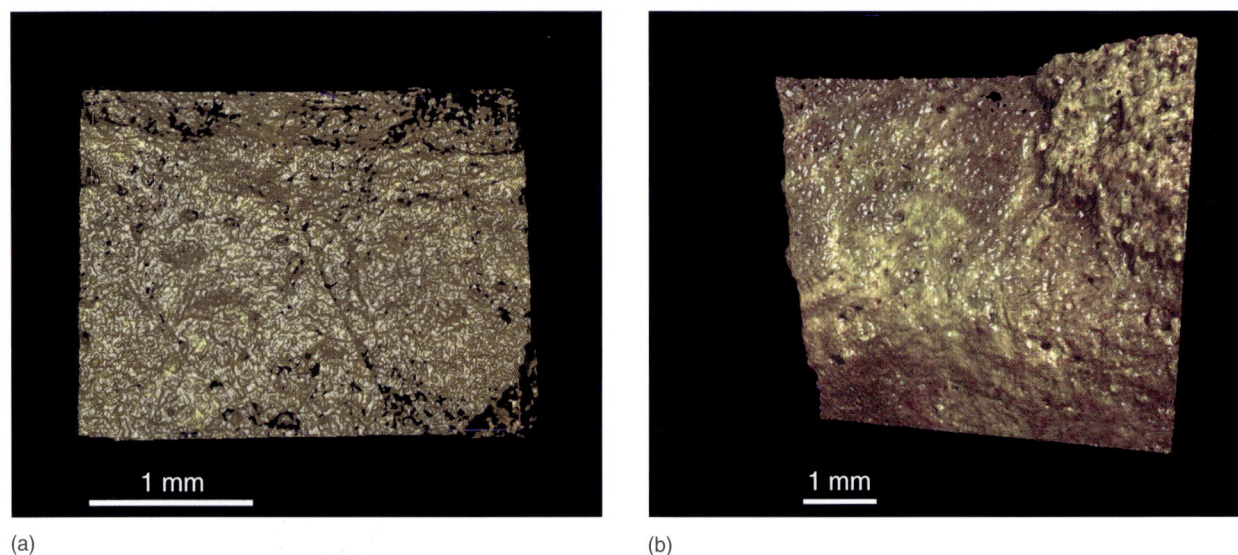

(a)

(b)

Figure 8.22 Probable cut marks on the phalanx of a medium-sized deer (Δ.43764). Oblique Alicona image of incisions: (a) first cluster of incisions (magnification 5x, vertical resolution 1.64μm, lateral resolution 7.82μm); (b) second cluster of incisions (magnification 2.5x, vertical resolution 9.46μm, lateral resolution 19.50μm)

Five bones were selected for more detailed microscopic study using an Alicona InfiniteFocus (AIFM) variation focus microscope. The AIFM is an optical microscope that integrates multiple scans to create a true-colour, three-dimensional surface model of an object. These images can then be manipulated to measure surface features (Bello *et al.* 2009; Bello 2011; see Bello and Soligo 2008 for methodology applied to measuring cut marks).

Two bones have fine incisions that are interpreted as slicing marks made during butchery (Figs 8.18; 8.20). Specimen Δ.43143 (40070) consists of the symphyseal part of the pubis of a medium sized deer. (Fig. 8.20a-b). The bone surface is well preserved and the dorsal face is marked with an isolated linear incision orientated along the cranial-caudal axis. Also apparent from the Alicona images is a second parallel, but much shorter, incision interpreted as 'shoulder effect'. Measurements taken at three points along main incision gave an average width of 124μm and the average depth

1μm. The measurements, presence of microscopic internal striations, and a possible shoulder effect, are consistent with a shallow cut mark (Andrews and Cook 1985; Behrensmeyer *et al.* 1986; Bello and Soligo 2008; Boulestin 1999; Domínguez-Rodrigo *et al.* 2009; Greenfield 1999; Shipman 1981; White 1992; Fig. 8.21a-b). The cross-sectional profile also resembles that of slicing-marks made with a stone tool (Fig. 8.21). The location and orientation of the cut mark indicates filleting or evisceration.

The second cut specimen (Δ.43764, context 40039/70) is the proximal half of a second phalanx of a medium sized deer (Fig. 8.20a and c). Macroscopically, the surface appears to be well preserved, with no obvious signs of weathering, trampling or post-depositional corrosion, but the breakage appears to have occurred during excavation. Two distinct clusters of striations were observed on the palmar surface. The first group consists of two short incisions, both with oblique orientations to the anatomical axis of the bone; the other group of two incisions has a perpendicular orientation with respect to the main axis (Fig. 8.22a and b). None of the incisions exhibits clear diagnostic features which would discriminate between trampling marks (Andrews and Cook 1985, Behrensmeyer *et al.* 1986; Domínguez-Rodrigo *et al.* 2009) and humanly induced butchery-marks. It may be significant that the average width (125μm) and depth (16μm) of these marks is compatible with equivalent dimensions of specimen Δ.43143. Although the microscopic morphology of the marks is difficult to interpret, their location is suggestive of human-derived butchery marks. Similar marks can be observed in butchered archaeological assemblages and in experimental butchery from skinning or cutting of tendons that hold the foot together.

Three bones (Δ. 40784, 40689, 41828) were selected to illustrate the morphology of grooves made accidentally with metal tools (trowels) during excavation and on-site cleaning of the bones (Figs 8.23-25).

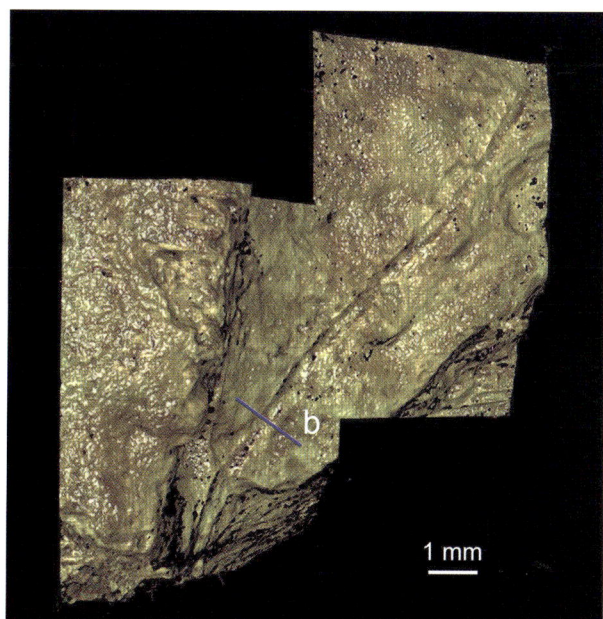

(a)

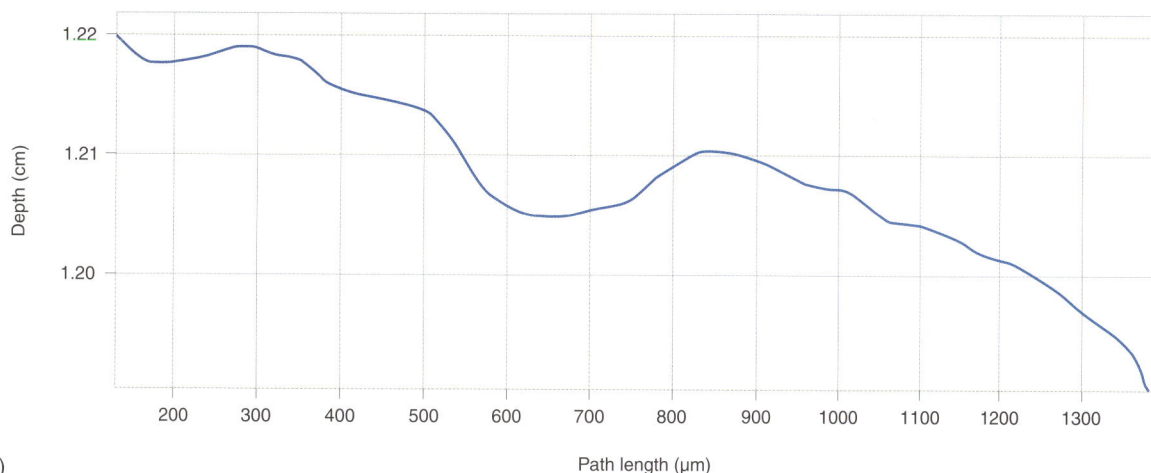

(b)

Figure 8.23 ?Elephant bone splinter (Δ.40784) with trowel mark. Oblique Alicona image (magnification 2.5x, vertical resolution 9.46μm, lateral resolution 19.50μm); (b) cross-sectional profile

Figure 8.24 Oblique Alicona image of groove on
?Elephant bone splinter (Δ.40689), showing typical
features of a trowel mark (magnification 2.5x, vertical
resolution 9.46µm, lateral resolution 19.50µm)

Although some bear a superficial resemblance to butchery cut marks, the trowel marks are usually lighter in colour than the surrounding cortical bone. Another indication that they were inflicted during the excavation is apparent because they cut-across, or 'smudged' manganese deposits and sediment adhering to the surface of the bone. Microscopically, they have smooth rounded profiles and open cracks or micro-faults perpendicular to the long-axis of the groove; they are typically longer and wider than cut marks. Specimen Δ.40784 (40070) is an indeterminate bone fragment, probably from an elephant. The bone is marked by an isolated long sinuous groove, which extends onto the broken edge of the fragment (Fig. 8.23a). Width and depth measurements were taken at four points along the groove and gave an average width of 377µm and an average depth of 64µm. The smooth U-shape cross-sectional is readily apparent in the AIFM profile (Fig. 8.3b). The second bone with excavator-damage (Δ.40689) is an indeterminate bone fragment that, because of its cortical thickness, is also likely to be from an elephant. The groove (Fig. 8.24) is similar in morphology to the previous mark, although it is somewhat broader (average measured at three points = 548.7µm) and deeper (average measured at three points = 103.9µm). The final piece is a fragment of large mammal bone (specimen 41828) which is impossible to identify either to element or taxon. There is a single relatively long curved groove (Fig. 8.25a). Although this feature is much narrower (average width measured at four points = 149.2µm) and shallower (average depth measured at four points = 10µm) than the other trowel marks, it is identical in having a smooth U-shaped profile (Fig. 8.25b).

DISCUSSION AND CONCLUSIONS

Southfleet Road is a rare Lower Palaeolithic example of a site with a single elephant skeleton found in association with an isolated scatter of *in situ* stone tools. Careful

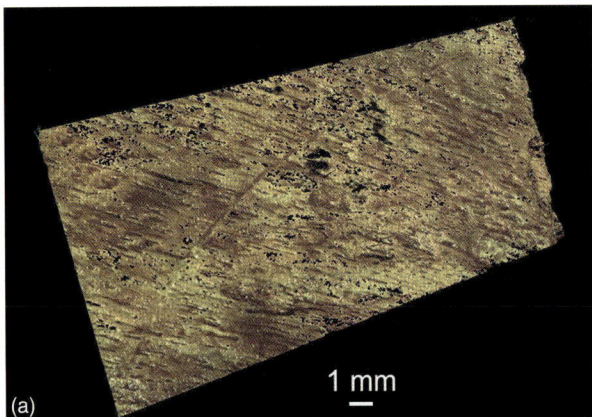

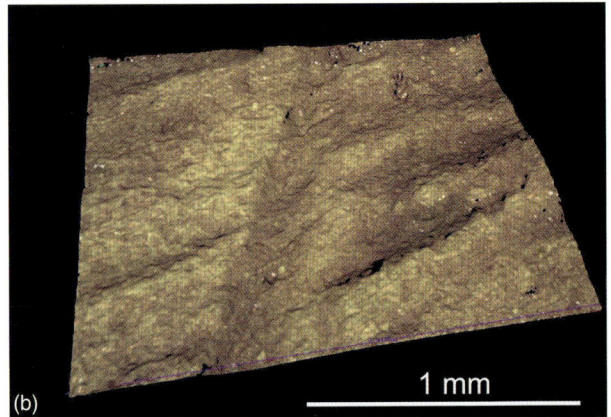

Figure 8.25 Indeterminate large mammal bone (Δ.41828) with trowel mark: (a) oblique Alicona images of the curved groove (magnification 2.5x, vertical resolution 9.46µm, lateral resolution 19.50µm) and (b) enlargement (magnification 10x, vertical resolution 407nm, lateral resolution 1.46µm), showing featureless U-shaped profile

study of the bones, however, has failed to identify any unequivocal evidence for modification of the elephant bones resulting from butchery or marrow extraction. This does not rule-out processing of the carcass by early humans, because not all butchering activities leave traces on bones (Haynes 1991). Beyond the concentration by the elephant skeleton, lithic artefacts are scattered at a low density throughout the Phase 6 deposits containing the other large mammal remains, without any apparent co-association of faunal and lithic remains. They are abundant in the lithic concentration south of Trench D, where however few faunal remains were found – although this is thought to be due to preservational issues rather than an archaeological pattern. Direct evidence for lithic use for large mammal butchery is, however, tenuous and based on only two probable cut-marked deer bones. In this respect Southfleet Road is comparable to many European Lower Palaeolithic sites where there is little or no evidence for human involvement in the accumulation and modification of associated large mammal remains (Gaudzinski 1999; Gaudzinski-Windheuser and Turner 1999). Whether this is an indication of minimal human involvement in the accumulation of the bones is unclear as detailed taphonomic analyses have not been undertaken at key sites (but see Gaudzinski-Windheuser *et al.*, 2010). At the Upper Palaeolithic-Mesolithic site of Three Ways Wharf (Uxbridge, UK), poor preservation of bones was implicated in the difficulty in recognising cut marks and other butchery evidence (Lewis and Rackham 2010). Although this site has yielded several thousand large mammal bones from open-air hunting camps, less than a dozen bones exhibit convincing cut marks, and no impact features were identified. This highlights the fact that cut marks made with stone tools are generally superficial features that can be obliterated by even minor weathering, flaking and abrasion of the bone surface. Cut marks and other impact features are also susceptible to post-depositional weathering and soil corrosion, which may erode microscopic features thus rendering analyses difficult. The generally poor state of preservation of the Southfleet bones must therefore be taken into account when considering the scant evidence for large mammal butchery at the site.

Other Clactonian sites with butchered faunal remains include Swanscombe (Phase I deposits), Clacton-on-Sea, Essex and Barnham, Norfolk (Parfitt 1998b). At the last site, the assemblage of large mammal bones is not particularly extensive, but does include a bovid femur shaft fragment with cuts and impact damage. Many more butchered bones are present in various collections from Clacton (Parfitt unpublished), but these often lack precise contextual information and the material recovered from archaeological excavations has yet to be studied in detail. Binford (1985) concluded that the Clactonian industry at Swanscombe, Kent, was associated with marginal scavenging of carnivore-ravaged carcasses for bone marrow. However, a re-analysis of butchered large mammal bones identified by Binford (ibid.) has established that the alterations in the Swanscombe sample include both natural and excava-

tion damage. Preliminary study of the cut marks on the Clacton and Swanscombe bones shows that they appear to relatively shallow and narrower than cuts made with handaxes. Although this observation has yet to tested with measurements (cf Bello *et al.* 2009), the shallowness of the cut-marks together with poor preservation of bones surfaces, may account for problems encountered with recognising and interpreting Clactonian butchery practices.

Establishing whether elephants were hunted and butchered during the Lower and Middle Palaeolithic periods is a particularly difficult problem (Clark and Haynes 1970; Shipman and Rose 1983a; Scott 1986; Jones and Vincent 1986; Binford 1987; Villa 1990; Piperno and Tagliacozzo 2001; Mazza *et al.* 2006; Mussi and Villa 2008; Waters *et al.* 2011; Slimak *et al.* 2011; Schreve 2006 vs Smith 2012). Some authors have suggested that early humans undertook planned hunting of elephants using thrusting or throwing spears (Weber 2000), whereas other studies have invoked scavenging at death sites (Binford 1987; Anzidei *et al.* 2012). More marginal utilisation may have involved the exploitation of bones to make handaxes and other cutting tools (Gaudzinski *et al.* 2005; Boschian and Saccà 2009). Other studies have suggested that associations between Proboscidean bones, even when found as an isolated skeleton, and stone tools can be entirely fortuitous (Byers 2002). Consideration of hominin involvement with megafaunal, and particularly Proboscidean, remains must take account of results from modern elephant butchery experiments, which have shown that it is possible to strip meat from the carcass without marking the bones (Frison 1989; Frison and Todd 1986; Haynes 1991). Elephant bones are unusual in that the diaphyses are encased in thick periosteum and articular surfaces have a thick layer of cartilage, both of which protect the bones from accidental contact with cutting tools. The sheer quantity of potentially edible tissue on an elephant carcass may also be an important factor that could have resulted in only limited butchery. This may have involved partial skinning and filleting to gain access to largest muscle blocks or removal of the internal organs. Palaeolithic evidence for partial exploitation of a carcass may include the butchered elephant skull from Gesher Benot Ya'aqov, Israel (Goren-Inbar *et al.* 1994), where cut marks indicate the removal of the trunk, and breakage of the skull has been implicated in the removal of the brain. Another example of partial exploitation may have taken place at Áridos 2, Spain (Yravedra *et al.* 2010). Here, cut marks are found on the ventral surfaces of the ribs indicating evisceration and removal of internal organs, which can only have occurred shortly after death and before scavenging carnivores had attacked the carcass (Yravedra *et al.* 2012). Another important factor determining the extent of human utilisation of elephant carcasses was the speed of soft-tissue decay and the loss of edible tissue to scavengers. Although carcasses may have remained in an edible condition for several months during winters in northern latitudes (or even longer in the permafrost zone), those in the tropics and temperate situations would have decayed rapidly through the action of microbes and maggots. This would have limited the opportunities for fully exploiting the carcass, making it less likely that bones were marked during butchery. There is also evidence for Palaeolithic evidence for the exploitation of elephant bones for oils and bone grease. As elephant bones do not have marrow cavities, the fat and grease contained within the spongy and cancellous bone can be extracted by hanging broken bones in the sun, or by heating and boiling. Although bone breakage suggestive of bone fat extraction has been recorded in Palaeolithic contexts (Yravedra *et al.* 2012), there is currently no evidence to suggest that bone grease extraction involved the use of fire.

Non-dietary utilisation of elephant carcasses could have taken place long after any edible tissue remained on the carcass. In areas where lithic raw materials are scarce, there is evidence that elephant bones were knapped to make cutting tools. Dried hides could have been used for a variety of purposes (eg ethnographic examples of foot pads used as bowls) and other useful soft tissues such as tendons would have been suitable for making bindings. In treeless landscapes, elephant bones with its high fat content may have provided the only reliable source of fuel. At La Cotte de St. Brelade, Jersey,

the abundance of burned bones in the Middle Pleistocene cold stage levels has been suggested to indicate use of bones for fuel (Stringer 2006). Bones may also have been used to make simple structures or windbreaks. The archaeological signal for many of these activities would be extremely weak. The current consensus is that exploitation of elephants and other large mammals, such as rhino and hippos, was more than just a marginal practice before the Upper Palaeolithic (Yravedra *et al.* 2012).

Although there is no direct evidence from cut marks or impact damage to indicate that the Ebbsfleet elephant was butchered, the tight spatial association of lithic artefacts and elephant bones may be sufficient to justify the assumption. This is supported by the vertical distribution of the stone tools, which are found at the same level as the bones and by use-damage on some of the artefacts, which has been interpreted as resulting from their use as butchery tools (Chapter 17). Although the patterns in bone distribution and the taphonomic alterations observed on the Ebbsfleet elephant bones can be explained by natural (ie non-human) processes alone, the spatial associations and lithic use-wear evidence provide compelling evidence that the carcass was exploited by early humans (Chapter 22).

Chapter 9

Mammalian biostratigraphy

by Simon A. Parfitt

INTRODUCTION

One of the key lines of evidence for dating the Clactonian archaeology at Southfleet Road is the associated mammalian fauna. Mammalian remains from Pleistocene deposits have long been widely employed for correlation and dating of Palaeolithic sites (Schreve and Thomas 2001). More recently, attempts have been made to combine biostratigraphical information from palynology, molluscs and mammal remains with amino acid geochronology to establish a more robust chronology for the British Quaternary (Penkman *et al.* 2011). For the British late Middle and Late Pleistocene, the most complete sequence of stratified molluscan and vertebrate faunas is associated with ancient fluvial deposits in the lower Thames Valley. In this area a flight of terraces reflects the aggradation by the river following its diversion by Anglian ice into its present-day course through central London. The major aggradations are believed to have formed in response to uplift and major cyclical climatic fluctuations between glacial and interglacial stages (Bridgland 2000; Bridgland and Maddy 2002). In this model, downcutting between terraces occurred during cold stages, and temperate sediments in each terrace represent different interglacials. Four pre-Holocene interglacials are recognised within the staircase of terraces in the Lower Thames:

1. The oldest occurs at Swanscombe in the highest (Boyn Hill/Orsett Heath Gravel) terrace and was clearly formed immediately after the southward diversion of the Thames to its present course during the Anglian glacial stage (Gibbard 1979). The interglacial associated with this terrace is widely correlated with the Hoxnian (MIS 11).

2. The second interglacial occurs at a number of sites in the terrace formed by the Lynch Hill/Corbets Tey Gravel. Sites associated with this aggradation include Hackney Downs, Belhus Park, Purfleet and Grays Thurrock (Bridgland 1994).

3. The penultimate interglacial occurs in the terrace formed by the Taplow/Mucking Gravel and has been recognised at Ilford (Uphall Pit), Crayford, Aveley and West Thurrock (Bridgland 1994).

4. The last interglacial (Ipswichian) occurs beneath the Kempton Park terrace and is exemplified by sites with *Hippopotamus* at Trafalgar Square and Brentford (Stuart 1982).

The mammalian faunas from these terrace deposits have been documented extensively (Bridgland 1994; Schreve 2006a; 2006b). This work has identified important differences in mammalian faunal composition between interglacials. These differences reflect fluctuations between glacials and interglacials, with a succession of extinctions and immigration events, as well as morphological changes within taxa. These factors have combined to produce unique combinations of mammalian taxa in the different terraces, which can been used as a basis for differentiating interglacial stages and for making correlations with isolated deposits elsewhere in southern England. The mammalian evidence has also been used to identify shorter duration climatic oscillations within interglacials that have been linked to isotope substages (Schreve 2001b); these attempts at fine-resolution correlation, however, remain highly controversial (Ashton *et al.* 2008; Pettitt and White 2012, 230–1). Work in progress at key British Middle Pleistocene localities, for example at Hoxne, Ebbsfleet Valley, is beginning to challenge the current mammal biozonation of the Middle and Late Pleistocene in Britain (Ashton *et al.* 2008; Preece and Parfitt 2000; 2008). The evidence from the Ebbsfleet Valley (Wenban-Smith *et al.* forthcoming) and the extensive mammalian fauna from Southfleet Road in particular, provide important cornerstones in this revision.

This chapter presents the mammalian biostratigraphical evidence for the age of the Southfleet Road sequence. The climatic implications of the large mammal taxa are discussed, as this evidence is important in establishing an interglacial context. The main focus, however, is a metrical and morphology comparison of the Southfleet Road large mammals with samples from British Middle and Late Pleistocene sites, particularly those from the Lower Thames Valley. At Southfleet Road, the mammalian fauna provides compelling evidence that the sediments containing the Clactonian artefacts were deposited during the early part of the Hoxnian interglacial. The final part of the chapter attempts to integrate this evidence into the emerging picture of a more complex British sequence, which can in turn be related to climatic oscillations and substages of MIS 11.

BIOSTRATIGRAPHICAL AND PALAEO-ECOLOGICAL SIGNIFICANCE OF THE SOUTHFLEET ROAD LARGE MAMMALS

Primates

Macaca sylvanus (L.), Barbary macaque

A notably exotic element of the Southfleet road fauna is macaque, represented by a well-preserved first phalanx (Fig. 9.1) from the tufa channel.

Today, there are small isolated populations of Barbary macaque in the Rif and Atlas Mountains of western Morocco and north-eastern Algeria, but the species was more widespread during the Pleistocene (Delson 1980), extending into northern Europe during interglacials. Remains of macaque have been reported from six other Pleistocene sites in Britain. The earliest record is from the Upper Freshwater Bed at West Runton, Norfolk (Hinton 1908; Stuart 1996) and dates to the early Middle Pleistocene, Cromerian Interglacial Stage. Currently, the youngest known material is from late Middle Pleistocene (MIS 9) deposits at Cudmore Grove (Essex) and Grays Thurrock and Purfleet (Schreve *et al.* 2002) in the Thames Valley. At Swanscombe, Barbary macaque was identified on the basis of a fragmentary ulna from the Lower Loam. Somewhat younger is the material from Hoxne Suffolk (late MIS 11). This sample includes several isolated teeth (Singer *et al.* 1982) from fluvial sediments that overlie the Hoxnian lake muds (Ashton *et al.* 2008). Earlier Pleistocene macaques exhibit morphological and metrical differences from extant Barbary macaque, implying subspecific status (Alba *et al.* 2011); Singer *et al.* (1982) showed that the Hoxne sample is indistin-

guishable morphologically from extant North African *Macaca sylvanus*.

Currently, the preferred habitat of the Barbary macaque is mountainous slopes in forests of cedar, oak and pine (Fa 1984) with nearby water. Marginal habitats occupied by Barbary macaque include coastal scrub and

Figure 9.1 *Macaca sylvanus*. First phalanx (Δ.50014, context 40070), anterior view

Table 9.1 List of *Oryctolagus cuniculus* specimens with measurements. Measurements (in mm) taken according to the method of von den Driesch 1974

Phase	Context	Sample	Element	Measurements
5	40025	40343	Upper cheek-tooth frag	
			Lower cheek-tooth frag	
			2 terminal phalanges	
		40348	2 upper cheek-tooth frags	
			R innominate frag	
			R astragalus	GL 11.7
			R calcaneus frag	
			1st phalanx frag	
			3 terminal phalanges	
		43860	Upper cheek-tooth frag	
		40380	L upper cheek-tooth frag	
			Lower cheek-tooth frag	
			L lower incisor	MD 2.93, BL 1.94
			R ulna (proximal end)	
6a	40039	*	L M_{2-3}	
6b	40070	40329	Upper & lower cheek-teeth (digested)	
		40330	R tibia (distal end)	Bd 10.7; Dd 6.29
	40144	40336	R M_2 (digested)	

Astragalus: GL = greatest length; Tibia: Bd = breadth of distal end, Dd = depth of distal end. Incisor: MD = mesiodistal diameter, BL = buccolingual diameter

* Δ. 41079

rocky slopes with vestigial vegetation. They feed primarily on seeds, fruits, roots and leaves of cedar and oak, with other foods such as insects and other small animals forming a minor component. As well as providing food, trees are also important for sleeping and providing safe refuges from predators.

The environment associated with the Barbary macaque from Hoxne was probably mixed woodland dominated by conifers, during a cooler phase near the end of MIS 11. At Swanscombe, however, macaque occurred during the early temperate part of the interglacial, when the dominant regional vegetation was deciduous woodland. The presence of Barbary macaque at Southfleet Road reinforces other faunal indications of woodland at the site.

Lagomorphs (Lagomorpha)

Oryctolagus cuniculus (L.), rabbit

A total of 24 bones and teeth were identified as rabbit (Table 9.1). These remains represent a minimum of eight individuals (assuming elements from the same individual do not occur in separate samples), with most coming from context 40025 (Phase 5). The morphology of the teeth and postcrania compare closely with modern *Oryctolagus cuniculus*.

The natural range of the rabbit during the early Holocene was confined to Iberia and the southern coastline of France. Very few Pleistocene records are known beyond this region (Donnard 1981; Rogers *et al.* 1994), with the notable exception of southern England, where several Middle Pleistocene sites have yielded rabbit remains. Here, rabbit was first shown to have been a genuine element of the British Pleistocene fauna by its presence in Hoxnian deposits at Ingress Vale and Barnfield Pit (Mayhew 1975). At Swanscombe, rabbit remains are abundant in the Lower Loam, but it is also present at several levels within the underlying gravels. According to Mayhew (ibid.), the rabbit bones and teeth from Swanscombe are indistinguishable skeletally from those of modern rabbits from southern England. More recently, rabbit remains have been found in Hoxnian contexts at East Farm, Barnham (Parfitt 1998b) and Beeches Pit (Preece *et al.* 2006; 2007). Besides these Hoxnian sites, it has been identified at Boxgrove (Currant 1986b; Parfitt 1999) and at Westbury-sub-Mendip (Parfitt unpublished), probably from different temperate stages within the 'Cromerian Complex' (Preece and Parfitt 2000; 2008).

The cluster of Hoxnian sites with rabbit in southern England suggests that optimal conditions for its survival existed during the peak of this interglacial. Important environmental factors may have included a combination of an oceanic climate with mild winters and suitable vegetation of open woodland and expanses of interconnecting grassland (Bjärvall and Ullström 1986; Mitchell-Jones *et al.* 1999). Rabbits may have favoured coastal situations (eg Boxgrove), but its presence at other sites suggests that it made incursions inland, perhaps along the grassy floodplains of large rivers.

Ecologically, the presence of rabbit at Southfleet Road is indicative of warm temperate conditions and reinforces other evidence for grassland with nearby cover provided by patchy scrub or from trees along the woodland edge.

Rodents (Rodentia)

Castor fiber (L.), beaver

This species is represented by a single complete upper fourth premolar (Fig. 9.2), which is indistinguishable in size and morphology from modern comparative European beaver.

Figure 9.2 *Castor fiber*. Right P⁴ (Δ.40899, context 40100): (a) buccal view; (b) occlusal view

At Swanscombe and Ingress Vale, two species of beaver are represented in the assemblages. The extinct 'giant' beaver *Trogontherium cuvieri* occurs at both sites, but European Beaver is known only from Swanscombe; both are represented by very few remains (Mayhew 1975). The beaver fits well with an interpretation of freshwater habitats bordered by deciduous woodland in the vicinity of these sites.

Carnivores (Carnivora)

Panthera leo (L.), lion

Lion is the only large carnivore represented at Southfleet Road. The identification is based on a complete astragalus from context 40025 (Fig. 9.3). The specimen is indistinguishable from that of modern *P. leo*, but the greatest length indicates a large and robust animal, consistent with other British Pleistocene samples (Table 9.2).

In northern Europe, lion is recorded from interglacials with deciduous woodland and cold stages, in association with more open herbaceous vegetation. Fossil evidence (Burger *et al.* 2004; Turner 1984; 2009) suggests that the lion first colonised Europe about 600,000 years ago. Genetic analysis of Late Pleistocene fossil material suggests that the European populations were isolated from those in Asian and African. European Pleistocene lions include individuals of an exceptionally large size (Ballesio 1980; Parfitt 1998b; Turner 2009), which are often referred to a separate species *P. spelaea* (or subspecies *P. leo fossilis – P. leo spelaea*). However, as Turner (1984) has shown, they are indistinguishable osteologically from the modern species, to which they are now referred (Lister and Brandon 1991; Turner 2009).

Remains of lion occur in small numbers in Hoxnian assemblages from Clacton, Swanscombe (Sutcliffe 1964) and Barnham (Parfitt 1998b). The somewhat younger large proximal humerus from Hoxne (Stuart *et al.* 1993) is from fluvial sediments that overlie the Hoxnian lake beds. In terms of size, all of the Hoxnian

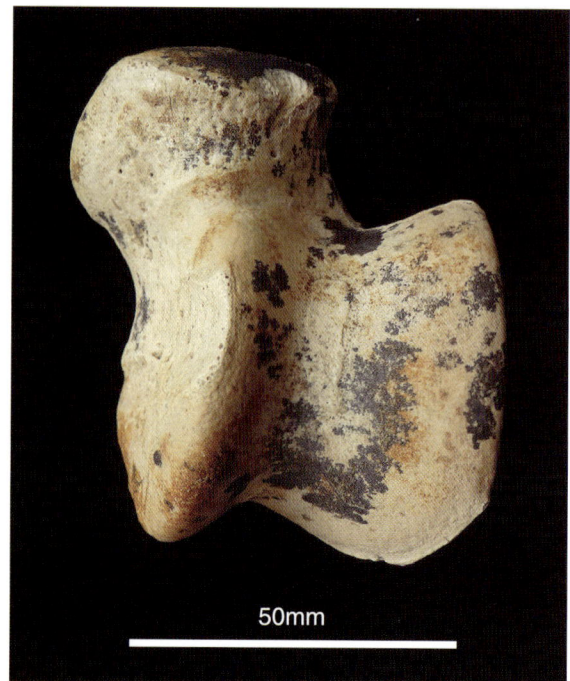

Figure 9.3 *Panthera leo.* Astragalus (Δ.43845, context 40025), dorsal view

specimens (including the astragalus from Southfleet Road) are larger and more robust on average than those of modern African lions (Parfitt 1998b).

At Southfleet Road, lions would have hunted in grassland amongst scattered trees in open parkland. Their principal prey would have included herbivores, such as fallow and red deer as well as bovids.

Elephants (Elephantidae)

Palaeoloxodon antiquus Falconer and Cautley, straight-tusked elephant

Straight-tusked elephant is represented by parts of a dispersed skeleton from context 40078, Phase 6, and a

Table 9.2 Greatest length (in mm) of the astragalus in Recent African and European Pleistocene *Panthera leo*

	n	*Mean + SD*	*Observed range*
Recent			
Africa[a]	8	53.1 ± 3.59	47.5–59.2
Africa[b]	22	53.8 ± 4.3	46.5–61.5
Last Cold Stage			
Gailenreuth, Germany	2		67.1, 67.4
Jaurens, France[c]	5	60.9 ± 5.18	56.0–67.0
Middle Pleistocene			
Southfleet Road	1	64.7	
Barnham, Suffolk (Hoxnian)	1	70.1	
Westbury-sub-Mendip, Somerset (early Middle Pleistocene)	1	73.5	

[a] Modern comparative sample (Natural History Museum, London) comprising four males and four females
[b] Modern comparative sample (Gross 1992) comprising 22 males and eight females
[c] Ballesio 1980
Other measurements, Parfitt (unpublished)

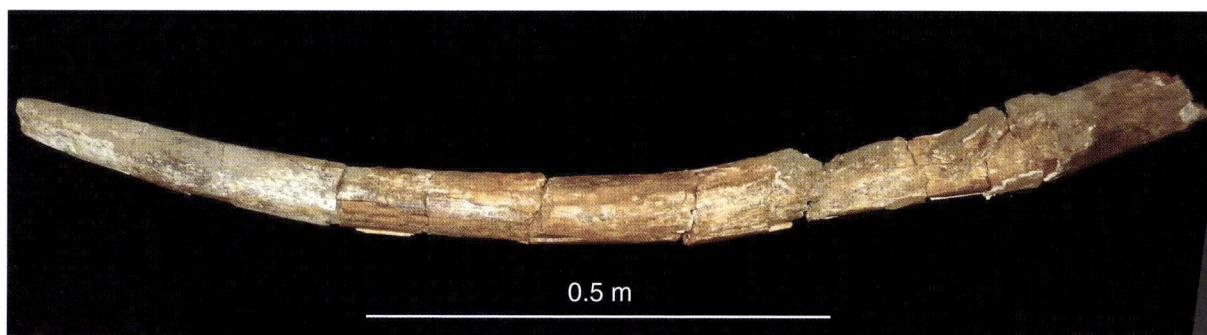

Figure 9.4 *Palaeoloxodon antiquus*. Tusk (Δ.43788, context 40025). The relatively small size of the tusk (which is almost complete) suggests that it is from either a female or young individual

tusk from context 40025, Phase 5 (Fig. 9.4). Most of the bones from the dispersed skeleton are poorly preserved, but identifiable pieces include both (badly crushed) tusks, the third upper molars, several vertebrae, ribs and carpal bones (see Chapter 8). An isolated cuneiform (Δ.43378) from context 40144 probably belongs to the partial skeleton from context 40078.

At Swanscombe, straight-tusked elephant remains occur throughout the fossiliferous sequence. These include numerous isolated teeth, but also post-cranial bones and a complete tusk from the Middle Gravel (Sutcliffe 1964). Straight-tusked elephant has also been reported from Hoxnian Thames-Medway deposits at Clacton and in the lake muds at Hoxne (Stuart 1982). As with most records from Britain, these finds are from interglacial contexts associated with palaeobotanical evidence for temperate deciduous woodland. Its appearance in at least two horizons at Southfleet Road is not unexpected given other faunal evidence of parkland and forested environments at the site.

Rhinoceroses (Rhinocerotidae)

Rhinoceroses were an important part of the fauna of Britain during the Pleistocene. Several species have been identified in the early Middle Pleistocene, most of which are assigned to the genus *Stephanorhinus*. These include *S. hundeshemensis*, two un-named but closely related forms (*Stephanorhinus* sp. A and *Stephanorhinus* sp. B, of Breda *et al.* 2010), and *S.* cf. *megarhinus*, the latter currently known only from Boxgrove (Breda *et al.* 2010). A major phase of faunal turnover occurred during the Anglian Glacial Stage with the extinction *S. hundesheimensis* and its allies and their replacement by narrow-nosed rhinoceros *S. hemitoechus* and Merck's rhinoceros *S. kirchbergensis*. These were the dominant rhinoceroses in northern Europe during temperate stages in western Europe. At present, the earliest known entry of the woolly rhinoceros (*Coelodonta antiquitatis*) into Britain was in the late Middle Pleistocene. This is attested by a skull and a humerus from the Ebbsfleet Valley, which were found in cold stage deposits attributed to MIS 8. In central Europe it has been found in securely dated and well-stratified interglacial contexts. One such site is Neumark-Nord in Germany, where the remarkable association of three rhinoceros species (*S. hemitoechus*, *S.*

kirchbergensis, *Coelodonta antiquitatis*) has been found in lake sediments containing a wealth of palaeobotanical evidence. The lake appears to have been infilled during the Last Interglacial MIS 5e (Sier *et al.* 2011, but see Van der Made 2010 for an alternative view) and the rhinoceros fossils are associated with open steppe forest, interspersed with steppe meadows, heath and shrub (Van der Made and Grube 2010).

Currant (1996) has raised the possibility that three rhinoceros species were present at Swanscombe during the early part of the Hoxnian. He identifies three associated upper cheek teeth from the Lower Gravel, which he ascribes to the Early Pleistocene rhinoceros species *S. etruscus*. Although this identification is contested (see below), another 'archaic' rhinoceros, *S. megarhinus*, is present in even younger contexts in Europe, thus adding a further level of complexity to the history of European Pleistocene rhinoceroses (Breda *et al.* 2010).

Stephanorhinus hemitoechus (Falconer), narrow-nosed rhinoceros

Narrow-nosed rhinoceros is represented by a badly crushed skull, with most of the upper dentition, from context 40039 (Phase 6a). The teeth (Fig. 9.5) are identified as *S. hemitoechus* on the basis of size (Fig. 9.6; Table 9.3), their high crowns, reduced anterior and enlarged posterior teeth, and rough enamel. A further 18 specimens are tentatively assigned to this species, including fragmentary teeth and skull fragments assigned to this taxon based on their close proximity to the skull (Table 9.4).

In northern Europe, narrow-skulled rhinoceros has been recorded in the early Middle Pleistocene (Fortelius *et al.* 1993), but it is unknown from deposits of this age in Britain (Breda *et al.* 2010). The material from Clacton and Swanscombe (eg Basal Gravel) are currently the oldest records from Britain (Breda *et al.* 2010). The dentition of *S. hemitoechus* is relatively high-crowned, suggesting that it fed mainly on abrasive herbaceous vegetation. Climatically, it encompassed both peak interglacials and interstadials in Britain (Lister and Brandon 1991).

Stephanorhinus kirchbergensis (Jäeger), Merck's rhinoceros

This rhinoceros is present in the assemblage from context 40100, where it is represented by a poorly-

Table 9.3 List of rhinoceros bones and teeth from Southfleet Road

Taxon	Phase	Context	Find no. Δ	Element
S. hemitoechus	6a	40039	41812	R P^3
			42047	R P^4
			41845	R M^1 frag (ectoloph)
			42048	R M^2
			41785	R M^3
			41786, 41875, 42939	L P^4 (refitting frags)
			41691[a], 41843, 41874	L M^1 (refitting frags)
			42491	L M^2
			42492	L M^3
S. cf. *hemitoechus*[b]	6a	40039	41921	L upper premolar frag
			41780	Cheek tooth frag
			41817	Cheek tooth frag
			42335	Tooth frag
			41690	Tooth frag
			41839	Tooth frag
			42238	L occipital condyle & skull frag
			41920	Petrosal
			42420/42421	Petrosal
			43833	R lunar
	6	40078	42003	L scapula frag (part of glenoid and neck)
			40632	L humerus (distal and shaft, distal fused)
			40629	L femur (proximal end, fused)
	6	40100	41475	L radius frag (proximal & shaft)
			41588	L radius frag (distal end)
			41589	L radius frag (distal end)
	6b	40144	43652	R innominate frag (acetabulum and pubis)
			43672a	L innominate frag (pubis)
S. kirchbergensis	6	40078?	43712	L ulna (distal end)
	6	40100	40844	R M$_3$
S. cf. *kirchbergensis*[b]			41007	Cheek tooth frags
			41026	R lower cheek tooth (crushed)
			40857	Tooth frag
			40861	Mandible frag
			41028	Mandible frag

[a] Associated with crushed skull   [b] In the absence of diagnostic characters, identifications are based on size or close spatial association with the skull (context 40039) and mandible (context 40100)

Table 9.4 Measurements (in mm) of rhinoceros bones and teeth from Southfleet Road, taken according to the method of Fortelius *et al.* 1993

Taxon	Phase	Context	Find no. Δ	Element	BL	LL	MB	DB
S. hemitoechus					BL	LL	MB	DB
	6a	40039	41812	P^3	37.6	32.8	48.4	48.6
			42047	P^4	41.2	37.1	56.0	52.6
			42048	M^2	57.9	47.8	62.0	
			41785	M^3	59.1		55.9	
S. cf. *hemitoechus*	6	40100	41475	Radius	Dpr	Bsr		
					c 54.5	*c* 52.0		
	6	40078	40629	Femur	LC			
					80.6			
S. kirchbergensis					BL	LL	MB	DB
	6	40100	40844	M^3	53.0	52.6	30.2	32.5
	6	40078?	43712	Ulna	BDu	BDau		
					77.3	73.5		

Teeth: BL = buccal length, LL = lingual length, MB = mesial width, DB = distal width; Radius: Dpr = proximal depth, Bsr = smallest breadth of the shaft of the radius; Femur: LC = length of caput femoralis; Ulna: BDu = distal breadth; BDau = breadth of distal articular surface

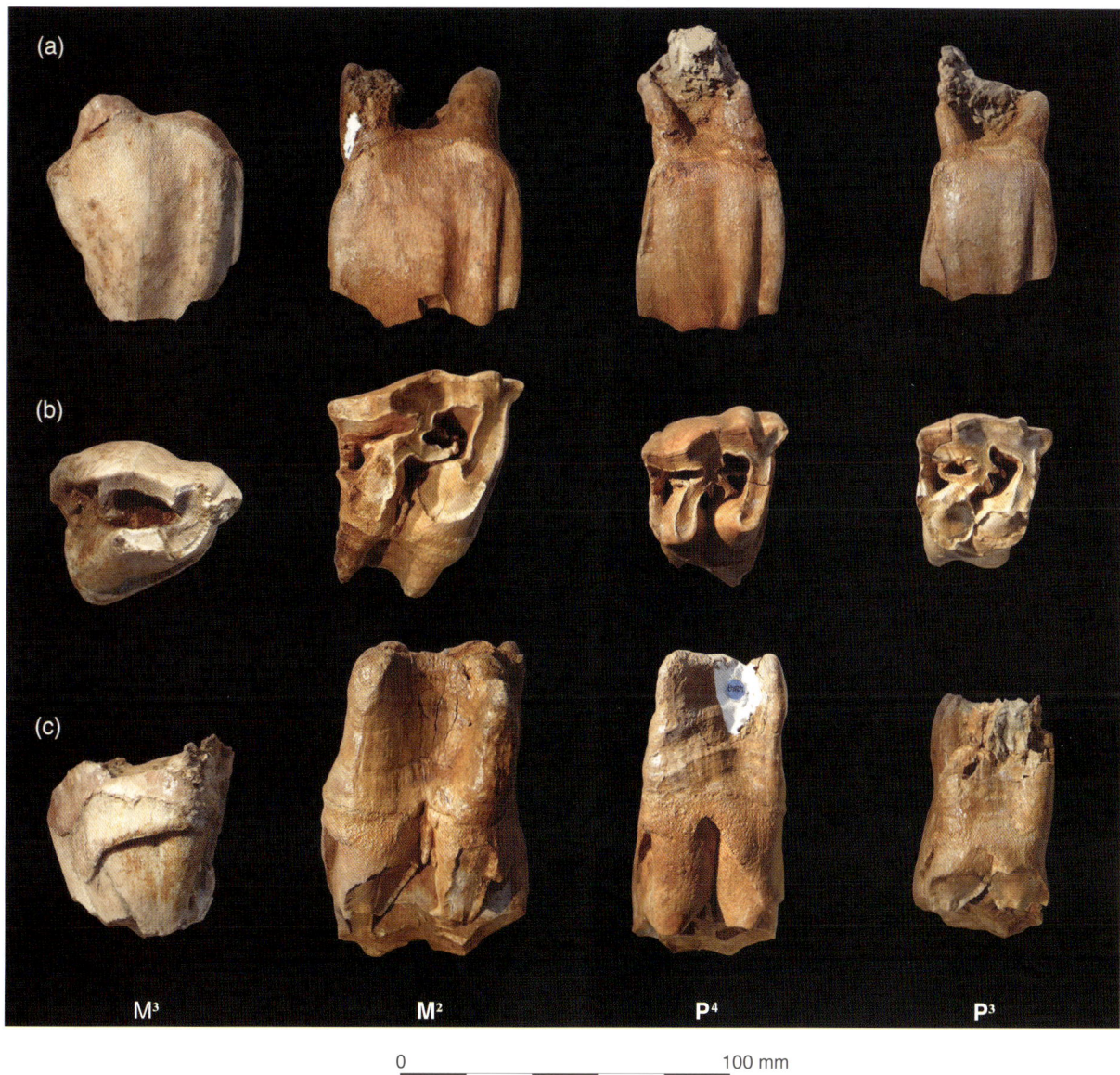

Figure 9.5 Associated right upper cheek teeth P³-M³ of *Stephanorhinus hemitoechus*: (a) labial view; (b) occlusal view; (c) lingual view

preserved mandible with the associated lower third molar (Fig. 9.7) and second badly crushed molar. A distal ulna from context 40078 is the only postcranial element that can be ascribed with certainty to Merck's rhinoceros (Fig. 9.8). The identification is based on its exceptionally large size compared to the ulna of *S. hemitoechus* (Table 9.5); woolly rhinoceros (*Coelodonta*) can be excluded on the basis of morphology.

At present, the earliest well-dated records of Merck's rhinoceros in Britain are from the Hoxnian at Clacton and Swanscombe. It may have appeared earlier in other parts of western Europe (eg Mosbach, Germany), but these finds should now be reassessed given the morphological similarities between teeth of *S. kirchbergenis* and *S. megarhinus* (Breda *et al.* 2010). Merck's rhinoceros was characteristic of temperate, wooded interglacials in the European Middle and Late Pleistocene, and has not been recorded from any securely dated cold stage mammalian assemblages in Europe (Van der Made 2010; Billia 2011). During cold

Table 9.5 Dimensions of the *S. kirchbergensis* distal ulna from Southfleet Road, and comparison with German and British samples of *S. kirchbergensis* and *S. hemitoechus* (Fortelius *et al.* 1993).

	BDu (mm)	*BDau (mm)*
S. kirchbergensis		
Southfleet Road	77.3	73.5
Stockstadt, Rhineland	75.5	65.0
S. hemitoechus		
Ilford		40.0
Maastricht-Belvédère, The Netherlands	36.0	33.0

BDu = distal breadth; BDau = breadth of distal articular surface

stages, its range contracted to refugia in Asia, from where it repeatedly recolonised Europe at the start of successive interglacials (Van der Made 2010).

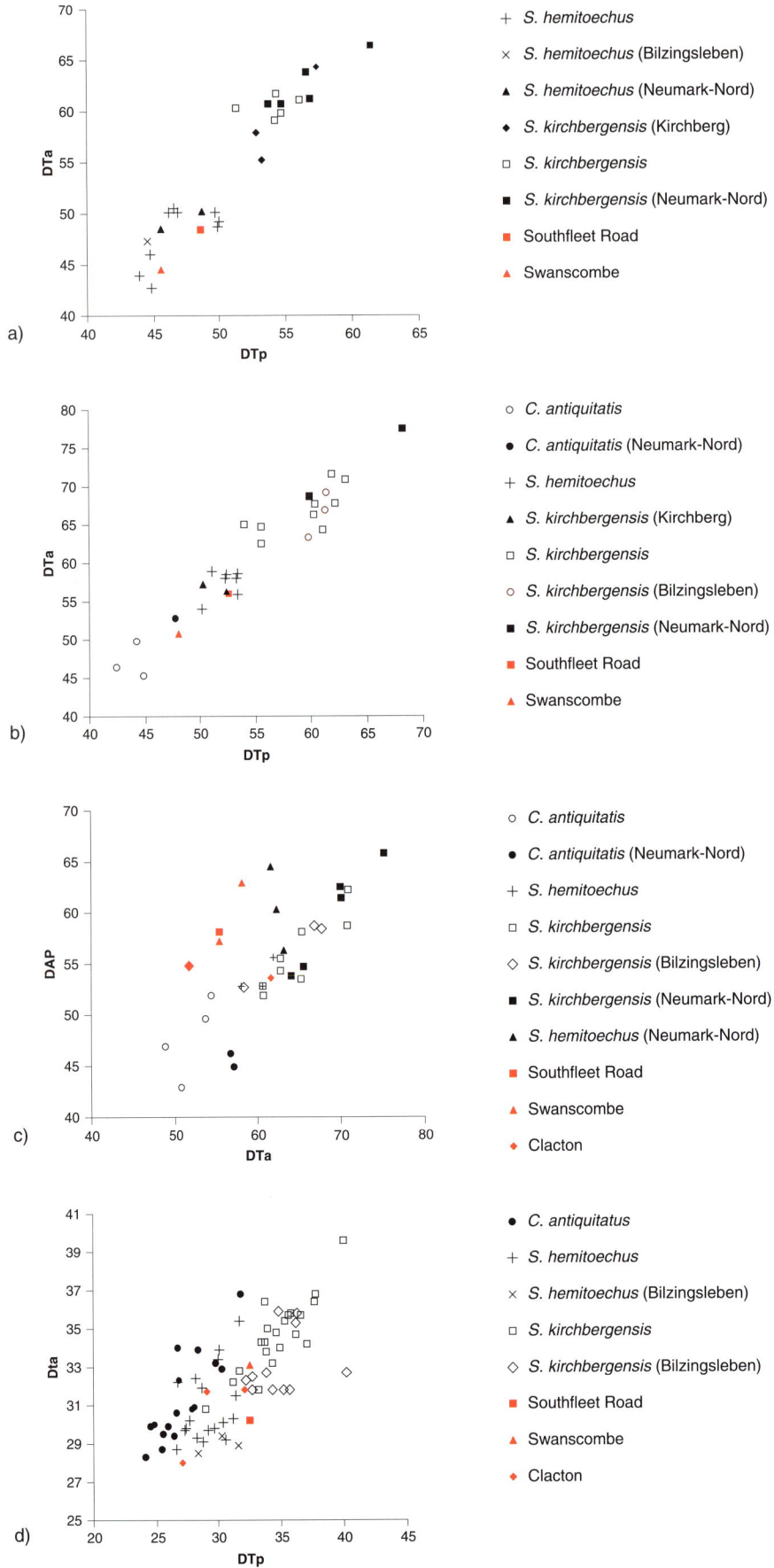

Figure 9.6 Biometric comparisons of rhinoceros cheek teeth. (a) P³, (b) P⁴, (c) M³, (d) M₃. Comparative samples and measurement abbreviations from Van der Made (2010)

Morphological adaptations (such as low-crowned molars, horizontal carriage of the skull) suggest that Merck's rhinoceros fed on bushes and other browse in woodland. This conclusion is supported by the remarkable preservation of a diverse assemblage of non-grass taxa and woody plant remains in fossas of upper cheek teeth of this species from Neumark-Nord, Germany (Van der Made and Grube 2010). Environmental reconstructions of Neumark-Nord suggest extensive forest, broken by patchy open habitats of meadows, heath and shrub; a similar environment can be envisaged at Southfleet Road.

Pigs (Suidae)

Sus scrofa L., wild boar

Only two elements could be ascribed with certainty to boar (Fig. 9.9). The specimens are an unworn tip of an upper canine from context 40039 (Phase 6a) and a worn upper second incisor from context 40100 (Phase 6).

The fossil history of British Pleistocene wild boar has been reviewed by Parfitt (in Lister *et al.* 2010). Measurements show an interesting pattern of size change during the Middle Pleistocene, with a marked reduction in body size after the Anglian Glacial Stage (Lister *et al.* 2010, figure 20). Hoxnian wild boar from Swanscombe and Ingress Vale seem to have been of unusually small body size, matching those living in northern Europe at the present day. Unfortunately, the sample from Southfleet Road does not include measurable specimens.

In terms of ecology, Pleistocene wild boar was

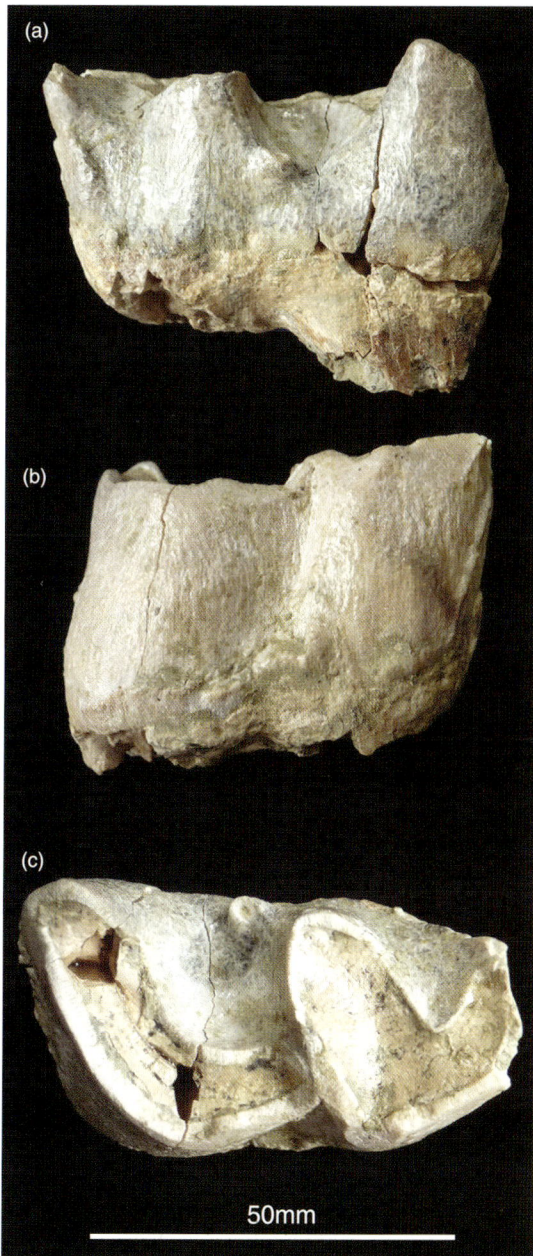

Figure 9.7 *(left)* *Stephanorhinus kirchbergensis*. Right M$_3$ (Δ.40884, context 40100): (a) lingual view; (b) buccal view; (c) occlusal view

Figure 9.8 *(below)* Ulna of *Stephanorhinus kirchbergensis* from Southfleet Road (Δ.43712, context 40078)

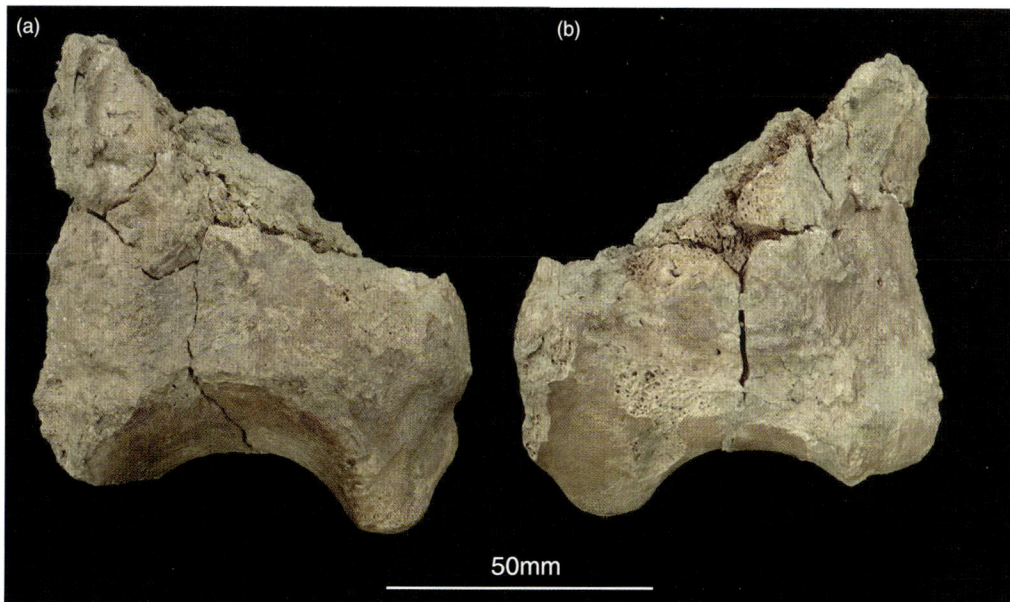

Figure 9.9 *Sus scrofa* : (a) canine (Δ.41876, context 40039); (b) incisor (Δ.41207, context 40100) from Southfleet Road

Deer (Cervidae)

Deer were the most numerous of the large mammals in most contexts from Southfleet Road. The sample comprises a total of 349 specimens and includes antlers, some complete postcranial bones and numerous isolated teeth. Of these red deer (*Cervus elaphus*) is represented by 33 specimens, fallow deer (*Dama dama*) by 11 specimens, while only 3 bones were assigned to roe deer (*Capreolus capreolus*). The assemblage includes a high number of indeterminate medium-sized cervid remains. These are likely to represent fallow or red deer, but are too fragmentary or damaged to identify fully. The distinction between Hoxnian fallow deer and red deer is further complicated due the overlap presence of relatively small red deer and exceptionally large fallow deer that occurred together at this time in southern England (Lister 1993; 1996).

At Swanscombe medium-sized cervids are also the most common large mammals in each of the main fossiliferous strata (Sutcliffe 1964; Lister 1986). Giant deer (*Megaloceros giganteus*) and roe deer are also present, but occur in very low numbers (Lister 1986). Lister (1981; 1986; 1994) has shown that Hoxnian deer species are particularly distinctive with a combination of morphological features in their antlers and teeth that make them particularly useful biostratigraphic indicators.

Cervus elaphus L., red deer

A total of 33 bones and teeth are identified to red deer (Table 9.6), making this the most common cervid in the Southfleet Road large mammal assemblage. The sample includes 18 postcranial bones (Fig. 9.10), six cheek teeth, eight shed antlers (Figs 9.11-12) and one portion of antler beam. Two antlers were almost complete, but

strongly associated with temperate woodland. They were absent from cold stage mammalian assemblages, with only rare occurrences during interstadials. The occurrence of wild boar at Southfleet Road is consistent with temperate conditions inferred from other faunal indicators. Local habitats must have included substantial woodland and dense thickets with close proximity to marsh and reed-beds (Bjärvall and Ullström 1986; Mitchell-Jones *et al*. 1999).

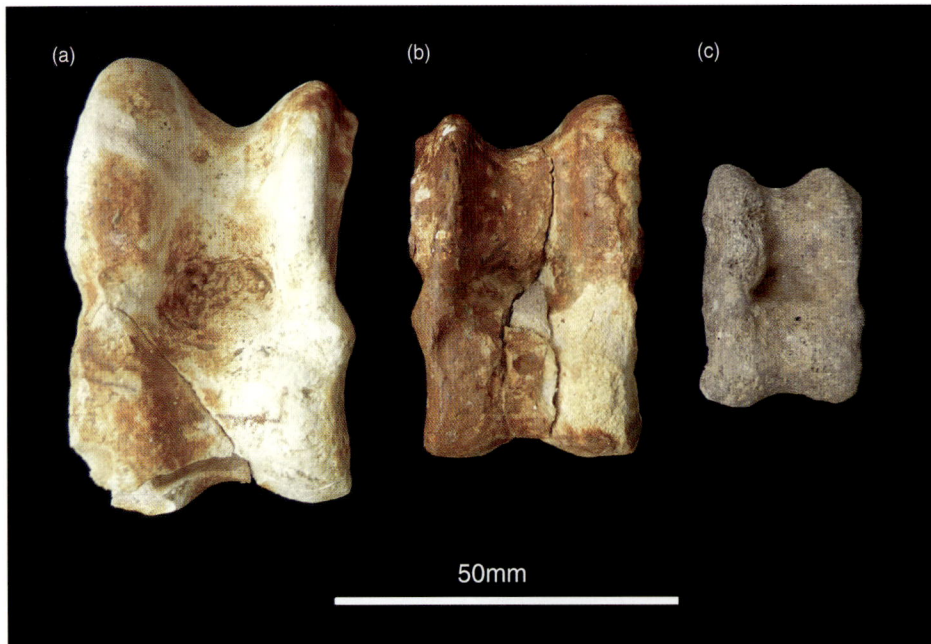

Figure 9.10 Three deer astragali from Southfleet Road: (a) *Cervus elaphus* (Δ.41407, context 40100); (b) *Dama dama* (Δ.41034, context 40039); (c) *Capreolus capreolus* (Δ.5002, context 40070)

Table 9.6 List of *Cervus elaphus* bones and teeth from Southfleet Road

Phase	Context	Find no. Δ	Element	Comments
6b	40070	43165	L antler (shed) – basal region with broken brow and bez tines	
6	40078	40741	Antler (shed) – basal region with brow and bez tines, part of beam	
6	40078	42945	R antler (shed) – basal region with brow and bez tines, part of beam	
6	40078	42992	L antler (shed) – basal region with brow and bez tines, part of beam	
6	40078	43848	L antler (shed)	Almost complete, but badly damaged
6	40078(?)	42282	L antler – beam and trez tine	
6	40100	40898	L antler (shed)	Almost complete, but badly damaged. Fig. 9.11
6	40100	41104	L antler base (shed)	
6a	40103	43787	R antler (shed) – basal region with brow tine	
6a	40039	42130	Axis frag	
6b	40070	43076	R P^2	Early-mid wear
6b	40070	43145	R M$^{1\text{ or }2}$	Early wear
6b	40070	43188	L P$_2$	Early-mid wear
6	40100	41722	L P$_3$	
6b	40070	43140	L P$_4$	Mid-late wear
6	40100	41142	R M$_3$	Early-mid wear
6a	40103/039	43663	L scapula frag	cf. *C. elaphus*
6a	40039	42128	R scapula frag	
6	40100	40626	L scapula frag	
6	40100	41327	L scapula frag	cf. *C. elaphus*
6	40100	40613	L radius frag	Prox. fused
6b	40144	50006	R radius frag	Unfused distal epiphysis
6b	40070	50001	L scaphoid	
6b	40070	50010	R lunate	Juvenile (forming bone)
6	40078	42147	L tibia frag	Dist. fused
6	40100	43917	L tibia frag	Dist. fused
6	40100	41377	R distal fibula	cf. *C. elaphus*
6a	40039	42287	R astragalus	
6a	40039	40076	L astragalus	cf. *C. elaphus*
6	40100	41407	R astragalus	
6a	40103	42476	L metatarsal frag	
6	40100	41305	1st phalanx	
6b	40070	43823a	2nd phalanx	

these proved too fragile to lift intact. The sample includes remains from juveniles as well as adults.

Red deer is one of the most common mammals recorded from the British Pleistocene and also one of the most variable, exhibiting great morphological and size variation through the Middle and Late Pleistocene and Holocene (Lister 1981; 1989; 1995; Grigson and Mellars 1987). The evolution of British Pleistocene red deer was studied in detail by Lister (1981). He noted that the Swanscombe collection includes two antler-tops with several points (Lister 1986, figure 3) that show that the Swanscombe red deer had the capacity to form a 'crown'. This feature is not found in red deer antlers from pre-Anglian/Elsterian contexts, but is dominant in modern adult red deer (Lister 1986; 1993). There is only one antler top from Southfleet Road (Fig. 9.12b). This also has three points forming a simple 'cup', and closely resembles the antler top from the Basal Gravel at Swanscombe (Fig. 9.12a).

The method adopted for assessing the skeletal size of the Southfleet Road red deer was developed by Lister (1993) in his analysis of the red deer from Hoxne. In this percentage method, each fossil specimen is standardised in relation to the corresponding element of a standard skeleton or a sample of individuals from a 'standard population'. An advantage of this approach is that it is particularly well-suited for combined statistical analysis of different bone elements. Table 9.7 gives measurements for the Southfleet Road sample, which are plotted on Figure 9.13 as percentage deviations from the standard Star Carr mean for each bone. From Figure 9.13 it can be seen that the Southfleet Road points are all smaller than the Star Carr mean, with most falling beyond one Standard Deviation of the Star Carr mean. Pooling the different measurements, the mean deviation of the Southfleet Road sample from the Star Carr mean is -9.2% (n = 4). Although the plotted sample is small, other specimens of red from Southfleet Road are

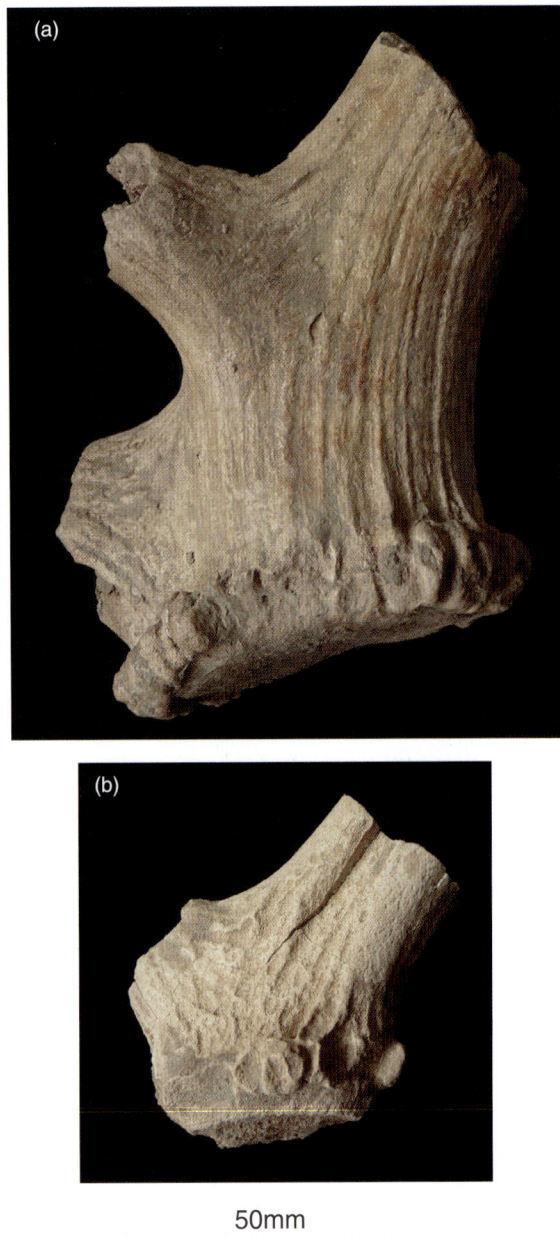

Figure 9.11 Two shed antler bases from Southfleet Road:
(a) *Cervus elaphus* (Δ.42992, context 40078); (b) *Dama
dama* (Δ..42198, context 40070)

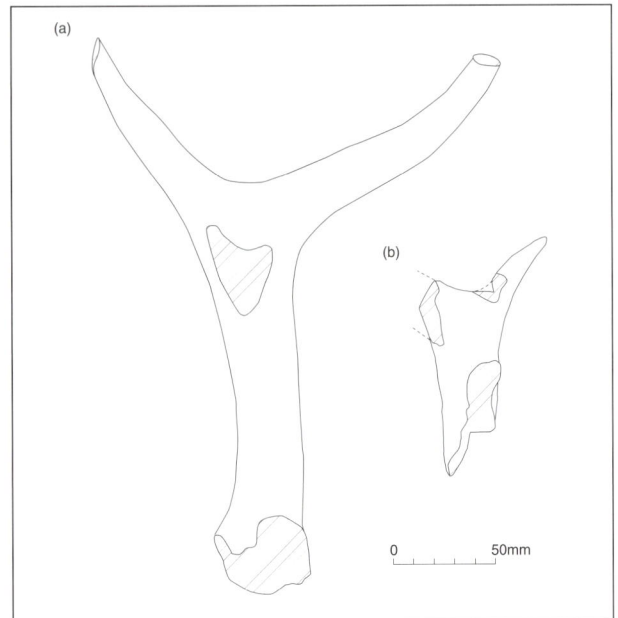

Figure 9.12 Drawing of antler crowns of *Cervus elaphus*
from (a) Swanscombe (Natural History Museum
(A.S. Kennard collection) M49726, Basal Gravel), and (b)
Southfleet Road (Δ. 40898, context 40100)

that the length of the single M_3 (L = 29.8 mm) from
Southfleet Road is smaller than the Star Carr sample
(31.6-39.0, mean 34.55, n = 30).

The red deer is a highly adaptable herbivore with a
distribution covering diverse environments and climatic
zones. It can subsist on both browse when shrubs and
tree shoots are available, but also grazes in open habitats.
Optimal red deer habitats include woodland and
woodland edge, but it can thrive in open habitats
(moors, open mountain areas and steppe), only avoiding
taiga with deep winter snow (Bjärvall and Ullström
1986; Mitchell-Jones *et al.* 1999). The relative abundance of red deer at Southfleet Road suggests that the
Ebbsfleet Valley may have been better suited to red deer
than the nearby floodplain of the Thames, where
remains of fallow deer are dominant and red deer are
less common in the lower part of the succession.

Dama dama (L.), fallow deer

Only 11 specimens of fallow deer have been recovered
(Table 9.8). Of these, four are shed antler bases (Fig.
9.11). The sample includes three isolated teeth, a broken
mandible with molars, the distal end of a tibia and two
astragali (Fig. 9.10). Table 9.7 gives measurements for
the Southfleet Road sample, which are plotted in Figure
9.14 as percentage deviations from a standard sample of
modern fallow deer from Richmond Park, London
(Lister 1993). From Figure 9.14 it can be seen that the
Southfleet Road fallow deer was considerably larger
than those in Britain today, and within the size-range of
Hoxnian large-bodied fallow deer from Swanscombe
and Clacton. In terms of body size, this corresponds to
a mean body weight of around 80kg for the Swanscombe

similarly small, indicating that the mean size of these
deer was smaller than that of red deer from Star Carr.
Turning to the other Hoxnian samples, it can be seen
that the Southfleet Road red deer are similar in size to
those from Swanscombe (percentage deviation, -5.6%)
and Clacton (percentage deviation, -10.5%), which are
all significantly smaller than the larger-bodied animals
from Hoxne (Lister 1993). At Hoxne, Lister (1993) has
shown that the red deer combined small dental size with
somewhat larger body size. This contrasts with the
situation at Swanscombe and Clacton animals, where
small dental size corresponds to small body size.
Unfortunately, it is not possible to test this relationship
for the Southfleet sample, where only six measurable
teeth are represented. Nevertheless, it may be significant

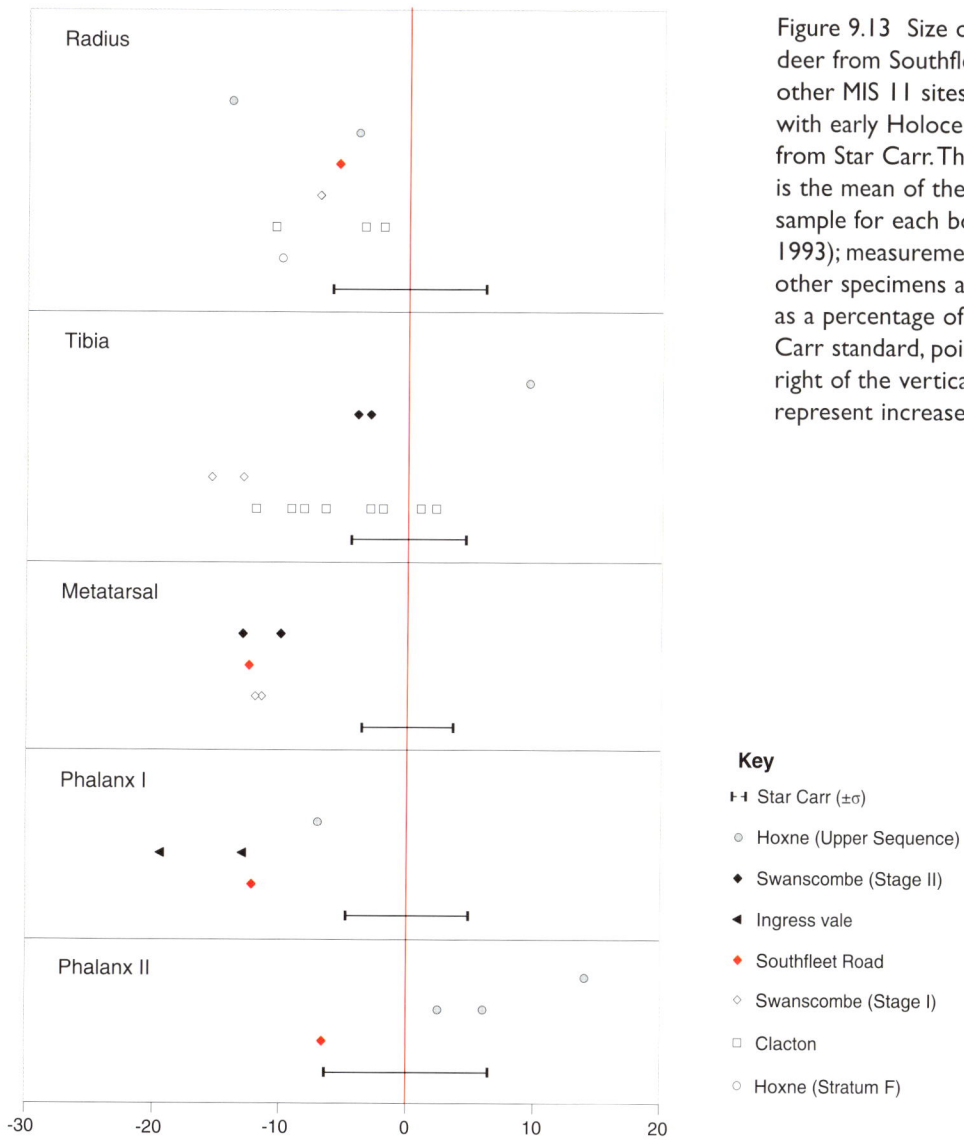

Figure 9.13 Size of the red deer from Southfleet Road and other MIS 11 sites compared with early Holocene material from Star Carr. The zero line is the mean of the Star Carr sample for each bone (Lister 1993); measurements of the other specimens are calculated as a percentage of the Star Carr standard, points to the right of the vertical line represent increased size

Key

⊢⊣ Star Carr (±σ)
○ Hoxne (Upper Sequence)
◆ Swanscombe (Stage II)
◀ Ingress vale
◆ Southfleet Road
◇ Swanscombe (Stage I)
□ Clacton
○ Hoxne (Stratum F)

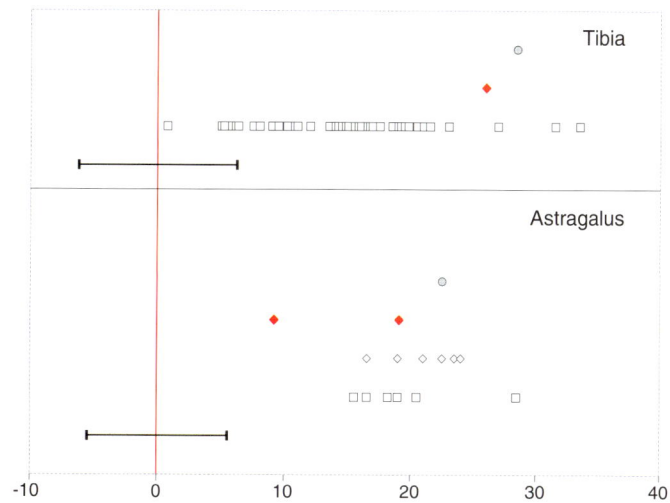

Figure 9.14 Size of the fallow deer from Southfleet Road, Clacton, Swanscombe (Stage I), Hoxne (Upper Sequence), compared with modern fallow deer. The zero line is the mean of a modern sample from Richmond Park, London (Lister 1993); measurements of the other specimens are calculated as a percentage of the modern standard. For key to symbols see Fig. 9.13

Table 9.7 Measurements (in mm) of cervid bones and teeth from Southfleet Road. Measurements (in mm) taken according to the method of von den Driesch 1974

Taxon	Sub-phase	Context	Find no. Δ	Element					Comments
Cervus elephus					Basal circumference				
	6	40078	42945	Antler	93.0				
	6	40078	42992		144.0				
	6	40100	40898		145.0				
					L	W			
	6b	40070	43076	P²	13.1	-			
	6b	40070	43145	M¹ or ²	20.1	21.5			
	6b	40070	43188	P₂	10.0	6.6			
	6	40100	41722	P₃	15.7	9.8			
	6b	40070	43140	P₄	15.4	10.8			
	6	40100	41142	M₃	29.8	14.1			
					GLP	LG	BG	SLC	
	6a	40103/039	43663	Scapula	-	-	-	30.3	cf. *C. elaphus*
	6	40100	40626		58.4	45.0	39.9		
					Bp	BFp			
	6	40100	40613	Radius	56.0	53.1			
					Bd	Dap			
	6	40078	42147	Tibia	52.1	38.5			
	6	40100	43917		48.5	35.8			
					GL				
	6a	40039	42287	Astragalus	c 59.4				
	6a	40039	40076		c 49.1				cf. *C. elaphus*. Minimum dimension due to damage and erosion
	6	40100	41407		56.6				
					Bp				
	6a	40103	42476	Metatarsal	c 38.3				
					GL	Bp	SD	Bd	
	6	40100	41305	1st Phalanx	57.2	19.7	16.1	18.5	
	6b	40070	43823a	2nd Phalanx	38.3	19.6	16.3	15.7	
Dama dama					Basal circumference				
	6b	40070	42198	Antler	95.0				
					L	W			
	6a	40039	41665	M¹ or ²	21.2	20.7			cf. *D. dama*
	6a	40039	42239	M³	19.6	21.8			cf. *D. dama*
	6	40100	41114	P₃	12.5	8.4			
	6	40100	41341	M₁	17.1	12.1			
				M₃	26.7	12.3			
					Bd				
	6a	40039	41046	Tibia	42.8				cf. *D. dama*
					GL	Bd			
	6a	40039	41034	Astragalus	46.5	28.7			
	6	40100	40727			c 26.3			cf. *D. dama*

animals, as compared to 50kg in Recent British fallow deer (Lister 1986).

Hoxnian fallow deer are generally referred to the subspecies *Dama dama clactoniana*. This form is characterised by an unusual morphology of the antlers, which are less palmated than modern fallow deer, with an extra anterior tine between second tine and palmation (Sutcliffe 1964; Leonardi and Petroni 1976; Lister 1986). This form is best known from the Hoxnian and broadly contemporaneous sites in Italy. Unfortunately the distal parts of fallow deer antlers are not preserved in the Southfleet assemblage.

Other distinctive morphological features of Hoxnian fallow deer are found in the dentition. Lister (1981; 1986) has studied the enamel patterns of the lower third premolar of Pleistocene fallow deer and has observed that Hoxnian specimens often have a metaconid that is fused with the entoconid high in the crown (Lister 1986, figure 2). This enamel configuration is not found in Last Interglacial and modern British fallow deer, but is present in the single worn example from Southfleet Road (Fig. 9.15). Although

Table 9.7 (continued)

Taxon	Sub-phase	Context	Find no. Δ	Element			Comments
Cervid					L	W	
	6b	40070	43007	Upper molar	19.1		
	6	40100	41113	$M^{1\ or\ 2}$		21.1	
	6	40100	41126	$M_{1\ or\ 2}$	18.0	12.5	
					BT		
	6b	40070	43093	Humerus	50.3		
	6	40100	41391	Humerus	c 45.9		
					BPC		
	6	40100	40875	Ulna	33.0		
					Bd		
	6	40078	41908	Metacarpal	37.6		
					LAR		
	5	40025	43827	Innominate	54.8		
					DC		
	6a	40039	41985	Femur	36.4		
					Bd		
	6	40078	41762	Tibia	47.9		

L = length; W = width; Bd = distal breadth; GL = greatest length; GLP = greatest length of glenoid process; LG = length of glenoid cavity; BG = breadth of glenoid cavity; SLC = smallest length of neck of scapula; BP = breadth of proximal end; BFp = breadth of humeral articular surface; Dd= depth of distal end; SD = smallest breadth of diaphysis; BPC = breadth of proximal articular surface of the ulna; LAR = length of the acetabulum on the rim; DC = depth of caput femoralis

Table 9.8 List of *Dama dama* bones and teeth from Southfleet Road

Phase	Context	Find no. Δ	Element	Comments
6b	40070	42198	L antler (shed) base and part of brow tine & beam	
6b	40070	50004	Antler base (shed)	
6a	40103	43784	L antler (shed) base and brow tine	
5	40025	43868	Antler (shed) base	cf. *D. dama*
6a	40039	41665	L $M^{1\ or\ 2}$	cf. *D. dama*. Mid wear
6a	40039	42239	L M^3	cf. *D. dama*. Mid wear
6	40100	41114	L P_3	Late wear (Fig. 9.15)
6	40100	41341	R mandible with M_{1-3}	Late-mid wear
6a	40039	41046	L tibia (distal and shaft)	cf. *D. dama*. Distal fused
6a	40039	41034	L astragalus	
6	40100	40727	R astragalus	cf. *D. dama*

widely used as a 'population marker' for Hoxnian fallow deer, metaconid fusion is also present at high frequency in the post-Hoxnian fallow deer from Grays, Essex. Although the dating of Grays has been contentious, there is now a growing consensus that the Grays deposits date to MIS 9 (Bridgland 1994; Penkman *et al.* 2011). The fallow deer from Grays are also unusually large, with an average percentage deviation of 14.2 % from the modern standard sample (compared with 16.5 % for Swanscombe and 19.5 % for Clacton). The Hoxne fallow deer (late MIS 11) are of comparable size. These data suggest that fallow deer maintained large body size throughout MIS11 and into the succeeding MIS 9 interglacial. Whether the unusual antler form of the Hoxnian fallow deer persisted into the following MIS 9 interglacial is unknown as no antler remains have been found at Grays.

From an ecological perspective, fallow deer is less adaptable than red deer and more closely tied to woodland, favouring parkland with ample herbaceous foods and relatively warm climates (Bjärvall and Ullström 1986; Mitchell-Jones *et al.* 1999). In the British Pleistocene, fallow deer is restricted to interglacials and is one of the most reliable indicators of interglacial conditions in a British Pleistocene fauna (Stuart 1982).

Capreolus capreolus (L.), roe deer

Only two bones can be ascribed with certainty to roe deer. These are a complete astragalus (Fig. 9.10, context 40070) and a fragment from the midshaft of a metacarpal (context 40144); both are from the tufa channel. A splinter of a metapodial diaphysis from context 40100 is also probably from roe deer.

Very few Pleistocene sites in Britain have so far produced remains of roe deer (Stuart 1982). To date, the only other Hoxnian specimens come from Swanscombe, where it is represented by a distal end of tibia from the

Basal Gravel (Sutcliffe 1964; Lister 1986). A small number of roe deer remains have been recovered from each of the Middle Pleistocene interglacials (Boxgrove, Ostend, Westbury-sub-Mendip and West Runton – early Middle Pleistocene; Hoxne – late MIS 11, Grays Thurrock – MIS; Selsey – MIS 7). Individual samples are generally too small to allow detailed discussion of body size, although data from Boxgrove clearly indicate

that some early Middle Pleistocene roe deer were exceptionally large (Lister 1993).

The length and distal breadth of the astragalus from Southfleet Road are given in Table 9.9 and plotted in Figure 9.16. Here they are compared with measurements of early Holocene roe deer from Star Carr (Legge and Rowley-Conwy 1988) and Cherhill (Grigson, in Evans and Smith 1983) and modern

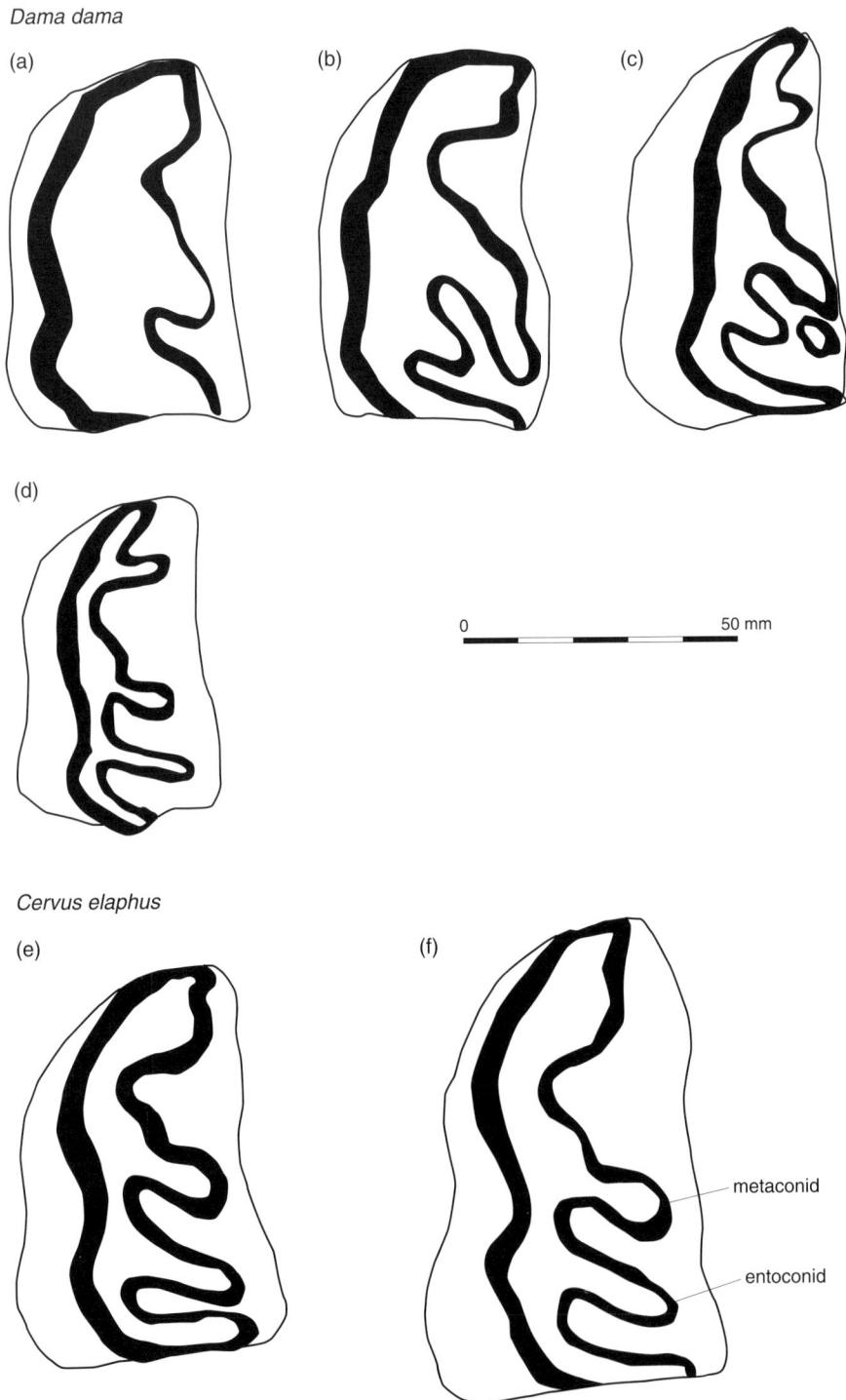

Figure 9.15 Comparison of modern and fossil fallow deer (*Dama dama*) and red deer (*Cervus elaphus*) showing states of metaconid-entoconid fusion in the lower third premolar: (a) Southfleet Road (Δ.41114, context 40100); (b) Swanscombe (M49724. Image inverted); (c) Swanscombe (M49716. Image inverted); (d) England, modern reference specimen (Department of Palaeontology, Natural History Museum, London); (e) Grays, Essex (M21643); (f) Southfleet Road (Δ.41722, context 40100)

comparative material. In these comparisons, the Southfleet astragalus falls within the range of the Star Carr measurements and is slightly larger than the modern west European comparative material. Mesolithic and Neolithic roe deer are large in comparison to modern roe deer from the same region, indicating that roe deer underwent a very marked decrease in size during the Holocene. Similar size reduction in other taxa is thought to be the result of a greater level of human interference and habitat destruction. Lister (1993) has noted that the teeth of late MIS 11 roe deer from Hoxne may have been slightly smaller than from Star Carr, whereas the tibia from Swanscombe is slightly larger than the modern comparative sample. This hints at size change in roe deer during MIS 11, but larger samples are needed to test this hypothesis.

The present-day ecology of the roe deer suggests a mosaic landscape existed in the vicinity of the site, with woodland or thick scrub interspersed with more open

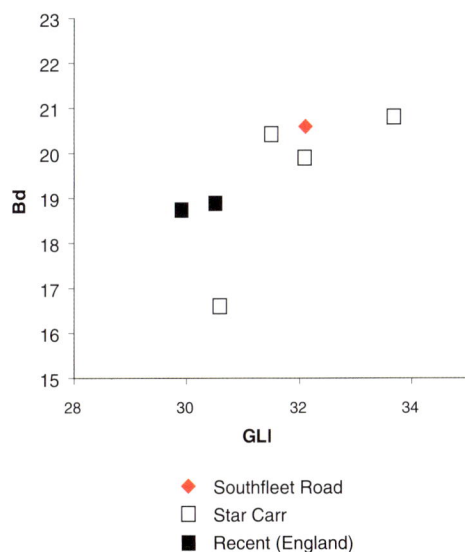

Figure 9.16 Distal breadth (Bd) plotted against greatest length (GLl) of the astragalus to illustrate the proportions of the Southfleet Road astragalus in comparison with samples of modern and early Holocene (Starr Carr) roe deer

areas supporting herbaceous vegetation (Bjärvall and Ullström 1986; Mitchell-Jones *et al.* 1999).

Bovids (Bovidae)

Bos primigenius Bojanus, aurochs

A substantial portion of an aurochs skull comprising both frontals with horncores and the basicranium was found in fluvial sands of Phase 3 (Fig. 9.17, Table 9.10). A second individual is represented by part of the nasal and maxillary regions of a skull with most of the cheek teeth from the right side (Fig. 9.18, Table 9.11). In addition to these cranial remains, several fragmentary bones and teeth were recovered from other contexts (Tables 9.10 and 9.12). It is not possible to identify these to species or genus level as they are either incomplete or otherwise poorly preserved.

Several species of large bovid have been recorded from Middle Pleistocene in northern Europe. In the late Middle Pleistocene, these include the ubiquitous *Bison priscus* (steppe bison) and *Bos primigenius*. Also represented is water buffalo (*Bubalus murrensis*), which has been found in deposits correlated with the Hoxnian in Germany (eg Steinheim, Schönebeck). According to Sutcliffe (1964) only *Bos primigenius* was present in the Swanscombe assemblage. However, careful reanalysis of the Swanscombe bovid material by Gee (1991; 1993) has identified a few bones of bison. The genus *Bos*, which arose in Africa at the end of the Early Pleistocene (Martínez-Navarro *et al.* 2007) is first recorded in Europe at Venosa-Notarchirico, Italy (*c* 0.5–0.6 million years ago). The Swanscombe and Clacton aurochs material is significant as they represent the earliest records of this species from north-western Europe. These aurochs are notable for their large size and distinctive horncores, which are less curved, somewhat flatter in cross-section and show less torsion than in more recent populations (Gee 1991).

The Southfleet Road aurochs skull is notable because it is unusually complete. Skulls from Swanscombe and Clacton are generally more fragmentary, but the sample includes a number of horncores and two frontlets from Clacton (one of which is restored). Measurements of the

Table 9.9 Comparison of measurements of the *Capreolus capreolus* astragalus from Southfleet Road and other samples

		GLl (mm)			Bd (mm)	
	n	*Observed range*	*Mean + SD*	*n*	*Observed range*	*Mean + SD*
Southfleet Road	1		32.1	1		20.4
Holocene						
Starr Carr[a]	4	30.6–33.7	32 ± 1.30	4	18.6–20.8	19.9 ± 0.96
Cherhill[b]	1		32.8			
Recent						
England[c]	2	29.9, 30.5		2	18.8, 18.9	

[a] Legge and Rowley-Conwy 1988
[b] Grigson 1983
[c] Parfitt own data

Table 9.10 List of bovid remains from Southfleet Road. Several of the bones could only be identified as from *Bos* or *Bison* owing to lack of diagnostic features. The proximal metacarpal resembles *Bison*, but the specimen is too incomplete and eroded to provide a firm attribution to this taxon

Taxon	Phase	Context	Find no. Δ	Element
Bos primigenius	3	40062	50186	Partial skull (basicranium, frontlet with both horncores)
	5	40025	43298	R premaxilla – maxilla with P⁴ (roots), M¹⁻³ (mid wear)
	5	40025	43813	L nasal (probably part of *B. primigenius* skull, Δ. 43298)
	6	40078	42561	R metatarsal frag (proximal lateral & medial facets)
Bos or *Bison*	6a	40039	42546	L pisiform & lunate (rearticulates to Δ. 43454)
	6a	40039	43454	L scaphoid (rearticulates to Δ. 42546)
	6	40158	43861	R tibia, shaft & distal end (fused)
	?	40089	40062	R metacarpal frag (part of proximal end & shaft)
Large bovid	5	40025	43842	Skull frag
	6	40078	41909	R metacarpal frag (part of distal end, fused)
	6	40078	40711	Metapodial frag (distal condyle)
	7	40167	50168	Cheek tooth frags

Table 9.11 Measurements (in mm) of *Bos primigenius* teeth from context 40025, Southfleet Road

Phase	Context	Find no. Δ	Length of molar row[a]	M¹L	M¹W	M²L	M²W	M³L	M³W
5	40025	43298	107	30.2	27.5	34.9	29.1	36.2	26.9

Note: All dental measurements taken at the occlusal surface   [a] Length of molar row, measured along alveoli on the buccal side

Table 9.12 Measurements (in mm) of the *Bos primigenius* skull from context 40062, Southfleet Road and two Hoxnian examples from Clacton

	Southfleet Road	*Clacton*	
		M15624	*M2342*
Least breadth between bases of horncores (vdD 31)	340	210	245
Least distance between horncore tips (vdD 42)	-	520	965
Distance between horncore tips (vdD 42a)	-	1970	1890
Greatest tangential distance between the outer curves of the horncores (vdD 43)	*c* 1090	952	1140
Length of outer curvature of horncore (vdD 47)	-	860	805
Horncore basal circumference (vdD 44)	-	475	465
Greatest (oro-basal) diameter of the horncore base (vdD 45)	*c* 164	174	173
Least (dorso-basal) diameter of the horncore base (vdD 46)	-	118	123

VdD = von den Driesch measurement number

Table 9.13 Biometric comparisons of the *B. primigenius* horncores from Southfleet Road with Hoxnian and Holocene samples. Statistics for the Holocene samples are: Minimum-Mean-Maximum (sample size)

	Length (mm)	*Basal circumference (mm)*	*Greatest (oro-basal) diameter (mm)*	*Least (dorso-basal) diameter (mm)*
Holocene (♂ & ♀)[a]	335–564.76–845 (90)	180–301.69–395 (100)	-	-
Southfleet Road	-	-	*c* 164	-
Clacton				
M15624	860	475	174	118
M27871	*c* 840	470	173	111
M2342	805	465	173	123
Swanscombe				
M20583, Basal Gravel	730	435	159	104
No number	-	400	138	113
No number	-	485	168	137

[a] from Grigson 1978

Swanscombe and Clacton skull are compared with the Southfleet Road example in Table 9.13. Although somewhat crushed, the Southfleet Road skull displays similar flattening of the horncore base. In addition, the Southfleet Road horncore appear less curved and with less torsion than those of later Pleistocene and Holocene horncores.

Although the aurochs is now extinct, details of its ecology can be reconstructed from its fossil record as well as historical descriptions of the last surviving European populations. It seems to have been a highly adaptable species that was able to shift its diet from grasses to browse, depending on local circumstances. While aurochs was ubiquitous during wooded inter-glacial conditions in northern Europe, there is ample evidence that it adapted to a variety of ecological conditions, including interstadials during the Last Cold Stage (Currant 1986a).

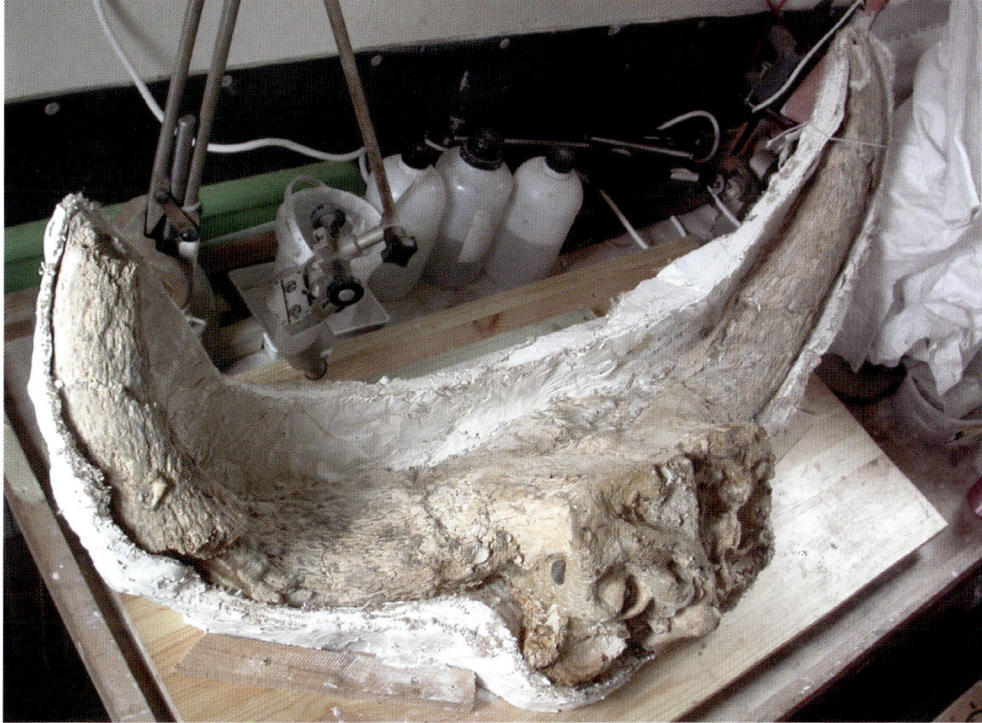

(a)

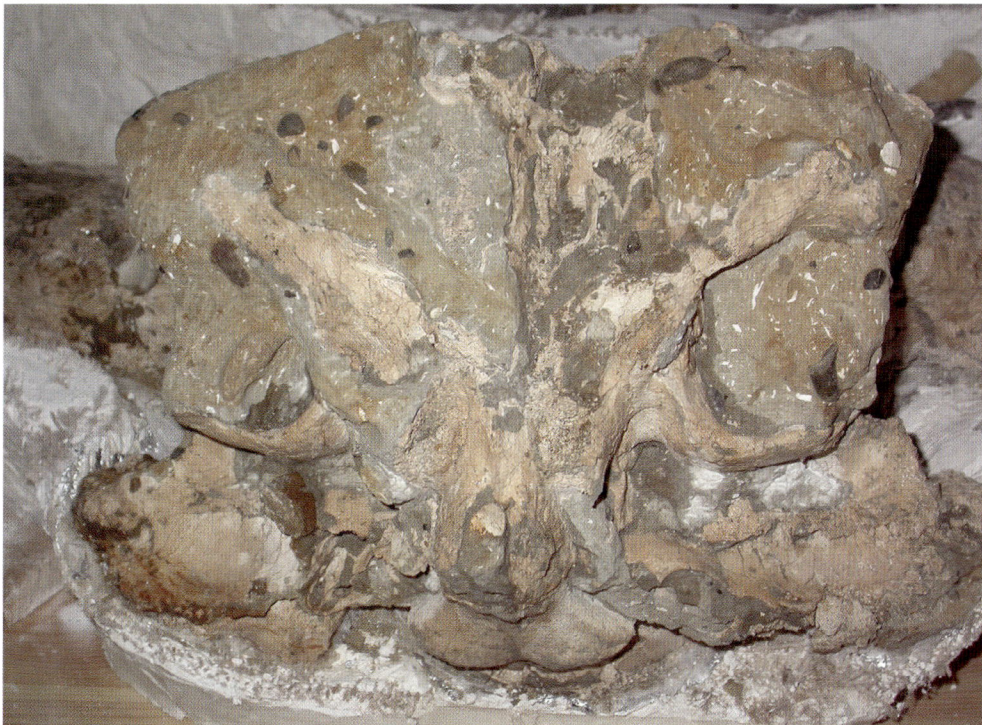

(b)

Figure 9.17 *Bos primigenius* skull (Δ.50186, context 40062) from Southfleet Road during conservation: (a) oblique frontal view; (b) basicranium, with occipital condyles at base [Photographs by N. R. Larkin]

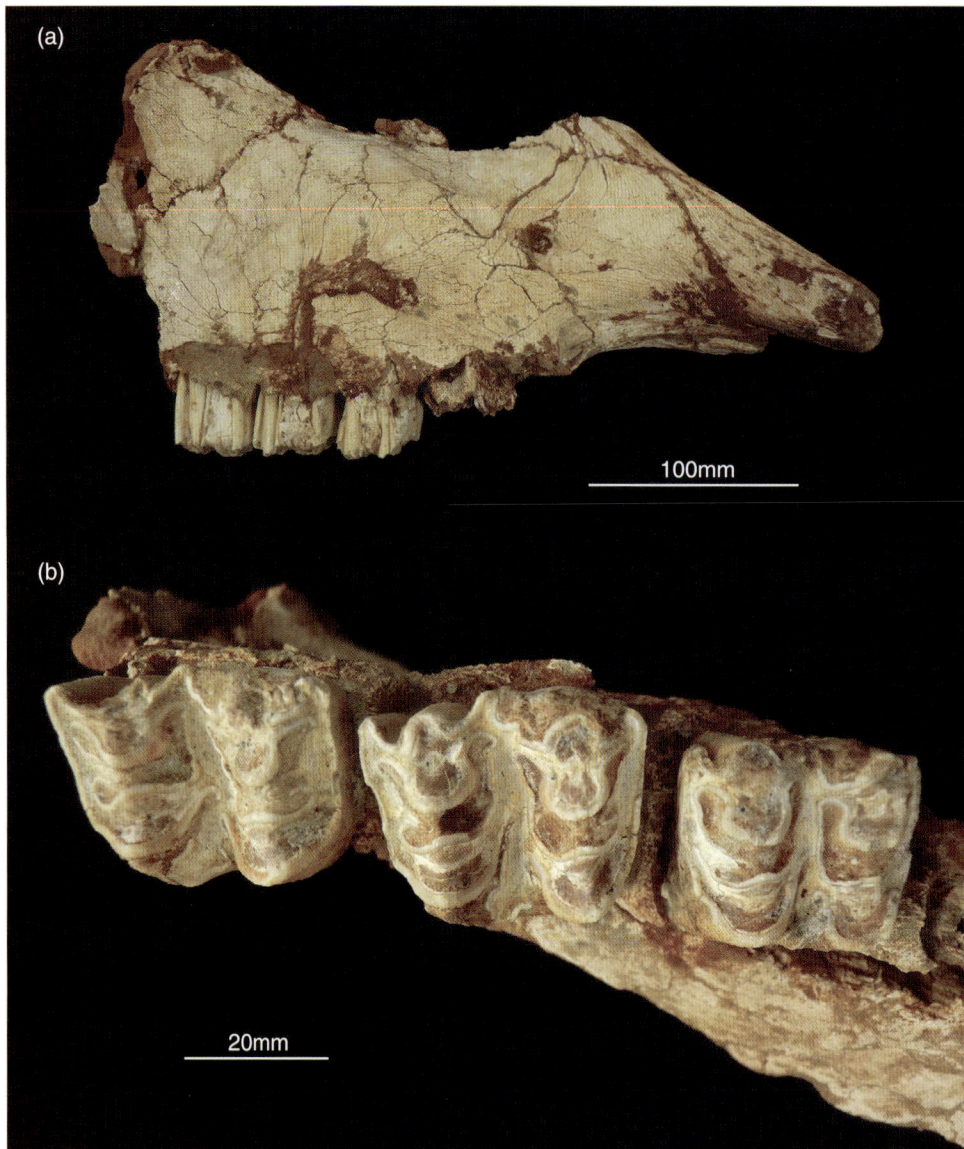

Figure 9.18 *Bos primigenius* maxilla (Δ.43298, context 40025): (a) lateral view, (b) occlusal view of M$^{1-3}$

MAMMALIAN EVIDENCE FOR THE AGE OF THE SOUTHFLEET ROAD INTERGLACIAL SEDIMENTS

A chart showing the occurrence of mammalian species at Southfleet Road is given in Table 9.14 and Figure 9.19. In sum, the mammalian fauna is entirely consistent with an interglacial phase providing regional deciduous woodland as well as locally open vegetation. The fauna includes a number of extant species found today in the European temperate zone. The same species (or their close relatives) are also important elements in interglacial faunas throughout NW Europe. Although several of these taxa are also known from interstadial contexts (for example straight-tusked elephant, aurochs), the Southfleet Road assemblage includes thermophilic species, such as rabbit and Merck's rhinoceros that are known only from interglacial contexts in Britain.

There appears to be no significant, abrupt break, but rather a succession of subtle changes in the composition of the faunal through the sequence, reflecting changes in the local vegetation and proximity to water bodies. Importantly, there is no evidence from the vertebrate fauna suggesting a substantial hiatus between the basal fluvial sediments (Phases 3-5) and the overlying fine-grained sequence of Phase 6, reinforcing similar indications from the lithological and ostracod records (Chapters 4 and 11 respectively). Both faunal groups suggest interglacial conditions, although the basal fluvial sediments have also yielded ground squirrel, a rodent more typical of cold stage faunas in Britain (Stuart 1982). Another small mammal that is more commonly found in cold stage faunas is the northern vole. This vole is found together with ground squirrel in Phase 5, but it is also present in Phase 6b, where it is associated with a fully interglacial woodland molluscan fauna. The presence of ground squirrel and northern vole in an otherwise interglacial context at Southfleet Road may suggest continuity with the preceding cold stage and survival in suitable microhabitats of dry grassland favoured by ground squirrels and marshland habitats favouring northern vole. The presence of a fully inter-glacial ostracod fauna in underlying deposits attributed to Phases 3 and 4 appears to rule out any possibility that this small mammal assemblage is associated with cold conditions in Phase 5.

Two lines of evidence help to constrain the age of the sediments containing the Clactonian artefacts, namely: (a) their geomorphological position in relation to other mammal-bearing fluvial sequences in the lower Ebbsfleet Valley (Wenban-Smith *et al.* forthcoming); and (b) mammalian biostratigraphy. In terms of altitude, the Southfleet Road fluvial sediments (occurring at *c* 20-25m OD) are higher (and therefore older) than the fluvial sequences (occurring at *c* 8-12m OD) in the vicinity of the famous 'Bakers Hole' site. These lower-level deposits have yielded interglacial faunas with horse and mammoth, which have been correlated with the penulti-mate (MIS 7) interglacial.

Species of particular biostratigraphic interest include the rhinos *Stephanorhinus hemitoechus* and *S. kirchber-gensis* and the aurochs (Table 9.14, Fig. 9.19). These ungulates are often common in late Middle Pleistocene temperate faunas, but unknown in Britain before the Anglian stage. Although aurochs is a persistent element in British post-Anglian faunas, Hoxnian aurochs are distinguished by their exceptional size combined with bulky horncores of a distinctive shape (Gee 1991). The morphology of the aurochs horncore from Southfleet Road is of the same distinctive form as Clacton and Swanscombe and provides strong evidence that the site dates to the Hoxnian stage. The detailed morphology and size of the red deer and fallow deer is also of biostrati-graphical interest. Lister (1993) has highlighted the relatively small size of red deer at Clacton and Swans-combe (spanning Basal Gravel to Upper Middle Gravel); in comparison the animals from Hoxne are much larger bodied. He further noted (ibid.) that the Hoxne red deer combined relatively large body size with small dental size. In terms of skeletal and dental size the Southfleet red deer match those from Swanscombe and Clacton. Similarly, fallow deer from Swanscombe and Clacton are distinctive, being typified by unusual antler form, relatively large size and features of the premolar dentition that define the late Middle Pleistocene subspecies *Dama dama clactoniana* (Lister 1981; 1986). The stratigraphical range of *D. d. clactoniana* is poorly known and the taxonomic status of the fallow deer from Hoxne as well as those from the succeeding (MIS 9) interglacial is currently uncertain. Although antlers have yet to be found from these sites, the fallow deer from Grays and Hoxne are similarly large-bodied and, on the basis of size alone, indistinguishable from *D. d. clactoniana*. Another feature potentially linking these populations is the high frequency of metaconid fusion in the third lower premolar found in samples from Clacton/Swanscombe and Grays. The foregoing discussion illustrates that the fallow deer and red deer from Southfleet Road are entirely consistent with a Hoxnian age. However, in the absence of well-preserved fallow deer antlers from Southfleet Road and

Grays, it is not currently possible to discriminate between MIS 11 and MIS 9 using evidence from the cervids in isolation. Stronger evidence for an upper age limit is provided by the presence of pine vole *Microtus (Terricola)* cf. *subterraneus*. This vole is unknown in Britain from the Holocene and has not been found in any Late Pleistocene fauna; during the later Middle Pleistocene it is known only from sites correlated with MIS 11 such as Hitchin, Beeches Pit, Barnham, Hoxne and Swanscombe. The absence of pine voles in post-MIS 11 faunas in Britain is unlikely to be due to insufficient sampling, as the British small mammal fauna is now rather well known, with rich samples from successive warm and cold stages spanning this time period.

The significance of other biostratigraphic 'indicator species' (*sensu* Schreve 2001a, b) is less clear. This group includes the mole *Talpa minor* and rabbit, which Schreve (2001a) has included in her list of taxa that define the 'Swanscombe Mammal Assemblage Zone'. In Britain, both taxa are unknown from deposits younger than the 'Upper Sequence' at Hoxne (Fig. 9.19; Table 9.15). However, these are rare taxa that are of dubious biostratigraphical significance because their absence could represent sampling bias due to small sample size. Virtually nothing is known of the fossil species of mole present in Britain between the Hoxnian and the Holocene. Similarly, the fossil record of lagomorphs is extremely poor for this time period (Mayhew 1975).

The problem of range extension is highlighted by the unexpected presence at Southfleet Road of the large and distinctive shrew *Sorex (Drepanosorex)* sp. This shrew has long-been regarded as a pre-Anglian 'indicator species' (Bishop 1982; Parfitt 1999; Schreve 2001a). It is a common element in many European Early and early Middle Pleistocene faunas, and in Britain it is well-represented in the early Middle Pleistocene at West Runton, Westbury and Boxgrove (Maul and Parfitt 2010). The identification of *Sorex (Drepanosorex)* sp. at Southfleet Road is based on a mandibular ramus from the tufaceous channel-fill; although broken, this specimen is identical in preservation to other small mammal bones from the same context, ruling out reworking from earlier deposits. In terms of morphology, the Southfleet Road mandible is indistinguishable from mandibles in the type series of *Sorex (Drepanosorex) savini* from West Runton. Given the weight of evidence suggesting a post-Anglian age for the Southfleet Road sequence, the most parsimonious explanation for the presence of *Sorex (Drepanosorex)* sp. at Southfleet Road is that its stratigraphical range extended beyond the Anglian. Additional evidence for a range extension of *Sorex (Drepanosorex)* sp. comes from poorly described post-Elsterian (?Holstenian) specimens of *Sorex (Drepanosorex)* sp. (sometimes assigned to *Drepanosorex postsavini*) from sites in central Europe (van Kolfschoten, pers. comm.). That such a characteristic species could have been overlooked in post-Anglian/Elsterian context warns against placing too much reliance on species presence or absence in mammalian biostratigraphy (cf. Lister 1992).

Table 9.14 The Southfleet Road Middle Pleistocene mammal fauna. Group A includes occurrences in the basal fluvial sediments (Phase 5) and those from Group B are found in the overlying fine-grained sediments and tufa-filled channel (Phase 6b). # Taxa first appear in the British sequence in Hoxnian; † extinct taxa; * taxa are known only in the British Pleistocene interglacials and are missing from cold stage assemblages (excluding interstadials)

Group	A	B	
Phase	5	6	6b
MAMMALIA			
Lipotyphla			
Sorex minutus, pygmy shrew	+		+
Sorex sp. 1, shrew (smaller than *S. araneus*)	+	+	+
Sorex (*Drepanosorex*) sp., shrew (large) †			+
Neomys sp., water shrew	+	+	+
Soricidae gen. et sp. indet., shrew		+	+
Talpa minor, mole †		+	+
Chiroptera			
Myotis daubentonii, Daubenton's bat *			+
Primates			
Macaca sylvanus, Barbary macaque			+
Homo sp., hominin (C = Clactonian)	C	C	C
Lagomorpha			
Oryctolagus cuniculus, rabbit *	+	+	+
Rodentia			
Sciurus sp., squirrel *			+
Spermophilus sp., ground squirrel	+		
Clethrionomys glareolus, bank vole	+	+	+
Castor fiber, beaver		+	
Arvicola cantianus, water vole †	+	+	+
Microtus (*Terricola*) cf. *subterraneus*, common pine vole	+		+
Microtus agrestis, field vole			+
Microtus agrestis / *M. arvalis*, field or common vole	+	+	+
Microtus oeconomus, northern vole	+		+
Microtus sp., vole	+	+	+
Apodemus maastrichtiensis, mouse †			+
Apodemus sylvaticus, wood mouse	+	+	+
Apodemus sp., mouse	+	+	+
Carnivora			
Mustela cf. *putorius*, polecat			+
Mustela sp., mustelid			+
Panthera leo, lion	+		
Proboscidae			
Palaeoloxodon antiquus, straight-tusked elephant †	+	+	+
Elephantidae gen. et sp. indet., indeterminate elephant		+	
Perissodactyla			
Stephanorhinus hemitoechus, narrow-nosed rhinoceros # †		+	cf
Stephanorhinus kirchbergensis, Merck's rhinoceros* # †		+	
Artiodactyla			
Sus scrofa, wild boar		+	
Dama dama, fallow deer	cf	+	+
Cervus elaphus, red deer	+	+	+
D. dama or *C. elaphus*, fallow or red deer	+	+	+
Capreolus capreolus, roe deer		+	+
Cervidae gen. et sp. indet, indeterminate deer		+	+
Bos primigenius, aurochs # †	+	+	
Bos primigenius or *Bison priscus*, aurochs or bison	+	+	

Climatostratigraphy	Ipswichian						Hoxnian		Anglian	'Cromerian Complex'		
Marine Isotope Stage	5	6	7	8	9	10	11a	11c	12	13	14	?
British Interglacial Site	Joint Mitnor / Trafalgar Square		Aveley / West Thurrock / Selsey		Purfleet / Grays Thurrock / Cudmore Grove		Hoxne (B2-B1)	Swanscombe / Clacton / Barnham / Beeches Pit		Boxgrove		West Runton
Equus sussenbornensis												●
Equus altidens												●
Mimomys savini												●
Macroneomys brachygnathus												●
Megaloceros savini												●
Pliomys episcopalis										●		●
Praemegaloceros dawkinsi										●		●
Praemegaloceros verticornis										●		●
Ursus deningeri										●		●
Sorex (Drepanosorex) sp.										○		○
Talpa minor										○		○
Microtus (Terricola) subterraneus							○	○		○		○
Trogontherium cuvieri							○	○				●
Macaca sylvanus							●	○				○
Crocuta crocuta							○	●		●		●
Mammuthus sp.	●		●		●					○		●
Oryctolagus cuniculus			●		○			○		○		
Ursus spelaeus			○		○			●		●		○
Equus ferus			●		●		●	●		●		●
Arvicola sp.	○		○		○		○	○		○		○
Stephanorhinus kircherbergensis			○		○			○				
Palaeoloxodon antiquus	○		○		○			○				
Megaloceros giganteus	●		●		●		●	●				
Bos primigenius	○		○		○			○				
Stephanorhinus hemitoechus	○		○		○			○				
Ursus arctos	●		○		●							
Hippopotamus sp.	●		●									

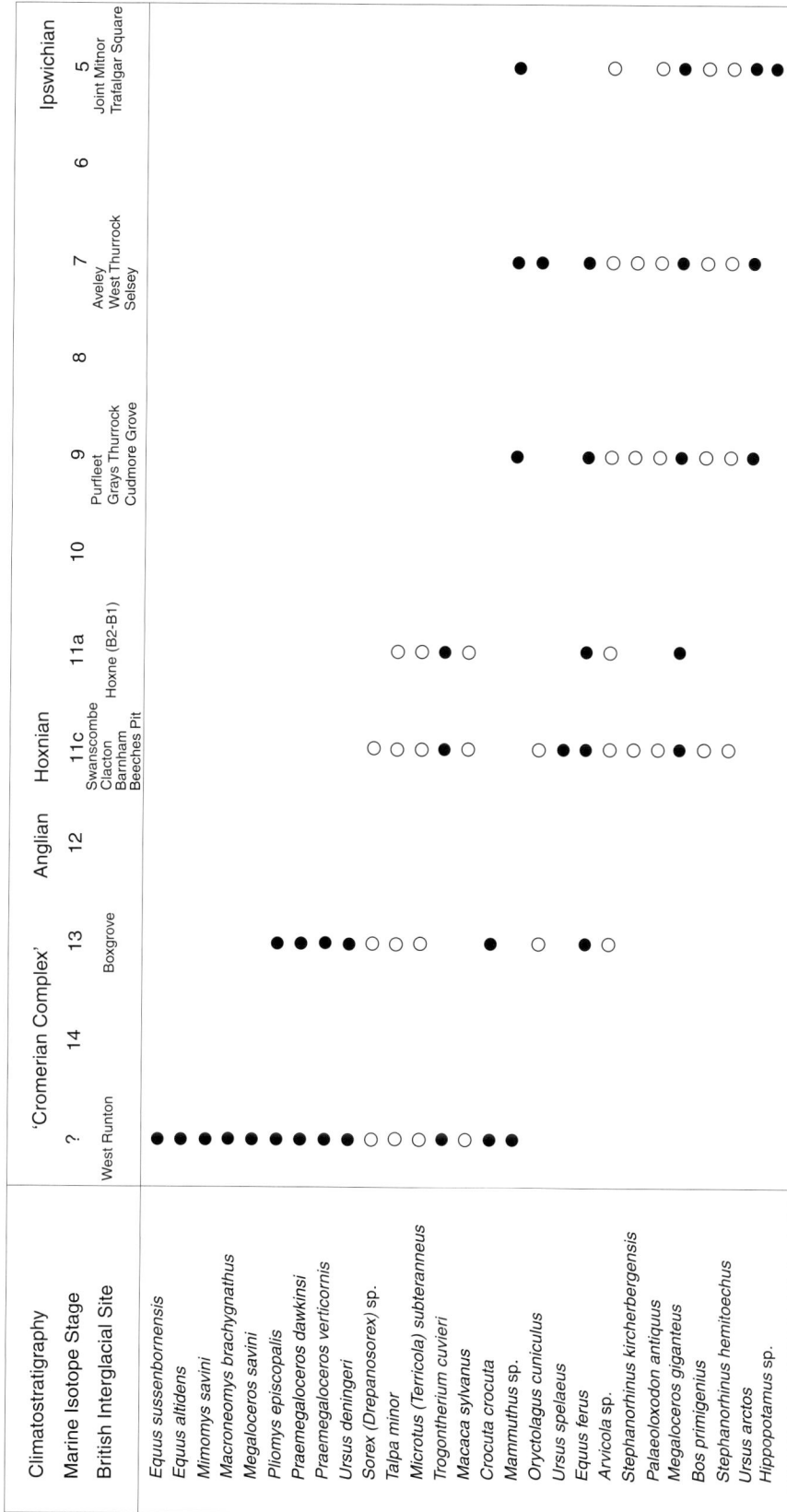

Figure 9.19 Summary of mammalian faunal change during the Middle and Late Pleistocene in Britain, showing the stratigraphic ranges of selected mammal taxa. O = taxon present at Southfleet Road; ● = taxon present at other sites. Many species have longer ranges in continental Europe and their absence from contemporaneous sites in Britain may be due to sampling failure or climatic, environmental and palaeogeographical factors.

Table 9.15 Comparison of Southfleet Road mammal fauna with other MIS 11 sites. With the exception of Clacton, where pollen is well preserved, attribution of the assemblages to the substages of the Hoxnian is based on lithostratigraphical and/or biostratigraphical inferences (Preece et al. 2007).

Site	Hoxne	South-fleet Rd.	Swans-combe	Clacton	Barn-ham	Beeches Pit	Ingress Vale	Swans-combe	Beeches Pit	Hoxne	Hoxne
Horizon	Stratum F	5-6	LG-LL	II	?II	Lower fauna II-III	III	UMG	Upper fauna	Stratum C	Strata B2-B1
Pollen substage	Late Anglian	?I-II	?I-II	Hoxnian	?II	Hoxnian	III	?III-IV	?	?	?
Stage	Anglian	Hoxnian	Hoxnian	Hoxnian	Hoxnian	Hoxnian	Hoxnian	Hoxnian	Hoxnian	Hoxnian	Hoxnian
Marine Isotope Stage	12	11c	11c	11c	11c	11c	11c	11c	11c	11b	11a
Lipotyphla											
Sorex minutus, pygmy shrew	–	+	–	–	+	+	–	–	+	–	–
Sorex sp. 1, shrew (smaller than *S. araneus*)	–	+	–	–	+	+	–	–	+	–	+
Sorex (*Drepanosorex*) sp., shrew (large)	–	+	+	–	+	+	–	–	+	–	+
Neomys sp., water shrew	–	+	–	–	+	+	–	–	+	–	+
Crocidura sp., white-toothed shrew	–	–	+	–	+	–	–	–	–	–	+
Talpa minor, mole	–	+	+	–	+	+	–	–	+	–	+
Desmana sp., Russian desman	–	–	–	–	+	–	–	–	–	–	+
Chiroptera											
Myotis daubentonii, Daubenton's bat	–	+	–	–	–	–	–	–	–	–	–
Plecotus sp., long-eared bat	–	–	–	–	+	–	–	–	–	–	–
Pipistrellus pipistrellus, common pipistrelle	–	–	–	–	+	–	–	–	–	–	–
Primates											
Macaca sylvanus, Barbary macaque	–	+	+	–	–	+	–	–	–	–	–
Homo sp., hominin (C = Clactonian, A = Acheulian)	–	–	C	C	C/A	A	A	A	A	–	A
Lagomorpha											
Oryctolagus cuniculus, rabbit	–	+	+	–	+	+	–	–	–	–	–
Lepus timidus, mountain hare	–	–	–	–	–	–	–	+	–	–	–
Rodentia											
Sciurus sp., squirrel	–	+	–	–	+	+	–	–	+	–	–
Spermophilus sp., ground squirrel	–	+	–	–	–	–	–	–	–	–	+
Trogontherium cuvieri, beaver-like rodent	–	–	–	+	–	–	+	–	+	–	+
Castor fiber, beaver	–	–	+	+	–	–	–	–	–	+	–
Dicrostonyx sp., collared lemming	–	–	–	–	–	–	–	–	–	+	+
Lemmus lemmus/Myopus schisticolor, Norway/wood lemming	–	–	–	–	–	–	+	+	+	–	+
Clethrionomys glareolus, bank vole	–	+	+	+	+	+	+	–	+	–	+
Arvicola cantianus, water vole	–	+	+	+	+	+	+	–	+	–	+
Microtus (*Terricola*) cf. *subterraneus*, common pine vole	–	+	+	–	+	+	–	cf.	+	–	+
Microtus agrestis, field vole	–	+	+	–	+	–	–	–	+	–	+
Microtus arvalis, common vole	–	cf.	cf.	–	+	–	–	cf.	+	–	cf.
Microtus agrestis/M. arvalis, field or common vole	–	+	+	+	+	+	–	+	+	–	+

Table 9.15 (continued)

Microtus oeconomus, northern vole	–	+	–	–	+	+	–	–	+	–	–	–	+	–
Apodemus maastrichtiensis, mouse	–	+	–	+	+	+	+	–	–	–	+	–	–	–
Apodemus sylvaticus, wood mouse	–	+	+	+	+	–	+	+	–	cf.	+	+	+	–
Eliomys quercinus, garden dormouse	–	–	–	–	–	–	+	–	–	–	–	–	–	–
Cetacea														
Tursiops cf. *truncatus*, bottlenose dolphin	–	–	–	–	–	+	–	–	–	+	–	–	–	–
Carnivora														
Canis lupus, wolf	–	+	+	+	–	–	+	+	+	+	–	–	–	sp.
Ursus spelaeus, cave bear	–	+	+	–	–	–	–	–	–	–	–	–	–	–
Ursus sp., bear	–	–	–	+	+	+	–	–	–	–	–	+	–	+
Mustela nivalis, weasel	–	–	+	+	–	+	–	–	+	–	–	–	–	–
Mustela cf. *putorius*, polecat	+	+	–	+	–	–	–	–	–	–	–	–	–	–
Mustela sp., mustelid	+	+	–	–	–	+	–	–	–	+	–	–	–	–
Martes martes, pine marten	–	+	–	–	–	–	–	–	–	–	–	–	–	–
Lutra lutra, otter	–	+	–	–	–	–	–	–	–	–	–	+	–	+
Felis sylvestris, wild cat	–	–	–	–	–	–	–	–	–	–	–	+	–	–
cf. *Lynx lynx*, lynx	–	–	–	–	–	–	–	–	–	–	–	+	–	+
Panthera leo, lion	–	+	+	+	–	+	+	+	+	+	–	+	–	+
Proboscidae														
Palaeoloxodon antiquus, straight-tusked elephant	–	+	+	+	+	+	+	+	–	+	–	+	–	–
Perissodactyla														
Equus ferus, horse	–	+	+	+	–	+	+	+	+	+	–	+	–	+
Equus hydruntinus, equid (at Swanscombe, horizon unknown)	–	?	–	–	–	–	–	–	–	?	–	–	–	–
Stephanorhinus kirchbergensis, Merck's rhinoceros	+	+	+	+	–	+	+	+	+	+	–	–	–	–
Stephanorhinus hemitoechus, narrow-nosed rhinoceros	+	+	+	+	–	+	+	+	+	+	–	+	–	+
Stephanorhinus sp., rhinoceros	–	+	+	+	+	+	+	+	+	+	–	–	–	+
Artiodactyla														
Sus scrofa, wild boar	–	+	+	+	+	+	+	+	–	+	–	–	–	–
Megaloceros giganteus, giant deer	–	+	+	+	–	+	–	–	–	+	–	+	–	+
Dama dama clactoniana or *D. dama* ssp. indet., fallow deer	–	+	+	+	+	+	+	+	–	+	+	+	–	+
Cervus elaphus, red deer	+	+	+	+	+	+	+	+	+	+	+	+	–	+
Capreolus capreolus, roe deer	–	+	+	+	–	+	+	+	–	+	–	+	–	+
Bos primigenius, aurochs	–	+	+	+	–	+	+	+	+	+	–	+	–	+
Bison priscus, bison	–	+	+	–	–	–	–	–	–	+	–	–	–	–
Bos primigenius or *Bison priscus*, aurochs or bison	–	+	+	+	–	+	+	+	+	+	–	+	–	–

Where the genus is certain, but not the species, the genus is followed by 'sp'; the term 'cf' is used to indicate that the specimen resembles the named species or genus very closely, but is too incomplete to show all diagnostic features.   (LG = Lower Gravel; LL = Lower Loam; UMG = Upper Middle Gravel), Barnham Beeches Pit (Lower fauna = beds 3–5; Upper fauna = beds 6–7) and Ingress Vale (Dierden's Pit).

In summary, there is a striking resemblance between the Southfleet Road assemblage and those from Clacton and Swanscombe (Basal Gravel-Lower Loam), as well as other more-or-less contemporary interglacial mammalian faunas from Beeches Pit and Barnham, which have been assigned to the first post-Anglian interglacial (Hoxnian, MIS 11). Although the faunal similarities would appear to imply that the deposits are of the same age, recent work on the Hoxne sequence has shown that the main faunal horizon was deposited during a second (post-Hoxnian) temperate episode. These deposits are separated from the Hoxnian lake sediments by a significant phase of climatic deterioration as represented by the 'Arctic Bed'. The second interglacial stage appears to fall within the latter part of MIS 11, based on the close similarity in its mammalian fauna to that from Swanscombe (Ashton *et al.* 2008). Work is still ongoing to establish whether there are any features of the mammalian fauna which allow differentiation of the two temperate episodes. One important difference between these two faunal groups is in the size of the red deer (Lister 1993). Those from Swanscombe and Clacton are significantly smaller than the red deer from Hoxne, which combine larger body-size with relatively small dental size. The correlation of Southfleet Road with the first post-Anglian interglacial is supported by the red deer, which show a close affinity to the samples from Swanscombe and Clacton.

MAMMALIAN FAUNAS AND A REVISED BIOSTRATIGRAPHICAL SCHEME FOR MIS 11 IN BRITAIN
by Simon A Parfitt and Antony J. Sutcliffe[1]

The emerging picture of climatic complexity within MIS 11 observed in ice and deep-sea records provide the template for this stage (Tzedakis *et al.* 2001; EPICA Community Members 2004). In terms of duration, the ice and deep-sea records show that MIS 11 spans about 65 000 years (425-360 ka). Climatically, the first half of MIS 11 includes a particularly warm interglacial of long-duration, which is terminated by a pronounced and rapid cooling at about 390 000 ka. The latter part of MIS 11 consists of probably several stadials and 'interstadials', one of which may represent the second warm interval observed in the Hoxne sequence.

Anglian Glacial Stage (MIS 12)

The British Anglian ice sheet was the most extensive glaciation of the Pleistocene and at its maximum extent it covered most of the British Isles, with the ice sheet extending as far south as north London and the River Severn. The glaciated areas were subject to considerable erosion, and as a consequence fossiliferous deposits of Anglian age are extremely rare in Britain. There are few records of mammalian remains that may be of Anglian age. The Mundesley (Norfolk) ground squirrel (identified as *Spermophilus undulatus* by Mayhew 1975) is usually assigned to the beginning of this time period. Similarly, the occurrence of remains of the ground squirrel and arctic lemming (*Dicrostonyx* sp.) together with other boreal/arctic biota has been recorded in the 'Arctic Fresh-water Bed' (Parfitt *et al.* 2010b) beneath glacial deposits (Happsiburgh Till) at Ostend (Norfolk). At Boxgrove (West Sussex), cold stage mammals (including grey-sided vole *Clethrionomys rufocanus* and Norway lemming *Lemmus lemmus*) assigned to the early part of the Anglian cold stage occur in chalky slope deposits and 'brickearth' immediately above the interglacial palaeosol horizon. The cavern-infill deposits at Westbury-sub-Mendip (Somerset) also include beds with cold stage mammal faunas (Stringer *et al.* 1996), but this sequence is no longer thought to be equivalent in age to Boxgrove and probably dates to earlier stages within the 'Cromerian Complex' (Preece and Parfitt 2000; Breda *et al.* 2010; Lister *et al.* 2010). At Hoxne, three red deer foot bones found resting on Anglian till represent the only known instance of mammals from the Anglian late-glacial stage (Stuart *et al.* 1993).

Hoxnian Interglacial Stage (MIS 11c)

Conditions immediately following the retreat of the ice sheet are recorded in late Anglian lacustrine sediments that formed on the newly deglaciated landscape in East Anglia. Of the localities with deposits attributed to the Hoxnian stage only a few have yielded any abundance of mammalian remains, the most important being Swanscombe, Kent, Clacton, Essex, Beeches Pit and Barnham East Farm, both in Suffolk (Tables 9.15 and 9.16). At Hoxne, Norfolk (stratotype of the Hoxnian), the mammalian remains are mainly confined to deposits overlying the Hoxnian lacustrine sediments and separated from them by a major cold stage. All of these localities are also important archaeological sites. Unfortunately, at none of these sites (even at the Hoxnian type locality) is there a complete pollen record, although a full sequence, ranging from Anglian till, through pollen zones I-IV and into the beginning of the following cold episode is known from Marks Tey, Essex (Turner 1970). The sequence of faunal changes during this period has therefore been reconstructed from geological sequences that are correlated primarily on the basis of biostratigraphy.

The site with the longest sequence of deposits is Barnfield Pit, Swanscombe, part of the Boyn Hill/Orsett Heath Terrace of the River Thames. This section has been discussed in detail by Conway *et al.* (1996) who group the deposits into three phases. Mammals are common at Swanscombe in all deposits up to and including the Upper Middle Gravels (Phase II), source of the famous human

---

[1] This section represents an updated version of an unpublished collaborative review paper with the late Antony Sutcliffe (1927-2004).

skull remains (Stringer and Hublin 1999). Two noteworthy features suggest a break in deposition at the base of the Middle Gravels: an abrupt typological change in the flint artefacts from Clactonian in the Lower Gravel and Lower Loams (Phase I) to Acheulian in the Middle Gravel (Phase II), and a horizon covered with animal footprints on the weathered surface of the Lower Loam. A striking feature of the Swanscombe sequence is the change in the character of the freshwater fauna that occurs between the Lower Loam and Middle Gravels. Assemblages from the upper part of the aggradation record an influx of so-called 'Rhenish' molluscan species, which Kennard (1942) suggested indicated a link between the Thames and Rhine. A similar molluscan fauna occurs at Clacton and Tillingham, where it is associated with brackish conditions indicating high sea level. A further important mammal locality in the Swanscombe terrace complex is Dierden's Pit, Ingress Vale, situated less than half a kilometre from Barnfield Pit. Most of the mammal remains probably derive from a shell bed, which has yielded the same 'Rhenish Suite' of molluscs found in the Lower Middle Gravel at Barnfield Pit. Kerney (1971) placed the shell bed above the Lower Loam of Barnfield Pit but slightly earlier than the main body of the Middle Gravel, implying that the hiatus in the Barnfield Pit is a

local phenomenon resulting from a shift in the position of the river.

Clacton is situated on the coast far out in the estuary of the Thames where it grades into the North Sea. The site stratigraphy, described by Warren (1923), was elaborated by Pike and Godwin (1953), Warren (1955), Kerney (1971), Turner and Kerney (1971) and more recently by Bridgland et al. (1999). A lower series of freshwater deposits with abundant mammalian remains and Clactonian artefacts, assigned to pollen zones IIb-IIIa, is overlain by estuarine clays and silts of IIIb age with an eroded surface at about 10m OD overlain by hillwash and soil. The Clactonian deposits were the focus of major archaeological excavations by Mary Leakey and John Wymer (Singer et al. 1973).

The second group of sites with mammalian remains is located in the Breckland region of East Anglia and includes Barnham, Beeches Pit and Elveden. The glacial deposits (including the chalky Lowestoft Till) in this region provide a useful lithostratigraphic marker that extends to the Thames catchment (as at Hornchurch where fluvial deposits of the first post-diversion terrace directly overly till). The interglacial faunas from the Breckland sites are found in lacustrine and fluvial deposits that directly overly the glacial deposits.

Table 9.16 Stratigraphic subdivision of the early part of the late Middle Pleistocene. The climatic stages in the deep-sea record are numbered back from the Holocene (Marine Isotope Stage (MIS) 1) such that odd-numbered marine isotope stages represent warm intervals and even-numbered stages represent cold episodes. Substages of MIS 11 follow Tzedakis et al. 2001

MIS	Pollen substage	Climatostratigraphy and suggested correlation with British sites	
11a		Un-named temperate stage	Hoxne (Strata B2-B1)
11b		Un-named cold stage	Hoxne (Stratum C)
11c	Ho IV	Hoxnian Interglacial Stage	Swanscombe (UMG)
	Ho III		Dierden's Pit (Ingress vale)
	Ho II		Clacton / Barnham / Swanscombe (LG, LL) / Southfleet Road
	Ho I		Hoxne (Stratum F)
12		Anglian Glacial Stage	Ostend ('Arctic Fresh-water Bed')

At Barnham, sedimentation commenced during the closing stages of the Anglian Glacial Stage with the deposition of glaciofluvial sand and gravel and Lowestoft till in a steep-sided channel incised into the Chalk bedrock (Ashton *et al.* 1998; Preece and Penkman 2005). Rich assemblage of small vertebrates, rare large mammal remains and an associated molluscan assemblage occurs in the upper fluvio-lacustrine part of the channel fill. The principal fossiliferous deposits date to the early temperate stage of the interglacial and have yielded southern thermophiles (European pond terrapin *Emys orbicularis* and Aesculapian snake *Zamenis longissimus*) indicating that summers were warmer than the present day.

The interglacial deposits at Beeches Pit rest directly upon glacial deposits (outwash gravels and till, referable to the Anglian Lowestoft Formation), with no apparent break in the sequence (Preece *et al.* 2007). The interglacial sediments consist of limnic, tufaceous and colluvial silts containing rich vertebrate, molluscan (including tufa with a highly distinctive 'Lyrodiscus fauna') and ostracod faunas. Faunal evidence from the upper levels provides clear indications for climatic deterioration. Both the molluscan and vertebrate faunas suggest correlation with the Hoxnian and the faunal changes seen at Beeches Pit have clear parallels with the succession at Swanscombe (Preece *et al.* 2007). A correlation with MIS 11 is further supported by uranium series dates from the tufa (*c* 455 ka BP), thermoluminescence dates from burnt flints (414 ± 30 ka BP) and amino acid racemization data from *Bithynia* opercula (Penkman *et al.* 2011).

The climatic significance of these various faunas has been discussed by a series of writers. Oakley (1952), from information by Bate and Hinton, noted a similarity of the faunas of the Lower Gravel and Middle Gravel of Barnfield Pit. These were indicative of woodlands interspersed by grasslands, although an increase in the relative abundances of horse and bison in the Middle Gravels suggested an increase in the proportion of grassland. He assigned the Upper Middle Gravels with *Lemmus* and an overlying Upper Loam to the closing stages of the 'Great' (Hoxnian) Interglacial. Sutcliffe (1964), who re-examined the collections, also noted a similarity of the mammalian faunas of the Lower Gravel and Middle Gravels, and he independently found support for Oakley's claim for an increase in the abundance of remains of non-sylvan species in the Middle Gravel. He observed a general similarity between the mammalian faunas of Swanscombe and Steinheim an der Murr, Germany, a site commonly referred to as the Holsteinian of the chronology of the European continent. Cook *et al.* (1982), from the evidence of the previously unstudied finds from the Lower Loam excavated by Waechter during 1969-72, linked this deposit (in which they observed and abundance of woodland species) with the Lower Gravel. They drew attention to the apparently considerable magnitude of depositional break between the deeply weathered footprint surface to the Lower Loam and the bottom of the Lower Middle Gravel and they interpreted the rodents of the Upper Middle Gravel as indicative of the development of cooler more continental conditions. Stuart (1982) found the mammalian faunas of the Lower Loam and Lower Gravel generally consistent with regional mixed forest, although with some taxa suggestive of, probably local, more open habitats. He drew attention to the occurrence of the pond terrapin (*Emys orbicularis*) at Ingress Vale, showing that at some stage during the Hoxnian the climate was warmer than today (Stuart 1979).

The faunas of all the sites described above have been widely quoted in the literature as being Hoxnian in age (Schreve 2001a), representing various parts of an interglacial cycle. Thus far, on the evidence of the few fossiliferous localities which have been studied, the story

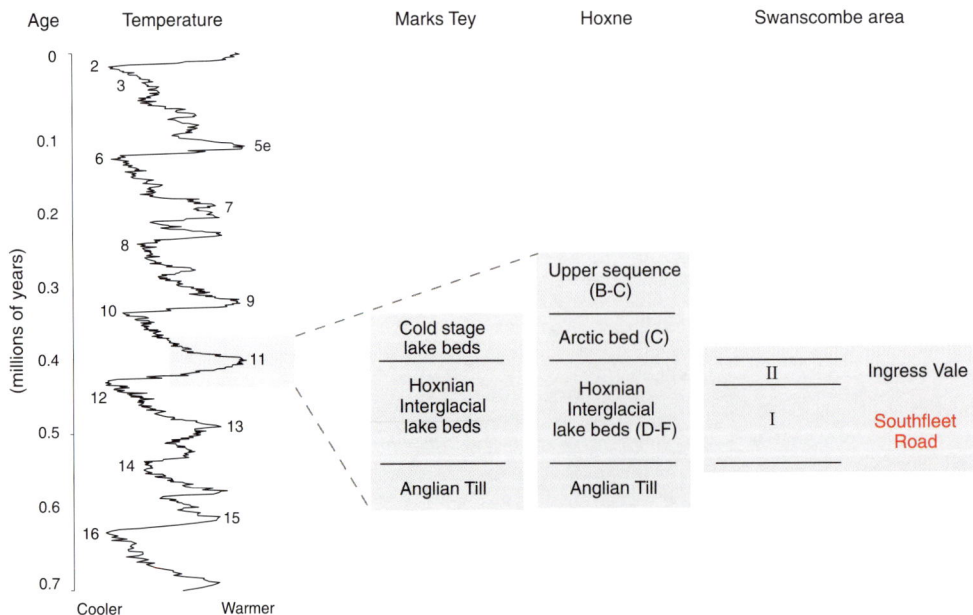

Figure 9.20 Suggested correlation between key Hoxnian sequences and the deep sea oxygen isotope record

of the mammals up to about pollen substages III and IV of the Hoxnian appears to be relatively straightforward. With the palynological evidence at Hoxne, however, truncated after pollen zone IIIa, the further continuation of the sequence has been far from clear. Schreve (2001a, b) has suggested that the main faunal horizons at Hoxne can be correlated with the Swanscombe Upper Middle Gravel. Other authors have suggested a more complex climatostratigraphy involving subdivision into different marine isotope stages (Bowen et al., 1989) or correlation with substages within MIS 11 (Schreve 2001a; b). None of these models is entirely satisfactory. New interpretations of the geological and environmental succession at Hoxne, combined with a reassessment of the biostratigraphical evidence, have helped to clarify this issue. This revision suggests that all of the above mentioned deposits can be accommodated within the first prolonged temperate substage (= Hoxnian Integlacial) equivalent to MIS 11c in the marine record (Table 9.16). The climatic deterioration of the Upper Middle Gravels being unlikely to have occurred earlier than pollen zone IV, more probably representing the early substage of the succeeding cold stage.

Combining these findings in the most simple possible way, the Clacton – Swanscombe (up to the Middle Gravels), Southfleet Road – Beeches Pit (beds 3 to 6) – Barnham sequence is interpreted as a single interglacial episode with woodland fauna, followed by an increase in the amount of grassland and finally the arrival of lemming (*Myopus* or *Lemmus*) at the end of the interglacial or beginning of the following cold stage. On mammalian evidence alone, there is no indication of the major temporal hiatus widely believed to be present between the Lower Loam and Middle Gravels of Swanscombe (Cook et al. 1982; Schreve 2001b).

Mid-MIS 11 cold episode (?MIS 11b)

A return to fully arctic conditions is recorded at Hoxne in the lake sediments that unconformably overly the detritus mud of Stratum D (Ho IIIa). The palaeobotanical remains from this horizon were studied by Reid, who identified dwarf birch (*Betula nana*) and three species of dwarf willow (*Salix* spp.). Recent bulk sieving of the 'Arctic Bed' (Stratum C, Ashton et al. 2008) has recovered rare teeth of Arctic lemming *Dicrostonyx* sp., together with a beetle fauna indicating severely cold winter temperatures.

Late MIS 11 mammalian faunas from Hoxne (?MIS 11a)

The chronological relationship of the upper layers at Hoxne to the orthodox chronology has been widely discussed. Stuart et al. (1993) suggested that the archaeological industries and their associated faunas were of about the same age as the Swanscombe Upper Middle Gravel, possibly within the range of pollen zones IIIb-IV. This interpretation was modified by Schreve (2001a, b), who also correlated Hoxne with the Upper Middle

Gravel, illogically assigning them to the same (Hoxnian) interglacial as the lower part of the Swanscombe sequence, but to a later (post-Hoxnian) temperate substage within MIS 11. The uncertainties surrounding the correlation and age of the Hoxne sequence have been clarified to some extent by recent fieldwork (Ashton et al. 2008). This has established that the Hoxnian lake muds and the temperate deposits containing the archaeological industries are demonstrably separated in time by a major cold stage as represented by the 'Arctic Bed'. The stratigraphy of Reid (Evans et al. 1896), West (1956) and Singer et al. (1993) can now be accommodated into a more comprehensive framework (Fig. 9.12; Table 9.16). Although correlation of the Hoxne deposits with the oxygen isotope record must remain a matter of conjecture for the time being, the separation of the Hoxne mammalian faunas from those of the Hoxnian interglacial (such as Swanscombe (phases I and II), Clacton, Barnham) by an intervening cold episode of arctic severity suggests that the mammalian fauna from Strata B2 and B1 at Hoxne belong to the latter part of MIS 11, possibly MIS 11a.

The sequence at Hoxne is contained within a basin in the Anglian till and comprises lacustrine muds overlain by a series of fluviatile and solifluction deposits (West 1956; Gladfelter and Singer 1975; Wymer 1974; 1983). The basal lake deposits (F-D) have been assigned to pollen zones I-IIIa of the Hoxnian, a hiatus in deposition truncating the palynological records of the interglacial sequence, leaving zone IV unproved. Concentrations of mammals and Acheulian artefacts occur at two levels in the succeeding part of the sequence; a lower assemblage in the basal part of Stratum C, and an upper assemblage and industry in bed 5. These archaeological horizons (Strata B2 and B1) have yielded abundant mammalian remains, both large and small; only a few mammalian remains have been found in the interglacial lake muds, so little is known about the mammalian fauna of the Hoxnian *sensu stricto* at the type locality. The vertebrate faunas from Strata B2 and B1 are very similar in composition (Stuart et al. 1993. The large-mammal fauna is dominated by herbivores: horse *Equus ferus*, red deer *Cervus elaphus*, fallow deer *Dama dama*), together with occasional remains of macaque *Macacca sylvanus*, bear *Ursus* sp., otter *Lutra lutra*, lion *Panthera leo*, rhinoceros *Stephanorhinus* sp., a small bison *Bison* sp. (Parfitt, unpublished observation) and roe deer *Capreolus capreolus*. The red deer can be distinguished from those at Swanscombe and Clacton by their combination of relatively large body size combined with small dental size (Lister, in Stuart et al. 1993). Two species of beaver (European beaver *Castor fiber* and the extinct beaver-like *Trogontherium cuvieri*) were also present. Sieving has recovered a rich small-mammal fauna that includes the mole *Talpa minor*, a soricid *Sorex* sp., pine vole *Microtus* (*Terricola*) cf. *subterraneus* and lemming (identified as *Lemmus lemmus* by Stuart et al. 1993). Stuart et al. (ibid.) conclude that the vertebrate fauna indicate interglacial conditions that supported forest dominated by deciduous trees.

Conclusion

The discovery of mammalian remains associated with the Clactonian archaeology has important archaeological implications. A detailed environmental and climatic picture can be reconstructed from the vertebrate evidence, which fits best into an interglacial landscape with at least regional deciduous woodland. The mammalian and pollen data suggest that the earliest Clactonian artefacts occur early in the interglacial cycle and extended into the early temperate substage. The upper age limit for this interglacial stage is constrained by the presence of the pine vole *Microtus* (*Terricola*) cf. *subterraneus*, which suggests that the sediments are earlier than those on the north bank of the Thames around Purfleet and Grays Thurrock (Lynch Hill/ Corbets Tey Gravel) correlated with MIS 9 (Penkman *et al.* 2011). The presence of other biostratigraphically significant mammals (*Stephanorhimus hemitoechus*, *S. kirchbergensis* and *Bos primigenius*), together with morphometric data, indicate a close correspondence with the mammalian faunas from Swanscombe and Clacton.

The archaeological significance of this correlation is that it establishes a link between the Clactonian archaeology in the tributary valley at Southfleet Road and the main Thames sequence at Barnfield Pit, Swanscombe, where the Clactonian artefacts occur in the lower part (phase I, Lower Gravel and Lower Loam) of the Boyn Hill/Orsett Heath Gravel. In the present state of knowledge, the Clactonian occupation was probably more-or-less contemporaneous at both sites and most likely occurred during the Hoxnian interglacial (MIS 11c).

Chapter 10

Molluscan analyses

by Tom S. White, Richard C. Preece and Francis Wenban-Smith

INTRODUCTION

The preliminary report on the Southfleet Road site, Swanscombe, recorded the discovery of a deep sequence of Pleistocene deposits with various faunal, floral and archaeological remains (Wenban-Smith *et al.* 2006). These included an elephant skeleton and a substantial Clactonian lithic assemblage from Phase 6 of the site's stratigraphical sequence, as presented in this volume (Chapter 4). Amongst the faunal remains, molluscan remains were identified in Phase 6b of the sequence, the 'tufaceous channel-fill' that occurred towards the base of the grey clay in the centre of the main east-facing section (Fig. 4.29). These deposits were also rich in small vertebrate (Chapter 7) and (in some horizons) ostracod remains (Chapter 11). Therefore, various series of samples were taken from these deposits for subsequent analysis. By far the richest samples for molluscan (as well as small vertebrate) remains were from sediments attributed to 40070, the main (and basal) tufaceous deposit of Phase 6b. The deposition of the main body of the clay is thought to have been in a lacustrine or backwater environment, with colluvial/sheetwash overbank input and periodic desiccation. The 'tufaceous channel-fill' probably represents a small, fast-flowing stream which cut into the clay when it was exposed as a short-lived land-surface during one these periods of desiccation. Although not a true *in situ* tufa, the high carbonate content suggests this stream was possibly spring-fed, and includes micritic tufaceous particles, reworked from true tufaceous sediments in the close vicinity. The calcareous nature of 40070 has clearly been important for the preservation of molluscan remains, since few Pleistocene shells have been recovered from other sediments.

In the preliminary site report, molluscs from a single sample from the middle of context 40070 were analysed to provide an impression of the climate and local environment (Wenban-Smith *et al.* 2006). For this more detailed report, columns of samples through the tufaceous channel-fill sequences have been analysed in order to investigate changes in the composition of the molluscan faunas through 40070 and into the overlying contexts 40143 and 40144, which also contain (albeit much sparser) molluscan remains.

Two columns of 20-30L bulk samples (Columns 1 and 2) were taken at different points along the longitudinal section 40075 of the tufaceous channel (Fig. 10.1a).

These were sub-sampled for ostracods and molluscs before the remainder was processed for small vertebrate recovery, leading to complementary molluscan and small vertebrate records from the same sample series (see Chapter 7). A vertical series of sub-samples from two monoliths parallel to Column 1, <40321> and <40326>, were also analysed. Between them, these samples covered the thickest parts of the tufaceous channel-fill, and went well into the underlying Phase 6a deposits (40039 and 40103). At Column 1, the top of 40070 was directly overlain by context 40144, lacking the intervening (and intermittently occurring) context 40143. Therefore a vertical series of sub-samples was also examined from monolith <40282>, from Trench XIII (Fig. 10.1b), which contained both 40143 and 40144 in sequence above the top of 40070. Finally, a third column of bulk samples with molluscan and ostracod sub-sampling (Column 3) was constructed from the east-west transverse section 40079 across the east side of the tufaceous channel (Fig. 10.1c). This was taken to assess palaeo-environmental differences between the channel-edge and the channel-centre. In many cases shells from contexts adjacent to 40070 have probably been derived from that context or, in the case of underlying deposits, the result of sampling across stratigraphical boundaries, where sediment contortions and intrusions made collection of absolutely stratigraphically pure samples impossible.

Many samples from other sediments were also assessed for molluscan remains (Appendix 1) but none were found, apart from in one deposit (Phase 3, 40062). Here, numerous *Bithynia* opercula but no other molluscan remains (apart from derived Tertiary fragments) were recovered in conjunction with a relatively rich ostracod fauna (see Chapter 11). Although this assemblage does not make a particularly informative contribution to the general palaeoenvironmental reconstruction, these Phase 3 *Bithynia* were submitted for amino acid dating (Chapter 13); the significance of the results depends on careful consideration of their taphonomic history, and the possibility of derivation.

Several samples intended primarily for recovery of small vertebrate remains were also scanned for fossil molluscs. Most of these contained large quantities of reworked Tertiary shell, with only a trace of Pleistocene material. Two rich Pleistocene faunas were picked from samples <40295> (40144) and <40284> (40143). The former duplicates the mollusc sub-sample of the same number in Column 1, although the additional material

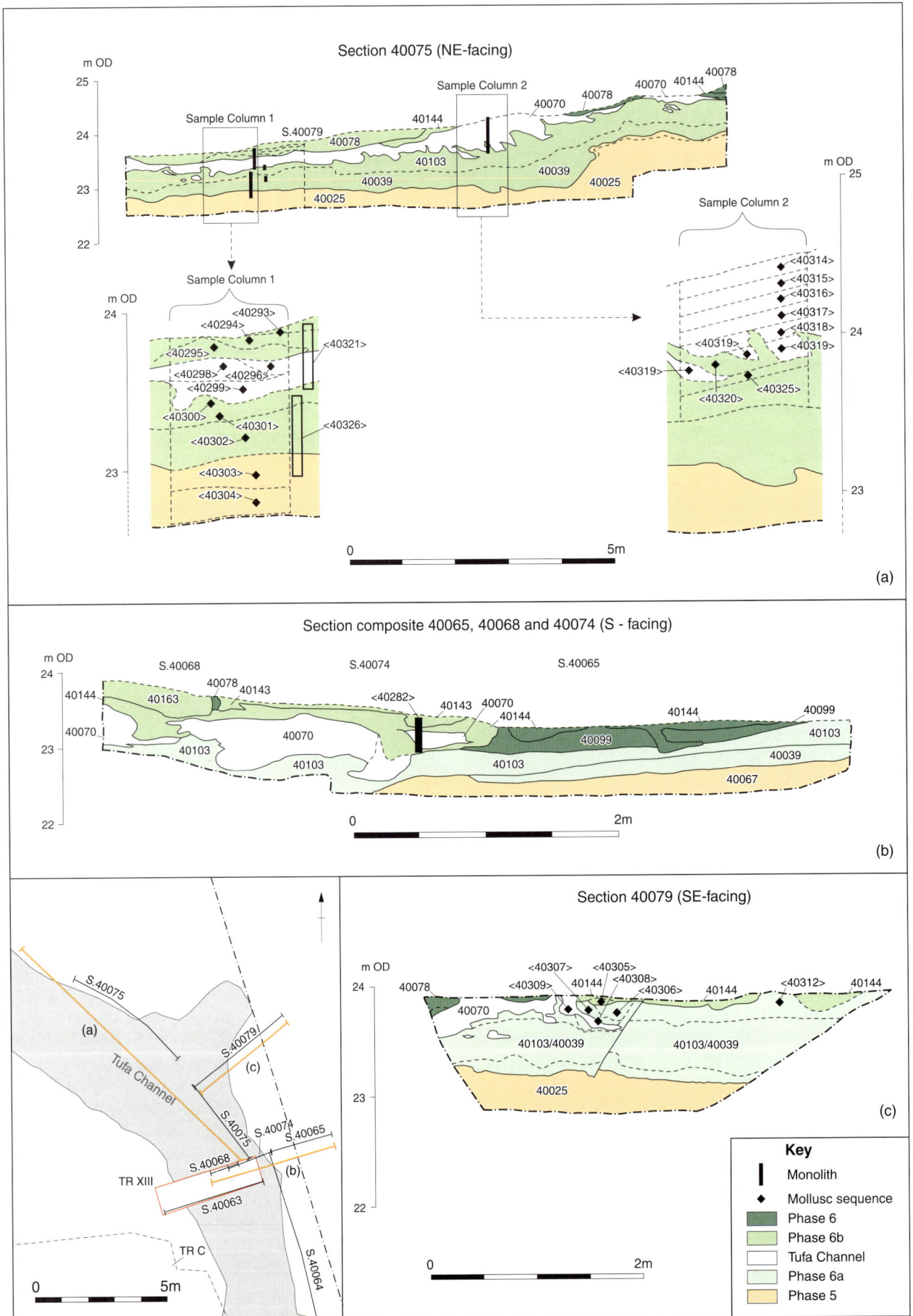

Figure 10.1 Locations of mollusc sequences through tufaceous channel fill, Phase 6b: (a) longitudinal Section 40075, showing Column 1, Column 2, monolith <40321> and monolith <40326>; (b) east-west composite section across tufaceous channel showing monolith <40282> from Trench XIII; (c) east-west Section 40079 across east side of tufaceous channel showing sample Column 3

provides a more statistically viable record of the proportions of different species at this level. The latter was of uncertain location in relation to the column samples, but the material recovered broadly duplicates the other data from context 40143, and likewise provides a more complete record.

METHODS AND TAPHONOMIC BIASES

Samples were weighed before being wet-sieved. For some samples it was necessary to use small quantities of dilute hydrogen peroxide to break up the clayey matrix before sieving. Residues above 500 μm were picked for mollusc shell fragments. Small vertebrate remains, including fish otoliths, were retained and passed on to the faunal specialist, Simon Parfitt (see Chapter 7). Residues below 500 μm were retained but were not requested for ostracod analysis as dedicated samples were available.

In many samples, the mollusc shells have been badly crushed and are represented only as fragments of body whorl. The absence of apical and apertural fragments means that it is impossible to provide a minimum number of individuals for some species. This problem particularly affects the land taxa *Discus ruderatus* and the Clausiliidae. When apical or apertural fragments are preserved they are often of juvenile individuals, suggesting that post-depositional processes have preferentially fragmented large shells. This preservational bias means that although many of the samples are rich in shell fragments, the number of quantifiable individuals is low.

Some shells were heavily encrusted with carbonate and could not be satisfactorily identified. All of the samples discussed below for which countable molluscan remains were present came from the tufaceous channel-fill sequence (Phase 6b), which without doubt represents a single short-lived stream development and channel-fill event in MIS 11. They can thus be considered both *en masse* as a representative assemblage from the part of MIS 11 represented (with due consideration of their landscape context) (see below). They can also be considered in terms of the changing local environment represented by variations in the proportions and ecological preferences of species at different levels through the sequences investigated (see below).

SPECIES LIST AND NOTES ON SIGNIFICANT SPECIES

The full list of species from the tufaceous channel-fill sequence, Phase 6b, is given at the end of this chapter (Table 10.7) compiled from all samples and all the three constituent contexts 40070, 40143 and 40144. In total 49 species were recognised, with 24 of them aquatic and 25 primarily terrestrial (including Succineidae). The ecological tolerances of some of the more interesting occurrences are given below:

Ena montana. Species of old woodland with a modern distribution mainly in Central Europe and Alpine areas, although pockets remain in SW England (Kerney and Cameron 1979). Requires warm summer temperatures.

Clausilia pumila. Species of damp woods now extinct in Britain, Central and Eastern European distribution (Kerney and Cameron 1979).

Arianta arbustorum. Widespread species, found in meadows, woods and hedgerows, but always in damp places – restricted in areas with dry climate and good drainage.

Azeca goodalli. Found in moss and ground litter in open woods, often in rocky places. Prefers light shade, avoiding both the extremes of open ground and dense woodland (Kerney 1999); the mossy edges of woodlands are often its preferred habitat (Paul 1974).

Vertigo angustior. Restricted to moist places affected neither by periodic dessication nor by flooding (Kerney 1999). Requires warm places vegetated by grasses, mosses or low herbs. This species is currently in decline, having been common at the beginning of the postglacial period after which suitable habitats were replaced by forest. This might also have been the case at Southfleet Road, as several woodland indicators are present in the assemblage; the lack of the more shade-demanding species evident at Hitchin suggests that woodland was not sufficiently well-developed for these taxa, allowing some more open species like *V. angustior* to remain.

SEQUENCE INTERPRETATIONS

Sample <40035> (context 40070)

This sample was scanned for a preliminary assessment of the molluscan remains by R. Preece; results are published in Wenban-Smith *et al.* (2006). Although doubling the number of species known from the site, subsequent analyses have not significantly altered the climatic and local environment interpretations offered by Preece in that publication. Residues from this initial assessment sample contained 26 taxa, 13 aquatic and 13 terrestrial, although quantitatively the aquatics are dominant (Table 10.1). The most common species were *Valvata piscinalis*, *Bithynia tentaculata* and bivalves of the genus *Pisidium*. This suite is typical of small hard-water streams, supported by the presence of the river limpet, *Ancylus fluviatilis*, which favours fast-flowing water and stony substrates. Many specimens were encrusted in carbonate. Marshland environments are indicated by the presence of *Vertigo antivertigo*, *Zonitoides nitidus* and indeterminate Succineidae. *Arianta arbustorum*, a large species, is common as fragments and more abundant than *Cepaea* sp. Fully temperate conditions are reflected by the presence of *Ena montana*, the modern range of which is thought to be limited by lack of summer warmth

Table 10.1 Molluscan remains from assessment sample <40035> (context 40070)

Southfleet Road Assessment Sample

Sample	<40035>	Notes
Context	40070	
Dry weight	Not rec.	

Valvata piscinalis Müller	14	
Bithynia tentaculata L. opercula	44	
Stagnicola palustris agg.	4	
Radix balthica Müller	1	
Anisus leucostoma (Millet)	2	
Gyraulus sp.	1	
Ancylus fluviatilis (Müller)	1	
Pisidium amnicum (Müller)	9	
Pisidium moitessierianum Paladilhe	1	
Pisidium nitidum Jenyns	11	
Pisidium subtruncatum Malm	12	
Pisidium tenuilineatum Stelfox	11	
Total aquatic taxa	67	
Succineidae	3	
Cochlicopa lubrica	1	
Vertigo antivertigo (Drapernaud)	1	
Vallonia costata (Müller)	1	
Ena montana (Drapernaud)	5	
Discus ruderatus (Férusac)	+	Fragments only
Zonitoides nitidus (Müller)	2	
Deroceras / Limax spp.	4	
Clausiliidae	17	
Trochulus hispidus (L.)	1	
Arianta arbustorum (L.)	+	
Cepaea sp.	+	
Total land taxa	35	
Total shells	102	
Tertiary shells	+	

(Kerney 1968). The Clausiliidae are well represented but highly fragmented; these including *Clausilia pumila*, a central European forest species no longer living in Britain (Kerney and Cameron 1979). *Discus ruderatus*, another species now extinct in Britain, was also present; its congener, *D. rotundatus*, which replaces *D. ruderatus* later in the interglacial, was not recorded.

Column 1

The lowermost samples in this sequence, from contexts 40025 and 40103, were devoid of Pleistocene shells. Sample <40300> incorporated material from contexts 40103 and 40070, suggesting that the shells were derived from the latter. Above 40070, some shells were obtained from the basal sample <40295> from 40144; however, the fact that the overlying samples from this context were devoid of Pleistocene shell remains suggests that the material from the base of context 40144 was probably derived from 40070.

The species represented are detailed in Table 10.2, and represented as a histogram in Figure 10.2a. The aquatic taxa are, like the assessment sample, dominated by *Valvata piscinalis* and members of the *Pisidium*. *Bithynia tentaculata* is also well represented, predominantly by opercula. Land taxa include a large number of fragments of *Discus ruderatus*, although none of these were apices. The terrestrial assemblage is very similar to that of sample <40035> with the notable addition of *Helicella itala*, a species of dry calcareous grassland.

The range of aquatic species indicates deposition of context 40070 in an aqueous environment, probably a small stream. The terrestrial species indicate a surrounding damp woodland environment, but relatively open rather than densely wooded. Figure 10.2a does not illustrate any significant changes in the fauna or environmental conditions through the Column 1 sequence; some of the apparent increases in species such as *Valvata piscinalis* are a function of the better recovery of shells from these levels.

Column 2

Unlike Column 1, this sample column does not go into contexts above 40070, which were truncated by machine excavation, but provides the deepest record of recovery (and possible change) within 40070. Again, Pleistocene shell was only recovered from samples within context 40070, the basal two samples from context 40103 being barren. The samples from this column were much richer in countable individuals and the shells were generally less fragmented. The species represented are detailed in Table 10.3 and represented as a histogram in Fig. 10.2b. This sequence is almost identical to Column 1. The general interpretation of both the aquatic and terrestrial taxa is the same as for Column 1, being a small fast-flowing stream with nearby woodland. Traces of dry grassland species, such as *H. itala*, are again present.

Again, the impression of increasing abundance in Fig. 10.2b is somewhat artificial due to the low recovery of shells from the basal samples. The fauna does not change significantly through this sequence, although the appearance of *Valvata cristata* in the uppermost sample might suggest shallowing waters and more weedy conditions (mirrored in Column 3, see below). The consistency of the mollusc fauna through the sequence suggests that the local environment was relatively stable during the deposition of 40070, lending further support to the hypothesis that the tufaceous channel was a relatively short-lived feature

Column 3, Section 40079

The basal samples from this sequence, from context 40103, yielded no Pleistocene shells. Four samples from 40070 and a single sample from context 40144 contained a molluscan assemblage (Table 10.4; Fig. 10.2c). The assemblages from context 40070 were of

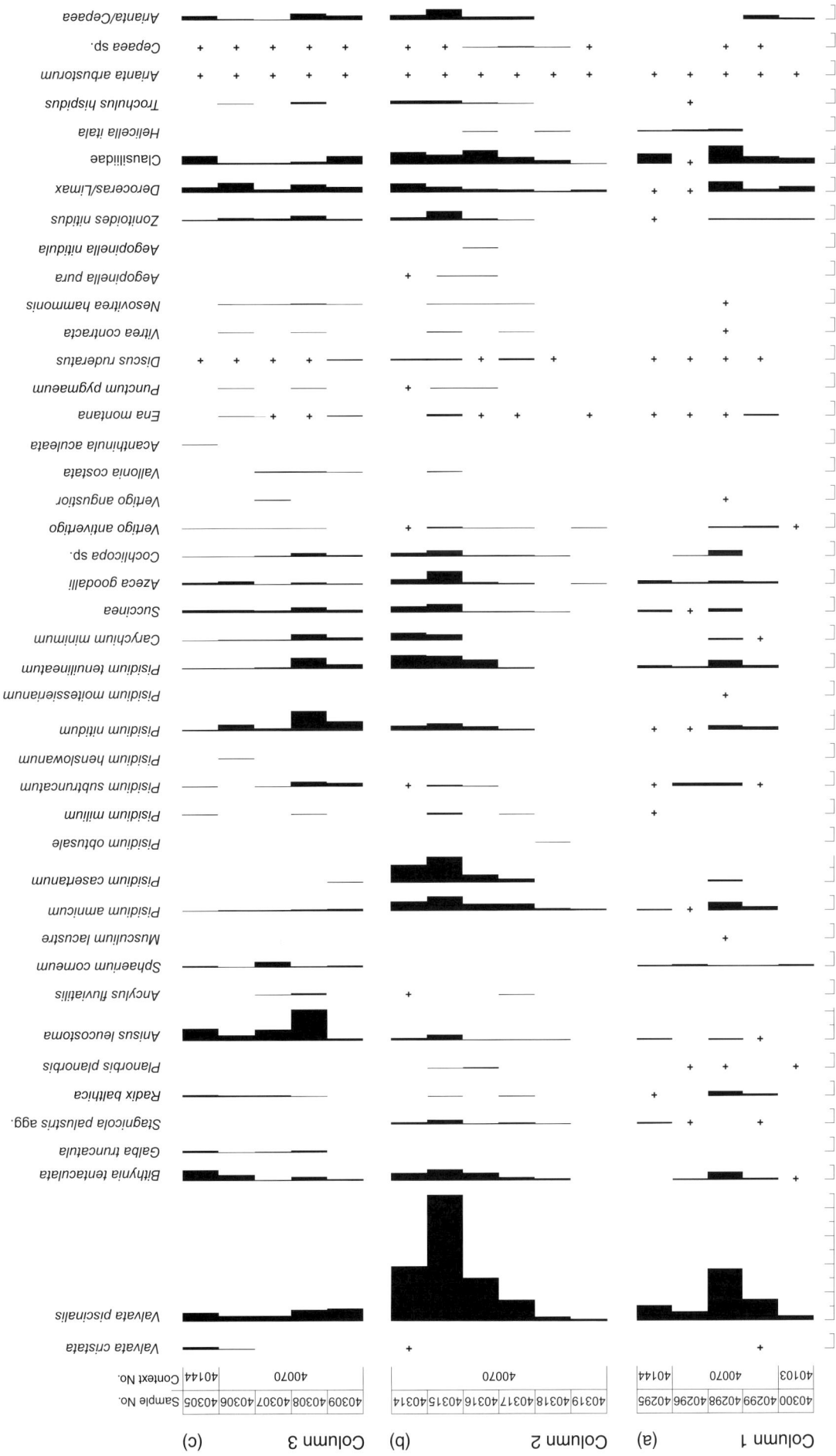

Figure 10.2 Mollusc histograms showing proportional changes (as %) of molluscs through tufaceous channel: (a) Column 1; (b) Column 2; (c) Column 3, Section 40079

Table 10.2 Molluscan remains from Column 1

Column 1					*Monolith*	*Bulk sample*
Sample	<40300>	<40299>	<40298>	<40296>	<40295>	<40295>
Context	*40070/*					
	40103	*40070*	*40070*	*40070*	*40144*	*40144*
Dry weight (g)	*1061.5*	*997.5*	*1301.1*	*797.8*	*1272.7*	*c 20 kg*
Aquatic taxa						
Valvata cristata Müller	-	1	-	-	-	-
Valvata piscinalis (Müller)	9	38	93	17	28	12
Bithynia tentaculata (L.)	1	3	14	2	-	1
Bithynia opercula	12	25	49	12	3	12
Stagnicola palustris (Müller) agg.	-	1	-	1	3	4
Radix balthica (Müller)	-	3	9	-	1	4
Planorbis planorbis (L.)	1	-	1	1	-	-
Anisus leucostoma (Millet)	-	1	2	-	3	4
Sphaerium corneum (L.)	3	2	2	4	3	-
Musculium lacustre (Müller)	-	-	1	-	-	-
Pisidium amnicum (Müller)	-	6	15	1	2	+
Pisidium casertanum (Poli)	-	-	4	-	-	-
Pisidium milium Held	-	-	-	-	1	-
Pisidium nitidum Jenyns	-	6	9	1	1	1
Pisidium subtruncatum Malm	-	1	6	6	1	-
Pisidium moitessierianum Paladilhe	-	-	1	-	-	-
Pisidium tenuilineatum Stelfox	-	5	14	3	6	3
Total aquatic (minus opercula)	14	67	171	36	49	41
Terrestrial taxa						
Carychium minimum Müller	-	2	3	-	-	-
Succineidae	-	-	6	2	4	-
Azeca goodalli (Férusac)	-	4	5	3	6	7
Cochlicopa sp.	-	-	10	1	-	-
Vertigo antivertigo (Draparnaud)	1	3	2	-	-	-
Vertigo angustior Jeffreys	-	-	1	-	-	-
Ena montana (Draparnaud)	-	3	1	1	+	-
Discus ruderatus (Férusac)	-	+	+	+	+	+
Vitrea contracta Westerlund	-	-	1	-	-	-
Nesovitrea hammonis (Ström)	-	-	1	-	-	-
Zonitoides nitidus (Müller)	2	2	2	-	1	-
Deroceras / Limax spp.	10	5	18	2	1	1
Clausiliidae	10	13	32	1	18	17
Helicella itala (L.)	-	-	4	3	2	-
Trochulus hispidus (L.)	-	-	-	1	-	-
Arianta arbustorum (L.)	3	7	+	+	+	+
Cepaea / Arianta	-	1	+	-	-	-
Total terrestrial	26	40	86	14	32	25
Total Mollusca	40	107	257	50	65	66

similar character to those from the same context in Columns 1 and 2. The transition from 40070 to 40144 is marked by the appearance of *Valvata cristata* (Fig. 10.2c), which suggests increasingly shallow and weedy environments towards the top of the sequence (also noted, albeit less convincingly, at the top of Column 2). A single shell of *Acanthinula aculeata* was recovered from the uppermost sample. This species is characteristic of deciduous woodland environments. Fewer grassland taxa were recorded from this sequence, with only *V. angustior* being notable. Otherwise, this

sequence shows very little change in the terrestrial and aquatic faunas.

Monoliths <40321> and <40326 >

As with other sequences, the only samples which contained Pleistocene shells from the deposit sequence represented in these monoliths relate to context 40070, all from the upper monolith <40321>. Shells were picked from 5 samples (Table 10.5) but were not very numerous by comparison with the three main sequences detailed

Table 10.3 Molluscan remains from Column 2

Column Sample 2						
Sample	*<40319/C>*	*<40318/C>*	*<40317/C>*	*<40316/C>*	*<40315/C>*	*<40314/C>*
Context	*40070*	*40070*	*40070*	*40070*	*40070*	*40070*
Dry weight (g)	*928.1*	*1142.7*	*955.6*	*1051*	*964.1*	*880*
Aquatic taxa						
Valvata cristata Müller	-	-	-	-	-	2
Valvata piscinalis (Müller)	3	8	38	77	226	99
Bithynia tentaculata (L.)	-	2	5	13	20	14
Bithynia opercula	11	9	24	36	52	37
Stagnicola palustris (Müller) agg.	-	1	3	2	8	4
Radix balthica (Müller)	-	-	1	-	1	-
Planorbis planorbis (L.)	-	-	-	2	1	-
Anisus leucostoma (Millet)	-	1	1	1	10	4
Gyraulus laevis (Alder)	-	-	-	-	3	-
Ancylus fluviatilis (Müller)	-	-	1	-	-	1
Pisidium amnicum (Müller)	3	4	13	13	26	17
Pisidium casertanum (Poli)	-	-	7	14	46	32
Pisidium obtusale (Lamarck)	-	1	-	-	-	-
Pisidium milium Held	-	-	1	-	3	-
Pisidium nitidum Jenyns	-	-	3	8	13	9
Pisidium subtruncatum Malm	-	-	-	1	3	2
Pisidium tenuilineatum Stelfox	2	-	2	16	23	24
Pisidium spp.	-	1	-	-	-	-
Total aquatic (minus opercula)	8	18	75	147	383	208
Terrestrial taxa						
Carychium minimum Müller	-	-	-	-	11	14
Succineidae	-	1	2	2	15	11
Azeca goodalli (Férusac)	1	-	2	4	24	9
Cochlicopa sp.	-	1	2	2	11	7
Vertigo antivertigo (Draparnaud)	1	-	1	1	3	2
Vertigo angustior Jeffreys	-	-	-	1	1	-
Vallonia costata (Müller)	-	-	-	-	2	-
Ena montana (Draparnaud)	+	-	+	+	4	-
Punctum pygmaeum (Draparnaud)	-	-	-	1	1	1
Discus ruderatus (Férusac)	-	+	3	+	3	3
Aegopinella pura (Alder)	-	-	-	1	1	1
Aegopinella nitidula (Draparnaud)	-	-	-	1	-	-
Vitrea contracta Westerlund	-	-	1	-	2	-
Nesovitrea hammonis (Ström)	-	-	1	1	1	-
Zonitoides nitidus (Müller)	1	-	1	3	17	6
Deroceras / Limax spp.	4	2	5	6	10	16
Clausiliidae	1	7	13	25	17	22
Helicigona lapicida (L.)	-	-	-	1	1	1
Helicella itala (L.)	-	1	-	1	-	-
Trochulus hispidus (L.)	-	-	1	2	6	6
Arianta arbustorum (L.)	+	+	+	+	+	+
Cepaea sp.	+	1	2	1	+	+
Cepaea / Arianta	-	-	6	6	19	9
Total terrestrial	8	13	40	59	149	108
Total Mollusca	16	31	115	206	564	316

above. The general environmental interpretation is the same as for other samples. The lowermost sample from which Pleistocene shell fragments were recovered (360–390mm) was notably iron-stained and contained a significant quantity of worn Tertiary shell fragments. *Bithynia* opercula were most common, together with a single fragment of *Sphaerium corneum* and an apex of a clausiliid. An additional aquatic species in this monolith was *Gyraulus crista*, from context 40070, which has a preference for weedy environments and is intolerant of desiccation, further reinforcing that the stream was a permanent waterbody during deposition of 40070.

Table 10.4 Molluscan remains from Column 3 (Section 40079)

Column 3, Section 40079					
Sample	<40309>	<40308>	<40307>	<40306>	<40305>
Context	40070	40070	40070	40070	40144
Dry weight (g)	1018	1123.8	1125.6	1024.9	1088.7
Aquatic taxa					
Valvata cristata Müller	-	-	-	1	5
Valvata piscinalis (Müller)	23	21	10	10	16
Bithynia tentaculata (L.)	3	7	2	11	20
Bithynia opercula	14	27	15	18	58
Galba truncatula (Müller)	-	4	2	1	4
Radix balthica (Müller)	-	1	3	3	4
Anisus leucostoma (Millet)	4	56	20	10	22
Ancylus fluviatilis (Müller)	-	4	1	-	-
Sphaerium corneum (L.)	4	2	10	1	3
Pisidium amnicum (Müller)	4	3	2	2	1
Pisidium casertanum (Poli)	2	-	-	-	-
Pisidium milium Held	-	2	-	-	1
Pisidium subtruncatum Malm	7	9	1	-	-
Pisidium henslowanum (Sheppard)	-	-	1	-	-
Pisidium nitidum Jenyns	17	36	5	12	3
Pisidium tenuilineatum Stelfox	8	20	3	2	2
Total aquatic (minus opercula)	72	165	59	54	82
Terrestrial taxa					
Carychium minimum Müller	6	12	2	2	1
Succineidae	5	10	4	5	5
Azeca goodalli (Férusac)	2	4	1	6	4
Cochlicopa sp.	3	7	2	1	1
Vertigo antivertigo (Draparnaud)	1	2	2	-	-
Vertigo angustior Jeffreys	-	1	1	1	1
Vallonia costata (Müller)	-	-	2	-	-
Ena montana (Draparnaud)	2	+	+	1	-
Acanthinula aculeata (Müller)	-	-	-	-	1
Punctum pygmaeum (Draparnaud)	-	1	-	1	-
Discus ruderatus (Férusac)	2	+	+	+	+
Vitrea contracta Westerlund	-	1	-	1	-
Nesovitrea hammonis (Ström)	1	2	1	1	-
Zonitoides nitidus (Müller)	3	9	4	5	2
Deroceras / Limax spp.	10	15	6	18	10
Clausiliidae	15	5	3	3	16
Arianta arbustorum (L.)	8	12	+	2	8
Trochulus hispidus (L.)	-	4	-	1	-
Total terrestrial	58	85	28	48	49
Total Mollusca	130	250	87	102	131

Monolith <40282 >

The sequence from this sample was unique in containing both contexts 40143 and 40144 overlying context 40070. It also provided an (albeit sparse) ostracod sequence to complement the molluscan record. Four relatively small samples contained Pleistocene shells. Additional molluscan material from 40143 also came from bulk sample <40284>, which also provided a complementary small vertebrate fauna from the same horizon. The species represented are detailed in Table 10.6, but there are insufficient shells from most of the samples in this monolith to make a histogram worthwhile.

The assemblages from contexts 40070 and 40144 are very similar in composition to those discussed above, showing minimal change upwards through the sequence. Those from the uppermost context (40143) show little difference in the range of aquatic species, but a marked reduction in the quantity and diversity of terrestrial species, perhaps reflecting a quiet episode of relatively still water.

Additional species in the sequence from this monolith include, from 40070, the aquatic *Bathyomphalus contortus*, a fairly catholic planorbid, which has a preference for weedy environments and is intolerant of desiccation, further reinforcing that the stream was a

Table 10.5 Molluscan remains from monolith <40321>

Sample	40321/C	40321/C	40321/C	40321/C	40321/C
Context	40070	40070	40070	40070	40144?
Depth	39-36cm	36-33cm	33-31cm	31-29cm	27-18cm
Dry weight (g)	170	145	100	150	555
Aquatic taxa					
Valvata piscinalis (Müller)		2		13	6
Bithynia tentaculata (L.)		1	1	+	1
Bithynia tentaculata (opercula)	5	9	2	18	5
Galba truncatula (Müller)					
Gyraulus crista (L.)				1	
Sphaerium corneum (L.)	1			1	1
Pisidium amnicum (Müller)				1	
Pisidium nitidum Jenyns		2		3	1
Pisidium subtruncatum Malm				2	2
Pisidium moitessierianum Paladilhe				+	1
Pisidium tenuilineatum Stelfox				3	2
Pisidium spp.			1		1
Total aquatic (minus opercula)	1	5	2	24	15
Terrestrial taxa					
Carychium minimum Müller				1	
Azeca goodalli (Férusac)				4	3
Cochlicopa sp.				1	1
Vertigo antivertigo (Draparnaud)				1	
Ena montana (Draparnaud)		+		+	+
Discus ruderatus (Férusac)		+	+	+	+
Vitrea contracta Westerlund					1
Zonitoides nitidus (Müller)				1	
Deroceras / Limax		1	1		
Clausiliidae	1	+	1	+	+
Arianta arbustorum (L.)		+			+
Cepaea / Arianta					2
Total terrestrial	1	1	2	8	7
Total Mollusca	2	6	4	32	22
Vertebrates	+	+	+	+	+
Tertiary shell fragments	+	+			

permanent waterbody. A single fragment of *Helicigona lapicida*, a calciphile landsnail characteristic of rocky environments and deciduous woodland, also from 40070, is the only additional terrestrial taxon.

CLIMATE AND ENVIRONMENT THROUGH THE SEQUENCE

The molluscan faunas from Southfleet Road have been recovered from relatively short sequences through context 40070 of the tufaceous channel-fill, and provide few indications of significant environmental changes. This is partly due to preservation, the highly fragmented nature of the material preventing accurate quantification of many species. It is probably also a reflection of the relatively short time span represented by the tufaceous channel deposits. The quantity and diversity of aquatic species through the tufaceous channel-fill sequence (40070, 40144 and 40143) indicates an aqueous deposi-

tional environment, and there are faint indications in the sequences of Column 2 and 3 (duplicated in the small vertebrate record (Chapter 7) of shallower and more weedy conditions at the top of 40070 and associated with deposition of 40144. The assemblage contains many species indicative of fully-temperate conditions, and a damp local woodland environment seems to have prevailed throughout the period of time represented, although there are sparse indications of drier and more open grassland in the vicinity.

CORRELATIONS WITH OTHER SITES

Preece (in Wenban-Smith *et al.* 2006) compared the fauna with that from the Lower Loam at Barnfield Pit, Swanscombe, where *Ena montana* is also a common constituent (Kerney 1971). The Lower Loam fauna represents a floodplain environment of a much larger river; it was therefore considered possible on purely molluscan

Table 10.6 Molluscan remains from monolith <40282> and bulk sample <40284>

Sample	<40282/C>	<40282/C>	<40282/C>	<40282/C>	<40284>
Context	40070	40070	40144	40143	40143
Depth	22-15cm	15-8cm	8-2cm	2-0cm	-
Dry weight (g)	305.3	211.9	393.8	74.2	c 750
Aquatic taxa					
Valvata piscinalis (Müller)	36	64	10	-	6
Bithynia tentaculata (L.)	4	18	6	5	40
Bithynia tentaculata (opercula)	12	35	26	5	177
Galba truncatula (Müller)	2	4	3	-	-
Stagnicola palustris (Müller) agg.	2	-	1	-	3
Radix balthica (Müller)	4	1	3	3	32
Lymnaea stagnalis (L.)	-	-	-	-	1
Gyraulus crista (L.)	-	1	2	1	1
Planorbis planorbis (L.)	+	+	2	7	
Anisus leucostoma (Millet)	2	4	15	-	
Planorbarius corneus (L.)	-	-	1	1	
Bathyomphalus contortus (L.)	2	-	3	-	
Ancylus fluviatilis (Müller)	1	-	-	-	
Acroloxus lacustris (L.)	-	-	-	1	
Sphaerium corneum (L.)	6	7	5	1	3
Pisidium amnicum (Müller)	-	+	-	-	
Pisidium nitidum Jenyns	3	8	1	1	1
Pisidium subtruncatum Malm	-	+	2	2	
Pisidium henslowanum (Sheppard)	-	-	1	-	
Pisidium moitessierianum Paladilhe	1	+	-	-	
Pisidium tenuilineatum Stelfox	7	11	3	-	1
Pisidium spp.	-	-	-	-	
Total aquatic (minus opercula)	70	118	58	22	88
Terrestrial taxa					
Carychium minimum Müller	2	2	3	-	-
Carychium tridentatum Risso	-	5	-	-	-
Succineidae	2	9	2	-	-
Azeca goodalli (Férusac)	3	12	4	-	1
Cochlicopa sp.	4	3	2	-	-
Vallonia costata (Müller)	2	1	-	-	-
Ena montana (Draparnaud)	+	+	+	-	-
Discus ruderatus (Férusac)	1	1	+	1	+
Vitrea contracta Westerlund	2	-	-	-	-
Nesovitrea hammonis (Ström)	1	-	-	1	-
Zonitoides nitidus (Müller)	8	-	1	-	-
Deroceras / Limax	9	11	7	-	1
Clausilia pumila Pfeiffer	1	-	-	-	-
Clausiliidae	7	4	1	-	-
Helicigona lapicida (L.)	-	+	-	-	-
Trochulus hispidus (L.)	+	2	1	1	-
Cepaea sp.	+	-	-	-	-
Arianta arbustorum (L.)	1	2	+	-	7
Total terrestrial	43	52	21	3	9
Total Mollusca	113	170	79	25	97
Ostracods	+	+	+	+	-
Otoliths	-	+	+	+	-
Tertiary shell fragments	-	-	1	-	-

grounds that the Southfleet Road tufaceous channel-fill was deposited by a small tributary of this river. Comparison can also be drawn with Hitchin, a MIS 11 tufa site in Hertfordshire (Kerney 1959). Here, *Ena montana* and *Azeca goodalli* are both common. The Southfleet Road assemblages lack the more exotic elements of the Hitchin tufa, perhaps suggesting they represent a slightly earlier part of the interglacial before characteristic woodland species such as *Platyla polita* had colonised Britain. The Southfleet Road material also shows similarities with the assemblage from Bed 3b at Beeches Pit, which is also considered to be a correlative of the Lower Loam (Preece *et al.* 2007). The most important taxa present at both sites are *Discus ruderatus*, which is replaced in later parts of the Beeches Pit sequence by *D. rotundatus*, and *Vitrea contracta*, which occurs only in Beds 3a and 3b of the Beeches Pit sequence (Preece *et al.* 2007).

CONCLUSIONS

The 'tufaceous channel-fill' at Southfleet Road appears to represent a right-bank tributary of the MIS 11 Thames and may well be a correlative of the Lower Loam deposits at Barnfield Pit, Swanscombe. Pollen evidence indicates that the Phase 6 grey clay belongs to the early-temperate phase of the Hoxnian (Chapter 12). This is supported by the molluscan evidence (Table 10.7), which indicates a local terrestrial environment of temperate woodland with areas of swamp and dry grassland nearby. The damp grassland environment, characterised by species such as *Vertigo angustior*, might be receding as the woodland develops; certainly the most shade-demanding taxa evident at sites such as Hitchin are not present at Southfleet Road.

The most common aquatic constituent of the Southfleet Road assemblage is *Valvata piscinalis*, although none of the specimens approach the *antiqua* form that first appears at the top of the Lower Loam at Barnfield Pit (Kerney 1971) and continues through the Lower Middle Gravels at Dierden's Pit (White *et al.* 2013). This lends further support to the Southfleet Road deposits being contemporary with an earlier part of the Barnfield Pit sequence, namely the lower/middle part of the Lower Loam, or the Lower Gravel. Similarly, *Ena montana*, presently a rare species in Britain that predominantly lives in old woodland, is only known from the Lower Gravel and Lower Loam at Barnfield Pit and *Vertigo angustior* is not known from Middle Gravels and above.

The proportion of land snails is high, attaining levels of around 50% in some samples. This is in keeping with

Table 10.7 Full list of species from tufaceous channel fill sequence, Phase 6b

Aquatic taxa:

Valvata cristata	*Valvata piscinalis*
Bithynia tentaculata	*Galba truncatula*
Stagnicola palustris	*Radix balthica*
Planorbis planorbis	*Anisus leucostoma*
Bathyomphalus contortus	*Gyraulus laevis*
Gyraulus crista	*Planorbarius corneus*
Ancylus fluviatilis	*Acroloxus lacustris*
Sphaerium corneum	*Musculium lacustre*
Pisidium amnicum	*Pisidium casertanum*
Pisidium obtusale	*Pisidium milium*
Pisidum subtruncatum	*Pisidium nitidum*
Pisidium moitessierianum	*Pisidium tenuilineatum*

Land taxa:

Carychium minimum	*Carychium tridentatum*
Succineidae	*Azeca goodalli*
Cochlicopa lubrica	*Vertigo antivertigo*
Vertigo angustior	*Vallonia costata*
Acanthinula aculeata	*Ena montana*
Punctum pygmaeum	*Discus ruderatus*
Vitrea crystallina	*Nesovitrea hammonis*
Zonitoides nitidus	*Deroceras / Limax*
Clausilia pumila	*Clausilia* cf. *bidentata*
Aegopinella nitidula	*Aegopinella pura*
Helicella itala	*Trochulus hispidus*
Arianta arbustorum	*Helicigona lapicida*
Cepaea cf. *nemoralis*	

the tufaceous nature of the mollusc-bearing channel deposits, which suggests that the water body was a small spring-like stream with woodland environments located nearby. The stream was a permanent water-body, however, as several aquatic taxa that require fast-flowing water, such as *Ancylus fluviatilis*, are present, as well as some intolerant of desiccation (eg *Gyraulus crista* and *Bathyomphalus contortus*).

The assemblages are very consistent, which suggests that the molluscan evidence from the tufaceous channel-fill records a relatively short time interval during which the local environment did not change significantly. There is also little variation in the proportions of species represented through the channel-fill sequences (see Figs 10.2a,b,c) through this unit, although the low countable numbers of shells from some contexts might be masking this information. It has therefore not been possible to do much more than characterise the environments represented mainly by context 40070, since there are no dynamic environmental changes recorded in the molluscan fauna.

Chapter 11

Ostracods and other microfossils

by John E. Whittaker, David J. Horne and Francis Wenban-Smith

INTRODUCTION

As with many other categories of biological remains, ostracods have potential not only to reconstruct climate and palaeoenvironment, but also to contribute to biostratigraphic dating. Ostracods are microscopically small, bivalved aquatic crustaceans (their diagnostic shells mostly being between 0.1 and 2mm long), usually completely invisible to the naked eye *in situ* in unprocessed sediment. Therefore, their presence and prevalence in sediment sequences needs to be investigated to a certain extent blindly, relying on their tendency to be preferentially preserved in fine-grained calcareous sediments to guide selection of samples for evaluation.

Several phases of evaluation for ostracods thus took place whilst fieldwork was in progress, as potentially promising new sedimentary horizons were revealed during the excavation. As discussed below, most of these evaluations were negative. However, a relatively rich and diverse ostracod fauna was identified in context 40143, the white silt capping the tufaceous channel-fill sequence (Phase 6b) and this was consequently sampled more intensively, leading to the results presented below.

Following fieldwork, a much more systematic post-excavation assessment for ostracods was carried out, based on sub-sampling from the extensive monolith and bulk sample archive, to investigate for ostracod presence throughout all phases of the site sequence. Although mostly negative, this did however reveal the completely unexpected presence of a rich interglacial ostracod fauna in context 40062, from Phase 3 of the site sequence, interpretation of which is discussed below.

Finally, a very similar ostracod fauna to that from Phase 3 was recovered from test pits (TP) 31 and 33 from the 2006 Station Quarter South field evaluation immediately to the east of the elephant site (Wessex Archaeology 2006b). As discussed previously (Chapter 4), the contexts from which these samples came are here mostly interpreted as periodically desiccated lacustrine sediments equivalent to Phase 4 of the site sequence, and the results from this ostracod analysis are also presented here.

MATERIALS AND METHODS

Samples were either provided straight from fieldwork in a small plastic bag with provenance information written on the outside, or were provided post-excavation as sub-samples collected from monoliths or larger bulk samples. In the former case, each sample had its own unique number (with the exception of four early samples from the December 2003 field evaluation, which had duplicate numbers with pollen evaluation samples from the same sediments). In the latter case, samples were given the same number as the source sample suffixed by '/B' to indicate sub-sampling for ostracod analysis, and further suffixed by a depth range in centimetres for sub-samples collected from monoliths.

For each sample, the whole sample was processed, up to a maximum weight of 225g. Where over 225g was provided, the remainder was left in the bag, unprocessed. After weighing, all the samples were, if not completely dry to begin with, first placed in ceramic bowls and dried in an oven. When dry, a teaspoon of sodium carbonate was added to each sample to help in the subsequent removal of the clay fraction, and hot water poured on them. They were left to soak overnight. Washing was with warm/hot water through a 75 µm sieve, the resulting residues being returned to the bowls for final drying. Good breakdown was invariably achieved, but in a few cases this procedure had to be repeated to achieve a full breakdown. After drying, the sample residues were stored in plastic labelled bags and later picked of their microfaunal content under a binocular microscope. Each sample was dry-sieved through a nest of sieves (>500µm, >250µm, >150µm, and pan), the fractions being sprinkled, one at a time, onto a grid-lined picking tray. Useful organic remains, such as fish and small mammal remains, were picked out, and the presence of others (for example *Bithynia* opercula or other molluscan remains) merely recorded on a presence (x)/ absence basis. The ostracods, on the other hand, were picked out into 3x1" multi-squared faunal slides and the abundance of each species recorded semi-quantitatively (present/several specimens, common, etc).

EVALUATION DURING FIELDWORK

Four episodes of sample collection and processing for ostracod assessment took place during fieldwork (Table 11.1). Firstly, four samples were collected from the Phase 4 deposits at the central logged location (Log 40011, Fig. 11.2b) of the main west-facing section during the preliminary field investigation of December 2003; these were

Table 11.1 Evaluations for ostracods and other microfauna carried out during fieldwork, December 2003 to July 2004

Date	Section	Phase	Context	Sample <>	Weight processed	Ostracods	Molluscs	Bithynia op.	Fish/ amphibians	Small mammals	Slug plate	Charophyte oogonia	Notes
Dec 2003	Log 40011	6	40039	40007	225g	-	-	-	-	-	-	-	Totally decalcified with iron mineralisation
Dec 2003	Log 40011	5	40025	40009	225g	-	-	-	-	-	-	-	Totally decalcified with iron mineralisation
Dec 2003	Log 40011	4	40026	40012	225g	-	-	-	-	-	-	-	Totally decalcified with iron mineralisation
Dec 2003	Log 40011	4	40027	40015	225g	-	-	-	-	-	-	-	Totally decalcified with iron mineralisation; a few moulds of derived Tertiary gastropods
Mar 2004	Sec 40016	6	40068	40086	225g	-	-	-	-	-	-	-	
Mar 2004	Sec 40016	6b	40070	40087	225g	-	xx	-	xx	xx	-	-	Abundant molluscs and small vertebrates, but no ostracods
Mar 2004	Sec 40016	6b	40070	40088	225g	-	xx	-	xx	xx	-	-	Abundant molluscs and small vertebrates, but no ostracods
Jun 2004	Sec 40063	6b	40143	40248	134g	x	x	x	x	-	-	-	
Jun 2004	Sec 40063	6b	40143	40249	108g	x	x	x	x	-	-	x	
Jun 2004	Sec 40063	6b	40143	40250	128g	x	x	x	x	-	-	-	
July 2004	Sec 40068	6b	40143	40264	330g	-	x	x	x	x	-	-	
July 2004	Sec 40068	6b	40143	40265	330g	x	x	x	x	x	x	-	
July 2004	Sec 40068	6b	40143	40266	326g	x	x	x	x	x	x	-	
July 2004	Sec 40063	6b	40143	40268	246g	x	x	x	x	x	-	-	
July 2004	Sec 40063	6b	40143	40269	150g	x	x	x	x	-	-	x	
July 2004	Sec 40063	6b	40143	40270	158g	x	x	x	x	x	-	x	
July 2004	Sec 40063	6b	40143	40271	276g	xx	x	x	x	x	x	x	

x - present; xx - abundant

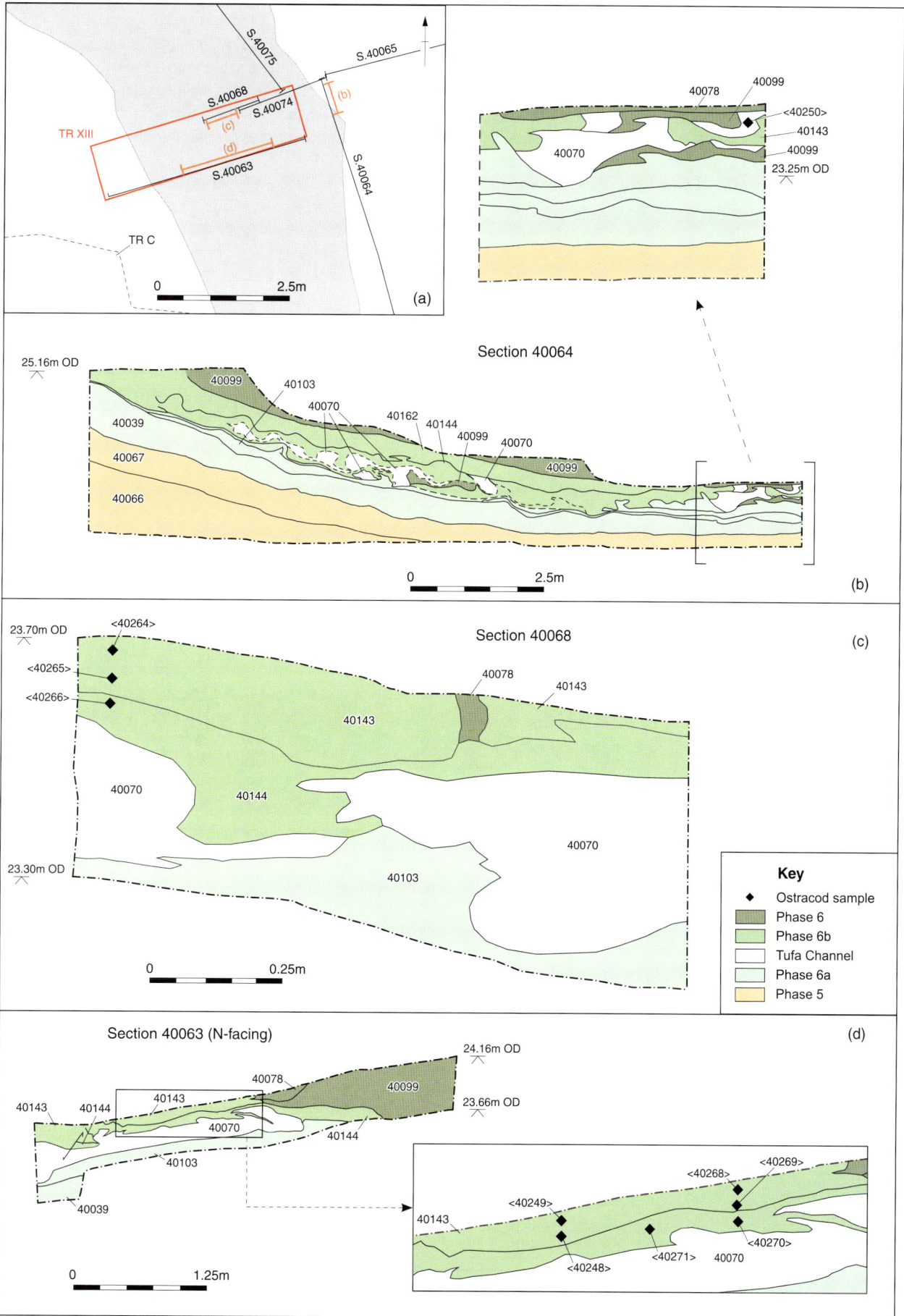

Figure 11.1 Ostracod sampling locations for context 40143: (a) thumbnail of sampling locations; (b) main east face of site (Section 40064); (c) Trench XIII, south-facing Section 40068; (d) north-facing Section 40063

found to be barren not only of ostracods, but also of all other Pleistocene remains. Secondly, three samples were collected from the Phase 6 sediments in the main east-facing section in March 2004, including two from context 40070, the main shell-rich bed of the tufaceous channel (Phase 6b). No faunal remains were present in the first of these and, somewhat surprisingly, the last two samples also proved to be barren of ostracods, despite their calcareous nature and the rich preservation of molluscan and small vertebrate remains. Thirdly, three samples: <40248>, <40249> and <40250>, were collected from the white silt (context 40143) capping the tufaceous channel-fill sequence, and assessed for ostracods in June 2004 (Fig. 11.1b, d). These, in pleasant contrast to the previous assessments, proved to contain a reasonably abundant and diverse ostracod fauna. Therefore, a fourth episode of sample collection and processing took place in July 2004, focusing on more intensive sampling of context 40143 in the sections of Evaluation Trench XIII, where there were the best and thickest exposures of this intermittently occurring horizon (Fig. 11.1c, d). Identification and interpretation of this ostracod fauna are presented below.

POST-EXCAVATION ASSESSMENT AND ANALYSIS

Once excavation was completed, a major programme of environmental assessment was carried out (Appendix 1), including an extensive investigation for the presence of ostracod remains throughout the finer-grained horizons of the site sequence. In total 57 samples were initially submitted for assessment (Table 11.2). These were then supplemented by nine additional samples from monolith <40068> (Fig. 11.2a), following the unexpected discovery in its lower part (context 40062, Phase 3) of a rich interglacial ostracod fauna, discussed in more detail below. Finally, analysis also took place of the residual ostracod fauna from bulk sample <40042>, also from context 40062 (Fig. 11.2a). This had been sieved through a mesh of 0.5mm to investigate for small vertebrate remains, but which it was later realised also contained remains of some of the larger ostracods present in this deposit.

Apart from this evidence from Phase 3, and sparse further remains from Phase 6b sediments (the tufaceous channel-fill), the post-excavation assessment proved entirely negative for ostracods. However some other micro-fauna were noted in some samples, and various other observations were made of interpretive potential, such as the occurrence of charophyte oogonia and fossilised rootlet hollows (Table 11.2). These allow some interpretation of depositional and post-depositional environments, and are also discussed where appropriate below.

Phase 3

The deposits investigated here as Phase 3 were equivalent to Unit 1 of the preliminary report (Wenban-Smith

et al. 2006). As discussed above, on the west side of the site they comprised grey clay-silty sand with chalk pebbles, rich in derived Tertiary shell material (context 40028). They became thicker and more differentiated eastwards, with finer-grained more clay-silty deposits (contexts 40062 and 40063) forming the top of the Phase 3 sequence in the east-facing site section 40016 (Fig. 4.6a).

Figure 11.3 shows the results from analysis of the eight samples from monolith <40068>, together with sample <40042>, covering contexts 40063 and 40062 (Fig. 11.2a). Context 40062 proved to contain the richest freshwater ostracod faunas found during the present survey, and following the initial examination of two samples from samples 50-52cm and 56-58cm this was expanded to an analysis of the entire monolith, including the overlying context 40063. Surprisingly, the climatic/ecological results from the initial assessment (herein) proved to be at variance with the initial thinking that this phase of the site sequence represented cool/cold conditions (as proposed in Wenban-Smith et al. 2006). Another set of sub-samples from the same monolith provided for mollusc assessment seemed to be almost barren (Chapter 10). It was therefore considered possible that the two assessed ostracod samples (or their source monolith) might have been mis-labelled at some point in time. The finest grade of residue from an extra bulk sample <40042> was then provided via Simon Parfitt (who had sorted it for small mammal and fish/amphibian remains). This sample was undoubtedly from context 40062. Even though it had been put through a 500 µm sieve, its organic remains, including ostracods (although some of the smaller species were lost), were so near-identical to those recorded as from monolith <40068> (Fig. 11.3) that it has been concluded that the samples and their microfaunas described here from monolith <40068> are genuinely ascribed.

The uppermost two samples, from context 40063, are unfortunately weathered and decalcified, as is much of the material generally from other horizons at the site. However, context 40062 below (monolith <40068> 48-60cm; and sample 40042) provided an extremely rich, fully interglacial freshwater ostracod faunas (with at least 8 species found), together with molluscs, Bithynia opercula and charophyte oogonia (Fig. 11.3). Furthermore, they were identical to the faunas from Station Quarter South TPs 31 and 33 excavated a short distance to the east in 2006, reported on by Whittaker (2006) and discussed below. The ostracod fauna from TP 31 was recovered from sediments attributed to Phase 4, and that from TP 33 was from the top of a body of sediment attributed to Phase 3 (see Fig. 11.5). There is little difference between the finer-grained upper Phase 3 deposits at the site and the presumed Phase 4 lithology further east; in fact, they may interdigitate with each other, or the more developed eastern Phase 4 deposits may be lateral equivalents of the Phase 3 sediments.

The ostracod faunas indicate a shallow, swampy but clean, slow-moving waterbody rich in vegetation or a weedy shallow lake. This is supported by additional

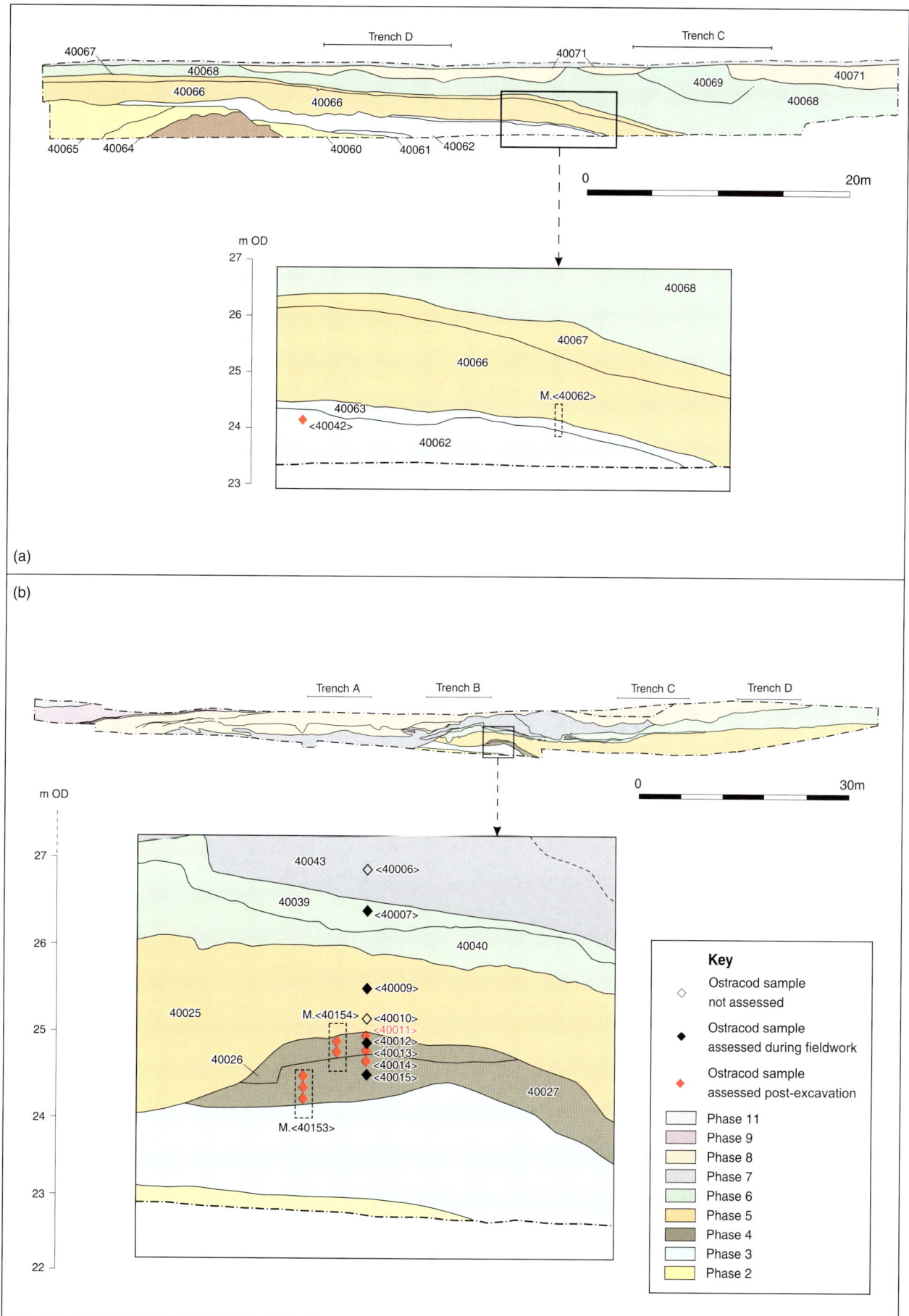

Figure 11.2 Ostracod sampling locations: (a) for Phase 3, in east face of site (Section 40016), monolith <40068> and bulk sample <40042>; (b) in west face of site, Log 40011, Section 40015

Table 11.2 Post-excavation assessment for ostracods and other microfauna, 2009–2010

Sample <>	Context	Phase	Weight processed	freshwater ostracods	molluscs	Bithynia opercula	fish/amphibian remains	small mammal teeth	charophyte oogonia	iron + iron tubes (rootlets)	tufa	Assessment notes
40420/B	40166	7	225g									Completely barren
40414/B	40166	7	225g							x		Iron mineral (?limonite and goethite) and iron tubes with imprints of plant stems and rootlets
40415/B	40166	7	225g							x		Iron mineral (?limonite and goethite) and iron tubes with imprints of plant stems and rootlets
40416/B	40166	7	225g							x		Iron mineral (?limonite and goethite) and iron tubes with imprints of plant stems and rootlets
40417/B	40166	7	225g							x		Iron mineral (?limonite and goethite) and iron tubes with imprints of plant stems and rootlets
40361/B	40164	7	225g									Completely barren
40418/B/0-2cm	40166	7	185g							x		Iron mineral (?limonite and goethite) and iron tubes with imprints of plant stems and rootlets
40418/B/5-7cm	40158	6	70g					x		x		
40418/B/13-15cm	40158	6	125g					x				
40311/B	40144	6b	225g				x					
40310/B	40070	6b	225g	x	x	x	x				x	
40313/B	40070	6b	225g	x	x	x	x		x		x	
40312/B	40103	6a	225g				x					
40282/B/0-2cm	40143	6b	50g	x	x	x	x		x		x	
40282/B/2-5cm	40144	6b	135g	x	x	x	x		x		x	
40282/B/5-8cm	40144	6b	160g	x	x	x	x				x	
40282/B/11-13cm	40070	6b	55g	x	x	x	x		x		x	
40282/B/17-19cm	40070	6b	80g		x	x	x	x			x	
40282/B/22-24cm	40103	6a	55g		x		x			x		
40321/B/1-5cm	40144	6b	80g									
40321/B/6-9cm	40144	6b	110g									
40321/B/10-14cm	40144	6b	110g									
40321/B/15-18cm	40144	6b	105g									
40321/B/19-22cm	40144	6b	110g				x					
40321/B/23-27cm	40144	6b	105g				x					
40321/B/29-31cm	40070	6b	c 80g			x	x	x			x	
40321/B/31-33cm	40070	6b	70g			x	x	x			x	
40321/B/33-36cm	40070	6b	95g	x		x	x	x			x	
40321/B/36-39cm	40070	6b	95g			x	x	x	x	x	x	
40321/B/39-42cm	40070	6b	75g			x	x	x	x	x	x	
40321/B/42-45cm	40103	6a	80g					x	x			
40157/B/3-5cm	40040	6a	95g									
40157/B/18-20cm	40040	6a	125g									
40157/B/31-33cm	40040	6a	105g							x		
40326/B/1-3cm	40103?	6a	45g							x		
40326/B/17-19cm	40103	6a	160g							x		
40326/B/31-33cm	40103	6a	135g							x		
40326/B/40-42cm	40025	5	105g							x		Iron tubes formed around plants or rootlets and suggests a shallow water-body or riverine setting whose sediments have subsequently been subject to weathering

Table 11.2 (continued)

Sample <>	Context	Phase	Weight processed	freshwater ostracods	molluscs	Bithynia opercula	fish/amphibian remains	small mammal teeth	charophyte oogonia	iron + iron tubes (rootlets)	tufa	Assessment notes
40154/B/26-28cm	40026	4	110g							x		
40154/B/41-43cm	40026	4	105g							x		
40153/B/0-2cm	40027	4	105g							x		
40153/B/12-14cm	40027	4	45g							x		
40153/B/27-29cm	40027	4	50g							x		
40011	40026	4	55g							x		
40013	40026	4	80g							x		
40014	40027	4	75g						x	x		
40068/B/10-12cm *	40066	5	90g							x		Iron tubes formed around plants or rootlets and suggests a shallow water body or riverine setting whose sediments have subsequently been subject to weathering
40068/B/23-25cm *	40066	5	110g							x		Iron tubes formed around plants or rootlets and suggests a shallow water body or riverine setting whose sediments have subsequently been subject to weathering
40068/B/32-34cm *	40066	5	80g							x		Iron tubes formed around plants or rootlets and suggests a shallow water body or riverine setting whose sediments have subsequently been subject to weathering
40068/B/38-40cm *	40063	3	115g							x		De-calcified upper part of context
40068/B/44-46cm *	40063	3	115g							x		De-calcified upper part of context
40068/B/48-50cm *	40062	3	125g	x	x	x			x	x		
40068/B/50-52cm	40062	3	105g	x	x	x			x	x		
40068/B/52-54cm *	40062	3	95g	x	x	x			x	x		
40068/B/54-56cm *	40062	3	115g	x	x	x			x			
40068/B/56-58cm	40062	3	80g	x	x	x			x			
40068/B/58-60cm *	40062	3	115g	x	x	x			x			
40095/B/10-15cm	40065	2	175g							x		Black iron mineral, probably goethite, suggests weathering
40078/B/46-51cm	40065	2	175g							x		Black iron mineral, probably goethite, suggests weathering
40076/B/33-38cm	40064	2	180g							x		Black iron mineral, probably goethite, suggests weathering
40073/B/4-9cm	40064	2	180g							x		Black iron mineral, probably goethite, suggests weathering
40098/B/16-22cm	40064	2	175g							x		Scraps of reworked Tertiary molluscs
40097/B/10-15cm	40064	2	175g									Completely barren
40066	40065	2	225g							x		Black iron mineral, probably goethite, suggests weathering
40062	40059	1	225g									Completely barren
40060	44057	1	225g							x		Iron (?) mineral, possibly reworked from Thanet Sand

* additional samples from monolith 40068, following initial assessment

Organic remains

CONTEXT	40063		40062						40042
SAMPLE	40068/B/38-40cm	40068/B/44-46cm	40068/B/48-50cm	40068/B/50-52cm	40068/B/52-54cm	40068/B/54-56cm	40068/B/56-58cm	40068/B/58-60cm	40042
iron mineral + iron tubes	x	x	x	x					
molluscs			x	x	x	x	x	x	x
Bithynia opercula			x	x	x	x	x	x	x
charophyte oogonia			x	x	x	x	x	x	x
freshwater ostracods			x	x	x	x	x	x	x

Ecology	40063		40062						
	weathered; decalcified		weedy shallow, but clean waterbody						

Freshwater Ostracods

CONTEXT	40063		40062						40042
SAMPLE	40068/B/38-40cm	40068/B/44-46cm	40068/B/48-50cm	40068/B/50-52cm	40068/B/52-54cm	40068/B/54-56cm	40068/B/56-58cm	40068/B/58-60cm	40042
Fabaeformiscandona balatonica			xxx	xxx	xx	xx	xx	xx	xx
Herpetocypris reptans			xx	xx	xx	xx	xx	xx	xxx
Candona neglecta			x	x	x	x	x		x
Ilyocypris bradyi			x	x	x	x	x	x	
Cypridopsis vidua			x		o	x	o	x	
Ilyocypris quinculminata			o			o		o	
Cyclocypris sp.			x	x	x		x	x	
Potamocypris zschokkei				o					

Organic remains are recorded on a presence (x) / absence basis only
Ostracods are recorded: o - one specimen; x - several specimens; xx - common; xxx - abundant

Freshwater ostracods
Extinct, not known in sediments younger than MIS11 (Hoxnian)

Figure 11.3   Phase 3: ostracods from monolith <40068> and bulk sample <40042>

evidence from the charophytes (see discussion under Phase 4 sediments, below) and it is unfortunate that no associated molluscan evidence was recovered to complement the ostracod data. Ecological preferences of the three most common ostracods are given below, adapted from Meisch (2000).

Herpetocypris reptans (Baird, 1835) Waterbodies with rich vegetation and a muddy bottom; slow waters; swampy waters; springs.

Fabaeformiscandona balatonica (Daday, 1894) Shallow pools and swampy, shallow fringes of waterbodies that can dry up in summer.

Ilyocypris bradyi (Sars, 1890) Slow waters, swamps, etc. Waters flowing from springs, ponds fed from springs.

In the Station Quarter South assemblages (see below), the occurrence of *Ilyocypris quinculminata* was reported, so far the only age-diagnostic ostracod found on the site or in the vicinity; it is also present here, albeit rarely, in samples from 48-50, 54-56 and 58-60cm in monolith <40068>. This distinctive species is present in several Cromerian (MIS 13) sites: Little Oakley and Sidestrand, East Anglia; Waverley Wood, Warwickshire (MIS1 5); Boxgrove, West Sussex (MIS 13) but is not known from sediments younger than the Hoxnian (MIS 11). Its occurrences however in MIS 11 are curious (Whittaker and Horne 2009). In full, these are: Trysull (type-locality) and Froghall, Staffordshire; Hatfield, Hertfordshire; Copford, Marks Tey and Rivenhall, in Essex; and Elveden and Hoxne, in Suffolk. There is not a single occurrence, apart from this Southfleet Road site, from the Thames environs of this age. It has not been found during work in progress with Richard Preece and Tom White (Cambridge University) from sediments at Dierden's Pit, Swanscombe that are thought most likely equivalent to the Lower Middle Gravel of the Barnfield Pit sequence, slightly younger in MIS 11 than the suggested age of Phases 3-6 of the Southfleet Road site sequence (see Chapter 22). Nor has it been found in the Lower Loam at Barnfield Pit, which may be thought to be more of a correlative and the lower part of which was examined at Swanscombe for ostracods in the 1980s by Robinson (1996). This species, of course, may have been restricted to a particular facies or niche: the Southfleet Road site would have been a minor backwater compared to the major river represented at Barnfield Pit, but there is also the possibility that it may only occur within a part (or parts) of the Hoxnian interglacial.

That this may be early in the Hoxnian is evidenced both by the position of these Phase 3 deposits towards the base of the Southfleet Road sequence, and also by the amino acid ratios obtained from *Bithynia* opercula from context 40062 (see Chapter 13). It was initially thought during excavation (Oxford Archaeology 2005; Wenban-Smith *et al.* 2006) that the Phase 3 sediments were soliflucted or fluvio-glacial in their origin, and thus by implication, of late Anglian or very early Hoxnian age. However, there are no cold ostracod indicators

whatsoever in the microfaunas of context 40062, and such markers are now very well-known from cold climate deposits, not only within MIS 11 (as at Beeches Pit, Suffolk, for example (Whittaker, *in* Preece *et al.* 2007)), but also from widespread younger, Devensian, solifluction deposits in the Warblington-Bognor-Worthing region on the South Coast of England (see Bates *et al.* 2009). A mutual ostracod temperature range (MOTR) analysis of the ostracod species by one of us (DJH) from context 40062 (below) indicates a fully temperate environment, albeit probably with greater seasonality than today. The only molluscan remains recovered in context 40062 were *Bithynia* opercula, which were common. These are considered to be confined to interglacials and interstadials (Richard Preece, pers. comm.) so, assuming they are not reworked, these too suggest warm climatic conditions. Reworking must be considered a possibility though, since *Bithynia* would generally prefer cleaner flowing water than the quiet/stagnant, periodically desiccated, small lake/pond thought to be the dominant Phase 3 environment.

These Phase 3 deposits contain the oldest undoubted Pleistocene microfaunas in the Southfleet Road succession. The initial suggestion (Wenban-Smith *et al.* 2006) that they were formed by solifluction in a periglacial climate must be rejected, it also being highly improbable that the ostracod faunas are derived, or the sediments in some way rafted in as a whole block. Under these circumstances the microfaunas reported here would have undergone destructive damage and probably lost their ecological/climatic integrity as an assemblage. The ostracod *Herpetocypris reptans* in particular is very large and thin-shelled and would be expected to be totally destroyed under those circumstances.

There is however some reworking in the lower part of context 40062 as it contains much derived Tertiary molluscan material (mostly scraps) as well as the distinctive reworked ostracod, *Vetustocytheridea lignitarum*. The distribution of the latter is well known, being found in the Blackheath Beds (Harwich Formation) and the Woolwich Formation of the Late Palaeocene and in the Upnor Formation of earliest Eocene age (Lord *et al.* 2009). It occurs in abundance in the 'Lower and Upper Shelly Clays' and the 'Laminated Beds' (of the Woolwich Formation) in the Swanscombe sequence. Consequently it is therefore probable that the Phase 3 deposits at Southfleet Road contain derived material from this same deposit, washed downslope from the high ground previously (pre-quarrying) to the west of the site (Chapter 2).

As reported above, the two samples subsequently examined from monolith 40068, context 40063 (interval 38-46cm), directly above these fossiliferous deposits, were completely barren of ostracods, or any other Pleistocene fauna for that matter. Their only content was iron mineral (limonite and goethite) and iron tubes formed around plant tissue. This would appear to indicate weathering and decalcification of the upper Phase 3 sedimentary sequence.

Mutual Ostracod Temperature Range (MOTR) Analysis: Phase 3 sediments, monolith <40068>, context 40062

This type of analysis (Horne 2007) is similar to the Mutual Climate Range (MCR) method of Russell Coope for Coleoptera. It also uses proxy data (in this case, a large present-day ostracod database called NODE) to estimate past air temperature ranges. The July and January means for the Phase 3, Monolith <40068>, ostracods have thus been calculated as follows:

	Calibrated July Mean	Calibrated January Mean
Herpetocypris reptans	+10 to +25°C	-8 to +15°C
Fabaeformiscandona balatonica	+17 to +21°C	-4 to –1°C
Cypridopsis vidua	+9 to +28°C	-8 to +17°C
Cyclocypris ovum	+7 to +26°C	-17 to +4°C
Prionocypris zenkeri	+14 to +26°C	-4 to +6°C
Candona neglecta	+7 to +27°C	-10 to +13°C
Potamocypris zschokkei	+6 to +24°C	-10 to +9°C

Ilyocypris bradyi cannot be calibrated due to taxonomic uncertainties.

Ilyocypris quinculminata cannot be used, either, as it is extinct.

The calibration of *Fabaeformiscandona balatonica* is new and is based on only five records in NODE, so its reliability is rather questionable.

Application of the MOTR method gives:
July mean temperature range: +17 to +21°C
January mean temperature range: -4 to -1°C

If *F. balatonica* is omitted, results are:
July mean temperature range: +14 to +24°C
January mean temperature range: -4 to +4°C

Present day values for Ebbsfleet, by comparison, are July +17.5°C, January +4°C. Hence, July temperatures during the deposition of the Phase 3 sediments were essentially the same as today, although possibly they could have been a few degrees warmer. January temperatures were essentially the same as today, but possibly a few degrees colder. The presence of *F. balatonica*, however, might be taken to indicate winter temperatures

at least 5 degrees colder than today. The MOTR analysis therefore indicates a fully temperate climate with perhaps greater seasonality than today.

Phase 4

Fig. 11.4 shows the results of the eight samples examined from contexts 40026 and 40027, from Phase 4 sediments in the central part of the main west-facing Section 40015, at or in the vicinity of Log 40011 (Fig. 11.2b). All eight samples contain iron mineral and iron tubes (formed around plant stems and rootlets). A significant find occurred in sample 40014 (context 40027) which contains charophyte oogonia (oospores), or rather their cortex preserved in iron. This gives some indication of the nature of the waterbody that once occurred at the site at this level, whose evidence has been mostly destroyed by subsequent weathering. Charophytes are water plants. They grow completely submerged in the water of wetlands, rivers, streams, lakes and swamps, in fact all sorts of non-marine watery habitats. They reproduce sexually via gametes produced by male and female reproductive organs (antheridia and oogonia, respectively), and they have seed-like oospores that germinate into the new plant. Charophytes are useful components of aquatic ecosystems. Many species require high water quality and clarity for survival, so their presence can indicate the existence of a healthy ecosystem. They decline when water becomes polluted, murky or eutrophic.

In contrast to this minimal evidence from the Phase 4 deposits at the site, rich ostracod faunas were recovered in 2006 from deposits thought to be equivalent in TPs 31 and 33 a short distance to the east, excavated as part of the Station Quarter South field evaluation (Wessex Archaeology 2006b; Chapter 4). These faunas were recovered from four samples in clayey silts with polygonal cracking indicating periodic drying out of a very quiet waterbody (Fig. 11.5). As discussed above, these deposits may directly overlie the Phase 3 sediments on the eastern side of the site, they may interdigitate with them, or they may in fact be direct lateral equivalents, continuing an eastward dipping/thickening and fining trend.

Samples <3> and <4> (contexts 3304 and 3306, respectively) came from towards the top of the sequence of TP 33 and have virtually identical faunas to each other (Fig. 11.6). The higher sample <3> was interpreted as coming from Phase 4 deposits, and the lower sample <4>

Organic remains

CONTEXT	40026		40027				40026		40027
SAMPLE	40154/B/26-28cm	40154/B/41-43cm	40153/B/0-2cm	40153/B/12-14cm	40153/B/27-29cm		40011	40013	40014
iron mineral + iron tubes	x	x	x	x	x		x	x	x
charophyte oogonia									x

Ecology	?shallow waterbody; weathered	shallow waterbody with high water quality; subsequently weathered

Organic remains are recorded on a presence (x) / absence basis only

Figure 11.4 Microfossil assessment: Phase 4

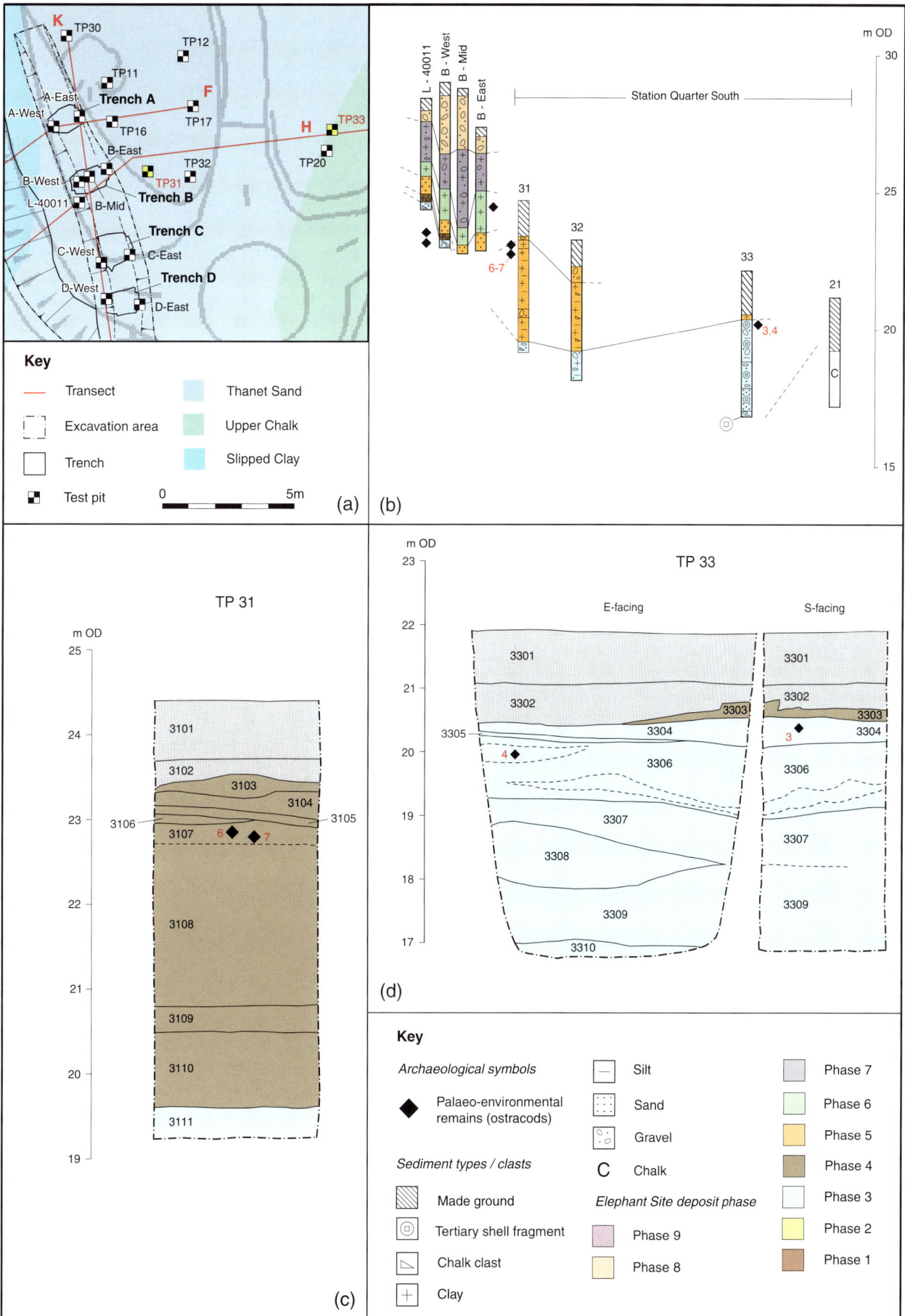

Figure 11.5 Ostracod sampling locations, Station Quarter South: (a) thumbnail of Test Pits 31 and 33 locations; (b) stratigraphic context of ostracod samples <3>, <4>, <6> and <7>; (c) Test Pit 31; (d) Test Pit 33

as from the top of Phase 3 deposits. Their organic remains are listed in the upper part of Fig. 11.6 and the ostracods below. Molluscan fragments occur again but this time with *Bithynia* opercula (in large numbers) and Pleistocene freshwater ostracods comprising the following species: *Herpetocypris reptans, Fabaeoformiscandona balatonica, Ilyocypris bradyi, Cypriopsis vidua, Prionocypris zenkeri* and *Ilyocypris quinculminata*.

Samples <6> and <7> (context 3107) from TP 31 have a near-identical ostracod fauna to this, but in even greater numbers (Fig. 11.6); and the very same six species, with the addition of *Cyclocypris ovum*. The organic remains now include charophyte oogonia (fruiting bodies of the stonewort), slug plates and calcareous tubes clearly formed around the stems and rootlets of plants. Rather than this signifying samples <6> and <7> are necessarily coeval with <3> and <4>, it probably indicates nothing more than both were laid down in very similar environments.

The ecological requirements of the three commonest ostracods in the assemblages (*Herpetocypris reptans, Fabaeoformiscandona balatonica* and *Ilyocypris bradyi*) are listed above (Phase 3, discussion); the less common species having similar preferences. Together, they indicate a shallow, swampy, slow-moving water body rich in vegetation, context 3107 even more so, in that it provides additional evidence of charophytes and the rootlets and stems of aquatic plants. All the ostracod species, save one, are found living in Britain today, so there is nothing to suggest that these deposits indicate anything other than a full interglacial. The exception, *Ilyocypris quinculminata*, however, is of great biostratigraphic value and immediately places them in the period MIS 13 (late Cromerian complex) to MIS 11 (the Hoxnian). This species is previously known from about 12 sites (all in South and Central England), listed above in the Phase 3 discussion. Curiously, however, the species has yet to be found in any of the classic Swanscombe 100-ft terrace deposits, such as at Barnfield Pit or Dierden's Pit.

Ostracods from the tufaceous channel-fill, Phase 6b

No ostracods were found in any of the samples from Phase 6 (undifferentiated clay, usually brecciated and/or browner with fragmentary rotted plant material) or 6a (the specific lower clay contexts, in ascending order 40040, 40039 and 40103 in the central part of the site). Many of the Phase 6a samples show evidence of weathering by the presence of iron minerals and iron tubes formed around plant tissue: contexts 40039, 40103 and 40158 (Table 11.2), or were completely barren (top of context 40040). All that can be said is that these sediments were probably deposited in a semi-terrestrial environment, with associated shallow waterbodies that subsequently were subject to desiccation and/or weathering.

In Phase 6b, the tufaceous channel-fill, evaluation while fieldwork was in progress failed to find any ostracods in the main basal mollusc-rich context 40070, but established that a relatively rich ostracod fauna was present in context 41043, the white silt capping the tufaceous channel-fill sequence. Ten samples from this deposit were analysed from three localities, all in or near Trench XIII, where the best exposures of context 40143 were present (Fig. 11.1). At Section 40068, a sequence of three samples (<40264>-<40266>) was taken down through the full thickness of the deposit (Fig. 11.1c). At section 40063, on the opposite face of Trench XIII, another sequence of three samples (<40268>-<40270>) was taken down through the full thickness of the deposit. This was supplemented here by samples <40249> and <40248> from towards the top of the deposit, and sample <40271> from the middle of the deposit (Fig. 11.1d). Finally, a single sample (<40250>) was taken at the third locality, the north end of Section 40064 (Fig. 11.1b).

Table 11.3 summarises all the microfauna found in the ten samples from Sections 40068, 40063 and 40064. There are slight differences in their associations, and more detailed reports on the small vertebrate and molluscan material are provided separately (Chapter 7 and Chapter 10, respectively). The material is in excellent condition and is comparable to that of the underlying tufa, all suggesting temperate conditions. Although fish remains are ubiquitous, the stratigraphic sequence from Sections 40068 and 40063 seem to show an increase in amphibian, reptile and small mammal remains towards the top, with a concomitant decrease in fish remains, possibly indicating drying out of the waterbody through time. This is considered further in the concluding discussion (Chapter 22), in conjunction with other lines of evidence.

All but one sample contains freshwater ostracods (Table 11.3). None is extinct and all are living in Britain today. Of these only one candonid is ubiquitous. It is usually

		TP 31		TP 33	
	PHASE	4		3	
	CONTEXT	3107		3304	3306
ORGANIC REMAINS	SAMPLE NO.	<6>	<7>	<3>	<4>
molluscs		x	x	x	x
reworked microfossils					
fish/amphibian bone/teeth					
Bithynia opercula		x	x	x	x
charophyte oogonia		x	x		
freshwater ostracods		x	x	x	x
slug plates		x	x		
calcareous tubes (rootlets)		x	x		

OSTRACOD SPECIES

Herpetocypris reptans	xx	xx	x	x
Fabaeoformiscandona balatonica	xx	xx	x	x
Ilyocypris bradyi	xx	x	x	x
Cypridopsis vidua	x	x		x
Cyclocypris ovum	x	x		
Ilyocypris quinculminata	x		x	
Prionocypris zenkeri		x	x	x

Organic remains are recorded on a presence (x) / absence basis
x - present; xx - common/abundant
Extinct, not known in sediments younger than MIS11 (Hoxnian)

Figure 11.6 Ostracods from Station Quarter South Test Pits 31 and 33 (Phases 3 and 4 deposits)

Table 11.3 Distribution of ostracods through context 40143 (tufaceous channel, Phase 6b) in samples from Sections 40063, 40068 and 40064

Section		40068			40063						40064
Sample <>		40264	40265	40266	40249	40268	40248	40269	40271	40270	40250
Vertical zone of context 40143		Upper	Middle	Lower	Upper	Middle	Lower	All			
Fabaeformiscandona balatonica			x(j)	x(j)	x(j)	x(j)	x(j)	x(j)	x	x(j)	o(j)
Ilyocypris bradyi			x	x		x		x	x		x
Pseudocandona rostrata				o(j)	o	x(j)		x	xx		
Herpetocypris reptans								x	x	x	
Pseudocandona longipes					x				x	x	
Cyclocypris sp.					o		o			o	
Potamocypris zschokkei							x				

o - one specimen; x - several specimens; xx - common; (j) juveniles only

represented only by juveniles, but in sample <40271> a few adults show the species to be *Fabaeformiscandona balatonica*. In the three sections another six ostracod species are also found in varying associations, which if *in situ* could indicate slightly different niches being represented. Overall, however, almost all of the ostracods live either in shallow waterbodies fed by springs (eg *Pseudocandona rostrata, P. longipes*), slow waters flowing from springs (eg *Ilyocypris bradyi*) or in the seepages/springs themselves (*Potamocypris zschokkei*). The occurrence (only) in Section 40063 of *Herpetocypris reptans* (usually in the same samples as charophyte oospores) evidences well vegetated, unpolluted waters, but the reason for its restriction is unclear. The most common ostracod, *Fabaeformiscandona balatonica*, today prefers shallow and swampy pools that dry up in summer, often also in shady conditions, in woodland. The evidence of the small vertebrate fauna, as we have seen, could indeed indicate a pool drying out over time. However, the ostracod evidence points to a seasonal phenomenon, whereas the latter must have taken place over a much longer period of time. The presence of mainly juvenile candonids could also be a seasonal factor, but could also indicate some redeposition (supported by the occurrence of much molluscan shell debris and concentrated *Bithynia* opercula) as the waterbody overflowed from time to time. This may possibly have been dependent on season, but was never catastrophic as the sediments are fine grained and other fossil preservation (for example of fragile gastropods) is usually good.

Following completion of fieldwork, the more intensive post-excavation assessment investigated various sequences through the tufaceous channel-fill (Table 11.2), and recovered sparse ostracod assemblages in a number of places, including material from contexts 40070 and 40144, as well as some further material from context 40143 (Fig 11.7). Considering that the thin and intermittent remnants of context 40143 were often highly contorted and intermingled with the underlying sediments, it is questionable whether any of this material genuinely belongs to any context other than 40143. The numbers of ostracods were quite low and the broken nature of much of the associated molluscan fauna also indicates that there may have been some reworking or redeposition. The ostracod species that occur live today in shallow waterbodies fed by springs, slow waters flowing from springs or in the seepages/springs themselves. Because abundant tufa is found throughout most of these sediments, the ostracods may have been living in the actual tufa-springs and associated pools, but the ostracod numbers are surprisingly sparse if everything is *in situ*, as also indicated by the micritic nature of the tufa.

Phase 7 sediments

These sediments directly, conformably (despite the contorted geometry of the junction) overlie the dark brown brecciated clay of the Phase 6 deposits in the central part of the site (Chapter 4). A series of samples from the fine-grained context 40166 forming the base of the syncline infill in the centre of the 'skateboard ramp' feature were investigated for microfaunal remains. None was found, but mineralised imprints of plant stems and rootlets allow some interpretation.

Figure 11.8 shows the results of the seven samples examined from contexts 40164 and 40166. All were barren of ostracods and anything else calcareous. Two were completely barren but 5 of the 7 samples contained iron mineral (?limonite and goethite) and iron tubes. The tubes showed distinct imprints of plant stems and rootlets. According to Candy (in Ashton *et al.* 2005) this iron is associated with weathering or near-surface groundwater, formed prior to the onset of fully terrestrial conditions, or pedogenic activity. The coarse nature of some of these Phase 7 deposits may indicate a high energy setting at times, or more likely the mass movement of gravel-rich sediments, but these samples were totally barren. Samples <40420> and <40361> were gravel-rich, and also examined for clast lithology (see Chapter 6 and Fig. 6.1c; d). The occurrence of mineralised plant tissue fossils, in association with the finer-grained sediment of the other samples, indicates the former presence of wetland plants in a shallow waterbody or a much quieter riverine setting. The actual plant remains and any once-present faunal remains having been destroyed by post-depositional oxidation and weathering.

The Ebbsfleet Elephant

Organic remains

CONTEXT	40144	40070	
SAMPLE	40311/B	40310/B	40313/B
fish bone and teeth	x	x	x
tufa		x	x
molluscs		x	x
Bithynia opercula		x	x
slug plates		x	x
freshwater ostracods		x	x
small mammal teeth			x
amphibian bone			
iron			

CONTEXT	40143	40144		40070	
SAMPLE	40282/B/0-2cm	40282/B/2-5cm	40282/B/5-8cm	40282/B/11-13cm	40282/B/17-19cm
fish bone and teeth	x	x	x	x	x
tufa	x	x	x	x	x
molluscs	x	x	x	x	x
Bithynia opercula	x	x	x	x	x
slug plates				x	x
freshwater ostracods	x	x	x	x	
small mammal teeth	x	x		x	x
amphibian bone		x	x		
iron					

CONTEXT	40144			
SAMPLE	40321/B/1-5cm	40321/B/6-9cm	40321/B/10-14cm	40321/B/15-18cm
fish bone and teeth				
tufa				
molluscs				
Bithynia opercula				
slug plates				
freshwater ostracods				
small mammal teeth				
amphibian bone				
iron				

Ecology	Actual tufa springs and associated pools, or redeposited from tufa springs

Ecology	Actual tufa springs and associated pools, or redeposited from tufa springs

	weathered

Freshwater ostracods

CONTEXT	40144	40070	
SAMPLE	40311/B	40310/B	40313/B
Ilyocypris gibba/bradyi		xx	x
Prionocypris zenkeri		o	
Potamocypris zschokkei		o	
Pseudocandona sp. (juv.)		o	
Fabaeformiscandona balatonica			

CONTEXT	40143	40144		40070	
SAMPLE	40282/B/0-2cm	40282/B/2-5cm	40282/B/5-8cm	40282/B/11-13cm	40282/B/17-19cm
Ilyocypris gibba/bradyi	x	x		o	
Prionocypris zenkeri					
Potamocypris zschokkei			x		
Pseudocandona sp. (juv.)					
Fabaeformiscandona balatonica	o	o			

CONTEXT	40144			
SAMPLE	40321/B/1-5cm	40321/B/6-9cm	40321/B/10-14cm	40321/B/15-18cm
Ilyocypris gibba/bradyi				
Prionocypris zenkeri				
Potamocypris zschokkei				
Pseudocandona sp. (juv.)				
Fabaeformiscandona balatonica				

Ostracods are recorded: o - one specimen; x - several specimens; xx - common; xxx - abundant

Organic remains are recorded on a presence (x) / absence basis only

Completely barren
Freshwater ostracods

Figure 11.7 *(above and right)* Phase 6b: microfossil assessment and ostracod analysis

Organic remains

CONTEXT	40166					40164	40166
SAMPLE	40420/B	40414/B	40415/B	40416/B	40417/B	40361/B	40418/B/0-2cm
iron mineral + iron tubes		x	x	x	x		x

Ecology	Weathered. Originally contained plants in either wetland or quiet riverine setting.

Organic remains are recorded on a presence (x) / absence basis only

Completely barren

Figure 11.8 Microfossil assessment: Phase 7 sediments

CONTEXT	40144		
SAMPLE	40321/B/19-22cm	40321/B/23-27cm	40321/B/29-31cm
fish bone and teeth	x	x	x
tufa			x
molluscs			x
Bithynia opercula			x
slug plates			x
freshwater ostracods			
small mammal teeth			
amphibian bone			
iron			

CONTEXT	40070			
SAMPLE	40321/B/31-33cm	40321/B/33-36cm	40321/B/36-39cm	40321/B/39-42cm
fish bone and teeth	x	x	x	x
tufa	x	x	x	x
molluscs	x	x	x	x
Bithynia opercula	x	x	x	x
slug plates	x	x	x	x
freshwater ostracods		x		
small mammal teeth			x	x
amphibian bone			x	
iron			x	x

Ecology	Actual tufa springs and associated pools, or redeposited from tufa springs

Ecology	Actual tufa springs and associated pools, or redeposited from tufa springs

CONTEXT	40144		
SAMPLE	40321/B/19-22cm	40321/B/23-27cm	40321/B/29-31cm
Ilyocypris gibba/bradyi			
Prionocypris zenkeri			
Potamocypris zschokkei			
Pseudocandona sp. (juv.)			
Fabaeformiscandona balatonica			

CONTEXT	40070			
SAMPLE	40321/B/31-33cm	40321/B/33-36cm	40321/B/36-39cm	40321/B/39-42cm
Ilyocypris gibba/bradyi		x		
Prionocypris zenkeri		o		
Potamocypris zschokkei				
Pseudocandona sp. (juv.)				
Fabaeformiscandona balatonica				

DISCUSSION AND CONCLUSIONS

The main value of the ostracods, where they occur, has been to assist in the palaeoenvironmental interpretation. Climatic information, made available through the MOTR method of Horne (2007), has also proven particularly useful and is emphasising the interglacial status of the Phase 3 sediments. The species *Ilyocypris quinculminata*, whose biostratigrapical distribution is now very well known, also suggests that the Phase 3 sediments at the site belong to the (?earlier) Hoxnian. Sediments further east (in Station Quarter South, interpreted here as equivalent to Phases 3 and 4 of the site sequence), have a virtually identical ostracod assemblage and likewise contain *Ilyocypris quinculminata*. They are not only formed under similar climatic/ecological conditions in the same time period, but are probably on lithostratigraphic grounds a lateral continuation of the same deposit.

Chapter 12

Pollen

by Charles Turner, Barbara Silva and Francis Wenban-Smith

INTRODUCTION

There were three main phases of pollen investigation at the site: (1) as part of the initial field evaluation; (2) during the main fieldwork programme; and (3) post-excavation assessment and analysis. As mentioned previously (Chapter 3), several samples were evaluated during the initial field evaluation of December 2003, some of which were found to apparently contain moderately abundant and well-preserved pollen remains, although concerns about the taphonomy were raised (Oxford Archaeology 2004). This positive result was one of the factors in recognising the site as important and justifying more thorough investigation, with a focus on pollen analysis through the sequence. Therefore, one strand of investigation in the following larger-scale excavation was recovery of monolith series through the deposits that were thought to be pollen-bearing, with the intention of constructing detailed pollen diagrams through the site sequence.

The site was visited by a range of specialists in late March 2004, following discovery of the elephant skeleton, and it was decided then to carry out some on-the-spot sampling to further evaluate for pollen preservation: firstly, in the vicinity of the newly discovered elephant; secondly, in other clayey horizons in the Phase 5 clay-laminated sands that were at that time thought to be polleniferous; and thirdly, in as-yet-uninvestigated horizons of Phases 6 and 9. Analysis of this second set of samples established reasonable preservation and countable quantities of pollen in the same brown organic-rich deposit (context 40078) that contained the elephant skeleton. However pollen remains were not found in samples from any other deposits, including those from Phase 5, where pollen had initially been identified in the original phase of evaluation. Therefore, a set of duplicate analyses was carried out, processing and analysing parts of the same samples as used for the initial pollen evaluation. The results of this re-analysis again came back negative, suggesting that some of the positive results of the initial evaluation, particularly those that had apparently shown pollen in excellent condition, were probably due to modern contamination rather than being ancient pollen. Further pollen investigation took place towards the end of the excavation in July-August 2004, when the very dark and organic-rich context 40158 was exposed in the central part of the site, forming the Phase 6 deposits at the base of the

synclinal 'skateboard ramp' (Fig. 3.22). The results of these analyses are reported on together (Turner, below).

Following completion of fieldwork, a much more systematic investigation took place, assessing for pollen remains throughout previously unevaluated horizons, namely Phases 1, 2 and 7 (lower part) of the site sequence. In addition, a vertical series of samples was taken from a monolith series through the full thickness of the Phase 6 clay and assessed for pollen. Although pollen presence had been confirmed in context 40158 and in the particular thin brown organic-rich bed (context 40078) associated with the elephant skeleton, the prevalence and preservation of pollen through most of the (several metres thick) Phase 6 deposits remained uncertain. These Phase 6 assessment samples were closely spaced in the nearest monoliths to the elephant skeleton, samples <40364> and <40365>, where the sequence above and below the elephant horizon was preserved, hoping to put the pollen profile from the elephant horizon in context within a longer pollen sequence. The results of this more detailed assessment are discussed further by Silva (below). This chapter on pollen analysis then concludes with a review of the more significant results of the various pollen analyses carried out, and discussion of the implications for climate, environment and dating.

INITIAL FIELD EVALUATION

Five samples were selected for initial evaluation in December 2003 from the central part of the main west-facing section at Log 40011, where the greatest range of phases of the site sequence (as understood at that time) was exposed (Fig 12.1a). Two samples were assessed from Phase 4, which was particularly clayey here, and which was thought to perhaps be a basal lake clay from the start of an interglacial period, and one from each of the overlying Phases 5, 6 and 7. The samples were analysed at Oxford Archaeology North, and this summary is taken from the evaluation report (Oxford Archaeology 2004a, Appendix 1).

Samples were prepared using the standard techniques of Potassium Hydroxide, acetolysis and hot Hydro-fluoric Acid treatment (Faegri and Iversen 1989). The residues were mounted in silicone oil and examined with an Olympus BH-2 microscope using x400 magnification routinely and x1000 for critical grains. Counting

Table 12.1 Initial pollen evaluation results from December 2003 fieldwork (Oxford Archaeology 2004a, Appendix 1), all counts based on examination of two slides

Phase [original deposit group]	Context	Sample <>	Sampling notes	Results
Phase 7 [6]	40043	40006	-	A single pollen grain (Poaceae)
Phase 6 [5]	40039	40007	From upper dark brown/purple brecciated level, the presumed palaeo-landsurface	Eight grass (Poaceae) pollen grains and 29 indeterminate grains
	40100 [originally 40029]	40008	From purer, thicker blue-grey clay to south of Log 40011	Not analysed
Phase 5 [4]	40025	40009 *	From olive-grey clay lamination	Eighty six pollen grains were identified on the two slides from this sample. The pollen preservation was excellent. Pollen from herbaceous plants dominated the assemblage and included grass (Poaceae), nettle (*Urtica*), goosefoot family (Chenopodiaceae) and stitchwort family (Caryophyllaceae) grains. Some birch (*Betula*) alder (*Alnus*) and hazel (*Corylus*) pollen
		40010	From olive-grey clay lamination	Not analysed
Phase 4 [3]	40026	40011	Top of grey/orange clay	Not analysed
		40012 *	Middle of grey/orange clay	Nineteen pollen grains, mainly from herbaceous taxa, were identified on the two slides from this sample. These included pollen from grasses (Poaceae), nettles (*Urtica*), plantains (*Plantago*) and Ericales. Single grains of birch (*Betula*) and oak (*Quercus*) were also recorded. Pollen preservation was again excellent
		40013	Bottom of grey/ orange clay	Not analysed
	40027	40014	Middle of lower, more sandy/ pebbly clay	Not analysed
		40015 *	Bottom of lower, more sandy/ pebbly clay	Fifty two pollen grains, mainly from herbaceous taxa, were identified on the two slides from this sample. They included pollen from grasses (Poaceae), nettles (*Urtica*), dead nettle family (Lamiaceae) and rock-rose (*Helianthemum*). Some birch (*Betula*), elm (*Ulmus*) and pine (*Pinus*) pollen was also recorded. Pollen preservation was again excellent

Subsequent re-analysis of the samples marked * established that the apparently positive results might be due to modern contamination (see Table 12.2)

continued until a sum of at least 50-100 pollen grains from land pollen types had been reached on two or more complete slides, to reduce the possible effects of differential dispersal under the coverslip (Brooks and Thomas 1967). If pollen was very sparse, counting continued until two complete slides had been assessed. Pollen identification was carried out using the standard keys of Faegri and Iversen (1989) and Moore *et al.* (1991) and a small reference collection. Because the samples were only being assessed, pollen grains not identified rapidly were recorded in either larger categories eg Asteraceae (Daisy-type) and Lactuaceae (Dandelion-type) or as undifferentiated grains. Indeterminate grains were recorded using groups based on those of Birks (1973). The results (Table 12.1) indicated low pollen frequency, as might be expected at the start of the climatic amelioration after a period of glaciation, but excellent preservation in three particular samples: <40009> (Phase 5), <40012>

(Phase 4) and <40015> (Phase 4), where pollen grains were also more abundant. Contamination was ruled out because the samples were taken from freshly exposed faces several metres below the present day land surface, the laboratory equipment was carefully cleaned, and it was thought unlikely for any atmospheric contamination to have taken place in late December, when the samples were collected, or in January, when the slides were prepared. It was therefore recommended that more detailed sampling should take place.

ANALYSES DURING FIELDWORK
by Charles Turner

Following from discovery of the elephant skeleton, a specialist meeting was convened on-site on 23rd of March 2004, during which several samples were collected

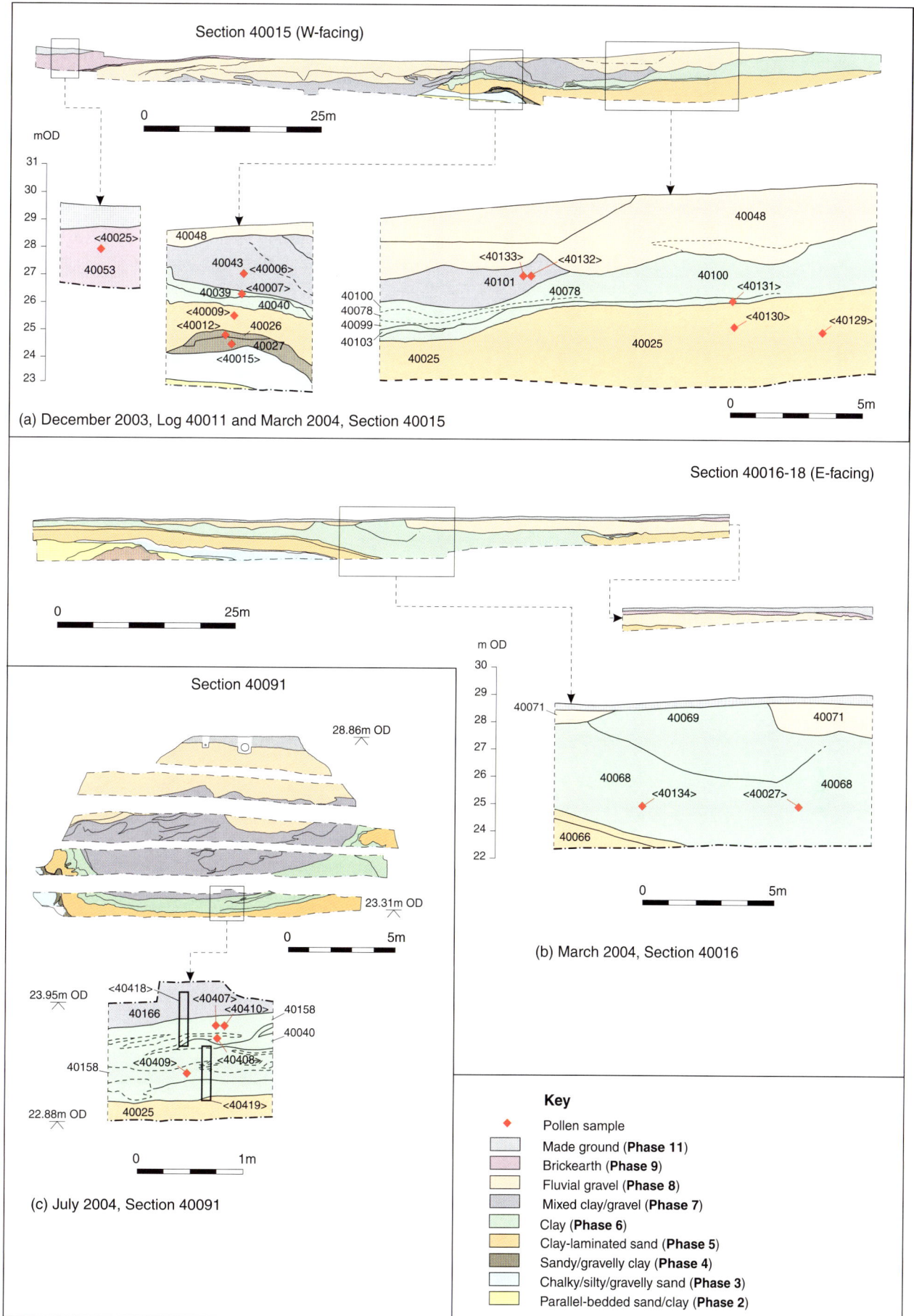

Figure 12.1 Pollen sampling locations during fieldwork

for immediate pollen analysis (Table 12.2; Fig. 12.1). Two samples, <40132> and <40133>, were collected from the same dark brown organic-rich clayey bed (context 40078) that contained the elephant, from amongst the scatter of the elephant bones to remove any doubt as to their association with the elephant (Fig. 12.1a). Three more samples, <40129>, <40130> and <40131>, were collected from particularly clayey beds *c* 10-20mm thick in the upper part of the Phase 5 deposits (context 40025) in the main west-facing section 40015 (Fig. 12.1a). These appeared in the field to have been subjected to fairly rigorous oxidation processes and not to have good potential for pollen preservation, and so it was decided to verify the apparently positive results from the initial evaluation. Two samples were collected from the main east-facing section 40016 (Fig. 12.1b), one from the tufaceous channel (<40134>, from context 40070 and the other from the upper part of the overlying grey clay of Phase 6 (<40027>, from context 40069. The former of these was highly calcareous, with relatively abundant molluscan and small vertebrate remains and, when a sample of this material was sieved, it also produced discrete pebbles of grey silty clay, which were then also prepared for pollen analysis. One sample, <40025>, was collected from the

Phase 9 brickearth (context 40053) capping the sequence at the north end of the site (Fig. 12.1a). Standard methods were used to prepare the samples for pollen analysis, including treatments with 10% NaOH, 7% HCl, 70% HF and Erdtman acetolysis (Faegri and Iversen 1989) as well as treatment with sodium pyrophosphate (NaP_2O_7) (Bates *et al.* 1978).

The results of these analyses were mostly negative (Table 12.2). All the samples from the Phase 5 clay-laminated sands were devoid of any trace of fossil pollen. Indeed, apart from the exotic *Lycopodium* spores routinely added during sample preparation, the slides prepared from these samples contained no, or virtually no, organic residues (ie nothing that absorbed safranin stain). This finding was entirely consistent with the field evidence that these thin silt layers had been subjected to oxidation processes incompatible with pollen survival. Likewise, the samples from Phase 9 (context 40053), the upper part of Phase 6 (context 40069) and the clay pebbles from the tufaceous channel-fill (context 40070) also contained no pollen whatsoever.

Since the negative results from Phase 5, samples <40129>-<40131> conflicted with the positive result from the initial evaluation of the same deposit, sample

Table 12.2 Pollen analyses during the 2004 excavation; samples marked * were re-analysed from the initial evaluation (see Table 12.1)

Phase of analysis	Site sequence Phase	Context	Sample/s <>	Sampling notes	Results
March-April 2004	9	40053	40025	From thicker brickearth at north end of main west-facing section 40015	No pollen
	6	40069	40027	From upper part of thick grey clay in central part of main east-facing section 40016	No pollen
	6b	40070 (clayey pebbles)	40134	Clayey pebbles sieved from context 40070 were investigated for pollen	No pollen
	6	40078	40132 40133	From darker brown organic-rich clay bed containing *Palaeoloxodon* skeleton	Countable quantities of fossil pollen and spores, although assemblages have been affected by differential destruction (Table 12.3)
	5	40025	40129 40130 40131	From grey clay-silt laminations within otherwise sandy/gravelly deposits at south end of main west-facing Section 40015	No pollen
May-July 2004	5 4	40025 40026 40027	40009* 40012* 40015*	Re-analysis of further sediment from same samples as initially assessed – see Table 12.1	No pollen
July-August 2004	6	40158	40407 40408 40409	Three small pollen samples from very dark organic-rich bed in Section 40091	Some amorphous non-cellular organic detritus; no pollen apart from some very corroded conifer grains, probably *Pinus*, and possibly of derived Tertiary origin
			40410	Slightly larger sample for plant macro remains from same level as <40407>	Single *Azolla* spore, but no Pleistocene pollen or other identifiable plant macro-fossils

<40009> (Table 12.1), a duplicate analysis was carried out on surviving sediment from the original sample. This was conducted alongside re-analysis of surviving sediment from the two other original samples that had also produced apparently positive results, samples <40012> and <40015> (Table 12.1). All three duplicate analyses came back entirely negative, suggesting that the initial positive results may have been due to laboratory contamination at some stage in the preparation process.

The only samples that produced any Pleistocene pollen were the two samples <40132> and <40133> from immediately beside the elephant. Here there were countable quantities of fossil pollen and spores (Table 12.3), although it is clear that in both cases the assemblages have been affected by differential destruction. Also present on the slides were a few fragments of cypselas (ie achenes) of the hemp agrimony *Eupatorium cannabinum*. The most important thing to realise about both these samples is that their palynomorph assemblages have undergone considerable differential destruction, through corrosion probably caused by both biological oxidation of pollen grain walls by bacteria and by direct chemical oxidation when drying out of the sediments introduced aerobic conditions. Differential destruction/preservation of pollen grains and spores has been extensively studied by Havinga (1964; 1967; 1985). The general conclusions of this and other work are summarised by Birks and Birks (1980). Resistance to corrosion depends substantially on the sporopollenin content of the outer walls (exines) of pollen grains and spores. It is clear that taxa noted by Havinga as having a high percentage content of sporopollenin, namely fern spores, pine *Pinus*, lime *Tilia*, alder *Alnus* and hazel *Corylus*, are precisely those which are best represented in the Southfleet assemblages. Other taxa, which would also be expected to be present, even abundant, in temperate forest assemblages, such as oak *Quercus*, elm *Ulmus*, ash *Fraxinus* and some herb pollen types, are barely represented or absent; and these are taxa which contain lower sporopollenin percentages in their pollen exines. It is also pertinent that the pollen and spore taxa that survive and are countable also have distinctive features or shapes (eg, the wings of *Pinus*, the arci of *Alnus*, the sunken pores of *Tilia* and the size and heavy sculpture of *Polypodium* spores). This tends to make them instantly recognisable, even when somewhat corroded.

It follows that, although the Southfleet pollen assemblages (Table 12.3) have been calculated as conventional percentages based on the sum of total land pollen, excluding spores (though corroded grains *etc*, are given as percentages of all grains) it is meaningless to make direct comparisons with percentage data shown in pollen diagrams from interglacial sites such as Hoxne (West 1956) or Marks Tey (Turner 1970), where the pollen was well-preserved. However, the Southfleet pollen assemblages do make it clear that context 40078 was deposited (and the *Palaeoloxodon* was active) under fully temperate vegetational and climatic conditions probably in or immediately adjacent to a swampy alder carr. This is suggested not only by the presence of

relatively abundant *Alnus* pollen, but also the preservation of frequent *Alnus* sieve plates derived from the breakdown of wood fragments within 40078. Larger fragments thought to be derived from rotted wood were also noted in the sediments during excavation but did not prove to retain sufficient cellular structure for identification (see Appendix 4). The few surviving traces of *Quercus*, *Ulmus*, *Tilia* and *Corylus* pollen, as well as *Pinus*, suggest that higher and drier ground in the wider site vicinity supported the typical mixed, largely deciduous forest that normally characterises the early-temperate substage of interglacial periods. However, the less frequent shrub taxa of the forest, such as holly *Ilex*, ivy *Hedera* or Type X are not represented, being either too poorly preserved or too sparsely produced to survive and be found in these weathered assemblages.

Given the preservation of pollen and spore assemblages in samples <40132> and <40133>, it was thought likely that further samples from the same stratum in other nearby faces of the excavation would also be likely to yield such assemblages. Therefore, four extra samples, <40407>-<40410>, were taken on a field visit in late July 2004, from the very dark organic-rich bed of context 40158, in the south-facing section 40091 of Trench B (Table 12.2; Fig. 12.1c). This context was thought to be equivalent stratigraphically to the elephant horizon of context 40078, and also appeared to have good potential for pollen preservation. This hope was strengthened by the presence of a distinctive fragment of a megaspore of the water fern *Azolla filiculoides* in sieved residues from sample <40410>. All four samples yielded varying amounts of organic detritus, mostly amorphous rather than cellular, and occasional, often crumpled and/or corroded, bisaccate conifer pollen grains. Most might be ascribed to *Pinus* (of undeterminable age), but some certainly appear to be reworked Tertiary palynomorphs, as do a few small, indeterminate tricolpate pollen grains. However, none of these samples yielded clear, let alone countable, Pleistocene pollen assemblages. Even pollen taxa like *Alnus* or Pteropsida (Filicales) spores, both relatively frequent in samples <40132> and <40133>, relatively resistant to destruction and easily identifiable even when poorly preserved, appeared to be absent from these pollen preparations. Sample <40410> had originally been collected for processing for plant macrofossils, but even when a much larger sample was sieved and examined, only two minute fragments of wood were seen and no further sign of *Azolla*.

The presence of *Azolla filiculoides*, however sparse the evidence, is nevertheless of stratigraphical importance. If we discount the early Middle Pleistocene ('Cromer Complex') interglacials as too old to be taken into account here, *A. filiculoides* apparently persisted in the European fossil record beyond the Elsterian/Anglian glacial Stage for a further two interglacial intervals: the Hoxnian/Holsteinian Interglacial, usually correlated with Marine Isotope Substage 11c in the deep-sea stratigraphy, and the Dömnitz Interglacial (Menke 1968; Erd 1970) which corresponds to part of MIS 9. In

Table 12.3 Pollen spectra from two samples from brown organic-rich band (context 40078) associated with *Palaeoloxodon* remains

	Sample <40132>		Sample <40133>	
	No. of grains counted	*Percentage of total land pollen*	*No. of grains counted*	*Percentage of total land pollen*
Trees & shrubs				
Betula	1	1%		
Pinus	8.5	10%	4.5	20%
Ulmus	1	1%		
Quercus	1	1%	1	4%
Tilia	3	3.5%	1	4%
Alnus	64	73%	12	45%
Corylus	8	9%	5	20%
Herbs				
Poaceae	1	1%	3	11%
Total land pollen	87.5		26.5	
Spores				
Pteropsida (Filicales)	130	149%	16	60%
Osmunda			1	4%
Polypodium	1	1%		
Crumpled & broken grains	40	(14%)	9	(15%)
Corroded grains	19	(3%)		
Exotic	86		40	
***Eupatorium cannabinum* cypsela fragments**	2			
Sieve plates derived from *Alnus* wood	70	3		

Britain *A. filiculoides* has been regularly recorded from Hoxnian sites such as Marks Tey (Turner 1970) and Hoxne itself (West 1956). More recently it has been recorded from Cudmore Grove, East Mersea (Roe 1999; Schreve 2001a) and Hackney Downs, London (Green *et al.* 2006), both sites that are generally believed to correlate with MIS 9 (Roe *et al.* 2009). After that interval *A. filiculoides* appears to have become extinct in Europe (though its megaspores are sometimes reworked into Eemian or Holocene deposits) until it was reintroduced from North America back into Europe in historical times, so that it is again widespread today.

It is possible to speculate a little further on the age of these deposits, from the pollen profiles of the two polleniferous samples, <40132> and <40133>, from context 40078. The absence of any trace of fir *Abies* and also hornbeam *Carpinus*, though the latter but not the former might be relatively susceptible to corrosion, make an early-temperate rather than a late-temperate substage of an interglacial, as defined by Turner and West (1968), more probable. In south-east England *Alnus* expands during the Hoxnian Interglacial to its highest percentages in pollen diagrams in subzone Ho IIb, but remains abundant also in Ho IIc and then through zone Ho III, the late-temperate substage. By contrast, at Cudmore Grove, thought to correlate with MIS 9 (Roe 1999), *Alnus* appears to have low values in the early-temperate zone II and much higher ones in the late-temperate (zone III), although this may simply be the result of a hiatus in the stratigraphy, cutting out the later part of zone II. On the balance of the palaeobotan-

ical data, it is therefore probable that the lower parts of Phase 6 at Southfleet, where sampled for pollen in context 40078 by the elephant skeleton, represent the early-temperate zone of the Hoxnian Interglacial, Ho II (MIS 11c).

POST-EXCAVATION ASSESSMENT AND ANALYSIS
by Barbara Silva

Once the excavation was finished, a much more systematic investigation for pollen preservation was carried out, focusing on horizons which had not yet been fully investigated. It was hoped that the darker bed (context 40057) of the Phase 1 'tilted block' sequence might contain pollen evidence, and therefore that pollen analysis could provide some information on this otherwise enigmatic sequence. Consequently eleven samples were chosen through the Phase 1 sequence, as well as seven through the overlying Phase 2 deposits (Table 12.4; Fig. 12.2a, b). Fifteen samples were assessed from the monolith sequence through the Phase 6 grey clay south of Trench D, where the major flint concentration had been found (Table 12.5; Fig. 12.2c). Eleven samples were assessed through deposits of Phase 6b, the tufaceous channel-fill, focusing upon the slightly browner and more organic-looking fine silty sand of context 40144, which looked more promising than other deposits of the channel-fill sequence for pollen preservation (Table 12.6; Fig. 12.3). Fifty-seven samples were

Table 12.4 Pollen assessment: Phases 1–2, Section 40016

Phase	Site area	Context	Source sample <>	Type	Site sample/ sub-sample <>	Pollen sample no.	Pollen concen- tration	Pollen preservation	Pollen taxa
2	South (E)	40065	40095	Monolith	40095/A/12-13cm	91	n/a	n/a	Not present
2	South (E)	40065	40067	Spot	40067	92	n/a	n/a	Not present
2	South (E)	40065	40078	Monolith	40078/A/49-50cm	93	n/a	n/a	Not present
2	South (E)	40064	40076	Monolith	40076/A/34-35cm	94	n/a	n/a	Not present
2	South (E)	40064	40073	Monolith	40073/A/17-18cm	95	n/a	n/a	Not present
2	South (E)	40064	40098	Monolith	40098/A/18-19cm	96	n/a	n/a	Not present
2	South (E)	40064	40097	Monolith	40097/A/12-13cm	97	n/a	n/a	Not present
1	South (E)	40059	40063	Spot	40063	98	n/a	n/a	Not present
1	South (E)	40059	40094	Monolith	40094/A/25-26cm	99	n/a	n/a	Not present
1	South (E)	40059	40093	Monolith	40093/A/24-25cm	100	n/a	n/a	Not present
1	South (E)	40059	40092	Monolith	40092/A/45-46cm	101	n/a	n/a	Not present
1	South (E)	40058	40085	Monolith	40085/A/5-6cm	102	very low	moderate	*Rumex*
1	South (E)	40058	40084	Monolith	40084/A/26-27cm	103	n/a	n/a	Not present
1	South (E)	40058	40083	Monolith	40083/A/57-58cm	104	n/a	n/a	Not present
1	South (E)	40057	40082	Monolith	40082/A/40-41cm	105	n/a	n/a	Not present
1	South (E)	40057	40082	Monolith	40082/A/57-58cm	106	n/a	n/a	Not present
1	South (E)	40057	40061	Spot	40061	107	n/a	n/a	Not present
1	South (E)	40057	40081	Monolith	40081/A/20-21cm	108	n/a	n/a	Not present

Table 12.5 Pollen assessment: Phase 6 (grey clay, south of Trench D), Section 40015

Phase	Site area	Context	Source sample <>	Type	Site sample/ sub-sample <>	Pollen sample no.	Pollen concen- tration	Pollen preservation	Pollen taxa
6	South (W)	40100	40196	Monolith	40196/A/5-6cm	1	n/a	n/a	Not present
6	South (W)	40100	40196	Monolith	40196/A/21-22cm	2	n/a	n/a	Not present
6	South (W)	40100	40196	Monolith	40196/A/41-42cm	3	n/a	n/a	Not present
6	South (W)	40100	40195	Monolith	40195/A/0-1cm	4	very low	moderate	*Ranunculus*
6	South (W)	40100	40195	Monolith	40195/A/16-17cm	5	n/a	n/a	Not present
6	South (W)	40100	40195	Monolith	40195/A/32-33cm	6	n/a	n/a	Not present
6	South (W)	40100	40194	Monolith	40194/A/4-5cm	7	n/a	n/a	Not present
6	South (W)	40100	40194	Monolith	40194/A/25-26cm	8	n/a	n/a	Not present
6	South (W)	40100	40194	Monolith	40194/A/45-46cm	9	n/a	n/a	Not present
6	South (W)	40100	40193	Monolith	40193/A/5-6cm	10	n/a	n/a	Not present
6	South (W)	40100	40193	Monolith	40193/A/15-16cm	11	n/a	n/a	Not present
6	South (W)	40100	40193	Monolith	40193/A/30-31cm	12	n/a	n/a	Not present
6	South (W)	40039	40193	Monolith	40193/A/33-34cm	13	n/a	n/a	Not present
6	South (W)	40039	40193	Monolith	40193/A/41-42cm	14	n/a	n/a	Not present
6	South (W)	40039	40193	Monolith	40193/A/49-50cm	15	n/a	n/a	Not present

Table 12.6 Pollen assessment: Phases 6a–6b (tufaceous channel), Trench XIII and Section 40064

Phase	Site area	Context	Source sample <>	Type	Site sample/ sub-sample <>	Pollen sample no.	Pollen concen- tration	Pollen preservation	Pollen taxa
6b	Central (E)	40144	40311	Bulk	40311/A	23	very low	poor	Poaceae
6b	Central (E)	40144	40321	Monolith	40321/A/5-6cm	24	n/a	n/a	Not present
6b	Central (E)	40144	40321	Monolith	40321/A/14-15cm	25	n/a	n/a	Not present
6b	Central (E)	40144	40321	Monolith	40321/A/23-24cm	26	n/a	n/a	Not present
6a	Central (E)	40103	40321	Monolith	40321/A/44-45cm	27	n/a	n/a	Not present
6b	Central (E)	40143	40250	Spot-sed	40250/A	85	n/a	n/a	Not present
6b	Central (E)	40143	40249	Spot-sed	40249/A	86	n/a	n/a	Not present
6b	Central (E)	40143	40248	Spot-sed	40248/A	87	n/a	n/a	Not present
6b	Central (E)	40070	40281	Monolith	40281/A/4-5cm	88	n/a	n/a	Not present
6b	Central (E)	40070	40281	Monolith	40281/A/12-13cm	89	n/a	n/a	Not present
6a	Central (E)	40103	40281	Monolith	40281/A/22-23	90	very low	moderate	*Pinus*, Poaceae

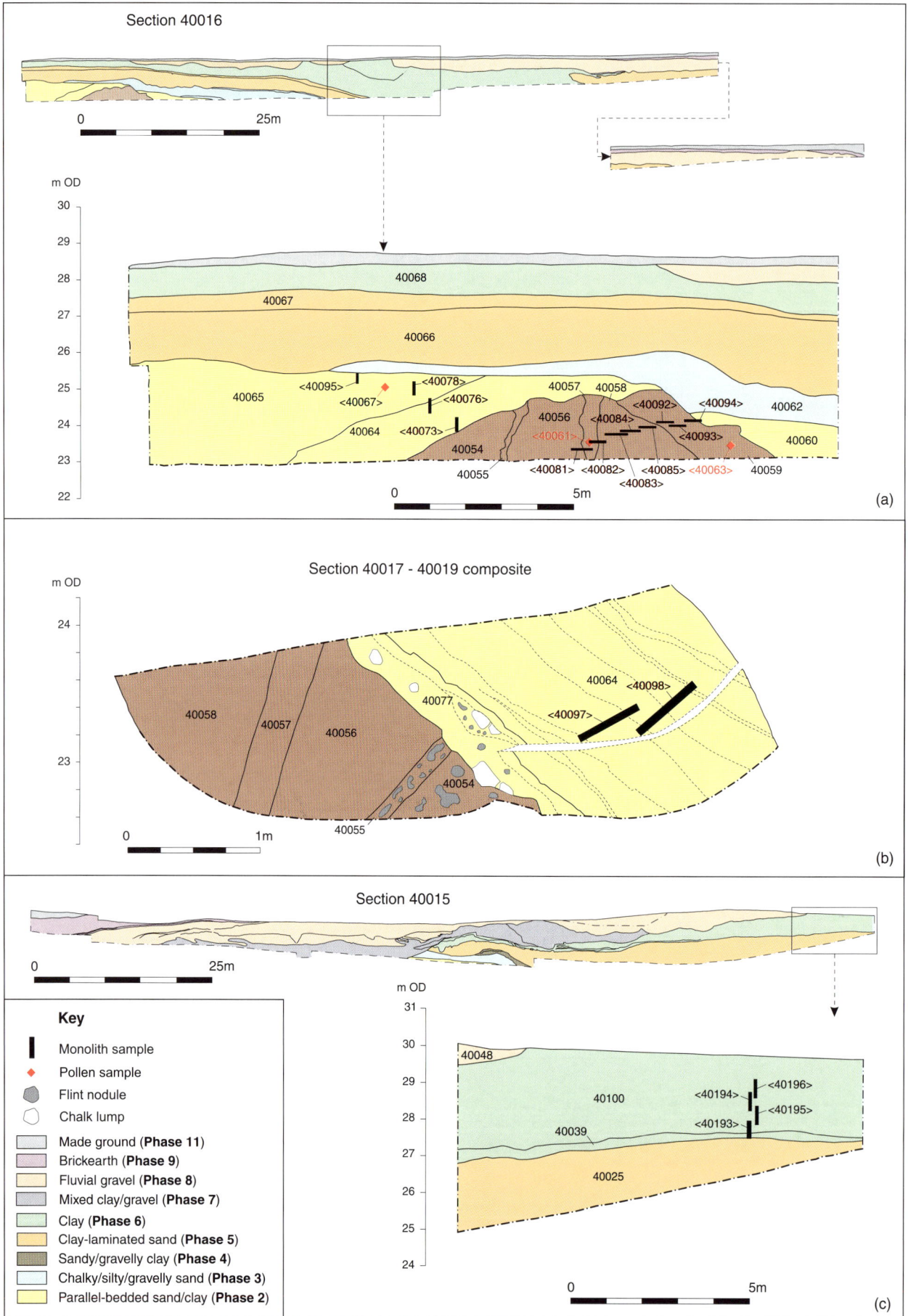

Figure 12.2 Pollen sampling locations, post-excavation assessment and analysis: (a) Phases 1–2, Section 40016; (b) Phase 2, Section 40017; (c) Phase 6, south of Trench D, Section 40015

Table 12.7 Pollen assessment: Phases 6a–6 (elephant horizon and surrounding grey clay), Section 40085

Phase	Site area	Context	Source sample <>	Type	Site sample/ sub-sample <>	Pollen sample no.	Pollen concen-tration	Pollen preservation	Pollen taxa
6	Central (E)	40162	40340	Monolith	40340/A/0-1cm	28	n/a	n/a	Not present
6	Central (E)	40162	40340	Monolith	40340/A/12-13cm	29	n/a	n/a	Not present
6	Central (E)	40162	40340	Monolith	40340/A/24-25cm	30	n/a	n/a	Not present
6	Central (E)	40162	40340	Monolith	40340/A/39-40cm	31	very low	poor	*Betula*
6	Central (E)	40162	40340	Monolith	40340/A/54-55cm	32	n/a	n/a	Not present
6	Central (E)	40162	40340	Monolith	40340/A/69-70cm	33	n/a	n/a	Not present
6	Central (E)	40162	40342	Monolith	40342/A/3-4cm	34	n/a	n/a	Not present
6	Central (E)	40162	40342	Monolith	40342/A/18-19cm	35	n/a	n/a	Not present
6a	Central (E)	40103	40342	Monolith	40342/A/30-31cm	36	n/a	n/a	Not present
6a	Central (E)	40103	40342	Monolith	40342/A/33-34cm	37	n/a	n/a	Not present
6a	Central (E)	40103	40342	Monolith	40342/A/48-49cm	38	n/a	n/a	Not present
6a	Central (E)	40103	40342	Monolith	40342/A/66-67cm	39	n/a	n/a	Not present
6	Central (E)	40162	40344	Monolith	40344/A/0-1cm	40	n/a	n/a	Not present
6	Central (E)	40162	40344	Monolith	40344/A/9-10cm	41	n/a	n/a	Not present
6	Central (E)	40162	40344	Monolith	40344/A/18-19cm	42	n/a	n/a	Not present
6	Central (E)	40162	40344	Monolith	40344/A/27-28cm	43	very low	poor	*Plantago*
6	Central (E)	40162	40344	Monolith	40344/A/39-40cm	44	n/a	n/a	Not present
6	Central (E)	40162	40344	Monolith	40344/A/48-49cm	45	n/a	n/a	Not present
6	Central (E)	40162	40344	Monolith	40344/A/60-61cm	46	very low	poor	*Alnus*, Poaceae
6	Central (E)	40162	40345	Monolith	40345/A/0-1cm	47	n/a	n/a	Not present
6	Central (E)	40162	40345	Monolith	40345/A/12-13cm	48	n/a	n/a	Not present
6	Central (E)	40162	40345	Monolith	40345/A/24-25cm	49	n/a	n/a	Not present
6	Central (E)	40162	40345	Monolith	40345/A/36-37cm	50	n/a	n/a	Not present
6	Central (E)	40162	40345	Monolith	40345/A/48-49cm	51	n/a	n/a	Not present
6	Central (E)	40162	40345	Monolith	40345/A/63-64cm	52	n/a	n/a	Not present
6	Central (E)	40100	40352	Monolith	40352/A/3-4cm	53	very low	poor	Poaceae
6	Central (E)	40099	40352	Monolith	40352/A/12-13cm	54	n/a	n/a	Not present
6	Central (E)	40099	40352	Monolith	40352/A/18-19cm	55	very low	poor	Poaceae
6	Central (E)	40099	40352	Monolith	40352/A/21-22cm	56	n/a	n/a	Not present
6	Central (E)	40099	40352	Monolith	40352/A/27-28cm	57	n/a	n/a	Not present
6	Central (E)	40099	40352	Monolith	40352/A/33-34cm	58	n/a	n/a	Not present
6	Central (E)	40099	40352	Monolith	40352/A/42-43cm	59	very low	moderate	*Picea*
6	Central (E)	40099	40352	Monolith	40352/A/51-52cm	60	n/a	n/a	Not present
6	Central (E)	40099	40353	Monolith	40353/A/0-1cm	61	n/a	n/a	Not present
6	Central (E)	40099	40353	Monolith	40353/A/9-10cm	62	n/a	n/a	Not present
6	Central (E)	40099	40353	Monolith	40353/A/21-22cm	63	n/a	n/a	Not present
6	Central (E)	40099	40353	Monolith	40353/A/36-37cm	64	n/a	n/a	Not present
6a	Central (E)	40103	40353	Monolith	40353/A/44-45cm	65	very low	poor	Cyperaceae, Poaceae
6a	Central (E)	40103	40353	Monolith	40353/A/53-54cm	66	n/a	n/a	Not present
6	Central (El)	40100	40364	Monolith	40364/A/0-1cm	67	low	moderate	Pinus, *Corylus, Alnus, Tilia,* Poaceae
6	Central (El)	40100	40364	Monolith	40364/A/6-7cm	68	low	moderate	*Alnus, Picea, Corylus,* Poaceae, Polypodiaceae
6	Central (El)	40100	40364	Monolith	40364/A/12-13cm	69	n/a	n/a	Not present
6	Central (El)	40100	40364	Monolith	40364/A/18-19cm	70	low	moderate	*Betula*
6	Central (El)	40078	40364	Monolith	40364/A/27-28cm	71	n/a	n/a	Not present
6	Central (El)	40078	40364	Monolith	40364/A/30-31cm	72	very low	moderate	Asteraceae
6	Central (El)	40078	40364	Monolith	40364/A/33-34cm	73	n/a	n/a	Not present
6	Central (El)	40078	40364	Monolith	40364/A/36-37cm	74	n/a	n/a	Not present
6	Central (El)	40078	40364	Monolith	40364/A/39-40cm	75	n/a	n/a	Not present
6	Central (El)	40078	40364	Monolith	40364/A/42-43cm	76	n/a	n/a	Not present
6	Central (El)	40078	40364	Monolith	40364/A/45-46cm	77	n/a	n/a	Not present
6	Central (El)	40099	40364	Monolith	40364/A/49-50cm	78	n/a	n/a	Not present
6	Central (El)	40099	40364	Monolith	40364/A/52-53cm	79	n/a	n/a	Not present
6	Central (El)	40078	40365	Monolith	40365/A/0-1cm	80	n/a	n/a	Not present
6	Central (El)	40078	40365	Monolith	40365/A/6-7cm	81	n/a	n/a	Not present
6	Central (El)	40099	40365	Monolith	40365/A/17-18cm	82	n/a	n/a	Not present
6a	Central (El)	40103	40365	Monolith	40365/A/29-30cm	83	n/a	n/a	Not present
6a	Central (El)	40103	40365	Monolith	40365/A/38-39cm	84	n/a	n/a	Not present

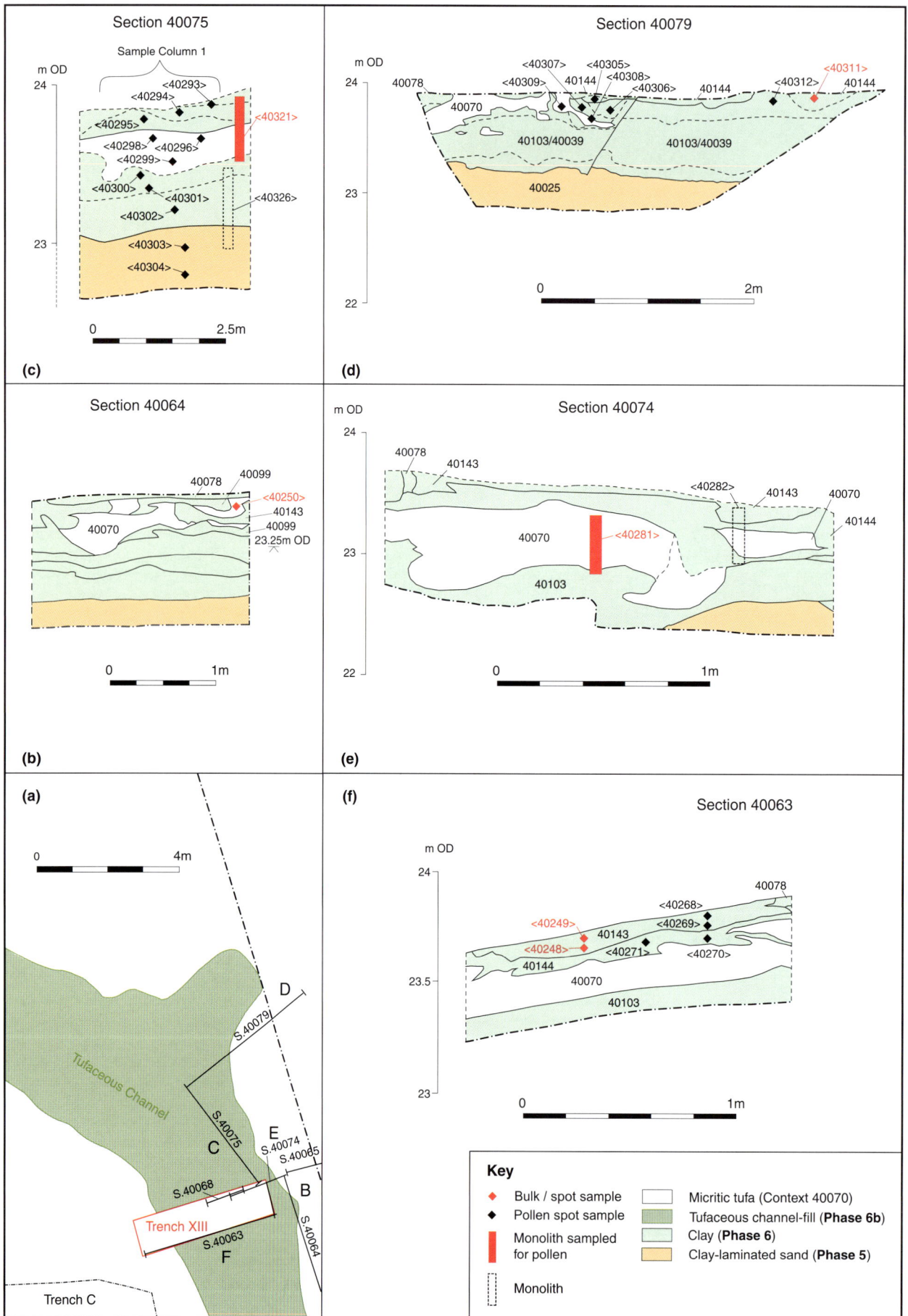

Figure 12.3 Pollen sampling locations, post-excavation assessment and analysis, Phases 6a–6b: (a) thumbnail of sampling locations; (b) Section 40064; (c) tufaceous channel, Section 40075; (d) tufaceous channel, Section 40079; (e) Trench XIII, Section 40074 (south-facing); and (f) Trench XIII, Section 40063 (north-facing)

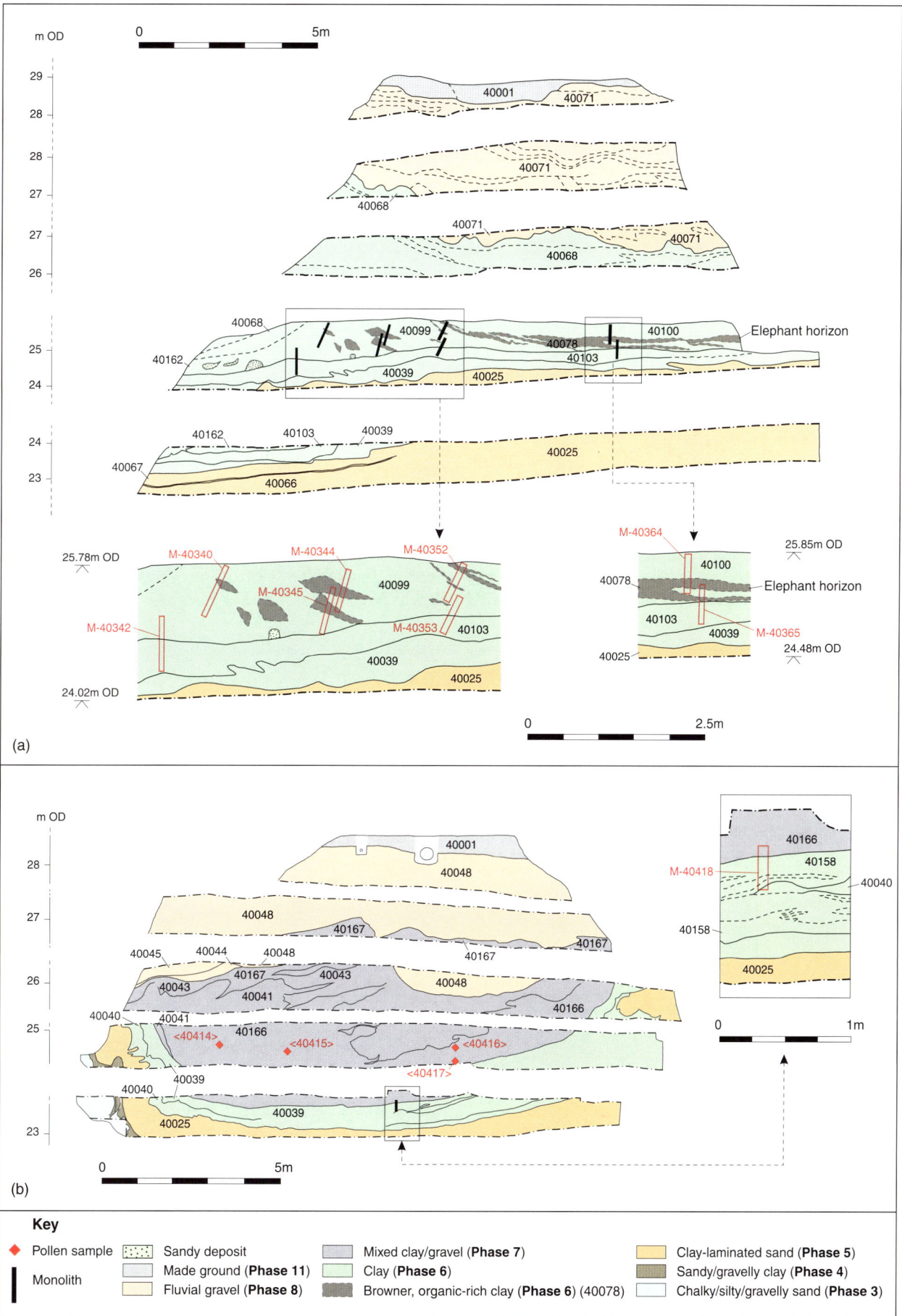

Figure 12.4 Pollen sampling locations, post-excavation assessment and analysis, Phases 6–7, including elephant horizon: (a) Trench C, Section 40085 (monoliths <40340>, <40342>, <40344>, <40345>, <40352>, <40353>, <40364> and <40365>); (b) Trench B, Section 40091

Table 12.8 Pollen assessment: Phases 6–7 (Trench B), Section 40091

Phase	Site area	Context	Source sample <>	Type	Site sample/ sub-sample <>	Pollen sample no.	Pollen concen- tration	Pollen preservation	Pollen taxa
7	Central (N)	40166	40414	Bulk	40414/A	16	n/a	n/a	Not present
7	Central (N)	40166	40415	Bulk	40415/A	17	n/a	n/a	Not present
7	Central (N)	40166	40416	Bulk	40416/A	18	n/a	n/a	Not present
7	Central (N)	40166	40417	Bulk	40417/A	19	n/a	n/a	Not present
7	Central (N)	40166	40418	Monolith	40418/A/0-1cm	20	n/a	n/a	Not present
6	Central (N)	40158	40418	Monolith	40418/A/5-6cm	21	very low	poor	*Pinus*, *Pediastrum*
6	Central (N)	40158	40418	Monolith	40418/A/14-15cm	22	low	moderate	Cheno-podiaceae, *Ulmus*, *Alnus*, Poaceae, *Betula*, Polypodiaceae

assessed from the monolith sequence through the Phase 6 clay in section 40085. These included samples from the lower parts of the clay (Phase 6a) below the elephant horizon, a deep sequence of samples through the clay above the elephant horizon, and two monoliths through a lateral equivalent of the elephant horizon, less than 5m from the elephant skeleton (Table 12.7; Fig. 12.4a). Finally, seven samples were assessed through the very dark context 40158 (Phase 6) seen in the south-facing

section 40091 of Trench B and the overlying fine silty sand of context 40166 (Table 12.8; Fig. 12.4b).

In total, 108 samples were assessed for pollen preservation. Preparation of the samples was carried out in a clean laboratory environment, with every care taken to minimise cross-contamination and the incorporation of modern pollen. The methodology used is summarised in the accompanying figure (Fig. 12.5). The slides were scanned to provide an assessment of the key pollen and

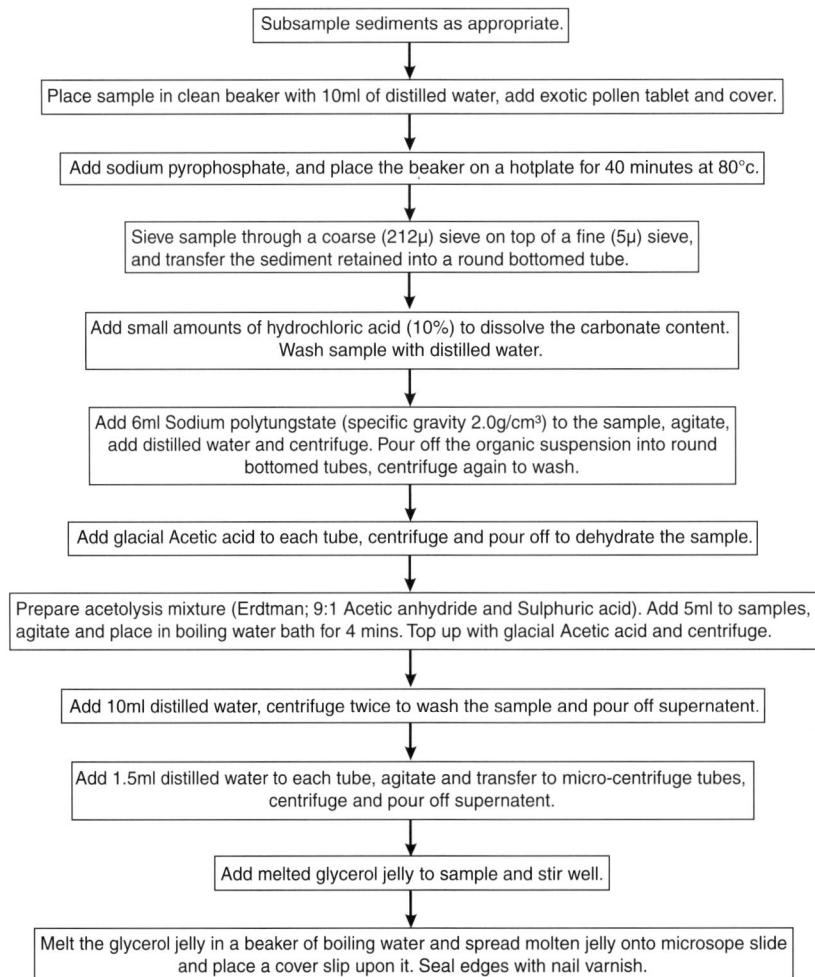

```
┌────────────────────────────────────────────┐
│ Subsample sediments as appropriate. │
└────────────────────────────────────────────┘
                    ↓
┌────────────────────────────────────────────────────────────────┐
│ Place sample in clean beaker with 10ml of distilled water, add   │
│ exotic pollen tablet and cover. │
└────────────────────────────────────────────────────────────────┘
                    ↓
┌────────────────────────────────────────────────────────────────┐
│ Add sodium pyrophosphate, and place the beaker on a hotplate for │
│ 40 minutes at 80°c. │
└────────────────────────────────────────────────────────────────┘
                    ↓
┌────────────────────────────────────────────────────────────────┐
│ Sieve sample through a coarse (212µ) sieve on top of a fine (5µ) │
│ sieve, and transfer the sediment retained into a round bottomed  │
│ tube. │
└────────────────────────────────────────────────────────────────┘
                    ↓
┌────────────────────────────────────────────────────────────────┐
│ Add small amounts of hydrochloric acid (10%) to dissolve the     │
│ carbonate content. Wash sample with distilled water. │
└────────────────────────────────────────────────────────────────┘
                    ↓
┌────────────────────────────────────────────────────────────────┐
│ Add 6ml Sodium polytungstate (specific gravity 2.0g/cm³) to the  │
│ sample, agitate, add distilled water and centrifuge. Pour off the│
│ organic suspension into round bottomed tubes, centrifuge again to│
│ wash. │
└────────────────────────────────────────────────────────────────┘
                    ↓
┌────────────────────────────────────────────────────────────────┐
│ Add glacial Acetic acid to each tube, centrifuge and pour off to │
│ dehydrate the sample. │
└────────────────────────────────────────────────────────────────┘
                    ↓
┌────────────────────────────────────────────────────────────────┐
│ Prepare acetolysis mixture (Erdtman; 9:1 Acetic anhydride and    │
│ Sulphuric acid). Add 5ml to samples, agitate and place in boiling│
│ water bath for 4 mins. Top up with glacial Acetic acid and       │
│ centrifuge. │
└────────────────────────────────────────────────────────────────┘
                    ↓
┌────────────────────────────────────────────────────────────────┐
│ Add 10ml distilled water, centrifuge twice to wash the sample and│
│ pour off supernatent. │
└────────────────────────────────────────────────────────────────┘
                    ↓
┌────────────────────────────────────────────────────────────────┐
│ Add 1.5ml distilled water to each tube, agitate and transfer to  │
│ micro-centrifuge tubes, centrifuge and pour off supernatent. │
└────────────────────────────────────────────────────────────────┘
                    ↓
┌────────────────────────────────────────────────────────────────┐
│ Add melted glycerol jelly to sample and stir well. │
└────────────────────────────────────────────────────────────────┘
                    ↓
┌────────────────────────────────────────────────────────────────┐
│ Melt the glycerol jelly in a beaker of boiling water and spread  │
│ molten jelly onto microsope slide and place a cover slip upon it.│
│ Seal edges with nail varnish. │
└────────────────────────────────────────────────────────────────┘
```

Figure 12.5 Pollen slide preparation method (Branch 2000, modified from Moore, Webb & Collinson 1991)

spore taxa, pollen and spore concentrations and preservation, and results are given for each group of assessed samples in the summary tables (Table 12.4 – Table 12.8). Pollen and spores were absent, or only present in very low quantities, in almost all of the samples processed.

Only three samples contained an interpretable presence of pollen grains, generally moderately preserved albeit in low concentrations. One of these was sample <40418/A/14-15cm> from context 40158, from the same horizon as two earlier samples <40407> and <40410> taken and studied while fieldwork was in progress, the latter of which produced the *Azolla* spore (see above). The other two were both from different depths within monolith <40364> (0-1cm and 6-7cm) both of these representing grey clay of context 40100 that overlay the browner, more organic-rich clay of context 40078 that contained the elephant skeleton, and which had previously produced countable pollen data (see above). Curiously, the assessment samples thought to be equivalent to the elephant horizon, between 27 and 47cm depth in monolith <40364>, and between 1 and 5cm depth in monolith <40365>, produced minimal pollen remains in this phase of investigation. Nonetheless, there were general similarities in the taxa present in the polleniferous assessment samples, with similar indications of mixed woodland and alder carr in the landscape. Despite the low concentrations noted, making it unlikely that a statistically significant pollen count (ie 300 Total Land Pollen) would be practically obtainable from these samples, further analyses (and counting) were carried out for the polleniferous horizons in monoliths <40364> and <40418>.

Thus, eleven samples were analysed in more detail. These comprised seven from the upper part of monolith <40364> between 0 and 25cm from its top, representing context 40100 directly above the elephant horizon (see Table 12.9) and four from the middle part of monolith <40418> between 5 and 20cm from its top, representing context 40158. This is thought to be a broad lateral equivalent of the elephant horizon (see Table 12.10).

Overall the pollen concentrations in these samples were low (between 62 and 2051 grains/cm^3) (Table 12.11) and consequently the inferences made here are tentative. Generally the pollen grains identified were in fair to good condition suggesting that they had not been subject to extensive re-working. It is however likely that selective preservation of pollen grains has taken place in favour of the more robust pollen grains such as pine *Pinus*, lime *Tilia* and alder *Alnus*.

Monolith <40364> (context 40100)

Pollen concentrations were moderate to very low (62-1672 grains/cm^3) from this context. A summary of the pollen counts is given here (Table 12.9), with a graphic diagram of the profile through the sequence (Fig. 12.6) and details of the changing concentrations through the profile (Table 12.11). This sequence is characterised by a diverse arboreal assemblage including fir *Abies*, alder *Alnus*, oak *Quercus*, lime *Tilia* and elm *Ulmus* as well as coniferous elements such as pine *Pinus* and spruce *Picea*. The presence of other taxa including hazel *Corylus*, daisy family Asteraceae and grasses *Poaceae* is also noted. Aquatic taxa are barely represented, with only *Pediastrum* (a green algae) and water lily *Nuphar*.

The similarity of the pollen data throughout the sequence suggests that the sediments were deposited relatively quickly with little evidence for vegetation succession, although a dominance of Poaceae at 21-22cm in monolith <40364> suggests a short-lived phase of local more open conditions. The pollen taxa identified tentatively suggest deposition under temperate conditions, with the regional presence of mixed woodland and perhaps alder carr growing in wetter areas. This interpretation accords with the previous investigations by Turner of the horizon of the elephant skeleton (see above; also in Wenban-Smith *et al.* 2006), apart from in the presence of *Abies*. The presence of herbaceous

Table 12.9 Summary table of the raw pollen counts from monolith <40364>

Taxa	0-1cm	3-4cm	6-7cm	9-10cm	15-16cm	18-19cm	21-22cm
Abies	0	2	0	0	8	0	0
Alnus	53	14	7	0	24	1	0
Betula	0	0	0	0	1	0	0
Picea	12	11	4	0	0	0	0
Pinus	7	17	5	0	34	1	1
Quercus	1	1	1	0	0	0	0
Tilia	6	3	1	0	4	0	0
Ulmus	1	0	0	0	0	0	0
Corylus type	11	10	1	0	14	0	0
Asteraceae undiff	0	0	0	0	1	0	0
Caryophyllaceae	2	0	0	0	0	0	0
cf *Limonium* type	0	1	0	0	0	0	0
Cyperaceae	1	0	0	0	5	0	0
Poaceae	20	0	0	0	15	0	19
Nuphar	1	0	0	0	1	0	0
Pediastrum	0	0	0	0	1	0	0
Polypodium	3	24	9	0	10	2	2

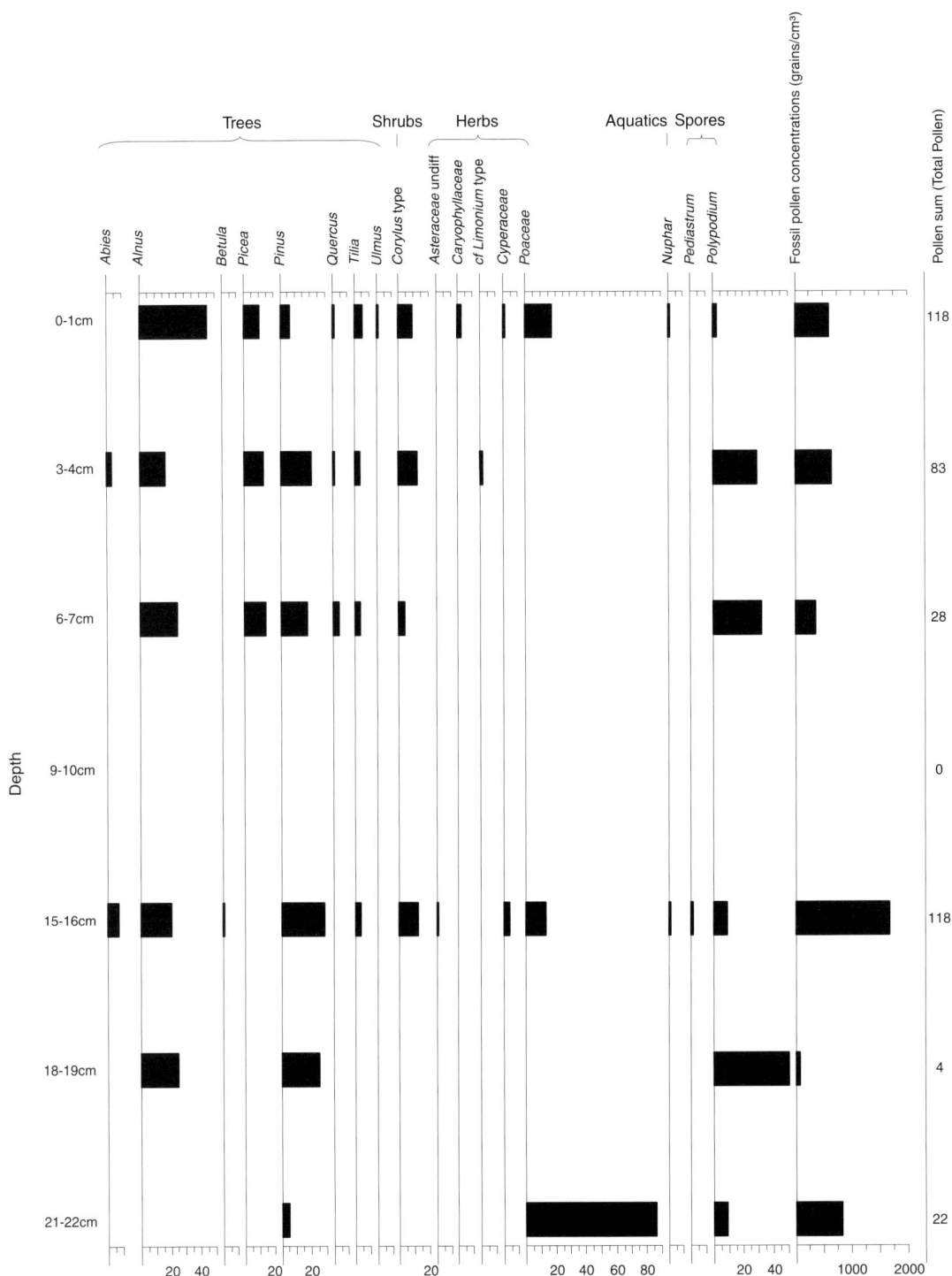

Figure 12.6 Total pollen percentage diagram for monolith <40364> (context 40100, upper part of monolith)

elements suggests that open, and possibly disturbed,areas were also present, as well as some open water inferred from the presence of *Pediastrum* and *Nuphar*.

Monolith <40418>

Pollen concentrations were moderate to very low (102-2051 grains/cm^3) for the samples from this context. A summary of the pollen counts is given here (Table 12.10), with a graphic diagram of the profile through the sequence (Fig. 12.7) and details of the changing concentrations through the profile (Table 12.11). The discussion

that follows is based on the pollen analysis of the sub-sample from 14-15cm depth, as the other levels were characterised by very low pollen content. This level is characterised by a diverse arboreal assemblage including *Alnus*, *Ulmus* and *Pinus*. The presence of other taxa including *Corylus*, Asteraceae and Poaceae is also noted.

The pollen taxa identified suggest deposition under temperate conditions, with mixed woodland and perhaps alder carr present in the landscape. The presence of Poaceae and other herbs suggests that open areas, possibly subject to disturbance, would also have characterised the landscape.

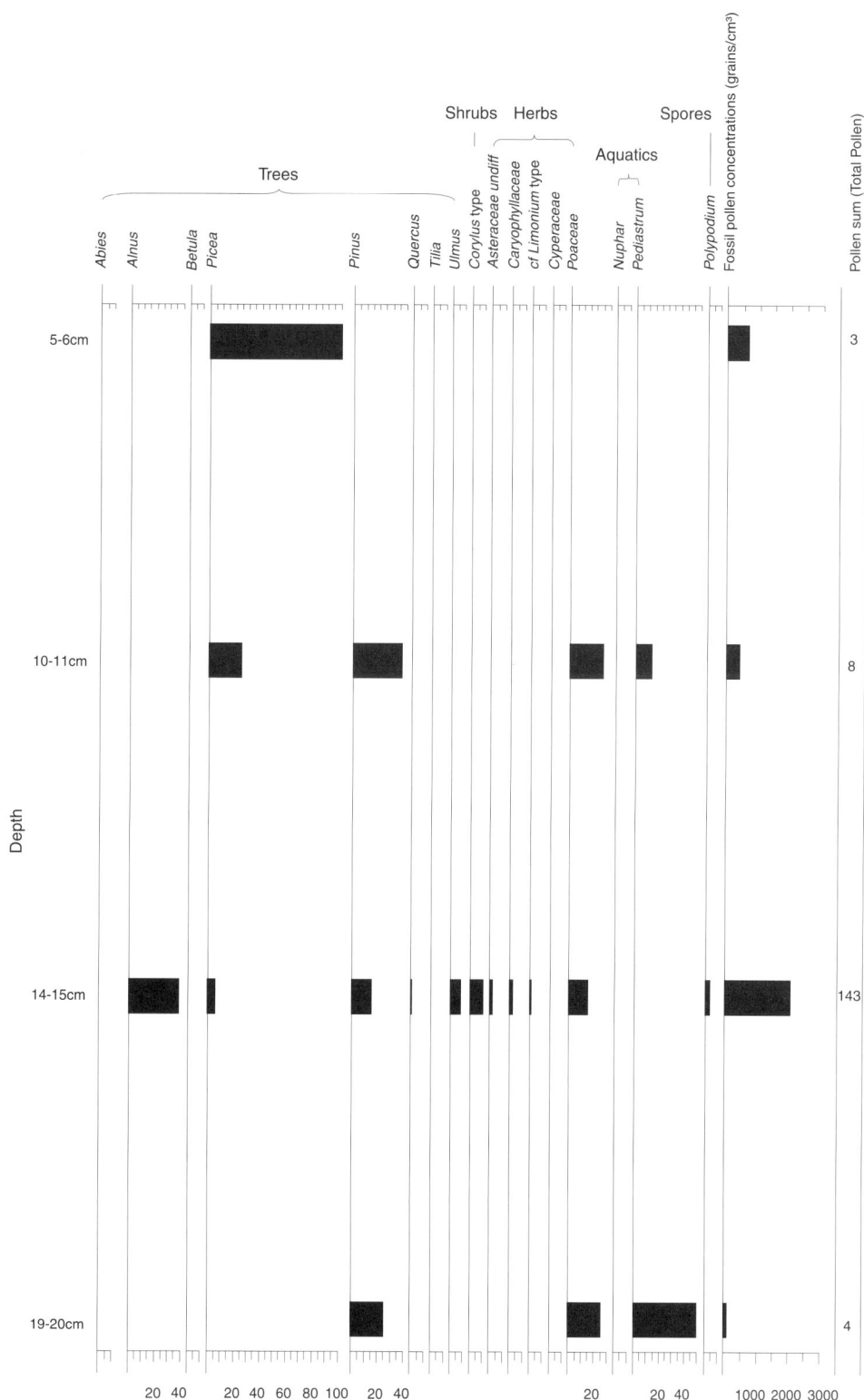

Figure 12.7 Total pollen percentage diagram for monolith <40418> (all samples from context 40158)

DISCUSSION AND CONCLUSIONS

The interpretative potential of the pollen remains from the site is, unfortunately, constrained by its poor preservation, generally low abundance and differential preservation. Contra the results of the initial field evaluation, it is clear that there are only two horizons where pollen (other than occasional Tertiary derivations) is preserved. These are, firstly, the lower part of Phase 6,

in the area of the elephant skeleton, where the grey clay of context 40100 interdigitates at its base with occasional darker brown more organic-rich beds, including context 40078. Secondly, a little (*c* 10-15m) to the north-east of the elephant, where the full thickness of Phase 6 becomes conflated to the very dark brown (almost black) organic-rich clay-silt beds *c* 0.2-0.3m thick of context 40158. The profiles from these two horizons are very similar, and indeed there is direct

Table 12.10 Summary table of the raw pollen counts from monolith <40418>

Taxa	5-6cm	10-11cm	14-15cm	19-20cm
Abies	0	0	0	0
Alnus	0	0	55	0
Betula	0	0	0	0
Picea	3	2	8	0
Pinus	0	3	22	1
Quercus	0	0	1	0
Tilia	0	0	0	0
Ulmus	0	0	11	0
Corylus type	0	0	14	0
Asteraceae undiff	0	0	3	0
Caryophyllaceae	0	0	3	0
cf *Limonium* type	0	0	1	0
Cyperaceae	0	0	0	0
Poaceae	0	2	20	1
Nuphar	0	0	0	0
Pediastrum	0	1	0	2
Polypodium	0	0	5	0

Table 12.11 Pollen concentrations, analysed samples from monoliths <40364> and <40418>

Monolith	Sub-Sample <>	Fossil pollen concentration (grains/cm³)
M-<40364>	40364/A/0-1cm	591
	40364/A/3-4cm	647
	40364/A/6-7cm	361
	40364/A/9-10cm	No pollen
	40364/A/15-16cm	1672
	40364/A/18-19cm	62
	40364/A/21-22cm	826
M-<40418>	40418/A/5-6cm	643
	40418/A/10-11cm	399
	40418/A/14-15cm	2051
	40418/A/19-20cm	102

lateral continuity between them so they can be considered together.

The general similarity between the pollen evidence from the two samples from context 40078 associated with the elephant skeleton, and that from monolith <40364> from context 40100, the lower part of the grey clay sealing the elephant horizon, suggests that this part of the sequence formed under reasonably stable climatic and local environmental conditions. The evidence indicates a local environment of swampy alder carr, with a mixed, mostly deciduous forest on surrounding higher and drier ground, characteristic of the early temperate substage of an interglacial. The presence of various shrubs and grasses indicates that this forest is not unbroken, but contains open areas probably maintained by herbivore grazing (see Chapter 7). A marked abundance of grasses at the base of the sequence from

monolith <40364> may, however, reflect a short-lived phase of more open local environment. Then, the presence of the distinctive grain of *Abies* higher up the sequence from this monolith suggests that this part of the sequence is not contemporary with the elephant, since this distinctive and robust grain would have been recognised in the samples from beside the elephant, had it been present.

There is nothing in the pollen evidence on its own to confirm that the interglacial concerned is the Hoxnian, or MIS 11. However, all the indications support this attribution. From the elephant horizon, the presence of *Azolla*, the absence of *Abies* and *Carpinus*, and the abundance of *Alnus* are all suggestive of the early-temperate substage, zone Ho II of MIS 11. It may in fact be towards the end of zone Ho II considering that just above, in monolith <40364>, the scarce presence of *Abies* provides a faint indication that this part of the sequence may date to the very end of zone Ho II or the start of zone Ho III, when this species starts to develop a greater presence in the Hoxnian.

Chapter 13

Amino acid dating

by Kirsty Penkman and Francis Wenban-Smith

INTRODUCTION

A new technique of amino acid racemization (AAR) analysis has been developed over the last 10 years (Penkman 2005; Penkman *et al.* 2007, 2008) that has proved robust and consistently reliable in geo-chrono-logical dating of molluscan remains at the sub-Pleistocene timescale. It has successfully distinguished material to the level of separate marine isotope stages over the last 500,000 years, MIS 1 to MIS13 (Penkman 2010; Penkman *et al.* 2011). This new technique (further explained below) combines a new Reverse-Phase High Pressure Liquid Chromatography method of analysis (Kaufman and Manley 1998) with the isolation of an 'intra-crystalline' fraction of amino acids by bleach treatment (Sykes *et al.* 1995). This combination of techniques results in the analysis of D/L values for multiple amino acids from the chemically-protected protein within the biomineral, enabling both decreased sample sizes and increased reliability of the analysis. The intra-crystalline fraction of the calcitic opercula of the fluvial gastropod *Bithynia tentaculata* has been found to be a particularly robust repository for the original protein, making these opercula the prime (although not the only) source of amino acid data for reliable geochronolog-ical determinations.

At the Southfleet Road site other lines of evidence, notably biostratigraphy (Chapters 9, 10 and 11) and geological correlation (Chapter 4), combine to suggest strongly that the interglacial sediments (Phase 6) associ-ated with the elephant skeleton and the dense lithic scatter south of Trench D are attributable to MIS 11, the Hoxnian interglacial. The presence of molluscan remains, including abundant *Bithynia* opercula at certain horizons in the site sequence (Chapter 10), provided the potential to confirm this attribution with the most reliable independent chronometric technique currently available for this time period. Molluscan remains were most abundant in Phase 6b, in the main basal context 40070 of the tufaceous channel sequence and in the overlying contexts 40144 and 40143. Although the more fragile shells of molluscs were absent through the rest of the site sequence, small quantities of the more robust *Bithynia* opercula were also recovered from a number of other deposits (Table 13.1). These include:

1. Context 40078 (Phase 6, though in a part of the deposit more equivalent to context 40158, rather than at the exact horizon of the elephant skeleton);

2. Context 40103 (Phase 6a, in the lowest part of the Phase 6 clay, beneath the tufaceous channel-fill sequence);

3. In the top part of context 40025 (Phase 5), where a few *Bithynia* opercula were recovered from some of the large bulk samples taken for small vertebrates;

4. Context 40062 (Phase 3) where, surprisingly, *Bithynia* opercula were abundant in bulk sample <40042>, taken for small vertebrate assessment, in association with a rich interglacial ostracod fauna (Chapter 11).

Clearly, the possibility of derivation has to be consid-ered for the opercula in these last deposits where they are not in association with other Pleistocene molluscan remains, and this is returned to below when considering the dating results.

A programme of amino acid dating analysis was undertaken that focused on these remains. It had the aims of:

1. Confirming the attribution of Phase 6b to MIS 11;

2. Investigating whether underlying and overlying deposits with *Bithynia* opercula were also attribut-able to MIS 11, or whether they should be attributed to other MIS stages;

3. Investigating whether it was possible to achieve intra-MIS 11 sub-stage distinction between material from different horizons based on the amino acid dating results;

4. Trying to establish correlations with key horizons from other English MIS 11 sites with significant Palaeolithic archaeological remains (Table 13.2), in conjunction with which the evidence from Southfleet Road is later considered (Chapter 22).

The results of this programme are summarised and discussed below; full details of the methods and individual analyses are also presented as Appendix 9. Analyses were carried out at the University of York amino acid laboratory (NEaar) between 2005 and 2010.

Table 13.1 Southfleet Road samples from which *Bithynia* opercula were picked for amino acid dating
[samples marked * represent material picked from small vertebrate samples]

Phase	Context	Sample <>	Figure	Sample source	Quantity	NEaar lab code/s	Result code/s	Context grouping [Fig. 13.3]
6	40078 [=40158]	40275*	-	Bulk sample	2	6433-4	SFR-4	6 - 40078
6b	40143	40282/C/0-2	Fig. 13.1d	Mollusc sub-sample from monolith	3	6430-2	SFR-3	6b - 40143
6b	40144	40295/C	Fig. 13.1b	Mollusc sub-sample from bulk sample	3	6435-7	SFR-5	6b - 40144
6b	40144	40333*	-	Bulk sample	3	6438-40	SFR-6	
6b	40070	40315/C	Fig. 13.1b	Mollusc sub-sample from bulk sample	3	6424-26	SFR-1	6b - 40070
"	"	"	"	"	8	6606-13	SFR-1	
6b	40070	40317/C	Fig. 13.1b	Mollusc sub-sample from bulk sample	3	6427-29	SFR-2	
6b	40070	40162*	Fig. 13.1a	Bulk sample	7	2041-7	Eb-A	-
6a	40103	40320*	Fig. 13.1b	Bulk sample	2	6441-2	SFR-7	6a - 40103
5	40025	40286*	Fig. 13.1c	Bulk sample	2	6443-4	SFR-8	5 - 40025
5	40025	40343*	Fig. 13.1c	Bulk sample	2	6445-6	SFR-9	
3	40062	40042*	Fig. 13.1a	Bulk sample	2	6166-7	SFR-10	3 - 40062
"	"	"	"	"	3	6447-9	SFR-10	
"	"	"	"	"	5	6534-38	SFR-10	
"	"	"	"	"	3	6614-16	SFR-10	

Table 13.2 MIS 11 comparator sites contributing to UK amino acid dating framework [based on *Bithynia* opercula]

Site	Location within site	Context, bed	Sample <>	Sample source	No. analysed	Identifier for Fig. 13.5	Reference
Hoxne	-	Stratum B2	'(64)' 2-8 mm	BM/AHOB excavations, 2001	4	Ho-B2	West 1956; Ashton *et al.* 2008
Hoxne	-	Stratum B2	'(50)' 2-8 mm	BM/AHOB excavations, 2001	4		
Swanscombe, Barnfield Pit	-	Lower Loam	Uncertain depth in sediment	Sampled by J. Rose, EuroMam 2004 (10-14 May)	6	Sw	Ovey 1964; Kerney 1971; Conway *et al.* 1996
Barnham	Pit 4	Bed 5c	118-128cm	BM excavation, 1993 (BEF 93)	6	Ba	Preece and Penkman 2005; Ashton *et al.* 1998
Hoxne	-	Stratum E	Sq 1,spit 5 (40-50), 'Sample 5'	BM/AHOB excavations, 2000	4	Ho-E	West 1956; Ashton *et al.* 2008
Beeches Pit, West Stow	Cutting 2	Bed 7	-	'1998 series'	4	BP	Preece *et al.* 2007

AMINO ACID DATING: THEORY AND METHODS

A detailed review of the theory and methods behind the new approach to amino acid dating applied here is provided in Appendix 9. In summary, amino acids, the building blocks of proteins that occur within mollusc shells and opercula, occur as two isomers that are chemically identical, but optically different. These isomers are designated as either D or L, depending upon whether they rotate plane-polarised light to the right or left respectively. In living organisms, the amino acids are almost exclusively L and the D/L value approaches zero. The potential application to geochronology arises from the fact that after death, amino acid isomers start to interconvert (or 'racemise'), trending over sufficient time towards parity with the D/L value approaching one. The proportion of D to L amino acids can therefore be taken

as a proxy for time passed. Once scaled by an independent framework based on lithostratigraphy, biostratigraphy and other geochronological methods, can be used to estimate the age of a fossil sample, alongside other indications of protein decomposition, such as the degradation of unstable amino acids.

Over the last 30 years, various attempts have been made to exploit this natural time-dependent property to allow the D/L values of Pleistocene molluscan remains to be used for Pleistocene dating. It was demonstrated in the 1980s that D/L values correlate sufficiently strongly with independently established dating frameworks for the technique to have value (eg Bowen *et al.* 1989). However, variable results between different species and wide error margins have restricted the reliability of the technique, and compromised the confidence with which it can be applied to deposits that are otherwise undatable. In the new technique applied here, attention is focused on a variety of different amino acids (and their decay products) which decay at varying rates. In combination these provide a more robust general dating result and greater sensitivity in certain time ranges depending upon which protein/acid cycle is analysed. In addition, analysis is focused on the protected intra-crystalline fraction, which (in contrast to the majority of the organic shell/opercula material) is relatively protected from contamination and external factors during burial (such as temperature changes and aqueous saturation). It therefore provides a closed system from which one can expect D/L values that are more closely aligned to nothing other than the passage of time.

It has been established in the recent work at NEaar, during development of this new technique, that the calcitic opercula of *Bithynia* are a particularly robust repository of protected proteins and amino acids, making these commonly occurring molluscan remains of particular value for dating purposes. A dating framework for the last 500,000 years of the British terrestrial Pleistocene archive has therefore been established on the basis of data derived primarily from opercula of *Bithynia tentaculata* (Penkman *et al.* 2011), and all the analyses discussed in this chapter were carried out on opercula of this species.

Shells selected for amino acid dating analysis were examined under a low powered microscope and any adhering sediment removed. The shell samples were then sonicated and rinsed several times in HPLC-grade water, after which they were crushed to <100µm. They then underwent a bleaching process to remove contaminants and organic materials other than the desired intra-crystalline fraction. This resulted in a tiny amount of dried powder derived from the original opercula, or from an individual operculum, if just one was used for a particular analysis. For each sample, this powder then underwent a series of further chemical treatments, designed to release the various amino acids present, the D and L quantities of which could then be isolated and measured. The details of this process are described in Appendix 9, and the results are discussed below. A key element of the analytical approach is the isolation of both 'Free' and 'Hydrolised' amino acid fractions, the D/L values of which should both give independent dating results and also correlate in closed and uncompromised systems. A lack of correlation between these fractions can therefore be a valuable indicator that a sample has been compromised, and its result is unreliable.

In the analyses conducted on the Southfleet Road material, the extent of racemization (D/L) was established for five amino acids and their decay products (Asparagine/aspartic acid – Asx; Glutamine/Glutamic acid – Glx; Serine – Ser; Alanine – Ala; and Valine – Val), along with the ratio of the concentration of Serine to Alanine ([Ser]/[Ala]), for both the Free and Hydrolised – henceforth, 'Hyd' – fractions. These indicators of protein decomposition have been selected as their peaks are cleanly eluted with baseline separation and they cover a wide range of rates of reaction. It has been demonstrated that with increasing age, the extent of racemization (ie the values of the D/L ratios) will increase, whilst the

[Ser]/[Ala] value will decrease due to the decomposition of the unstable serine. Therefore the results from the Southfleet Road analyses can be interpreted in light of the framework now established for the British Pleistocene (Penkman *et al.* 2011).

ANALYSIS PROGRAMME AND SAMPLING LOCATIONS

In total, 208 analyses were carried out, on 52 separate samples (Table 13.1; Fig. 13.1). The only horizon of the site where molluscan remains were abundant was the basal context 40070 of the tufaceous channel-fill (Phase 6b). As soon as this was recognised, a preliminary selection of *Bithynia* opercula from one of the early bulk small vertebrate assessment samples (<40162>) was analysed to verify that material from the site was suitable for AAR analysis. This having proved successful (NEaar 2041-2044), a more comprehensive programme was established, in which *Bithynia* opercula were collected from Phase 3 (context 40062), Phase 5 (context 40025, upper part) and from throughout the Phase 6 sequence (from the base: contexts 40103, 40070, 40144, 40143 and 40078). Following this main phase of analysis, an additional phase took place in which more analyses were carried out on material from Phase 3 (context 40062) and from Phase 6b (context 40070) in order to have sufficient data for comparison of the results between these horizons to be statistically meaningful.

The majority of material analysed was recovered from mollusc or bulk small vertebrate samples, the locations of which were recorded on drawn sections. The Phase 3 material was recovered from bulk sample <40042>, taken from the bottom level of the main east-facing section 40016 early in the excavation (Fig. 13.1a).

The Phase 5 material was recovered from two bulk samples taken from sediment surrounding larger vertebrate remains found within the top 0.2m of context 40025 in the central part of the site, between Trenches B and C, beside the east-west transverse section 40080 (Fig. 13.1c). Bulk sample <40286> was recovered from around the bovid maxilla (Δ.43298 – see Chapter 7), and sample <40343> was recovered from around the single, poor condition elephant tusk (Δ.43788) found near the north end of the tufaceous channel.

The Phase 6 material was all recovered from samples in and around the tufaceous channel, mostly from the two mollusc sampling columns in the longitudinal section 40082 (Fig. 13.1b). The preliminary sample <40162> from which *Bithynia* opercula were collected and analysed came from a bulk small vertebrate assessment sample from one of the early southern exposures of the tufaceous channel in the main east facing section, where it was both thin and heavily contorted (Fig. 13.1a). The later samples (particularly the material from mollusc sub-samples <40315/C> and <40317/C> from Column 2) came from more tightly constrained horizons within the greatest thickness of the main tufaceous channel-fill context 40070 (Fig. 13.1b), so these are

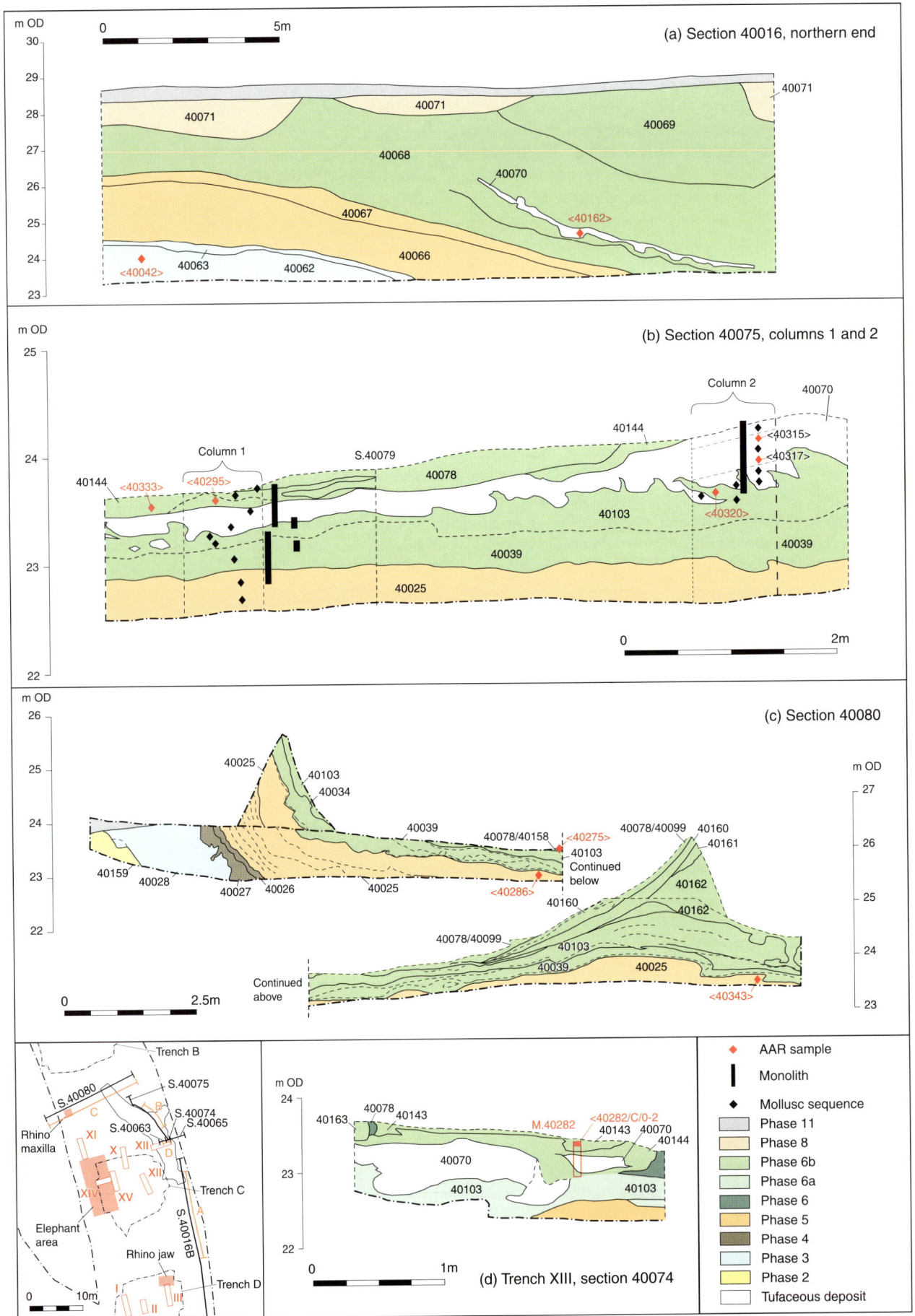

Figure 13.1 Mollusc sampling locations from which *Bithynia* opercula were recovered for amino acid dating: (a) Section 40016, northern end; (b) Section 40075, Columns 1 and 2; (c) Section 40080; (d) Trench XIII, Section 40074

regarded as having marginally higher stratigraphic integrity.

Deposits in and around the tufaceous channel were in places contorted and interdigitated, particularly the thin overlying beds 40143 and 40144. Consequently there is a legitimate question mark over whether the sparse molluscan evidence from these contexts is derived from context 40070, rather than being independent evidence from later phases of deposition. However, the molluscan and small vertebrate analyses of these overlying contexts 40144 and 40143 show

continuing local environmental and ecological trends which suggest good integrity of the biological remains (Chapters 7 and 10). Context 40144 was well-defined and reasonably widely distributed, so there is high confidence that the samples and analysed opercula attributed to this context are reliably provenanced. The *Bithynia* opercula analysed from this context came from two samples (Fig. 13.1b): firstly, from molluscan subsample <40295/C> from Column 1 through the tufaceous channel; and secondly, from bulk sample <40333> which was part-sieved on site for small

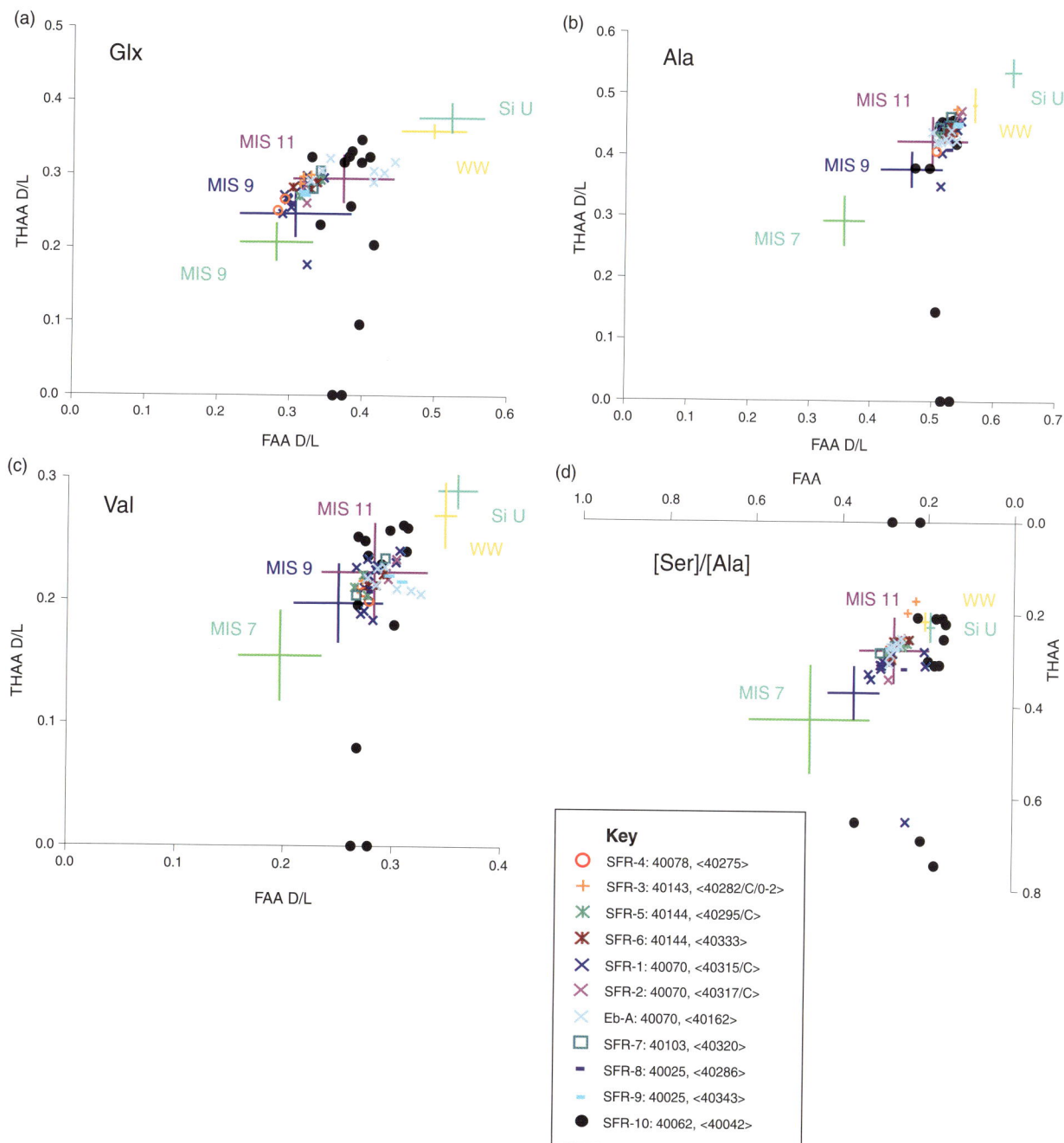

Figure 13.2 D/L Hyd vs D/L Free for *Bithynia* opercula from Southfleet Road, compared with UK Middle Pleistocene framework for: (a) Glutamic Acid/Glutamine Glx; (b) Alanin – Ala; (c) Valine – Val; and (d) [Serine]/[Alanine] – [Ser]/[Ala] [The bars of each cross represent two standard deviations about the mean for data obtained from opercula from sites correlated with MIS 7, MIS 9, MIS 11, Waverley Wood (WW) and Sidestrand 'Unio bed' (SiU) – see Penkman *et al.* 2011]

vertebrate recovery, but the precise location of which was not recorded

Context 40143, the top level of the tufaceous channel-fill sequence, was very intermittent and often highly contorted, making it difficult to collect large enough samples for mollusc and small vertebrate analysis. It was, however, very distinctive, making it easy to recognise, and likewise leading to high confidence in the provenance of the samples and opercula attributed to it. The best exposures were in Trench XIII, and a subsample for molluscan analysis was taken from monolith <40282> in section 40074 (Fig. 13.1d). Three *Bithynia* opercula were recovered for AAR dating from this monolith.

Finally, two *Bithynia* opercula were recovered from bulk small vertebrate sample <40275>, attributed in the site archive to context 40078, but which came from the very dark, organic-rich Phase 6 deposits in Trench B that were later reattributed as context 40158 (Fig. 13.1c). This sample was originally taken in the hope of recovering insect and/or plant macro-remains, but these being lacking, it was sorted for small vertebrate remains, in course of which two *Bithynia* opercula were recovered, which were then incorporated in the AAR dating program. Despite its similar original context number, this material is therefore not from the same horizon as the elephant skeleton, but can probably be regarded as stratigraphically slightly younger, being from the top part of the Phase 6 sequence.

All processed samples were assigned a NEaar lab code and a shorthand result code (Eb-A and SFR1-10), as shown (Table 13.1), the result code being used in the various graphics and figures used to illustrate the results.

During hydrolysis the vials of two of the samples from context 40062, sample <40042> (SFR-10) cracked, and so no Hyd data is available for these two samples. The material from this sample was in general slightly problematic; the *Bithynia* opercula were particularly friable, with several disintegrating during the initial rinsing step to clean them. As Figure 13.2 shows, besides the two with no Hyd value (plotted on the x-axis, ie with a notional y-axis value of 0), some of the remaining opercula analysed from this horizon also showed much lower than expected THAA D/L values. This is clear evidence of a compromised closed system and an inaccurate dating result for those specimens. The [Ser]/[Ala] plot (Fig. 13.2d) shows three of these Phase 3 samples as clearly compromised, with excessively high values of THAA. One of the opercula from context 40070, <40315/C> (SFR-1) also showed this behaviour. In previous analyses of nearly 500 opercula samples, less than 2% of opercula were compromised in this way, so it is extremely unusual to have so many from one horizon. The friability of the opercula from context 40062 indicates that mineral diagenesis has occurred in at least some of them. This may merely reflect a poor preservational environment; although, if that were the case, it would be puzzling for the smaller and more delicate ostracod fauna to have survived (Chapter 11). This is therefore perhaps an indication that these opercula may

have been derived, and one should therefore be wary to argue that their slightly older results (see below) relate to the deposit from which they were recovered. The total data set for these samples is shown in the Free vs Hyd plots below (Fig. 13.2, SFR-10), but the compromised samples were removed from the more detailed bar charts (Fig. 13.3) and the comparative statistical analyses to avoid skewing the data.

SOUTHFLEET ROAD ANALYSES COMPARED WITH THE UK MIDDLE PLEISTOCENE MIS FRAMEWORK

As outlined above, analyses were carried out for five amino acids of differing racemization rates. One of these, Aspartic acid/Asparagine (Asx), is one of the fastest racemising of the amino acids discussed here (due to the fact that it can racemise whilst still peptide bound, Collins *et al.* 1999). This enables good levels of resolution at younger age sites, but decreased resolution beyond MIS 7. The results showed clearly that all the Southfleet Road samples were substantially older than MIS 7 (see Appendix 9), so the results from this particular set of analyses are not presented here.

The data obtained from Glutamic Acid/glutamine (Glx), serine (Ser), alanine (Ala) and valine (Val) are discussed in detail below. If the amino acids were contained within a closed system, the relationship between the Free and the Hyd fractions should be highly correlated, with non-concordance enabling the recognition of compromised samples (Preece and Penkman 2005). The plot of Free versus Hyd data from each sample can also be used as a relative timescale, with younger samples falling towards the bottom left corner of the graph and older samples falling towards the upper right corner, along the line of expected decomposition.

The data from the Southfleet Road samples have been plotted in this way for each of the amino acids (Fig. 13.2), with crosshairs representing the data obtained from other MIS 7, MIS 9, MIS 11 and (probable) MIS 13 sites from the UK with independent geochronology, as listed in Penkman *et al.* (2011). There is unfortunately a lack of comparative *Bithynia* opercula from independently dated MIS 13 sites. The data here are based on analyses of material from two sites: the Unio bed at Sidestrand, Norfolk (shown in Fig. 13.2 as 'Si-U') and Waverley Wood (shown in Fig. 13.2 as 'WW'). At the former interglacial sediments are sealed beneath Anglian glacial till and are also dated to MIS 13 by water vole biostratigraphy, although an MIS 15 date is also possible (Preece *et al.* 2009). The Waverley Wood material is more reliably dated to MIS 13 (Shotton *et al.* 1993), but the opercula come from *Bithynia troschelii* rather than *B. tentaculata*; the racemization rates for these two species are, however, sufficiently comparable to be used interchangeably (Penkman *et al.* 2013). These plots show each Southfleet Road analysis as a separate point, with the colour/shape varying to reflect each separate sample, as indicated on the figure key. The data are thus split to

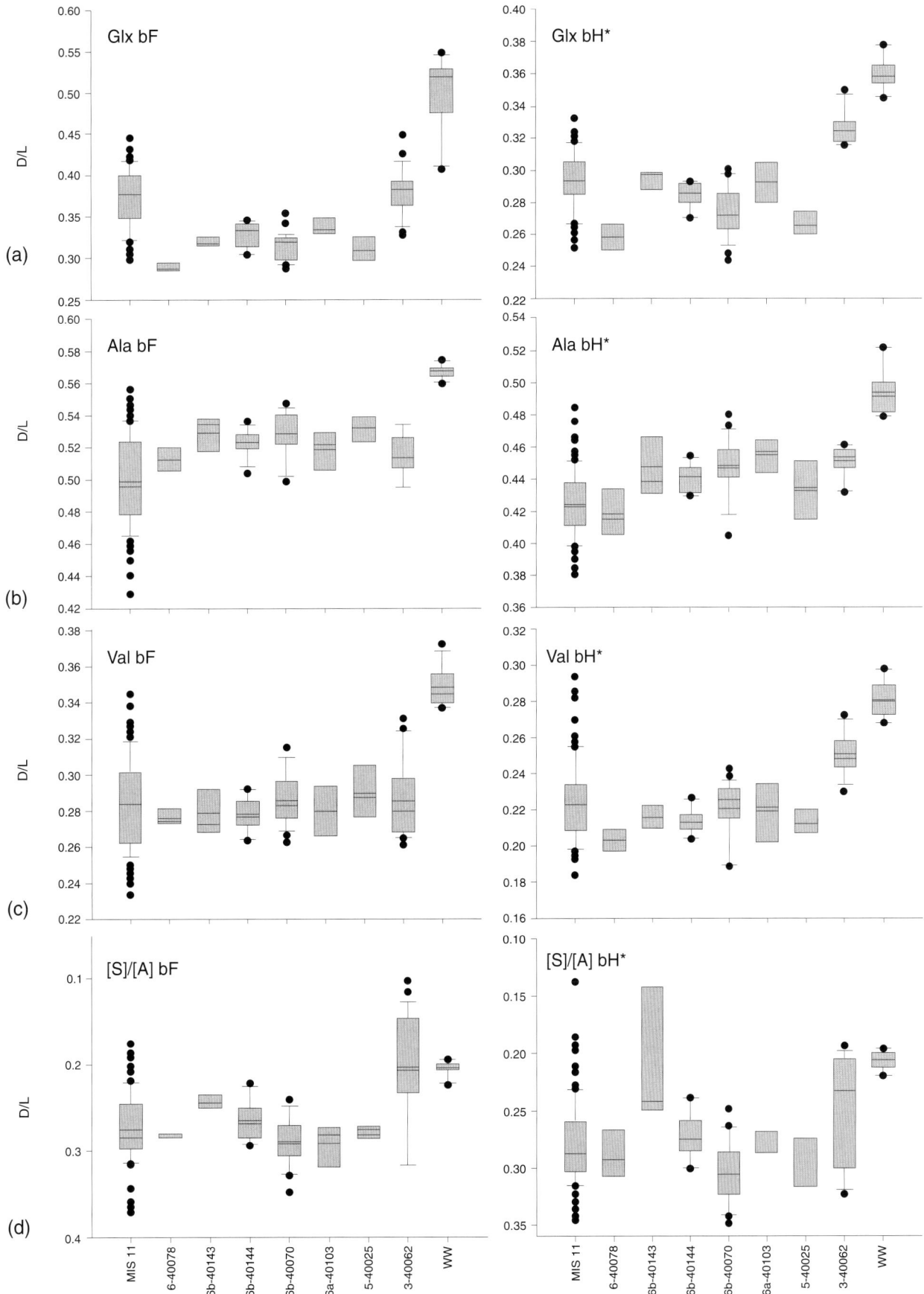

Figure 13.3 Free (left) and Hyd (right) D/L values for uncompromised *Bithynia* opercula from Southfleet Road, plotted/ grouped by context in stratigraphic order, and compared with data from MIS 11 and MIS 13 [Waverley Wood (WW)], for: (a) Glutamic Acid/Glutamine Glx; (b) Alanin – Ala; (c) Valine – Val; and (d) [Serine]/[Alanine] – [Ser]/[Ala]

the maximum level of division, since several of these samples are stratigraphically equivalent (as shown in Table 13.1), and could reasonably be lumped together. This diagram, the interpretation of which is discussed in more detail below, establishes a broad correlation of the data with the MIS framework. It clearly shows that most, if not all, of the analyses indicate association with MIS 11, defined here on the basis of comparative material from the sites of Barnham, Hoxne, Beeches Pit, Woodston, Swanscombe, Clacton, Elveden and Dierden's Pit (Penkman *et al.* 2011).

This diagram is complemented by another (Fig. 13.3), which combines data from the same context, as shown in the right-hand column of Table 13.1, in an attempt to investigate whether there is any discernible trend in the amino acid data that correlates with the intra-site stratigraphy. If so, this might indicate that the deposits formed over sufficient time for intra-MIS 11 AAR resolution to be present. In this latter diagram, the Free and Hyd datasets for each analysis within each context group are plotted as box-plots, with top and bottom of the boxed area defined by the 25th and 75th percentiles. Within the box, the solid line indicates the median and the dashed line the mean. Where more than 9 data points are available, the 10th and 90th percentiles can be calculated (shown by lines below and above the boxes respectively). The results of duplicate analyses are included to provide a statistically significant sample size. The outside columns on each box-plot diagram provide overall comparisons with MIS 11 and MIS 13 [Waverley Wood] datasets, and between these, site stratigraphic order descends from left to right.

Considering Fig. 13.2, the first plot (Fig. 13.2a) shows data for Glutamic Acid/glutamine (Glx), which is one of the slower racemising amino acids discussed here. This makes it less useful for distinguishing younger Pleistocene material, but the low levels of racemization do help discriminate between material of Middle Pleistocene age. The Glx D/L values from Southfleet Road show values within the range of those expected from sites of MIS 9 and MIS 11 age. The plot also shows four sample points from Phase 3 (SFR-10 – context 40062, sample <40042>) and one point from Phase 6b (SFR-1 – context 40070, sample <40315/C) where the fall away from the expected line of correlation between the Free and the Hyd data indicates compromised samples. The results of these are therefore unreliable. Apart from the compromised data, two datasets (SFR-10, from sample <40042>, Phase 3; and Eb-A, from sample <40162>, Phase 6b) cluster in a slightly older part of the plot. Whilst this potentially makes sense for the Phase 3 material, it is anomalous for the other group, which should be the same age as other material from context 40070 (samples SFR-1 and SFR-2).

The second plot (Fig. 13.2b) shows data for Alanine (Ala). This is a hydrophobic amino acid, whose concentration is partly contributed from the decomposition of other amino acids (notably serine). Ala racemises at an intermediate rate, so is one of the most useful amino acids for distinguishing samples at Middle-Late Pleisto-

cene timescales. The Ala data shows a tight clustering of data, consistent with a correlation with MIS 11 and clearly enabling discrimination from both sites of MIS 9 age and the two comparator sites of probable MIS 13 age. Aside from the three data points where the very low values of the Hyd dataset (THAA D/L) clearly show compromised results, the two samples from SFR-10 and that from SFR-1 which fall within the MIS 9 cluster show clear evidence in the other amino acids of being compromised, so can be disregarded. These leaves a tight cluster of uncompromised data points corresponding with the crosshair for the comparative MIS 11 dataset.

The third plot (Fig. 13.2c) shows data for Valine, which has extremely low rates of racemization. As the concentration of Val is quite low, the difficulty of measuring the D/L accurately results in higher variability. It does however still prove useful for age discrimination within material of Middle Pleistocene age. The D/L values for Val in the Free and the Hyd fractions again support the other amino acid data, centring on the crosshair for MIS 11. There is also, however, again a preferential clustering of the Phase 3 dataset (SFR-10) in the slightly older part of the plot, although still by no means approaching the comparative MIS 13 values; this is discussed further below.

The fourth plot (Fig. 13.2c) shows data for the ratio of the concentrations of serine and alanine, which provides an extremely useful tool for age estimation. Serine is a very unstable amino acid, and it can degrade via dehydration into alanine (Bada *et al.* 1978). As the protein within a sample breaks down, the concentration of serine will decrease with an increase in the concentration of alanine, thus the [Ser]/[Ala] value will decrease with increasing time. In order to ease the interpretation, the axes are plotted in reverse, so that the age-related direction of increase in protein degradation is the same as for the racemization plots. The [Ser]/[Ala] of the Southfleet Road samples are again consistent with an age in MIS 11, but the level of discrimination is not particularly high. As with the other plots, the Phase 3 dataset (SFR-10) is grouped in the slightly older part of the plot, although with a high proportion of compromised data. Also in the slightly older part of the plot, curiously, are two data points from SFR-3 (Phase 6b, context 40143) which on stratigraphic grounds should be amongst the youngest material.

In combination, the data presented above provide robust confirmation that Phases 5 and 6 of the site sequence were laid down in MIS 11. They also suggest that Phase 3 of the site sequence was probably also laid down in MIS 11. However, the consistent clustering of dataset SFR-10 from Phase 3 as slightly older within MIS 11 raises the possibility that the AAR data are demonstrating intra-MIS 11 distinction within the aggradational sequence. In order to investigate intra-sequence variability in more detail, the Free and Hyd data are here shown as box-plots (Fig. 13.3), as described above, with material grouped by context (but omitting material from sample <40162>, where the tufaceous pocket sampled was relatively thin and contorted, near the edge of the tufaceous channel). As before, there are four datasets, one for each of the three amino acids Glx (Fig. 13.3a), Ala (Fig. 13.3b) and Val (Fig. 13.3c), and a fourth showing [Ser]/[Ala] (Fig. 13.3d). And for each of these fours datasets there is a separate set of plots for the Free and Hyd D/L values. These data show several consistent trends.

Firstly, the group of material from Phase 3 (SFR-10) appears (statistically significantly at the 10% level in the box plot for Glx, Fig. 13.3a) slightly older than the other material from the elephant site in half of the analyses – Glx (Free and Hyd); Val (Hyd); and [Ser]/[Ala] (Free). This group of material is nonetheless clearly aligned with other data from MIS 11 rather than MIS 13 in the three individual amino acid analyses (Fig. 13.3a-c). In the fourth analysis [Ser]/[Ala] (Fig. 13.3d), it appears better aligned with the MIS 13 comparator Waverley Wood for the Free dataset, and the situation is similar for the Hyd dataset, although the wide variability in the sample dataset make the separation from MIS 11 less clear-cut. The top of Phase 3 is marked by a zone of decalcification (context 40063), perhaps reflecting a depositional hiatus between the top of Phase 3 and the base of Phase 5, although there is no evidence of a major depositional unconformity. Any depositional hiatus is likely to be minor and of minimal chrono-stratigraphic significance, considering the interglacial attribution of Phases 3 and 5, and the lack of any intervening evidence for climatic deterioration.

Secondly, there is almost complete consistency for the results from Phase 5 (context 40025), Phase 6a (context 40103) and Phase 6b, within the tufaceous channel (contexts 40070, 40144 and 40143). In all these datasets, (apart from one: [S]/[A] Hyd – where the results for context 40143 appear anomalously old), these groups are indistinguishable both from each other, and from the MIS 11 comparator dataset, confirming the MIS 11 attribution throughout this stretch of the site sequence.

Thirdly and finally, at the top of the sequence, the data from the very dark organic rich upper part of the clay (context 40158, sample SFR-4) appears very slightly younger in several of the analyses: Glx, Free and Hyd (Fig. 13.3a); Ala, Hyd (Fig. 13.3b); and Val, Hyd (Fig. 13.3c). The number of data points is, however, too low for this result to be statistically significant.

In general, the data consistently show an MIS 11 date for Phases 5 and 6, and there is also a suggestion of a recognisably earlier MIS 11 date for Phase 3. This latter result is worthy of some discussion, because to-date AAR dating has not achieved sub-stage precision. It should be remembered that this result was not consistently replicated in all the datasets. There was, for instance, no distinction represented in the Free or the Hyd data from Alanine (Fig. 13.3b), which would normally be one of the most useful for distinguishing between different MIS stages within the Middle Pleistocene. It should also be remembered that there was a high degree of compromised chemical systems in the opercula from Phase 3 and that many of them were in poor condition, perhaps

indicating preservational and post-depositional problems that might have affected the results. No other molluscan remains were recorded in association with the Phase 3 opercula (see Chapter 10), suggesting conditions for preservation were not ideal. On the other hand, they were associated with a rich interglacial ostracod fauna that was undoubtedly an *in situ* autochthonous assemblage (see Chapter 11). This fauna might be expected to be less resistant to decay than any larger molluscs, which are, however, entirely lacking apart from the (quite abundant) *Bithynia* opercula.

On balance, it appears possible that the opercula from Phase 3 have been dated by AAR to an earlier part of MIS 11 than Phases 5 and 6. However, the evidence that the opercula were poorly preserved and often chemically compromised slightly undermines the reliability of this result. The lack of preservation of associated mollusc shells, despite the presence of smaller and more delicate ostracods, also suggests that the opercula may have been reworked. This indicates that, even if they are dating to an earlier part of MIS 11, or perhaps a late MIS 13 or an intra-MIS 12 interstadial, this dating does not relate to deposition of Phase 3. Phase 3 is therefore probably of very similar age to Phases 5 and 6, supporting the zoological and geological evidence for climatic and depositional continuity through Phases 3 to 6. Further investigation is, however, required to try and establish whether preservational factors are having an influence, and whether the result is replicable with larger datasets from deep sequences at other MIS 11 and MIS 12–13 comparator sites.

PHASE 6B ANALYSES (TUFACEOUS CHANNEL, CONTEXT 40070) COMPARED WITH KEY MIS 11 HORIZONS

Having confirmed that Phase 6, which contained the elephant skeleton and the lithic concentration south of Trench D, can be securely dated by AAR to MIS 11, and investigated the AAR dating of the site sequence more generally, the final stage of the AAR dating programme was to investigate how the dating results from the site compared with a number of other significant MIS 11 comparators, particularly those that also have important archaeological remains (Table 13.2). Two more charts were plotted, analogous to those used to investigate the intra-site data. The first of these (Fig. 13.4) shows scatter plots of Free vs Hydrolised D/L values for the same datasets as discussed above, in other words (a) Glx; (b) Ala; (c) Val; and (d) [Ser]/[Ala], for the most heavily analysed sample (sample <40315/C>, SFR-1) from the middle of context 40070 in the tufaceous channel, alongside datasets from: Barnfield Pit, Swanscombe (Lower Loam); Barnham (Bed 5c); Hoxne (Stratum E); Hoxne (Stratum B2); and Beeches Pit, West Stow (Cutting 2, Bed 7). The dataset from Phase 3 of the elephant site (sample <40042>, SFR-10) was also included, to continue investigation of whether this could be related to other specific sites and/or a specific part of MIS 11. The second of these shows separate box-plots for Free and Hydrolised data separately (Fig. 13.5), for the same set of comparator sites, and the same range of racemization data. These data fail to show any significant inter-site groupings or correlations, and this is probably currently beyond the precision of the technique for the Middle Pleistocene, although something to be aspired to in future research. The scatter-plots (Fig. 13.4) and the histograms (Fig. 13.5) all show that, for almost all the datasets, the elephant site Phase 6b data are indistinguishable from any of the MIS 11 comparator sites, particularly for Val and [Ser]/[Ala] (Fig. 13.4c, d; Fig. 13.5c, d). There are occasional minor and statistically insignificant

deviations from this similarity, but these do not correspond conisistently with known stratigraphic order or postulated relative dating (ie the Glx (Hyd) values for Stratum B2 at Hoxne are slightly greater than those for the lower Stratum E) and there is no consistency in which particular site/horizon appears slightly older or younger. The data from Barnham have marginally lower D/L values (Free and Hyd) for Val and [Ser]/[Ala] but are conversely marginally higher for Free Ala and Glx. Likewise, the data from the Lower Loam at Swanscombe have marginally lower D/L values than the Phase 6b data for Ala (Free and Hyd) but are otherwise marginally higher than the other MIS 11 comparator sites.

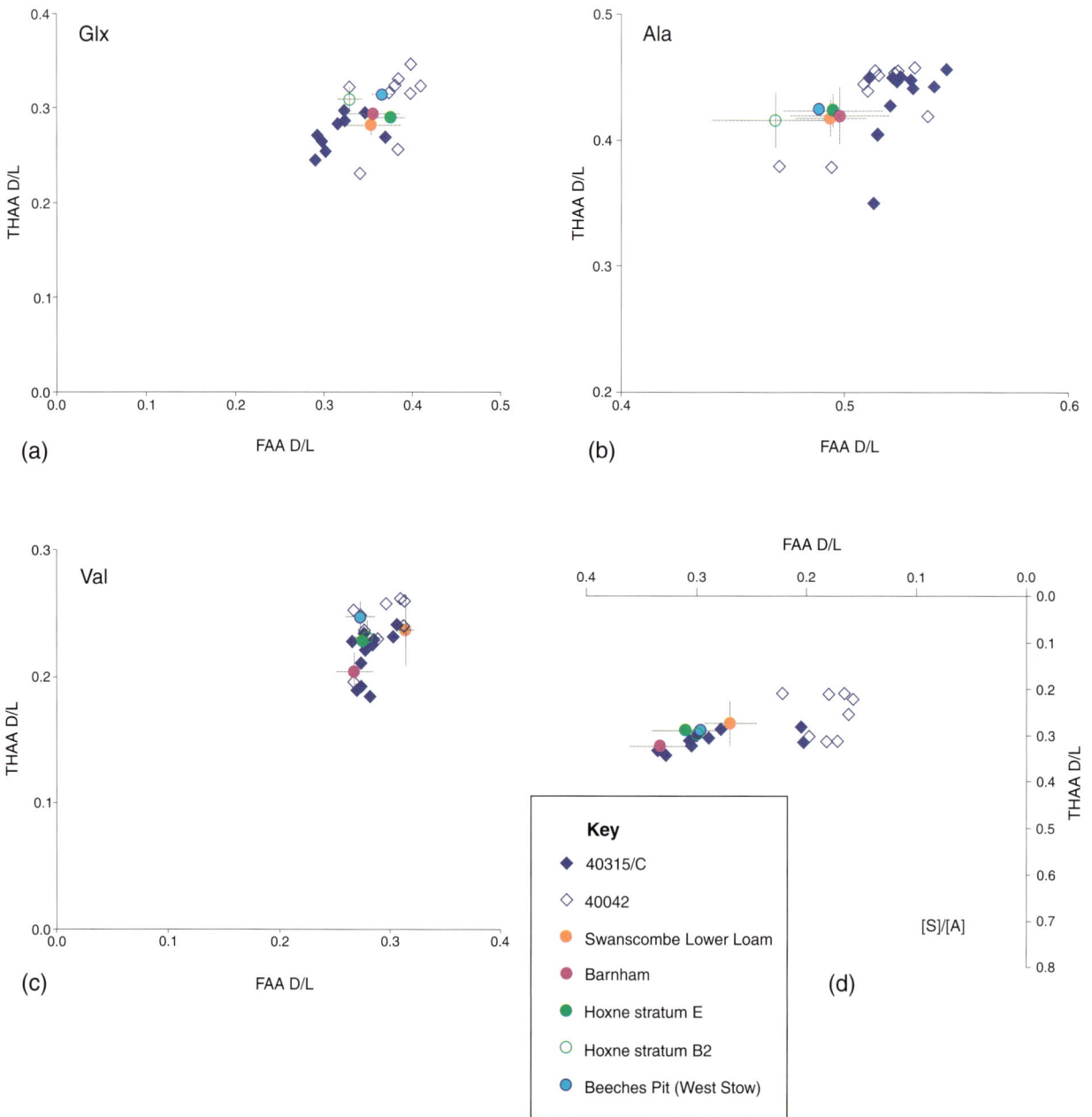

Figure 13.4 D/L Hyd vs D/L Free for uncompromised *Bithynia* opercula from Southfleet Road, Phase 6b tufaceous channel (sample <40315C>, context 40070, SFR-1) and Phase 3 (sample <40042>, context 40062, SFR-10) compared with key MIS 11 horizons for: (a) Glutamic Acid/Glutamine Glx; (b) Alanin –Ala; (c) Valine – Val; and (d) [Serine]/[Alanine] – [Ser]/[Ala] [See Table 13.2 for MIS 11 comparator horizons]

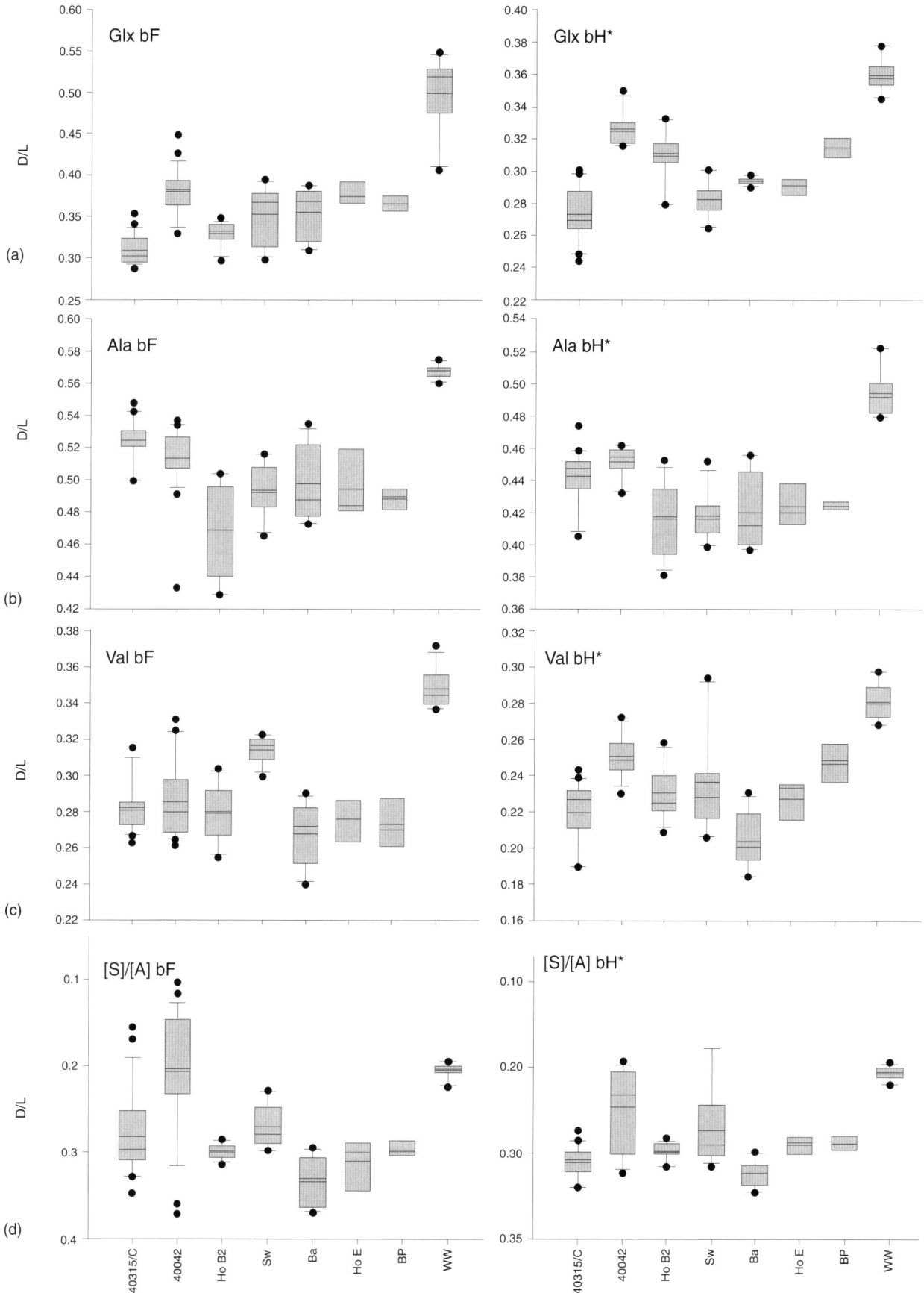

Figure 13.5 Free (left) and Hyd (right) D/L for uncompromised *Bithynia* opercula from Southfleet Road, Phase 6b tufaceous channel (sample <40315/C>, context 40070, SFR-1) and Phase 3 (sample 40042, context 40062, SFR-10) compared with key MIS 11 horizons and Waverley Wood (WW) for: (a) Glutamic Acid/Glutamine Glx; (b) Alanin – Ala; (c) Valine – Val; and (d) [Serine]/[Alanine] – [Ser]/[Ala] [See Table 13.2 for MIS 11 comparator horizons]

DISCUSSION AND CONCLUSIONS

The amino acid dating programme has convincingly confirmed the attribution of Phase 6 of the site sequence to MIS 11. Although the evidence from Phase 5 is much sparser, this too seems securely provenanced and reliably dated to MIS 11. Below this, the data from Phase 3, from where abundant opercula were recovered, appear to cluster slightly earlier within MIS 11. There are, however, possible indications of reworking and an unusually high proportion of chemically compromised specimens in the material from Phase 3, which is generally in poor condition. These factors may be affecting the results, and it is suggested that, if the dating is nonetheless correct, the dated opercula are reworked from an earlier horizon, rather than the Phase 3 deposits having formed in a significantly earlier part of MIS 11 to Phases 5 and 6.

Comparisons with other MIS 11 sites were investigated, particularly those with significant archaeological horizons, namely: Swanscombe, Hoxne, Barnham and Beeches Pit. However, no consistent pattern or groupings emerged from the data that allowed any more detailed inter-site correlation of specific archaeological horizons. This appears to currently be beyond the precision of the AAR technique, but is certainly an area for further future investigation.

Chapter 14

Optically stimulated luminescence (OSL) dating

by Jean-Luc Schwenninger and Francis Wenban-Smith

INTRODUCTION

A small programme of OSL dating was carried out at the site. Although we had some preconceptions that the probable age of the site was *c* MIS 11, it was important not to let this prior expectation over-ride independent chronometric investigation of the date. Furthermore, it was not known until fieldwork was substantially in progress how much other vertebrate and molluscan material (including *Bithynia* opercula) that could contribute to dating was present. Therefore, since the site was rich in sand/silt horizons sedimentologically suitable for OSL dating, a series of tube-samples for OSL dating was taken throughout the sequence (Table 14.1).

Later, once fieldwork and some preliminary faunal analysis had been completed, it was clear that the most significant archaeological horizon of the site, the Phase 6 clay, which included the elephant skeleton and the flint scatter south of Trench D, was reliably dated to MIS 11 by both independent chronometric means (amino acid dating, Chapter 13) and biostratigraphy (Chapter 9). Therefore, when it came to determining the post-excavation programme, there was evidently little benefit in devoting resources to carrying out OSL dating on sediments that: (a) were already confidently dated to MIS 11 and (b) given this, were in any case at the upper limit of the viability of the method.

Nonetheless, a small programme was carried out on some of the horizons at the top of the sequence, for which there was no other dating evidence. The highest level with molluscan and/or other biological remains allowing direct dating was the Phase 6 clay. Above this, the stratigraphic superposition of deposits of Phases 7, 8 and 9 established their relative ages, although it was uncertain whether this sequence was laid down in relatively quick succession following Phase 6, or whether it included significant chronological hiatuses. It was suspected on geo-morphological grounds that the Phase 8 gravels were probably of similar age to the Lower Middle Gravel at Barnfield Pit (ie also MIS 11). However there was no other evidence to support this (apart from, perhaps, the typology of the handaxes from the gravel (see Chapter 20), although dating deposits from their artefactual content should never be relied upon). Also, there was no evidence of a major stratigraphic unconformity between the Phase 8 gravel and the overlying brickearth of Phase 9, nor was there any indication of the date of the brickearth or of the likely passage of time between its deposition and the cessation of fluvial activity associated with the underlying Phase 8 gravels.

Therefore OSL dating was initially focused on the top of the sequence, with sand-rich beds from Phases 8 and 9 being analysed (see below). When these results proved surprisingly young, it was decided to carry out a further phase of analysis that: (a) investigated a sand-rich bed at the base of the Phase 8 gravels; and (b) analysed as a control a sand-rich sample from the base of the Phase 5 gravels, well below the Phase 6 deposits that are confidently attributed to MIS 11. As discussed in more detail below, this control analysis provided a dating result significantly younger than compatible with the Phase 6 dating evidence, leading to the suspicion that some of the other OSL results might also be questionable. Nonetheless this programme of OSL work is presented in full here, and serves as a useful and thought-provoking case-study, demonstrating that, when independently verified by other chronometric, biostratigraphic and geomorphological means, potentially misleading results have been obtained that would otherwise have been unsuspected. As discussed below, it is therefore perhaps now necessary to treat OSL dating results with more caution, and to seek additional independent and stratigraphic controls when applying the technique.

METHODS AND SAMPLING LOCATIONS

OSL has been well-established over at least the last 20 years as one of the main approaches to dating sediments younger than *c* 350,000 years, which is near the effective limit of the technique as currently practiced. Developed from the thermo-luminescence (TL) method of the 1990s (Wintle 1991), current OSL methods use a single aliquot regenerative dose (SAR) protocol as described by Murray and Wintle (Murray and Wintle 2000; Wintle and Murray 2006). In the simplest terms, the technique measures the time since sand grains were last exposed to daylight. It is thus particularly suitable for Quaternary sequences, where sand-rich beds are ubiquitous, and which often lack other means of dating.

Samples are taken in the field by hammering a light-opaque sampling tube of *c* 6-8cm diameter into a freshly cleaned face, then carefully removing this and covering the ends with light-opaque caps. A small additional sample of *c* 5-10cc is also taken for analysis of moisture content; this latter sample does not need to be kept away from light, but needs to be sealed with tape and double-bagged to avoid evaporation of natural soil moisture prior to analysis. Ideally, a measurement is then taken in the sample hole of the background radiation levels using a portable gamma-ray spectrometer; however, this step is not essential and the background dose rate can also be measured in the laboratory, extrapolating from the sediment forming the tube sample.

Sample preparation and optically stimulated luminescence measurements were performed at the Luminescence Dating Laboratory at the Research Laboratory for Archaeology and the History of Art, University of Oxford. The dating results are based on luminescence measurements of sand-sized quartz grains (180-255mm) extracted from the samples and mounted onto aluminium discs as small sized (3-4mm) multigrain aliquots and using the weighted mean of repeat measurements performed on multiple aliquots. All samples were measured in automated Risø luminescence readers (Bøtter-Jensen 1988 and 1997; Bøtter-Jensen *et al.* 2000) using a SAR post-IR blue OSL measurement protocol (Murray and Wintle 2000; Banerjee *et al.* 2001; Wintle and Murray 2006).

Dose rate calculations are based on the concentrations of radioactive elements (potassium, thorium and uranium) within the sediment. The beta dose rate was calculated from the concentrations of radioisotopes by fusion ICP-MS analyses and with the exception of sample X-1967, discussed below, the external gamma-dose rate was derived from the *in situ* radioactivity measurements. The final OSL age estimates include an additional 2% systematic error to account for uncertainties in source calibration. Dose rate calculations are based on Aitken (1985). These incorporated beta attenuation factors (Mejdahl 1979), dose rate conversion factors (Adamiec and Aitken 1998) and an absorption coefficient for the water content (Zimmerman 1971). The contribution of cosmic radiation to the total dose rate was calculated as a function of latitude, altitude, depth and average over-burden density based on data by Prescott and Hutton (1994).

INITIAL ANALYSES

The samples chosen for the initial phase of analysis were the highest three in the sampled sequence of deposits (Table 14.1). At the top of the sequence, sample <40254> (RLAHA lab code X-2060) came from the higher west end of Transect 2; one of the longitudinal strips cleared to investigate deposits to the north of the main site, where a handaxe and flint debitage scatter were found on the stripped surface of the brickearth bank (see Chapter 3). The sample came from a sand-rich bed in the upper part of the Phase 9 brickearth, close beside the location of the handaxe find (Fig. 14.1a). The second sample chosen for analysis was sample <40056> from context 40051 (RLAHA lab code X-1966), an undulating sand bed that occurred at the north end of section 40015, between the uppermost gravel bed of Phase 8 and the base of the brickearth (Fig. 14.1b). Although the upper and lower boundaries of this sand bed were sharply defined, there was no evidence that

Table 14.1 Southfleet Road OSL samples, in stratigraphic order

Phase	Context	Trench	Section	Sample <>	Initial analysis (RLAHA lab code)	Further analysis (RLAHA lab code)
9b	40087	Transect 2	Plan 40008	40254	X-2060	-
9a	40051	-	40015	40056	X-1966	-
8c	40049	-	40015	40057	X-1967	-
8b	40047	-	40015	40058	-	-
8a	40098	A	40066	40244	-	X-2056
8a	40098	A	40066	40245	-	-
8a	40045	-	40015	40243	-	-
6b	40144	-	40064	40247	-	-
6b	40070	-	40064	40246	-	-
6b	40070	-	40063	40255	-	-
5	40066	-	40016	40052	-	-
5	40066	-	40016	40053	-	X-1963
5	40066	-	40016	40054	-	-
5	40066	-	40016?	40358	-	-
5	40066	-	40016?	40359	-	-
5	40163	-	40086	40354	-	-
5	40163	-	40086	40355	-	-
5	40163	-	40086	40356	-	-
5	40163	-	40086	40357	-	-
2	40065	-	40016	40047	-	-
2	40065	-	40016	40048	-	-
2	40064	-	40016	40049	-	-
2	40060	-	40016	40051	-	-
1	40056	-	40016	40050	-	-

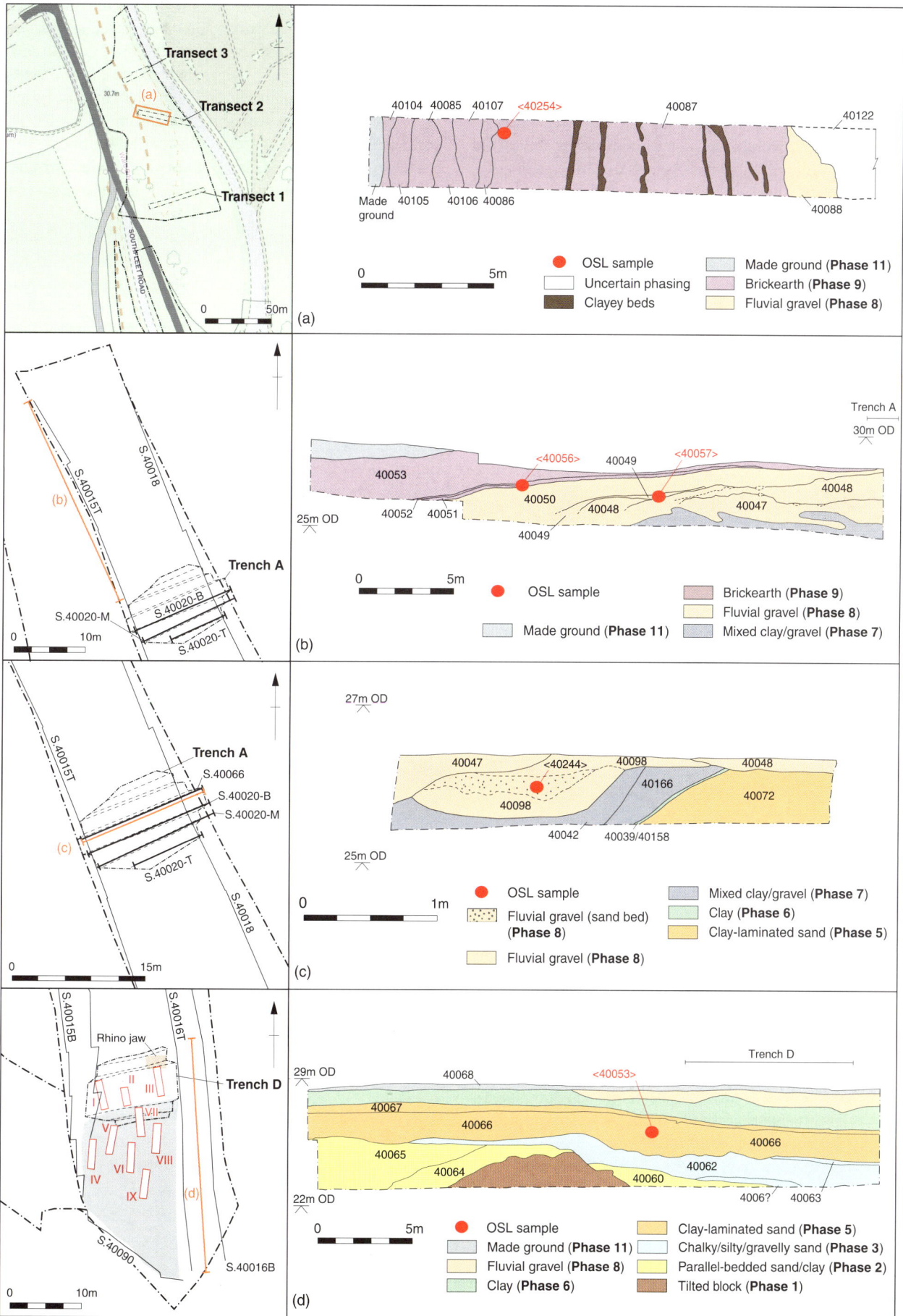

Figure 14.1 OSL sampling locations: (a) Transect 2, sample <40254>; (b) Section 40015, samples <40056> and <40057>; (c) Trench A, Section 40066, sample <40244>; (d) Section 40016, sample <40053>

either was associated with erosional truncation, so it was not initially thought that a major period of time was represented by the transition from Phase 8 to Phase 9. The third sample chosen at this stage came from slightly lower in the sequence in the same part of the site, and was sample <40057> from context 40049 (RLAHA lab code X-1967), which was a sand bed within the upper part of the Phase 8 gravels (Fig. 14.1b).

All samples were collected as described above, supported by collection of *in situ* readings of the background sedimentary radiation dose rate with a portable gamma-ray spectrometer.

The quartz OSL signal characteristics of all three samples were generally good featuring high signal intensity, negligible recuperation and good recycling ratios. Only a small number of aliquots showed a clear infrared stimulated signal (which indicates the presence of contaminant feldspar mineral grains) thereby confirming suitable sample preparation and purity of the quartz extract. The adoption of a post-IR blue OSL measurement protocol (Banerjee *et al.* 2001) in which each OSL measurement is preceded by an IR measurement in order to deplete the contribution of feldspathic minerals to the OSL signal further enabled to reduce the IR/OSL ratio to negligible levels.

Repeat measurements on the same sample showed that in the case of samples X-1966 and X-1967 there is a wide degree of scatter between aliquots with some clear outliers and this directly accounts for the rather large error of ~55ka obtained for sample X-1966. This variability may be attributed to problems of incomplete bleaching which commonly affect mineral grains from fluvial sediments. For this reason obvious high outliers were omitted from these calculations.

Although, this study only includes a limited set of three samples, the luminescence measurements seemed to provide a stratigraphically consistent set of results (Table 14.2). There is good agreement between samples X-1966 and X-1967 which provided OSL age estimates of respectively *c* 281 and 287 ka [ka = thousands of years ago]. If accepted as correct, these dates would place the gravel and overlying sand in MIS 8. The date calculated for X-2060 is substantially younger (*c* 58 ka) as a result of a reduced palaeodose and a higher environmental dose rate. This date would place the upper part of the brickearth in the later part of the Devensian, at the end of MIS 4 or the start of MIS 3.

Although the close agreement of samples X-1966 and X-1967 corresponds with the sedimentological observations suggesting no major depositional hiatus between the sampled deposits, the actual date suggested was

substantially younger than the expected date of *c* 380,000-360,000 BP, even allowing for the margin of error accompanying the dating result.

Even more striking, was the dating result for X-2060 from the upper part of the brickearth, which was very much younger than anticipated. The dated sand-rich bed was definitely within the main body of the brickearth, which contained no evident sedimentary boundaries in the exposure seen, right down to its basal junction with gravel deposits that were thought to be equivalent to the top of the Phase 8 gravels.

The accuracy of this result should not, however, be ruled out. It is becoming increasingly well-established from numerous OSL investigations carried out recently in the Ebbsfleet Valley (Wenban-Smith and Bates 2011a; Wenban-Smith *et al.* forthcoming) and the Dartford area (Wessex Archaeology 2008c; Wenban-Smith *et al.* 2010; Wenban-Smith and Bates 2011b) that the Devensian glacial (taken as starting with the post-Ipswichian climatic deterioration of MIS 5d, *c* 115,000 BP) was a period of major colluvial/slopewash activity in north-west Kent, with repeated evidence of massive slope movement and redeposition of substantial bodies of sand/silt and brickearth. Even if the precision of this result is questionable, it still suggests that there is a major, unsuspected depositional hiatus between context 40051 (Phase 9a) and the main body of the overlying brickearth (Phase 9b). And it also seems possible that this brickearth may well be associated, at least in its upper part, with the Devensian, rather than MIS 11 or 10, as previously suspected.

FURTHER ANALYSES

Following from the results of the first phase of OSL analysis, two further samples were dated (Table 14.1). The first of these, sample <40244> (RLAHA lab code X-2056) came from the base of the Phase 8 gravels, from a substantial sand bed in context 40098 at the bottom of the south-facing section 40066 of Trench A (Fig. 14.1c). Even though the dating result from the upper part of the gravel was considered to be too young, the possibility that it was correct was not ruled out, and it was therefore decided to investigate further down the gravel sequence to see if a similar, or slightly earlier, result was obtained.

The second additional sample analysed, sample <40053> (RLAHA lab code X-1963), came from the Phase 5 clay-laminated sands in the southern part of the site in the main east-facing section 40016 (Fig. 14.1d).

Table 14.2 Southfleet Road OSL results: initial analyses

Phase	Context	Sample <>	Lab code	Palaeodose (Gy)	Dose rate (Gy/ka)	Age estimate (ka)
9b	40087	<40254>	X-2060	125.5 ± 23.2	2.15 ± 0.11	58.5 ± 11.3
9a	40051	<40056>	X-1966	304.7 ± 56.9	1.09 ± 0.07	281 ± 56
8c	40049	<40057>	X-1967	298.5 ± 26.5	1.04 ± 0.07	287 ± 33

This sand was stratigraphically sealed beneath the Phase 6 clay, which was securely dated to MIS 11 by amino acid racemization (Chapter 13) and on biostratigraphic grounds (Chapter 9). This sample therefore serves as a control for the application of OSL as a dating technique at the site, to establish the apparent dating result for a deposit that is firmly believed to be associated with early MIS 11, *c* 400,000 BP in age.

The quartz OSL signal characteristics of both samples were generally good, featuring high signal intensity, negligible recuperation and good recycling ratios. Some aliquots showed a small but clear infrared stimulated signal (>5%), indicating the presence of contaminant feldspar mineral grains. The adoption of a post-IR blue OSL measurement protocol (Banerjee *et al.* 2001) in which each OSL measurement is preceded by an IR measurement in order to deplete the contribution of feldspathic minerals to the OSL signal enabled reduction of the IR/OSL ratio to negligible levels (<3%).

Repeat measurements on the same sample revealed a wide degree of scatter between aliquots (especially for sample X-2056, from the base of the Phase 8 gravel) which can lead to rather large errors on the final dates. The presence of high outliers is often encountered in rapidly deposited fluvial sediments such as this, and this is attributed to incomplete bleaching of the OSL signal at the time of deposition. The same issue affected some of the samples in the first phase of analysis and clear outliers were omitted from the age calculations presented here (Table 14.3).

The results of both these additional analyses were incompatible with both site stratigraphy and the independent dating of Phase 6 to MIS 11. For sample X-2056 from the base of the Phase 8 gravel, the apparent age of *c* 211 ka is substantially younger than the previous result from towards the top of the gravel. Taking account of the (albeit substantial) error margins of these dates, the stratigraphically higher date has a range of 320-254 ka, and the stratigraphically lower one a range of 255-167 ka. If the gravel was very rapidly deposited, one could interpret the combined results as suggesting a date of *c* 255 ka, towards the end of MIS 8, based on the overlap of the error margin range. This date would still, however, the substantially younger than the MIS 11 date of *c* 380,000 BP expected on geological grounds.

For sample X-40053, the result of *c* 142 ka was completely at odds with all other dating information, and has worrying implications for the widespread and uncritical application of OSL dating, particularly in conjunction with the relatively narrow error margin of ± 18 years, suggesting a relatively accurate and precise result. The

sample was taken in the field under optimum conditions, with *in situ* measurement of the sedimentary background radiation dose rate using a portable gamma ray spectrometer, and there was nothing in the laboratory analysis to indicate that the result was problematic. However, the apparent result would not only place the deposit in MIS 6, despite being securely sealed beneath the Phase 6 deposits that are securely dated to MIS 11, but also directly contradicts the other OSL dating results from the site (Table 14.2; Table 14.3). These (with the exception of X-2060 from the very top of the site sequence) have all provided older dates for stratigraphically higher deposits.

DISCUSSION AND CONCLUSIONS

The incompatibility of the additional results with: (a) previous results from higher in the stratigraphic sequence, (b) site stratigraphy and (c) independent dating evidence, demonstrates that many of the OSL dating results must be wrong. Since there is no means of judging which ones are least unreliable, they must therefore all be regarded as suspect – the only possible exception being the Devensian date of *c* 58 ka for X-2060, from the top of the Phase 9 brickearth, the highest natural deposit in the site sequence. This result not only matches the stratigraphic sequence, but it is also compatible with a substantial body of evidence from the region in and around Swanscombe: the Darent Valley (Wenban-Smith *et al.* 2010; Wenban-Smith and Bates 2011b); the Ebbsfleet Valley (Wenban-Smith and Bates 2011a; Wenban-Smith *et al.* forthcoming) and the Dartford area (Wessex Archaeology 2008c), for slope instability and substantial colluvial/slopewash redeposition throughout the Devensian from *c* 115,000 to 10,000 BP (MIS 5d-2). The quantity and consistency of these results lend them credence. They also tally with the likely depositional processes associated with this predominantly cold period and with available lithostratigraphic data, in that the proposed Devensian colluvial sediments are always near the top of the Quaternary sequence and never buried by deposits for which independent dating suggests an earlier age.

Besides this one Devensian date, the rest of the OSL results reported here are regarded as unreliable, and the other evidence of geomorphological position and lithostratigraphic relationships with deposits dated by other means is given greater credence when considering the age of the deposits for which OSL dating was attempted. The results suggest that, although we should always keep

Table 14.3 Southfleet Road OSL results: further analyses

Phase	Context	Sample <>	Lab code	Palaeodose(Gy)	Dose rate(Gy/ka)	Age estimate(ka)
8a	40098	<40244>	X-2056	271.06 ± 54.06	1.28 ± 0.07	211 ± 44
5	40066	<40053>	X-1963	221.75 ± 25.03	1.56 ± 0.07	142 ± 18

an open mind that our preconceptions may not be right, the Quaternary community in general should be cautious about uncritically accepting OSL dating results that are not independently supported and that significantly conflict with prior expectation. It should also be a priority for the OSL dating community to develop the technique further, perhaps with increased application of approaches such as single grain dating. Also, crucially, with further analytical controls which, especially when a date cannot be supported by independent means, provide better indications of when a particular dating result is reliable.

Chapter 15

Lithic artefacts: overview and approach to analysis

by Francis Wenban-Smith

INTRODUCTION AND APPROACH TO ANALYSIS

The focus of the analytical approach was not only to make a technological and typological record of the collection that allowed its comparison with the wider British and NW European Lower/Middle Palaeolithic record, but also to investigate the behaviour and cognitive processes behind the lithic remains by analysing the *chaîne opératoire* and the spatial organisation of production. Complementing these cultural and archaeological goals, and as a necessary prelude to them, site formation and taphonomic processes were also investigated, as it was necessary to consider how these might have affected and/or distorted the lithic remains, disguising (or perhaps falsely presenting) patterns relating to hominin activity.

There is a danger in lithic analysis of indiscriminate recording of an over-abundance of superfluous empirical data. This may happen for various reasons. Partly, there is a long history in lithic analysis of untheorised empiricism, whereby the lithic analytical chapter just starts along the lines: 'The artefacts from xxx [site/layer/period] can be grouped into xxx main groups ..' followed by, for each group, sections on raw material, handaxes, cores, flake-tools and perhaps flakes, without any preceding discussion of the basis and objectives of classification. Partly perhaps, the act of measurement brings a reassuring empiricism and scientific control to the otherwise alien world of Palaeolithic technology, allied to an unconscious equation of analysis with measurement. Finally, also perhaps partly because certain data have previously been measured or recorded, and, although no longer participating in any interpretive debate, have become embedded in the intellectual/processual DNA of lithic analysis. However, it was attempted here to adopt a more focused approach to the recording of lithic data. The analysis of the lithic collection was undertaken with a number of clear objectives in mind (paragraph 1, above), and all recorded observations and measurements were chosen as relevant to these objectives. The methods of analysis and the data chosen for recording are outlined below, complemented by a lithic methodological appendix with more detailed descriptions of the technological and typological categories used and the measurement protocols for quantitative attributes (Appendix 6).

The sequence at the site included at least nine different major depositional phases, Phases 1-9 (Chapter 4) probably all from the Pleistocene, the majority of which contained lithic remains. The collection was initially divided into assemblages by stratigraphic phase. Consideration was then given as to the integrity of the assemblage from each phase, and whether further subdivision into smaller assemblages for analysis was useful. This was done on the basis of artefact condition, spatial concentration and (for the artefact-rich horizon of the Phase 6 clay) refitting and microdebitage distribution.

METHODS OF ANALYSIS

Study of the artefact collection initially involved going through each artefact in turn, checking and recording its provenance, and recording the range of data identified at the outset as relevant to the analysis. Full details of the analytical process and the data recorded are presented in Appendix 6. In summary, five groups of data were recorded, plus miscellaneous notes (Table 15.1):

- recording reference
- packing/storage information
- site provenance data
- lithic technological/typological data (categorical)
- lithic analytical data (quantitative)

The main lithic technological categories identified are summarised here (Table 15.2), and the more detailed typological subdivisions, for instance of handaxe shape and flake tool type, are given in Appendix 6 (Table A6.4; Table A6.9). Although a bipartite technological classification was initially applied during recording, following the categories and subcategories C1 and C2 specified in Table 15.2, this was simplified for subsequent analysis with each technological category regarded as of potential interpretive significance being allocated a single numeric code (Table 15.3). These categories and codes are used in the assemblage summary tables in the subsequent lithic chapters. Most of the technological and typological categories applied are uncontroversial, although distinguishing between some of the categories requires a (usually not discussed) emic engagement with the mind of the knapper, for instance distinguishing between flake-tools and flakes/irregular waste used as cores, or between cores and core-tools. At an etic level, both these pairs of categories could be subsumed under, respectively, the two categories 'worked flakes' and 'worked nodules'. Readers can, if they wish, perform this sleight-of-mind for themselves when considering the results, and distance themselves from my interpretation of knapping desire and purpose. However, part of this analysis has been to make these interpretations, founded ultimately on not just many years of looking at early Palaeolithic flint artefacts, but most importantly upon many years of experimental flint working leading to a reasonable basis for understanding and interpreting prehistoric engagements with the same material. Also, to develop an understanding of the site, and the hominins who inhabited it, based on the accumulated interpretations of the artefactual collection and the lithic *chaîne opératoire* in its landscape and environmental context.

In some cases, it seemed very obvious that, for instance, a large flake with a blunt cortical side opposed to a sharp edge, out of which a single notching flake had been struck, should be regarded as a flake-tool. In others, there was great difficulty in attempting to decide whether several quite small and chunky pieces of flint debitage or irregular waste with several small removals should be regarded as tools or cores. Although not very convincing as tools, it was hard to imagine that the even smaller and lightweight removals were themselves more desirable as tools. Consequently these were mostly categorised as miscellaneous notched tools, although one must also of course be aware that the indestructibility of flint means that juvenile knapping results enter the archaeological record along with the adult products, so this is always a potential (although another generally undiscussed) source of confusion in the attempt to

interpret lithic collections. It is perhaps particularly applicable to small-scale and hard-to-make-sense-of unstructured reduction episodes.

Another technological problem that became apparent, and this was applicable at the straight forward etic level, without any concern over emic engagement, was the distinction between flakes, irregular waste and natural unworked flint clasts that were a background part of the sediment. Flakes were attributed by being clearly recognisable as individual removals (or broken parts of) with striking platforms and ventral surfaces. Irregular waste was applied as a general category for irregular pieces of knapping debitage that did not conform to the definition of flake. The particular problem here was that, since much of the raw material contained frost-fractures, a knapping blow could easily lead to the breaking-up of a flint nodule due to pre-existing flaws. Many of the resulting pieces would not themselves always exhibit any clear evidence of hominin interference, but would result from it, and would thus conceptually be 'debitage' as much as the finest flake. This problem can in principle be addressed by total recovery and refitting, and in fact is demonstrated by some of the refitting results of the project (Chapter 18). However, this is not a practical solution to the problem at the stage of initial analysis, nor is it particularly useful since total recovery of all frost-fractured pieces is impractical, and would in any case need to matched by complete and time-consuming refitting.

Table 15.1 Lithic analysis recording proforma

Type of data	Name	Description
Recording reference	Rec sht	Recording sheet, in number order of recording
	Sht #	Line number on recording sheet
Packing/storage	OA box no.	Box number as originally received
Site provenance data	Δ ID	Unique lithic identifier, small find number
	Context	Taken from finds bag, cross-checked with paper archive
	Area	Area of site: Trenches A-D; Transects 1-3, Strips A-D
	Trench	Evaluation trenches I-XV
	Sample <>	Sample number, for lithic bulk spit-sieved samples
	Spit	Spit-number, for lithics from spits in evaluation trenches I-XV
Categorical data	Cnd	Condition
	C1	Main technological category
	C2	Secondary technological category
	T1	Technology/typology, sub-category 1 (varies acc. C1, C2)
	T2	Technology/typology, sub-category 2 (varies acc. C1, C2)
	T3	Technology/typology, sub-category 3 (handaxes, flake-tools)
	T4	Technology/typology, sub-category 3 (handaxes, flake-tools)
	WhL	Completeness, wholeness (varies acc C1, C2)
Quantitative data	%Cx	Percentage remnant cortex, on dorsal surface of flakes
	DSC	Dorsal scar count, scars from debitage estimated as ≥20mm [not including striking platform, for flakes]
	ML	Maximum length, measured along ventral surface for flakes from point of percussion, mm [1]
	MW	Maximum width mm, orthogonal to ML[1]
	MT	Maximum thickness mm, orthogonal to ML
	WtG	Weight grammes
Notes	N	Notes, not usually entered on database but useful on paper record

[1] ML, MW for debitage – estimate extra for damage/abrasion <20mm

Table 15.2 Lithic artefact technological categories, as originally recorded

C1	C2	Description
0 - Natural	-	Not humanly worked, can be interpreted as raw material, can be excluded from database, but if so needs to be quantified
1 - Raw material	-	No sign of working, but clearly a manuport
2 - Tested nodule	-	Nodule with only a couple of flakes off, no sign of whether a core or core-tool
3 - Chunk	-	Knapped chunk. Uncertain whether core or core-tool, poss. because broken, or not very knapped, or just very ugly
4 - Core	1 - Conventional	Flakes removed, generally reasonably large, from natural lump of raw material and no sign of preferential edge/part for use
	2 - On flake	Debitage used as a core
	3 - On core-tool	Eg, if re-used or after breakage
5 - Debitage	1 - Irregular waste	Lump, fragment or shatter; piece bigger than 20 mm but not otherwise classifiable, often resulting from knapping frost-fractured pieces; usually show some sign of percussive impact, but in principle can apply to pieces that look completely natural, but are interpreted as resulting from hominin knapping
	2 - Flake, blade	Flakes, or parts of flakes, must have signs of being part of a single removal, else classified as C2=1
	3 - Chip/spall	Flake/irregular waste less than 20mm
	4 - Flake-flake	Debitage from flaking a flake
6 - Tool	1 - Handaxe (core-tool)	Usually evidence of preferential edge/part for use and bifacially worked; attention to straightening, to opposing handle, removal of small shaping flakes of no use in themselves
	2 - Handaxe (on flake)	When a handaxe is made on a blank that shows definite evidence of originally having been a piece of debitage
	3 - Flake-tool	Worked/utilised flake; working can be backing (eg, possible interpretation as backed knife), retouching (eg, to form scraping edge) or notching
	4 - Percussor	Evidence of focused battering, can appear on cores/core-tools, can have some working to facilitate handling
	5 - Anvil	Battering on very large pieces, usually would be interpreted as percussors

Table 15.3 Simplified lithic technological categories, with numeric codes used in analysis

Artefact category	Numeric code	C1, C2	Details
Percussor	5	6, 4	Localised battering on rounded protrusions on flint nodules
Tested nodules	10	2, -	Tested/abandoned nodule; single/failed flake removals
Cores	20	7, 1	Cores; numerous flake removals, no apparent bifacial shaping or edge creation
Cores-on-flakes	30	7, 2	Cores-on-flakes; large, chunky flakes or irregular waste with several removals
Handaxes	40	6, 1	Handaxes, including very simple core-tools and 'proto' handaxes
Handaxes-on-flakes	50	6, 2	Clear evidence that bifacial shaping applied to a flake
Flake-tools	61	6, 3 (T3=20)	Flake-tool 'Utilised flakes' – flakes without secondary working, but showing macro-wear interpreted as from use
	62	6, 3 (T3=21)	Flake-tool 'Knives' – flakes with secondary working opposing a sharp cutting edge, often also with macro-wear to indicate use
	63	6, 3 (T3=10)	Flake-tool 'Single notch' – classic Clactonian notch
	64	6, 3 (T3=12)	Flake-tool 'Linear/double notch' – two (usually, v. occ more) notches beside each other on one edge of a flake
	65	6, 3 (T3=11)	Flake-tool 'Multiple notch' – more than one secondary notch scattered around a flake blank
	66	6, 3 (T3=30, etc)	Flake-tool 'Miscellaneous' – other secondarily worked flakes
Flake-flakes	80	5, 4	Debitage removals from secondary flaking, ie knapping of flakes
Flakes	90	5, 2	Debitage with clear striking platforms and ventral surfaces
Irregular waste	100	5, 1 (and 8)	Pieces of knapping waste that are not proper flakes, and natural-looking pieces that are thought to result from shattering of frost-fractured raw material during knapping
Chips	110	5, 3	Chips – pieces of knapping waste <20mm maximum dimension
Natural	120	0, -	Natural – pieces of flint thought to be wholly natural in origin

In practice, many pieces that lacked clear artificially-induced fracture planes were categorised as irregular waste. This was on the basis of sometimes very faint indications that hominin interference had precipitated their fragmentation, such as a contrast in condition between different frost-fracture planes, or a slight impression of more directional conchoidal ripples on a fracture plane. Although this categorisation is unlikely to have provided a 100% accurate record of the genuine situation, it is hoped that mis-attributions may have been equally made, so that the overall composition of the assemblage is reasonably accurate. Even with the possibility that some pieces of natural flint have been categorised as irregular debitage, it is better that this should happen (both at the excavation stage, as well as the analytical stage) than that they should be totally omitted. The refitting results demonstrated that pieces that superficially appeared natural did on occasion derive from knapping reduction sequences. A particularly important instance of this was the broken percussor from around the elephant skeleton (see Chapter 17), for which most of the pieces showed no obvious sign of hominin interference. They were nonetheless collected during mechanical excavation of the clay beyond the immediate vicinity of the skeleton, where they were out-of-place sedimentologically in relation to the surrounding homogenous clay.

In addition to this recording process, which was applied to every lithic object recovered, the substantial part of the lithic collection from the Phase 6 clay was studied in more detail. Refitting and investigation of the microdebitage distribution was used to investigate taphonomy and site formation, and ultimately it was hoped, to reveal details of intra-site organisation of behaviour. It was clear during the excavation that there was a substantial concentration of lithic artefacts south of Trench D, another concentration surrounding the elephant skeleton and otherwise a sparse scatter of relatively isolated artefacts in the other areas of the clay. It was decided during excavation that investigation of the distribution of microdebitage associated with these two lithic concentrations could help in establishing whether they were undisturbed. Therefore a sampling programme was carried out to recover microdebitage from these concentrations (details in Chapter 3), the results of which are discussed subsequently in the respective chapters on the material from around the elephant skeleton (Chapter 17) and the concentration south of Trench D (Chapter 18).

A refitting programme was also carried out on the collection from Phase 6. This programme contributed to a number of analytic objectives. In the first place, it addressed the issue of taphonomy and disturbance, since one could expect high proportions of refitting material for undisturbed knapping scatters. Secondly, it allowed investigation of the significance, or otherwise, of finer stratigraphic subdivisions recognised during excavation process. Thirdly, dependent upon the degree of disturbance, it had the potential to investigate intra-site movement of the end-products of knapping sequences, and the general spatial organisation of lithic production within the site. It was also possible that the distance and directions of refitting material relative to each other might alternatively reflect site formation processes. And fourthly, when refitting was successful in reconstructing knapping sequences, it provided a much better view of the knapping *chaîne opératoire* than possible from separate artefacts.

Carrying out the refitting programme required that all artefacts involved were marked with their site find number, so that they could then be arranged on tables for refitting. Exceptional care was taken to avoid loss of provenance information, with the initial marking checked before re-bagging against the original finds bag marked with the original site provenance, and then checked again when subsequently removed for refitting. A few marking errors were found and corrected during both checking stages, indicating that both were necessary. Refitting was carried out at the British Museum, Department of Early Prehistory, Franks House, which was the only available venue with adequate space to lay out the Phase 6 collection, which included more than 2200 artefacts, many of them of substantial size. Particular gratitude is due to Nick Ashton of the British Museum for facilitating this.

Three different assemblages of material were defined within Phase 6: the concentration south of Trench D (assemblage 6.1), the sparsely distributed material from the rest of the Phase 6 clay (assemblage 6.2), and the material from around the elephant skeleton (assemblage 6.3). For assemblage 6.1, the large quantity of material (n=2010) was initially divided into three sub-groups (North, Middle and South) based on subsidiary spatial concentrations within the scatter. Each sub-group was also initially divided into upper and lower phases, based on site stratigraphy. The approach to the refitting programme was, for each of these sub-groups, to lay out the artefacts together and attempt to refit material within them. These initial sub-groupings were then gradually amalgamated. There was a general drift in the refitting process towards arranging the amalgamated assemblage 6.1 material on the basis of raw material type and technological category (core, flake, flake-flake and irregular waste) in the attempt to find further refits that crossed stratigraphic boundaries and that were separated by greater distances.

For the much smaller quantity of lithic material in assemblages 6.2 (n=135) and 6.3 (n=93), there was no need to create spatial sub-groups. Otherwise the same process was followed, with initial separation of the material stratigraphically, followed by amalgamation on the basis of raw material and technological category. This phase of amalgamation also, incidentally, included attempts to find refits between the different Phase 6 assemblages of 6.1, 6.2 and 6.3; however, no inter-assemblage refits were achieved.

The results of the refitting programme are discussed subsequently, in the respective chapters on the material from around the elephant skeleton (Chapter 17) and the concentration south of Trench D (Chapter 18).

Table 15.4 Overview of lithic collection by stratigraphic phase and analysis groups

Phase	Context/s	Analysis assemblage	n flints (total)	Notes and relevant chapter
11 – Not *in situ*	0 40001 40039? 40100?	Group 11.2	12	Technologically distinctive group attributed to 18th century gunflint manufacture (Chapter 21)
	0 40001 40012 40039? 40048? 40069? 40100? 40100?? 40133	Group 11.1	43	Various out-of-context Palaeolithic material (Chapter 21)
T1 – Transect 1	40080 40081 40082 40083 40084	Group T-1	8	Uncertain how these deposits relate to main site sequence (Chapter 21)
9–10	40176	Group 9-10	10	Rejected by Hugo Anderson-Whymark as Holocene, so perhaps mystery later L/M Pal assemblage (Chapter 21)
9 – Brickearth bank	40053 40076	Group 9.1	20	Brickearth at main site, and brickearth bank to N of main site (Chapter 21)
8 – Sandy gravel (palaeo-Ebbsfleet)	40014 40048 40050 40071 40102	8c	160	Upper beds of fluvial gravel; recovered by a variety of means: stray finds, bulk sieving and during machining (Chapter 20)
	40047	8b	22	More widespread bed of more gravelly gravel, overlying basal bed (Chapter 20)
	40098	8a	27	Basal, sandy bed of fluvial gravel (Chapter 20)
7 – Mixed clay gravel (syncline infill)	40045	-	1	A natural flint; part of group of contexts at top of syncline infill sequence (40044-40046) that interdigitate with base of Phase 8 gravels (Chapter 19)
	40042 40043 40164 40166 40167	7	90	Recovered by a variety of means: stray finds, bulk sieving and during machining (Chapter 19)
6 – Grey clay, with organic-rich beds and tufaceous channel deposits	40039 40078 40078? 40099 40100 40103	6.3	93	Flints from near elephant skeleton; recovered by hand-excavation (Chapter 17)
	40039 40068 40069 40070 40078 40099	6.2	135	More dispersed flints that were not near the elephant skeleton or part of the main scatter south of Trench D, including some from within the Phase 6b tufaceous channel fill; mostly recovered during machine-reduction (Chapter 18)

continued overleaf

Table 15.4 (continued)

Phase	Context/s	Analysis assemblage	n flints (total)	Notes and relevant chapter
	40100 40103 40144 40158 40036 40039 40039? 40078 40100	6.1	2010	Flints from main concentration south of Trench D; mostly recovered by hand excavation (Chapter 18)
5 – Clay-laminated sand	40025 40072	5	21	(Chapter 16)
3 – Chalky/silty/gravelly sand	40028 40061 40062 40159	3	8	(Chapter 16)
2 – Parallel-bedded sand/clay	40060	2	1	Probably natural (see Chapter 16)
1 – Tilted block	40056	1	1	Probably natural (see Chapter 16)
Total			2662	

As an adjunct to the refitting programme, and as part of the general objective of investigating/representing the rich Phase 6 lithic collection, a 3-D GIS model was constructed. Each artefact was given a different symbol according to its technological category, sized according to its weight, and coloured according to its stratigraphic provenance within Phase 6. Once the model was constructed, it was then possible to examine the 3-D distribution of the artefacts, focusing on any selected combination of artefact types. It was also possible to make a direct visual assessment of whether there were any trends in, for instance, size distribution spatially across the site or vertically within the Phase 6 clay. Refitting connection lines were also added into the model, likewise making it easy to visualise/explore in three dimensions the vertical and spatial connections between refitting material. This model is available online in the Archaeology Data Service (ADS) archive http://dx.doi.org/10.5284/1018062, and interested readers are encouraged to investigate it.

OVERVIEW OF THE LITHIC COLLECTION

In total, there were 2662 lithic items in the collection resulting from the Southfleet Road excavation (Table 15.4), including a quite substantial number (n=202) of wholly natural unworked pieces. This reflects a deliberate policy that excavators who were uncertain whether a lithic object was worked should treat it as if worked, and record it as a small find for future consideration. This hopefully has meant that as-many-as-

possible genuine artefacts were recovered. The deposits at the site included numerous natural flint clasts for which it was obviously impractical to attempt recovery, and their presence was recorded in context sedimentological descriptions. In the Phase 6 clay, from which the greatest part of the lithic collection came (n=2238), there was also the greatest number of natural flints recovered (n=157). In this case they provide a useful sample, supplementing memory and written notes, of the natural clasts that were present but not collected in this predominantly fine-grained deposit. Their presence also has site formation implications because it is necessary to either regard them as hominin manuports, which is considered unlikely for the great majority that appear to have no value as raw material or as a tool in their own right, or to consider by what natural process they became incorporated in the sediment. Then to consider whether this also has implications for accumulation of the associated artefactual content (see Chapter 18).

As can be seen from the summary table (Table 15.4), there were very few lithic finds from Phases 1, 2 and 3. Those from Phases 1 and 2 were almost certainly of natural origin, although a few undoubted artefacts were recovered from Phase 3. The material from these phases is discussed in the following chapter (Chapter 16).

There were slightly more lithic finds in the sand of Phase 5 (n=21), although three of these were of natural origin. These finds mostly came from towards the top of the Phase 5 sands, and, particularly for some of the larger ones, it is possible that some of them more properly belong with the overlying Phase 6 material, from which they are technologically and typologically indistinguishable (Chapter 16).

Above the Phase 6 clay, from which the great majority of lithic finds were recovered as discussed above, a reasonably substantial assemblage (n=90) was recovered from the Phase 7 deposits of the syncline infill. Most of these were recovered by sieving of bulk samples, once it was discovered during mechanical excavation that artefacts were present in these deposits; the details of the lithic assemblage from Phase 7 are considered in Chapter 19.

Above the Phase 7 deposits, another reasonably substantial lithic collection was recovered from the Phase 8 gravel (n=209). These artefacts were mostly recovered by a combination of sieving of bulk spit samples and during careful monitoring of machine reduction, so the majority are securely provenanced not just to the gravel in general, but to specific beds within it. This allowed the collection to be divided into three assemblages from different levels within the gravel for more detailed analysis (Chapter 20).

A single artefact was recovered *in situ* from the Phase 9 brickearth during the watching brief phase of machine reduction. Otherwise, all the artefacts from this phase of the site sequence where recovered either from a bulk spit

sieve sample from Trench A (n=3), or from the stripped brickearth surface *c* 75m to the north of the main site (n=16). The Phase 9 collection is discussed in Chapter 21, together with other material for which the context was not certain, namely:

1. The small collection from Transect 1 (n=8), which could not be phased stratigraphically in relation to the main site sequence.

2. Material recovered from late prehistoric features that was deemed to be derived Palaeolithic material on account of its condition and lack of similarity with known late prehistoric lithic material by the late prehistoric lithic analyst (Hugo Anderson-Whymark).

3. Material recovered from what was thought to be recently made ground or out-of-context loose on the ground surface in various parts of the site. This latter collection included a fresh condition and technologically distinctive assemblage, apparently from recently-made ground close beneath the asphalt road surface that capped the site sequence; this was initially puzzling, but was subsequently attributed to 18th-century gunflint manufacture (Chapter 21).

Chapter 16

Lithic artefacts: Phases 1, 2, 3 and 5

by Francis Wenban-Smith

INTRODUCTION

This chapter presents the lithic material from the earlier phases of the site, prior to the abundant evidence of Phase 6, discussed subsequently (Chapter 17; Chapter 18). The tiny assemblages (n=1, in both cases) from Phase 1, the 'tilted block' (see below), and Phase 2, the 'primary sump infill' deposits to the west of the tilted block, are almost certainly non-artefactual. However their existence needs to be noted, and the possibility considered that they might have been worked. The assemblage from Phase 3 is not much larger (n=8), and includes some natural material, but it also includes three indisputably worked lithic artefacts. This makes it of importance, despite its paucity, as being the first definite evidence of hominin presence in the site sequence. No lithic material was recovered from Phase 4, which was only represented at the site by about 200mm thickness of silty clay in one small width (Log 40011) of the main west-facing section (see Chapter 4). Above this, there was a small assemblage (n=21) from the Phase 5 sands, which seems to represent another phase of hominin presence at the site prior to the main phase of occupation in Phase 6. It does, however, need to be considered whether the evidence that appears to be from Phase 5 is merely intrusive from the basal deposits of the overlying deposits, namely either from Phase 6 in the central and southern parts of the site, or from Phase 8 in the northern part of the site where these gravels directly cut into the Phase 5 sands.

PHASE I

There is just one item forming the lithic assemblage from Phase 1, which is Δ.40063 from context 40056. This context is one of the intermediate beds in the tilted block sequence (Fig. 4.6a; Fig. 4.16; Chapter 4), comprising a grey sand that was equated with Thanet Sand. The artefact is a small, thin flint chip 12mm long in absolutely mint condition. A small area of dark green cortex on its butt suggests that it originates from a piece of Bullhead Bed flint. It was found whilst scraping the section clean for recording and sampling, and it is uncertain whether it was indisputably *in situ*, or whether it is intrusive, perhaps forced into the sediment in the course of the original mechanical excavation that created the face being

cleaned. If not intrusive, tiny flint chips and pebbles are occasionally present in Thanet Sand, although they are usually not quite so fresh as in this instance. It is also not recorded how close the chip was found to the adjacent context 40055, which consisted of a bed about 100mm thick of freshly fractured Bullhead Bed flint nodules with many very sharp chips and fragments. This chip is certainly insufficient evidence on which to argue for hominin occupation in Phase 1; it is probably intrusive from the context 40055, or was perhaps forced into the otherwise fine-grained sediments in course of mechanical excavation.

PHASE 2

The assemblage from Phase 2 also contains just one item. It (Δ.43943) is a medium-sized nodule of Bullhead Bed flint, *c* 150mm long and weighing almost 1kg, bearing a number of what look like scars from flake removals. It was found on the stripped floor of the site (Fig. 16.1), after mechanical excavators had cleared the path for the new roundabout link road. This was in the bottom context, 40077, of the Phase 2 sequence, which was a chalk rubble with occasional flint nodules that directly overlay the 'tilted block' (see Fig. 4.16). The nodule was damaged by heavy machine plant movement, with its upper exposed ridge having undergone obvious modern crushing, leading to formation of several flake scars up to 50mm long, the chips and flakes from which were still present in the ground, and which were also recovered. However, in addition to this obvious modern damage there was also several other larger flake scars, so the nodule was collected for further consideration as a possible artefact. Although originating from various directions, all of the larger flake-scars were on the upper exposed face of the flint nodule, when found, and some of them originated from the ridge with obvious modern crushing. Perhaps the least unconvincing evidence for ancient hominin interference is a battered patch at one end, the bottom-most end as shown in the photo (Fig. 16.1). This shows impact fractures from repeated blows by a hard object on the flattened edge of the nodule, leading to detachment of some cortical chips. This must either be from a hammerstone in the Palaeolithic, or by a modern attempt to shift the nodule by striking it with a shovel or a mattock. Perhaps the most likely possibility is that this battering results from the shovel-cleaning of the

Figure 16.1 Phase 2 assemblage: flint Δ.43943 *in situ* in ground

surface on which it was found, carried out prior to its discovery. On balance, it seems most likely that this is not a genuine Palaeolithic artefact, but a much-abused result of modern interference.

PHASE 3

In contrast to the earlier phases of the site, this is the first horizon where there is indisputable evidence of hominin presence. The lithic collection from the Phase 3 deposits contains eight finds (Table 16.1). More than half of these are ambiguous as to whether they are of natural or hominin origin, but at least three of them are definite artefacts, namely: Δ.50034 (context 40028), a cylindrical flint nodule from one side of which a single flake has been removed and with intense battering at one end; Δ.43219 (context 40061), a small mint condition flake, technologically undiagnostic, but bearing scars of earlier removals struck from the same direction (Fig. 16.2a); and Δ.43941 (also context 40061), a larger, chunkier flake struck across a cylindrical flint nodule, which shows the dorsal scar of a preceding removal struck from the opposite side of the nodule (Fig. 16.2b).

Within the sequence of contexts attributed to Phase 3, context 40061 was at the base, and comprises a chalk and flint gravel with frequent broken Tertiary shell fragments. This gravel was overlain on the eastern side of the site (see Fig. 4.6a) by a more sandy and clayey

deposit (context 40062), also with chalk and flint pebbles, and broken derived Tertiary shell fragments. This was equivalent to context 40028 on the west side of the site (Fig. 4.5a; b). These contexts are thought to represent slope-wash deposits, where they descend into the west edge of a rising water body.

The cylindrical flint nodule is battered on all of its protrusions, apart from on the end were the flake scar is present, where there are several very sharp and unabraded edges. The battering is particularly heavy at the other end of the nodule, where several cortical chips have flaked off, making it look very much like it has been used as a knapping percussor. The only doubt over whether this is a genuine percussor is because the battering extends over every protrusion of the flint nodule, and not just at its end. However, the battering is more intense at the end, and the scars from the cortical chip removals are fresh, supporting its interpretation as a percussor, and not just a simple core or tested nodule.

The two flakes are both technologically undiagnostic, although the larger one (Fig. 16.2b) has a natural cutting edge that would have made it a useful tool in its own right. Both flakes are unstained and unpatinated, and in mint or fresh condition, suggesting firstly, a minimum of post-depositional movement and therefore secondly, that they result from hominin activity on the banks of the water body.

The Phase 3 deposits produced a fully interglacial ostracod fauna (Chapter 11), so this appears to be

Table 16.1 Phase 3 lithic collection

Context	Find ID - Δ	Length (mm)	Weight (g)	Description and interpretation
40028	50034 *		954	Cylindrical nodule with a flake off; battering at one end possibly due to use as a percussor
40061	43219 *	40	16	A definite flint flake; unstained, unpatinated and in mint condition
	43220		80	Natural lump? Possibly irregular waste
	43241	38	5	Small thin flake; lacks clear striking platform but has dorsal scar; slightly patinated and abraded; probably natural
	43242		92	Natural lump
	43941 *	64	116	Another definite flint flake; medium-size, quite chunky flake in fresh condition, from cylindrical flint nodule; unstained and unpatinated
40062	43447		2	Small chip, probably natural
40159	43630		460	Cylindrical piece of flint nodule, probably natural

* definite artefacts

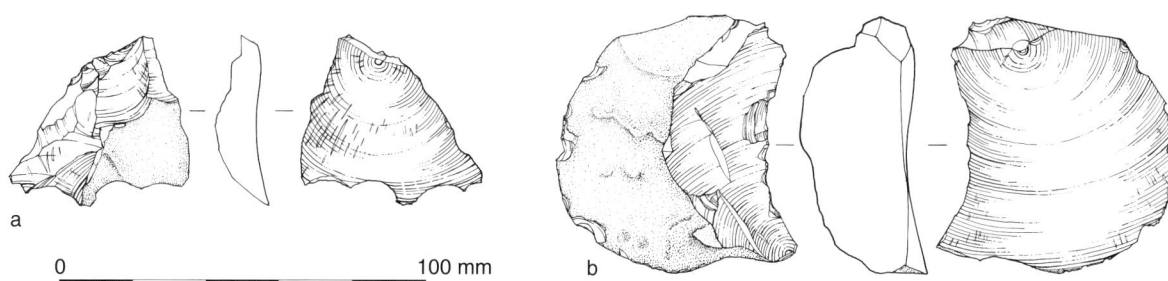

Figure 16.2 Phase 3 lithics: (a) flake Δ.43219 (context 40061); (b) flake Δ.43941 (context 40061) [ill. B. McNee]

evidence of hominin activity that pre-dates, by an uncertain time interval, the main evidence of activity in the Phase 6 clay. Despite the thickness of the deposits, about 3m, between the artefactual horizons of Phases 3 and the junction of the top of Phase 5 with the bottom of Phase 6, there is no sign of a major depositional hiatus or cold climatic conditions. This suggests that Phases 3 through to 6 were all laid down in the same temperate interglacial episode. The Phase 3 assemblage is far too small to consider its technological characteristics as representative of an industrial tradition. There is, however, nothing in the meagre assemblage to distinguish it from the slightly larger Phase 5 assemblage, or the very much larger Phase 6 assemblages.

PHASE 5

Disregarding three natural pieces, the worked flint collection from Phase 5 contained 18 artefacts, 13 of them in mint or fresh condition, and 5 of them in abraded or very abraded condition (Table 16.2).

The abraded assemblage all came from the northern end of the site, where the Phase 5 sands are directly, and unconformably, overlain by the Phase 8 gravels. More specifically, they also all came from the top part of the sands, based on the recorded positions of the artefacts and the recorded position of the gravel/sand junction in various section drawings near their find spots. There is unfortunately no more precise record of how close to

the gravel/sand junction the artefacts were found. Nonetheless, they are thought more likely to be from the sand rather from the base of the gravel, as they are much less abraded than the majority of artefacts recovered from the gravel (Chapter 20). The Phase 5 sand is a fluvial deposit, although the increase in clayey laminations in its top part suggests the start of the stagnating and drying-up trend that is continued in the overlying Phase 6 clay. The artefacts were probably washed into the positions where they were found by episodes of higher energy water-flow, although it is unlikely they have travelled far due to their low level of abrasion. Thus they probably represent evidence of contemporary hominin activity near the edge of the fluvial water body, or on temporary dry surfaces within the water channel.

The abraded assemblage comprises one possible percussor and four flakes. The 'percussor' is a battered flint pebble, with battering on every protruding part. The fact that it has also clearly been naturally abraded makes it hard to distinguish any marks of hominin knapping from natural battering. It looks very natural, but was collected because it was an unusually large clast for the predominantly fine-grained Phase 5 deposits. The flakes are all technologically undiagnostic, hard-hammer struck, quite chunky and of medium size, and all with varying amounts of cortex, ranging from *c* 15% to, in one case, 95%. They thus seem to represent sporadic knapping activity applied to freshly and locally collected raw material.

Table 16.2 Phase 5 lithic collection

	5 – Percussor (?)	10 – Tested nodule	20 – Core	30 – Core-on-flake	40 – Core-tools	50 – Handaxe-on-flake	60s – Fl-tools	80 – Fl-flakes	90 – Flakes	100 – Irreg. waste	Sub-total (n)
Phase 5 - Fresh	-	2	2	-	-	-	1	-	5	3	13
%	-	15.4	15.4	-	-	-	7.7	-	38.5	23.1	
Phase 5 - Abraded	1	-	-	-	-	-	-	-	4	-	5
%	20.0	-	-	-	-	-	-	-	80.0	-	
Phase 5 - All	1	2	2	-	-	-	1	-	9	3	18
%	5.6	11.1	11.1	-	-	-	5.6	-	50.0	16.7	

The fresher assemblage from Phase 5 (Table 16.3) mostly comes from the central part of the site, between Trenches B and C, and was found during machine excavation in this area, which was continued into the top *c* 0.5m of the Phase 5 sands. A few artefacts in fresh condition were also found on the east side of the site, north of Trench A, and a single one was found in the bottom spit of Evaluation Trench IV, south of Trench D. As with the abraded assemblage, they were all recovered from the top part of the Phase 5 sands, without precise records of how close to the junction of the overlying deposit. In this case this was usually the Phase 6 clay.

Despite the possibility that some of this assemblage is intrusive from the overlying Phase 6, all of it was well-provenanced to horizons well below the main archaeological levels of Phase 6, so it nonetheless represents evidence from a different, and earlier, phase of occupation. The presence of hominin artefactual evidence is wholly compatible with the depositional environment in the top of the Phase 5 sands, where a sharp increase in thickness and frequency of clayey beds suggests cessation of fluvial deposition and development of quiet conditions.

The fresh condition of the artefacts suggests minimal post-depositional disturbance and reworking, so they probably represent intermittent activity on temporarily exposed dry surfaces, at a time when the fluvial regime associated with deposition of the Phase 5 sands was coming to an end, with the water-body stagnating and drying-up. At this time there was perhaps some colluvial input from the valley sides, which may have introduced some of these artefacts into their location of discovery.

Table 16.3 Phase 5, fresh assemblage (see Table 15.4 for description of 'tech group' codes)

Tech group	Find ID - Δ	Context	Area, Trench	Whl	%Cx	DSC	WtG	Notes
10 –Tested nodules	43836	40025	B-C	-	-	-	2015	Split slab of tabular flint; heavily frost-fractured
	43837	40025	B-C	-	-	-	1150	Rounded, cylindrical part of nodule, smooth cortex
20 –Cores	43839	40025	B-C	1	6	6	485	Globular core with three main removals; remaining flint looks accessible and good quality (Fig. 16.3a)
	50144	40025	A-B	1	0	13	890	Globular core with numerous removals; remaining flint looks inaccessible and frost-fractured
63 –Notched flake-tool	43808	40025	B-C	1	5	1	335	Large flake with single notch (Fig. 16.3b)
90 - Flakes	43834	40025	B-C	1	5	2	99	Quite large flake, migrated platform
	43856	40025	B	1	0	4	18	Small flake, migrated platform
	50109	40025	A-B	1	6	2	8	Small flake on derived Tertiary pebble; stable platform
	50060	40072	N of A	1	7	2	33	Small flake, stable platform
	50166	40025	A-B	1	4	6	575	Large thick flake, stable platform; several well-developed ring-cracks and Hertzian cones indicate very violent knapping
100 – Irregular waste	40818	40025	Tr IV (spit 11)	-	-	-	25	Small lump with combination of natural frost-fracturing and scars from deliberate knapping
	43824	40025	B-C	-	-	-	794	Large globular lump, heavily frost-fractured but some conchoidal rippling suggests knapping removals
	43835	40025	B-C	-	-	-	85	Small frost-fractured lump; possibly entirely natural

The raw material in the assemblage is very varied, although all being typical flint of the Swanscombe area. The cortex on one of the tested nodules is abraded smooth and a relatively dark blueish-grey, suggesting a history of derivation and transport prior to collection for attempted knapping. The cortex of most of the other artefacts is white, and in a fresh rough condition, suggesting a minimum of transport and reworking after the original derivation from Chalk bedrock. One of the smaller flakes (Δ.50109) has a smooth rounded dark grey battered outer surface, indicating the raw material used was a derived Tertiary flint pebble.

Technological and typological details of the assemblage are provided in the summary table (Table 16.3). As a group, the assemblage is entirely compatible with a typical Clactonian industrial tradition (see Chapter 22 for discussion of the definition and recognition of 'Clactonian'). There is a combination of firstly, flake production by a simple unstructured approach to core reduction, with a combination of episodes of repeated flaking from the same striking platform ('stable platform') and *ad hoc* changing of the striking platform to one of the scars from a previous removal ('migrating platform'); and secondly,

the secondary working of one flake to make a notched flake tool. However, the small size of the assemblage means that it cannot be unequivocally accepted as representing a Clactonian industrial tradition.

The raw material shows a high incidence of frost-fracturing, particularly in the larger tested nodule and the larger of the two cores, which has probably influenced the cessation of reduction. In contrast, reduction of the smaller of the two cores (Fig. 16.3a) was stopped when it still had plenty of good quality flint available, and which was easily accessible from a range of possible striking points. Most of the flakes had more than 50% cortex (Table 16.3), indicating they were from reasonably early in their respective reduction sequences.

There is one secondarily worked flake in the fresh assemblage, a large flake from which a single notch has been struck at the distal end (Fig. 16.3b). Prior to this removal, the remainder of the perimeter of the parent flake would have been entirely cortical and not much use for cutting. After it, there was a slightly concave and very robust sharp edge formed, and it is reasonable to suggest, therefore, that this artefact can be regarded as a deliberately formed flake-tool, to facilitate cutting.

Figure 16.3 Phase 5 lithics:
(a) core Δ.43839 (context 40025);
(b) notched flake-tool Δ.43808 (context 40025)

DISCUSSION AND CONCLUSIONS

The assemblages discussed in this chapter are of small size, but represent occupation at, or nearby, the site at an earlier stage than the main occupation evidence of Phase 6 (Chapters 17 and 18). The tiny assemblages from Phases 1 and 2 are probably not of Palaeolithic hominin origin. However, the small assemblages from Phases 3 and 5 include sufficient indisputable artefacts to demonstrate hominin presence. The ostracod fauna from Phase 3 (Chapter 11) indicate prevailing interglacial conditions, and there is nothing in the sequence between Phases 3, 5 and 6 to indicate a major depositional hiatus or evidence of climatic cooling. It therefore seems likely that the lithic material from Phases 3 and 5 represents earlier hominin activity in the same interglacial episode as represented in Phase 6.

Although the assemblages are too small to make reliable generalisations, both assemblages are technologically and typologically indistinguishable from the much larger Phase 6 collection. There is evidence of on-the-spot production of flakes from a range of lithic raw material, applying a relatively unstructured knapping strategy based on removal of one or more flakes from one platform before migrating to another. The presence of a high proportion of cortical flakes reflects the early stages of production, in turn suggesting that the raw material was locally acquired, unless minimally worked flint nodules were being moved around the landscape, which would be surprising. The rough condition of the cortex on the majority of artefacts suggests the raw material had not been affected by fluvial transport or by major slope-wash abrasion. However, the high incidence of frost-fracturing (which clearly predates the knapping, as many of the knapping fracture planes are visibly affected by pre-existing frost-fracture planes) reflects prolonged exposure to the elements, presumably the glacial cold of MIS 12.

Establishing the raw material source is an important part of reconstructing patterns of mobility in relation to knapping activity and site usage. This is considered in more detail for Phase 6 (Chapters 17, 18), where there is much more evidence, including the recovery of abundant non-artefactual flints in the Phase 6 collection.

Chapter 17

Lithic artefacts: Phase 6, the elephant area

by Francis Wenban-Smith

INTRODUCTION

The great majority of the lithic collection from the site came from the Phase 6 clay, which produced 2,238 finds in total, although these included 157 natural pieces. The Phase 6 collection was sub-divided into three subsidiary assemblages for analysis (Table 17.1):

- Assemblage 6.1, the dense concentration south of Trench D;
- Assemblage 6.2, the generally more dispersed material north of Trench D, excepting the cluster around the elephant skeleton; and
- Assemblage 6.3, the specific cluster of artefacts around the elephant skeleton (Fig. 17.1).

Assemblages 6.1 and 6.2 are discussed subsequently (Chapter 18); the remainder of this chapter focuses on assemblage 6.3 from around the elephant skeleton. After a brief review of the quantity and provenance of this assemblage (below), the following section considers its taphonomy and site formation processes, taking account of results from sampling for microdebitage, the clustering of the lithic material and the degree of refitting. After this is an analysis of the technological *chaîne opératoire*, focusing on the source and transport of raw material, the evidence from refitting sequences of reduction and the typology of artefacts recognised as tools. The final section considers the interpretation of the elephant area as a whole, covering what might have been the circumstances of the elephant's death and the hominin interaction with its carcass, and whether the lithic evidence can provide any wider insights into hominin behaviour of the era.

Table 17.1 Phase 6 lithic collection, subsidiary assemblages

Assemblage	Context/s	Natural pieces	Total flints	Notes
6.3	40039 40078 40078? 40099 40100 40103	12	93	Flints from near elephant skeleton
6.2	40039 40068 40069 40070 40078 40099 40100 40103 40144 40158	6	135	More dispersed flints that were not near the elephant skeleton or part of the main scatter south of Trench D, including some from within the Phase 6b tufaceous channel fill
6.1	40036 40039 40039? 40078 40100	139	2010	Flints from main concentration south of Trench D
Total		157	2238	

Table 17.2 Assemblage 6.3, artefact collection from around the elephant: provenance and stratigraphic phasing (excluding natural pieces)

Phase	Context	Artefacts (n)	Notes
6	40100	1	-
	40078?	2	From bulk spoil samples taken after first discovery of elephant; uncertain context
	40078	62	Darker brown organic-rich bed in clay with elephant bones
	40099	14	Grey clay under the elephant bone horizon
6a	40103	1	Grey clay divided from overlying context 40099 by thin Fe-rich horizon, which was however faint immediately below the elephant skeleton
	40039	1	The bottom bed of the clay, sealed beneath the base of context 40103 in the vicinity of the elephant
Total		81	

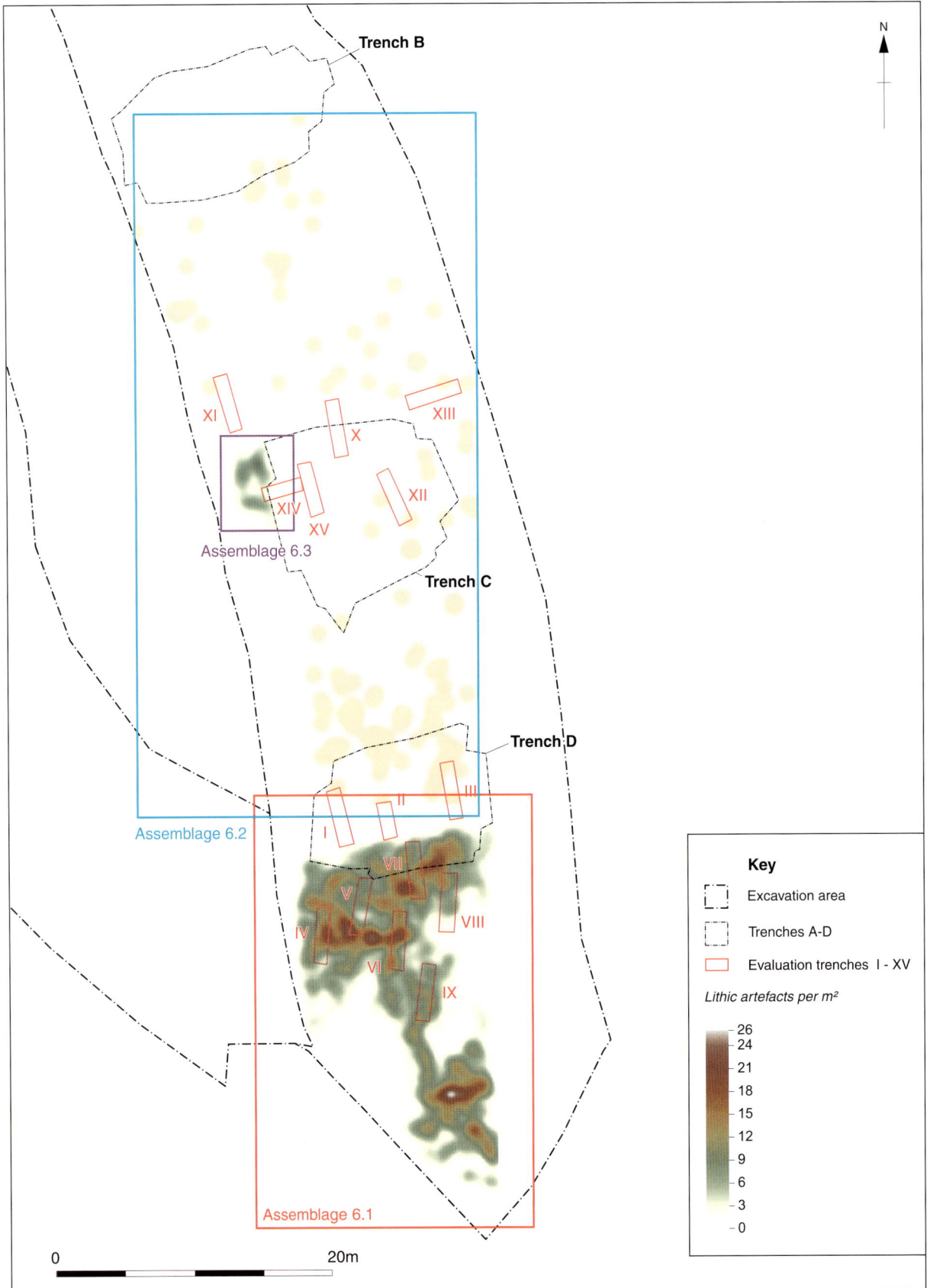

Figure 17.1 Phase 6 lithic distribution overview

PROVENANCE AND QUANTIFICATION

In total, there were 81 lithic artefacts recovered from the immediate vicinity of the elephant skeleton (Table 17.2; Fig. 17.2). Most of these were excavated, but two came from the bulk samples of machine spoil taken immediately

after initial discovery of the skeleton. These were allocated context '40078?' as, although most of the sediment was from the same general horizon as the elephant bones, contamination from other horizons was also likely. The great majority of artefacts came from the same specific dark brown organic-rich context 40078 as the elephant

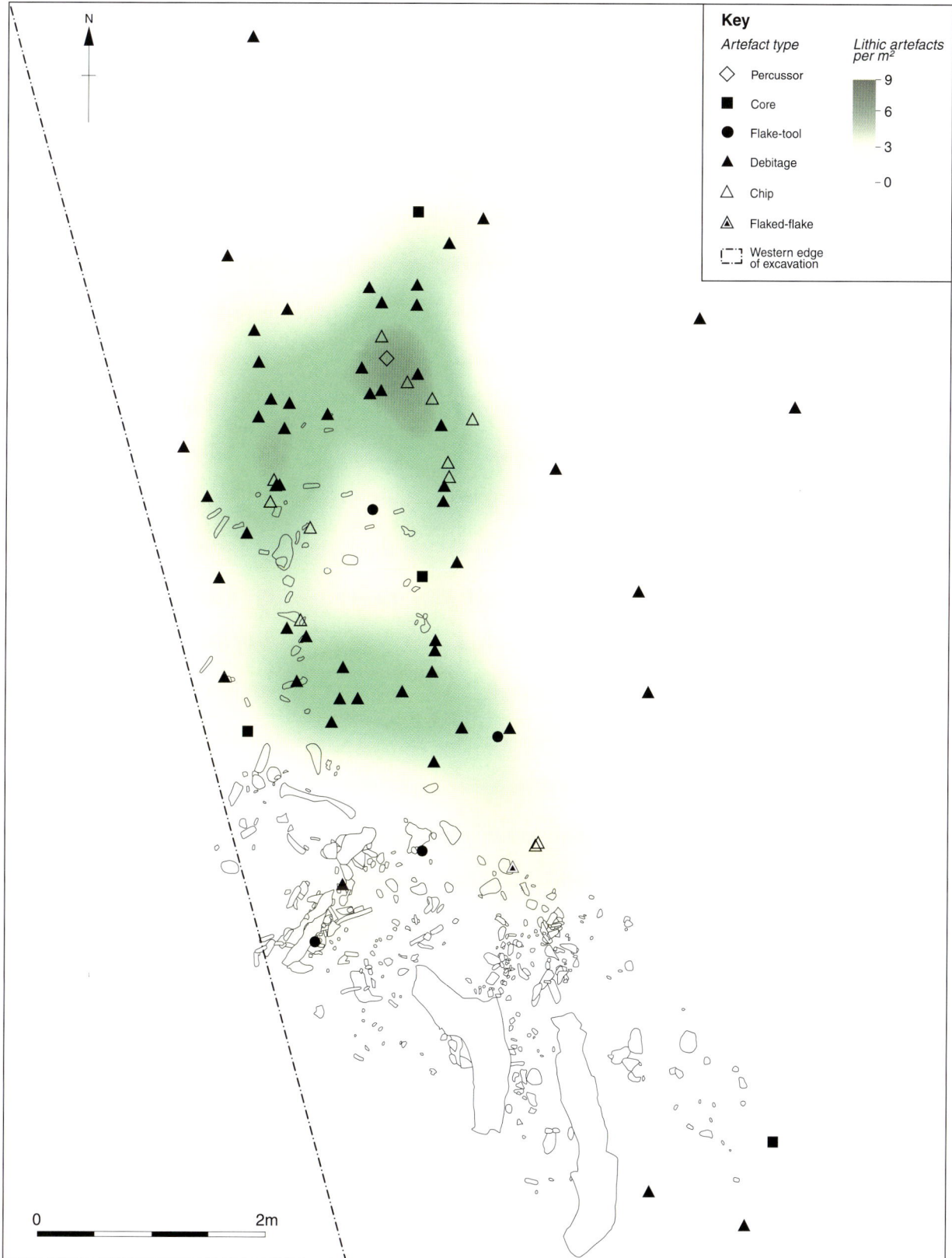

Figure 17.2 Lithic artefacts in relation to the elephant skeleton

bones, and thus are securely associated with them. The association of a few of the artefacts with the elephant skeleton is, however, questionable, particularly the two from the bulk samples, and the single artefact from context 40039 which was recovered from a clearly distinct bed well below the elephant horizon. All these artefacts were, however, initially included for consideration as part of the assemblage potentially associated with the elephant skeleton, prior to refitting studies to investigate the integrity of their stratigraphic provenance. Likewise, although almost all of the artefacts in the assemblage were in absolutely mint condition, as if freshly knapped, a few artefacts that were merely in 'fresh' condition and a single artefact (from context 40078) that was in slightly abraded condition were also all included in this initial consideration. This was prior to the refitting study and consideration of site formation processes.

TAPHONOMY AND SITE FORMATION

Almost all the artefacts from the assemblage around the elephant skeleton came from two contexts, 40078 and 40099; a few others came from overlying or underlying contexts, or were of uncertain provenance (Table 17.2). The mint or fresh condition of most artefacts and their clustering around the elephant skeleton provided an initial indication that they probably represent undisturbed evidence from hominin activity associated with the carcass. However, it remained to be established whether the different context provenance distinguished different episodes/phases of artefact deposition, or whether they perhaps reflected post-depositional artefact movement within the sediment, post-depositional development of sedimentary boundaries or imprecise field recording. To investigate the taphonomy

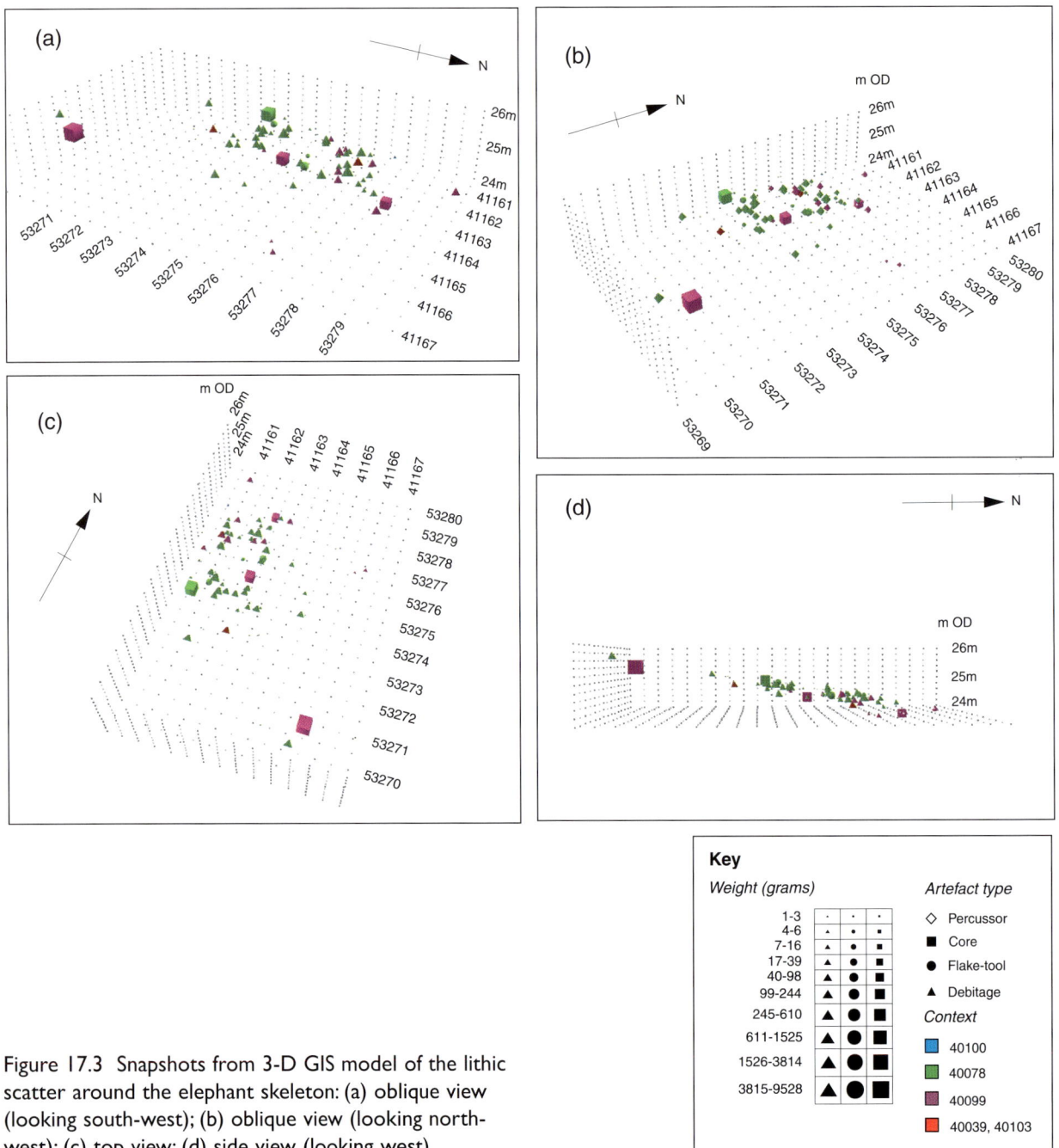

Figure 17.3 Snapshots from 3-D GIS model of the lithic scatter around the elephant skeleton: (a) oblique view (looking south-west); (b) oblique view (looking north-west); (c) top view; (d) side view (looking west)

and site formation process of the assemblage further, three approaches were adopted. Firstly, a 3-D GIS model was created with artefacts represented by different symbols for different technological categories, with the symbols sized according to their weight and colour-coded by context. This allowed initial investigation of the 3-D distribution of the artefacts, and of whether there was any immediate visual evidence of size-sorting, either laterally or vertically. Some snapshots from the model are illustrated here (Fig. 17.3), showing various views of the scatter around the elephant skeleton. This is however a poor second to viewing the 3-D model for real, which is available in the ADS online resources accompanying this monograph (ADS 2013).

The 3-D modelling showed that almost all of the artefacts were contained in a narrow band *c* 10-15cm thick that dipped northwards at an angle of *c* 20° from horizontal (Fig. 17.3d). There was just one exception (apart from those lacking precise coordinates, such as those recovered from the two bulk samples). This was the find from context 40039, which was a piece of irregular waste found at least 10cm lower than the rest of the lithic material; it was therefore excluded from subsequent

analyses of the material around the elephant. There was no evidence of size-sorting, either horizontally or vertically.

Secondly, a refitting study was carried out. This included all the material from close to the elephant skeleton, designated as assemblage 6.3, which was examined in conjunction with the more sparsely distributed flints from the wider area of *c* 10m around the elephant skeleton. The results demonstrated a high proportion of refitting material near the skeleton, with seven distinct refitting groups A-G ranging from 2 to 24 artefacts in size (Table 17.3; Fig. 17.4). The single abraded flake in the assemblage from around the elephant (Δ.40659) did not refit, and so was excluded from subsequent analysis of the elephant assemblage, along with the two flakes of uncertain provenance from the bulk samples and the artefact from context 40039, none of which refitted. Of the 77 artefacts now comprising the elephant assemblage, including 12 tiny chips, 52 of them were indisputably refittable (ie 68% of the assemblage), and an additional two were thought likely to be part of the Group B sequence. The maximum spatial separation for any material within the same refitting group was also measured (Table 17.3). The greatest ranges (of between 4

Table 17.3 Assemblage 6.3: refitting groups and non-refitting artefacts summary [total excludes: natural pieces; two insecurely provenanced pieces Δ.44020 and Δ.44021 from bulk samples; debitage Δ.43843 from context 40039; and abraded debitage Δ.40659 from context 40078]

Refitting group	Description	Artefacts (n)	Context	Maximum separation (m)	Notes
A	'Piebald nodule'	7	40078	2.7	Sequence of flakes without core, from distinctive banded grey/white flint with green cortex
B	'Large cylindrical core' Δ.40494	7	40078	4.0	Sequence of flakes with core from early in its reduction
B?	Additional debitage, possibly related to large cylindrical core	1	40099	4.0	Cortical irregular waste from end of a cylindrical nodule, that is probably start of Group B sequence
	Δ.40494?	1	40078	5.1	Small flake-flake that could be from secondary working of missing flake early in Group B reduction sequence
C	'Main core' flaking sequence, Δ.40871	18	40078	5.0	Reasonably complete reduction sequence from
		6	40099		initial decortication of nodule through to core
D	Broken percussor	5	40078	4.7	Broken flint percussor
		2	40099		
E	Broken core, 'Shattered nest'	3	40099	2.6	Core that has broken into three pieces from one blow, one of these pieces then knapped further
F	Broken flake	2	40078	0.4	Medium size flake, partly cortical that has broken on knapping
G	Broken cortical flake	2	40078	0.3	Small cortical flake that has split during percussion
-	Core	1	40099	-	Large core on southern fringe of elephant lithic concentration
-	Flake-tools	2	40078	-	Two utilised flakes and one notched tool
		1	40103		
-	Flakes	3	40078	-	Mostly from edge of elephant lithic concentration
		2	40099		
		1	40100		
-	Irregular waste	2	40078	-	-
-	Chips	11	40078	-	-
Total		77			

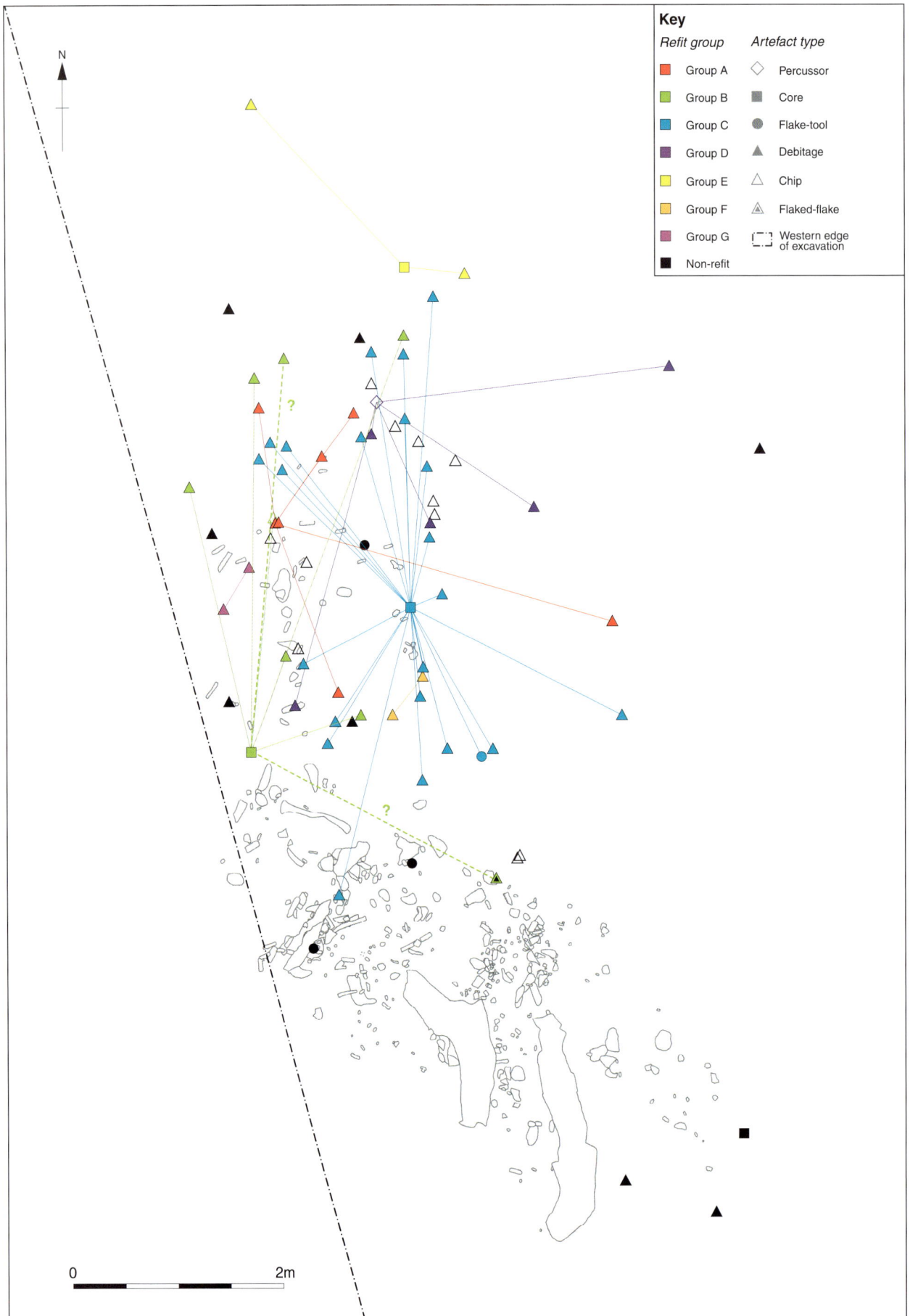

Figure 17.4 Orthogonal view (from above) of refitting groups A–G around elephant skeleton [lines linked to core or heaviest piece for groups where n>2]

and 5m) were recorded for the larger refitting groups B, C and D. The other, smaller groups had much more reduced refitting ranges, particularly the broken flakes represented by refitting groups F and G, which were less than 0.5m apart in both cases. The high proportion of refitting material (rising to 79% if one excludes the 11 small chips that were not refitted, but of which many appear likely to belong to the same raw material as the refitting groups), the internal clustering of the refitting groups and the generally small distances of separation all suggest a minimum of post-depositional disturbance.

The refitting summary table (Table 17.3) also shows the contexts of the refitting material within each of the refitting groups. This demonstrates that, although most of the groups contain material from just one context, two of the larger groups (C and D) contain material from both contexts 40078 and 40099. This establishes that there is no basis for regarding the material from these two contexts as meaningfully divisible. Although the two artefacts from contexts 40103 and 40100 did not refit to any other material, they are included in subsequent analyses as part of the elephant assemblage as they were shown in the 3-D GIS model to come from the same narrow band of artefacts as the rest of the

elephant assemblage. Also, the distinction of these contexts from contexts 40078 and 40099 in the vicinity of the elephant was often unclear.

The third and final element of the taphonomic investigation was the microdebitage study. Experimental studies of technologically similar simple flake-core reduction sequences (see Wenban-Smith 1985 and 1996; Toth 1982) have established that even uncomplicated knapping strategies without platform preparation produce abundant microdebitage in the size ranges characterised here as chips and spalls, that is less than 20mm long; and also that all sizes of debitage are tightly spatially co-clustered for undisturbed knapping scatters. Based on the experimental work of Wenban-Smith (1985), which provides the most relevant comparison with similar reduction sequences applied to similar nodular flint raw material, small flakes in the size range 10-50mm typically remain within a circle of radius *c* 0.75m around the centre of the tight cluster of larger flakes (Fig. 17.5). Smaller chips in the size range 2.5-10mm typically remain within a slightly larger arc of up to 1m. The quantities of material produced in these size ranges were also collated in this study. Data from four experiments carried out in 1985 are shown, along with a fifth experiment carried out in 2011 as a control (Table

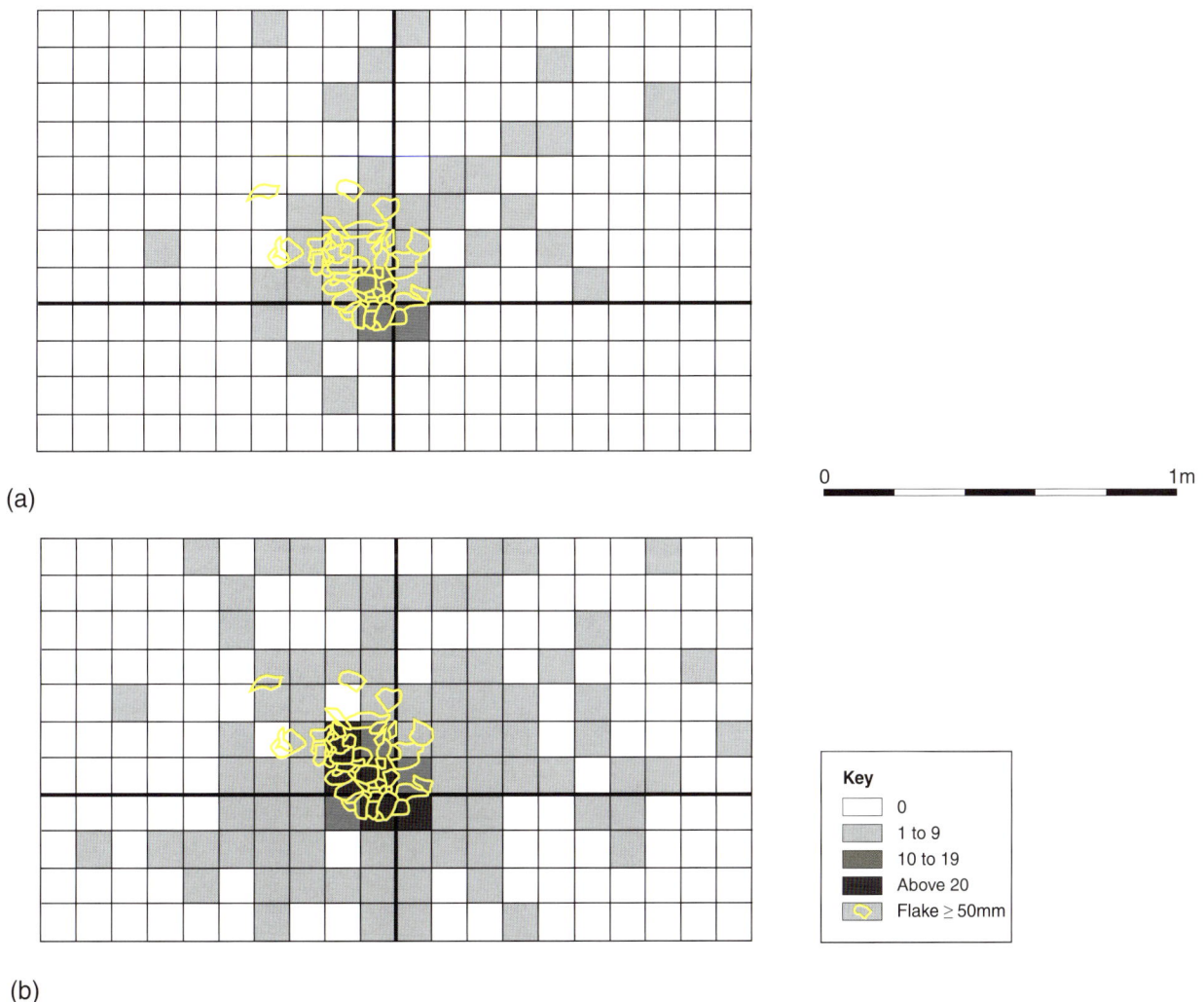

(a)

(b)

Figure 17.5 Spatial distribution of experimental flake/core (Clactonian) debitage: (a) flakes 10–50mm; (b) chips 2.5–10mm (from Wenban-Smith 1985)

17.4). All five experiments gave very consistent results, which indicated that for a knapping sequence long enough to produce at least 10 flakes ≥50mm, significant quantities of debitage (approximately four times as many) would be produced in the size range 10-50mm. There would also be even greater quantities of microdebitage in the size range 2.5-10mm (approximately ten times as many).

In order to investigate microdebitage from the lithic concentration around the elephant, an L-shaped pattern of sample squares was dug, with the long axis approximately NNW-SSE through the centre of the scatter, and the shorter axis orthogonal to this at the southern end of the scatter (Fig. 17.6). Each sample square was 0.5 x 0.5m wide and about 5cm deep. Unfortunately, rather

Table 17.4 Debitage size-profiles: comparative experimental data, assemblage 6.3 and microdebitage sampling strip

Size-range	Ex 1		Ex 2		Ex 3		Ex 4		Ex 5 (2011)		All exps		Ass. 6.3		Microdebitage strip	
	n	%	n	%	n	%	n	%	n	%	n	%	n	%	n	%
≥ 50mm	14	8	32	6	27	7	39	6	10	6	136	7	27	41	10	53
≥10-50mm	53	30	115	21	100	25	128	20	46	26	440	23	38	58	6	32
2.5 - 10mm	108	62	403	73	270	68	469	74	121	68	1359	70	1	1	3	15

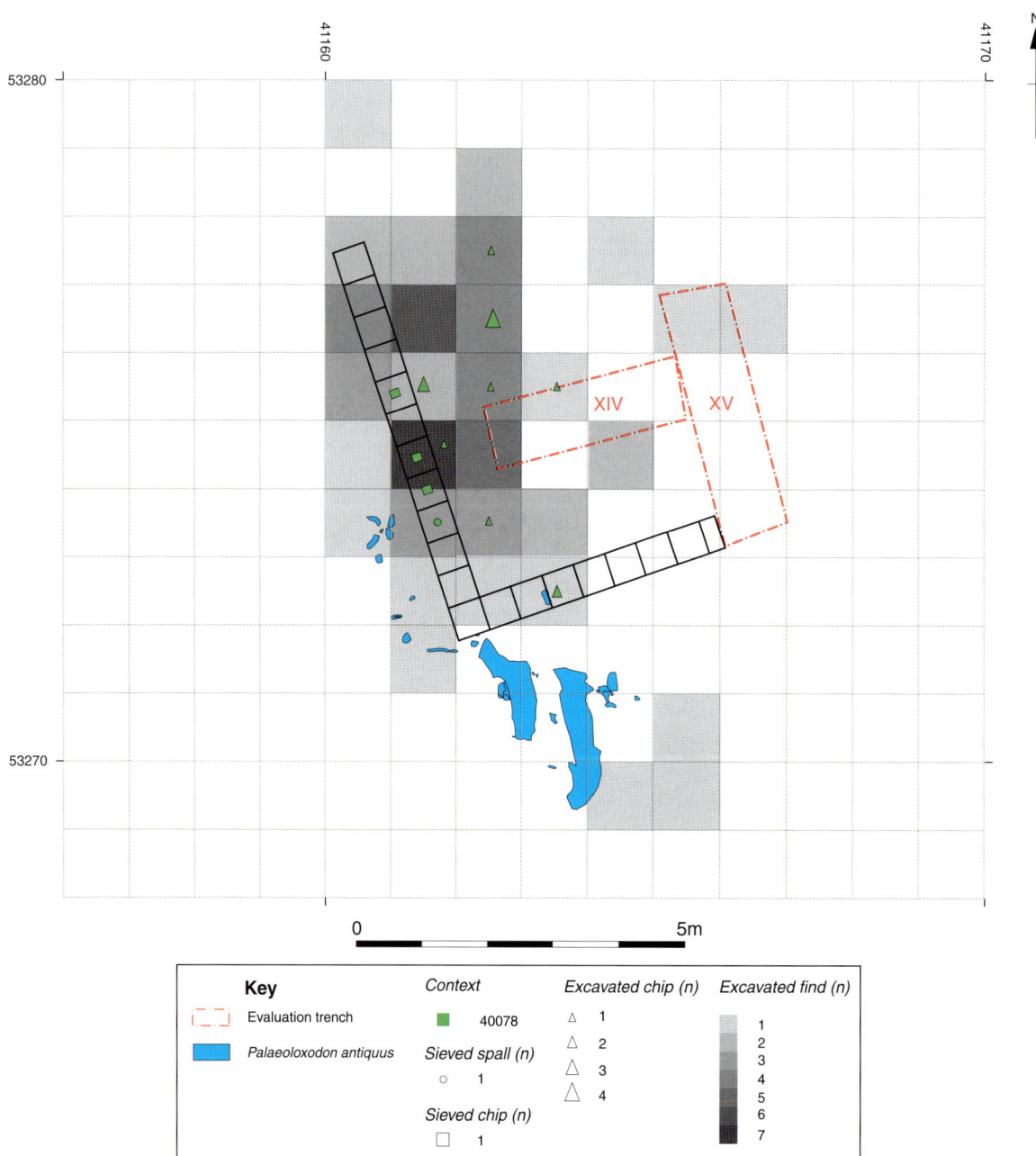

Figure 17.6 Microdebitage sampling and lithic recovery around the elephant skeleton

than the total sediment recovery that was intended within these squares, a sediment subsample of approximately only 2kg was taken from within each square, compromising meaningful comparison of the quantity of microdebitage recovered from the sampling with that from the comparative experimental work. Nonetheless, the results are presented here. Comparative quantitative data for the elephant scatter generally, and the longer axis of the L-shaped sample strip, are presented in the tabular data summary (Table 17.4). The quantities of microdebitage recovered in the sample strip are shown in relation to the distribution of larger artefacts in the elephant concentration (Fig. 17.6). Even allowing for the incomplete sampling, there seems to be an anomalous lack of the smaller chips and spalls that one would expect to be present in an entirely undisturbed and freshly knapped lithic scatter.

This impression was reinforced during hand excavation in the scatter area. Even though one cannot expect to reliably recover very small debitage when trowelling, one is generally aware when small pieces are encountered, particularly when flint artefacts are contained in a very clayey matrix as was the case here, as they scrape against the metallic trowel. From both personal experience of trowelling by the skeleton and the anecdotal reports of other excavators, this was not a common occurrence in the elephant area, and it was thought that a very high proportion of the lithic remains that were present were found and recovered. All flint pieces detected during trowelling were recorded as small finds, leading to the assemblage from around the elephant including 12 chips less than 20mm long. If there had been a greater abundance of small debitage, many more would have been noted and recorded during trowelling.

Taken together, the evidence seems to indicate a slight spatial disturbance of the larger lithic material, as well as of the elephant skeleton parts, alongside a sedimentary burial process that has led to the dispersal of the very small and more transportable lithic microdebitage. The distribution arc of each of the larger refitting groups B, C and D is, at *c* 5m, slightly larger than in the comparative experimental models of single reduction sequences (Fig. 17.5). It should of course be remembered that the Palaeolithic knappers may not have completed their reduction sequences on the same spot, and, as discussed below, Group D is a broken percussor rather than a reduction sequence. In addition, the elephant bones themselves are distributed more widely than, and have lost the patterning of, an entirely undisturbed skeleton (Chapter 8). Nevertheless, the fact that the elephant bones are still relatively concentrated in a scatter, with the tusks for instance still identifiably parallel, indicates minimal disturbance of faunal material in the area of the elephant skeleton. The larger lithics too are evidently minimally disturbed, given the high degree of refitting with the constituents of all the lithic refitting groups contained within such restricted areas. It seems inconceivable that the co-occurrence of the elephant skeleton and the undisturbed refitting scatters in the

same location, and the same narrow stratigraphic horizon, should be a coincidence within the wider context of the sparsely distributed lithic and faunal remains in the surrounding Phase 6 clay. Consequently it seems inescapable that the knapping activity reflects hominin engagement with the carcass.

As discussed previously (Chapter 4), the Phase 6 clay is thought to have been mostly introduced into a standing/fluctuating water body by slopewash from high ground to the west. The elephant carcass, which was originally lying on a firm dry surface rather than mired in soft sediment, as indicated by the lack of deformation of the sedimentary sequence under the horizon where the bones were recovered, would have had substantial thickness when fresh. It seems that as it collapsed (or perhaps exploded) with decay, and perhaps also slightly spread out laterally due to water movement and slopewash input, some parts of the skeleton became incorporated in organic-rich peaty/clayey sediment during a phase of raised water level. This would have partially submerged parts of the decaying skeleton, leading to their preservation. The larger lithic artefacts that were associated with hominin activity in the area of the skeleton seem to have remained almost undisturbed during this process, perhaps also subject to very slight lateral displacement. However the smaller microdebitage that would almost certainly also have been present have been dispersed during this process of burial.

The final and most unfortunate aspect of the taphonomy and site formation process of the elephant area, is that its western part was bulk-excavated by machine in summer 2003 and taken away in the back of a lorry without anyone knowing of its presence. It is therefore now entirely uncertain whether there were many, or few, faunal and/or lithic remains lost during this process. There is thus forever a fog of uncertainty over whether the lithic collection is a relatively complete representation of activity in the vicinity of the elephant skeleton, or whether certain potentially behaviourally significant gaps and absences in the lithic collection (discussed below) would have been filled by recovery and analysis of the lost part of the site.

TECHNOLOGY, TYPOLOGY AND THE CHAÎNE OPÉRATOIRE

Introduction and raw material

The assemblage regarded as reliably associated with the elephant skeleton contained 77 artefacts, of which 12 were chips <20 mm long, and there were also four other artefacts from the area that were excluded from the elephant assemblage on grounds of their provenance, as discussed above. Each artefact was analysed separately, prior to refitting, and the initial technological breakdown of the assemblage is given without consideration of the refitting results (Table 17.5; Fig. 17.7). Besides two broken pieces of percussor (which were subsequently shown by refitting to belong to the same single

Table 17.5 Assemblage 6.3: technological categories, excluding natural pieces

	5 - Percussor (broken)	10 - Tested nodule	20 - Core	30 - Core-on-flake	40 - Core-tools	50 - Handaxe-on-flake	60s - Fl-tools	80 - Fl-flakes	90 - Flakes	100 - Irreg. waste	110 - Chips	Sub-total (n)
Elephant scatter	2	-	4	-	-	-	4	1	37	17	12	77
% - inc chips	2.6	-	5.2	-	-	-	5.2	1.3	48.1	22.1	15.6	
% - excl chips	3.1	-	6.2	-	-	-	6.2	1.5	56.9	26.2	-	65
Non-elephant material	-	-	-	-	-	-	-	-	2	2	-	4
% - inc chips	-	-	-	-	-	-	-	-	100.0	100.0	-	
% - excl chips	-	-	-	-	-	-	-	-	100.0	100.0	-	4

percussor), there were four cores and four flake-tools, each representing approximately 6% of the assemblage. The remainder was waste debitage, comprising about 57% flakes and 26% irregular waste, if the chips are discounted. In general, it was regarded as preferable to omit chips from the quantitative assemblage summaries, as their recovery was very patchy, depending upon the care of individual excavators and the methods of excavation. Omitting the chip-counts therefore makes quantitative intra-site assemblage comparisons more meaningful, particularly when expressed as percentages, both between the different assemblages from Phase 6, and between assemblages from different phases of the site's sequence.

The raw material represented in the assemblage is very varied. The cortical remnants on the debitage and cores are in a range of conditions, from fresh and unabraded suggesting a raw material source reasonably fresh from Chalk, to highly smoothed suggesting a history of substantive transport and reworking prior to collection for knapping. There is a reasonably high incidence of frost-fracturing, most of which demonstrably precedes knapping, where possible to tell. Most of the frost-fracturing is relatively minor, without major implications for flake production or technological potential. However quite a few pieces are very severely frost-fractured, which would have severely limited their potential for producing sharp-edged flakes or for carrying out any more ambitious reduction scheme such as handaxe manufacture or Levalloisian production. The flint itself is typical grey nodular flint of the Swanscombe area. Some pieces show evidence of knots

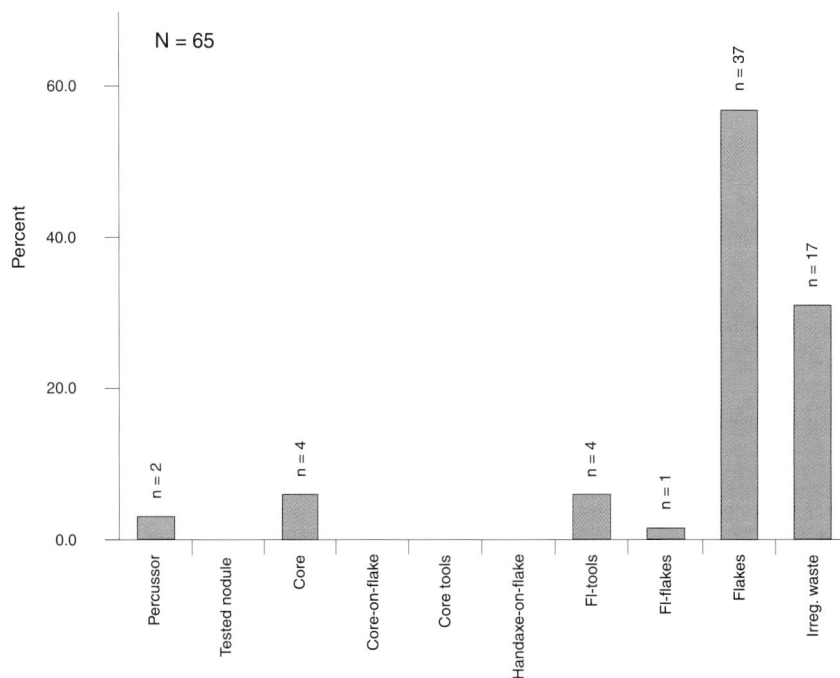

Figure 17.7 Technological summary of assemblage 6.3, from around the elephant skeleton

Table 17.6 Assemblage 6.3: cores

Find ID	Refitting group	Whl	%Cx	DSC	ML	WtG	Notes
Δ.40494	B	1	9	6	158	1386	Cylindrical core, with alternating flaking at one end [Fig. 17.10]
Δ.40871	C	1	2	12	92	324	Globular core, with small remnant cortical patch, and migrating flaking [Fig. 17.8a]
Δ.41100	-	1	5	6	217	1856	Broken tabular slab of flint, with one very large flake removal, and several smaller ones
Δ.41376	E	0	3	8	84	186	Broken piece of larger core, subsequently flaked further [Fig. 17.8b]

of clearer, more glossy flint surround by more opaque flint – the so-called Devil's Eye of local folklore – and much of it is quite pale, cherty and coarse-grained in the interior, with only a narrow band near the nodule's exterior being relatively translucent and glossy. The refitting work demonstrates at least seven different flaking sequences carried out on different pieces of raw material (Groups A-G); in addition to the refitting material, there is one large core (Δ.41100), three flake-tools and two pieces of larger debitage which all appear to come from additional distinct pieces of raw material, making a total of at least thirteen distinct episodes of reduction not including various smaller pieces of debitage and irregular waste.

Four cores and a percussor

A certain amount of technological information is interpretable from the separate artefacts, particularly the cores, prior to the more detailed discussions based on the refitted reduction sequences. The cores all represent apparently unstructured, or very simply structured, flaking strategies (Table 17.6).

Core Δ.41100 (not illustrated, and for which no refitting flakes were found) was found slightly apart from the main group, at the extreme south-west end of the scatter, near the tips of the tusks (Fig. 17.2). It was a thick, broken (possibly deliberately) tabular slab of flint, with several smaller removals at one end, and then, on the opposite face, one huge flake removal (c 140 x 120 x 40mm). It was attempted to find this removal in the material from around the elephant skeleton and in the wider surrounding area, but this was unsuccessful. Such a large flake would have been easily recognisable, so it clearly is not present in the Phase 6 collection. There are a few minor frost-fracture flaws visible, but there is obviously a substantial quantity of good quality flint remaining in the core, which was easily accessible if desired.

Core Δ.40494 was also quite large (Fig.17.13a); its cortex was blueish-grey, and moderately abraded, suggesting a history of derivation, and the interior flint was a quite rare combination of glossy, fine-grained and at the same time a pale opaque mottled grey. It had undergone a short sequence of flaking at one end, including removal of flakes from alternating platforms. There were also some other impact marks on the core

which caused cortical chips to be detached, but not successful flakes. The core was abandoned with a substantial amount of good quality flint easily accessible and unexploited. Most of the debitage from this core was found nearby and refitted to it (as Group B), and more details of the reduction sequence are discussed further below.

Core Δ.40871 was relatively small (Fig. 17.8a), weighing less than a quarter of the two above-mentioned cores. It had, however, been subject to a much longer sequence of reduction, with scars from 12 flakes on the core, and 23 pieces of refitting debitage present in the assemblage from around the elephant. The core itself retains a small area of cortex opposing a very obtuse edge, from which a series of alternating flakes had been struck. Most of the removals from this core were present in the scatter around the elephant skeleton, including large pieces of irregular waste from the start of its reduction, and the more detailed reduction sequence revealed by the refitting of this material is discussed below under Group C (below). As with core Δ.40494, the raw material is dense, fine-grained and an opaque mottled pale grey. The cortex is, however, relatively unabraded suggesting a lesser history of derivation prior to collection for knapping.

The last core, or core remnant, associated with the elephant skeleton is Δ.41376, which joined to two other pieces as Group E (Fig. 17.8b). In contrast to the other cores, it is made of a more typical type of local flint raw material, dark grey and slightly glossy with some large pale grey coarser inclusions, with the interior stained slightly orange/ochre by iron oxidisation; the cortex is white and relatively unabraded. When considered in isolation, the main core remnant piece is relatively uninformative. It shows a few small flake removals from one platform, before a further attempted removal deeper back into the same platform struck on a pre-existing fracture-plane that has led to the core breaking up. When considered as a refitted group (see below) there is, however, more technological information.

None of the debitage or flake-tools shows any distinctive morphology or scar patterning that represents anything other than these simple reduction strategies. Many of the flakes show clear ring-cracks at the point of percussive impact, indicating striking with a hard-hammer percussor. And there are also many ring-cracks on the surviving cores and flakes representing hard-hammer

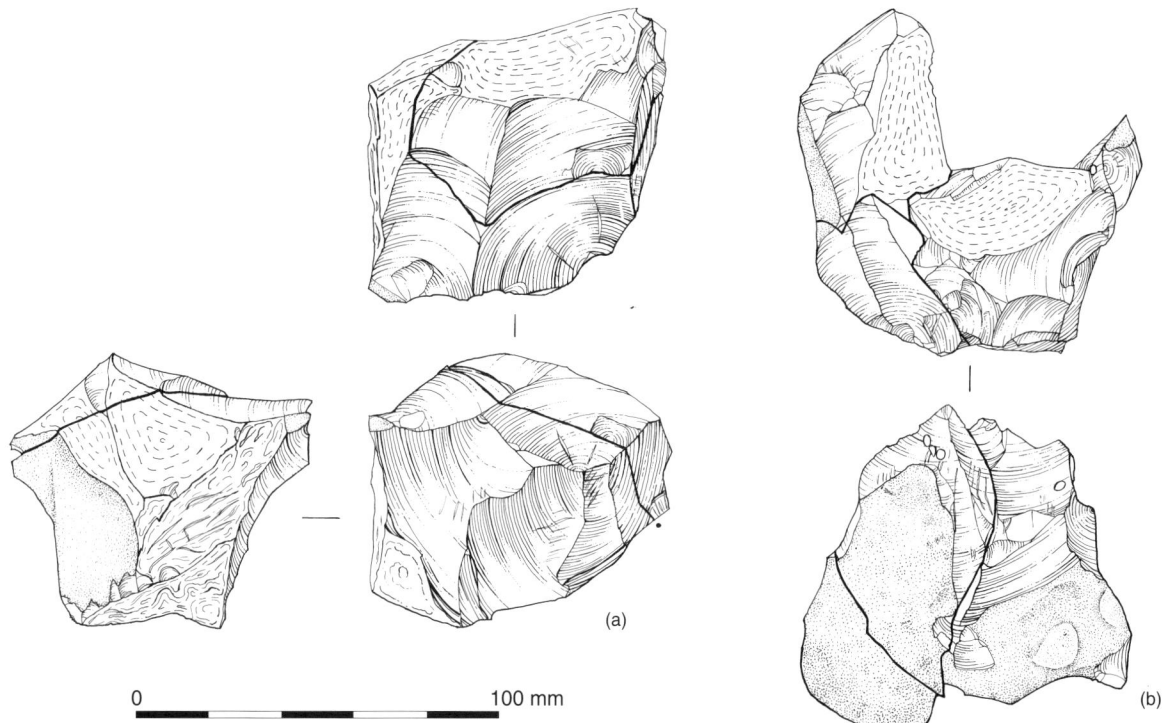

Figure 17.8 Cores from around elephant skeleton: (a) Δ.40871, with last refitting flake Δ.41040 [Group C]; (b) refitted broken core pieces Δ.40834, Δ.41351 and Δ.41376 [Group E] [ill. B. McNee]

percussive blows that failed to detach flakes. When refitting was undertaken, it quickly became clear that one of the refitting groups [Group D] did not represent a flaking reduction sequence, but the broken parts of a nodular flint percussor. Seven broken pieces of the percussor, including a tiny interior chip, were found in the northern half of the lithic scatter associated with the elephant skeleton, within an area *c* 4.5 x 2m (Fig. 17.9a).

This percussor is a small flint nodule, about 120 x 80 x 60mm in size, and weighing *c* 750g, including all the refitting pieces. It is flat on one side, and slightly domed on the other, with a broad ridge extending most of its longer axis (Fig. 17.9b). It has grey cortex worn smooth by natural working and abrasion, and its perimeter and protrusions are also heavily battered and abraded by natural processes. Superimposed on this background of natural abrasion, there are distinct areas of fresh percussion impact marks on several protrusions, particularly on two distinct protrusions on its upper ridge, from one of which also emanate several small flake scars. This evidence of fresh and localised battering underpins interpretation of the nodule as a knapping percussor. Its interior is heavily frost-fractured, and it seems that the act of percussion has caused it to break up. Some of the fresh battering scars cross from one broken piece to another, proving that this battering took place prior to its disintegration. Other small scars do not however cross between broken pieces, suggesting an attempt to continue using the percussor after it had begun to break up. The messy splintering of the highest protrusion on its domed surface seems to reflect an attempt to hold the broken

pieces of the percussor together, and to continue knapping with it in its broken state.

The following section looks in more detail at the sequences of reduction and *chaînes opératoires*, as revealed by the refitting sequences. The subsequent section considers certain debitage products as potential tools, and, in the one instance of secondary working, considers how this was applied to turn the flake into a presumably more useful cutting tool.

Refitted reduction sequences

This section focuses on the knapping strategies applied at the elephant area as revealed by the longer refitting sequences of Groups A, B, C and E (Fig. 17.10). Even more than the preceding analysis of the cores, these allow characterisation of the lithic material culture not just as types of differently shaped outcomes, but as patterns of knapping procedure. Furthermore, within the context of an essentially undisturbed activity area, the stages of reduction present reflect the spatial organisation of the production process. Absences in the flaking sequence may represent the selection and export of certain flaking products as desired or chosen, contributing both to a tentative understanding of the intentions of the knapper and to a behavioural interpretation of the site.

As well as illustrating each reduction sequence with photos and drawings, it was found useful to represent each sequence as a tabulated order of flaking events (Table 17.7–17.10). Even when certain flakes were absent, the sequences were all sufficiently complete that continuous sequences of reduction could be recon-

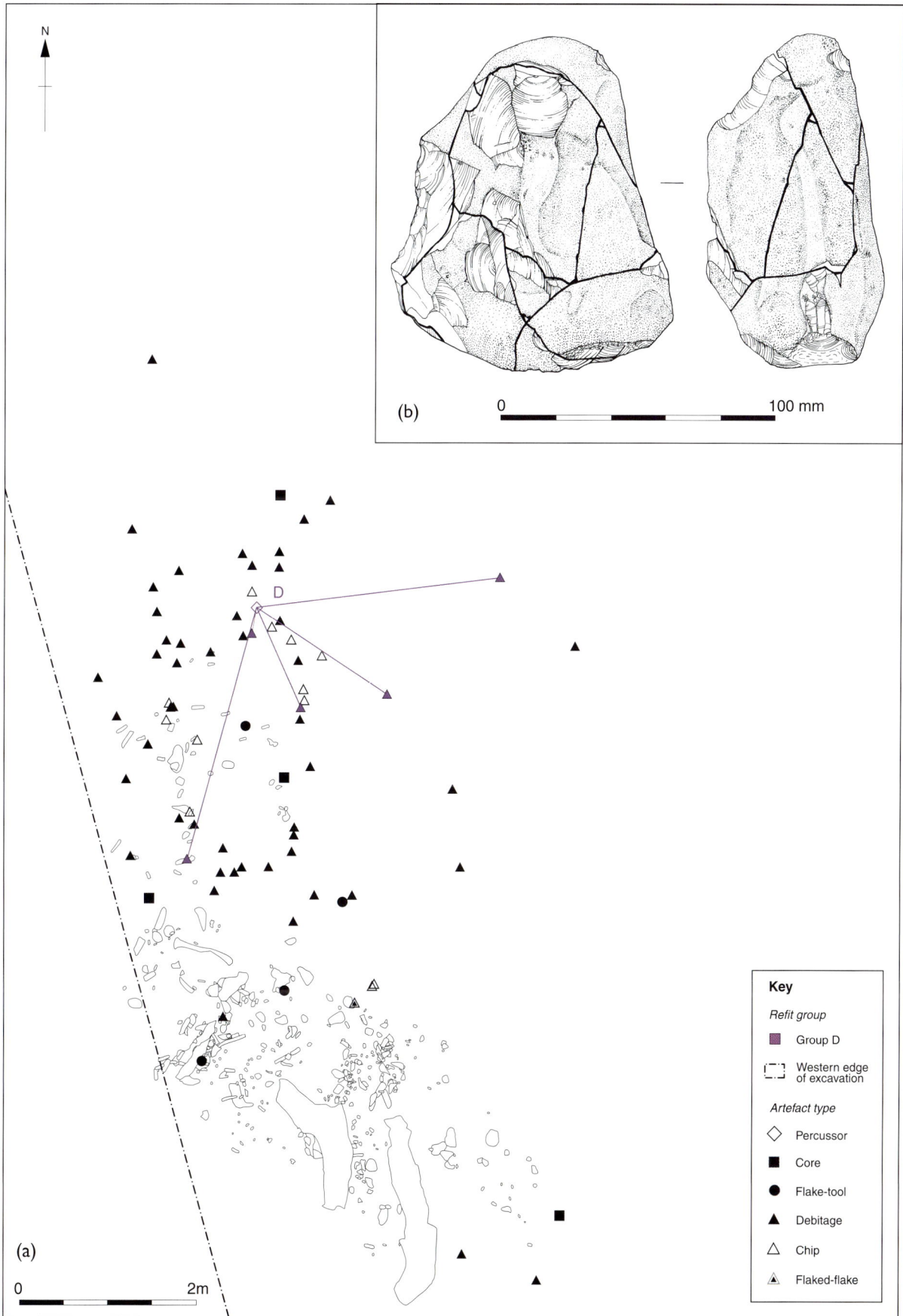

Figure 17.9 Refitted percussor from near elephant skeleton [Group D]: (a) distribution of Group D refitting pieces; (b) the refitted percussor [ill. B. McNee]

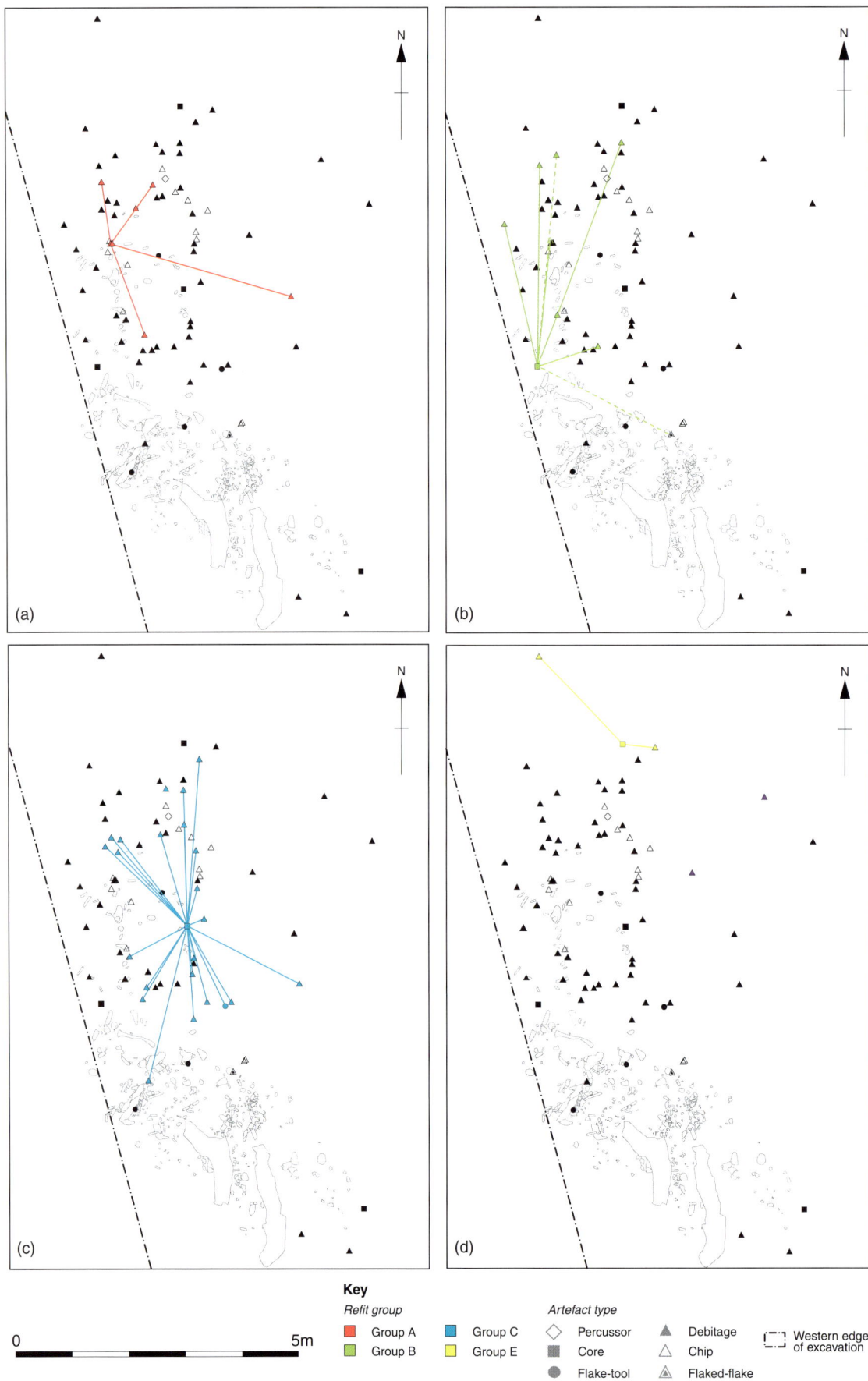

Figure 17.10 Distribution of main refitting flake groups around the elephant skeleton: (a) Group A; (b) Group B; (c) Group C; (d) Group E

structed from the combination of the surviving material and the scars of missing removals. In these tables, certain descriptive conventions were followed. As specified in the captions, different symbols were used for different technological categories. Solid filled symbols were used to represent material present in the sequence, with its find number. Hollow symbols were used for missing material, with a notional letter suffixed in a series starting at 'a' for each reduction sequence. Each row in the summary tables represents a separate flaking event. Each time the same striking platform was used for a subsequent removal (or attempted removal), the following row is separated by a dashed line. Each time the striking platform changes, the row following is separated by a solid line. When the striking platform changes, there are also specific comments on this re-positioning. When the new platform is struck on the scar of the immediately preceding removal, this is regarded as an 'alternate' platform; when struck on the scar of an earlier removal, it is regarded as merely a 'new' platform, even if it superficially appears alternate.

Finally, one technical aspect that was investigated in course of this part of the study was handedness of the knapper, following the work of Toth (1985; Schick and Toth 1993: 140-142). Toth suggested that, when striking a series of flakes from a single platform, handedness is reflected in the debitage by a tendency for a preferential migration of the striking impact point away from the knapper, leading to a recognisable bias in the distribution of cortex on the right dorsal side of the resulting flakes (when oriented with the butt upwards). This was investigated here, not by the examination of the dorsal cortical distribution on flakes (or, as is equally possible, of the side/order of sequential dorsal scars), but by direct examination of the direction of rotation of the striking impact point when sequential refitting flakes were struck from the same platform. This rotation is described here as movement of the percussion point relative to the flat, stationary platform, when viewed from above. Thus, for

typical right-handed knapping, the striking point would be described as migrating away from the knapper *anticlockwise*, whereas for a left-handed knapper, it would be described as a *clockwise* rotation (Fig. 17.11).

Group A, 'Piebald nodule'

This group of refits included seven artefacts, all of them categorised as flakes when studied separately. Three of them were broken, but none of the broken flakes joined with each other. The raw material was an opaque fine-grained flint with distinctive piebald banding near the cortex. The cortex was a distinctive pale mottled greenish/greyish white, and there was an intermittent orange iron-stained band beneath the cortex in places, suggesting the raw material originated from the Bullhead flint bed. The raw material was very distinctive, and although a search was made through the rest of the Phase 6 collection, no other pieces of this refitting sequence were found.

The sequence appears to represent the initial stages of reduction of a core, with the core and any later removals absent. The constituent flints are mostly distributed within an area of about 3 x 2m, with a single outlier (Δ.40247) a little further away on the eastern side of the scatter (Fig. 17.10a). The five flints from the middle part of the sequence are shown here fitted together (Fig. 17.12a), together with the first Δ.41024 (Fig. 17.12b) and the last Δ.41025 (Fig. 17.12c); and the sequence is described in more detail in the accompanying tabular summary (Table 17.7).

The sequence starts with two 100% cortical flakes (events # 1-2), one of which is missing, and the other of which (Δ.41024) creates a wide clear striking platform which is then used for the following sequence of four consecutive removals (events # 3-6). There is no sideways migration of the striking point between events # 3 and # 4; between events # 4 and # 5 there is a 60mm anticlockwise migration; and then between events # 5 and # 6 there is a matching clockwise migration back towards the earlier striking points. The initial anti-clockwise move

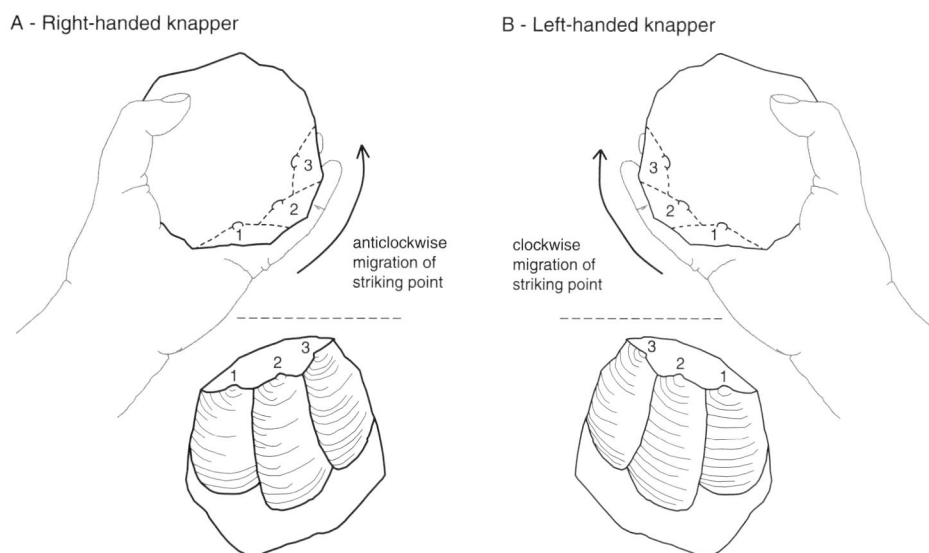

A - Right-handed knapper   B - Left-handed knapper

anticlockwise migration of striking point

clockwise migration of striking point

Figure 17.11 Suggested preferential striking point migrations for right- and left-handed knappers when removing consecutive flakes from the same platform

between events # 4 and # 5 could perhaps tentatively be construed as matching the preference of a right-handed knapper, but is inconclusive on its own.

The last flake of this sequence breaks in two, and the quite substantial distal piece 'Δe' is missing. After this, there is removal of a small chunky flake 'Δf' that appears to represent preparation of a new striking platform, followed by removal of a substantial flake Δ.40957 (event # 8) across the direction of the earlier sequence of consecutive removals, using the scar of flake 'Δf' as an alternate platform. A number of incipient cones from failed flake removals on the surface of Δ.40957 perhaps

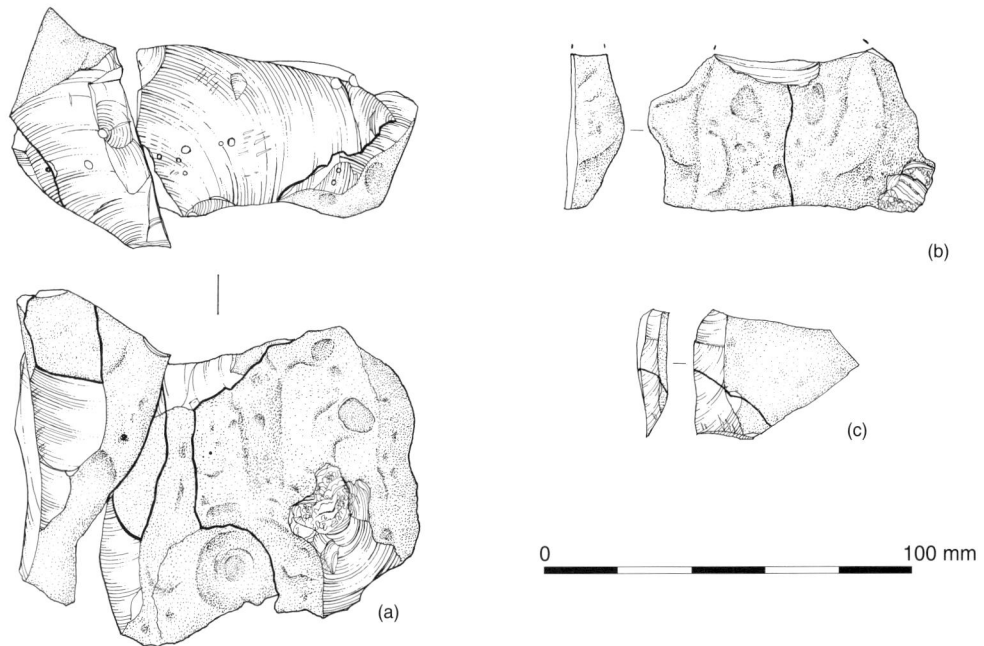

Figure 17.12 Refitting from around elephant skeleton: Group A: (a) five flints from middle of the sequence; (b) first flake in the sequence, Δ.41024; (c) last flake in the sequence, Δ.41025 [ill. B. McNee]

Table 17.7 Elephant area, Group A 'Piebald nodule': reduction *chaîne opératoire*

Event order #	Core/platform positioning	Flake removal	Comments
1-2?	On protruding cortical contour	△a⇩	Sequence starts with two 100% cortical flakes
1-2?	On different protruding cortical contour	▲41024=△b⇩	△a, and ▲41024=△b, their relative order unknown; distal part △b is missing
ò			
3	New platform, alternate (▲41024=△b scar)	▲40733=△c=△d⇩	Butt of flake present, but missing two small 100% cortical mesial and distal pieces △c and △d
4	Same platform (directly behind)	▲40248⇩	Quite solid flake; at least two blows taken to remove it, reflected in incipient cones on striking platform
5	Same platform (other end, *c* 60mm anti-clockwise)	▲41189⇩	Quite large 100% cortical flake
6	Same platform (back to original end, *c* 50mm clockwise)	▲40247=△e⇩	Quite substantial broken distal piece △e is missing
7 *	New platform (▲41189 scar)	△f⇩	Small chunky flake △f, possibly deliberate platform preparation
8	New platform, alternate (△f scar)	▲40957⇩	Some visible damage on one obtuse edge, interpreted as excavation damage, not use-wear
9	New platform, alternate (▲40957 scar)	▲41025⇩	Small flake, mostly cortical — last part of surviving Group A sequence
10+?	Uncertain how/whether knapping continued	△g+, □h+	Core and possible subsequent removals missing – do not believe present in rest of Phase 6 assemblage, as raw material is very distinctive

△ - missing debitage from refitted sequence; □ - missing core from refitted sequence; ▲ - debitage present in reduction sequence
* No evidence of intervening flakes in the surviving sequence, but possible invisible reduction episodes on absent material

attest to an intervening stage of attempted continuation of use of that platform after event # 6, before moving onto creation of a new platform by flake 'Δf'. The scar from flake Δ.40957 is then in turn used as an alternate platform for the final removal of the sequence, Δ.41025 which is a small, mostly cortical flake with a sharp edge down one side (Fig. 17.12c).

The only potentially significant gap within the sequence is the missing distal part of flake Δ.40247, which would have been of reasonable size, and would have had a sharp edge down one side opposed by a cortical side that would have been relatively comfortable to hold. The other potentially significant flake with potential for use as a cutting tool is Δ.40957 which has a robust sharp edge at its distal end, but which did not, however, show any sign of macro use-damage – although it did have some abrasion interpreted as excavation damage on one obtuse edge.

It is interesting that one of the two initial decortication flakes is absent, as is the core and any subsequent removals. Unfortunately, due to the loss of part of the site, it is entirely uncertain whether these would have been close at hand, or whether they represent larger-scale mobility, and a more extended spatial scale of the organisation of production.

Group B, 'Large cylindrical core'

This group of material (Fig. 17.13) comprises one large core Δ.40494, six pieces of refitting debitage, and an additional piece of irregular waste (Δ.41259) that does not refit, but has such similar raw material and cortical characteristics that it is confidently believed to represent part of the same reduction sequence. The core was found in amongst the spread of bones (Fig. 3.17), and the rest of the refitting group was found within an area of approximately 2 x 4m to the north of the core (Fig. 17.10b).

This refitting group appears to represent the progression from the very first cortical removal from a virgin core, to abandonment of the core after removal of relatively few flakes (<10) and with a substantial amount of good quality flint remaining and accessible. There are, however, some missing flakes from the recovered sequence, which is summarised in the accompanying table (Table 17.8), and discussed in more detail below.

The first removal is thought to be the non-refitting piece of irregular waste Δ.41259 (Fig. 17.13b). This clearly represents the initial removal of the cortical end of a cylindrical flint nodule; the raw material and cortical characteristics are so similar to core Δ.40494 that it is confidently believed that this piece represents the start of its reduction. There is then a gap in the sequence, with removal of at least two further flakes Δa and Δb of reasonable size by so called 'salami-slice' technique, whereby the convex cortical surface of a cylindrical nodule is used directly as a striking platform, with some rotation of the core between flake removals. After these, the next two flakes Δ.40725 (Fig. 17.13c) and Δ.41188 (Fig. 17.13e) are removed from successive alternating platforms, starting with the scar of removal 'Δb'. After this, event # 6 is the removal of flake Δ.40732 = Δ.41279 from the same platform as event # 5, with the striking point rotated slightly clockwise (contra the presumed preference of a right-hander). This flake broke in two as a *siret* fracture

Table 17.8 Elephant area, Group B 'Large cylindrical core': reduction *chaîne opératoire*

Event order #	Core/platform positioning	Flake removal	Comments
1	Striking onto cortex at one end of cylindrical nodule	▲41259 *⇩	Start of sequence with knocking off irregular piece rom one end of cylindrical nodule
2, 3	Rotating core, and other 'salami slice' flakes off same end, struck on cortex	△a, △b⇩	Probably at least two flakes missing from sequence (one of which might be the blank for the flake-flake ▲40658)
4	New platform, alternate (△b scar)	▲40725⇩	100% cortical flake, with hinge termination
5	New platform, alternate (▲40725 scar)	▲41188⇩	Small flake with narrow platform and hinge termination
6	Same platform (c 25mm clockwise)	▲40732=▲41279⇩	Medium-size flake, broken during removal; not apparently used, although useful sharp edge
7	New platform, alternate (▲40732=▲41279 scar)	△41005⇩	Small flake, hinge termination; can be interpreted as failure due to incorrect striking too close to platform edge
8	New platform, on cortex further down core, without altering its orientation	△c (failed)⇩	Impact marks on cortical protrusion show failed attempt to strike a flake, leading to a few tiny cortical chips (not recovered)
9	Same cortical platform (further up core surface, c 35mm behind)	▲41023, ■40494 **	Impact marks from at least two heavy blows fail to remove large flake, but only small cortical flake ▲41023

△ - missing debitage from refitted sequence; ▲ - debitage present in reduction sequence; ■ - core left at end of sequence

* This piece does not refit, but is confidently grouped with the other material from this sequence on the basis of its similar cortex and flint texture, and its clear match with the missing end of the surviving core ■40494

** One of the mysteries of this core is that it was abandoned while still containing a substantial mass of good-quality flint; here, this abandonment is tied in with the broken hammerstone (refitting Group D), which is suggested to have broken at Event #9 of this sequence, leading to abandonment of the core without further knapping

(Fig. 17.13d); the large piece still retained what seems to be a useful sharp edge opposed by a blunt cortical edge, but there is no sign of macro-damage suggesting use.

The following removal (Δ.41005, shown fitted to the core – Fig. 17.13a), struck on the alternate platform resulting from the removal of flake Δ.40732 = Δ.41279, was a small flake that seems to represent a failure due to striking too close to the edge of the platform, leading to a short, thin flake with a hinge termination. This seems to have been the beginning of the end of this core's life. Although the precise order of subsequent events is slightly speculative, there are two distinct areas of impact directly onto the cortex of the main body of the core, identified here as events # 8-9. Event # 8 is presumed to take priority, because it does not require any re-orientation of the core after event # 7, and leaves several impact marks on a cortical protrusion, without detachment of a flake. Event # 9 is then represented by further batter-marks *c* 35 mm higher up the core,

behind another protrusion that might serve as an (ambitious) striking platform to remove a substantial flake from the main body of the core. However, no major removal was forthcoming, although a small chip was detached (D.41023, shown here fitted to the core – Fig. 17.13a).

The impacts associated with events # 8-9 would have been pretty heavy, particularly in conjunction with failure to remove a flake, and would have sent substantial shock-waves through the percussor. Although speculative, it seems reasonable to suggest that the abandonment of core Δ.40494 whilst still containing a substantial quantity of accessible and high quality fine-grained flint (it weighs almost 1400g) might be associated with breakage of the percussor resulting from events # 8-9, and that the broken percussor is present at the site, fitted together as Group D (Fig. 17.9).

It is also very tentatively suggested that flake-flake Δ.40658 (Fig 17.13f) might be from the same raw

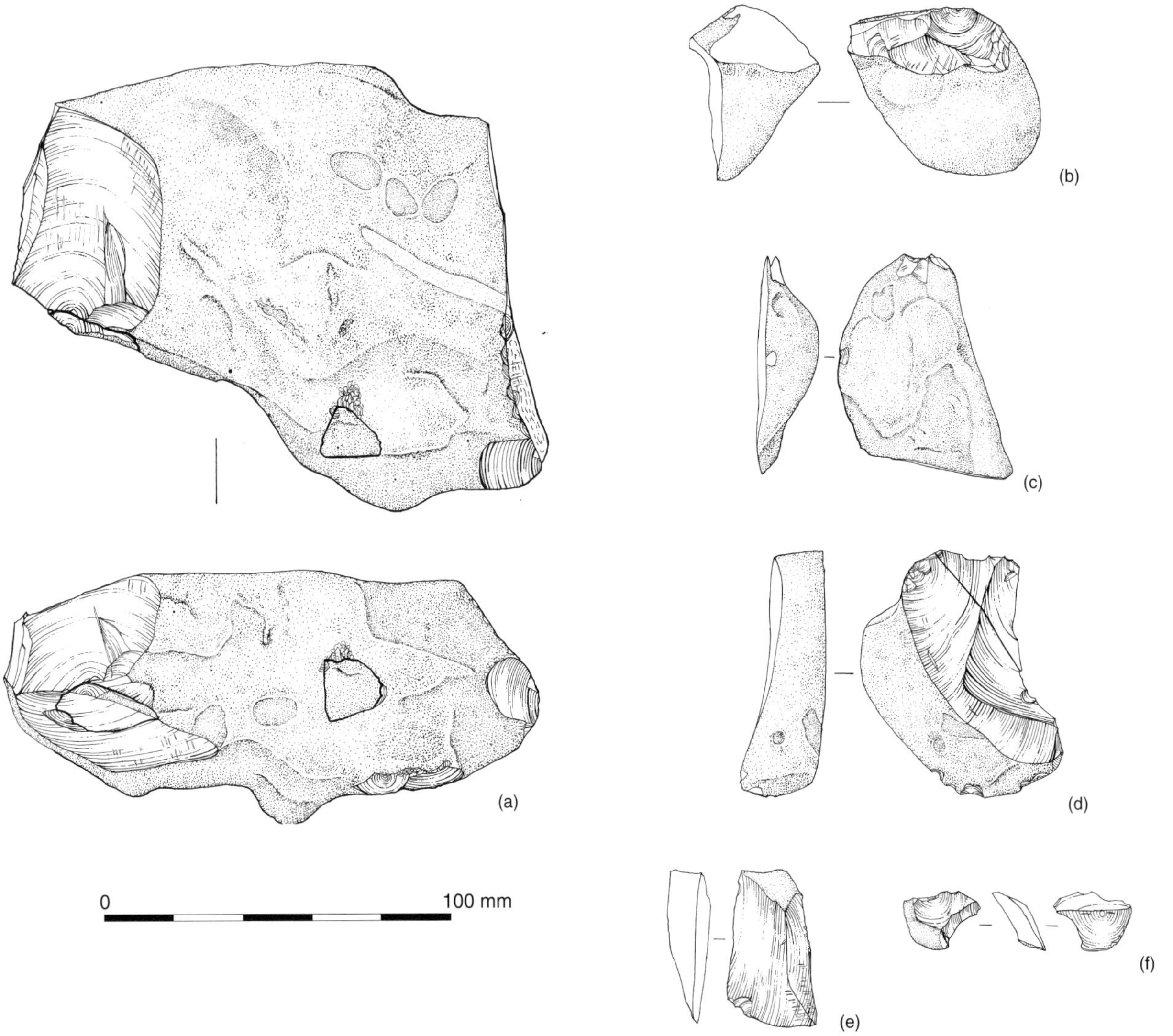

Figure 17.13 Refitting from around elephant skeleton: Group B: (a) core Δ.40494 with refitting flakes Δ.41005 and Δ.41023; (b) irregular waste Δ.41259, thought to be from the start of the sequence; (c) flake Δ.40725; (d) joined flake Δ.40732=Δ.41279; (e) flake Δ.41188 ; (f) flake-flake Δ.40658 from making notched flake-tool, thought to perhaps be of same raw material as Group B [ill. B. McNee]

material as the rest of Group B, and that it therefore represents secondary working of one the missing flakes Δa or Δb to form a notched cutting tool.

In summary, Group B represents a flaking sequence that was started and finished by the elephant carcass, with the reduction strategy being a combination of salami-slice and alternating platform technique applied to one end of a cylindrical flint nodule. Some of the resulting debitage is absent, and it is therefore thought that these may have been either collected for use unmodified as cutting tools, or for secondary working to make a more designed notched flake-tool.

Group C, 'Main core'

The most spectacular of the refitting sequences from around the elephant was Group C, which comprised 23 pieces of debitage fitted back to core Δ.40871 (Fig. 17.8a). All these pieces were contained within an area approximately 5 x 3m, with the core roughly in the middle (Fig. 17.10c). The raw material for Group C was a quite substantial flint nodule, probably initially weighing c1800g, allowing for some of the missing pieces of debitage from amongst the recovered refitting material. Its cortex was generally white, with some pale greyish/blueish patches, and rough and unabraded suggesting a minimal history of derivation and transport. The interior flint was dense, fine-grained and mostly a pale, mottled opaque grey; in places there was, however, a thin band of glossy more translucent dark grey flint near the cortex.

As discussed above, when merely considering the core on its own, the reduction approach can be described as one of broadly alternating reduction on a thick edge opposed by a cortical face. When the full refitted sequence was considered, it became clear that there is more structure and pattern in its reduction, with clear evidence of a preferential orientation of the core, interspersed with episodes of transverse striking across the preferred orientation as certain platforms became exhausted or less viable. A literally blow-by-blow account of the core's reduction is presented in the tabular summary (Table 17.9) and the photographic montage of key stages of its reduction (Fig. 17.14) so this is not repeated here in text form. Rather, the overall strategy of reduction is summarised, with certain key points highlighted.

Although 23 pieces of refitting debitage were recovered, there are several gaps in the sequence. The full sequence is thought to have almost 30 removal events, before abandonment of the core, with several small pieces of cortical waste debitage missing along with three reasonably sized flakes Δi, Δj and Δm, which are candidates for having been selected and extracted as desirable end-products.

As with Groups A and B, very early stages of reduction seem to have taken place at the spot, after some initial testing (event # 1, the evidence of which was not recovered). This included preliminary removal of protruding cortical lumps to reduce the less usable bulk of the raw material and create striking platforms to start its reduction (events # 2-5).

After this, the preferred orientation of the core becomes established, with one aspect generally being maintained as the upper striking platform surface (Fig. 17.14). The general pattern of reduction is one of occasional platform alternation interspersed with episodes of repeated flake removals from the same platform. There is regular evidence on the striking platforms that a hard-hammer percussor was used, and the repeated incidence of similarly wide ring-cracks seems to indicate consistent use throughout reduction of the same percussor with a relatively wide hemispherical spread at the point of knapping impact. For series of flakes struck from the same platform, there is a marked preference for the striking point to migrate anti-clockwise through the series, suggestive of a right-handed knapper. Of the four instances where there are such series, three of them (events # 6-9, # 16-17 and # 19-21 – see Fig. 17.15a) include an anti-clockwise striking point migration, and the fourth (events # 10-12) has no sideways migration. In addition, event # 24 represents an anti-clockwise migration on essentially the same striking plane, but on a different flake scar, so technically a new platform following the nomenclatural convention applied. There are hints of what could be interpreted as personal knapping idiosyncrasies in the sequence. There are several instances of deliberately striking on dihedral platforms, usually avoided by modern and prehistoric knappers (although note the comments of Bradley and Sampson 1986), within the context of Levalloisian core-surface trimming. There is also a regular habit of striking on the furthest edge of plain platforms (assuming a right-handed knapper; or conversely, if left-handed, on the nearest edge).

Apart from one or two small preliminary decortication flakes, the bulk of the early decortication of the nodule took place at the spot, and the last removal (event # 27, flake Δ.41040) was also recovered less than 1.5m from the core. This latter flake (illustrated attached to the core, Fig. 17.8a) showed some faint macro use-wear on one of its sharp protrusions, as well as a larger chip on one edge, possibly a deliberate secondary notch removal, and so is regarded as probably having been used as a tool. It was found at the southern edge of the lithic concentration, nearer the elephant bone spread, in close proximity to two other flakes interpreted as having been utilise, supporting its interpretation as a tool.

Group E, 'Shattered nest'

This group of material comprises just three pieces of flint, two of them initially classified as irregular waste, and one of them as a broken core (Fig. 17.8b). All three were found at the northern edge of the scatter by the elephant skeleton, within an area of roughly 2 x 2m (Fig. 17.10d). Rather than representing a sequence of reduction, the refitting pieces represent the shattering of a core into three pieces simultaneously, caused by a percussion blow struck directly onto a frost-fracture. However, the various flake scars on the refitted group still allow reconstruction of a narrative of reduction (Table 17.10).

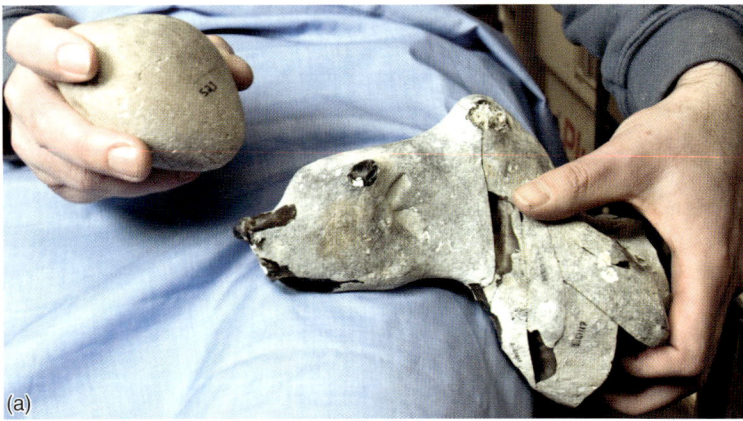

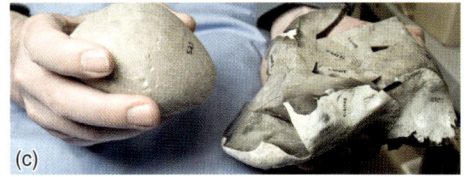

(m)

Figure 17.14 *(facing page and this page)* Refitting Group C from around elephant skeleton, progression of reduction and final core: (a) event #2, Δ.41068; (b) event #4, Δ.40885=Δ.40791; (c) event #5, Δ.41096; (d) event #6 Δ.40250; (e) event #7 Δ.41103; (f) event #10, Δ.41094=Δ.41067; (g) event #14, Δ.41208; (h) event #15 , Δ.40787; (i) event #18, Δ.41212; (j) event #19, Δ.40771, (k) event #22, Δ.41162; (l) event #23, Δ.40786; (m) final core Δ.40871, with last flake Δ.41040 attached; (n) refitted nodule, with all pieces attached.

0 —————————————— 100 mm

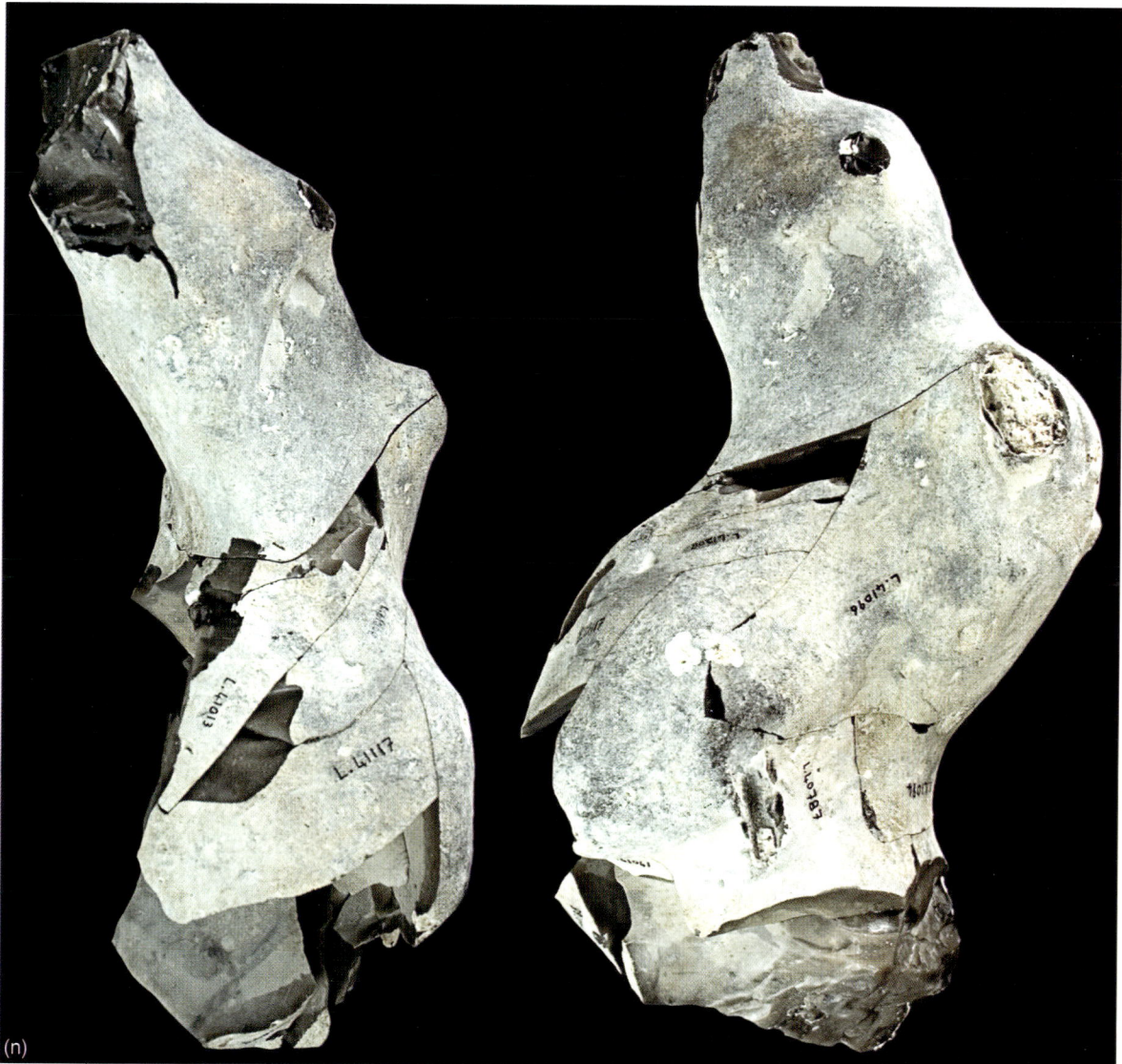

(n)

Table 17.9 Elephant area, Group C 'Main core': reduction *chaîne opératoire*

Event order #	Core/platform positioning	Flake removal	Comments
1	Start of sequence, cortical platform at tip of cylindrical protrusion	△a, △b⇓	At least one, and probably two, pieces of irregular cortical waste △a, △b knocked off tip of cylindrical protrusion (possibly likewise at other end of core)
2	Same platform (directly behind)	▲41068⇓	Cortical platform
3	New platform, alternate (▲41068 scar)	△c⇓	Small cortical flake △c knocked off tip of cylindrical protrusion
4	Same platform (directly behind △c scar, but on cortex)	▲40885=▲40791=△d⇓	Cylindrical protrusion snaps off as irregular waste, together with ▲40791 and a quite substantial third piece △d
5	New platform, alternate (▲40885 scar)	▲41096=△e⇓	Large flake, hinged profile and 100% cortical; prob. simultaneous detachment of missing small chunk of irregular waste re from platform overhang
6	New platform, alternate (▲41096 scar)	▲40250⇓	
7	Same platform (directly behind)	▲41103⇓	
8	Same platform (*c* 45 mm anti-clockwise)	▲40628=▲41108=△⇓	Removal ▲40628=▲41108 is missing proximal end △f; uncertain relative order of this flake in relation to previous two from same platform
9	Same platform (directly behind)	▲41013⇓	Quite a wide flake, successfully removes remnant ridge left from #3
10	New platform, alternate (▲41013 scar)	▲41067=▲41094⇓	Deep platform, leads to thick, hinged flake which nonetheless tidies core to inverted pyramidal form
11	Same platform (directly behind)	△f⇓	Small triangular waste from resulting platform overhang
12	Same platform (directly behind)	△41229 (failed)⇓	Failed detachment of ▲41229 leaves internal fracture-plane defining proximal end of flake's ventral surface
13	New platform, natural flint surface on left dorsal ridge of ▲41229	▲41229⇓	Flake eventually removed by striking sideways onto left side of its dorsal ridge
14	New platform, near-alternate (ridge between ▲41067=▲41094 and ▲41096=△e scars)	▲41208=△g⇓	Flake struck directly on dihedral ridge between two flake scars; distal end △g of flake snapped during removal and missing
15	New platform, alternate (▲41208=△g scar)	▲40787=△h⇓	Chunky, 50% cortical flake, hinged at distal end
16	New platform, alternate (▲40787=△h scar)	▲41117⇓	
17	Same platform (c 25mm anti-clockwise)	△i⇓	Solid flake, with straight, naturally sharp distal end – a good candidate for use/modification as a flake-tool
18	New platform, alternate (△i scar)	▲41212⇓	
19	New platform, alternate (▲41212 scar)	▲40771⇓	Short triangular flake
20	Same platform (directly behind)	▲41095⇓	Slightly longer, elongated triangular flake
21	Same platform (c 35 mm anti-clockwise)	▲40869⇓	Squat transverse rectangular flake, struck on shallow ridge between ▲41212 and ▲40787 scars; platform preparation, shifts angle to bypass hinge protrusions at bottom end of car
22	New platform, alternate (▲40869 scar)	▲41162⇓	
23	New platform, alternate (▲41162 scar)	▲40786⇓	
24	New platform (anticlockwise c 30 mm, past edge of ▲41162 scar)	△j⇓	
25	New platform, alternate (△j scar)	▲41041=△k=△l⇓	Small siret-fractured piece of proximal end present; rest of flake missing, probably represented by at least two pieces
26	New platform, alternate (▲41041=△k=△l scar)	△m⇓	Mod. small flake, but would have had sharp edge down one side and at distal end good for cutting use
27	New platform (clockwise 100°)	▲41040, ■40871	Final flake of Group C is ▲41040 (which may have been used as a flake-tool), and is found 1.55m from core ■40871

[△ - missing debitage from refitted sequence; ▲ - debitage present in reduction sequence; ■ - core left at end of sequence]

Table 17.10 Elephant area, Group E 'Shattered nest' core: reduction *chaîne opératoire*

Event order #	Core/platform positioning	Flake removal	Comments
1, 2, 3	Uncertain history of migrating and/or alternating platform use prior to start of surviving sequence	△a, △b, △c⇩	Probably about. three previous removals, one of which exposes crack from internal frost-fracture
4	New platform, directly on frost-fracture crack	▲40834=▲41351=▲41376⇩	Knapping blow on frost-fracture crack between ▲41351 and ▲41376 leads to shattering of nodule into these three pieces
5	New platform, pre-existing scar	△d⇩	Knapping continues on largest piece ▲41376, with removal of a medium-size flake △d (not present)
6	New platform, alternate (△d scar)	△e⇩	Tiny, failed flake (not present), crushing edge of platform
7	Same platform (directly behind)	▲41376=□f	Attempted removal fails and core breaks again along pre-existing frost-fracture, leaving visible point of impact and start of cone of percussion; remnant core □f absent

[△ - missing debitage from refitted sequence; □ - missing core from refitted sequence; ▲ - debitage present in reduction sequence]

Unlike the other refitting sequences associated with the elephant carcass, the earliest stages of its reduction are absent from the recovered assemblage. There seem to have been at least three removals – Δa, Δb and Δc – prior to start of the surviving sequence, one of which exposes a pre-existing frost-fracture. The next blow is applied directly onto the frost-fracture, and it is possible that this was with the deliberate intention of breaking the core into smaller pieces of non-flawed flint with which to continue knapping. The two smaller pieces (Δ.40834, weighing 62g; and Δ.41351, weighing 90g) are both then abandoned without further reduction. The larger piece (Δ.41376, weighing 186g) is however flaked further, with removal of a flake Δd which was not recovered. The scar from flake Δd was then used as an alternate platform for attempted removal of another flake Δe. This latter flake was however mis-struck, and the edge of the striking platform was crushed, with only the removal of a tiny hinged flake. The subsequent blow is slightly deeper into the same platform, but the fracture-plane intersects a pre-existing frost-fracture, and the core breaks again, leaving the surviving broken piece Δ.41376 and another broken piece □f that is missing.

This sequence contrasts with the other three refitting sequences in that firstly, the flint is poorer quality, strongly affected by frost-fracturing and secondly, the earliest stages of its reduction were not recovered. Even though the flint was of poor quality, knapping persisted after the nodule broke for the first time, suggesting in this instance some pressure to maximise its exploitation.

Refitted flakes

In addition to the four refitted flaking sequences of Groups A, B, C and E, and the broken percussor of Group D, there were two other refitting groups, F and G, both of them comprising two joining pieces of a broken flake. The former of these, Group F (Fig.

17.15b), was quite a large flake (weighing *c* 120g when whole). Both pieces were found only about 0.40m apart at the southern side of the scatter by the elephant. The cortex is fresh, white and unabraded, and the flint is fine-grained, glossy and mottled light/dark grey. Its dorsal scars show at least two previous knapping episodes with intervening core rotation prior to its own removal. When struck, it instantly broke into two due to internal fracture-planes caused not by frost-fracture, but by previous failed knapping blows. There is no evidence of the core from which this flake came or of any later removals from the same piece of raw material.

The last refitting group, Group G, is a relatively tiny exterior chip that has split in two as a *siret* fracture. Both pieces were found close together at the western side of the lithic scatter (Fig. 17.15c). It is not thought that this chip belongs with any of the other refitting groups, although this cannot definitively be ruled out because of its small size. There are no scars from previous removals, so it provides no technological information, other than that, being wholly from the exterior of a flint nodule (naturally polished, weathered flint, rather than cortex) it probably represents an early stage of nodular reduction. It is also possible that it has chipped off a percussor, although it does not seem to join with the broken percussor of Group D.

Flake-tools and miscellaneous debitage

There were just four artefacts identified as tools in the assemblage by the elephant. One of these was quite a small secondarily worked flake with a single notch on one side (Fig. 17.16a). The other three were unmodified flakes (or in one case, irregular waste) with sharp edges that showed visible signs of minor chipping/scaling interpreted as macro-use wear. One of these is flake Δ.41040 (Fig. 17.8a) which was the last flake detached from core Δ.40871, from the Group C refitting

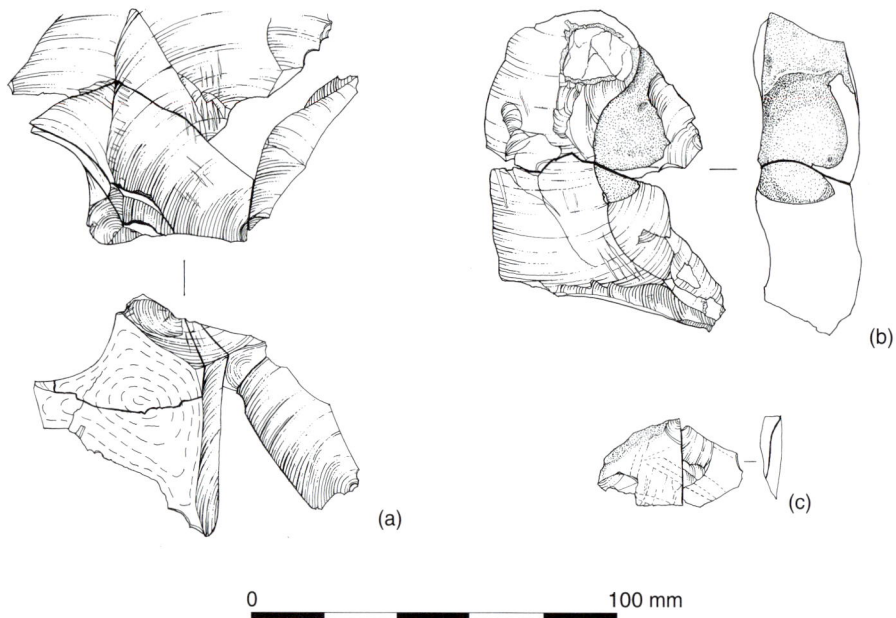

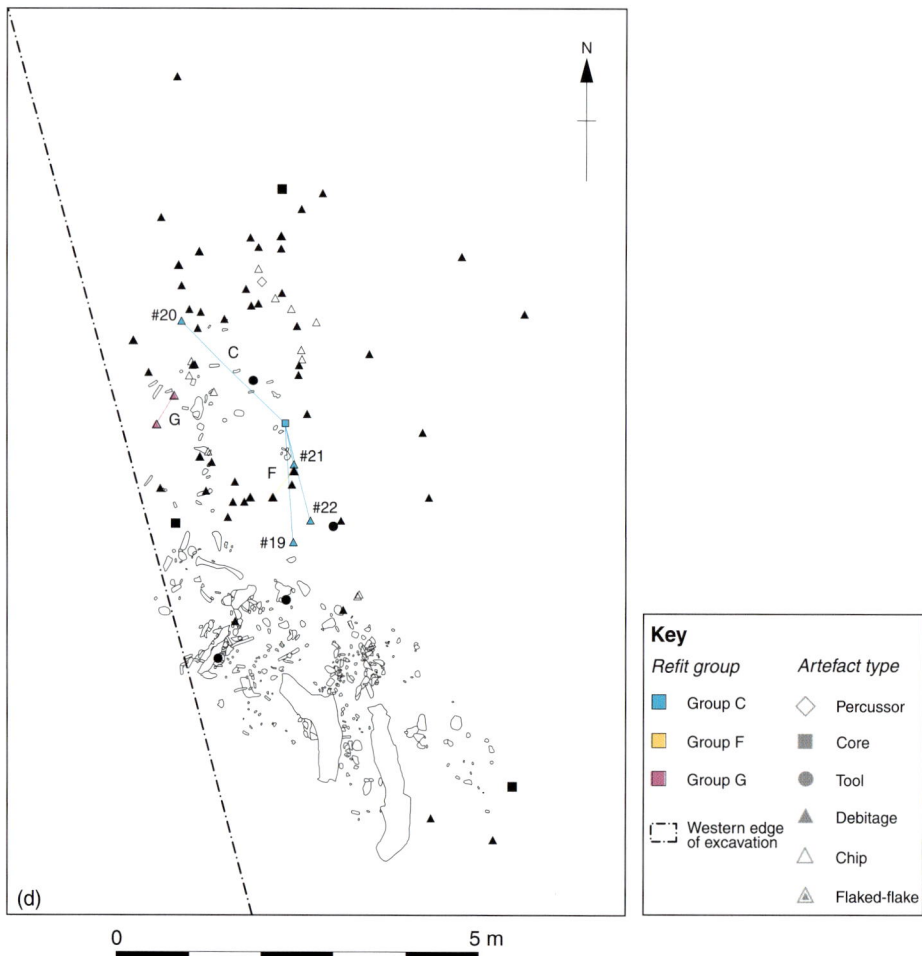

Figure 17.15 Refitting from around elephant skeleton: (a) Group C, events #19–22 - showing anticlockwise migration of striking point on platform; (b) Group F, broken flake Δ.40657=Δ.40792 ; (c) Group G, broken flake Δ.40799= Δ.40955; (d) thumbnail of illustrated flint locations and refitting connections [ill. B. McNee]

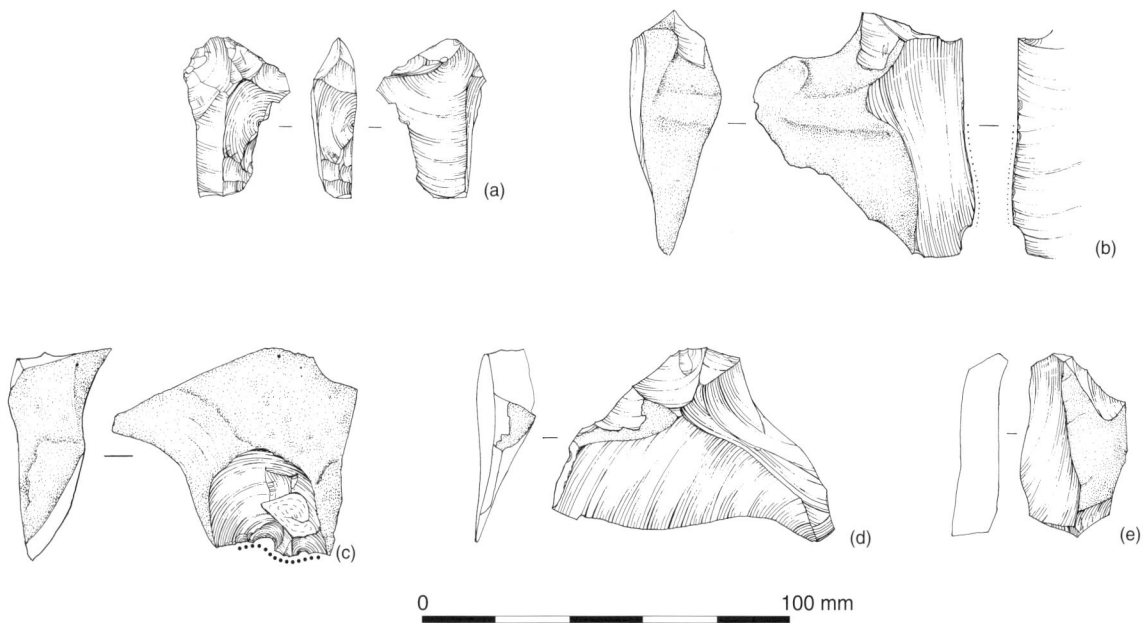

Figure 17.16 Non-refitting flints from around elephant skeleton: (a) single-notched flake, Δ.41066; (b) utilised flake, Δ.40883; (c) utilised flake, Δ.40357; (d) flake, Δ.41069; (e) flake, Δ.40723 [ill. B. McNee]

sequence. The other two are shown here (Δ.40883 – Fig. 17.16b; Δ.40357 – Fig. 17.16c). Although the notched flake-tool was found in the centre of the scatter, the other utilised pieces were all found at its southern edge, in amongst the surviving elephant bones (Fig. 17.2). In addition to these flake tools, one small flake Δ.40658 (Fig. 17.13f) was interpreted as a flake-flake from making a notched flake-tool.

Apart from these tools, and the flake-flake, the remainder of the non-refitting elephant lithic assemblage comprised six flakes and two pieces of irregular waste. Two of the larger flakes are also illustrated (Fig. 17.16d; e). Both have long sharp edges, but there is no visible evidence of them having been used. All of the aforementioned larger artefacts seem to be of distinct raw material both from each other, and from the six refitting reduction sequences, indicating that the site retains evidence of working of a minimum of twelve separate pieces of raw material. Of the remaining six pieces of debitage, several were small and stained, unlike the refitting and other material discussed above, and it is possible that some or most of them are coincidental background noise, rather than being genuinely evidence of activity associated with the elephant carcass. One of them (Δ.42139) was a small piece of the distal end of a flake, made of distinctive orange semi-translucent flint that was definitely not present elsewhere in the assemblage. The others had the raw material obscured by staining, and, if not coincidental background noise, could perhaps be associated with the same raw material pieces as already represented by the rest of the assemblage. Therefore, in total, there are probably thirteen pieces of raw material represented in the assemblage, four of them by refitting reduction sequences, and the other nine by single artefacts.

The chaîne opératoire and organisation of production

Although opinion differs on this (cf. Bar-Yosef and van Peer 2009) the approach taken here is that discussion of the *chaîne opératoire* involves a deeper, more emic engagement with the cognitive and physical landscape of the knapper than the mere description of the reduction sequence. Thus, rather than merely describing the progression of knapping reduction and the apparent stages of production present wholly as perceived from a modern perspective, it attempts to imbue these with ancient purposive intent and integrate them into the organisation of the lithic technological system, from initial raw material procurement, through transformation by knapping and selective appropriation of certain lithic products for use, up to the point of their loss/ abandonment/discard.

One of the first points to arise from analysis of the assemblage is that the three longer refitting sequences (Groups A, B and C) all include the early stages of reduction. This suggests that the raw material was locally obtained, unless heavy raw material was brought a substantial distance before most of it was abandoned as waste debitage after knapping. As indicated by the geological investigation to the west of the site (Chapter 4) and the quantity of natural flint nodules in the assemblage from the Phase 6 clay south of Trench D (Chapter 18), it seems clear that flint raw material was available near the site, on, and at the foot of, the slope that would have been rising up to the west. Therefore it seems likely that at least three pieces of raw material were collected locally and knapped on the spot as an immediate expedient response to the presence of the elephant carcass. Work done previously by Wenban-Smith (1996) investigated a range of flake attrib-

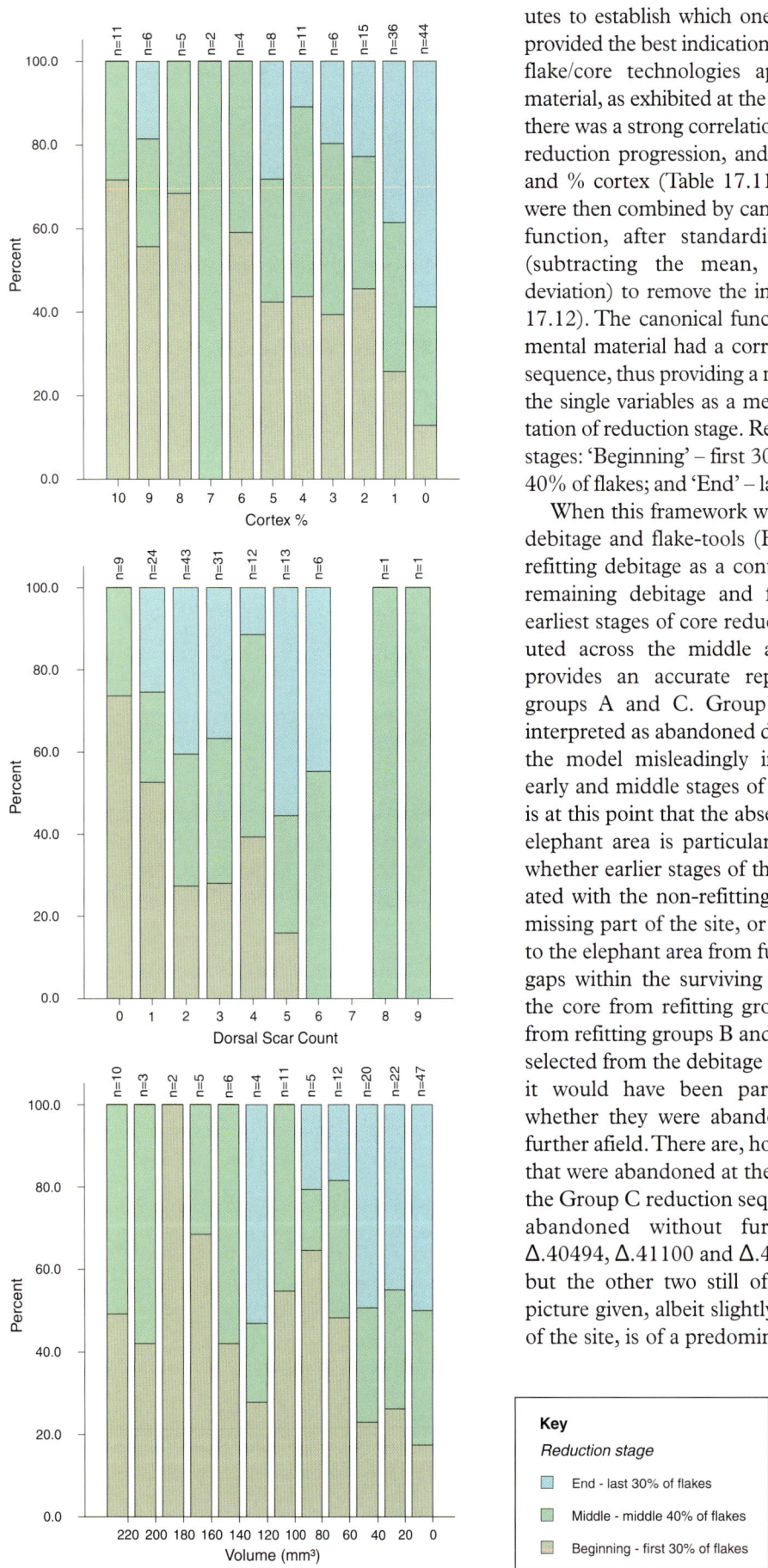

utes to establish which ones, and in which combination, provided the best indication of reduction order. For simple flake/core technologies applied to nodular flint raw material, as exhibited at the elephant site, it was found that there was a strong correlation of the dorsal scar count with reduction progression, and an inverse correlation of size and % cortex (Table 17.11; Fig. 17.17). These variables were then combined by canonical correlation into a single function, after standardising the different variables (subtracting the mean, and dividing by standard deviation) to remove the impact of scale variation (Table 17.12). The canonical function for the combined experimental material had a correlation of 0.65 with reduction sequence, thus providing a marked improvement on any of the single variables as a means of modeling the representation of reduction stage. Reduction was divided into three stages: 'Beginning' – first 30% of flakes; 'Middle' – middle 40% of flakes; and 'End' – last 30% of flakes (Fig. 17.18a).

When this framework was applied to the non-refitting debitage and flake-tools (Fig. 17.18b), as well as to the refitting debitage as a control, it could be seen that the remaining debitage and flake-tools seem to lack the earliest stages of core reduction and to be evenly distributed across the middle and later stages. The model provides an accurate representation of the refitting groups A and C. Group B is somewhat anomalous, interpreted as abandoned due to percussor breakage, and the model misleadingly indicates debitage from both early and middle stages of reduction as being present. It is at this point that the absence of the western side of the elephant area is particularly frustrating. It is uncertain whether earlier stages of the reduction sequences associated with the non-refitting material were present in the missing part of the site, or whether these were imported to the elephant area from further afield. There are certain gaps within the surviving refitting assemblage, such as the core from refitting group A and some of the flakes from refitting groups B and C. These have probably been selected from the debitage produced for use as tools, and it would have been particularly useful to establish whether they were abandoned at the site or exported further afield. There are, however, a number of flake tools that were abandoned at the site, including Δ.41040 from the Group C reduction sequence. Also present at the site, abandoned without further reduction, are cores Δ.40494, Δ.41100 and Δ.40871, the latter slightly small, but the other two still of substantial size. The overall picture given, albeit slightly hazy due to the missing part of the site, is of a predominantly expedient lithic techno-

Figure 17.17 Experimental model for relationship of selected flake attributes to Clactonian reduction stage: (a) percentage of dorsal cortex; (b) dorsal scar count; (c) flake volume

Table 17.11 Correlation coefficients of flake attributes with reduction order for three Clactonian experiments, independently and combined (Wenban-Smith 1996)

Attribute	Ex. 1	Ex. 2	Ex. 3	Ex. 1-3	Sig.level
Max Length	-0.29	-0.36	-0.45	-0.36	0.000
Max Width	-0.40	-0.15	-0.36	-0.30	0.000
Max Thickness	-0.28	-0.35	-0.42	-0.34	0.000
Platform length	-0.27	-0.24	-0.36	-0.29	0.000
Dorsal scar count	0.49	0.34	0.44	0.41	0.000
Percentage cortex	-0.48	-0.29	-0.38	-0.37	0.000
Flake volume (cm^3)	-0.39	-0.31	-0.45	-0.38	0.000
Cortical area (cm^2)	-0.51	-0.36	-0.45	-0.43	0.000

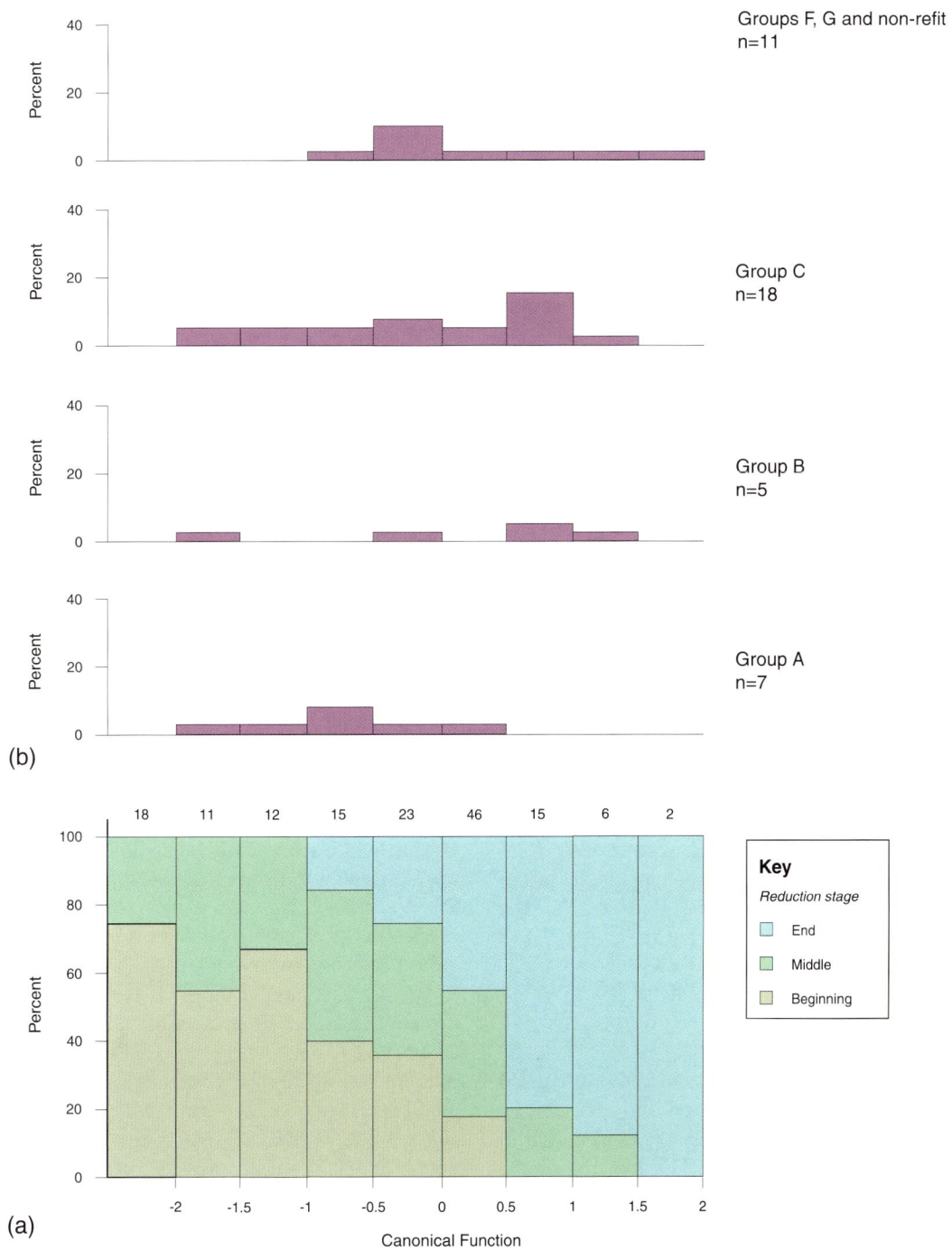

Figure 17.18 Canonical correlation function for flake reduction order: (a) experimental model; (b) applied to debitage from elephant area

Table 17.12 Canonical correlation functions of selected flake attributes with reduction order for three Clactonian experiments, independently and combined (Wenban-Smith 1996)

Attribute	Ex. 1	Ex. 2	Ex. 3	Ex. 1-3
Max Length	-0.31	-0.45	-0.85	-0.42
Dorsal scar count	0.79	0.79	0.89	0.80
Percentage cortex	-0.14	-0.10	-0.37	-0.39
Flake volume (cm³)	-0.37	-0.59	-0.37	-0.39
Cortical area (cm²)	-0.13	0.33	0.74	-
Correlation coeff.	0.76	0.53	0.75	0.65

logical *chaîne opératoire*, whereby raw material for knapping and adequate percussors are both present in the surrounding landscape, and are collected, used and abandoned as required. However some of the flakes and flake-tools could have been brought to the site having been knapped elsewhere.

The technological strategy seems to be based around the production of sharp-edged flakes of sufficient size to be convenient for handling, some of which were used unmodified as cutting tools, and some of which were secondarily worked with addition of a notch. Although superficially the knapping strategy is an entirely unstructured *ad hoc* reduction approach, with flakes removed in series from a platform until it is exhausted and then a new platform found, often using the alternate scar of the previous removal, there are some flint hints of a more structured volumetric conception of the nodule being worked. In the longer reduction sequence of Group C, rather than random *ad hoc* rotation of the core to continually find newly orientated striking platforms there is a consistent upper side to the core from which most flakes are struck, with occasional deviations when flakes are struck sideways across the primary flaking direction prior to continuation of flaking in the primary direction. Most of the flakes produced seem to be abandoned forthwith as waste, although some are selected for use and modification. It is, unfortunately, entirely uncertain whether some of the larger isolated debitage and utilised flakes found at the site were parts of reduction sequences that were nearby, but were lost before archaeological excavation took place, or whether they were imported to the elephant carcass area from further afield.

DISCUSSION

This chapter has presented the results of more detailed analysis of the lithic collection from the area of the elephant skeleton than in the previously published interim report (Wenban-Smith *et al.* 2006). This has not altered the headline interpretation of that report, namely that there is clear evidence of hominin knapping activity associated with the carcass, most likely focused on its butchery for meat; on the contrary, this interpretation has if anything been reinforced. However, some details

presented in the interim report have been revised, and it is important to spell these out. As reported previously, the lithic artefact concentration associated with the elephant was contained within an area of approximately 6 x 4m immediately adjacent to the main concentration of bones (Fig. 17.2). However, the lithic artefact assemblage was smaller than originally thought, once natural pieces and unreliably associated material were excluded, with only 77 artefacts forming the definitive elephant assemblage, of which 12 were chips <20mm long. Furthermore, although the lithic concentration was clearly juxtaposed to the bone concentration, the previous report of an area of lithic concentration between the tusks was found to be incorrect, although there was a mini-concentration of three outlying artefacts from the main scatter beyond the tip of the more south-easterly tusk.

It is uncertain how the elephant arrived at the location. The underlying sediment was not deformed, suggesting it first came to rest on a dry ground surface, and was definitely not mired in swampy ground. However, the general environment of sedimentary deposition associated with its burial is one of swampy/peaty conditions at the edge of a fluctuating water-body. This probably partially submerged the area of the skeleton for a sustained period, sufficient to ensure deposition of peaty clay-silts that helped preserve it until the present day. There is no direct evidence to indicate whether or not the elephant was actively hunted and killed, whether it died naturally on the spot, or whether perhaps the carcass floated in. The latter seems unlikely, as the carcass would have been pretty rancid prior to being sufficiently bloated to become waterborne, and therefore not a very appetizing prospect for even a very hungry hominin. Concerning the possibility of hunting, the elephant was a prime adult, probably male based on its large size. Such animals are less likely, in the modern era, to die in the wild of natural causes than are juveniles or elderly individuals (Haynes 1988). Although it may seem unlikely, considering its size, there is documented evidence of adult elephants being killed by modern humans using simple technology based around a spear (eg Zwilling 1942), and it is known from the finds at Clacton (Wymer 1985), Schöningen (Thieme 1997) and Boxgrove (Pitts and Roberts 1997) that wooden spears were part of the technological capacity of hominins of this era, so hunting cannot be ruled out, particularly if the animal was perhaps already injured. This is considered further in the concluding discussion (Chapter 22).

However it came to the spot, the juxtaposition of the lithic remains, their mint condition and the high degree of refitting strongly suggest a genuine causative association between the lithic remains and the elephant carcass. The lithic clustering and their refitting indicate a lack of disturbance. Within an area of the Phase 6 clay which generally has a sparse presence of lithic artefacts and faunal remains, the association of both the elephant skeleton and the lithic evidence suggests that flint knapping activity and lithic artefact deposition has developed in relation to the presence of the carcass. This interpretation is reinforced by the fact that the main concentration of lithics is immediately beside the main concentration of bones, rather than directly superimposed, suggesting the main area of knapping activity respected the position of the carcass, as shown in the site layout figure (Fig. 17.2).

Interpretation of activity around the carcass is frustratingly hampered by the missing part of the site. In the surviving evidence, there are three longer sequences of reduction (Groups A, B and C), all of which start with early stages of reduction of a (very likely) locally obtained flint nodule. There is evidence of at least one percussor in use, represented by the broken percussor of Group D. There are four flake-tools present in the assemblage, as well as evidence of a missing notched flake-tool represented by a flake from secondary working. There are gaps of perhaps 5-6 flakes in the refitted reduction sequences that it could be suggested represent the selection and movement of tools. When the rest of the assemblage is considered, the total number of pieces of raw material represented is *c* 12-13. There are clearly various imponderables, not only to do with the missing part of the site. It is also uncertain whether the elephant should be regarded as having been exploited on a single occasion, or whether it was subject to repeated episodes of butchery, assuming that exploitation of the elephant's meat or some other body part for food was the prime focus of the hominin activity.

It is likely that the meat, or other body parts, would have remained fresh enough to eat for at least several days after its death, perhaps as long as a week depending upon the prevailing environmental temperature. Once discovered, it would probably therefore have been subject to repeated visits by the hominins in the vicinity. Under this scenario, it is possible on post-discovery return visits that suitable tools for any desired butchery tasks would be obtained *en route* to the carcass, so would not leave such substantial evidence of reduction. It is therefore possible to tentatively speculate that the group-size of the hominins who were included in the initial discovery of, and knapping activity around, the carcass may have been as low as 3-4, based on the number of longer reduction sequences (and without consideration of the unknown missing evidence).

It is also noteworthy that, although the tusks and upper molars of the elephant were present, the jaw and lower molars were absent; the latter in particular would have been likely to survive had they been present despite the poor preservation of many bone elements. As seems to be the case at other elephant butchery sites, for instance Notarchirico in Italy (Piperno and Tagliacozzo 2001), one of the typical hominin exploitations of an elephant carcass is the overturning of its skull and removal of the mandible to get at nutritious soft parts such as the brain, tongue and trunk. The absence of the mandible from the rest of the skeleton provides an additional indication of hominin exploitation for meat.

The relative paucity of lithic remains around the elephant does not make a good basis for recognition of a wider industrial tradition, since they relate to perhaps one specific fortnight of intermittent activity at one location in the landscape. They therefore provide a very specific snapshot rather than a more general view. However, the evidence that there is shows a very clear picture of a simple flake/core technology, with certain flake-products selected for use without modification, and others slightly modified by simple secondary working to form a notch in the cutting edge. Although the primary exploitation of the carcass seems to reflect an expedient approach to raw material provisioning, tool manufacture, use and discard, there is also some indication that there is an accompanying element of more logistically organised behaviour, with tools moved around the landscape in anticipation of use. This is evidenced by the absence of potential cutting tools in the refitted reduction sequences and from the presence of other tools that are not part of any refitting sequences.

Chapter 18

Lithic artefacts: Phase 6, the concentration south of Trench D and other more scattered pieces

by Francis Wenban-Smith

INTRODUCTION

This chapter presents the lithic collection from south of Trench D, and the various lithic artefacts that were more sparsely distributed throughout the rest of the Phase 6 clay. As discussed previously (Chapter 17; Table 17.1; Fig 17.1), the lithic collection from the Phase 6 clay was divided into three primary assemblages: 6.1, 6.2 and 6.3, for analysis based on their spatial clustering within the excavated area. The great majority of lithic material from the Phase 6 clay was contained within the concentration south of Trench D (Fig 18.1), and this material was designated as assemblage 6.1 (n=2010, including 139 natural pieces). The northern edge of this concentration is defined by the southern edge of Trench D, which was dug by machine before discovery of the concentration to its south, so it could be suggested that the sharp northern cut-off of this concentration is artificially created. A number of flints were recovered during machining of Trench D, and if as many had been found as were later discovered to its south, machining would have been halted in Trench D itself. Three Evaluation Trenches I, II and III were dug within Trench D, and these produced very few artefacts. Therefore, although it is probable that the machine excavation of Trench D has slightly enhanced the sharpness of the northern edge of the concentration later found to its south, it is still believed that this concentration had a well-defined northern boundary approximately where it appears to do so. The collection from the rest of the

Phase 6 clay (excluding the assemblage associated with the elephant) was far more dispersed and far less numerous (Fig 18.1), and this material was designated as assemblage 6.2 (n=135, including 6 natural pieces). The material from around the elephant skeleton, discussed in the previous chapter (Chapter 17), was designated assemblage 6.3.

The remainder of this chapter follows the same broad structure as Chapter 17, although with some minor variations and changes of emphasis which reflect some important differences in the nature of the material from assemblages 6.1 and 6.2. After a brief review of the quantities and provenances of these assemblages (below), the following section considers their taphonomy and site formation processes, based on the combination of artefact condition, stratigraphy, spatial clustering, microdebitage sampling and refitting. The outcome of these analyses was that there is little benefit in stratigraphically sub-dividing either of the assemblages, and that a certain amount of more abraded lithic material within them should be excluded and analysed separately. This more abraded material does, however, represent important reworked evidence of an earlier phase of occupation.

Having established the most valid assemblages for analysis, technology, typology and the *chaîne opératoire* are looked at, including the evidence from refitted sequences of primary flaking and secondary flake modification. Finally, in the concluding section the evidence as a whole is considered in terms of how/

Table 18.1 Assemblage 6.1: lithic collection from Phase 6 grey clay south of Trench D

Phase	Context	Natural pieces (n)	Chips (n)	Other artefacts (n)	Total (n)	Notes
6	40036	-	-	2	2	Flints found on stripped surface of west bank of new road cutting during preliminary field evaluation in December 2003
	40078	-	-	2	2	Found in a slightly reddish-brown stained band, at northern side of concentration; discoloration not thought to be of any stratigraphic significance
	40100	119	110	1612	1841	-
6a	40039?	1	-	-	1	Contradictory provenance information in archive
	40039	19	5	140	164	Brown-stained bed at base of Phase 6 clay
Total		139	115	1756	2010	

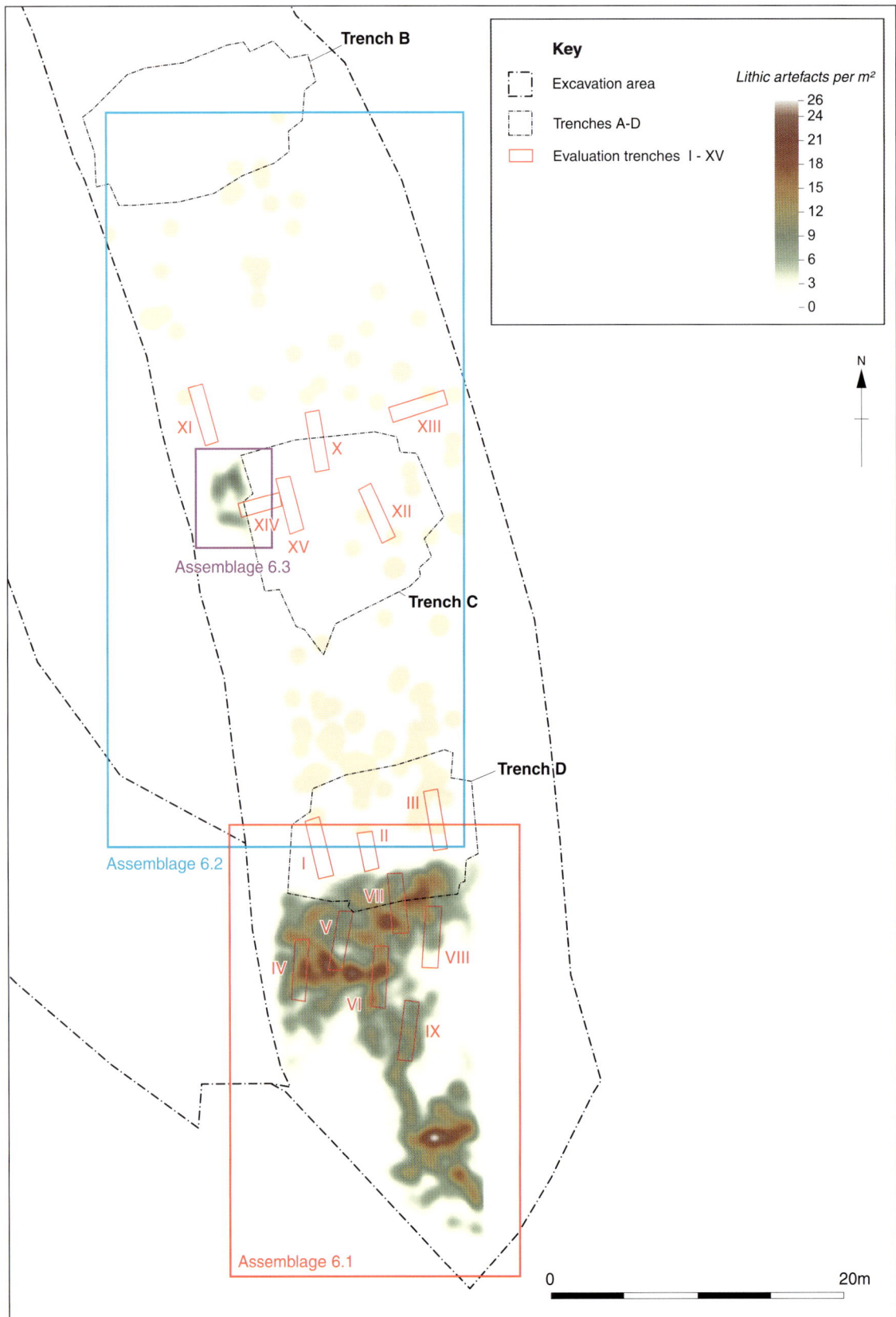

Figure 18.1 Assemblages 6.1 and 6.2: distribution of lithic finds (excluding natural pieces)

whether it provides any insights to hominin behaviour at the site and how it relates to the evidence from other sites in the Swanscombe area and further afield in south-east England. The extent to which it contributes to debate over the existence or otherwise of a Clactonian industrial tradition in south-east England in the earlier temperate part of the Hoxnian interglacial is also addressed.

PROVENANCE AND QUANTIFICATION

For assemblage 6.1 (Table 18.1), almost all of the constituent artefacts came from just two contexts. These were context 40100, the bottom 0.75m of the main body of the grey Phase 6 clay in the southern part of the site, and context 40039, the slightly sandy yellowish-brown stained clay-silt bed about 0.1-0.15m thick that underlay context 40100 in this part of the site. As discussed previously (Chapter 3), the Phase 6 clay in the area south of Trench D was being stripped by machine when flint artefacts began to be recognised in the lower part of the clay, at which point machine excavation ceased, and hand excavation began. The more northerly part of the area south of Trench D was trowelled and several evaluation trenches were also excavated with a combination of trowelling and careful mattocking. These methods probably led to a greater recovery of smaller chips <20mm long in the area where they were applied. In the central and southern parts of the area south of Trench D, excavation was carried out solely by mattocking, probably leading to a lesser recovery of smaller debitage. Therefore, aside from their specific inclusion in the microdebitage studies and their attempted use (rarely successfully) within the refitting programme, the chips were excluded from the quantitative analyses carried out. In general, smaller debitage of this size would not contain useful technological information for Lower/Middle Palaeolithic flaking sequences, although all were examined as potential flake-flakes from tool manufacture, and if thought to be so, were included as such in the quantitative analyses.

Assemblage 6.2 (Table 18.2) was recovered from the thicker and stratigraphically more complex part of Phase 6, between Trenches B and D. The majority of artefacts (*c* 63%) came from contexts generally attributed to the main body of the Phase 6 clay (mostly from context 40100). However, a reasonable proportion (*c* 35%) was recovered from Phase 6a, from the bottom two contexts of the Phase 6 clay (contexts 40039 and 40103). In addition to these, a few (n=3) artefacts were recovered from the Phase 6b tufaceous channel-fill, context 40070. Most of the artefacts from assemblage 6.2 were recovered from careful monitoring of machine excavation, so many of them were slightly damaged by the process of discovery, and there is probably a bias in this assemblage against recovery of smaller artefacts. Nevertheless, the assemblage still includes 11 chips <20mm long.

TAPHONOMY AND SITE FORMATION

Almost all of the artefacts from assemblages 6.1 and 6.2 were in mint or very fresh condition. For assemblage 6.1, there were just 15 artefacts in slightly-moderately abraded condition and two artefacts in very abraded condition, ie less than 1% of the total assemblage was not in mint or fresh condition. The more abraded specimens must be intrusive into the assemblage, representing evidence of older phases of occupation reworked from the higher ground to the west of the site. Some of the slightly abraded specimens may belong with the main assemblage, but it was decided to exclude them as they contributed nothing additional to the analytical results, and it was preferred to base these wholly on the fresh material. For assemblage 6.2, there were only two artefacts not in mint or fresh condition (approximately 1.5% of the assemblage), both of them technologically undiagnostic debitage in slightly-moderately abraded condition, and these were likewise excluded from the more detailed quantitative analyses of the material.

The first objective of the lithic analysis was to consider whether there was any interpretive importance

Table 18.2 Assemblage 6.2: lithic collection from Phase 6 grey clay, excluding concentrations by elephant skeleton and south of Trench D

Phase	Context	Natural pieces (n)	Chips (n)	Other artefacts (n)	Total (n)
6	40158	-	-	1	1
	40100	-	8	54	62
	40099	-	-	1	1
	40078	3	-	11	14
	40069	-	-	4	4
	40068	-	-	2	2
6b	40144	1	-	-	1
	40070	-	-	3	3
6a	40103	-	-	5	5
	40039	2	3	37	42
Total		6	11	118	135

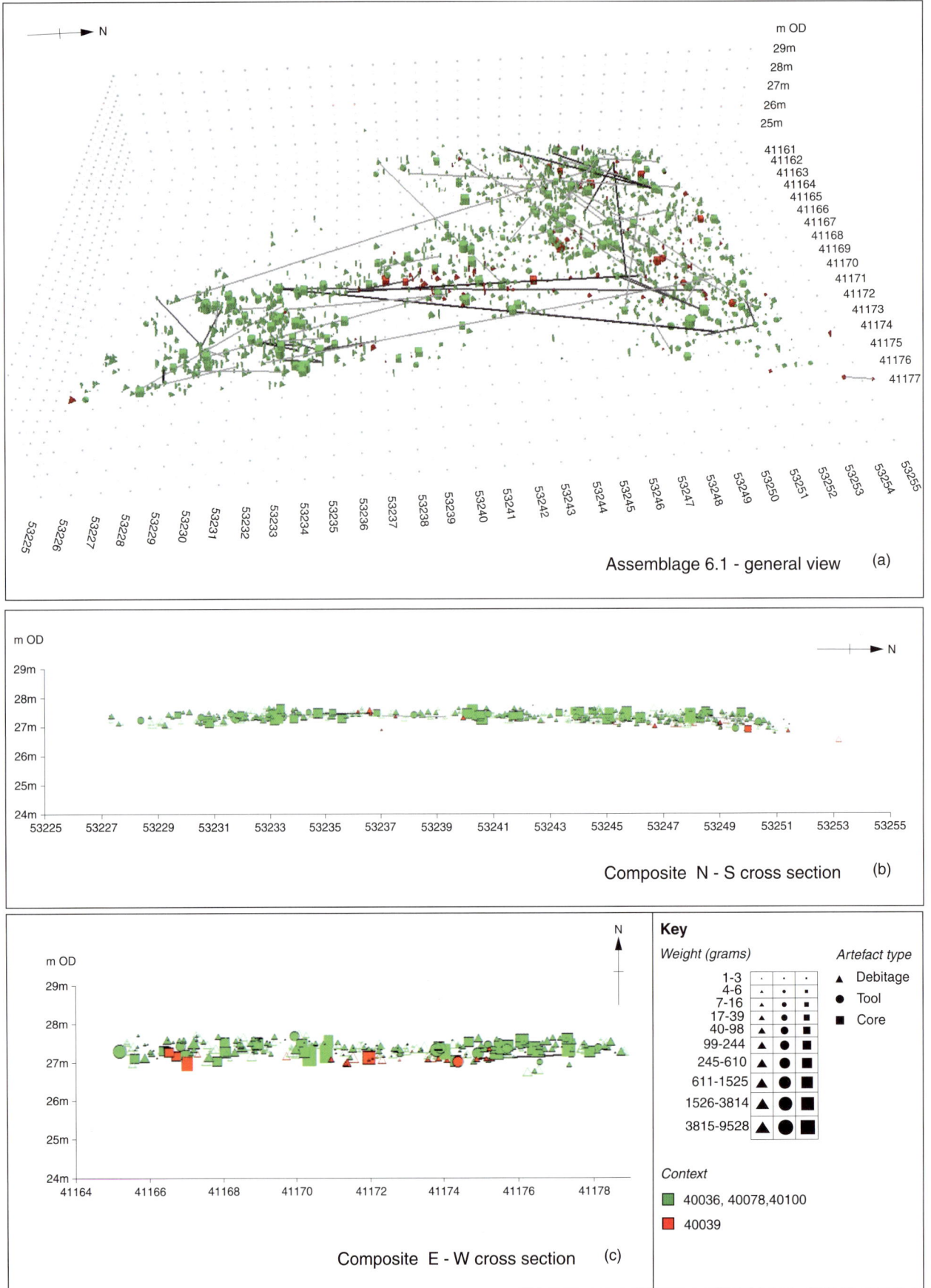

Figure 18.2 Assemblage 6.1: snapshots from 3-D GIS model, with refit connections: (a) general view; (b) conflated side-view (looking west); (c) conflated end view (looking north)

to the different stratigraphic horizons from which the artefacts were recovered, or whether the material would be better conflated for interpretive purposes. The same three approaches were adopted as for the investigation of the lithics from around the elephant skeleton. Firstly, an analogous 3-D GIS model was created for assemblage 6.1 with artefacts represented by different symbols for different technological categories, with the symbols sized according to their weight and colour-coded by context. This allowed initial investigation of the 3-D distribution of the artefacts, and of whether there was any immediate visual evidence of size-sorting, either laterally or vertically. Some snapshots from the model are illustrated here (Fig 18.2), showing the overall view of the concentration south of Trench D (assemblage 6.1), and cross-sections through the scatter. This is however a poor second to viewing the model for real, and as discussed previously, it is intended that the 3-D model is available for general investigation as part of the online resources accompanying this monograph.

The model showed that, within the area of lithic concentration, artefacts of all sizes were evenly distributed, both laterally and vertically. The slight increase in smaller debitage at its northern end is almost certainly due to the greater use of more careful hand-trowelling rather than a general attribute of the assemblage. It was also apparent that while there were distinct areas of greater artefact concentration (also see Fig. 18.1), these did not match the discrete concentrated clusters that would be expected from a palimpsest of undisturbed knapping activity, but were more homogenous. Therefore, although the lithic material was not looking like the result of undisturbed knapping activity, there was no sign that it was size-sorted within the sediment.

Secondly, a refitting study was carried out. All of the material from assemblage 6.1 (including the more abraded material, none of which was however found to refit) was laid out together. It was initially arranged to correspond with the on-site spatial clustering, and several weeks were spent attempting to fit it back together, in conjunction with the material from assemblages 6.2 and 6.3 (from near the elephant) since it was also hoped to use refitting evidence to link material from different parts of the site, and perhaps to investigate intra-site movement of artefacts in the event that it was found to represent a wider area of undisturbed activity. Refits were broken down into five main types:

B   *in situ* 'Break'; identified by unstained fresh joining surfaces, and almost zero distance between refitting pairs, for example for RG #35 and for RG #11, where Δ.43407 and Δ.43418 represent the eventual *in situ* detachment of a failed flake removal (see Fig. 18.14a).

J   knapping break, 'Join'; when a single flake has snapped during knapping, as distinct from subsequently in the ground or by deliberate breakage.

Jc   'Complex join'; when a piece has been deliberately broken, or when a piece broken during knapping

has undergone subsequent knapping.

R   separate debitage 'Removals', but with scars from intervening removals that show them not to be directly consecutive.

Rc   consecutive debitage 'Removals', so far as can be established.

Complementing this more descriptive classification of the refits, the last four types (that is apart from those thought to have been broken *in situ*) were also categorised into one of three groupings for taphonomic investigation. This was based on an interpretation of the likelihood that any piece of a refitting group would have been of any interest to a hominin, and therefore likely to have been moved by behaviour. Alternatively, whether it was regarded as likely to have been of minimal interest, and so any movement beyond the normal range of undisturbed material is likely to reflect sedimentary site formation processes. The three taphonomic groupings were:

1. 'T' – broken flakes and irregular waste, thought to have broken during knapping, and unlikely to have been used as tools, and therefore any movement/ separation most likely to relate to sedimentary 'Taphonomy'.

2. 'T?' – sequential flake removals, with no suspicion that either piece was likely to have been useful as a tool and affected by behavioural movement, therefore probably, although not definitely, any movement/ separation relating to sedimentary 'Taphonomy'.

3. 'B/B?' – sequential flake removals, with the possibility/ expectation that either piece might have been moved by 'Behaviour'. This category includes flakes that might have been removed with intervening hominid movement, for instance flakes from different stages of reduction of the same core and more than one flake-flake from the same flake-tool.

The refitting study was moderately successful, with 64 refitting groups found, including three from assemblage 6.2, and with 144 artefacts refitted in total, representing about 8% of the total fresh material from assemblages 6.1 and 6.2, excluding chips. Each refitting group was given a unique identifying number, RG #1-64, and the refits were tabulated as pairs of connecting artefacts (Table 18.3). Most of the refitting groups comprised just two artefacts, whether broken pieces or separate removals, but there were also a few groups with greater numbers of refitting pieces (Table 18.4). The greatest number of refitting pieces in a single group was 6, for RG #11, which was an intermittent sequence of five pieces of debitage fitted back to a core (discussed in more detail below). These results contrast with those from around the elephant where much longer reduction sequences were refitted, and almost 70% of the assemblage, including chips, was refitted (see Chapter 17). This immediately suggests a contrast between the level of disturbance of the material from assemblage 6.1

Table 18.3 Assemblages 6.1 and 6.2: collation of refits

Ref grp #	Ref type	Taph-onomy grouping	Flint 1 of group (or pair within group)				Flint 2 of group (or pair within group)				XYZ distance - m	Vertical distance - m	Bearing from N (lighter from heavier)
			Find ID - Δ	Context	C1	C2	Find ID - Δ	Context	C1	C2			
1	Rc	2	41329	40100	5	2	41211	40100	5	2	1.75	0.13	10.8
2	R	2	42145	40100	6	3	42436	40100	5	4	2.84	0.34	178.75
3	Rc	2	43178	40070	5	2	43062	40070	5	2	1.23	0.31	184.9
4	J	1	42544	40039	6	4	42572	40039	5	1	1.91	0.4	26.92
5	R	2	40812	40039	5	2	40813	40039	5	2	1.03	0.1	3.9
6	Rc	2	41747	40100	5	4	41934	40100	6	3	1.4	0.07	-86.65
6	R	2	41934	40100	6	3	41940	40100	5	2	2.17	0.22	163.52
6	R/Rc	2	41940	40100	5	2	41747	40100	5	4	2.96	0.29	136.98
7	Jc	3	41294	40100	5	1	41671	40100	5	4	1.77	0.24	-80.95
8	J	1	41345	40100	5	1	41517	40100	5	1	1.05	0.06	120.71
9	R	3	42470	40100	6	3	42468	40100	5	4	0.24	0.01	6
9	R	3	42468	40100	5	4	41371	40100	7	1	3.39	0.1	223
9	R	3	41371	40100	7	1	42470	40100	6	3	3.59	0.1	220.73
10	J	1	42682	40100	5	1	42689	40100	5	1	1.54	0.44	172.82
11	R/J	2	41993	40100	5	2	41855	40039	5	2	2.43	0.1	-86.04
11	R/J	2	43418	40100	7	1	41855	40039	5	2	15.87	0.29	3.96
11	R/J	2	43407	40100	5	2	41993	40100	5	2	16.13	0.24	12.67
12	J	1	41300	40100	5	2	41805	40100	5	2	1.62	0.21	56.24
13	R	2	41663	40100	7	1	42409	40100	5	2	2.42	0.19	232.26
14	J	1	41959	40100	5	1	42257	40100	5	1	1.67	0.01	119.85
15	J	1	42289	40100	5	1	43024	40100	5	1	2.09	0.3	264.53
16	J	1	40205	40100	5	1	40385	40100	5	1	1.08	0.08	229.46
17	R	3	42021	40100	5	2	42022	40100	7	1	1.17	0.04	105.62
18	R	3	40264	40100	6	3	42258	40100	5	2	2.69	0.24	119.29
19	J	1	42242	40100	5	2	42278	40100	5	2	0.49	0.01	140.3
20	R	2	42418	40100	5	2	42568	40100	5	2	4.09	0.14	57.66
21	R	2	41154	40100	5	2	41882	40100	5	2	11.4	0.66	51.68
22	J	1	40374	40100	5	2	42079	40100	5	1	2.75	0.14	26.55
23	R	2	41851	40100	5	2	41957	40100	5	2	1.84	0.24	-65.13
24	J	1	41779	40100	5	1	41891	40100	5	1	0.73	0	202.74
25	J	1	41516	40100	5	2	42946	40100	5	1	11.24	0.14	9.01
26	R	2	40300	40100	5	2	42353	40100	5	2	2.32	0.25	130.59
27	J	1	42594	40100	5	2	42812	40100	5	2	3.36	0.06	124.06
28	R	3	43182	40100	7	1	43217	40100	5	2	0.49	0.13	79.87
29	B	0	43081	40100	5	1	43082	40100	5	1	0.05	0.02	-45.49
30	R/J	3	43399	40100	7	1	43267	40100	7	1	0.57	0.02	-24.54
30	R	3	43198	40100	6	4	43267	40100	7	1	1.36	0.15	193.91
30	R/J	3	43267	40100	7	1	43262	40100	5	1	1.44	0.01	177.49
30	R/J	3	43262	40100	5	1	43399	40100	7	1	1.98	0.02	-8.7
31	R	2	43227	40100	5	1	43202	40100	5	2	0.24	0.19	201.74
31	R	1	43388	40100	7	1	43227	40100	5	1	1.62	0.17	26.86
32	R	2	43370	40100	5	2	43478	40100	5	2	1.5	0.05	175.63
33	J	1	43547	40100	5	1	43735	40100	5	1	1.6	0.01	88.73
36	R/B	0	43636	40100	5	2	43635	40100	5	2	0.02	0	128.54
36	R/B	2	43635	40100	5	2	43542	40100	5	2	0.49	0.21	88.19
37	J	1	42913	40039	5	1	40817	40039	5	1	0.49	0.05	-50.05
37	J	1	40817	40039	5	1	43051	40039	5	1	2.12	0.14	163.91
37	J	1	43051	40039	5	1	42913	40039	5	1	2.54	0.09	-22.26
38	J	1	43512	40100	5	1	43666	40100	5	1	2.3	0.24	-78.53
38	J	1	43666	40100	5	1	43609	40100	5	1	2.4	0.32	189.47
38	J	1	43609	40100	5	1	43512	40100	5	1	3.25	0.08	234.26
39	J	1	43411	40100	5	1	43398	40100	5	1	0.47	0.09	215.66
39	J	1	43570	40100	5	1	43411	40100	5	1	1.14	0.06	6.7
39	J	1	43398	40100	5	1	43570	40100	5	1	1.57	0.15	195.01
40	R	2	42783	40100	5	1	43523	40100	5	2	9.07	0.35	168.04
41	R	3	43111	40100	7	2	43307	40100	5	2	1.82	0.2	186.81
42	Jc	1	43476	40100	5	1	43734	40100	5	1	6.2	0.21	-41.98

Table 18.3 (continued)

Ref grp #	Ref type	Taphon- omy grouping	Flint 1 of group (or pair within group)				Flint 2 of group (or pair within group)				XYZ distance - m	Vertical distance - m	Bearing from N (lighter from heavier)
			Find ID	Context	C1	C2	Find ID	Context	C1	C2			
43	R	2	41797	40100	5	1	43604	40100	5	2	19.12	0.33	158.06
44	J	1	41256	40100	5	1	42358	40100	5	1	4.99	0.12	61.22
45	J	1	43458	40100	5	1	43517	40100	5	1	0.97	0.15	20.08
46	Rc	3	42789	40100	5	2	43469	40100	7	1	8.72	0.21	-15.9
47	R	2	41155	40100	5	2	42170	40100	5	2	3.17	0.12	256.27
48	J	1	43327	40100	5	1	43328	40100	5	1	0.14	0.03	95.2
49	Rc	2	40616	40100	5	2	44018	40100	5	2	0.37	0.16	94.82
50	R	2	40274	40100	5	2	43584	40100	5	2	16.88	0.02	-10.7
51	Rc	2	41539	40100	5	2	42573	40039	5	2	5.37	0.06	231.12
52	R	2	40717	40100	5	2	40553	40100	5	2	6.12	0.16	-1.64
52	R	3	40553	40100	5	2	43102	40100	5	2	4.94	0.07	178.47
52	R	3	43102	40100	5	2	40717	40100	5	2	11.05	0.23	-1.59
53	J	1	42894	40100	5	1	43059	40100	5	1	2.37	0.05	-12.79
54	J	1	41282	40100	5	2	41480	40100	5	2	5.7	0.03	58.14
55	Rc	2	42467	40100	5	2	41938	40100	5	2	4.33	0.03	258.45
55	R/Rc	3	41938	40100	5	2	42619	40039	5	2	4.1	0.07	-61.93
55	R	3	42619	40039	5	2	42467	40100	5	2	7.93	0.11	-82.3
56	B	0	42457	40100	5	2	42458	40100	5	2	0.13	0.01	8.73
57	Rc	2	43594	40100	5	1	43697	40100	5	2	3.01	0.07	167.37
58	R	2	40468	40100	5	1	42624	40100	5	2	5.09	0.18	54.25
59	J	1	40203	40100	5	2	42465	40100	5	1	2.06	0.21	183.42
59	J	1	42465	40100	5	1	41554	40100	7	1	5.16	0.13	40.84
59	J	1	41554	40100	7	1	40203	40100	5	2	6.89	0.09	210.45
60	J	1	42090	40100	5	1	42723	40100	5	2	1.42	0.27	-5.86
61	R	3	40429	40100	7	1	42009	40100	5	2	7.24	0.01	-73.36
62	Jc	3	40618	40039	5	2	42972	40100	5	2	9.97	0.11	62.21
63	R	3	40425	40100	6	3	41537	40100	5	2	2.03	0.12	-55.94
64	J	1	41669	40100	5	1	42378	40100	5	1	3.57	0.06	-15

Table 18.4 Assemblages 6.1 and 6.2: refitting group sizes

Refitting group size (n pieces)	Assemblage 6.1	Assemblage 6.2	Total	Refitting groups #
2	49	3 (RGs: #1, #3 and #4)	52	All the rest
3	10	-	10	RGs: #6, #9, #31, #36-39, #52, #55 and #59
4	1	-	1	RG #30
5	-	-	-	-
6	1	-	1	RG #11
Total	61	3	64	

Table 18.5 Assemblage 6.1: distribution of refitting flints between 'upper' and 'lower' stratigraphic horizons ['Both-Upp.' represents the number of flints from upper contexts in refitting groups with flints from both upper and lower horizons; 'Both-Low.' represents the number of flints from lower horizons in the same groups]

Horizon	All flints		Refitting flints		Refitting groups	
	n	%	n	%	n	%
'Upper'	1602	92	120	87	55	90
Both-Upp.	-	-	9	6.5	4	6.5
Both-Low.	-	-	4	3		
'Lower'	139	8	5	3.5	2	3.5
Total	1741		138	7.9	61	

Table 18.6 Assemblage 6.2: refitting flints by phase and context

Horizon	All flints		Refitting flints		Refitting groups		
	n	*%*	*n*	*%*	*n*	*%*	*Context*
Phase 6	71	61.2	2	33.33	1	33.33	40100
Phase 6b	3	2.6	2	33.33	1	33.33	40070
Phase 6a	42	36.2	2	33.33	1	33.33	40039
Total	116		6	5.2	3		

with that by the elephant, which is thought to be essentially undisturbed.

In assemblage 6.1, the refitting results were initially used to investigate the stratigraphic integrity between Phase 6 (context 40100) and Phase 6a (context 40039) (Table 18.5). There were 138 refitting flints in total, comprising nearly 8% of the total (fresh) assemblage, 92% of which came from context 40100, and 8% of which came from context 40039. Most of the refitting groups did not cross between these two contexts, which is statistically to be expected even if they have no stratigraphic significance. However, four of the refitting groups (RGs #11, #51, #55 and #62) included pieces from both contexts 40100 and 40039, leading to the conclusion that there was little stratigraphic integrity between the material from these two contexts. Therefore, all of the fresh condition material from Phases 6 and 6a south of Trench D was combined into a single assemblage (n=1854, including 113 chips) for subsequent analyses.

In assemblage 6.2 (Table 18.6), there were only three refitting groups of flints (RGs #1, #3 and #4), all of them comprising just two artefacts. The overall proportion of flints refitted was approximately 5%, which was broadly comparable to the figure from assemblage 6.1. The numbers of flints are too low for any meaningful statistical inferences, although each of these pairs represented a different stratigraphic phase, and none of the refitting pairs crossed any context boundary. The fresh material from this assemblage was likewise amalgamated (n=127, including 11 chips) to develop a quantitative technological and typological profile, since the material from the separate contexts would have been too low in number for meaningful comparisons, and there was not thought to have been any significant time depth between formation of the different contexts. It does, however, seem likely that the material from the different phases in this assemblage represents discrete parcels of material.

When the spatial connections between the refitting material were plotted (Fig. 18.2a; Fig. 18.3; Fig. 18.4) it can be seen not only that the overall quantities of refits are much lower than would be expected for undisturbed knapping debitage from the flake/core reduction sequences represented in the assemblage, but also that the clustering and distribution of the refitting material does not correspond with undisturbed material: for instance in contrast to four experimental scatters of the same flake/core technology (Fig. 18.5) and to the

material by the elephant skeleton (Fig. 17.10). However, the overall quantity of refitting material is still quite significant, at approximately 8% of the total fresh assemblages, and suggests an only-slightly-spatially-disturbed accumulation of at least some material, and high stratigraphic integrity within the Phase 6 clay.

The data for refit direction and distance were then subject to further analyses, to try and identify any directional trends that might relate to site formation processes, and any residual behavioural information. All of the refitting material was analysed as pairs of connecting points. For groups of three, there were therefore usually three pairs of points, although in some instances (for example RG #31, Fig 18.7b) the sequence could be reduced to one removal, and then one break. Three statistics were calculated for each refitting pair: (1) their distance of separation in three dimensions XYZ, (2) their vertical separation regardless of spatial distance apart, and (3) the direction of the lighter piece from the heavier, as a bearing from North. These data are presented for each refitting pair in the collated refitting data table (Table 18.3). The directional data for each of the three different taphonomic refit groups are summarised on separate rose diagrams (Fig. 18.6). Three concentric circles represent increasing distance of separation, ranging from 'close' (< 2.5m, well within the typical spread range of undisturbed knapping remains), through 'medium' (2.5 – 7.5m, at the far end of the typical spread range of undisturbed knapping remains) and 'far' (>7.5m, well beyond the typical spread range of undisturbed knapping remains). Complementing the rose diagrams for each of the 3 taphonomic groupings are summary tables of the basic statistics 'Maximum value', the 3rd/4th quartile boundary 'Q3', the mean, the 1st/2nd quartile boundary 'Q1', minimum value and the population's sigma value 'SD' (Table 18.7).

It is immediately clear from the rose diagrams (Fig. 18.6) that the directions of refitting connections are essentially random for all three of the taphonomic groups of refit, and within each group for the different distance ranges. On the orthogonal plan-view of the refits from assemblage 6.1 (Fig. 18.3 – which also labels some of the longer refitting connections), there appears to be a slight predominance of longer north-south connections. However, this is a misleading illusion due to the lithic concentration being more extended north-south so there is an inevitable potential for more, and longer, refits in this direction. Also, three of the longer refitting connections (from RG #11 and RG #52) showing as darker lines in

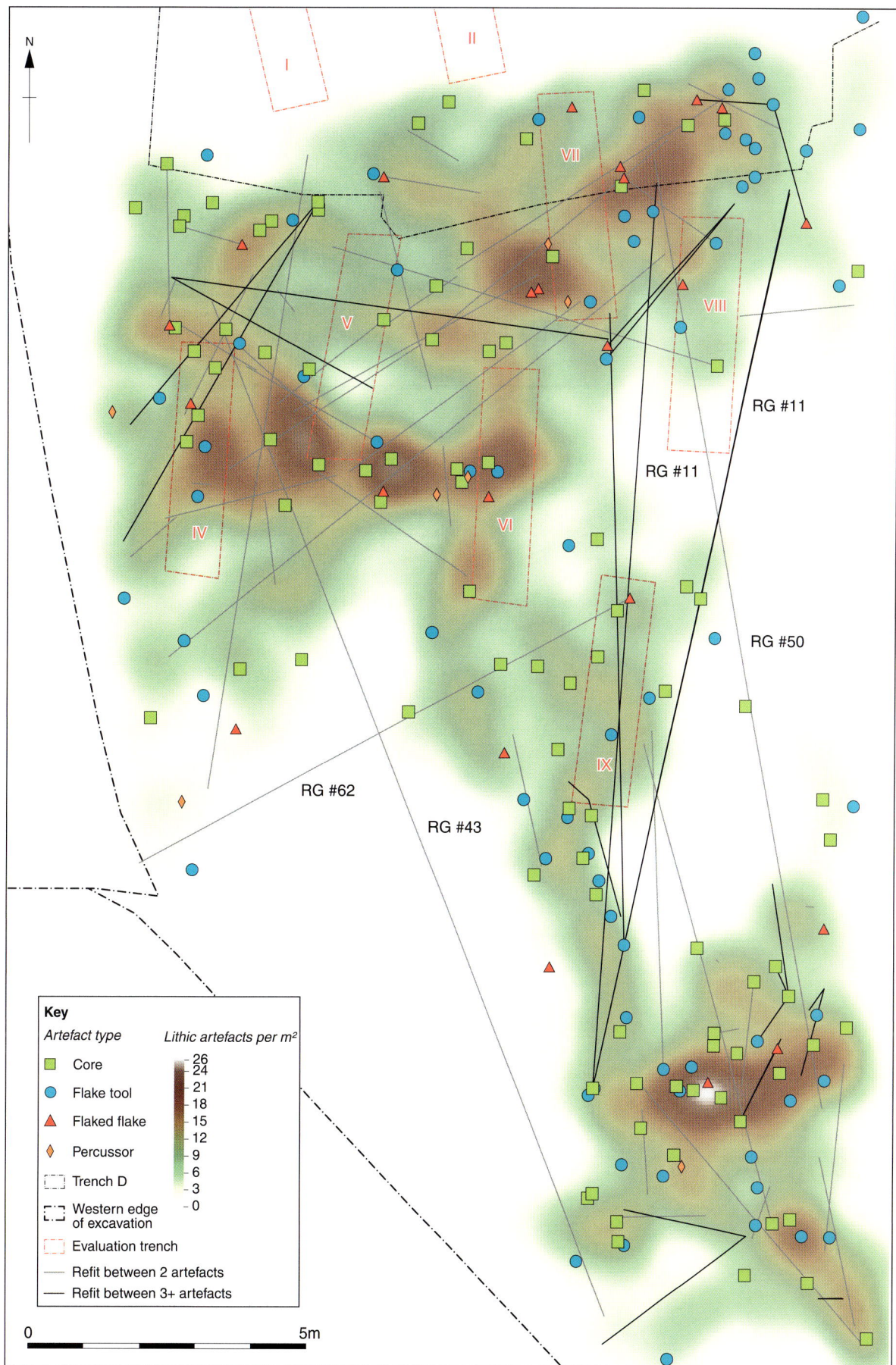

Figure 18.3 Assemblage 6.1: orthogonal view (from above) of refitting connections [lines linked to core or heaviest piece for groups where n>2]

Figure 18.4 Assemblage 6.2: orthogonal view (from above) of refitting connections [lines linked to core or heaviest piece for groups where n>2]

the north-south direction. There is certainly no trend for more 'behavioural' refits to be further apart than, or in different directions from, 'taphonomic' ones (Table 18.7). Aside from some higher outliers in the intermediate group 2 'T/T?' the majority of refits in all three groups occur with similar separations, between approximately 1m and 6m apart. The Q3 point is slightly lower for the presumed-

more-taphonomic group 1, consisting of broken material, but this may be influenced by the intrusion of unrecognised material that broke *in situ* in the sediment, rather than being broken during knapping.

The most separated refit found was RG #43. This consisted of a small flake struck off a larger piece of irregular waste (Fig. 18.7a), with the irregular waste

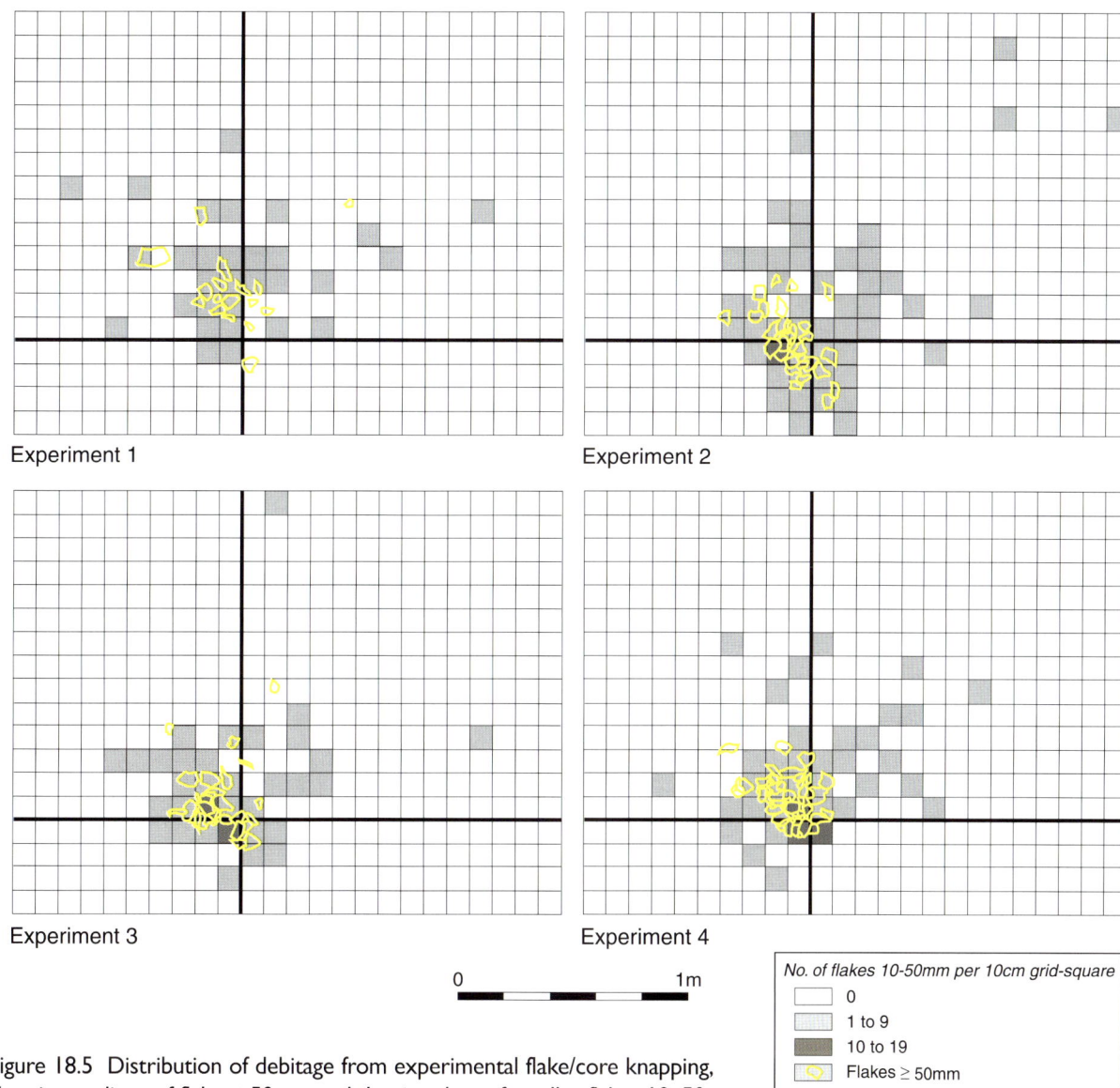

Experiment 1

Experiment 2

Experiment 3

Experiment 4

0 ▬▬▬▬ 1m

No. of flakes 10-50mm per 10cm grid-square
- ☐ 0
- ▨ 1 to 9
- ▦ 10 to 19
- ⬡ Flakes ≥ 50mm

Figure 18.5 Distribution of debitage from experimental flake/core knapping, showing outlines of flakes ≥50mm and density plots of smaller flakes 10–50mm

Table 18.7 Assemblage 6.1: refitting separation summaries, by taphonomic group

	Taph. group 1 - T (n=34)		Taph. group 2 - T? (n=28)		Taph. group 3 - B/B? (n=20)		All Taph. groups (n=82)	
	Distance XYZ	*Vertical separation*	*Distance XYZ*	*Vertical separation*	*Disancet XYZ*	*Vertical separation*	*Distance XYZ*	*Vertical separation*
Max	11.24	0.44	19.12	0.66	11.05	0.24	19.12	0.66
Q3	3.12	0.20	5.56	0.26	5.51	0.16	4.79	0.21
Average	2.60	0.14	5.17	0.20	3.83	0.11	3.78	0.15
Q1	1.21	0.06	1.69	0.10	1.42	0.03	1.42	0.06
Min	0.14	0.00	0.24	0.02	0.24	0.01	0.14	0.00
SD (Pop)	2.24	0.11	5.43	0.13	3.27	0.08	4.00	0.12

1. Taphonomy (n=34)

2. Taphonomy? (n=28)

3. Behaviour (n=20)

Key

Frequency

Distance

A   0 - 2.5m

B   2.5 - 7.5m

C   +7.5m

found at the north-west corner of the main lithic concentration and the smaller flake found almost at the southern limit of the excavated area, almost 20m away (Fig. 18.3). Other notably separated refitting groups were RGs #11, #50 and #62 (Fig. 18.3; Fig. 18.7c,d,e). The first two of these were both attributed to the intermediate taphonomic group 2 'T?' and the third, which involved a flake-flake was in group 3 'B/B?', more likely to have been affected by hominin mobility. Group RG #11 was the longest sequence found, and represented a failed core with its last two small removals, and one earlier one (Fig. 18.7c; Fig. 18.18a). Unfortunately, the XYZ location of the early removal was lost, but the core was found in the southern part of the concentration and the two last failed removals approximately 15m to the north, about 2.5m from each other (Fig. 18.3).

Refitting group #50 was represented by two moderately small, quite nondescript flakes, one of which was found in the southern part of the concentration and the other almost 17m away towards its north-east corner. Refitting group #62 (Fig. 18.7e; Fig. 18.27b) was more interesting technologically. It comprised a large, long flake that had broken in two (during knapping presumably, judging by the absence of any sign of deliberate breakage). The broken distal end was then further knapped at one corner, by the break. The refitting pieces comprise the proximal end, and the flake-flake from the secondary working of the distal end; the worked piece is absent. The small flake-flake was found in the central part of the lithic concentration, and the larger proximal end some 10m away to the west.

An overall conclusion from the refitting work is that assemblages 6.1 and 6.2 do not represent wholly undisturbed knapping activity, in contrast to assemblage 6.3 from around the elephant. There is nonetheless a reasonably high degree of refitting, and the presence of quite numerous refits separated by less than 4m (n=58, roughly 70% of the total population of refit separations) suggests that at least some of the material is minimally disturbed. Finally, it is concluded that there is no evidence of trends of refit separation or direction that can be attributed to distinct sedimentary site formation processes or hominin behaviour and mobility.

The refitting work was complemented by an investigation of microdebitage quantities and distribution. As discussed previously (Chapter 17), experimental replication of similar flake/core knapping strategies applied to nodular flint raw material has provided a comparative data set for expected quantities of microdebitage in

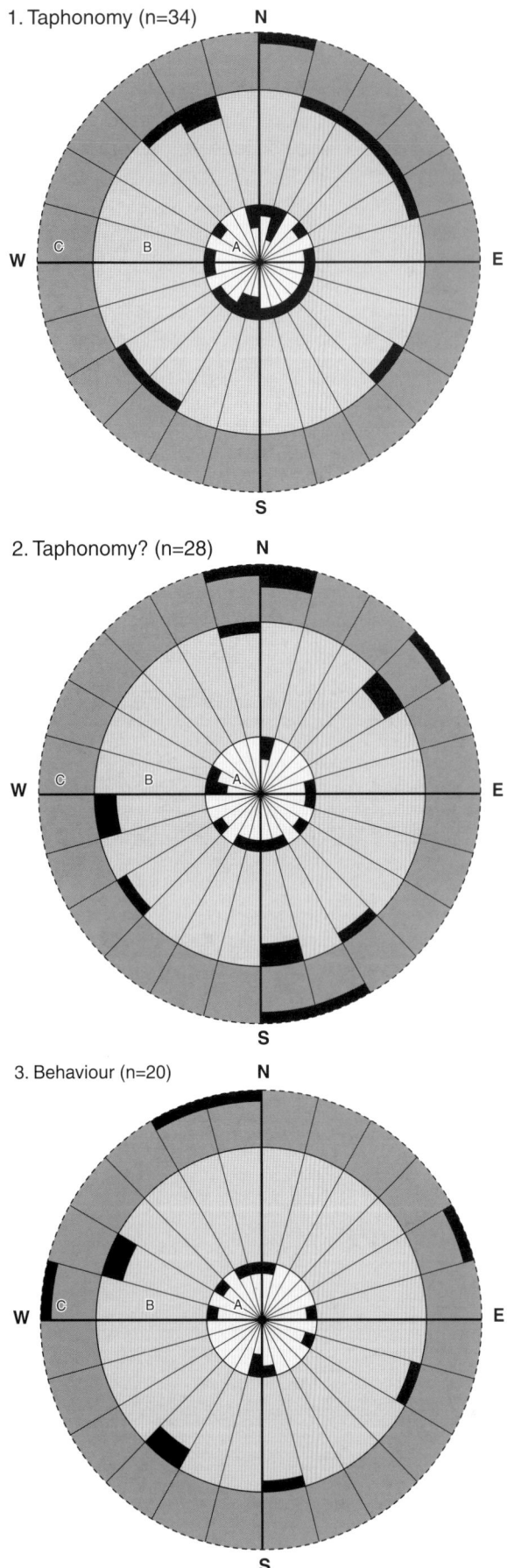

Figure 18.6 Rose diagrams showing distances and directions of refitting connections for the three taphonomic groups: (1) T - Taphonomy, (2) T? - Taphonomy? and (3) B/B? - Behaviour

relation to larger flakes and irregular waste. It is clear that, even when simple flake/core knapping strategies without deliberate platform preparation or fine surface-trimming are carried out, large quantities of microdebitage are produced. Furthermore, for undisturbed knapping scatters from single moderate reduction episodes (Fig. 17.5; Fig. 18.5), the microdebitage is concentrated in a tight cluster around the knapping point. Typically there are about 100 small flakes 10–50mm and over 250 chips in the size range 2.5–10mm within a circle of radius 0.5m, tailing off more than 1m from the central cluster. A sampling programme for microdebitage was therefore carried out in the area south of Trench D, in Evaluation Trenches IV, V and VII, and in the area of the *S kirchbergensis* rhino jaw (Group Δ.40843) just to the north of the main lithic concentration.

The programme is described in detail in the earlier chapter on excavation methods (Chapter 3). In summary, within each of the evaluation trenches, samples were taken in a vertical and horizontal grid of contiguous squares, normally 0.5 x 0.5m, as hand excavation progressed down in spits of 50mm thick.

At Trench IV a total of 216 samples were taken from nine excavation spits (Fig. 18.9). At Trench V a total of 248 samples were taken from eight excavation spits (Fig. 18.10). At Trench VII a total of 312 samples were taken from nine excavation spits (Fig. 18.11) and at the rhino jaw, 24 samples were taken from a single excavation spit (Fig. 18.12). All of the samples were sieved, and any small sharp flint pieces thought to be microdebitage were sorted from the resulting residues. These were divided into one of three categories: (1) ≥ 20mm, (2) 4–20mm (ie did not go through a 4mm sieve-mesh), and (3) 2–4mm (ie went through 4mm sieve-mesh, but not a 2mm mesh). The results for each trench and the rhino jaw group are shown in the respective figures (Figs 18.9 – 18.12) and the quantitative data for each sampled area are also summarised (Table 18.8) with the total microdebitage

Figure 18.7 Photos of selected refitting groups: (a) RG #43; (b) RG #31; (c) RG #11; (d) RG #50; (e) RG #62

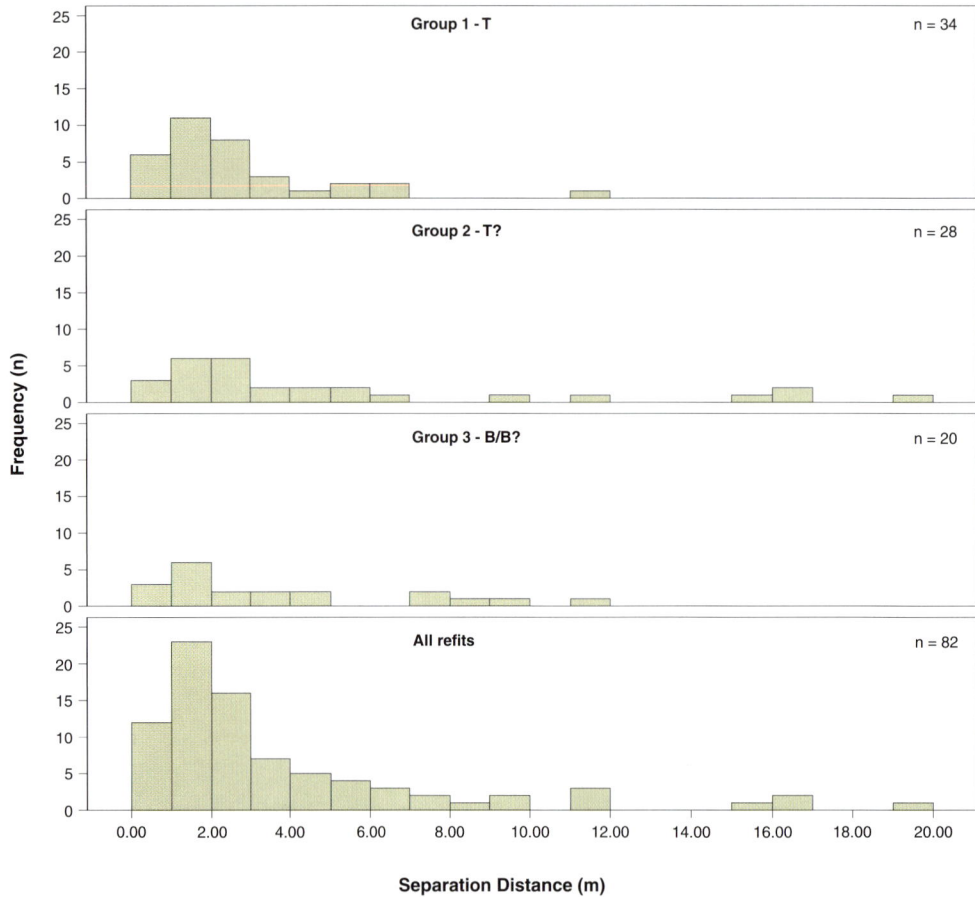

Figure 18.8 Histogram of refit separations

counts compared with the quantities of artefacts ≥ 20mm in the sampled areas.

Apart from those by the rhino jaw, these data present interesting and apparently contradictory results. At the rhino jaw, there is a sparse spread of all three artefact-size categories, with no sign that microdebitage is concentrated in a patch and associated with larger debitage so as to represent undisturbed knapping evidence. Although there are greater quantities of microdebitage than larger flakes, there are not as many as would be expected for undisturbed material. In conjunction with the absence of any refitting of the artefacts in the area of the rhino jaw microdebitage sampling, it seems evident that there is no association of undisturbed lithic evidence with the rhino jaw.

In Trenches IV, V and VII, the data are more complex. The sampling patterns allow investigation of both the spatial distribution of microdebitage, and a vertical profile through the sampled sequence in each trench. The

same trends are visible in all three diagrams. Firstly, while there are some localised areas richer in microdebitage, for instance in the centre of Trench V, spit 3, these are never so densely concentrated as the experimental models for undisturbed knapping remains, and there are no instances of a rapid drop-off from a high concentration to a low concentration, as would be the case for an undisturbed scatter. Secondly, the microdebitage is quite evenly distributed through the sequence vertically. Thirdly, there seems to be no correspondence between the areas of concentration of the larger artefacts ≥ 20mm in size, and the areas with greater quantity of micro-debitage. And fourthly, there is no particular correspon-dence between areas richer in chip-size microdebitage (in other words 4–20mm) and areas richer in spalls (2–4mm). All of these factors indicate that, particularly in conjunction with the above-mentioned refitting results, that this area of the site cannot be regarded as containing undisturbed lithic remains.

Table 18.8 Assemblage 6.1, debitage and microdebitage quantitative comparisons: Trenches IV, V, VII and rhino jaw (group Δ.40843) compared to experimental data (Exp.)

Size-range	Exp.		Tr IV		Tr V		Tr VII		Rhino jaw	
	n	%	n	%	n	%	n	%	n	%
≥ 20mm	24	14	15	5	23	4	25	9	5	14
4–20mm	44	25	33	12	92	17	32	11	14	40
2–4mm	109	61	230	83	424	79	223	80	16	46

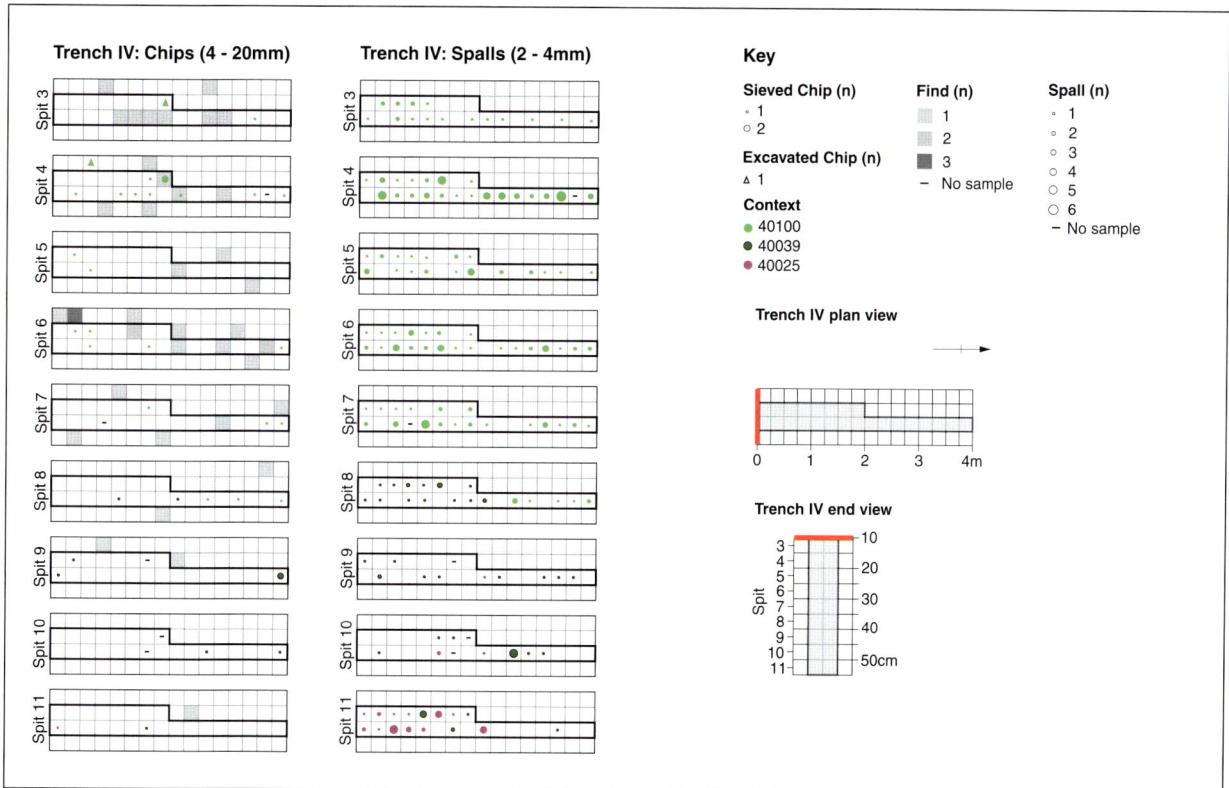

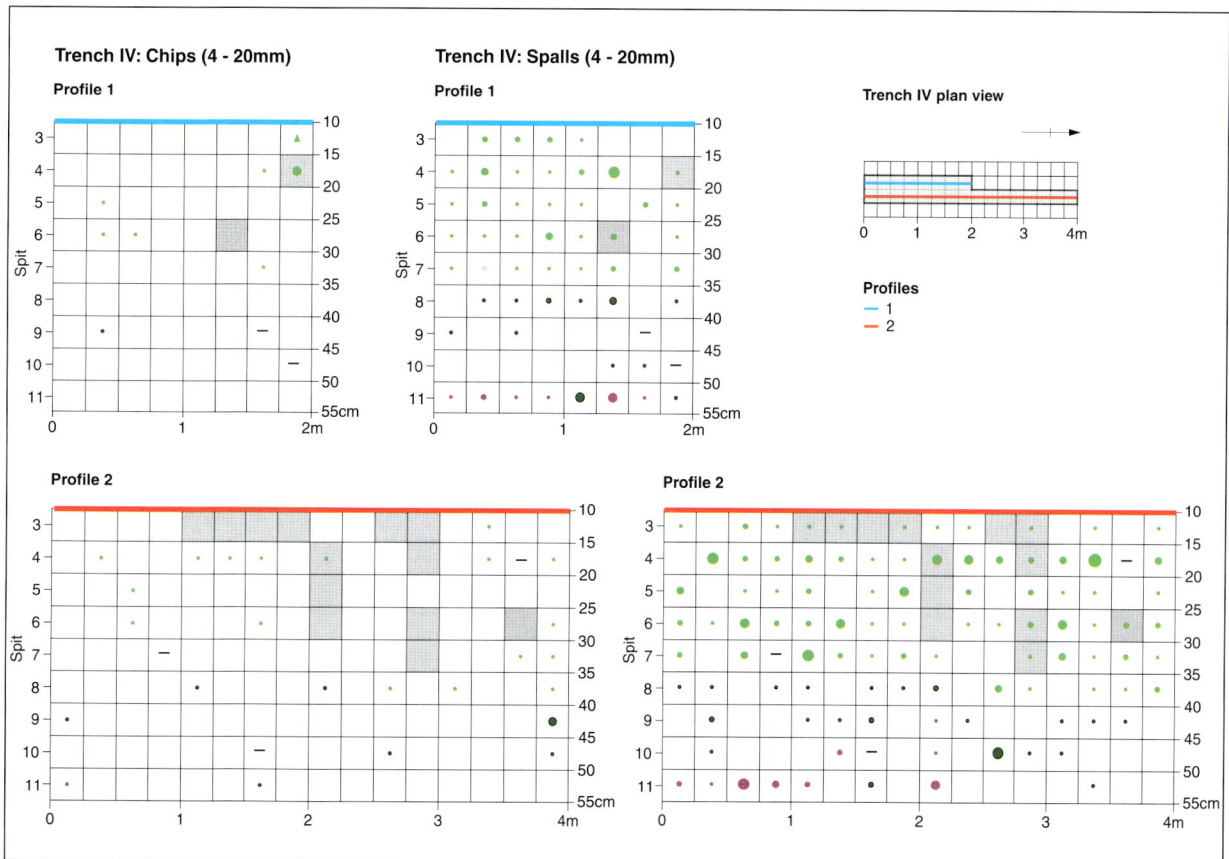

Figure 18.9 Microdebitage sampling and recovery, Trench IV

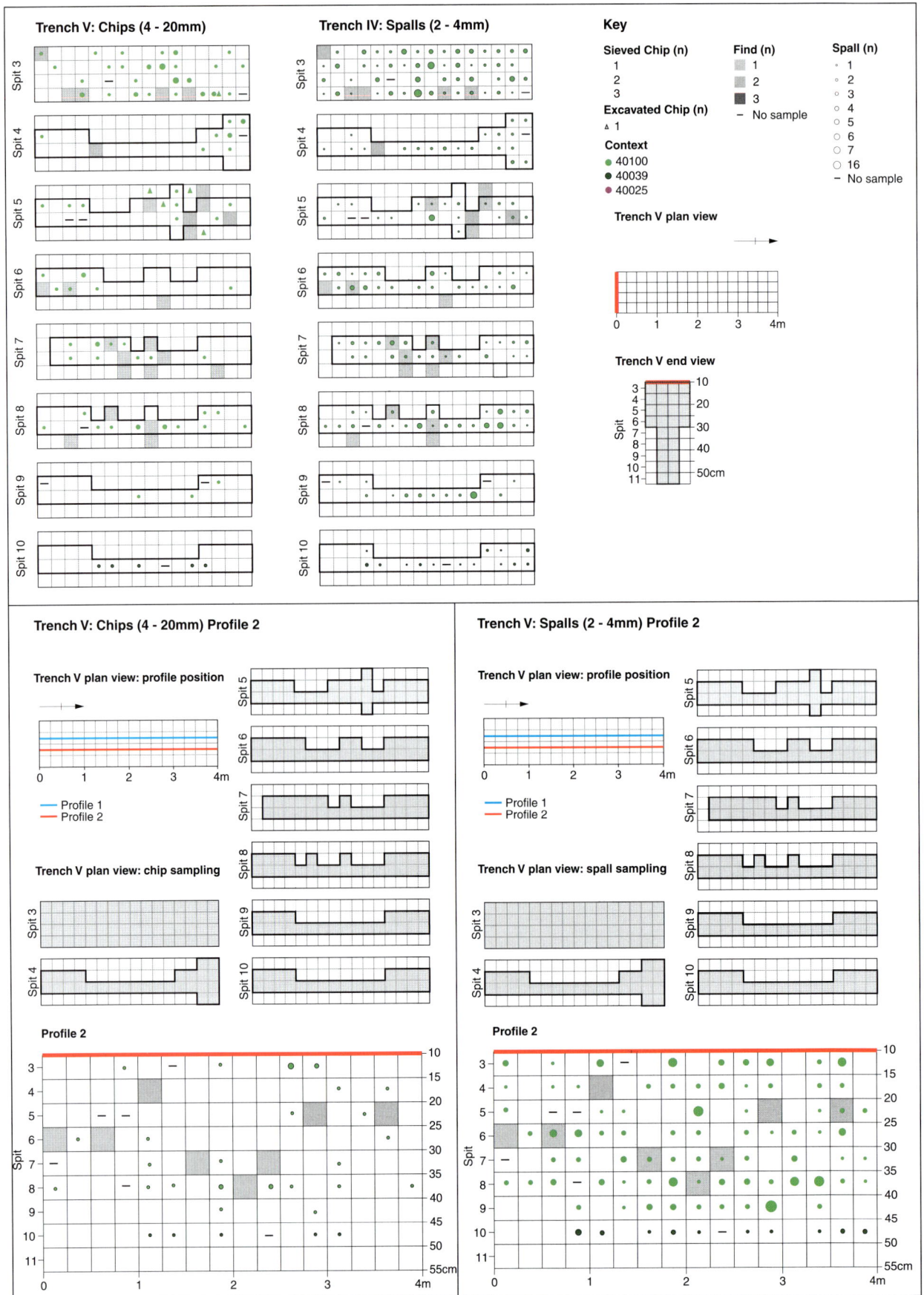

Figure 18.10 Microdebitage sampling and recovery, Trench V

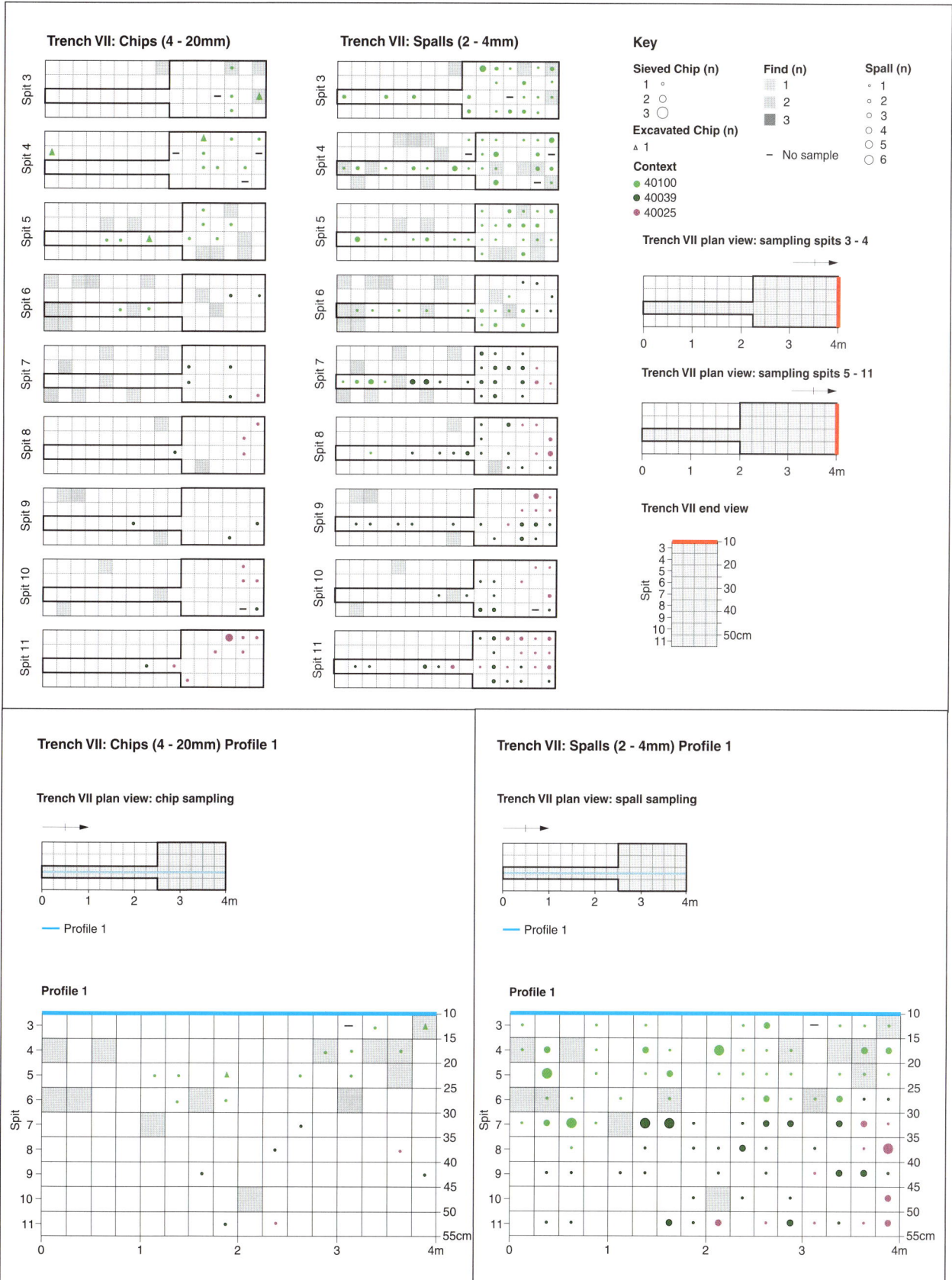

Figure 18.11 Microdebitage sampling and recovery, Trench VII

Rhinoceros jaw area: chip (4–20mm) and spall (2–4mm) plan

Section 40060.b

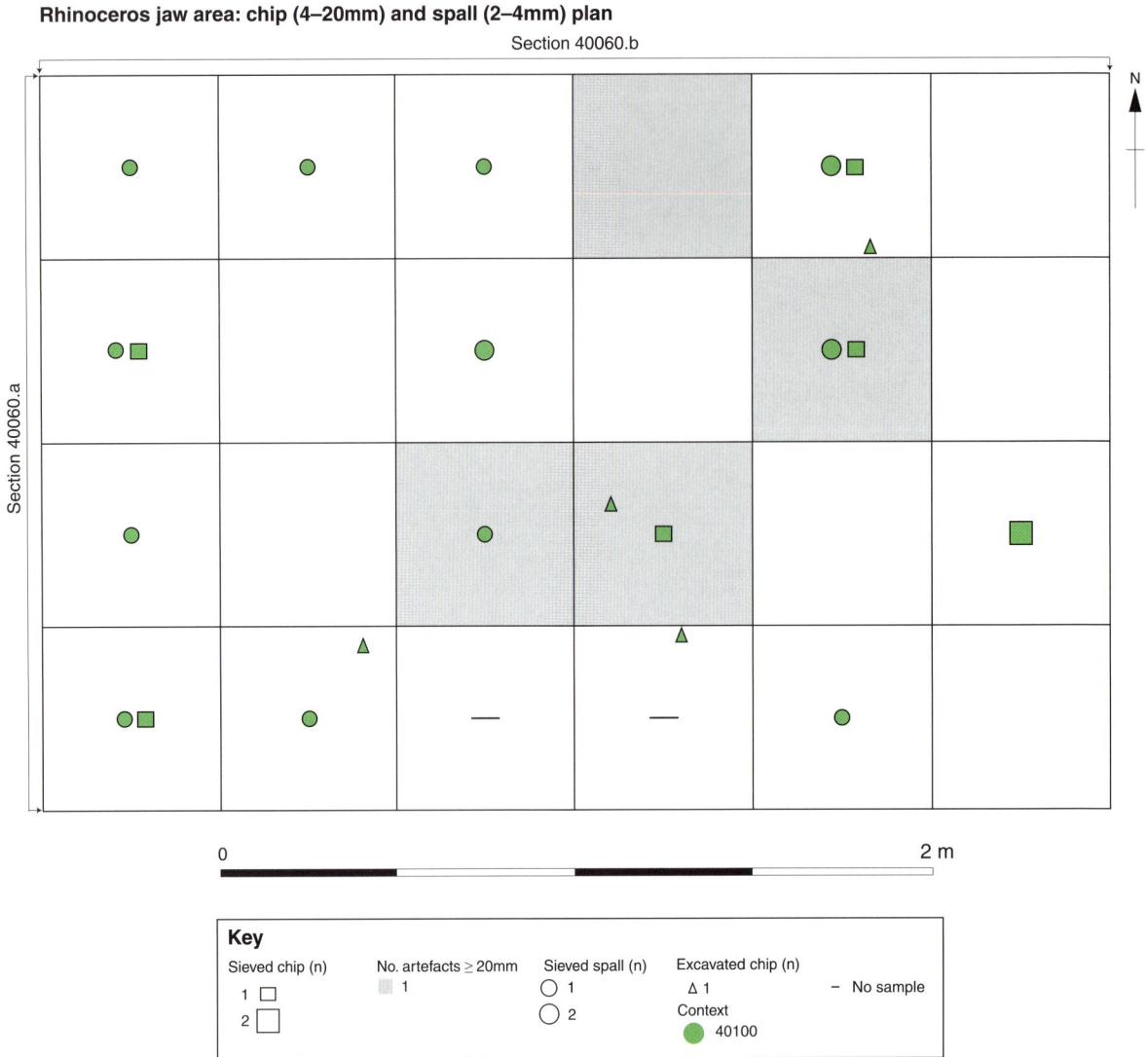

Figure 18.12 Microdebitage sampling and recovery, rhinoceros (*S. kirchbergensis*) jaw group Δ.40843

However, there are nonetheless high quantities of microdebitage (Table 18.8). In fact, when the total quantities of microdebitage are considered in relation to the number of larger lithic artefacts from the sampled volume of sediment, there are higher proportions of the smallest spall-size pieces than in the experimental models, typically approximately 80% versus 60%. Bearing in mind the proposed sedimentary depositional process for this deposit (see Chapter 4), which is one of clay-rich sheetwash from the west into a water-body of fluctuating level, periodically desiccating, it seems likely that knapping activity is taking place on the spot or the near vicinity. It is likely that the microdebitage is being dispersed and more evenly distributed by low-energy fluvial activity, perhaps with occasional slopewash episodes where sediment is mobilised *en masse*. These would carry with them, and slightly mix/disperse, larger artefacts from activity on the western slope and at the edge of the water-body, at the same time as rearranging and dispersing the microdebitage.

TECHNOLOGY, TYPOLOGY AND THE *CHAÎNE OPÉRATOIRE*

Introduction and raw material

The lithic artefact assemblage south of Trench D (assemblage 6.1) comprised 1871 artefacts, including 115 chips (see Table 18.1, Fig. 18.13a). That from the remainder of the Phase 6 clay (assemblage 6.2), except from around the elephant skeleton, comprised 129 artefacts with 11 of them chips (see Table 18.2; Fig. 18.13b). The raw material used was always flint, but other than this commonality, was extremely varied in every possible aspect: exterior condition, internal condition, texture, size, shape and colour, although this last property mostly reflects post-depositional burial history, rather than being a factor at the time of collection and knapping.

The great majority of the flint is nodular chalk flint with a very thin cortex, as is typical of much flint from the Upper Chalk of the Swanscombe area, which contains numerous seams with nodules of varying sizes, from small to huge. Some seams are almost tabular, or form as a network of connected nodules, which then break up on derivation into nodules with a solid central node and cylindrical projections. All these shape variants are present in assemblages 6.1 and 6.2. The cortical condition is never completely fresh and white, as if collected freshly derived from Chalk bedrock, but it is usually off-white or pale blueish/greyish white and slightly-moderately abraded, suggesting some degree of reworking before collection for knapping. Much of the material also has a smooth, weathered and well-abraded cortex, often darker blue-grey in colour, reflecting a greater degree of pre-collection exposure and reworking.

There is also a reasonably high incidence of frost-fracturing in the raw material, ranging from minor pot-lids and frost-fractures that would have had minimal effect on knapping potential, through to nodules/pieces that were so riven by frost-fractures that they were completely un-knappable. One of these (Δ.42777) was recovered, and fell into 59 pieces after excavation (Fig. 18.14c). After these were pieced back together, it was clear that at least one flake removal had been attempted, after which it was presumably abandoned as a lost cause; this piece was amongst those classified as a 'tested nodule'. The majority of the evidence of frost-fracture clearly pre-dates use for knapping, as many percussion fracture-planes are evidently influenced by pre-existing frost-fracture flaws. However, there is also some evidence of *in situ* temperature changes having had a post-knapping impact, for instance flake Δ.43751 has two small 'pot-lids' developed on its ventral surface, one of them developed from the point of percussion and found *in situ* in its hollow, and the other not recovered (Fig. 18.14b).

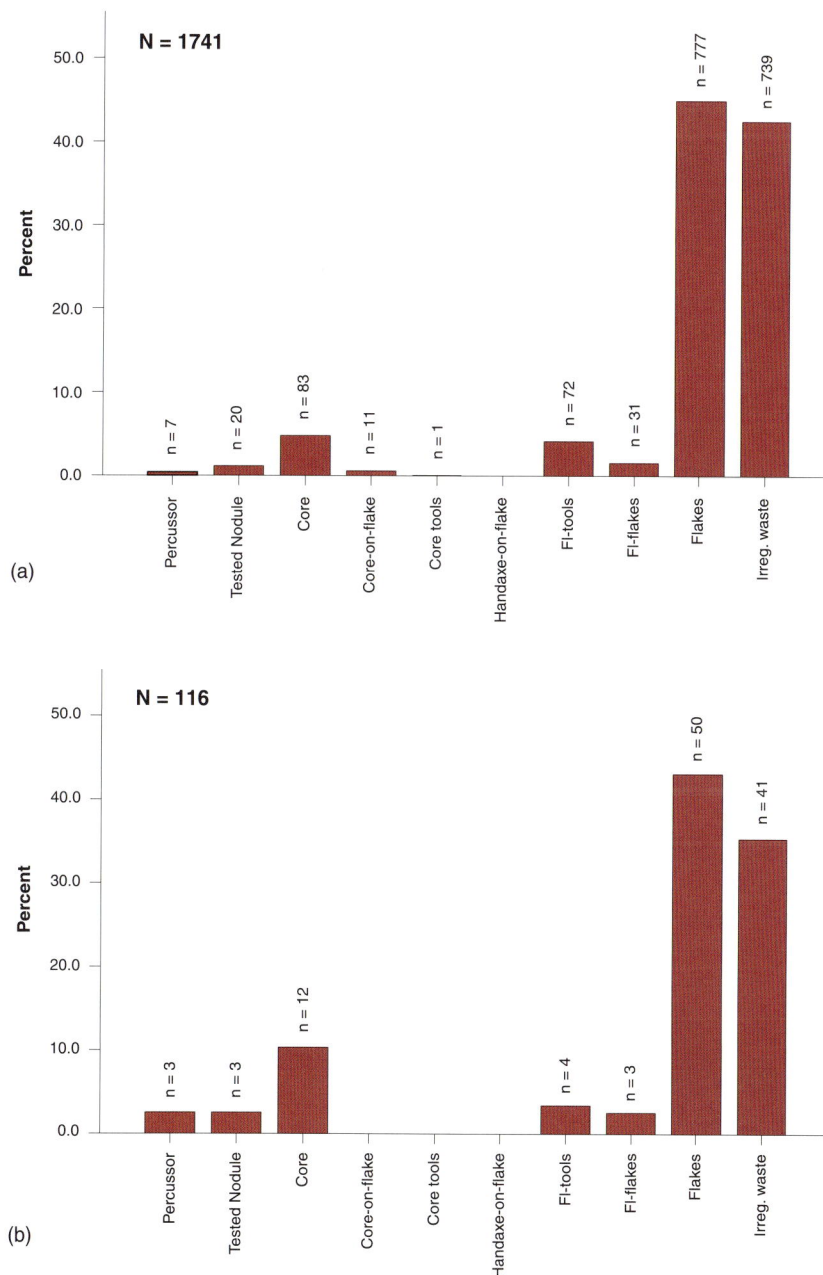

Figure 18.13 Histograms of technological categorisation: (a) assemblage 6.1; (b) assemblage 6.2

NOT TO SCALE

Figure 18.14 Photos of selected artefacts: (a) *in situ* break, Δ.43407=43418, RG #11; (b) *in situ* break, Δ.44017=43497, RG #35; (c) pot-lids on ventral surface, flake Δ.43751, RG #34; (d) shattered tested nodule, Δ.42777; (e) percussor, Δ.42644; (f) core-tool, Δ.42377

In addition to the nodular chalk flint, there is a reasonable quantity (approximately 2-5%) of Bullhead Bed flint, with its distinctive dark green cortex and sub-cortical orange-stained band, the cortex typically being moderately abraded. There is also some use of well-rounded Tertiary pebbles/cobbles as raw material for knapping. These are clearly recognisable from their well-rounded, chatter-marked and heavily abraded exteriors, normally stained dark grey, brown or pale ochre. Their interior flint is always coarse with very poor flaking properties and prone to break up on knapping, so it is hard to imagine that this was a very desirable raw material. Nonetheless, it was knapped, although the only artefacts found were a few pieces of debitage and irregular waste, so there is no evidence that it was ever successfully incorporated into a *chaîne opératoire* for tool manufacture and use.

It seems likely that the raw material was all very locally available. The Phase 6 clay sediments included a fair number of sizeable natural flint clasts, 145 of which were collected as finds in assemblages 6.1 and 6.2, and these are included in the site archive. Many more natural pieces were not recovered, but it is worth noting anecdotally that they were most common in the area south of Trench D, and included reasonably numerous very well-rounded derived Tertiary cobbles up to about 100–120mm long. Whether the concentration of natural clasts precisely matched the spatial clustering of the artefact distribution is, however, uncertain.

Test pit investigations in the area to the west of the site (Chapter 4) have confirmed the presence there, albeit a puzzling and difficult-to-explain presence, of chalk-rich sediments rich in flint nodules on what would have been slightly higher ground to the west of the site. These are presumed to equate broadly with Phase 3 of the site sequence, and thus to have formed flint-rich valley-side deposits that would have extended down to the water-body at the floor of the valley. They would have provided a ready source of flint raw material for hominins in the vicinity during Phases 4, 5 and 6. The evidence of abrasion and frost-fracturing in the raw material confirms its reworking and exposure to the elements, and considering the dating of the site to the earlier Hoxnian interglacial, it seems likely that the flint raw material had been exposed to the cold of the preceding Anglian glaciation. The presence of derived Tertiary material suggests input not only from chalk-rich sediments on the lower slopes, but also from Tertiary deposits that would have been higher up the slope to the west of the site, capping the now-quarried Swanscombe Hill (Fig. 2.3).

Technological summary and knapping methods

The breakdown of assemblages 6.1 and 6.2 into the main technological categories used in this analysis is given below (Table 18.9; Table 18.10), and the results for both assemblages are also summarised as bar-charts (Fig. 18.13). Assemblage 6.1 included 17 abraded artefacts, mostly in the category 'slightly-moderately abraded', but two of them were 'well-abraded'. The latter group included one medium-size technologically undiagnostic waste-flake, together with a hard-to-interpret bifacially flaked artefact (Δ.42862) that seems to be a broken part of a biface that has suffered a deep plunging flake transversely across it, which has then been further flaked. This artefact and the equally well-abraded flake are clearly intrusive into the assemblage and must represent derived evidence of earlier hominin occupation in the region, probably pre-Anglian. The former group are all technologically undiagnostic waste debitage comprising a mixture of small-medium flakes and irregular waste. They may also represent pre-Anglian evidence that has been subject to less severe reworking, or they may represent earlier Hoxnian activity on the valley side above the site that has then been transported down into the site area by slopewash, to mingle with relatively undisturbed evidence of activity at the foot of the slope.

Assemblage 6.2 has just two abraded artefacts, both in slightly-moderately abraded condition. One of them is a small piece of irregular waste. The other is a large but technologically undiagnostic waste flake (Δ.40352) with no cortex, several dorsal scars and a long sharp edge with visible scaling all along it. This was interpreted as natural damage, rather than use-wear, although this was possibly wrong, and the artefact could alternatively be interpreted

Table 18.9 Assemblage 6.1: technological categories, excluding natural pieces

	5 – Percussor	10 – Tested nodule	20 – Core	30 – Core-on-flake	40 – Core-tools	50 – Handaxe-on-flake	60s – Fl-tools	80 – Fl-flakes	90 – Flakes	100 – Irreg. waste	110 – Chips	Sub-total (n)
Fresh material	7	20	83	11	1	–	72	31	777	739	113	1854
% - inc chips	0.4	1.1	4.5	0.6	0.1	–	3.9	1.7	41.9	39.9	6.1	
% - excl chips	0.4	1.1	4.8	0.6	0.1	–	4.1	1.8	44.6	42.4		1741
Abraded material	–	–	–	–	1	–	–	–	8	6	2	17
% - inc chips	–	–	–	–	5.9	–	–	–	47.1	35.3	11.8	
% - excl chips	–	–	–	–	6.7	–	–	–	53.3	40.0		15

Table 18.10 Assemblage 6.2: technological categories, excluding natural pieces

	5 – Percussor	10 – Tested nodule	20 – Core	30 – Core-on-flake	40 – Core-tools	50 – Handaxe-on-flake	60s – Fl-tools	80 – Fl-flakes	90 – Flakes	100 – Irreg. waste	110 – Chips	Sub-total (n)
Fresh material (n)	3	3	12	-	-	-	4	3	5	41	11	127
% - inc chips	2.3	2.3	9.0	-	-	-	3.0	2.3	37.6	30.8	8.3	
% - excl chips	2.5	2.5	9.8	-	-	-	3.3	2.5	41.0	33.6		116
Abraded material (r)	-	-	-	-	-	-	-	-	1	1	-	2
% - inc chips	-	-	-	-	-	-	-	-	50.0	50.0	-	
% - excl chips	-	-	-	-	-	-	-	-	50.0	50.0	-	2

Table 18.11 Assemblages 6.1-6.2: percussors

Assemblage	Find ID	Whl	WtG	Notes
6.1	Δ.40719	1	211	Abraded flint pebble with patch of fresher battering at one end
	Δ.41235	1	395	Heavy battering at one end
	Δ.41622	1	292	Small nodular lump with one protrusion showing numerous impact marks
	Δ.42555	1	646	Localised impact marks at one end, but possibly natural rather than percussion
	Δ.42644	1	344	Definitely a percussor; clear, localised impact marks (Fig. 18.14e)
	Δ.42935	1	259	Localised impact marks suggesting use as percussor, but also damaged by mattock when found
	Δ.43060	1	221	Small core remnant with numerous flake removals and patch of localised battering on rounded cortical protrusion, suggesting additional use as a percussor (Fig. 18.20b)
	Δ.43442	1	559	Lump of irregular waste with localised batter-marks that suggest use as a percussor rather than attempted flake removal
6.2	Δ.40394	1	152	Fractured Tertiary pebble, poss. impacts from percussion?
	Δ.42544	0	227	Possibly broken during use, refits with irregular waste Δ.42572
	Δ.42571	0	82	Broken piece; heavily battered, possibly two phases of use

as a heavily utilised flake-tool. As with assemblage 6.1, these abraded elements probably represent intrusive evidence of pre-Anglian or earlier Hoxnian hominin activity on the high ground to the west of the site, introduced to the site sequence by slopewash from the west.

For the fresh assemblages, it is immediately clear (Fig. 18.13) that, despite assemblage 6.1 being 15x larger than assemblage 6.2, the relative proportions of different categories of artefact are virtually identical. The only difference is the slightly increased number of pieces identified as 'percussor' in assemblage 6.2, which is easily explained as a disproportionate statistical influence of a chance variation in a small assemblage. The majority of both assemblages 6.1 and 6.2 was formed of flakes (approximately 40–45%) and irregular waste (approximately 35–40%), with the relative quantity of flakes slightly higher in assemblage 6.1. This probably reflects the increased recovery of smaller, thinner flakes in the area south of Trench D due to the increased use of hand excavation. Cores constitute a higher proportion of assemblage 6.2 than 6.1 (about 10% versus 5%), and this likewise probably reflects the increased recovery of

smaller debitage by hand excavation in the area south of Trench D. Tested nodules are a very small element of both assemblages (about 1–2%). There is just one example of an artefact classifiable as a core-tool rather than a core, in assemblage 6.1; this is discussed in more detail below, but suffice it to say that it is not bifacially worked and could not be classified as even the most proto of proto-handaxes.

The rest of both assemblages consists of flake-tools (about 3–4%), that is flakes and irregular waste that have been subject to secondary working (when not interpreted as cores) and flake-flakes caused by this secondary working (about 1-2%). The latter are probably under-represented since they are not always clearly identifiable, and were only classified as such when there was clear evidence that they had been struck on a pre-existing flake removal. All these elements are discussed in more detail below.

The whole assemblage seems to have been created by hard-hammer percussion. Most flakes have typical hard-hammer diagnostic features such as circular ring-cracks indicating the point of percussion, the absence of a lip between the point of percussion and the ventral surface,

Table 18.12 Assemblages 6.1 and 6.2: core statistics (all cores)

Assemblage	Statistic	ML	DSC	WtG	Notes
6.1 (n=83)	Max	146	24	901	Includes 12 cores classified as 'broken'; excluding these makes negligible difference to the figures, and it is often problematic to establish that a core was broken after its last removal; nonetheless, comparative data for whole cores are provided in Fig. 18.15
	Q4/Q3	83.5	8	224.5	
	Mean	73.85	7.01	177.73	
	Q2/Q1	62	5	79	
	Min	42	1	31	
	Sd (pop)	19.65	3.84	154.49	
6.2 (n=12)	Max	275	20	6350	Includes one very large core, Δ.42916 (Fig. 18.16)
	Q4/Q3	95	9	397.75	
	Mean	111.55	7.5	1143.67	
	Q2/Q1	77	4.75	245	
	Min	62	1	62	
	Sd (pop)	64.20	4.77	1991.66	

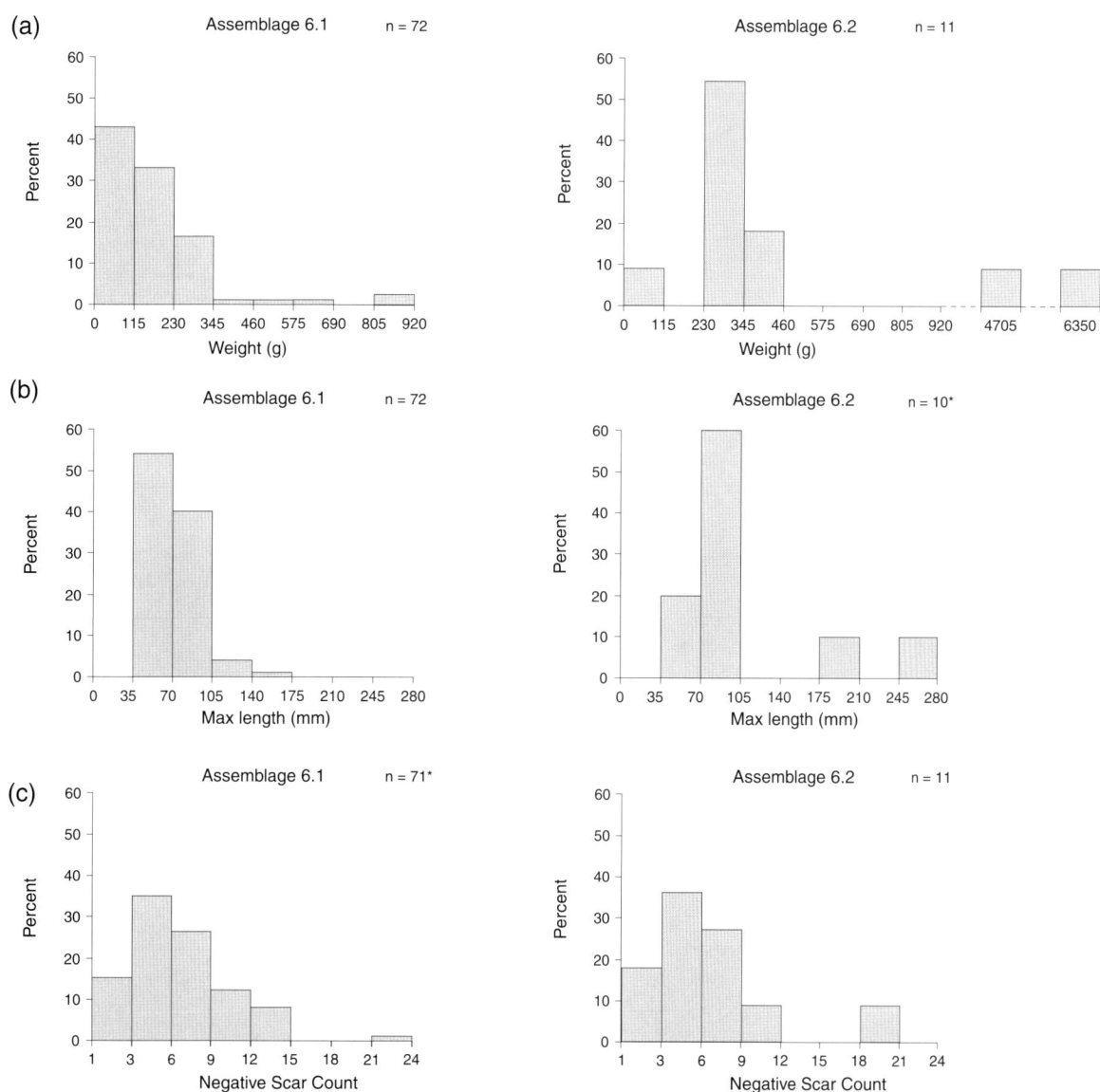

Figure 18.15 Histograms of core size and reduction attributes (unbroken cores only) from assemblages 6.1 and 6.2: (a) weight; (b) maximum length; and (c) negative scar count [* missing one measurement of negative scar count from Ass. 6.1 and one measurement of max. length from Ass. 6.2]

and a higher incidence of visible conchoidal rippling on the ventral bulb of percussion. The last probably reflecting the harsher vibrations introduced by a hard-hammer blow than a soft-hammer one.

In addition to this indirect evidence of percussor type, a number of flint pieces were found with localised patches of battering suggesting their use as hard-hammer percussors, including one (Δ.43060) that also served as a core (Table 18.11; see also Fig. 18.20b, below). On some of these, it was hard to distinguish natural battering from hominin percussion, but on others there was very clear and localised fresh battering clearly indicating use as a percussor (Δ.42644, Fig. 18.14d). The average weight of the more complete percussors was a little under 350g, which corresponds well with the preference of modern knappers in some recent experiments (pers. observation).

A core-tool, cores and flaking strategies

Out of all the material comprising (the fresh elements of) assemblages 6.1 and 6.2, there was just one artefact classifiable as a core-tool. There was not one single example of a handaxe on a flake and there was not one single piece of debitage suggestive of bifacial thinning and/or shaping. The 'core-tool' (Δ.42377) was a cylindrical flint nodule about 120–150mm long, weighing 520g (Fig. 18.14e). One end was diagonally truncated by a natural frost-fracture, and this frost-fracture had been used as the platform for removal of a flake to form a transverse sharp

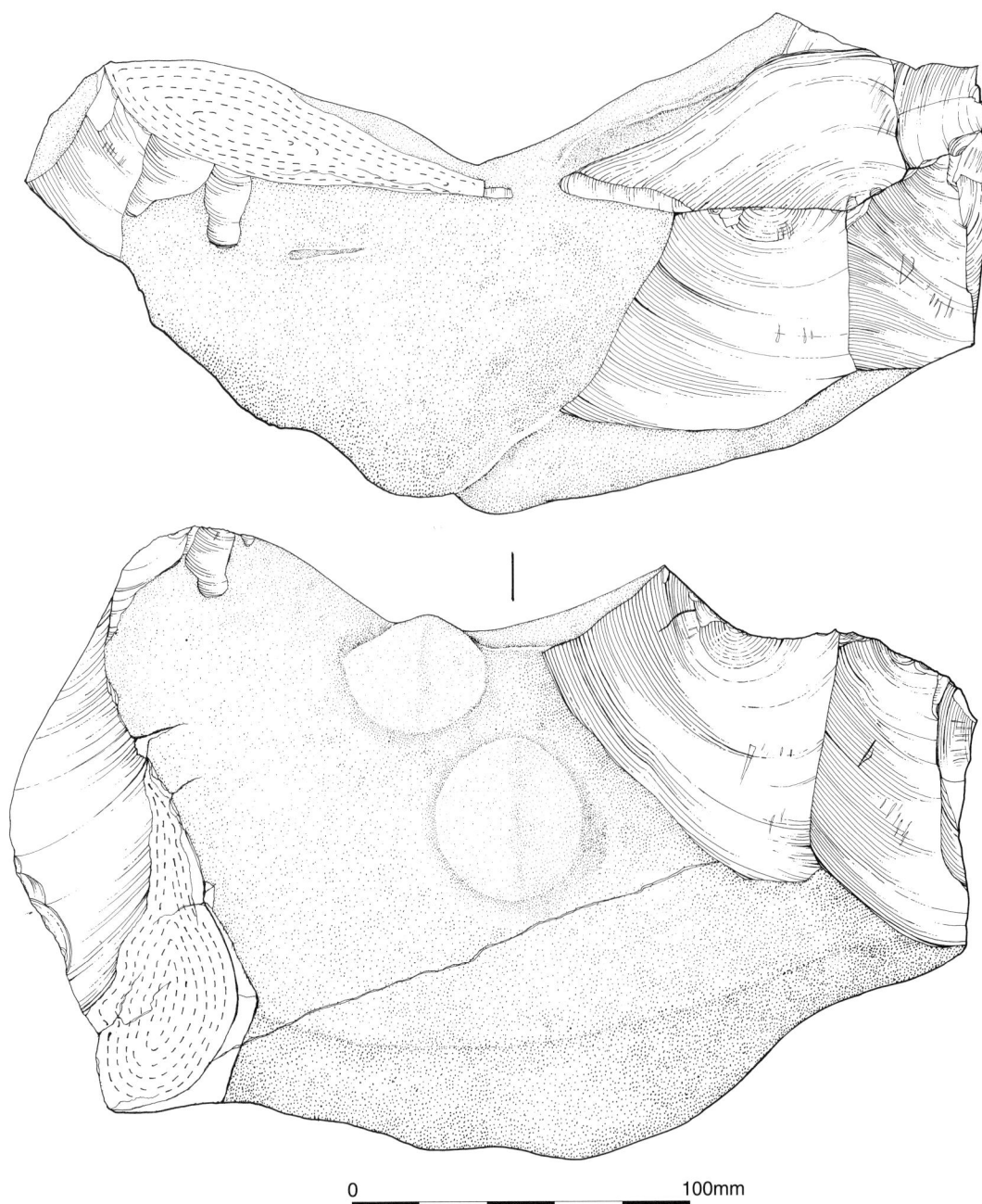

0   100mm

Figure 18.16 Assemblage 6.2: core Δ.42916 [ill. B. McNee]

edge across the end of the nodule. This edge had some small, invasive chips on it, which look like macro use-wear rather than natural abrasion. There is no hint of bifacial working. So one thing that is clear about the site is that there is no evidence of bifacial tool manufacturing in any of the Phase 6 assemblages 6.1, 6.2 or 6.3, aside from the single, very abraded intrusion discussed above.

As well as nodules from which only single small removals were made, classified here as 'tested nodules', there were 83 cores in assemblage 6.1, and 12 in assemblage 6.2. The basic size statistics of the cores from these assemblages are given (Table 18.12). It can be seen that they occur in a wide range of sizes from about 40mm to 275mm long, weighing from about 30g to 6350g and with widely varying intensities of reduction with from 1–24 countable negative flake scars. The few instances of cores with single flake scars were, incidentally, distinguished from 'tested nodules' due to their small size, the higher quality flint (tested nodules were often interpreted as abandoned due to frost-fracturing or being an awkward shape to knap) and the relatively large size of the flake removed compared to the core. Apart from one huge core (Δ.42916) which somewhat skewed the statistics, the sizes and degree of reduction of cores were broadly similar in both assemblages 6.1 and 6.2. Histograms were prepared showing the distributions of these three size variables (Fig. 18.15a–c). The collection of cores from assemblage 6.1 is probably more representative as it is substantially larger, and was recovered by hand-excavation, and so is less biased towards larger specimens. It shows that most cores were abandoned at quite a small size (between approximately 30 and 100mm maximum length and weighing less than 200g). It also shows that most cores had quite low dorsal scar counts, predominantly in the range 4–6, although this is

an unreliable indication of the actual number of flakes produced, as many of the earlier removals would not leave scars that remained visible at the end of the reduction sequence.

All the cores seemed to result from simply structured approaches to producing series of flakes of varying sizes. A representative selection of 15 cores was chosen for illustration (Figs 18.16–18.21), including three with refitting flakes: RGs #11, #30 and #59. Some cores were abandoned after removal of several, or sometimes very few, large flakes (for example core Δ.42916 – Fig. 18.16) and others were quite intensively worked and showed the removal of many small-medium flakes (Fig. 18.20). In terms of the knapping strategies adopted, it is a moot point whether the *post hoc* modern imposition of concepts such as 'approaches' and 'strategies' is valid, or whether it is more appropriate to describe the reduction sequence carried out as neutrally as possible and observe groupings and repetitions without any suggestion that these reflect deliberately applied strategies. The approach taken here is that it is not thought that there was any intention to create a core in a final shape, so the repeated occurrence of similarly shaped cores must be the unintentional outcome of repeated approaches to reduction. Also, that it is possible to describe the sequence of reduction without casting it as a pre-conceived strategy, and then have a subsequent discussion of whether it is possible to cross into the territory of considering the repetition of any particular reduction pathways as deliberately conceived strategic approaches.

Four main pathways of reduction were observed in the core collection, as well as numerous short episodes of single platform or alternate flaking, and removal of flakes from randomly migrating platforms in what seemed an

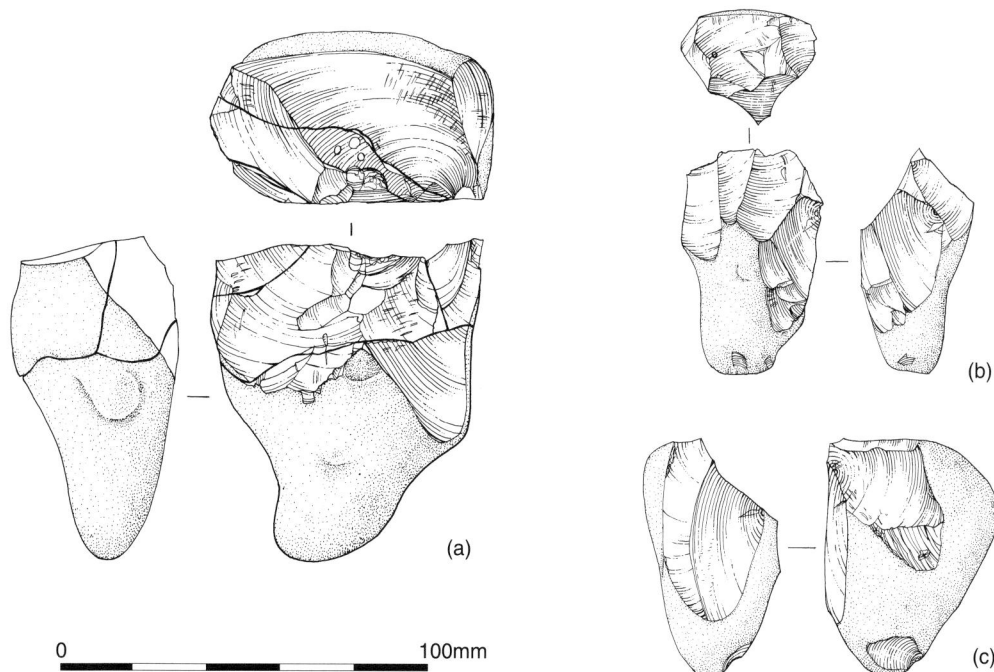

Figure 18.17 Assemblage 6.1 cores, alternately flaked around part of nodule: (a) RG #59, Δ.40203=Δ.41554=Δ.42465; (b) Δ.42775; (c) Δ.43713 [ill. B. McNee]

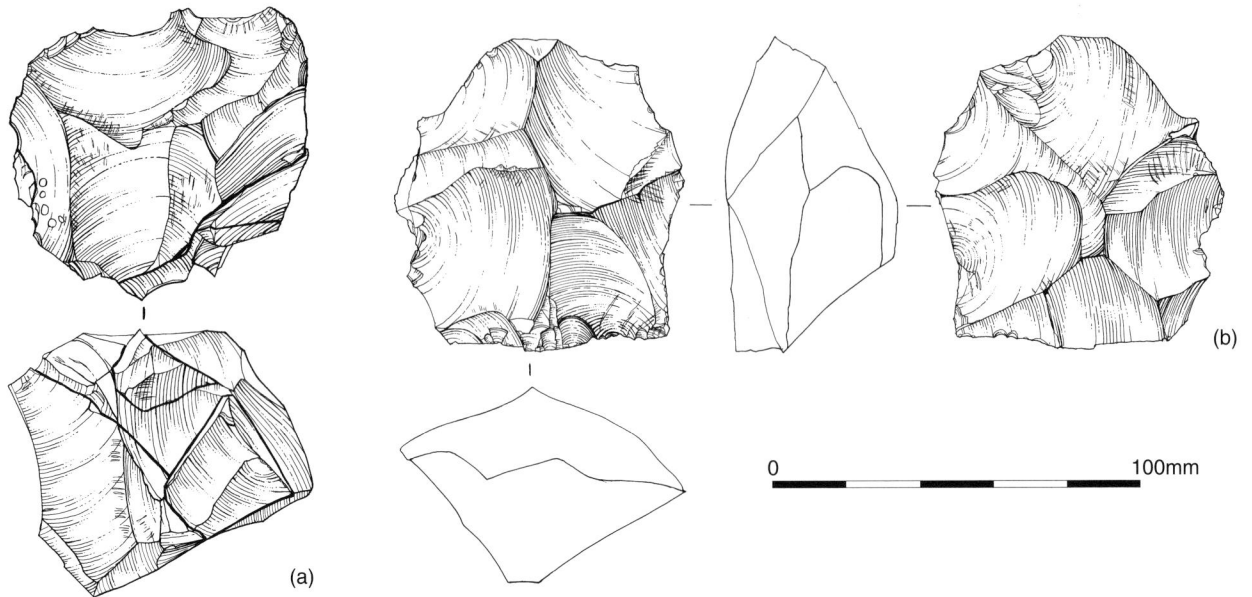

Figure 18.18 Assemblage 6.1, bi/uni-pyramidal cores: (a) RG #11, Δ.43418=Δ.43407, Δ.41993=Δ.41994, Δ.41855; (b) Δ.43718 [ill. B. McNee]

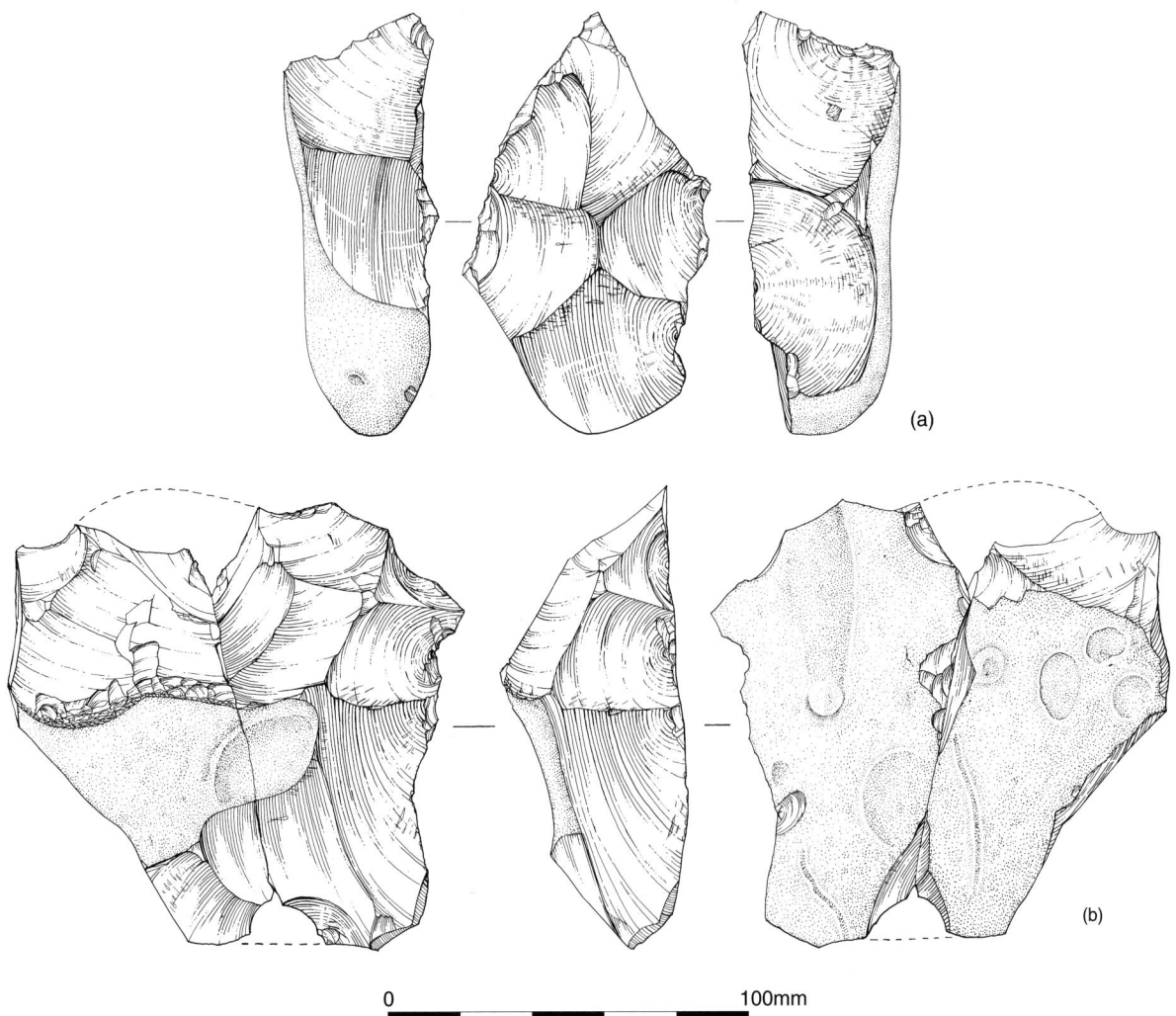

Figure 18.19 Assemblage 6.1, unifacial/single platform cores: (a) Δ.43309; (b) Δ.42437 [ill. B. McNee]

entirely *ad hoc* manner. Firstly, several cores were alternately flaked at one end, with this pattern of flaking sometimes ceasing after only a few removals, and sometimes continuing further around a nodule (Fig. 18.16 – Δ.42916; Fig. 18.17 – Δ.42775, Δ.43713 and RG #59, although this latter sequence is interrupted by breakage of the core).

Another regular occurring pattern of reduction (also represented in the main core Δ.40871 of refitting Group C from around the elephant skeleton) was for some flakes to be removed from a flatter 'top' surface, but for this 'top' surface to be mostly used as a striking platform for flakes that were struck around most of its perimeter, resulting in a uni-pyramidal end-form of the core (Fig. 18.18b – Δ.43718). There were also instances of repeated flaking from just one platform all around a core (Fig. 18.19b – Δ.42437), or conversely for most of the flakes to come from one flat surface of a core, and only a few from around the perimeter (Fig. 18.19a – Δ.43309). A variation on this was for the distinction between 'top' surface and main flaking direction to be less clear-cut, and for pseudo-bifacial alternate flaking to proceed around the perimeter of a core, leaving a pseudo-bifacial or bi-pyramidal form (Fig. 18.18a – RG#11; Fig. 18.20d,

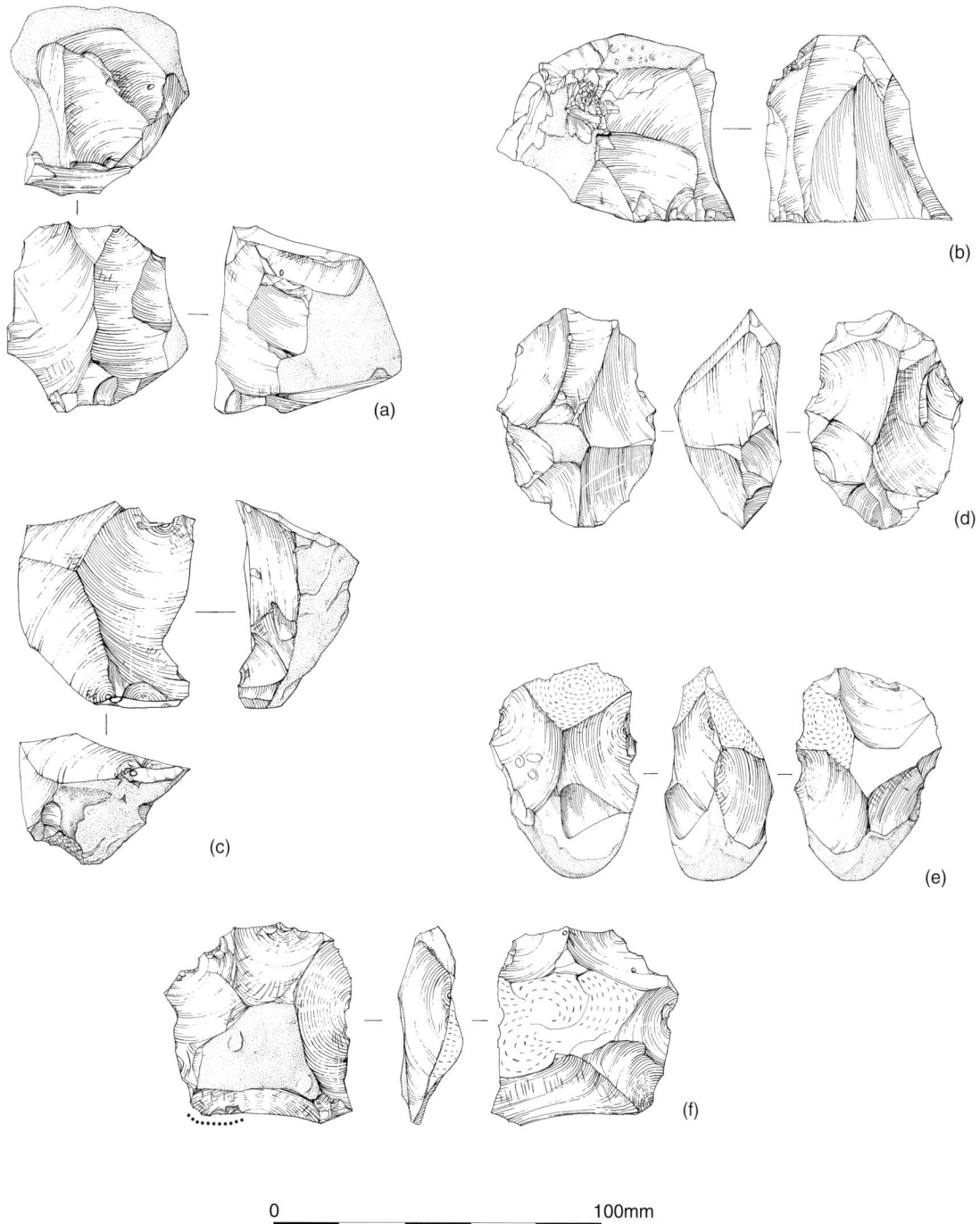

Figure 18.20 Assemblage 6.1, small globular cores (a)–(c) and pseudo-bifacial forms (d)–(f): (a) Δ.42801; (b) Δ.43060, with marks from use as percussor; (c) Δ.43235, on flake; (d) Δ.40742; (e) Δ.40835; (f) Δ.42628, on irregular waste, with faint macro use-wear shown as dots [ill. B. McNee]

e, f – Δ.40742, Δ.40835 and Δ.42628), with very sinuous edges and no indication of thinning/straightening the edge as a bifacial cutting edge, although one of these illustrated cores was quite thin (Δ.42628), and had signs of possible macro use-wear on one of its sharp edges, leading to ambiguity over whether it should be regarded as a core or a flake-tool.

There were quite a few small lumps with no apparent structure to the reduction pathway, which had been flaked via a combination of new and alternating platforms, two of which are shown here, the latter of which also has batter-marks suggesting it was also used as a percussor (Fig 18.20a, b – Δ.42801; Δ.43060).

Several of the cores (n=11, in assemblage 6.1) were made on large flakes or pieces of irregular waste (see for example Fig. 18.20c, f – Δ.43235, Δ.42628). As discussed below, one of the problems in interpretation was to try and distinguish between chunky flake-tools and small cores on flakes or waste debitage. This was an impossible task, although it was attempted since some of the single and double-notched flakes seemed so clearly to be deliberately made tools that it seemed negligent to exclude them from consideration as such. This, however, had the knock-on effect of having to force inappropriate categorical boundaries on other material. Aside from the possibility that some of these smaller cores are purposeless juvenile knapping, their interpretation as cores presumes that some quite small flakes were useful for certain light duty tasks.

The following section considers the reduction pathways applied further, based on the small number of slightly longer refitted flaking sequences found.

Refitted reduction sequences

In contrast to the material from around the elephant, there were no long refitted reduction sequences that demonstrated significantly more detailed pathways of flake reduction than could be observed in the cores themselves. The most informative sequences were RG #11 (Fig. 18.18a; Table 18.13), RG #30 (Fig. 18.21a, b; Table 18.14) and RG #52 (Fig. 18.21c; Table 18.15; Table 18.16), which provided slightly more information on sequences of core position, and exemplify some of the variety of approaches.

RG #11: (Δ.43418 = Δ.43407; Δ.41993 = Δ.41994; Δ.42511; Δ.41855)

In the first of these, RG #11, also considered above purely as a core, there were six constituent pieces (Fig. 18.22). However, these only represent three removals and the core, since the core is formed of two pieces (Δ.43407 = Δ.43418) that were almost certainly broken *in situ* when frost-action and sediment heaving exacerbated the pre-existing fracture-plane of a failed flake removal. Likewise, one of the three removals is formed of two joining pieces thought to have broken *in situ* (Δ.41993 = Δ.41994). The other two flakes completing

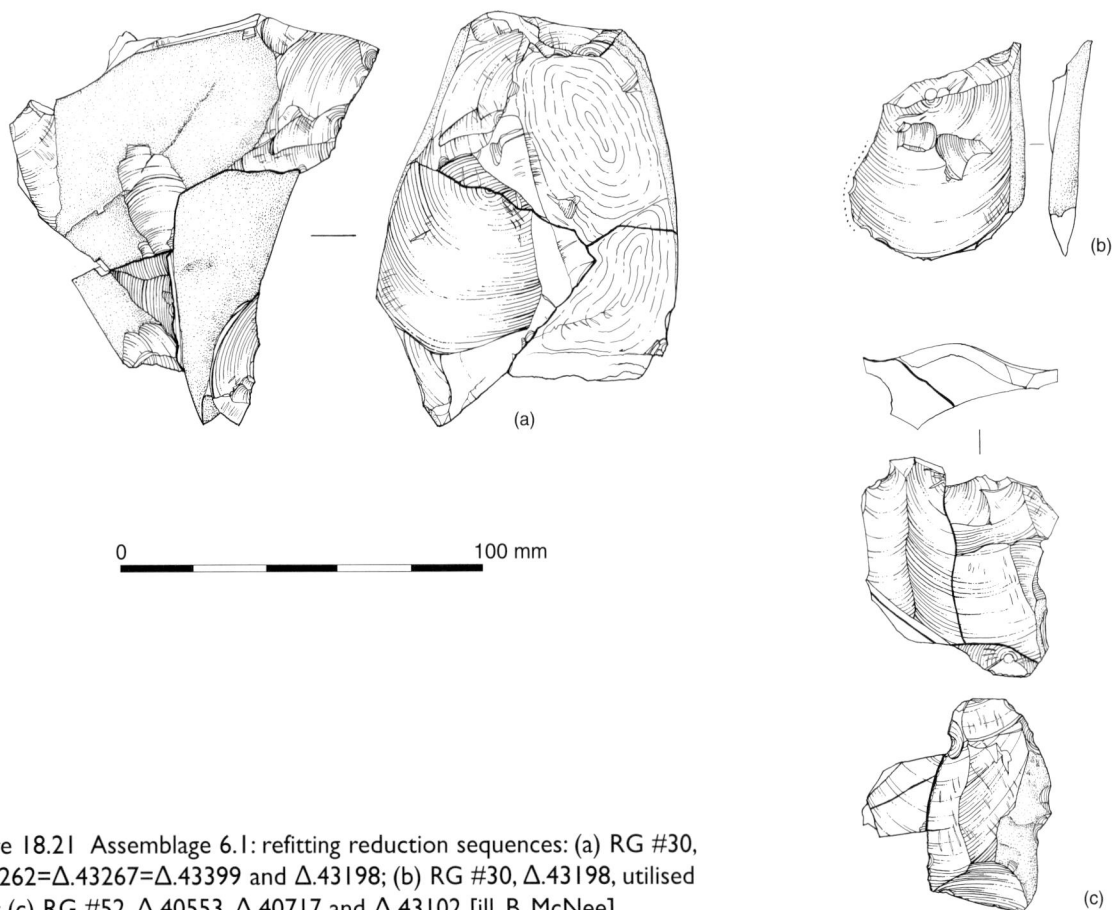

0   100 mm

Figure 18.21 Assemblage 6.1: refitting reduction sequences: (a) RG #30, Δ.43262=Δ.43267=Δ.43399 and Δ.43198; (b) RG #30, Δ.43198, utilised flake; (c) RG #52, Δ.40553, Δ.40717 and Δ.43102 [ill. B. McNee]

the refitting group are Δ.42511 (not illustrated) and Δ.41855 (illustrated attached to the core).

When considered as a core, this was a good example of a core with a bi-pyramidal end-form resulting from pseudo-bifacial alternate flaking around its perimeter. The overall strategy of flaking represented in the core is very clear, at least as represented in the surviving evidence of its negative scars and refitting removals.

There are also clear signs of the failures that led to its abandonment. The earliest flake in the sequence is Δ.42511 (which unfortunately lacks XYZ spatial provenance). This flake has approximately 10 dorsal scars indicating previous flaking. At least one of these appears to have been from a very sizeable flake that ended in a step fracture, part of the ridge from which is preserved on the dorsal surface of Δ.42511 and the other part on the

Table 18.13 Assemblage 6.1, RG#11 'bi-pyramidal core': reduction *chaîne opératoire*

Event order #	Core/platform positioning	Flake removal	Comments
-	General alternate platform flaking around perimeter of core	△n+⇩	At least ten flakes missing from the early sequence, probably many more; most of them with a tendency towards hinge terminations
1	New platform, scar of one of the earlier removals	▲42511⇩	Flake struck from perimeter of core, towards centre of one pyramidal face
2	New platform, alternate (▲42511 scar) – core turned upside down	△a ⇩	Squat flake with hinge termination
3	Same platform, (c 15mm clockwise)	△b ⇩	Another, even squatter flake with step termination
4	Same platform, (c 5mm anti-clockwise)	▲41993=▲41994=△c⇩	Another, even more squat flake with step termination; breaks with Siret fracture, one side of which is missing
5	Same platform, (c 5mm clockwise)	▲41855⇩	Small flake with hinge termination; fails to clear steps on core surface resulting from events #2–4
6	New platform, alternate (▲41855 scar) – core turned upside down again	△d ⇩	Small flake (c 50mm long) that travels across what was previously the top surface of the core, guided by the left ridge of the earlier removal ▲42511
7	New platform, opposite side of core's median perimeter	△e⇩	Another small flake that also successfully crosses the core surface
8	Same platform, (c 30mm clockwise)	△f⇩	Squat flake that ends with a hinge termination
9?	New platform – core turned upside down again	▲43407 (failed), ■ 43418	There are several percussion points on the platform, one of which has caused initiation of a fracture plane towards the centre of the core; [this event could alternatively have happened between events #5 and 6]; ▲43407 was not detached at the time, but it was later split off by frost action that extended the fracture plane, probably during burial

△ - missing debitage from refitted sequence; ▲ - debitage present in reduction sequence; ■ - core left at end of sequence

Table 18.14 Assemblage 6.1: refitting group RG#30

Find ID	ML	MW	MT	WtG	Notes
Δ.43262	-	-	-	101	Irregular waste - no flake removals ≥20mm
Δ.43267	72	56	27	77	Core - one flake removal (Δ.43198)
Δ.43399	97	68	50	381	Core - scars from several flake removals, but all of them small and/or failed with hinge/step terminations, and affected by internal flaws that persist within the piece
Δ.43198	44	43	11	31	Utilised flake - has use macro-wear on sharp edge

Table 18.15 Assemblage 6.1: refitting group RG#52, basic statistics

Find ID	%Cx	DSC	ML	MW	MT	WtG	Notes
Δ.40553	1	7	52	53	15	41	Flake
Δ.40717	0	5	46	25	10	13	Flake
Δ.43102	3	3	54	33	17	27	Utilised flake? Has faint micro-chipping and one larger mini-notch along its sharp edge

Table 18.16 Assemblage 6.1, RG#52 'migrating platform' *chaîne opératoire*

Event order #	Core/platform positioning	Flake removal	Comments
-	General migrating platform flaking, from two platforms opposed to each other	△n+⇓	At least three flakes missing from the early sequence, probably many more
1	New platform, migrated round *c* 180° from previous removal	▲43102⇓	Flake has cortex down right-hand side, and scars from three previous removals,
2	New platform, migrated round *c* 90° from previous removal	△a⇓	Small but quite solid flake *c* 45 x 25mm crosses full length of core surface
3	Same platform, (*c* 15mm clockwise)	▲40717⇓	Smaller flake that also crosses full length of core surface
4	Same platform, (*c* 10mm anticlockwise)	△b ⇓	Short, squat flake *c* 15 x 25mm that only travels a short distance, with slight hinge at end
5	Same platform, (directly behind)	▲40553⇓	Small-medium flake that thickens towards distal end, crossing full length of what would have been quite a small core

△ - missing debitage from refitted sequence; ▲ - debitage present in reduction sequence

core. There is no remnant of cortex on the core or on any of its refitting flakes, and it was clearly of much more substantial size when its knapping commenced. It retains scars of some 20 removals and could have produced considerably more, although the evidence of this is not preserved. The result is a roughly bi-pyramidal core with a sinuous median perimeter from which flakes were struck off one or other of the two pyramidal faces, depending which way up the core was held.

After removal of Δ.42511, the core was turned upside down and the scar from this removal was used as an alternate platform for two consecutive flakes off the opposite face, events #2-3, Δa and Δb (neither recovered). These two flakes were followed by two further removals struck from the same platform, events #4-5, Δ.41993 = Δ.41994 = Δc and Δ.41855. All four flakes were short and squat, with hinge/step terminations and failed to travel along the surface of the core. Although one has to be wary in ascribing success or failure to prehistoric knappers, this is unlikely to have been thought a satisfactory outcome. It is only in much later prehistory, such as the Neolithic or Bronze Age that flake-production strategies for arrowhead blanks seem to deliberately aim at producing squat flakes with hinge terminations.

After event #5, there is a slight ambiguity in the remainder of the reduction sequence. It is possible that the event shown as #9? in the tabular summary (Table 18.13) was interspersed between events #5 and #6, in which case it would have been a clockwise migration around the perimeter of the core, and a failed attempt to detach a substantial flake from the same face of the core as was being flaked by events #2-5 and would have removed the surface blockage caused by the hinge/step fracturing. As it is, it is guessed as slightly more likely, considering the predominance of alternate platform flaking, that the core was turned over after event #5 and the scar of flake Δ.41855 was used as an alternate platform for the removal of flake Δd (event #6). After this, the following two flakes (Δe and Δf, events #7-8) were struck from the same face, but from the opposite side of the median perimeter of the core. None of these

three flakes were recovered; Δd and Δe were successful in crossing the core surface, but Δf finished in a steep hinge termination, leaving a protruding step on this core-surface too.

It is suggested here that the final event (#9?) of the core's life was then turning it over again, and making another attempt to remove a flake from the opposite face. There are several percussion marks on the surface of the core (Fig. 18.18a), one of which has initiated a fracture plane towards its centre, but which has failed to remove a flake. This fracture plane has subsequently become extended due to *in situ* frost-fracturing (Fig. 18.14a), leading to recovery of part of the core with the proximal end of the failed flake as a separate find (Δ.43407) from the main part of the core (Δ.43418). It is, however, also possible that this (proposed final) event was interspersed between events #5 and #6, and that the failure of this side of the core instigated its inversion. The sequence of events #6-8 then represented the end-game of its reduction and the failure of the other face with the hinge termination resulting from flake Δf was the final event of the sequence.

Despite the surface failures that have apparently led to its abandonment, it would have been possible to continue flaking the core, removing flakes of size approximately 40–50 x 20–30mm. These would have had quite uneven edges, however, and would have had messy dorsal surfaces with various ridges and bumps caused by old step-fractures and steep ridges from scar intersections. This perhaps presents a useful indication of the size and type of flake that were, on at least one occasion, regarded as not worth bothering with.

RG #30: (Δ.43262 = Δ.43267 = Δ.43399; Δ.43198)

The four flints comprising this refitting group were all found reasonably close together (within an area of about 2.5 x 1m) in the southern part of the concentration south of Trench D (Fig. 18.22). The group exemplifies a completely different approach to flake production than nicer cores and longer sequences such as RG #11. In this group, a nodule that has been heavily affected by internal frost-fracturing has been broken into irregular

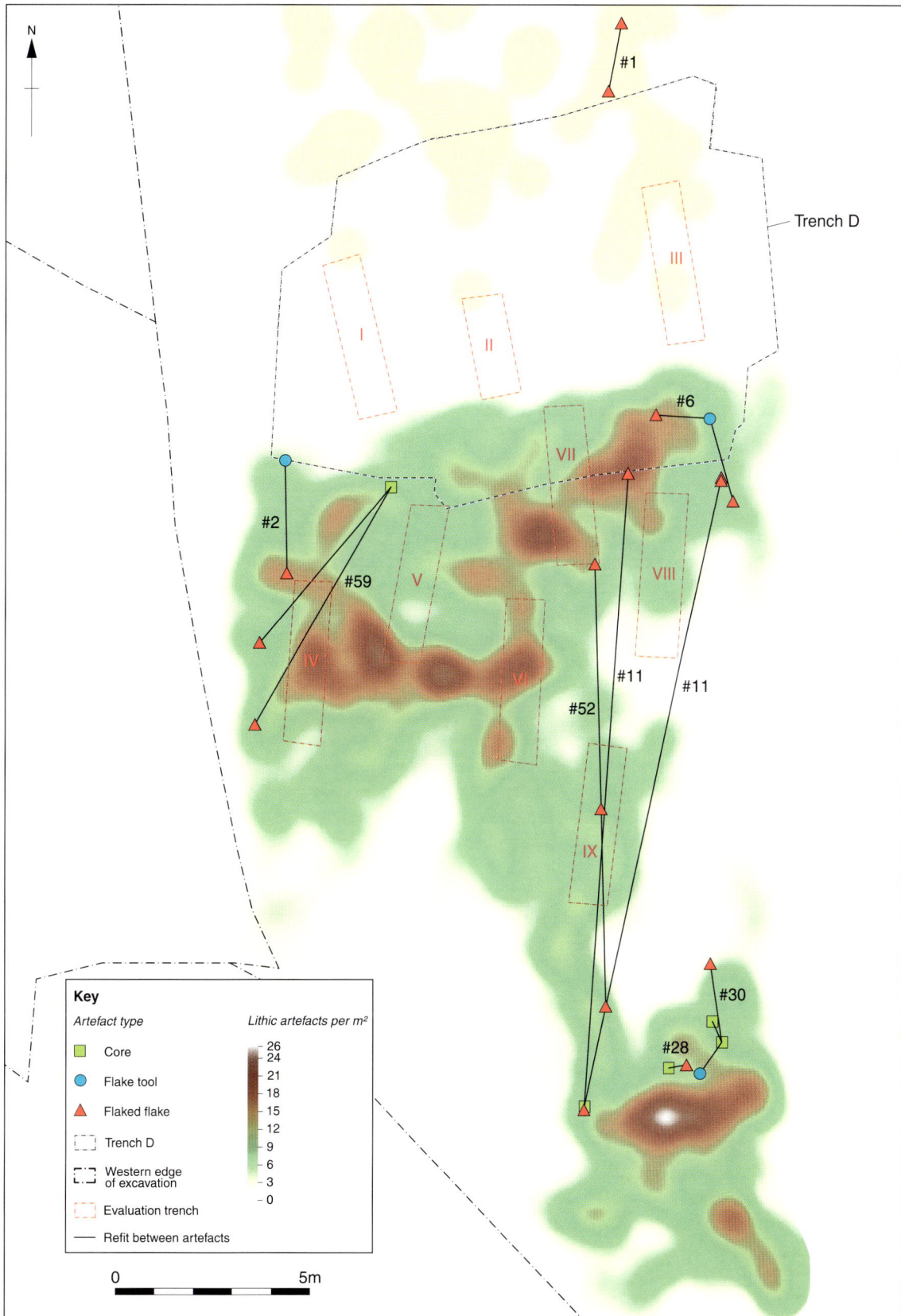

Figure 18.22 Spatial distribution of selected refitting groups and key artefacts: RG #1, RG #2, RG #6, RG #11, RG #28, RG #30, RG #52, RG #59

waste pieces, and then at least two of these waste pieces have served as cores for the attempted removal of flakes.

The group comprises three pieces of irregular waste (Table 18.14) that join together, and appear to have broken apart simultaneously along internal frost-fracture planes of the parent nodule (Fig. 18.21a). The pieces combine to form one end of a much larger flint nodule, the rest of which was not recovered. A number of flake removals were attempted from the largest piece (Δ.43399, weighing about 380g) but all were unsuccessful due to the persistence within the piece of internal flaws caused by frost-fracturing. One of the two smaller pieces (Δ.43267, weighing about 80g) was, however, used as a core for the removal of a single flake (Δ.43198) which was recovered only approximately 1.5m away. This flake (Fig. 18.21b) had a blunt cortical side opposed by a straight convex sharp edge, and this edge had slight chipping and nibbling suggesting use macro-wear. As a whole, this refitting group reflects a much more expedient technological approach, with the production of a flake for immediate use, followed by its prompt abandonment.

RG #52: (Δ.40553; Δ.40717; Δ.43102)

This refitting group comprises three separate small-medium flake removals from what would probably have been a small globular core (Fig. 18.21c; Table 18.15; Table 18.16). They were found quite widely separated (Fig. 18.22), with Δ.40717 in the north part of the concentration south of Trench D, Δ.43102 in the southern part and Δ.40553 approximately halfway between the other two.

The first removal of the surviving sequence was Δ.43102, which bears the scars of at least three previous removals, the last of which crosses its distal end struck from an opposing direction. After this, the core is rotated through about 90° and a flake (event #2, flake ra) struck off, which was not recovered. This was followed by removal from the same platform, with clockwise striking point migration of approximately 15mm, of another flake that was recovered (event #3, Δ.40717). Then, from the same platform, another small flake was struck and this too was not recovered (event #4, rb). Finally, still from the same platform, a further flake was struck (event #5, Δ.40553). The core was not found in the excavated collection, although a careful search was made.

The distribution of the artefacts along their north-south connecting line does not match their removal order. The first removal (Δ.43102) is the most southerly of the pieces, then the second removal (Δ.40717) is the most northerly, and the final removal (Δ.40553) was found roughly halfway between. Considering the evidence of artefact dispersal thought to be by non-hominin sedimentary processes, these separations can not be reliably interpreted as reflecting intra-site movement, although this remains a possibility. This sequence, in contrast to RG #11 (discussed above) demonstrates that core reduction was often continued when the maximum potential length of any flakes produced would be around 50mm, so although flakes this small may not have been desirable, they were often produced. Whether or not this reflects a mindless engagement with lithic material – if it doesn't move, knap it – or whether it reflects a more considered and thoughtful engagement whereby it was actively decided to produce flakes of a certain size, remains a conundrum for wider consideration in the interpretation of lithic material culture of this era. It is also of course possible that the missing so-called 'core' would be better regarded as a tool, and that these flakes should be viewed as secondary waste from tool production, rather than potential tools or blanks; this issue is considered further below.

Secondary flake modifications and flake-tools

A significant technological element of the assemblage was the quantity of flakes and small pieces of irregular waste that were subject to the removal of further, secondary flake-flakes. It is accepted that there is an unknowable grey and subjective area in attempting to distinguish secondary modifications aimed at creating flake-tools, from secondary flaking of larger debitage pieces where the flake-flake product is not waste, but the desired end-product in its own right. Distinctions have been attempted here based on the size and shape of the piece subject to secondary working, the size of the secondary flake-flakes (beneath about 30mm they were not regarded as having been desirable end-products in their own right), the distribution and outcome of the secondary flaking and whether the flaked piece looked to have any viability as a tool in terms of its handling and cutting potential, or for some other potential use. Nonetheless, there remained a residual rump of ambiguous pieces, some of which are presented below.

It is postulated here that the majority of secondary working was aimed at transforming debitage into tools, mostly more useful cutting tools. The secondary working often produced a sharp concave notch that would have formed an ideal tool for cutting through substances such as animal skin, much like the hook on the end of a present-day box-knife. Conversely, the secondary working on a minority of pieces seems to have been aimed at deliberately smoothing and blunting an irregular edge opposed to a naturally straight and sharp edge. It is suggested here that, rather than being con-strued as a 'scraper' which would be the immediate interpretation of many lithic analysts, focusing on the secondary working, these pieces represent the creation of comfortably blunt handling facets, to facilitate use of the opposing sharp edge of the piece as a 'knife' for cutting. Finally, quite a few pieces of debitage that are unaffected by secondary flaking show localised areas of micro-chipping and damage on otherwise pristine sharp edges; these are interpreted here as macro use-wear, and these pieces are classified as 'utilised flakes'.

It is also necessary to recognise that, while an interpretive assumption has been made that many secondarily worked pieces should be construed as 'tools', and these are grouped below into 'types' according to the nature and distribution of secondary working, this is not the same as asserting that these tool-types were intentionally formed and conceived as distinct groups by their prehistoric makers. Rather, it is suggested that most tools should be understood purely as a flexible and plastic combination of handle and a working edge, usually for cutting, with the working edge usually formed by a secondarily flaked notch. The beauty of flint as a raw material for this technological approach is that it is highly flexible; most flakes can receive a notch, and any flake in use (whether notched or not) can be rejuvenated with replacement or additional notches if the immediate needs change, or if a particular cutting edge becomes damaged. One of the features of both the notched tools and the flake-flakes, discussed in more detail below, is that they often show signs of macro use-wear on what would have been a cutting edge; so it seems clear that individual cutting tools had a biography of use and transformation, perhaps progressing from un-notched through single-notched to multiple-notched forms, with an accompanying legacy of

flake-flake debitage. Whether the timescale of this life-cycle should be thought of in terms of a short period such as a few hours, a longer period of punctuated abandonment and reclamation, or a combination of the two is another factor to be considered and integrated into an understanding of the lithic *chaîne opératoire*.

As shown in the earlier tabular summaries (Table 18.9 and Table 18.10) there were 76 pieces identified as flake-tools in assemblages 6.1 and 6.2, 34 pieces identified as flake-flakes from secondary working and eleven pieces identified as cores-on-flakes. In the remainder of this section a representative selection of these finds is presented, highlighting certain types of end-product and patterns of secondary working that seem to regularly recur, and also presenting some refitting evidence of secondary debitage modification. Five main groupings of flake-tools were identified, all of them variations on notched and un-notched cutting tools (Table 18.17). The quantities of each of these groups are given for

Table 18.17 Technological categories of secondarily modified pieces, subsidiary types for flake-tools and cross-references to figures

Technological category	Tool-type	Analysis code	Description	Figure/s
Flake-tool	Utilised flake	61	Use-damaged, evidence of macro use-wear but no secondary flaking	Fig. 18.21a Fig. 18.23
	Flake-knife	62	Blunting/backing retouch opposite/beside natural cutting edge, which can show macro-wear, to facilitate handling and use	Fig. 18.23a,b
	Single notch	63	Clear single notch, can be backed by natural cortical handle or blunting/backing retouch	Fig. 18.24
	Double/linear notch	64	Two or more notches, aligned on one edge to form crude denticulate	Fig. 18.25
	Multiple notch	65	Two or more notches, scattered around; for instance orthogonal to each other, or on different sides of the same flake	Fig. 18.26
	Miscellaneous other	66	Any other secondary working that does not fit into the other categories, often when broken	-
Core-on-flake	-	30	Debitage used as a core	Fig. 18.20c
Flake-flake	-	80	Debitage from flaking a flake	Fig. 18.27
Ambiguous worked pieces	Cores? Flake-tools?	-	Solid pieces of debitage or irregular waste, with small-medium secondary removals	Fig. 18.28

Table 18.18 Assemblages 6.1 and 6.2: flake-tools, flake-flakes and cores-on-flakes

	61 – Utilised flakes	62 – Knives	63 – Single notches	64 – Double/linear notches	65 – Multiple notches	66 – Misc flake-tools	60s – All Fl-tools	80 – Fl-flakes	30 – Cores-on-flakes
Ass 6.1	15	5	17	17	8	10	72	31	11
Ass 6.2	2	-	2	-	-	-	4	3	-
Ass 6.1 & 6.2	17	5	19	17	8	10	76	34	11

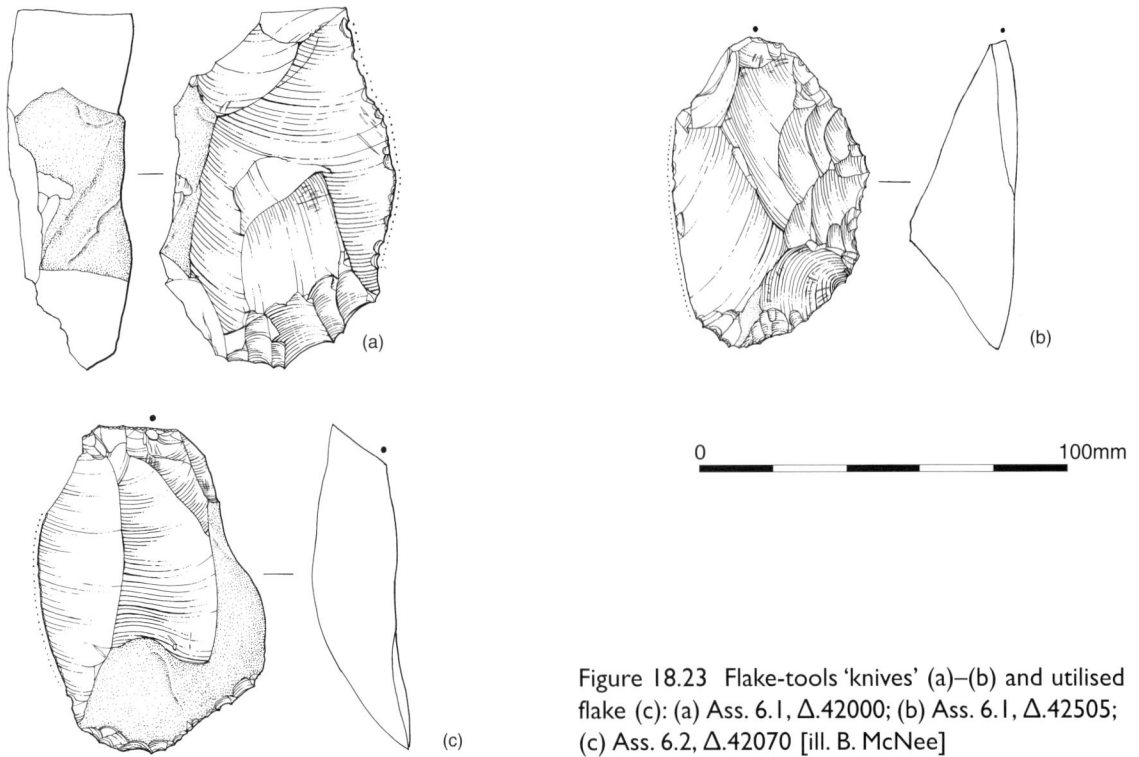

Figure 18.23 Flake-tools 'knives' (a)–(b) and utilised flake (c): (a) Ass. 6.1, Δ.42000; (b) Ass. 6.1, Δ.42505; (c) Ass. 6.2, Δ.42070 [ill. B. McNee]

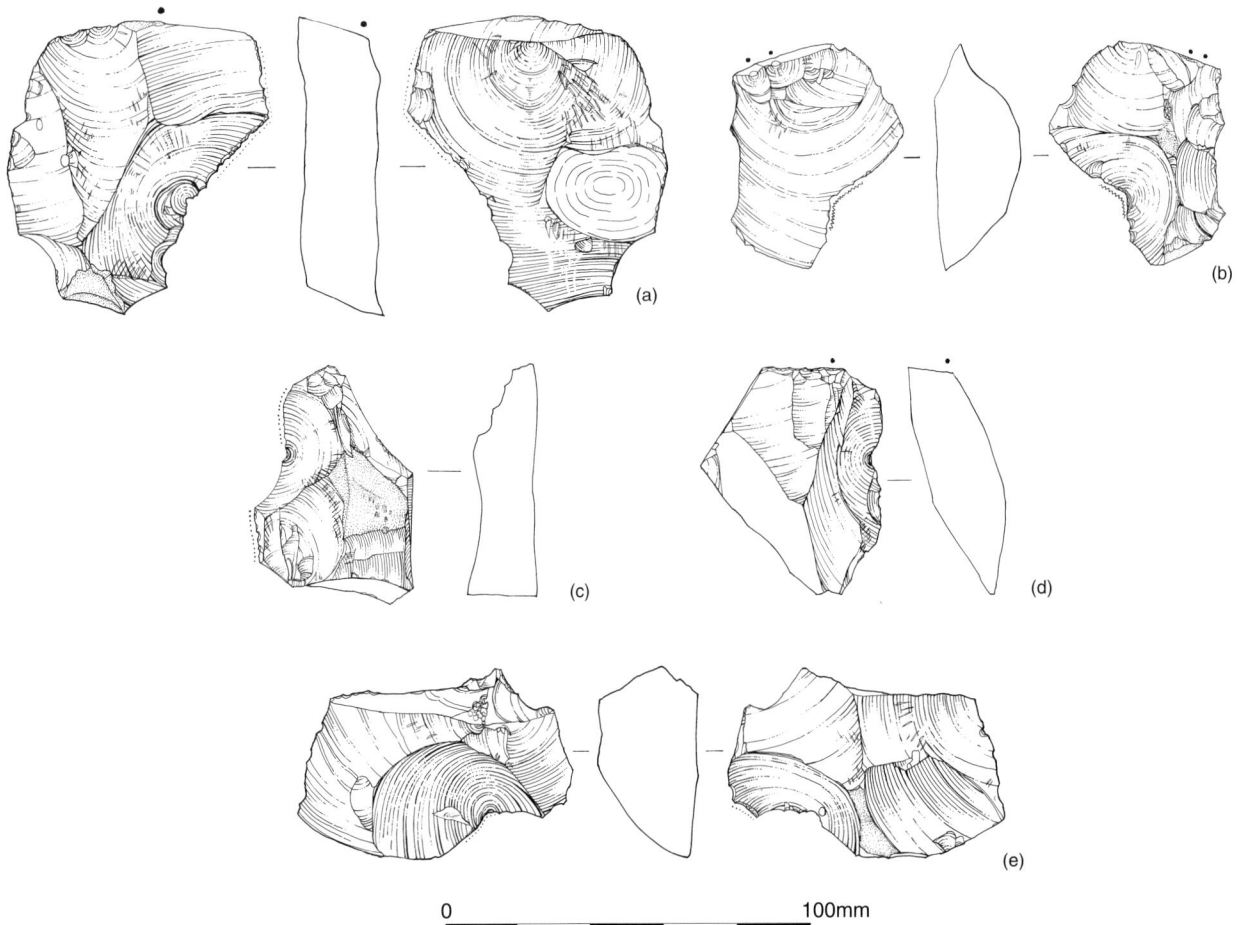

Figure 18.24 Flake-tools, single notches: (a) Ass. 6.2, Δ.40211; (b) Ass. 6.1, Δ.41577; (c) Ass. 6.1, Δ.42805; (d) Ass. 6.1, Δ.42842; (e) Ass. 6.1, Δ.43809 [ill. B. McNee]

assemblages 6.1 and 6.2, together with the quantities of flake-flakes and cores-on-flakes (Table 18.18). Various investigations were made exploring the spatial distribution of flake-tools and other technological categories within assemblages 6.1 and 6.2. Although there were some minor variations in relative proportions of different categories in different parts of the site, as is statistically inevitable, the overall impression was of remarkable homogeneity, so the remainder of this section focuses on their technological and typological characteristics without consideration of their spatial distribution.

Three of the pieces identified as utilised flakes are illustrated, RG #30 Δ.43198 (Fig. 18.21b), RG #52, Δ.43102 (Fig. 18.21c) and Δ.42070 (Fig. 18.23c). Clearly there is some difficulty in the reliable differentiation of macro use-wear from incidental damage to delicate sharp flake edges, whether during knapping, burial or post-depositionally *in situ*. However, several pieces had localised damage that had a slightly more invasive and regular character than the isolated and evenly distributed small chips that were regarded as incidental natural damage. Much use may have left no

visible trace, and it is suspected that many flakes classified as waste debitage may in fact have been used. Two of the five tools classified as 'flake-knives' are illustrated (Δ.42000 and Δ.42505, Fig. 18.23a, b). Both of these seem particularly clear examples of instances where secondary retouch has been applied, not to form a working part of the tool, but to facilitate handling and use of the unworked sharp edge of the tool for cutting. For Δ.42000 there is a natural blunt cortical facet down one side of quite a large flake, opposed by a regular slightly convex sharp edge. The distal end of the flake seems to have been lightly trimmed to follow the convexity of the cutting edge, and clear it for use. Furthermore, there is marked 'nibbling' along this edge that seems a very clear example of macro use-wear. For Δ.42505, secondary working has formed a convex blunt face opposing a regular slightly convex sharp edge, which has occasional small invasive chips suggestive of macro-use wear. It seems highly likely that this working was aimed at facilitating handling of the tool with use of the sharp edge for cutting, rather than for use of the opposite blunt face as a scraping facet. If the latter was the mode of use, then

Table 18.19 Assemblage 6.1: quantitative comparison and size statistics for flake-tools and debitage (only whole pieces included)

	61 – Utilised flakes	62 – Knives	63 – Single notches	64 – Double/linear notches	65 – Multiple notches	66 – Misc flake-tools	60s – All Fl-tools	90 – Flakes
(n=)	7	4	12	8	2	235	566	
Max L - Max	115	80	95	112.00	60.00	40.00	115.00	125
- Q4/Q3	90.5	66.5	63.25	74.5	55.75	35.25	70	51
- Mean	74.57	59.75	59.25	62.88	51.50	30.50	61.11	40.44
- Q2/Q1	56.5	49.75	51.25	46.25	47.25	25.75	47	28
- Min	50	46	24	29	43	2121	12	
- SD pop	23.80	13.05	19.17	25.01	8.50	9.50	22.54	16.64
Max W - Max	56	65	77	53.00	45.00	47.00	77.00	132
- Q4/Q3	52	56	48.25	46.25	43.5	38.75	48.5	41
- Mean	44.71	43.75	45.92	40.63	42.00	30.50	43.11	33.35
- Q2/Q1	39.5	30.25	40.25	35.5	40.5	22.25	37	23
- Min	28	25	31	26	39	1414	10	
- SD pop	9.45	16.02	11.12	9.03	3.00	16.50	11.75	14.54
Max T - Max	38	32	36	47.00	26.00	16.00	47.00	75
- Q4/Q3	28.5	20	26.25	24.75	25	13.75	26	17
- Mean	21.86	18.50	22.67	22.50	24.00	11.50	21.43	12.97
- Q2/Q1	15	14	19	15.75	23	9.25	16	8
- Min	7	11	13	11	22	77	2	
- SD pop	10.15	8.02	6.07	10.45	2.00	4.50	8.65	7.43
Weight g - Max	170	115	263	276.00	48.00	39.00	276.00	543
- Q4/Q3	119	79.75	80.75	81.25	47.25	30.5	81.5	34
- Mean	81.14	54.75	74.25	80.13	46.50	22.00	70.17	28.27
- Q2/Q1	43	18.5	43	29	45.75	13.5	29.5	7
- Min	11	17	18	11	45	55	2	
- SD pop	54.04	40.34	63.20	80.46	1.50	17.00	62.29	42.04

the tool would have been very awkward to handle, with the sharp cutting edge pressing right into the hand.

Five of the 20 pieces identified as single notches are illustrated (Fig. 18.24). Of these five, four have visible use macro-wear on part of the sharp edge created by the notch. Although prevalence of this property was not formally quantified, it was commonly found on notched tools, whether single, double or multiple. The notches are usually, although not always, placed at one side or other of a flake towards its distal end. As one of the more numerous categories of flake tool, the size statistics are worth considering (Table 18.19; see also Figs 18.31–32, below). These show that the average size of single-notch flake tools (and indeed, of all flake-tools) in their present state after secondary flaking, is approximately 60 x 45 x 22mm, with an average weight of roughly 75g. This is significantly larger than the average size of unworked flakes. It therefore perhaps gives an indication of the size of flake blank that was regarded as preferred for tool-use.

Another reasonably common group of secondarily-worked notched tools was double/linear notches, of which there were 16 in total, of which five are illustrated (Fig. 18.25). One of the illustrated specimens (Δ.42973 – Fig. 18.25d) is quite large (maximum length 88mm; weight 115g) and has two notches side-by-side opposing

a cortical ridge, so it seems pretty clear that the notches were the functional part of the tool, although there is also some apparent abrasion at the proximal end, possibly minor trimming of a sharp edge to facilitate handling, rather than use-wear. Another one (Δ.43276 – Fig. 18.25e) likewise has two notches in a row, but here they are on a thicker flake with an opposing sharp edge so it is much less clear which bit of the artefact was the working part, if indeed just one part of it was, rather than different edges being used as appropriate to specific tasks. Δ.43716 (Fig. 18.25f) is quite similar to Δ.42973, although a little smaller.

In contrast to the other illustrated specimens, Δ.42866 (Fig. 18.25c) is substantially smaller, although its maximum length of 41mm and weight of 14g are by no means the smallest in this flake-tool category. One of the enduring mysteries of Lower/Middle Palaeolithic flint artefacts is the regular occurrence of tiny versions of lithic tool forms. Since lithic artefacts are virtually indestructible once formed, and since there must have been a stage of juvenile knapping experimentation and learning that would have contributed to the archaeological record, it is not entirely fanciful to suggest that some, or most, of these might relate to juvenile flint-knapping, emulating surrounding adult practices.

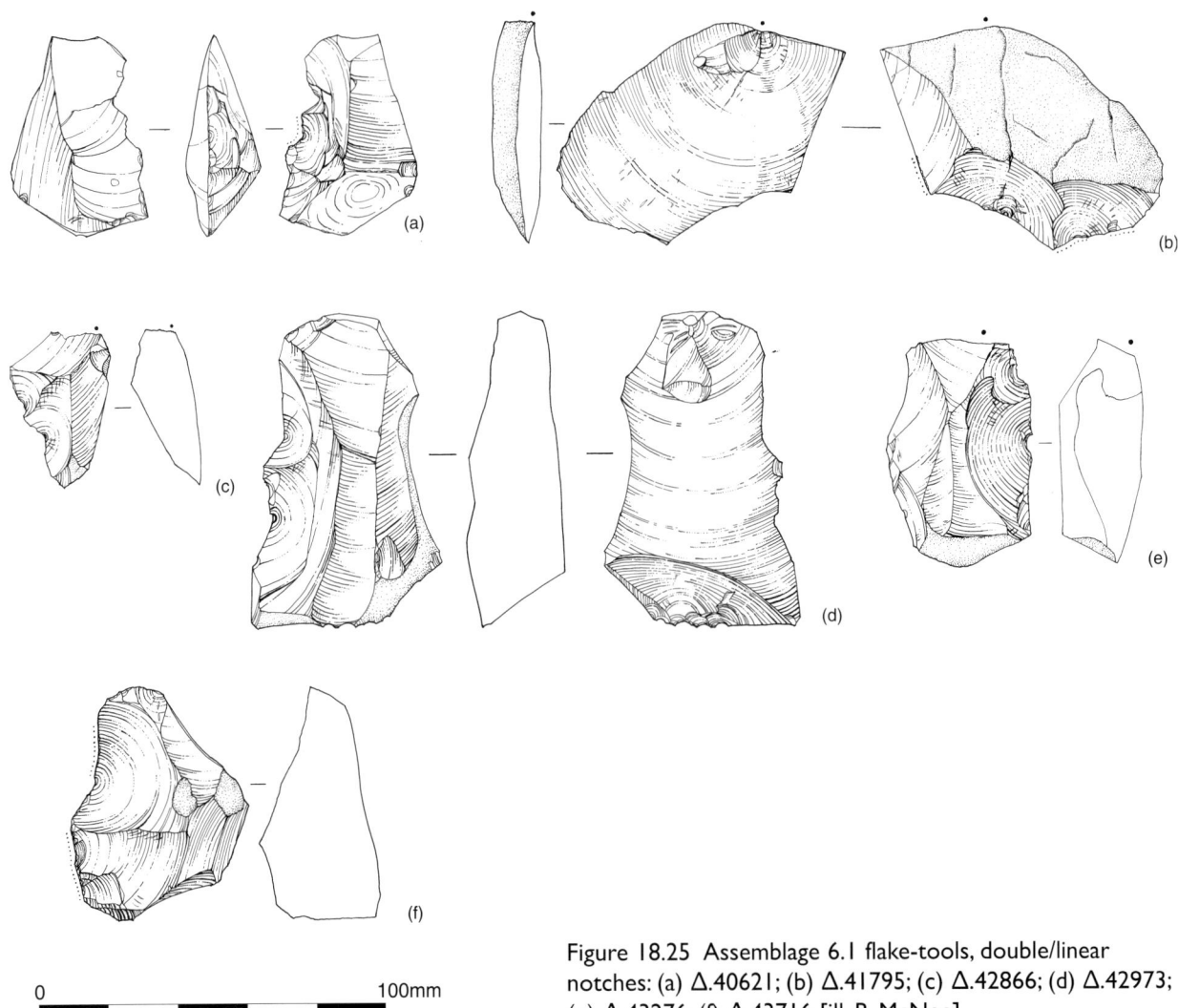

0   100mm

Figure 18.25 Assemblage 6.1 flake-tools, double/linear notches: (a) Δ.40621; (b) Δ.41795; (c) Δ.42866; (d) Δ.42973; (e) Δ.43276; (f) Δ.43716 [ill. B. McNee]

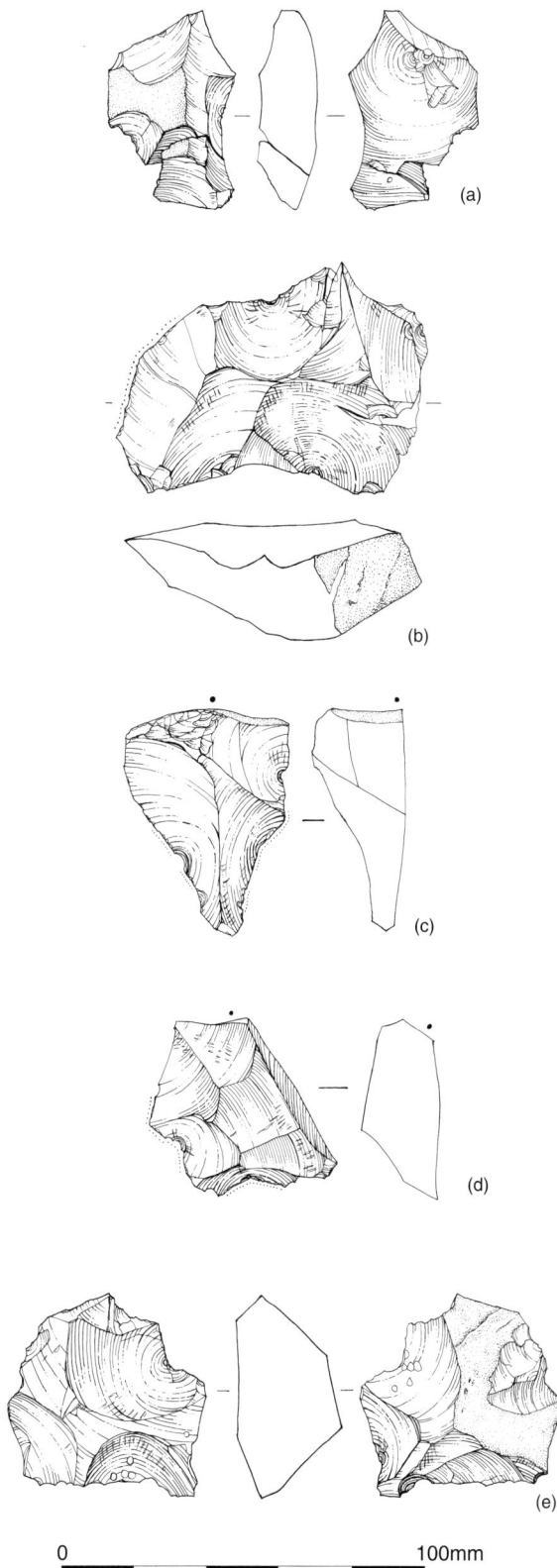

Finally, the fifth illustrated artefact in this group is Δ.40621 (Fig. 18.25a), which, rather than having two large notches in a row, has three smaller ones to create a coarsely denticulated edge opposite a reasonably sharp edge. It is not immediately clear how this piece should be understood in terms of its likely mode of use and handling, but there is micro-chipping in two places: within the middle notch, and on the end of one of the end notches, that perhaps reflects macro use-wear.

There were only eight multiple-notched tools that were whole and could not be attributed to one of the other notched-tool categories, five of which are illustrated (Fig. 18.26), including one refitting group with both the parent tool and the secondary flake: RG #63 (Fig. 18.26a). On one of these (Δ.42824, which was one of the larger flake-tools recovered, with a maximum length of 85mm and weighing 120g; Fig. 18.26b), there are three medium-size secondary flakes that have removed the proximal and distal ends of what was once a substantial flake, leaving one sharp edge on what has become the putative working end of the tool, but on what was originally one of the flake's sides. This sharp edge has tiny chipping indicative of macro use-wear, so conceptually this tool could equally have been categorised as a 'flake-knife'. The other tools with multiple notches are less easy to make a sense of. The notches on the larger tools (Δ.42873 – Fig. 18.26c; Δ.42967 – Fig. 18.26d; and Δ.43473 – Fig. 18.26e) would all have had sharp edges that were potentially useful for cutting, although none of these pieces seem especially convenient to handle. The slightly chaotic and repeated ring-cracks from failed percussion blows on Δ.43473 again bring to mind juvenile emulation of adult behaviour, perhaps applied to a discarded tool.

Finally, piece Δ.40425 (Fig. 18.26a) is the proximal end of a flake that is now reduced to a maximum length of 35mm and weighs only 20g. It has had a radial series of notches flaked around its distal end, all struck on the ventral surface. One of these secondary flake-flakes (Δ.41537) was recovered, only about 2m away, both artefacts being found towards the northern edge of the lithic concentration south of Trench D. Again, it is hard to suggest a mode of use and handling for this small piece of flint, although it was clearly deliberately flaked to leave it in its current form. The recovery of both pieces in such proximity suggests a minimum of disturbance, although there were other secondary removals both preceding and subsequent to the one recovered, that were not found in the excavated assemblage. They may have been missed due to their small size, or they perhaps have been lost during machining of Trench D, a short distance to the north. It seems very unlikely that the missing flake-flakes, which would have been about 10mm long, were selected for use elsewhere. It is also unlikely that the last (missing) two flake-flakes of the reduction sequence were knapped elsewhere, before the tool was moved back and abandoned at the exact spot of the earlier (recovered) flake-flake.

Most of the miscellaneous flake-tools were either broken, and therefore uncertainly classifiable to any of

Figure 18.26 Assemblage 6.1 flake-tools, multiple notches: (a) RG #63 - Δ.40425-Δ.41537; (b) Δ.42824; (c) Δ.42873; (d) Δ.42967; (e) Δ.43473 [ill. B. McNee]

the other forms, or had minor secondary working in odd places that could not be easily construed as blunting/ backing or as a notch. Some of them were in the category of ambiguous knapped pieces for which it could not be decided whether they should be categorised as a core or a flake-tool, and one of these (Δ.40611) is illustrated and discussed further below (Fig. 18.28e).

The inevitable technological counterpart of the high incidence of working of parent flakes was the production of secondary flake-flakes, a representative selection of which is illustrated here including two that refit to each other (Fig. 18.27). There is a clear imbalance in the assemblage between the proven quantities of notches from secondary working (a minimum of approximately 75) and the recovered quantity of flake-flakes (n=31). However, this almost certainly does not reflect organisational structure of the lithic production. Rather, it probably reflects that firstly, the secondary flake-flake products are small, and therefore their recovery is likely to have been less than complete, and secondly, rarely clearly identifiable as distinct from normal flakes. Of the four flake-flakes that were found to refit to parent debitage: RG #6, Δ.41940 and Δ.41746 (Fig. 18.28b); RG #2, Δ.42436 (Fig. 18.28a); and RG #63, Δ.41537 (Fig. 18.26a), all were originally classified as conventional

flakes prior to their refitting, after which the original records were revised. This suggests that flake-flakes are almost certainly under-represented in the overall results of the technological analysis. Consequently, the average dimensions of conventional flake products are probably slightly larger than indicated in the summary table and figures (Table 18.19; see also Figs 18.31-32, below), since the data undoubtedly includes measurements from a number of flake-flakes, which would generally be smaller than normal flakes.

Several of the flake-flakes show signs of macro use-wear (eg. Δ.42319, Fig. 18.27f) and/or previous flake-flake removals, emphasising that secondary working is carried out in conjunction with, and as part of an ongoing process of, tool-use. Another (Δ.43252, Fig. 18.27h) shows heavy use-wear on an unmodified flake-edge, lending credence to a model of a life-history of cutting tools involving a progressive transformation from unmodified utilised flakes to single notches, and then perhaps further on to double and multiple notches.

Two refitting flake-flakes were found that were of particular interest, shown here (Fig. 18.27a, RG #1 – Δ.41329 and Δ.41211). These were found only about 1.75m away from each other, just to the north of Trench D (Fig. 18.22), and only about 2m away from the

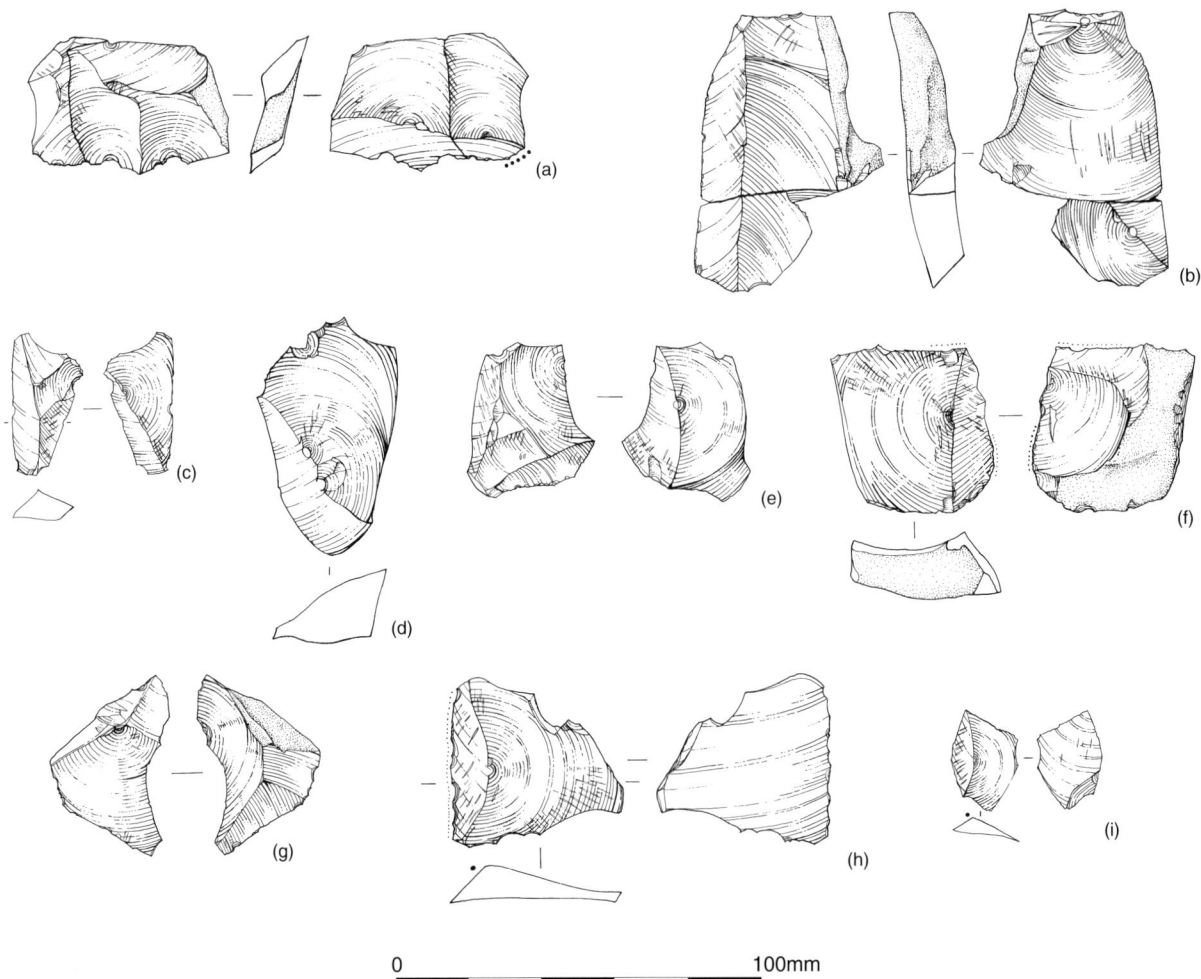

Figure 18.27 Assemblages 6.1 (b)–(i) and 6.2 (a), secondary flake-flakes: (a) RG #1 – Δ.41329-Δ.41211; (b) RG #62 – Δ.40618-Δ.42972; (c) Δ.40486; (d) Δ.40502; (e) Δ.41676; (f) Δ.42319; (g) Δ.43203; (h) Δ.43252; (i) Δ.43347 [ill. B. McNee]

S kirchbergensis rhino jaw (Group Δ.40843 – see Chapter 7). There is however no reliable basis for associating these finds with the rhino jaw, as there is a general sparse distribution of lithics in this area, without any apparent focus on the area where the jaw was found. Besides from the fact that they refit and were found close together, suggesting a minimum of disturbance, and that they therefore represent activity carried out at the site, these flake-flakes exemplify the proposed progressive re-sharpening tool-use model. When refitted, they retain as dorsal scars the evidence of two earlier flake-flake removals, with signs of macro use-wear. They also perhaps provide a hint that the linear double notch might not just be a progression of more intensive resharpening from a single notch, but might in fact be a deliberately conceived type, with the previous double linear notch re-sharpened to a new linear double notch as a single event.

The final group of secondarily worked flakes to consider is the ambiguous forms that could not be reasonably categorised as either cores or flake-tools with any confidence. A selection of these is shown (Fig. 18.28), including three refitting groups (RG #2, RG #6 and RG #28 – Fig. 18.22) where both the secondary flakes and the parent pieces were recovered to aid in their attempted classification. In RG #2 for instance (Fig. 18.28a), the parent piece (Δ.42145) was originally a very large flake, and has been subject to the removal of numerous chunky flake-flakes, just one of which (Δ.42436) was recovered, a little less than 3m away, not the last secondary flake incidentally, two later removals were not found. The parent piece was classified as a 'core-on-flake' when first analysed, but closer examination revealed some possible macro use-wear on one sharp protrusion, so this should perhaps be reconsidered as a 'flake-knife', with the large secondary

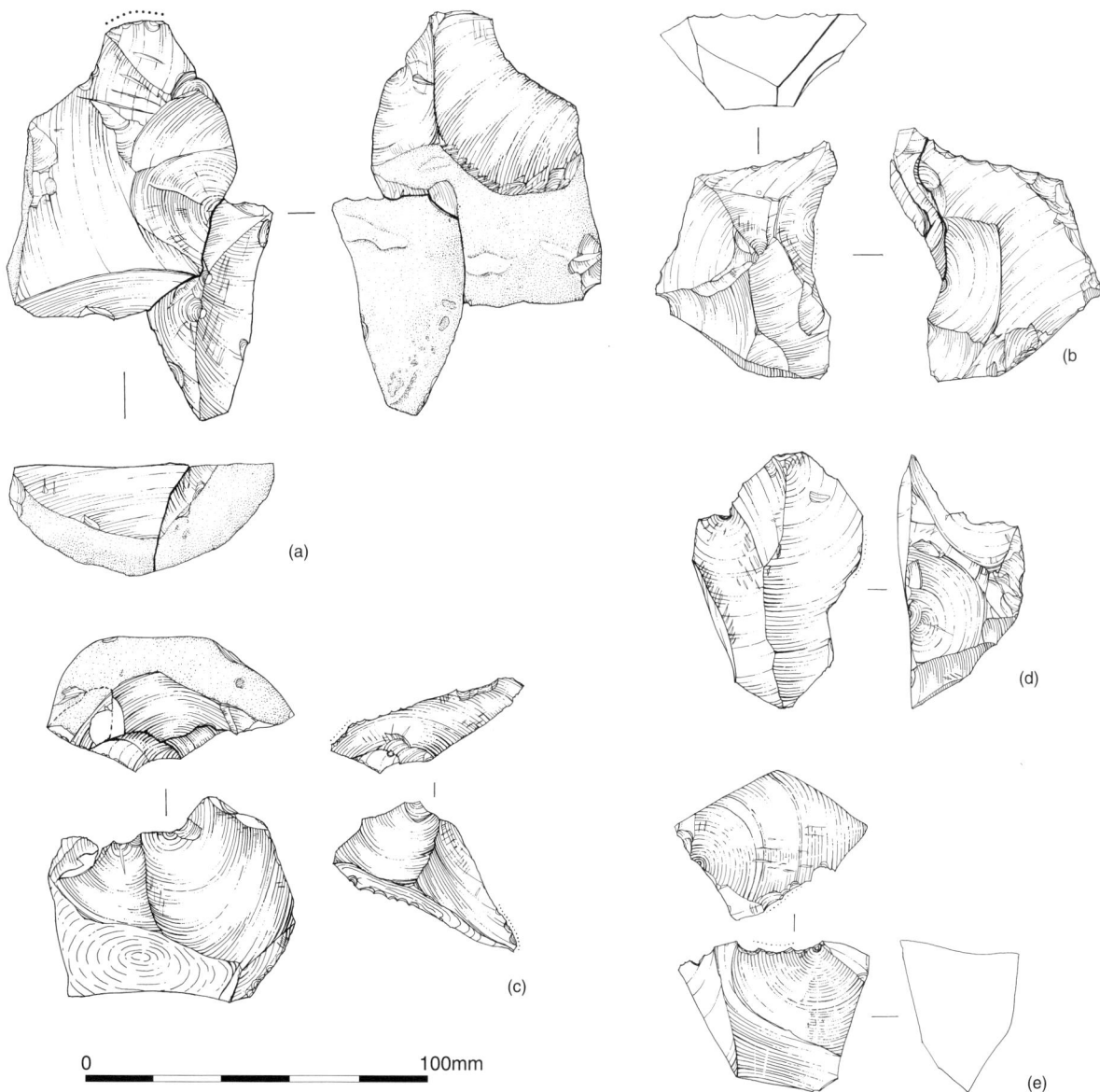

Figure 18.28 Assemblage 6.1, chunky secondarily worked flakes that are ambiguous as to whether 'core-on-flake' or 'flake-tool': (a) RG #2 – Δ.42145–Δ.42436; (b) RG #6 – Δ.41934–Δ.441747–Δ.41940; (c) RG #28 – Δ.43182–Δ.43217; (d) Δ.40517; (e) Δ.40611 [ill. B. McNee]

flakes merely removed to make it fit the hand more comfortably.

Pieces Δ.40517 (Fig. 18.28d), Δ.40611 (Fig. 18.28e) and Δ.41934, from RG #6 (Fig. 18.28b), were all classified as flake-tools in the original analysis and this remains a reasonable possibility, although they do not exhibit the obvious functionality and handling convenience of the majority of the other flake-tools. It is hard to imagine that the refitting flake-flakes of RG #6 (Δ.41940 and Δ.41747) would have had much useful functionality, particularly the former, which was 23mm long and weighed 5g. The latter, which was slightly larger (28mm long and weighing 10g), did however have one sharp edge approximately 30mm long which showed some tiny denticulations, the negative scars from which were stained pale grey, in contrast to the deep red staining of the rest of the piece. These scars did not appear, however, to be from damage during excavation. They may represent *in situ* post-depositional damage or they may perhaps represent macro use-wear reflecting use of this small piece as a cutting tool subsequent to its removal from the parent piece. The position of these denticulated chips means, incidentally, that it is not possible that they were formed while attached to the parent piece.

One final point to make about the flake-tools found, is the anecdotal observation that there is a disproportionate focus for secondary flaking on flakes from the slightly coarser opaque interior of flint nodules. Many flint nodules have translucent glossier flint nearer their outside cortex, but a progressively coarser and more opaque texture deeper into their interior. These pieces are now often stained on a gamut from pale greenish-yellow through to deep ochre following their prolonged burial, probably associated with a greater capacity to absorb minerals or moisture. It seems that the Palaeolithic knappers at the site actively preferred for flake-tools the slightly tougher opaque flint that most would today regard as less satisfactory. This is contra modern preference, whereby it is generally assumed that more glossy, translucent flint is of higher quality and would be preferred, for instance proposed as a universal pan-cultural law of lithic technology by Rhys Jones (Jones and White 1988). It may have been less slippery to hold, perhaps, or it may have maintained a tougher edge with better cutting properties for longer than the brittle, ostensibly sharper edge from more glossy and translucent flint.

The chaîne opératoire *and organisation of production*

Unlike the lithic concentration around the elephant (Chapter 17), where the refitting confirmed that the assemblage was essentially undisturbed and long refitting reduction sequences therefore directly represented the *chaîne opératoire* and the organisation of production, assemblages 6.1 and 6.2 seem to have been more disturbed. They do not contain long refitting sequences representing activity that was happening at or near the site. However, when compared with representative datasets for complete experimental flaking sequences (Fig. 18.29a; Fig. 18.30a) – using the data from Wenban-Smith's (1996) Clactonian experiments, which involved reduction of three nodular flint cores, with each sequence producing *c* 50 flakes – the assemblage of refitting flakes (n=56) corresponded well with the datasets for dorsal scar counts and percentage cortex (Fig. 18.29b; Fig. 18.30b), two of the variables shown (ibid.) to correlate best with reduction order, with correlation coefficients of 0.41 and -0.37 respectively, both significant at the 1 in a 1000 level. Since these attributes correlate so well with reduction order, it is suggested that the refitting flake assemblage, although of small size and consisting of numerous short sequences, nonetheless matches an overall pattern of evenly balanced flake production at the site, without a bias towards either early or late stages of core reduction.

The rest of assemblage 6.1, although not refitting and not representing undisturbed knapping remains from on-the-spot activity, is nonetheless thought to represent evidence of minimally disturbed activity in the near vicinity. Consequently some quantitative analyses were also carried out on this much larger quantity of material in an attempt to investigate the *chaîne opératoire* and the spatial organisation of production. Some of these are necessarily crude, since they are vulnerable to uncertainties. These are, firstly, the influence of the high degree of frost-fracturing and the possibly misleading quantitative effects of the high proportion of irregular waste consequently produced. Secondly, the extent of the invisibility of early flakes from cores with longer sequences of reduction, where the scars of early flake removals were not retained on the eventual core.

Nonetheless a simple comparison was carried out to investigate how the quantity of debitage in assemblage 6.1 compared with the overall number of scars on the cores from the assemblage. Added to the scars from tested nodules, which were all allocated a provisional count of one scar since this data was not recorded when analysed, the combined total of debitage scars represented in the cores is 618. This is based on the total dorsal scar count: 20 from tested nodules, 568 from cores and 30 from cores-on-flakes. The total number of pieces of debitage represented by the flakes and flake-tools is 749, counting only whole debitage and proximal pieces: 694 from flakes, and 55 from flake-tools. However, it needs to be remembered that cores-on-flakes are themselves pieces of debitage, so a further ten pieces need to be added to the debitage total, bringing it to nearly 760. Considering that there is likely to be some invisibility of flake scars from early in reduction, and that the quantity of debitage probably includes quite a few flake-flakes, these quantities look broadly comparable. This suggests that the pattern of flake production correspond with locally obtained raw material that was exploited on the spot. It is assumed for this analysis that debitage from the cores-on-flakes would have been more likely to have been categorised as 'flakes' rather than 'flake-flakes' when originally recorded.

A slightly more detailed quantitative analysis was also carried out on the flake assemblage, comparing the dorsal percentage of cortex and the number of dorsal scars with complete experimental sequences of flake production using what was thought to be a similar *ad hoc* knapping approach utilising the data from Wenban-Smith's (1996) Clactonian experiments. This analysis (Fig. 18.29; Fig. 18.30) established that, to a slightly lesser degree than the refitting flake data, there was still a good correspondence between the profile for the complete set of experimental data and the archaeological data. The most significant area of comparison is between the experimental data and the large hand-excavated dataset of assemblage 6.1. The main points of contrast are the much lower proportion of flakes with less than 10% cortex in the excavated assemblage (about 34% versus about 54%), and the more equal numbers of flakes with two rather than one dorsal scars (24%:22% versus 29%:16%), although the total proportion of flakes with either one or two dorsal scars is virtually identical. In general these data seem to likewise suggest that all stages of reduction are represented pretty equally in the excavated collection. The relative lack of flakes with minimal cortex probably reflects the relatively high incidence of much shorter reduction sequences in the archaeological assemblage (see Fig. 18.15), compared to the experimental dataset which was based on three long reduction sequences, of approximately 50 flakes each.

These analyses do not necessarily indicate an expedient lithic technological *chaîne opératoire*, with raw material

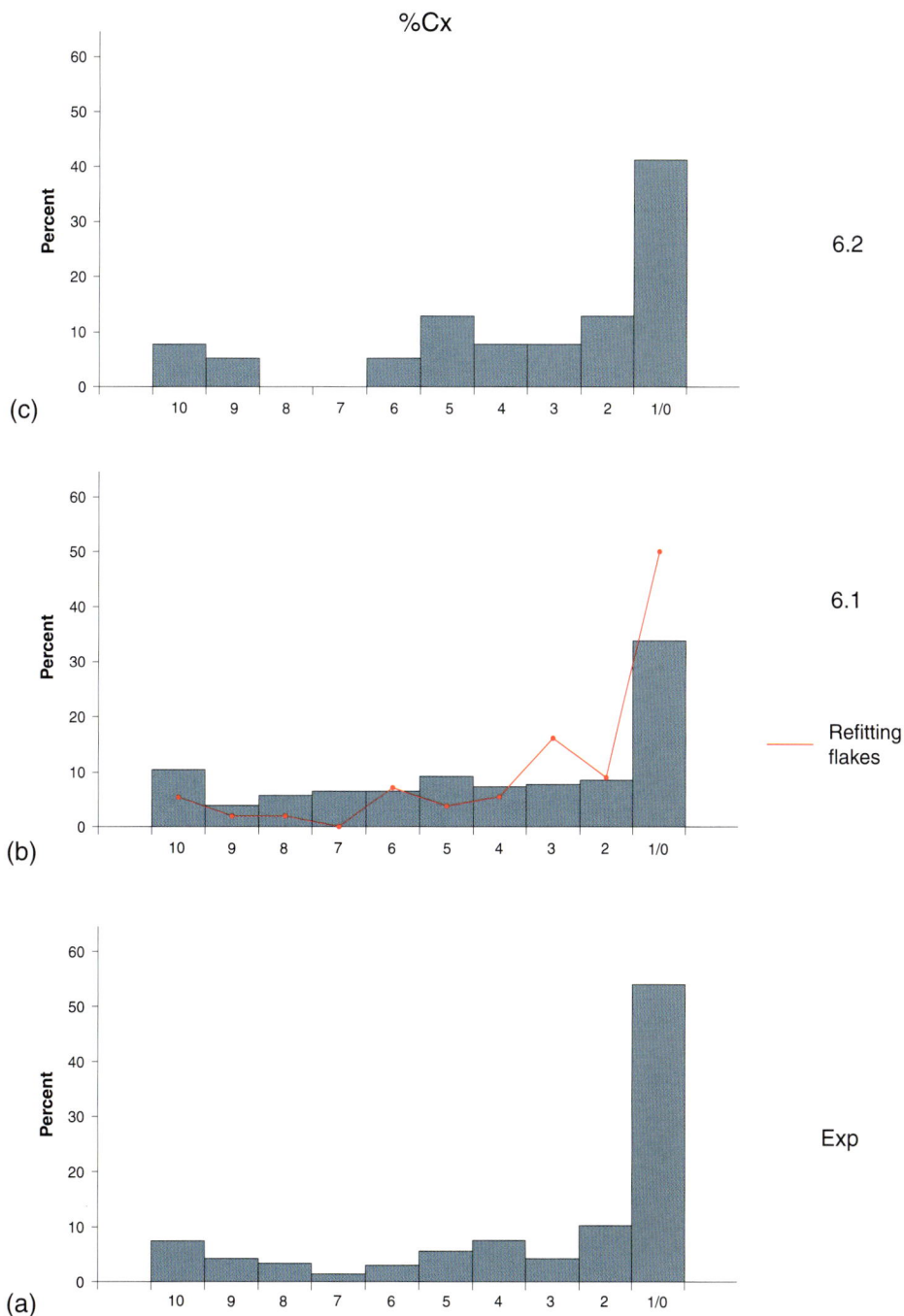

Figure 18.29 Histogram of percentage dorsal cortex (Cx%) on whole debitage: (a) comparative experimental data for complete reduction sequence; (b) assemblage 6.1 with line for refitting flakes; (c) assemblage 6.2

locally obtained, reduction carried out on the spot, and certain flake products immediately subject to secondary flaking to form notched cutting tools that are then used, with resharpening if needed, and discarded on the spot. This is because they are based on an assemblage of flakes that are mostly all from different individual reduction sequences. It therefore remains uncertain whether the missing parts of individual sequences are merely due to sampling and post-depositional mixing, or whether there was temporal separation and/or spatial mobility associated with the progression of individual *chaîne opératoires*. There is clear evidence of some behavioural mobility in the least disturbed, refitting element of the assemblage. This includes evidence of tool rejuvenation, but no tool (RG #1) and many flake-sequences without their core (Table 18.3), although their temporal/spatial extensions are unknown. However, the balanced profile of the assemblage indicates a lack of pattern in the organisation of production around the local landscape,

with all stages seeming equally likely to occur in the area represented by the excavated assemblage, which is also an area rich in the raw material used. There is therefore certainly no indication of a preferential export of flake-blanks or part-worked cores as might be expected if the raw material source was exploited in a more logistically organised way, for instance as with the handaxe-manufacturing locale at Red Barns, Hampshire (Wenban-Smith *et al.* 2000; Wenban-Smith 2004b), where the marked predominance of distinctive handaxe-manufacturing debitage clearly indicates a pattern of the manufacture and export of handaxes at the site (ibid.)

DISCUSSION

This section recapitulates the main conclusions of the preceding analyses and investigations of the lithic material in assemblages 6.1 and 6.2, and considers them

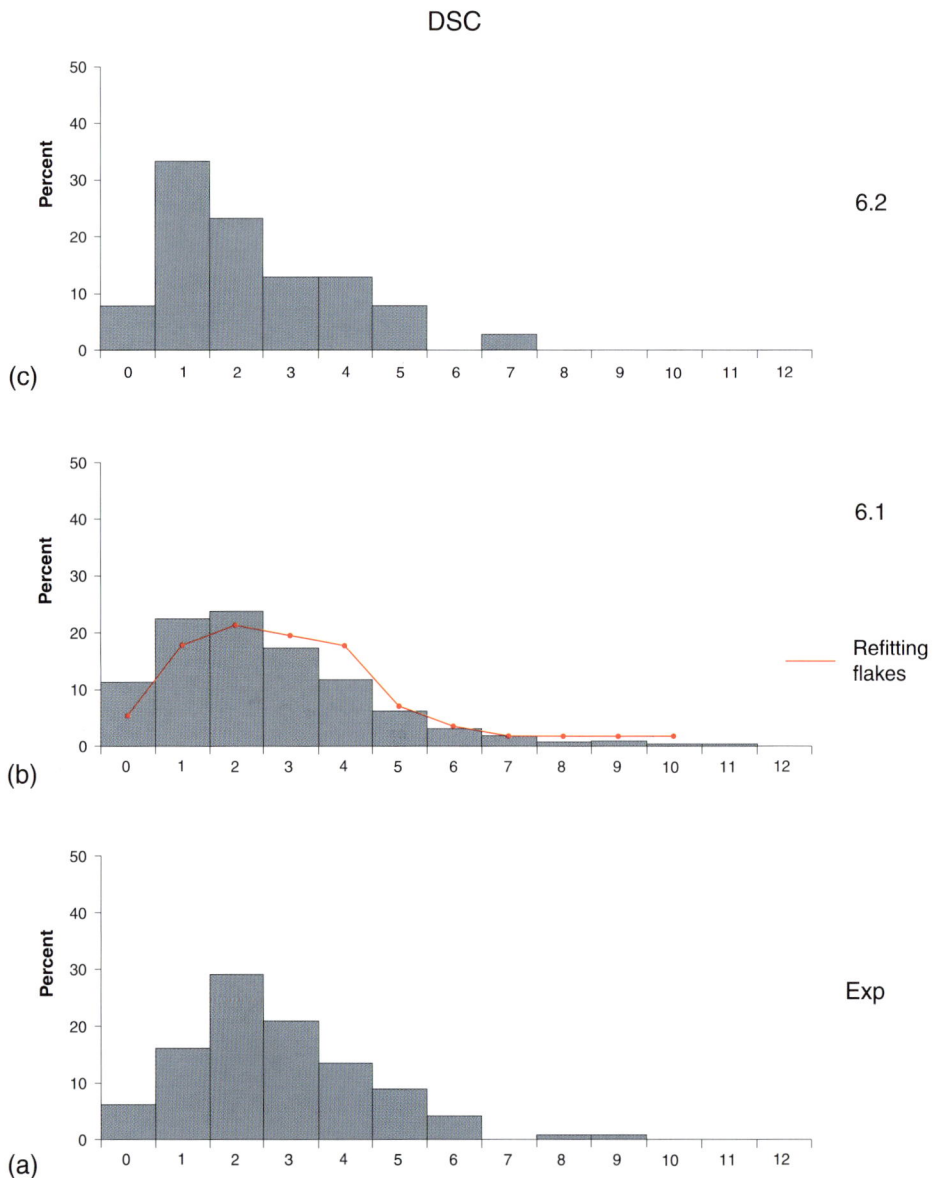

Figure 18.30 Histogram of dorsal scar count DSC on whole debitage: (a) comparative experimental data for complete reduction sequence; (b) assemblage 6.1 with line for refitting flakes; (c) assemblage 6.2

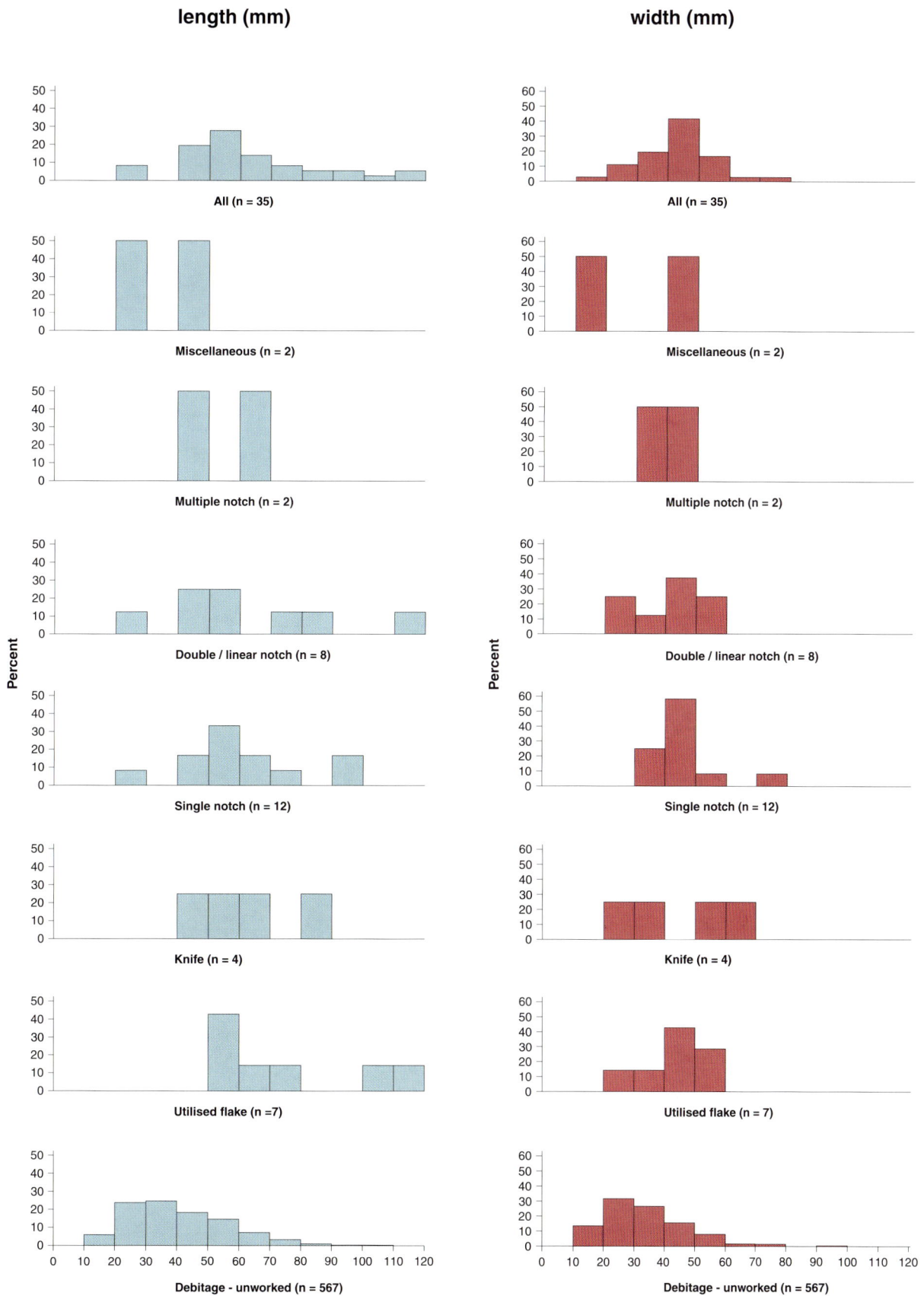

Figure 18.31 Histograms of length and width distributions for flakes and flake-tools from Assemblage 6.1 (whole artefacts only)

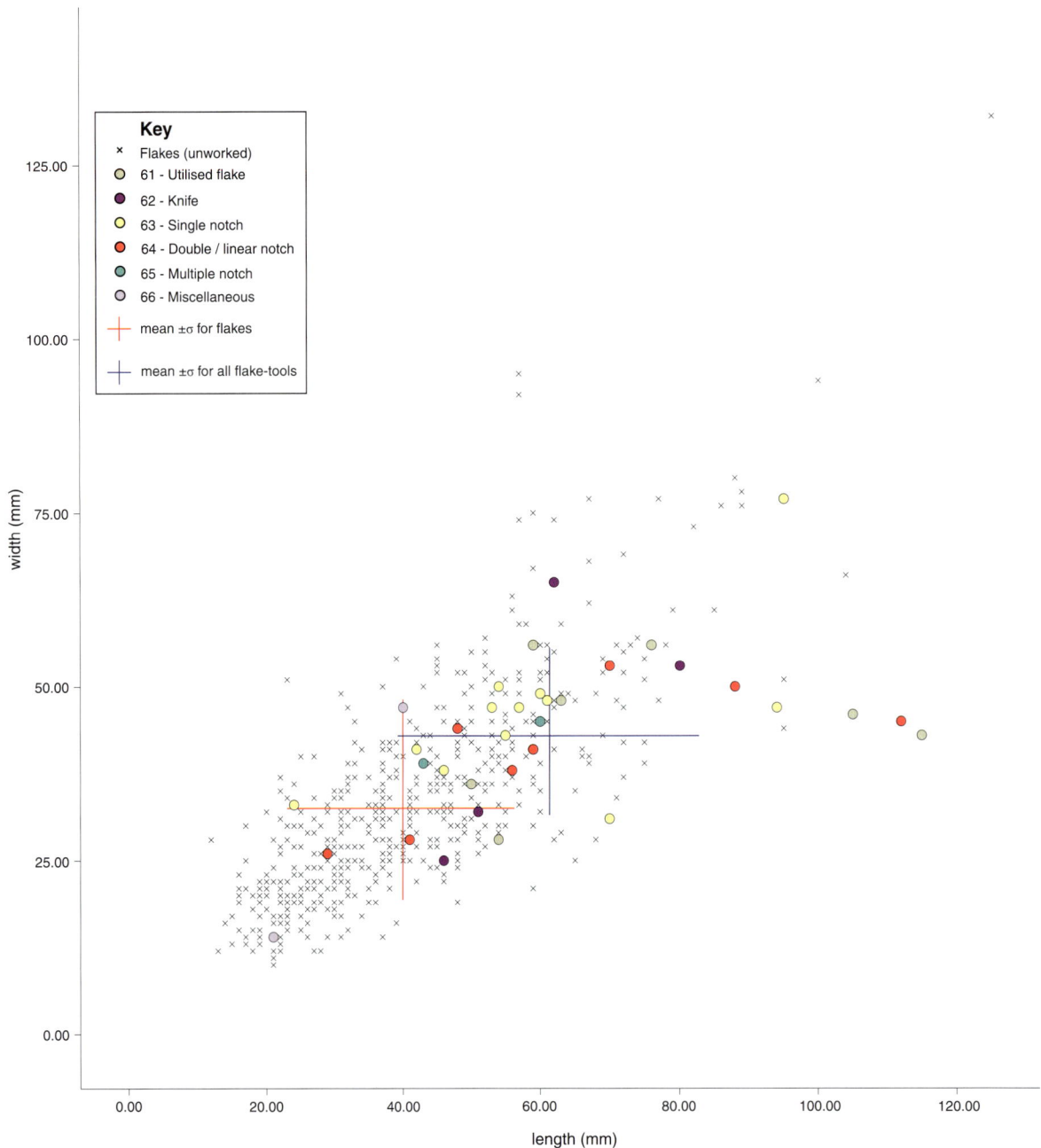

Figure 18.32 Bivariate scatterplot of length versus width for flakes and flake-tools from assemblage 6.1 (whole artefacts only)

in the wider contexts of the Lower/Middle Palaeolithic of (a) the Swanscombe area and (b) south-east England.

Site formation and integrity

The refitting and microdebitage studies have firmly established that assemblages 6.1 and 6.2, here focusing on the parts of them that were in mint or fresh condition, do not consist of entirely undisturbed lithic remains. However, nor do they seem to represent a particularly mixed collection. The great majority are in mint condition, and there is a reasonable amount of refitting material, with about 5% of the artefacts from assemblage 6.2 and about 8% of those from assemblage 6.1 refitting. Although there is quite a homogenous mixture of material of different sizes, shapes and techno-

logical categories in the areas where lithic material is present, the overall distribution of lithic material south of Trench D is strongly clustered (see Fig. 18.1; Fig. 18.3). There are patches of high lithic concentration occurring next to patches with virtually no lithic remains. Furthermore, of the refits that were found, many of them represent technological refits (such as flaking sequences, and secondary flaking) that were found only a short distance apart, suggesting a minimum of disturbance. It is suggested here that assemblages 6.1 and 6.2 represent a complementary combination of two contrasting site formation processes. The majority of the material is thought to represent the remains of knapping activity that was taking place in the close vicinity, to the west of the water-body that would have been present, over perhaps many hundreds of years. It has probably become slightly conflated by removal of fine-grained clay sediments by sheetwash processes, and fed en masse into the water-body at the foot of the slope, as minor fans or tongues of sediment with rich concentrations of artefacts. Superimposed upon the landscape of this process, are the undisturbed remains of occasional bouts of activity that have become incorporated with minimal disturbance. It is therefore postulated that the lithic material represents a combination of slightly disturbed material from activity over a reasonable period of time, in the local area and a little upslope to the west, with almost entirely undisturbed material from activity over the same period of time but at the site itself. One might expect the proposed mass movement to have caused more abrasion to many of the artefacts, but their concentration seems to have been too low for this to have happened; hardly any were found in touching distance of each other, apart from those thought to have broken *in situ*.

Organisation of production and the chaîne opératoire

Various methods were applied to investigate the *chaîne opératoire* and the organisation of production. The direct evidence of refitting was rather unsatisfactory, as no complete sequences of reduction were recovered. Those that were found did not, however, provide direct evidence that any pieces of raw material were collected, worked and abandoned at the spot, in contrast to the evidence from around the elephant skeleton (see Chapter 17). However, when the assemblage was considered as a whole, all avenues of analysis seemed to suggest that this was nonetheless the case. The overall quantities of debitage and negative scars on cores broadly matched each other. The characteristics of the flake assemblage that are most closely linked to stage of reduction (percentage of cortex, and dorsal scar count) broadly corresponded with comparative experimental material representing complete sequences of production, from flaking similar nodular flint cores from start to finish. Likewise, the presence in the assemblages of fairly numerous (about 4%) secondarily worked flakes, mostly interpreted as flake-tools, together with secondary flake-flakes from their working (including some refitting examples, see Figs 18.26 – 18.28) indicate that this technological aspect was also being carried out at the site, or in its near vicinity. Some of the refitting evidence did, however, indicate that the technological *chaîne opératoire* was not wholly expedient, but that there were at least some occasions where, for instance, a flake-tool was re-sharpened, and then exported for use elsewhere (RG #1, Fig 18.27a). The refitting evidence and, in particular, the presence of macro use-wear both in the notches of notched flake-tools and in the secondary flake-flakes from their rejuvenation, also suggests that these notched flake tools were not used and abandoned immediately after being made. Rather it seems that they went through a cycle of use and maintenance before their discard. It is uncertain how extended this

cycle might have been temporally and spatially, and gaining further insight into this should be a key priority for improving understanding of the behaviour and cognitive capabilities of these hominins.

In general, in conjunction with the presumed slightly-transported mode of site formation for the majority of the lithic material, the lithic evidence seems to represent the time-averaged accumulation of a palimpsest of activity on the west side of the water-body, or periodically desiccated plain. This would have been present at the foot of the slope up to the west. There is no spatial structure to the flint knapping/use activity within the excavated area, nor evidence of a spatial organisation to different stages of the technological *chaîne opératoire*. It seems that over the (probably reasonably substantial) period of time represented by the lithic remains, the same overall *chaîne opératoire* has been enacted in slightly different places, depending upon where resources were encountered. This means that a snapshot of one place (such as represented by the excavation) provides a balanced representation of the full *chaîne opératoire* without necessarily containing all the elements of the same individual *chaîne opératoires*. However this apparent lack might reflect the slight lack of integrity of the lithic material, rather reflecting the organisational structure of hominin behaviour.

Finally, the flaking methods and the types of flake-tool represented closely match the characteristics of the lithic assemblages from a number of other horizons at other sites from the same period – ie the early part of MIS 11, the early temperate stage of the Hoxnian interglacial) – identified as 'Clactonian', in particular from (a) the Lower Loam and Lower Gravel at nearby Barnfield Pit (Conway *et al.* 1996), (b) unit 5 (pale grey silt) at Barnham (Ashton *et al.* 1998) and (c) the marl/gravel at the golf course excavation at Clacton-on-Sea (Wymer 1985: 264-284). However, what is of particular importance about the collection represented by assemblages 6.1 and 6.2 is (a) that so much of the material found exhibits such technological and typological consistency and (b) that although some of this material is thought to be entirely undisturbed, the majority of it is thought to be slightly time and space averaged, and therefore represents a sample of stable and consistent material cultural practice from a community of hominins through a sustained time period. There is not one shaped/thinned bifacially-edged tool, let alone one that could be dignified by the term 'handaxe', and nor is there any instance of distinctive debitage from handaxe manufacture. The excavated material from Phase 6 at Southfleet Road, particularly assemblage 6.1, therefore provides the most substantial assemblage, and one with the most suitable taphonomic history, to represent and validate the Clactonian as an early MIS 11 industrial tradition in Britain. This is considered further in the concluding chapter (Chapter 22), in conjunction with the lithic material from Phases 7 and 8 of the site's sequence, discussed in the following chapters (Chapters 19 and 20).

Chapter 19

Lithic artefacts: Phase 7, the syncline infill

by Francis Wenban-Smith

INTRODUCTION

Phase 7 of the site sequence is represented by the series of deposits that conformably overlie the Phase 6 clay. These deposits (see Chapter 4 for a full description and interpretation) are absent in the southern part of the site, where the Phase 6 clay is unconformably truncated by the Phase 8 gravel. However, in the central part of the site the surface of the Phase 6 clay dips northward and at the same time takes on a marked synclinal structure, the 'skateboard ramp', as seen in Trench B (Fig. 19.1a). The hollow of the syncline is filled with a sequence of, from the base: a thin intermittent gravel layer (40164); a yellowish-brown sand/silt body with fine clay-silt laminations (40166); a contorted, chaotically structured clayey gravel (40167) and a slightly gravelly clay (40043). The upper parts of this sequence (40167 and 40043) overlap the western side of the synclinal basin and also extend across the west side of the northern half of the site (Fig. 19.1b). Here, context 40042 was recognised and can be regarded as the lateral equivalent of context 40167. These sediments are thought to mostly be colluvial/slopewash deposits, originating from higher ground to the west.

PROVENANCE AND QUANTIFICATION

In total, 82 flint artefacts were recovered from the Phase 7 deposits (Table 19.1). No hand-excavation took place in this part of the sequence. The deposits were machined away, initially in the area between Trenches B and C to allow investigation of the Phase 6 clay and the elephant area, and subsequently as part of the general ground reduction north of Trench B. As described previously (Chapter 3), when it became clear that there were artefacts in the Phase 7 synclinal infill in Trench B, a 500L bulk sample <40197> was taken from the gravel-rich context 40167 in which they were being found. This was sieved for artefacts on site through a 12mm mesh, which led to the recovery of 19 artefacts. The remainder of the Phase 7 collection was either recovered during the rest of the machining between Trenches B and C, or during the later Watching Brief on the ground reduction north of Trench B. Thus, apart from the assemblage from the bulk sample <40197>, the Phase 7 collection is probably biased towards larger artefacts. Those that were found are liable to have sustained some damage from the mechanical excavator.

ASSEMBLAGE INTEGRITY: CONDITION, STAINING AND PATINATION

The first question to consider for the Phase 7 material is whether there is meaningful stratigraphic integrity between the assemblages from the different contexts. The basal context in the sequence (40164) produced just one artefact, which was a medium-sized chunky flake, dark-grey stained and moderately abraded, suggesting a history of derivation and reworking. The overlying context (40166), which was finer-grained, produced six artefacts, four of them flakes and two of them flake-tools (these latter described in more typological detail below).

Table 19.1 Phase 7, lithic artefacts from the syncline infill: provenance and stratigraphic phasing (excluding natural pieces)

Phase	Context	Artefacts (n)	Context notes
7	40043	4	Greenish-grey, variably gravelly silty clay towards top of syncline infill sequence
	40167	62	Non-fluvial gravel in Trench B, truncated by contexts 40098 and 40048, overlying context 40166; source of bulk sample <40197>, sieved on site for lithic artefacts (Fig. 19.1a)
	40042	9	Gravelly bed between Trenches B and C, see W-facing section 40015 (Fig. 19.1b)
	40166	6	Sand/silt infilling main centre of syncline (Fig. 19.1a)
	40164	1	Gravel bed right at base of syncline infill group (Fig. 19.1a)
Total		82	

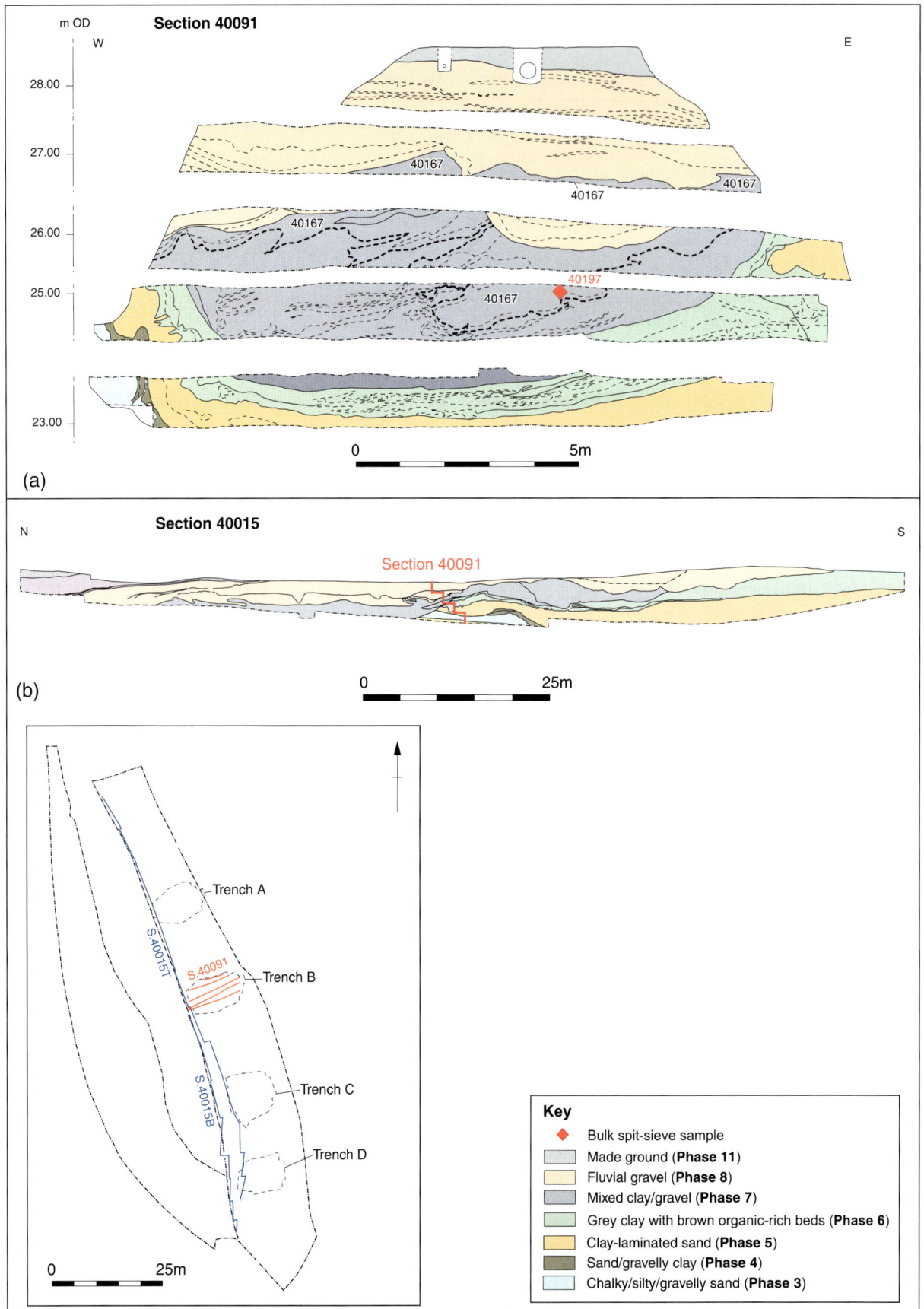

Figure 19.1 Stratigraphic distribution of Phase 7 sediments: (a) Section 40091, Trench B south-facing; (b) Section 40015, main west-facing section

Both of the flake-tools were moderately abraded; one of them is in places quite strongly patinated and the other is strongly stained ochre. Of the four pieces of debitage from context 40166, two are in mint condition and two are moderately abraded; all are very lightly stained with a faint greenish-brown superimposed upon the natural grey of the flint.

The majority of the Phase 7 collection (n=71) comes from 40167 and 40042. In this part of the collection, only eight artefacts (11%) are in mint condition; 24 of them (34%) are in fresh condition; 31 (44%) are in slightly-moderately abraded condition; and eight (11%) are in very abraded condition. Furthermore, they exhibit a wide range of patination and staining, with some being strongly white patinated, some moderately blue-white patinated, and some unpatinated. Likewise, there is the full gamut of staining, from none to deep ochre. The final element of the Phase 7 collection came from 40043, which can be regarded as a gravel-free bed within the upper part of context 40167(=40142). This produced just four artefacts; however they were all at least moderately abraded and one of them was in the extremely abraded category, a rarity in the site's collection.

Overall, the Phase 7 collection shows little sign that its context provenance reflects site formation integrity, especially considering the fact that that all of the contexts are interpreted as colluvial/slopewash deposits originating from the high ground to the west. The impression is given of a mixed assemblage that has been caught up from general artefactual debris in the landscape to the west, some parts of which have suffered more abrasion than others during the deposition process. The fact that the Phase 7 deposits do not erosionally truncate the Phase 6 deposits suggests the Phase 7 collection is not directly derived from the latter. Rather, it is suggested here that they share a common source, with artefacts in the Phase 7 deposits originating from the artefact-rich landscape contemporary with Phase 6 to the west of the site. Therefore there seems little rationale for sub-dividing the Phase 7 assemblage into separate horizons based on their precise strati-graphic context. Consequently it is treated here as a single entity, representing a sample of lithic material culture broadly contemporary with the assemblages from Phase 6, but derived from a slightly wider landscape of occupation to the west of the site.

LITHIC ANALYSIS

Introduction and technological overview

As a whole, the technological and typological profile of the Phase 7 collection (Table 19.2; Fig. 19.2) is indistinguishable from the material from Phase 6. The relative frequency of different technological categories is also broadly similar, with the one exception that there is a much-reduced recovery of irregular waste. This probably reflects that the Phase 6 assemblage was recovered by hand-excavation in fine clay sediment, with every flint clast being carefully scrutinised for signs of working and with recovery taking place if there was any doubt. This led to the recovery of large quantities of irregular waste, much of it exhibiting only frost-fractures, but still interpreted as having been caused by hominin knapping (cf. Chapter 15). In contrast, during machining of the predominantly gravelly Phase 7 deposits, only artefacts that were clearly distinguishable as knapped were spotted and recovered, leading to a reduced recovery of irregular waste. More details of the different technological elements of the Phase 7 collection are given below.

Percussor

The collection included one percussor, Δ.40331 from context 40167, and it was a very fine example (Fig. 19.3a). It consisted of the rounded end of a flattened cylindrical flint nodule, and weighed 560g, heavier than most of the percussors identified at the site. Firstly, there was evidence of at least one heavy impact on the more sharply rounded tip of the piece, which had caused a substantial chip to flake off, ruining the potential of this part of the piece as a percussor. Secondly, on the central high point of the more smoothly rounded dome on the side of the piece, there was a tightly focused area of

Table 19.2 Phase 7 lithics: quantities in artefact categories

	5 – Percussor	*10 – Tested nodule*	*20 – Core*	*30 – Core-on-flake*	*40 – Core-tools*	*50 – Handaxe-on-flake*	*60s – Fl-tools*	*80 – Fl-flakes*	*90 – Flakes*	*100 – Irreg. waste*	*110 – Chips*	*Sub-total (n)*
Abraded	-	1	3	-	-	-	3	1	35	5	-	48
%	-	2.1	6.3	-	-	-	6.3	2.1	72.9	10.4	-	
Fresh	1	-	7	-	-	-	3	2	12	9	-	34
%	2.9	-	20.6	-	-	-	8.8	5.9	35.3	26.5	-	
All	1	1	10	-	-	-	6	3	47	14	-	82
%	1.2	1.2	12.2	-	-	-	7.3	3.7	57.3	17.1	-	

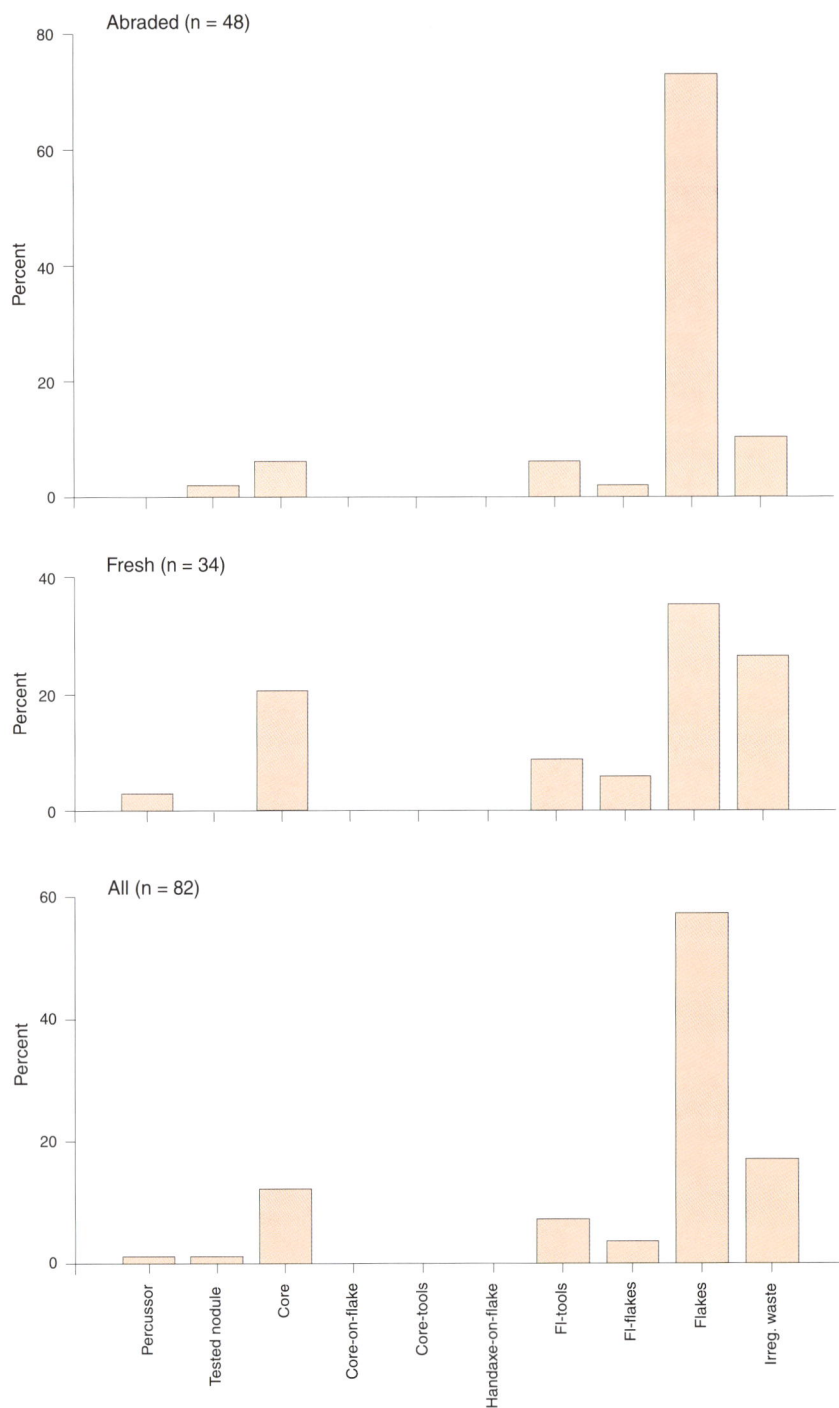

Figure 19.2 Phase 7 artefacts, histogram of technological categories

marks from repeated percussion blows that had completely removed the cortex, leaving a prominent central Hertzian cone. A few small flakes had also been removed from the piece, possibly to facilitate its handling as a percussor.

Cores

Ten cores were found in the Phase 7 deposits, three of them in moderately abraded condition, and the other seven in mint or fresh condition, mostly the latter (Table 19.3). Their dimensions and other key characteristics such as the number of negative flake scars and the remaining amount of cortex are all described in the summary table. The range of flaking approaches represented was indistinguishable from Phase 6, with a combination of generally migrating platforms leading to globular core end-products, and a stronger focus on either one face or a single platform leading to flatter or more angular core products. Four of the cores are shown in the photographic plate (Fig. 19.3b-e): Δ.50023, an angular core that is strongly white-patinated (Fig. 19.3b); Δ.40187, another angular core, greyish-white stained/patinated showing the internal node of dark glossy flint colloquially known as the 'Devil's Eye' (Fig. 19.3c); Δ.40213, a flat core from which flakes have been prefer-

(a)

(b)

(c)

0 ──────────── 100mm

Figure 19.3 Selected Phase 7 artefacts: (a) percussor Δ.40331; (b) core Δ.50023; (c) core Δ.40187

(d)

(e)

(f)

0 _____ 100mm

(g)

(h)

Figure 19.3 *(continued)* Selected Phase 7 artefacts: (d) core Δ.40213; (e) core Δ.40239; (f) miscellaneous flake-tool Δ.40212; (g) flake-knife Δ.50139; (h) single-notch flake-tool Δ.50054

Table 19.3 Phase 7 lithics: cores

Context	Find ID	Cnd	%Cx	DSC	ML	WtG	Notes
40043	Δ.50023	3	2	9	73	144	Angular core; strongly white-patinated [Fig. 19.3b]
40167	Δ.40187	1	1	14	84	235	Angular lump of flint with 'Devil's Eye'; greyish-white stained/patinated [Fig. 19.3c]
	Δ.40213	3	4	13	92	345	Flakes preferentially removed from one face of a flat nodule, using platforms around its perimeter [Fig. 19.3d]
	Δ.40215	2	1	8	84	325	Globular lump; orange-brown stained
	Δ.40238	3	2	7	61	80	Small angular lump
	Δ.40239	2	0	13	51	65	Small remnant globular lump of what was once probably a large core; this piece is the glossy 'Devil's Eye', which normally is surrounded by a mass of more opaque flint [Fig. 19.3e]
	Δ.41649	2	5	8	65	132	Small angular lump, with fair amount cortex remaining
	Δ.50142	2	1	8	102	405	Piece of irregular waste, single platform flaked around perimeter
	Δ.50149	2	3	9	104	265	Slightly flattened lump, alternating platform flaking around part of its perimeter
	Δ.50159	2	0	9	95	217	Angular lump, orange-brown stained

entially removed from one face, using platforms created around the perimeter, which slightly presages later, more organised Levalloisian and radial Mousterian approaches (Fig. 19.3d); and Δ.40239, a small heavily-reduced lump that is the remnant 'Devil's Eye' central node of what was probably originally a quite substantial flint nodule (Fig. 19.3e).

Flake-tools and flake-flakes

Six flake-tools were found in the Phase 7 assemblage, three of them in abraded condition, and three of them in mint or fresh condition (Table 19.4). It is hard to generalise from such a small assemblage as to whether the overall range of tool-types matches the larger Phase 6 flake-tool collection, but none of them would have been out of place in Phase 6, particularly the single-notched flake-tool Δ.50054 and the flake-knife Δ.50139. The relatively heavy abrasion on one of the pieces (Δ.50183) has obscured the possible signs of secondary trimming; and the presumed secondary flake on another (Δ.50145) seems to serve no useful purpose, being relatively flat along the ventral surface, and may be the residual result of a historic fracture-plane from a failed earlier attempt to remove a flake. Key metrical data and other technological attributes are given in the summary table, and three of the flake-tools are shown in the photographic plate: Δ.40212, a naturally segmental lump of irregular waste, which has had its convex (cutting?) edge straightened and slightly sharpened by unifacial trimming (Fig. 19.3f); Δ.50139, a flake which has had secondary flakes removed along one side from both ventral and dorsal surfaces, leaving a straight, sharp edge on the opposite side, which has some tiny invasive

Table 19.4 Phase 7 lithics: flake-tools

Context	Flake-tool type	Find ID	Cnd	%Cx	DSC	ML	WtG	Notes
40167	66 - Misc	Δ.40212	2	8	2	95	258	Segmental lump of irregular waste, with edge slightly straightened by unifacial trimming [Fig. 19.3f]
	62 - Knife	Δ.50139	1	3	3	84	99	Machine-damaged and stained faint greenish/brownish-grey; elongated piece with at least three secondary removals down one side (off both ventral and dorsal surfaces) opposite sharp straight edge that has faint macro use-wear [Fig. 19.3g]
40042	63 - Single notch	Δ.50054	2	8	1	75	116	Medium-size flake with clear single notch [Fig. 19.3h]
	62 - Knife?	Δ.50183	4	0	2	60	51	Blue-white patinated, with straight edge opposed to blunt cortical back; distal end looks like trimmed to facilitate handling as a knife
40166	66 - Misc?	Δ.50145	3	7	1	95	311	Large, thick flake, orange-brown stained, with one removal off ventral surface, possibly due to fracture plane from previous knapping rather than a deliberate secondary removal
	66 - Misc	Δ.50167	3	2	7	86	252	Partly blue-white patinated; thick flake with at least one notch, and possibly two, on one side, and also possibly trimmed at distal end

scaling at one end that looks like macro use-wear (Fig. 19.3g); and Δ.50054, a medium-size flake with a clear single notch (Fig. 19.3h).

In addition to the flake tools, three flake-flakes from secondary working of debitage were recovered, all from context 40167. Two of these (Δ.40231 and Δ.40294) were recovered from the sieved bulk-sample <40197>, and the third (Δ.50146) (a much larger piece of a brutally truncated flake) was recovered from the general monitoring of machine excavation. Smaller debitage such as secondary flake-flakes were clearly under-represented in the assemblage, due to the fact that it mostly results from monitoring of machine excavation rather than sieving or hand-excavation. Nonetheless, the fact that three were found suggests that they might have been relatively abundant.

Debitage

Finally, the bulk of the assemblage from Phase 7 consisted of flakes, of which there were 47 in total, about 60% of them in abraded condition (Table 19.2). Some quantitative statistics were calculated for the unbroken ones (Table 19.5) to investigate how the fresh and abraded elements compared, and how they compared with the Phase 6.1 and 6.2 flake assemblages. It was not possible to meaningfully compare flake sizes between different horizons within the Phase 7 sequence, as there were so few flakes in contexts other than 40167/40042.

The fresh flakes were slightly smaller than the abraded ones, although not statistically significantly, with average dimensions of c 53 x 43 x 18mm, and an average weight of c 70g, as against c 62 x 50 x 20mm, and an average weight of c 100g. There was, however, a more significant contrast with the overall data from assemblages 6.1 and 6.2 (see Table 18.18), where the respective figures are c 40 x 33 x 13mm, and an average weight of 28g. This probably reflects the combination of firstly, the more complete recognition and recovery of smaller debitage during the hand-excavation of the fine-grained Phase 6 clay; and secondly, the formation process of context 40167, from which most of the Phase 7 artefacts were recovered. This is predominantly gravel-rich with frequent flint clasts of small-cobble size, and so

Table 19.5 Phase 7 lithics: size statistics for flakes (only whole flakes included)

Context		Fresh (n=8)				Abraded (n=30)				All (n=38)			
		ML	MW	MT	WtG	ML	MW	MT	WtG	ML	MW	MT	WtG
40043	Max	-	-	-	-	100	65	24	206	100	65	24	206
(n=2, all	Q4/Q3	-	-	-	-	89.5	56.25	21	159.75	89.5	56.25	21	159.75
abraded)	Mean	-	-	-	-	79.00	47.50	18.00	113.50	79.00	47.50	18.00	113.50
	Q2/Q1	-	-	-	-	68.5	38.75	15	67.25	68.5	38.75	15	67.25
	Min	-	-	-	-	58	30	12	21	58	30	12	21
	SD pop	-	-	-	-	21.00	17.50	6.00	92.50	21.00	17.50	6.00	92.50
40042,	Max	100	75	37	252	103	90	45	466	103	90	45	466
40167	Q4/Q3	48.25	47.25	14.25	43	75	66	25	137	75	64	25	120.5
(n=31, 6	Mean	45.67	38.17	15.17	57.67	60.48	49.56	20.20	99.16	57.61	47.35	19.23	91.13
of them	Q2/Q1	31.25	23.25	9	9	41	29	12	30	38.5	28	11	17
fresh)	Min	16	16	7	4	22	15	4	4	16	15	4	4
	SD pop	26.51	19.91	10.12	88.26	21.75	20.57	10.60	103.91	23.49	20.93	10.69	102.39
40166	Max	81	68	29	158	96	80	24	228	96	80	29	228
(n=4, 2	Q4/Q3	77.5	62.8	27.3	136.8	82.0	67.5	20.5	174.5	84.8	71.0	25.3	175.5
of them	Mean	74.0	57.5	25.5	115.5	68.0	55.0	17.0	121.0	71.0	56.3	21.3	118.3
fresh)	Q2/Q1	70.5	52.3	23.8	94.3	54.0	42.5	13.5	67.5	60.3	42.8	19.0	58.3
	Min	67	47	22	73	40	30	10	14	40	30	10	14
	SD pop	7.0	10.5	3.5	42.5	28.0	25.0	7.0	107.0	20.6	19.2	7.0	81.5
40164	Max	-	-	-	-	62	66	37	126	62	66	37	126
(n=1,	Q4/Q3	-	-	-	-	-	-	-	-	-	-	-	-
abraded)	Mean	-	-	-	-	62	66	37	126	62	66	37	126
	Q2/Q1	-	-	-	-	-	-	-	-	-	-	-	-
	Min	-	-	-	-	62	66	37	126	62	66	37	126
	SD pop	-	-	-	-	-	-	-	-	-	-	-	-
All	Max	100	75	37	252	103	90	45	466	103	90	45	466
	Q4/Q3	70.5	55.3	23.8	94.3	78.0	66.0	25.0	160.3	78.0	66.0	25.0	152.8
	Mean	52.75	43.00	17.75	72.13	62.27	50.33	20.40	102.47	60.26	48.79	19.84	96.08
	Q2/Q1	33.8	27.8	11.0	11.0	42.5	30.0	12.0	23.3	41.0	30.0	12.0	18.3
	Min	16	16	7	4	22	15	4	4	16	15	4	4
	SD pop	26.27	19.87	10.00	83.19	22.35	20.62	10.47	101.93	23.55	20.69	10.43	99.06

is likely to have crushed or otherwise dispersed many of the smaller and more lighter weight flint flakes that would probably originally have been present in the source area to the west of the site.

Technologically, none of the flakes exhibit any diagnostic features that reflect any flaking strategy other than those reflected in the cores and known to have occurred in Phase 6. These are, firstly, unstructured migrating platforms, episodes when several flakes are removed in sequence from the same platform before moving onto another and, secondly, episodes of platform alternation, using the scar of one removal as the platform for the next. When features relating to the percussor were preserved (which was quite often), they were always indicative of hard-hammer percussion; there was no evidence of soft-hammer use.

DISCUSSION AND CONCLUSIONS

Although the Phase 7 assemblage is small within the context of the overall artefact collection from the site, and particularly in relation to the substantial collection from Phase 6, it is nonetheless informative in several areas. Firstly, based on the interpretation that the Phase 7 sediments were laid down by colluvial/slopewash processes, and therefore that the artefacts themselves were gathered from an area of probably at least several hectares to the west of the site, they provide a sample of lithic material culture from a wider space/time envelope than the much more abundant material from assemblages 6.1 and 6.2. Nonetheless, the Phase 7 assemblage is indistinguishable technologically and technologically, suggesting a stable lithic technological adaptation over a sustained period of time. Furthermore, although gathered from a wider area and probably representing a more extended time envelope, there is still not a hint of either a bifacial core-tool, or of debitage representing bifacial shaping and/or thinning characteristic of handaxe manufacture.

Secondly, the Phase 7 assemblage contains the same technological categories in similar proportions to assemblages 6.1 and 6.2, apart from debitage, explicable

as due to contrasts in site formation processes and excavation methodology (discussed above, Chapter 19). This suggests that there is a minimum of organisational structure to the *chaîne opératoire* in the part of the landscape represented by the catchment area of the assemblage, and that all stages of production are equally likely to have occurred at any point in the sampled landscape.

Thirdly, even though stratigraphically higher, the Phase 7 assemblage is thought to represent hominin lithic manufacturing activity broadly contemporary with the Phase 6 assemblage. This is an instructive case-study which demonstrates that stratigraphically higher does not necessarily mean chronologically later. A similar situation pertains, for instance, at Boxgrove, where the lithic collection from the thick solifluction gravels (unit 11) that ultimately buried the occupation horizon of unit 4c is probably broadly contemporary with the lithic evidence from this horizon, although derived from a much wider downland landscape above the cliff-line to the north (Roberts and Parfitt 1999).

Fourthly and finally, the fact that a colluvial/slopewash deposit can gather material from the wider landscape, and through the coincidence of the survival of parts of the sedimentary archive and its present-day excavation provide a sample of material in a specific location and at a particular altitudinal level (here, between 24 and 27m OD), does not mean that this height has any significance for integrating the remains into a fluvial terrace chrono-stratigraphic framework. This is despite the fact that these remains were sealed beneath fluvial gravels and could well have been regarded as part of the same fluvial sequence without the wide sedimentary exposures and intensive investigations carried out at the site. This point will be returned to in the concluding discussion (Chapter 22). It is germane to a discussion of the possible post-Hoxnian cyclical recurrence of similar core/flake dominated industries attributed to the Clactonian in MIS 9 at (perhaps), Purfleet (Schreve *et al.* 2002), and in MIS 10/9 (perhaps) at Globe Pit, Little Thurrock (Wymer 1968, 314-317; Bridgland and Harding 1993; Bridgland 1994, 228-236).

Chapter 20

Lithic artefacts: Phase 8, the overlying fluvial gravel

by Francis Wenban-Smith

INTRODUCTION

Phase 8 of the site sequence consisted of fluvial gravels that unconformably truncated the underlying sequence, variously overlying deposits of Phases 5, 6 and 7 in different parts of the site (Fig. 20.1a). The Phase 8 gravels extended across the whole excavated area of the site (see Chapter 4 for a full description), with their base dipping from 27m OD at the southern end to 25.5m at Trench A and also slightly sloping down to the west along the length of the site.

The gravels were divided into three stratigraphic sub-phases, 8a-8c (Table 20.1). The basal Phase 8a was represented by context 40098, and was only seen at the base of Trench A (Fig. 20.1b), which was excavated more deeply than the main sections either side of the site. Overlying this, and unconformably truncating it, was the more extensive gravel bed of context 40047, identified as Phase 8b, which was visible in the central part of the main west-facing section 40015 (Fig. 20.1a). This bed was slightly sandy in places, and wedged out against the synclinal infill deposits of Phase 7, interdigitating with them in places. This bed was then overlain, and unconformably truncated, by the much more extensive main gravel beds of Phase 8c. These extended the full length of the site, and were variously allocated the different context numbers (40014, 40048, 40071 and 40102) in different parts of the site (see Chapter 4, Table 4.3). At the northern end of the site, there were small exposures of higher gravel/sand beds (contexts 40049 and 40050) that conformably overlay context

40048, dipping north; these two contexts were included within Phase 8c, but did not produce any artefacts.

PROVENANCE AND QUANTIFICATION

In total, 184 artefacts were recovered from the Phase 8 gravels (Table 20.1). Two flakes were found during preliminary field investigation of the site in December 2003, loose on an exposed sloping section of the gravel. It was thought likely that they originated from the gravel, but they were not found *in situ* within it, and despite a careful search of the extensive gravel exposures, no flint artefacts were found *in situ* in this phase of work. Therefore there was no expectation at the outset of the main excavation programme that the gravel would be an important source of artefactual material. However, it was nonetheless agreed that the gravel be subject to a systematic sampling programme, with a vertical series of samples of at least 100L taken through the gravels in Trenches A, B, C and D, and sieved on-site for lithic artefacts.

This process commenced in Trench A, and did not prove very productive in the upper levels of the gravel, although a few artefacts were found. However, a fresh condition pointed handaxe was recovered from the bottom sample in the lowest level of the gravel, context 40098 (Phase 8a). At the same time, machine-clearance and hand-cleaning of the main west-facing section in the southern part of the site was leading to recovery of other handaxes from the gravels. Therefore, since it was now

Table 20.1 Phase 8, artefacts from the fluvial gravel (excluding natural pieces): provenance, stratigraphic phasing and condition

Phase	Context	Total artefacts (n)	1 -Mint	2 - Fresh	3 - Sl-mod abraded	4 -Very abraded	5 – Extr. abraded	Notes
8c	40014	2	-	-	1	1	-	
	40048	74	3	18	39	14	-	
	40102	47	10	14	14	9	-	Including one chip
	40071	17	-	6	8	3	-	Including two chips
8b	40047	21	1	7	11	2	-	Including one chip
8a	40098	23	-	14	8	1	-	Basal gravel bed at north end of site, in main W-facing section
Total		184	14	59	81	30	-	

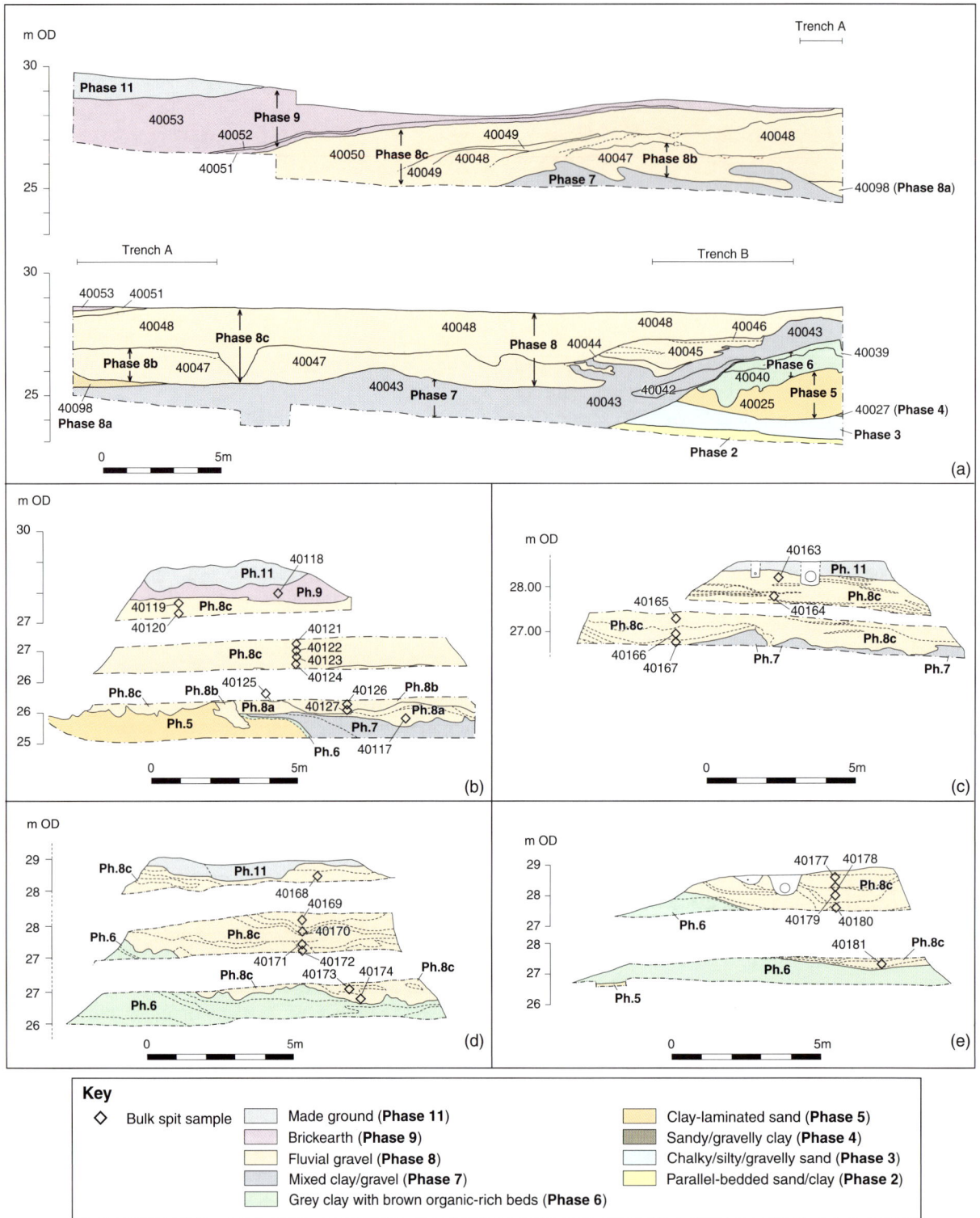

Figure 20.1 (a) Gravel phasing and stratigraphy at north end of west-facing Section 40015; (b) Trench A, stratigraphy and bulk spit sampling; (c) Trench B, stratigraphy and bulk spit sampling; (d) Trench C, stratigraphy and bulk spit sampling; (e) Trench D, stratigraphy and bulk spit sampling

clear that the gravels not only contained flint artefacts, but that they were in fact relatively rich in handaxes, spit sample sizes were increased from 100L to 250L for Trenches B, C and D. Full details of sample sizes and artefact recovery through Trenches A, B, C and D are given in the tabular summary (Table 20.2). The locations within the sequence of all the gravel samples are also showing in the summary trench section diagrams (Fig. 20.1b–e).

In total, 39 artefacts were recovered from the controlled sieve-sampling programme, derived from nearly 5300L of sieved gravel (Table 20.2). This part of the Phase 8 collection therefore represents a complete and unbiased sample of the lithic artefactual content of the gravel. The remainder of the artefact collection from Phase 8 was recovered either directly *in situ* during hand-cleaning of the various sections through the gravel, whether in Trenches A–D or from the main east-facing and west-facing sections of the site, or from monitoring of machine excavation through the gravel. Once it was realised that the gravel contained artefacts, it was removed in shallow machine spits, with both the machine driver and an archaeologist monitoring for discovery of flint artefacts. All artefacts found were recovered, with precise 3-D recording of their position when possible, although several were found in freshly excavated spoil of certain stratigraphic provenance but without their precise locations being known. This recovery process has probably resulted in a bias towards larger and more easily recognisable artefacts, particularly handaxes.

There were two main stages to the machine excavation of the gravel. The first stage was early in the excavation, when it was necessary to strip off the gravel to allow investigation of the underlying Phase 6 deposits. Approximately 95 artefacts were found in the gravel during this initial stage (including the 39 recovered from the systematic sieving program), when work was focused on the central and southern parts of the site; artefacts from this phase of work have find numbers in the sequence starting Δ.40000. The second stage took place after the main excavation was completed, as part of the watching brief on the bulk ground reduction north of Trench B. Approximately 90 artefacts were recovered during this stage, all from the gravels in the northern part of the site, which have find numbers in the sequence starting with Δ.50000.

ASSEMBLAGE INTEGRITY: CONDITION, STAINING AND PATINATION

The different Phases, 8a-8c, of the gravel were divided by unconformable junctions and the whole artefact collec-

tion was reliably provenanced to one of them. Consequently it was decided to maintain these stratigraphic subdivisions for initial analysis, and then to amalgamate the data at a later stage if desired. The summary table of lithic recovery (Table 20.1) also shows the breakdown of artefacts by condition for each context. There is some indication of a progression through the sequence of increasing abrasion, with about 60% of the (relatively small, admittedly) Phase 8a assemblage being in fresh condition, 38% of the (slightly smaller) Phase 8b assemblage being in mint or fresh condition, and 36% of the (relatively large) Phase 8c assemblage being in mint or fresh condition. The majority of artefacts were in slightly/moderately or very abraded condition, although none were in the category of extremely abraded.

The model adopted for site formation in relation to condition is that the condition of an artefact reflects firstly the environment and rapidity of its initial burial, and then the degree to which it is subsequently disturbed and reworked by new channelling within the braided gravel floodplain environment. Thus it is possible that the increasing prevalence of abrasion between Phase 8a and Phase 8b/8c reflects some reworking of material from Phase 8a, when truncated by development of the overlying Phase 8b and 8c. However, it is thought more likely that this mostly represents increased fluvial activity and reworking within Phase 8b and 8c. Therefore no distinction is made in subsequent analyses between parts of assemblages in differing condition, apart from in the discussion of handaxe typology (below) when there is a faint indication that twisted ovate/cordate forms are only found in fresh condition in Phase 8a, in contrast to more pointed forms which are found in fresh condition through Phases 8a–8c. Phase 8a (context 40098) was richest in sand/silt beds reflecting episodes of quieter deposition, which probably helped preservation of any artefacts that were abandoned in this environment. Likewise, the mint and fresh parts of the Phase 8c assemblage mostly came from

Table 20.2 Trenches A-D: artefact recovery from bulk spit-sampling of Phase 8 gravel

Phase	Context	Trench A		Trench B		Trench C		Trench D	
		Vol - lit	Finds	Vol - L	Finds	Vol - L	Finds	Vol - L	Finds
8c	40048, 40102	140	1 x irr w	250	3 x flks	250	None	250	2 x flks
		100	None	250	None	250	None	250	1 x h-axe, 3 x flk
		100?	None	250	None	250	1 x flk	250	None
		100	1 x irr w	250	2 x flks	250	2 x flks	250	1 x h-axe, 3 x flk
		100?	1 x flk	250	None	250	1 x flk	250	None
		100	1 x flk	-	-	250	3 x h-axes	-	-
							1 x fl-tool		
							1 x flk		
		100?	1 x flk	-	-	250	3 x flks	-	-
8b	40047	100	None	-	-	-	-	-	-
		100	1 x chip	-	-	-	-	-	-
8a	40098	100	1 x h-axe	-	-	-	-	-	-
			5 x flks						
Total		1040?	12	1250	5	1750	12	1250	10

irr w - irregular waste; flk - flake; h-axe - handaxe; fl-tool - flake-tool

the southern part of the site, where there were more numerous sand/silt lenses reflecting quieter deposition, perhaps in an environment of exposed gravel bars on an abraded floodplain between quiet channels. The presence of a fair amount of material in absolutely mint condition, particularly in context 40102, indicates rapid burial and a minimum of disturbance. It is likely that this material is probably found where it was discarded, and therefore, although being found in a fluvial gravel environment mostly containing slightly disturbed material, it represents *in situ* evidence of hominin activity on the braided floodplain. The material in fresher condition is often unstained and unpatinated, but the majority of material is at least slightly stained, ranging from slightly yellowish, greyish or greenish, to orange-

brown or strong ochre. A few pieces are also patinated to varying degrees, from partially mottled to completely blue-white, although usually with additional staining.

ANALYSIS

Introduction and technological overview

The quantities of artefacts attributed to different technological categories for each of Phases 8a–8c are tabulated, with both absolute counts and percentages (Table 20.3), although the latter are subject to disproportionate fluctuations due to their low absolute values for the smaller assemblages from Phase 8a and Phase

Table 20.3 Phases 8a-8c: technological categories, excluding chips and natural pieces

	5 - Percussor	10 - Tested nodule	20 - Core	30 - Core-on-flake	40 - Core-tools	50 - Handaxe-on-flake	60s - Fl-tools	80 - Fl-flakes	90 - Flakes	100 - Irreg. waste	Sub-total (n)
Phase 8c	-	4	4	-	23	2	8	-	76	20	137
%	-	2.9	2.9	-	16.8	1.5	5.8	-	55.5	14.6	
Phase 8b	-	3	1	1	2	1	2	-	6	4	20
%	-	15.0	5.0	5.0	10.0	5.0	10.0	-	30.0	20.0	
Phase 8a	-	-	-	-	4	-	1	-	12	6	23
%	-	-	-	-	17.4	-	4.3	-	52.2	26.1	
All	-	7	5	1	29	3	11	-	94	30	180
%	-	3.9	2.8	0.6	16.1	1.7	6.1	-	52.2	16.7	

Table 20.4 Phase 8: cores, two (marked *) also used as percussors

Phase	Context	Find ID	Cnd	%Cx	DSC	ML	WtG	Notes
8c	40102	Δ.40488	1	6	8	80	209	Unstained/unpatinated; pointed end of cylindrical nodule with alternating flake removals at other end; fresh chip from possible use as a percussor on rounded cortical tip
	40048	Δ.50019	2	1	10	57	103	Small broken irregular waste with scars from several removals; loamy adherents suggest possibly intrusive from modern or Late Prehistoric deposits
		Δ.50085	2	5	7	122	966	Large lump; end of sub-cylindrical nodule with alternating flaking around opposing end and part of sides to form sinuous bifacial edge; could perhaps just be failed attempt at a handaxe; fresh chip on rounded cortical end suggests use as a percussor
		Δ.50124	3	5	6	87	206	Small flattish lump from end of a nodule, with a few opposing alternating flakes to leave a flattened bifacial form with a sinuous edge; but no sign of edge-straightening to suggest intention to make a handaxe
8b	40047	Δ.50093	3	2	3	99	443	Unstained; ugly globular frost-fractured chunk with a few removals from opposing ends of one platform; slightly lenticular in profile, could perhaps be central part of broken biface, abandoned early in its reduction
		Δ.50099	2	-	-	-	345	Stained deep ochre; large flake, broken along frost-fracture, then used as core for some medium flakes

8b. Nonetheless, the technological profiles for all three assemblages are broadly similar, consisting mostly of flakes (30–55%), followed by irregular waste (14–26%), handaxes (15–18%), flake-tools (4–10%) and low proportions of the other technological categories, when present. Although no specific percussors were identified, chipping due to impact on the rounded cortical parts of two of the cores (see below) was interpreted as due to their use as knapping percussors. The most notable features of these assemblages are firstly, the presence of a significant proportion of handaxes; and secondly, the high proportion of handaxes relative to the quantity of debitage. This is discussed in more detail further below, in the section on debitage.

Cores and percussors

Six cores were found, individual details of which are given in the accompanying table (Table 20.4); although no separate percussors were identified in their own right in the Phase 8 assemblages, two of the cores (Δ.40488

and Δ.50085) had fresh chips on their rounded cortical protrusions, suggesting short-lived use for knapping.

The cores from Phase 8 were, to be frank, a pretty sorry group. One of them (Δ.50019, from Phase 8c) had greyish-brown loamy adherents, suggesting it was probably a modern or late prehistoric intrusion. As discussed subsequently (Chapter 21), there is evidence of late prehistoric activity and post-medieval flint knapping at the site from deposits directly overlying the Phase 8 gravels, and cut into them in places. Two of them (Δ.50085, from Phase 8c; and Δ.50093, from Phase 8b) could possibly just be broken parts of failed attempts at handaxe manufacture; neither of them showed any sign of removals other than small flakes, and both were broken due to frost fracture and had a crude tendency towards creation of a bifacial perimeter.

The other three cores showed similar technological approaches to knapping reduction as those from Phase 6. Two of them (Δ.40488 and Δ.50124, both from Phase 8c) were small remnants of flint nodules retaining a cortical nub opposed by simple alternating platform flake

Table 20.5 Handaxe (core-tool) and handaxe (on-flake) shape categories

Shape category	Wymer type	Description
0 - Unspecific	-	Indeterminate, eg when broken or unclassifiable to other categories
1 - Rough-out/abandoned	-	Pieces which appear to have been abandoned before completion, for instance because of frost-fracturing, persistent failure to achieve thinning, or breakage
2 - Simple	Proto	Includes McNabb and Ashton's (1992) 'non-classic' handaxes, simple bifacial or unifacial edges opposed to natural handles
31 - Crude pointed (large)	D	Large (≥100mm) pointed/sub-pointed biface, no soft-hammer, thick, wavy edges, thicker and heavier at butt
32 - Crude pointed (small)	E	Small (<100mm) pointed/sub-pointed biface, no soft-hammer, thicker and heavier butt, thick, wavy edges
4 - Classic pointed	F	Well-made pointed handaxe with clear butt, straightish sides and thinned towards tip, can be any size; butt can be unworked or crudely worked
50 - Sub-cordate	G	Progression from type F with convex sides, often more rounded point, thick/heavy butt, widest part of handaxe well towards butt; butt can be unworked, crudely worked
51 - Sub-cordate (plano-convex)	G	Similar to above but with clear plano-convex profile, cf. Wolvercote Channel
52 - Sub-cordate (twisted)	-	Sub-cordate plan shape, but tip distinctly twisted relative to butt
60 - Sub-ovate	GK	Much more ovate version of sub-cordate; tip is smoothly rounded without any well-defined point, widest part of handaxe is nearer middle of long axis, clear working to shape/thin butt and sides as convex curve, although not as much as for true ovate or cordate
7 - Cordate	J	Cutting edge all round tool with thinning and shaping around butt, centre of gravity near middle, bit more rounded than sub-cordate, but still has clear tip, with widest part of handaxe towards butt
71 - Twisted cordate	-	Cordate plan shape, but tip distinctly twisted relative to butt
80 - Ovate	K	Cutting edge and thinning/shaping all round, centre of gravity near middle, more rounded at base than cordate with widest part of handaxe towards middle, usually one end recognisable as tip by being more elongated from widest part of handaxe and often tranchet sharpened
81 - Twisted ovate	K	Ditto above, but clearly twisted tip
9 - Side-chopper	L	Segmental chopping tool, one knapped bifacial edge or sharper edge opposed by flat edge or natural backing; crucial distinction with cleaver is that business edge is parallel with main longitudinal axis rather than transverse
10 - Classic ficron	M	Very pointed with symmetrical concave sides and well-defined heavy butt, cf. Furze Platt, Cuxton (Wenban-Smith 2004)
11 - *Bout coupé*	N	Flat-butted cordate, trimmed all round butt, but with distinct corners between gently convex base and sides
12 - Cleaver	H	Key characteristic is straight cutting edge at tip end, transverse to main longitudinal orientation of tool, cf. Cuxton (Wenban-Smith 2004)

removals. And the final core (Δ.50099, from Phase 8b) was a very thick flake, which appeared to have broken along a pre-existing frost fracture during knapping, and then to have subsequently had a few further medium-size flakes removed, using the frost fracture as a striking platform.

Handaxes

In contrast to the cores, the Phase 8 deposits produced a fine collection of over 30 handaxes (Table 20.6), all of them stratigraphically provenanced to specific contexts within the Phase 8 gravels, and most of them with their precise location surveyed. The handaxe collection (including handaxes made on flakes) was classified by shape broadly following the groups established by Wymer in his analysis of Swanscombe handaxes (1968, 45-68). The main difference between the shape categories applied in this analysis (Table 20.5) and Wymer's typology is the recognition of twisted-profile versions of ovates and cordates as distinct types (shape categories 52 and 81) for ease of tabulation. Handaxes are also grouped into 'pointed' and 'ovate' forms; as can be seen, the majority of handaxes are in the pointed categories, although about 15% of the assemblage is of more ovate character, all except one of which have twisted profiles.

Considering that Phase 8a was restricted in extent, it provided a relatively substantial collection of artefacts, including four handaxes. All are in very fresh condition and two of them are illustrated (Δ.50055, a strongly twisted sub-cordate, Fig. 20.2a; and Δ.40106, a quite crudely made, but sharply pointed form, Fig. 20.2b). The twisted specimen had its tip twisted anticlockwise relative to its butt, when viewed from above, creating a Z-shaped rather than an S-shaped side profile. The other two handaxes from Phase 8a were both broken; one of them (Δ.40067) was entirely unclassifiable, but the other retained a sharp pointed tip, so was classified as a 'classic pointed' form (type 4).

Phase 8b was less restricted in extent than Phase 8a, but produced a roughly similar size assemblage, of both handaxes and other artefacts, suggesting that artefacts were generally less abundant within it (although there are unfortunately no precise quantitative data of sediment volumes to support this observation). It produced three bifacial core tools (not illustrated), one of which (Δ.50169, in fresh condition) had a long blunt cortical edge opposed to a bifacially flaked edge and so was classified as a side-chopper (type 9). The other two handaxes from Phase 8b are both in moderately abraded condition. One of them (Δ.50031) is a straight-sided, convex-pointed and thick-butted sub-cordate form (type 50); the other (Δ.50095) is very similar, although its butt was more trimmed and thinned, and it has a twisted Z-profile. However, it was not classified as a twisted type because of its general crudity and the fact that various step and hinge fractures on its faces reflect problems in its manufacture. It is therefore not possible to be confident that its twist was deliberate, rather than being the accidental outcome of thinning/shaping failures.

By far the largest assemblage of the Phase 8 collection was that from Phase 8c, from which 25 handaxes were recovered (Table 20.6). Deposits of Phase 8c were extensive across the site. No controlled record was made of the volume of excavated/investigated sediment compared with Phases 8a and 8b. Therefore it is not possible to quantify the relative abundance of artefacts in these phases; however, it is estimated that they were generally slightly less abundant in Phase 8c, and that the larger quantity of lithic material recovered is due to the much larger volume of sediment investigated.

The Phase 8c assemblage includes a wide diversity of handaxe shapes, including both pointed and ovate forms, a representative selection of which are illustrated (Fig. 20.3). The illustrated specimens are: an absolutely mint condition, unstained/patinated sharply pointed handaxe with slightly concave sides and very slightly plano-convex in side profile, which was classified as a ficron (Δ.40480, Fig. 20.3a); a thick-butted pointed form in fresh condition which was classified as type 4 'classic pointed' (Δ.40057, Fig. 20.3b); a bluntly pointed form with a thick, flat butt, with convex sides in plan-view and a very straight side profile, which was classified as sub-cordate (Δ.50008, Fig. 20.3c); and a bluntly pointed cordate form, trimmed all around the butt, which had a distinct Z-twisted profile and so was classified as a twisted cordate, type 71 (Δ.50071, Fig. 20.3d).

Table 20.6 Phases 8a-8c: handaxes and handaxes-on-flakes, shape categories

	Unspecific			Pointed					Ovate				Illustrations
	0 – Unspecific	*1 – Rough-out, abandoned*	*9 – Side-chopper*	*31 – Crudely pointed (large)*	*32 – Crudely pointed (small)*	*4 – Classic pointed*	*10 – Ficron*	*50 – Sub-cordate*	*7 – Cordate*	*71 – Twisted cordate*	*81 – Twisted ovate*	*Total (n)*	
Phase 8c	-	4	-	2	4	*5	*1	*5	1	*1	2	25	Fig. 20.3
Phase 8b	-	-	1	-	-	-	-	2	-	-	-	3	-
Phase 8a	1	-	-	*1	-	1	-	-	-	*1	-	4	Fig. 20.2
All	1	4	1	3	4	6	1	7	1	2	2	32	

* illustrated examples

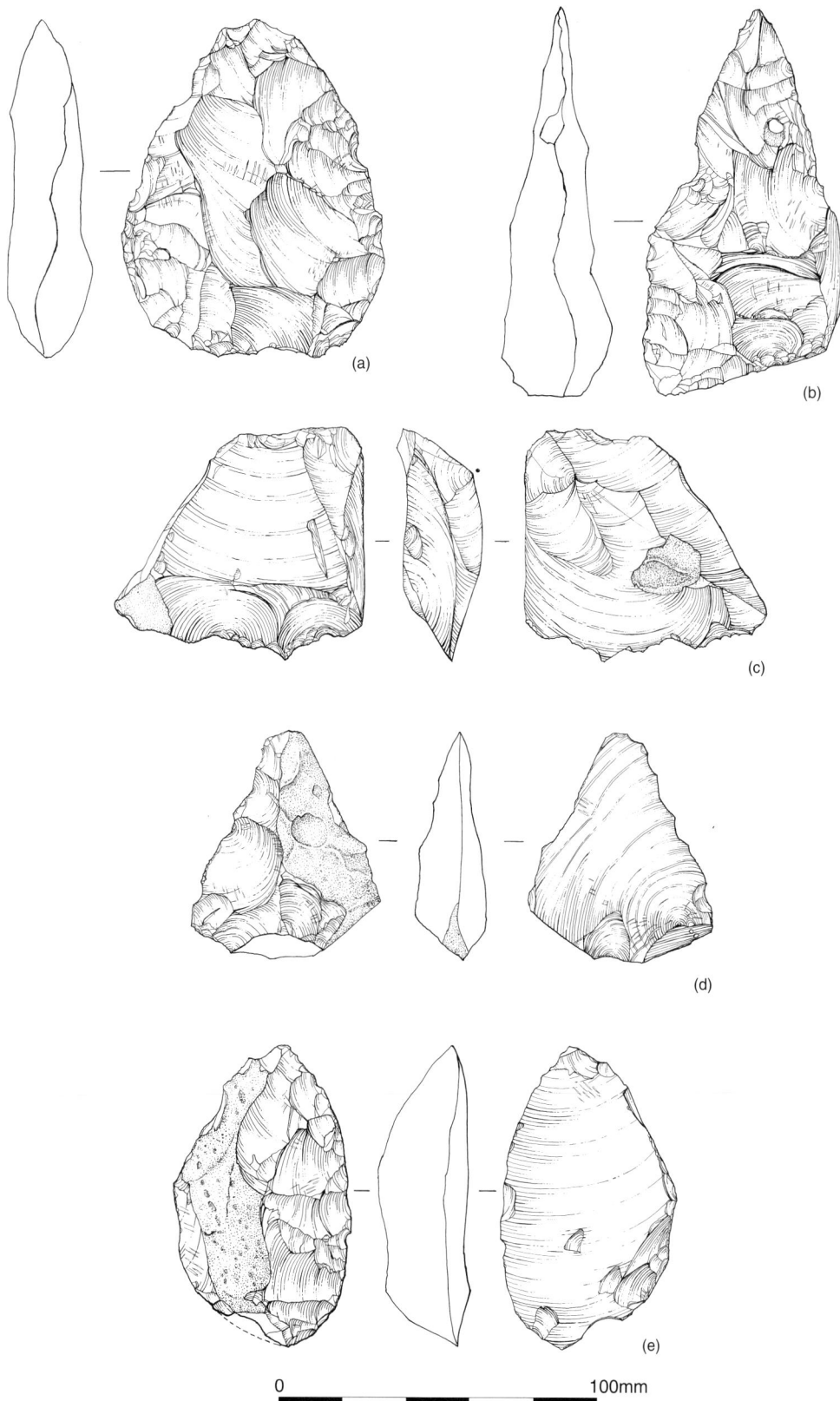

Figure 20.2 Phases 8a and 8b, handaxes and flake-tools: (a) Phase 8a – twisted cordate handaxe Δ.50055; (b) Phase 8a – pointed handaxe Δ.40106; (c) Phase 8a – double/linear notched flake-tool Δ.50030; (d) Phase 8b – double/linear notched flake-tool Δ.50022; (e) Phase 8b – miscellaneous flake-tool, 'Quina-type' scraper Δ.50113 [ill. B. McNee]

In general, the Phase 8 handaxe collection shows great diversity, with pointed and ovate forms both being present from top to bottom of the sequence. There are, however, some faint trends in the relationship of shape to degree of abrasion that hint at a subtler pattern. In Phase 8a, all of the handaxes are in fresh condition, and they include a combination of pointed and ovate forms, suggesting both types of handaxe shape were being manufactured in the close vicinity contemporary with deposition of the basal fluvial gravel bed. The evidence from Phase 8b is too limited to contribute to this discussion, with one atypical form in fresh condition and two sub-cordate forms in abraded condition. In Phase 8c, of the twelve straight-sided pointed forms (types 10, 31, 32

and 4), eight of them are in mint or fresh condition and four of them are in abraded condition. Of the five sub-cordate forms (type 50), two of them are in fresh condition and the other three are in abraded condition; of the four ovate forms (types 7, 71 and 81), all of them are in abraded condition. This suggests that there is at least some manufacture of pointed forms contemporary with deposition of the Phase 8c gravel beds, but that with all of the ovate forms being abraded, there is the possibility that they were reworked from earlier sediments rather than contemporary with Phase 8c.

Although (a) the numbers are too low for statistical significance and (b) there is also uncertainty over whether the degree of abrasion reflects reworking from

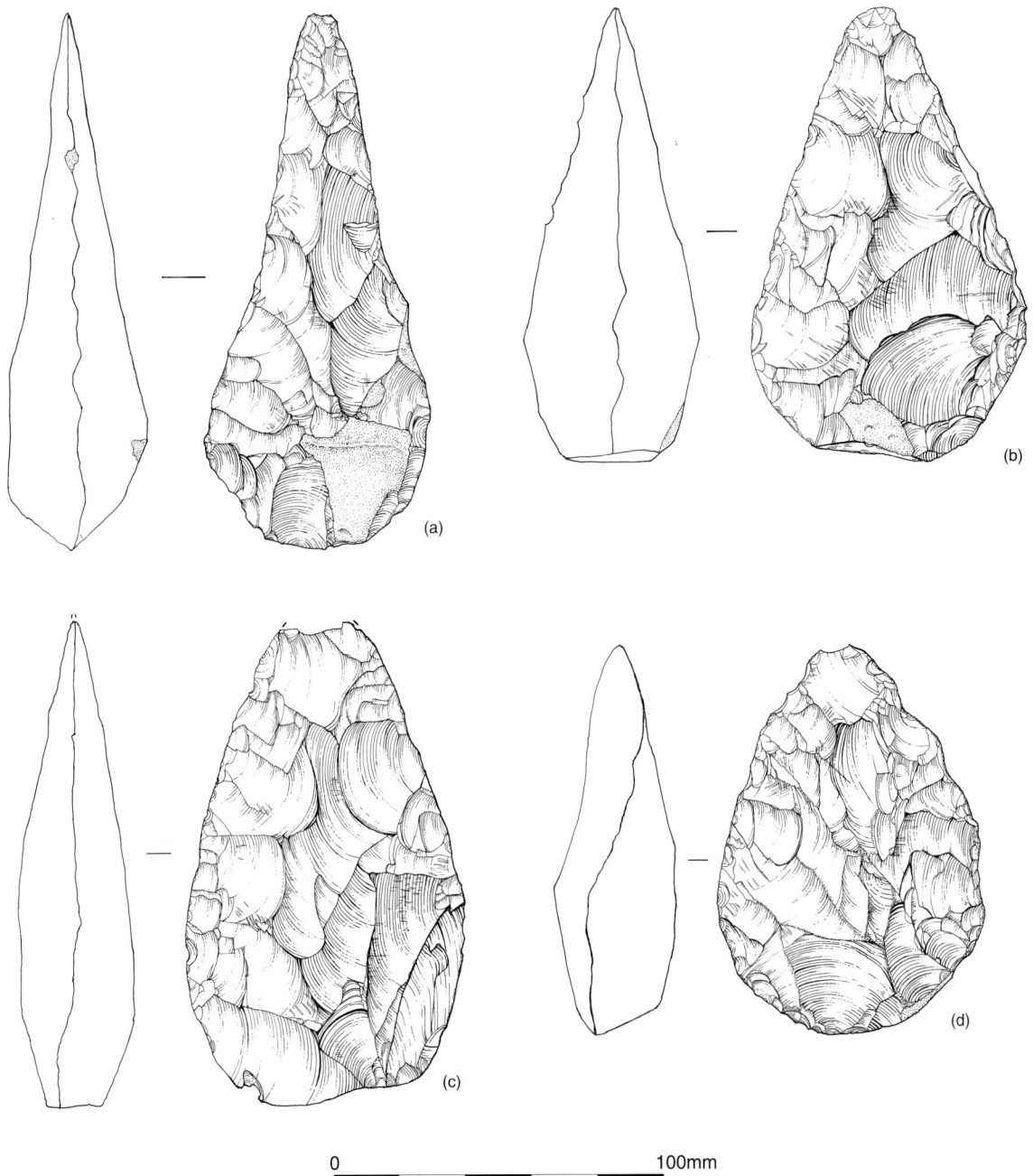

Figure 20.3 Phase 8c, handaxes: (a) ficron Δ.40480; (b) classic pointed Δ.40057; (c) sub-cordate Δ.50008; (d) twisted-profile cordate Δ.50071 [ill. B. McNee]

earlier gravel beds or whether it is merely due to the haphazard chance of varying degrees of reworking within the same gravel bed, it is possible that these trends reflect a decline in the manufacture of ovate handaxe forms upwards through the sequence. This suggestion is easily reinforced or falsified by further investigations; even though the gravel at the site was mostly removed in course of the archaeological investigation, substantial parts of it survive immediately to the north of the investigated area where it could be researched further.

Flake-tools

Complementing the core-tool element of the Phase 8 assemblage, there were also several (n=11) flake-tools recovered, most of them from Phase 8c (Table 20.7). It was not always possible to clearly differentiate some of these from simple handaxes when they had a pointed shape in plan. The interpretation here of two ambiguous specimens (Δ.50030, Fig. 20.2c; and Δ.50022, Fig. 20.2d) as linear notched flake-tools rather than handaxes on flakes was based upon the regularity of the secondary flaking, the fact that it was only carried out on one edge and one face of the parent flake, and because there was no attempt to straighten the secondarily flaked edge with more delicate flaking away of the ridges between the secondary flake scars.

The only flake-tool found in Phase 8a was a flake with two notches side-by-side transversely across its distal end (Δ.50030, Fig. 20.2c). This piece was in abraded condition, and was of similar type to several of the double/linear notched flake-tools from Phase 6. Hence it is possible that it was derived, as the Phase 8 gravels unconformably truncate the Phase 6 deposits and so must incorporate a reasonable proportion of reworked artefacts from them.

Two flake-tools were found in Phase 8b, both of which are illustrated (Fig. 20.2). One of them (Δ.50022, Fig. 20.2d) was very similar to the double/linear notched flake-tool from Phase 8a, and was likewise abraded, so could equally be derived from the Phase 6 clay. The other (Δ.50113, Fig. 20.2e) was quite different, and in fresh condition, making it less likely that it was derived. If found in a cave in south-west France it would probably be classified as a Mousterian 'Quina-type' side-scraper (Bordes 1979); one side of quite a large, thick flake has been carefully and methodically flaked with very numerous secondary flake removals to leave a steep, scaled convex edge. Nothing like this was present amongst the large selection of flake tools in the lower Phases 3, 5 and 6 at the site, all of which contained a range of much more simply made flake-tools with less-structured secondary flaking.

In Phase 8c, eight flake-tools were found, a representative selection of three of which are illustrated here (Fig. 20.4). Two utilised flakes were recovered, one described in the tabular summary (Δ.50047, Table 20.7) and the other both described and illustrated (Δ.50016, Fig. 20.4a). The latter also has some possible secondary flake-scars that could represent trimming of the opposing

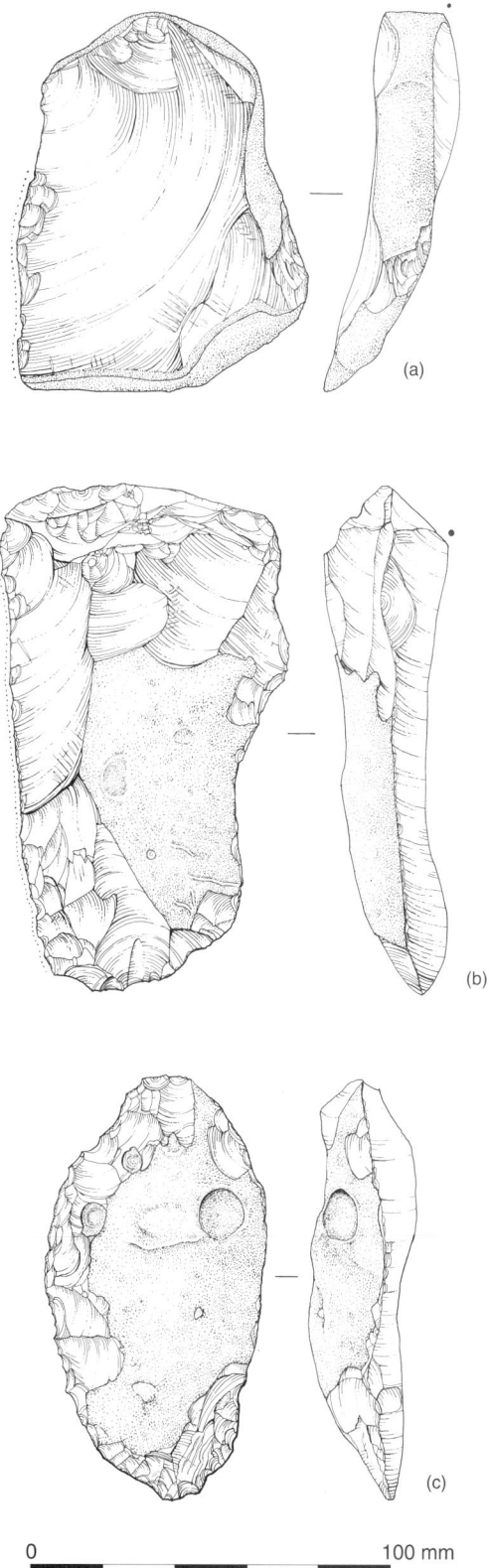

Figure 20.4 Phase 8c, flake-tools: (a) utilised flake Δ.50016; (b) 'flake-knife' Δ.50007; (c) miscellaneous flake-tool, 'Quina-type' scraper Δ.50003 [ill. B. McNee]

Table 20.7 Phase 8: flake-tools

Phase	Flake-tool type	Find ID	Whl	Cnd	%Cx	DSC	ML	WtG	Notes
8c	61 – Utilised flake	Δ.50016	1	2	3	2	94	202	Large flake with long straight edge showing clear macro use-wear; also some minor possible secondary flaking to facilitate handling [Fig. 20.4a]
		Δ.50047	1	2	-	-	125	495	Huge, thick flake, broken down middle with one heavily battered/chipped straight edge, some of which also may be invasive secondary flaking
	62 – Flake-knife	Δ.40135	1	4	0	3	87	115	Heavy abrasion obscures possible secondary flaking, but row of small flakes across one blunt edge opposite a straight and sharp edge suggests deliberate trimming to facilitate handling as a knife
		Δ.40145	1	2	0	9	74	110	Distal edge and one side of large bifacial thinning flake seem to have been slightly trimmed, leaving one sharp edge that could have been used for cutting
		Δ.50007	1	2	4	6	118	292	Large flake with natural straight sharp edge down two-thirds of one side, with the distal part of the edge trimmed to straighten it; clear macro use-wear [Fig. 20.4b]
		Δ.50092	-	4	5	0	62	55	Segmental scrap of flint with thick cortical back opposing sharp straight edge; several small flakes removed across cortical back to regularise its convex curve
	66 – Miscellaneous	Δ.40493	4	2	3	2	72	99	Unpatinated, slightly greenish-stained flake with numerous secondary removals, including: scalar 'scraper' retouch; invasive thinning across ventral surface; and two notches struck on ventral surface
		Δ.50003	1	2	9	0	98	200	'Mousterian Quina-type' side-scraper [Fig. 20.4c]
8b	64 – Linear/double notch	Δ.50022	1	3	5	2	61	68	Shape looks like small pointed handaxe in plan, but the only secondary flaking is notching along straight edge opposing cortical back [Fig. 20.2d]
	66 – Miscellaneous.	Δ.50113	1	2	-	-	90	144	'Mousterian Quina-type' side-scraper [Fig. 20.2e]
8a	64 – Linear/double notch	Δ.50030	20	3	2	3	91	136	Medium-size flake with two notches side-by-side across distal end [Fig. 20.2c]

blunt cortical edge. It also has very clear scaling that is probably macro use-wear on the straight edge, although it could also perhaps reflect deliberate fine-flaking of the edge to make a straight and finely denticulated cutting edge. Four flake-knives were recovered, one of which is described and illustrated (Δ.50007; Table 20.7 and Fig. 20.4b). The others are described in more detail in the accompanying table. Finally, there were also two miscellaneous flake-tools from Phase 8c. One of these (Δ.50003, Fig. 20.4c) was another Mousterian 'Quina-type' scraper, virtually identical to that from Phase 8a. The other (Δ.40493, not illustrated but described in Table 20.7) combined the remnants of what might once have been a thick convex scraping edge with two much more invasive secondary flakes that have left two deep notches in the ventral surface, orthogonal to each other.

There are too few flake-tools to consider whether they are any trends through Phases 8a to 8c. The two

double/linear notched flake-tools may be derived from the Phase 6 deposits. Besides these, there is a significant element of utilised flakes and flake-knives, tool-types that exhibit a considerable overlap, with the only difference being recognition of secondary flaking to facilitate handling of the tool for cutting. These tool-types also occur in Phase 6, although the specimens found in the Phase 8 gravels are often bigger than any from Phase 6, and the fresh condition of the majority indicates their contemporaneity with the gravel. The most interesting type of flake-tool recovered was the 'Quina-type' scrapers, two of which were found, both almost identical to each other, Δ.50030 from Phase 8a (Fig. 20.2c), and v.50003 from Phase 8c (Fig. 20.4c). The similarity of these tools, and the carefully positioned secondary flaking suggest that these tools were deliberately conceived following a mental template. As such, they provide key points of contrast with the flake-tools from

Phase 6, in the intensity of secondary flaking that has shaped them, in their final form, and in the suggestion that this final form was pre-conceived as a type.

Debitage

The largest element of the Phase 8 collection was waste debitage, with almost 100 pieces recovered (Table 20.3): 12 from Phase 8a (five of them mint/fresh), 6 from Phase 8b (one of them fresh), and 76 from Phase 8c (20 of them fresh). Some basic size statistics were calculated for these assemblages, using only the whole flakes (Table 20.8). These were compared with size statistics for 210 flakes from complete reduction sequences of three experimental handaxes of similar pointed/sub-cordate shape to the majority of those in the Phase 8 collection (data from Wenban-Smith 1996). The size data from Phases 8a and 8c were remarkably similar, with average size of *c* 55mm long x 40-45mm wide x *c* 15mm thick, and an average weight of *c* 65-80g. Abraded flakes in both these assemblages were very slightly larger, on

average. Flakes in Phase 8b were much smaller, but their low quantity makes it inappropriate to generalise.

In general the flakes from Phase 8 were slightly smaller than the comparative experimental data. It is likely that the smaller elements of the archaeological material were either present but not recovered, or had been winnowed out of the gravel by fluvial action, or perhaps both these factors have occurred. Most of the excavated debitage was technologically undiagnostic, but there were also a reasonable number that were clearly identifiable as from handaxe manufacture, including one tranchet-thinning flake (Δ.50133, not illustrated). The distribution of whole flakes in different size ranges was recorded, and also for each size-range the proportion of flakes that were identifiable as from handaxe manufacture (Table 20.9). These data were compared with the experimental model for complete reduction sequences. Rather than exclude broken debitage in the excavated collection that was clearly from handaxe manufacture, the likely lengths of the broken pieces were estimated.

Table 20.8 Phase 8: size and weight statistics for flakes (only whole pieces included), compared with combined experimental dataset from manufacture of three handaxes (from Wenban-Smith 1996)

Phase		Fresh (n=16)				Abraded (n=44)				All (n=60)			
		ML	MW	MT	WtG	ML	MW	MT	WtG	ML	MW	MT	WtG
8c (n=47,	Max	87	80	33	381	119	105	75	328	119	105	75	381
13 of them	Q4/Q3	71.0	59.0	15.3	109.0	73.3	58.0	22.0	93.0	72.0	59.0	22.0	97.5
fresh)	Mean	54.2	39.7	13.3	66.8	57.5	45.9	18.0	68.1	56.6	44.1	16.8	67.8
	Q2/Q1	39	24	7.75	9	36.75	30	11	16.25	37.5	26.5	9	13.5
	Min	28	16	3	3	21	10	5	3	21	10	3	3
	SD pop	19.4	20.4	9.6	103.5	24.4	21.0	12.9	79.2	23.2	21.0	12.3	86.6
8b (n=5,	Max	37	31	4	6	40	36	15	24	40	36	15	24
1 of them	Q4/Q3	-	-	-	-	37.8	34.5	15.0	21.8	37.0	34.0	15.0	21.0
fresh)	Mean	37	31	4	6	34.5	30.8	12.8	17.0	35.0	30.8	11.0	14.8
	Q2/Q1	-	-	-	-	32.75	29.75	12.3	14.75	35	31	7	6
	Min	37	31	4	6	26	20	7	5	26	20	4	5
	SD pop	-	-	-	-	5.2	6.3	3.3	7.2	4.8	5.6	4.6	7.8
8a (n=8,	Max	81	58	16	77	85	75	40	411	85	75	40	411
2 of them	Q4/Q3	66.8	49.8	14.3	58.8	64.0	44.3	17.8	53.3	72.8	48.3	16.8	65.8
fresh)	Mean	52.5	41.5	12.5	40.5	53.5	41.7	15.8	89.5	53.3	41.6	14.9	77.3
	Q2/Q1	38.25	33.25	10.8	22.25	41.25	29.75	7.38	14.25	38.75	28.5	8.38	11.25
	Min	24	25	9	4	35	27	5	6	24	25	5	4
	SD pop	28.5	16.5	3.5	36.5	17.9	16.3	11.8	144.9	21.0	16.3	10.5	128.6
All	Max	87	80	33	381	119	105	75	411	119	105	75	411
	Q4/Q3	71.5	58.3	14.5	85.0	70.3	55.0	21.3	68.3	71.0	55.8	20.0	78.8
	Mean	52.9	39.4	12.6	59.8	54.8	43.9	17.2	66.4	54.3	42.7	16.1	64.6
	Q2/Q1	35.5	24	7.5	8	35.75	29.75	10	15.5	35.75	27	9	12.25
	Min	24	16	3	3	21	10	5	3	21	10	3	3
	SD pop	20.6	19.4	9.0	95.6	23.4	20.0	12.3	89.5	22.7	19.9	11.7	91.2
Exp data	Max	-	-	-	-	-	-	-	-	122	109	47	-
(n=210)	Q4/Q3	-	-	-	-	-	-	-	-	59.5	50.75	14	-
	Mean	-	-	-	-	-	-	-	-	46.67	39.44	10.39	-
	Q2/Q1	-	-	-	-	-	-	-	-	30	27	5	-
	Min	-	-	-	-	-	-	-	-	2	10	1	-
	SD pop	-	-	-	-	-	-	-	-	21.91	18.88	8.33	-

Table 20.9 Phase 8: identifiability of debitage from handaxe manufacture

Flake-length (mm)	Exp. data		Phase 8a		Phase 8b		Phase 8c		All Phase 8	
	Total	%HA	Total	%HA	Total	%HA	Total	%HA	Total	%HA
20-40	90	24	2	50	4	0	14	7	20	10
40-60	62	30	⋆ 4	25	1	0	⋆⋆ 15	13	20	15
60-80	33	12	2	0	-	-	⋆ 12	33	14	29
80-100	14	21	⋆ 2	20	-	-	⋆⋆ 9	30	11	36
>100	6	0	-	-	-	-	2	0	2	0

Total = total number of flakes in size-range; %HA = the percentage of them that are identifiable as from handaxe manufacture

⋆ Includes estimated length of one broken flake recognisable as from handaxe manufacture

⋆⋆ Includes estimated length of two broken flakes recognisable as from handaxe manufacture

This analysis demonstrates at least three points of interest. Firstly, the size distribution profile of the archaeological material is quite different to that of the experimental model, with a marked absence of material less than about 50mm long. Secondly, of the archaeological debitage that was present, the smaller pieces are less likely to be identifiable as from handaxe manufacture than in the experimental model; whereas conversely, the large pieces are (slightly) more likely to be identifiable as such. This is perhaps an indication that later stages of handaxe reduction are not typically represented in the smaller debitage, as the experimental work clearly showed that smaller and more identifiable pieces were much more commonly produced in the end stages of handaxe manufacture (Wenban-Smith 1996). Thirdly, whatever the finer interpretation of variations in the percentage of identifiability for debitage of different size ranges, the key take-home message must be that there is a near-ubiquitous presence (apart from in the small assemblage of five whole flakes from Phase 8b) of a reasonable proportion of flakes identifiable as from handaxe manufacture. Disregarding the size of flakes and whether they are broken or whole, out of the whole debitage collection of 94 flakes (26 of them in fresh condition), thirteen of them were clearly identifiable as from handaxe manufacture (two of which are in fresh condition).

The final analysis carried out on the debitage collection was a more detailed attempt to model the expected composition of the assemblage if the debitage from the handaxes found was fully represented in the recovered collection. As pointed out in the interim report (Wenban-Smith *et al.* 2006) it is instantly clear to anyone with some experience of handaxe manufacturing that the proportion of debitage recovered was a gross under-representation of what would be expected from full reduction sequences of the handaxes found. This general assertion is supported here by quantitative data derived from experimental work. Numerous experiments have consistently demonstrated that manufacture of a handaxe typically leads to around 50-70 flakes ≥ 20mm being produced (Newcomer 1971; Wenban-Smith 1996 and 2004b). A crude estimate of the expected debitage composition in different size-ranges was then calculated, based on the three experiments reported in Wenban-Smith (1996) and assuming (which we already know to be wrong) complete representation of all categories of debitage size. The results of this model (Table 20.10), even allowing for the proven under-representation of smaller debitage, emphasise the staggeringly great under-representation of debitage in relation to the number of handaxes found. Another recognised source of error in the model is that it is based on the manufacture of sub-cordate and thick-butted handaxes, whereas the archaeological collection also includes cordate and ovate forms. However, these would probably have produced even greater numbers of more recognisable debitage, so their absence only serves to emphasise the disproportionate under-representation of debitage in the archae-

Table 20.10 Phase 8: expected quantities of debitage from handaxe manufacture in Phase 8 assemblages

Flake-length (mm)	Phase 8a Handaxes=4		Phase 8b Handaxes=3		Phase 8c Handaxes=25		All Phase 8 Handaxes=32	
	Exp.⋆	Act.	Exp.⋆	Act.	Exp⋆	Act.	Exp.⋆	Act.
20-40	80	2	75	4	465	14	614	20
40-60	56	4	51	1	321	15	423	20
60-80	29	2	27	-	171	12	225	14
80-100	12	2	12	-	73	9	95	11
>100	5	-	5	-	31	2	41	2

Exp. = expected number of flakes in size-ranges based on experimental model (data from Wenban-Smith 1996)

Act. = the actual quantities of debitage recovered in each size-range

⋆ Expected flakes model reduced to match proportions of broken flakes of uncertain length in archaeological assemblages: Phase 8a, by 33%; Phase 8b, by 17%; Phase 8c, by 38%; Phase 8, by 36%.

ological assemblages. This point is even further reinforced by the fact that they probably include some debitage from non-handaxe reduction, based on the presence of some cores.

Consequently, two important points seem very clear. Firstly, even allowing for some bias in the non-recovery of debitage, there is a spatially organised pattern for the technological *chaîne opératoire* in relation to handaxe manufacture that reflects raw material collection and handaxe-manufacture *elsewhere* in the landscape, beyond the catchment of the Phase 8 gravels, and then their movement to, and abandonment at, the site, or at least within the catchment of the Phase 8 deposits. Secondly, even when handaxe manufacture is apparently not taking place at the spot (at least nowhere near to the degree that matches the number of handaxes found) there is still a significant representation of handaxe-manufacturing debitage in the lithic collection. This would in turn suggest that where there is a consistent absence of any debitage evidence for handaxe manufacture in a fluvial context, such as in the Lower Gravel at Barnfield Pit, one can be confident about the lack of handaxe manufacture not just in the close vicinity, but in the much wider neighbourhood.

DISCUSSION AND CONCLUSIONS

The Phase 8 gravels represent a return to fluvial conditions after the interlude of swampy stagnation and slopewash deposition represented by the Phase 6 deposits, and the more substantial slopewash and mass slope movement represented by Phase 7. Although the date, climate and local environment of Phase 6 are reliably established to the early temperate sub-stage Ho II of the first, main part of the Hoxnian interglacial (MIS 11c), the climate and environment associated with the overlying deposits of Phase 7 and Phase 8 are entirely unknown as are the time periods represented by firstly, the transition from one depositional phase to the next, and secondly, the deposition of the sediments associated with Phases 7 and 8 respectively.

The Phase 8 gravels are thought to most likely also date to MIS 11, on the basis primarily of their geomorphological relationship to the Lower Middle Gravels of the Swanscombe 100-ft terrace (see Chapter 4). However, both this dating, and also a specific correlation with the Lower Middle Gravel, are by no means certain. The gravels could represent quite a different stage within the grand sweep of the later parts of MIS 11, which include both a stadial of arctic cold and a return to interglacial conditions, as represented at Hoxne (Ashton *et al.* 2008). Alternatively, they could perhaps represent a post-MIS 11 phase of deposition.

Whatever their precise date, the Phase 8 gravels contain abundant evidence of a handaxe-dominated lithic material culture that is present in the lowermost deposits of Phase 8a, and which at this level includes production of both pointed and ovate handaxe forms, with the ovate forms (technically a cordate in the classi-

ficatory scheme applied here, due to its flattened butt) being pronounced Z-twisted in side-profile. Although the Phase 8 deposits directly truncate the Phase 6 deposits in places, and therefore create an illusion of a sharp cultural transition, it is important to remember that they are separated by Phase 7 which (a) produced no archaeological material apart from that thought to have been derived from Phase 6, and (b) that the time period separating the occupation of Phase 6 from the occupation of Phase 8 is entirely unknown. Although both events are thought likely to have occurred within MIS 11, this still leaves around 40,000 years to play with; plenty of time for even the slowest drift in material cultural adaptation to lead to radically different techno-logical and typological practices.

Within the Phase 8 sequence there is a slight suggestion that the more ovate handaxe forms (most of which had a Z-twisted profile) are more commonly produced in the bottom part of the sequence. Those found higher up are in abraded condition and therefore potentially reflect a history of derivation from earlier deposits. From a material cultural point of view, although there was clearly an emphasis on handaxe manufacture, there was also a small but significant element of flake-tool manufacture. These include both simple utilised flakes and partly-trimmed flake-knives, and also more apparently designed convex unifacially flaked side-scrapers, similar in appearance to the Mousterian 'Quina-type' scrapers of south-west France.

Complementing these contrasts in technology and typology between Phase 6 and Phase 8, there is also a radically different structure to the spatial organisation of the *chaîne opératoire*. In Phase 6, all the evidence suggests that the *chaîne opératoire* was not spatially organised around the landscape. Rather, allowing for some small-scale mobility of flake-tools and partly-worked cores, it was (a) generally started and finished in the same part of the landscape, stimulated by an encounter with a resource such as a dead elephant, and (b) the distribution of lithic remains across the landscape was homogenised by the random nature (at least within the part of the landscape contributing to the horizon investigated at the site) of the distribution of resources requiring the application of lithic technology.

In contrast, there is a clear pattern in the evidence from the Phase 8 gravels of a consistent spatial organisation of the technological *chaîne opératoire*. The handaxes that were the predominant tool-type were mostly made elsewhere in the landscape, before becoming abandoned at, or in the vicinity of, the site. This is a complementary pattern to that represented at, for instance, the site at Red Barns in Hampshire. There it appears that an exposure of flint-rich chalky slope-wash deposits served as a location for the manufacture of handaxes that were then mostly taken away, leaving a disproportionate amount of debitage in relation to the small number of handaxes found (Wenban-Smith *et al.* 2000; Wenban-Smith 2004b).

Finally, there are some important points to make about the archaeological remains from the Phase 8 gravels concerning their recognition and value as a heritage resource. When first encountered, they were immediately recognised as fluvial (although there was some uncertainty about whether or not they were Pleistocene, or significantly earlier). However, there was no immediate indication that they contained any artefactual remains. Furthermore, even if there had been, there are many in the heritage curatorial world who would regard any lithic (or other archaeological) remains from river gravels as *de facto* 'disturbed' and therefore not of value for the business of investigating ancient societies, for which only '*in situ*' remains from undisturbed occupation surfaces are deemed worthy of investigation.

It was only after prolonged and systematic investigation that it began to become clear that there was a persistent artefactual presence within the Phase 8 gravel. The resulting relatively substantial collection was the result of systematic sieving of more than $5m^3$ of sediment and, as a very rough guess, the careful monitoring of the machining of several hundred cubic metres. This demonstrates that an initial apparent absence of artefacts in a deposit might prove to be misleading after more thorough investigation.

Furthermore, concerning the value of the artefacts that were eventually recovered, these prove to range in condition from absolutely mint (for example handaxe Δ.40480, Fig. 20.3a) to well-abraded. The mint and fresh specimens, even if not found on a recognisably intact landsurface, are nonetheless likely to have hardly moved since their deposition and to have been rapidly buried.

Hence, even though found within fluvial gravel, they can be regarded as essentially representing undisturbed evidence of activity contemporary with formation of the gravel. However, even the more abraded remains also have value as an interpretive resource. Although it would be very beneficial to understand their taphonomic history more fully, they nonetheless represent a sample of material cultural activity from a catchment area in the vicinity of their discovery location and from a not-necessarily-huge time period up to and including the deposition of their eventual context of discovery. Such evidence can in fact provide a more useful and reliable sample of broad cultural trends through the long vistas of the Lower/Middle Palaeolithic than the instinctively appealing undisturbed remains of short-lived episodes of activity perhaps only lasting one afternoon. Therefore it is necessary to start to take Pleistocene fluvial gravels more seriously as a potential Palaeolithic resource, and to investigate them in advance of development. On the occasions where no artefacts are found, it should be remembered that they may be present in reasonable numbers, but not necessarily commonly enough to have been identified in a limited evaluation. Additionally, it should be considered that a negative result in one part of a body of gravel, or even in the entirety of a substantial body, should not tarnish the potential of gravel in general as an archaeological resource. After all, just because one field doesn't contain a Roman villa, it doesn't mean that one shouldn't look in the next one!

Chapter 21

Lithic artefacts, miscellaneous collections from outside the main sequence: Phases T-1, 9, 9-10 and 11

by Francis Wenban-Smith

INTRODUCTION

The lithic remains presented in this chapter represent some disparate collections from the later phases of the site sequence, as well as a collection of out-of-context Palaeolithic material (Table 21.1). These collections are reviewed in turn below, in broad stratigraphic/dating order working towards the present-day.

The material from Transect 1 ('Group T-1', below) is probably broadly contemporary with the main Phases 3-8 of the site sequence. The material from the brickearth and its stripped surface to the north of the site ('Group 9.1', below) is all regarded as equivalent to Phase 9. A small group of material was recovered from the highest Pleistocene context in the site sequence, context 40176. This was a sand bed capping the brickearth at the north end of the site; this is studied below as 'Group 9-10' (below), since it is unclear how much of a hiatus, if any, occurs depositionally between context 40176 and the top of the brickearth.

Above this, the final two assemblages, attributed to Phase 11, represent material that was out-of-context or came from modern made ground, including late prehistoric features. One of these assemblages, designated as 'Group 11.1' (below), contains lithic artefacts that appear Lower/Middle Palaeolithic on technological/typological grounds or by their staining and patination. The other group, designated as 'Group 11.2', contains a technologically distinct subset of the Phase 11 material that was unstained and unpatinated, and also in mint condition;

this group is interpreted here as an 18th-century gunflint manufacturing industry.

GROUP T-1: TRANSECT 1

This group of material (Table 21.2) was all recovered from the stripped surface in and near Transect 1, about 50m north-east of the site (Fig. 21.1a). The transect surface was stripped by machine using a toothless bucket, and then cleaned by hand prior to recording of the exposed deposits (Fig. 21.1b). Seven artefacts were recovered from the transect during cleaning, and an additional artefact was recovered from the ground surface near the east end of the transect. It was not possible to integrate the sediments from which artefacts were recovered into the phased sequence of the main site, but they were clearly broadly contemporary with Phases 3–8, and suggestions for possible correlations are given in the artefact summary table (Table 21.2).

The artefacts from contexts 40083 and 40084 comprised two flakes and a core, all of them in mint condition. The two flakes, both of which came from context 40083, were technologically undiagnostic. The core from context 40084 was quite a large angular core reflecting removal of flakes from a migrating platform. As such, it fitted in with the technological character of the material recovered from Phases 3–7 at the main site. The two flakes from contexts 40081 and 40082 were also both technologically undiagnostic; the former was in

Table 21.1 Miscellaneous collections examined in Chapter 21

Assemblage group	Details of artefact context/s	Artefacts (n)
11.2	A technologically distinctive group that was mostly found in what was thought to be modern made ground at the base of Southfleet Road, and above the Pleistocene sequence	12
11.1	Various obviously-Palaeolithic material (eg bits of handaxes) that was found out-of-context, in modern made-ground or in Late Prehistoric features	42
9-10	From context 40176, a sand lens at the top of the Phase 9 brickearth, which was cut into by Late Prehistoric features and overlain by the made ground underlying the original Southfleet Road surface	10
9.1	From stripped surface of brickearth bank to north of site, and a few in situ from Phase 9 brickearth at site	18
T-1	From stripped Transect 1, to north of main site	8

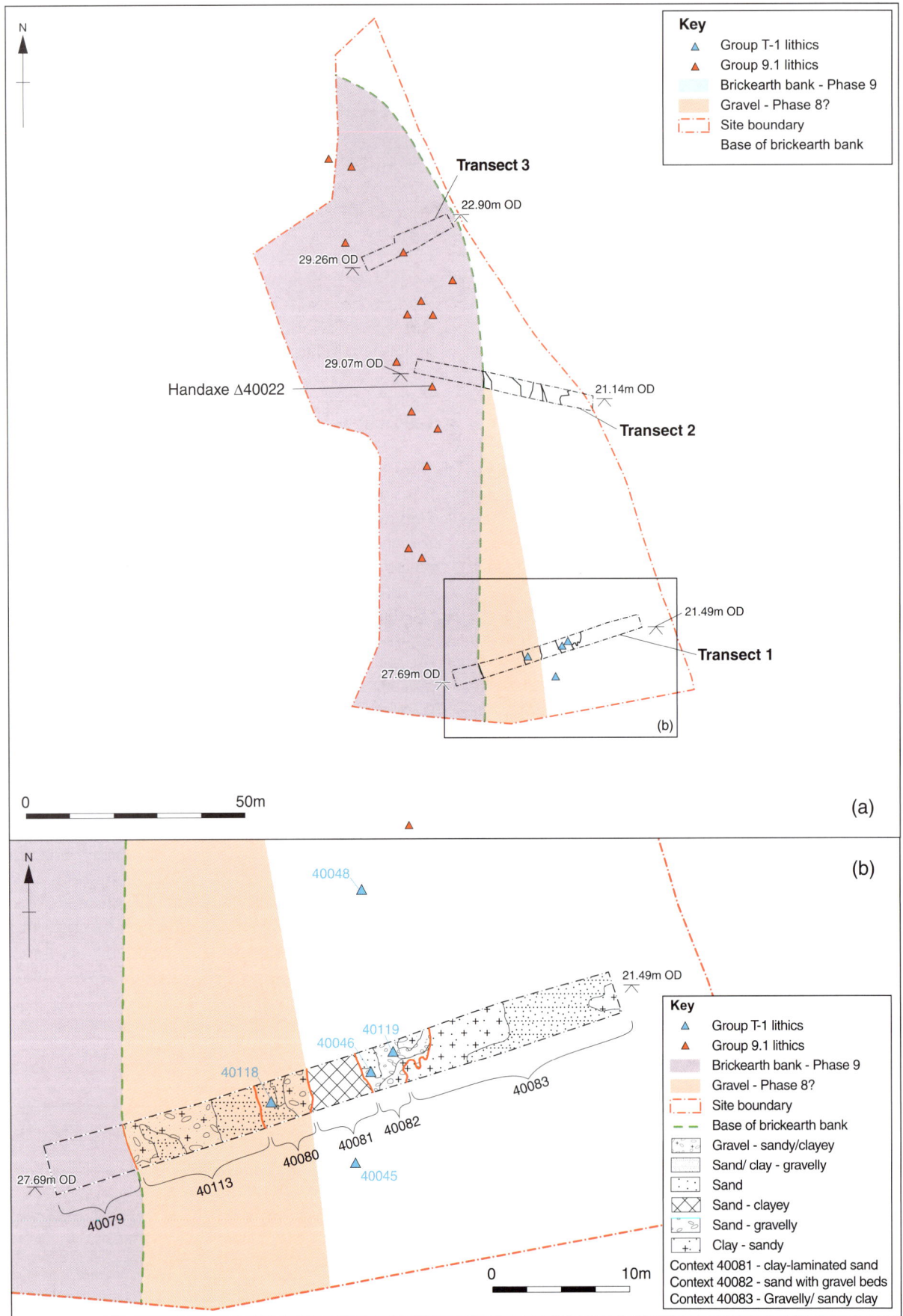

Figure 21.1 Northern part of site: (a) Transects 1, 2 and 3, showing exposed sediments and lithic findspots for Groups 9.1 and T-1; (b) closer view of Transect 1, showing internal stratigraphy and Group T-1 lithic findspots (for finds with XYZ co-ordinates available)

mint condition, and the latter was in moderately abraded condition. And context 40080, thought quite likely to be equivalent to Phase 8, contained the broken proximal end of a flake that appeared to have been crudely bifacially worked and therefore was interpreted as a broken handaxe-on-flake; it was in abraded condition. Context 40080 also produced two flakes, both of which were technologically undiagnostic; one was in abraded condition and the other in fresh condition.

Although this small assemblage does not add anything to the understanding of the main site, its existence, with the presence of mint condition artefacts, emphasises the continuing potential Palaeolithic significance of the uninvestigated deposits that survive in a north-south strip approximately 250m long by 50m wide to the north of the site, between Southfleet Road and the access road to Ebbsfleet International Station.

GROUP 9.1: BRICKEARTH

The majority of this group of material (Table 21.3) was collected from the machine-stripped surface of the brickearth bank (context 40076) to the north of the site, between the west ends of Transects 1 and 3. One artefact was also recovered directly from the brickearth (context 40053) at the north end of the site, and three were recovered from the top bulk sieve-sample <40118> into the brickearth at Trench A (Fig. 20.1b).

The artefacts collected from the surface of the brickearth bank were quite widely distributed (Fig. 21.1a), and comprised one handaxe and thirteen pieces of debitage. This assemblage did not appear to represent a specific scatter disturbed by the machine as they were in a wide variety of conditions. The handaxe (Δ.40022), which was a large, very bluntly pointed sub-cordate (Fig. 3.10), was in fresh condition and was unstained/unpatinated on one face; it was however moderately patinated and light brown stained on the other face, suggesting a substantial period of exposure lying flat on a palaeolandsurface before its eventual incorporation in buried sediments. The majority of the debitage was in mint or fresh condition; most pieces were unstained and unpatinated, or lightly stained/patinated, but a few were more strongly patinated and/or strongly brown/ochre stained. The majority of the debitage was technologically undiagnostic. However, one of the pieces in mint condition (Δ.40027) was clearly from thinning/shaping a large handaxe. Another piece (Δ.40037) also in mint condition, showed numerous parallel blade-like dorsal

Table 21.2 Group T-1, from Transect 1 to north of site: technological overview

Context	5 – Percussor	10 – Tested nodule	20 – Core	30 – Core-on-flake	40 – Core-tools	50 – Handaxe-on-flake	60s – Fl-tools	80 – Fl-flakes	90 – Flakes	100 – Irreg. waste	110 – Chips	Sub-total (n)
40080 - sandy/clayey gravel [Phase 8??]	-	-	-	-	-	1	-	-	2	-	-	3
40081 - clay-laminated sand [Phase 5??]	-	-	-	-	-	-	-	-	1	-	-	1
40082 - sand/gravel [Phase 3??]	-	-	-	-	-	-	-	-	1	-	-	1
40083 - gravelly clay [Phase 3??]	-	-	-	-	-	-	-	-	2	-	-	2
40084 - [Phase 3??]	-	-	1	-	-	-	-	-	-	-	-	1
All	-	-	1	-	-	1	-	-	6	-	-	8

Table 21.3 Group 9.1, from brickearth bank: technological overview

Context	5 – Percussor	10 – Tested nodule	20 – Core	30 – Core-on-flake	40 – Core-tools	50 – Handaxe-on-flake	60s – Fl-tools	80 – Fl-flakes	90 – Flakes	100 – Irreg. waste	110 – Chips	Sub-total (n)
40053 - brickearth (Phase 9) capping Pleistocene sequence at north end of main site	-	-	-	-	-	-	-	-	2	2	-	4
40076 - collected from machine-stripped surface of brickearth bank to north of main site	-	-	-	-	1	-	-	-	10	3	-	14
All	-	-	-	-	1	-	-	-	12	5	-	18

scars, suggesting perhaps a Levalloisian blade-like sequence, well-known in the MIS 8/7 deposits of the Ebbsfleet Valley and at nearby Crayford (Wenban-Smith 1995a and 2007).

The artefacts recovered directly from the brickearth were all in mint or fresh condition. Three of them were technologically undiagnostic, but one of them – Δ.40094, from Trench A sample <40118> – was from the later stages of thinning/shaping a handaxe.

Overall, although the majority of this assemblage was not found directly *in situ*, it draws attention to the potential of the brickearth as a possible source of lithic remains in good condition. Whether these might represent undisturbed material on undisturbed palaeo-landsurfaces within brickearth, or whether they represent colluvially transported material, remains uncertain. However, as was proven by the spectacularly fine handaxe recovered from Station Quarter South Test Pit 25, dug in 2006 (Wessex Archaeology 2006b; this volume Chapter 4 *Deposits in the site vicinity*) the brickearth is a poorly understood deposit that nonetheless produces lithic archaeological remains, and merits further investigation.

GROUP 9–10: SAND CAPPING BRICKEARTH

This group of material (Table 21.4) was all recovered from a sand bed (context 40176) overlying the brickearth at the north end of the site, towards the end of the Watching Brief. All of the assemblage is in moderately to very abraded condition, and moderately stained/patinated, apart from one flake that is in mint condition and unstained/unpatinated. The core (Δ.50180) is small (maximum length = 52mm; weight = 75g) and intensively worked. It is approximately pyramidal in shape, and each face has numerous parallel removals of small blade-like flakes; it would not be out of place in a Mesolithic or Neolithic assemblage. One of the flake-tools (Δ.50178) is a large, thick flake with one area of edge-crushing that is interpreted as either a concave scrapping edge or macro use-wear. The other two are medium-size flakes approximately 60–70mm long and 50–60mm wide that have a crude convex scraping-type edge made by 2–3 small secondary removals. One of these retains a facetted butt that is

slightly indicative of having been removed from the surface of a radially flaked Levalloisian/Mousterian-type core, but the dorsal scar pattern is uni-directional rather than radial. Alternatively, this flake could be a 'core tablet' representing rejuvenation of the platform of a core similar to that recovered here and what looks like facetting could merely be proximal ends of a series of parallel flake removals. The remainder of the debitage is technologically undiagnostic.

As a whole, this assemblage is not particularly inform-ative, and its date is uncertain. It is most likely Palaeolithic, despite the small pyramidal blade-flake core. Such pieces are not unknown in the Lower/Middle Palaeolithic (Pradel 1944), although often disregarded as intrusive when encountered; for instance, there is a similar one in the APCM Baker's Hole collection at the British Museum. It probably represents residual evidence of the later MIS 8/7 occupation of the Ebbsfleet Valley, remains from which were so abundantly found in the deposits that (prior to extraction by quarrying) were present in the more central part of the Ebbsfleet Valley to the north-east of the site (Wenban-Smith *et al.* forthcoming).

GROUP 11.1: DERIVED PALAEOLITHIC MATERIAL

This group of material (Table 21.5) mostly represents artefacts that appear obviously Palaeolithic on grounds of typology/technology and size/condition, but were found out-of-context or in the modern made ground. Several artefacts from the excavation for which their provenance was misplaced are also included in this group. Many of the pieces are handaxes, or broken bits of the same. They are of little interpretative value in themselves and add nothing to the excavated collection due to their lack of provenance. They do, however, represent a useful case-study of the type of collection that might be recovered prior to any detailed investigations in an area of rich Lower/Middle Palaeolithic archaeology, which would highlight an area as meriting further investigation for better provenanced material. They also have value for teaching/handling/display as a relatively expendable material representation of this very distant past, for which minor physical damage and loss of some items would cause no reduction in the future research potential of the

Table 21.4 Group 9-10, from context 40176: technological overview

Context	5 – Percussor	10 – Tested nodule	20 – Core	30 – Core-on-flake	40 – Core-tools	50 – Handaxe-on-flake	60s – Fl-tools	80 – Fl-flakes	90 – Flakes	100 – Irreg. waste	110 – Chips	Sub-total (n)
40176 - sand lens at the top of the Phase 9 brickearth at north end of main site	-	-	1	-	-	-	3	-	6	-	-	10
Total	-	-	1	-	-	-	3	-	6	-	-	10

Table 21.5 Group 11.1, derived and out-of-context Palaeolithic material: technological overview

Context	5 – Percussor	10 – Tested nodule	20 – Core	30 – Core-on-flake	40 – Core-tools	50 – Handaxe-on-flake	60s – Fl-tools	80 – Fl-flakes	90 – Flakes	100 – Irreg. waste	110 – Chips	Sub-total (n)
0 – out-of-context, not in situ	-	-	-	-	9	-	1	-	1	-	-	11
40001 – modern made ground	-	-	-	-	1	1	-	-	-	-	-	2
40012 – modern stripped ground surface, above bank to west of site	-	-	-	-	1	-	-	-	-	-	-	1
40133 – fill of Late Prehistoric feature	-	-	-	-	-	-	1	-	-	-	-	1
40039? – uncertain provenance	-	-	-	-	-	-	-	-	1	-	-	1
40048? – uncertain provenance	-	-	-	-	-	-	-	-	1	-	-	1
40069? – uncertain provenance	-	-	-	-	-	-	-	-	2	1	-	3
40100? – uncertain provenance	-	-	-	-	-	-	-	-	-	-	1	1
40100?? – uncertain provenance	-	1	1	-	-	-	2	-	12	5	-	21
Total	-	1	1	-	11	1	4	-	17	6	1	42

site archive. This is, of course, in contrast to the better provenanced material, which needs to be more carefully curated and preserved for future research.

GROUP 11.2: AN 18th CENTURY GUNFLINT INDUSTRY

In amongst the wider collection of Group 11, a very curious lithic assemblage instantly stood out during initial analysis. Despite being provenanced to the 'Made Ground' capping the sequence of the main site, which occurred between the top of the Pleistocene deposits (usually the Phase 8 gravel, but the brickearth at the north end of the site) and the asphalt surface of the old Southfleet Road, there were several clear lithic artefacts that were all in absolutely mint condition. These were unstained and unpatinated, on typical local slightly coarse Swanscombe-area grey flint with inclusions. These were initially regarded by myself as most likely something late prehistoric, or perhaps some by-product of 19th or early 20th century road construction. They were

therefore passed to Hugo Anderson-Whymark, who was dealing with later prehistoric flint work from the HS1 projects in the Ebbsfleet Valley. However, he rejected them and passed them back to me! There are not that many flints involved (only nine initially from context 40001, although three others were later added on grounds of their condition and technological similarity, Table 21.6). This probably resulted from the minimal attention given to recovery of material from the made ground overlying the Pleistocene sequence, which was the focus of the archaeological work.

The assemblage is very coherent technologically. It mostly comprises quite large and very chunky flint flakes, violently struck, with notches from flake-flakes struck sideways across the ventral surface, sometimes a single notch, sometimes double opposing notches (often with the distal ends of their flake scars intersecting). Of the twelve artefacts in the assemblage, nine are secondarily worked debitage of this nature, two are elongated flakes with opposing notches across the ventral surface (Fig. 21.2b; c), three of them chunky flakes with single notches (Fig. 21.2d; e), and the remainder chunky

Table 21.6 Group 11.2, 18th century gunflint industry from 'made ground': technological overview

Context	5 – Percussor	10 – Tested nodule	20 – Core	30 – Core-on-flake	40 – Core-tools	50 – Handaxe-on-flake	60s – Fl-tools	80 – Fl-flakes	90 – Flakes	100 – Irreg. waste	110 – Chips	Sub-total (n)
0 – out-of-context, not in situ	-	-	-	1	-	-	-	-	-	-	-	1
40001 – modern made ground	-	-	-	7	-	-	1	1	-	-	-	9
40039? – uncertain provenance	-	-	-	1	-	-	-	-	-	-	-	1
40100? – uncertain provenance	-	-	-	-	-	-	1	-	-	-	-	1
Total	-	-	-	9	-	-	2	1	-	-	-	12

debitage with various less structured notching. In addition to these, there is one secondary flake-flake from this notching, which has itself then been struck to leave the notch scar from what might be termed a Tertiary flake, that is a flake from what was already a flake-flake (Fig. 21.2f). The notches (excepting the latter small notch on the flake-flake) are constantly about 20mm long (parallel with their axis of percussion) and 40–50mm wide (transverse to their axis of percussion). Finally, there are two secondarily worked flakes with a few small areas where minor retouching seems to have taken place; these were classified as flake tools, but the apparent retouch may merely be damage, or an incidental part of the debitage removal process.

Having been rejected as a late prehistoric industry, and having consulted with Alan Saville and Frances Healy, who mentioned gunflints, I remembered there had once been a couple of articles in the Lithic Studies Society journal *Lithics* on this topic. After a quick trawl through back-issues I re-encountered an article by McNabb and Ashton (1990) that describes and illustrates a technologically identical assemblage (Fig. 21.2a), found by A. T. Marston near Dartford on the Thames foreshore. It describes the so-called 'Wedge technique' for gunflint manufacture, which apparently died out in about 1780 when it was superseded by the better-known blade-based approach exemplified by the 19th century Brandon gunflint tradition (Forrest 1983;

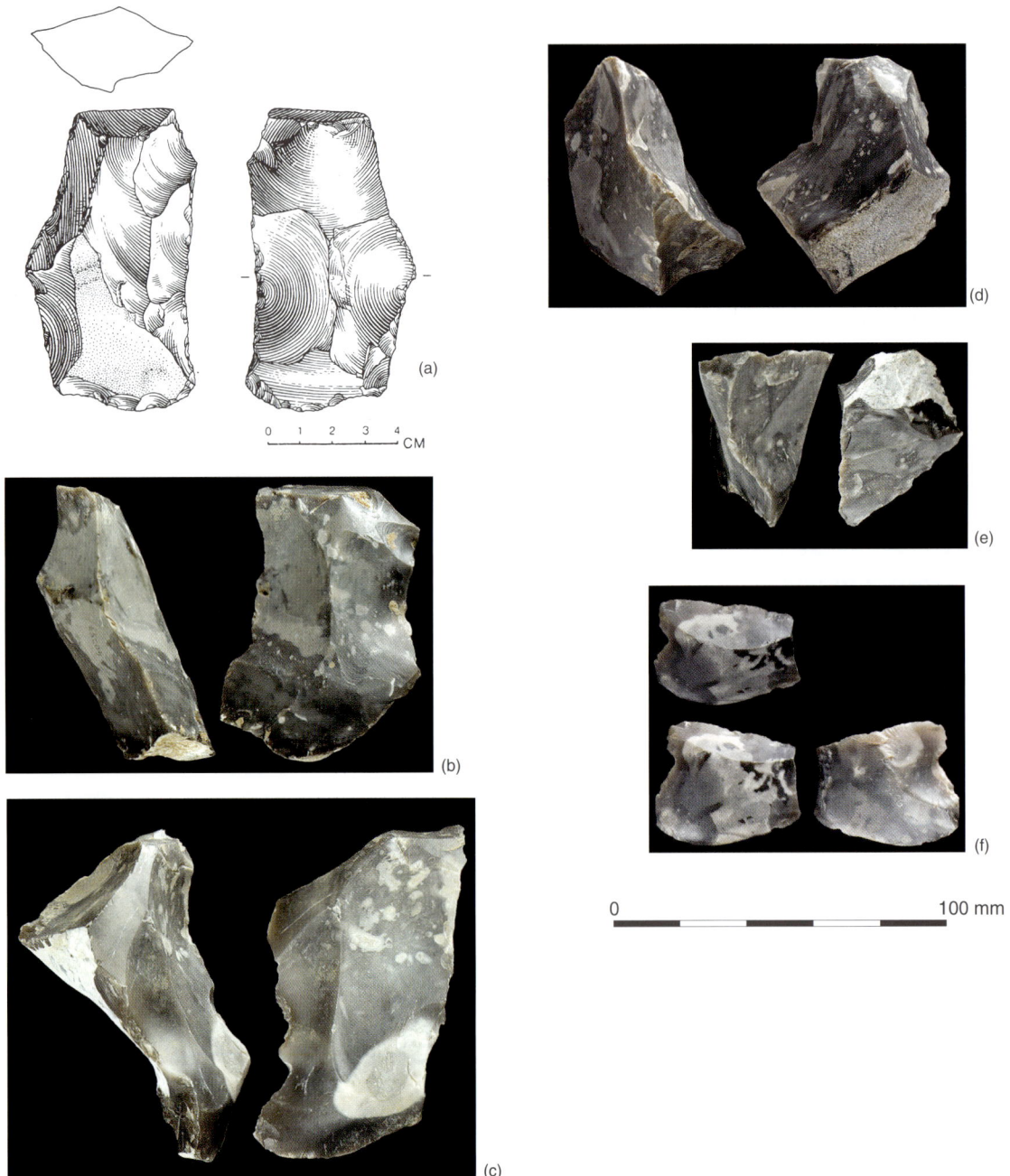

Figure 21.2 Group 11.2, 18th century gunflint industry: (a) Wedge core (double opposed) example from Thames foreshore (McNabb and Ashton 1990); (b) Wedge core (double opposed), Δ.40257; (c) Wedge core (double opposed), Δ.40259; (d) Wedge core (single), Δ.40251; (e) Wedge core (single), Δ.40253; (f) Wedge flake-flake blank, Δ.40256

Karklins 1984). They also, quoting Lotbinière (1977), mention by name a gunflint manufacturer W. Levett who was active in the Northfleet area in the mid-late 18th century. They suggested as an irony that the typically Clactonian approach of creating notched flake-tools from flake blanks was not the cause of Marston's original attribution to Clactonian. However, there is in fact quite a contrast between the typical genuine Clactonian approach to notching, as exemplified in Phase 6 (see, for example, Chapter 18 *Secondary flake modifications and flake-tools*) – which involves striking notch removals on the ventral surface of a flake – and the gunflint approach, as exemplified here (Fig. 21.2), which involves striking on the dorsal surface with the secondary removal therefore coming off the ventral surface, parallel with it rather than orthogonal to it. Even more ironic in this case is that, not only does the gunflint assemblage display the classic wedge technique, using thick chunky flakes as cores, but also that it was incorporated in the uppermost deposits of a *bona fide* Clactonian site. One can therefore surmise that perhaps Levett himself, or one of his contemporaries, must have taken advantage of the outcropping exposure of the same deposit rich in nodular flint that provided the source of flint raw material for early hominin tool manufacture 400,000 years earlier, and sat knapping gunflints by the roadside leaving the waste to confuse later archaeologists.

Chapter 22

Discussion and conclusions: Clactonian elephant hunters, north-west European colonisation and the Acheulian invasion of Britain?

by Francis Wenban-Smith

INTRODUCTION

The preceding chapters have presented detailed studies of different aspects of the site: stratigraphy, sedimentology, palaeontology, palaeobotany and archaeology. In this final chapter, the disparate specialist analyses are integrated to (a) provide an overview of the history of landscape development and climatic/environmental change through the parts of the Middle Pleistocene represented in the deposit sequence, (b) present the evidence of hominin occupation and activity within this framework and (c) consider some of the wider implications of the evidence.

The site presents narratives and resonances at different temporal and spatial scales. At one level, it provides a 'just-so' story about a locale, with deep explanatory roots for the present-day geomorphology, which in turn has structured the pattern of land usage, aggregate extraction, road networks and urban development, providing a fundamental connection between the present environment and the geological past. The archaeological and palaeontological evidence populate this ancient landscape with hominin and other faunal and floral life, likewise providing roots for the contemporary experience. At a wider level, the evidence from the site contributes to our wider understanding of the British Lower Palaeolithic, complementing information from broadly contemporary horizons at nearby Barnfield Pit (Conway *et al.* 1996) and from further afield in East Anglia. In particular: East Farm Pit, Barnham (Ashton *et al.* 1998); Beeches Pit, West Stow (Hallos 2004; Gowlett *et al.* 2005) and Clacton-on-Sea (Oakley and Leakey 1937; Singer *et al.* 1973). Additionally, the evidence from Southfleet Road integrates into the sparse web of similarly undisturbed sites from across the Lower/Middle Pleistocene Old World, for example: Boxgrove, UK (Roberts and Parfitt 1999); Aridos, Spain (Villa 1990); Notarchirico, Italy (Piperno *et al.* 1998); Gesher Benot, Israel (Goren-Inbar *et al.* 1994) and Mwanganda's Village, Malawi (Clark and Haynes 1970), to provide a series of snapshots of early hominin life across the grand sweep of Palaeolithic time.

At a deeper level, the investigation and analysis of the Southfleet Road site provides a series of case-studies. In the first case, it is a multi-disciplinary Palaeolithic investigation, with the lithic analysis that is the long-standing focus of Palaeolithic archaeology supported by Quater-nary Earth Science and zoological work. Second, it is a case-study of a lithic analytical approach, incorporating traditional technological and typological characterisation, but complementing these with analysis of the *chaîne opératoire* and the spatial organisation of lithic production. Third, it is a case-study of the contrasting approaches to interpretation of the Lower/Middle Palaeolithic record that are crystallised in the specific 'Clactonian debate' that has persisted since the 1930s (Oakley and Leakey 1937; Singer *et al.* 1973; Ohel 1979; Ashton *et al.* 1994; Wenban-Smith 1998; White 2000) and to which the evidence from the Southfleet Road site is highly germane. Like other great debates of Old World early Palaeolithic archaeology, such as the so-called 'Mousterian' debate (Binford and Binford 1966; Binford 1973; Bordes 1981; Dibble and Rolland 1992), the Clactonian debate concerns not just the story derived from the lithic evidence, but a clash of more fundamental perspectives on the nature of the early archaeological record, and in particular the lithic record, and how to approach its interpretation.

The archaeological evidence through the sequence takes a variety of forms. Sparse lithic evidence incorporated into the lower deposits, formed by active processes such as landslips, slopewash or river torrents (Phase 3), provides evidence of occupation in the general area at, or before, the horizons from which the evidence was recovered. The paucity of this evidence makes it impossible, however, to develop an idea of the wider lithic cultural/industrial tradition.

At other horizons, most notably in clayey deposits (Phase 6) thought to have been formed at the edge of a fluctuating waterbody at the foot of a slope forming its western bank, undisturbed remains of hominin activity have been found, representing flint tool manufacture and exploitation of a single elephant carcass over a restricted period, perhaps between one day and a week. This evidence thus provides a clear snapshot of hominin activity involving flint tool manufacture at the site of a single large food resource. While the space-time envelope of this undisturbed evidence is restricted, making it of questionable utility for determining a wider industrial tradition, it is complemented by a much larger accumulation of technologically similar material from the same stratigraphic horizon, in the concentration south of Trench D. While some of this latter material may likewise be undisturbed, the majority is thought to have been slightly mixed and transported by slopewash processes, perhaps only a few metres, and to represent a sample from a more sustained accumulation of lithic technological events in the bank margin zone between the floodplain and the clayey slope rising to the west of the site. It thus provides a significantly more robust basis for identifying the presence of a distinct lithic industrial tradition sustained throughout the period represented by the persistence of the floodplain edge as a place of activity. The tradition that emerges is based on the loosely structured reduction of locally obtained flint nodules to produce flake blanks of various sizes and shapes. The larger (or more conveniently shaped) of these were then selected for use or secondary flaking and transformed into a variety of simple flake-tools, often with clear notch removals, thought to have been particularly efficacious for cutting meat and skin.

A slightly wider temporal and spatial scale of interpretation is provided by the lithic material from Phase 7, thought to represent mass movement of deposits downslope from the high ground to the west, and thus containing reworked evidence of lithic activity broadly contemporary with that from the underlying Phase 6, but from a much wider landscape catchment. The evidence is nonetheless technologically and typologically indistinguishable from that from Phase 6, further reinforcing the notion of the sustained production by a hominin group inhabiting the Swanscombe locale during the period represented by deposition of the Phase 6 sediments (and probably also the upper Phase 5 sediments). Of a lithic industrial tradition based upon cores, flakes and simple flake-tools, but entirely lacking in bifacial handaxes.

Although the tight time-space envelope of the activity around the elephant skeleton makes it, paradoxically, less useful for determination of lithic industrial tradition – *contra* the thinking of the 1960s and 1970s, whereby one of the main drivers of the importance of finding undisturbed sites was to achieve culturally pure lithic artefact samples without reworked contamination from older occupations – the undisturbed horizon of elephant exploitation at the site can be securely linked with the rich palaeo-environmental remains of the tufaceous channel (Phase 6b) by means of the recovery within it of foot bones originating from the elephant skeleton. The association therefore provides a firm, and exceptionally rare, instance where Middle Pleistocene hominin presence and activity can be placed, not only in a landscape context, but also in a specific climatic and environmental context. In this instance, it can be placed firmly in the fully temperate interglacial climatic optimum of MIS 11, probably Ho-II of the Hoxnian interglacial. At this time the environment of south-east Britain would have been predominantly forested, although without doubt with various more open grassy spaces maintained by grazing herbivores, perhaps mostly in riparian riverbank situations as in this instance. Thus, although clearly contradicting the position of Gamble (1986) which was based primarily on the atypical record

of the Late Pleistocene, when Britain was not apparently inhabited in the peak interglacial maximum of MIS 5e, questions still remain about the degree of hominin penetration into major tracts of more densely forested parts of the landscape, away from the river systems that have also (perhaps misleadingly) provided the depositional conditions for preservation of most archaeological evidence from the period.

Higher in the sequence, there is a stratigraphic unconformity between Phase 7 and Phase 8. The fluvial gravels of Phase 8 truncate the underlying sequence, representing a depositional hiatus of uncertain duration and a major change in the activity of the depositional environment, which was relatively placid throughout Phases 4-7. The archaeological remains of Phase 8 present a major contrast to those of the underlying sequence. They include both mint condition and abraded material, thus providing evidence of both activity at the site contemporary with deposition of the river gravels and activity in the wider catchment area. Not only are they technologically and typologically very different, being dominated by the presence of often finely made handaxes of a range of forms, often finely made (sharply pointed, flat sub-cordate and twisted cordate examples are all found) alongside a range of flake-tools including large unifacial side-scrapers, but they are also characterised by being part of a wholly different organisational approach to the lithic technological system.

In the earlier deposits, there was no evidence of the lithic technology being organised in the landscape. At all the different spatial/temporal scales at which the evidence survived, there was a uniform picture of lithic production as a floating *ad hoc* response to immediate need. Reduction sequences were often started and finished at the same spot using readily available raw material to make tools for immediate use and discard, with no sign of any spatial patterning of activity in the landscape. In Phase 8, there is in contrast a preponderance of finished, intensely worked handaxes, but a distinct lack of waste debitage commensurate with their manufacture. This reflects a significant re-alignment of the organisation of the lithic technological system in relation to mobility and the encountering of resources. It does not necessarily reflect increased mobility, but it reflects both a greater spatial separation of tool use and discard from the place of manufacture, and the spatially structured repetition of this process in the landscape. These are archaeological characteristics that could be taken as reflecting a greater degree of cognitive anticipation and logistic planning, and which represent a more modern human style of technological practice in a 'cultural' geographic framework as opposed to the less actively constructed 'niche geography' of much non-human animal behaviour, in Binford's (1987b) terms. Thus, this contrast raises the question of whether not only is there a straightforward technological/typological industrial/cultural contrast between the archaeological remains of Phase 8 and the underlying deposits. But does this contrast (if accepted) also reflect a more deep-rooted difference in cognitive capabilities and behavioural practices, one that perhaps could relate to different hominin species, or evolutionary lineages; this is considered further below.

In the remainder of this chapter, different aspects of the results of the work at the site are recapped in more detail, starting with an overview of the basic results and progressing through a series of discussions of the interpretive implications within the wider context of current Palaeolithic research. It culminates in some thoughts about methods of investigation of Palaeolithic sites and particular issues with carrying out Palaeolithic archaeology in advance of development, as opposed to on a purely research-led basis.

OVERVIEW: STRATIGRAPHY, DATING, ENVIRONMENT AND ARCHAEOLOGY

The complete sequence of the site is summarised here (Table 22.1). Geomorphologically, the site was situated throughout its depositional history in the base of a north-south trending valley, with a high gentle slope rising to the west, and possibly a steeper cliff or bank to the east.

Phase 1 ('Tilted Block') and Phase 2 (Parallel-bedded sand/clay)

The bottom two Phases (1-2) of the stratigraphic sequence, present at the southern end of the main site, are of uncertain date, and lack any biological or archaeological evidence. Phase 1 (the 'Tilted Block') attests to significant disruption of the underlying Chalk bedrock in the site area, reinforced by data from test pits and boreholes in the surrounding area (see Chapter 4). This disruption, although poorly understood and to-date little investigated, is probably critical to the site's existence, providing a locale both locally rich in flint raw material and prone to depositional aggradation, thus ensuring both hominin activity and subsequent preservation of the resulting evidence in conjunction with various zoological and floral remains.

Phase 2 (Parallel-bedded sand/clay) seems to represent a phase of infilling of a very uneven local landscape, whereby sumps and local depressions within a landscape characterised by cliffs and jagged pinnacles of chalk became infilled by silt/sand deposited by quiet water or slopewash, interspersed with episodes of standing water represented by clay bands typically 10-20mm thick. This phase of the sequence is unconformably truncated by deposits of Phase 3 (Chalky/silty/gravelly sand), and these phases of the site sequence may be separated by a significant hiatus. However, the inferred depositional environments of Phases 2 and 3 are remarkably similar in parts, perhaps indicating a not-so-great time separation.

Phase 3

The basal sediments of Phase 3 are medium-coarse silty/sandy flint gravel beds rich in chalk pebbles and comminuted Tertiary shell fragments (context 40061)

that seem to have been fluvially deposited in their lower parts, although probably including significant slopewash input. These deposits dip and thicken to the east of the main site, extending below 18m OD, with their base not reached. Clast lithological analysis of these deeper eastern gravelly sediments (Wessex Archaeology 2006b) confirmed a south-bank Thames tributary origin, namely an early course of the Ebbsfleet. The upper sediments of Phase 3 are characterised by a much-reduced flint gravel component and a lack of sedimentary structure. They comprise (contexts 40028 and 40062) clayey/silty sand with occasional small flint and chalk pebbles and Tertiary shell fragments, with these inclusions becoming smaller and less common eastward across the site. The upper parts of the Phase 3 sequence are thought to have been deposited in much quieter water, again with significant slopewash input having a stronger influence further west within the site area, towards the western valley side. The top of these sediments is decalcified in places (context 40063), perhaps indicating a break in deposition or a period of emergence as a short-lived landsurface, although there is no sign of soil development. The Phase 3 sediments are both distorted by post-depositional ground movement (Chapter 4) and unconformably truncated by the Phase 5 sediments, so their original geometry is unknown.

Ostracod remains within the upper part of Phase 3 (Chapter 11) consisted of a rich freshwater fauna typical of fully temperate interglacial conditions, with a range of species typical of quiet swampy waterbodies and springs, surrounded by rich vegetation, and prone to periodic drying. The ostracod assemblage indicated a mean July temperature of 17-21°C and a mean January temperature of –4 to –1°C, indicating a similar climate to the present day, but with slightly greater seasonality. It also included one particular species, *Ilyocypris quinculminata*, that is of biostratigraphic significance, not being known from any sites younger than MIS 11. Its two occurrences in the area of the Hoxnian stratotype, Coleman's Farm, Rivenhall and Hoxne itself, are both, incidentally, in deposits attributed to the 'true' Hoxnian of MIS 11c, in fully temperate sediments stratigraphically underlying the Arctic Bed of stratum C (cf. Ashton *et al.* 2008). A few vertebrate remains were recovered from Phase 3 sediments (Chapter 7), the only identifiable specimen being the well-preserved skull of a wild aurochs, *Bos primigenius*, supporting the ostracod evidence for a temperate climate. The only molluscan remains recovered from the Phase 3 sediments were quite numerous *Bithynia* opercula, which provided an AAR result suggesting a date within MIS 11 that was statistically separable as earlier within MIS 11 than Phases 5 and 6 of the site sequence (Chapter 13). Although tempting to take this result at face value, it was concluded that the Phase 3 *Bithynia* opercula were probably derived and had also perhaps been compromised by poor preservation in light of (a) the abundant biostratigraphic and AAR dating evidence that Phases 5 and 6 correlated with the main earlier temperate phase of MIS 11 (the Hoxnian) and (b) the absence of a cool episode and major depositional hiatus between Phases 3 and 5. It was

Table 22.1 The full sequence at the Southfleet Road site: archaeological, palaeontological and dating summary

Site phase	Archaeolgical remains	Industry	Biological remains	Climate	Palaeo-environment	Dating, zonation
11 - Not *in situ* (out-of-context, or from modern made ground)	Various flint artefacts	18th C gunflint cores; derived Palaeolithic material	-	-	-	Post-Medieval
10 - Holocene (various Late Prehistoric features)	Various pottery and flint artefacts	Neolithic?	-	-	-	MIS 1
9 - Brickearth bank (colluvial, including reworked aeolian deposits?)	Handaxes, debitage	Acheulian	-	?	?	MIS 2-5d? MIS 6-10? MIS 11a-b?
8 - Sandy gravel (fluvial, palaeo-Ebbsfleet)	Common handaxes, pointed and twisted cordate forms; large scrapers and other flake-tools; occasional debitage	Acheulian	-	Temperate/cooling?? (no direct evidence, but inferred from models of fluvia sequence deposition where return to gravel deposition accompanies cooling conditions)	Becoming more open? (no direct evidence, but inferred from models of fluvial sequence deposition where return to gravel deposition accompanies cooling conditions)	MIS 10?? MIS 11a-c? Ho III-IV??
7 - Mixed clay/gravel (syncline infill - slopewash, landslip)	Cores/flakes (varied condition)	Clactonian (derived)	Wood pieces (mostly fragmentary and decayed)	Temperate	Woodland in vicinity Ho II-III?	MIS 11c
6 - Grey clay, with organic-rich beds and tufaceous channel-fill deposits (quiet water, periodically drying: with slopewash input)	Abundant cores, flakes, notched tools and flake-knives (all mint); some refitting Elephant carcass with refitting lithic scatters	Clactonian	Woodland fauna (bank vole, wood mouse, land snails - including *Discus ruderatus*) Freshwater aquatic fauna Straight-tusked elephant Narrow-nosed rhinoceros	Fully temperate	Local shady, damp wood-land (alder carr) near river or stream, with open grass-land in area	MIS 11c Ho IIb-c
5 - Clay-laminated sand (fluvial)	Cores, flakes, notched tool	Clactonian?	Grassland fauna with woodland elements (Straight-tusked elephant, lion, aurochs, rabbit, ground squirrel, deer) Freshwater fish	Fully temperate	Grassy, herbaceous floodplain with patchy reed beds and woodland	MIS 11c Ho II

Table 22.1 (continued)

Site phase	Archaeological remains	Industry	Biological remains	Climate	Palaeo-environment	Dating, zonation
4 – Sandy/gravelly clay (lacustrine, periodically drying)	-	-	Ostracods [in nearby test pits for Station Quarter South - Wessex Archaeology 2006b]	Fully temperate	- Ho II	MIS 11c
3 – Chalky/silty/gravelly sand (quiet water, with slopewash input)	Flaked nodule, two flakes	Clactonian??	Aurochs skull Ostracods	Fully temperate	Patches of herbaceous meadows and open grassland	MIS 11c Ho II
2 – Parallel-bedded sand/ clay (lacustrine/slopewash?)	-	-	-	-	-	Earlier Middle Pleistocene?
1 – Tilted block (bedrock upheaval/collapse?)	-	-	-	-	-	Early-Middle Pleistocene?

therefore concluded that the Phase 3 sediments were of similar age to the overlying sediments of Phases 5-6, and likewise attributable to the early temperate stage of the Hoxnian interglacial.

Phase 3 is the lowest level of the site sequence with firm evidence of hominin activity. Lithic artefacts indisputably of hominin origin are present in the basal deposit (context 40061), as well as in the overlying context 40028 (Chapter 16). The fresh condition of these artefacts suggests occupation in the site vicinity contemporary with formation of the deposits, probably on the bank of the waterbody thought to have been present. The artefact assemblage comprises two technologically undiagnostic flint flakes and a nodule from which at least one flake was been struck off. Although compatible with the rich flake/core industrial traditions of higher site levels (Phase 6 in particular), there are far too few lithic remains to postulate any attribution of industrial/cultural affinity.

Phase 4 (Sandy/gravelly clay)

Phase 4 sediments were virtually absent at the main site, but were extensive and well-developed a short distance to the east, seen in test pits dug in 2006 (Chapter 4). The western edge of the Phase 4 sediments was seen in the narrow exposure of Log 40011 on the western side of the main site, where they were only about 200mm thick, and where they included sandy and gravelly beds/patches. To the east of the site, they consisted of fine homogenous clayey silts up to three metres thick, present between about 20 and 23m OD, with trace colour patterns from polygonal cracking due to periodic desiccation. They contained a distinctive fully temperate interglacial ostracod fauna indistinguishable from that of Phase 3, including the biostratigraphically significant *Ilyocypris quinculminata* (Chapter 11). These sediments are therefore thought to represent a still, muddy lake or pond, periodically drying up and probably butting up against, and perhaps interdigitating with, sediments of Phase 3. Molluscs were also present in the sediments seen in the 2006 test pits, suggesting a low-energy freshwater habitat (Chapter 4). No other fauna or flora are known; nor have Phase 4 sediments produced any artefactual remains.

Phase 5 (Clay-laminated sand)

Phase 5 sediments extend as a significant sheet of deposits about two metres thick across the whole length of the main site, dipping gently northward and unconformably truncating the underlying sediments of Phases 2 and 3. They predominantly consist of fine-medium sand, interspersed with undulating clayey/silty laminations and occasional gravel lenses; the former becoming more developed and more closely spaced towards the top of the sequence, and the latter better developed towards the bottom of the sequence. Wavy and cross-cutting bedding structures with thick homogenous sand beds in the bottom part of Phase 5 attest to very rapid fluvial deposition. This continues up through the sequence with episodes of higher energy marked by gravel lenses and quieter episodes marked by the clay/silt laminations. The uppermost part of the Phase 5 sequence (distinguished as context 40067) is marked by a significant increase in the frequency and thickness of the clay/silt laminations, as well as a marked increase in their waviness, to the point of being significantly contorted. This is thought to represent both a change in the environment of deposition, with an increase in the prevalence of quiet conditions, and perhaps occasional desiccation, and *in situ* deformation due to loading by overlying sediments whilst saturated. Gravity folding (*boudinage*) was also seen in the clay/silt laminations, and this is thought to have been associated with post-depositional re-arrangement of the sediments into the synclinal basin they now form.

Some faunal remains were recovered from the upper parts of the Phase 5 sediments, comprising occasional large mammal remains (Chapter 7), small vertebrate remains from a few bulk samples taken where decalcification was less pronounced (perhaps due to lenses rich in derived Tertiary shell fragments: Chapter 7) and occasional *Bithynia* opercula from these same bulk samples, which were used for amino acid dating (Chapter 13). The small vertebrates included numerous fish and amphibians, confirming the waterlain nature of the upper Phase 5 sediments, likewise supported by the presence of *Bithynia* opercula, which would have been more resistant to decay than other Pleistocene molluscan remains, which were absent. The larger mammal remains included lion, aurochs and deer; the smaller mammals included rabbit, ground squirrel (*Spermophilus* sp.), and a range of shrews and voles including water shrew (*Neomys* sp.), water vole (*Arvicola cantianus*), northern vole (*Microtus oeconomus*) and bank vole (*Clethryonomys glareolus*). As a whole, the vertebrate assemblage indicates a fully temperate climate and a well-vegetated waterside habitat with local grassland, scrubby woodland and some drier meadows/grassland with sandy substrate suitable for rabbit burrowing. Some of the large mammalian bones show cracking and splitting from exposure on a landsurface, indicating at least episodic drying up of the waterbody in this part of the sequence.

A small collection of 18 lithic artefacts was recovered from the upper parts of the Phase 5 deposits, 13 of them in mint/fresh condition and the remainder in more abraded condition. The latter assemblage (if not intrusive from the overlying Phase 8 gravel) is thought to represent hominin activity in the vicinity of the watercourse, transported by fluvial activity . The former group is thought to represent minimally disturbed evidence of intermittent activity on the spot, coinciding with temporary drying up of the watercourse and the exposure of short-lived landsurfaces. Technologically and typologically the collection (although of small size) conforms to the classic Clactonian industrial tradition, as also represented in Phase 6 (see below) with (a) the production of flakes by a simple, unstructured approach

to core reduction and (b) the secondary working of flakes to produce simple notched flake-tools.

Phase 6 (Grey clay, with organic-rich beds and tufaceous deposits)

The Phase 6 sediments were mostly grey brecciated clay, with occasional angular flint pebbles and cobbles of nodular flint, and were present in the central and southern parts of the main site, conformably overlying the Phase 5 fluvial sands. They dipped gently and thickened towards the central part of the site, where their surface formed a steep-sided synclinal basin, thought to result from post-depositional deformation of the sediments. The Phase 6 clays then thinned in the northern part of the site, becoming vestigial, and being unconformably truncated by the Phase 8 fluvial gravels. Where they were higher in the southern part of the site, the Phase 6 clays were internally relatively homogenous with faint sandy and silty facies. Where lower in the central part of the site, they were thicker and more complex in their lower parts, with brown organic-rich beds (context 40078), well-defined iron-pans dividing the basal clayey beds (Phase 6a) and including a small channel-cutting filled with tufaceous sediments (Phase 6b). Taking into account the geomorphological context of the site, the nature of the sediments, their geometry and the soil micromorphological analyses (Chapters 4 and 5), the Phase 6 sediments are thought to mostly represent slopewash deposits entering from the west into a stagnant, swampy waterbody that periodically dried up exposing temporary landsurfaces. The investigated site is thought to represent the west shore of this waterbody, at its junction with the slope of the western valley side, in a zone that thus oscillated as water levels fluctuated between saturation, and thus prone to peat formation, and exposure as drier land.

The Phase 6 sediments contained the main archaeological horizon of the site, with (a) the elephant skeleton (extinct straight-tusked elephant, *Palaeoloxodon antiquus*) and its associated lithic scatter in the central western part of the site and (b) the lithic concentration in the southern part of the site, south of Trench D. It also produced the majority of biological remains. Large vertebrate remains were found scattered throughout the grey clays that constituted the majority of Phase 6. These were often tiny weathered scraps, although occasional larger and better-preserved remains were also present in the lower-lying parts of the deposit in the central part of the site, notably the remains of the elephant skeleton, a rhinoceros skull (*Stephanorhinus hemitoechus*) and a rhinoceros jaw (*S. kirchbergensis*). The elephant skeleton was associated with a thin dark-brown horizon within the basal part of the Phase 6 clay (context 40078), on the west side of the central part of the site. This horizon was rich in fragments of rotted organic material, which produced a sparse and poorly preserved pollen assemblage attributable to the early temperate zone Ho-II of the Hoxnian (Chapter 12). The small channel of a short-lived stream filled with tufaceous and other calcareous sediments (Phase 6b) was found in the central part of the site, incorporated within the lower part of the Phase 6 clay approximately 15m to the north-east of the elephant skeleton. The Phase 6b sequence was rich in a range of palaeo-environmental remains, including molluscs, ostracods and small vertebrates, as well as larger mammalian fossils. Crucially, these included foot bones from the elephant, thus establishing precise contemporaneity of the tufaceous channel and its environmental evidence with the elephant skeleton and its pollen assemblage.

The faunal remains from the tufaceous channel provide a clear picture of a fully temperate interglacial climate, perhaps slightly warmer than the present day, with a local mosaic of habitats that included closed-canopy woodland, grassland and wetland habitats bordering a stream. Biostratigraphic indications from the mammalian evidence (Chapter 9) provide firm support for dating the Phase 6 deposits to MIS 11, the Hoxnian (*sensu* Swanscombe, Barnfield Pit phases I and II; and *sensu* Strata D-E at the Hoxne type site). In particular these include the co-occurrence of species not known in the UK after MIS 11 (pine vole *Microtus* (*Terricola*) cf *subterraneus*, shrew *Sorex* (*Drepanosorex*) and mole *Talpa minor*), with species known in Britain only from MIS 11 or later in the UK (narrow-nosed rhino *S. hemitoechus*, Merck's rhino *S. kirchbergensis*, aurochs *Bos primigenius*, water vole *Arvicola cantianus*). The morphology of the aurochs horn cores, the fallow deer and the red deer are also similar to the specific forms known from securely dated MIS 11 horizons at sites such as Clacton and Swanscombe. Further confirmation of an MIS 11 date is provided by the amino acid analyses on numerous *Bithynia* opercula from the tufaceous channel-fill, which clearly show levels of racemization matching accepted MIS 11 horizons at Barnham, Barnfield Pit, Hoxne and Beeches Pit (Chapter 13).

There are two quite distinct main areas of lithic artefact recovery within the Phase 6 clay. Firstly, there is a tight cluster of approximately 80 lithic artefacts immediately beside the elephant skeleton, categorised as 'Assemblage 6.3' in this study (Chapter 17). Their spatial association with the elephant remains, their high degree of refitting and their tight clustering combine to indicate a completely undisturbed assemblage reflecting manufacture of flint tools on the spot for butchery of the elephant carcass. Interpretation of these remains is hampered by the incomplete survival of the site, the elephant having been chopped in half by mechanical bulk ground extraction before its discovery, leading not only to loss of some of the skeleton, but also to loss of whatever archaeological evidence was surrounding its western parts. However, based on the recovered remains, the assemblage around the elephant comprised one percussor, four cores, four reduction episodes (Groups A-C and E), four presumed flake-tools and a minimum of thirteen separate pieces of raw material, nine of them represented by single pieces of debitage. The percussor was found in seven pieces (Group D), and is thought to have broken during attempted

reduction of one of the larger cores found incompletely reduced with several refitting flakes (Group B). The longer reduction sequences all represent the early stages of reduction, by simple unstructured flaking, of pieces of nodular flint, presumably locally obtained. The flake-tools comprise a single-notched flake and three flakes that were not secondarily flaked, but were interpreted as tools on the basis of visible use-wear on suitable sharp edges. Technologically and typologically the material conforms to the classic Clactonian industrial/cultural repertoire, discussed further below. It is suggested that the site represents a combination of knapping activity associated with initial discovery of the elephant's carcass (or following its killing), expediently producing flint tools from locally-obtained raw material to butcher it for meat or other nutritious tissue, in conjunction with discarded tools from subsequent visits (perhaps during a period of around 1 week when the meat retained freshness), that were not made at the spot, but were brought to it in the knowledge of the awaiting carcass.

The elephant was not mired in soft sediment, but died at the spot when the ground was exposed as a dry land surface, judging by the lack of disturbance to the underlying sediments. The immediate area would probably have been mostly wooded, densely in patches, although with some more open areas with shrubs and grasses; and the ground would have sloped up to the west from the elephant carcass. The edge of a quiet waterbody would probably have been close by to the east, where there would probably have been soft clayey sediments and swampy conditions. The rise and westward expansion of this waterbody not long after the death of the elephant (and its butchery) led to its preservation due to its submergence and the consequent growth of peaty horizons in conjunction with burial by slopewash sediments.

The second main area of lithic artefact recovery is the much denser concentration of nearly 1900 flint artefacts found in the Phase 6 clay at the southern end of the site, to the south of Trench D, approximately 30m to the south of the elephant skeleton. These are grouped for analysis here as 'Assemblage 6.1' (Chapter 18). Although a few were in abraded condition, and so regarded as older reworked intrusions, the great majority (more than 1850 artefacts, of which approximately 110 were chips < 20mm long) were in very fresh condition, and regarded as a single assemblage. The artefacts forming the assemblage were found within the lower part of the Phase 6 clay in a stratigraphically equivalent horizon to the elephant skeleton and the tufaceous channel, and are presumed to be broadly (and probably in at least part, exactly) contemporary with Assemblage 6.3 from around the elephant. However, with the exception of the poorly preserved and fragmented jaw of a Merck's rhinoceros found right at the northern edge of the concentration, there were no identifiable faunal (or other palaeoenvironmental) remains found in association with them. The assemblage was altitudinally higher (having been recovered in a sub-horizontal band between *c* 27 and 27.5m OD) than that from by the elephant skeleton (which occurred in a narrow band approximately 150mm thick that sloped between *c* 24 and 24.5m OD). This may reflect a higher and drier position less conducive to biological preservation and more conducive to hominin occupation. However, in light of the gross post-depositional sedimentary deformation represented by the synclinal basin in the central part of the site, it is uncertain what the relative topography of these locales would have been in MIS 11, and what impact this might have had on the distribution of hominin activity.

The artefacts of Assemblage 6.1 mostly comprised waste debitage (90%), cores (5%), and flake tools (4%). The assemblage was strongly spatially clustered with northern and southern areas of high concentration (up to 25 artefacts per m^2) interspersed with areas of very low concentration. However, the clustering did not conform to patterns reflecting undisturbed knapping scatters; likewise, the lack of microdebitage and the refitting results – only about 8% of the assemblage refitted, without long and closely-spaced refitting sequences – did not suggest a palimpsest of entirely undisturbed artefactual remains from activity on the spot. There were, however, several instances of refitting material with short separations suggesting a minimum of disturbance. It was concluded that most of the artefactual material had been mixed and transported a short distance by slopewash processes, but that it still represented contemporary activity from the immediate vicinity, in the bankside area of the waterbody. It was also concluded that, in amongst the general mass of slightly moved material, was a small element of entirely undisturbed material representing artefacts recovered from the precise position where they had been discarded.

Assemblage 6.1 represents a consistent picture of the minimally structured (or perhaps, entirely unstructured) reduction of locally obtained pieces of nodular flint raw material, producing flake blanks that were then either used without further modification as cutting tools, or secondarily worked into simple flake-tools. The most common form of secondary working was the striking of one (or more) small flakes from one edge of a flake, leaving a sharp concave notch (or more than one notch, often linearly aligned), that would have provided excellent cutting edges. The presence of macroscopic use-wear on the sharp edges of these notches, as well as on sharp edges of some unworked flakes, supports their interpretation as cutting tools. There was just one example in the un-derived fresher condition material of a simple core-tool, and no examples of bifacially-shaped handaxes or debitage from their manufacture. Typologically and technologically, Assemblage 6.1 exactly conforms to the Palaeolithic industrial tradition labelled as 'Clactonian' since the 1920s (Breuil 1926; Warren 1926; Wymer 1968, 34-44; Wymer 1985, 277-283; Roe 1981, 70), as represented in the Lower Loam and Lower Gravel at Barnfield Pit, Swanscombe (Ashton and McNabb 1996), the pale silt (unit 5) at Barnham (Ashton 1998) and at Clacton-on-Sea itself (Warren 1951and 1958; Oakley and Leakey 1937; Singer *et al.* 1973). Subject to subsequent debate, the

status of the Clactonian is discussed separately below, along with the industrial/cultural attribution of Assemblage 6.1. The lithic material represented an even balance of all stages of production of the *chaîne opératoire*, suggesting lithic production as mostly an expedient response to need, leading to a homogenous distribution of knapping remains around the occupied landscape.

In addition to these two main concentrations of lithic remains, there was a general background noise within the Phase 6 grey clay of a low density of scattered isolated lithic artefacts, categorised as 'Assemblage 6.2' in this study. Totalling only about 125 artefacts, this assemblage mirrored Assemblages 6.1 and 6.3 in its typological and technological characteristics. Having been recovered by watching mechanical excavation rather than by hand excavation, Assemblage 6.2 is probably slightly biased towards larger artefacts. It includes three refitting pairs of artefacts, each pair with a separation distance of less than two metres, suggesting a minimum of disturbance, if any. The assemblage is thought to have formed in the same way as Assemblage 6.1, with a combination of the entirely undisturbed remains of activity discarded during periodic exposure of a dry surface, and material gently transported in by slopewash from the bankside area to the west. Various isolated large mammal remains were also found dispersed throughout the same area of the Phase 6 grey clay as Assemblage 6.2, including some with evidence of hominin interference (Chapter 7). There are however, apart from around the elephant skeleton, no spatial associations between any specific lithic and faunal remains that suggest a causative connection between them.

Phase 7 (Mixed clay/gravel)

Phase 7 sediments were well-developed in the central part of the site, filling the U-shaped synclinal basin of the 'skateboard ramp'. The basal junction with the top of Phase 6 sediments was generally sharp, but was nowhere erosive and unconformable, suggesting no major depositional hiatus. Phase 7 deposits were absent at the northern and southern ends of the site, where the Phase 8 gravel that unconformably overlay them had its base cut down to the Phase 6 clay. The Phase 7 sediments mostly consisted of variably gravelly brecciated clay, with a thick basal sand/silt bed in the trough of the synclinal basin. They were mostly structureless, although there were some thin parallel sand/silt/clay beds in the synclinal trough. Their upper, more gravelly parts contained occasional concentrations of what looked like rotted and charred plant macro-remains and produced an assemblage of variably stained/abraded lithic artefacts (Chapter 19). Sediments of this phase continue to the west and north-west of the site, seen in various nearby investigations extending upslope as a substantial clayey/gravelly mass, and interpreted as a local mass movement deposit, originating from west or north-west of the site (Chapter 4). This is thought to have overridden the site, perhaps causing the synclinal trough by lateral pressure due to a major landslip event when the ground conditions were saturated, or by the weight of slipped sediments compressing softer organic-rich sediments in the centre of the trough.

The Phase 7 sediments thus continue the depositional trends of the underlying sequence of Phases 3-6, representing a phase when slopewash processes became more pronounced. A more extensive and significant downslope movement of deposits to the east caused much greater sediment accumulation at the site, completely over-riding the swampy valley floor, rather than providing minor input to its western side. Despite their great thickness, they probably formed very rapidly, and represent an insignificant portion of Pleistocene time. No biological remains were found in the basal sand/silt beds filling the synclinal trough to provide information on climate and environment. The rotted wood and other plant remains in the overlying clayey gravel suggest a continuation of temperate conditions, and there is no sedimentary evidence of climatic deterioration between Phases 6 and 7.

The lithic remains contained in the Phase 7 deposits are thus probably broadly contemporaneous with those from the underlying Phase 6, and may include material gathered from higher up the valley sides, as well as from closer at hand on the valley floor at the foot of its western flank. They therefore provide an important record of hominin activity and lithic production from a wider catchment area than represented in the Phase 6 deposits. Technologically and typologically, the lithic material is identical to that from Phase 6. It also likewise represents an even balance of the different stages of lithic production, suggesting that expedient tool making was being carried out across the slightly wider landscape, and not just the bankside occupation area of the valley floor.

Phase 8 (Ebbsfleet gravel)

The Phase 8 deposits comprise gravels, sandy/clayey in places, that cut unconformably across the underlying sequence. They dip gently from south to north, directly overlying deposits of Phase 6 at *c* 27.5m OD at the south end of the site, and cutting into deposits of Phase 5 at *c* 25.5m OD at the north end. In between, they cut into the deposits of Phase 7 filling the synclinal trough of the 'skateboard ramp' in the central part of the site.

Clast lithological analysis has established that these are fluvial gravels, representing an early course of the Ebbsfleet (Chapter 6). Numerous investigations in unquarried areas to the north and north-west of the site have revealed the continuation of the gravel body, heading towards a confluence with the Thames some 300m further north (Chapter 4). They contain no biological remains to help with dating or palaeo-climate/environment, but on geomorphological grounds they can be broadly correlated with the Lower Middle Gravel of the Swanscombe 100-ft terrace at Barnfield Pit, widely accepted as dating to the Hoxnian, MIS 11 (see for example Bridgland 1994).

It was not immediately clear that the Phase 8 gravels were also archaeologically important. Initial examination of the extensive exposures in the main east-facing and west-facing site sections failed to produce any artefacts, apart from two flakes that were found loose on the gravel surface, and were therefore not indisputably from the deposit. However, it then became clear through systematic sieving and careful watching of mechanical excavation both that the gravels were a reasonably rich source of lithic artefactual remains, and also that there is a major technological and typological contrast between the lithic material from Phase 8 and that from underlying deposits.

The Phase 8 gravels produced an assemblage of 180 artefacts, including more than 30 handaxes. The majority of artefacts were in fresh (32%) or slightly abraded (44%) condition, with the remainder mint (8%) or very abraded (16%). Thus, although some of the assemblage probably represents undisturbed remains of activity on the spot, on temporarily exposed gravel bars on the floodplain, most were probably transported a short distance from a slightly wider catchment area, or had been subject to reworking within the gravel. None was sufficiently abraded to be considered as signicantly transported or reworked from significantly older deposits, so the assemblage was treated as a whole for analysis, sub-divided into three depositional phases based on internal stratigraphy within the gravel.

Although not supported by quantitative data, the impression was formed that artefacts were most abundant in the lower parts of the gravel and became scarcer higher up. This is contrary to the sizes of the recovered assemblages, but allows for the fact that much greater volumes of sediment were excavated from the higher parts of the gravel. In general, the Phase 8 handaxe collection shows great diversity, with both pointed and ovate forms present from top to bottom of the sequence. The more ovate handaxe forms (most of which have a Z-twisted profile) are more common in the bottom part of the gravel, with those found higher up invariably in abraded condition, potentially reflecting a history of derivation from the basal gravel layers. Although there was clearly an emphasis on handaxe manufacture, there was also a small but significant element of flake-tools, comprising simple utilised flakes and partly-trimmed flake-knives, and also convex unifacially flaked side-scrapers, similar in appearance to the Mousterian 'Quina-type' scrapers of southwest France.

Complementing these contrasts in technology and typology between Phases 6 and 8, there is also a radically different structure to the spatial organisation of the *chaîne opératoire*. In Phase 6, all the evidence suggests firstly: that the *chaîne opératoire* was not spatially organised around the landscape, but generally started and finished in the same part of the landscape in relation to encounters with resources leading to lithic production; and secondly, that that the distribution of lithic remains across the landscape was homogenised by the spatially unpatterned distribution of these encounters.

In contrast, there is a clear pattern in the evidence from the Phase 8 gravels of a consistent spatial organisation of the technological *chaîne opératoire*. The quantity of debitage recovered is far less than commensurate with manufacture of about 30 handaxes, indicating that they were mostly made elsewhere in the landscape, before becoming abandoned at, or in the vicinity of, the site. This is a complementary pattern to that represented at, for instance, the site at Red Barns in Hampshire (Wenban-Smith *et al.* 2000; Wenban-Smith 2004b). There it appears that an exposure of flint-rich chalky slope-wash deposits served as a location for the manufacture of handaxes that were then mostly taken away, leaving a disproportionate amount of debitage in relation to the small number of handaxes found.

There thus seem to be both technological/typological and behavioural/organisational contrasts between the hominin behaviour of Phases 6 and 8, as reflected in the lithic remains; possible implications of this are further considered below.

Phase 9 (Brickearth bank)

The Phase 9 brickearth deposits were present at the northern end of the main site, conformably overlying the northward-dipping surface of the Phase 8 gravels, with a base level of *c* 26m OD. They continue to the north, where they were substantially cut into by groundworks, with the truncated surface forming a sloping bank covering an area approximately 100 x 25m. Numerous test pits were dug above the higher western side of this bank in 2006, allowing further examination of the brickearth sediments and establishing that their uppermost surviving parts reached at least 30m OD.

Although with thin sand and gravel beds at their base, the Phase 9 deposits mostly comprise a thick homogenous body of reddish-brown sandy/clayey silt, colloquially known as 'brickearth'. The main brickearth body contains occasional lenses and patches of fine gravel, and occasional faint parallel clayey/sandy beds dipping gently to the east, transverse to the north-south axis of the Ebbsfleet valley. These deposits are therefore presumed to be primarily colluvial in origin, reflecting slopewash from higher ground to the west, probably with an aeolian component.

They lack any biological remains that could contribute to dating them, or give any indication of climate or palaeo-environment. There was no evidence for a major depositional hiatus or climatic deterioration between the top of the Phase 8 gravel and the base of the brickearth. Hence the most likely age of the basal part of the deposit is late MIS 11 or MIS 10, if one accepts the Phase 8 gravels as belonging to MIS 11. However, the Phase 9 brickearth body could be of younger age, or could contain a series of colluvial deposits of different ages without clear lithostratigraphic junctions between them. An OSL dating result of *c* 60k BP was obtained from towards the top of the brickearth, and a result of *c* 280k BP from its base (Chapter 14). Although these dating results should be treated with great caution in light of the anomalous results from lower in the sequence, they suggest that the brickearth, despite the lack of visible internal lithostratigraphic boundaries, might include different phases of deposition.

A collection of 14 lithic artefacts was recovered from the truncated surface of the brickearth, exposed by mechanical excavation of a sloping bank through it. These artefacts, which included a substantial subcordate handaxe, were scattered across the sloping bank, over an area approximately 100 x 20m. None of them was found *in situ* in the brickearth, and they were in varied condition and with different degrees of patination. All however, appeared to be of Lower/Middle Palaeolithic origin. The handaxe was strongly stained and patinated on one side, but not the other, suggesting prolonged exposure prior to burial.

A few pieces of waste debitage were recovered *in situ* from the brickearth at the main site, establishing that the deposit does contain lithic material. These were mostly technologically undiagnostic, apart from one flake that represented the later stages of thinning/shaping a handaxe. Finally, a magnificent pointed handaxe in mint condition was recovered from the upper part of the brickearth in one of the 2006 test pits (Fig. 4.42). This artefact was also stained brown on one face, with the other entirely unstained, likewise suggesting a period of exposure prior to burial.

The Phase 9 brickearth is without doubt equivalent to the 'ferruginous loam' reported by Carreck (1972, 61) exposed in the quarry faces to the east of Southfleet Road. He suggested on the basis of its general height above OD that it might be related to 'the Boyn Hill Terrace', but was unable to examine it closely. This interpretation can now be ruled out, since the deposit is underlain in many places by Ebbsfleet gravels, and is now thought to be mostly a mixed colluvial/aeolian slopewash deposit lining the west flank of the Ebbsfleet valley; it remains an incompletely understood deposit, of uncertain date and depositional origin. It does, however, contain mint condition Lower/Middle Palaeolithic artefacts, and merits further investigation where it survives to the north of the site and to the east of Southfleet Road.

Phase 10 (Holocene features)

Holocene features cut into the Pleistocene deposits produced various Late Prehistoric lithic and ceramic material, not considered in this volume. In amongst this material was found a single, secondarily-worked flake from the lower fill (context 40133) of a pit (context 40129), which had staining, patination and abrasion suggested derivation from Palaeolithic times. It was not recognisable as any of the types used in the analysis, so was classified as a 'miscellaneous flake-tool' and adds nothing to the understanding of the site. It was included with other derived Palaeolithic material from modern made ground, under assemblage group 11.1, discussed below.

Phase 11 (Modern made ground; not in situ)

The lithic collection from the site included about 50 artefacts that were not provenanced to Pleistocene contexts. Apart from those whose Pleistocene provenance had become misplaced during the excavation or post-excavation process, this collection included artefacts found loose around the site and from deposits thought to be modern made ground from above the Pleistocene sequence. The Phase 11 collection mostly comprised a variety of derived Lower/Middle Palaeolithic material, including 12 handaxes (or broken parts of), a core, and a selection of flake-tools and waste debitage, all similar to material from the known main artefact-bearing deposits of Phases 6-8.

However, in amongst this material was a distinctive assemblage of 12 artefacts (grouped as 'Assemblage 11.2') that immediately stood out during analysis because of its mint condition and unpatinated/unstained appearance, and its curious technological characteristics. It mostly comprises quite large and very chunky flint flakes, violently struck, with notch scars from secondary flake-flakes struck sideways across the ventral surface, sometimes a single notch and sometimes double opposing notches (often with the distal ends of their flake scars intersecting). This assemblage was identified as an 18th century gunflint industry, at least one major practitioner of which (William Levett) is known to have been active in the Northfleet area (Chapter 21).

It seems likely that the same local availability of flint raw material as stimulated hominin activity at the site in the Hoxnian interglacial, outcropping in the late 18th century on the side of the Southfleet Road, led to its exploitation some 400,000 years later for a wholly different purpose, although by remarkably similar means.

THE ELEPHANT: SITE FORMATION AND HOMININ EXPLOITATION

Hominin exploitation of megafauna in the Pleistocene is a major topic of current debate (Gaudzinski and Turner 1999; Gaudzinski *et al.* 2005; Delagnes *et al.* 2006; Yravedra *et al.* 2012; Rabinovich *et al.* 2012; Sacca 2012; and subsidiary references), evolving from critiques of the later 20th century (such as Binford 1981; Isaac and Crader 1981; Isaac 1983; Gamble 1987; Nitecki and Nitecki 1987; Villa 1990). Questions began to be asked about whether the co-occurrence of megafaunal remains and lithic artefacts in the same depositional horizon implied hominin megafaunal exploitation, and if there was hominin exploitation, what was its nature? Was it primary targeted hunting, systematic scavenging or marginal expedient scavenging? And what was the importance of megafaunal remains for purposes other than nutrition? The discussion below is focused upon Proboscidean remains, and particularly sites with *Palaeoloxodon antiquus*, but is also broadly applicable to other megafauna such as hippopotamus and rhinoceros,

apart from in relation to specifically Proboscidean behaviour and anatomical characteristics.

There are two main types of site pertinent to these questions (cf. Gaudzinski *et al.* 2005). Firstly, there are widely spread scatters, sometimes of high density, where megafaunal remains and lithic artefacts both occur in the same horizon, but with no strong co-association of spatial clustering, and several (or numerous) individuals represented. Here (for example at Cotte St. Brelade – Scott 1986; Torralba – Binford 1987a; Lynford – Boismier *et al.* 2012; Castel di Guido – Sacca 2012; Revadim Quarry – Rabinovich *et al.* 2012), debate focuses on the presence/absence of direct signs of hominin interference with the faunal remains, and upon indirect signs of hominin exploitation in the profile of the faunal assemblage, in particular: skeletal elements, age and seasonality. Secondly, there are undisturbed single-carcass sites, where remains of a single individual are found, often in association with lithic artefacts. For these sites (for example Mwanganda's Village – Clark and Haynes 1970; Olduvai FLK North – Leakey 1971; Barogali – Berthelet and Chavaillon 2001; Lehringen and Gröbern – Weber 2000; Aridos 1 and 2 – Santonja and Villa 1990, Yravedra *et al.* 2010) the spatial association of lithic artefacts provides a strong *a priori* indication of associated hominin activity. Once alternative non-hominin taphonomic factors have been duly considered, debate can focus on the nature of the hominin exploitation, rather than whether there is a hominin role at all. There are also other, more intermediate sites (such as Benot Ya'aqov – Goren-Inbar *et al.* 1994; Ariendorf 2 – Gaundzinski *et al.* 2005; Notarchirico – Cassoli *et al.* 1999; PRERESA – Yravedra *et al.* 2012; and perhaps also Nadung'a 4 – Delagnes *et al.* 2006) where remains of single carcasses, perhaps associated with clusters of lithic remains, are disguised against a background of other lithic and faunal remains.

The Southfleet Road site exemplifies the second site type, providing clear evidence of the co-association of an undisturbed cluster of lithic artefacts with the carcass of a large adult male elephant, in the prime of its life (at an estimated 35-40 years old) and weighing perhaps 8000-10,000kg. The presence of several distinct reduction episodes, the exceptionally high degree of artefact refitting (79% of artefacts >20mm), the tight clustering of the refitting flints (weighing from 1 to *c* 1400g), their presence in the same narrow band of sediment as the elephant bones and the fine-grained nature of the containing sediment combine to indicate that the lithic remains represent on-the-spot knapping and discard (Chapter 17). Within the context of a surrounding lithic and faunal find density of well below 1/m², the close juxtaposition of the cluster of lithic remains with the elephant skeleton is unlikely to be coincidental.

Considering the vanishingly small likelihood that either the elephant coincidentally died immediately beside a freshly made knapping scatter, involving several reduction episodes on the same spot, or that the knapping activity occurred immediately beside a fresh elephant carcass in a bankside location otherwise lacking obvious resources, but had no connection with it, it is beyond reasonable doubt to assume that the lithic remains at the spot represent undisturbed evidence from hominin exploitation of the elephant carcass. Even though there is no evidence of cut marks or deliberate breakage, it is well-established that elephant remains can be butchered without leaving many recognisable marks (Haynes 1991; Haynes and Krasinski 2010). Also, the poor preservation of many of the surviving elephant bones has destroyed/obscured many elements from which cut marks or breakage might have been identified (Chapter 8).

The available quantity of meat and other edible remains would depend upon whether the hominins had first access to the fresh carcass, or whether they were lower down the exploitation chain. This in turn depends partly upon whether the beast was hunted or scavenged. This is considered further below. It is intuitively likely that one of the prime benefits of an elephant carcass would be its meat. Without doubt, the fresh carcass would have had a huge amount of meat, and this would probably have been desirable to the Southfleet Road hominins and exploited by them. However, an even more desirable resource would have been its fat, and more fatty and nutritional parts of the carcass such as brain, trunk, tongue, offal and pads within the feet. Other instances of Middle Pleistocene elephant carcass butchery show breakage of the skull to get at the brain (for example Gesher Benot Ya'aqov – Goren-Inbar *et al.* 1994), and removal of the jaw, probably to get at the tongue (for example Notarchirico – Cassoli *et al.* 1999). At Southfleet Road, the skull remnants are too fragmentary and poorly preserved to consider whether or not the skull might have been broken into. The mandible is, however, missing and no sign of it, or of the robust molars that it would have contained, was found. This seems likely to represent hominin behaviour as it would not easily have become naturally disarticulated. In addition, one of the feet was found about 20m to the east of the rest of the elephant, in the upper part of the tufaceous channel sequence. Although feet are one of the body parts of dead elephants that are most prone to natural detachment and carnivore scavenging, the well-preserved bones of this foot showed no sign of carnivore interference, suggesting that this too may reflect hominin activity.

Although a systematic functional assessment of the associated lithic assemblage was not attempted, lithic production seemed (as with the much greater lithic concentration in the southern part of the site) to be focused on the manufacture of sharp-edged flake blanks of medium-large size. These could either be used unmodified as cutting tools, or could be used as blanks for creation of simple notched cutting tools. Four flake-tools were found near the elephant, one of them being a notched tool, and the other three being unworked flakes with signs of macro-wear on their main sharp edge suggesting minor damage during use for cutting. A fifth tool was also evidenced by the flake-flake from creation of a notched tool (which was itself absent). And three gaps in the reduction sequence of refitting Group C suggest a further three flakes selected and extracted as

tools. Refitting Group D was interpreted as a broken knapping percussor; and marks of percussion on core D.40494 were interpreted as failed flake removals, perhaps leading to breakage of the percussor represented by Group D. Although this interpretation is preferred, these could also represent evidence of a heavier duty tool component, used for bashing at solid parts of the elephant skeleton, particularly the skull. It is however thought unlikely that any attempt at marrow extraction was carried out, as elephant bones do not have easily accessible marrow, but require special processing to extract it from hollows within the bones (Sacca 2012). It is also thought unlikely that any use was made of elephant bone as a raw material for tool manufacture. Flint raw material was locally abundant, and the use of elephant bone as a knapping raw material is only known – in the European Middle Pleistocene – from Italy (Gaudzinski *et al.* 2005; Sacca 2012), and is thought to be a response to scarcity of good lithic raw material.

Unfortunately, the absence of part of the site compromises consideration of the size of the hominin group potentially involved in exploitation of the elephant. Based on the surviving evidence, there were four cores reduced at the site from their early stages, and there is evidence of 13 separate pieces of raw material, including individual flakes. There is evidence of one percussor; and of perhaps eight flake-tools in use. It is suggested here that exploitation of the elephant might have involved an initial episode of lithic production from local raw material when the fresh carcass was first exploited, followed by further visits bringing tools made elsewhere to the carcass while it retained sufficient freshness to remain edible. On this basis, one could postulate a band of between four and 13 individuals, with the low end of this range being preferred on the balance of probability and the evidence of four core reduction episodes from first exploitation of the beast.

Comparative data on elephant and other megafaunal butchery sites is hard to obtain, particular as most comparator sites include a greater potential degree of unrelated background material. At Aridos 1, eight cores were found, with three quartzite percussors, two biface tips and 39 flake-tools (Santonja and Villa 1990). At Aridos 2 (less completely preserved), four cores were found, together with five tools of various types, including a cleaver, a biface and three flake-tools (ibid.). At Notarchirico, one percussor was found in the area of the main elephant, together with 4 four handaxes, five 'chopper/cores' and three flake-tools (Cassoli *et al.* 1999). At Gröbern, a smaller group of flakes was recovered which could be divided into about six different original raw material pieces (Weber 2000). And at Gesher Benot Ya'aqov, there were nine handaxes found in close association with one elephant skull (Goren-Inbar *et al.* 1994), although it is uncertain to what extent these were associated with its exploitation, or are part of a general background of archaeological material in the sediments. Perhaps the best and most completely recovered comparator site is the horse butchery site GTP 17 at Boxgrove (Pitts and Roberts 1997; Pope 2002, 95). Here, eight distinct refitting scatters were found, representing the manufacture of eight handaxes, all of which were absent from the excavated site, and so were therefore taken away after use. Taken together, and notwithstanding a host of uncertainties over how these tool kits relate to hominin band numbers, a consistent picture emerges of about 6-10 as the number of individuals represented by the key data of core quantity, handaxe quantity or raw material pieces represented. Bearing in mind that some of the Southfleet Road site was lost to mechanical excavation, this corresponds well with the earlier band-size estimate of '4-13 and probably at the low end of the range' of those directly involved in exploitation of the elephant. Other studies (such as Gamble and Steele 1999) suggest 20-40 as the overall size of Lower Palaeolithic groups in north-west Europe at this time. Therefore we can perhaps envisage that the elephant was exploited by part of a larger hominin group, and that meat and other nutritious elements were transported back to other group members.

More problematic, however, is consideration of whether the elephant was found dead and scavenged after other carnivores had first access, whether it was found freshly dead (or in a wounded/disabled state, making its final dispatch an easy task) allowing first access to it, or whether it was hunted and killed. The possibility that it was found stuck in muddy sediments at the edge of the waterbody can be discounted. The sediment layer under the skeleton was undisturbed, and the bones were found in a narrow band at one horizon suggesting it died and initially decomposed on a dry ground surface. It is suggested here that this surface was later submerged by rising water level and slopewash deposits, leading to burial and preservation of the skeleton and its associated archaeological remains. There is also no local landscape feature such as a cliff or gully that would have led the elephant to become trapped in this location.

Although it should be emphasised that there is no direct evidence to support the notion that it was hunted, there are however certain factors that make this a likely possibility. Firstly, it is wrong to assume that hominins lacking technology such as guns, nets, metal-working and bows would be incapable of hunting a healthy elephant, despite the great size disparity. There is extensive ethnographic evidence that elephants could be successfully hunted with nothing more than a sturdy wooden thrusting spear (Zwilling 1942; Movius 1950; Adam 1951). Secondly, there is clear evidence that wooden spears were part of European Lower/Middle Palaeolithic technology. This includes the puncture wound on the Boxgrove GTP 17 horse scapula (Pitts and Roberts 1997); the spear point from Clacton-on-Sea (Wymer 1985); and the spears from Schöningen (Thieme 1997). There is also evidence that these were successfully used for elephant hunting in the early Late Pleistocene at Lehringen (Movius 1950; Adam 1951). Thirdly, there is no evidence of other carnivore activity affecting the Southfleet Road skeleton. Although much of the bone is in poor condition and thus does not preserve the crucial evidence, the well-preserved foot bones found in the

tufaceous channel show no sign of carnivore activity, and these would be one of the first pieces of the elephant that they would have scavenged. This suggests that the hominins had first access to the fresh carcass and may well have protected it for a period thereafter.

Finally (and perhaps paradoxically), the great size of the elephant and the fact that it was a male in its prime makes it more, rather than less, likely that it was hunted and killed by hominins. Such beasts are less likely to die of natural causes in the present day (Conybeare and Haynes 1984; Haynes 1991), and would have been more able to withstand other carnivore predators such as lions. Furthermore, if one considers other *Palaeoloxodon* single carcass archaeological sites in the Lower/Middle Palaeolithic there is a disproportionately high presence of larger adult males (for example at Lehringen, Gröbern and Aridos 2; not to mention Upnor, although no hominin association has been demonstrated for this latter). Together with the large adult male from Southfleet Road, these rare discoveries are perhaps the faint archaeological echo of a pattern of Proboscidean exploitation by hunting across the European Lower/Middle Palaeolithic.

ELEPHANT HUNTING AND THE ECOLOGY OF HOMININ ADAPTATION IN THE NORTH-WEST EUROPEAN MIDDLE PLEISTOCENE

Another point of continuing debate since the 1970s is the ecology of hominin adaptations in Europe during the Middle Pleistocene and the implications for hominin behaviour and patterns of colonisation and settlement into more northerly latitudes, with their reduced growing seasons and greater seasonality (Geist 1978; Gamble 1986, 1987 and 1993; Roebroeks *et al.* 1992; Roebroeks 2001; 2007). The joy (or perhaps, the curse) of palaeoanthropology is that there are few enough firm facts, and great enough imprecision in those we do have, to have free rein in imagining the past in a variety of ways, and from diverse intellectual perspectives. Thus there has been a pincer movement, whereby from one direction the Palaeolithic past is deduced from ecological principles, with suitably supportive facts highlighted from the archaeological record (perhaps Gamble 1987). From the other, a vast collection of facts (particularly environmental and lithic data) has been accumulated as building blocks for the overall picture (cf. Roebroeks 2007), without necessarily being integrated into a coherent vision of ecologically viable adaptations, and so without recognition/consideration of any behavioural implications. What of course one hopes from a pincer movement, is that at some point the prey is cornered; and perhaps we are reaching that point in the problem of the Middle Pleistocene settlement of north-westerly Europe, particularly that part north of the Pyrenees and north-west of the Alps.

As summarised by Roebroeks (2001), the intensity of investigation in north-west Europe since the mid-19th century provides a robust basis for accepting the broad pattern of occupation revealed. It indicates that (with the exception of a few earlier incursions, such as at Pakefield and Happisburgh on the Norfok coast – Parfitt *et al.* 2005; 2010) it is only in MIS 13, *c* 500,000 years BP, that hominin settlement appears to have become more sustained as far north as Great Britain. Gamble (1986; 1987) suggested, based on the combination of ecological theory and the occupational history of the last interglacial/glacial cycle, that Middle Pleistocene hominins were not able to survive in the dense peak-interglacial forests of north-west Europe. Particular difficulties were presumed to be posed by the seasonality of the more abundant plant resources, and the locking up of the less abundant (but nutritionally essential) animal biomass in either small parcels, not viable and too difficult to hunt in forested conditions, or in large dangerous herbivores that bred too slowly to form the basis of a hominin diet, even if they could be hunted. Then Gamble (ibid.) emphasised the material cultural evidence of widening social networks in the upper Palaeolithic and Mesolithic as the key to overcoming this ecological bind. However, apart from the fact that it is questionable whether this last interglacial/glacial cycle provides a valid model for the earlier Middle Pleistocene, there are as pointed out by Roebroeks *et al.* (1992) and confirmed by subsequent discoveries (Caours, France – Antoine *et al.* 2006; Beeches Pit, England – Preece *et al.* 2007) numerous records of Middle Pleistocene occupation in association with fully interglacial forested environments. There are also numerous records for occupation in temperate but not densely forested environments where there would have been the same seasonality, although a greater proportion of nutritious biomass in herds of medium-large herbivores.

So, *was gebt*? Primarily, hunting; at the time of Gamble's initial work in this area (Gamble 1986; 1987) there was no good evidence for successful large herbivore hunting, and, as summarised by Binford (1985), in a reaction to the long-standing trope of 'Man the Hunter', early hominin occupation of more northerly latitudes (and indeed also tropical ones) was widely thought to be underpinned by marginal tool-assisted scavenging of carcasses resulting from carnivore predation and natural death. Since then, there has been a growing body of evidence that Middle Pleistocene hominins were in fact successful hunters of large-medium size mammals. At Boxgrove *c* 500,000 BP, apart from the evidence of horse-hunting and butchery at GTP 17 (Pitts and Roberts 1997), cut-marks and skeletal elements at the main Q1B excavation area suggest preferential hunting and exploitation of other larger mammals such as rhinoceros and the larger species of deer (SA Parfitt, pers. comm.). At Schöningen, besides the spear itself (Thieme 1997), the associated faunal assemblage dominated by horse bones reflects their hunting (Voormolen 2008). And, returning to the Proboscidean megafaunal theme, there is of course the Lehringen elephant with the spear embedded in its ribs, provocatively claimed as a 'snow-probe' by Gamble (1987). As discussed by Roebroeks (2001), the practice of hunting large herbivores in the Middle Pleistocene can

also be associated with social and behavioural developments such as language, co-operation, strategic planning and the development of gender-based social structures. All of these contribute to the ability to maintain a viable adaptation in more challenging northern latitudes, whether in more open cool or mild conditions, or in densely-wooded peak interglacial conditions. It does, however, still seem to be only anatomically modern humans in the Upper Palaeolithic who solved the problem of surviving in peak glacial conditions; even then, not in the more northerly parts of Europe such as Great Britain, northern France and 'Benelux' (Otte 1990).

Complementing the evidence of hunting, more detailed evidence for environmental conditions at several sites associated with peak interglacial forested environments suggests that these were not necessarily situated within unbroken forest tracts, but were within a mosaic landscape that might support a greater density of herbivores, and might likewise make them easier to locate and hunt. This is for instance the case at Beeches Pit (Preece *et al.* 2007), Barnham (Ashton *et al.* 1998) and the occupation horizons associated with the Lower Loam at Barnfield Pit, Swanscombe (Conway *et al.* 1996), as well as here at Southfleet Road.

Building on Roebroeks' (2001) assessment, and Geist's remarkably prescient analysis – 'The only niche open [for Middle Pleistocene hominin northward colonisation] is that of a supercarnivore, who can despatch very large, slow, dangerous herbivores, and also the [other] large carnivores [lions, wolves *etc.*].' (Geist 1978, 281-282) – it is suggested here not only that large mammal hunting was an important aspect of more northerly adaptations in the Middle Pleistocene, but more specifically that (1) elephant hunting in particular was a crucial part of this adaptation, facilitating viable adaptations in more-wooded environments and (2) that elephant hunting may have underpinned northward colonisation at the start of the Middle Pleistocene, based on the substantive history of megafaunal exploitation in the Lower and early Middle Pleistocene in Africa and the near East. As discussed above, although just one instance of a single elephant carcass in the northern European Middle Pleistocene can be directly interpreted as hunted on the basis of firm evidence – that of Lehringen, based on the wooden spear between its ribs – the age/size profile of the assemblage of *c* 6-8 individuals found as single carcasses with archaeological associations is markedly different from a natural death profile and indicative of targeted hunting. Part of the ecological thinking of Gamble's initial model was that megafauna such as rhino and elephant were too slow reproducing to underpin a hominin adaptation, besides presumed difficulties in hunting and locating them. However, this is unlikely to have been an important factor considering the tiny hominin populations of the time, and the much larger megafaunal populations. Prime adult males would in fact have been the most expendable element of the megafaunal adaptation, having been ejected from the matriarchal herd at maturity and then roaming in bachelor groups, with occasional reconnection with the matriarchal herd for mating (Haynes 1991). Thus, the loss of some prime males would have had a negligible effect on the overall stability of the megafaunal population, but would have conversely been crucially important in aiding the survival of a small hominin group, considering the quantity of meat and other nutritional parts on a single fresh carcass.

LITHIC TECHNOLOGY, MOBILITY AND THE ORGANISATION OF PRODUCTION

Turning to lithic technology, there are two main archaeological horizons at the site, with deeply contrasting lithic material culture, in terms of both (a) technological pathways and resulting tools, and (b) the spatial organisation of lithic production. In Phases 6 and 7 (the latter regarded as containing derived evidence from the former), there is a consistent representation of the same lithic material cultural signature at three different spatial/temporal scales: short-term activity beside the elephant carcass, medium-term bankside activity, and medium/long-term activity in the slightly wider local landscape represented in the assemblage from Phase 7. The technological *chaîne opératoire* typically involves the expedient collection and use of local flint raw material in response to an encounter with a resource. The piece of flint was then rapidly reduced to a collection of medium-large flakes and irregular waste by a minimally structured reduction pathway involving a combination of episodes of alternate flaking, new platform selections and repeated flake removal from single platforms. Selected larger pieces with sharp edges and convenient handling properties were then either used without further modification, or turned into simple tools, mostly by single or double notching, and/or removal of sharp projections to facilitate handling. Although these tools can be grouped into 'types' *post hoc*, it is entirely uncertain whether these types would have been meaningful to the makers, and it is suggested here that they may represent stages of progression along a continuum of reduction/use intensity. The overall functionality of the lithic industry is clearly aimed at a cutting capability, although heavier cores and rounded nodular flint pieces could have been used for various heavy-duty percussion tasks. The nature and location of battering on several pieces clearly suggests flint knapping, but these pieces could also have been used for other tasks such as breaking bones for marrow, or breaking into an elephant skull to get the brain.

Although there is some suggestion of tool movement around the landscape, leading to some tools being resharpened and abandoned separately from their debitage scatters, there is no apparent spatial structuring of the stages of lithic production within the local landscape. All stages of production are equally represented in the various assemblages studied, even though completed individual reduction sequences are mostly not present. Although reduction sequences were therefore not necessarily always started and finished on the same spot, or even within the site area, there was not, at this scale, any spatial structure to the knapping activity. The technological *chaîne opératoire* was either entirely completed within the site area, or was equally likely to have been started as finished. This matches a model of primarily expedient exploitation of a resource whose precise location was unpredictable within a site catchment area. Over time, this has created a homogenous lithic signature, with an equal representation of early and late reduction stages and discarded tools, similar at different spatial/temporal scales.

This contrasts greatly with the signature from Phase 8. Here we have a lithic material culture dominated by bifacial handaxes, and with virtually no evidence of flake production from simple cores. In terms purely of reduction pathway and implicit cognitive capability behind their manufacture, these core tools present a deeply significant contrast. Compared to the (relatively easy, but not as simple as one might think) task of producing a series of flakes in an unstructured manner, the production of handaxes involves the skilful (try it!) removal of an (often long) series of thinning/shaping flakes. The whole sequence of knapping is aimed at producing a single bifacial tool, generally symmetrical in both plan and cross-section, with a sinuous and moderately sharp edge around most of its perimeter. TThere is thus a much greater investment of time and effort in producing a single tool. There is much debate about the cognitive implications of handaxe manufacturing and symmetry (Lycett 2008; Kohn and Mithen 1999; Machin *et al.* 2005; Spikins 2012), and also over the extent to which the range of finished forms we find in the archaeological record were deliberately intended from the outset, or result from factors such as raw material type or resharpening intensity (White 1998). Although no consensus has been reached in this debate, most with practical experience of experimental handaxe replication regard it a skilled craft requiring advanced cognitive capabilities to plan the reduction pathway to achieve an intended end product (see discussion in Wenban-Smith 2004b). Whether intentionally shaped or not, the handaxes found in Phase 8 show a variety of forms, ranging from sharply pointed to bluntly sub-cordate and cordate, many of the latter with markedly twisted profiles.

Of much greater import, however, is that there is also a significant contrast in the organisation of the production of the lithic assemblage from the Phase 8 gravel. There is a major imbalance between the quantity of handaxes present (most of them well-worked with scars from numerous flake removals) and the quantity of debitage. It is clear that handaxes have been preferentially brought to the site and discarded (even if not necessarily used at the site), having been manufactured elsewhere, outside the catchment zone of the Phase 8 fluvial gravel. This is unlikely to be a taphonomic or recovery bias, since much of the debitage from the handaxe manufacture would have been larger and less mobile than the dominant small-medium flint pebble clasts of the containing gravel. Much of the debitage would have been of similar size to the smaller handaxes, and equally as recognisable as the handaxes during machine excavation. Much of the lithic assemblage is also in fresh or moderately fresh condition suggesting not too much disturbance and transport. Furthermore, in the comparative situation of the Swan Valley School site (Wenban-Smith and Bridgland 2001), a much higher proportion of debitage was recovered from slightly coarser gravel, suggesting that fluvial action does not typically winnow out handaxe-making debitage. Therefore it seems inescapable that the dominance of handaxes is a behavioural organisational signal, reflecting a structured use of the landscape, with repeated discard at the site over the medium term represented by the formation of the gravels.

This complements evidence from other sites, such as Red Barns in Hampshire (Wenban-Smith *et al.* 2000; Wenban-Smith 2004b), where the reverse seems to be the case, and where there is strong evidence of a raw material source used for handaxe manufacture on-the-spot, from which finished tools were then exported prior to use/discard elsewhere. Likewise, further evidence for the association of handaxe-making adaptations with a more structured relationship between mobility and lithic production is provided by the GTP 17 horse butchery site and Q1B at Boxgrove. At the former location, raw material has been brought a short distance from its source, knapped, and then all the handaxes taken away from the location. At the latter, about 700m to the west, there is a disproportionate concentration of handaxes in relation to the associated debitage, reflecting a general pattern of import/discard of handaxes, as in the Phase 8 gravel at Southfleet Road. What of course would be desirable is to connect through refitting some of the handaxes at Q1B with the debitage from GTP 17. Regardless of whether this latter can be achieved, there is a consistent pattern of structured lithic production across the landscape. This can be equated to a logistically organised adaptation, with technological production integrated into pattern of mobility in a manner that suggests deliberate operational planning, and anticipation of tool-using encounters with key resources, rather than a more expedient adaptation dominated by tool manufacture and use as a response to such encounters. This also ties in with interpretations of more advanced cognitive capability based on the manufacture of handaxes, and, as pointed out by Roebroeks (2001), with the notion that survival in the more challenging seasonal environments of NW Europe would also mandate greater planning and anticipatory capabilities.

Of course, the absence of the evidence for these capabilities in the ostensibly simpler technology of Phase 6 is not in itself definitive that these capabilities were not present. The fact of a viable adaptation in the fully temperate conditions of Phase 6, probably supported by the hunting of large dangerous herbivores such as elephant, in itself suggests a high degree of capability. Nonetheless, the lithic evidence shows a greater degree of technological skill and cognitive capability and a more structured approach to lithic production in Phase 8, suggesting a more logistically planned adaptation. As discussed immediately below, this could be pertinent to current discussion of the continental distribution of on-the-one-hand core/flake industries, and on-the-other handaxe industries, and of the apparently rapid replacement of the former by the latter in southern Britain early in the interglacial of MIS 11, and more debatably perhaps also in MIS 9.

THE CLACTONIAN: CLASSIFICATORY FICTION, EVOLVED TRADITION OR HOMININ PHYLUM?

One of the major debates of Lower/Middle Palaeolithic archaeology over the last 50 years has been over the existence or otherwise of a distinctive Clactonian industrial tradition (Wenban-Smith 1998; White 2000; and quoted references). The analysis of lithic artefacts has been a fundamental part of Palaeolithic research since its beginning. Predicated upon the apparently straightforward notion that lithic material cultural practices, and in particular tool types and knapping techniques, were in some way shared between a hominin group, and then transmitted down to younger generations, the European Palaeolithic has been constructed as a story of cultural development and interaction through the Middle and Late Pleistocene, with specific cultural traditions occurring within a framework of Lower, Middle and Upper Palaeolithic stages. By the start of the 20th century, the NW European Lower and Middle Palaeolithic was divided into three epochs, Chellean, Acheulian and Mousterian (G and A de Mortillet 1900), presumed to be applicable at a pan-European scale. In this framework the first Chellean developmental stage was represented by large simple handaxes, the second Acheulian stage by more finely-made handaxes in conjunction with flake-tools and the third Mousterian stage purely by flake-tools.

In the first part of the 20th century more controlled research led to a more detailed framework for the Pleistocene of NW Europe incorporating four glacial-interglacial cycles (Penck and Brückner 1909), and contradictions in the existing linear developmental framework began to be exposed. Smith and Dewey (1913) recovered a large assemblage consisting exclusively of cores and large flakes, some of them interpreted as retouched into tools, from the Lower Gravel at Swanscombe, stratified beneath levels (the Middle Gravels) containing Chellean and Early Acheulian types of handaxe. Additionally, Warren (1922) made a large collection of similar material from the Elephant Bed at Clacton-on-Sea between 1911 and 1916. The Clacton material was attributed by Breuil (in Warren 1922) to a Mesvinian industry, based on its similarity to material recovered from Mesvin in Belgium. Breuil subsequently divided the Belgian Mesvinian into early and later stages, and having retained the term Mesvinian for the later material, suggested the term Clactonian for the earlier material from Clacton, Swanscombe and Mesvin, after the site at Clacton-on-Sea (Warren 1926; Breuil 1926; 1932).

Breuil also noted that, during the climatic oscillations accompanying the Palaeolithic, there seemed to be a repeated pattern in different interglacials of an early and sharp replacement of flake/core-based industrial traditions with handaxe-based traditions. Therefore Breuil (1926; 1932) re-wrote the Lower Palaeolithic as a story of the parallel and contemporaneous development of two separate industrial traditions practised by culturally separate human groups, one using predominantly bifacial technology in warmer conditions, and the other using exclusively core and flake tool technology in colder conditions. These groups moved north and south following climatic zones and their associated fauna in conjunction with the climatic oscillations of the Pleistocene, leading to the occasional superposition of interglacial Acheulian industries above late glacial or early interglacial Clactonian industries, for instance as at Swanscombe.

Breuil's (1932) definition of the Clactonian lithic industry specified that it was based on flakes and cores, and lacked handaxes. The cores were worked in an *ad hoc* fashion, to produce large flakes with wide striking platforms. The resulting cores could be large or small, and ended up in a variety of shapes with differing degrees of reduction. The flakes were either left unretouched or made into crude scrapers with a minimum of retouch. There were also occasional partly bifacially worked tools, but never regular and symmetrical handaxes, such as found in Chellean and Acheulian industries.

Acheulian industries, in contrast, were characterised primarily by the presence of numerous well made symmetrical handaxes. Breuil and Koslowski (1931; 1932; 1934) divided the Acheulian industrial tradition into seven stages, based on their assessment of changing handaxe shapes and presumed dating of the Somme valley sequence, in northern France. All these stages include pointed and ovate forms, with and without features such as tranchet sharpening and worked butts. It was recognised that flake-tools were an integral, although subsidiary, element of these Acheulian industries. These were presumed to have been mostly made on debitage from handaxe manufacture, although it was recognised that some unstandardised core and flake technology was also practised.

This framework was then substantially reinforced, at least in Britain, by work at several sites in the ensuing decades. Firstly, work by Paterson (1937) at the East Anglian site of Barnham identified the same sequence of events as at Barnfield Pit. Levels with exclusively flake-core technology were overlain by a level with handaxes, all in a sequence thought to post-date the Anglian glaciation and therefore broadly contemporary with the Barnfield Pit sequence. Further work at the Clacton-on-Sea type site by Oakley and Leakey (1937), put a greater focus on secondarily worked types of flake-tools in Clactonian assemblages, with recognition of four types of scrapers (nosed, trilobed hollow, discoidal/quadrilateral and butt-end) and two types of points (triangular and beaked). This, perhaps undue, focus on secondary working as a means of creating 'definable types' (ibid. p 226) of

scraping edges was also compromised by the difficulty of distinguishing secondary working from natural abrasion, and many of these scraper types would today be regarded as abraded, unworked waste debitage. Oakley and Leakey provide clear illustrations of types interpreted here as a 'single notch' (ibid. p. 229, fig 3, no. 8) and a 'knife' (ibid. p. 229, fig 3, no. 3), classified by them as an 'end-scraper' and a 'trimmed flake, with nibbling from use' respectively. Subsequently, Warren (1951) provided a definitive typological overview of the Clactonian from Clacton-on-Sea. It both emphasised the importance of a range of simple core-tools, including cores used as heavy-duty tools, and also tidied the range of flake-tools into groups approaching the more technologically-based classifications used in this volume, although still with a strong emphasis on the shape of the outcome rather than its technological basis or presumed functionality. Warren still retained at least six different types of 'trimmed flakes', including: side-scrapers, bill-hook forms, sub-crescent forms, flake-points and piercers. Amongst these, 'sub-crescent forms' were essentially the type of flake-tool classified here as 'single notch', described by Warren (ibid., p. 117) as 'flakes with one deep and broad secondary facet struck out of one of the side edges'.

At the Globe Pit, Little Thurrock, Wymer (1957) and Snelling (1964) recovered substantial Clactonian assemblages from gravels thought (then) to underlie brickearths attributed to the Hoxnian interglacial. At Swanscombe, excavations by Wymer (Ovey 1964) and Waechter (Conway *et al.* 1996) confirmed a sharp transition within the Hoxnian interglacial from a flake/core industrial tradition (in the Lower Gravel and Lower Loam) to a handaxe-based Acheulian tradition in the overlying Lower Middle Gravel. At Clacton, Kerney and Turner used a combination of molluscan and pollen analysis to tie in the Clacton sequence with the Barnfield Pit deposits and the now-well-understood sequence of pollen zones through the Hoxnian (Turner 1970; Kerney 1971; Turner and Kerney 1971). This established that Warren's Clactonian industry at Clacton-on-Sea was broadly contemporary with that in the Lower Loam at Barnfield Pit, both dating to the early temperate pollen zone HoIIb-c. Supported by the attribution of a handaxe industry at Hoxne to zone HoIIc, Wymer (1974) then reiterated Breuil's original framework of parallel cultural phyla, suggesting a rapid replacement in SE England early in the Hoxnian of groups with a Clactonian industry by Acheulian groups with handaxe-based industrial traditions.

This culture-historical orthodoxy (cf. Trigger 1989) has since been subject to numerous challenges. As articulated by Singer *et al.* (1973, 8) as long ago as the early 1970s, there are three main possible explanations of the apparent archaeological pattern:

1. *Separate groups* – the two industries represent separate, contemporary hominin groups, with different stone-working traditions and possibly also physical origins

2. In situ *evolution* – Clactonian and Acheulian industries are the chronologically sequential products of a single hominin group, with the latter developing out of the former

3. *Technological facies* – the Clactonian industry is essentially a classificatory fiction, merely a contemporary facies of the Acheulian produced under particular circumstances by the same group.

Although recovering a substantial well-provenanced Clactonian assemblage from their excavations on the Golf Course site near Jaywick Sands, these authors did not, however, take a view on which of these explanations they preferred.

Option 3 was substantially aired in the 1990s, primarily by McNabb and Ashton, in conjunction with new investigations at the critical site of Barnham (McNabb and Ashton 1992; 1995; Ashton *et al.* 1994; 1998) and publication of Waechter's work at Swanscombe (Conway *et al.* 1996). Although not addressing the apparent chronological ordering of the two industrial traditions within the British Hoxnian, the essence of this position is that the long-standing distinction between Clactonian and Acheulian industries is a theoretical misconception based on the inappropriate pigeon-holing into one or other industry of a fundamentally similar technological record whose main element of variability is the proportion of handaxes in assemblages. Thus, the absence of handaxes at some sites does not represent a non-handaxe cultural tradition – the Clactonian – but the non-handaxe end of a technological continuum involving the differential distribution across the landscape of different lithic assemblages by a single hominin group with a varied technological repertoire. Factual evidence presented in support of this argument is: (a) typically 'Clactonian' core and flake technology in many Acheulian industries; (b) handaxes and their distinctive manufacturing debitage in Clactonian industries; and (c) evidence of handaxe manufacture contemporary with the main *in situ* Clactonian horizon at Barnham. It is also suggested that one reason why Clactonian industries may have been produced was a lack of suitable raw material for handaxe manufacture.

This option has, however, been rejected by most British Palaeolithic archaeologists, even before discovery and reporting of the Southfleet Road site (Wenban-Smith 1995b and 1998; Wymer 1998; Roe 1996; White 2000; Pope 2002. M. B. Roberts, pers. comm.). There are several key factors in this rejection:

1. The non-bifacial element of Acheulian industries is uncontroversial and was recognised both in the original definition of Acheulian industries and by the Abbé Breuil when he first distinguished Clactonian industries.

2. There is in fact no unequivocal evidence of handaxe manufacture in any well-provenanced Clactonian horizon. All of the claimed instances are from out-of-context collections lacking reliable stratigraphic

provenance: the beach at Clacton-on-Sea; talus slopes in Rickson's Pit; and loose material on the surface of Waechter's Swanscombe excavations. Concerning the beach finds at Clacton-on-Sea, we now know from finds at Happisburgh amongst other places that handaxe-making took place in East Anglia prior to the Anglian glaciation (Parfitt et al. 2010), not to mention subsequently (Wymer 1985), so occasional stray handaxe finds in amongst the more abundant derived Clactonian material on the beach foreshore are to be expected. At Rickson's Pit: (a) it is very unclear how the recorded sections (Dewey 1932) relate to the Clactonian horizons of the Lower Gravel and Lower Loam at Barnfield Pit; (b) there is in any case no record of where the handaxe-manufacturing debitage was recovered; and (c) as demonstrated by more recent excavations immediately to the west at Swan Valley School (Wenban-Smith and Bridgland 2001), it would almost certainly have been recovered loose from the surface of a section capped by, or consisting of, Lower Middle Gravel (or reworked material from it) which would have definitely and uncontroversially have contained evidence of handaxe manufacturing. At Waechter's excavation, the exposed surface of the Lower Loam was at the foot of sections containing Lower Middle Gravel, and it is highly likely that these would have contained handaxes, and that a handaxe might have dropped onto the Lower Loam surface.

3. There are no handaxes, nor any debitage from their manufacture, amongst the thousands of well-provenanced artefacts recovered from throughout the Lower Gravel at Swanscombe. Considering that this is a fluvial deposit, and therefore incorporates a time-averaged collection from the surrounding catchment area over what is likely to have been a substantial time period, it seems inconceivable that if handaxes were being made by the same hominin group as responsible for the abundant Clactonian material, no evidence of this would have been found from the Lower Gravel. Although not all debitage from handaxe manufacture is recognisable as such, experimental replication (Wenban-Smith 1996: 94-100 and fig 3.29, p 131) has suggested that between 5% and 20% (depending upon size, shape and reduction intensity) can be identified, greatly multiplying the archaeological visibility of handaxe manufacture beyond the implement itself. Even if every single handaxe had been exported from the catchment area, it is certain that some distinctive debitage from their manufacture would have become incorporated in the contemporary fluvial deposits and been subsequently recovered.

4. At Barnham, it has been claimed by Ashton *et al.* (1994 and 1998) that the main horizon with evidence of handaxe manufacture (in Area IV-4) can be shown to be contemporary with the main undisturbed Clactonian horizon, on the surface of the Cobble

Layer in Area I. Prior to more critical discussion, it is accepted that there is good evidence of handaxe manufacture at Area IV-4 and that it comes from the upper part, and surface of, a gravel-rich bed, overlain by fine-grained deposits. The earlier of these two publications (Ashton *et al.* 1994, 586, fig 2), showed the sequence at Area I as comprising the Cobble Layer (with the Clactonian material) overlain by a thin bed of grey silt-and-clay, overlain in turn by a thin band of dark brown/black clay (which contained a handaxe). The sequence at Area IV-4 showed the horizon with handaxe-manufacturing debitage (which was interpreted as equivalent to the Cobble Layer at Area I) as being directly overlain by the same dark brown/black clay as contained the handaxe at Area I. Thus, rather than proving contemporaneity, these observations were equally compatible with handaxe-making taking place later than Clactonian flake/core production. In the subsequent publication, the stratigraphic sequence at Area IV-4 was reinterpreted (Lewis 1998, 35, fig 4.8) so that the top of the gravel there was divided from the dark brown/black clay (now 'unit 6') by the same grey silt-and-clay as at Area I (now 'unit 5e'), providing stronger evidence of apparent contemporaneity. However, it seems clear from synthesis of the other published sections in the 1998 volume, and particularly the sections from Area IV-5 which link Area I with Area IV-4 (Fig. 22.1), that the gravel bed associated with the bifacial evidence at Area IV-4 is not in fact equivalent to the Cobble Layer (now 'unit 4') at Area I. The pale silt/sand of unit 5 can be traced down into the centre of the pit from Area I to the west end of the long section of the Area IV-5 extension (Lewis 1998, 40, fig 4.11). Then, it can clearly be seen in this section that this pale silt/sand becomes overlain to the east by a new gravel layer, which in turn becomes overlain further to the east by a new silt/sand layer. These two new layers can then be traced south and east, via sections 1 and 3 of Area IV-5 (Lewis 1998, 36, fig 4.9), to correlate with the sequence at Area IV-4. Thus the gravel layer at Area IV-4 containing the handaxe-making evidence in its upper part is *not* directly equivalent to the Cobble Layer at Area I, but occurs in the upper part of a quite distinct gravel bed, divided from the Cobble Layer at Area I by further silt/sand deposition. There is, therefore, a chronological hiatus of uncertain duration between the Clactonian and handaxe-manufacturing horizons.

5. Investigations at Barnham (Wenban-Smith and Ashton 1998) have shown that the raw material available to the knappers at the Area I Clactonian horizon (in the cobble layer) was perfectly suitable for handaxe manufacture. Furthermore, work at Red Barns (Wenban-Smith *et al.* 2000) has shown that the poor quality frost-fractured raw material at Red Barns was nonetheless used for a handaxe-dominated industry, even in an area of Chalk bedrock where fresher flint nodules were probably abundant nearby. These observations refute poor raw material availability/quality as an explanation for why a handaxe-manufacturing hominin group should choose not to do so in certain places.

6. One aspect of the archaeological record not addressed by Ashton and McNabb is the consistent stratigraphic superposition of Acheulian horizons above Clactonian horizons in the British Hoxnian. Generally, biostratigraphic and chronometric approaches to dating are insufficiently precise to allow relative date ordering within the Hoxnian of distinct Clactonian and Acheulian horizons between different sites, so the key facts here are the repeated instances of stratigraphic superposition of Clactonian above Acheulian at Swanscombe, Barnham and now Southfleet Road. These all point one way, suggesting that the latter follows the former, rather than contemporaneity. Furthermore, the molluscan evidence from Swanscombe allows comparison with the Hoxnian pollen framework at Clacton, and confirms that the handaxe-bearing Middle Gravels post-date the Clactonian horizons at Clacton, as well as at Swanscombe (Kerney 1971). The only possible dating anomaly is at Beeches Pit (Preece *et al.* 2007). Here an attempt has been made to use another aspect of the molluscan record, the relative proportions of *Discus ruderatus* and *Discus rotundatus*, to suggest that the deposits associated with handaxe manufacturing at locality AH (layer 3) are broadly contemporary with the Clactonian horizons at Swanscombe, and Barnham. However, there is an (admittedly minor) presence of *D ruderatus* in the only known mollusc record from the lower Middle Gravels (Kerney 1971, fig 3), suggesting that the presence of this species is not incompatible with Acheulian horizons that are known to overlie Clactonian ones. Also, as accepted by Preece *et al.* (2007, 1281), the temporal resolution of the proposed Hoxnian replacement of *D ruderatus* by *D rotundatus* is insufficient to distinguish between genuine contemporaneity of Clactonian and Acheulian, or rapid change from the former to the latter. Furthermore, as discussed further below, the apparent stratigraphic suddenness of this change need not indicate particular rapidity in calendar terms.

The evidence from Southfleet Road provides an important addition to the British Lower/Middle Palaeolithic record. The size and technological/typological consistency of the artefactual assemblages from Phases 6 and 7 validate identification of the Clactonian as a genuine industrial tradition in south-east England in the Hoxnian. Crucially, the Southfleet Road collection includes assemblages representing three complementary spatial/temporal scales: short-term activity at one spot (Assemblage 6.3); medium-term activity in a bankside locale (Assemblages 6.1 and 6.2); and medium-term activity over a slightly wider valley-side landscape (Phase 7 material). The quantity of artefacts in Assemblage 6.1

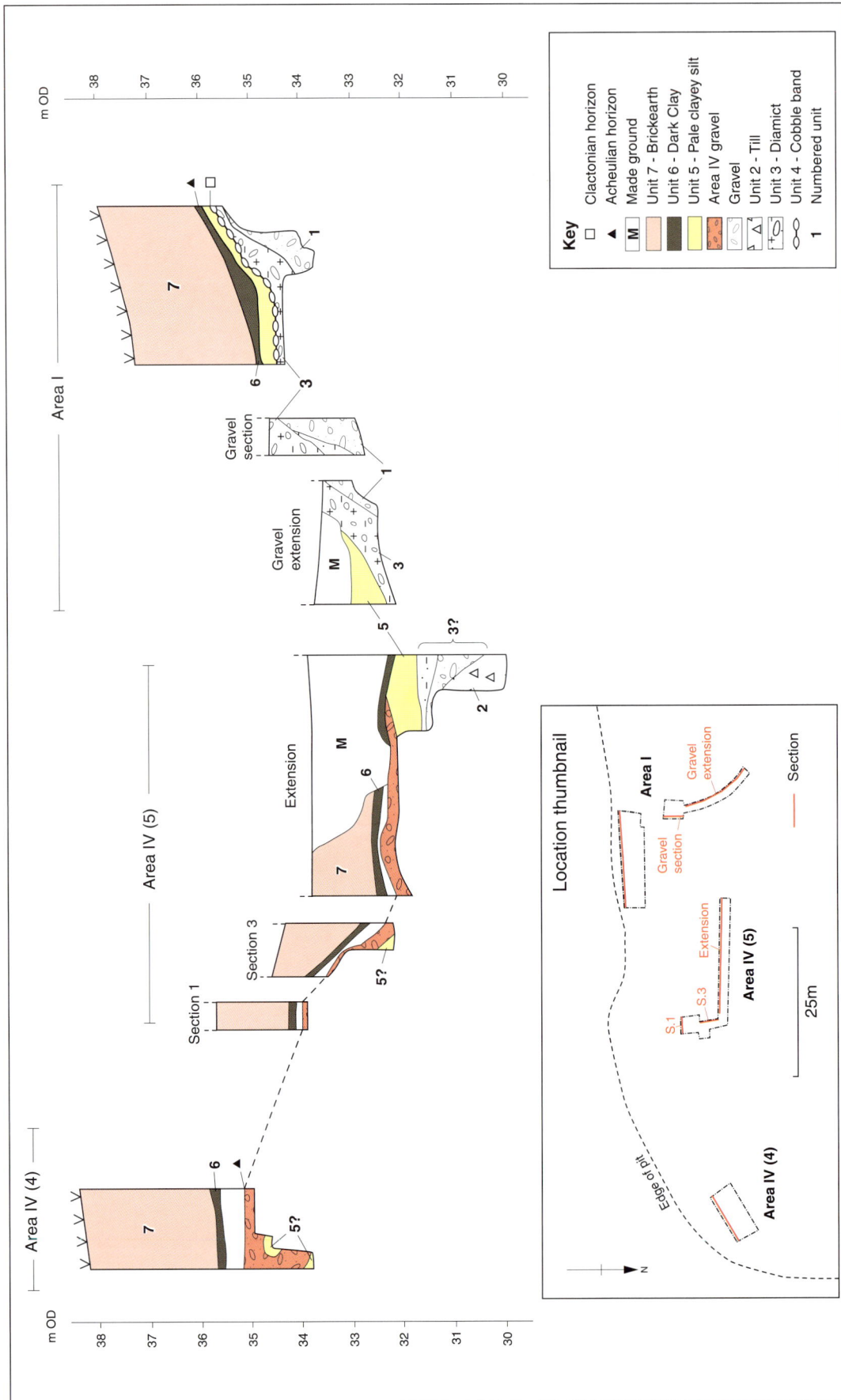

Figure 22.1   Clactonian and Acheulian horizons at East Farm Pit, Barnham

is greater than the other well-provenanced assemblages of Barnham (unit 5, Area I – Ashton 1998), Clacton-on-Sea (the fresh condition material from the Marl and the Gravel at the golf Course excavation – Singer *et al.* 1973) and Swanscombe (assemblages from Waechter's excavations in the Lower Loam and the Lower Gravel – Conway *et al.* 1996). Taken together, these facts seem to reflect a consistent Clactonian lithic tradition in a variety of situations over a sustained period across south-east England, rather than a variably applied technological facies.

The molluscan, pollen and vertebrate evidence combine to suggest the Clactonian occupation in Phase 6 of the site sequence was in the early-temperate substage, zone Ho II, broadly contemporary with Clactonian horizons at Barnfield Pit, Barnham and Clacton-on-Sea. With a dated sequence of technological change duplicating that from Barnfield Pit and Barnham, this adds to the evidence suggesting replacement of Clactonian industries by Acheulian ones in Britain in the Hoxnian interglacial of MIS 11. This information is summarised here, attempting to illustrate the imprecision of the relationship of the key archaeological horizons to the Hoxnian pollen zone framework (Fig. 22.2). The only slight anomaly in this consistent pattern, now that the archaeological evidence from Hoxne is regarded as from a much later temperate episode in MIS 11 (Ashton *et al.* 2008), is the apparent presence of handaxe-manufacturing at area AH at Beeches Pit in late zone HoII/early zone HoIII, based on the slight preponderance of *Discus ruderatus* (Preece *et al.* 2007). This does not however undermine the proposed sequence of lithic industrial change due to both the imprecision of dating this molluscan faunal attribute, and uncertainty over the duration of occupation represented in area AH and its precise relationship with the mollusc sequence of layer 3.

So, having ruled out the suggestion that the Clactonian and Acheulian are contemporary facies of a single technological complex, and accepting them as distinct and chronologically successive cultural/industrial entities, at least within the British Hoxnian, is it possible to prefer either of options 1 or 2, or some more subtle explanation as discussed by White and Schreve (2000)? Pertinent to this, is a brief consideration of some evidence from beyond the British Hoxnian, from later periods and from the continental European mainland. Firstly, it is often suggested that there are other instances of 'Clactonian' occupation in Britain early in MIS 9 at Globe Pit, Purfleet and Cuxton (for example Bridgland 1994; White and Schreve 2000; Schreve *et al.* 2002; Stringer 2006). There is thus a repeated pattern in successive interglacials of early Clactonian colonisation of Britain followed by a later wave of colonisation by 'different regional populations with different technological repertoires' (White and Schreve 2000, 18). This apparent repetition of a similar sequence of colonisation in separate interglacials irresistibly raises the spectre of Breuil's original model of parallel cultural phyla (Breuil 1926; 1932), implying at least two unchanging material cultural traditions that

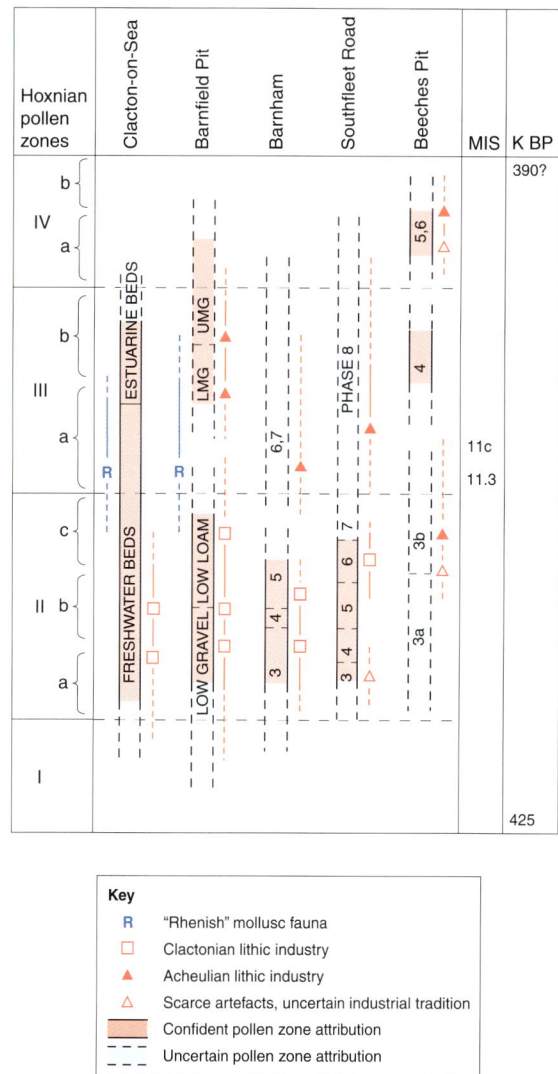

Figure 22.2 Relationship of Clactonian and Acheulian horizons at key localities to Hoxnian pollen zones

somehow persist in northern Europe between early MIS 11 and MIS 9. The suggestion is that this persists in a Central-Eastern European cultural refugium, possibly associated with a distinct hominin evolutionary lineage. However, critical to this whole line of discussion is the identification of Clactonian in MIS 10/9 in Britain, and this is in fact by no means robustly established.

At Purfleet (Palmer 1975; Schreve *et al.* 2002) and Cuxton (Cruse 1987), the claimed Clactonian presence is based upon far too few artefacts to be accepted as representative of a sustained cultural tradition, and also lacks the notched flake-tool component that is a typical element of the MIS 11 Clactonian assemblages. Furthermore, the age of the Cuxton sequence is at present very uncertain; the latest OSL results (Wenban-Smith *et al.* 2007) suggest a date in MIS 7, although this result is now under more careful scrutiny (Wenban-Smith *et al.* in prep.). So, even though the sparse material from the lower deposits of these two sites does not show evidence of handaxe manufacture, they must on present evidence be discounted as contributing to this debate.

At Globe Pit (Wymer 1957; Bridgland and Harding 1993), there is indeed a substantial lithic assemblage that is technologically and typologically sufficiently similar to the Clactonian as recognised in MIS 11, to be regarded as attributable to it. Also, it comes from a gravel deposit that has unanimously (in recent times, although not through most of the 20th century, cf. Hinton and Kennard 1900; West 1969; and Evans 1971) been regarded as part of the MIS 10-9-8 Corbets Tey/Lynch Hill Formation, and indeed is now suggested as the stratotype (the Little Thurrock Gravel) for the lower part of this Formation (Schreve *et al.* 2002). Although the gravel containing the lithic material dips down to the south (underpinning the earlier slopewash interpretations) rather than resting on a sustained sub-horizontal bench, it is firmly regarded by Bridgland (1994; in Schreve *et al.* 2002; and pers. comm. 2013) as the valley-side feather-edge of an indisputably fluvial deposit on the basis of sub-horizontal bedding within it. Although this seems like strong evidence, examination of much wider exposures of Pleistocene deposits on Kentish valley sides (ie, in association with the improvement of the M25/A2 junction near Dartford, see Wenban-Smith *et al.* 2010), has shown that what might look like clear fluvial bedding within an *in situ* terrace deposit in a small exposure can be revealed as part of a much wider slopewash deposit when more fully examined. Therefore, there is still some scope for continuing debate over whether the artefact-bearing gravels at Globe Pit are *in situ* fluvial deposits of late MIS 10, as generally accepted. Notwithstanding this quibble, of greater import is that the artefacts themselves do not represent *in situ* occupation, but are a reworked/transported assemblage. Assessing the degree of reworking and transportation depends upon the extent to which edge-nibbling on various flakes is regarded as natural abrasion or hominin flake-tool manufacture. Considering their gravel context, the former is more likely, which suggests a lesser degree of fresh condition material than reported by Wymer (1957). Bearing in mind these observations, and even accepting the gravel as part of the Corbets Tey/Lynch Hill formation, it seems quite possible that as suspected by West (1969) the contained Clactonian material is derived from older MIS 11 deposits, rather than contemporary with the gravel formation. Even though no artefacts were found in the higher Orsett Heath Formation terrace gravels a short distance to the north during section cleaning in the 1990s (Bridgland and Harding 1993), this was by no means an exhaustive investigation. In any case, an absence of lithic material in the surviving deposits would not establish that none was ever present; it is quite possible that the putative Clactonian-artefact-bearing part of this Formation has now disappeared during the downcutting and gravel reworking associated with the transition from MIS 11 to MIS 9. Thus there is currently insufficient evidence to be confident that there is a repetition of genuinely Clactonian occupation in Britain in the early part of the MIS 10-9-8 cycle, so most of the remainder of this discussion focuses upon the Clactonian/Acheulian transition in Britain in MIS 11.

There is however one point to be made before continuation of this discussion. If one did regard the Clactonian industrial tradition as persisting in a central-eastern European refuge area, and perhaps associated with a distinct hominin lineage, then one could point to the clear contrasts in the organisation of lithic production between Clactonian and Acheulian adaptations, the more spatially structured pattern of landscape use in Acheulian adaptations and the evidence of more logistically organised behaviour, to suggest a fundamentally different adaptation that could perhaps reflect the different cognitive capabilities of a different hominin lineage. In the almost total absence of a skeletal record we are presently in the dark over the association (or otherwise) of hominin physiologies with lithic material culture through the Middle Pleistocene. However, the increasing evidence from DNA analyses, such as the discovery of the distinct eastern European Denisovans regarded as a sister group of Neanderthals (Krause *et al.* 2010; Reich *et al.* 2010), suggests that there might be surprises in store. One such surprise could be the association of non-handaxe industries from Eastern Europe with a distinct hominin lineage, possibly even Denisovans.

Back in Britain, and discounting the above possibility, it has long been suggested in support of option 1 that the apparently rapid replacement of Clactonian by Acheulian in the British Hoxnian reflects the direct replacement of one hominin group by another (Breuil 1926; 1932; Wymer 1974; White and Schreve 2000; Stringer 2006). Although this might appear rapid at the Pleistocene geological scale, it must however be emphasised that the period during which this change took place, between HoIIb and HoIIIa, spans at least 5,000 years (Turner 1970), and probably more. This is therefore ample time for an isolated hominin population, on what would probably at that time have been the island of Britain (Preece 1995) to have developed a different lithic industrial tradition that involved both different techno-logical/typological practices and a wholly different organisation of production integrated with a more logisti-cally organised adaption. Crucial to this debate is the history of Britain through the Hoxnian as an island or a peninsula. It is widely accepted (Gibbard 1995; White and Schreve 2000) that the Straits of Dover were initially formed in the Anglian. The subsequent onset of peak MIS 11 interglacial conditions and accompanying sea level rise happened slightly more slowly than in some other interglacials (eg MIS 9) allowing a period of peninsularity early in the Hoxnian during which hominins and other fauna were able to colonise Britain before it was isolated as an island due to rising sea level. The puzzlingly late influx of 'Rhenish' species in the Swanscombe and Clacton-on-Sea MIS 11 sequences (Meijer and Preece 1995) could be taken as indicating an early period of insularity. Despite the late advent of Rhenish species, it is hard to see from other evidence of temperature and sea-level history through MIS 11c (Tzedakis *et al.* 2001; Waelbroeck *et al.* 2002), which can be regarded as the true Hoxnian of the Swanscombe, Clacton and Barnham sequences (Ashton *et al.* 2008), how peninsularity could have been re-established before the climatic dip of MIS 11b that would have allowed colonisation by new hominin groups. Consequently it is hard to imagine firstly, how a new hominin group could reach Britain at this time, secondly, if there were separate groups in Britain and north-west Europe at this time how they could maintain their cultural separation and distinct industrial traditions through the early Hoxnian, or thirdly, if that was the case, why there has been no evidence of the handaxe-manufacturing tradition from the first part of the Hoxnian.

Therefore option 2 is preferred here, namely that the initial Hoxnian population of south-east Britain had a Clactonian lithic industrial tradition, based on a minimally spatially organised *chaîne opératoire*, and mostly associated with the expedient use of locally found raw material in response to tool-using need. This then developed over a period of maybe 5,000 years into a handaxe-based Acheulian industrial tradition, integrated within a more spatially and logistically organised *chaîne opératoire*, and associated with the transport and use of bifacial tools, provisioned and carried in advance of anticipated tool-using need. As previously discussed by both Wenban-Smith (1998) and White and Schreve (2000), there are a multiplicity of factors that could underpin this development. The Clactonian could, for instance, work well as the adaptation of a small, mobile pioneer population, but more stable adaptations in interglacial environments could then encourage popula-tion growth and the development of more logistically organised adaptations that could also support social transmission of the more difficult skills of handaxe manufacture (cf. Mithen 1994). Alternatively, changing raw material or animal resource availability could stimulate transition from a primarily expedient tradition of tool-use to one in which it became more important to carry a tool in anticipation of use, and this could have encouraged development of a handaxe-based lithic industrial tradition. The Clactonian assemblages of Phases 6 and 7 contain the seeds of this behavioural transition, with some evidence of interruption of *chaîne opératoires*, and movement of tools around the landscape in anticipation of use. Likewise, the simple core-tool and the ambiguous chunky flake-core/flake-tool elements of Clactonian assemblages provide the seed of the techno-logical transition toward more intensively worked and deliberately shaped handaxes.

Whichever interpretive option is preferred, the key point here is that interpreting the Lower/Middle Palaeolithic material cultural record is by no means a simple matter of data collection and application of an agreed theoretical framework. Every stage of the interpretive process is highly contested. The selection of lithic attributes as 'data' is highly integrated with theoret-ical preconceptions, and these in turn underpin subsequent interpretations. Thus, although debate over the interpretation of 'the Clactonian' is at one level a debate about what was happening in Britain in the Hoxnian, it is also a debate about perspectives on the Lower/Middle Palaeolithic record. It involves preconcep-tions on: (a) time-depth and continuity, ie. whether we see the record as representing tiny snippets or substantial

slices of Pleistocene time; (b) hominin groups and adaptations, how we model their density, distribution and networks, both across a region and down through time; and (c) material cultural traditions and change, how variable we think might be their material cultural output, how new generations of hominins maintain and acquire material cultural practices, and what factors might stimulate change. Whatever perspectives a modern worker takes on these matters, will then inevitably inform both the selection/characterisation of relevant data and also the resulting interpretation. Therefore, what is required is not just uncritically to inherit a dataset, or a conception of relevant data, but to try and explore the foundations of the data, and to articulate the perspectives underpinning its interpretation. Here, it has been argued that the Pleistocene depositional record artificially conflates vast stretches of prehistoric time, that deposi-tional and taphonomic processes need to be more carefully considered and that the term 'Clactonian' has been misapplied to some small lithic assemblages. Although unarticulated, there are also widely shared preconceptions, unchallenged in this analysis, that the hominin groups of this era are very low density, maintaining stable lithic industrial traditions over substantial time periods, transmitted by observation and practice from one generation to the next.

At a wider scale, this explanation for the British situation does not directly contradict the notion of an eastern European non-handaxe province, and a southern/western European province with a handaxe-making industrial tradition in the Middle Pleistocene. It merely argues that, in the island of Britain in the Hoxnian, it seems more likely that *in situ* evolution of the lithic industrial tradition has taken place rather than a second wave of colonisation. It does however put an increased emphasis on appreciating the long time-depth and intermittence of the depositional archive of this period, leaving a highly punctuated and conflated Palaeolithic record, with vast periods of time entirely absent and short periods disproportionately represented in rapidly formed sediment accumulations, leading to a potentially misleading record of periods of stability interspersed by rapid change. It also emphasises (a) the potential, in conjunction with short-term stability, for long-term flexibility and cyclical drift within the overall context of lithic-assisted adaptations as an aid to survival, and (b) the importance of investigating not only the technological/typological characterisation of the lithic products, but also the spatial organisation of the *chaîne opératoire*. Thus bearing in mind, for instance, the record of pre-Anglian handaxe-based occupations in Britain (for example at Boxgrove and Happisburgh) and the unifacial flake-tools of High Lodge (Ashton *et al.* 1992), it implies that handaxe-based and flake-tool-based lithic technological adaptations are liable to fade and recur over vast timescales, as changing parts of an integrated social and behavioural solution to the problem of survival in the always slightly different, but climatically and environmentally also broadly cyclical, Pleistocene world.

PALAEOLITHIC ARCHAEOLOGY AND DEVELOPMENT CONTROL: SITES, METHODS AND THE RESEARCH FRAMEWORK

The history of discovery, fieldwork and post-excavation analysis at Southfleet Road provides a case-study from which some valuable lessons can be learnt. In the first place, the site was very nearly missed altogether, only being discovered at a late stage of the HS1 development programme in the Ebbsfleet Valley, after the archaeological programme was thought to have been completed. This occurred for a number of reasons. Firstly, at the very outset of the HS1 project in the early 1990s, I prepared an overview (Wenban-Smith 1992) of the Palaeolithic priorities and acceptability of different rail link routes through the Ebbsfleet Valley for inclusion in the overall survey of cultural effects along the HS1 corridor (Oxford Archaeology 1994). This incorporated two mistakes. First, I took the available geological mapping of the area (British Geological Survey 1977) at face value as an accurate representation of the distribution and nature of the Pleistocene deposits, and this underpinned my assessment of Palaeolithic potential for uninvestigated areas. Second, I was focused on deposits surviving within the quarried parts of the Ebbsfleet Valley affected by the HS1 route, rather than those around the edge of the main quarried area. Thus, although the Swanscombe 100-ft terrace deposits on the west side of the valley were identified as a Palaeolithic concern the area where the elephant was later found at the far south-west edge of the HS1 area (which at that time had not been defined as such) was overlooked. This area was at that time mapped as 'Thanet Sand below Head' deposits and, had it been on my radar, I would have regarded it as the lowest category of Palaeolithic importance, 'Palaeolithic priority Category 5' (Pleistocene deposits of low Palaeolithic potential; monitoring of works with option of rescue excavation recommended). Likewise, the stretch of 'ferruginous loam' identified by Carreck (1972, 61) to the north of the elephant site (Chapter 2) was not deemed significant as it had been seen at the top of the quarry face below the east side of Southfleet Road, and so was not going to be affected by the rail link works. It was only after subsequent work in the area, particularly investigations in advance of the Swan Valley Community School in the late 1990s (Wenban-Smith and Bridgland 2001) and in Eastern Quarry in 2005 (Wessex Archaeology 2006a) that the presence of extensive deep spreads of unmapped Pleistocene sediments with high Palaeolithic potential was recognised extending south of Swanscombe towards the elephant area. Had I been aware of this in the early 1990s, then the unquarried area where the elephant was later found would have been higher on my agenda, even though it was far away from the different route options being considered at that time.

This emphasises that while BGS mapping is an essential starting point for the consideration of Palaeolithic potential, since any important Palaeolithic remains will be contained in Pleistocene sediments, it cannot be regarded as precisely accurate for the distribution, thickness or nature of Pleistocene sediments. A key task for any desk-based archaeological assessment in advance of development must be, therefore, to look at the geomorphological context of a site in relation to Pleistocene mapping in the general region, and to consider the possible presence and likely nature/potential of any unmapped deposits. Other similar examples include the sites of Red Barns, Hampshire (Wenban-Smith et al. 2000) and Harnham, Wiltshire (Bates et al. in prep.) where nationally important Lower/Middle Palaeolithic discoveries have been made in areas mapped as Chalk bedrock.

Secondly, it is important for assessment (and mitigation) of the archaeological impact of major development projects to take a wide view of potential impact, taking account of related infrastructural development and not just the headline project. In this case, focus at the beginning, and at every subsequent stage of the archaeological process, was on the route of the HS1 track, the link to the existing North Kent line, the new pylons (ZR3A and ZR4) and the footprint of the Ebbsfleet International station. Less attention was paid to field evaluation in other areas where there would be ancillary landscape remodelling and other impacts such as roads and services. The elephant was found at the south-western edge of the overall area of HS1 development in the Ebbsfleet Valley, well away from the track and the station. Despite being an area of substantial landscape remodelling, this area received no archaeological attention prior to the December 2003 investigations (Chapter 3), leading to the elephant (and the other Palaeolithic and later prehistoric archaeological remains) only being discovered after substantial groundworks had already taken place.

Thirdly, the investigation of the site highlights and exemplifies some of the issues concerning the importance, or otherwise, of fluvial gravel deposits as a Lower/Middle Palaeolithic resource. Many (perhaps most) archaeologists and curators continue to regard such gravel deposits as essentially 'disturbed', lacking in situ occupation horizons and therefore not meriting investigation in advance of development, whether or not artefactual (or faunal) remains are suspected, or known, to be present. In this instance, it was not apparent at first that the Phase 8 gravel capping the site contained any flint artefacts. It was only after a high volume of systematic sieving, extensive section cleaning and the start of mechanical reduction that artefacts began to be recovered. By the end of the project an assemblage of 180 artefacts had been recovered, including more than 30 handaxes, emphasising that apparently sterile gravel bodies may in fact contain an abundant archaeological resource when more thoroughly investigated.

Furthermore, many of the artefacts from the gravel were in mint or fresh condition, reflecting a minimum of disturbance. The likelihood was that they were recovered close to where they had been discarded during activity on the gravelly braided floodplain of the palaeo-Ebbsfleet; thus gravel deposits can contain minimally disturbed evidence, as also established at other sites such as Lynford, Norfolk (Boismier *et al.* 2012). However, an even more important point to take on board is that the more-disturbed artefactual evidence from fluvial gravels provides an important complement to the undisturbed evidence, giving information on lithic industries in the slightly wider landscape over the medium-term period during which a fluvial gravel body was formed. At the wider Pleistocene timescale, the evidence from individual gravel bodies is the most suitable unit from which to develop a wider picture of hominin presence across Britain, and of the broad trajectories of lithic material cultural change and regional variation. Therefore, far from being unworthy of targeted investigation, fluvial gravel bodies are a key resource for investigating the Lower/Middle Palaeolithic and addressing current research priorities.

Excavations at the site also highlight a number of practical matters concerning field methods for Palaeolithic/Pleistocene investigations. Firstly, the concurrent programme of environmental assessment allowed important progression in understanding the palaeoenvironmental potential of different horizons, leading to modification of the sampling programme and targeting of the most appropriate deposits while fieldwork was in progress. While a balance must always be struck, this obviated the need for a massively redundant field sampling programme involving a large volume of precautionary sampling from deposits of uncertain potential.

Secondly, the exposure of long, deep Pleistocene sequences on both sides of the main site spine illustrated how complex Pleistocene stratigraphy can be, with beds drastically dipping, thinning, thickening or disappearing over short distances, and important stratigraphic junctions being almost indistinguishable at some points. Many beds exhibited complex 3-dimensional geometry, and the spectacular synclinal basin in the central part of the site was only revealed when transverse cross-sections were dug across the main site spine, to link the two long north-south faces. These long exposures and the unexpected synclinal geometry in the centre of the site demonstrate the difficulty (and the likely inaccuracy) of modelling sub-surface Pleistocene sequences between widely spaced test pits across a site. This has also been repeatedly demonstrated at many other major continuous sections in the HS1 works in the Ebbsfleet Valley, for instance at Trench 3776 TT at the ZR4 pylon site and in Trenches 3971 TT and 3972 TT (Wenban-Smith *et al.* forthcoming). Clearly, any test pits are better than none; and many are better than few. However, even with many it is important to bear in mind (a) that the sub-surface deposit model may not be especially accurate, (b) that important stratigraphic boundaries may only become apparent when wider exposures are seen and (c) that site areas where particularly important remains or deposits are present may be restricted in extent and unpredictably located. This last point is well-exemplified at Southfleet Road not only by the restricted areas of the Phase 6 clay containing the elephant skeleton and the larger concentration south of Trench D, but also by the

restricted extent of the tufaceous channel and the very patchy preservation of ostracod and small vertebrate remains in the deposits of Phases 3 and 5 respectively.

Finally, it is worth recapping that, although the main north-south spine of the site that underlay the old Southfleet Road is now gone, having been throughly investigated in the work reported throughout this volume, Pleistocene deposits of high potential still survive in the immediate vicinity. First, the curving and sloping bank to the west of the new Southfleet Road cutting is formed by *in situ* Pleistocene deposits equivalent to Phases 6-9 of the main site sequence. These survive close beneath the current ground surface and continue further west into the arable field within the perimeter of the National Grid Northfleet West Substation property. Field evaluation here has demonstrated the continuation of some areas of high Palaeolithic importance (Museum of London Archaeology 2011). Secondly, the Phase 8 palaeo-Ebbsfleet gravels and underlying deposits of Phases 3-5 are still present directly under the new Southfleet Road surface and to its east, north of a line through Trench B. Thirdly, deposits of Phases 3-6 are still present to the east of the old site spine, south of the same line through Trench B, where they have been investigated by test pits dug for the Station Quarter South field evaluation (Wessex Archaeology 2006b). Finally, undisturbed deposits are still present close beneath the current ground surface in the area of Transects 1-3 to the north-east of the main site (Fig. 3.11; Fig. 21.1). The western half of this area contains deposits of Phases 8 and 9, the palaeo-Ebbsfleet gravel and the brickearth; the eastern half contains deposits of uncertain correlation to the site sequence (Fig. 4.34; Fig. 21.1), but which are thought to be broadly equivalent to Phases 3-6. These latter deposits have been shown to contain fresh condition lithic artefacts and faunal remains; they are thus of high Palaeolithic potential and require targeted field evaluation prior to any further proposed development impact.

CONCLUDING THOUGHTS

by Francis Wenban-Smith and Stuart Foreman

So, a little over nine years after the site was recognised and fieldwork began, and after a three-year-programme of painstaking specialist analyses, how might we sum up its importance? This major investment of time and resources, leading to production of this volume, is in itself testament to the recognition by Stone Age specialists and curatorial authorities that the site was of undoubted national and international importance. The discovery of any undisturbed remains of this great age (roughly 400,000 years ago) is an incredibly rare event in itself, valued for the insight provided into the life of early hominins. To recover the specific evidence of the butchery of a single large animal, and in particular of the evocative extinct beast the straight-tusked elephant *Palaeoloxodon antiquus*, with the flint tools lying where they were dropped beside the carcass, takes it onto a higher plane; such recoveries are noted and celebrated across the globe, feeding into academic research and public consciousness. In Britain, the only comparable discovery is the horse-butchery site GTP 17 at Boxgrove (Pitts and Roberts 1997; Pope 2002). In Europe, there are maybe 6-7 sites of this nature known from the full earlier Stone Age period prior to the advent of modern humans roughly 40,000 years ago, and a similar number are known from the wider Old World and Middle East (discussed in the preceding sections of this chapter, and see also Gaudzinski *et al.* 2005; Surovell *et al.* 2005; and Yravedra *et al.* 2012).

The Ebbsfleet Elephant will take its place in this pantheon of great sites, contributing to academic debate in future years. In the words of Mark Roberts (Institute of Archaeology, University College London), Director since the early 1980s of the seminal Boxgrove project that did so much to kick-start modern Palaeolithic multidisciplinary investigations, and can be said to have trained the current generation of Palaeolithic researchers:

"The site has an extremely important bearing on some of the major research questions and debates of the British Palaeolithic. The geological sequence and palaeoenvironmental evidence links it with other internationally important archaeological sites in the Swanscombe area and south-east England, addressing the long-standing debate over post-Anglian Clactonian and Acheulian industries. This volume includes a comprehensive synthesis of the current state of knowledge, and the site's contribution to the debate".

Its impact and importance go beyond the academic domain, however. This is the first time that a major Palaeolithic site has been brought to publication entirely through developer-funding, and the authors hope that the standard has been set at a suitably high level. By demonstrating what can be achieved, the HS1 project in the Ebbsfleet Valley has helped to raise the profile of Palaeolithic archaeology as a mainstream concern in developer-funded archaeology. Palaeolithic specialists, perhaps more than archaeologists from other periods, benefit most greatly from large-scale development that exposes deep-lying deposits to reveal important new discoveries. The history of archaeological discovery in the Ebbsfleet Valley begins with the start of quarrying in the late 19th century, and will no doubt continue in the 21st century as 'Ebbsfleet' is reinvented as a new settlement on very ancient foundations. Exceptionally important Palaeolithic discoveries, where they are accompanied by an appropriate level of research and publication, can be regarded as a positive benefit arising from the development, rather than merely as 'mitigation' of an environmental impact. The Southfleet Road site certainly falls into this category. Apart from the headline discovery of the elephant butchery area, numerous other aspects of the site combine to make it of foremost importance, amongst which the rich faunal remains from the tufaceous channel and the abundant lithic artefacts from the rest of the site.

A common complaint from developers is that the most important finds often seem to emerge late in the day of a planned excavation, when all assigned resources have been expended. The discovery of the Southfleet Road site certainly conformed to this perceived pattern, which is equally frustrating for archaeologists. Archaeological work in a developer-funded environment is firstly about identifying and avoiding important remains, wherever possible. And where avoidance is not possible, it then becomes about managing the risk of significant discoveries and ensuring that available time and resources are targeted effectively at identifying and understanding important sites before they are lost. Since the methods do not yet exist to reliably detect ephemeral prehistoric archaeological sites in deep and complex sediment sequences, even in a locale with a long history of previous investigation such as the Ebbsfleet Valley, identifying important Palaeolithic sites is like looking for a needle in a haystack. We have to accept that even the best-managed projects will potentially affect significant sites.

It is to the great credit of those involved in managing the HS1 construction project that, once the importance of the elephant site became clear, generous additional time and resources were made available to conduct a thorough investigation. This was completed to the demanding standards expected by the HS1 archaeological Statutory Consultees (formed from English Heritage, Kent County Council Heritage Environment Conservation) in spite of the extreme pressure on the construction team to finish building the access road to the new Ebbsfleet International station. The subsequent investigation of the site, more than any other excavated along the HS1 route, is a testament to the exceptionally high standards of environmental management adopted by the HS1 project and its Statutory Consultees, deservedly recognised in Heritage and Construction industry awards.

Given the extremely valuable results it is to be hoped that the Southfleet Road site does not remain for long a one-off example. Sites with Palaeolithic potential are perhaps more at risk than most of being covertly bypassed as 'too-difficult-to-deal-with' in the developer-funded environment. The characteristic combination of deep deposit sequences and very sparse and irregularly distributed remains places the Palaeolithic period among the most troublesome 'final frontiers' of British Archaeology. It remains a work-in-progress to educate all participants in the development-related curatorial process (curators, consultants and contractors) in how to manage the risk of disturbing important Palaeolithic remains most effectively, and how to investigate them most appropriately when they are encountered. The undoubted international importance of the Southfleet Road site, and the research results presented in this volume, will we hope provide encouragement on the curatorial side that it is worth persevering with the mysteries of the Palaeolithic. Even more importantly, we hope that they will engage and inspire developers, engineers and construction workers that archaeological work produces results of interest and importance, and that the British archaeological community has the experience, methods and capacity to investigate and record such sites highly effectively. In particular it is hoped that the high standards set through the work on High Speed 1 will be adopted in development of the new High Speed 2 line linking London to northern Britain, and ultimately lead to discovery and investigation of the archaeology along HS2 for the benefit of future generations.

Critical to the success of this project has been the close interaction between university- and museum-based specialists and archaeologists from the local government/ commercial sector. The involvement of academic specialists was regarded as essential in this case because of the intrinsically difficult and multi-disciplinary nature of Palaeolithic archaeology. High-level management decisions were made by the archaeologists within the HS1/ RLE Environment Team, advised by their external Palaeolithic specialist Mark Roberts in discussion with the Statutory Consultees (in particular Lis Dyson of Kent County Council), with input from the archaeologists involved on the site (Francis Wenban-Smith, and Richard Brown and Darko Maricevic of Oxford Archaeology). Oxford Archaeology brought to the team the organisational capacity needed to deal with a substantial detailed excavation in the face of imperative construction deadlines, and long experience of working with engineers and planners. However as major Palaeolithic excavations are a rare occurrence in the world of developer-funded archaeology, the RLE and OA archaeological teams were more than usually reliant on the depth of expertise provided by the specialist team. OA were able to deploy a number of excavators with previous Palaeolithic excavation experience, Mark Roberts' Boxgrove project having featured as a training ground for several of the project team. However many of the team had no prior period-specific experience; guidance and training in procedures specific to the Palaeolithic period was therefore essential and was provided by members of the specialist team working on-site in a supervisory role, in particular Francis Wenban-Smith (general excavation strategy, lithics recovery and recording), Martin Bates (geoarchaeological recording), and Simon Parfitt and colleagues from the Natural History Museum (faunal remains recovery and recording). This level of collaboration is a model that can and should be applied more widely to other periods, particularly in the case of the most significant discoveries. Hopefully the various segments of the archaeological profession can learn valuable lessons from the experience, as in this case.

Finally, this volume provides a record of Palaeolithic and other Quaternary investigations into a substantial volume of Pleistocene sediments that were bulk-extracted as part of the major infrastructural development of the HS1 track and Ebbsfleet International station. Although some material remains, mostly faunal fossils and lithic artefacts, were recovered and are now saved for posterity in museum collections, supported by an archive of paper records, drawings, survey records and photographs, it is salutary to remember that everything reported in this volume is a remembrance of something vanished. That which was, is gone, and it is ultimately for this reason that archaeological investigations take place in advance of development, to discover, and preserve, the story of the land and the people of it; a story that establishes and maintains connection with the physical environment and its past occupants. Thus this story is presented here; a history of landscape development integrated with the climatic and environmental fluctuations of the Pleistocene, during which at various points hominin ancestors lived and knapped, and encountered an elephant. Then much later their distant descendants quarried, built and knapped some more, and finally dug and built again, at a vaster scale with modern mechanisation wreaking rapid and drastic change unimaginable to the earliest inhabitants.

Perhaps the most surprising aspect of this story is, besides any more spiritual connection it might support, the evident physical connection between the deep Pleistocene landscape history and the present day pattern of occupation, use and infrastructural development. Thus, the town of Swanscombe is exactly situated on the level spread of the local outcrop of the Boyn Hill/ Orsett Heath terrace of the Thames. The chalk hills around the town have mostly been quarried away, but the surviving network of roads likewise echoes the deep-

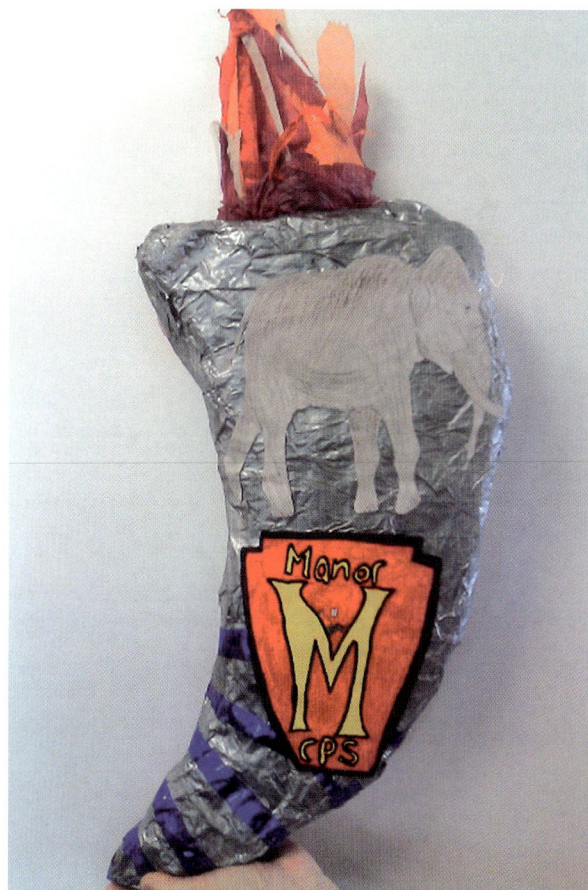

Figure 22.3 Manor Community Primary School, Swanscombe, Olympic torch

rooted Ice Age landscape structure, mapping onto ancient drainage pathways and terrace remnants. At Southfleet Road, it is perhaps not entirely coincidental that the elephant was preserved beneath the road: this road following a fold in the landscape; this fold directly resulting from the influence of the synclinal basin on the subsequent fluvial landscape; and the synclinal basin probably fundamentally related to the depositional environment and landscape topography that led to preservation of the elephant skeleton, and perhaps also to its demise on the spot where it was (much) later rediscovered.

Besides the contribution to academic research and and public understanding of the Stone Age in Britain, the most rewarding aspect of this story has, however, been its impact on the next generation. At the time of writing, the Olympic torch is criss-crossing the land; the Dartford area having been omitted from the route, the local schools designed and paraded their own torches. Inspired by both this new discovery and the earlier discovery of an isolated tusk and Palaeolithic flint tools during construction of the neighbouring Swan Valley school, the children of Manor Community Primary School, Swanscombe, have looked to this engagement with the evidence of the past in their torch (Fig. 22.3), adopting the shape of the tusk and the colour and texture of the associated flint tools, and fronted by the great beast *Palaeoloxodon antiquus* itself. What more appropriate legacy could there be of our archaeological endeavour?

Appendix I

Environmental sample inventory and assessment

Table A1.1 Sample inventory

Sample type	Phase	Context	Sample <no.>	Sample size	Sampled for	Assessment programme	Analysis	Comments
Bulk-sed	9b	40053	40023	30L	Small verts	Prelim assessment April 2004 – nothing found	-	
Bulk-sed	8c	40018	40004	10L	Clast lithology	-	Clast lithology & angularity/roundness	
Bulk-sed	8c	40018	40005	10L	Clast lithology	-	Lost	
Bulk-sed	8c	40071	40001	10L	Clast lithology	-	Clast lithology & angularity/roundness	
Bulk-sed	8c	40071	40002	10L	Clast lithology	-	Clast lithology & angularity/roundness	
Bulk-sed	8c	40071	40044	30L	Small verts	Prelim assessment April 2004 – nothing found	-	
Bulk-sed	6	40078	40100	220L	Elephant bits	Assessed for small vertebrates, larger identifiable elephant bones and lithic artefacts, 2009	-	
Bulk-sed	6	40078	40101	110L	Elephant bits	Assessed for small vertebrates, larger identifiable elephant bones and lithic artefacts, 2009	-	
Bulk-sed	6	40078	40128	100L	megafauna	Assessed for small vertebrates, 2009	-	
Bulk-sed	6	40078	40159	110L	Elephant bits	Assessed for small vertebrates, larger identifiable elephant bones and lithic artefacts, 2009	-	
Bulk-sed	6	40078	40194	10L	megafauna	-	-	
Bulk-sed	6	40078	40275	40L	Small verts	Assessed for small vertebrates, 2009	-	
Bulk-sed	6	40078	40276	10L	Insects	-	-	
Bulk-sed	6	40078	40327	20L	Small verts	Assessed for small vertebrates, 2009	Small vertebrate material contributes to 2010 analysis	
Bulk-sed	6	40100	40176	30L	Small verts	Prelim assessment April 2004 – nothing found	-	
Bulk-sed	6c	40039	40238	0.2L	Small verts	Assessed for SVs & molluscs	Small vertebrate material contributes to 2010 analysis	
Bulk-sed	6	40068	40026	30L	Small verts	Prelim assessment April 2004 – nothing found	-	
Bulk-sed	6	40068	40032	30L	Small verts	Prelim assessment April 2004 – a few frags of large animal bone	-	
Bulk-sed	6	40068	40203	10L	Charred remains	Assessed 2009	Wood ID and C14 by Oxford Archaeology	
Bulk-sed	6	40069	40029	30L	Small verts	Prelim assessment April 2004 – nothing found	-	
Bulk-sed	6	40162	40346	20L	Insects	Assessed for insects, 2009	-	
Bulk-sed	6b	40070	40035	30L	Small verts	Prelim assessment April 2004 – abundant molluscs and small vertebrates	Material forms basis of mollusc and SV analyses in J Quaternary Sci paper (Wenban-Smith et al. 2006); small vertebrates analysed in 2010	
Bulk-sed	6b	40070	40160	50L	Small verts	Prelim assessment April 2004 – abundant molluscs and small vertebrates	-	
Bulk-sed	6b	40070	40162	50L	Small verts	Prelim assessment April 2004 – abundant molluscs and small vertebrates	Small vertebrate material contributes to 2010 analysis	
Bulk-sed	6b	40070	40182	20L	Small verts	Prelim assessment April 2004 – scarce molluscs and common small vertebrates	-	
Bulk-sed	6b	40070	40183	50L	Small verts	-	-	
Bulk-sed	6b	40070	40184	40L	Small verts	-	-	

Table A1.1 (continued 1)

Sample type	Phase	Context	Sample <no.>	Sample size	Sampled for	Assessment programme	Analysis	Comments
Bulk-sed	6b	40070	40185	10L	Small verts	-	-	
Bulk-sed	6b	40070	40186	20L	Small verts	Prelim assessment April 2004 - no molluscs, but common small vertebrates	-	
Bulk-sed	6b	40070	40237	190L	Small verts	-	-	
Bulk-sed	6b	40070	40241	300L	Small verts	-	-	Sieved, not picked; residue retained for archive
Bulk-sed	6b	40070	40251	150L	Small verts	-	-	
Bulk-sed	6b	40070	40274	80L	Small verts	-	-	
Bulk-sed	6b	40070	40277	10L	Small verts	Taken away by SA Parfitt	Small vertebrate material contributes to 2010 analysis	
Bulk-sed	6b	40070	40280	10L	Small verts	-	-	
Bulk-sed	6b	40070	40288	100L	Small verts	-	-	
Bulk-sed	6b	40070	40289	100L	Small verts	-	-	
Bulk-sed	6b	40070	40310	10L	Small verts	Assessed for ostracods & small vertebrates, 2009	Small vertebrate material contributes to 2010 analysis; also, a few ostracods	
Bulk-sed	6b	40070	40313	10L	Small verts	Assessed for ostracods & small vertebrates, 2010	Small vertebrate material contributes to 2010 analysis	
Bulk-sed	6b	40070	40324	30L	Small verts	-	-	
Bulk-sed	6b	40070	40412	100L	Small verts	-	-	Sieved, not picked; residue retained for archive
Bulk-sed	6b	40143	40267	100L	Small verts	Assessed for small vertebrates, 2009	Small vertebrate material contributes to 2010 analysis	
Bulk-sed	6b	40143	40284	10L	Small verts	Assessed for small vertebrates, 2009	Small vertebrate & molluscan material contributes to 2010 analysis	
Bulk-sed	6b	40144	40311	10L	Small verts	Assessed for small vertebrates, pollen and ostracods, 2009	Small vertebrate material contributes to 2010 analysis	
Bulk-sed	6b	40144?	40297	30L	Small verts	Straight-to-analysis	Small vertebrate material contributes to 2010 analysis	
Bulk-sed	6a	40039	40022	30L	Small verts	Prelim assessment April 2004 - no molluscs, but common small vertebrates	-	
Bulk-sed	6a	40039	40261	60L	Small verts	Assessed for small vertebrates, 2009	Small vertebrate material contributes to 2010 analysis	
Bulk-sed	6a	40040	40021	30L	Small verts	Prelim assessment April 2004 -	-	
Bulk-sed	6b	40070	40283	30L	Small verts	-	-	
Bulk-sed	6a	40103	40175	30L	Small verts	Prelim assessment April 2004 -	-	
Bulk-sed	6a	40103	40312	20L	Small verts	Small vertebrates, ostracods & molluscs (tufa channel, Column 3 - eastern offset)	Small vertebrate material contributes to 2010 analysis	
Bulk-sed	6	40158	40262	30L	Small verts	Assessed for small vertebrates, 2009	-	
Bulk-sed	6	40158	40263	20L	Insects	Assessed for insects, 2009	-	

Table A1.1 (continued 2)

Sample type	Phase	Context	Sample <no.>	Sample size	Sampled for	Assessment programme	Analysis	Comments
Bulk-sed	6	40158	40413	20L	Insects	Assessed for insects, 2009	-	
Bulk-sed	7	40043	40350	30L	Small verts	Prelim assessment Aug 2004 -	-	
Bulk-sed	7	40164	40361	40L	Small verts; PLUS clast lithology	Assessed for ostracods & small vertebrates, 2009	Clast lithology & angularity/roundness	
Bulk-sed	7	40166	40349	30L	Small verts	Prelim assessment Aug 2004 -	-	
Bulk-sed	7	40166	40414	40L	Small verts	Assessed for pollen, ostracods & small vertebrates, 2009	-	
Bulk-sed	7	40166	40415	10L	Small verts	Assessed for pollen, ostracods & small vertebrates, 2009	-	
Bulk-sed	7	40166	40416	40L	Small verts	Assessed for pollen, ostracods & small vertebrates, 2009	-	
Bulk-sed	7	40166	40417	10L	Small verts	Assessed for pollen, ostracods & small vertebrates, 2009	-	
Bulk-sed	7	40167	40003	10L	Clast lithology	-	Clast lithology & angularity/roundness	
Bulk-sed	7	40167	40420	50L	Clast lithology	Assessed for ostracods and small vertebrates, 2009	Clast lithology & angularity/roundness	
Bulk-sed	6	40039	40242	10L	Small verts	-	-	
Bulk-sed	6	40100	40161	30L	Small verts	Prelim assessment April 2004 -	-	
Bulk-sed	5	40025	40020	30L	Small verts	Prelim assessment April 2004 -	-	
Bulk-sed	5	40025	40285	10L	Small verts	Assessed for small vertebrates, 2009	-	
Bulk-sed	5	40025	40286	100L	Small verts	Assessed for small vertebrates, 2009	Small vertebrate material contributes to 2010 analysis	
Bulk-sed	5	40025	40343	100L	Small verts	Assessed for small vertebrates, 2009	Small vertebrate material contributes to 2010 analysis	
Bulk-sed	5	40025	40348	100L	Small verts	Assessed for small vertebrates, 2009	Small vertebrate material contributes to 2010 analysis	
Bulk-sed	5	40025	40380	280L	Small verts	Assessed for small vertebrates, 2009	Small vertebrate material contributes to 2010 analysis	
Bulk-sed	5	40025	40411	200L	Small verts	Assessed for small vertebrates, 2009	Small vertebrate material contributes to 2010 analysis	
Bulk-sed	5	40066	40040	30L	Small verts	Prelim assessment April 2004 -	-	
Bulk-sed	5	40067	40041	30L	Small verts	-	-	
Bulk-sed	3	40028	40292	20L	Small verts	Prelim assessment July 2004 -	-	
Bulk-sed	3	40062	40042	30L	Small verts	Prelim assessment April 2004 -	Small vertebrates, molluscs and ostracods	
Bulk-sed	3	40062	40043	30L	Small verts	Prelim assessment April 2004 -	-	
Bulk-sed	3	40062	40045	30L	Small verts	Prelim assessment April 2004 -	-	
Bulk-sed	3	40062	40291	100L	Small verts	Assessed for small vertebrates, 2009	-	
Bulk-sed	2	40064	40064	30L	Small verts	Prelim assessment April 2004 -	-	
Bulk-sed	2	40064	40114	20L	Small verts	Prelim assessment April 2004 -	-	
Bulk-sed	2	40065	40065	30L	Small verts	Prelim assessment April 2004 -	-	
Bulk-sed	2	40077	40116	30L??	Small verts	Prelim assessment April 2004 -	-	
Bulk-sed	1	40056	40115	30L	Small verts	Prelim assessment April 2004 -	-	
Bulk-sed	1	40057	40059	30L	Small verts	Prelim assessment April 2004 -	-	

Table A1.1 (continued 3)

Sample type	Phase	Context	Sample <no.>	Sample size	Sampled for	Assessment programme	Analysis	Comments
Bulk-sed (p-sieved)	1	40058	40055	30L	Small verts	Prelim assessment April 2004 -	-	
Bulk-sed (p-sieved)	1	40059	40046	30L	Small verts	Prelim assessment April 2004 -	-	
Bulk-sed (p-sieved)	6b?	40144?	40336	10L	Small verts	Assessed for small vertebrates, 2009	Small vertebrate material contributes to 2010 analysis	
Bulk-sed (p-sieved)	6b	40070	40290	100L	Small verts	-	Sieved and residues sorted for larger identifiable remains	
Bulk-sed (p-sieved)	6b	40070	40329	??	Small verts	-	Sieved and residues sorted for larger identifiable remains	
Bulk-sed (p-sieved)	6b	40070	40330	??	Small verts	-	Sieved and residues sorted for larger identifiable remains	
Bulk-sed (p-sieved)	6b	40070	40331	??	Small verts	-	Sieved and residues sorted for larger identifiable remains	
Bulk-sed (p-sieved)	6b	40070	40332	20L	Small verts	-	Sieved and residues sorted for larger identifiable remains	
Bulk-sed (p-sieved)	6b	40070	40335	80L	Small verts	-	Sieved and residues sorted for larger identifiable remains	
Bulk-sed (p-sieved)	6b	40070	40337	??	Small verts	-	Sieved and residues sorted for larger identifiable remains	
Bulk-sed (p-sieved)	6b	40070	40338	10L	Small verts	-	Sieved and residues sorted for larger identifiable remains	
Bulk-sed (p-sieved)	6b	40070	40339	1L	Small verts	-	Sieved and residues sorted for larger identifiable remains	
Bulk-sed (p-sieved)	6b	40070	40347	75L	Small verts	-	Sieved and residues sorted for larger identifiable remains	
Bulk-sed (p-sieved)	6b	40070	40351	10L	Small verts	-	Sieved and residues sorted for larger identifiable remains	
Bulk-sed (p-sieved)	6b	40070	40381	40L	Small verts	-	Sieved and residues sorted for larger identifiable remains	
Bulk-sed (p-sieved)	6b	40144	40333	60L	Small verts	Assessed for small vertebrates, 2009	Small vertebrate material contributes to 2010 analysis	
Bulk-sed (p-sieved)	6b	40144	40334	80L	Small verts	Assessed for small vertebrates, 2009	-	
Bulk-sed (p-sieved)	5	40025	40382	??	Small verts	Assessed for small vertebrates, 2009	Small vertebrate material contributes to 2010 analysis	
Bulk-inc	6b	40070	40314	30L	Small verts	Straight-to-analysis	Small vertebrates & molluscs (tufa channel, Column 2)	
Bulk-inc	6b	40070	40315	30L	Small verts	Straight-to-analysis	Small vertebrates & molluscs (tufa channel, Column 2)	
Bulk-inc	6b	40070	40316	30L	Small verts	Straight-to-analysis	Small vertebrates & molluscs (tufa channel, Column 2)	

Table A1.1 (continued 4)

Sample type	Phase	Context	Sample <no.>	Sample size	Sampled for	Assessment programme	Analysis	Comments
Bulk-inc	6b	40070	40317	30L	Small verts	Straight-to-analysis	Small vertebrates & molluscs (tufa channel, Column 2)	
Bulk-inc	6b	40070	40318	30L	Small verts	Straight-to-analysis	Small vertebrates & molluscs (tufa channel, Column 2)	
Bulk-inc	6b	40070	40319	20L	Small verts	Straight-to-analysis	Small vertebrates & molluscs (tufa channel, Column 2)	
Bulk-inc	6a	40103	40320	20L	Small verts	Straight-to-analysis	Small vertebrates & molluscs (tufa channel, Column 2)	
Bulk-inc	6a	40103	40325	30L	Small verts	Straight-to-analysis	Small vertebrates & molluscs (tufa channel, Column 2)	
Bulk-inc	6	40078	40293	30L	Small verts	Straight-to-analysis	Small vertebrates & molluscs (tufa channel, Column 1)	
Bulk-inc	6b	40144	40294	10L	Small verts	Straight-to-analysis	Small vertebrates & molluscs (tufa channel, Column 1)	
Bulk-inc	6b	40144	40295	20L	Small verts	Straight-to-analysis	Small vertebrates & molluscs (tufa channel, Column 1)	
Bulk-inc	6b	40070	40296	10L	Small verts	Straight-to-analysis	Small vertebrates & molluscs (tufa channel, Column 1)	
Bulk-inc	6b	40070	40298	20L	Small verts	Straight-to-analysis	Small vertebrates & molluscs (tufa channel, Column 1)	
Bulk-inc	6b	40070	40299	20L	Small verts	Straight-to-analysis	Small vertebrates & molluscs (tufa channel, Column 1)	
Bulk-inc	6a	40070	40300	20L	Small verts	Straight-to-analysis	Small vertebrates & molluscs (tufa channel, Column 1)	
Bulk-inc	6a	40103	40301	40L	Small verts	Straight-to-analysis	Small vertebrates & molluscs (tufa channel, Column 1)	
Bulk-inc	6a	40039	40302	40L	Small verts	Straight-to-analysis	Small vertebrates & molluscs (tufa channel, Column 1)	
Bulk-inc	5	40025	40303	20L	Small verts	Straight-to-analysis	Small vertebrates & molluscs (tufa channel, Column 1)	
Bulk-inc	5	40025	40304	20L	Small verts	Straight-to-analysis	Small vertebrates & molluscs (tufa channel, Column 1)	
Bulk-inc	6b	40144	40305	30L	Small verts	Straight-to-analysis	Small vertebrates & molluscs (tufa channel, Column 3 - eastern offset)	
Bulk-inc	6b	40070	40306	20L	Small verts	Straight-to-analysis	Small vertebrates & molluscs (tufa channel, Column 3 - eastern offset)	
Bulk-inc	6b	40070	40307	20L	Small verts	Straight-to-analysis	Small vertebrates & molluscs (tufa channel, Column 3 - eastern offset)	
Bulk-inc	6b	40070	40308	20L	Small verts	Straight-to-analysis	Small vertebrates & molluscs (tufa channel, Column 3 - eastern offset)	
Bulk-inc	6b	40070	40309	30L	Small verts	Straight-to-analysis	Small vertebrates & molluscs (tufa channel, Column 3 - eastern offset)	
Spot-sed	9b	40053	40024	100g	Ostracods	-	-	
Spot-sed	9b	40053	40025	100g	Pollen	Prelim assessment July 2004	-	
Spot-sed	7	40023	40006	100g	Ostracods	-	-	
Spot-sed	7	40023	40006	100g	Pollen	Prelim assessment January 2004	-	
Spot-sed	6a	40024	40007	100g	Ostracods	Prelim assessment January 2004	-	
Spot-sed	6a	40024	40007	100g	Pollen	Prelim assessment January 2004	-	
Spot-sed	6a	40039	40278	0.5	Molluscs	Assessed for SVs & molluscs	Small vertebrate material contributes to 2010 analysis	
Spot-sed	6	40158	40407	75cc	Pollen	Prelim investigation July 2004 (C Turner)	-	
Spot-sed	6	40158	40408	75cc	Pollen	Prelim investigation July 2004 (C Turner)	-	
Spot-sed	6	40158	40409	75cc	Pollen	Prelim investigation July 2004 (C Turner)	-	

Table A1.1 (continued 5)

Sample type	Phase	Context	Sample <no.>	Sample size	Sampled for	Assessment programme	Analysis	Comments
Spot-sed	6	40158	40410	250cc	Plant macro remains	Prelim investigation July 2004 (C Turner) -	-	Azolla spore; amorphous non-cellular organic detritus
Spot-sed	6	40078	40132	100g	Pollen	Prelim investigation July 2004 (C Turner)	Countable pollen	
Spot-sed	6	40078	40133	100g	Pollen	Prelim investigation July 2004 (C Turner)	Countable pollen	
Spot-sed	6a	40039	40279	0.5L	Charred/rotted plant remains??	Lost	-	
Spot-sed	5	40025	40130	100g	Pollen	Prelim investigation July 2004 (C Turner) -	-	
Spot-sed	5	40025	40131	100g	Pollen	Prelim investigation July 2004 (C Turner) -	-	
Spot-sed	6	40069	40030	100g	Ostracods	-	-	
Spot-sed	6	40069	40031	100g	Pollen	-	-	
Spot-sed	6	40068	40027	100g	Pollen	Prelim investigation July 2004 (C Turner) -	-	
Spot-sed	6	40068	40028	100g	Ostracods	-	-	
Spot-sed	6	40068	40033	100g	Ostracods	-	-	
Spot-sed	6	40068	40034	100g	Pollen	-	-	
Spot-sed	6	40068	40086	100g	Ostracods	Prelim assessment March 2004 -	-	
Spot-sed	6b	40143	40248	200g	Ostracods	Prelim ostracod assessment June 2004; pollen assessment 2009	Ostracods present and analysed	
Spot-sed	6b	40143	40249	200g	Ostracods	Prelim ostracod assessment June 2004; pollen assessment 2009	Ostracods present and analysed	
Spot-sed	6b	40143	40250	200g	Ostracods	Prelim ostracod assessment June 2004; pollen assessment 2009	Ostracods present and analysed	
Spot-sed	6b	40143	40253	1L	Molluscs	Taken away by SA Parfitt	Small vertebrate material contributes to 2010 analysis	
Spot-sed	6b	40143	40271	200g	Ostracods	Prelim assessment July 2004 -	Ostracods present and analysed	
Spot-sed	6b	40070	40036	100g	Ostracods	-	-	
Spot-sed	6b	40070	40037	100g	Pollen	-	-	
Spot-sed	6b	40070	40038	2L	Molluscs	-	-	
Spot-sed	6b	40070	40039	100g	Ostracods	-	-	
Spot-sed	6b	40070	40087	100g	Ostracods	Prelim assessment March 2004 -	-	
Spot-sed	6b	40070	40088	100g	Ostracods	Prelim assessment March 2004 -	-	
Spot-sed	6b	40070	40134-1	100g	Pollen	Prelim investigation July 2004 (C Turner) -	-	
Spot-sed	6b	40070	40134-2	25 g	Plant macro remains	-	-	
Spot-sed	6b	40070	40252	1L	Molluscs	Taken away by SA Parfitt	Small vertebrate material contributes to 2010 analysis	
Spot-sed	6b	40070	40422		Small verts	-	-	
Spot-sed	6	40029	40008	100g	Ostracods	-	-	
Spot-sed	6	40029	40008	100g	Pollen	-	-	
Spot-sed	5	40025	40129	100g	Pollen	Prelim investigation July 2004 (C Turner) -	-	

Table A1.1 (continued 6)

Sample type	Phase	Context	Sample <no.>	Sample size	Sampled for	Assessment programme	Analysis	Comments
Spot-sed	6	40100	40257	10L	Molluscs	Assessed for molluscs	-	
Spot-sed	2	40065	40066	100g	Ostracods	Ostracod assessment 2009	-	
Spot-sed	2	40065	40067	100g	Pollen	Pollen assessment 2009	-	
Spot-sed	1	40059	40062	100g	Ostracods	Ostracod assessment 2009	-	
Spot-sed	1	40059	40063	100g	Pollen	Pollen assessment 2009	-	
Spot-sed	1	40057	40060	100g	Ostracods	Ostracod assessment 2009	-	
Spot-sed	1	40057	40061	100g	Pollen	Pollen assessment 2009	-	
Spot-sed	6	40100	40256	10L	Molluscs	-	-	
Sed-increment	5	40025	40009	100g	Ostracods	Prelim assessment January 2004	-	
Sed-increment	5	40025	40009	100g	Pollen	Prelim assessment January 2004; Re-assessed July 2004	-	
Sed-increment	5	40025	40010	100g	Ostracods	-	-	
Sed-increment	5	40025	40010	100g	Pollen	-	-	
Sed-increment	4	40026	40011	100g	Ostracods	Ostracod assessment 2009	-	
Sed-increment	4	40026	40011	100g	Pollen	-	-	
Sed-increment	4	40026	40012	100g	Ostracods	Prelim assessment January 2004 -	-	
Sed-increment	4	40026	40012	100g	Pollen	Prelim assessment January 2004; Re-assessed July 2004	-	
Sed-increment	4	40026	40013	100g	Ostracods	Ostracod assessment 2009	-	
Sed-increment	4	40026	40013	100g	Pollen	-	-	
Sed-increment	4	40027	40014	100g	Ostracods	Ostracod assessment 2009	-	
Sed-increment	4	40027	40014	100g	Pollen	-	-	
Sed-increment	4	40027	40015	100g	Ostracods	Prelim assessment January 2004 -	-	
Sed-increment	4	40027	40015	100g	Pollen	Prelim assessment January 2004; Re-assessed July 2004	-	
Sed-increment	6	40100	40366	20L	Small verts	-	-	
Sed-increment	6	40100	40367	20L	Small verts	-	-	
Sed-increment	6	40078	40368	20L	Small verts	-	-	
Sed-increment	6	40099	40369	20L	Small verts	-	-	
Sed-increment	6a	40103	40370	20L	Small verts	-	-	
Sed-increment	6a	40103	40371	20L	Small verts	-	-	
Sed-increment	6a	40039	40372	20L	Small verts	-	-	
Sed-increment	6b	40143	40264	200g	Ostracods	Prelim assessment July 2004	Ostracods present and analysed	
Sed-increment	6b	40143	40265	200g	Ostracods	Prelim assessment July 2004	Ostracods present and analysed	
Sed-increment	6b	40143	40266	200g	Ostracods	Prelim assessment July 2004	Ostracods present and analysed	
Sed-increment	6b	40143	40268	200g	Ostracods	Prelim assessment July 2004	Ostracods present and analysed	

Table A1.1 (continued 7)

Sample type	Phase	Context	Sample <no.>	Sample size	Sampled for	Assessment programme	Analysis	Comments
Sed-increment	6b	40143	40269	200g	Ostracods	Prelim assessment July 2004	Ostracods present and analysed	
Sed-increment	6b	40143	40270	200g	Ostracods	Prelim assessment July 2004	Ostracods present and analysed	
Sed-increment	6	40162	40398	20L	Small verts	-	-	
Sed-increment	6	40162	40399	20L	Small verts	-	-	
Sed-increment	6	40162	40400	20L	Small verts	-	-	
Sed-increment	6	40162	40401	20L	Small verts	-	-	
Sed-increment	6	40162	40402	20L	Small verts	-	-	
Sed-increment	6a	40103	40403	20L	Small verts	-	-	
Sed-increment	6a	40103	40404	20L	Small verts	-	-	
Sed-increment	6a	40039	40405	20L	Small verts	-	-	
Sed-increment	6	40162	40393	20L	Small verts	-	-	
Sed-increment	6	40162	40394	20L	Small verts	-	-	
Sed-increment	6	40162	40395	20L	Small verts	-	-	
Sed-increment	6	40162	40396	20L	Small verts	-	-	
Sed-increment	6	40162	40397	20L	Small verts	-	-	
Sed-increment	6	40100	40387	20L	Small verts	-	-	
Sed-increment	6	40078	40388	20L	Small verts	-	-	
Sed-increment	6	40078	40389	20L	Small verts	-	-	
Sed-increment	6	40099	40390	20L	Small verts	-	-	
Sed-increment	6	40099	40391	20L	Small verts	-	-	
Sed-increment	6a	40103	40392	20L	Small verts	-	-	
Sed-increment	6	40100	40373	20L	Small verts	-	-	
Sed-increment	6	40100	40374	20L	Small verts	-	-	
Sed-increment	6	40100	40375	20L	Small verts	-	-	
Sed-increment	6	40078	40376	20L	Small verts	-	-	
Sed-increment	6	40078	40377	20L	Small verts	-	-	
Sed-increment	6	40099	40378	20L	Small verts	-	-	
Sed-increment	6	40099	40379	20L	Small verts	-	-	
Monolith	6b	40070	40328	63cm	Pollen/ micro-pal/ Soil m-morph	-	-	
Monolith	7	40043	40018	135cm	Pollen	-	-	
Monolith	6b	40143	40282	24cm	Pollen/ micro-pal/ Soil m-morph	Assessed for molluscs & ostracods, 2009	Molluscs analysed; Small vertebrate material contributes to 2010 analysis; also, a few ostracods	

Table A1.1 (continued 8)

Sample type	Phase	Context	Sample \<no.\>	Sample size	Sampled for	Assessment programme	Analysis	Comments
Monolith	6b	40070	40281	27cm	Pollen/micro-pal/Soil m-morph	Pollen assessment 2009	-	
Kubiena	6	40100	40229		Soil m-morph		-	
Kubiena	6	40100	40230		Soil m-morph		-	
Kubiena	6	40100	40231		Soil m-morph		-	
Kubiena	6	40099	40232		Soil m-morph		-	
Kubiena	6a	40103	40233		Soil m-morph		-	
Kubiena	6a	40039	40234		Soil m-morph		-	
Kubiena	6a	40039	40235		Soil m-morph		-	
Kubiena	6a	40040	40236		Soil m-morph		-	
Kubiena	6b	40070	40322	10cm	Soil m-morph		-	
Kubiena	6a	40039	40323	12cm	Soil m-morph		Soil micro-morphology	
Kubiena	6	40100	40260	1 foil tin (15cm)	Soil m-morph		-	
Kubiena	6	40100	40259	1 foil tin (18cm)	Soil m-morph		-	
Kubiena	6	40039	40258	1 foil tin (17cm)	Soil m-morph		-	
Kubiena	6	40100	40225	12cm	Soil m-morph		-	
Kubiena	6	40100	40226	10cm	Soil m-morph		-	
Kubiena	6	40039	40227	10cm	Soil m-morph		-	
Mon-inc	7	40166	40418	50cm	Pollen/micro-pal/Soil m-morph	Assessed for pollen & ostracods, 2009	Soil micro-morphology; some pollen	
Mon-inc	6	40158	40419	50cm	Pollen/micro-pal/Soil m-morph	-	-	
Mon-inc	7	40043	40158	50cm	Pollen	-	-	
Mon-inc	6a	40040	40157	45cm	Pollen	Assessed for molluscs & ostracods, 2009	-	
Mon-inc	6a	40040	40156	45cm	Pollen	Sieved for small vertebrates	-	
Mon-inc	5	40025	40155	45cm	Pollen	Sieved for small vertebrates	-	
Mon-inc	5	40025	40154	45cm	Pollen	Assessed for ostracods, 2009	-	
Mon-inc	4	40027	40153	50cm	Pollen	Assessed for ostracods, 2009	-	

Table A1.1 (continued 9)

Sample type	Phase	Context	Sample <no.>	Sample size	Sampled for	Assessment programme	Analysis	Comments
Mon-inc	7	40101	40148	55cm	Pollen	-	-	
Mon-inc	6	40100	40149	50cm	Pollen	-	Soil micro-morphology	
Mon-inc	6	40078	40150	45cm	Pollen	-	Soil micro-morphology	
Mon-inc	6a	40103	40151	45cm	Pollen	-	Soil micro-morphology	
Mon-inc	6a	40039	40152	50cm	Pollen	-	-	
Mon-inc	6	40099	40142	45cm	Pollen	-	-	
Mon-inc	6	40099	40143	40cm	Pollen	-	-	
Mon-inc	5	40025	40144	45cm	Pollen	Sieved for small vertebrates	-	
Mon-inc	5	40025	40145	45cm	Pollen	Sieved for small vertebrates	Small vertebrate material contributes to 2010 analysis	
Mon-inc	5	40025	40146	45cm	Pollen	Sieved for small vertebrates	-	
Mon-inc	5	40025	40147	50cm	Pollen	Sieved for small vertebrates	-	
Mon-inc	6a	40039	40135	45cm	Pollen	-	-	
Mon-inc	5	40025	40136	45cm	Pollen	Sieved for small vertebrates	-	
Mon-inc	5	40025	40137	45cm	Pollen	Sieved for small vertebrates	-	
Mon-inc	5	40025	40138	45cm	Pollen	Sieved for small vertebrates	-	
Mon-inc	5	40025	40139	45cm	Pollen	Sieved for small vertebrates	-	
Mon-inc	5	40025	40140	50cm	Pollen	Sieved for small vertebrates	-	
Mon-inc	5	40025	40141	60cm	Pollen	Sieved for small vertebrates	-	
Mon-inc	6	40100	40364	55cm	Pollen/ micro-pal/ Soil m-morph	Assessed for pollen, 2009	Some pollen	
Mon-inc	6	40078	40365	55cm	Pollen/ micro-pal/ Soil m-morph	Assessed for pollen, 2009	Soil micro-morphology	
Mon-inc	6	40100	40362	54cm	Pollen/ micro-pal/ Soil m-morph	-	-	
Mon-inc	6	40078	40363	53cm	Pollen/ micro-pal/ Soil m-morph	-	-	
Mon-inc	6	40099	40352	58cm	Pollen/ micro-pal/ Soil m-morph	Assessed for pollen, 2009	-	
Mon-inc	6	40099	40353	54cm	Pollen/ micro-pal/ Soil m-morph	Assessed for pollen, 2009	-	

Table A1.1 (continued 10)

Sample type	Phase	Context	Sample <no.>	Sample size	Sampled for	Assessment programme	Analysis	Comments
Mon-inc	6	40162	40344	67cm	Pollen/ micro-pal/ Soil m-morph	Assessed for pollen, 2009	-	
Mon-inc	6	40162	40345	67cm	Pollen/ micro-pal/ Soil m-morph	Assessed for pollen, 2009	-	
Mon-inc	6	40162	40340	75cm	Pollen/ micro-pal/ Soil m-morph	Assessed for pollen, 2009	-	
Mon-inc	6	40162	40341	75cm	Pollen/ micro-pal/ Soil m-morph	-	-	
Mon-inc	6	40162	40342	75cm	Pollen/ micro-pal/ Soil m-morph	Assessed for pollen, 2009	-	
Mon-inc	6	40078	40321	40cm	Pollen/ micro-pal/ Soil m-morph	Assessed for pollen, molluscs & ostracods, 2009	Molluscs analysed; also, a few ostracods	
Mon-inc	6a	40103	40326	50cm	Pollen/ micro-pal/ Soil m-morph	Assessed for molluscs & ostracods, 2009	-	
Mon-inc	5	40067	40072	75cm	Pollen	-	-	
Mon-inc	5	40066	40071	65cm	Pollen	-	-	
Mon-inc	5	40066	40070	70cm	Pollen	-	-	
Mon-inc	5	40066	40069	55cm	Pollen	-	-	
Mon-inc	5	40062	40068	50cm	Pollen	Assessed for molluscs & ostracods, 2009	Ostracods present and analysed	
Mon-inc	6	40100	40196	50cm	Pollen	Assessed for pollen, 2009	Soil micro-morphology	
Mon-inc	6	40100	40195	50cm	Pollen	Assessed for pollen, 2009	Soil micro-morphology	
Mon-inc	6	40100	40194	50cm	Pollen	Assessed for pollen, 2009	Soil micro-morphology	
Mon-inc	6	40100	40193	50cm	Pollen	Assessed for pollen, 2009	Soil micro-morphology	
Mon-inc	6	40039	40192	50cm	Pollen	Sieved for small vertebrates	-	
Mon-inc	5	40025	40191	45cm	Pollen	Sieved for small vertebrates	-	
Mon-inc	5	40025	40189	55cm	Pollen	Sieved for small vertebrates	-	
Mon-inc	5	40025	40188	55cm	Pollen	Sieved for small vertebrates	-	
Mon-inc	5	40025	40187	55cm	Pollen	Sieved for small vertebrates	-	

Table A1.1 (continued 11)

Sample type	Phase	Context	Sample <no.>	Sample size	Sampled for	Assessment programme	Analysis	Comments
Mon-inc	6	40068	40113	55cm	Pollen	-	-	
Mon-inc	6	40068	40112	55cm	Pollen	-	-	
Mon-inc	6	40068	40111	55cm	Pollen	-	-	
Mon-inc	6	40068	40110	55cm	Pollen	-	-	
Mon-inc	6	40068	40109	55cm	Pollen	-	-	
Mon-inc	6	40068	40108	55cm	Pollen	-	-	
Mon-inc	6	40068	40102	60cm	Pollen	-	-	
Mon-inc	5	40067	40103	45cm	Pollen	-	-	
Mon-inc	5	40066	40104	55cm	Pollen	Sieved for small vertebrates	-	
Mon-inc	5	40066	40105	65cm	Pollen	Sieved for small vertebrates	-	
Mon-inc	5	40066	40106	60cm	Pollen	Sieved for small vertebrates	-	
Mon-inc	5	40066	40107	55cm	Pollen	Sieved for small vertebrates	-	
Mon-inc	2	40065	40095	45cm	Pollen	Assessed for pollen & ostracods, 2009	-	
Mon-inc	2	40065	40079	55cm	Pollen	Assessed for pollen, 2009	-	
Mon-inc	2	40065	40078	60cm	Pollen	Assessed for pollen & ostracods, 2009	-	
Mon-inc	2	40065	40077	55cm	Pollen	-	-	
Mon-inc	2	40065	40076	65cm	Pollen	Assessed for pollen & ostracods, 2009	-	
Mon-inc	2	40064	40075	50cm	Pollen	-	-	
Mon-inc	2	40064	40074	50cm	Pollen	-	-	
Mon-inc	2	40064	40073	60cm	Pollen	Assessed for pollen & ostracods, 2009	-	
Mon-inc	2	40064	40099	57cm	Pollen	-	-	
Mon-inc	2	40064	40098	55cm	Pollen	Assessed for pollen & ostracods, 2009	-	
Mon-inc	2	40064	40097	50cm	Pollen	Assessed for pollen & ostracods, 2009	-	
Mon-inc	2	40064	40096	63cm	Pollen	Assessed for pollen & ostracods, 2009	-	
Mon-inc	1	40059	40094	50cm	Pollen	Assessed for pollen, 2009	-	
Mon-inc	1	40059	40093	45cm	Pollen	Assessed for pollen, 2009	-	
Mon-inc	1	40059	40092	50cm	Pollen	Assessed for pollen, 2009	-	
Mon-inc	1	40058	40085	50cm	Pollen	Assessed for pollen, 2009	-	
Mon-inc	1	40058	40084	55cm	Pollen	Assessed for pollen, 2009	-	
Mon-inc	1	40058	40083	65cm	Pollen	Assessed for pollen, 2009	-	
Mon-inc	1	40058	40082	50cm	Pollen	Assessed for pollen, 2009	Soil micro-morphology	
Mon-inc	1	40057	40081	60cm	Pollen	Assessed for pollen, 2009	-	
Mon-inc	1	40056	40080	60cm	Pollen	-	-	
Mon-inc	1	40056	40089	50cm	Pollen	-	-	

Table A1.1 (continued 12)

Sample type	Phase	Context	Sample <no.>	Sample size	Sampled for	Assessment programme	Analysis	Comments
Mon-inc	1	40056	40090	50cm	Pollen	-	-	
Mon-inc	1	40056	40091	60cm	Pollen	-	-	
OSL-tube	9b	40087	40254		OSL dating	-	1st batch	
OSL-tube	9a	40051	40056		OSL dating	-	1st batch	
OSL-tube	8c	40049	40057		OSL dating	-	1st batch	
OSL-tube	8b	40047	40058		OSL dating	-	-	
OSL-tube	8a	40098	40244		OSL dating	-	2nd batch	
OSL-tube	8a	40098	40245		OSL dating	-	-	
OSL-tube	8a	40045	40243		OSL dating	-	-	
OSL-tube	6b	40144	40247		OSL dating	Assessed for molluscan remains	-	Assessed for molluscan remains
OSL-tube	6b	40070	40246		OSL dating	-	-	
OSL-tube	6b	40070	40255		OSL dating	-	-	
OSL-tube	5	40163	40354		OSL dating	-	-	
OSL-tube	5	40163	40355		OSL dating	-	-	
OSL-tube	5	40066	40052		OSL dating	-	-	
OSL-tube	5	40066	40053		OSL dating	-	2nd batch	
OSL-tube	5	40066	40054		OSL dating	-	-	
OSL-tube	5	40066	40358		OSL dating	-	-	
OSL-tube	5	40066	40359		OSL dating	-	-	
OSL-tube	5	40163	40356		OSL dating	-	-	
OSL-tube	5	40163	40357		OSL dating	-	-	
OSL-tube	2	40065	40047		OSL dating	-	-	
OSL-tube	2	40065	40048		OSL dating	-	-	
OSL-tube	2	40064	40049		OSL dating	-	-	
OSL-tube	2	40060	40051		OSL dating	-	-	
OSL-tube	1	40056	40050		OSL dating	-	-	
Other	8c	40018	40016	na	Clast orientation study	-	-	
Other	8c	40018	40017	na	Clast orientation study	-	-	
Other	6b	40070	40421	??	Molluscs	-	-	
Other	6	40078	40240	??	Wood species ID	-	-	
Other	6	40100	40190	??	Wood species ID Assessed 2009	-	-	

Appendix 2

Loss-on-ignition and magnetic susceptibility of the sedimentary sequences at Southfleet Road

by John Crowther

INTRODUCTION

Loss-on-ignition (LOI) and low frequency mass-specific magnetic susceptibility (χ) determinations were made on 105 bulk samples from the sediments at Southfleet Road in the hope that they might provide evidence of possible soils/land surfaces within the sequences of deposits. As the sediments accumulated, it would be anticipated that former soils/land surfaces would have had a relatively high organic matter content (as estimated by LOI) as a result of plant growth and inputs of organic litter. An enhanced magnetic susceptibility would also be expected as a consequence of natural fermentation processes within soils (Le Borgne, 1955). It should be noted, however, that both properties may have been significantly affected by post-depositional processes. Organic matter content is likely to have diminished as a result of decomposition processes, and magnetic susceptibility may have been affected by the mobilisation (through gleying), leaching and reprecipitation of iron (Fe) compounds as a result of waterlogging. Also, χ is affected both by the degree enhancement and the Fe content – and where (as is likely to be the case in these sedimentary sequences) the latter is quite variable then χ may poorly reflect the levels of enhancement. The LOI and χ data do therefore need to be interpreted with caution.

METHODS

Analysis was undertaken on the fine earth fraction (ie < 2mm) of the samples. LOI was determined by ignition at 375°C for 16hrs (Ball 1964), previous studies having established that there is no significant breakdown of carbonates at this temperature. In addition to χ, determinations were made of χ_{max} (maximum potential magnetic susceptibility, which generally closely reflects

the Fe content) on 20 samples, representative of the range of χ values recorded, by subjecting a sample to optimum conditions for susceptibility enhancement in the laboratory. χ_{conv} (fractional conversion), which is expressed as a percentage, is a measure of the extent to which the potential susceptibility has been achieved in the original sample, viz: (χ/χ_{max}) x 100.0 (Tite 1972; Scollar *et al.* 1990). In many respects this is a better indicator of magnetic susceptibility enhancement than raw χ data, particularly in cases where sediments have widely differing χ_{max} values (Crowther and Barker 1995; Crowther 2003). A Bartington MS2 meter was used for magnetic susceptibility measurements. χ_{max} was achieved by heating samples at 650°C in reducing, followed by oxidising conditions. The method used broadly follows that of Tite and Mullins (1971), except that household flour was mixed with the soils and lids placed on the crucibles to create the reducing environment (after Graham and Scollar 1976; Crowther and Barker 1995).

RESULTS (Tables A2.1-A2.6)

The analytical data are presented in Table A2.6, summary statistics relating to samples form particular contexts and sequences in Tables A2.1–A2.5, a plot of the fractional conversion data in Fig. A2.1, and plots of variations in LOI and χ down individual sediment sequences in Figs. A2.2–2.3. Here, a broad overview of the individual properties is presented, before consideration of the results from individual contexts and sequences.

Overview of individual properties

1. Loss-on-ignition. The samples are predominantly minerogenic (Table A2.1), with 99 of the 105 samples

Table A2.1 Summary statistics for all samples

	No.	Minimum	Maximum	Mean	Std dev
LOI (%)	105	0.375	6.69	1.88	0.776
χ (10^{-8} m³ kg⁻¹)	105	2.0	27.4	10.7	4.48
χ_{max} (10^{-8} m³ kg⁻¹)	20	22.0	7390.0	1230	2030
χ_{conv} (%)	20	0.08	37.7	6.60	10.1

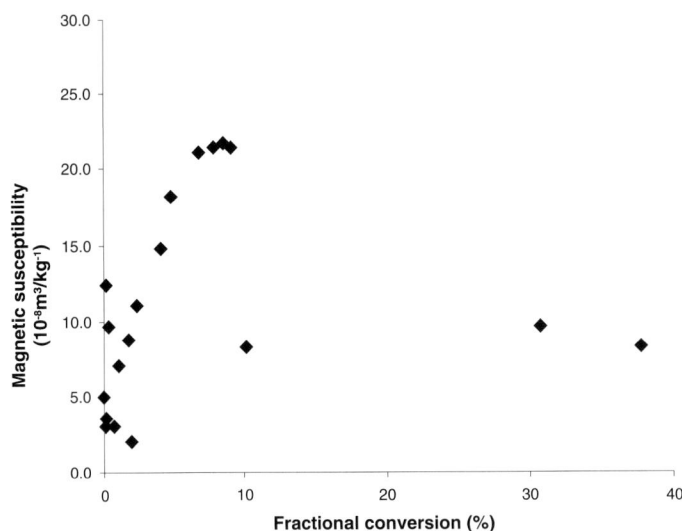

Figure A2.1. Plot of χ against χ$_{conv}$ for 20 representative samples

having a LOI < 3.00%. Of the remaining samples, only one sample from context 40158 stands out as having a notably higher LOI of 6.69%. Because of post-depositional organic decomposition, these data inevitably underestimate and may well poorly reflect the original organic matter content of the sediments. It should also be noted that a number of samples which appeared (by virtue of their darker colour) to be more organic, did not necessarily have relatively high LOI values. The samples were found to vary quite markedly in texture and, while there is unlikely to be any breakdown of clays at the temperature (375°C) used for ignition, it is possible that some of the variability in organic matter content may be directly related to texture in that finer sediments are more likely to contain more resistant clay-humus complexes and will tend to be less well aerated and therefore less vulnerable to decomposition. It would be interesting to investigate whether there is an underlying correlation between LOI and texture.

2. Magnetic susceptibility. The most striking feature of the magnetic susceptibility data is the extremely high variability in χ$_{max}$ with values ranging from 22.0–7390 x 10^{-8} m^3 kg^{-1} (Table A2.1). Given this exceptionally high variability, and the relatively low variability in χ (range, 2.0–27.4 x 10^{-8} m^3 kg^{-1}), it is highly unlikely that there will be a strong relationship between χ and χ$_{conv}$ (Crowther 2003), and this is borne out by Fig. A2.1. Furthermore, since such variability in χ$_{max}$ is likely to be largely attributable to variations in Fe content, which in these sedimentary sequences could well have been subject to post-depositional change through gleying and associated leaching/reprecipitation, the χ and χ$_{max}$ data may poorly reflect the characteristics of the sediments at the time of deposition. Thus, little reliance can be placed on the magnetic susceptibility data – with samples with a high χ$_{conv}$ (maximum, 37.7%) not necessarily being indicative of significant enhancement in the original sediments. In these circumstances, the χ data need to be interpreted with extreme caution.

Table A2.2 Summary of LOI (%) data for each context

	No.	Minimum	Maximum	Mean	Std dev
40025	5	0.375	0.844	0.530	0.187
40039	11	1.18	3.04	1.68	0.538
40040	4	0.511	1.98	1.18	0.605
40043		0.912	1.44	1.18	0.373
40056	5	1.39	1.91	1.62	0.188
40057	7	1.18	1.55	1.37	0.136
40058	3	0.972	1.09	1.05	0.065
40070	1	3.23	3.23	3.23	-
40078	7	2.02	2.30	2.17	0.087
40099	6	1.95	2.35	2.15	0.164
40100	20	1.21	2.36	1.87	0.254
40103	8	0.820	3.19	2.38	0.753
40158	5	1.75	6.69	3.49	1.87
40160	2	0.946	1.03	0.988	0.059
40162	18	1.78	2.52	2.19	0.176
40166	1	1.38	1.38	1.38	-

Comparison of different contexts

1. Loss-on-ignition. Many of the individual contexts display quite marked variability in LOI (Table A2.2). Of the 10 contexts for which ≥ 5 samples were analysed, three have a standard deviations of ≥ 0.500%, which are high for contexts with such low mean LOI values. Heterogeneity of this magnitude within individual contexts suggests that significant changes in environmental conditions occurred as each context developed, with the more organic-rich samples being likely associated with periods of soil development/surface exposure.

Table A2.3 Summary of $\chi(10^{-8}$ m^3 kg$^{-1})$ data for each context

	No.	Minimum	Maximum	Mean	Std dev
40025	5	2.9	5.1	3.9	0.873
40039	11	6.6	14.3	9.1	2.79
40040	4	3.0	9.6	6.6	2.74
40043	2	7.6	9.3	8.5	1.20
40056	5	19.5	21.7	20.6	0.887
40057	7	14.7	21.6	19.2	2.34
40058	3	15.9	27.4	19.7	6.64
40070	1	12.3	12.3	12.3	-
40078	7	9.0	10.4	9.8	0.544
40099	6	9.2	12.5	10.9	1.48
40100	20	8.1	12.3	10.0	0.851
40103	8	2.0	13.9	10.4	3.82
40158	5	7.6	10.0	8.8	0.950
40160	2	4.8	4.9	4.9	0.071
40162	18	7.7	10.4	9.4	0.775
40166	1	8.3	8.3	8.3	-

Table A2.4 Summary of LOI (%) data for each sequence

	No.	Minimum	Maximum	Mean	Std dev
1	5	0.912	1.98	1.47	0.453
2	10	0.511	6.69	2.24	1.83
3	23	0.375	3.19	1.76	0.809
4	20	1.21	2.36	1.82	0.287
5	4	0.82	3.23	1.70	1.06
6	7	1.89	2.46	2.19	0.199
7.1	5	1.78	2.52	2.18	0.276
7.2	4	2.13	2.42	2.30	0.132
7.3	7	2.06	2.27	2.17	0.084
7.4	5	1.93	3.07	2.43	0.412
8.1	5	0.972	1.37	1.14	0.149
8.2	10	1.23	1.91	1.51	0.193

Table A2.5 Summary of $\chi(10^{-8}$ m^3 kg$^{-1})$ data for each sequence

	No.	Minimum	Maximum	Mean	Std dev
1	5	6.5	9.6	8.3	1.27
2	10	3.0	10.0	7.2	2.26
3	23	2.9	14.3	8.9	3.39
4	20	7.0	12.3	9.8	1.15
5	4	2.0	12.3	7.5	4.29
6	7	9.3	12.5	11.3	1.37
7.1	5	7.7	9.6	8.8	0.783
7.2	4	8.7	10.1	9.6	0.638
7.3	7	8.9	10.4	9.9	0.516
7.4	5	8.3	11.8	9.8	1.31
8.1	5	14.7	27.4	18.4	5.18
8.2	10	19.2	21.7	20.5	0.972

Table A2.6 Analytical data

Sequence	Sample	Depth (cm)	Sequence depth (cm)	Context (rev)	LOI (%)	χ $(10^{-8}\,m^3kg^{-1})$	χ_{max} $(10^{-8}\,m^3kg^{-1})$	$\chi_{conv\,(\%)}$ $(10^{-8}\,m^3kg^{-1})$
1	40158	7-8	7.5	40043	0.912	7.6		
1	40158	17-18	17.5	40043	1.44	9.3		
1	40158	27-28	27.5	40039	1.85	8.4		
1	40158	37-38	37.5	40040	1.98	9.6	2200	0.44
1	40158	46-47	46.5	40040	1.15	6.5		
2	40418	2-3	2.5	40166	1.38	8.3		
2	40418	12-13	12.5	40158	6.69	8.2	81	10.1
2	40418	25-26	25.5	40158	1.75	7.6		
2	40418	36-37	36.5	40160	1.03	4.9	6310	0.08
2	40418	44-45	44.5	40158	3.12	10.0		
2	40419	4-5	49	40160	0.946	4.8		
2	40419	11-12	56	40158	2.81	8.7	454	1.92
2	40419	21-22	66	40158	3.10	9.4		
2	40419	31-32	76	40040	1.09	7.3		
2	40419	42-43	87	40040	0.511	3.0	2080	0.14
3	40148	4-5	4.5	40100	1.76	10.0		
3	40148	11-12	11.5	40100	2.11	10.3		
3	40148	21-22	21.5	40078	2.30	10.3		
3	40148	30-31	30.5	40078	2.23	10.4		
3	40148	43-44	43.5	40078	2.13	10.1		
3	40149	15-16	40.5	40078	2.14	9.8		
3	40149	22-23	47.5	40078	2.17	9.0		
3	40149	32-33	57.5	40078	2.17	9.2		
3	40149	44-45	69.5	40099	2.06	9.3		
3	40150	7-8	77.5	40099	1.95	9.2		
3	40150	17-18	87.5	40099	2.01	10.4		
3	40150	32-33	102.5	40103	3.19	13.8		
3	40150	42-43	112.5	40103	2.76	13.9		
3	40151	8-9	111	40039	3.04	14.3		
3	40151	17-18	120	40039	1.63	14.0		
3	40151	24-25	127	40039	1.70	7.3		
3	40151	34-35	137	40039	1.31	7.6		
3	40151	44-45	147	40039	1.18	6.6		
3	40152	8-9	148.5	40025	0.493	3.5	1610	0.22
3	40152	15-16	155.5	40025	0.844	5.1		
3	40152	24-25	164.5	40025	0.534	4.5		
3	40152	33-34	173.5	40025	0.405	3.6		
3	40152	43-44	183.5	40025	0.375	2.9	399	0.73
4	40196	4-5	4.5	40100	2.36	10.7		
4	40196	14-15	14.5	40100	2.01	10.1		
4	40196	24-25	24.5	40100	2.19	10.9		
4	40196	34-35	34.5	40100	1.84	9.9		
4	40196	44-45	44.5	40100	1.69	10.6		
4	40195	6-7	44	40100	1.84	10.2		
4	40195	16-17	54	40100	1.21	8.1		
4	40195	26-27	64	40100	2.05	10.2		
4	40195	36-37	74	40100	2.05	10.3		
4	40195	46-47	84	40100	2.02	9.8		
4	40194	6-7	81.5	40100	1.72	10.3		
4	40194	9-10		40100	Not sampled			
4	40194	16-17	91.5	40100	1.70	12.3		
4	40194	26-27	101.5	40100	1.62	9.3		
4	40194	36-37	111.5	40100	1.70	9.4		
4	40194	46-47	121.5	40100	1.99	10.5		
4	40193	7-8	122.5	40100	1.82	9.4		
4	40193	17-18	132.5	40100	1.62	9.0		
4	40193	27-28	142.5	40100	2.01	9.4		
4	40193	37-38	152.5	40039	1.73	8.2		
4	40193	47-48	162.5	40039	1.21	7.0	652	1.07
5	40322	4-5	4.5	40070	3.23	12.3	7390	0.17

Table A2.6 (continued)

Sequence	Sample	Depth (cm)	Sequence depth (cm)	Context (rev)	LOI (%)	χ $(10^{-8}\,m^3kg^{-1})$	χ_{max} $(10^{-8}\,m^3kg^{-1})$	$\chi_{conv\ (\%)}$ $(10^{-8}\,m^3kg^{-1})$
5	40322	11-12	11.5	40103	0.82	2.0	97	2.06
5	40323	2-3	22.5	40039	1.45	8.7		
5	40323	11-12	31.5	40039	1.28	6.8		
6	40365	4-5	4.5	40078	2.02	9.5		
6	40365	9-10	9.5	40099	2.22	12.5		
6	40365	15-16	15.5	40099	2.30	12.4		
6	40365	23-24	23.5	40099	2.35	11.6		
6	40365	30-31	30.5	40103	1.89	9.3		
6	40365	40-41	40.5	40103	2.46	12.4		
6	40365	50-51	50.5	40039	2.11	11.1	455	2.44
7.1	40344	9-10	9.5	40162	2.30	9.4		
7.1	40344	24-25	24.5	40162	1.78	7.7		
7.1	40344	29-30	29.5	40162	2.52	9.6		
7.1	40344	43-44	43.5	40162	2.24	8.8		
7.1	40344	55-56	55.5	40162	2.07	8.3	22	37.7
7.2	40345	4-5	4.5	40162	2.25	8.7		
7.2	40345	11-12	11.5	40162	2.42	10.1		
7.2	40345	43-44	43.5	40162	2.38	10.0		
7.2	40345	55-56	55.5	40162	2.13	9.6		
7.3	40340	7-8	7.5	40162	2.27	10.2		
7.3	40340	17-18	17.5	40162	2.08	10.4		
7.3	40340	27-28	27.5	40162	2.12	10.2		
7.3	40340	37-38	37.5	40162	2.06	10.0		
7.3	40340	47-48	47.5	40162	2.20	9.5	31	30.7
7.3	40340	57-58	57.5	40162	2.26	9.8		
7.3	40340	67-68	67.5	40162	2.19	8.9		
7.4	40342	6-7	6.5	40162	2.29	9.1		
7.4	40342	18-19	18.5	40162	1.93	8.3		
7.4	40342	38-39	38.5	40103	2.43	10.2		
7.4	40342	54-55	54.5	40103	2.45	9.8		
7.4	40342	70-71	70.5	40103	3.07	11.8		
8.1	40082	8-9	8.5	40058	0.972	15.9		
8.1	40082	18-19	18.5	40058	1.09	27.4	1000	2.74
8.1	40082	28-29	28.5	40058	1.08	15.9		
8.1	40082	38-39	38.5	40057	1.18	14.7	348	4.22
8.1	40082	51-52	51.5	40057	1.37	18.2	373	4.88
8.2	40081	5-6	5.5	40057	1.36	19.2		
8.2	40081	8-9	8.5	40057	1.23	19.6		
8.2	40081	15-16	15.5	40057	1.52	19.7		
8.2	40081	18-19	18.5	40057	1.35	21.6	239	9.04
8.2	40081	25-26	25.5	40057	1.55	21.5	270	7.96
8.2	40081	28-29	28.5	40056	1.39	21.7	255	8.51
8.2	40081	35-36	35.5	40056	1.62	21.2	311	6.82
8.2	40081	38-39	38.5	40056	1.56	20.5		
8.2	40081	45-46	45.5	40056	1.91	20.0		
8.2	40081	48-49	48.5	40056	1.64	19.5		

The extreme case is context 40158 (std dev., 1.87%), for which individual LOI values range from 1.75–6.69%. Allowing for post-depositional organic decomposition, then it seems likely that the most organic rich of these samples is from a sediment that was originally very humic (possibly peaty) in composition. Other contexts, by comparison, are much more uniform in terms of LOI (e.g. the 18 samples from context 40162 have a range of 1.78–2.52%), and in these cases there is therefore less evidence for changing environmental conditions as the sediments were deposited.

In addition to within-context variability, there are also differences in mean LOI between the contexts. Clearly, care needs to be exercised when comparing mean values based on small numbers of samples. Of the contexts with ≥ 5 samples, the mean values range from 0.530% (40025) to 3.49% (40158). Such differences probably reflect significant differences in the sedimentary environment at the time of deposition, with those contexts with a higher mean LOI being most likely to be associated with periods of soil development/surface exposure.

2. Magnetic susceptibility. As with LOI, the χ data display quite marked within- and, in this case particularly between-context, variability (Table A2.3). For example, contexts 40056 and 40057 stand out as having relatively high mean χ values (20.6 and 19.2 x 10^{-8} m^3 kg^{-1}, respectively), whereas the remaining contexts for which ≥ 5 samples were analysed have means of ≤ 10.9 x 10^{-8} m^3 kg^{-1}, with a minimum of 3.9 x 10^{-8} m^3 kg^{-1} (40025). As noted above, these data do need to be interpreted with caution because of the difficulties of distinguishing between enhancement through organic fermentation processes and the effects of Fe content, the latter of which may well have been subject to post-depositional change.

Comparison of different sequences

Summary LOI and χ data for each of the sequences are presented in Tables A2.4 and A2.5, respectively. Since many of the sequences include samples from more than one context, there is inevitably both within- and between-sequence variability in the data sets. Of potentially greater interest are the patterns of variation down individual sedimentary sequences, and these are presented in Figs. A2.2–A2.13. Interestingly, despite the serious reservations concerning the interpretation of the χ data, in several of the sequences there is a very close correlation between χ and LOI. This is particularly well illustrated by Sequence

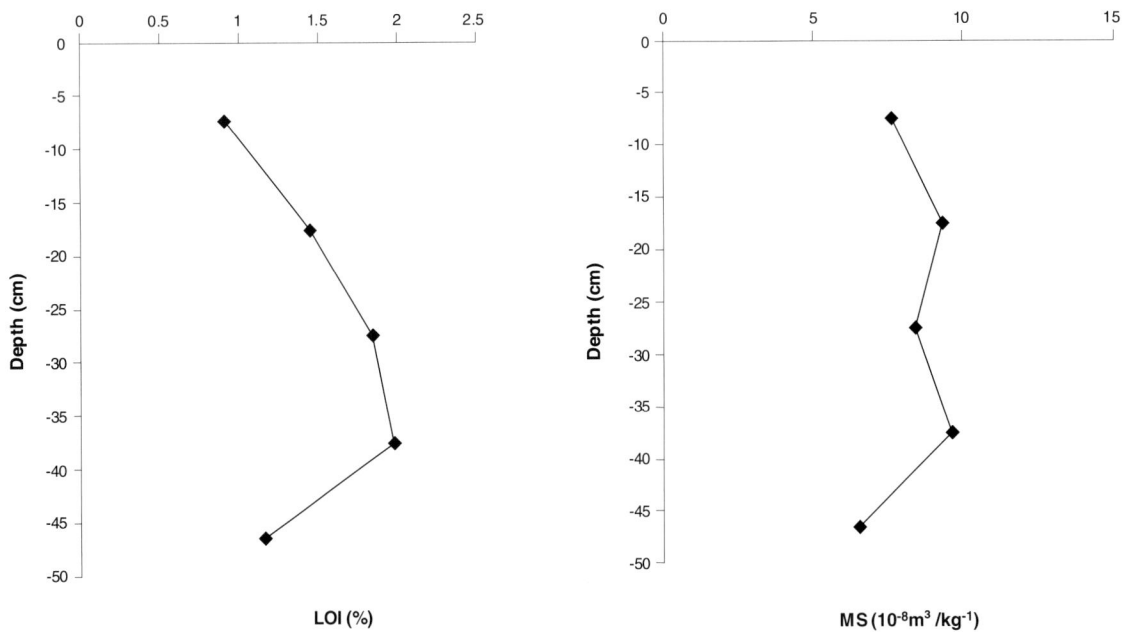

Figure A2.2 Variations in LOI (%) and $\chi(10^{-8}$ m^3 kg^{-1} down Sequence 1

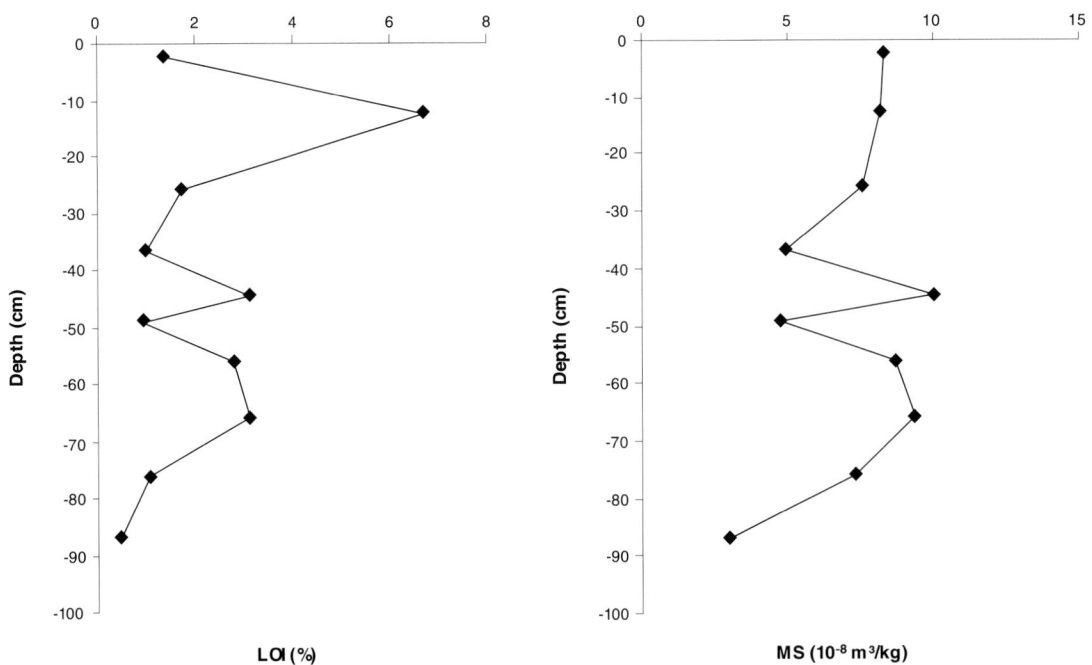

Figure A2.3 Variations in LOI (%) and $\chi(10^{-8}$ m^3 kg^{-1}) down Sequence 2

3 (Fig. A2.4), and in cases like this, where χ supports the LOI evidence, the χ data are perhaps more likely to reflect the degree of enhancement. Clearly, the detailed patterns will need to be examined in light of other evidence (colour, texture, composition, etc.). However, principal points to emerge from the LOI and χ plots are as follows (all depths refer to sequence depth):

Sequence 1 (Fig. A2.2): Generally higher LOI at 27.5–37.5cm, though not especially high (maximum, 1.98%). Sample at 37.5cm (context 40040) would appear from LOI and χ to be more similar to sample at 27.5cm (context 40039) than the underlying sample from context 40040.

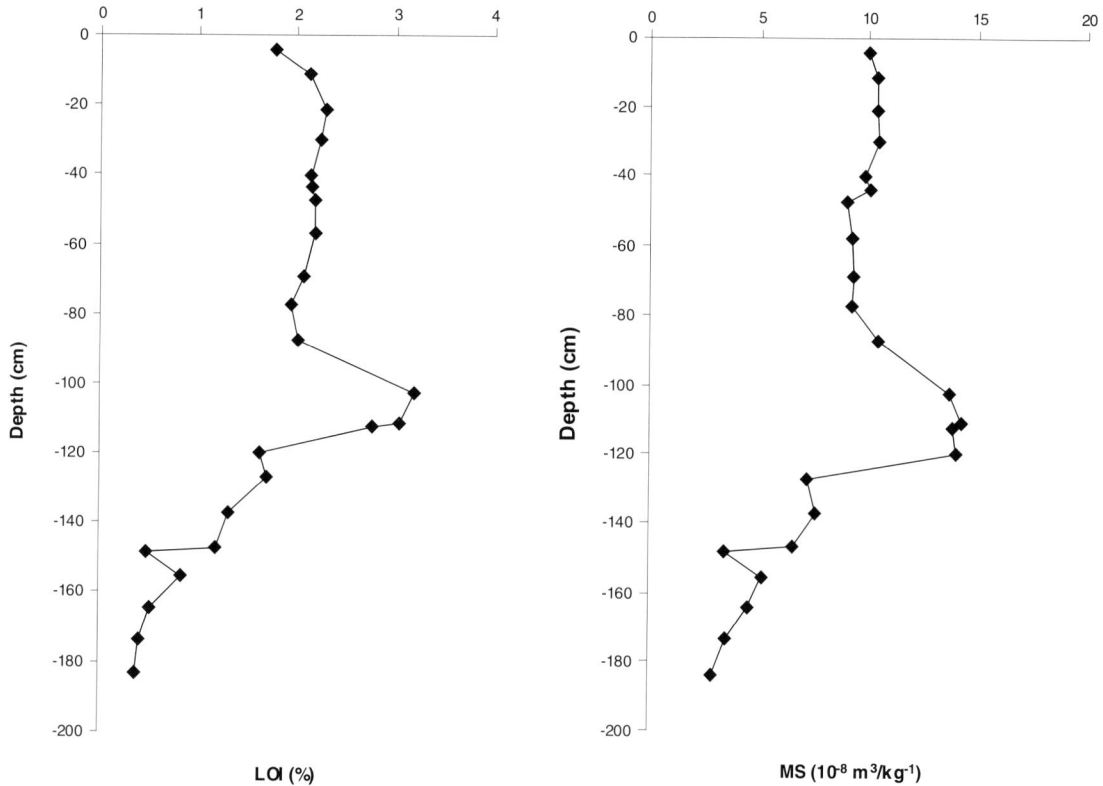

Figure A2.4 Variations in LOI (%) and χ(10⁻⁸ m³ kg⁻¹) down Sequence 3

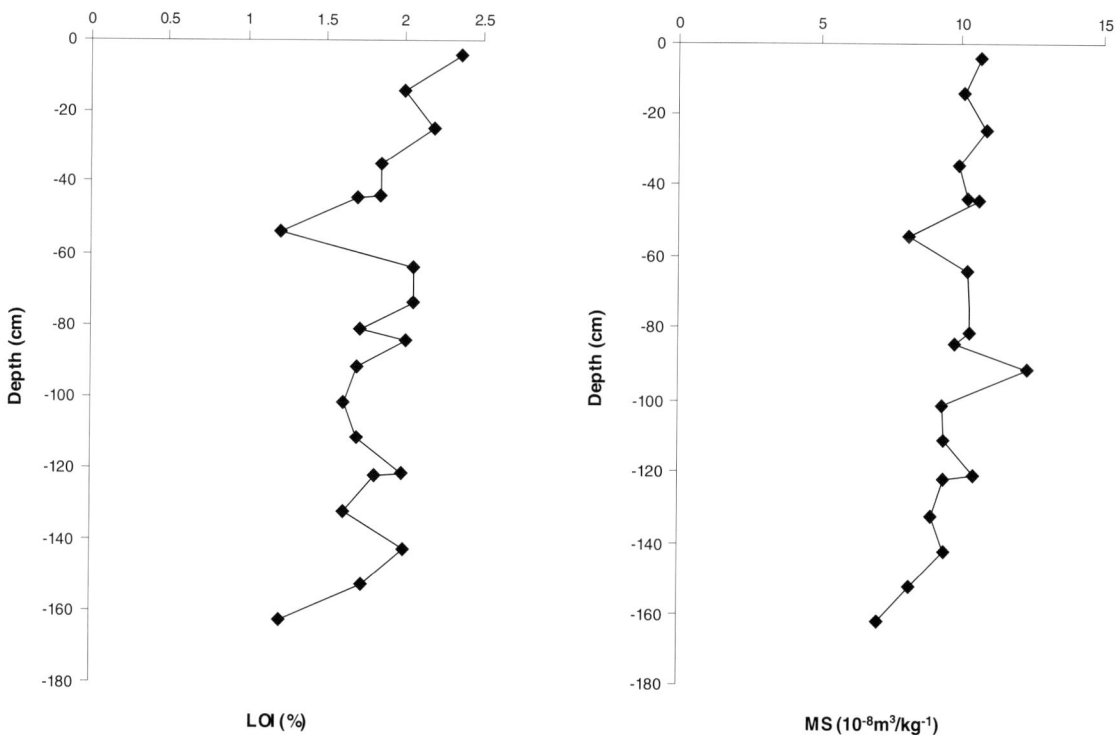

Figure A2.5 Variations in LOI (%) and χ(10⁻⁸ m³ kg⁻¹) down Sequence 4

Sequence 2 (Fig. A2.3): The higher LOI and, to some extent, χ values are associated with interdigitated context 40158, with only the sample at 25.5cm within this context having a notably lower LOI (1.75%). The very high peak in LOI at 12.5cm (6.69%) is almost certainly associated with soil (possibly peat?) formation, and the secondary peaks at 44.5 and 56–66cm also seem likely to be associated with periods of soil development/ surface exposure.

Sequence 3 (Fig. A2.4): Clear peak in LOI and χ at 102.5–111cm (i.e. context 40103) which is likely to be associated with soil development/surface exposure. Sediments become increasingly minerogenic towards base of sequence (contexts 40039 and, especially, 40025).

Sequence 4 (Fig. A2.5): Context 40100, which extends down to *c* 145cm, displays mostly relatively minor variations in LOI, with several possible minor peaks (at 4.5, 24.5, 64–74, 84, 121.5 and 142.5cm) and one sample (at 54cm) showing a notably lower LOI (1.21%). Several of the peaks in LOI correspond with peaks in χ.

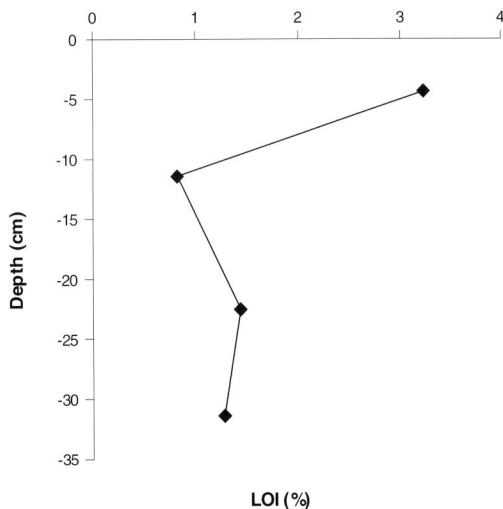

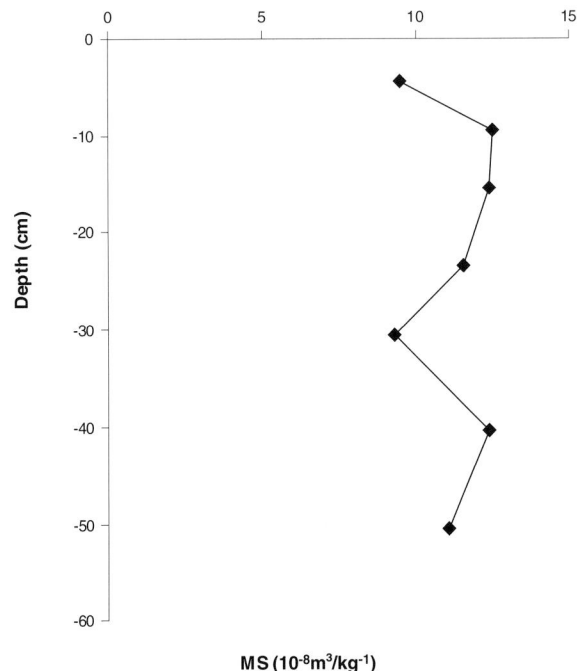

Figure A2.6 Variations in LOI (%) and χ(10^{-8} m^3 kg^{-1}) down Sequence 5

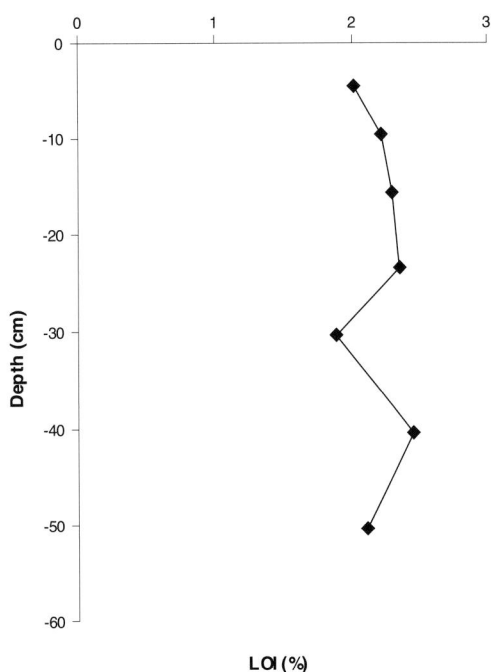

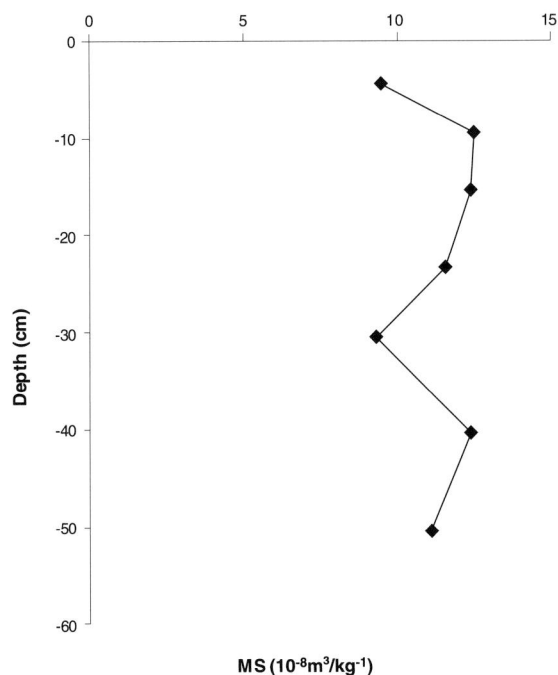

Figure A2.7 Variations in LOI (%) and χ(10^{-8} m^3 kg^{-1}) down Sequence 6

Sequence 5 (Fig. A2.6): This sequence reveals a close correlation between LOI and χ, and a clear distinction between context 40070, which has a higher LOI and χ, and the underlying contexts (40103 and 40039).

Sequence 6 (Fig. A2.7): This sequence also reveals a close correlation between LOI and χ, with both showing a minor peak towards the base of context 40103 (at 40.5cm).

Sequence 7.1 (Fig. A2.8): A single context (40162) showing a close correlation between LOI and χ, with both showing a minor peak at 29.5cm.

Sequence 7.2 (Fig. A2.9): A single context (40162) showing a close correlation between LOI and χ, though the range of variation down the sequence is relatively small.

Sequence 7.3 (Fig. A2.10): A single context (40162) showing a possible minor peak in LOI and χ at 57.5cm, though the range of variation down the sequence is relatively small.

Sequence 7.4 (Fig. A2.11): A close correlation between LOI and χ, and, though the range of variation down the sequence is relatively small, the lower context (40103)

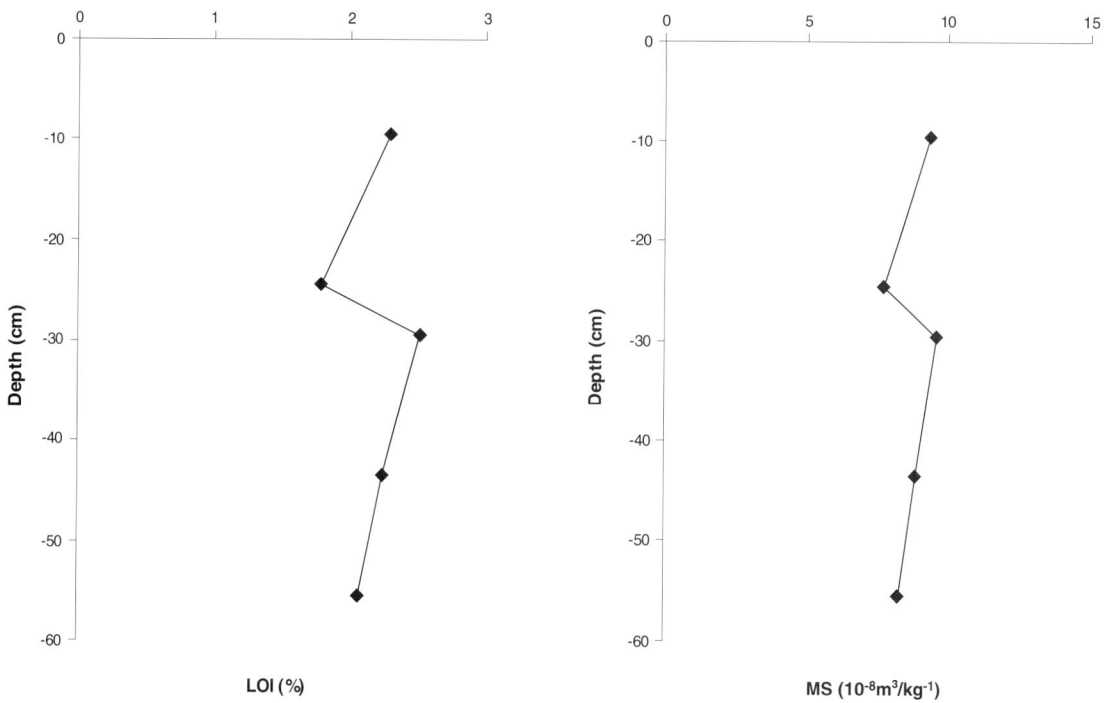

Figure A2.8 Variations in LOI (%) and χ(10^{-8} m^3 kg^{-1}) down Sequence 7.1

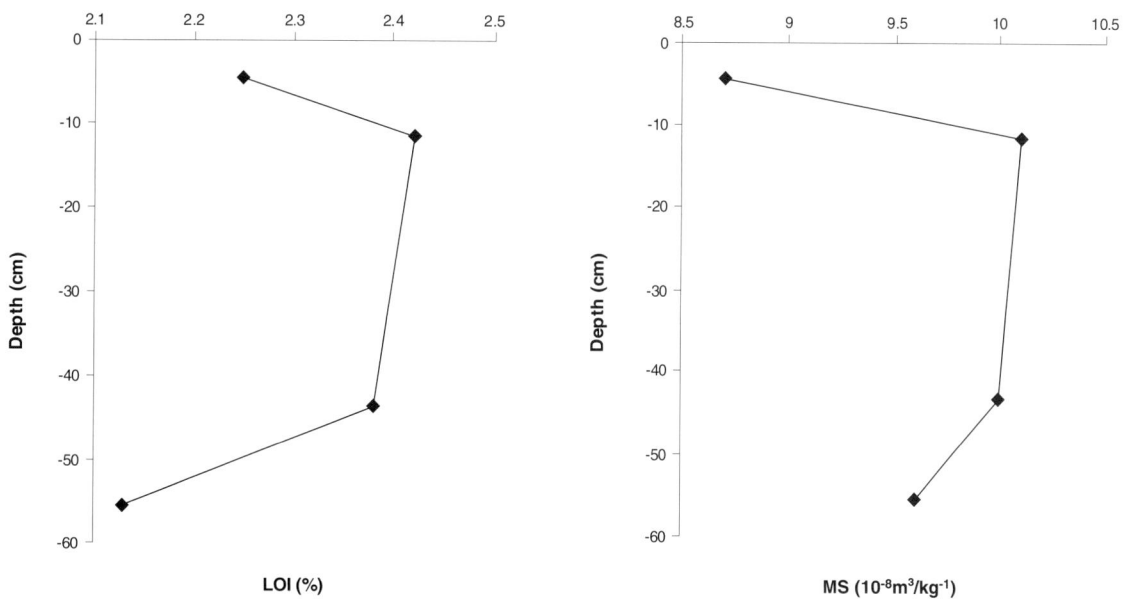

Figure A2.9 Variations in LOI (%) and χ(10^{-8} m^3 kg^{-1}) down Sequence 7.2

has a somewhat higher LOI and χ than the upper (40162).

Sequence 8.1 (Fig. A2.12): A general increase in LOI down the section from context 40058 to 40057, and a

notably high χ (27.4×10^{-8} m³ kg⁻¹) at 18.5cm. It should be noted that the latter sample has a relatively high χ_{max} (1000×10^{-8} m³ kg⁻¹) and needs therefore to be interpreted with caution – ie the higher χ may well simply reflect a higher Fe content.

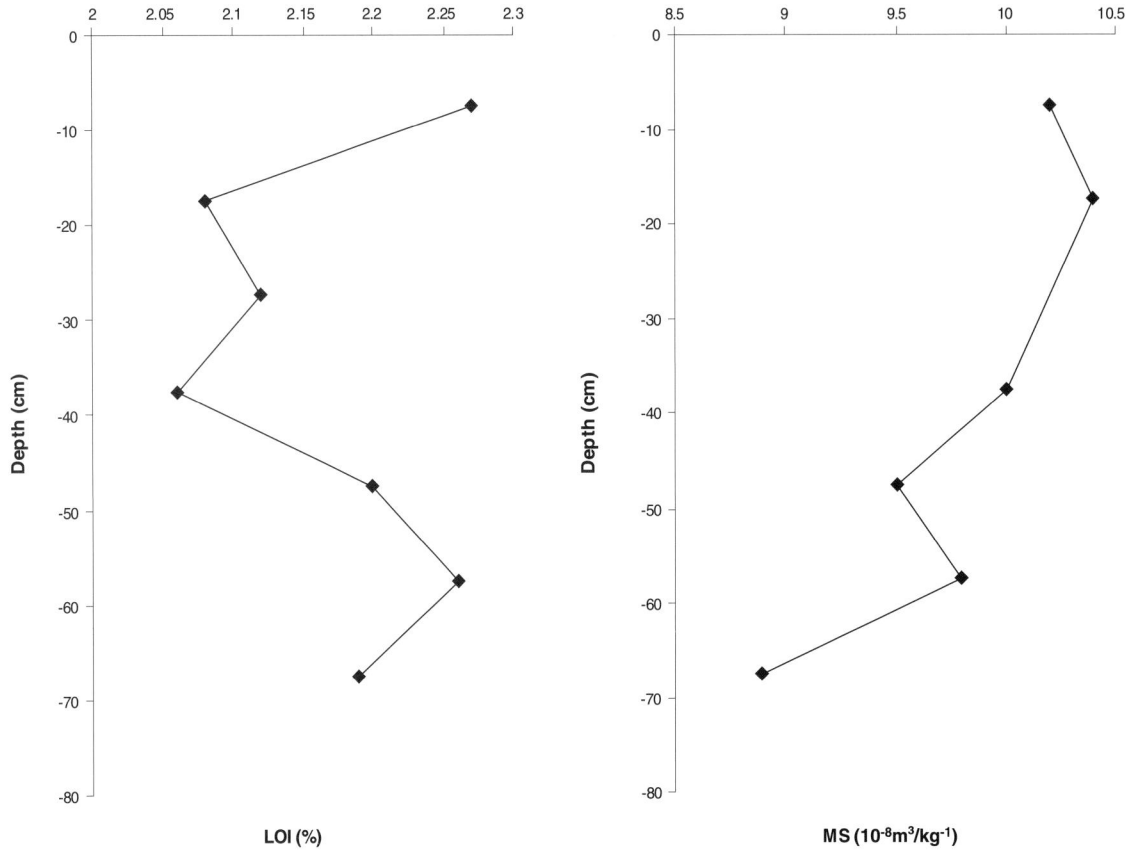

Figure A2.10 Variations in LOI (%) and χ(10^{-8} m³ kg⁻¹) down Sequence 7.3;
[horizontal scales do not start at 0.0 (ie plots exaggerate differences between samples]

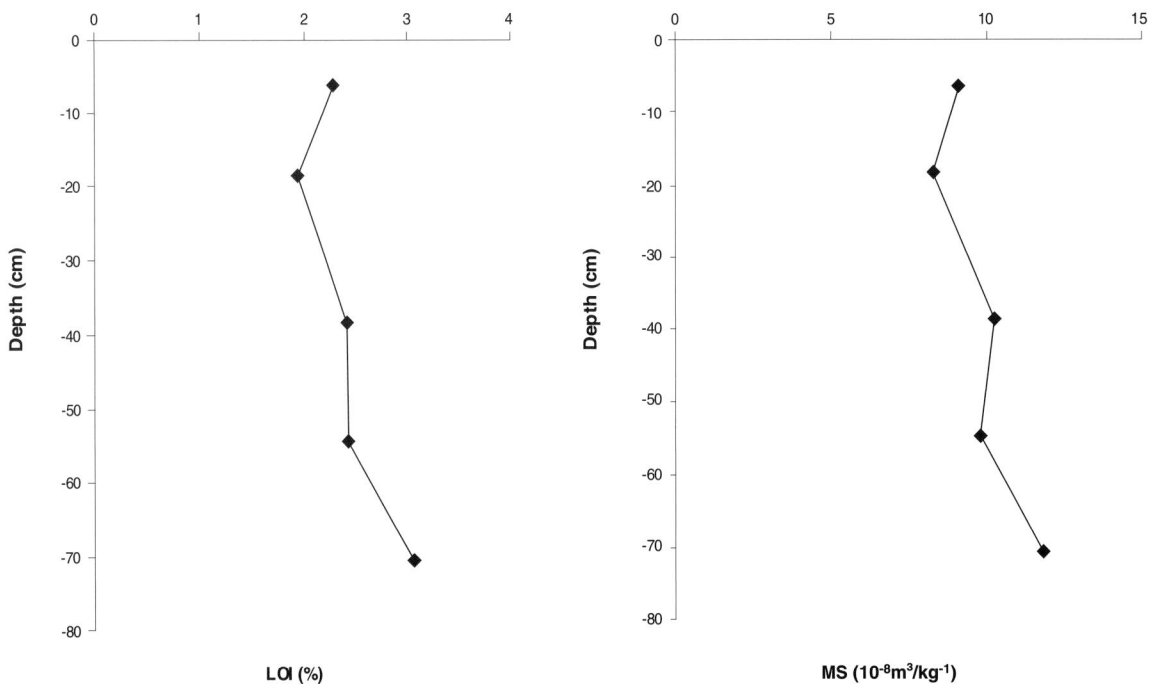

Figure A2.11 Variations in LOI (%) and χ(10^{-8} m³ kg⁻¹) down Sequence 7.4

Sequence 8.2 (Fig. A2.13): A general increase in LOI down the section from context 40057 to 40056, with a possible minor peak at 45.5cm in context 40056. The peak in χ shown in the plot is relatively small and is unlikely to be of significance.

CONCLUSIONS AND RECOMMENDATIONS

As noted in the introduction, LOI and χ data from these sequences may well poorly reflect the character of the initial sediments, because of likely post-depositional organic decomposition and mobilisation of Fe. As the investigations of the relationship between χ and $χ_{conv}$ have shown, interpretation of the χ data is further complicated by what would appear to be wide variations in Fe content through the sequences of deposits. Nonetheless, the results have revealed quite marked variability both within and between contexts. One major peak in LOI (probably the remains of a humic or even peaty deposit) has been identified in context 40158 (Sequence 2), which is very likely to be associated with

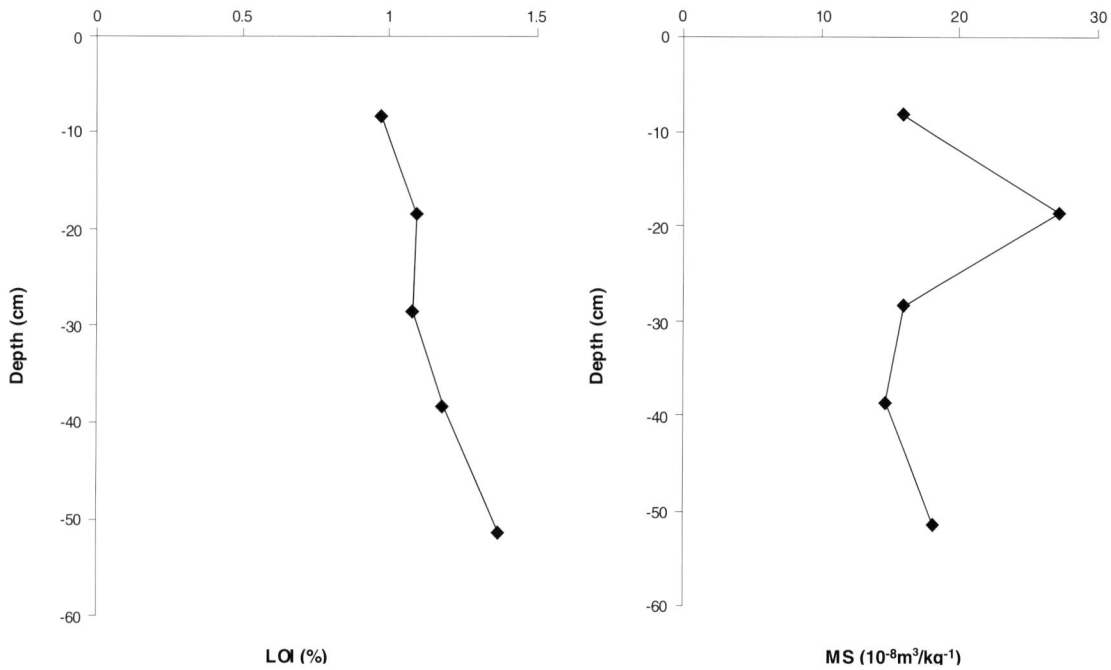

Figure A2.12 Variations in LOI (%) and χ(10^{-8} m^3 kg^{-1}) down Sequence 8.1

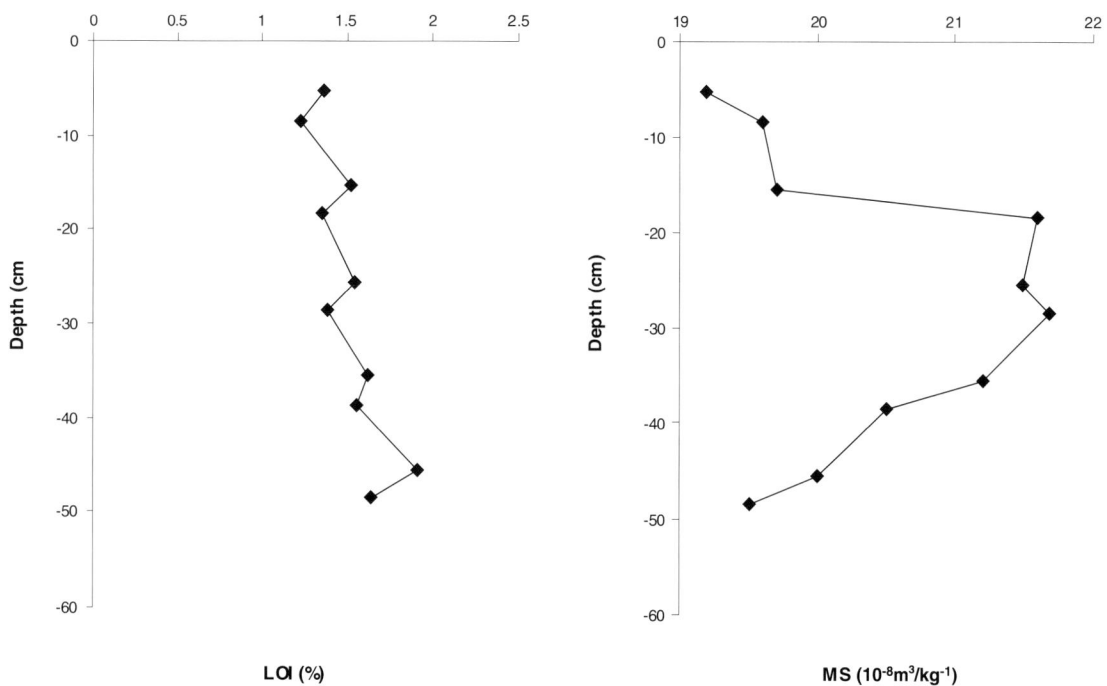

Figure A2.13 Variations in LOI (%) and χ(10^{-8} m^3 kg^{-1}) down Sequence 8.2
[horizontal scale for χ does not start at 0.0 (ie plot exaggerates differences between samples)]

a significant period of soil development/surface exposure. Elsewhere, various minor peaks in LOI have been identified, which may also be indicative of periods of soil development/surface exposure, particularly where there is a close correlation with χ.

It is recommended that the present data are examined in the light of other field and post-excavation evidence – eg textural and compositional variations (which may affect both LOI and χ), and the results of thin section investigations.

Appendix 3

Diatom assessment of samples from Southfleet Road

by Nigel Cameron

INTRODUCTION

The aim of the diatom assessment was to evaluate the potential to use diatom analysis of the Southfleet Road sediments for environmental reconstruction and in particular to determine the types of aquatic environment in which the sediments were deposited.

A total of 32 samples for diatom assessment were sub-sampled from monolith and bulk sediment samples taken from the Lower Palaeolithic site at Southfleet Road, Ebbsfleet (Wenban-Smith 2009; Wenban-Smith *et al.* 2006) The sample details are listed below (Table A3.1) along with the laboratory diatom sample number that each sample was given at UCL. Note in particular that

the order of groups of diatom sample numbers does not follow the order in which sediment samples are listed in the original sample tables supplied by F. Wenban-Smith, B. Silva and E. Stafford. The order of the diatom sample numbers was determined by the priority for assessment of the samples and the order in which groups of samples arrived. In addition, samples of lower priority are listed as possible samples for diatom evaluation in the table, however, sediments for some these were not sent for diatom analysis (Diatom sample number recorded here as 'none'). Further, there is no diatom sample 'D25' because the sub-sample from Sample 40340, 12-13cm was not sent for analysis (see Table 1. Samples assessed for diatoms from the site at Southfleet Road).

Table A3.1 Samples/sub-samples selected for diatom assessment

Diatom Sample No.	Source sample/sub-sample <>	Context	Phase
D1	40250/D	40143	6b
D2	40249/D	40143	6b
D3	40248/D	40143	6b
D4	40281/D/nn-nn	40070	6b
D5	40281/D/nn-nn	40070	6b
D6	40281/D/nn-nn	40103	6a
D7	40364/D/6-7cm	40100	6d
D8	40364/D/33-34cm	40078	6d
D9	40364/D/52-53cm	40099	6c
D10	40365/A/0-1cm	40078	6d
D11	40365/D/29-30cm	40103	6a
D12	40068/D/10-12cm	40066	5
D13	40068/D/44-46cm	40063	3
D14	40068/D/50-52cm	40062	3
D15	40068/D/56-58cm	40062	3
D16	40342/D/3-4	40162	6
D17	40352/D/3-4	40100	6
D18	40352/D/51-52	40099	6
D19	40352/D/18-19	40099	6
D20	40353/D/36-37	40099	6
D21	40321/D/5-6cm	40144	6b
D22	40321/D/14-15cm	40144	6b
D23	40321/D/23-24cm	40144	6b
D24	40321/D/44-45cm	40103	6a
D26	40340/D/39-40cm	40162	6c
D27	40342/D/48-49cm	40103	6a
D28	40344/D/9-10cm	40162	6c
D29	40344/D/39-40cm	40162	6c
D30	40344/D/60-61cm	40162	6c
D31	40345/D/12-13cm	40162	6c
D32	40345/D/48-49cm	40162	6c
D33	40353/D/9-10cm	40099	6c

Table A3.2 Summary of diatom evaluation results for Southfleet Road

Diatom Sample	Diatoms assemblage	Diatom numbers	Quality of preservation	Diversity	Assemblage type	Potential for % count
D1	absent	-	-	-	-	none
D2	see text	one fragment	extremely poor	-	unknown	none
D3	absent	-	-	-	-	none
D4	absent	-	-	-	-	none
D5	absent	-	-	-	-	none
D6	absent	-	-	-	-	none
D7	absent	-	-	-	-	none
D8	absent	-	-	-	-	none
D9	absent	-	-	-	-	none
D10	absent	-	-	-	-	none
D11	absent	-	-	-	-	none
D12	absent	-	-	-	-	none
D13	absent	-	-	-	-	none
D14	absent	-	-	-	-	none
D15	absent	-	-	-	-	none
D16	absent	-	-	-	-	none
D17	absent	-	-	-	-	none
D18	absent	-	-	-	-	none
D19	absent	-	-	-	-	none
D20	absent	-	-	-	-	none
D21	absent	-	-	-	-	none
D22	absent	-	-	-	-	none
D23	absent	-	-	-	-	none
D24	absent	-	-	-	-	none
D26	absent	-	-	-	-	none
D27	absent	-	-	-	-	none
D28	absent	-	-	-	-	none
D29	absent	-	-	-	-	none
D30	absent	-	-	-	-	none
D31	absent	-	-	-	-	none
D32	absent	-	-	-	-	none
D33	absent	-	-	-	-	none

METHODS

Diatom preparation followed standard techniques (Battarbee 1986, Battarbee et al. 2001). Two coverslips of different concentrations were made from each sample and fixed in Naphrax for diatom microscopy. A large area of the coverslips on each slide was scanned for diatoms at magnifications of x200, x400 and x1000 under phase contrast illumination.

RESULTS AND DISCUSSION

The results of the diatom evaluation are shown in Table A3.2. A single, small striate fragment from an indeterminate pennate diatom species was recorded in D2. Diatom assemblages were absent from all thirty-two samples. It is not therefore possible to comment on the aquatic environments in which these sediments were deposited.

Given the ubiquity of diatoms in natural water bodies, the absence of their remains from waterlain sediments is likely to be the result of taphonomic processes rather than absence of diatoms from the water. In particular this can be the result of silica dissolution caused by factors such as high sediment alkalinity, the under-saturation of sediment pore water with dissolved silica, cycles of prolonged drying and rehydration, exposure of sediment to the air or the rapid accumulation of flood deposits (eg, Flower 1993; Ryves et al. 2001). However, these factors do not preclude the preservation of diatoms. Unfortunately because of the absence of diatoms here it is not possible to comment further on the nature of sediment deposition or changes in water quality.

CONCLUSIONS

Diatoms assemblages were absent from all thirty-two samples. The loss of diatom assemblages from these deposits may have been the result of one or more taphonomic factors leading to diatom valve breakage and silica dissolution, and consequently there is no further potential for diatom analysis of these samples.

Appendix 4

Plant macrofossils and wood charcoal

by Denise Druce and Francis Wenban-Smith

INTRODUCTION

Visible fragments of what looked like soft, black remnants of plant material or wood were ubiquitous in contexts 40078, 40158 and 40167, as well as occurring as occasional sparse patches in contexts 40039, 40068 and 40100. Samples from 40158 were taken and processed for charcoal and plant macrofossil remains in conjunction with those taken for pollen analysis (Chapter 12) and assessment of insect remains (Appendix 5). Nothing woody was found during the insect assessment, but a megaspore of the water fern *Azolla filiculoides* was found during sieving before pollen analysis. No other identifiable plant remains were recovered, merely some amorphous non-cellular organic detritus. Four samples were specifically collected with charcoal and wood remains analysis in mind, focusing on patches found during excavation that were thought to be charcoal-rich (Table A4.1).

ASSESSMENT

Unfortunately, of these four samples, all except one was misplaced once taken off-site, so the only sample investigated was <40203>. It produced a very sandy flot. A couple of very small charcoal pieces (<2mm) were picked out, as well as several pieces of black vitrified material which looked plant-derived, although poorly preserved. Several more pieces of this black material were recovered from the heavy residue. This organic material was thought possibly suitable for radiocarbon dating. There was much modern root material in the flot, so there is a possibility that the small quantity of charcoal which was present may be from more recent contamination. Considering the possibility of modern

intrusion, balanced against the significance of any wood identification, it was therefore decided to proceed with identification of the wood remains and also to carry out radiocarbon dating to establish whether or not they were likely to be of Pleistocene origin.

RESULTS

The majority of charcoal fragments from <40203> were highly comminuted and measured less than 0.5mm. A few, however, were large enough to warrant a reasonably clear transverse section. Six fragments were identified as coniferous wood; however, they were too small to enable positive identification to species level. The flot also contained modern plant remains and rare molluscan remains, as well as common heat-affected vesicular material, which probably equates to the indeterminate organic material observed during the initial assessment and also the sieving of samples <40407>-<40410> carried out during pollen analysis. Given the presence of the coniferous wood in the sample, it is tempting to interpret this material as the remains of burnt resin; however this is by no means conclusive.

DISCUSSION

The identified wood charcoal could represent any of the native British and northern European coniferous woods, which includes yew *Taxus baccata*, pine *Pinus sylvestris*, silver fir *Abies alba*, and spruce *Picea abies*. All four have been recorded in middle Pleistocene deposits at Clacton-on-Sea (Bridgland *et al.* 1999), and both pine and yew pollen have been recorded in early-to-mid Holocene deposits from other sites in the

Table A4.1 Samples taken for charcoal/plant macro-fossil assessment

Sample No.	Context	Phase	Weight/volume collected	Collection notes
40190	40100	6	Unrecorded	Wood remains in clay between Trenches B and C
40240	40078	6	Unrecorded	A piece of degraded wood within clay 40078 near *Palaeoloxodon* spread
40203	40068	6	*c* 10L	A small patch of charcoal-rich clay identified during machining between Trenches B and C
40279	40039	6a	0.5L	Fragments of charcoal (?) found within 40039 near rhino maxilla group Δ.42477

Thames Valley including Ebbsfleet HS1 and STDR4.

Although the sample of coniferous charcoal from sample <40203> was sent to the Scottish Universities Environmental Research Centre AMS Facility for dating, unfortunately the sampled failed due to insufficient carbon yield (GU-24990).

Appendix 5

Assessment of samples for insect remains

by Russell Coope

INTRODUCTION

Although the majority of the sediments at the site did not appear especially promising for insect preservation, the darker brown, more organic-rich brecciated clay of contexts 40158 and 40162 (in the central part of the site, near the elephant skeleton) did appear to have some potential for insect preservation, particularly in light of the confirmed pollen preservation, the presence of rotted plant material and the suggestion (see Chapter 5) that the sediment would previously have been more 'peaty' in nature. Therefore three samples were selected for assessment for insect remains.

METHODS

Half of each sample was washed over a 300μm sieve, and any resulting residue dried and examined.

RESULTS AND CONCLUSIONS

No vestige of insect remains, or indeed any other floral or faunal remains, was found from any of the samples investigated, listed below.

Table A5.1 Samples/sub-samples selected for insect assessment

Sample No.	Context	Phase	No. of boxes	Weight processed
40263	40158	6	1 – 10L	11.5kg
40413	40158	6	1 – 10L	11kg
40346	40162	6	1 – 10L	12kg

Samples selected for assessment

The three samples selected are listed below (Table A5.1). Two (samples <40263> and <40413>) came from context 40158, the very dark brown brecciated clay rich in fragmentary organic material that formed the full thickness of the Phase 6 sequence at the foot of the synclinal 'skateboard ramp', in the central part of the site, south of Trench B. The third (sample <40346>) came from context 40162, a slightly browner and sandier facies within the lower part of the Phase 6 clay in the central part of the site. This deposit was considered possibly to have some organic preservation.

Sample <40346>
Sample consisted of grey clay with lighter patches. Many 'rootlets' but no seeds, insects or molluscs.

Sample <40263>
Sample consisted of dark brown clay. Nothing retained on the sieve.

Sample <40413>
Sample consisted of dark brown clay. Nothing retained on the sieve.

Appendix 6

Worked flint post-excavation analysis methods

by Francis Wenban-Smith

INTRODUCTION

The underlying philosophy of the post-excavation lithic analysis was to categorise the artefacts on a technological basis as neutrally as possible, identifying categories such as: core; debitage; flake-tool; and core-tool. This was supplemented by a restricted range of qualitative and quantitative data that it was anticipated would contribute to analysis of the collection. There is a danger in lithic analysis of indiscriminate recording of an overabundance of superfluous empirical data. This analysis was undertaken with a number of clear objectives in mind, and the recording was focused upon data that were chosen as relevant to these objectives, which were: investigation of knapping strategies and primary technological pathways; tool typology; and investigation of the organisational structure of lithic production and the *chaîne opératoire*.

These data, recorded at the post-excavation stage of analysis, were then combined with site provenance and locational data recorded during excavation to provide the basis for the overall lithic analytical project.

THE LITHIC COLLECTION

The lithic collection was initially divided into two primary groups, each of which subject to fundamentally different approaches to excavation, recording and analysis, namely: (1) artefacts ≥ 20 mm maximum length; and (2) microdebitage < 20mm long. During the excavation, it was not specifically decided to omit recovery and recording of artefacts < 20mm long, but it is certain that the majority of artefacts this size would not have been recovered by the trowelling and mattocking methods applied (Chapter 3), although a few were recovered. Furthermore, it was thought that artefacts of this small size would not usually have been deliberately made in this early period (in contrast, for instance, with the Mesolithic or Neolithic), so there was less value in spending time studying them in detail. However, analysis of the quantity and distribution of microdebitage is of interpretive value in investigating taphonomy and disturbance, so part of the excavation was focused upon a systematic recovery of microdebitage in a selection of the evaluation trenches south of Trench D, by the collection of sediment samples on site that were subsequently sieved through a fine mesh.

Lithic artefacts (≥20mm)

Artefacts ≥ 20mm maximum length were subject to detailed recording, in six distinct data groups:

- Recording referencing
- Packing/storage information
- Site provenance information
- Categorical lithic data
- Quantitative lithic data
- Notes

The details of different data recorded in each of these groups are given below:

Microdebitage (< 20mm)

The microdebitage was divided into two categories: 'chips' and 'spalls'. They are not included in the more detailed technological and typological analyses. There is, however, useful information on the integrity and post-depositional disturbance of an assemblage from: (a) their relative quantities in relation to each other and larger debitage; and (b) their spatial distribution in relation to larger debitage. For ease and consistency of comparison, artefacts <20mm maximum dimension were put through stacked 4mm and 2mm sieves, and those retained in the 4mm sieve were recorded as chips, and those in the 2mm sieve were recorded as spalls.

Comparative data from experimental knapping and sites that are confidently believed to represent complete undisturbed knapping scatters then give a baseline for identifying complete and undisturbed debitage distributions.

RECORDED DATA

There were six main groups of data for artefacts ≥20mm, each of which associated with a range of information to be recorded (Table A6.1), described in turn below. The data were recorded by hand onto a paper recording proforma. This took the form of a landscape format A4 paper sheet with columns for each piece of data to be recorded and 20 blank rows for different artefacts on each sheet. After recording, the paper record was typed into an Excel spreadsheet, which could then be linked with the digital survey data

Table A6.1 Lithic analysis recording proforma

Type of data	Name	Description
Recording reference	Rec sht	Recording sheet, in number order of recording
	Sht #	Line number on recording sheet
Packing/storage	OA box no.	Box number as originally received
Site provenance data	Δ ID	Unique lithic identifier, small find number
	Context	Taken from finds bag, cross-checked with paper archive
	Area	Area of site: Trenches A-D; Transects 1-3, Strips A-D
	Trench	Evaluation trenches I-XV
	Sample <>	Sample number, for lithic bulk spit-sieved samples
	Spit	Spit-number, for lithics from spits in evaluation trenches I-XV
Categorical data	Cnd	Condition
	C1	Main technological category
	C2	Secondary technological category
	T1	Technology/typology, sub-category 1 (varies acc. C1, C2)
	T2	Technology/typology, sub-category 2 (varies acc. C1, C2)
	T3	Technology/typology, sub-category 3 (handaxes, flake-tools)
	T4	Technology/typology, sub-category 3 (handaxes, flake-tools)
	WhL	Completeness, wholeness (varies acc C1, C2)
Quantitative data	%Cx	Percentage remnant cortex, on dorsal surface of flakes
	DSC	Dorsal scar count, scars from debitage estimated as ≥20mm [not including striking platform, for flakes]
	ML	Maximum length, measured along ventral surface for flakes from point of percussion, mm [1]
	MW	Maximum width mm, orthogonal to ML[1]
	MT	Maximum thickness mm, orthogonal to ML
	WtG	Weight grammes
Notes	N	Notes, not usually entered on database but useful on paper record

[1] ML, MW for debitage — estimate extra for damage/abrasion <20 mm

thus providing the potential to generate site plans and statistics based on the lithic interpretations.

Recording reference

Each sheet was numbered incrementally as it was used, and each row on each sheet (representing separate artefacts) was also numbered from 1-20 down the sheet. Once this data was entered into the digital record alongside the provenance and lithic analytical data, it meant that it was then easy at a later point to go back and investigate the original paper record for any particular artefact. Without this procedure, this would have been a virtually impossible task considering the original paper archive constituted 134 sheets, since the artefacts were not examined in a particular order, other than the context order in which they were provided originally.

Packing/storage

A record was made during the lithic analysis of the box that material was received in. Lithic artefacts were initially put back into the same box they came from after the first examination. This original packing order gradually became more muddled, as it became clear that several artefacts were initially incorrectly numbered or attributed

to the wrong context, leading to them being correctly attributed and placed in different boxes. Maintenance of this record of where artefacts were being stored was useful while study was taking place. Later, the material was entirely reorganised to match significant stratigraphic groupings, so this part of the record became redundant.

Site provenance data

Provenance data is usually collected on site and before analysis, and stored in a separate database, before being added later to the analysis data, but there has to be a key field to link the two such as Δ.ID (individual lithic ID) or <SN> (Sample number, eg. for artefact collections from large sieving programmes). At Southfleet Road, stratigraphic provenance and 3-D spatial location data were collected during excavation for the great majority of flint artefacts. A few were recovered from bulk spit-sieve samples, or from excavated spoil of known stratigraphic provenance but uncertain precise location, so did not have full XYZ locational data. The XYZ data was collected digitally using a total survey station, tied in with the artefact find number. The stratigraphic provenance data was collected manually, with the context number (and, if applicable, site area, trench number and spit/sample number) written on the plastic bag into which lithic finds were placed, together with the unique individual find number. This was supplemented by maintenance of a paper find register, which listed the finds numbers used, the stratigraphic context of each find along with notes on the location of the find and its material (at Southfleet Road, almost invariably 'Flint' or 'Bone').

When the artefacts were examined, the provenance information written on the finds bag was recorded on the paper lithic recording proforma. This was subsequently cross-checked with the paper record (which was typed up separately into a different Excel spreadsheet) and any discrepancies investigated and resolved. Six different aspects of provenance were recorded, as specified in the proforma (Table A6.1).

Categorical data

The first categorical attribute recorded was the condition of each artefact, 'Cnd', which was classified as one of five different degrees of damage/abrasion (Table A6.2), ranging from absolutely mint condition, as freshly knapped, to so heavily abraded that virtually a beach pebble. The next two attributes, C1 and C2, relates to the technological category. Seven basic technological categories were recognised (Table A6.3), represented by C1, ranging from natural unworked flint (C1=0) to artefacts interpreted as tools (C1=6). For cores, debitage and tools, the second category C2 represents further more specific subdivisions, for instance irregular waste debitage would be coded as C1,C2 = '5,1', a 'flake-flake' representing a flake removed in course of trimming a flake blank to form a flake-tool would be coded as C1,C2 = '5,4' and a handaxe on a flake would be C1,C2 = '6,2'. The use of numerical codes was found useful as saving time in data entry, and also easier for coding queries for subsequent quantitative analyses investigating the proportions of different types of artefact in different assemblages.

Table A6.2 Lithic condition categories

Grade	Category	Description
1	Mint	As freshly knapped, razor sharp
2	Sharp/fresh	Sharp to handle, ridges unaffected, but slight abrasion on parts of edges
3	Slightly abraded/rolled	Ridges slightly abraded, edges lightly–moderately battered, smooth to touch
4	Well-abraded	Ridges very abraded, all edges moderately–heavily battered
5	Extremely abraded	Almost a beach pebble, ridges non-existent or vestigial, heavily battered surfaces

Table A6.3 Lithic artefact technological categories

C1	C2	Description
0 - Natural	-	Not humanly worked, can be interpreted as raw material, can be excluded from database, but if so needs to be quantified
1 - Raw materia	-	No sign of working, but clearly a manuport
2 - Tested nodule	-	Nodule with only a couple of flakes off, no sign of whether a core or core-tool
3 - Chunk	-	Knapped chunk. Uncertain whether core or core-tool, poss. because broken, or not very knapped, or just very ugly
4 - Core	1 - Conventional	Flakes removed, generally reasonably large, from natural lump of raw material and no sign of preferential edge/part for use
	2 - On flake	Debitage used as a core
	3 - On core-tool	Eg if re-used or after breakage
5 - Debitage	1 - Irregular waste	Lump, fragment or shatter; piece bigger than 20mm but not otherwise classifiable, often resulting from knapping frost-fractured pieces; usually show some sign of percussive impact, but in principle can apply to pieces that look completely natural, but are interpreted as resulting from hominin knapping
	2 - Flake, blade	Flakes, or parts of flakes, must have signs of being part of a single removal, else classified as C2=1
	3 - Chip/spall	Flake/irregular waste less than 20mm
	4 - Flake-flake	Debitage from flaking a flake
6 - Tool	1 - Handaxe (core-tool)	Usually evidence of preferential edge/part for use and bifacially worked; attention to straightening, to opposing handle, removal of small shaping flakes of no use in themselves
	2 - Handaxe (on flake)	When a handaxe is made on a blank that shows definite evidence of originally having been a piece of debitage
	3 - Flake-tool	Worked/utilised flake; working can be backing (eg possible interpretation as backed knife), retouching (eg. to form scraping edge) or notching
	4 - Percussor	Evidence of focused battering, can appear on cores/core-tools, can have some working to facilitate handling
	5 - Anvil	Battering on very large pieces, usually would be interpreted as percussors

The next four attribute categories T1–T4 reflect more detailed technological and typological attributions. These varied according to the basic technological category, so for instance, handaxes have a different selection of options than flake tools, and other technological categories lack more detailed T1–T4 options.

For handaxes, a classificatory scheme was applied based upon that developed by Wymer (1968) in his review of the Lower/Middle Palaeolithic archaeology of the Thames Valley, with a different numeric code entered as T1 according to the handaxe shape (Table A6.4). Other handaxe attributes such as the degree of butt-trimming, the presence and knapping direction of tranchet sharpening and the peasants and orientation of a twisted profile are recorded as attributes T2–T4, as shown and described in the respective tables (Table A6.5 – A6.7).

For flake tools, the first two technological/typological attributes T1 and T2 are used to categorise the flake-blank, whether it is a natural scrap (and not realistically

Table A6.4 Handaxe shape categories (T1)

Shape category - T1	Wymer type	Description
0 - Unspecific	-	Indeterminate, eg. when broken or unclassifiable to other categories
1 - Rough-out/abandoned	-	Pieces which appear to have been abandoned before completion, for instance because of frost-fracturing, persistent failure to achieve thinning, or breakage
2 - Simple	Proto	Includes McNabb and Ashton's 'non-classic' handaxes, simple bifacial or unifacial edges opposed to natural handles
31 - Crude pointed (large)	D	Large (≥100mm) pointed/sub-pointed biface, no soft-hammer, thick, wavy edges, thicker and heavier at butt
32 - Crude pointed (small)	E	Small (<100mm) pointed/sub-pointed biface, no soft-hammer, thicker and heavier butt, thick, wavy edges
4 - Classic pointed	F	Well-made pointed handaxe with clear butt, straightish sides and thinned towards tip, can be any size; butt can be unworked or crudely worked
50 - Sub-cordate	G	Progression from type F with convex sides, often more rounded point, thick/heavy butt, widest part of handaxe well towards butt; butt can be unworked, crudely worked
51 - Sub-cordate (plano-convex)	G	Similar to above but with clear plano-convex profile, cf. Wolvercote Channel
52 - Sub-cordate (twisted)	-	Sub-cordate plan shape, but tip distinctly twisted relative to butt
60 - Sub-ovate	GK	Much more ovate version of sub-cordate; tip is smoothly rounded without any well-defined point, widest part of handaxe is nearer middle of long axis, clear working to shape/thin butt and sides as convex curve, although not as much as for true ovate or cordate
7 - Cordate	J	Cutting edge all round tool with thinning and shaping around butt, centre of gravity near middle, bit more rounded than sub-cordate, but still has clear tip, with widest part of handaxe towards butt
71 - Twisted cordate	-	Cordate plan shape, but tip distinctly twisted relative to butt
80 - Ovate	K	Cutting edge and thinning/shaping all round, centre of gravity near middle, more rounded at base than cordate with widest part of handaxe towards middle, usually one end recognisable as tip by being more elongated from widest part of handaxe and often tranchet sharpened
81 - Twisted ovate	K	Ditto above, but clearly twisted tip
9 - Side-chopper	L	Segmental chopping tool, one knapped bifacial edge or sharper edge opposed by flat edge or natural backing; crucial distinction with cleaver is that business edge is parallel with main longitudinal axis rather than transverse
10 - Classic ficron	M	Very pointed with symmetrical concave sides and well-defined heavy butt, cf. Furze Platt, Cuxton (Wenban-Smith 2004)
11 - Bout coupé	N	Flat-butted cordate, trimmed all round butt, but with distinct corners between gently convex base and sides
12 - Cleaver	H	Key characteristic is straight cutting edge at tip end, transverse to main longitudinal orientation of tool, cf. Cuxton (Wenban-Smith 2004)

Table A6.5 Handaxe butt trimming categories (T2)

T2	Description
0	Inapplicable — indeterminate or unknown, eg. when broken
1	Untrimmed butt — entirely cortex or natural fracture
2	Slightly trimmed butt — over 50% cortex or natural fracture
3	Mostly trimmed butt — less than 50% cortex or natural fracture
4	Wholly trimmed butt — all butt and corners trimmed

classifiable as a core-tool), piece of irregular waste or a normal flake (Table A6.8). The last two attributes T3 and T4 then reflect interpretation of the type of flake-tool, with T3 representing the general category of flake-tool, and T4 representing more detailed subdivision (Table A6.9). These subdivisions are not intended to be an exhaustive typological list of the Lower/Middle Palaeolithic, merely to represent basic groupings identified during analysis of the material from the Southfleet Road site. In general, it was not attempted to apply/develop a detailed typology, in light of serious doubts as to whether this would be interpretively meaningful. Rather, it was just attempted to reflect

basic technological groupings, and then the range of technical detail exhibited by different technical categories in different assemblages were described and drawn on a case-by-case basis (see Chapters 17, 18 and 20 in particular).

The final categorical attribute recorded was the completeness/brokenness of each artefact. This was coded from 0–4 (Table A6.10), with minor variations of criteria depending upon the different technological categories. As described in the table, broken-off pieces estimated as less than 20mm maximum length were disregarded. For debitage, flake-tools and handaxes, different coding was used to reflect whether the

Table A6.6 Handaxe tranchet sharpening categories (T3)

T3	Description
0	Inapplicable — indeterminate or unknown, eg. when broken
1	Absent — no tranchet
2	Left tranchet — struck from left (removal underneath and tip away; can be on both faces, as long as consistently from left)
3	Right tranchet — struck from right (removal underneath and tip away; can be on both faces, as long as consistently from right)
4	Complex tranchet — struck from both left and right (removals from both faces, with face underneath and tip away)

Table A6.7 Handaxe twisting categories (T4)

T4	Description
0	Inapplicable — indeterminate or unknown, eg. when broken
1	S-twist, anticlockwise tip-twist when viewed from above, looking down on tip, with tip twisting anticlockwise relative to butt; bifacial edge descending to R across middle of handaxe when profile viewed from side, with tip to R
2	Z-twist, clockwise tip-twist when viewed from above, looking down on tip, with tip twisting clockwise relative to butt; bifacial edge ascending to R across middle of handaxe when profile viewed from side, with tip to R

Table A6.8 Flake-tool blanks, technological categories (T1 and T2)

T1	T2	Description
0	0	Made on a scrap of natural, eg. frost-fractured, flint
5	1	Made on a piece of irregular waste
5	2	Made on a normal flake

Table A6.9 Flake-tools, typological categories (T3 and T4)

T3	T4	Details
1 - Notches	10 Single notch	Clear single notch, can be backed by natural cortical handle or blunting/backing retouch
	11 Multi-notched	Two or more notches, scattered around
	12 Linear notches	Two or more notches, aligned on one edge (sometimes to form crude denticulate)
2 - Flake-knives	20 Utilised flake	Use-damaged, evidence of macro use-wear but no retouch
	21 Knife	Blunting/backing retouch opposite/beside natural cutting edge, which can show macro-wear, to facilitate handling and use
3 - Scrapers	30 Gen scraper	General scraping edge/s
	31 "Mousterian" scraper	Convex unifacial scraping edge down one long side of a medium-large flake
4 - Saws	40 Gen saw	Unifacial/bifacial sharpening of edge/edges of flake to form sawing edge on flake
5	50	Point/awl?
6	60	Misc. other; describe on case-by-case basis

Table A6.10 Completeness/breakage codes for different lithic technological categories
[Irregular waste is by definition always whole]

Code	Core	Debitage or flake-tool	Handaxe
0	Broken	-	Broken: piece of core-tool; not clearly butt, middle or tip remaining
1	Whole	Whole — missing bits <20mm	Whole: missing bits <20mm
2	-	20 - Proximal present: missing proximal bit <20mm and distal bit ≥20mm 21 Proximal present: Siret fracture (left), when looking at ventral surface, with strik. plat. upward 22 Proximal present: Siret fracture (right) , when looking at ventral surface, with strik. plat. upward	Butt present: missing butt <20mm and tip ≥20mm
3	-	Mesial present: missing proximal and distal bits both ≥20 mm	Middle present: missing butt and tip, both ≥20mm
4	-	Distal present: missing proximal bit ≥20mm and distal bit <2 mm	Tip present: missing butt ≥20 mm and tip <20mm

proximal, middle part or distal end was present. And for flakes and flake-tools, a slightly more detailed coding was used when the proximal end was present, to identify and distinguish between left and right sides of Siret fractures.

Quantitative data

A relatively restricted range of six quantitative attributes was recorded (Table A6.1), chosen as being particularly relevant to taphonomic/post-depositional interpretation and investigation of the *chaîne opératoire* and organisation of lithic production. Experimental work by Wenban-Smith (1996) has firmly established that two attributes in particular – percentage cortex and dorsal scar count – are especially useful for categorising the stage of the reduction sequence represented by waste debitage. Therefore these two attributes were recorded here for flakes and flake-tools. The percentage of cortex (or unknapped natural flint surface) on the dorsal surface was categorised by eye as one of 11 grades 0–10 as specified in the accompanying table (Table A6.11). The dorsal scar count was based on the estimated size of the complete removal represented by a dorsal scar, not just the size of the visible remnant, and scars representing removals thought to be less than 20mm maximum length were disregarded. For flake tools, it was attempted to focus on counting scars from flakes removed during the original knapping process rather than from secondary working to form the flake-tool.

The remaining four quantitative attributes recorded were the basic size measurements of length, width and thickness (ML, MW and MT), as well as weight (WTG). There are innumerable variations on the precise way of measuring these apparently straightforward flake-size attributes. Those applied here are illustrated (Fig. A6.1), with the additional factor that broken-off parts less than 20mm were ignored, and measurements estimated to allow for their presence if necessary. This was so that the size data were directly comparable with experimental models based on complete flakes.

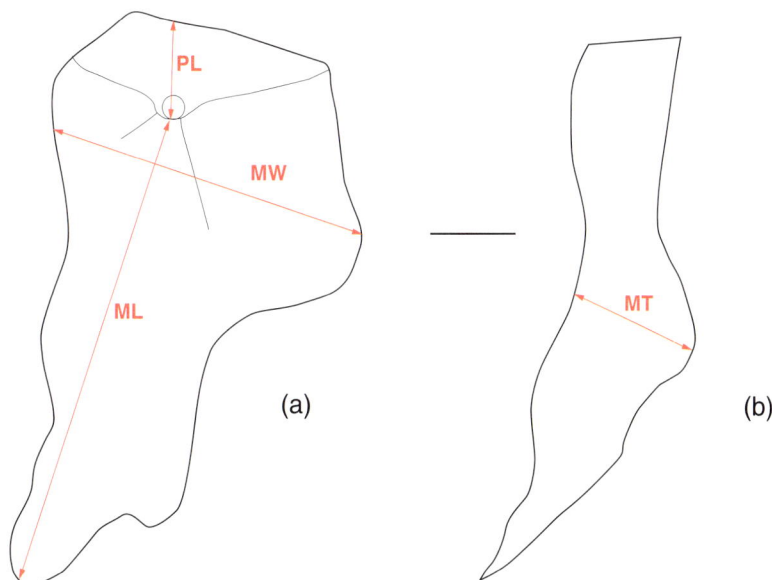

Figure A6.1 Measurement of flake size: (a) maximum length ML and maximum width MW; (b) maximum thickness MT

Table A6.11 Codes for recording amount of cortex present

Code	Description
0	0%
1	1–10%
2	11–20%
3	21–30%
4	31–40%
5	41–50%
6	51–60%
7	61–70%
8	71–80%
9	81–90%
10	91–100%

Notes

Finally, various notes and sketches were recorded on the paper archive, drawing attention to anything of interest seen, such as distinctive raw material, unusual knapping strategies or the distribution of fine macro-wear traces on the sharp edges of otherwise unworked flakes. To avoid ambiguity when making these descriptive notes, core-tools are by default oriented butt down, and flakes with striking platform up. When describing working on flakes, 'struck off' refers to the surface bearing the scar/s of any retouch/removals, and 'struck on' refers to the surface hit by the percussor. Hence the normal situation is struck on the ventral surface and off the dorsal surface.

Appendix 7

Thin section data

by Richard Macphail

Reproduced overleaf

Table A7.1 Soil micromorphology: descriptions and preliminary interpretations

Microfacies type (MFT)/ Soil microfabric type (SMT)	Sample No.	Depth (relative depth) Soil Micromorphology (SM)	Preliminary Interpretation and Comments
MFT A1/SMT 1a1, 1a2, 2a1	M40082	290-370mm SM: Heterogeneous with dominant fine and medium sandy SMT 1a1 in lower half, mixing with common fine sandy SMT 1a2 and frequent (fragments/fills) of SMT 2a1 upwards; *Microstructure:* massive with fissures (1.5-2mm wide), 25% voids, fine to coarse fissures, vughs with simple packing voids, (root?) channels traces (0.5-0.7mm). *Coarse Mineral –* C:F (Coarse:Fine limit at 10μm), SMT 1a1=85:15: moderately poorly sorted angular coarse silt/very fine sand-size quartz, feldspar, micas (and opaques including probable haematite), with medium sand-size mainly moderately weathered sub-rounded glauconite (some little-weathered glauconite), fine and medium sandy clay clasts (colourless, 1st order grey birefringence – palygorskite?, relict of Cretaceous), and rare very coarse sand-size rounded clasts of SMT 2a1; SMT 1a2: C:F=60:40, very dominant moderately poorly sorted angular coarse silt/very fine sand-size quartz, etc (as SMT 1a1), with few to frequent medium sand-size glauconite. *Coarse Organic and Anthropogenic: Fine Fabric –*SMT 1a1: cloudy dusty pale brownish grey (PPL), moderately low interference colours (close porphyric to coated grain, speckled and patchily crystallitic, XPL), very pale brown (OIL), rare trace of once-humic iron-staining; SMT 1a2: as SMT 1a1, with close porphyric c/f distribution; SMT 2a1: cloudy and finely speckled pale brown (PPL), moderately low interference colours (open porphyric, stipple speckled b-fabric, XPL), greyish brown (OIL), possible thin humic staining and occasional very fine relict amorphous organic matter. *Pedofeatures: Textural –*occasional void coatings/infillings of SMT 2a1; *Amorphous:* rare ~200μm-thick iron void hypocoatings, associated with thin relict root(?) channels; rare of iron impregnations of possibly once-humic sediment? *Fabric –*many fabric mixing, including v-shaped burrows, some 15mm wide, with broad 2mm thick burrows?? – a long line of later(?) root channel(?)	Context 40057 (Phase 1) South (E) 'Tilted block' Lowermost sequence of site Boundary zone between non-calcareous moderately poorly sorted, coarse silty and fine and medium sandy, glauconite medium sand-rich sediments with little fine material (40057), which upwards become mixed with moderately sorted coarse silts and fine sands, sometimes with fewer glauconite, and mixed with clasts and burrowed-in weakly humic clay (LOI=1.22%). Glauconite is moderately weathered with some little-weathered grains, and opaques include haematite. A relict thin (1.5-2mm) probable root channel is marked by iron hypocoatings; iron staining also affects traces of organic matter. *This is a moderately mixed junction between fine and medium non-calcareous glauconitic sands, and overlying silty glauconitic sands, and clayey silts, probably deposited as alluvium and recording diminishing energy. There has probably been some burrowing and rooting, the latter affected by secondary iron staining, which marks a minor amount of sediment ripening/soil formation. Clayey deposits were probably weakly humic originally. Context 40058 could not be sampled because the sediment monolith was not intact, but apparently clayey inwash into 40057 probably derives from overlying soliflucted 40058.*
MFT E4/SMT 3a3, (3a1)	M40418A	45-125mm 45-70mm SM: Heterogeneous with poorly sorted sands and fine sands (SMT 3a3), with coarse silt and coarsely mixed clay (depleted SMT 3a1); *Microstructure:* massive, finely laminated; 10% voids, fine packing voids and fissures; *Coarse Mineral –*C:F, 80:20, mainly well sorted coarse silts, with patches of less well-sorted fine to medium sands; few glauconite and very few coarse silt-size clay papules present. *Coarse Organic and Anthropogenic: Fine Fabric –* as SMT 3a1 and 3a3. *Pedofeatures – Amorphous:* trace amounts of iron staining and possible Fe-Mn fine nodules. *Fabric –* occasional broad burrows.	Context 40166 (Unit 4?)-40158 (Phase 7/6a) Context 40158 Sediments become less humic and contain less fine detrital organic matter (OM), being clayey deposits, but with marked iron-staining of organic traces.

Table A7.1 (continued 2)

Microfacies type (MFT)/ Soil microfabric type (SMT)	Sample No.	Depth (relative depth) Soil Micromorphology (SM)	Preliminary Interpretation and Comments
MFT E3/SMT 6a2	M40418B	70-115mm: as MFT C1/SMT 5a1 (M40150B), with few sandy SMT 3a3 115-125: as MFT E3/SMT 6a2 (M40418B) 170-250mm SM: Heterogeneous; *Microstructure:* massive, partially laminated, 15% voids, fissures. *Coarse Mineral* – C:F, 60:40 (sandy) and 40:60 humic clays. *Coarse Organic and Anthropogenic* – examples of 5mm long horizontally oriented blackened plant material (eg, 3 thin [20-30μm-thick] monocot? leaf fragments?; occasional fragments and patches of reddish amorphous OM 1-3mm in size; many fine charcoal (100-250 μm). *Fine Fabric* – SMT 6a2: speckled and dotted brown to dark reddish brown (PPL), moderately low to moderate interference colours (open to close porphyric, grano-striate b-fabric, XPL), brown to dark brown (OIL), humic with very abundant amorphous and very fine charred OM. *Pedofeatures: Textural* – very abundant clayey pans (sedimentary muds) and channel infills (1+mm wide), and associated intercalations and void/matrix coatings. *Fabric:* –many thin burrows. 215-250mm As MFT E2/SMT 3a3, M40418C, below, but with increasing amounts of SMT 6a2 (humic clay containing very fine charcoal) BD (40158): 6.68% LOI	Contexts 40158-40040 (Phase 7/6a) Mainly laminated very humic clayey fine sands and humic clays, with abundant included fine amorphous organic matter and fine charcoal (100-250μm), and few fragments of amorphous peat. An example of a horizontally oriented 5mm long very thin blackened monocotyledonous leaf/leaves fragment occurs. Humic clays sometimes occur as clayey pans and infills. Minor thin burrowing has occurred. Markedly high LOI of 6.68%. *These are laminated peaty clays and peaty sands, horizontally deposited in low to moderately low energy conditions, allowing horizontal deposition of detrital leaves, fragments of pure peat and ubiquitous fine and very fine charcoal. Organic matter shows no sign of being ferruginised. Low energy (seasonal) minerogenic peat formation associated with burned landscape – wildfires??*
MFT E2/SMT 3a3	M40418C	400-480mm 400-450mm (40158) SM: Homogeneous fine to medium sands (clayey channel fills); *Microstructure* –massive with some relict laminae (1-5mm); 40% voids, open vughs and channels; *Coarse Mineral* – C:F, 90:10, moderately well sorted very fine, fine sands with medium angular and subangular sands and very coarse silt-size quartz, quartzite and feldspar, very few glauconite. *Coarse Organic and Anthropogenic* – many thin ferruginised amorphous OM stringers? *Fine Fabric* – as SMT 3a3 (iron-stained). *Pedofeatures* – occasional ~1mm clayey 'circular' channel fills. *Amorphous* –abundant ferruginisation, thin ironpan formation, some possibly pseudomorphic of OM. *Fabric* – many 0.5-1mm size thin burrows. 450-480mm (40039)	Contexts 40040-40025-40158 (Strat. Phase 7/6a) Moderately well sorted massive and sometimes laminated very fine, fine sands containing coarse silt and medium sand. Many thin burrows, examples of clayey inwash down fine (1mm) root(?) channels and very thin stringers of ferruginised OM (??) inwash. *Channel fine sands mainly, that were burrowed and rooted, and thus episodically/seasonally exposed. Possibly humic matter from above filtered down and became ferruginised. Clayey sediments above were introduced downprofile along empty relict root channels.*
MFT E1 /SMT 3a3, 3a4		SM – Homogeneous coarse silt-very fine fine sands (SMT 3a3), with clayey infills (SMT 3a4). *Microstructure* – laminated (0.5-2mm), 35%, simple packing voids, some vughs and fissures associated with laminae. *Coarse Mineral* – coarse silt-very fine fine sands, with C:F of 90:10. *Coarse Organic and*	Context 40039 Laminated moderately well sorted coarse silts and very fine fine sands, with patchy 'layer'/infills of clay with occasional ferruginised very fine organic matter and phytoliths present. Very thin ironpans associated with sandy laminae appear to be relict

Table A7.1 (continued 3)

Microfacies type (MFT)/ Soil microfabric type (SMT)	Sample No.	Depth (relative depth) Soil Micromorphology (SM)	Preliminary Interpretation and Comments
			amorphous OM and can involved very thin relict excrement pseudomorphs.

Fluvial coarse silts and very fine sands; perhaps seasonal (spring) alluviation with winter? period of clay deposition forming infills. Sand laminae also sometimes associated with thin peat formation and its partial working by small invertebrates. Coolish climate? Near channel alluviation. |
| MFT D3/SMT 3a1, 3a2, 3a3 | M40365A | 50-130mm (context 40078-40099 boundary at 110mm)
SM – Heterogeneous with mainly poorly iron-stained SMT 3a1, and frequent silty and fine sandy SMT3a3. *Microstructure* –massive (relict laminae/pans), with fine fissuring (medium prisms), 25% voids, fine fissuring and fine to medium channels. *Coarse Mineral*– as M40151B, with example of 2.5mm-size flint and 1.5mm-size quartz. *Coarse Organic and Anthropogenic* – very abundant fine root traces, some ferruginised and some as blackened remains (monocot?); occasional very fine to fine fragments of peat/plant (black and red in PPL, eg 1mm in size) ; *Fine Fabric*– as SMT 3a1 with very thin humic staining, rare to occasional ferruginised and blackened very fine OM. *Pedofeatures: Textural* – many pans, textural intercalations, associated with embedded grains and void coatings and infills (matrans); Depletion: occasional strongly iron depleted zones. *Amorphous*– rare amounts of weak yellowish staining and ferruginised OM; some as poor pseudomorphs of plant material/roots, possibly relict iron/sodium carbonate? *Fabric* –very abundant coarse fabric mixing (rooting disturbance as in monocot peat, mineral junctions, inwash of silty material etc). BD (40078): 2.44% LOI. | Context 40078-40099 transition (Phase 6c/6a)
Weakly humic clayey sediment with fine organic fragments and (monocotyledonous?) root traces and common patches of partially bedded silty clay loam (40078) over iron-depleted clayey 40099. 40078 includes blackened relict peat/mocot plant/root fragments, up to 1mm in size (red and black under PPL); overall OM content reflected in 2.44% LOI. Textural intercalations and associated matrix coatings and fabric mixing associated with relict channels and fissures.

Weakly humic remains of putative junction between monocotyledonous peat and minerogenic sediments, with flood wash bringing in silts; hence mixing down along old root channels. Rooting down through waterlogged clayey sediment caused fabric disruption, mixing and intercalations. A period of moderate stasis, with flooding of inwash silts and peat growth? |
| MFT B1/SMT 3a1, 3a2 | M40365B | 220-300mm
SM– Moderately heterogeneous with very dominant iron-stained and iron-depleted clayey SMT 3a1 and very few fine and medium sandy SMT 3a3. *Microstructure* – massive, 30% voids (20% intrapedal), fine fissures. *Coarse Mineral* – C:F, as M40151B; more sandy SMT 3a3 can be focused along possible relict root channels. *Coarse Organic and Anthropogenic* – many traces of medium (max 5mm) roots. *Fine Fabric* – SMT 3a1 and 3a3, as M40151B. *Pedofeatures: Textural* – very abundant relict panning (~1mm thick diffuse laminae). *Depletion* – many iron-depleted areas, some | Contexts 40099 (Phase 6c/6a)
Mainly iron-depleted clay, characterised by ~1mm-thick clay panning (sedimentation), with silt-rich clay loam along possible relict root channel fills, which are marked by iron staining and relict likely hypocoatings. Relict root features are 3-5mm wide; some show ferruginised traces/oxidised traces of probable pyrite framboids associated with roots.

Waterlogged clay sediments deposited as muddy pans, very low energy alluvial events/flooding. These were rooted by plants, possibly shrubs/ |

Table A7.1 (continued 4)

Microfacies type (MFT)/ Soil microfabric type (SMT)	Sample No.	Depth (relative depth) Soil Micromorphology (SM)	Preliminary Interpretation and Comments
		especially associated with old rooting. *Amorphous* – abundant moderate to strong iron staining.	*woodland(?) and more silty clay loam infilled these on decay.*
MFT B1/SMT 3a1, 3a2	M40365C	310–390mm Sample quite fragmented and also was probably part insect burrowed previously. SM – Heterogeneous with grey clayey SMT 3a1 and iron-stained SMT 3a2. *Microstructure* – massive?; fine fissuring and channels (10% voids?). *Coarse Mineral* – C:F (Coarse:Fine limit at 10μm), as M40151A and B. *Coarse Organic and Anthropogenic* – many traces of (sometime possible woody) roots (now ferruginised) up to 3mm in size. *Fine Fabric* – as M40151A; *Pedofeatures* – as M40151A.	Contexts 40103 (Phase 6c/6a) Central part of site Generally, iron depleted, very weakly humic clay, but with many ferruginised traces of roots, some possibly once-woody(?) up to 3mm in diameter. *Waterlogged lacustrine clay sediments acting as rooting medium for wetland and possible woodland plants.*
D1b/SMT 10a, 10b	M40196	85–165mm 1) Brownish layer: 85-135(155)mm (dominantly SMT 10a) 2) Greyish layer: 135(155)-165mm (dominantly SMT 10b) SM – Heterogeneous. *Microstructure* – massive, with curved horizontal fissuring, 10% intrapedal voids, fine channels and closed vughs. *Coarse Mineral* – as M40195. *Coarse Organic and Anthropogenic* – 1) 2 fire-cracked flint (5mm and 10mm-in size; calcined), strongly rubefied clay ~1mm (opaque, reddish brown under OIL, compared to yellowish orange surrounding iron stained soil), and rare fine rubefied clay/mineral and calcined grains. *Fine Fabric* – SMT 10a: speckled darkish brown (PPL), moderately low interference colours (open and close porphyric, speckled and grano-striate and uni-striate b-fabric, XPL), orange (OIL), trace amounts of very fine blackened OM/charcoal(?); SMT 10b: speckled greyish brown (PPL), XPL as SMT 10a, grey to pale yellow (OIL). *Pedofeatures* – very abundant textural intercalations, muddy laminae and void (sometimes closed vugh, polyconcave vugh) matrix coatings, also embedding large grains such as clacined flint fragment. *Amorphous* – very abundant iron staining in upper half.	Context 40100 (Phase 6) brownish over greyish This is a brownish clay loam, partially mixed with grey clay loam, and becoming more dominantly grey downwards. The micro-fabric is characterised by very abundant textural pedofeatures (intercalations, pans and matrix void coatings – and associated closed vughs) and iron staining. Fine channelling and fissuring affected the massive soil-sediment. Two fire-cracked flints (5mm and 10mm-in size; calcined), and an enigmatic embedded strongly rubefied clay clast (~1mm) occurs. Flints are also embedded in the soil-sediment matrix. *This is a muddy mixed clay loam, with mainly iron stained brownish sediment in the upper half of the slide and iron-depleted clay loam below. The upper part also shows mixed grey and brown microfabrics, and the inclusion of probable fire-cracked flints, and possible rubefied clay material, as two ~mm-size fragments and as a scatter of very fine rubefied material. These materials may be relict of a combustion zone, but have been eroded and fragmented by colluvial processes.*
MFT D1a/SMT 9a1 and 9a2 Over MFT D1b/SMT 9a1	M40195	130–210mm SM – Heterogeneous/broadly layered but with SMT 9a throughout, 1) 130-160mm: dusty clayey, 2) 160-185mm: fine clayey with clay clasts (SMT 9a2), 3) 185-210: sandy and gravelly clay. *Microstructure* – massive; 20% with curved planar voids/ fissures, collapsing vughs and channels. *Coarse Mineral* – C:F, 1) 30:70, with coarse silt; 2) 50:50, with coarse silt and fine sand and coarse sand to gravel-size angular clay clasts (from CwF?); 3) 80-90:20-10, with coarse silt to very coarse sand and few gravel (max	Context 40100 (sand lens boundaries (Phase 6) This thin section is located along layered junctions between sandy microfacies (coarse silt, fine to very coarse sands with fine [max 5mm flint] gravel) and overlying a moderately, sandy clayey layer containing many fine gravel-size brownish clay clasts (as in M40196 [Context 40100] or Clay-with-Flints like?), and with upwards, a greyish clayey layer where textural intercalations, matrix void coatings and clayey pans are abundant. There is weak

Table A7.1 (continued 5)

Microfacies type (MFT)/ Soil microfabric type (SMT)	Sample No.	Depth (relative depth) Soil Micromorphology (SM)	Preliminary Interpretation and Comments
		5mm)(ironstone, iron-stained clay also present); both weathered and freshish glauconite is present. *Coarse Organic and Anthropogenic* – possible inclusion of fine angular flakes. *Fine Fabric* – SMT 9a: dusty to speckled greyish to dark greyish brown (PPL), moderately low interference colours (open to close porphyric, speckled, grano- and striate b-fabric, XPL), pale greyish yellowish brown (OIL), rare very fine blackened/charred OM. SMT 9b (clay clast material): Contains 30% coarse silt and fine-medium sand, dark yellowish brown (PPL), low to moderately low interference colours (close porphyric, speckled, grano- and striate b-fabric, XPL), yellowish orange (OIL), and as 9a. *Pedofeatures: Textural* – occasional textural intercalations, muddy laminae and void (sometimes closed vugh, polyconcave vugh) matrix coatings (150-300µm-thick), increasing to very abundant upwards. *Amorphous* – abundant iron staining, sometimes possibly picking out relict organic fabrics(?).	to moderate iron staining throughout, possibly sometimes picking out relict amorphous organic matter. Both weathered and moderately fresh glauconite is present. Rare fine blackened/charred very fine organic matter occurs. *The thin section records a relatively high energy event/wash compared to the clayey deposits generally, with first sands and fine gravels being deposited, followed by sandy clays containing gravel-size clay clasts (eroded 40100 material [from elsewhere]/ a Clay-with-Flints -like material (Avery et al. 1959)), suggesting a lowering of energy and more muddy colluvial(?) deposition upwards. Lastly, muddy clayey sediments are deposited which are characterised by textural intercalations, matrix void coatings and clayey pans. Generally, this appears to be an upward-fining/decreasing energy sequence.*
MFT C4a/SMT 7a1, 7a3 (7a2)	M40194	320-400mm SM– Heterogeneous with dominant clayey SMT 7a1 (with silty variants), common (pale iron stained) SMT 7a3 and very few SMT 7a2 (speckled with trace amounts of very fine charcoal?). *Microstructure* – massive (coarse prisms?), 10% voids, fine channels and fissures. *Coarse Mineral* – as below; 2.1mm-size quartzite. *Coarse Organic and Anthropogenic* – trace amounts of very fine charcoal; very fine sand size burnt? mineral grains/flint? 2mm-size cracked flint. *Fine Fabric* – SMT 7a1: cloudy and dusty pale grey (PPL), moderate interference colours (open porphyric, speckled, uni-and grano-striate, XPL), pale grey (OIL), very poorly humic with rare very fine amorphous, blackened and charred OM, phytoliths present; SMT 7a2: finely speckled grey (PPL), moderate low interference colours (as 7a1), pale grey (OIL), very poorly humic with rare to occasional very fine amorphous, blackened and charred OM, phytoliths present; SMT 7a3: pale cloudy (ochreous speckled and dotted) yellowish grey brown (PPL), moderate interference colours (open porphyric, speckled, uni-and grano-striate, XPL), pale orange to orange (OIL), many to abundant fine (5-10µm) ferruginised amorphous OM/pyrite pseudomorphs? *Pedofeatures: Textural* –very abundant clayey intercalations; occasional silt concentrations. *Depletion* – occasional strong iron depletion, sometimes including fine fabric SMT 7a2. *Amorphous* – probable oxidised pyrite spheroids?; abundant weak Fe staining and many ferruginised OM/oxidise pyrite? *Fabric* – very abundant mixing of weakly stained/ferruginous microfabric SMT 7a3 (oxidised pyrite) BD (40100): 1.73% LOI, 0.179 mg g 1 phosphate-P.	Context 40100 (stones/artefact inclusions?) (Phase 6) Heterogeneous, massive, compact iron depleted clay and iron-stained clay (containing coarse silt and fine sand) probable 2mm-size fire-cracked flint. Characterised by textural intercalations (associated with uni- and grano-striate b-fabric), sometimes silty clay in nature, and patches and infills of yellow-stained clay with fine (5-10µm) ferruginised amorphous OM? (very fine nodules – possible pyrite pseudomorphs). Trace amounts of very fine charcoal, blackened detrital OM and phytoliths present; examples of possible burned mineral grains. Fabric mixing of two major microfabric types. *Muddy and likely physically disturbed sediment, which was generally iron-depleted (gleyed) but which once had organic sediment mixed in. The latter has been affected by possible pyrite formation (associated with relict OM) and ferruginisation of this material or the fine OM. Possible dung inputs and animal tallow (cf Unit 4u at Boxgrove freshwater pond?).*

Table A7.I (continued 6)

Microfacies type (MFT)/ Soil microfabric type (SMT)	Sample No.	Depth (relative depth) Soil Micromorphology (SM)	Preliminary Interpretation and Comments
MFT C4b/SMT 8a1 and 8a2	M40193A	290–365mm SM – as below. *Microstructure* – as below, with prismatic fissuring – 30% voids (currently); closed vughs. *Coarse Mineral* – C:F, 40-60:60-40, poorly sorted coarse silt to medium, coarse, very coarse sand and example of rounded flint gravel (2.5mm) with bleached rim. *Coarse Organic and Anthropogenic* – root trace. *Fine Fabric* – as below. *Pedofeatures: Textural* – very abundant intercalations and associated closed vughs, for example. *Amorphous* – many moderate iron staining, with ferruginised organic very thin excrements in root trace/channel. *Excrements* – trace of organic very thin excrements in root trace.	Context 40100 (Phase 6) (southern end of site?) Similar to below, but more poorly sorted fine to coarse sands and sandy concentrations. Very abundant textural intercalations, with associated closed vugh formation. Fewer (many) and more weakly iron stained fabrics. Example of fine gravel and root trace with ferruginised once-organic very thin excrements were noted. *Muddy trampled(?) water-saturated sediments containing higher amounts of poorly sorted sands, compared to below; perhaps as the result of slightly increasing episodic fluvial activity(?).*
MFT C4b/SMT 8a1 and 8a2	M40193B	370–450mm SM – Heterogeneous, as M40194 with SMT 8a1 and 8a2, with three broad layers of clayey fine sand, centre 20mm characterised by textural pedofeatures. *Microstructure* – massive, 15% voids, fine channels, partially collapsed (?) fissures, vughs. *Coarse Mineral* – C:F, 40:60. *Coarse Organic and Anthropogenic* – 400μm-size calcined flint fragment; 8mm-size angular fire-cracked(?) flake with traces of rubefication. *Fine Fabric* – SMT 8a1 (as 7a1) and 8a2 (as SMT 7a3), both with 40% sand-size coarse mineral material. *Pedofeatures* – *Textural*: very abundant clayey infills eg in 3mm wide infills along junction between upper and lower layers, and intercalations throughout. *Amorphous* – very abundant iron staining – weak to strong – especially concentrated in central layer.	Context 40039 lower (Phase 6) (southern end of site?) Similar to M40194 but more sandy, with three c 20-25mm thick layers of clayey fine sand (40% fine sand), with textural clayey intercalations and Fe staining throughout, but concentrated in central layer where there are clay infills up to 3mm wide and an irregular boundary. A 400μm-size calcined flint fragment and a 8mm-size angular fire-cracked(?) flake (traces of rubefication were noted. *Layered muddy fine sand deposition, with for example marked iron staining, clayey intercalations and clayey infills, possibly indicative of trampling. Artefact inclusions indicate human presence.*
MFT C3b/SMT 6a1	M40149	350–430mm As 40099 in M40150A, below, with 2mm-size angular flint fragment (flake?) occurring as an embedded grain; trace amounts of very fine fragments of amorphous OM (max 150μm) in fine channels towards upper part of thin section. SM – Heterogeneous; *Microstructure*: as below, fissured, 30% voids. *Coarse Mineral* –as below. *Coarse Organic and Anthropogenic* – trace amounts of amorphous OM; possible flint flake; *Fine Fabric*: SMT 6a1. *Pedofeatures* – as below. *Textural* – abundant sedimentary clayey intercalations, including 4mm-thick curved layer/fill at 410mm. *Amorphous* – trace amounts of void hypocoatings and OM pseudomorphs. *Fabric* – very broad burrow (?) with collapsed thin excrements.	Context 40078-40099 interface (Phase 6d/6c/6a) 40099 – 6a Iron depleted clay, as below, with abundant sedimentary clayey intercalations, including a 4mm-thick layer/fill at 410mm depth. A possible 2mm-size flint flake is present, as an 'embedded grain'. Near the top, trace amounts of amorphous organic matter inclusions occur in channels. Very weak trace amounts of iron hypocoatings occur. *Continued muddy lacustrine sediment accumulation, with again a possible flint flake fragment occurring as an embedded grain suggesting it sank down under muddy conditions. Inclusions of preserved amorphous organic matter may stem from inwash from overlying contexts (40078?).*
MFT C3a/SMT 6a1	M40150A	220–295mm 220–240(275)mm	Context 40099-40103 transition (Phase 6d/6c/6a); 6c over 40099 – 6a

Table A7.1 (continued 7)

Microfacies type (MFT)/ Soil microfabric type (SMT)	Sample No.	Depth (relative depth) Soil Micromorphology (SM)	Preliminary Interpretation and Comments
		SM – Homogeneous SMT 6a1. *Microstructure* – massive, fissured, poor prisms, 15% voids, very fine fissures and channels, to medium-size. *Coarse Mineral* – as below, but with 12mm-size angular fragment of flint (flake?). *Coarse Organic and Anthropogenic* – 12mm-size flint flake?; rare blackened root traces (monocot wetland plants?). *Fine Fabric* –cloudy and finely speckled greyish yellow (PPL), moderate interference colours (very open porphyric, speckled and weakly mosaic and striated b-fabric (grano-striate/embedded grains), XPL); pale grey with white areas (clay depleted/weather coatings and infills); very weakly humic with occasional very fine detrital blackened and brown OM, possible trace amounts of charcoal. *Pedofeatures:Textural* – occasional intercalations and rare matrix infills (of collapsed voids?). *Depletion* – very abundant iron depletion throughout; leached/iron depleted fissure and channel hypocoatings (very acid/strong gleying effect). BD (40099): 2.26% LOI, 0.194 mg g 1 phosphate-P.	Iron depleted, very weakly humic clay containing rare detrital fine OM, blackened (organic) monocot? root traces and traces of charcoal. It includes a possible flint flake (12mm). Occasional intercalations and matrix infills, and strongly leached fissure and channel hypocoatings. Bulk analyses record 2.26% LOI, 0.194 mg g 1 phosphate-P; Fe-depleted (SEM/EDS). *Massive iron-depleted lacustrine clay sediments, showing rooting (by wetland plants – blackened organic monocotyledonous root traces), and inclusions of detrital OM including trace amounts of very fine charcoal. Anomalous 12mm size flint fragment and character suggest it may be a flint flake/of anthropogenic origin; possibly naturally sunk down in muddy sediments. Strong leaching has affected the sediment, hence moderate OM preservation and markedly low phosphate content.*
MFT C2/SMT 5a1		240(275)-295mm Very similar to 40103 in M40150B, with marked concentration of iron staining in uppermost 10mm, as a series and concentrations of 1mm sub-parallel bands (sloping at 45 degrees in thin section sample). These bands and underlying sediment also include very abundant poorly pseudomorphic root traces. In topmost 5mm, 1mm-size rounded clay (soil?) clasts occur with darkish reddish brown finely ganular clay/once-humic soil infilling the voids (often more ferruginous). Iron-stained root traces and mottles possibly pseudomorphic of once-humic (now-ferruginised) soil (burrows and root channels) is present. An atypical 6mm round flint gravel clast is present; trace amounts of very fine charcoal. Pedofeatures – as M40150B, with trace amounts of void matrans forming from textural intercalations. BD (40103 – Fe pan): 3.63% LOI, 1.77 mg g 1 phosphate-P. SEM/EDS : 10 analyses of ironpan and iron depleted zones (Table 5.2)	40103 - 6a As below, but with a (sloping?) concentration of iron staining along uppermost 10mm – forming an iron pan. This is partially made up of a concentration once-humic burrows and channel (fills), and in places 1mm-size 'soil' clasts cemented by infills of sometimes strongly ferruginised (once-humic) fine soil (?). Trace amounts of very fine charcoal and a 6mm example of rounded flint gravel. Marked high LOI (3.63%) and enriched phosphate-content (1.77 mg g 1 phosphate-P); Fe and P-enriched (SEM/EDS). *Possible ripened sediment/soil surface, showing 1) slight truncation in places, 2) traces of rooting and small mesofauna burrowing, 3) very locally eroded and transported ripened soil/sediment clasts, and 4) local invash of humic fine soil. This appears to be a possible buried ripened soil which may have had short-lived woodland (?) cover, before being affected by inundation and renewed sedimentation (40099). (Gentle inundation caused slaking and structural collapse/formation of intercalations and void matrans, and erosion of local soil/sediments, and invash.) Buried 'topsoil' interpretation is consistent with LOI and enriched phosphate probably also reflects this and 'geogenic' concentration of P due to groundwater movement.*
MFT C1/SMT 5a1	M40150B	300-375mm SM – Homogeneous SMT 5a1. *Microstructure* – massive, crack and relict prismatic, 25% voids, very fine to medium (1mm) curved planar voids and fissures, fine to coarse (2.5mm) extant and relict channels. *Coarse Mineral* –	Context 40103 (Phase 6d/6c/6a); 6a Massive clay containing small amounts of coarse silt, fine and medium sand. Clay has fissures and relict channels, some with

Table A7.1 (continued 8)

Microfacies type (MFT)/ Soil microfabric type (SMT)	Sample No.	Depth (relative depth) Soil Micromorphology (SM)	Preliminary Interpretation and Comments
		C:F (Coarse:Fine limit at 10 μm), 20:80; moderately sorted coarse silt to fine to medium sand, as below. *Coarse Organic and Anthropogenic* – occasional (to possibly very abundant?) ferruginised broad and fine woody(?) root traces (0.5-2.5mm). *Fine Fabric* – SMT 5a1: typically pale yellowish brown (PPL; cf. depleted and iron-stained), moderate interference colours (very open porphyric, speckled and occasionally grano-striate ('embedded grain'), XPL), grey to pale yellow (OIL), very weakly humic with trace amounts of phytoliths. *Pedofeatures: Textural* – occasional textural intercalations, rare channel infill of iron depleted clay (plus partial clay destruction). *Depletion* – very abundant iron depletion in matrix and as channel hypocoatings associated with 'roots'. *Amorphous* – very abundant iron impregnation affecting relict roots and forming parallel lines/possible root-like stains. *Fabric* – many apparent slickenslide-like features (clay orientation); *Excrements*: possible rare traces of iron pseudomorphs of very thin excrements in root channels. BD (40103 – Fe mottling): 3.42% LOI, 0.538 mg g 1 phosphate-P	likely ferruginised (woody) root pseudomorphs (0.5-2.5mm; with traces of very thin ferruginised excrements). Sediment is mottled with iron depleted and iron stained areas and other possible root stains. Some channels have leached/iron-depleted hypocoatings; example of channel with leached/iron-depleted clay fill. *'Basal' lacustrine clay sediments, containing scattered small amounts of coarse silt, fine and medium sand – possibly locally blown-in/washed-in. This clay acted as a rooting substrate, possibly for woodland, as suggested by the presence of woody roots (some being dominated by small invertebrate mesofauna). The sediment was also affected by shrinking and swelling (an effect of woodland growth?), marked iron depletion (and possibly clay break down at times) and iron staining – hydromorphic/gleying effects. Enigmatic high measurements of LOI possibly relate to woody root traces, while enhanced P is probably secondary and linked to iron-staining.*
MFT B2/SMT 3a1(3a3)	M40151A	70-150mm (finely cracked dried out sample) SM – Moderately heterogeneous with very dominant iron-stained and iron-depleted clayey SMT 3a1 and very few fine and medium sandy SMT 3a3. *Microstructure* – massive (currently very finely fissured), 30% voids, fine fissures. *Coarse Mineral* – C:F, as M40151B, with 3a1, C:F=30:70 containing fine and medium sand. *Coarse Organic and Anthropogenic* – possible traces of rare thin roots. *Fine Fabric* – SMT 3a1 and 3a3, as M40151B. *Pedofeatures – Depletion*: many iron-depleted areas; *Amorphous*: very abundant moderate to strong iron staining, strongest staining describes curved 'fill'(?); possible root stains. *Fabric* – 30mm deep by 4.5mm wide curved 'fill'; very broad burrow fills (focus of sands); possible thin weak iron stained burrows.	Upper-context 40103 (Sub-phase 6a) (Overlying 40103 partially occurs as strongly iron-depleted clay in uppermost 20mm) Massive non-calcareous clay sediments (with few sand inclusions), with marked iron-staining, with a infill features, burrows, once humic peds and excrements, demarcated by strong iron-staining. Sands are only present in rare broad burrow fills. Clays seemingly less humic compared to lower down, and only very fine charred/blackened detrital OM noted. *A minerogenic wetland clayey sediment appears to have developed as a topsoil (hydroseral succession?), with burrows, peds, roots and excrements of relatively humic soil showing marked iron impregnation. Overlying iron-depleted 40103 is probably a flood clay.*
MFT B1/SMT 3a1, 3a2, 3a3	M40151B	320-410mm SM – Heterogeneous with dominant clayey (SMT 3a1), common ferruginous clayey (3a2) and sandy (3a3). *Microstructure* –massive with fissures/crack microstructure (residual bedding and fine laminae); 30% voids, including coarse fissures, but only 20% with fine cracks in matrix; fine channels and vughs. *Coarse Mineral* – C:F, SMT 3a1=10:90 or 30:70 (once laminated silts and clay?), well sorted coarse silt-fine sand quartz, feldspar, mica (opaques include limonite); 3a2, C:F=0/10:100/90 (once-humic?); SMT 3a3, C:F=80/90:20/10, moderately well sorted fine and medium sands; 700 μm –size mica fragment. *Coarse Organic and*	40039 (Sub-phase 6a)(Basal clay etc) Part layered, part coarsely mixed non-calcareous clays, silty clay, fine and medium sands and ferruginised once-organic sediment; silty clays are composed of mixed silt and clay fine laminae. Charcoal as both rare fine (eg. max 800 μm-size monocot charcoal) and occasional very fine material occurs alongside very fine blackened (detrital OM), and phytoliths are present; patches of totally ferruginised organic matter including plant fragment

Table A7.1 (continued 9)

Microfacies type (MFT)/ Soil microfabric type (SMT)	Sample No.	Depth (relative depth) Soil Micromorphology (SM)	Preliminary Interpretation and Comments
		Anthropogenic – ~350 μm-size charcoal fragments (eg of 800 μm-size monocot charcoal?) and finer, trace in lower part becoming rare alongside occasional upwards alongside very fine blackened detrital material (uppermost clay layer); trace amounts of blackish and reddish brown humified peat including 1mm-size fragment. *Fine Fabric* – SMT 3a1: cloudy very pale brownish grey (PPL), moderately low interference colours (open porphyric, stipple speckled with striated b-fabric, XPL), grey (OIL – iron-stained areas are orange to brown), occasional to many very fine blackened detrital, and charred OM, trace of brownish amorphous OM; phytoliths present. *Pedofeatures:Textural* – many textural intercalations, associated with embedded grains and void coatings and infills (matrans 150-200 μm thick; some stained yellowish – see amorphous). *Depletion* – occasional strongly iron depleted zones; *Amorphous:* rare amounts of weak yellowish staining and possibly associated sand size yellowish, isotropic nodules embedded within same matrix (FeP?); very abundant moderate to strong iron impregnation of fine clayey matrix and once-organic sediment, with fine iron nodules (~30 μm), granular or pseudomorphs? some as poor pseudomorphs of plant material/roots, possibly relict iron/sodium carbonate? *Fabric* – very abundant coarse fabric mixing. BD (40039): 1.88% LOI, 0.556 mg g 1 phosphate-P SEM : 17 analyses of iron-depleted and iron-stained areas (Table 5.2).	pseudomorphs occur (current LOI is 1.88%). Sediments also characterised by textural intercalations and associated matrix embedded grains and matrix void coatings, some stained yellow; rare fine sand-size yellow isotropic nodules/infills are probably iron phosphate formations (see EDS Table 5.2; sediment characterised by slight phosphate enrichment, 0.556 mg g 1 phosphate-P). *Once bedded and laminated weakly humic clays, silts, and occasionally medium sands, with patches (inclusions) of peat(?), now ferruginised with plant pseudomorphs; wetland (marsh) locally burned monocotyledonous vegetation(?). Fabric mixing and associated intercalations, matrix coatings, and yellow staining and nodule formation (FeP), alongside slight phosphate enrichment may infer animal trampling and defecation.*
D2/SMT 4a2	M40323A	0-70mm SM – Homogeneous SMT 4a2. *Microstructure* – massive, diffuse <1mm thick laminae upwards, with embedded/cemented 0.5mm-size subrounded clasts in uppermost 5mm; 10% voids, very thin fissures, channels and vughs. *Coarse Mineral* – as below, but very coarse sand and gravel-free. *Coarse Organic and Anthropogenic – Fine Fabric:* SMT 4a2: speckled and cloudy darkish yellowish brown (PPL), moderate interference colours (close porphyric, speckled and grano-striate b-fabric, XPL), very pale yellowish grey (OIL), generally very weakly humic with rare very fine OM, but upwards 'humic pans'(?) of now-ferruginised; trace amounts of phytoliths. *Pedofeatures:Textural* – as below, with matrix pans upwards associated with laminae; *Depletion* – strongly iron-depleted fine matrix. *Amorphous* – abundant iron impregnation associated with relict fine rooting and once-humic <1mm-thick pans (thin iron panning).	(Context 40039 Upper (Tufaceous channel), Phase 6b/6a); 6a Massive, moderately well sorted clayey fine and medium sand, with in the main strongly iron-depleted fine fabric, which is generally very weakly humic, but appears, upwards, to have a series of thin (<1mm) once-humic (now-iron-replaced) pans. There are trace amounts of phytoliths present. In uppermost 5mm a scatter of 0.5mm-size clay clasts occur. Iron also picks out many traces of probable fine rooting and voids around these clasts. The diffuse boundary to 40070 is marked by intercalated and curved thin ironpans and iron stained burrows; clay infills some burrows and channels. *Moderately low becoming low energy fine and medium sandy colluvial deposition, marked by thin rooting and periodic organic matter accumulation (seasonal very thin peat formation?) Lastly inwash of eroded clay clasts is recorded. This all occurred under waterlogged conditions (hence iron depletion and organic matter panning), perhaps as the channel silted up(?). The sediment was affected by secondary but penecontemporaneous iron staining.*

Table A7.I (continued 10)

Microfacies type (MFT)/ Soil microfabric type (SMT)	Sample No.	Depth (relative depth) Soil Micromorphology (SM)	Preliminary Interpretation and Comments
			The ironpans along the diffuse boundary pick out micro gullying and erosion, typical of some Pleistocene brickearth deposits; clay has washed down from overlying contexts into later burrows and channels.
			Context 40039 Middle as below in M40323B
		Diffuse boundary from 40039 Upper to 40039 Middle at ~60mm.	
MFT D1/SMT 4a1	M40323B	70-130 mm	Context 40039 Middle (Tufaceous channel), (Phase 6b/6a); 6a
		SM – mainly homogeneous fine and medium sandy SMT 4a1, with few subhorizonal intercalations of clayey SMT 4a1. *Microstructure* – massive, with fine bedding laminations and clayey intercalations; 20% voids, fine vughs and channels (root traces?). *Coarse Mineral* – C:F, 80:20, moderately sorted fine and medium sand-size quartz, quartzite, feldspar and mica, with very few coarse sand and gravel (2-2.5mm; quartzite and 3mm flint – trace of clay-coating/embedded grain, traces of rubefication/calcination); very few glauconite. *Coarse Organic and Anthropogenic* – possible traces of burning on flint gravel; Fine Fabric: SMT 4a1, as SMT 3a1 – see M40151B, granostriate b-fabric, with only rare very fine amorphous OM and trace amounts of very fine charcoal. *Pedofeatures: Textural* – occasional clayey intercalations (cf grano-striate), void infills. *Amorphous* – rare iron staining of relict rare very broad burrows/channels (edge only in thin section) and sediment layer hypocoatings. *Fabric* – abundant traces of very broad burrowing (6+mm), with curved infills associated with penecontemporaneous clay inwash.	Massive, moderately well sorted clayey fine and medium sand (very few fine gravel), with relict sedimentary laminae, traces of very broad burrows and penecontemporaneous clayey inwash and intercalation and grano-striate formation. Clayey intercalations may show evidence of rooting – preferential channel formation. Rare iron staining (of edge of very broad burrow?) and sedimentary laminae hypocoatings. *Moderately low energy muddy fine and medium sandy colluvial deposition, with penecontemporaneous very broad burrowing and clayey infilling. Minor secondary iron staining.*

Site sequence geometry and sub-surface structural geology

by John Hutchinson

The Site

The site is a rescue dig associated with roadworks for the HS1 rail network. It is situated around Grid Ref. TQ 613 732 on the Upper Chalk. The classic Swanscombe Pleistocene exposures lie between 1.2 and 2.5km to the north and north-west.

The 'solid' strata in the vicinity are, in upwards order:

1. The Upper Chalk, which everywhere underlies the site (dip within a couple of degrees of the horizontal)
2. The Thanet Beds, with the Bullhead Bed at their base
3. The Woolwich Beds
4. The Blackheath Beds
5. The London Clay that caps the hill in Swanscombe Park 1km west of the site.

The Quaternary here has not yet been fully logged and worked out. The interim names of units currently used on site (very broadly in upwards order) are:

A. Soliflucted Chalk (Coombe Rock)
B. Reworked Thanet Beds
C. Lacustrine deposits
D. Reworked fine sands and silty clays with a few pebbles, probably derived largelyfrom the Thanet, Woolwich and Blackheath Beds
E. Gravels and sands (apparently little disturbed by the disturbances describedbelow), possibly from a braided river
F. Soliflucted London Clay and associated head

The 1:50,000 geological map shows a terrace of Boyn Hill Gravel about 1.5km NNW of the site and Alluvium at its foot. The relationships of these and the classic sections at Swanscombe to the geological exposures on the Southfleet Road site have yet to be established.

The main exposures seen, from the site office, northwards towards the newly diverted access road are:

a. Small exposure of soliflucted Chalk, overlain by B (above)
b. Rotated block of intact Chalk with contiguous Bullhead Bed and Thanet Beds. Dip of contact (bedding) very steeply (probably 70 to 80O northwards.
c. In cut slope to W, various (unlogged) reworked materials, probably including London Clay

d. In area to the east of c) excellent exposures of a double basin structure, rather fully exhumed , with additional cross trenches. The best exposed basin (to the W) is delineated by a layer of C. Its infill, chiefly D overlain by E, has been largely removed by archaeological excavations. This basin is about 16m across, with a depth (amplitude) of about 3m. The axes of this and the other basin run roughly nort–south. The basins are slightly asymmetrical, with steeper dips on their western limbs. The maximum dips in these approach the sub-vertical. The southern trench allows a good view of the basins in cross section. They are very regular, apart from some secondary undulations, and are virtually unbroken by faults. The deformed clayey layer, C, is continuous across each basin and is underlain by a compact fine sand, probably B. Each basin appears to be elongated in roughly a N-S direction and to peter out at each end. Towards the north end of the eastern basin apparent loading structures were seen. These are of very small scale, around 2m across, in comparison with the two basin structures.
e. We also looked at a new road cutting to the SW of the main site. Soliflucted London Clay is exposed its south face.

No plans, face logs or sections, or drawings showing their general inter-relationships, were seen during the visit; these were currently being produced.

Possible origins of the above features

Dips in the solid geology of the central and eastern parts of the London Basin are very low, generally not exceeding a few degrees. Against this background, the high dips observed in some of the Southfleet Road excavations, both in the Chalk and in the overlying Pleistocene deposits, are remarkable and apparently anomalous.

The main processes in southern England which can produce steep dips are reviewed briefly below:

Tectonics

Dips in the Chalk as high as 55° occur on the southern limb of the London Basin at the Hog's Back (Sumbler 1996) and, further south, approach the vertical on the south side of the Hampshire Tertiary Basin in the Isle of Wight.

The chief features against the double basin at our site being of structural origin is its location in an area of sub-horizontal dip, its very local nature and the apparent absence of hinge structures between the two basins.

Superficial valley disturbances

A wide variety of superficial valley disturbances exists, which is reviewed by Hutchinson (1991; 1992). These can produce local dips of 70° or more in valley bulges and are widely distributed in the Jurassic outcrop of central England and in the Weald (Hutchinson 1992). However, they require the valley to be underlain by a thick argillaceous stratum.

The absence of such a stratum beneath the Southfleet Road site, its location away from the axis of the local valley and the very local nature of the observed anomalous features rule out an explanation depending upon the above mechanisms.

Pingos

Relict pingos, chiefly of open type and generally (but not always) more or less circular or oval in plan, are common in southern England and Wales. Some are reported to have diapiric structures in their lower parts that can result in steep dips. Pingos are frequently found in clusters towards the foot of slopes, particularly if artesian groundwater pressure is present.

While some of these criteria may be fulfilled at Southfleet Road, the continuity of the beds, especially B and C, across the basins and the absence of boggy, peaty infills argue strongly against a pingo origin.

Landsliding

Pressure exerted by the toe of a landslide can produce deformation, folding and faulting of its lower parts and of the resisting ground. An example at Lyme Regis is described by Hutchinson and Hight (1987), but the effects extend no more than about 5m below ground level.

To judge better whether a slide sufficiently deep-seated to affect the area of the basins could have occurred at our site, one should draw a W – E section through areas c and d, picking up the basin structures and the geology and showing as far as possible the original natural ground surface. My impression on site was that, apart from the London Clay, most of the deposits overlying the Chalk are firm and granular and difficult to develop a major slide in which could penetrate down to or below the level of the basins. Furthermore, no slip surfaces were seen there and the

basin structures themselves do not correspond to the passive style that one would expect in a slide toe area.

Frost-heave and periglacial solifluction

While both these processes can result in considerable deformation of the ground, they are too shallow to explain the Southfleet Road structures.

Solution of the Chalk

Solution pipes, swallow holes and solution dolines are numerous in the Chalk, especially around the margin of London Clay capped hills (the umbrella effect). The work of Higginbottom and Fookes (1970) and Jones (1981) indicate that the extent and magnitude of solution effects in the Chalk have been underestimated. The former authors report an apparent concentration of solution features on the route of the M40 motorway through the Chilterns, which in some cases are underlain by linear depressions in the Chalk surface as deep as 15m below the general Chalk surface level. The latter author reports particularly severe solution beneath a cover of Blackheath Beds around the southern outskirts of London and a concealed, irregular rockhead on the North Downs and, to a lesser extent, the Chiltern Hills. He concludes that the high, steep-sided pinnacles of Chalk often found beneath the Blackheath Beds and the well-known pipes exposed at Lenham represent rather dramatic examples of solutional forms widely developed on chalklands with a thin cover of superficial deposits. Good reviews of karst development and subsidence on the Chalk outcrop in England are provided by Edmonds (1983; 1988).

Both this background and the detailed local geology point to the solution of Chalk as by far the most likely mechanism producing the exposures b and d. In the former case it is not abnormal to have two solution features adjoining each other or for them to have a linear tendency. It will be interesting to see if this parallels a regional joint direction in the Chalk. In the latter case it is suggested that the exposure represents a fallen karstic pinnacle of Chalk capped by the lower Thanet Beds. The intact nature of the Chalk/Thanet beds contact may indicate that the ground was frozen at the time of collapse (the maximum depth of permafrost here is estimated by Hutchinson (1991) to lie between about 30 and 60m).

If the above conclusions are correct, the surface of the Chalk bedrock should be highly irregular. This should be checked by geophysics and/or closely spaced borings.

Appendix 9

Amino acid dating

by Kirsty Penkman, Victoria Morris and Richard Allen

INTRODUCTION

Advances in Quaternary Science during the past few decades, particularly with respect to absolute dating methods and the correlation of terrestrial stratigraphic sequences with oxygen isotope records (ie major climatic signals) interpreted from the geochemical analysis of deep sea sediment cores, have provided an opportunity to develop a high resolution chronology of human occupation and activity over the past 600,000 years or so. This report details attempts to obtain age estimates using amino acid racemization (AAR).

Amino acid analyses were undertaken at the York Laboratory (NEaar) from key horizons. This involves isolating the intra-crystalline protein fraction of two biominerals: the calcitic opercula from the fluvial gastropod *Bithynia tentaculata*, for which an excellent and growing database of protein degradation data has recently been assembled (Penkman *et al.* 2008b). This laboratory has been developing an improved methodology for the technique, and studies are ongoing to calibrate the amino acid data with reference to other dating techniques.

Amino acid racemization geochronology

A new technique of amino acid analysis has been developed for geochronological purposes (Penkman 2005; Penkman *et al.* 2007; 2008), combining a new Reverse-Phase High Pressure Liquid Chromatography method of analysis (Kaufman and Manley 1998) with the isolation of an 'intra-crystalline' fraction of amino acids by bleach treatment (Sykes *et al.* 1995). This combination of techniques results in the analysis of D/L values of multiple amino acids from the chemically-protected protein within the biomineral; enabling both decreased sample sizes and increased reliability of the analysis. Amino acid data obtained from the intra-crystalline fraction of the calcitic *Bithynia* opercula has been found to be a particularly robust repository for the original protein. This has enabled an increased level of resolution and therefore this material has been focused on in this study.

Theory

Amino acids, the building blocks of proteins, occur as two isomers that are chemically identical, but optically different. These isomers are designated as either D (dextrorotary) or L (laevorotary) depending upon whether they rotate plane polarised light to the right or left respectively (Fig. A9.1). In living organisms the amino acids in protein are almost exclusively L and the D/L value approaches zero. D-amino acids are synthesised by some organisms; they are found free in invertebrate body fluids where they play a role in osmoregulation and can occur peptide bound in bacterial peptidoglycan, where part of their function is resistance to proteases. The potential application to geochronology arises from the fact that after death amino acid isomers start to interconvert. This process is commonly termed racemization. In time the D/L value approaches one. The proportion of D to L amino acids is therefore an estimate of the extent of protein degradation, and if this is assumed to be predictable over time can be used to estimate the age of a sample. Other indications of protein decomposition, such as the degradation of unstable amino acids, can also be used to estimate age.

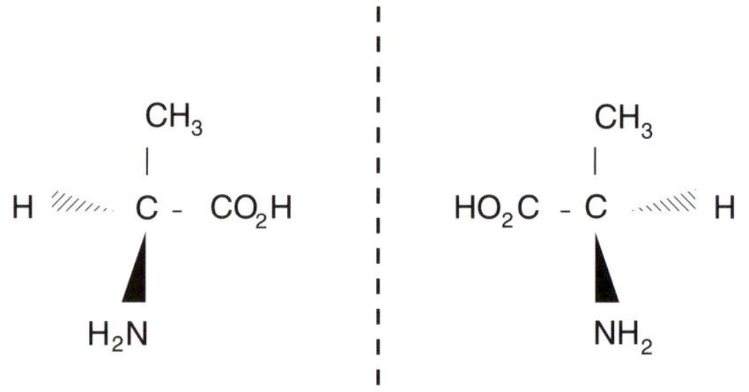

Figure A9.1   Figure 1: L- and D- amino acid structure

Mechanisms of racemization

The rate of racemization is governed by a variety of factors, most of which have been studied in detail only for free amino acids. North East amino acid racemization (NEaar) analyse the intra-crystalline amino acid fraction and in this way, within a closed environment in which other factors (water content, concentration of cations, pH) are constant, the extent of racemization is a function of time and temperature. Over a small geographical area, such as that represented in this study, it can be assumed that the integrated temperature histories are effectively the same. Any differences in the extent of decomposition of protein within the sample are therefore age-dependent.

Intra-crystalline protein decomposition

The organic matter existing within individual crystals (intra-crystalline fraction) is believed to be a more reliable substrate for analysis than the whole shell (Sykes *et al.* 1995; Penkman 2005; Penkman *et al.* 2008). The initial bleaching step in the recovery of the intra-crystalline fraction removes both secondary contamination and the organic matrix of the shell. This organic matrix degrades and leaches at an unpredictable rate over time, leading to variation in the concentration and D/L of the amino acids. Thus, as appears to be the case in ostrich eggshell (Miller *et al.* 2000), the D/L values of amino acids in the intra-crystalline fraction of shells have been analysed; in the case of ostrich eggshell no bleaching step was used. The molluscan racemization data reported therefore contrasts with previous work that examined D/L values from whole mollusc shells containing both intra- and inter-crystalline material.

This isolation of the intra-crystalline fraction is believed to provide a closed system repository for the amino acids during the burial history of the shell. Only the amino acids within this fraction are protected from the action of external rate-affecting factors (except temperature), contamination by exogenous amino acids and leaching. Amino acids within the whole shell are not protected and can be leached out into the environment. Figure A9.2 shows a schematic of the intra-crystalline fraction with respect to the whole shell. The low level of Free amino acids observed in the inter-crystalline fraction of unbleached samples (Penkman *et al.* 2007) indicates that these have been lost through diagenesis, and as these tend to be more highly racemised than the Total fraction, this loss would lead to a lower than expected D/L for the Total fraction of the whole shell.

Traditionally AAR studies targeted a single amino acid racemization reaction, that of L-isoleucine to D-alloisoleucine (A/I), due to the technical ease of separation and its slow rate of racemization. The approach used in this study diverges from this, as dates are derived from the analysis of multiple amino acids. Whilst racemization rates differ between individual amino acids, they should be highly correlated in a closed system. By linking together different amino acids, and then linking this to a temperature driven model of decay, which includes hydrolysis, racemization and degradation, the extent of protein degradation can be derived. The pattern of decomposition appears to be different between mollusc genera, requiring separate models for each genus or species studied.

Once a closed system inside mollusc shells has been isolated, then the kinetics of protein decomposition are much simpler to predict. In this laboratory the concept of age estimation using the extent of overall Intra-crystalline Protein Degradation (IcPD) has been devised, which links the hydrolysis, racemization and decomposition of all the amino acids isolated by this method. The concept behind the IcPD is to combine multiple information from a single sample to derive an overall measure of the extent of diagenesis of the protein in that fossil. Similar ideas have been used before, although not in such a comprehensive way (Wehmiller *et al.* 2010). Divergence from the normal in a plot of A/I vs Gly/Ala is thought to indicate leaching in molluscs (Murray-Wallace and Kimber 1987). Kaufman (2000)

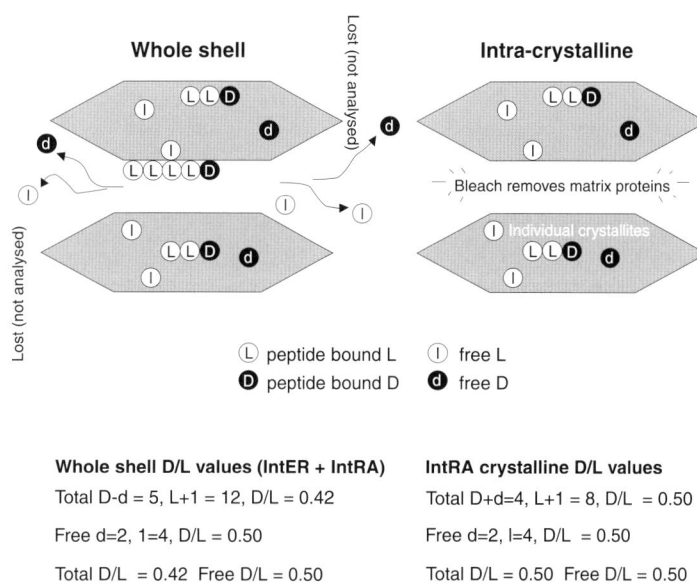

Whole shell D/L values (IntER + IntRA)

Total D-d = 5, L+1 = 12, D/L = 0.42

Free d=2, 1=4, D/L = 0.50

Total D/L = 0.42 Free D/L = 0.50

IntRA crystalline D/L values

Total D+d=4, L+1 = 8, D/L = 0.50

Free d=2, l=4, D/L = 0.50

Total D/L = 0.50 Free D/L = 0.50

Figure A9.2 Schematic of intra-crystalline amino acids entrapped within carbonate crystallites. Unlike the proteins of the organic matrix between the crystallites, which leach from the shell with time, in a closed 'intra-crystalline' system the amino acids are entrapped. Thus the relationship between the DL ratios of different amino acids and between free (non-protein bound) and total (both free and originally protein-bound amino acids, released by acid hydrolysis) amino acids is predictable. Analysis of the whole shell would result in lower than expected D/L for the total fraction, due to the loss of the more highly racemised frees.

used ratios of Asx to Glx to screen out samples with any unusual values.

In a closed system, it should be possible to predict the relationship between geological time and IcPD increase, using not just racemization but other measures of protein decomposition, such as total and relative concentrations. It follows from the innovations above that, assuming sampling is from an idealised closed system, the pattern of protein decomposition governs the observed racemization of (a) free amino acids and (b) the total system, (c) the percentage of free amino acids and (d) the total concentration of amino acids.

This model can also be used as a method of assessing the internal reliability of each biomineral used and to determine how closely these substrates approximate to a closed system. Subsequently palaeotemperature information can be included and estimates made of the link between degradation and absolute age in environments with fluctuating temperatures. If an accurate temperature model is used, then age estimates can be derived directly from the IcPD data, although the results presented here do not incorporate any palaeotemperature information and are presented simply as a relative dating tool.

MATERIALS AND METHOD

Amino acid racemization (AAR) analyses were undertaken on *Bithynia tentaculata* opercula:

- Two individual opercula from Southfleet Road, context 40078, <40275>, Bulk SV sample (NEaar 6433-4; SFR4Bto1-2)
- Three individual opercula from Southfleet Road, context 40143, <40282/C/0-2>, mollusc subsample from monolith (NEaar 6430-2; SFR3Bto1-3)
- Three individual opercula from Southfleet Road, context 40144, <40295/C>, mollusc subsample from bulk SV sample (NEaar 6435-7; SFR5Bto1-3)
- Three individual opercula from Southfleet Road, context 40144, <40333>, bulk SV sample (NEaar 6438-40; SFR6Bto1-3)
- Eleven individual opercula from Southfleet Road, context 40070, <40315/C>, mollusc subsample from bulk SV sample (NEaar 6424-6, 6606-6613; SFR1Bto1-11)
- Three individual opercula from Southfleet Road, context 40070, <40317/C>, mollusc subsample from bulk SV sample (NEaar 6427-9; SFR2Bto1-3)
- Seven individual opercula from Southfleet Road, context 40070, <40162>, bulk SV sample (NEaar 2041-7; EbABto1-7)
- Two individual opercula from Southfleet Road, context 40103, <40320>, bulk SV sample (NEaar 6441-2; SFR7Bto1-2)
- Two individual opercula from Southfleet Road, context 40025, <40286>, bulk SV sample (NEaar 6443-4; SFR8Bto1-2)
- Two individual opercula from Southfleet Road, context 40025, <40343>, bulk SV sample (NEaar 6445-6; SFR9Bto1-2)
- Fourteen individual opercula from Southfleet Road, context 40062, < 40042>, bulk SV sample (NEaar 6166-7, 6447-9, 6534-6538, 6614-6616; SRA1Bto1-2 and SFR10Bto1-12)

Sample Preparation

Shells were examined under a low powered microscope and any adhering sediment removed. The shell samples were then sonicated and rinsed several times in HPLC-grade water. The shells were then crushed to <100µm. Only bleached samples were analysed.

Bleaching

50µl of 12% solution of sodium hypochlorite at room temperature was added to each milligram of powdered sample and the caps retightened. The powders were bleached for 48 hours with a shake at 24 hours. The bleach was pipetted off and the powders were then rinsed five times in HPLC-grade water and a final rinse in HPLC-grade methanol (MeOH) to destroy any residual oxidant by reaction with the MeOH. The bulk of the MeOH was pippetted off and the remainder left to evaporate to dryness.

Hydrolysis

Protein bound amino acids are released by adding an excess of 7 M HCl to the bleached powder and hydrolysing at 110°C for 24 hours (H★).

20µl per milligram of sample of 7 M Hydrochloric Acid (HCl) was added to each Hydrolysis ('Hyd', H★, THAA) sample in sterile 2ml glass vials, were flushed with nitrogen for 20 seconds to prevent oxidation of the amino acids, and were then placed in an oven at 110°C for 24 hours. After 10 minutes in the oven, the caps of the 2ml vials were re-tightened to prevent the escape of vapour. After 24 hours, the samples were dried in a centrifugal evaporator overnight.

Demineralisation

Free amino-acid samples ('Free', F, FAA) were demineralised in cold 2M HCl, which dissolves the carbonate but minimises the hydrolysis of peptide bonds, and then dried in the centrifugal evaporator overnight.

Rehydration

When completely dry, samples were rehydrated with 10µl per mg of Rehydration Fluid: a solution containing 0.01 mM HCl, 0.01 mM L-homo arginine internal standard, and 0.77 mM sodium azide at a pH of 2. Each vial was vortexed for 20 seconds to ensure complete dissolution, and checked visually for undissolved particles.

Approximately 20µl of rehydrated sample was then placed in a sterile, labelled 2 ml autosampler vial containing a glass insert, capped and then placed on the autosampler tray of the HPLC.

For each set of sub-samples a blank vial was included at each stage to account for any background interference from the bleach, acid, or rehydration fluid added to the samples.

Analysis of Free and Hydrolysed Amino Acids

Amino acid enantiomers were separated by Reverse Phase High Pressure Liquid Chromatography (RP-HPLC). NEaar uses the method of Kaufman and Manley (1998) using an automated RP-HPLC system. This method achieves separation and detection of L and D isomers in the sub- picomole range.

Samples (2µl) were derivitised with 2.2µl *o*-phthaldialdehyde and thiol *N*-isobutyryl-L-cysteine automatically prior to injection. The resulting diastereomeric derivatives were then separated on Hypersil C_{18} BDS column (sphere d. 5µm; 250 x 3mm) using a linear gradient of a sodium acetate buffer (23 mM sodium acetate, 1.3 mM Na_2EDTA; pH6), methanol, and acetonitrile on an integrated HP1100 liquid chromatograph (Hewlett-Packard, USA).

Individual amino-acids are separated on a non-polar stationary phase according to their varied retention times: a function of their mass, structure, and hydrophobicity. A fluorescence detector is used to determine the concentrations of each amino-acid and record them as separate peaks on a chromatogram. A gradient elution programme was used to keep the retention time to below 120 minutes.

The fluorescence intensity of derivitised amino acids was measured (Ex = 230 nm, Em = 445 nm) in each sample and normalised to the internal standard. All samples were run in duplicate. Quantification of individual amino acids was achieved by comparison with the standard amino acid mixture.

External standards containing a variety of D- and L-amino acids, allowing calibration with the analyte samples, were analyzed at the beginning and end of every run, and one standard was analyzed every ten samples. Blanks which had been subjected to identical preparation procedures were randomly interspersed amongst the standards.

The L and D isomers of 10 amino acids were routinely analysed. During preparative hydrolysis both asparagine and glutamine undergo rapid irreversible deamination to aspartic acid and glutamic acid respectively (Hill 1965). It is therefore not possible to distinguish between the acidic amino acids and their derivatives and they are reported together as Asx and Glx.

RESULTS AND DISCUSSION

In total we conducted 208 analyses, all of which were on bleached samples. The extent of racemization in five amino acids (D/L of Asx, Glx, Ser, Ala and Val), along with the ratio of the concentration of Ser to Ala ([Ser]/[Ala]), are reported for both the Free and Hyd fractions. These indicators of protein decomposition

have been selected as their peaks are cleanly eluted with baseline separation and they cover a wide range of rates of reaction. It is expected that with increasing age, the extent of racemization (D/L) will increase whilst the [Ser]/[Ala] value will decrease, due to the decomposition of the unstable serine.

The data obtained from Asx, Glx, serine (Ser), alanine (Ala) and valine (Val) are discussed in detail below. If the amino acids were contained within a closed system, the relationship between the Free and the Hyd fractions should be highly correlated, with non-concordance enabling the recognition of compromised samples (Preece and Penkman 2005). The plot of Free to Hyd data from each sample can also be used as a relative timescale, with younger samples falling towards the bottom left corner of the graph and older samples falling towards the upper right corner, along the line of expected decomposition. The data from the Southfleet Road samples have been plotted in this way below for each of the amino acids, with crosshairs representing the data obtained from other MIS 5e, MIS 7, MIS 9 and MIS 11 sites from the UK with independent geochronology. During hydrolysis the Hyd vials of two of the samples from context 40062, <40042> cracked, and so no Hyd data is available for these two samples

From the majority of horizons, 2-3 individual opercula were analysed, but two horizons were analysed in more detail: Phase 3 (context 40070, <40162>) and Phase 6 (context 40070, < 40315/C>) to test the ability of the opercula samples to resolve between the different phases of the site.

The samples from context 40062, <40042>, were particularly friable, with several disintegrating during the initial rinsing step to clean the opercula. As the figures below show, four of the eleven samples analysed from this horizon showed much lower than expected THAA D/Ls, clear evidence of a compromised closed system. One of the context 40070, <40315/C> samples also showed this behaviour. In the analysis of nearly 500 opercula samples, less than 2% of opercula analysed were compromised, so it is extremely unusual to have so many from one horizon. The friability of the opercula from this horizon indicates that mineral diagenesis has occurred in at least some of these samples. The total data set is shown in the Free vs Hyd plots below, but the compromised samples are removed from the statistical analysis to avoid skewing the data.

Aspartic acid / Asparagine (Asx)

Asx is one of the fastest racemizing of the amino acids discussed here (due to the fact that it can racemise whilst still peptide bound; Collins *et al.* 1999). This enables good levels of resolution at younger age sites, but decreased resolution beyond MIS 7.

The D/L Asx data from Southfleet Road have very similar values to each other, and to those from sites correlated with MIS 9 and MIS 11 (Fig. A9.3). Other than the compromised samples from context 40062, sample 40042, there is one outlying sample from context

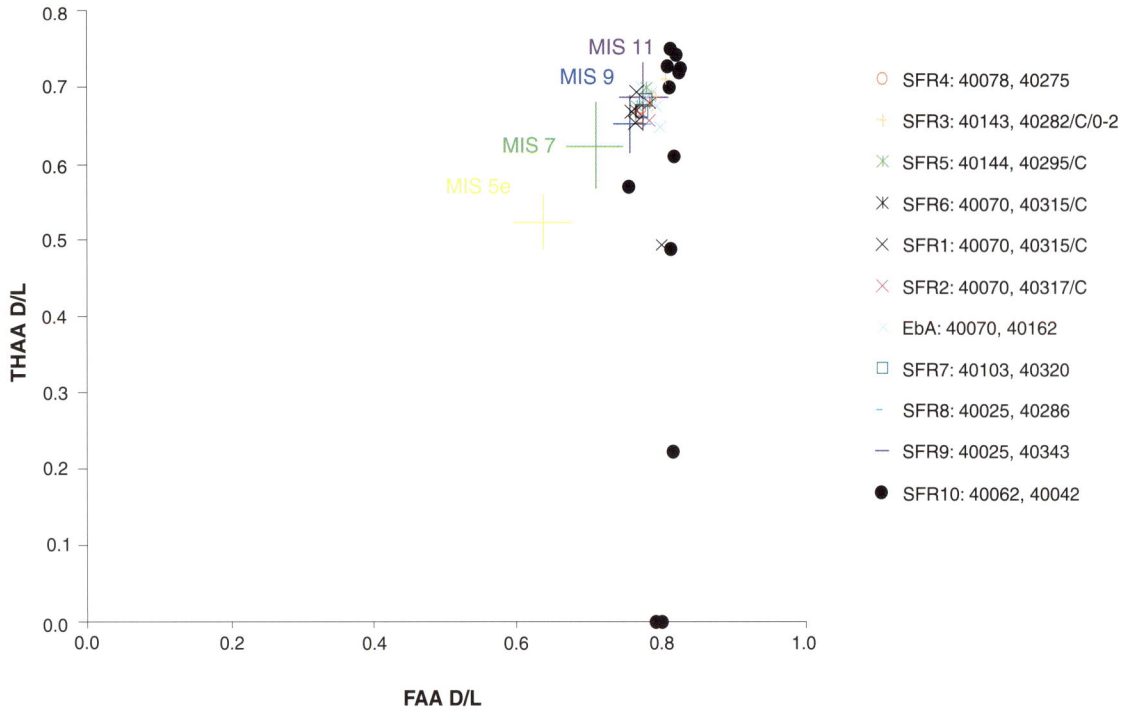

Figure A9.3 D/L Hyd vs D/L Free for Asx in *Bithynia tentaculata* opercula from Southfleet Road. The error bars represent two standard deviations about the mean for data obtained from opercula from sites correlated with MIS 5e, MIS 7, MIS 9 and MIS 11.

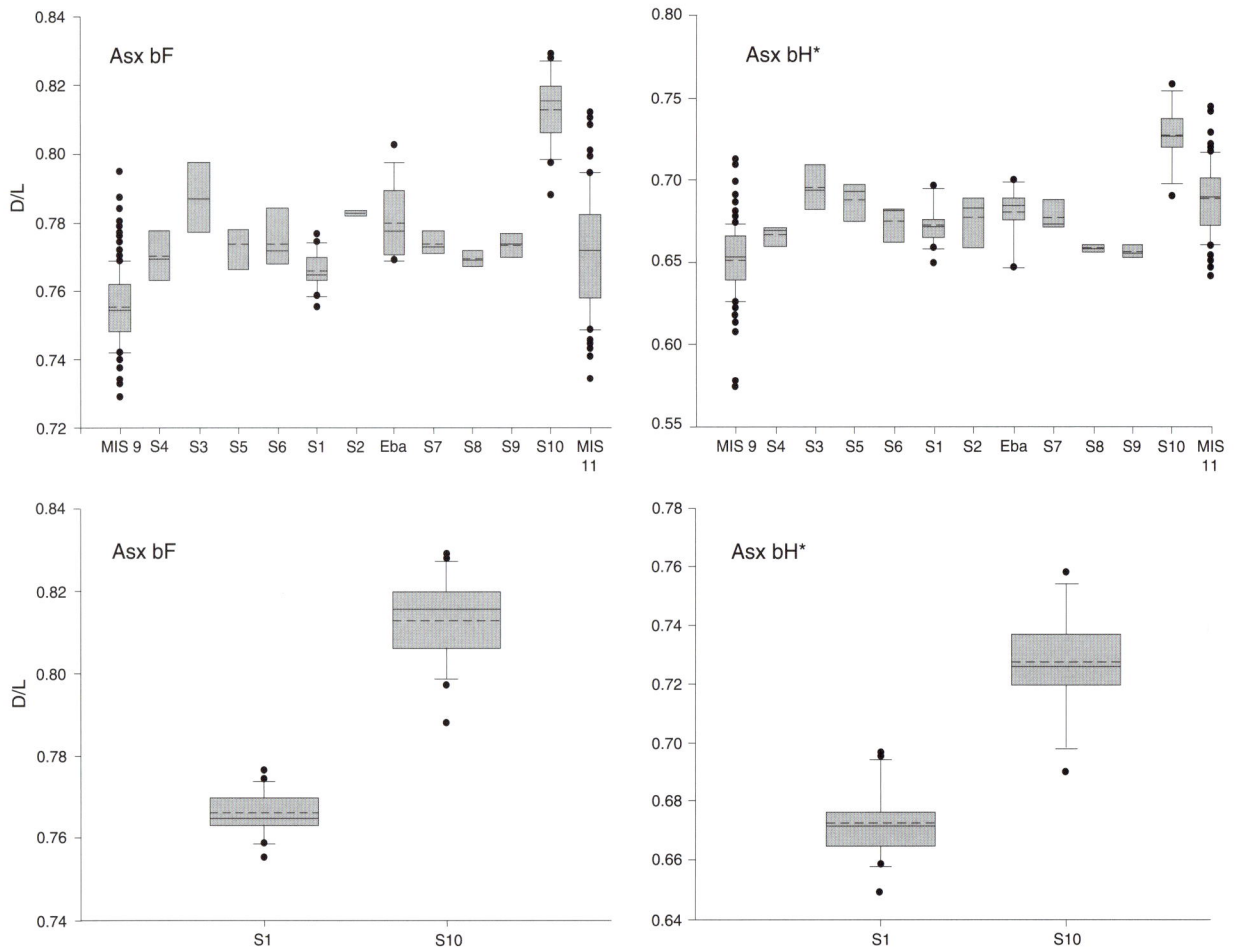

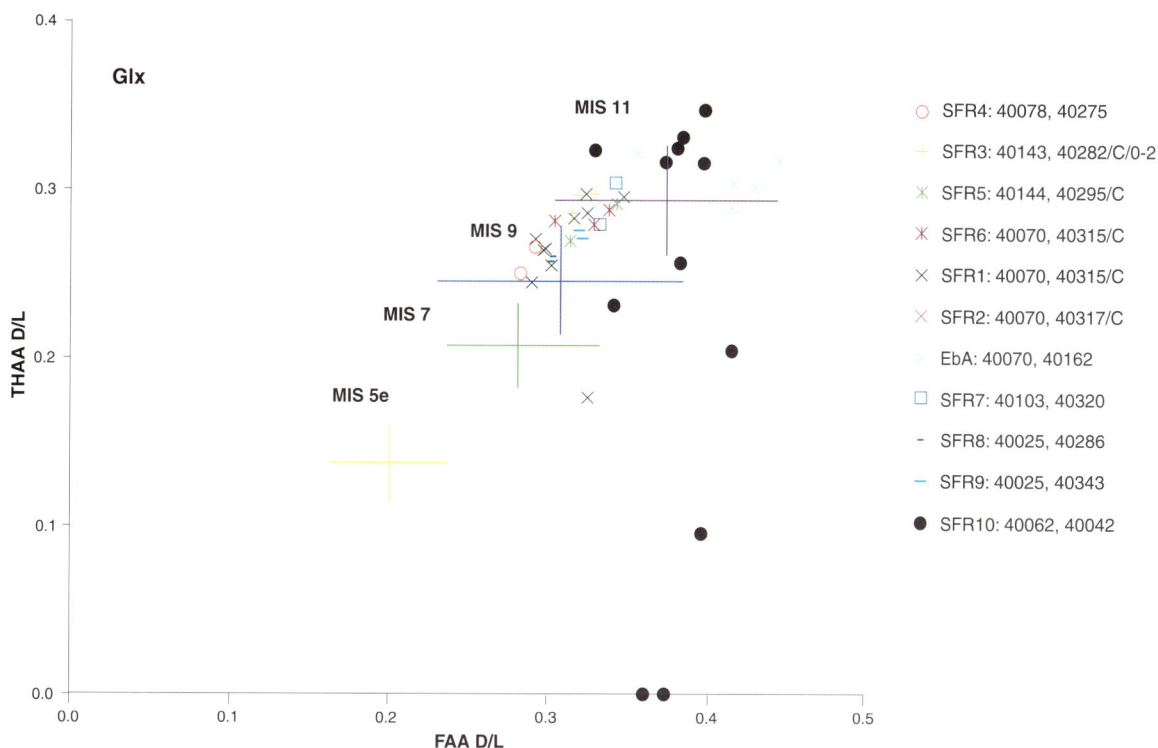

Figure A9.5 D/L Hyd vs D/L Free for Glx in *Bithynia tentaculata* opercula from Southfleet Road. The error bars represent two standard deviations about the mean for data obtained from opercula from sites correlated with MIS 5e, MIS 7, MIS 9 and MIS 11.

40070, <40315/C> (NEaar 6612, SFR1Bto10), which falls away from the expected line of decomposition in the Free vs Hyd plot. The amino acid composition of this sample was also divergent, and this sample does not therefore show closed system behaviour.

As the Asx racemization is so near equilibrium, it is surprising that a degree of behaviour consistent with the stratigraphic order is apparent, with the youngest (SFR4; context 40078, <40275>) and oldest (SFR10; context 40062, <40042>) distinguishable (Fig. A9.4). As can be seen from the lower figures, the samples from Phase 3 (context 40062, <40042>) are significantly higher than those from Phase 6 (context 40070, <40315/C>).

Glutamic Acid / Glutamine (Glx)

Glx is one of the slower racemizing amino acids discussed here and so the level of resolution from young sites is less than that seen with faster racemizing amino acids such as Asx. However, the low levels of racemization do help discriminate between material of Middle Pleistocene age. It is noteworthy that Glx has a slightly unusual pattern of racemization in the free form, due to the formation of a lactam (see Walton 1998). This results in difficulties in measuring Glx in the Free form, as the lactam cannot be derivitised and is therefore unavailable for analysis.

The Glx D/L values from Southfleet Road show values within the range of those expected from sites of MIS 9 and MIS 11 age (Fig. A9.5). Sample SFR1Bto10 also shows divergent behaviour in this amino acid.

While the Free Glx D/L is very variable in the EbA samples (context 40070, <40162>), this amino acid does show some increase in extent of racemization in concordance with the stratigraphy (Fig. A9.6), again enabling resolution between S1 (Phase 6) and S10 (Phase 3).

Figure A9.4 *(facing page)* Free (left) and Hyd (right) D/L for Asx in *Bithynia tentaculata* opercula from Southfleet Road, plotted in stratigraphic order. For each group, the base of the box indicates the 25th percentile. Within the box, the solid line plots the median and the dashed line shows the mean. The top of the box indicates the 75th percentile. Where more than 9 data points are available, the 10th and 90th percentiles can be calculated (shown by lines below and above the boxes respectively). The results of each duplicate analysis are included in order to provide a statistically significant sample size The lower figure shows the data for the two horizons studied in greater detail: S10 from Phase 3 (Context 40062, sample 40042) and S1 from Phase 6 (Context 40070, sample 40315/C). Note different scales on the y-axes.

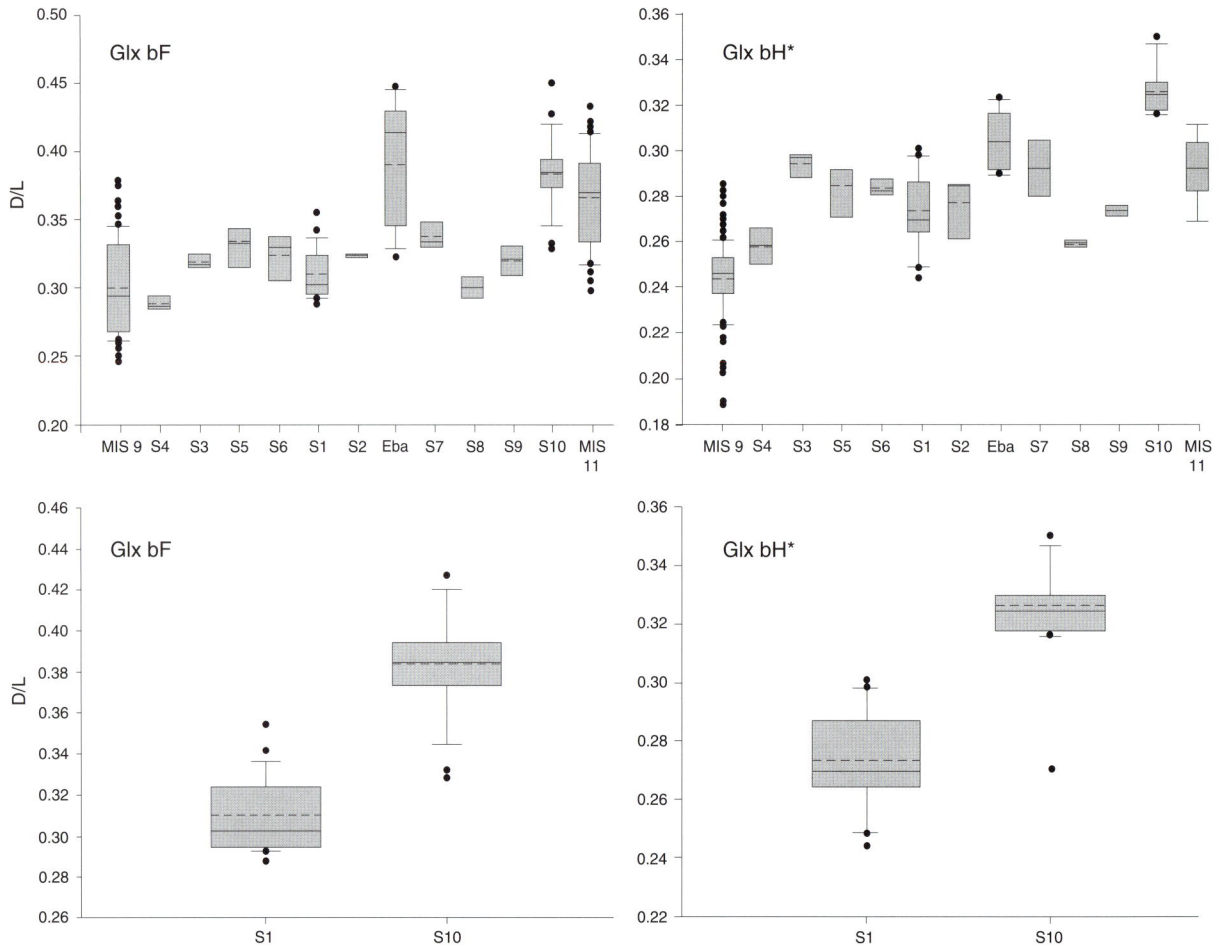

Figure A9.6 Free (left) and Hyd (right) D/L for Glx in *Bithynia tentaculata* opercula from Southfleet Road, plotted in stratigraphic order. For the full legend see Figure A9.4. Note different scales on y-axes

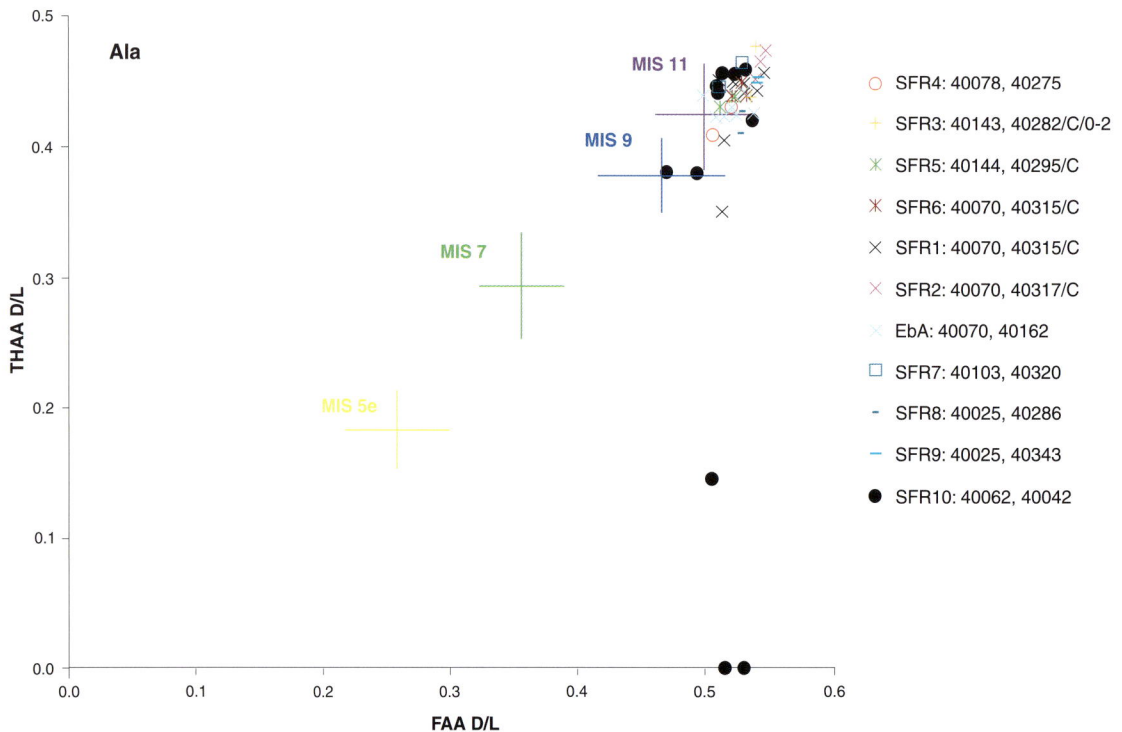

Figure A9.7 D/L Hyd vs D/L Free for Ala in *Bithynia tentaculata* opercula from Southfleet Road. The error bars represent two standard deviations about the mean for data obtained from opercula from sites correlated with MIS 5e, MIS 7, MIS 9 and MIS 11.

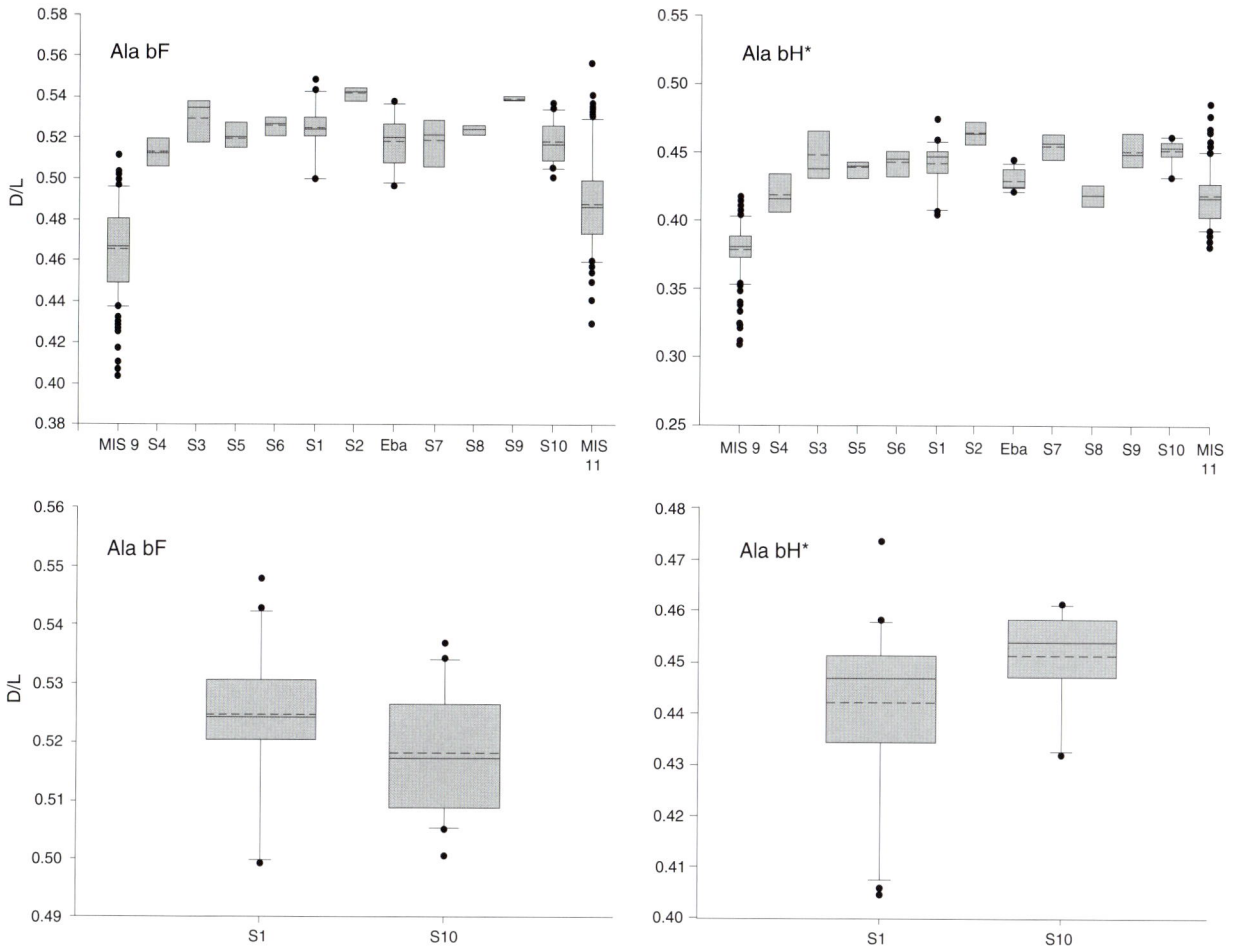

Figure A9.8 Free (left) and Hyd (right) D/L for Ala in *Bithynia tentaculata* opercula from Southfleet Road, plotted in stratigraphic order. Note different scales on y-axes

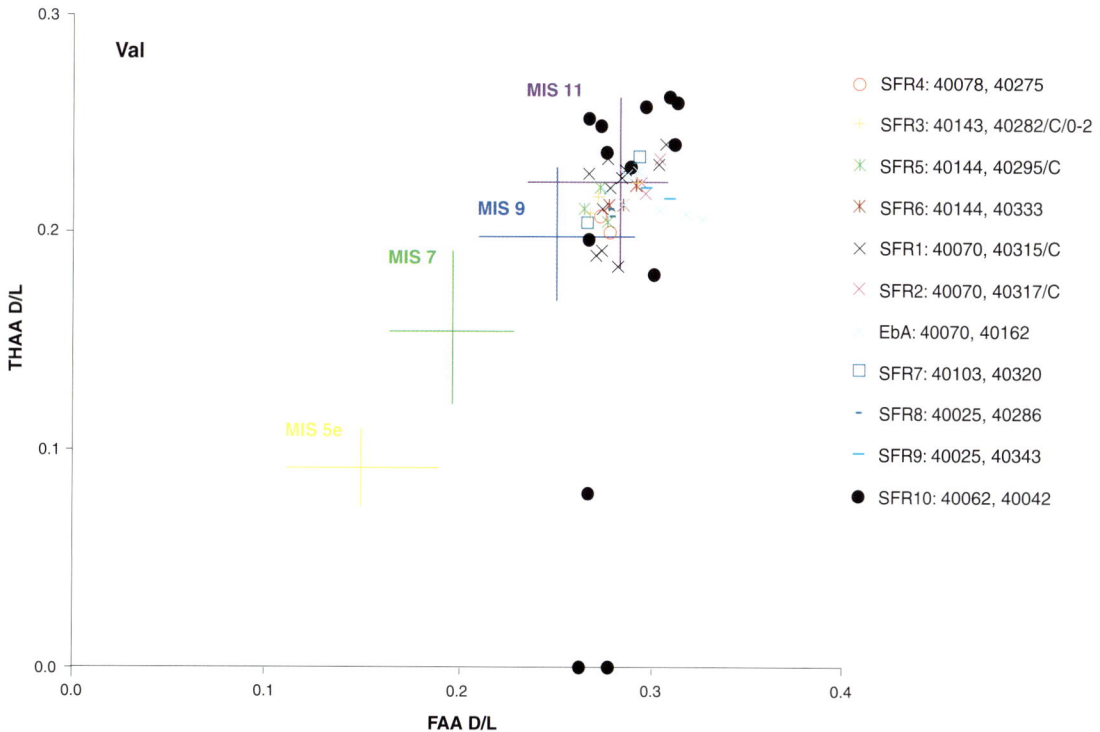

Figure A9.9 D/L Hyd vs D/L Free for Val in *Bithynia tentaculata* opercula from Southfleet Road. The error bars represent two standard deviations about the mean for data obtained from opercula from sites correlated with MIS 5e, MIS 7, MIS 9 and MIS 11

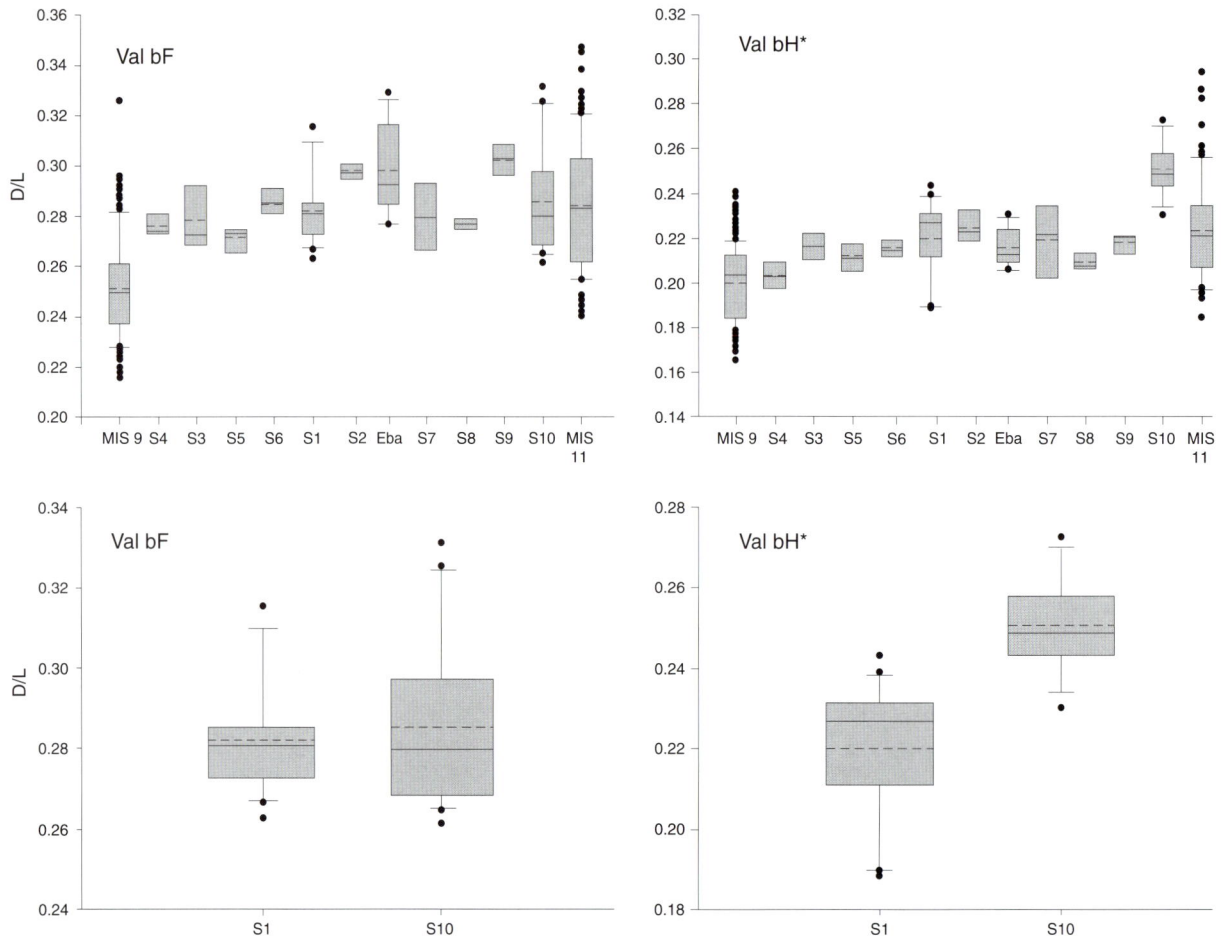

Figure A9.10 Free (left) and Hyd (right) D/L for Val in *Bithynia tentaculata* opercula from Southfleet Road, plotted in stratigraphic order. Note different scales on y-axes

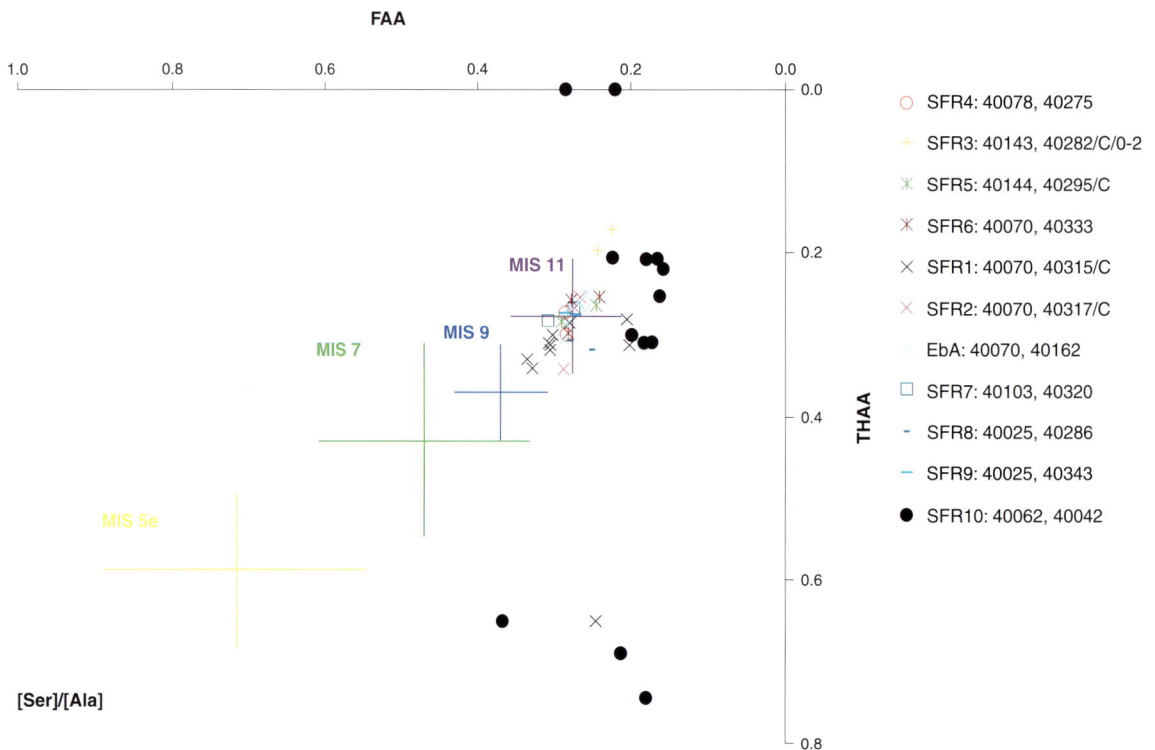

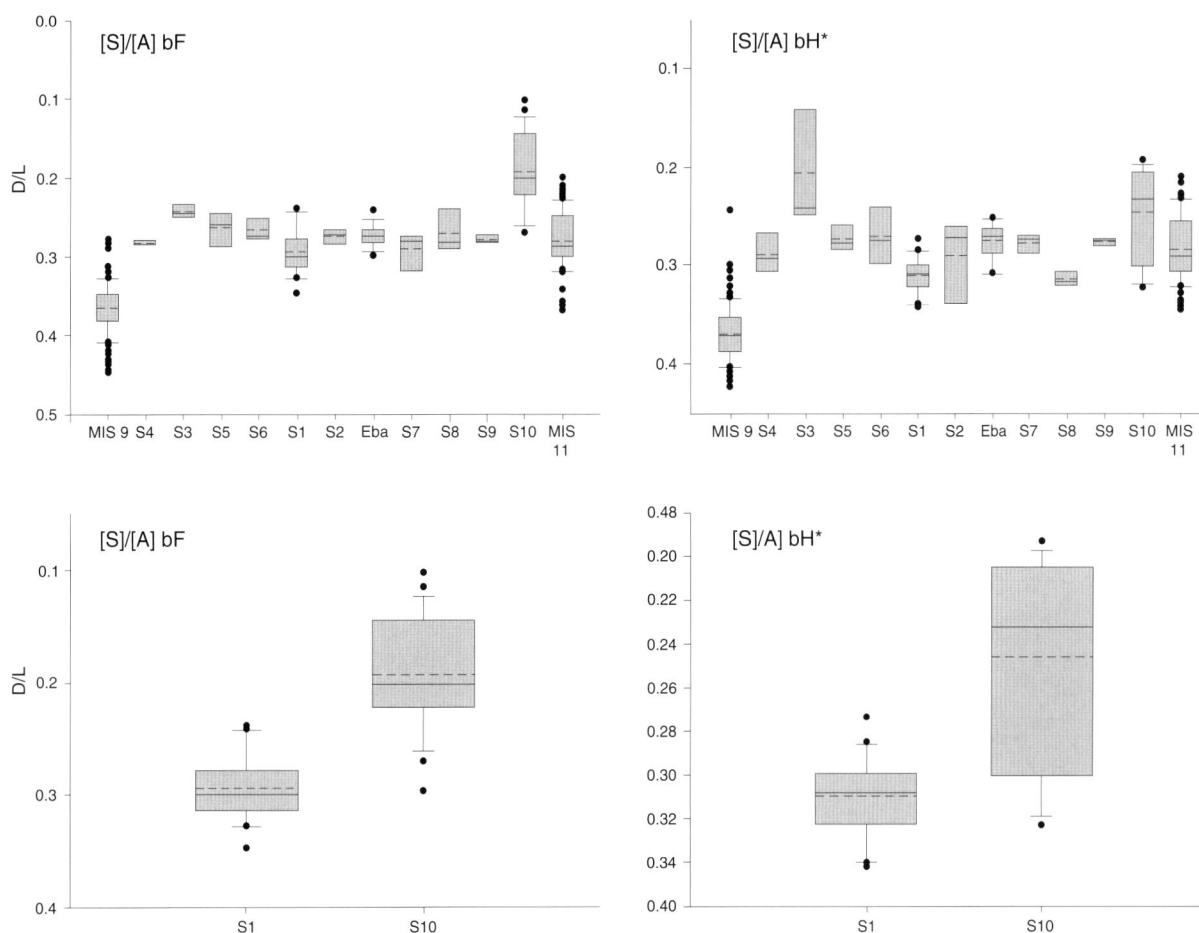

Figure A9.12 Free (left) and Hyd (right) [Ser]/[Ala] in *Bithynia tentaculata* opercula from Southfleet Road, plotted in stratigraphic order. As the [Ser]/[Ala] value decreases with increasing protein decomposition, the axes of this plot has been reversed so that the direction of protein decomposition is the same as that for the D/L graphs. Note different scales on y-axes

Alanine

Alanine (Ala) is a hydrophobic amino acid, whose concentration is partly contributed from the decomposition of other amino acids (notably serine). Ala racemises at an intermediate rate, so is one of the most useful amino acids for distinguishing samples at these timescales. The Ala data shows a tight clustering of data, consistent with a correlation with MIS 11 (Fig. A9.7) clearly enabling discrimination from sites of MIS 9 age. The two samples from S10 which fall within the MIS 9 cluster are those that show clear evidence in the other amino acids of being compromised. There is however little discrimination within the site using this amino acid (Fig. A9.8).

Valine (Val)

Valine has extremely low rates of racemization, and as the concentration of Val is quite low, the difficulty of measuring the D/L accurately results in higher variability. It does however still prove useful for age discrimination within material of Middle Pleistocene age. The Val D/L in the Free and the Hyd fractions again support the other amino acid data (Fig. A9.9). Only in the Hyd fracton is discrimination within the site possible using Val (Fig. A9.10).

[Serine]/[Alanine]

The ratio of the concentrations of serine and alanine provides an extremely useful tool for age estimation.

Figure A9.11 *(facing page)* [Ser]/[Ala] Hyd vs Free in *Bithynia tentaculata* opercula from Southfleet Road. The error bars represent two standard deviations about the mean for data obtained from opercula from sites correlated with MIS 5e, MIS 7, MIS 9 and MIS 11. As the [Ser]/[Ala] value decreases with increasing protein decomposition, the axes of this plot has been reversed so that the direction of protein decomposition is the same as that for the D/L graphs, with younger samples falling to the bottom left corner and older samples falling to the top right corner of the graph

Serine is a very unstable amino acid, and it can degrade via dehydration into alanine (Bada *et al.*, 1978). As the protein within a sample breaks down, the concentration of serine will decrease with an increase in the concentration of alanine, thus the [Ser]/[Ala] value will decrease with increasing time. In order to ease the interpretation, the y-axes in Figure 11 are plotted in reverse, so that the direction of increase in protein degradation is the same as for the racemization graphs.

The [Ser]/[Ala] of the Southfleet Road samples are again consistent with an age in MIS 11, but the level of discrimination is not particularly high (Fig. A9.11). The variability in the data precludes any definitive within-site stratigraphy from the amino acids alone, but the samples from context 40062, <40042> (SFR10) do show the highest levels of protein breakdown.

DISCUSSION

Comparison with other sites

The analysis of the closed system of protein within shells allows a new concept of age estimation to be developed, which incorporates multiple amino acid data to give a single measure of the overall extent of protein breakdown within a sample. This measurement, the Intra-crystalline Protein Degradation value (IcPD, formerly DMK) simplifies the presentation of the data to two compound values for each sample, one for the Free and one for the Hydrolysed amino acids. As these should be highly correlated, they can be cross-plotted, giving an aminostratigraphic framework with younger samples lying at low values and older samples with higher values, given a similar temperature history for all the sites. A study has been undertaken of interglacial sites within the UK that has allowed the tentative correlation of the aminostratigraphic framework to the marine oxygen isotope stage (MIS) record (Penkman 2005; Penkman *et al.* 2013).

On the basis of the relative D/L values and concentrations, the amino acid data from the opercula from Southfleet Road are very similar to the IcPD from UK sites correlated with MIS 11, including Hoxne, Barnham, Swanscombe, Beeches Pit, Elveden and Clacton.

CONCLUSIONS

In this study the amino acid data has been used as a relative dating technique to present an aminostratigraphy for the area in question. The conversion of relative sequences into absolute dates and accurate correlation between different areas is currently being undertaken, but preliminary correlations have been made to the MIS record.

The samples from Southfleet Road are consistent with an assignment within MIS 11. The samples from Phase 3 (context 40062, <40042>) show the highest

levels of protein breakdown within the sequence and are statistically distinguishable from those of the Phase 6 deposits. The samples from context 40078, <40275> (SFR4) generally shows the lowest levels of IcPD.

GLOSSARY

18MΩ water: The water has a resistivity of 18MΩ/cm, indicating a lack of ions.

HPLC grade water: In addition to low ion content, HPLC grade water has a low organic content (typically < 2 ppb).

Amino acids: the building blocks of proteins and consist of an alpha carbon atom (C) which has four different groups bonded to it: an amino group (-NH$_2$), a carboxyl group (-COOH), a hydrogen atom (-H), and a side chain, (often called an R group). About 20 amino acids normally occur in nature and some of these can undergo further modification (eg, the hydroxylation of proline to hydroxyproline). The amino acids are commonly known by three letter codes (see below, Abbreviations). They exist free in the cell, but are more commonly linked together by **peptide bonds** to form proteins, peptides, and sub-components of some other macromolecules (eg bacterial peptidoglycan).

Amino acid isomers: amino acids occur as two stereoisomers that are chemically identical, but optically different. These isomers are designated as either D (dextro-rotary) or L (laevo-rotary) depending upon whether they rotate plane polarised light to the right or left respectively (Fig 6). In living organisms the amino acids in protein are almost exclusively L and the D/L ratio approaches zero. Two amino acids, isoleucine and threonine, have two chiral carbon atoms and therefore have four stereoisomers each. As well as racemization, these two amino acids can undergo a process known as epimerization. The detection of the L-alloisoleucine epimer (derived from L-isoleucine) is possible by conventional ion-exchange chromatography, and was thus the most commonly used reaction pathway in geochronology.

Asx: Measurements of aspartic acid following hydrolysis also include asparagines, which decomposes to Asx. This combined signal of aspartic acid plus asparagine (Asp +Asn) is referred to as Asx (Collins *et al*, 1999).

D-amino acid: dextrorotary amino acid, formed following synthesis of the protein as it degrades over time (remember as 'dead amino acid').

IcPD: Conventional racemization analysis tends to report an allosioleucine / isoleucine (A/I or D/L ratio). This amino acid ratio has the advantage of being relative easy to measure and also sufficiently slow to be used to 'date' sediments in the European Quaternary.

Our IcPD approach utilises multiple amino acids. However we have avoided trying to give a whole series of D/L values for each amino acid in each sample. Instead we are using a theoretical model of protein degradation. The model outputs are then used to compare observed D/L values of any amino acid against the A/I value at the same stage of protein decomposition. The relative rate of racemization of any amino acid (its DL ratio) is then reported as an A/I equivalent – which as a working title we have named the Intra-crystalline Protein Degradation value (or IcPD) (Penkman, Collins and Kaufman in prep).

Instead of getting a single A/I ratio we obtain a series of (IcPD) values, currently IcPD $_{Asx}$, IcPD $_{Glu}$, IcPD $_{Phe}$, IcPD $_{Ala}$, IcPD $_{Val}$, and a (pretty unreliable) A/I ratio (IcPD $_{A/I}$ = A/I). Other ratios, notably IcPD $_{Ser}$, are not currently of implemented in the model – ie we don't have a good degradation model for this amino acid yet.

Because each amino acid has its own particular characteristics, only in a well behaved system will IcPD $_{Asx}$ = IcPD $_{Glu}$ = IcPD $_{Phe}$ = IcPD $_{Ala}$ = IcPD $_{Val}$ = A/I. If an amino acid has an unusually low ratio (due to modern contamination) or unusually high racemization (due to inclusion of bacterial cell wall contaminants) either some or all of the amino acids will no longer fit to the idealised degradation model. Indeed we can use elevation of IcPD $_{Asx}$ = IcPD $_{Glu}$ and = IcPD $_{Ala}$ to provide a bacterial contamination index. We have not done so in this case as there was no evidence of contamination.

IcPD values: Intra-crystalline Protein Degradation value, a summary value obtained from multiple amino acid D/L values from a single sample all normalised to a common model of protein degradation and racemization.

Enantiomers / optical isomers: mirror image forms of the same compound that cannot be superimposed on one another.

Epimerisation: the inversion of the chiral -carbon atom.

Free amino acid fraction: The fraction of amino acids directly amenable to racemization analysis. Only amino acids which have already been naturally hydrolysed (over time) are measured. These are the most highly racemised

Hydrolysis: A chemical reaction involving water leading to the breaking apart of a compound (in this case the breaking of peptide bonds to release amino acids).

L-amino acid: levorotary amino acid, the constituent form of proteins (remember as 'living amino acid').

Peptide bond: an amide linkage between the carboxyl group of one amino acid and the amino group of another.

Racemization: the inversion of all chiral carbon atoms, leading to the decrease in specific optical rotation. When the optical rotation is reduced to zero, the mixture is said to be racemised.

Stereoisomers: molecules of the same compound that have their atoms arranged differently in space.

Total amino acid fraction: The extent of racemization of all amino acids in a sample, determined following aggressive high temperature hydrolysis with strong mineral acid, which has the effect of breaking apart all peptide bonds so that the total extent of racemization in all amino acids both free and peptide bound are measured.

Zwitterion: A dipolar ion containing ionic groups of opposite charge. At neutral pH the ionic form of amino acids which predominates is the zwitterions

IcPD = Glx not alle / Ile?

Due to the problem of being unable to accurately measure A/I in our current system, we have switched to a version IcPD which is normalised for Glutamic acid. Although D/L Glu A/I, we have not yet fully established this relationship.

What does the date estimated from IcPD mean?

The date is our best estimate based upon the temperature history of the site. If we wanted to constrain this further we would need reliable independent dates. There are considerable differences in racemization rates between different molluscs. This reflects differences in rates of decomposition of proteins within the shell – the so-called species effects (Lajoie et al, 1980).

Past Use of Amino Acid Racemization Dating.

The presence of proteins in archaeological remains has been known for some time. Nearly fifty years ago Abelson (1954) separated amino acids from subfossil shell. He suggested the possibility of using the kinetics of the degradation of amino acids as the basis for a dating method (Abelson 1955). In 1967 Hare and Abelson measured the extent of racemization of amino acids extracted from modern and sub-fossil *Mercenaria mercenaria* shells (edible clam). They found that the total amount of amino acids present in shell decreased with the age of the shell. The amino acids in recent shell were all in the L configuration and over time the amount of D configuration amino acid increased (Hare and Abelson 1967). However, even after 35 years this method of dating is still subject to vigorous debate, with the application of AAR to date bone being particularly controversial (Bada 1990; Marshall 1990). Major reviews of AAR include: Johnson and Miller (1997), Hare, von Endt, and Kokis (1997), Rutter and Blackwell (1995), Murray-Wallace (1993), Bada (1991) and Schroeder and Bada (1976). Racemization is a chemical reaction and a number of factors influence its rate (Rutter and Blackwell, 1995). These include: amino acid structure, the sequence of amino acids in peptides, pH, buffering effects, metallic cations, the presence of water and temperature. To establish a dating method the kinetics and mechanisms of the racemization (and epimerization) reaction of free and peptide bound amino acids need to be established. To this end various workers in the late 1960s and the 1970s studied free amino acids in solution and carried out laboratory simulations of post mortem changes in the amino acids in bone (Bada 1972) and shell (Hare and Abelson 1967; Hare and Mitterer 1969). Attempts have also been made to relate the kinetics of free amino acids, with those in short polypeptides and the proteins in various archaeological samples (Smith and Evans 1980; Bada 1982).

The ability of this technique to be used as a geochronological and geothermometry tool has led to its use in many environmental studies. Goodfriend (1991; 1992) analysed terrestrial gastropods. Other studies have looked at bivalves (Goodfriend and Stanley 1996), foraminifera (Harada *et al* 1996), ostrich egg shells (Miller *et al.* 1992; 1997) and speleothems (Lauritzen 1994). Early methods of chemical separation, using Ion-Exchange liquid chromatography, are able to separate the enantiomers of one amino acid found in proteins, L-isoleucine (L-Ile, I), from its most stable diastereoisomer alloisoleucine (D-aile, A). By analysing the total protein content within non-marine mollusc shells from UK interglacial sites, an amino acid geochronology was developed using the increase in A/I, with correlations made with the marine oxygen isotope warm stages (Bowen *et al.* 1989).

Abbreviations used in this report

Abbrev	1-letter code	number of chiral centres	
Ala	A	1	Alanine
Arg	R	1	Arginine
Acn			acetonitrile
AA			Amino acid(n)
Asn	N	1	Asparagine
Asp	D	1	Aspartic acid
Asx			Asparagine + Aspartic acid + succinimide
Asu			Succinimide
Cys	C	1	Cysteine
DCM			Dichlormethane
GABA			γ-Aminobutyric acidγ
Gln	Q	1	Glutamine
Glu	E	1	Glutamic acid
Gly	G	0	Glycine
His	H	1	Histidine
HPLC			High-Performance Liquid Chromatography
Hyp			Hydroxyproline
IBD(L)C			N-Isobutyryl-D(L)-Cysteine
Ile	I	2	Isoleucine
Leu	L	1	Leucine
Lys	K	1	Lysine
MeOH			Methanol
Met	M	1	Methionine
Nle			Norleucine
OPA			ortho-Phthaldialdehyde
Orn			Ornithine
Phe	F	1	Phenylalanine
Pro	P	1	Proline
Ser	S	1	Serine
Thr	T	2	Threonine
Trp	W	1	Tryptophan
Tyr	Y	1	Tyrosine
Val	V	1	Valine

Appendix 10

The bird remains

by John R. Stewart

INTRODUCTION

The bird remains were identified by means of modern comparative material of the author. Avian anatomical description follows the terminology described in Baumel (1979). Stewart and Hernández Carrasquilla (1997) published a review of literature which aids in the identification of bird skeletal remains. Woolfenden (1961) has been consulted for the identification of the Anseriformes as a whole, Woelfle (1967) for ducks, Stewart (1992) for the Turdidae. Comparative metrical data used in the identifications and measurement protocols come from the above publications. The size categories follow Harrison and Stewart (1999).

DESCRIPTION OF THE SOUTHFLEET ROAD BIRD REMAINS

Context 40070

Sample <40162>
1. Os carpi ulnare
   Passeriformes (< Blackbird size)
2. 1. Os carpi ulnare
   Passeriformes (< House sparrow size)

Sample <40035>
1. Proximal (L) ulna
   Small bird – cf. Passeriformes (House sparrow size)

Sample <40307>
1. Phalange, proximal end with shaft, distal end damaged.
   Small bird – cf. Passeriformes (Blackbird – House sparrow size)

Sample <40314>
1. Proximal (R) ulna.
   Passeriformes (Blackbird size)
2. Coracoid (R) fragment – region of the procoracoid
   Passerine (Blackbird size)

Sample <40315>
1. Humerus (R) shaft
   Passeriformes (< Blackbird size)
2. Distal (L) tibiotarsus
   Small bird – cf. Passeriformes (< House sparrow size)

Sample <40317>
1. Proximal (L) carpometacarpus
   Passeriformes (House sparrow size)
2. Proximal (L) tarsometatarsus
   Passeriformes (House sparrow size)
3. Distal tibiotarsus (L/R?) – Immature as no articulation
   Small bird – cf. Passeriformes (< House sparrow size)

Sample <40318>
1. Distal (L) tibiotarsus
   Passeriformes (Blackbird size)

Sample <40329>
1. Tibiotarsus (L) shaft (broken into 2 fragments)
   Anatidae (Mallard size)
2. Scapula (R) articular end with ca. 2cm of corpus present (broken into 2 fragments)
   Anatidae (Mallard size)
3. Distal (R) humerus
   Passeriformes (House sparrow size)
4. Distal (L) humerus
   Passeriformes (House sparrow size)
5. Distal (L) tarsometatarsus
   Passeriformes (House sparrow size)
6. Distal (L) tarsometatarsus – trochlea missing
   Passeriformes (House sparrow size)

Sample <40033X>
1. Distal (R) ulna
   Small bird cf. Passeriformes (Blackbird size)
2. Phalange of pes
   Not Anatidae? (Mallard size)
3. Distal tibiotarsus (R)
   Small bird (House sparrow size)
4. Proximal (R) ulna
   Small bird cf. Passeriformes (House sparrow size)

Sample <40330>
1. Distal humerus (L)
   Passeriformes (Backbird size)
2. Distal (L) ulna
   Small bird – cf. Passeriformes (House sparrow size)
3. Thoracic vertebra
   Small bird (Blackbird size)
4. Possibly a bird distal carpometacarpus fragment

Sample <40331>
1. Proximal (L) humerus (Bp: 8.1 mm).
   Turdus cf. *philomelos / iliacus*. The morphological
   characters that can be used to distinguish *Turdus*
   from *Sturnus* (Stewart 1992, 2007) were applied.
2. Carpometacarpus (R) (GL: 18.0)
   Turdus / Sturnus. The morphological characters that
   can be used to distinguish *Turdus* from *Sturnus*
   (Stewart 1992; 2007) were not applied due to
   damage to proximal articulation. Based on modern
   measurements this would be *T. philomelos / iliacus*
   (Stewart 1992; 2007).
3. Distal (R) ulna
   Anatidae (Mallard size?)
4. Sternal extremity of coracoid (L)
   Small bird – cf. Passeriformes (Blackbird size)

Sample <40335>
1. Proximal ungual phalanx fragment – tip broken off

Small bird (House sparrow size)
2. Distal tibiotarsus (L/R?) – Immature as no
   articulation
   Small bird (House sparrow size)
3. Phalange of pes
   Medium sized bird?

Sample <40347>
1. Tarsometatarsus (L) shaft
   Anatidae (Mallard size) – immature as has grainy
   texture to shaft
2. Proximal (L) carpometacarpus
   Passeriformes (House sparrow size)

Sample< 40381>
1. Os carpi radiale (?)
   Anatidae (Mallard size)
2. Ungual phalanx
   Small bird – cf. Passeriformes (House sparrow size)

Bibliography

Note: Where journal abbreviations are used they follow the ISSN List of Title Word Abbreviations and (for archaeological publications) those published by the Council for British Archaeology (1991).

Abelson, P H, 1954 Amino acids in fossils, *Science* **119**, 576

Abelson, P H, 1955 *Organic Constituents of Fossils*, Carnegie Institution of Washington Year Book **54**, 107–9

Adam, von K D, 1951 Der Waldelefant von Lehringen, eine Jagdbeute des diluvialen Menschen, *Quartär* **5**, 79–92

Adamiec, G, and Aitken, M J, 1998 Dose-rate conversion factors: new data, *Ancient TL* **16**, 37-50

ADS, 2013 HS1 Ebbsfleet Elephant Digital Archive, Archaeology Data Service, University of York http://archaeologydataservice.ac.uk/archives/view/ebbsfleet_oa_2013/landing.cfm

Aitken, M J, 1985 *Thermoluminescence Dating*, Academic Press, London

Alba, D M, Carlos Calero, J A, Mancheno, M Á, Montoya, P, Morales, J, and Rook, L, 2011 Fossil remains of *Macaca sylvanus florentina* (Cocchi, 1872) (Primates, Cercopithecidae) from the Early Pleistocene of Quibas (Murcia, Spain), *J Hum Evol* **61**, 703–18

Allen, P, 1991 Deformation structures in British Pleistocene sediments, in *Glacial Deposits in Great Britain and Ireland* (eds J Ehlers, P L Gibbard and J Rose), 455-469, Balkema, Rotterdam

Andrews P, Biddulph E, Hardy H, and Brown, R, 2011a. *Settling the Ebbsfleet Valley, High Speed 1 Excavations at Springhead and Northfleet, Kent; the Late Iron Age, Roman, Saxon and Medieval Landscape – Volume 1: The Sites*, Oxford Wessex Archaeol Monogr **1**, Oxford and Salisbury

Andrews P, Mepham L, Schuster J, and Stevens C J, 2011b *Settling the Ebbsfleet Valley, CTRL Excavations at Springhead and Northfleet, Kent; the Late Iron Age, Roman, Saxon and Medieval Landscape – Volume 4: Post-Roman Finds and Environmental Reports*, Oxford Wessex Archaeol Monogr **4**, Oxford and Salisbury

Andrews, C W, 1928 *On a Specimen of Elephas antiquus from Upnor*, Brit Mus (Nat Hist), London

Andrews, P, 1990 *Owls, Caves and Fossils*, Brit Mus (Nat Hist), London

Andrews, P, and Cook, J, 1985 Natural modifications to bones in a temperate setting, *Man* **20**, 675–691

Antoine, P, Limondin-Lozouet, N, Auguste, P, Locht, J-L, Ghaleb, B, Reyss, J-L, Escude, E, Carbonel, P, Mercier, N, and Bahain, J-J, 2006 The Caours tufa (Somme, France): evidence from an Eemian sequence associated with a Palaeolithic settlement, *Quaternaire* **17**, 281–320

Anzidei, A P, Bulgarelli, G M, Catalano, P, Cerilli, E, Gallotti, R, Lemorini, C, Milli, S, Palombo, M R, Pantano, W, and Santucci, E, 2012 Ongoing research at the late Middle Pleistocene site of La Polledrara di Cecanibbio (central Italy), with emphasis on human-elephant relationships, *Quat Internat* **255**, 171-187

Ashton, N M, 1998 The technology of the flint assemblages, in Ashton *et al.* 1998, 205–235

Ashton, N M, and McNabb, J, 1994 Bifaces in perspective, in *Stories in Stone* (eds N Ashton and A David), 182–191, Lithic Stud Soc Occ Pap **4**, London

Ashton, N M, and McNabb, J, 1996 The flint industries from the Waechter excavations, in Conway *et al.* 1996, 201–236

Ashton, N M and Lewis, S G, 2002 Deserted Britain: declining populations in the British late Middle Pleistocene, *Antiquity* **76**, 388-396

Ashton, N M, Cook, J, Lewis, S G, and Rose, J (eds), 1992 *High Lodge: excavations by G. de G. Sieveking 1962–68 and J. Cook 1988*, Brit Mus Press, London

Ashton, N M, McNabb, J, Irving B, Lewis S, and Parfitt, S, 1994 Contemporaneity of Clactonian and Acheulian flint industries at Barnham, Suffolk, *Antiquity* **68**, 585–589

Ashton, N M, Lewis, S G, and Parfitt, S 1998 *Excavations at the Lower Palaeolithic site at East Farm, Barnham, Suffolk 1989-94*, Brit Mus Occ Pap **125**, London

Ashton, N, Lewis, S, Parfitt, S, Candy, I, Keen, D, Kemp, R, Penkman, K, Thomas, G, Whittaker, J, and White, M, 2005 Excavations at the Lower Palaeolithic site at Elveden, Suffolk, UK, *Proc Prehist Soc* **71**, 1–61

Ashton, N, Lewis, S G, Parfitt, S, and White, M, 2006 Riparian landscapes and human habitat preferences during the Hoxnian (MIS 11) Interglacial, *J Quat Sci* **21**, 497–505

Ashton N, Lewis S G, Parfitt S A, Penkman K E H, and Coope G, 2008 New evidence for complex climate change in MIS 11 from Hoxne, Suffolk, UK, *Quat Sci Rev* **27**, 652–668

Avery, B W, 1990 *Soils of the British Isles*, CAB International, Wallingford

Avery, B W, Stephen, I, Brown, G, and Yaalon, D H, 1959 The origin and development of brown earths on Clay-with-Flints and Coombe Deposits, *J Soil Sci* **10**, 177–195

Bada, J L, 1972 The dating of fossil bones using the racemization of isoleucine, *Earth Planet Sci Let* **15**, 223–31

Bada, J L, 1982 Racemization of amino acids in nature, *Interdiscip Sci Rev* **7** **(1)**, 30–46

Bada, J L, 1990 Racemization dating, *Science* **248**, 539–40

Bada, J L, 1991 Amino acid cosmogeochemistry, *Phil Trans Royal Soc, London* **B333**, 349–58

Bada J L, Shou M-Y, Man E H, and Schroeder R A, 1978 Decomposition of hydroxy amino acids in foraminiferal tests: Kinetics, mechanism and geochronological implications, *Earth Planet Sci Let* **41**, 67–76

Ball, D F, 1964 Loss-on-ignition as an estimate of organic matter and organic carbon in non-calcareous soils, *J Soil Sci* **15**, 84–92

Ballesio, R, 1980 Le gisement Pléistocène supérieur de la grotte de Jaurens á Nespouls, Corrèze, France: les Carnivores (Mammalia, Carnivora). II Felidae, *Nouvelles Archives du Muséum d'Histoire Naturelle de Lyon* **18**, 61–102

Banerjee, D, Murray, A S, Bøtter-Jensen, L, and Lang, A, 2001 Equivalent dose estimation using a single aliquot of polymineral fine grains, *Radiat Meas* **33**, 73–94

Bar Yosef O, and van Peer P, 2009 The *chaîne opératoire* approach in Middle Paleolithic archaeology, *Curr Anthropol* **50** (1), 103–131

Barnett C, McKinley J I, Stafford, E, Grimm, J M, and Stevens C J, 2011 *Settling the Ebbsfleet Valley, High Speed 1 Excavations at Springhead and Northfleet, Kent; the Late Iron Age, Roman, Saxon and Medieval Landscape – Volume 3: Late Iron Age to Roman human remains and environmental reports*, Oxford Wessex Archaeol Monogr **3**, Oxford and Salisbury

Bates, C D, Coxon, P, and Gibbard, P L, 1978 A new method for the preparation of clay-rich sediment samples for palynological investigation, *New Phytologist* **81**, 459–463

Bates, M R, 1998 Pleistocene deposits at Portfield Pit, Westhampnett East, Chichester, in *The Quaternary of Kent and Sussex. Field Guide* (eds J B Murton, C A Whiteman, M R Bates, D R Bridgland, A J Long, M B Roberts and M P Waller), 178–186, Quat Res Ass, London

Bates, M R, Bates, C R, Bates, S, Jones, S, Schwenninger, J-L, Walker, M J C, and Whittaker, J E, 2009 Cold stage deposits of the West Sussex Coastal Plain: the evidence from Warblington, Hampshire, in *The Quaternary of the Solent Basin and West Sussex Raised Beaches* (eds R M Brian, M R Bates, R T Hosfield and F F Wenban-Smith), 60–72, Quat Res Ass, London

Bates, M R, Wenban-Smith, F F, Bridgland, D R, Collins, M C, Keen, D H, Parfitt, S A, Rhodes, E, and Whittaker, J, in prep. Late Middle Pleistocene human occupation and palaeoenvironmental reconstruction at Harnham, Salisbury (UK), submitted to J Quat Sci

Battarbee, RW, 1986 Diatom analysis, in *Handbook of Holocene Palaeoecology and Palaeohydrology* (ed. B E Berglund), 527–571, J Wiley, Chichester

Battarbee, R W, Jones, V J, Flower, R J, Cameron, N G, Bennion, H B, Carvalho, L, and Juggins, S, 2001 Diatoms, in *Tracking Environmental Change Using Lake Sediments Volume 3: terrestrial, algal, and siliceous indicators* (eds J P Smol and H J B Birks), 155–202, Kluwer, Dordrecht

Baumel, J J (ed.), 1979 *Nomina Anatomica Avium*, Academic Press, London

Behrensmeyer, A K, 1993 The taphonomic record of faunal change in Amboseli Park, Kenya, *Research and Exploration* **9**, 402–421

Behrensmeyer, A K, Gordon, K, and Yanagi, G, 1986 Trampling as a cause of bone surface damage and pseudo-cut marks, *Nature* **319**, 768–771

Bello, S M, 2011 New results from the examination of cut-marks using 3-Dimensional imaging, in *The Ancient Human Occupation of Britain* (eds N M Ashton, S G Lewis and C B Stringer), 249–262, Developments in Quaternary Science **14**, Elsevier, Amsterdam

Bello, S M, and Soligo C, 2008 A new method for the quantitative analysis of cutmark micromorphology, *J Archaeol Sci* **35**, 1542–1552

Bello, S M, Parfitt, S A, and Stringer, C, 2009 Quantitative micromorphological analyses of cut marks produced by ancient and modern handaxes, *J Archaeol Sci* **36**, 1869–1880

Benninghof, W S, 1962 Calculation of pollen and spores density in sediments by addition of exotic pollen in known quantities, *Pollen et Spores* **4**, 332–333

Berthelet, A, and Chavaillon, J, 2001 The early Palaeolithic butchery sites of Barogali (Republik of Djibouti), in *The World of Elephants: proceedings of the 1st International Congress* (eds G. Caverretta, P. Gioia, M. Mussi and M.R. Palomba), 587–591, Consiglio Nazionale delle Richerce, Rome

Biddulph, E, Smith, R S, and Schuster J, 2011 *Settling the Ebbsfleet Valley, High Speed 1 Excavations at Springhead and Northfleet, Kent; the Late Iron Age, Roman, Saxon and Medieval Landscape – Volume 2: Late Iron Age to Roman find reports*, Oxford Wessex Archaeol Monogr **2**, Oxford and Salisbury

Billia, E M E, 2011 Occurrences of *Stephanorhinus kirchbergensis* (Jäeger, 1839) (Mammalia, Rhinocerotidae) in Eurasia – An Account, *Acta Palaeontologica Romaniae* **7**, 17–40

Binford, L R, 1973 Interassemblage variability – the Mousterian and the 'functional' argument, in *The Explanation of Culture Change: models in prehistory* (ed. C. Renfrew), 227–254, Duckworth, London

Binford, L R, 1981 *Bones: ancient men and modern myths*, Academic Press, New York

Binford, L R, 1985 Human ancestors: changing views of their behavior, *J Anthropol Archaeol* **4**, 292–327

Binford, L R, 1987a Were there hunters at Torralba? in *The Evolution of Human Hunting* (eds M.H.

Nitecki, and D.V. Nitecki), 47–105, Plenum Press, New York

Binford, L R, 1987b Searching for camps and missing the evidence? Another look at the Lower\Middle Palaeolithic, in *The Pleistocene Old World: regional perspectives* (ed. O Soffer), 17–31, Plenum Press, New York

Binford, L R and Binford, S R, 1966 A preliminary analysis of functional variability in the Mousterian of Levallois facies, *Am Anthropol* **68 (2)**, 238–295

Birks, H J B, 1973, *Past and Present Vegetation of the Isle of Skye: a palaeoecological study*, CUP, Cambridge

Birks, H J B, and Birks, H H, 1980 *Quaternary Palaeoecology*, Edward Arnold, London

Bishop, M J, 1982 The mammal fauna of the early Middle Pleistocene cavern infill site of Westbury-sub-Mendip, Somerset, *Spec Pap Palaeontol* **28**, 1–108

Bjärväll, A, and Ullström, S, 1986 *The Mammals of Britain and Europe*, Croom Helm, London

Blumenschine, R J, and Selvaggio, M M, 1988 Percussion marks on bone surfaces as a new diagnostic of homind behaviour, *Nature* **333**, 763–765

Boismier, W A, Gamble, C S, and Coward, F, 2012 *Neanderthals among Mammoths: excavations at Lynford Quarry, Norfolk*, English Heritage, Swindon

Booth, P, Champion, T, Foreman, S, Garwood, P, Glass, E, Munby, J, and Reynolds, A, 2011 *On Track: the archaeology of High Speed 1 Section 1 in Kent*, Oxford Wessex Archaeol Monogr **4**, Oxford

Bordes, F, 1979 *Typologie du Paléolithique Ancien et Moyen*, (3rd edition), Centre Nationale de Récherche Scientifique, Paris

Bordes, F, 1981 Vingt-cinq ans après: le complexe Moustérien revisité, *Bulletin de la Société Préhistorique Française* **78**, 77–87

Boschian, G, and Saccà, D, 2009 Ambiguities in human and elephant interactions? Stories of bones, sand and water from Castel di Guido (Italy), *Quat Internat* **214**, 3–16

Bøtter-Jensen, L, 1988 The automated Riso TL dating reader system, *Nuclear Tracks and Radiat Meas* **14**, 177–180

Bøtter-Jensen, L, 1997 Luminescence techniques: instrumentation and methods, *Radiat Meas* **27**, 749–768

Bøtter-Jensen, L, Bulur, E, Duller, G A T, Murray, A S, 2000 Advances in luminescence instrument systems, *Radiat Meas* **32**, 523–528

Boulestin, B, 1999 *Approche Taphonomique des Restes Humaines. le cas des Mésolithiques de la Grotte des Perrats et le problème du cannibalisme en préhistoire récente européenne*, Brit Archaeol Rep (Int Ser) **776**, Oxford

Bouma, J, Fox, C A, and Miedema, R, 1990 Micromorphology of hydromorphic soils: applications for soil genesis and land evaluation, in *Soil Micromorphology: A basic and applied science* (ed. L A Douglas), 257–278, Volume Developments in Soil Science **19**, Elsevier, Amsterdam

Bowen, D Q (ed.), 1999 *A Revised Correlation of Quaternary Deposits in the British Isles*, The Geol Soc, London, Spec Rep **23**

Bowen, D Q, and Sykes G A, 1988 Correlation of marine events and glaciations on the Northeast Atlantic Margin, *Phil Trans Royal Soc, London* **B318**, 619–35

Bowen, D Q, Hughes, S, Sykes, G A, and Miller, G H, 1989 Land-sea correlations in the Pleistocene based on isoleucine epimerization in non-marine molluscs, *Nature* **340**, 49–51

Bradley, B, and Sampson, C G, 1986 Analysis by replication of two Acheulian artefact assemblages from Caddington, England, in *Stone Age Prehistory: studies in memory of Charles McBurney* (eds G N Bailey and P Callow), 29–45, CUP, Cambridge

Branch, N P, 2000 Pollen preparation procedure, unpubl document, Dept Geography, Royal Holloway Univ London

Breda, M, Collinge, S, Parfitt, S A, and Lister, A M, 2010 Metric analysis of ungulate mammals in the early Middle Pleistocene of Britain, in relation to taxonomy and biostratigraphy. I: Rhinocerotidae and Bovidae, in *The West Runton Freshwater Bed and the West Runton Mammoth* (eds A J Stuart and A M Lister), *Quat Internat* **228**, 136–156

Breuil, H, 1926 Palaeolithic industries from the beginning of the Rissian to the beginning of the Würmian Glaciation, *Man* **116**, 176–179

Breuil, H, 1932 Les industries à éclats du Paléolithique ancien: I. Le Clactonien, *Préhistoire* **1**, 125–190

Breuil, H, and Koslowski, L, 1931 Etudes de stratigraphie Paléolithique dans le nord de la France, la Belgique et l'Angleterre: la vallée de la Somme, *L'Anthropologie* **41**, 449–488

Breuil, H, and Koslowski, L, 1932 Etudes de stratigraphie Paléolithique dans le nord de la France, la Belgique et l'Angleterre: V – Basse terrasse de la Somme, *L'Anthropologie* **42**, 27–47, 291–314

Breuil, H, and Koslowski, L, 1934 Etudes de stratigraphie Paléolithique dans le nord de la France, la Belgique et l'Angleterre: La Belgique, *L'Anthropologie* **44**, 249–290

Bridgland, D R, 1986 *Clast Lithological Analysis*, Tech Guide 3, Quat Res Assoc, Cambridge

Bridgland, D R, 1988 The Pleistocene fluvial stratigraphy and palaeogeography of Essex, *Proc Geol Ass* **99**, 291-314

Bridgland, D R, 1994 *Quaternary of the Thames*, Chapman and Hall, London

Bridgland, D R, 1995 The Quaternary sequence of the eastern Thames basin: problems of correlation, in *The Quaternary of the Lower Reaches of the Thames. Field Guide* (eds D R Bridgland, P Allen and B A Haggart), 35-52, Quat Res Ass, Cambridge

Bridgland, D R, 1999 Analysis of the raised beach gravel deposits at Boxgrove and related sites, in *A Middle Pleistocene Hominid Site at Eartham Quarry, Boxgrove, West Sussex, UK* (M B Roberts and S A Parfitt), 100–110, Engl Herit Archaeol Rep **17**, London

Bridgland, D R, 2000 River terrace systems in north-west Europe: an archive of environmental change, uplift and early Human occupation, *Quat Sci Rev* **19**, 1293–1303

Bridgland, D R, 2001 The Pleistocene evolution and Palaeolithic occupation of the Solent River, in *Palaeolithic Archaeology of the Solent River* (eds F F Wenban-Smith, and R T Hosfield), 15–25, Lithic Stud Soc Occ Pap **7**

Bridgland, D R, 2006 The Middle and Upper Pleistocene sequences in the Lower Thames: a record of Milankovitch climatic fluctuation and early human occupation of southern Britain, *Proc Geol Ass* **117**, 281-306

Bridgland, D R, and Harding, P, 1993 Middle Pleistocene Thames terrace deposits at Globe Pit, Little Thurrock, and their contained Clactonian industry, *Proc Geol Ass* **104**, 263–283

Bridgland, D R, and D'Olier, B, 1995 The Pleistocene evolution of the Thames and Rhine drainage systems in the southern North Sea Basin, in *Island Britain, a Quaternary Perspective* (ed. R C Preece), 27–45, Geol Soc Spec Publ **96**, The Geol Soc, London

Bridgland, D R, and Maddy, D, 2002 Global correlation of long Quaternary fluvial sequences; a review of baseline knowledge and possible methods and criteria for establishing a database, *Netherlands J Geosciences* **81**, 265–281

Bridgland, D R, Lewis, S G. and Wymer, J J, 1995 Middle Pleistocene stratigraphy and archaeology around Mildenhall and Icklingham, Suffolk: report on the Geologists' Association Field Meeting, 27 June 1992, *Proc Geol Ass* **106**, 57–69

Bridgland, D R, Field, M H, Holmes, J A, McNabb, J, Preece, R.C, Selby, I, Wymer, J J, Boreham, S, Irving, B G, Parfitt, S, Peglar, S and Stuart, A J 1999, Middle Pleistocene interglacial Thames-Medway deposits at Clacton-on-Sea, England: Reconsideration of the biostratigraphical and environmental context of the type Clactonian Palaeolithic industry, *Quat Sci Rev* **18**, 109–146

Bridgland, D R, Harding, P, Allen, P, Candy, I, Cherry, C, George, W, Horne, D, Keen, D H, Penkman, K E H, Preece, R C, Rhodes, E J, Scaife, R, Schreve, D C, Schwenninger, J.-L, Slipper, I, Ward, G, White, M J, White, T S, Whittaker, J E, 2012 An enhanced record of MIS 9 environments, geochronology and geoarchaeology: data from construction of the High Speed 1 (London – Channel Tunnel) rail-link and other recent investigations at Purfleet, Essex, UK, *Proc Geol Ass*, http://dx.doi.org/10.1016/j.pgeola.2012.03.006

British Geological Survey, 1977 *Dartford: England and Wales Sheet 271, Drift Geology,* 1:50,000 Series, Geol Survey of Great Britain, Southampton

British Geological Survey, 1998 *Dartford: England and Wales Sheet 271, Solid and Drift Geology,* 1:50,000 Series, Keyworth, Nottingham

Bromage, T G, Boyde, A, 1984 Microscopic criteria for the determination of directionality of cutmarks on bone, *Am J Phys Anthropol* **65**, 359–366

Brooks, D and Thomas, K W, 1967 The distribution of pollen grains on microscope slides. The non randomness of the distribution, *Pollen and Spores* **9**, 621–629.

Bull, C R, 1990 A Glimpse of Pre-Industrial Swanscombe, *Proc Dartford Hist and Antiq Soc* **27**, 4–13

Bullock, P, Fedoroff, N, Jongerius, A, Stoops, G, and Tursina, T, 1985 *Handbook for Soil Thin Section Description*, Waine Res Publs, Wolverhampton

Burchell, J P T, 1935 Some Pleistocene deposits at Kirmington and Crayford, *Geol Mag* **72**, 327–331

Burchell, J P T, 1957 A temperate bed of the last interglacial period at Northfleet, Kent. *Geol Mag* **94**, 212–214

Burger, J, Rosendahl, W, Loreille, O, Hemmer, H, Eriksson, T, Gotherstrom, A, Hiller, J, Collins, M. J, Wess, T. and Alt, K W, 2004 Molecular phylogeny of the extinct cave lion *Panthera leo spelaea. Mol Phylogenet Evol* **30**, 841–49

Byers, D A, 2002 Taphonomic analysis, associational integrity, and depositional history of the Fetterman Mammoth, Eastern Wyoming, U.S.A, *Geoarchaeology* **17**, 417–440

Candy, I., Coope, G R, Lee, J.R, Parfitt, S A, Preece, R C, Rose, J, and Schreve, D C, 2010 Pronounced warmth during early Middle Pleistocene interglacials: Investigating the Mid-Brunhes Event in the British Pleistocene sequence, *Earth Sci Rev* **103**, 183–196

Carreck, J N, 1972 Chronology of the Quaternary deposits of south-east England, with special reference to their vertebrate faunas, unpubl M. Phil thesis, Univ London

Cassoli P F, Fabbo G, Fiore I, Piperno M and Tagliacozzo A, 1999 Taphonomic analysis of some levels of the Lower Palaeolithic site of Notarchirico (Venosa, Basilicata, Italy), in *The Role of Early Humans in the Accumulation of European Lower and Middle Palaeolithic Bone Assemblages* (eds S Gaudzinski and E Turner), 153–173, Monographien des Römisch-Germanischen Zentralmuseums Mainz, Habelt, Mainz and Bonn

CgMs Consulting and Wenban-Smith, F F 2003 Springhead Quarter, Ebbsfleet Development: Palaeolithic/Pleistocene observations during geotechnical site investigations, unpubl client rep submitted to Kent Co Council

Chaplin, R E, 1971 *The Study of Animal Bones from Archaeological Sites*, Seminar Press, London and New York

Clark, J D and Haynes, C V, 1970 An elephant butchery site at Mwanganda's Village, Karonga, Malawi, and its relevance for Palaeolithic archaeology, *World Archaeol* **1**, 390–411

Coe, M, 1978 The decomposition of elephant carcasses in the Tsavo (East) National Park, Kenya, *J Arid Environ* **1**, 71–86

Coles, B, 2010 The European Beaver, in *Extinctions and*

Invasions. A Social History of British Fauna (eds T
O'Connor and N Sykes), 104–115, Windgather
Press, Oxford

Collins, M J, Waite, E R, and van Duin, E C T, 1999
Predicting protein decomposition: the case of
aspartic-acid racemization kinetics, *Phil Trans Royal
Soc London* **B354 (1379)**, 51–64

Conway, B W, McNabb, J, and Ashton, N, (eds) 1996
Excavations at Barnfield Pit, Swanscombe, 1968–72,
Brit Mus Occ Pap **94**, London

Conybeare, A, and Haynes, G, 1984 Observations on
elephant mortality and bones in water holes, *Quat
Res* **22**, 189–200

Cook J, Stringer, C B, Currant, A P, Schwarcz, H, and
Wintle, A G, 1982 A review of the chronology of the
European Middle Pleistocene hominid record,
Yearbook Physical Anthropol **25**, 19–65

Council for British Archaeology, 1991 *Signposts for
Archaeological Publication* (3rd edition), CBA, York

Corbet, G B, and Southern, H N (eds), 1977 *The
Handbook of British Mammals*, Blackwell, Oxford

Courty, M A, 2001 Microfacies analysis assisting
archaeological stratigraphy, in *Earth Sciences and
Archaeology* (eds P Goldberg, V T Holliday and C R
Ferring), 205–239, Kluwer, New York

Courty, M A, Goldberg, P, and Macphail, R I, 1989
Soils and Micromorphology in Archaeology, CUP,
Cambridge

Crowther, J, 2003 Potential magnetic susceptibility and
fractional conversion studies of archaeological soils
and sediments, *Archaeometry* **45**, 685–701

Crowther, J and Barker, P, 1995 Magnetic susceptibility: distinguishing anthropogenic effects from the
natural, *Archaeol Prospect* **2**, 207–215

Cruise, G M, Macphail, R I, Linderholm, J, Maggi, R,
and Marshall, P D, 2009 Lago di Bargone, Liguria,
N. Italy: a reconstruction of Holocene environmental and land-use history, *The Holocene* **19**,
987–1003

Cruse, R J, 1987 Further investigation of the Acheulian
site at Cuxton, *Archaeol Cantiana* **104**, 39–81

Cuenca-Bescós, G, Rofes, J and Garcia-Pimienta, J,
2005 *Environmental change across the Early-Middle
Pleistocene transition: small mammalian evidence from
the Trinchera Dolina cave, Atapuerca, Spain*, Geol Soc
Spec Publ **247**, 277–286

Currant, A P, 1986a The Lateglacial mammal fauna of
Gough's Cave, Cheddar, Somerset, *Proc Univ Bristol
Spelaeological Soc* **17**, 286–304

Currant, A P, 1986b Interim report on the small
mammal remains, in Excavation of a Lower
Palaeolithic site at Amey's Eartham Pit, Boxgrove,
West Sussex: a preliminary report (M B Roberts),
Proc Prehist Soc **52**, 215–245

Currant, A P, 1996 Notes on the mammalian remains
from Barnfield Pit, Swanscombe, in B Conway *et al.*
1996, 163–167

Davies, P, 2002 The straight-tusked elephant
(*Palaeoloxodon antiquus*) in Pleistocene Europe,

unpubl dissertation, Univ London

Davis, S M, 1992 *A Rapid Method for Recording
Information about Mammal Bones from Archaeological
Sites*, Ancient Monuments Lab Rep 19/92, English
Heritage, London

Delagnes A, Lenoble A, Harmand S, Brugal J-P, Prat
S, Tiercelin J-J and Roche H, 2006 Interpreting
pachyderm single carcass sites in the African Lower
and Early Middle Pleistocene record: a multi-
disciplinary approach to the site of Nadung'a 4
(Kenya), *J Anthropol Archaeol* **25**, 448–465

Delson, E, 1980 *Fossil Macaques*. Phyletic relationships
and a scenario of development, in *The Macaques:
Studies in Ecology, Behaviour and Evolution* (ed. D
Lindburg), 10–30, Van Nostrand Reinhold, New
York

Dennell W R, Martinón-Torres M, and Bermúdez de
Castro, J M, 2011 Hominin variability, climatic
instability and population demography in Middle
Pleistocene Europe, *Quat Sci Rev* **30**, 1511-1524

Department of the Environment, 1990 *Planning Policy
Guidance: Archaeology and Planning, PPG 16*,
HMSO, London

Dewey, H, 1932 The Palaeolithic deposits of the Lower
Thames Valley, *Q J Geol Soc London* **88**, 36–56

Dewey, H, Bromehead, C E N, Chatwin C P and
Dines, H G, 1924 *The Geology of the Country around
Dartford: explanation of sheet 271*, Memoirs of the
Geological Survey of Great Britain

Dibble, H L and Rolland, N, 1992 On assemblage
variability in the Middle Palaeolithic of western
Europe, in *The Middle Palaeolithic: adaptation,
behaviour and variability* (eds H L Dibble and P A
Mellars), 1–27, Univ Mus Press, Philadelphia

Dick, W A, and Tabatabai, M A, 1977 An alkaline
oxidation method for the determination of total
phosphorus in soils, *J Soil Sci Soc America* **41**, 511-14

Domínguez-Rodrigo, M, de Juana, S, Galán, A B, and
Rodríguez, M, 2009 A new protocol to differentiate
trampling marks from butchery cut marks, *J Archaeol
Sci* **36**, 2643–2654

Donnard, E, 1981 *Oryctolagus cuniculus* dans quelques
gisements quaternaries Française, *Quaternairia* **23**,
145–57

Drewett, P L, 1997 The Channel Tunnel Rail Link:
archaeological research strategy, unpubl rep for Rail
Link Engineering

Driesch, A von den, 1974 *A Guide to the Measurement of
Animal Bones from Archaeological Sites*, Peabody
Museum Bulletin **1**, Harvard Univ, Cambridge, Mass

Duchaufour, P, 1982 *Pedology*, Allen and Unwin, London

Edmonds, C N, 1988 The engineering geology of karst
development and the prediction of subsidence risk
upon the Chalk outcrop in England, unpubl Ph.D.
dissertation, Royal Holloway and Bedford New
College, Univ London

Edmonds, C N, 1983 Towards the prediction of
subsidence risk upon the Chalk outcrop, *Q J Eng
Geol* **16**, 261-266

English Heritage, 1991 *Exploring our Past: Strategies for the Archaeology of England*, HBMC England, London

English Heritage, 1998 *Identifying and Protecting Palaeolithic Remains: Archaeological Guidance for Planning Authorities and Developers*, English Heritage, London

English Heritage/Prehistoric Society, 1999 *Research Frameworks for the Palaeolithic and Mesolithic of Britain and Ireland*, English Heritage, London

EPICA Community Members, 2004 Eight glacial cycles from an Antarctic ice core, *Nature* **429**, 623–628

Erd, K, 1970 Pollen analytical classification of the Middle Pleistocene in the German Democratic Republic, *Palaeogeogr, Palaeoclimatol, Palaeoecol* **12**, 129–145

Essex County Council and Kent County Council, 2004 Archaeological Survey of Mineral Extraction Sites around the Thames Estuary: Aggregates Levy Sustainability Fund 1, Assessment Report. Project 3374, Report lodged with Archaeology Data Service, ADS Collection 774, DOI 10.5284/1000016

Evans, J, Morse, E, Reid, C, Ridley, E P, and Ridley, H N, 1896 The relation of Palaeolithic man to the glacial epochs, *Rep Brit Assoc Liverpool*, 400–416

Evans, J G, and Smith, I F, 1983 Excavations at Cherhill, North Wiltshire, 1967, *Proc Prehist Soc* **49**, 43–117

Evans, P, 1971 *Towards a Pleistocene time-scale, Part 2 of The Phanerozoic Time-scale – A Supplement*, The Geol Soc, London, Spec Publ **5**, 123-356

Fa, J E, 1984 *The Barbary Macaque. a case study in conservation*, Plenum Press, New York

Faegri, K, and Iversen, J, 1989 *Textbook of Pollen Analysis* (4th edn with K. Krzywinski), John Wiley, Chichester and New York

Fisher, P F, and Bridgland, D R, 1986 Analysis of pebble morphology, in *Clast Lithological Analysis* (ed. D R Bridgland), 43-58, Tech Guide 3, Quat Res Assoc, Cambridge

Flower, R J, 1993 Diatom preservation: experiments and observations on dissolution and breakage in modern and fossil material, *Hydrobiologia* **269/270**, 473–484

Forrest, A J, 1983 *Masters of Flint*, Terance Dalton, Lavenham

Fortelius, M, Mazza, P and Sala, B, 1993 *Stephanorhinus* (Mammalia: Rhinocerotidae) of the western European Pleistocene, with a revision of *S. etruscus* (Falconer, 1868), *Palaeontographia Italica* **40**, 63–155

Frison, G C, 1989 Experimental use of Clovis weaponry and tools on African elephants, *Am Antiq* **54**, 766–784

Frison, G C, and Todd, L C, 1986 *The Colby Mammoth Site. Taphonomy and Archaeology of a Clovis Kill in Northern Wyoming*, Univ of New Mexico Press, Albuquerque

Gabunia, L, Vekua, A, Lordkipanidze, P, Swisher, C C III, Ferring, R, Justus, A, Nioridze, M, Tralchrelidze, M, Anton, S C, Bosinski, G, Joris, O, de Lumley, M-A, Majsuradze, G, and Mashkhelishnili, A, 2000 Earliest Pleistocene hominid cranial remains from Dmanisi, Republic of Georgia: Taxonomy, geological setting and age, *Science* **288**, 1019–1025

Gamble, C S, 1986 *The Palaeolithic Settlement of Europe*, CUP, Cambridge

Gamble, C S, 1987 Man the shoveller: alternative models for Middle Pleistocene colonization and occupation in northern latitudes, in *The Pleistocene Old World Regional Perspectives* (ed. O Soffer), 81–98, Plenum Press, New York

Gamble, C S, 1993 *Timewalkers: the Prehistory of Global Colonization*, Sutton, Stroud

Gamble, C S, and Steele, J, 1999 Hominid ranging patterns and dietary strategies, in *Hominid Evolution: lifestyles and survival strategies* (ed. H. Ullrich), 396–409, Archea, Schwelm

Gaudzinski, S, 1999 The faunal record of the Lower and Middle Palaeolithic of Europe: remarks on human interference, in *The Middle Palaeolithic of Europe* (eds W Roebroeks and C. Gamble), 215–233, Univ Leiden, Leiden

Gaudzinski, S, and Turner, E, 1999 Summarising the role of early humans in the accumulation of European Lower and Middle Palaeolithic bone assemblages, in *The Role of Early Humans in the Accumulation of European Lower and Middle Palaeolithic Bone Assemblages* (eds S Gaudzinski and E Turner), 381–394, Monographien des Römisch-Germanischen Zentralmuseums Mainz. Habelt, Mainz and Bonn

Gaudzinski, S, Turner, E, Anzidei, A P, Álvarez-Fernández, E, Arroyo-Cabrales, J, Cing-Mars, J, Dobosi, V T, Hannus, A, Johnson, E, Münzel, S C, Scheer, A, and Villa, P, 2005 The use of Proboscidean remains in every-day Palaeolithic life, in *Studying Proboscideans: Knowledge, Problems and Perspectives* (eds M R Palombo, M Mussi, P Gioia and G Cavarretta), *Quat Internat* **126–128**, 179–194

Gaudzinski-Windheuser, S, Lindler, L, Rabinovich, R, and Goren-Inbar, N, 2010 Testing heterogeneity in faunal assemblages from archaeological sites, Tumbling and trampling experiments at the early-Middle Pleistocene site of Gesher Benot Ya'aqov (Israel), *J Archaeol Sci* **37**, 3170–3190

Gee, H E, 1991 Bovids in the Pleistocene of Britain, unpubl PhD thesis, Univ. Cambridge

Gee, H E, 1993 The distinction between postcranial bones of *Bos primigenius* Bojanus, 1827 and *Bison priscus* Bojanus, 1827 from the British Pleistocene and the taxonomic status of *Bos* and *Bison*, *J Quat Sci* **8**, 79–92

Geist, V, 1978 *Life Strategies, Human Evolution, Environmental Dsign. Toward a Biological Theory of Health*, Springer Verlag, New York.

Gibbard, P L, 1979 Middle Pleistocene drainage in the Thames valley, *Geol Mag* **116**, 87–90

Gibbard, P L, 1994 *Pleistocene History of the Lower Thames Valley*, CUP, Cambridge

Gibbard, P L, 1995 The formation of the Strait of Dover, in *Island Britain: a Quaternary Perspective* (ed. R C Preece), 15–26, The Geol Soc, London

Gladfelter, B G, and Singer, R, 1975 Implications of East Anglian glacial stratigraphy for the British Lower Palaeolithic, in *Quaternary Studies* (eds R P Suggate and M M Creswell),139–45, Royal Soc New Zealand, Wellington

Goldberg, P, and Guy, J, 1996 Micromorphological observations of selected rock ovens, Wilson-Leonard site, Central Texas, in *Paleoecology; Colloquium 3 of XIII International Congress of Prehistoric and Protohistoric Sciences* (eds L Castelletti and M Cremaschi), 115–122, ABACO, Forli

Goldberg, P, Weiner, S, Bar-Yosef, O, Xu, Q, and Liu, J, 2001 Site formation processes at Zhoukoudian, China, *J Hum Evol* **41**, 483–530

Goldberg, P, and Macphail R I, 2006 *Practical and Theoretical Geoarchaeology*, Blackwell, Oxford

Goodfriend, G A, 1991 Patterns of racemization and epimerization of amino acids in land snail shells over the course of the Holocene, *Geochim Cosmochim Acta* **55**, 293–302

Goodfriend, G A, 1992 Rapid racemization of aspartic acid in mollusc shells and potential for dating over recent centuries, *Nature* **357**, 399–401.

Goodfriend, G A, and Stanley, D J, 1996 Reworking and discontinuities in holocene sedimentation in the Nile Delta – documentation from amino-acid racemization and stable isotopes in mollusk shells, *Marine Geol* **129**, 271–83.

Goren-Inbar, N, Lister, A, Werker, E, and Chech, M, 1994 A butchered elephant skull and associated artifacts from the Acheulian site of Gesher Benot Ya'aqov, Israel, *Paléorient* **20 (1)**, 99–112

Gowlett J, Hallos J, Hounsell S, Brant V, and Debenham N, 2005 Beeches Pit: Archaeology, assemblage dynamics and early fire history of a Middle Pleistocene site in East Anglia, UK, *Eurasian Prehist* **3 (2)**, 3–38

Graham, I D G, and Scollar, I, 1976 Limitations on magnetic prospection in archaeology imposed by soil properties, *Archaeo-Physika* **6**, 1–124

Green, C P, Branch, N P, Coope, G R, Field, M H, Keen, D H, Wells, J M, Schwenninger, J-L, Preece, R C, Schreve, D C, Canti, M G, and Gleed-Owen, C P, 2006 Marine Isotope Stage 9 environments of fluvial deposits at Hackney, north London, UK, *Quat Sci Rev* **25**, 89–113

Greenfield, H J, 1999 The origins of metallurgy: distinguishing stone from metal cut-marks on bones from archaeological sites, *J Archaeol Sci* **26**, 797–808

Grigson, C, and Mellars, P A, 1987 The mammalian remains from the middens, in *Excavations on Oronsay. Prehistoric Human Ecology on a small Island* (ed P A Mellars), 243–289, Edinburgh Univ Press, Edinburgh

Hallos, J, 2004 Artefact dynamics in the Middle Pleistocene: implications for hominid behaviour, in *Lithics in Action: Proceedings of the Lithic Studies Society Conference held in Cardiff, September 2000* (eds E A Walker, F F Wenban-Smith and F Healy), 26–37, Lithic Stud Soc Occ Pap **8**, Oxbow Books, Oxford

Harada, N, Handa, N, Ito, M, Oba, T, and Matsumoto, E, 1996 Chronology of marine sediments by the racemization reaction of aspartic acid in planktonic foraminifera, *Organic Geochem* **24**, 921–30

Hare, P E, and Abelson, P H, 1967 Racemization of amino acids in fossil shells, *Carnegie Institution of Washington Year Book* **66**, 526–8

Hare, P E, and Mitterer, R M, 1969 Laboratory simulation of amino-acid diagenesis in fossils, *Carnegie Institution of Washington Year Book* **67**, 205–8

Hare, P E, von Endt, D W, and Kokis, J E, 1997 Protein and amino acid diagenesis dating, in *Chronometric Dating in Archaeology* (eds R E Taylor and M J Aitken), 261-96, Advances in Archaeological and Museum Science **2**, Plenum Press, New York

Harrison, C J O, and Stewart, J R, 1999 The birds remains, in *A Middle Pleistocene Hominid Site at Eartham Quarry, Boxgrove, West Sussex, UK* (M B Roberts and S A Parfitt), Engl Herit Archaeol Rep **17**, London

Havinga, A J, 1964 Investigation into the differential corrosion susceptibility of pollen and spores, *Pollen et Spores* **6**, 621–635

Havinga, A J, 1967 Palynology and pollen preservation, *Rev Palaeobot Palynol* **2**, 81–98

Havinga, A J, 1985 A 20-year experimental investigation into the differential corrosion susceptibility of pollen and spores in various soil types, *Pollen et Spores* **26**, 541–558

Haynes, G, 1988, Longitudinal studies of African elephant death and bone deposits, *J Archaeol Sci* **15**, 131–157

Haynes, G, 1991 *Mammoths, Mastodonts and Elephants: Biology, Behavior and the Fossil Record*, CUP, Cambridge

Haynes, G, and Krasinski, K E, 2010 Taphonomic fieldwork in Southern Africa and its application in studies of the earliest peopling of North America, *J Taphonomy* **8**, 181–202

Hays, J D, Imbrie, J, and Shackleton, N J, 1976 Variations in the earth's orbit: pacemaker of the ice ages, *Science* **194**, 1121–1132

Higginbottom, I E, and Fookes, P G, 1970 Engineering aspects of periglacial features in Britain, *Q J Eng Geol* **3**, 85–117

High Speed 1, 2011 *Tracks and Traces: The Archaeology of High Speed 1*, I C Art and Design Ltd

Hill, A, and Behrensmeyer, A K, 1984 Disarticulation patterns of some modern east African mammals, *Paleobiology* **10**, 366–76

Hill, R L, 1965 Hydrolysis of proteins, *Adv Protein Chem* **20**, 37–107

Hinton, M A C, 1908 Note on the discovery of the bone of a monkey in the Norfolk Forest Bed, *Geol Mag* **5**, 440–4

Hinton, M A C and Kennard, A S, 1900 Contributions to the Pleistocene geology of the Thames valley, I. The Grays Thurrock area, Part I, *Essex Naturalist* **11**, 336–370

Hinton, M A C and Kennard, A S, 1905 The relative ages of the stone implements of the Lower Thames Valley, *Proc Geol Ass* **19**, 76–100

Holman, J A, 1998 *Pleistocene Amphibians and Reptiles in Britain and Europe*, CUP, Cambridge

Horne, D J, 2007 A Mutual Temperature Range method for Quaternary palaeoclimatic analysis using European nonmarine Ostracoda, *Quat Sci Rev* **26**, 1398–1415

Hutchinson, J N, 1991 Periglacial and slope processes, *Geol Soc Engineering Geol Spec Publ* **7**, 283–331

Hutchinson, J N, 1992 Engineering in relict periglacial and extraglacial areas in Britain, in *Applications of Quaternary Research, Quaternary Proceedings* 2 (ed. J Gray), 49–65, Quat Res Ass Cambridge

Hutchinson, J N, and Hight, D W, 1987 Strongly folded structures associated with past permafrost degradation and periglacial solifluction at Lyme Regis, Dorset, in *Periglacial Processes and Landforms in Britain and Ireland* (ed. J Boardman), 245–256, CUP, Cambridge

Isaac, G L, 1983 Recent single-carcass bone scatters and the problem of 'butchery' sites in the archaeological record, in *Animals and Archaeology, Hunters and Their Prey* (eds J Clutton-Brock and C Grigson), 3–19, Brit Archaeol Rep (Int Ser) **163**, Oxford

Isaac, G L, and Crader, D C, 1981 To what extent were the early hominids carnivorous? An archaeological perspective, in *Omnivorous Primates* (eds R S O Hardinger and G Teleki), 37–103, Columbia Univ Press, New York

Johnson, B J, and Miller, G H, 1997 Archaeological applications of amino acid racemization, *Archaeometry* **39**, 265–87

Jones, D K C, 1981 *Southeast and Southern England*, Methuen, London and New York

Jones, P R, and Vincent, A S, 1986 A study of bone surfaces from La Cotte de St. Brelade, in *La Cotte de St. Brelade, 1961–1978* (eds P. Callow and J M Cornford), 185–192, Geo Books, Norwich

Jones, R M, and White, N, 1988 Point blank: stone tool manufacture at the Ngilipitji Quarry, Arnhem Land, in *Archaeology with Ethnography* (ed E Meehan and R Jones), 51–87, Aust Nat Univ, Canberra

Karklins, K, 1984 The gunflint industry at Brandon, *Canadian J Arms Collect* **22** (2), 51–59

Kaufman, D S, 2000 Amino Acid Racemization in Ostracodes, in *Perspectives in Amino Acid and Protein Geochemistry* (ed. G A Goodfriend, M J Collins, M L Fogel, S A Macko and J F Wehmiller),

145–160, OUP, Oxford

Kaufman D S, and Manley W F, 1998 A new procedure for determining DL amino acid ratios in fossils using Reverse Phase Liquid Chromatography, *Quat Sci Rev* **17**, 987–1000

Kemp, R A, 1985 The decalcified Lower Loam at Swanscombe, Kent: a buried Quaternary soil, *Proc Geol Ass* **96**, 343–354

Kennard, A S, 1942 Faunas of the High Terrace at Swanscombe, *Proc Geol Ass* **53**, 105

Kerney, M P, 1959 An interglacial tufa near Hitchin, Hertfordshire, *Proc Geol Ass* **70**, 322–337

Kerney, M P, 1968 Britain's fauna of land Mollusca and its relation to the the Post-glacial thermal optimum, *Symp Zool Soc London* **22**, 273–291

Kerney, M P, 1971 Interglacial deposits at Barnfield Pit, Swanscombe, and their molluscan fauna, *J Geol Soc London* **127**, 69–86

Kerney, M P, 1999 *Atlas of the Land and Freshwater Mollucs of Britain and Ireland*, Harley Books, Colchester

Kerney, M P, and Sieveking, G de G, 1977 Northfleet, in *South East England and the Thames Valley* (eds E R Shephard-Thorn and J J Wymer), 44–49, X INQUA Congress excursion guide A5, Geo Abstracts, Norwich

Kerney, M P, and Cameron, R A D, 1979 *A Field Guide to the Land Snails of Britain and North-West Europe*, Collins, London

Klein, R G, and Cruz-Uribe, K, 1984 *The Analysis of Animal Bones from Archaeological Sites*, Univ Chicago Press, Chicago and London

Klein, R, 2009, Hominin Dispersals in the Old World, in *The Human Past: World Prehistory and the Development of Human Societies* (ed. C Scarre), 84–123, Thames and Hudson, London

Kohn, M, and Mithen, S, 1999 Handaxes: products of sexual selection? *Antiquity* **73**, 518–26

Krause J, Fu Q, Good J M, Viola B, Shunkov M V, Derevianko A P, and Pääbo S, 2010 The complete mitochondrial DNA genome of an unknown hominin from southern Siberia, *Nature* **464**, 894–897

Kroll, W, 1991 Der Waldelefant von Crumstadt. Ein Beitrag zur Osteologie des Waldelefanten, *Elephas (Palaeoloxodon) antiquus* Falconer and Cautley (1847), unpubl dissertation, Ludwig-Maximilians-Universität, München

Langford, H, 1981 Discovery of straight-tusked elephant bones near Peterborough, *Quat Newslett* **81**, 25

Lajoie, K R, Peterson, E, and Gerow, B A, 1980 Amino acid bone dating: a feasibility study, South San Francisco Bay region, California, in *Biogeochemistry of Amino Acids* (ed. P E Hare, T C Hoering, and K J King), 477–89, Wiley, New York

Lauritzen, S E, Haugen, J E, Lovlie, R, and Giljenielsen, H, 1994, Geochronological potential of isoleucine epimerization in calcite speleothems, *Quat Res* **41**(1), 52–8

Le Borgne, E, 1955 Susceptibilité magnétique anormale du sol superficiel, *Annales de Geophysique* **11**, 399-419

Leakey, M D, 1971 *Olduvai Gorge, Excavations in Beds I and II, 1960–1963, Vol. 3.* CUP, Cambridge

Legge, A J, 2010 The aurochs and domestic cattle, in *Extinctions and Invasions. A Social History of British Fauna* (eds T. O'Connor and N. Sykes), 26–35, Windgather Press, Oxford

Legge, A J, and Rowley-Conwy, P A, 1988 *Star Carr Revisited. A Re-analysis of the Large Mammals*, Birkbeck College, London

Leonardi, G, and Petroni, C, 1976 The fallow deer of the European Pleistocene, *Geologica Romana* **57**, 1–67

Lewis, J S, Wiltshire, P, and Macphail, R I, 1992 A Late Devensian/Early Flandrian site at Three Ways Wharf, Uxbridge: environmental implications, in *Alluvial Archaeology in Britain* (eds S Needham and M G Macklin), 235–248, Oxbow Monogr **27**, Oxbow Books, Oxford

Lewis, J S C, and Rackham, J, 2010 *Three Ways Wharf, Uxbridge. A Lateglacial and Early Holocene Hunter-gatherer Site in the Colne Valley*, Mus London Archaeol Monogr **51**, London

Lewis, S G, 1998 Quaternary geology of East Farm brick pit, Barnham and the surrounding area, in Ashton *et al.* 1998, 23–78

Lister, A M, 1981 Evolutionary studies on Pleistocene deer, unpubl. PhD. thesis, Univ Cambridge

Lister, A M, 1986 New results on deer from Swanscombe, and the stratigraphical significance of deer in the Middle and Upper Pleistocene of Europe, *J Archaeol Sci* **13**, 319–338

Lister, A M, 1989 Rapid dwarfing of red deer on Jersey in the Last Interglacial, *Nature* **342**, 539–542

Lister, A M, 1992 Mammalian fossils and Quaternary biostratigraphy, *Quat Sci Rev* **11**, 329–344

Lister, A M, 1993 Cervidae (Deer), in *The Lower Paleolithic Site at Hoxne, England.* (eds R Singer, B G Gladfelter and J J Wymer), 176–190, Univ Chicago Press, Chicago

Lister, A M, 1994 The evolution of the giant deer, *Megaloceros giganteus* (Blumenbach), *Zool J Linnean Soc* **112**, 65–100

Lister, A M, 1995 Sea-levels and the evolution of island endemics: the dwarf red deer of Jersey, in *Island Britain: a Quaternary perspective* (ed. R C Preece), 151–172, Geol Soc Spec Publ **96**, London

Lister, A M, 1996 The morphological distinction between bones and teeth of fallow deer (*Dama dama*) and red deer (*Cervus elaphus*), *Int J Osteoarch* **6**, 119–143

Lister, A M, 2004 Ecological interactions of Elephantids in Pleistocene Eurasia: *Palaeoloxodon* and *Mammuthus*, in *Human Paleoecology in the Levantine Corridor* (eds N Goren-Inbar and J D Speth), 53–60, Oxbow, Oxford

Lister, A M, and Brandon, A, 1991 A pre-Ipswichian cold stage mammalian fauna from Balderton Sand and Gravel, Lincolnshire, England, *J Quat Sci* **6**, 139–75

Lister, A, and Agenbroad, L D, 1994 Gender determination of the Hot Springs Mammoths, in *The Hot Springs Mammoth Site* (eds L D Agenbroad and J I Mead), 208-260, The Mammoth Site of South Dakota, Hot Springs, South Dakota

Lister, A M, and Stuart, A J, (eds), 2010 The West Runton Freshwater Bed and the West Runton Mammoth, *Quat Internat* **228 (1-2)**, 1-248

Lister, A M, Parfitt, S A, Owen, F J, Collinge, S, and Breda, M, 2010 Metric analysis of ungulate mammals in the early Middle Pleistocene of Britain, in relation to taxonomy and biostratigraphy. II: Cervidae, Equidae and Suidae, in Lister and Stuart (eds) 2010, 157–179

Lord, A R, Whittaker, J E, and King, C, 2009 Paleogene, in *Ostracods in British Stratigraphy* (eds J E Whittaker and M B Hart), 373-409, The Micropalaeontol Soc Spec Publ, The Geol Soc, London

Lotbinière, S de, 1977 The story of the English gunflint, some theories and queries, *J Arms and Armour Soc* **9**, 18–53

Loveland, P J, and Findlay, D C, 1982 Composition and development of some soils on glauconitic Cretaceous (Upper Greensand) rocks in southern England, *J Soil Sci* **33**, 279–294

Lowe, J J, and Walker, M J C, 1997 (reprinted edition) *Reconstructing Quaternary Environments*, Longman Scientific and Technical, Harlow

Lycett, S J, 2008 Acheulean variation and selection: does handaxe symmetry fit neutral expectations, *J Archaeol Sci* **35**, 2640–2648

Machin, A, Hosfield, R, and Mithen, S J, 2005 Testing the functional utility of handaxes symmetry: fallow deer butchery with replica handaxes, *Lithics* **26**, 23–37

McNabb, J, and Ashton, N M, 1990 Clactonian gunflints? *Lithics* **11**, 44–47

McNabb, J, and Ashton, N M, 1992 The cutting edge: bifaces in the Clactonian, *Lithics* **13**, 4–10

McNabb, J, and Ashton, N M, 1995 Thoughtful flakers, *Camb Archaeol J* **5**, 289–298

Macphail, R I, 1999 Sediment micromorphology, in M B Roberts and S A Parfitt 1999, 118–148

Macphail, R I, 2001 The soil micromorphologist as team player: a multianalytical approach to the study of European microstratigraphy, in *Earth Science and Archaeology* (eds P Goldberg, V Holliday and R Ferring), 241–267, Kluwer Academic/Plenum, New York

Macphail, R I, 2009 Marine inundation and archaeological sites: first results from the partial flooding of Wallasea Island, Essex, UK, *Antiquity Project Gallery*, Vol 2009, http://antiquity.ac.uk/projgall/macphail/

Macphail, R I, and Cruise, G M, 2000 Soil micromorphology, in *Prehistoric Intertidal Archaeology in the Welsh Severn Estuary* (eds M Bell, A Caseldine and

H Neumann), 267-269 and CD-Rom, CBA Res Rep **120**, York

Macphail, R I, and Cruise, G M, 2001 The soil micromorphologist as team player: a multianalytical approach to the study of European microstratigraphy, in *Earth Science and Archaeology* (eds P Goldberg, V Holliday and R Ferring), 241-267, Kluwer Academic/Plenium, New York

Macphail, R I, and Crowther, J, 2009 A13 Thames Gateway, Movers Lane and Woolwich Manor Way Sites: soil micromorphology and bulk analyses, unpubl rep for Oxford Archaeology

Macphail, R I, Acott, T G, and Crowther, J, 2001 Boxgrove: Sediment microstratigraphy (soil micromorphology, image analysis and chemistry), unpubl rep, Institute of Archaeology, London

Macphail, R I, Crowther, J, and Cruise,G M, 2009 The Pilgrims' School, Winchester: soil micromorphology, pollen, chemistry and magnetic susceptibility, unpubl rep for Oxford Archaeology

Macphail, R I, Allen, M J, Crowther, J, Cruise, G M, and Whittaker, J E, 2010 Marine inundation: effects on archaeological features, materials, sediments and soils, *Quat Internat* **214**, 44–55

Maddy, D, Lewis, S G, Scaife, R G, Bowen, D Q, Coope, G R, Green, C P, Hardaker, T, Keen, D H, Rees-Jones, J, Parfitt, S, and Scott, K, 1998 The Upper Pleistocene deposits at Cassington, near Oxford, England, *J Quat Sci*, **13 (3)**, 205–231

Martínez-Navarro, B, Pérez-Claros, J A, Palombo, M R, Rook, L, and Palmquist, P, 2007 The Olduvai buffalo *Pelorovis* and the origin of *Bos*, *Quat Res* **68**, 220–6

Martinson, D G, Pisias, N G, Hays, J D, Imbrie, J, Moore, T C, and Shackleton, N J, 1987 Age dating and the orbital theory of the ice ages: Development of a high-resolution 0 to 300,000-year chronostratigraphy, *Quat Res* **27**, 1-30

Marshall, E, 1990 Racemization dating: great expectations, *Science* **247**, 799

Maul, L C, and Parfitt, S A, 2010 Micromammals from the 1995 Mammoth Excavation at West Runton, Norfolk, UK: morphometric data and taxonomic reappraisal, in *The West Runton Freshwater Bed and the West Runton Mammoth* (eds A Lister and A J Stuart), *Quat Internat* **228**, 91–115

Mayhew, D F, 1975 The quaternary history of some British Pleistocene rodents and lagomorphs, unpubl Ph.D. thesis, Univ Cambridge

Mazza, P P M, Martini, F, Sala, B, Magi, M, Colombini, M P, Giachi, G, Landucci, F, Lemorini, C, Modugno, F, and Ribechini, E, 2006. A new Palaeolithic discovery: tar-hafted stone tools in a European Mid-Pleistocene bone-bearing bed, *J Archaeol Sci* **33**, 1310–1318

Meijer T and Preece R C, 1995 Malacological evidence relating to the insularity of the British Isles during the Quaternary, in *Island Britain: a Quaternary Perspective* (ed R C Preece), 89–110, Geol Soc Spec Publ **96**, London

Meisch, C, 2000 Freshwater Ostracoda of Western and Central Europe in *Süßwasserfauna von Mitteleuropa, Band 8/3* (eds J Schwoerbel and P Zwick), Spektrum Akademischer Verlag, Heidelberg and Berlin

Mejdahl, V, 1979 Thermoluminescence dating: beta-dose attenuation in quartz grains, *Archaeometry* **21**, 61–72

Menke, B, 1968 Beiträge zur Biostratigraphie des Mittelpleistozäns in Norddeutschland (pollenanalytische Untersuchungen aus Westholstein), *Meyniana* **18**, 35–42

Miedema, R, Jongmans, A G, and Slager, S, 1974 Micromorphological observations on pyrite and its oxidation products in four Holocene alluvial soils in the Netherlands, in *Soil Microscopy* (ed. G K Rutherford), 772–794, The Limestone Press, Kingston, Ontario

Miller, G H, Beaumont, P B, Jull, A J T, and Johnson, B, 1992 Pleistocene geochronology and palaeothermometry from protein diagenesis in ostrich eggshells: implications for the evolution of modern humans, *Phil Trans Royal Soc London* **B337**, 149–57

Miller, G H, Magee, J W, and Jull, A J T, 1997 Low-latitude glacial cooling in the Southern Hemisphere from amino acid racemization in emu eggshells, *Nature* **385**, 241–4

Miller, G H, Hart, C P, Roark, E B, and Johnson, B J, 2000 Isoleucine epimerization in eggshells of the flightless Australian birds Genyornis and Dromaius, in *Perspectives in Amino Acid and Protein Geochemistry* (eds G A Goodfriend, M J Collins, M L Fogel, S A Macko and J F Wehmiller), 161–181, OUP, Oxford

Mitchell-Jones, A J, Amori, G, Bogdanowicz, W, Kryštufek, B, Reijnders, P J H, Spitzenberger, F, Stubbe, M, Thissen, J B M, Vohralík, V and Zima, J, 1999 *The Atlas of European Mammals*, T and A.D Poyser, London

Mithen, S, 1994 Technology and society during the Middle Pleistocene: hominid group size, social learning and industrial variability, *Cambridge Archaeol J* **4**, 3–32

Mol, D, de Vod, J, and van der Plicht, J, 2007 The presence and extinction of *Elephas antiquus* Falconer and Cautley, 1847, in Europe, *Quat Internat* **169–170**, 149–153

Moore, P D, Webb, J A, and Collinson, M E, 1991 *Pollen Analysis*, 2nd edn, Oxford

Mortillet, G de, 1869 Essai d'une classification des cavernes et des stations sous abris, fondée sur les produits de l'industrie humaine, *Comptes Rendus des Séances de l'Académie des Sciences* **68**, 553–555

Mortillet, G de, 1872 Classification des âges de la pierre, *Comptes Rendus: Congrès International d'Anthropologie et d'Archéologie Préhistorique, 6e session (Bruxelles)*, 432–44

Mortillet, G de, and A de, 1900 (3rd edn) *Le Préhistorique Origine et Antiquité de l'Homme*, Reinwald, Paris

Movius, H L, 1950 A wooden spear of Third Interglacial age from lower Saxony, *Southwestern J Anthropol* **6 (2)**, 139–142

Murphy, C P, 1986 *Thin section Preparation of Soils and*

Sediments, A B Academic Publishers, Berkhamsted

Murray, A S and Wintle, A G, 2000 Luminescence dating of quartz using an improved single-aliquot regenerative-dose protocol, *Radiat Meas* **32**, 57–73

Murray-Wallace, C V, 1993 A review of the application of the amino acid racemization reaction to archaeological dating, *The Artefact* **16**, 19–26

Murray-Wallace, C V, and Kimber, R W L, 1987 Evaluation of the amino acid racemization reaction in studies of Quaternary marine sediments in South Australia, *Aust J Earth Sci* **34**, 279–292

Museum of London Archaeology, 2011 Northfleet West substation, Southfleet Road, Swanscombe (County of Kent): geoarchaeological evaluation report, unpubl client rep (site code KT-SFL 03), submitted to Kent Co. Council

Mussi, M, and Villa, P, 2008 Single carcass of *Mammuthus primigenius* with lithic artifacts in the Upper Pleistocene of northern Italy, *J Archaeol Sci* **35**, 2606–2613

Newcomer, M.H. 1971 Some quantitative experiments in hand-axe manufacture, *World Archaeol* **3**, 85–94

Newton, W M, 1901 The occurrence in a very limited area of the rudest with the finer forms of worked stones, *Man* **66**, 81–82

Nitecki, M H, and Nitecki, D V (eds), 1987 *The Evolution of Human Hunting*, Plenum Press, New York and London

O'Connor, T P, 2003 *The Analysis of Urban Animal Bone Assemblages*: a handbook for archaeologists, The Archaeology of York **19/2**, CBA, York

Oakley, K P, 1952 Swanscombe Man, *Proc Geol Assoc* **63**, 271–300

Oakley, K P and Leakey, M D, 1937 Report on excavations at Jaywick Sands, Essex (1934) with some observations on the Clactonian industry, and on the fauna and geological significance of the Clacton channel, *Proc Prehist Soc* **3**, 217–260

Ohel, M, 1979 The Clactonian: an independent complex or an integral part of the Acheulian, *Current Anthropol* **20**, 685–726

Otte, M, 1990 The northwestern European Plain around 18,000 BP, in *The World at 18,000 BP: Vol 1, High Latitudes* (eds O. Soffer and C. Gamble), 54–68, Unwin Hyman, London

Ovey, C D (ed.), 1964 *The Swanscombe Skull: a survey of research on a Pleistocene site*, Royal Anthropol Inst Occ Pap **20**, 19–61

Oxford Archaeology, 1994 Channel Tunnel Rail Link: assessment of historic and cultural effects, final report, unpubl client rep

Oxford Archaeology, 2004a Channel Tunnel Rail Link contract URN/342/ARC/0018, archaeological works at Southfleet Road: method statement [rev AA, 20th January 2004], unpubl client rep

Oxford Archaeology, 2004b Channel Tunnel Rail Link contract URN/342/ARC/0018, archaeological works at Southfleet Road: method statement [rev AD, 20th May 2004], unpubl client rep

Oxford Archaeology, 2005 Channel Tunnel Rail Link contract URN/342/ARC/0018, Southfleet Road elephant site ARC 342 W 02: Preliminary post-excavation assessment report [342-EZR-SOXAR-00007], unpubl client rep

Palmer, S, 1975 A Palaeolithic site at North Road, Purfleet, Essex, *Trans Essex Archaeol Soc* **7**, 1–13

Parfitt, S A, 1998a Pleistocene vertebrate faunas of the West Sussex coastal plain: their stratigraphic and palaeoenvironmental significance, in *The Quaternary of Kent and Sussex: Field Guide* (eds J B Murton, C A Whiteman, M R Bates, D R Bridgland, A J Long, M B Roberts and M P Walker), 121–135, Quat Res Ass, London

Parfitt, S, 1998b The interglacial mammalian fauna from Barnham, in Ashton *et al.* 1998, 111–147

Parfitt, S A, 1999 Mammalia, in M B Roberts and S A Parfitt 1999, 197–290

Parfitt, S, 2008 A tree frog (*Hyla* sp.) from the West Runton Freshwater Bed (early Middle Pleistocene), Norfolk, and its palaeoenvironmental significance, *Quat Newslett* **114**, 20–27

Parfitt, S A, Barendregt, R W, Breda, M, Candy I, Collins, M J, Coope, G R, Durbidge, P, Field, M H, Lee, J R, Lister, M, Mutch, R, Penkman, K E H, Preece, R C, Rose, J, Stringer, C B, Symmons, R, Whittaker, J E, Wymer, J J, and Stuart, A J, 2005 The earliest record of human activity in northern Europe, *Nature* **438**, 1008–1012

Parfitt S A, Ashton N M, Lewis S G, Abel R L, Coope G R, Field M H, Gale R, Hoare P G, Larkin N R, Lewis M, Karloukovski V, Maher B, Peglar S M, Preece R C, Whittaker J E and Stringer C B, 2010a Early Pleistocene human occupation at the edge of the boreal zone in northwest Europe, *Nature* **466**, 229-233

Parfitt, S A, Coope, G R, Field, M H, Peglar, S M, Preece, R C P, and Whittaker, J E, 2010b Middle Pleistocene biota of the early Anglian 'Arctic Fresh-water Bed' at Ostend, Norfolk, UK, *Proc Geol Ass* **121**, 55–65

Paterson, T T, 1937 Studies in the Palaeolithic succession of England. No. 1, the Barnham sequence, *Proc Prehist Soc East Anglia* **3**, 87–135

Paul, C R C, 1974 *Azeca* in Britain, *J Conchological Soc* **28**, 155–172

Penkman, K E H, 2005 Amino acid geochronology: a closed system approach to test and refine the UK model, unpubl PhD thesis, Univ Newcastle

Penkman, K E H, 2010 Amino acid geochronology: its impact on our understanding of the Quaternary stratigraphy of the British Isles, *J Quat Sci* **25(4)**, 501–514

Penkman, K E H, Preece, R C, Keen, D H, Maddy, D, Schreve, D C and Collins, M J, 2007. Amino acids from the intra-crystalline fraction of mollusc shells, applications to geochronology, *Quat Sci Rev* **26**, 2958–2969

Penkman, K E H, Kaufman, D S, Maddy, D and
   Collins, M J, 2008a Closed-system behaviour of the
   intra-crystalline fraction of amino acids in mollusc
   shells, *Quat Geochronol* **3**, 2–25

Penkman, K, Collins, M, Keen, D, and Preece, R,
   2008b British Aggregates: An improved chronology
   using amino acid racemization and degradation of
   intra-crystalline amino acids (IcPD), Engl Herit Res
   Dept Rep ser 6-2008

Penkman, K E H, Preece, R C, Bridgland, D R, Keen,
   D H, Meijer, T, Parfitt, S A, White, T S, and Collins,
   M J, 2011 A chronological framework for the British
   Quaternary based on *Bithynia* opercula, *Nature* **476**,
   446–449

Penkman, K E H, Preece, R C, Bridgland, D R, Keen,
   D H, Meijer, T, Parfitt, S A, White, T S and Collins,
   M J, 2013 An aminostratigraphy for the British
   Quaternary based on *Bithynia* opercula, *Quat Sci
   Rev* **61**, 111–134

Pettitt, P, and White, M, 2012 *The British Palaeolithic*,
   Routledge, London and New York

Pickering, T R, and Hensley-Marschand, B, 2008
   Cutmarks and hominid handedness, *J Archaeol Sci*
   35, 310–315

Pike, K, and Godwin, H, 1953 The interglacial at
   Clacton-on-Sea, Essex, *Q J Geol Soc London* **108**,
   261–72

Piperno M, Lefèvre D, Raynal J-P and Tagliacozzo A,
   1998 Notarchirico. An early Middle Pleistocene site
   in the Venosa Basin, *Anthropologie* **36 (1-2)**, 85–90

Piperno M, Tagliacozzo A, 2001 The elephant butchery
   area at the Middle Pleistocene site of Notarchirico
   (Venosa, Basilicata, Italy), in *The World of Elephants,
   International Congress* (eds G Cavarretta, P Gioia,
   M Mussi and M R Palombo), 230–236, Consiglio
   Nazionale delle Ricerche, Rome

Pitts, M, and Roberts, M B, 1997 *Fairweather Eden: life
   in Britain half a million years ago as revealed by the
   excavations at Boxgrove*, Century, London

Polly, P D, and Eronen, J T, 2011 Mammal associations
   in the Pleistocene of Britain: implications of ecological niche modelling and a method for
   reconstructing palaeoclimate, in *The Ancient Human
   Occupation of Britain* (ed N. Ashton, S. Lewis and
   C. Stringer), 257–282, Elsevier, Amsterdam

Pope, M I, 2002 The significance of biface-rich
   assemblages: an examination of the behavioural
   controls on lithic assemblage formation in the Lower
   Palaeolithic, unpubl Ph.D. thesis, Univ
   Southampton

Powers, M C, 1953 A new roundness scale for
   sedimentary particles, *J Sediment Petrol* **23**, 117–119

Pradel, D L, 1944 Micro-nucléi du gisement acheuléo-
   moustérien de Fontmaure (Vienne), *Extrait du
   Bulletin de la Société Préhistoric Française* No **1-2-3**

Preece, R C (ed), 1995 *Island Britain: A Quaternary
   Perspective*, Geol Soc Spec Publ **96**, Geol Soc,
   London

Preece, R C, and Parfitt, S A, 2000 The Cromer
   Forest-bed Formation: new thoughts on an old

problem, in *The Quaternary of Norfolk and Suffolk.
   Field Guide* (eds S G Lewis, C A Whiteman and R C
   Preece), 1–27, Quat Res Ass, London

Preece, R C, and Penkman, K E H, 2005 New faunal
   analyses and amino acid dating of the Lower
   Palaeolithic site at East Farm, Barnham, Suffolk,
   Proc Geol Ass **116**, 363–377

Preece, R C, and Parfitt, S A, 2008 The Cromer
   Forest-bed Formation: some recent developments
   relating to early human occupation and lowland
   glaciation, in *The Quaternary of Northern East Anglia.
   Field Guide* (eds I Candy, J R Lee and A M
   Harrison), 60–83, Quat Res Ass, London

Preece, R C and Parfitt, S A, 2012 The Early and early
   Middle Pleistocene context of human occupation
   and lowland glaciation in Britain and northern
   Europe, *Quat Internat* **271**, 6-28

Preece, R C, Gowlett, J A J, Parfitt, S A, Bridgland, D
   R, and Lewis, S G, 2006 Humans in the Hoxnian:
   habitat, context and fire use at Beeches Pit, West
   Stow, Suffolk, UK, *J Quat Sci* **21**, 485–496

Preece, R C, Parfitt, S A, Bridgland, D R, Lewis, S G,
   Rowe, P J, Atkinson, T C, Candy, I, Debenham, N
   C, Penkman, K E H, Rhodes, E J, Schwenninger,
   J-L, Griffiths, H I, Whittaker, J E, and Gleed-Owen,
   C, 2007 Terrestrial environments during MIS 11:
   Evidence from the Palaeolithic site at West Stow,
   Suffolk, UK, *Quat Sci Rev* **26**, 1236–1300

Preece, R C, Parfitt S A, Coope, G R, Penkman, K E
   H, Ponel, P, and Whittaker, J E 2009 Biostratigraphic and aminostratigraphic constraints on the
   age of the Middle Pleistocene glacial succession in
   north Norfolk, UK, *J Quat Sci* **24**, 557–580

Prescott, J R, Hutton, J T, 1994 Cosmic ray contributions to dose rates for luminescence and ESR
   dating: large depths and long-term time variations,
   Radiat Meas **23**, 497–500

Pryor, W A, 1971 Grain shape, in *Procedures in
   Sedimentary Petrology* (ed. R E Carver), 131–150,
   John Wiley, New York

Pushkina, D, 2007 The Pleistocene easternmost
   distribution in Eurasia of the species associated with
   the Eemian *Palaeoloxodon antiquus* assemblage,
   Mammal Rev **37**, 224–245

Rabinovich, R, Ackermann, O, Aladjem, E, Barkai, R,
   Biton, R, Milevski, I, Solodenko, N, and Marder, O,
   2012 Elephants at the Middle Pleistocene Acheulian
   open-air site of Revadim Quarry, Israel, *Quat
   Internat* **276-277**, 183–197

Reich, D, Green, R E, Kircher, M, Krause, J,
   Patterson, N, Durand, E Y, Viola, B, Briggs, A W,
   Stenzel, U, Johnson, P L, Maricic, T, Good, J M,
   Marques-Bonet, T, Alkan, C, Fu, Q, Mallick, S,
   Li, H, Meyer, M, Eichler, E E, Stoneking, M,
   Richards, M, Talamo, S, Shunkov, M V, Derevianko,
   A P, Hublin, J J, Kelso, J, Slatkin, M, and Pääbo, S,
   2010 Genetic history of an archaic hominin group
   from Denisova Cave in Siberia, *Nature* **468** (7327),
   1053–1060

Rhodes, E J, and Schwenninger, J-L, 2007 Dose rates and radioisotope concentrations in the concrete calibration blocks at Oxford, *Ancient TL* **25**, 5–8

Roberts, M B, 2000 CTRL Section 2: Pepper Hill to St. Pancras Terminus, research strategy for Palaeolithic archaeology and Pleistocene geology, Annex 13, CTRL Minor Works Contract, Contract URN/342/ARC/0018, Rail Link Engineering

Roberts, M B, and Parfitt, S A, 1999 *A Middle Pleistocene Hominid Site at Eartham Quarry, Boxgrove, West Sussex, UK*, Engl Herit Archaeol Rep **17**, London

Roberts, M B, and Pope, M I, in press *Mapping the Early Middle Pleistocene Deposits of the Slindon Formation across the Coastal Plain of West Sussex and Eastern Hampshire*, Left Coast Press, London

Robinson, E, 1996 The ostracod fauna from the Waechter excavations, in Conway *et al.* 1996, 187–190

Rodríguez, J, Burjachs, F, Cuenca-Bescós, G, García, N, Van der Made, J, Pérez González, A, Blain, H-A, Expósito, I, López-García, J-M, García Antón, M, Allué, E, Cáceres, I, Huguet, R, Mosquera, M, Ollé, A, Rosell, J, Parés, J M, Rodríguez, X-P, Díez, C, Rofes, J, Sala, R, Saladié, P, Vallverdú, J, Bennasar, M L, Blasco, R, Bermúdez de Castro, J M, and Carbonell E, 2011 One million years of cultural evolution in a stable environment at Atapuerca (Burgos, Spain), *Quat Sci Rev* **30**, 1396–1412

Rodwell, J S, 1991, *British Plant Communities. Volume 2. Mires and Heaths*, CUP, Cambridge

Roe, D A, 1968 *A Gazetteer of British Lower and Middle Palaeolithic Sites*, CBA Res Rep **8**, London

Roe, D A, 1980 Introduction: precise moments in remote time, *World Archaeol* **12 (2)**, 107–108

Roe, D A, 1981 *The Lower and Middle Palaeolithic Periods in Britain*, Routledge and Kegan Paul, London

Roe, D A, 1996 The start of the British Palaeolithic: some old and new thoughts and speculations, *Lithics* **16**, 17–26

Roe, H M, 1999 Late Middle Pleistocene sea-level change in the southern North Sea: the record from eastern Essex, UK, *Quat Internat* **55**, 115–128

Roe, H M, Coope, G R, Devoy, R J N, Harrison, C J O, Penkman, K E H, Preece, R C, and Schreve, D C, 2009 Differentiation of MIS 9 and MIS 11 in the continental record: vegetational, faunal, amino-stratigraphic and sea-level evidence from coastal sites in eastern Essex, UK, *Quat Sci Rev* **28**, 2342–2373

Roebroeks, W, 2001 Hominid behaviour and the earliest occupation of Europe: an exploration, *J Hum Evolut* **41**, 437–461

Roebroeks W, 2007 Building blocks from an old brickyard, *Quat Sci Rev* **26**, 1194–1196

Roebroeks, W, Conard, N J and van Kolfschoten, T, 1992 Dense forests, cold steppes, and the Palaeolithic settlement of Northern Europe, *Curr Anthropol* **33**, 551–586

Rogers, P M, Arthur, C P, and Soriguer, R C, 1994 The rabbit in continental Europe, in *The European Rabbit: the history and biology of a successful coloniser* (eds H V Thomas and C M King), 22–63, OUP, Oxford

Rose, J, 1987 Status of the Wolstonian glaciation in the British Quaternary, *Quat Newslett* **53**, 1–9

Rose, J, 1991 Stratigraphic basis of the 'Wolstonian Glaciation', and retention of the term 'Wolstonian' as a chrnosstratigraphic stage name – discussion, in *Central East Anglia and the Fen basin. Field Guide* (eds S G Lewis, C A Whiteman and D R Bridgland), 15–20, Quat Res Ass, Cambridge

Rutter, N W, and Blackwell, B, 1995 Amino acid racemization dating, in *Dating Methods for Quaternary Deposits* (eds N W Rutter and N R Catto), 125–64, Geol Assoc Canada, St John's, Newfoundland

Ryves, D B, Juggins, S, Fritz, S C, and Battarbee, R W, 2001 Experimental diatom dissolution and the quantification of microfossil preservation in sediments, P*alaeogeog, Palaeoclimatol, Palaeoecol* **172**, 99–113

Sacca, D, 2012 Taphonomy of Palaeoloxodon antiquus at Castel di Guido (Rome, Italy): Proboscidean carcass exploitation in the Lower Palaeolithic, *Quat Internat* **276–277**, 27–41

Santonja, M, and Villa, P, 1990 The Lower Palaeolithic of Spain and Portugal, *J World Prehist* **4**, 45–94

Schick, K D, and Toth, N, 1993 *Making Silent Stones Speak*, Simon and Schuster, New York

Schneiderhöhn, P, 1954 Eine vergleichende Studie uber Methoden zur quantitativen Bestimmung von Abrundung und Form an Sandkornern, *Heidl Beitr Miner Petrogr* **4**, 172–191

Schreve, D, 1996 The mammalian fauna from the Waechter excavations, Barnfield Pit, Swanscombe, in Conway *et al.* 1996, 149–162

Schreve, D C, 2001a. Differentiation of the British late Middle Pleistocene interglacials: the evidence from mammalian biostratigraphy, *Quat Sci Rev* **20**, 1693–1705

Schreve, D C, 2001b. Mammalian evidence from fluvial sequences for complex environmental change at the oxygen isotope substage level, *Quat Internat* **79**, 65–74

Schreve, D, 2006 The taphonomy of a Middle Devensian (MIS 3) vertebrate assemblage from Lynford, Norfolk, UK, and its implications for Middle Palaeolithic subsistence strategies, *J Quat Sci* **21**, 543–556

Schreve, D C, and Thomas, G N, 2001 European Quaternary Biostratigraphy, *Quat Sci Rev* **20**, 1577–1764

Schreve, D C, Bridgland, D R, Allen, P, Blackford, J J, Gleed-Owen, C P, Griffiths, H I, Keen, D H, and White, M J, 2002. Sedimentology, palaeontology and archaeology of late Middle Pleistocene River Thames terrace deposits at Purfleet, Essex, UK, *Quat Sci Rev* **21**, 1423–1464

Schroeder, R A, and Bada, J L, 1976 A review of the

geochemical applications of the amino acid racemization reaction, *Earth-Sci Rev* **12**, 347–91

Scollar, I, Tabbagh, A, Hesse, A, and Herzog, I, 1990 *Archaeological Prospecting and Remote Sensing*, CUP, Cambridge

Scott, K, 1986 The bone assemblages of layers 3 and 6, in *La Cotte de St. Brelade, 1961–1978*. (eds P Callow and J M Cornford), 159–183, Geo Books, Norwich

Shackleton, N J, 1987 Oxygen Isotopes, ice volume and sea level, *Quat Sci Rev* **6**, 183–190

Shipman, P, 1981 Applications of scanning electron microscopy to taphonomic problems, *Ann NY Acad Sci* **276**, 357–385

Shipman, P, and Rose, J J, 1983a Evidence of butchery and hominid activities at Torralba and Ambrona: an evaluation using microscopic techniques, *J Archaeol Sci* **10**, 465–474

Shipman, P, and Rose, J, 1983b Early hominid hunting, butchering, and carcass processing behaviours: approaches to the fossil record, *J Anthrop Archaeol* **2**, 57–98

Shotton, F W, Keen, D H, Coope, G R, Currant, A P, Gibard, P L, Aalto, M, Peglar, S M, Robinson, J E, 1993 The Middle Pleistocene deposits of Waverley Wood Pit, Warwickshire, England, *J Quat Sci* **8**, 293–325

Sier, M J, Roebroeks, W, Bakels, C C, Dekkers, M J, Brühl, E, De Loecker, D, Gaudzinski-Windheuser, S, Hesse, N, Jagich, A, Kindler, L, Kuijper, W J, Laurat, T, Mücher, H J, Penkman, K E H, Richter, D, *et al.* 2011 Direct terrestrial-marine correlation demonstrates surprisingly late onset of the last interglacial in central Europe, *Quat Res* **75**, 213–218

Singer, R, Wymer, J J, Gladfelter, B G, and Wolff, R, 1973 Excavation of the Clactonian industry at the Golf Course, Clacton-on-Sea, Essex, *Proc Prehist Soc* **39**, 6–74

Singer, R, Wolff, R, Gladfelter, B G, and Wymer, J J, 1982 Pleistocene *Macaca* from Hoxne, Suffolk, England, *Folia Primatologica* **37**, 141–52

Singer, R, Gladfelter, B G, Wymer, J J (eds), 1993 *The Lower Paleolithic Site at Hoxne, England*, Univ Chicago Press, Chicago

Slimak, L, Svendsen, J I, Mangerud, J, Plisson, H, Heggen, H P, Brugère, A, and Pavlov, P Y, 2011 Late Mousterian persistence near the Arctic Circle, *Science* **332**, 841–845

Smith, G M, 2012 Middle Palaeolithic subsistence: The role of hominins at Lynford, Norfolk, UK, *Quat Internat* **252**, 68–81

Smith, R A, 1911 A Palaeolithic Industry at Northfleet, Kent, *Archaeologia* **62 (2)**, 515–532

Smith, R A, and Dewey, H, 1913 Stratification at Swanscombe: report on excavations made on behalf of the British Museum and H.M. Geological Survey, *Archaeologia* **64**, 177–204

Smith, G G, and Evans, R C, 1980 The effect of structure and conditions on the rate of racemization of free and bound amino acids, in *Biogeochemistry of Amino Acids* (eds P E Hare, T C Hoering, and

J King), 257–82, Wiley, New York

Snelling, A J R, 1964 Excavations at the Globe Pit, Little Thurrock, Grays, Essex 1961, *Essex Naturalist* **31**, 199–208

Spikins, P, 2012 Goodwill hunting? Debates over the 'meaning' of Lower Palaeolithic handaxe form revisited, *World Archaeol* **44 (3)**, 378–392

Spurrell, F C J, 1890 Excursion to Swanscombe, *Proc Geol Assoc* II, **cxlv–vi**

Stewart, J R, 1992 The Turdidae of Pin Hole Cave, Derbyshire, unpubl. MSc dissertation, City of London, North London and Thames Polytechnics

Stewart, J R, 1998 The avifauna, in Ashton *et al.*, 107–109

Stewart, J R, 2007 *The Evolutionary study of some Archaeologically Significant Avian Taxa in the Quaternary of the Western Palaearctic*, Brit Archaeol Rep (Int Ser) **1653**, Oxford

Stewart, J R, 2010 The bird remains from the West Runton Freshwater Bed, Norfolk, England, in *The West Runton Freshwater Bed and the West Runton Mammoth* (eds A J Stuart and A M Lister), *Quat Internat* **228**, 72–90

Stewart, J R, and Hernández Carrasquilla, F, 1997 The identification of extant European birds remains: a review of the literature, *Int J Osteoarch* **7**, 364–371

Stringer, C B, 2006 *Homo Britannicus: the Incredible Story of Human Life in Britain*, Allen Lane, London

Stringer, C, and Hublin, J-J, 1999 New age estimates for the Swanscombe hominid, and their significance for human evolution, *J Hum Evol* **37**, 873–877

Stringer, C B, Andrews, P, and Currant, A P, 1996 Palaeoclimatic significance of mammalian faunas from Westbury Cave, Somerset, England, in *The early Middle Pleistocene in Europe* (ed. C. Turner), 135–143, Balkema, Rotterdam

Stoops, G, 2003 *Guidelines for Analysis and Description of Soil and Regolith Thin sections*, Soil Science Society of America, Inc, Madison, Wisconsin

Stopes, H, 1900 On the discovery of *Neritina fluviatilis* with a Pleistocene fauna and worked flints in High Terrace gravels of the Thames valley, *J Royal Anthropol Inst* **29** (New Ser 2), 302–303

Stuart, A J, 1979 Pleistocene occurrences of the European pond tortoise (*Emys orbicularis* L.) in Britain, *Boreas* **8**, 359–371

Stuart, A J, 1982 *Pleistocene Vertebrates in the British Isles*, Longman, London

Stuart, A J, 1996 Vertebrate faunas from the early Middle Pleistocene of East Anglia, in *The early Middle Pleistocene of Europe* (ed. C Turner), 9–24, AA Balkema, Rotterdam

Stuart, A J, 2005 The extinction of the woolly mammoth (*Mammuthus primigenius*) and straight-tusked elephant (*Palaeoloxodon antiquus*) in Europe, *Quat Internat* **228**, 217–232

Stuart, A J, and Larkin, N M, 2010 Taphonomy of the West Runton mammoth, in *The West Runton Freshwater Bed and the West Runton Mammoth* (eds A Lister and A J Stuart), *Quat Internat* **228**, 217–232

Stuart, A J, Wolff, R G, Lister, A M, Singer, R, and Egginton, J M, 1993 Fossil Vertebrates, in *The Lower Paleolithic Site at Hoxne, England* (eds R Singer, B G Gladfelter and J J Wymer), 163–206, Univ Chicago Press, Chicago

Sumbler, M G, 1996 *London and the Thames Valley*, British Regional Geology (4ᵗʰ edn), BGS, Keyworth

Surrovell, T, Waguespack, N, and Brantingham, P J, 2005 Global archaeological evidence for proboscidean overkill, *Proc Nat Acad Sci* **102 (17)**, 6231–6236

Sutcliffe, A J, 1964 The mammalian fauna, in *The Swanscombe Skull: a survey of research on a Pleistocene site* (ed C D Ovey), 85–111

Sutcliffe, A J, 1995 The Aveley elephant site, Sandy Lane Pit (TQ 553807), in *The Quaternary of the Lower Reaches of the Thames* (eds D R Bridgland, P Allen and B A Hart), 189–199, Quat Res Ass, London

Swanscombe Committee, 1938 Report on the Swanscombe skull: prepared by the Swanscombe Committee of the Royal Anthropological Institute, *J Royal Anthrop Inst* **68**, 17–98.

Sykes, G A, Collins, M J and Walton, D I, 1995 The significance of a geochemically isolated intra-crystalline organic fraction within biominerals, *Organic Geochemistry* **23 (11/12)**, 1039–1065

Thieme, H, 1997 Lower Palaeolithic hunting spears from Germany, *Nature* **385**, 807–10

Thirly, M, Galbois, J, and Schmitt, J-M, 2006 Unusual phosphate concretions related to groundwater flow in a continental environment, *J Sediment Res* **76**, 866–877

Tite, M S, 1972 The influence of geology on the magnetic susceptibility of soils on archaeological sites, *Archaeometry* **14**, 229–236

Tite, M S, and Mullins, C, 1971 Enhancement of magnetic susceptibility of soils on archaeological sites, *Archaeometry* **13**, 209–19

Toth, N, 1982 The stone technologies of early hominids at Koobi Foora, Kenya: an experimental approach, unpubl Ph.D. dissertation, Univ California, Berkeley, Ann Arbor, University Microfilms

Toth, N, 1985 Archaeological evidence for preferential right-handedness in the Lower and Middle Pleistocene, and its possible implications, *J Hum Evol* **14,** 607–614

Trigger, B G, 1989 *A History of Archaeological Thought,* CUP, Cambridge

Turner, A, 1984 Dental sex dimorphism in European lions (*Panthera leo* L.) of the Upper Pleistocene: palaeoecological and palaeoethological implications, *Annales Zoologici Fennici* **21**, 1–8

Turner, A, 1997 *The Big Cats and their Fossil Relatives,* Columbia Univ Press, New York

Turner, A, 2009 The evolution of the guild of large Carnivora of the British Isles during the Middle and Late Pleistocene, *J Quat Sci* **24**, 991–1005

Turner, C, 1970 The Middle Pleistocene deposits at Marks Tey, Essex, *Phil Trans Royal Soc London* **B257**, 373–440

Turner, C, 1975a The correlation and duration of Middle Pleistocene interglacial periods in northwest Europe, in *After the Australopithecines: stratigraphy, ecology and culture-change in the Middle Pleistocene* (eds K W Butzer and G L Isaac), 259–308, Mouton, Le Hague

Turner, C, 1975b Der Einfluss grosser Mammalier auf die interglaziale Vegetation, *Quartärpaläontologie* **1**, 13–19

Turner, C, and Kerney, M P, 1971 A note on the age of the freshwater beds of the Clacton Channel, *J Geol Soc* **127**, 87–93

Turner, C and West, R G, 1968 The subdivision and zonation of interglacial periods, *Eiszeitalter und Gegenwart* **19**, 93–101

Tzedakis, P C, Hooshiemstra, H, Magri, D, Reille, M, Sadori, L, Shackleton, N, Wijmstra, T A, 2001 Establishing a terrestrial chronological framework as a basis for biostratigraphical comparisons, *Quat Sci Rev* **25**, 3416–3430

Van der Made, J, and Grube R, 2010 The rhinoceroses from Neumark-Nord and their nutrition, in *Elefantenreich – Eine Fossilwelt in Europe* (ed H Meller, H), 383–394, Landesmat für Denkmalpflege und Archäologie Sachsen-Anhalt and Landes-museum Vorgeschichte, Denmark

Van der Made, J, 2010 The rhinos of the Middle Pleistocene of Neumark-Nord (Saxony-Anhalt), in Neumark-Nord: Ein interglaziales Ökosystem des mittelpäläolithischen Menschen, *Veröffentlichungen des Landesamtes für Denkmalpflege und Archäologie* **62**, 433–527

Villa, P, 1990 Torralba and Aridos: elephant exploita-tion in Middle Pleistocene Spain, *J Hum Evol* **19**, 299–309

Villa, P, and Mahieu, È, 1991 Breakage patterns of human long bones, *J Hum Evol* **21**, 27–48

Voormolen B, 2008 Ancient hunters, modern butchers. Schöningen 13II-4, a kill-butchery site dating from the northwest European Lower Palaeolithic, *J Taphonomy* **6(2)**, 71–247

Waelbroeck, C, Labeyrie, L, Michel, E, Duplessy, J C, McManus, J F, Lambeck, K, Balbon, E, Labracherie, M, 2002 Sea-level and deep water temperature changes derived from benthic foraminifera isotopic records, *Quat Sci Rev* **21 (1-3),** 295–305

Walker, E A, 2001 Old collections – a new resource? The history of some English Palaeolithic collections in Cardiff, in *A Very Remote Period Indeed: papers on the Palaeolithic presented to Derek Roe* (eds S Milliken and J Cook), 249–259, Oxbow Books, Oxford

Walton D, 1998 Degradation of intracrystalline proteins and amino acids in fossil brachiopods, *Org Geochem* **28**, 389–410

Warren, S H, 1922 The Mesvinian industry of Clacton-on-Sea, *Proc Prehist Soc East Anglia* **3**, 597–602

Warren, S H, 1923 The *Elephas antiquus* bed of Clacton-on-Sea (Essex) and its flora and fauna, with reports by E M Reid, M E J Chandler, J Groves, C W Andrews, M A C Hinton, T H Withers, A S Kennard and B B Woodward, *Q J Geol Soc London* **79**, 606–34

Warren, S H, 1926 The classification of the Lower Palaeolithic with especial reference to Essex, *Trans South Eastern Union of Scientific Societies 1926*, 38–50

Warren, S H, 1951 The Clactonian flint industry: a new interpretation. *Proc Geol Ass* **62**, 107–135

Warren, S H, 1955 The Clacton (Essex) channel deposits, *Q J Geol Soc London* **111**, 283–307

Warren, S H, 1958 The Clactonian Flint Industry: a supplementary note, *Proc Geol Ass* **69**, 123–129

Waters, M R, Stafford, T W, McDonald, H G, Gustafson, C, Rasmussen, M, Cappellini, E, Olsen, J V, Sklarczyk, D, Jensen, L J, Gilbert, M T P, and Willerslev, E, 2011 Pre-Clovis mastodon hunting 13, 800 years ago at the Manis site, Washington, *Science* **334**, 351–353

Weber, T, 2000 The Eemian *Elephas antiquus* finds with artefacts from Lehringen and Gröbern: are they really killing sites? *Anthropologie et Préhistoire* **111**, 177–185

Wehmiller, J F, Thieler, E R, Miller, D, Pellerito, V, Bakeman Keeney, V, Riggs, S R, Culver, S, Mallinson, D, Farrell, K M, York, L L, Pierson, J, and Parham, P R, 2010 Aminostratigraphy of surface and subsurface Quaternary sediments, North Carolina coastal plain, USA, *Quat Geochronol* **5**, 459–494

Wenban-Smith, F F, 1985 Analysis of experimentally produced flint debitage and on-site application of results, unpubl BA diss, Institute of Archaeology, Univ London

Wenban-Smith, F F, 1992 Palaeolithic Priorities and Acceptability of Proposed Rail-link Routes in the Ebbsfleet Valley, unpubl rep submitted to Oxford Archaeology on behalf of Union Railways

Wenban-Smith, F F, 1995a, The Ebbsfleet Valley, Northfleet (Baker's Hole), in *The Quaternary of the Lower Reaches of the Thames. Field Guide* (eds D R Bridgland, P Allen and B A Haggart), 147–64, Quat Res Ass, Durham

Wenban-Smith, F F, 1995b Another one bites the dust. Review of *Excavations at Barnfield Pit, Swanscombe, 1968–72 (*eds B Conway, J McNabb and N Ashton), *Lithics* **16**, 99-108

Wenban-Smith, F F, 1996 The Palaeolithic archaeology of Baker's Hole: a case study for focus in lithic analysis, unpubl PhD thesis, Dept Archaeology, Univ Southampton

Wenban-Smith, F F, 1998 Clactonian and Acheulian industries in Britain: their chronology and significance reconsidered, in *Stone Age Archaeology: essays in honour of John Wymer* (eds N Ashton, F Healy and P Pettitt), 90–97, Oxbow Books, Oxford

Wenban-Smith, F F, 1999 Palaeolithic archaeology at Craylands Lane East: assessment report, unpubl client rep submitted to Kent Co Council

Wenban-Smith, F F, 2004a The Stopes Palaeolithic project: final report, unpubl rep for English Heritage

Wenban-Smith, F F, 2004b Bringing behaviour into focus: Archaic landscapes and lithic technology, in *Lithics in Action: Proceedings of the Lithic Studies Society Conference held in Cardiff, September 2000* (eds E A Walker, F F Wenban-Smith and F Healy), 48–56, Lithic Stud Soc Occ Pap **8**, Oxbow Books, Oxford

Wenban-Smith, F F, 2007 The Palaeolithic archaeology of Kent, in *The Archaeology of Kent to AD 800* (ed. J H Williams), 25–64, The Boydell Press, Woodbridge

Wenban-Smith, F F, 2009 Henry Stopes (1852–1902): engineer, brewer and anthropologist, in *Great Prehistorians: 150 years of Palaeolithic research, 1859–2009* (eds R T Hosfield, F F Wenban-Smith and M Pope), 65-85, Lithics **30** (spec edn), London

Wenban-Smith, F F, and Ashton, N, 1998 Raw material and lithic technology, in Ashton *et al.* 1998, 237–244

Wenban-Smith, F F, and Bridgland, D R, 2001 Palaeolithic archaeology at the Swan Valley Community School, Swanscombe, Kent, *Proc Prehist Soc* **67**, 219–259

Wenban-Smith, F F, and CgMs Consulting Ltd, 2002 Eastern Quarry, Swanscombe: preliminary Palaeolithic/Pleistocene field evaluation report, unpubl client rep submitted to Kent Co Council

Wenban-Smith, F F, and Bates, M R, 2011a Northfleet waste-water treatment works: archive report on Palaeolithic investigations, unpubl client report for Canterbury Archaeol Trust, submitted to Kent Co Council

Wenban-Smith, F F and Bates, M R, 2011b Palaeolithic and Pleistocene investigations, in *Excavations in North-West Kent 2005-2007: one hundred thousand years of human activity in and around the Darent Valley* (eds A Simmonds, F F Wenban-Smith, M R Bates, K Powell, D Sykes, R Devaney, D Stansbie and D Score), 17-59, Oxford Archaeol Monogr **11**, Oxford

Wenban-Smith, F F, Gamble C S, and ApSimon, A M, 2000 The Lower Palaeolithic site at Red Barns, Portchester, Hampshire: bifacial technology, raw material quality and the organisation of Archaic behaviour, *Proc Prehist Soc* **66**, 209–255

Wenban-Smith, F F, Allen, P, Bates, M R, Parfitt, S A, Preece, R C, Stewart, J R, Turner, C, and Whittaker, J E, 2006 The Clactonian elephant butchery site at Southfleet Road, Ebbsfleet, UK, *J Quat Sci* **21 (5)**, 471-483

Wenban-Smith, F F, Bates, M R, and Marshall, G, 2007 Medway Valley Palaeolithic project final report: the Palaeolithic resource in the Medway Gravels (Kent), unpubl rep for English Heritage, made available through the Archaeology Data Service,

Univ York (ADS) http://ads.ahds.ac.uk

Wenban-Smith F F, Bates M R, and Schwenninger J-L, 2010 Early Devensian (MIS 5d-5b) occupation at Dartford, southeast England, *J Quat Sci* **25** (**8**), 1193-1199

Wenban-Smith F F, Stafford E, and Bates, M R (eds) forthcoming *Prehistoric Ebbsfleet: Excavations and Research in Advance of High Speed 1 and STDR 4*, Monogr in Oxford/Wessex Archaeol HS1 series, forthcoming 2013

Wenban-Smith, F F, Schwenninger, J-L, Bates, M R, in prep. OSL dating and post-depositional sediment disturbance: single-grain results from the late Middle Pleistocene handaxe site at Cuxton, UK, for *J Archaeol Sci*

Wessex Archaeology, 1993 *The Southern Rivers Palaeolithic Project, Report No. 2 – The South West and South of the Thames*, Wessex Archaeol, Salisbury

Wessex Archaeology, 2004 Springhead Quarter, Ebbsfleet, Kent: archaeological evaluation report , unpubl client rep [WA 54924.01] for CgMs Consulting, submitted to Kent Co Council

Wessex Archaeology, 2006a Eastern Quarry (Area B and Additional Areas), Swanscombe, Kent: archaeological investigations, unpubl client rep [WA 61040.01] for CgMs Consulting, submitted to Kent Co Council

Wessex Archaeology, 2006b Station Quarter South, Ebbsfleet, Kent: archaeological test pits in the location of the Elephant Lake Site, unpubl client rep [WA 63542] for CgMs Consulting, submitted to Kent Co Council

Wessex Archaeology, 2007 Watching brief during geotechnical window sampling in Station Quarter South and a targeted (Palaeolithic) test pit evaluation in Eastern Quarry adjacent to Southfleet Road and in close proximity to the elephant lake site, Ebbsfleet, Kent, unpubl client report [WA 63544] for CgMs Consulting, submitted to Kent Co Council

Wessex Archaeology 2008a Weldon, Castle Hill (Eastern Quarry), Swanscombe, Kent: further Palaeolithic test-pitting along fastrack spine route and cross-site link road, unpubl client rep [WA 61041] for CgMs Consulting, submitted to Kent Co Council

Wessex Archaeology 2008b Weldon, Castle Hill (Eastern Quarry), Swanscombe, Kent: additional further Pal TPs , unpubl client rep [WA 61042] for CgMs Consulting submitted to Kent Co Council

Wessex Archaeology 2008c Plots A8 and A9 Crossways Business Park, Dartford, Kent: report on a programme of archaeological mitigation, unpubl client rep [WA 66223.02]

Wessex Archaeology 2009a Weldon, Castle Hill (Eastern Quarry), Swanscombe, Kent: Palaeolithic integrated deposit model and research framework, unpubl client rep [WA 61046] for CgMs Consulting on behalf of Land Securities, submitted to Kent Co Council.

Wessex Archaeology, 2009b Watching Brief on a Septic tank in Weldon, Castle Hill (Eastern Quarry), unpubl client rep [WA 68990] for CgMs Consulting, submitted to Kent Co Council

West, R G, 1956 The Quaternary deposits at Hoxne, Suffolk, *Phil Trans Royal Soc London* **B239**, 265–356

West, R G, 1969 Pollen analyses from interglacial deposits at Aveley and Grays, Essex, *Proc Geol Ass* **80**, 271–282

Westaway, R, 2011 A re-evaluation of the timing of the earliest reported human occupation of Britain: the age of sediments at Happisburgh, eastern England, *Proc Geol Ass* **122** (**3**), 383-396

Wheeler, W, 1977 The origin and distribution of the freshwater fishes of the British Isles, *J Biogeog* **4**, 1–24

White, M J, 1998 On the significance of Acheulean biface variability in southern Britain, *Proc Prehist Soc* **64**, 15–44

White, M J, 2000 The Clactonian question: on the interpretation of core-and-flake assemblages in the British Lower Palaeolithic, *J World Prehist* **14**, 1–63

White, M J, and Jacobi, R M, 2002 Two sides to every story: *bout coupé* handaxes revisited, *Oxford J Archaeol* **21**, 109–133

White, M J, and Schreve, D C, 2000 Peninsula Britain, palaeogeography, colonisation, and the Lower Palaeolithic settlement of the British Isles, *Proc Prehist Soc* **66**, 1–28

White, T D, 1992 *Prehistoric Cannibalism at Mancos 5Mtumr-2346*, Princeton Uni Press, Princeton

White, T S, 2012 Late Middle Pleistocene molluscan and ostracod successions, and their relevance to the Palaeolithic record, unpubl PhD thesis, Dept Zoology, Univ Cambridge

White, T S, Preece, R C, and Whittaker, J E 2013 Molluscan and ostracod successions from Dierden's Pit, Swanscombe: insights into the fluvial history, sea-level record and human occupation of the Hoxnian Thames, *Quat Sci Rev* **70**, 73-90

Whittaker, J E, 2006 Report on the ostracoda from 8 samples, in Station Quarter South, Ebbsfleet, Kent: archaeological test pits in the location of the Elephant Lake Site: Appendix 2 (Wessex Archaeol), unpubl client rep (WA 63542) for CgMs Consulting submitted to Kent Co Council

Whittaker, J E, and Horne, D J, 2009 Pleistocene, in *Ostracods in British Stratigraphy* (eds J E Whittaker and M B Hart), 447–467, Spec Publ of The Micropalaeontol Soc, Geol Soc London

Whittaker, K, Beasley, M, Bates, M R, and Wenban-Smith, F F, 2004 The lost valley, *Brit Archaeol* **74** (Jan 2004), 22–27

Wiltshire, P E J, Edwards, K J, and Bond, S, 1994 Microbially-derived metallic sulphide spherules, pollen, and the waterlogging of archaeological sites, *Proc Am Assoc of Sedimentary Palynologists* **29**, 207–221

Wintle, A G, 1991 Luminescence dating, in *Quaternary Dating Methods – a User's Guide* (eds P L Smart and

P D Frances), 108–127, Quat Res Ass Tech Guide **4**, Cambridge

Wintle, A G, and Murray, A S, 2006 A review of quartz optically stimulated luminescence characteristics and their relevance in single-aliquot regeneration dating protocols, *Radiat Meas* **41**, 369–391

Woelfle, E, 1967 *Vergleichend morphologische Untersuchungen an Einzelknochen des postcranialen Skeletts in Mitteleuropa vorkommender Enten, Halbgänse und Säger*, Ludwig-Maximilians-Universitat, Munchen

Woolfenden, G E, 1961 Postcranial osteology of the waterfowl, *Bull Florida State Mus, Biol Sci* **6** (1), 1–129

Wymer, J J, 1957 A Clactonian flint industry at Little Thurrock, Grays, Essex, *Proc Geol Ass* **68**, 159–177

Wymer, J J, 1968 *Lower Palaeolithic Archaeology in Britain as Represented by the Thames Valley*, John Baker, London

Wymer, J J, 1974 Clactonian and Acheulian industries in Britain – their chronology and significance, *Proc Geol Ass* **85**, 391–421

Wymer, J J, 1982 *The Palaeolithic Age*, Croon Helm, London

Wymer, J J, 1983 The Lower Palaeolithic site at Hoxne, *Proc Suffolk Inst Archaeol and Hist* **35**, 169–89

Wymer, J J, 1985 *The Palaeolithic Sites of East Anglia*, Geo Books, Norwich

Wymer, J J, 1988 Palaeolithic archaeology and the British Quaternary sequence, *Quat Sci Rev* **7**, 79–98

Wymer, J J, 1998 Review of 'Excavations at Barnfield Pit, Swanscombe, 1968–1972', *J Quat Sci* **13** (2), 179–181

Wymer, J J, 1999 *The Palaeolithic Occupation of Britain*, Wessex Archaeol, Salisbury

Yravedra, J, Domínguez-Rodrigo, M, Santonja, M, Pérez-González, A, Panera, J, Rubio-Jara, S, Baquedano, E, 2010 Cut marks on the Middle Pleistocene elephant carcass of Áridos 2 (Madrid, Spain), *J Archaeol Sci* **37**, 2469-2476

Yravedra, J, Rubio-Jara, S, Panera, J, Uribelarrea, D, and Pérez-González. A, 2012 Elephants and subsistence. Evidence of the human exploitation of extremely large mammal bones from the Middle Palaeolithic site of PRERESA (Madrid, Spain), *J Archeol Sci* **39**, 1063-1071

Zimmerman, D W, 1971 Thermoluminescent dating using fine grains from pottery, *Archaeometry* **13**, 29-50

Zwilling, E A, 1942 (4th edition) *Unvergessenes Kamerun: Zehn Jahre Wanderungen und Jagden: 1928-1938*, Paul Parey, Berlin

Index

wooded   137, 165, 172, 174, 179, 182, 187, 189,
      192, 194-5, 202-3, 235-6, 241-2, 247, 249,
      252, 261-2, 266, 452-3
Erith, London   201

fire   49, 123, 134, 137, 144-5, 194
fish   162, 172, 179-80, 191-2, 194, 202, 286, 452
   carp, Cyprinidae   202
   minnow *Phoxinus phoxinus*   180
   rudd *Scardinius erythrophthalmus*   202
   tench *Tinca tinca*   162, 202
flint – see lithics
Fonte Campanile, Italy   209-10
Froghall, Staffordshire   283

Gesher Benot Ya'aqov, Israel   231, 447, 458
Globe Pit, Little Thurrock, Kent   14-16, 423, 463,
      467-8
Grays, Thurrock, Essex   234, 247, 253
Gröbern, Germany   210, 225, 457-9
gunflint industry   68, 329, 331, 443-5, 457

Hackney Downs, London   296
Happisburgh, Norfolk   7-9, 460, 464, 469
Harnham, Wilts   9, 470
Hatfield, Herts   283
High Lodge, Suffolk   469
Hitchin, Herts   253, 273
hominin
   anatomically modern human   9
   association with Clactonian and Acheulian   468
   colonisation of Britain   460, 469
   Denisovans   468
   group size   459
   Homo antecessor   9
   Homo erectus   8-9
   Homo ergaster   8-9
   Homo heidelbergensis   9
   Neanderthal   9, 468
   occupation of forested environment   460
   skull, Swanscombe   13-15
Hoxne, Suffolk   194, 234-5, 237, 244
   amino acid dating   314-16, 536
   fauna   253, 258
   flora   296
   ostracods   283, 449
   stratigraphy   261
Hoxnian interglacial
   environment   165, 207, 437, 468
   fauna   160, 165, 179, 182, 190, 195, 203, 235-7,
      239, 241-2, 242, 246-7, 249-50, 253, 258,
      260-1
   hominin activity   192, 390, 457, 468
   lithics   22-3, 197, 233, 262, 371, 413, 463-5,
      467-9
   Marine Isotope Stages   260-1
   mollusca   307, 465
   ostracods   283, 286, 449
   pollen   34, 49, 144, 194, 197, 261, 295-6, 306,
      452, 463, 465, 467

sites   13, 56, 233-5, 261, 289, 453, 455
hunting   457, 460
   large mammals   366, 459-60
hydroseral succession   129, 136-7

Ingress Vale, Kent   149, 235-6, 241
Ipswichian interglacial   7, 9-10, 233

Jaywick Sands, Clacton, Essex   464
Jozwin, Poland   210

La Cotte de St Brelade, Jersey   232, 457
landsurface   29, 31-2, 34, 36, 42-3, 45, 48, 85,
      87-8, 90, 111, 119, 123, 452-3
Lehringen, Germany   457, 459-60
Levett, W   445, 457
lithics
   Acheulian   14-15, 23, 53, 462-3
   Bullhead flint   333, 353, 389
   chaîne opératoire   347, 363, 366, 386, 408-9, 413
      Acheulean   437, 461, 469
      Clactonian   454, 461, 469
      Clactonian and Acheulian compared   455-6
   Chellean   462
   Clactonian   13-15, 18, 22-3, 31, 43, 45, 53, 102,
      337, 452-4, 462-3, 467
      relationship with Acheulian   53, 55, 463, 465,
         467, 469
   classification   325-6, 328
   core   349-50, 353, 355, 357, 393, 396, 398, 400,
      418, 429, 458
   core and flake industry   8
   core-on-flake   401, 407
   core-tool   390, 392, 463
   cutting tool   400-1, 458
   emic and etic interpretation   325-6
   flake-flake   43, 406-7, 413, 421
   flake-knife   403, 405, 407, 421, 434
   flake-tool   361, 390, 396, 400-1, 403, 408, 413,
      421, 433-4, 454-5, 463, 510
   hammerstone   333
   handaxe   13-15, 17-20, 23, 40, 42-3, 105, 107,
      425, 429-30, 432, 455-6
      bout coupé   9
      classification   430, 509
      handaxe-on-flake   441
      manufacture   9, 23, 410, 429, 435-7, 456, 461
      ovate   13-15, 18-19, 31-2, 432-3, 437, 441, 455
      pointed   13, 15, 105, 432, 437, 455
      pre-Anglian   464, 469
   handedness (right and left)   353-4, 357
   juvenile flint-knapping   404
   Levalloisian   9, 14, 18, 20, 27
   microdebitage, distribution   345-7, 380-6
   miniature tools   404
   notched flake-tool   43, 45, 53, 326, 337, 357,
      400-1, 404-7, 410, 413, 421, 433-4, 458
   percussor   334-5, 350-1, 356, 390, 392, 396, 417-
      18, 429, 453, 458
   proto-handaxe   15